Christine M. Eckel
West Virginia School of Osteopathic Medicine

Theresa Stouter Bidle
Hagerstown Community College

Kyla Turpin Ross
Georgia State University

Anatomy & Physiology Laboratory Manual

Main Version

ANATOMY & PHYSIOLOGY LABORATORY MANUAL: MAIN VERSION

Published by McGraw-Hill, a business unit of The McGraw-Hill Companies, Inc., 1221 Avenue of the Americas, New York, NY 10020. Copyright © 2013 by The McGraw-Hill Companies, Inc. All rights reserved. Printed in the United States of America. No part of this publication may be reproduced or distributed in any form or by any means, or stored in a database or retrieval system, without the prior written consent of The McGraw-Hill Companies, Inc., including, but not limited to, in any network or other electronic storage or transmission, or broadcast for distance learning.

Some ancillaries, including electronic and print components, may not be available to customers outside the United States.

This book is printed on acid-free paper.

1 2 3 4 5 6 7 8 9 0 RMN/RMN 1 0 9 8 7 6 5 4 3 2

ISBN 978-0-07-732916-7
MHID 0-07-732916-3

Vice President, Editor-in-Chief: *Marty Lange*
Vice President, EDP: *Kimberly Meriwether David*
Senior Director of Development: *Kristine Tibbetts*
Publisher: *Michael S. Hackett*
Executive Editor: *James F. Connely*
Senior Developmental Editor: *Donna Nemmers*
Marketing Manager: *Chris Loewenberg*
Senior Project Manager: *April R. Southwood*
Senior Buyer: *Sandy Ludovissy*
Senior Media Project Manager: *Tammy Juran*
Senior Designer: *David W. Hash*
Interior Designer: *Christopher Reese*
Cover Anatomy Art: *Electronic Publishing Services Inc., NYC*
Cover Image: *Woman road running in the Santa Monica Mountains,*
  © *Ty Allison/Photographer's Choice/Getty Images*
Lead Photo Research Coordinator: *Carrie K. Burger*
Photo Research: *Danny Meldung/Photo Affairs, Inc.*
Compositor and Art Studio: *Electronic Publishing Services Inc., NYC*
Typeface: *9.5/12 Slimbach Std*
Printer: *R. R. Donnelley*

All credits appearing on page or at the end of the book are considered to be an extension of the copyright page.

Some of the laboratory experiments included in this text may be hazardous if materials are handled improperly or if procedures are conducted incorrectly. Safety precautions are necessary when you are working with chemicals, glass test tubes, hot water baths, sharp instruments, and the like, or for any procedures that generally require caution. Your school may have set regulations regarding safety procedures that your instructor will explain to you. Should you have any problems with materials or procedures, please ask your instructor for help.

www.mhhe.com

brief contents

PART I | **INTRODUCTION TO THE ANATOMY AND PHYSIOLOGY LABORATORY**

Chapter 1
The Laboratory Environment   1

Chapter 2
Orientation to the Human Body   27

Chapter 3
The Microscope   43

PART II | **ORGANIZATION OF THE HUMAN BODY**

Chapter 4
Cell Structure and Membrane Transport   57

Chapter 5
Histology   93

PART III | **SUPPORT AND BODY MOVEMENT**

Chapter 6
Integument   125

Chapter 7
The Skeletal System: Bone Structure and Function   145

Chapter 8
The Skeletal System: Axial Skeleton   165

Chapter 9
The Skeletal System: Appendicular Skeleton   199

Chapter 10
Articulations   233

Chapter 11
The Muscular System: Muscle Structure and Function   251

Chapter 12
The Muscular System: Axial Muscles   289

Chapter 13
The Muscular System: Appendicular Muscles   319

PART IV | **COMMUNICATION AND CONTROL**

Chapter 14
Nervous Tissues   355

Chapter 15
The Brain and Cranial Nerves   381

Chapter 16
The Spinal Cord and Spinal Nerves   427

Chapter 17
The Somatic and Autonomic Nervous Systems and Human Reflex Physiology   449

Chapter 18
General and Special Senses   465

Chapter 19
The Endocrine System   509

PART V | **MAINTENANCE AND REGULATION**

Chapter 20
The Cardiovascular System: Blood   529

Chapter 21
The Cardiovascular System: The Heart   551

Chapter 22
The Cardiovascular System: Vessels and Circulation   589

Chapter 23
The Lymphatic System and Immunity   641

Chapter 24
The Respiratory System   661

Chapter 25
The Digestive System   697

Chapter 26
The Urinary System   735

PART VI | **REPRODUCTION**

Chapter 27
The Reproductive System and Early Development   756

about the authors

With thanks to my mom and dad, my sister, and my best friend Travis.

CHRISTINE MARIE ECKEL is Associate Professor of Anatomy at West Virginia School of Osteopathic Medicine (WVSOM), where she has directed both the Medical Gross Anatomy and Medical Microanatomy courses and served as the Director of the Human Gift Registry. Christine received her B.A. and M.A degrees in Integrative Biology and Human Biodynamics, respectively, at the University of California, Berkeley, where she also received an Outstanding Graduate Student Instructor Award. She then moved to Utah and taught undergraduate Human Anatomy and Physiology at Salt Lake Community College (SLCC) for 11 years, where she received an Outstanding Teaching Award. While working full-time at SLCC, Christine began a Ph.D. program in Neurobiology & Anatomy at University of Utah School of Medicine (U of U SOM). At the U of U SOM, she was named the Betty Cook Karrh Endowed P.E.O. Scholar, and was awarded the Frank L. Christensen Endowed Fellowship. After earning her Ph.D. in Neurobiology and Anatomy, Christine began her current position at WVSOM where she teaches in an integrated curriculum and engages in teaching innovation and educational research projects. In 2011 she received the Golden Key Award for Science Faculty from the Atlas Club at WVSOM. Christine is a member of HAPS, AAA, AACA, and TAA. She is active in several committees within HAPS and the AAA, and served on the HAPS Board of Directors from 2004–2008. She is also a peer reviewer for the journals *Anatomical Sciences Education* and *Medical Education*. Christine's passions for anatomy and physiology, teaching, dissection, and photography are evident throughout the pages of this laboratory manual.

With love and thanks to my husband Jay and my daughter Stephanie for their support.

TERRI STOUTER BIDLE received her undergraduate degree from Rutgers University, her M.S. degree in Biomedical Science from Hood College in Maryland, and has completed additional graduate coursework in Genetics at the National Institutes of Health. She is a professor at Hagerstown Community College where she teaches Anatomy and Physiology and Genetics to pre-allied health students. Before joining the Hagerstown faculty in 1990, Terri was Coordinator of the Science Learning Center, where she developed study materials and a tutoring program for students enrolled in science classes. Terri has been a developmental reviewer, and has written supplemental materials for both textbooks and laboratory manuals.

To my husband Jim and daughter Ella: I treasure your constant love and unending support.

KYLA TURPIN ROSS received her undergraduate degree from Louisiana State University in Biological and Agricultural Engineering, and her Ph.D. in Biomedical Engineering from Georgia Institute of Technology and Emory University. Kyla then served as a postdoctoral fellow in the Fellowships in Research and Science Teaching (FIRST) program at Emory University, an NIH-funded program that provides training in both research and teaching. Kyla is now an academic professional at Georgia State University (GSU), where she teaches and manages the Human Anatomy and Physiology course. Kyla has extensive experience developing lecture and laboratory curricula, and incorporates active learning in the classroom as a method to reinforce difficult physiological concepts. In addition, Kyla plays an active role in mentoring GSU teaching assistants, and planning and hosting an annual teaching assistant workshop. She has served as a reviewer for numerous publications, and has authored a custom laboratory manual for GSU's Human Anatomy and Physiology course.

contents

Preface xv

PART I | INTRODUCTION TO THE ANATOMY AND PHYSIOLOGY LABORATORY 1

Chapter 1
The Laboratory Environment 1

Gross Anatomy 4

THE SCIENTIFIC PROCESS OF DISCOVERY 4
- EXERCISE 1.1 THE SCIENTIFIC METHOD 6
- EXERCISE 1.2 PRESENTING DATA 7

MEASUREMENT IN SCIENCE 8
- EXERCISE 1.3 UNITS OF MEASUREMENT 10

LABORATORY EQUIPMENT 10
- EXERCISE 1.4 IDENTIFICATION OF COMMON DISSECTION INSTRUMENTS 11
- EXERCISE 1.5 PROPER DISPOSAL OF LABORATORY WASTE 14

DISSECTION TECHNIQUES 14
- EXERCISE 1.6 PLACING A SCALPEL BLADE ON A SCALPEL BLADE HANDLE 16
- EXERCISE 1.7 DISSECTING WITH A SCALPEL 19
- EXERCISE 1.8 DISSECTING WITH SCISSORS 20
- EXERCISE 1.9 BLUNT DISSECTION TECHNIQUES 22

Chapter 2
Orientation to the Human Body 27

Gross Anatomy 31

ANATOMIC TERMINOLOGY AND THE ANATOMIC POSITION 31

ANATOMIC PLANES AND SECTIONS 31
- EXERCISE 2.1 DETERMINING ANATOMIC PLANES AND SECTIONS 32

DIRECTIONAL TERMS 33
- EXERCISE 2.2 USING DIRECTIONAL TERMS 33

REGIONAL TERMS 34
- EXERCISE 2.3 USING REGIONAL TERMS 35

BODY CAVITIES AND MEMBRANES 36
- EXERCISE 2.4 UNDERSTANDING BODY CAVITIES 36

ABDOMINOPELVIC REGIONS AND QUADRANTS 37
- EXERCISE 2.5 LOCATING MAJOR BODY ORGANS USING ABDOMINOPELVIC REGION AND QUADRANT TERMINOLOGY 37

Chapter 3
The Microscope 43

Histology 45

THE COMPOUND MICROSCOPE 45

CARING FOR THE COMPOUND MICROSCOPE 45
- EXERCISE 3.1 PARTS OF A COMPOUND MICROSCOPE 46

FOCUS AND WORKING DISTANCE 48
- EXERCISE 3.2 VIEWING A SLIDE OF THE LETTER e 48

DIAMETER OF THE FIELD OF VIEW 49
- EXERCISE 3.3 MEASURING THE DIAMETER OF THE FIELD OF VIEW 50
- EXERCISE 3.4 ESTIMATING THE SIZE OF A SPECIMEN 50

DEPTH OF FIELD 51
- EXERCISE 3.5 DETERMINING DEPTH OF FIELD 51

Gross Anatomy 52

THE DISSECTING MICROSCOPE 52
- EXERCISE 3.6 PARTS OF A DISSECTING MICROSCOPE 53

PART II | ORGANIZATION OF THE HUMAN BODY 57

Chapter 4
Cell Structure and Membrane Transport 57

Histology 61

STRUCTURE AND FUNCTION OF A TYPICAL ANIMAL CELL 61
- EXERCISE 4.1 OBSERVING CELLULAR ANATOMY 63

vi   Contents

MITOSIS   65
EXERCISE 4.2 OBSERVING MITOSIS IN A WHITEFISH EMBRYO   66

Gross Anatomy   66

MODELS OF A TYPICAL ANIMAL CELL AND STAGES OF MITOSIS   66
EXERCISE 4.3 OBSERVING CLASSROOM MODELS OF CELLULAR ANATOMY AND MITOSIS   67

Physiology   67

MECHANISMS OF MEMBRANE TRANSPORT   67
EXERCISE 4.4 DIFFUSION (WET LAB)   67

EXERCISE 4.5 OSMOSIS (WET LAB)   77

EXERCISE 4.6 FILTRATION (WET LAB)   79

EXERCISE 4.7 Ph.I.L.S. LESSON 1: OSMOSIS AND DIFFUSION: VARYING EXTRACELLULAR CONCENTRATION   81

Chapter 5
Histology   93

HISTOLOGY SLIDES   94

Histology   96

EPITHELIAL TISSUE   96
EXERCISE 5.1 IDENTIFICATION AND CLASSIFICATION OF EPITHELIAL TISSUE   98

CONNECTIVE TISSUE   104
EXERCISE 5.2 IDENTIFICATION OF EMBRYONIC CONNECTIVE TISSUE   105

EXERCISE 5.3 IDENTIFICATION AND CLASSIFICATION OF CONNECTIVE TISSUE PROPER   106

EXERCISE 5.4 IDENTIFICATION AND CLASSIFICATION OF SUPPORTING CONNECTIVE TISSUE   111

EXERCISE 5.5 IDENTIFICATION AND CLASSIFICATION OF FLUID CONNECTIVE TISSUE   114

MUSCLE TISSUE   114
EXERCISE 5.6 IDENTIFICATION AND CLASSIFICATION OF MUSCLE TISSUE   116

NERVOUS TISSUE   117
EXERCISE 5.7 IDENTIFICATION AND CLASSIFICATION OF NERVOUS TISSUE   118

PART III SUPPORT AND BODY MOVEMENT   125

Chapter 6
Integument   125

Histology   127

THE EPIDERMIS   127
EXERCISE 6.1 LAYERS OF THE EPIDERMIS   128

EXERCISE 6.2 PIGMENTED SKIN   129

THE DERMIS   130
EXERCISE 6.3 LAYERS OF THE DERMIS   130

EXERCISE 6.4 MEROCRINE (ECCRINE) SWEAT GLANDS AND SENSORY RECEPTORS   131

EXERCISE 6.5 THE SCALP—HAIR FOLLICLES AND SEBACEOUS GLANDS   133

EXERCISE 6.6 AXILLARY SKIN—APOCRINE SWEAT GLANDS   135

EXERCISE 6.7 STRUCTURE OF A NAIL   136

Gross Anatomy   137

INTEGUMENT MODEL   137
EXERCISE 6.8 OBSERVING CLASSROOM MODELS OF INTEGUMENT   137

Chapter 7
The Skeletal System: Bone Structure and Function   145

Histology   147

BONE TISSUE   147
EXERCISE 7.1 COMPACT BONE   148

EXERCISE 7.2 SPONGY BONE   149

EXERCISE 7.3 ENDOCHONDRAL BONE DEVELOPMENT   150

Gross Anatomy   152

CLASSIFICATION OF BONES   152
EXERCISE 7.4 IDENTIFYING CLASSES OF BONES BASED ON SHAPE   153

STRUCTURE OF A TYPICAL LONG BONE   153
EXERCISE 7.5 COMPONENTS OF A LONG BONE   154

EXERCISE 7.6 COW BONE DISSECTION   155

SURVEY OF THE HUMAN SKELETON   157
EXERCISE 7.7 THE HUMAN SKELETON   157

Chapter 8
The Skeletal System: Axial Skeleton 165

Gross Anatomy 169

THE SKULL 169
- **EXERCISE 8.1** ANTERIOR VIEW OF THE SKULL 171
- **EXERCISE 8.2** ADDITIONAL VIEWS OF THE SKULL 176
- **EXERCISE 8.3** SUPERIOR VIEW OF THE CRANIAL FLOOR 179
- **EXERCISE 8.4** BONES ASSOCIATED WITH THE SKULL 181

THE FETAL SKULL 181
- **EXERCISE 8.5** THE FETAL SKULL 182

THE VERTEBRAL COLUMN 183
- **EXERCISE 8.6** VERTEBRAL COLUMN REGIONS AND CURVATURES 184
- **EXERCISE 8.7** STRUCTURE OF A TYPICAL VERTEBRA 185
- **EXERCISE 8.8** CHARACTERISTICS OF INDIVIDUAL VERTEBRAE 186

THE THORACIC CAGE 190
- **EXERCISE 8.9** THE STERNUM 191
- **EXERCISE 8.10** THE RIBS 192

Chapter 9
The Skeletal System: Appendicular Skeleton 199

Gross Anatomy 202

THE PECTORAL GIRDLE 202
- **EXERCISE 9.1** BONES OF THE PECTORAL GIRDLE 203

THE UPPER LIMB 205
- **EXERCISE 9.2** BONES OF THE UPPER LIMB 207
- **EXERCISE 9.3** SURFACE ANATOMY REVIEW—PECTORAL GIRDLE AND UPPER LIMB 213

THE PELVIC GIRDLE 214
- **EXERCISE 9.4** BONES OF THE PELVIC GIRDLE 215

THE LOWER LIMB 218
- **EXERCISE 9.5** BONES OF THE LOWER LIMB 220
- **EXERCISE 9.6** SURFACE ANATOMY REVIEW—PELVIC GIRDLE AND LOWER LIMB 227

Chapter 10
Articulations 233

Gross Anatomy 236

FIBROUS JOINTS 236
- **EXERCISE 10.1** FIBROUS JOINTS 236

CARTILAGINOUS JOINTS 237
- **EXERCISE 10.2** CARTILAGINOUS JOINTS 238

SYNOVIAL JOINTS 239
- **EXERCISE 10.3** GENERAL STRUCTURE OF A SYNOVIAL JOINT 239
- **EXERCISE 10.4** STRUCTURAL CLASSIFICATIONS OF SYNOVIAL JOINTS 240
- **EXERCISE 10.5** PRACTICING SYNOVIAL JOINT MOVEMENTS 241
- **EXERCISE 10.6** THE KNEE JOINT 243

Chapter 11
The Muscular System: Muscle Structure and Function 251

Histology 254

SKELETAL MUSCLE TISSUE 254
- **EXERCISE 11.1** HISTOLOGY OF SKELETAL MUSCLE FIBERS 256
- **EXERCISE 11.2** CONNECTIVE TISSUE COVERINGS OF SKELETAL MUSCLE 257

THE NEUROMUSCULAR JUNCTION 257
- **EXERCISE 11.3** THE NEUROMUSCULAR JUNCTION 259

SMOOTH MUSCLE TISSUE 259
- **EXERCISE 11.4** SMOOTH MUSCLE TISSUE 260

CARDIAC MUSCLE TISSUE 260
- **EXERCISE 11.5** CARDIAC MUSCLE TISSUE 261

Gross Anatomy 261

GROSS ANATOMY OF SKELETAL MUSCLES 261
- **EXERCISE 11.6** NAMING SKELETAL MUSCLES 261
- **EXERCISE 11.7** ARCHITECTURE OF SKELETAL MUSCLES 264

ORGANIZATION OF THE HUMAN MUSCULOSKELETAL SYSTEM 265
- **EXERCISE 11.8** MAJOR MUSCLE GROUPS AND FASCIAL COMPARTMENTS OF THE LIMBS 266

Physiology 268

FORCE GENERATION OF SKELETAL MUSCLE 268

EXERCISE 11.9 MOTOR UNITS AND MUSCLE FATIGUE (HUMAN SUBJECT) 269

EXERCISE 11.10 CONTRACTION OF SKELETAL MUSCLE (WET LAB) 270

EXERCISE 11.11 Ph.I.L.S. LESSON 4: STIMULUS-DEPENDENT FORCE GENERATION 274

EXERCISE 11.12 Ph.I.L.S. LESSON 5: THE LENGTH-TENSION RELATIONSHIP 276

EXERCISE 11.13 Ph.I.L.S. LESSON 6: PRINCIPLES OF SUMMATION AND TETANUS 278

EXERCISE 11.14 Ph.I.L.S. LESSON 7: EMG AND TWITCH AMPLITUDE 279

Chapter 12
The Muscular System: Axial Muscles 289

Gross Anatomy 292

MUSCLES OF THE HEAD AND NECK 292

EXERCISE 12.1 MUSCLES OF FACIAL EXPRESSION 292

EXERCISE 12.2 MUSCLES OF MASTICATION 295

EXERCISE 12.3 EXTRINSIC EYE MUSCLES 296

EXERCISE 12.4 MUSCLES THAT MOVE THE TONGUE 297

EXERCISE 12.5 MUSCLES OF THE PHARYNX 298

EXERCISE 12.6 MUSCLES OF THE NECK 299

MUSCLES OF THE VERTEBRAL COLUMN 303

EXERCISE 12.7 MUSCLES OF THE VERTEBRAL COLUMN 303

MUSCLES OF RESPIRATION 307

EXERCISE 12.8 MUSCLES OF RESPIRATION 308

MUSCLES OF THE ABDOMINAL WALL 310

EXERCISE 12.9 MUSCLES OF THE ABDOMINAL WALL 310

EXERCISE 12.10 THE RECTUS SHEATH, INGUINAL LIGAMENT, AND INGUINAL CANAL 312

Chapter 13
The Muscular System: Appendicular Muscles 319

Gross Anatomy 323

MUSCLES THAT MOVE THE PECTORAL GIRDLE AND GLENOHUMERAL JOINT 323

EXERCISE 13.1 MUSCLES THAT MOVE THE PECTORAL GIRDLE AND GLENOHUMERAL JOINT 324

UPPER LIMB MUSCULATURE 326

EXERCISE 13.2 COMPARTMENTS OF THE ARM 326

EXERCISE 13.3 COMPARTMENTS OF THE FOREARM 329

EXERCISE 13.4 INTRINSIC MUSCLES OF THE HAND 334

MUSCLES THAT MOVE THE HIP JOINT 336

EXERCISE 13.5 MUSCLES THAT MOVE THE HIP 336

LOWER LIMB MUSCULATURE 339

EXERCISE 13.6 COMPARTMENTS OF THE THIGH 339

EXERCISE 13.7 COMPARTMENTS OF THE LEG 344

EXERCISE 13.8 INTRINSIC MUSCLES OF THE FOOT 347

PART IV | COMMUNICATION AND CONTROL 355

Chapter 14
Nervous Tissues 355

Histology 358

EXERCISE 14.1 GRAY AND WHITE MATTER 358

NEURONS 359

EXERCISE 14.2 GENERAL MULTIPOLAR NEURONS—ANTERIOR HORN CELLS 360

EXERCISE 14.3 CEREBRUM—PYRAMIDAL CELLS 360

EXERCISE 14.4 CEREBELLUM—PURKINJE CELLS 361

GLIAL CELLS 362

EXERCISE 14.5 ASTROCYTES 362

EXERCISE 14.6 EPENDYMAL CELLS 363

EXERCISE 14.7 NEUROLEMMOCYTES (SCHWANN CELLS) 363

EXERCISE 14.8 SATELLITE CELLS 364

PERIPHERAL NERVES 365

EXERCISE 14.9 COVERINGS OF A PERIPHERAL NERVE 366

Physiology 366

RESTING MEMBRANE POTENTIAL 366

EXERCISE 14.10 Ph.I.L.S. LESSON 8: RESTING POTENTIAL AND EXTERNAL [K^+] 367

EXERCISE 14.11 Ph.I.L.S. LESSON 9: RESTING POTENTIAL AND EXTERNAL [Na^+] 369

ACTION POTENTIAL PROPAGATION 371

EXERCISE 14.12 Ph.I.L.S. LESSON 10: THE COMPOUND ACTION POTENTIAL 372

EXERCISE 14.13 Ph.I.L.S. LESSON 11: CONDUCTION VELOCITY AND TEMPERATURE 374

EXERCISE 14.14 Ph.I.L.S. LESSON 12: REFRACTORY PERIODS 375

Chapter 15
The Brain and Cranial Nerves 381

Gross Anatomy 384

THE MENINGES 384
- **EXERCISE 15.1** CRANIAL MENINGES 384

VENTRICLES OF THE BRAIN 388
- **EXERCISE 15.2** BRAIN VENTRICLES 389

THE HUMAN BRAIN 390
- **EXERCISE 15.3** SUPERIOR VIEW OF THE HUMAN BRAIN 394
- **EXERCISE 15.4** LATERAL VIEW OF THE HUMAN BRAIN 395
- **EXERCISE 15.5** INFERIOR VIEW OF THE HUMAN BRAIN 396
- **EXERCISE 15.6** MIDSAGITTAL VIEW OF THE HUMAN BRAIN 397

CRANIAL NERVES 398
- **EXERCISE 15.7** IDENTIFICATION OF CRANIAL NERVES ON A BRAIN OR BRAINSTEM MODEL 398

THE SHEEP BRAIN 401
- **EXERCISE 15.8** SHEEP BRAIN DISSECTION 401

Physiology 408

TESTING CRANIAL NERVE FUNCTIONS 408
- **EXERCISE 15.9** TESTING SPECIFIC FUNCTIONS OF THE CRANIAL NERVES 410

Chapter 16
The Spinal Cord and Spinal Nerves 427

Histology 430

SPINAL CORD ORGANIZATION 430
- **EXERCISE 16.1** HISTOLOGICAL CROSS SECTIONS OF THE SPINAL CORD 430

Gross Anatomy 433

THE SPINAL CORD 433
- **EXERCISE 16.2** GROSS ANATOMY OF THE SPINAL CORD 434

PERIPHERAL NERVES 435
- **EXERCISE 16.3** THE CERVICAL PLEXUS 436
- **EXERCISE 16.4** THE BRACHIAL PLEXUS 437
- **EXERCISE 16.5** THE LUMBOSACRAL PLEXUS 440

Chapter 17
The Somatic and Autonomic Nervous Systems and Human Reflex Physiology 449

Gross Anatomy 452

SOMATIC NERVOUS SYSTEM 452
- **EXERCISE 17.1** IDENTIFYING COMPONENTS OF A REFLEX ON A CLASSROOM MODEL 452

AUTONOMIC NERVOUS SYSTEM 453
- **EXERCISE 17.2** PARASYMPATHETIC DIVISION 453
- **EXERCISE 17.3** SYMPATHETIC DIVISION 455

Physiology 459

SOMATIC REFLEXES 459
- **EXERCISE 17.4** PATELLAR REFLEX 460
- **EXERCISE 17.5** PLANTAR REFLEX 460
- **EXERCISE 17.6** CORNEAL REFLEX 461

AUTONOMIC REFLEXES 461
- **EXERCISE 17.7** PUPILLARY REFLEXES 462

Chapter 18
General and Special Senses 465

Histology 468

GENERAL SENSES 468
- **EXERCISE 18.1** TACTILE (MEISSNER) CORPUSCLES 468
- **EXERCISE 18.2** LAMELLATED (PACINIAN) CORPUSCLES 469

SPECIAL SENSES 470
- **EXERCISE 18.3** GUSTATION (TASTE) 470
- **EXERCISE 18.4** OLFACTION (SMELL) 472
- **EXERCISE 18.5** VISION (THE RETINA) 474
- **EXERCISE 18.6** HEARING 476

Gross Anatomy 479

GENERAL SENSES 479
- **EXERCISE 18.7** SENSORY RECEPTORS IN THE SKIN 479

SPECIAL SENSES 481
- **EXERCISE 18.8** GROSS ANATOMY OF THE EYE 481
- **EXERCISE 18.9** COW EYE DISSECTION 485
- **EXERCISE 18.10** GROSS ANATOMY OF THE EAR 487

Physiology 491

GENERAL SENSES 491
- **EXERCISE 18.11** TWO-POINT DISCRIMINATION 491
- **EXERCISE 18.12** TACTILE LOCALIZATION 492
- **EXERCISE 18.13** GENERAL SENSORY RECEPTOR TESTS: ADAPTATION 493

SPECIAL SENSES 494
- **EXERCISE 18.14** GUSTATORY TESTS 494
- **EXERCISE 18.15** OLFACTORY TESTS 495
- **EXERCISE 18.16** VISION TESTS 497
- **EXERCISE 18.17** HEARING AND EQUILIBRIUM TESTS 499

Chapter 19
The Endocrine System 509

Histology 512
- **EXERCISE 19.1** THE HYPOTHALAMUS AND PITUITARY GLAND 512
- **EXERCISE 19.2** THE PINEAL GLAND 515
- **EXERCISE 19.3** THE THYROID AND PARATHYROID GLANDS 516
- **EXERCISE 19.4** THE ADRENAL GLANDS 517
- **EXERCISE 19.5** THE ENDOCRINE PANCREAS—PANCREATIC ISLETS (OF LANGERHANS) 520

Gross Anatomy 521

ENDOCRINE ORGANS 522
- **EXERCISE 19.6** GROSS ANATOMY OF ENDOCRINE ORGANS 522

Physiology 524

METABOLISM 524
- **EXERCISE 19.7** Ph.I.L.S. LESSON 17: THYROID GLAND AND METABOLIC RATE 524

PART V | MAINTENANCE AND REGULATION 529

Chapter 20
The Cardiovascular System: Blood 529

Histology 533
- **EXERCISE 20.1** IDENTIFICATION OF FORMED ELEMENTS ON A PREPARED BLOOD SMEAR 533
- **EXERCISE 20.2** IDENTIFICATION OF MEGAKARYOCYTES ON A BONE MARROW SLIDE 536

Gross Anatomy 537
- **EXERCISE 20.3** IDENTIFICATION OF FORMED ELEMENTS OF THE BLOOD ON CLASSROOM MODELS OR CHARTS 537

Physiology 538

BLOOD DIAGNOSTIC TESTS 538
- **EXERCISE 20.4** DETERMINATION OF WHITE BLOOD CELL COUNTS 539
- **EXERCISE 20.5** DETERMINATION OF HEMATOCRIT 540
- **EXERCISE 20.6** DETERMINATION OF HEMOGLOBIN CONTENT 541
- **EXERCISE 20.7** DETERMINATION OF COAGULATION TIME 542
- **EXERCISE 20.8** DETERMINATION OF BLOOD TYPE 544
- **EXERCISE 20.9** DETERMINATION OF BLOOD CHOLESTEROL 545

Chapter 21
The Cardiovascular System: The Heart 551

Histology 554
- **EXERCISE 21.1** CARDIAC MUSCLE 554
- **EXERCISE 21.2** LAYERS OF THE HEART WALL 555

Gross Anatomy 556
- **EXERCISE 21.3** THE PERICARDIAL CAVITY 556
- **EXERCISE 21.4** GROSS ANATOMY OF THE HUMAN HEART 557
- **EXERCISE 21.5** THE CORONARY CIRCULATION 562
- **EXERCISE 21.6** SUPERFICIAL STRUCTURES OF THE SHEEP HEART 566
- **EXERCISE 21.7** CORONAL SECTION OF THE SHEEP HEART 567
- **EXERCISE 21.8** TRANSVERSE SECTION OF THE SHEEP HEART 568

Physiology 569

ELECTRICAL CONDUCTION WITHIN THE HEART 569
- **EXERCISE 21.9** Ph.I.L.S. LESSON 19: REFRACTORY PERIOD OF THE HEART 571
- **EXERCISE 21.10** Ph.I.L.S. LESSON 25: ELECTRICAL AXIS OF THE HEART 573
- **EXERCISE 21.11** ELECTROCARDIOGRAPHY USING STANDARD ECG APPARATUS 574
- **EXERCISE 21.12** Ph.I.L.S. LESSON 26: ECG AND HEART BLOCK 575
- **EXERCISE 21.13** Ph.I.L.S. LESSON 27: ABNORMAL ECGs 577

CARDIAC CYCLE AND HEART SOUNDS 578
- **EXERCISE 21.14** Ph.I.L.S. LESSON 23: THE MEANING OF HEART SOUNDS 578

EXERCISE 21.15 AUSCULTATION OF HEART SOUNDS 579

EXERCISE 21.16 Ph.I.L.S. LESSON 20: STARLING'S LAW OF THE HEART 579

Chapter 22
The Cardiovascular System: Vessels and Circulation 589

Histology 592

BLOOD VESSEL WALL STRUCTURE 592

EXERCISE 22.1 BLOOD VESSEL WALL STRUCTURE 592

ELASTIC ARTERIES 594

EXERCISE 22.2 ELASTIC ARTERY—THE AORTA 594

MUSCULAR ARTERIES 594

EXERCISE 22.3 MUSCULAR ARTERY 595

ARTERIOLES 595

EXERCISE 22.4 ARTERIOLE 595

VEINS 596

EXERCISE 22.5 VEIN 596

CAPILLARIES 597

EXERCISE 22.6 OBSERVING ELECTRON MICROGRAPHS OF CAPILLARIES 597

Gross Anatomy 600

PULMONARY CIRCUIT 600

EXERCISE 22.7 PULMONARY CIRCUIT 600

SYSTEMIC CIRCUIT 601

EXERCISE 22.8 CIRCULATION TO THE HEAD AND NECK 601

EXERCISE 22.9 CIRCULATION TO THE BRAIN 603

EXERCISE 22.10 CIRCULATION TO THE THORACIC AND ABDOMINAL WALLS 606

EXERCISE 22.11 CIRCULATION TO THE ABDOMINAL CAVITY 609

EXERCISE 22.12 CIRCULATION TO THE UPPER LIMB 614

EXERCISE 22.13 CIRCULATION TO THE LOWER LIMB 618

FETAL CIRCULATION 623

EXERCISE 22.14 FETAL CIRCULATION 623

Physiology 625

BLOOD PRESSURE AND PULSE 625

EXERCISE 22.15 Ph.I.L.S. LESSON 24: ECG AND FINGER PULSE 626

EXERCISE 22.16 BLOOD PRESSURE AND PULSE USING A STANDARD BLOOD PRESSURE CUFF 627

EXERCISE 22.17 BIOPAC LESSON 16: BLOOD PRESSURE 628

Chapter 23
The Lymphatic System and Immunity 641

Histology 644

LYMPHATIC VESSELS 644

EXERCISE 23.1 LYMPHATIC VESSELS 644

MUCOSA-ASSOCIATED LYMPHATIC TISSUE (MALT) 645

EXERCISE 23.2 TONSILS 646

EXERCISE 23.3 PEYER PATCHES 647

EXERCISE 23.4 THE VERMIFORM APPENDIX 648

LYMPHATIC ORGANS 648

EXERCISE 23.5 LYMPH NODES 648

EXERCISE 23.6 THE THYMUS 651

EXERCISE 23.7 THE SPLEEN 653

Gross Anatomy 655

EXERCISE 23.8 GROSS ANATOMY OF LYMPHATIC STRUCTURES 655

Chapter 24
The Respiratory System 661

Histology 664

UPPER RESPIRATORY TRACT 664

EXERCISE 24.1 OLFACTORY MUCOSA 665

LOWER RESPIRATORY TRACT 666

EXERCISE 24.2 THE TRACHEA 666

EXERCISE 24.3 THE BRONCHI AND BRONCHIOLES 668

LUNGS 669

EXERCISE 24.4 THE LUNGS 669

Gross Anatomy 671

UPPER RESPIRATORY TRACT 671

EXERCISE 24.5 SAGITTAL SECTION OF THE HEAD AND NECK 671

LOWER RESPIRATORY TRACT 673

EXERCISE 24.6 THE LARYNX 674

THE PLEURAL CAVITIES AND THE LUNGS 675

EXERCISE 24.7 THE PLEURAL CAVITIES 675

EXERCISE 24.8 THE LUNGS 675

EXERCISE 24.9 THE BRONCHIAL TREE 678

Physiology 679

RESPIRATORY PHYSIOLOGY 679

EXERCISE 24.10 MECHANICS OF VENTILATION 682

EXERCISE 24.11 AUSCULTATION OF RESPIRATORY SOUNDS 682

EXERCISE 24.12 (BIOPAC) PULMONARY FUNCTION TESTS 683

EXERCISE 24.13 Ph.I.L.S. LESSON 31: pH AND Hb-OXYGEN BINDING 686

EXERCISE 24.14 Ph.I.L.S. LESSON 35: EXERCISE-INDUCED CHANGES 688

Chapter 25
The Digestive System 697

Histology 700

SALIVARY GLANDS 700

EXERCISE 25.1 HISTOLOGY OF THE SALIVARY GLANDS 701

THE STOMACH 702

EXERCISE 25.2 WALL LAYERS OF THE STOMACH 702

EXERCISE 25.3 HISTOLOGY OF THE STOMACH 705

THE SMALL INTESTINE 707

EXERCISE 25.4 HISTOLOGY OF THE SMALL INTESTINE 708

THE LARGE INTESTINE 708

EXERCISE 25.5 HISTOLOGY OF THE LARGE INTESTINE 709

THE LIVER 710

EXERCISE 25.6 HISTOLOGY OF THE LIVER 710

THE PANCREAS 711

EXERCISE 25.7 HISTOLOGY OF THE PANCREAS 712

Gross Anatomy 712

THE ORAL CAVITY, PHARYNX, AND ESOPHAGUS 712

EXERCISE 25.8 GROSS ANATOMY OF THE ORAL CAVITY, PHARYNX, AND ESOPHAGUS 712

THE STOMACH 714

EXERCISE 25.9 GROSS ANATOMY OF THE STOMACH 714

THE DUODENUM, LIVER, GALLBLADDER, AND PANCREAS 716

EXERCISE 25.10 GROSS ANATOMY OF THE DUODENUM, LIVER, GALLBLADDER, AND PANCREAS 717

THE JEJUNUM AND ILEUM OF THE SMALL INTESTINE 718

EXERCISE 25.11 GROSS ANATOMY OF THE JEJUNUM AND ILEUM OF THE SMALL INTESTINE 719

THE LARGE INTESTINE 720

EXERCISE 25.12 GROSS ANATOMY OF THE LARGE INTESTINE 721

Physiology 722

DIGESTIVE PHYSIOLOGY 722

EXERCISE 25.13 DIGESTIVE ENZYMES 725

EXERCISE 25.14 Ph.I.L.S. LESSON 37: GLUCOSE TRANSPORT 727

Chapter 26
The Urinary System 735

Histology 738

THE KIDNEY 738

EXERCISE 26.1 HISTOLOGY OF THE RENAL CORTEX 740

EXERCISE 26.2 HISTOLOGY OF THE RENAL MEDULLA 741

THE URINARY TRACT 742

EXERCISE 26.3 HISTOLOGY OF THE URETERS 743

EXERCISE 26.4 HISTOLOGY OF THE URINARY BLADDER 744

Gross Anatomy 744

THE KIDNEY 744

EXERCISE 26.5 GROSS ANATOMY OF THE KIDNEY 744

EXERCISE 26.6 BLOOD SUPPLY TO THE KIDNEY 745

EXERCISE 26.7 URINE-DRAINING STRUCTURES WITHIN THE KIDNEY 748

THE URINARY TRACT 748

EXERCISE 26.8 GROSS ANATOMY OF THE URETERS 748

EXERCISE 26.9 GROSS ANATOMY OF THE URINARY BLADDER AND URETHRA 749

Physiology 752

URINE FORMATION 752

EXERCISE 26.10 URINALYSIS 754

ACID-BASE BALANCE 756

EXERCISE 26.11 A CLINICAL CASE STUDY IN ACID-BASE BALANCE 757

PART VI | REPRODUCTION 765

Chapter 27
The Reproductive System and Early Development 765

Histology 768

FEMALE REPRODUCTIVE SYSTEM 768
- **EXERCISE 27.1** HISTOLOGY OF THE OVARY 768
- **EXERCISE 27.2** HISTOLOGY OF THE UTERINE TUBES 772
- **EXERCISE 27.3** HISTOLOGY OF THE UTERINE WALL 774
- **EXERCISE 27.4** HISTOLOGY OF THE VAGINAL WALL 775

MALE REPRODUCTIVE SYSTEM 776
- **EXERCISE 27.5** HISTOLOGY OF THE SEMINIFEROUS TUBULES 776
- **EXERCISE 27.6** HISTOLOGY OF THE EPIDIDYMIS 778
- **EXERCISE 27.7** HISTOLOGY OF THE DUCTUS DEFERENS 779
- **EXERCISE 27.8** HISTOLOGY OF THE SEMINAL VESICLES 780
- **EXERCISE 27.9** HISTOLOGY OF THE PROSTATE GLAND 781
- **EXERCISE 27.10** HISTOLOGY OF THE PENIS 782

Gross Anatomy 783

FEMALE REPRODUCTIVE SYSTEM 783
- **EXERCISE 27.11** GROSS ANATOMY OF THE OVARY, UTERINE TUBES, UTERUS, AND SUSPENSORY LIGAMENTS 783
- **EXERCISE 27.12** GROSS ANATOMY OF THE FEMALE BREAST 786

MALE REPRODUCTIVE SYSTEM 788
- **EXERCISE 27.13** GROSS ANATOMY OF THE SCROTUM, TESTIS, SPERMATIC CORD, AND PENIS 789

Physiology 791

REPRODUCTIVE PHYSIOLOGY 791
- **EXERCISE 27.14** A CLINICAL CASE IN REPRODUCTIVE PHYSIOLOGY 795

FERTILIZATION AND DEVELOPMENT 797
- **EXERCISE 27.15** EARLY DEVELOPMENT: FERTILIZATION AND ZYGOTE FORMATION 797
- **EXERCISE 27.16** EARLY DEVELOPMENT: EMBRYONIC DEVELOPMENT 798

Appendix A-1

Credits C-1

Index I-1

preface

Human anatomy and physiology is a complex yet fascinating subject, and is perhaps one of the most personal subjects a student will encounter during his or her education. It is also a subject that can create concern for students because of the sheer volume of material, and the misconception that "it is all about memorization."

The study of human anatomy and physiology really comes to life in the anatomy and physiology laboratory, where students get hands-on experience with human cadavers and bones, classroom models, preserved and fresh animal organs, histology slides of human tissues, and discovering the process of scientific discovery through physiology experimentation. Yet, most students are at a loss regarding how to proceed in the anatomy and physiology laboratory. For example, students are often given numerous lists of structures to identify, histology slides to view, and "wet labs" to conduct, but are given comparatively little direction regarding how to recognize structures, or how to relate what they encounter in the laboratory to the material presented in the lecture. In addition, most laboratory manuals on the market contain little more than material repeated from anatomy and physiology textbooks, which provides no real benefit to a student.

This laboratory manual takes a very focused approach to the laboratory experience, and is intended to provide students with tools to make the subject matter more relevant to their own bodies and to the world around them. Rather than providing a recap of material from classroom lectures and the main textbook for the course, this laboratory manual is much more of an interactive workbook for students: a "how-to" guide to learning human anatomy and physiology through touch, dissection, observation, experimentation, and critical thinking exercises. Students are guided to first formulate a hypothesis about each experiment, before actually beginning exercises. Diagrams direct students in how to perform experiments, and don't just show the end results. The text is written in a friendly, conversational tone to put students at ease as they discover, organize, and understand the material presented in each chapter.

Organization

Because observation of histology slides, observation of human cadavers or classroom models, and "wet lab" experiments are usually performed in separate physical spaces or at specific times within each laboratory classroom, chapters in this laboratory manual are similarly separated into three sections: Histology, Gross Anatomy, and Physiology. Each exercise within these chapter sections has been designed with the student's actual experience in the anatomy and physiology laboratory in mind. Thus, each exercise covers only a single histology slide, classroom model, region of the human body, or wet lab experiment. At the same time, within-chapter "Concept Connection" and "Clinical View" boxes provide an opportunity to integrate the material from all three sections of each chapter. In addition, "Can You Apply What You've Learned?" and "Can You Synthesize What You've Learned?" questions in Post-laboratory Worksheets provide further opportunities for students to integrate the information and apply it to clinically relevant and practical situations. Organization of each chapter into a series of discreet exercises makes the laboratory manual easily customizable to any anatomy and physiology classroom, allowing an instructor to assign certain exercises, while telling students to ignore other exercises. Post-laboratory worksheets are also organized by exercise, which makes it easy for an instructor to assign questions that relate only to the exercises covered in their classroom.

the learning system

Features

The Eckel/Bidle/Ross: *Anatomy & Physiology Laboratory Manual* was developed as a complement to the McKinley/O'Loughlin/Bidle: *Anatomy & Physiology: An Integrative Approach* textbook. Each chapter opener includes an **outline** that lists a set of **learning objectives** for the chapter.

■ A chapter **introduction** opens with a real-life scenario that emphasizes the section of the body covered in the chapter, to connect the anatomy of our bodies with the physiology that helps us to perform day-to-day activities.

■ The laboratory manual exhibits the highest-quality **photographs and illustrations** of any laboratory manual on the market.

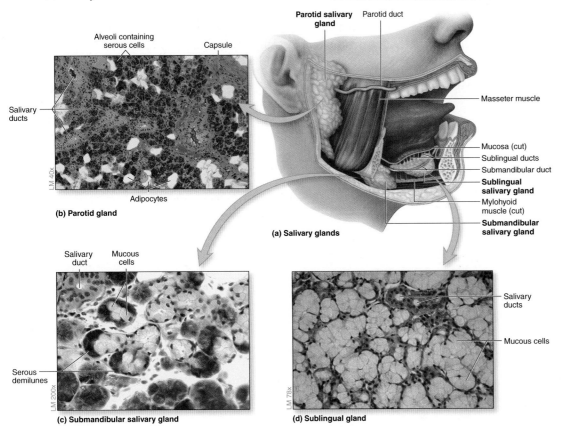

- The content of the laboratory manual is informed by the textbook, and both the textbook and the laboratory manual share similar pedagogic elements: **Concept Connection, Learning Strategy,** and **Clinical View** elements from the text are continued in the laboratory manual.

  - **Integrate: Concept Connection** boxes draw concepts from the classroom into the laboratory for a real-time review of how previously covered concepts relate to body systems.

  - **Integrate: Learning Strategy** boxes offer tried-and-tested learning strategies that consist of everyday analogies, mnemonics, and useful tips to aid understanding and memory.

  - **Integrate: Clinical View** sidebars reinforce facts through a clinical discussion of what happens when the body doesn't perform normally.

INTEGRATE

CONCEPT CONNECTION

Red bone marrow is the site of hemopoiesis, or blood cell formation. In young children, red bone marrow is found within most bones. However, by adulthood most red bone marrow is replaced by yellow bone marrow, which is mainly adipose tissue. In an adult, red bone marrow is limited to the ribs, sternum, vertebrae, pelvis, flat bones of the skull, and proximal epiphyses of the femur and humerus. Red bone marrow contains stem cells known as hemocytoblasts, which differentiate into *erythrocytes* (red blood cells), *leukocytes* (white blood cells), and *megakaryocytes*, which produce platelets. Chemical "factors," including hormones released in the blood, determine the ultimate fate of these stem cells. Bone marrow donation involves harvesting red bone marrow from either the iliac crest or the sternum (locations where red bone marrow is abundant and relatively easy to access). As with blood transfusions, donor and recipient bone marrow must "match" to prevent rejection. Properties of erythrocytes and leukocytes will be explored in chapters 20 (blood) and 23 (lymphatic and immune systems), respectively.

INTEGRATE

LEARNING STRATEGY

Here is a mnemonic that may help you remember the names of the carpal bones, and the order in which they are found: **S**o **L**ong **T**op **P**art, **H**ere **C**omes **T**he **T**humb.
So Long Top Part = Scaphoid, Lunate, Triquetrum, Pisiform
Here Comes The Thumb = Hamate, Capitate, Trapezium, Trapezoid
   Proximal row first, moving from lateral to medial (anatomic position).
Distal row second, moving from lateral to medial (anatomic position).

INTEGRATE

CLINICAL VIEW
Amyotrophic Lateral Sclerosis (ALS)

Amyotrophic lateral sclerosis (ALS), or Lou Gehrig's Disease, is so named for the famous baseball player, Lou Gehrig, who was diagnosed with this fatal disease at the age of 36 after suffering a dramatic decline in the quality of his game. ALS impacts somatic motor neurons, whose cell bodies are located in the posterior (ventral) horn of the spinal cord (see chapter 16). These neurons innervate skeletal muscle fibers. While the etiology of ALS is still unclear, it is known that ALS leads to the destruction of these motor neurons. As motor neurons die and retract from their respective muscle fibers, there is a loss in function in the nerve-muscle connection, or neuromuscular junction. With this loss comes an inability to excite the skeletal muscle fiber. Without excitation, there is no contraction, and if you don't use it, you lose it! That is, muscle fibers will begin to atrophy (*a-*, without + *-trophy*, nourishment) due to disuse. As you can imagine, this also impacts motor unit recruitment, as a loss of motor neurons results in atrophy of all of the fibers that the motor neurons innervate. As a compensatory mechanism, healthy motor neurons begin sprouting collateral processes to make connections with the fibers that are no longer receiving electrical signals from the central nervous system. Often, muscle fiber types will begin to "group" together, such that Type I fibers may be concentrated in one particular compartment of a muscle. This is in contrast to a healthy muscle where muscle fiber types are often distributed throughout a compartment. As the disease progresses, patients may begin to experience slurred speech, muscle fatigue, and clumsiness. Unfortunately, ALS is an aggressive disease that attacks all somatic motor neurons in the body, including those responsible for innervating skeletal muscles that aid in respiration, such as the diaphragm. Over time, often in as little as 3 to 5 years, there is so much denervation and muscle atrophy that patients are no longer able to effectively contract the diaphragm and die of respiratory failure.

Chapter 4: Cell Structure and Membrane Transport

Name: _____
Date: _____ Section: _____

PRE-LABORATORY WORKSHEET

Also available at www.connect.mcgraw-hill.com

connect plus+
ANATOMY & PHYSIOLOGY

1. Match the cell structures in column A with their functions in column B.

   Column A
   ___ 1. centrioles
   ___ 2. chromatin
   ___ 3. cytoplasm
   ___ 4. cytoskeleton
   ___ 5. rough endoplasmic reticulum (RER)
   ___ 6. Golgi apparatus
   ___ 7. lysosomes
   ___ 8. mitochondria
   ___ 9. nucleolus
   ___ 10. nucleus
   ___ 11. peroxisomes
   ___ 12. plasma membrane
   ___ 13. ribosomes

   Column B
   a. provides a selectively permeable barrier between the intracellular and extracellular environments of the cell
   b. includes cellular organelles and cytosol; cytosol contains enzymes that mediate many cytosolic reactions, such as glycolysis and fermentation
   c. contains the cell's genetic material (DNA)
   d. synthesizes rRNA and assembles ribosomes in the nucleus
   e. genetic material within the nucleus; consists of uncoiled chromosomes and associated proteins
   f. synthesizes new proteins destined for the plasma membrane, for lysosomes, or for secretion from the cell
   g. sites of protein synthesis; may be bound to the ER ("bound") or found within the cytoplasm ("free")
   h. a stack of flattened membranes that are the site where proteins from the ER are modified, packaged, and sorted for delivery to other organelles or to the plasma membrane of the cell
   i. membrane-enclosed sacs that contain digestive enzymes and function in the breakdown of intracellular structures
   j. membrane-enclosed sacs that contain catalase and other oxidative enzymes
   k. often referred to as the "powerhouse" of the cell, these organelles are the site of cellular respiration
   l. paired organelles composed of microtubules that are used to organize the spindle microtubules that attach to chromosomes during mitosis
   m. composed of protein filaments called microtubules, intermediate filaments, and microfilaments; provides the main structural support for the cell

2. Define *mitosis*: _____

3. List the four stages of mitosis (in order):
   a. _____
   b. _____
   c. _____
   d. _____

4. The stage of the cell cycle when cells are *not* undergoing mitosis is _____. During this phase, individual chromosomes <u>are/are not</u> (circle one) visible within the nucleus of the cell.

5. Define the following terms:
   a. *solute*
   _____

Chapter Four *Cell Structure and Membrane Transport* 59

- **Pre-laboratory worksheets** at the start of each chapter consist of important refresher points to provide students with a 'warm-up' before entering the laboratory classroom. Some questions pertain to previous activities that are relevant to upcoming exercises, while others are basic questions that students should be able to answer if they have read the chapter from their lecture text before coming into the laboratory classroom. The goal of completing these worksheets is to have students arrive at the laboratory prepared to deal with the material they will be covering, so valuable laboratory time isn't lost in reviewing necessary information.

- In-chapter **activities** offer a mixture of labeling exercises, sketching activities, table completion exercises, data recording and analysis, palpation of surface anatomy, and other sources of learning. In the gross anatomy exercises of this manual, structures such as cranial bones and muscles of the body are *not* always presented as labeled photos, since students already have labeled photos provided in their anatomy and physiology textbook. Instead, images are presented as labeling activities with a checklist of structures. The checklists serve two purposes: (1) they guide students to items they should be able to identify on classroom models, fresh specimens, or cadavers (if the laboratory uses human cadavers), and (2) they double as a list of terms students can use to complete the labeling activities. Answers to the labeling activities are provided in the Appendix. Thus, if a student does not know what a leader line is pointing to, or cannot remember the correct term, the Appendix serves as a resource for locating the correct answer.

- **Anatomy & Physiology Revealed® (APR)** correlations, indicated by the APR logo, direct students to related content in this cutting-edge software.

- Each chapter contains numerous **tables,** which concisely summarize critical information and key structures and serve as important points of reference while in the laboratory classroom. Most tables contain a column that provides word origins for each structure listed within the table. These word origins are intended to give students continual exposure to the origins of the language of anatomy and physiology, which is critical for learning and retention.

- Numerous **Physiology Interactive Lab Simulations© (Ph.I.L.S.) 3.0** exercises throughout the laboratory manual make otherwise difficult and expensive experiments a breeze, and offer additional opportunities to aid student understanding of physiology.

- **BIOPAC©** exercises are included in chapter 22 for blood pressure, and in chapter 24 for respiratory physiology.

- **Post-laboratory worksheets** at the end of each chapter serve as a review of the materials just covered, and challenge students to apply knowledge gained in the laboratory. The post-laboratory worksheets contain more in-depth, critical thinking question types than the pre-laboratory worksheets. Post-laboratory worksheets are perforated so they can be torn out and handed in to the instructor, if so desired. Assessment questions are organized by exercise, and are keyed to the Learning Objectives from the chapter opener outline.

  - **"Do You Know the Basics?"** questions quiz students on the material they have just learned in the chapter, using a variety of question formats including labeling, table completion, matching exercises, and fill-in-the-blank.

  - **"Can You Apply What You've Learned?"** questions are often clinically oriented and expose health-sciences students to problem solving in clinical contexts.

  - **"Can You Synthesize What You've Learned?"** questions combine concepts learned in the chapter to ensure student understanding of each chapter's objectives.

Teaching Supplements

Answers to the pre-laboratory and post-laboratory worksheets can be found within the **Instructor's Manual** for this Laboratory Manual at www.mhhe.com/mckinleyap. **Image files** for use in presentations and teaching materials are also provided for instructor use at this location.

The Best of Both Worlds

McGraw-Hill and Blackboard©

McGraw-Hill Higher Education and Blackboard have teamed up. What does this partnership mean for you? Blackboard users will find the single sign-on and deep integration of ConnectPlus within their Blackboard course an invaluable benefit. Even if your school is not using Blackboard, we have a solution for you. Learn more at www.domorenow.com.

ACKNOWLEDGMENTS

This laboratory manual is the product of the excellent work and dedication of a consummate group of talented professionals who have helped lead us through this publishing process. We are forever indebted to all of you for embarking on this journey with us.

First and foremost, we wish to thank Kelly Brown, Jim Connely, Marty Lange, Beva Lincoln, and Michelle Watnick for their belief in us as authors and for providing us the unique opportunity to share our enthusiasm for teaching anatomy and physiology through the pages of this laboratory manual.

From start to finish, this book has been carried through the able hands of developmental editors Kristine Queck, Kristine Tibbetts, and Donna Nemmers. Donna stepped in and managed to handle an enormous workload, a new book, and new authors with aplomb. All of you amaze us with your ability to juggle so many cards and have everything come out beautiful in the end. April Southwood, Senior Project Manager, is remarkable in her ability to keep everyone on track while also keeping a smile on everyone's faces. Beth Bulger (copy editor) and Patti Evers and Debbie Trilk (proofreaders) have keen eyes and a talent for paying attention to detail that was instrumental in catching things our tired eyes could not always see. There aren't enough ways to thank these professionals for putting up with our quirks, while keeping the hair on their heads intact!

The gorgeous design and line art that make this laboratory manual shine are products of a team of hugely talented designers and artists: David Hash, Eli Ensor, Libby Lamb, Lake Lloyd, Kim Moss, and the rest of the fantastic EPS team. Carrie Burger's direction and guidance was instrumental in bringing out the best of our photography program. Danny Meldung, of Photo Affairs, Inc. always managed to come up with wonderful photos, even on short notice. Jill Braaten (photographer) and Heald College provided excellent photos for several exercises in chapter 18 (general and special senses). Mike McKinley, Ph.D., extensively reviewed page proofs to ensure consistency between this laboratory manual and the McKinley/O'Loughlin/Bidle: *Anatomy & Physiology* textbook. His efforts are appreciated by us, and greatly enhance the relationship between this manual and the McKinley/O'Loughlin/Bidle text. Much of the art and photos have benefited from the input of Valerie O'Loughlin, Ph.D., who also deserves a special thank-you for being a wonderful colleague and friend. Shawn Weeks, M.S. and future D.O., stepped up to the plate for his first modeling gig and even donated his own blood for the good of the manual. Brian Kipp, Ph.D., authored the Ph.I.L.S. exercises for the manual. Thank you, all!

We thank all of the reviewers of the manual (listed below) for taking the time to review this manual and provide us with their insight and perspective, gained from years of experience in the classroom. We hope we have honored your suggestions for improvement, and we welcome continued feedback. We also thank the many students we have had the pleasure of interacting with over the years, for teaching *us* what works or does not work in the classroom.

The extent of our gratitude is limitless when it comes to the love, understanding, and support that our families, friends, and colleagues gave to us throughout this process. We are truly honored to live our lives in the presence of such wonderful people.

To the users of this laboratory manual: we sincerely hope we have created a learning resource that will not only excite you about the study of anatomy and physiology, but will also actively engage you in the laboratory as you learn about the wonders of the human body. We welcome your thoughts and suggestions for improvements.

Christine M. Eckel
Department of Biomedical Sciences
West Virginia School of Osteopathic Medicine
ceckel@osteo.wvsom.edu

Terri Stouter Bidle
Science Division
Hagerstown Community College
tsbidle@hagerstowncc.edu

Kyla Turpin Ross
Department of Biology
Georgia State University
kross@gsu.edu

Reviewers

Mike Aaron
  Shelton State Community College
Penny P. Antley
  University of Louisiana–Lafayette
Celina Bellanceau
  University of South Florida
James B. Crabbe
  County College of Morris–Randolph
Smruti A. Desai
  Lone Star College–Cyfair
Maria Florez
  Lone Star College–Cyfair
Lynn Gargan
  Tarrant County Community College–Northeast Campus
Alexander Ibe
  Weatherford College
Michael S. LaPointe
  Indiana University Northwest
Jerri K. Lindsey
  Tarrant County College–Northeast Campus
Michael P. McKinley
  Glendale Community College
Terrence Miller
  Central Carolina Community College
Greg Reeder
  Broward College
Dee Ann S. Sato
  Cypress College
William Stewart
  Middle Tennessee State University
Michele Iannuzzi Sucich
  SUNY Orange–Middletown

PART I | INTRODUCTION TO THE ANATOMY AND PHYSIOLOGY LABORATORY

CHAPTER 1

The Laboratory Environment

OUTLINE AND LEARNING OBJECTIVES

Gross Anatomy 4

The Scientific Process of Discovery 4

EXERCISE 1.1: THE SCIENTIFIC METHOD 6
1. Describe the steps involved in the scientific method
2. Define: independent variable, dependent variable, and confounding variables
3. Explain the importance of a good experimental control
4. Given a set of data, calculate the mean and range

EXERCISE 1.2: PRESENTING DATA 7
5. Describe the components of a typical graph and what each represents
6. Compose a graph using experimental data
7. Interpret a graph

Measurement in Science 8

EXERCISE 1.3: UNITS OF MEASUREMENT 10
8. Associate prefixes for metric untis of measurement with a power of ten that each prefix represents
9. List the metric unit used to measure each of the following: length, mass, temperature, and volume
10. Convert common measurements from English units to metric units

Laboratory Equipment 10

EXERCISE 1.4: IDENTIFICATION OF COMMON DISSECTION INSTRUMENTS 11
11. Identify commonly used dissection tools

EXERCISE 1.5: PROPER DISPOSAL OF LABORATORY WASTE 15
12. Describe the proper disposal methods for tissues, instruments, and paper waste

Dissection Techniques 14

EXERCISE 1.6: PLACING A SCALPEL BLADE ON A SCALPEL BLADE HANDLE 16
13. Demonstrate the proper technique for putting a scalpel blade on a scalpel handle
14. Demonstrate the proper technique for removing a scalpel blade from a scalpel handle

EXERCISE 1.7: DISSECTING WITH A SCALPEL 19
15. Demonstrate the proper technique for using a scalpel to cut tissues
16. Describe techniques used to prevent damage to underlying tissues when using a scalpel

EXERCISE 1.8: DISSECTING WITH SCISSORS 20
17. Demonstrate how to use scissors to dissect
18. Demonstrate "open scissors" technique

EXERCISE 1.9: BLUNT DISSECTION TECHNIQUES 22
19. Define "blunt dissection"
20. Demonstrate common blunt dissection techniques
21. Explain the importance of using blunt dissection techniques whenever possible

INTRODUCTION

Welcome to the human anatomy and physiology laboratory! If you are like most students, you are experiencing both excitement and anxiety about this course. The human body is a fascinating subject, and the study of human anatomy and physiology is an experience that you will surely never forget.

This laboratory manual is designed for an integrated, systems-based course that combines human gross anatomy, histology, and physiology. **Histology** is the study of tissues, and requires the use of a microscope. **Gross anatomy** is the study of structures that can be seen with the naked eye. This means any structure that you do not need a microscope to see. **Physiology** is the study of body functions. Once you have completed this course, it is our hope that you will have developed an understanding and appreciation for how tissue structure relates to gross structure, and how all levels of structure relate to function. However, in the laboratory itself, you may find yourself studying the three somewhat separately. That is, your laboratory studies in histology will likely involve observing histology slides with a microscope or using some sort of virtual microscopy system; your laboratory studies in gross anatomy will likely involve observing classroom models, dissecting animal specimens, or making observations of human bones and/or human cadavers; and your laboratory studies in physiology will involve performing wet ➡

lab or virtual (computer software–based) experiments. To assist you, the exercises in this manual are divided into three types of activities: histology, gross anatomy, and physiology activities. Where applicable, each chapter will begin with a section on histology and will end with a section on physiology. Although you will be performing the activities somewhat separately, your goal is to integrate what you are learning in each exercise and to associate structure with function. "Concept Connection" boxes and questions within exercises in each chapter will assist you with this task.

The purpose of this introductory chapter is to help you become familiar with the process of science, systems of measurement, common equipment and dissection techniques encountered in the anatomy and physiology laboratory, proper disposal of laboratory waste materials, and common dissection techniques.

INTEGRATE

CLINICAL VIEW
Use of Human Cadavers in the Anatomy and Physiology Laboratory

Where did that body lying on a table in your human anatomy and physiology laboratory come from? Typically, the body was donated by a person who made special arrangements before the time of death to donate his or her body to a body donor program so it could be used for education or research. Individuals who donate their bodies for these purposes made a conscious decision to do so. Such individuals have given us an incredible gift—the opportunity to learn human anatomy and physiology from an actual human body. It is important to remember that what that person has given you is, indeed, a gift. The cadaver deserves your utmost respect at all times. Making jokes about any part of the cadaver or intentionally damaging or "poking" at parts of the cadaver is unacceptable behavior.

The idea that you will be learning anatomy and physiology by observing structures on what was, at one time, a living, breathing human being might make you feel very uncomfortable at first. It is quite normal to have an emotional response to the cadaver upon first inspection. Would you want it any other way? You will need time and experience to become comfortable around the cadaver. Even if you think you will be just fine around the cadaver, it is important to be aware of your first response and of the responses of your fellow students. If at any time you feel faint or light-headed, sit down immediately. Fainting, though rare, is a possibility, and people can get hurt from falling down unexpectedly. Be aware of fellow students who appear to lose color in their faces or start to look sick—they might need your assistance.

Typically the part of the body that evokes the most emotional response is the face, because it is most indicative of the person that the cadaver once was. Because of this, the face of the cadaver should remain covered most of the time. This does not mean you are not allowed to view it. However, when you wish to do so, you should make sure that other students in the room know that you will be uncovering the face. If you have a particularly strong emotional response to the cadaver, take a break and come back to it later when you are feeling better.

Those of us with a great deal of experience around cadavers had a similar emotional response our first time as well. What happens with time is that you learn to disconnect your emotions from the experience. Certainly at one time the body that is the cadaver in your laboratory was the home of a living human being. However, now it is just a body. Eventually you will become comfortable using the cadaver and will find that it is an invaluable learning tool that is far more useful than any model or picture could ever be. You will literally see for yourself that there is nothing quite like the real thing to help you truly understand the structure of the human body. Make the most of this unique opportunity—and give thanks to those who selflessly donated their bodies so you could partake in the ultimate learning experience in anatomy and physiology.

Students who are curious about the uses of cadavers in science and research are encouraged to check out the following book from the library: Mary Roach, *Stiff: The Curious Lives of Human Cadavers* (New York: W.W. Norton, 2003).

INTEGRATE

CONCEPT CONNECTION

Homeostasis (*homoios-*, same kind + *-stasis*, standing still) is one of the most important concepts covered in an anatomy and physiology course. The premise is that stability of the internal environment (milieu) is maintained within limits despite a constantly changing internal and external environment. For each variable there is a desired physiological **set point**. Any control system required for maintaining homeostasis involves a stimulus, a receptor, a control center, and an effector. Any fluctuation from the set point constitutes a **stimulus**. Specialized **receptors** detect this change and relay the information to the **control center**. The role of the control center is to interpret the information and determine a course of action, which is then relayed to the **effector**. The effector then carries out the desired **response**, which is meant to bring the system back to the set point. Deviations from the set point initiate a physiological response that either reduces the deviation (negative feedback) or enhances the deviation (positive feedback). For instance, when you eat a meal containing carbohydrates, you digest the food and absorb the carbohydrates as glucose (blood sugar). As a result, your blood glucose levels rise. This stimulates the release of the hormone insulin by the pancreas, which works to reduce blood glucose levels. This is an example of **negative feedback,** because increasing levels of insulin work to **counteract** the increase in blood glucose. Negative feedback mechanisms are the feedback mechanisms most often encountered in physiology. However, you will also encounter a few examples of **positive feedback.** One example concerns uterine contractions during childbirth. As the uterine wall stretches, the hormone oxytocin is released. This hormone stimulates the uterus to contract, which further increases the tension (stretch) on the uterine wall. The increased stretching of the uterine wall, in turn, stimulates further release of oxytocin until the baby is delivered. Because the response to the stimulus results in an *increase* in the deviation from the set point, this is termed *positive* feedback. As you explore each system of the body, you will see several examples of these feedback mechanisms, which allow the body to maintain the dynamic equilibrium called homeostasis.

Chapter 1: The Laboratory Environment

Name: _____
Date: _____ Section: _____

PRE-LABORATORY WORKSHEET

Also available at www.connect.mcgraw-hill.com

1. Define *histology*: _____
   _____
   _____

2. Define *gross anatomy*: _____
   _____
   _____

3. Define *physiology*: _____
   _____
   _____

4. Define *hypothesis*: _____
   _____
   _____

5. List the five steps involved in the scientific method.

   a. _____
   b. _____
   c. _____
   d. _____
   e. _____

6. What is the *mean* with respect to a set of data points and how is the mean calculated?
   _____
   _____

7. The prefix *kilo-* means _____.

8. The prefix *deci-* means _____.

9. The metric unit used to report volume is _____.

10. The metric unit used to report mass is _____.

Chapter One *The Laboratory Environment* 3

Gross Anatomy

The Scientific Process of Discovery

What is science? **Science** is a way of knowing. A **scientist** is an individual who engages in research using the scientific method to learn facts about the world. The **scientific method** is a systematic approach to inquiry that allows us to study the world. The scientific method assumes that the answer to a question can be explained by phenomena that are *observable* and *measurable*. Specifically, that cause cannot be explained simply on the basis of faith or spirituality. In the anatomy and physiology laboratory, you will be using the scientific method to assist you in making observations and conclusions about the structure and function of the human body.

The scientific method is a rigorous and systematic approach to inquiry that requires certain steps be taken. However, the steps need not be followed precisely. That is, there is some flexibility in the order in which the steps take place. Very often several steps take place concurrently as the process of discovery evolves. The steps involved in the scientific method follow the general pattern shown below:

Observation → Hypothesis → Experiment → Data Collection → Data Analysis → Conclusions

Observation

The first step of the scientific method is **observation.** When you observe a phenomenon that you do not understand, you may make a guess as to the cause of that phenomenon. If you base your guess on previous knowledge, you are making an *educated* guess as to the cause for the phenomenon. In science we call this educated guess a **hypothesis.** For example, let's say you have observed that your body temperature seems to change during the day, and you are interested in knowing if your body temperature also changes during the course of the night while you are asleep. You have made an observation (body temperature changes during the day), and have followed up by asking a question: Does body temperature change during the night? The next step is for you to formulate a hypothesis.

Hypothesis

The second step of the scientific method is to **formulate a hypothesis.** One of the key features of a good hypothesis is that it must be testable. That is, you must be able to measure some aspect of the variable of interest. In our example, body temperature is the variable of interest, and can be measured using a thermometer. Another feature of a good hypothesis is that it must be specific, yet limited in scope. This is not to say that the explanation for a phenomenon is limited, nor that the questions you ask are limited. Instead, it means that you must limit your testable hypothesis to something that is measurable while all other conditions are controlled. An example of a simple, testable hypothesis is the following: Body temperature changes over time during the night. Once you have formulated a hypothesis, the next step is for you to design an experiment to test your hypothesis.

Experiment

One of the most creative and interesting aspects of the scientific method is to **design an experiment** to test a hypothesis. Designing a good experiment to test a hypothesis is a challenging task. The key feature of good experimental design is that you must try to predict any variables that may have an influence on the variables of interest and control for them. In our example, the variables of interest are body temperature and time during the night. Because we are interested in knowing if body temperature changes during the night, we need to design an experiment to measure body temperature at given time intervals during the night.

Variables

A **variable** is a characteristic that may or may not influence the outcome of an experiment. In our experiment we have identified two variables of interest: body temperature and time. Based on convention, we need to categorize the variables in our experiment as independent, dependent, or confounding variables.

The **independent variable** is a variable that is set at the outset of the experiment. It does not change as a result of the experimental procedure. Thus, it is said to be *independent* of the experimental procedure. In our example, time is the independent variable. Note that it is impossible for time to change as a result of any experimental procedure that we may perform. The **dependent variable** is the unknown variable that we are going to measure. In formulating our hypothesis, we often expect that the dependent variable will change as a result of the experimental procedure. Thus, it is said to be *dependent* on the experimental procedure. In our working example, body temperature is the dependent variable. **Confounding variables** are any variables that we think may affect our variable of interest. In our working example, we are only interested in the effect of time on body temperature. Thus, we must control all other variables that may affect our measured variable. Examples of some confounding variables are the amount and type of clothing the subject is wearing, the type of bedding the subject uses, how much and what the subject eats or drinks before going to sleep, and so on. In setting up an experiment, we generally try to do whatever we can to control as many confounding variables as possible. Finally, we also need a **control value** with which to compare our measured values of body temperature. In this example, we would likely use "normal" body temperature of 37 degrees Celsius as our control value.

Data Collection

Once we have designed a controlled experiment to test our hypothesis, we must then **collect the data.** In a true scientific experiment, before we collected the data we would perform a statistical test to determine the sample size necessary to get a meaningful result from our experiment. Why? If we measured body temperature for only one subject, it would be neither reasonable nor appropriate to extrapolate that data to include all individuals, because the person we measured may not be typical of most individuals. In the experiments you perform in your laboratory class, the study subjects will likely include most of your classmates. Thus, your data set will be limited in scope compared to the ideal situation. However, in most cases you will have collected enough data to obtain reasonable results.

Once the sample size has been determined, the next step is to collect the data. In this example, we would likely measure body temperature for each of our study subjects at specific time intervals during the night while also controlling for any confounding variables by having all study subjects, for example, wear the same clothing, use the same bedding, refrain from eating within a certain number of hours before bed, go to bed at a prescribed time. After data collection comes the fun part: data analysis!

Data Analysis

Once we have collected our data, we must find a way to **analyze the data** so that it makes sense both to us and to the rest of the scientific community. There are several ways to present data, and presentation depends somewhat on the variable in question. For any given data point, we first must calculate a mean, median, and standard deviation for that value, such that we get a value that represents the data for the group of study subjects as a whole. The **mean** is the average of all the data points, and is calculated by taking the sum of all the data points and dividing by the number of study subjects. The **median** is the middle value of all the data points. The **standard deviation** is a measure of the variability of the individual data points as compared to the mean. When the standard deviation is small, it means that individual data points are all very close to the mean. When the standard deviation is large, it means there is much variability between individual data points and the mean.

For the purposes of the exercises in this manual, you will not be reporting standard deviations for your experiments because this generally involves calculations that require the use of a computer program. An easier way of determining variability is to calculate and report the *range* of values. The **range** is simply the difference between the highest and lowest values, and is calculated by subtracting the lowest value from the highest value. If the range is small, then we know that there is very little variability of individual data points as compared to the mean. If the range is large, then we know there was quite a bit of variability of individual data points as compared to the mean. **Table 1.1** is a table showing hypothetical temperature data for five study subjects taken at two different times (12 a.m. and 6 a.m.). The mean, standard deviation, and range of the data are also shown in the table. Notice that even though the standard deviation and range are different for each time, they have the same pattern. That is, for the data taken at 6 a.m. both the standard deviation and the range are higher than for the data taken at 12 a.m. Thus, both of these measures tell us there was greater variability in the individual temperature values at 6 a.m. than there was at 12 a.m.

Conclusions

The final step in the scientific method involves **drawing conclusions** based on the results of your experiment. This requires reviewing your hypothesis in light of the data you have collected. In our example, we hypothesized that body temperature would change over the course of the night. **Figure 1.1** is a sample graph that is based on data from actual studies that looked at the variation in body temperature of a large number of study subjects during a daily cycle. Thus, we can use it as an estimate of data we might have obtained had we performed

Table 1.1	Body Temperature Data for Five Study Subjects	
Study Subject	**Body Temperature at 12 a.m. (°C)**	**Body Temperature at 6 a.m. (°C)**
Subject A	36.2	35.0
Subject B	36.8	34.0
Subject C	37.2	34.3
Subject D	37.0	35.9
Subject E	36.5	36.0
Mean	36.7	35.0
Standard Deviation	0.4	0.9
Range	1.0	2.0

our hypothetical experiment. Based on the results, we can draw the conclusion that body temperature does change as a result of time of the night. We can also go on to be more specific and say *how* body temperature changes during the night. That is, body temperature appears to decrease steadily throughout the night and reaches its lowest point just before waking (5 a.m.).

Often a hypothesis and related experiment will result in a few answers, but even more questions. For example, now that we know that body temperature changes during the night, we may ask *why* body temperature changes during the night. Using the scientific method, we can continue the scientific process by stating a new hypothesis. Our new hypothesis will lead us to pursue further experimentation, data collection, and analysis. As the process continues we will learn more and more about our area of interest.

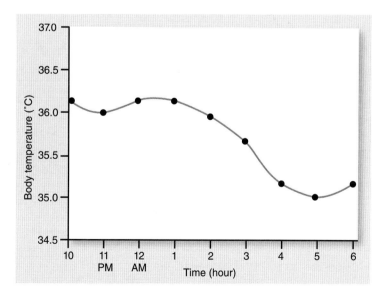

Figure 1.1 **Temperature Variance During the Course of the Night.** Each data point represents the mean for a number of study subjects.

EXERCISE 1.1

THE SCIENTIFIC METHOD

Let's say that you have observed that heart rate varies with activity level. That is, when an individual exercises heart rate increases, and when an individual is at rest heart rate decreases. Now, let's say that because you have observed that heart rate decreases when an individual is in a "rested" state, you are now interested in knowing if meditation, an activity that is designed to put the body in a calm/rested state, causes an individual's heart rate to decrease. You decide to perform a scientific experiment to figure out if this is accurate.

1. State your hypothesis: _____
   _____
   _____
   _____
   _____
   _____
   _____
   _____

2. What is the independent variable? _____

3. What is the dependent variable? _____

4. Design an experiment to test your hypothesis. In the explanation of your experimental design, describe what variable(s) you will control and what variable(s) you will measure.
   _____
   _____
   _____
   _____
   _____
   _____
   _____
   _____
   _____
   _____
   _____

5. Finally, describe any confounding variables that may affect the results of the experiment and explain how you might control for such variables.
   _____
   _____
   _____
   _____
   _____
   _____
   _____
   _____

EXERCISE 1.2

PRESENTING DATA

Once we have collected experimental data, we must present that data in a meaningful way so that others who view the data will be able to draw their own conclusions about the data. In the previous description of the scientific method, data was presented using a data table (table 1.1) and a graph (figure 1.1). These are two of the most common modes of data presentation, and are the modes you will most often use to present data for the experiments you perform in this manual. When presenting data using a **table,** it is common to include mean, number of study subjects, and standard deviation (or range) for each data point (see table 1.1 for an example). When presenting data using a **graph,** there are conventions to follow. Namely, the *independent* variable is plotted on the X-axis (horizontal) and the *dependent* variable is plotted on the Y-axis (vertical). By graphing data this way, you will be able to determine what effect, if any, variable X has on variable Y. For our sample experiment looking at the effect of time of the night on body temperature, the data have been plotted in the graph shown in figure 1.1.

1. Why was time plotted on the X-axis? _____

_____

2. Why was temperature plotted on the Y-axis? _____

_____

Other items to note when creating a graph are the following:

1. Each axis must be labeled with the appropriate numbers indicating the value of the measurement (in the graph in figure 1.1 these are the measured values for "hour" and "degrees Celsius").

2. Each axis must have a legend and provide the units that are used. In the graph in figure 1.1, the X-axis is labeled "Time" and the unit is "hour"; the Y-axis is labeled "Temperature" and the units are "degrees Celsius."

It is very important that you always take care to place labels on the graphs that you create. A common mistake many students make is to assume that their instructor—or whoever is grading their laboratory report—already knows what is supposed to be graphed, therefore the student does not need to put the units on the graph. Do not make this mistake! Always assume that the person who is going to read your graph has no idea what you are trying to present. If you fail to put units on your graph, the graph means nothing. For example, let's say you put "Time" on the X-axis, but did not provide units. The reader of your graph is left to wonder if your intervals indicate time in seconds, minutes, hours, or some other unit. Likewise, if you fail to indicate the units for "Temperature" on the Y-axis, the reader is left to wonder if the temperature is given in Celsius (°C) or Fahrenheit (°F).

Sample Graphing Exercise

1. Table 1.2 represents experimental data obtained by rolling a bowling ball on a track and measuring how far the ball rolls over time.

Table 1.2	Distance Traveled by a Bowling Ball
Time (sec)	Distance Traveled (meters)
0	0
1	6
2	8
3	9.5
4	10.5
5	11
6	11.5
7	11.75
8	11.8
9	11.9
10	12

Questions:

a. What is the independent variable? _____

_____

b. What is the dependent variable? _____

_____

2. Using the data in table 1.2, graph the data on the grid below. Be sure to label your axes and provide the appropriate units.

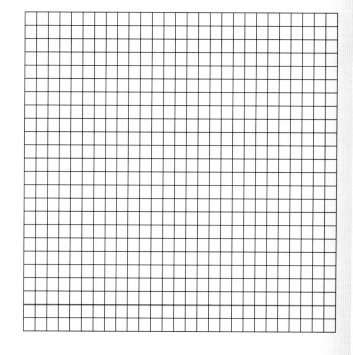

(continued on next page)

3. Now that you have graphed the data, write a brief paragraph in the space below explaining your interpretation of the data.

Measurement in Science

Systems of Measurement

If you have ever baked a cake or done any other activity that required making measurements, you are familiar with common units of measurement such as cups, gallons, teaspoons, liters, and the like. There are two systems of measurement that are most commonly used in the world: English and metric systems. Because of its uniformity and ease of use, the **metric system** is the system most widely used throughout the world. It is also the system of measurement that scientists use when reporting data. Thus, when you perform the laboratory activities in this manual, you will be reporting your results using metric units.

The metric system is very simple to use because all units are given as powers (multiples) of ten. To calculate a power of ten you simply take the superscript (i.e., 10^x, where x is the superscript) and multiply ten by ten that many times. Thus, $10^2 = 10 \times 10 = 100$; $10^3 = 10 \times 10 \times 10 = 1000$. The only unusual numbers are 10^0 and 10^1. When you see these numbers, just remember that ten to the zero power is 1 and ten to the 1st power is ten. In the metric system, specific prefixes denote the power of ten that is used in a measurement. For example, a kilometer is 10^3 meters, and thus represents 1000 meters. **Table 1.3** lists common metric prefixes and the power of ten that each represents.

When you need to convert one metric unit to another, you will move the decimal point to the right or to the left, depending on if you are converting a larger unit to a smaller unit (e.g., kilograms to grams) or vice versa (e.g., grams to kilograms). When converting from a larger unit to a smaller unit, move the decimal to the *right* the same number of units as the power of ten. This is the same as *multiplying* the larger measurement by the appropriate factor of ten. For example, to convert kilometers to meters (1 km = 1000 meters), move the decimal point to the right *three* positions because a kilometer is ten to the *third* power (10^3). Conversely, when converting from a smaller unit to a larger unit, move the decimal to the *left* the same number of units as the power of ten. This is the same as *dividing* the smaller measurement by the appropriate factor of ten. For example, to convert meters to millimeters (1 mm = 1/1000 meter), move the decimal point to the left *three* positions because a millimeter is ten to the negative *third* power (10^{-3}). Another way to remember this is that a power of ten that is positive moves the decimal to the right (forward) and a power of ten that is negative moves the decimal to the left (backward).

small unit → larger unit   MULTIPLY by appropriate power of ten

larger unit → smaller unit   DIVIDE by appropriate power of ten

Practice:

Convert 457 milligrams to grams

Convert 5698 centimeters to decimeters

Convert 4.3 kilometers to meters

Convert 0.5 liter to milliliters

Table 1.3 Metrics

Metric Prefix	Meaning	Symbol	Amount	Power of Ten	Length Measure	Mass Measure	Volume Measure	Word Origin
kilo-	one thousand	k	1000	10^3	kilometer (km)	kilogram (kg)	kiloliter (kL)	*chlioi*, a thousand
No prefix/base unit	Use the standard unit	–	1	10^0	meter (m)	gram (g)	liter (L)	NA
deci-	one-tenth	d	0.1	10^{-1}	decimeter (dm)	decigram (dg)	deciliter (dL)	*decimus*, tenth
centi-	one-hundredth	c	0.01	10^{-2}	centimeter (cm)	centigram (cg)	centiliter (cL)	*centum*, hundred
milli-	one-thousandth	m	0.001	10^{-3}	millimeter (mm)	milligram (mg)	milliliter (mL)	*mille*, thousand
micro-	one-millionth	μg	0.000001	10^{-6}	micrometer (μm)	microgram (μg)	microliter (μL)	*mikros*, small

Metric Conversions

If you make any measurements using the English system, you will need to convert English units to metric units for your reports. **Table 1.4** lists some common conversions between English and metric measurements. Some of these conversions are conversions that you should be able to make in your head easily. For example, 1 inch = 2.54 centimeters and 1 pound = 0.45 kilogram.

For example, to convert 16 ounces (English measurement) into milliliters (metric measurement), you will multiply the number of ounces (16) by 30 (see table 1.4) to get milliliters:

$$16 \text{ ounces} \times 30 = 480 \text{ milliliters}$$

Notice that the conversion in table 1.4 tells you how to convert ounces to milliliters. What if you want to convert ounces to liters instead? Recall from table 1.3 that one milliliter is one one-thousandth of a liter, and its power of ten is 10^{-3}. Thus, to convert ounces to liters, first convert to milliliters using the multiplier in table 1.4 (below), then convert milliliters to liters by moving the decimal point three places to the left: 480.0 milliliters = 0.48 liter.

Temperature Scales

Just as there are two systems of measurement for volume, length, and mass, there are also two systems of measurement for temperature. In the United States most of our temperatures (such as air temperatures reported by weather stations) are reported as degrees Fahrenheit (°F). However, often temperatures will instead be reported as degrees Celsius (°C, also known as centigrade). In science, it is appropriate to report temperature readings (such as body temperature) in degrees Celsius. To make conversions between these to temperature readings, use the following equations:

To convert degrees Fahrenheit to degrees Celsius:

$$°F = ((°C \times 9)/5) + 32$$

To convert degrees Celsius to degrees Fahrenheit

$$°C = ((°F - 32) \times 5)/9$$

Table 1.4 Common English-Metric Conversions

To Convert From:	To:	Multiply:	By:	Conversion
centimeters	inches	centimeters	0.39	1 cm = 0.39 inch
feet	centimeters	feet	30.48	1 foot = 30.48 cm
feet	meters	feet	0.3	1 foot = 0.3 m
fluid ounces	milliliters	fluid ounces	30	1 oz = 30 mL
gallons	liters	gallons	3.78	1 gallon = 3.78 L
grams	ounces	grams	0.035	1 g = 0.035 oz
inches	centimeters	inches	2.54	1 inch = 2.54 cm
kilograms	pounds	kilograms	2.2	1 kg = 2.2 lb
kilometers	miles	kilometers	0.62	1 km = 0.62 mile
liters	quarts	liters	1.06	1 L = 1.06 qt
liters	gallons	liters	0.26	1 L = 0.26 gallon
meters	yards	meters	0.9	1 m = 1.1 yd
meters	feet	meters	3.3	1 m = 3.3 feet
miles	kilometers	miles	1.61	1 mi = 1.61 km
milliliters	fluid ounces	milliliters	0.03	1 mL = 0.03 oz
millimeters	inches	millimeters	0.039	1 mm = 0.039 inch
fluid ounces	grams	fluid ounces	28.3	1 oz = 28.3 g
fluid ounces	milliliters	fluid ounces	29.6	1 oz = 29.6 mL
pounds	grams	pounds	453.6	1 lb = 453.6 g
pounds	kilograms	pounds	0.45	1 lb = 0.45 kg
quarts	liters	quarts	0.95	1 qt = 0.95 L
yards	meters	yards	0.9	1 yd = 0.9 m

EXERCISE 1.3

UNITS OF MEASUREMENT

1. Using tables 1.3 and 1.4, and the temperature conversion equations given in this chapter as references, make each of the following conversions:

   a. 5 millimeters = _____ centimeters

   b. 21 grams = _____ ounces

   c. 500 mL = _____ liters

   d. 5 inches = _____ centimeters

   e. 3 meters = _____ feet

   f. 3 gallons = _____ liters

   g. 3 liters = _____ mL

2. The set point for human body temperature is 98.6 °F. This means that 98.6 °F is the temperature set point that homeostatic mechanisms in the body strive to maintain. It is appropriate in science to report temperature values in degrees Celsius. Convert 98.6 °F to degrees Celsius in the space to the right. Be sure to show all of your work.

Laboratory Equipment

The typical human anatomy and physiology laboratory classroom consists of laboratory tables or benches that provide ample room for use of microscopes, laboratory equipment, and dissection materials. If your classroom uses human cadavers, it will also have a space dedicated to the tables where the cadavers are stored. When you enter the classroom, look around and familiarize yourself with your environment. Pay particular attention to the location of the sinks, eyewash stations, and safety equipment such as first-aid kits and fire extinguishers. Your instructor will give you a detailed introduction specific to your laboratory classroom, safety procedures, and accepted protocol. The main purpose of this chapter is to introduce you to common safety devices and dissection equipment, and to help you identify some of the items your laboratory instructor may be introducing to you. *Do not* use the information in this chapter as your sole source of information on laboratory safety, as it is not intended to be a safety manual for the laboratory.

Protective Equipment

The human anatomy and physiology laboratory poses few risks, although it is important to be aware of what these are. The main risks are damage to skin or eyes from exposure to laboratory chemicals (covered in the next section) and cuts from dissection tools. As a general precaution, whenever you are working with fresh or preserved specimens (animal or human), wear protective gloves to keep any potentially infectious or caustic agents from contacting your skin.

If there is a risk of squirting fluid, then wear protective eyewear (safety glasses or safety goggles). When wearing gloves, be sure to wear the correct size for your hands. If the gloves are too small, they may tear easily. If they are too big, you may not be able to handle instruments and tissues easily. When your gloves become excessively dirty, remove them and put on a new pair. When removing gloves, start at the wrist and pull toward your fingers, turning the glove inside-out as it is removed. This will prevent any potentially damaging fluids from contacting your skin during removal of the gloves.

When using dissecting tools there is always a risk of cutting yourself or others. First and foremost—*never* wear open-toed shoes to the laboratory. Dissecting tools are sometimes dropped and will cut your feet if you are not wearing protective footwear. When using sharp tools such as scalpels, always be aware of where you are pointing the blades of those instruments. They should always be pointed away from you *and* away from others in your laboratory. If you are dissecting, be aware of where others are standing or sitting, and consider the risk posed to yourself and others if your hand were to slip. *Never* put your hands in the dissecting field when someone else is dissecting. If another person asks for assistance holding tissues while they are dissecting, use forceps or some other device to hold the tissue so your hands are not within reach of the scalpel blade. Always be aware of the location of your scalpel, particularly when you are not using it. Individuals can be cut accidentally when they reach into a dissecting tray or table and unexpectedly find a scalpel sitting there. Whenever you are finished using dissecting pins, remove them from the specimen so they will not poke unsuspecting individuals.

EXERCISE 1.4

IDENTIFICATION OF COMMON DISSECTION INSTRUMENTS

Several dissection instruments are commonly found in the human anatomy and physiology laboratory classroom. **Table 1.5** describes each of these instruments and their uses.

1. Obtain a dissection kit from your laboratory instructor, or use your own dissection kit if you were required to purchase your own materials.

2. Using table 1.5 as a guide, identify the instruments listed in **figure 1.2**. Then label figure 1.2.

Table 1.5 Common Dissection Instruments

Tool	Description and Use	Photo	Word Origin
Blunt probe	An instrument with a blunt (not sharp) end on it. It is used to pry and poke at tissues when you do not want to damage them. Some probes come with a sharper point on the opposite end that can be used for "picking" at tissues.		*proba*, examination
Dissecting needles	Long, thick needles that have a handle made of wood, plastic, or metal. These needles are used to pick at tissues and to pry small pieces of tissue apart.		*dissectus*, to cut up
Dissecting pins	"T" shaped pins that are used to pin tissues to a dissecting tray, thus allowing you to see a particular area more easily		*dissectus*, to cut up
Dissecting tray	Metal or plastic tray used to hold a specimen. The tray is filled with wax or plastic. The wax and/or plastic is soft enough to allow you to pin tissues to it.		*dissectus*, to cut up
Forceps	Resemble tweezers, and are used for holding objects. Some are large and have tongs on the ends that assist with grabbing tough tissues. Some are small and fine (needle-nose) for picking up small objects. Forceps may also be straight-tipped or curve-tipped.		*formus*, form + *ceps*, taker
Hemostat	In surgery these are used to compress blood vessels and stop bleeding (hence the name). For dissection they are useful as "grabbing" tools. The handle locks in place, which allows you to pull on tissues without fatiguing your hand and forearm muscles.		*haimo-*, blood + *statikos*, causing to stop

(continued on next page)

(continued from previous page)

Table 1.5	Common Dissection Instruments *(continued)*		
Scalpel	A sharp cutting tool. Generally the blade and the blade handle will be separate, unless you are using a disposable scalpel. See specific directions in the text regarding proper use of a scalpel, as they can be dangerous!	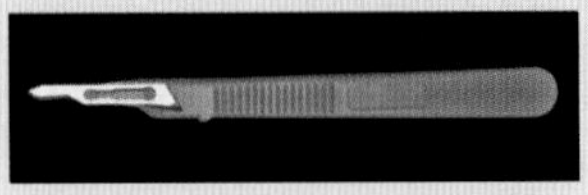	
Scalpel blade	Both the cutting part and the disposable part of a scalpel. The number of the blade indicates the size of the blade, and must be matched with an appropriately numbered blade handle. When the blades become dull, they may be removed and replaced with a new blade. Used blades must be disposed of in a sharps container.	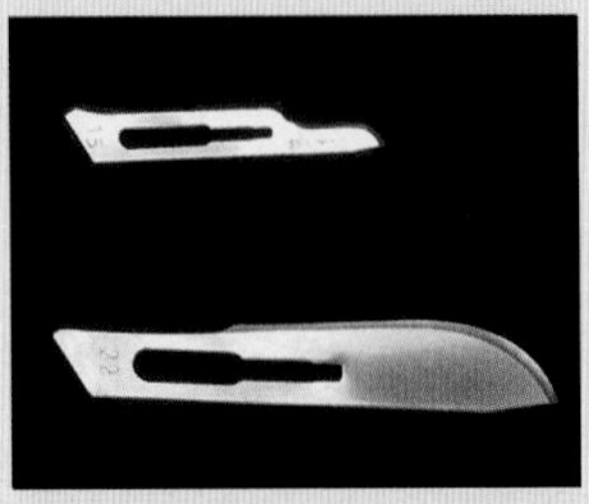	*scalpere,* to scratch
Scalpel blade handle	The nondisposable part of a scalpel that is used to hold the blade. The number on the handle indicates the size of the handle and is used to match it with a particular blade size. A scalpel blade handle can be a very useful tool for blunt dissection when used *without* a blade attached.	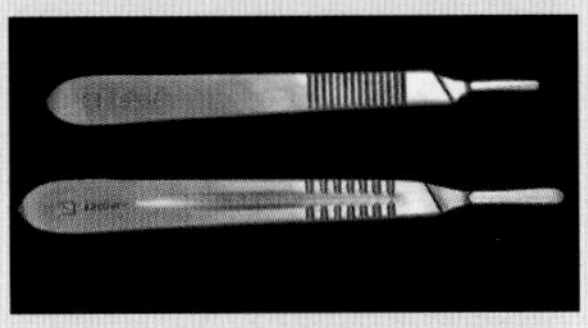	*scalpere,* to scratch
Scissors	Some scissors come with pointed blades and some have one curved (blunt) and one pointed blade. Scissors with the curved/blunt edge are used when you need to be careful not to damage some structures. To use them, direct the curved blade toward the structures you do not want to damage. Pointed-blade scissors are particularly helpful for using "open scissors" technique (see text).	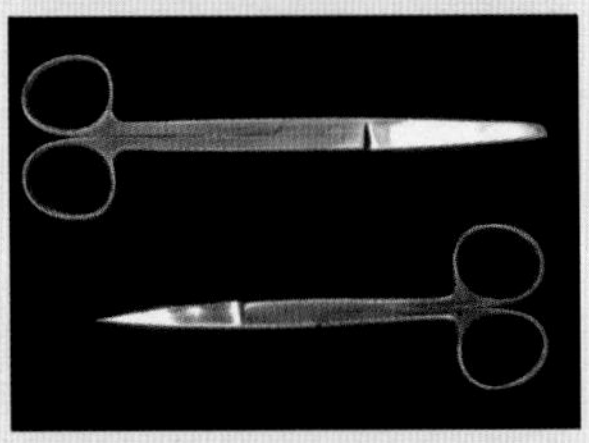	*scindere,* to cut

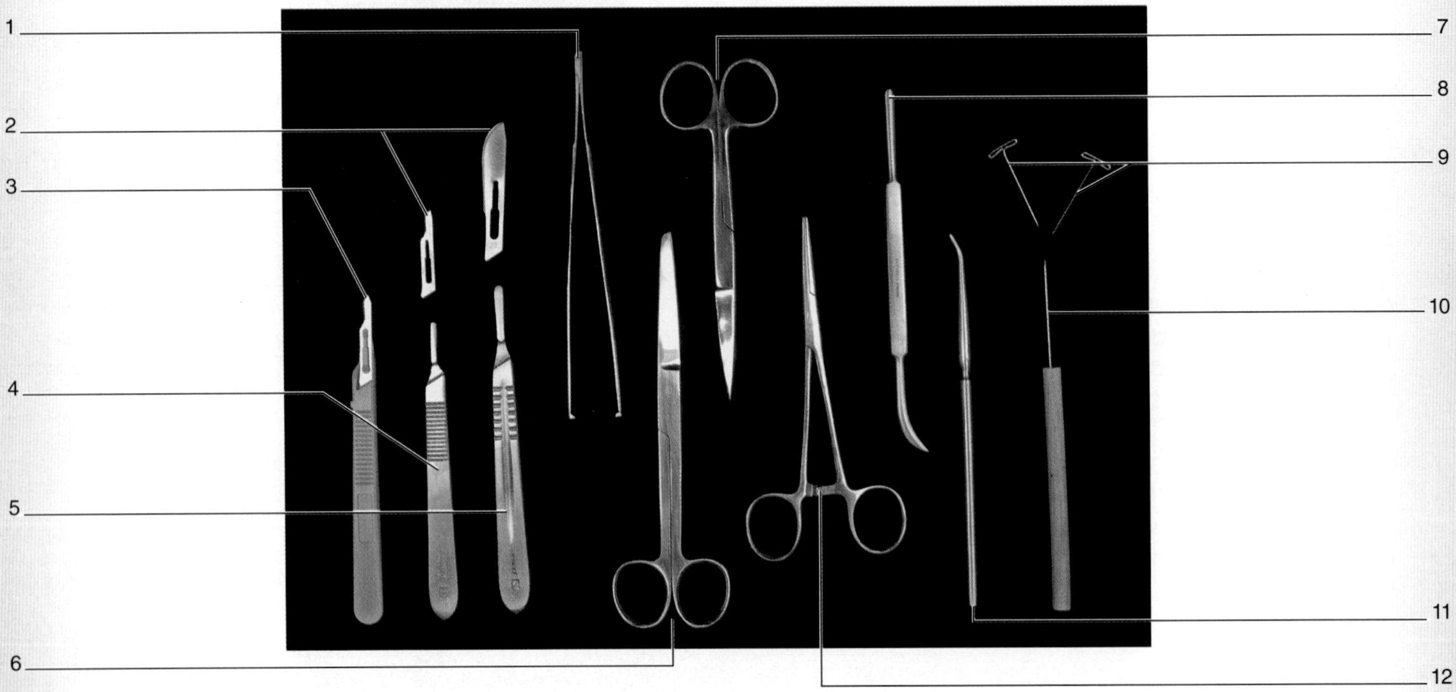

Figure 1.2 Identification of Common Dissection Instruments.

- ☐ blunt probe (used twice)
- ☐ dissecting needle
- ☐ dissecting pins
- ☐ forceps
- ☐ hemostat
- ☐ scalpel (disposable)
- ☐ scalpel blade handle (#3)
- ☐ scalpel blade handle (#4)
- ☐ scalpel blades
- ☐ scissors (curved)
- ☐ scissors (pointed)

Hazardous Chemicals

You will be using various chemicals from time to time in the human anatomy and physiology laboratory. The discussion in this chapter will be limited to chemicals that are used to preserve, or "embalm," animal specimens or human cadavers. Safety precautions for chemicals used in physiology "wet lab" experiments will be covered when these chemicals are encountered within specific laboratory exercises. As concerns embalming chemicals, generally you will not find any of them in their full-strength form in the laboratory. Instead, you will find tissues that have been previously injected with solutions containing these chemicals. Thus, safety measures in the laboratory are designed to protect you from the forms of these chemicals that you are most likely to encounter. The most common chemicals used for embalming purposes are formalin, ethanol, phenol, and glycerol.

Table 1.6 summarizes the uses and hazards of these chemicals. The majority of these chemicals are used to fix tissues and prevent the growth of harmful microorganisms, such as bacteria, viruses, and fungi. **Fixation** refers to the ability of the chemical to solidify proteins, thus preventing their breakdown. **Preservatives** both fix tissues and inhibit the growth of harmful microorganisms. Because most preservatives also dehydrate tissues, **humectants** are added to embalming solutions. Humectants, such as glycerol, attract water. When humectants act alongside preservatives, they help keep tissues moist. Other chemicals that may be added to embalming solutions are pigments, which make the tissues look more natural, or chemicals that mask the odors of the preservative chemicals. **Formalin** and **phenol** are the most toxic and odorous preservative chemicals. Luckily, your exposure to them in the anatomy laboratory will be very low. Although it may smell as if the concentrations of these chemicals are high, the odor is often misleading because these chemicals can be detected by odor in extremely small quantities. Although the concentrations of formalin and phenol that you are exposed to may be very low, if these chemicals have been used to preserve any specimens you are working with, then you need to wear protective clothing to prevent the chemicals from contacting your skin or eyes. Use gloves whenever you handle the specimens, and use protective eyewear whenever there is a risk of chemicals getting into your eyes. If your skin is exposed, rinse it immediately. If your eyes are exposed, use the eyewash station in your laboratory to rinse your eyes thoroughly. If you have experienced contact exposure to these chemicals and your skin or eyes continue to be irritated after rinsing, consult a medical doctor.

Table 1.6 Preservative Chemicals Encountered in the Human Anatomy & Physiology Laboratory

Chemical	Description	Use	Hazard	Preventing Exposure	Disposal
Ethanol	Inhibits growth of bacteria and fungi	Preservative	Flammable, so requires storage in a fire-safe cabinet. Generally safe in small quantities.	Gloves and eye protection. Rinse tissues immediately if exposed, particularly eyes. Seek medical attention if irritation persists.	Small amounts may be flushed down the sink along with plenty of water to dilute the solution
Formalin	Fixes tissues by causing proteins to cross-link (solidify). Destroys autolytic enzymes, which initiate tissue decomposition. Inhibits growth of bacteria, yeast, and mold.	Preservative	Flammable, so requires storage in a fire-safe cabinet. Toxic at full strength. Penetrates skin. Corrosive. Burns skin. Damages lungs if inhaled. May be carcinogenic.	Gloves and eye protection. Rinse tissues immediately if exposed, particularly eyes. Seek medical attention if irritation persists.	Do not pour into sinks
Glycerine (glycerol)	Helps control moisture balance in tissues. When used with formalin, it counteracts the dehydrating effects of formalin.	Humectant	Flammable, so requires storage in a fire-safe cabinet. Generally safe. Can pose a slipping hazard if spilled on the floor.	Gloves and eye protection. Rinse tissues immediately if exposed, particularly eyes. Seek medical attention if irritation persists.	Small amounts may be flushed down the sink along with plenty of water to dilute the solution
Phenol	Assists formalin in fixing tissues through protein solidification. Inhibits growth of bacteria, yeast, and mold.	Preservative	Flammable, so requires storage in a fire-safe cabinet. Extremely toxic at full strength. Rapidly penetrates the skin. Corrosive. Burns skin. Damages lungs if inhaled. NOTE: When used as embalming preservative, concentration (and thus toxicity) is extremely low.	Gloves and eye protection. Rinse tissues immediately if exposed. Use an eye wash station if solution gets in the eyes. Seek medical attention if irritation persists.	Do not pour into sinks

Proper Disposal of Laboratory Waste

There are several types of waste that must be disposed of in the human anatomy laboratory. Much of this waste is "normal" waste, such as tissues, paper towels, rubber gloves. Such waste should be disposed of in the regular garbage/waste container found in the classroom. However, it is likely that at times you will find yourself needing to dispose of potentially **hazardous waste,** which must be disposed of in a special container. The general rule for determining if something is potentially hazardous or not is this: If you think someone else may be injured *in any way* from handling this waste, it is hazardous. Follow this rule, and be sure to ask your instructor how to properly dispose of something any time you have a question as to whether it is hazardous or not. It is always better to err on the side of caution.

What is hazardous waste?

1. Any sort of fresh tissue and/or blood is potentially hazardous.
2. Laboratory chemicals
3. Broken glass, scalpel blades, or any other sharp item that may cut an individual who handles the waste

Sharps Containers

Sharps containers (figure 1.3) are plastic containers (often red or orange) that are used to dispose of anything "sharp," such as needles, scalpel blades, broken glass, pins, or anything else that has the potential to cut or puncture a person who handles it. Such items should NEVER go in the garbage, because they may injure anyone who handles the garbage thereafter. When in doubt, put it in the sharps container.

Biohazard Bags

Special **biohazard bags** may be available in your laboratory. These are used for biological materials such as blood or other fresh animal tissue that requires special disposal. When it comes to human blood, an item containing a small amount of blood (such as a band-aid) can be disposed of in a normal wastebasket. However, if you have a towel soaked with blood, then that must be disposed of in a biohazard bag. A biohazard bag is usually red or clear and has the symbol shown

Figure 1.3 Sharps Containers. Samples of two different models of sharps containers. Such containers allow you to place sharp objects into the container, but they cannot be removed once placed inside. Note the biohazard warning symbol on the containers.

Figure 1.4 Biohazard Waste Symbol.

in **figure 1.4** on it. When you are dealing with tissues that must be disposed of in a biohazard bag, your instructor will generally inform you of this. Again, when in doubt, always ask before disposing of something potentially hazardous. Important note: Human cadaveric tissues do *not* go into biohazard bags. They must be kept with the cadaver. Any piece of human tissue removed from a cadaver must eventually be returned to the cadaver to be cremated with the entire body.

Dissection Techniques

The word *dissect* literally means to cut something up. Most of us have been led to think that the first thing a surgeon or anatomist does when planning to dissect is to pick up a scalpel and cut. However, skilled dissection does not always involve actually cutting tissues. In fact, the dissector's best friend is a technique called "blunt dissection." Blunt dissection specifically involves separation of tissues without using sharp instruments (hence the term *blunt*). When dissecting tissues, always try using blunt dissection before picking up sharp instruments such as scissors and scalpels. Sharp instruments are very handy—as they are good at cutting things. However, often students will end up cutting many things they do not wish to cut, purely by accident. Thus, being sparse and prudent in your use of sharp tools is one of the most important things you can do to perform a good dissection.

For this exercise, the demonstration of techniques will be shown using a fresh chicken purchased from a grocery store. However, your instructor may choose another specimen for you to practice on. For now, the goal of the dissection is to separate the skin from the underlying tissues such as bones and muscle (the "meat") of the specimen.

Sharp Dissection Techniques

Sharp dissection techniques are the techniques most familiar to most of us. These techniques involve the use of sharp instruments such as scissors and scalpels. They are "cutting" techniques. They are advantageous in that you can use them to separate tough tissues from each other, or to remove pieces of tissue from a dissection specimen. The danger in using sharp techniques is that novice and experienced dissectors alike will often end up cutting things they do not wish to cut, such as blood vessels and nerves. Thus, sharp dissection techniques should be used with care.

EXERCISE 1.5

PROPER DISPOSAL OF LABORATORY WASTE

Circle the letter (a, b, or c) of the correct waste receptacle (shown in **figure 1.5**) for each item listed below.

1. broken scissors   A   B   C
2. cotton swab   A   B   C
3. dissecting pins   A   B   C
4. glass slide   A   B   C
5. paper towel   A   B   C
6. rubber glove   A   B   C
7. scalpel blades   A   B   C
8. fresh tissue   A   B   C
9. hypodermic needle   A   B   C

(a)

(b)

(c)

Figure 1.5 Common Waste Receptacles in the Laboratory. (a) Sharps container (b) Wastebasket (c) Hazardous waste bag

EXERCISE 1.6

PLACING A SCALPEL BLADE ON A SCALPEL BLADE HANDLE

Scalpels come in many forms. Some are of the disposable type, which typically means that the handle and blade come as one unit and the handle is made out of plastic **(figure 1.6)**. Often the blades and handles are separate items. Such items allow you to replace the blade whenever it becomes dull from use. In this exercise you will learn how to properly place a scalpel blade on a scalpel blade handle, and how to properly remove it once finished.

1. Obtain a **scalpel blade** and **scalpel blade handle** from your instructor. Scalpel blades and handles come in various sizes, and it is important to match the size of the blade to the size of the blade handle. Observe the scalpel handle and look for a number stamped on it, which will be a 3 or a 4 **(figure 1.7a)**. Next, observe the blade packet and note the number on it (figure 1.7b). A number 3 handle is used to fit number 10, 10A, 11, 12, 12D, or 15 blades. A number 4 handle is used to fit number 18, 20, 21, 22, 23, 24, 24D, or 25 blades. Larger handles and blades are generally used for making bigger, deeper cuts, whereas the smaller handles and blades are generally used for finer dissection. One of the most commonly used combinations in anatomy laboratories is a number 4 handle matched with a number 22 blade.

2. Once you have paired the scalpel handle and blade size, carefully open the scalpel blade packet halfway **(figure 1.8a-1)**. Note the bevel on the blade. This bevel matches the bevel on the blade handle, so that there is only one way to properly place the scalpel blade on the handle. The blade handle has a bayonet fitting that is matched to the opening on the scalpel blade (figure 1.8a-2), which will lock the blade in place on the handle. The safest way to place the blade on the handle is to first grasp the end of the blade using **hemostats** (figure 1.8a-2; table 1.5). Then, while matching the bevel on the blade to the bevel on the handle, slide the blade onto the handle until it clicks, indicating it is locked in place (figure 1.8a-3, a-4). If it does not go on easily, check to make sure you haven't placed it on incorrectly (example: figure 1.8b). Now it is ready to use!

3. The safest way to remove a blade from a handle is to use a device that is both a **blade remover** and a sharps container all in one (an example is shown in **figure 1.9**).

4. If a blade remover is not available, you will remove the blade using hemostats. Obtain a pair of hemostats. Pointing the blade away from you (but not toward someone else), clamp the part of the blade nearest the handle with the hemostats **(figure 1.10-1)**. Once you have a firm grip on the blade, slide it over the bayonet on the handle and away from you until the blade comes off of the handle (figure 1.10-2). Using the hemostats, transport the blade to a sharps container and dispose of it in the sharps container (figure 1.10-3).

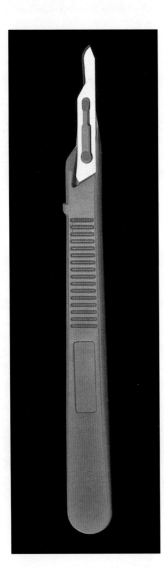

Figure 1.6 **Disposable Scalpel.** The scalpel blade and scalpel blade handle are both disposable. The entire unit must be disposed of in a sharps container.

Chapter One *The Laboratory Environment* 17

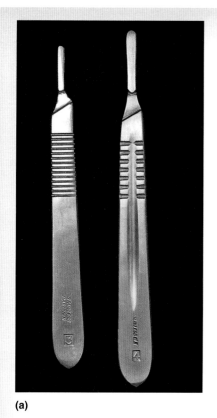
(a)

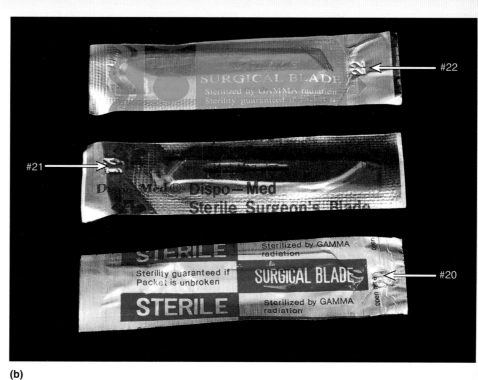
(b)

Figure 1.7 Scalpel Blade Handles and Blades. (a) The number on the scalpel blade handle indicates what size blades will fit on the handle. (b) The number on the blade wrapper indicates the size of the blade. See text for description of what size blades fit on what size blade handles.

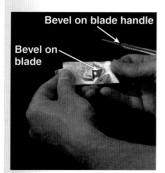
① Open the foil packet and note the bevel on the blade.

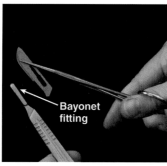
② Grasp the blade firmly using hemostat and line the blade up so that it matches the bevel on the blade handle.

③ Slide the blade onto the bayonet of the blade handle.

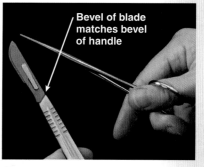
④ The blade should "click" as it locks in place on the blade handle.

(a)

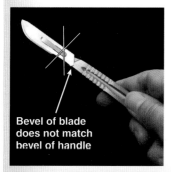
(b)

Figure 1.8 Scalpel Blade Placement. (a) Correct procedure. (b) Incorrect placement of a blade on a blade handle. Notice that the bevel on the blade does not match up with the bevel on the blade handle. If placed in this fashion, the blade will not be secure on the handle and may slip off the handle and injure someone.

(continued on next page)

(continued from previous page)

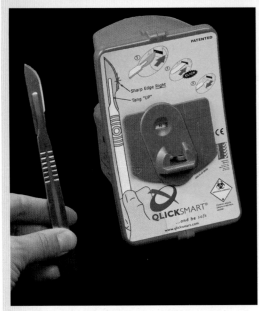
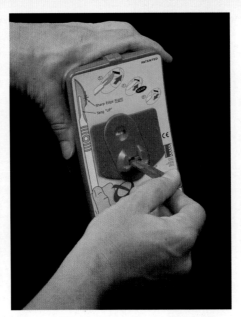
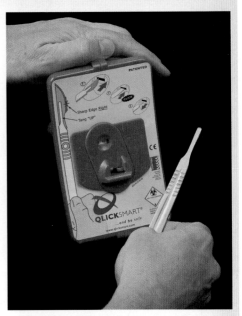

(1) Orient the blade and blade handle with sharp edge of the blade pointed to the right, as shown on the front of the device.

(2) Push the blade into the slot on the device until you hear and feel a distinct "click."

(3) While holding the removal device firmly with your free hand, pull the blade handle out of the device.

Figure 1.9 Removal of a Scalpel Blade from Handle Using All-in-One Blade Remover/Sharps Container.

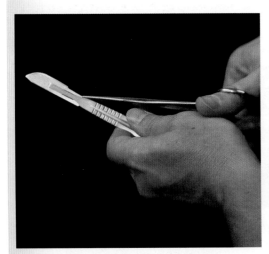

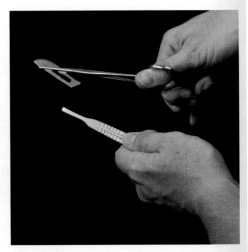

(1) With the blade pointed away from you and the bayonet surface of the handle also directed away from you, grasp the base of the blade with hemostats and lock the hemostats firmly to the blade.

(2) Slide the blade off of the bayonet on the blade handle. Again, push it away from you (and away from others in your vicinity as well).

(3) Once the blade has been removed from the handle, continue to grasp it firmly with the hemostats.

Figure 1.10 Removal of a Scalpel Blade from Handle Using Hemostats.

EXERCISE 1.7

DISSECTING WITH A SCALPEL

1. Obtain a dissection specimen from your instructor and place it on a dissecting tray.

2. Obtain a scalpel with a blade (see exercise 1.4) and some **tissue forceps** (see table 1.5). If the tissue is difficult to hold on to, you may want to use **hemostats** instead of forceps. Hemostats allow you to "lock" on to the tissue so the tissue is not dropped when you release your grip on the handle.

3. Using the forceps or hemostats, pull the skin away from the muscle on your dissection specimen **(figure 1.11-1)**. Carefully cut into the skin, using the tip of the scalpel blade (figure 1.11-2). Note how easily a new blade cuts into the tissue. When cutting with a scalpel, you want to take care not to cut too deep, or too aggressively, or you will damage tissues. Once you have cut a small slit in the skin, observe the stringy tissue that lies between the skin and the muscle. This tissue is a loose connective tissue called fascia (figure 1.11-3), which you will learn more about in chapter 5. Because our goal is to separate the skin from the muscle, this means we want to loosen the "grip" of the fascia that holds the skin and muscle together. One way to do this is to cut into the fascia using the scalpel.

4. Next, *without* holding the skin away from the muscle with forceps, just cut into the skin using a considerable amount of pressure. Note how easily you cut through the skin directly into the muscle. This is not desirable. You may ask yourself, "If I can't manage to use forceps to easily separate skin from muscle before cutting with the scalpel, how can I avoid damaging the underlying tissues?" One technique is to push a blunt probe or scalpel handle (*without* blade attached!) into the space between the skin and muscle, thus protecting the underlying tissues. Then cut with the scalpel superficial to the probe **(figure 1.12)**. This way the probe limits the depth at which the scalpel blade can cut, thus protecting the underlying tissues.

5. Once you have pulled back enough of the skin to grasp it easily with forceps or hemostats, you want to put as much tension on the skin as possible, thus stretching out the fibers in the fascia **(figure 1.13)**. Once the fascia is stretched, you can use the scalpel to cut the fascia and remove the skin from the specimen. When you cut with the scalpel, always point the sharp end of the blade toward the skin, not toward the underlying tissues, so as to protect those underlying tissues.

6. Practice using the forceps, hemostats, blunt probe, and scalpel to remove the skin from part of the specimen. Note areas where this is more difficult than others. As you are practicing, consider carefully whether or not the scalpel is the best instrument for the job, or if using it is causing you to damage too many tissues.

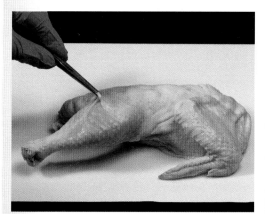

(1) Pull the skin away from the underlying tissues using tissue forceps.

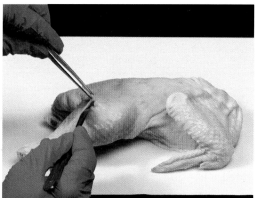

(2) Begin cutting the skin with the scalpel, taking care not to cut delicate tissues deep to the skin.

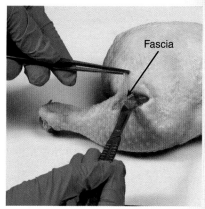

(3) To assist with removal of the skin, use the scalpel to gently cut away the fascia that loosely holds the skin to the muscle. Maintain as much tension on the skin as possible and always keep the sharp end of the blade pointed toward the skin, not the underlying tissues.

Figure 1.11 Dissecting with a Scalpel.

(continued on next page)

(continued from previous page)

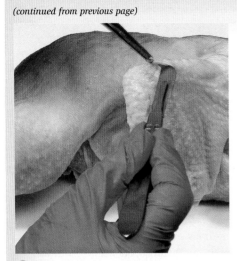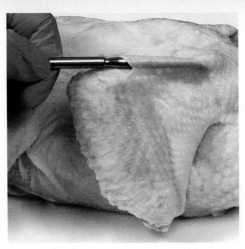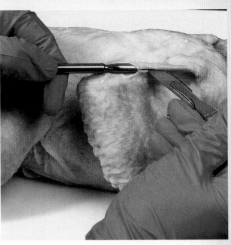

① Pull the skin away from the underlying tissues using tissue forceps and cut a small slit in the skin with the scalpel, taking care not to cut delicate tissues deep to the skin.

② Push the probe under the skin along the line where you wish to make your cut.

③ Cut the skin superficial to the probe with the scalpel. Notice how the blunt probe limits the depth at which the scalpel can cut, thus protecting underlying tissues.

Figure 1.12 Protecting Underlying Tissues with a Probe When Dissecting with a Scalpel.

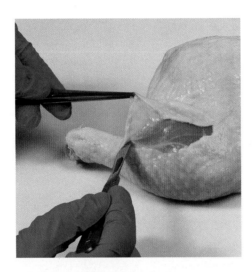

Figure 1.13 Removing the Rest of the Skin with the Scalpel. As you continue to cut, use the forceps to pull the skin away from the underlying tissues and be sure to keep as much tension as possible on the fascia. Cut the fascia with the scalpel, always keeping the sharp part of the scalpel blade pointed toward the skin. This way, if the blade slips accidentally, it will cut the skin, not the underlying tissues that you wish to preserve.

EXERCISE 1.8

DISSECTING WITH SCISSORS

1. Using the dissection specimen you began with in exercise 1.7, you will now practice using scissors to cut tissues.

2. Obtain a pair of **pointed scissors** and **forceps** (table 1.5). Using the forceps, grasp part of the skin covering part of the specimen that you have not already dissected and pull it away from the muscle. Next, cut a small slit into the skin until you can see the fascia beneath it. Continue to lengthen your cut until it is about 2 inches long **(figure 1.14)**.

3. In theory, you can begin to separate the skin from the muscle by cutting with the scissors in much the same way you did with the scalpel. However, there are a lot of tissues within that fascia that we may want to preserve, such as nerves and blood vessels. If we just cut away using "sharp" techniques, we may cut those vessels accidentally. For this reason, we prefer to use "blunt" dissection technique here to prevent ourselves from destroying potentially important structures. To do this we will use what is called an "open scissors" technique, so named because the dissecting action of the scissors is performed by starting with the scissors closed and then

actively opening them. This is exactly the opposite of how most scissors are used.

4. With the scissors closed, push the tip of the scissors into the space between the skin and the muscle so that it pierces the fascia **(figure 1.15-1)**. Once you have the tip of the scissors within the fascia, open the scissors (figure 1.15-2). Notice how this action causes the fibers within the fascia to separate from each other and loosens the hold between the skin and the fascia.

5. Continue to loosen the fascia using the open scissors technique. As you do so, notice small structures such as blood vessels and nerves that may run in the space between the skin and the underlying tissues. Notice how the fibers in the fascia easily separate from each other without damaging those structures when you use the open scissors technique. At times, the hold of the fascia is too tight, and "open scissors" technique will no longer work effectively. At those times, switch to "normal" scissors technique and cut away the tough tissue.

6. Practice using both open and normal scissors techniques to continue to remove skin from underlying tissues.

(1) Use tissue forceps to pull the skin away from underlying tissues.

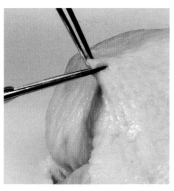
(2) Make a small cut in the skin with the scissors.

(3) Now you will have created a small hole in the skin with which to begin cutting directionally along the skin.

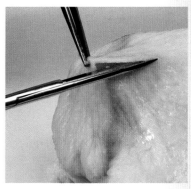
(4) Continue your cut along the skin, making sure to continue to use the tissue forceps to pull the skin away from underlying tissues before making each cut so as not to damage underlying structures.

Figure 1.14 Dissecting with Scissors.

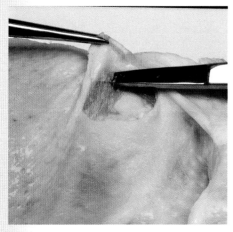
(1) After you have created a cut in the skin and can pull the skin away from underlying tissues using forceps, push the pointed end of a pair of scissors into the fascia deep to the skin, taking care not to pierce other underlying structures.

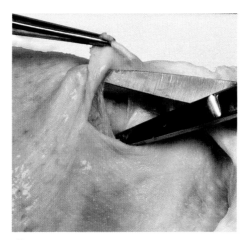
(2) Open the scissors, thus separating the fibers of the fascia.

Figure 1.15 Open Scissors Technique.

Blunt Dissection Techniques

There are times when sharp dissection technique is undesirable, because tissues might be damaged if we use sharp instruments. At these times it is best to switch to blunt dissection techniques. Blunt dissection is designed to separate tissues without damaging delicate structures.

EXERCISE 1.9

BLUNT DISSECTION TECHNIQUES

1. Using the dissection specimen you began with in exercise 1.7, you will now practice using blunt dissection techniques.

2. Obtain a pair of **pointed scissors** and **forceps** (table 1.5). Using the forceps, grasp part of the skin of the specimen where you have not previously dissected and pull it away from the muscle. Next, make a small cut into the skin until you can see the fascia beneath it **(figure 1.16-1)**. When you cut this way, you are using "sharp" dissection technique.

3. We have already learned that one way to separate skin from underlying tissues is to cut into the fascia. However, there are tissues within that fascia that we may want to preserve, such as nerves and blood vessels. If we just cut away using "sharp" techniques, we may cut those structures accidentally. For this reason, we prefer to use "blunt" dissection technique whenever possible so as to prevent ourselves from destroying potentially important structures. "Open scissors" is one blunt dissection technique that you learned in exercise 1.8. Use this technique to loosen the hold between the skin and the fascia on your specimen.

4. Once you have created a space between the skin and muscle that is large enough for you to put a finger in, set the scissors down. You will now proceed to separate the skin from the muscle using only your fingers (figure 1.16-2). Because you are not using any sharp instruments to perform this, it is referred to as "blunt" dissection.

5. Obtain a blunt probe (table 1.5). A blunt probe can be used in place of your fingers to separate structures when your fingers are too large (figure 1.16-2). Because it does not cut the tissue, this is also a "blunt" dissection technique.

6. Other items that can be used for blunt dissection are scalpel blade handles (*without* the blades on them!), or the rounded ends of forceps. Spend some time practicing the use of these tools by trying out different tools to separate skin from muscle in different regions of the dissection specimen. The general rule of thumb is that you want to start with sharp dissection techniques to cut slits in the skin, but then transition to blunt techniques whenever possible so as to prevent accidental damage to underlying tissues.

① Using tissue forceps and scissors, pull the skin away from the underlying tissues and make a cut in the skin. Use open scissors technique to loosen the fascia and to create a space where a blunt probe or your fingers may be pushed in.

② Using your fingers, pull the skin away from the underlying tissues. When necessary, use a sharp instrument to cut any fascia that is very tough and won't separate using blunt techniques.

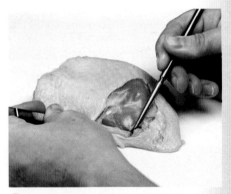
③ A blunt probe can be moved around under the skin to gently separate the connective tissue without damaging underlying structures.

Figure 1.16 **Blunt Dissection.** Blunt dissection techniques involve separating tissues with your fingers or blunt instruments such as a probe.

Chapter 1: The Laboratory Environment

Name: _____
Date: _____ Section: _____

POST-LABORATORY WORKSHEET

The ❶ corresponds to the Learning Objective(s) listed in the chapter opener outline.

Do You Know the Basics?

Exercise 1.1: The Scientific Method

1. Describe the steps involved in the scientific method. ❶

   _____
   _____
   _____
   _____

2. Define the following terms. ❷

   Independent variable:
   _____

   Dependent variable:
   _____

   Confounding variable:
   _____

3. Explain the importance of a good experimental control. ❸

   _____
   _____
   _____
   _____

4. Given the set of numbers below, calculate the mean and the range. Show your work in the box. ❹

25
98
32
46
22
87
34
26
15

Mean: Range:

Exercise 1.2: Presenting Data

5. Match the components of a typical graph in column A with what each component represents in column B. ❺

   Column A
   ____ label
   ____ units
   ____ values
   ____ X-axis
   ____ Y-axis

   Column B
   a. axis where the independent variable is plotted
   b. axis where the dependent variable is plotted
   c. description (name) of the variable in question
   d. description of what the numbers on the graph indicate
   e. numerical representation of the data points

6. The following data were obtained by measuring the distance traveled by a golf ball over time after it was hit with a putter. Graph the data using the grid provided below. Be sure to include all units and labels on your graph.

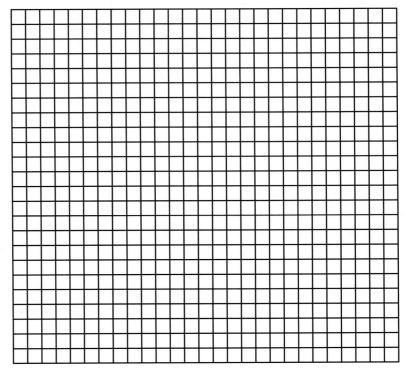

Time (min)	Distance (cm)
1	2
2	3.75
3	5.5
4	7
5	8.25
6	9.25
7	10
8	10.5
9	11
10	11.25

7. Observe the graph you created in question #6. In the space below, describe the relationship between distance traveled and time (e.g., is it linear/curved, does it increase/decrease?).

_____
_____
_____
_____

Exercise 1.3: Units of Measurement

8. Match each of the following prefixes in column A with the correct power of ten that it represents in column B.

Column A
____ centi-
____ deci-
____ kilo-
____ micro-
____ milli-
____ no prefix/base unit

Column B
a. 10^3
b. 10^0
c. 10^{-1}
d. 10^{-2}
e. 10^{-3}
f. 10^{-6}

9. List the metric unit used to measure each of the following:

Length: _____ Temperature: _____

Mass: _____ Volume: _____

10. Make the appropriate English-to-metric conversions described below.

a. You have a piece of dialysis tubing that is 8 inches long. How many centimeters long is the piece of dialysis tubing? _____ cm
Show your calculation in the space below.

b. You are weighing a patient, a 160-lb male. How many kilograms does your patient weigh? _____ kg
Show your calculation in the space below.

c. The boiling point of water is 100 °C. What is the boiling point of water in Fahrenheit? _____ °F
Show your calculation in the space below.

d. The freezing point of water is 0 °C. What is the freezing point of water in Fahrenheit? _____ °F
Show your calculation in the space below.

Exercise 1.5: Proper Disposal of Laboratory Waste

11. List the three items described in the text as "hazardous waste." ⓬

_____
_____
_____

12. Where do you dispose of used scalpel blades? ⓬

_____
_____
_____

13. Formalin and phenol are potentially hazardous chemicals. What steps should be taken to prevent exposure to these chemicals? ⓬

_____
_____
_____

14. What should you do if your skin or eyes are exposed to formalin or phenol? ⓬

_____
_____
_____

Exercise 1.6: Placing a Scalpel Blade on a Scalpel Blade Handle

15. Describe the proper technique for putting a scalpel blade on a scalpel handle. ⓭

_____
_____
_____

16. Describe the proper technique for removing a scalpel blade from a scalpel handle. ⓮

_____
_____
_____

Exercise 1.7: Dissecting with a Scalpel

17. List an example of a case when a scalpel is the preferred tool for dissection. ⓯

_____
_____
_____

Chapter One *The Laboratory Environment*

18. Describe one technique used to prevent damage to underlying tissues when dissecting with a scalpel.

Exercise 1.8: Dissecting with Scissors

19. List an example of a case when scissors are the preferred tool for dissection.

20. Describe *open scissors* technique.

Exercise 1.9: Blunt Dissection Techniques

21. Define *blunt dissection:*

22. Explain the importance of using blunt dissection techniques whenever possible.

Can You Apply What You've Learned?

23. An article reports data from an experiment that looked at the effect of body weight on risk of developing Type II diabetes mellitus.

    a. What is the independent variable in this experiment?

    b. What is the dependent variable in this experiment?

24. Briefly describe *fascia*.

25. You have just finished dissecting a fresh cow bone as part of the day's laboratory activities. What do you think is the most appropriate way to dispose of this waste?

26. You are dissecting the wing of a chicken, and the skin is held tight to the bones beneath it. You would like to remove the skin, but in doing so you would like to preserve the bone and muscle beneath the skin. What tools might you use, and how will you use them, to complete this task? (There is more than one correct answer to this question; use your best judgment.)

CHAPTER 2

Orientation to the Human Body

OUTLINE AND LEARNING OBJECTIVES

Gross Anatomy 31

Anatomic Terminology and the Anatomic Position 31

Anatomic Planes and Sections 31

EXERCISE 2.1: DETERMINING ANATOMIC PLANES AND SECTIONS 32
1. Demonstrate the anatomic position
2. Use correct anatomic terminology to define planes of the body
3. Identify the plane of section of four different views of the brain

Directional Terms 33

EXERCISE 2.2: USING DIRECTIONAL TERMS 33
4. Use directional terms to describe locations on the surface of the body

Regional Terms 34

Body Cavities and Membranes 34

EXERCISE 2.3: USING REGIONAL TERMS 35
5. Use regional terms to describe parts of the human body

EXERCISE 2.4: UNDERSTANDING BODY CAVITIES 36
6. List the major organs located within each of the major body cavities

Abdominopelvic Regions and Quadrants 37

EXERCISE 2.5: LOCATING MAJOR BODY ORGANS USING ABDOMINOPELVIC REGION AND QUADRANT TERMINOLOGY 37
7. Describe the abdominopelvic regions and quadrants
8. Describe the locations of the major abdominopelvic organs using abdominopelvic region and quadrant terminology

INTRODUCTION

The human body is both beautiful and complex. You are about to embark on an incredible journey in which you will develop an understanding of that beauty and complexity. To be successful in this venture, you must do a great deal of work. The good news is that if you put in the time, you *will* succeed. The study of anatomy and physiology requires an enormous time commitment.

To put things in perspective, consider this: A beginning student in human anatomy and physiology is typically asked to learn more new words in a one-semester anatomy and physiology class than a beginning student learns in the first semester of a foreign language class. In fact, you *will* be learning a new language. This language has its origins principally in Latin and Greek. One of your primary tasks will be to establish a firm understanding of the meanings of common word origins. You are encouraged to use the list of common word origins located on the inside front and back covers of this manual to look up the origins of all words that are new or unfamiliar. This will require a small amount of work each time you look up a word, but over time you will develop an impressive vocabulary of anatomical/medical terms. In addition, you will ➡

 **MODULE 1: BODY ORIENTATION**

establish a rich knowledge base that will allow you to begin to interpret the meanings of new words as you encounter them later in the course.

Let's practice by analyzing the origins of the words *anatomy* and *physiology*. The word *anatomy* can be broken down into two parts, *ana-* and *-tomé*. The word part *ana-* means "apart." The word part *-tomé* means "to cut." Thus, the word **anatomy** literally means "to cut apart." Though you might not literally be cutting up human bodies as medical students do and early anatomists did, you will at the very least be *conceptually* "cutting up" the body to understand its component parts. The word **physiology** consists of two word parts, *physis-*, nature and *-logos*, study. The word physiology means to study the nature of how the body functions.

Your primary goal as you complete the laboratory exercises in this chapter is to practice using correct anatomic directional, regional, and sectional terms to describe the body and its parts. You will also locate major body organs on a human torso model. If your laboratory uses human cadavers, you may also locate them on the cadaver. In addition, you will become familiar with the organ systems of the body and will use anatomic terminology to describe locations of the organs that compose these organ systems.

Chapter 2: Orientation to the Human Body

Name: _____
Date: _____ Section: _____

PRE-LABORATORY WORKSHEET

Also available at www.connect.mcgraw-hill.com

|ANATOMY & PHYSIOLOGY

1. In the space below, make a simple line drawing of an individual in the anatomic position. Next to your drawing, describe the anatomic position using your own words.

   _____
   _____
   _____
   _____
   _____
   _____
   _____
   _____
   _____
   _____
   _____

2. A plane that separates the body into superior and inferior portions is a _____ plane.

3. A plane that separates the body into right and left portions is a _____ plane.

4. A plane that separates the body into anterior and posterior portions is a _____ plane.

5. Match each directional term in column A with its definition in column B.

Column A

____ 1. anterior (ventral)
____ 2. caudal
____ 3. cranial
____ 4. deep
____ 5. distal
____ 6. inferior
____ 7. lateral
____ 8. medial
____ 9. posterior (dorsal)
____ 10. proximal
____ 11. superficial
____ 12. superior

Column B

a. in front of; toward the front surface
b. in back of; toward the back surface
c. above
d. below
e. at the tail end of the body
f. at the head end of the body
g. toward the midline of the body
h. away from the midline of the body
i. on the inside; beneath the surface of the body
j. on the outside; toward the surface of the body
k. closer to the attachment point of a limb to the trunk
l. farther from the attachment point of a limb to the trunk

Chapter Two *Orientation to the Human Body*

6. Match each regional name in column A with its description in column B.

Column A

____ 1. axillary
____ 2. brachial
____ 3. buccal
____ 4. carpal
____ 5. cephalic
____ 6. cervical
____ 7. digital
____ 8. femoral
____ 9. gluteal
____ 10. inguinal
____ 11. lumbar
____ 12. mental
____ 13. orbital
____ 14. plantar
____ 15. umbilical
____ 16. vertebral

Column B

a. arm
b. armpit
c. buttock
d. cheek
e. chin
f. eye
g. fingers or toes
h. groin
i. head
j. lower back; loin
k. navel
l. neck
m. sole of the foot
n. spinal column
o. thigh
p. wrist

7. The body cavity that encases the brain is the _____ cavity, and the body cavity that encases the spinal cord is the _____. These two cavities combined make up the _____.

8. The body cavity that surrounds the heart is the _____ cavity.

9. The body cavity that contains most of the digestive, reproductive, and urinary system organs is the _____ cavity.

10. The abdominopelvic cavity can be divided into a total of _____ quadrants or _____ regions. The central point of reference for dividing the abdominopelvic cavity into quadrants or regions is the _____.

Gross Anatomy

Anatomic Terminology and the Anatomic Position

Because the human body can be placed in numerous positions and each position has the potential to change the definition of terms such as *front* and *back*, in anatomy and medicine all such terms refer to the body when it is placed in the **anatomic position**. The anatomic position **(figure 2.1)** is the position of the body when one is standing up, facing forward (anterior), with the palms of the hands facing forward (anterior). In this position, no two bones of the body cross each other. The anatomic position is similar to the natural position of an individual when standing up, except for the position of the wrist and hand. Because the position of the wrist and hand is somewhat unusual in the anatomic position, always be extra careful when using directional terms that relate to the upper limbs.

Anatomic Planes and Sections

The study of anatomy that involves viewing sections (or slices) of an organ or the body is called **sectional anatomy**. An understanding of sectional anatomy is increasingly important in clinical settings, where medical imaging techniques such as CT (computed tomography) and MRI (magnetic resonance imaging) scans are used extensively. The ability to analyze such scans requires special skill that is developed over time. At this point in your education, it is important that you simply understand the terms related to the three major planes that divide the body into sections: the coronal plane, the transverse plane, and the sagittal plane **(table 2.1)**.

INTEGRATE

LEARNING STRATEGY

A **plane** (*planus*, flat) is an imaginary two-dimensional flat surface. One way of visualizing a plane is to imagine a transparency film passing through the body; a **section** (*sectio*, a cutting) is a slice made along one of these two-dimensional planes.

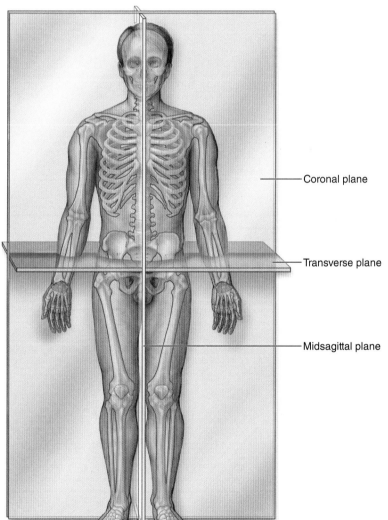

Figure 2.1 Anatomic Position and Body Planes.
In the anatomic position, no two bones are crossed.

Table 2.1	Anatomic Planes
Body Plane	**Description**
Coronal (frontal)	Separates anterior portions from posterior portions
Transverse (horizontal)	Separates superior portions from inferior portions
Sagittal (parasagittal)	Separates right portions from left portions
Midsagittal (median)	Separates right and left portions equally; runs down the midline of the body
Oblique	Runs at an angle to any of the three main planes of the body (coronal, transverse, or sagittal)

EXERCISE 2.1

DETERMINING ANATOMIC PLANES AND SECTIONS

1. **Figure 2.2** shows photographs of a brain that has been sectioned along different planes. Determine which plane the brain was sectioned along for each photo, then enter the information in the appropriate spaces in figure 2.2 below. Answer choices are provided in the figure legend, and each answer may be used only once. Refer to the chapter on the brain in your main textbook if you need help getting oriented.

2. *Optional Activity:* **APR 1: Body Orientation**—Review all dissections in this module to familiarize yourself with general anatomic terminology.

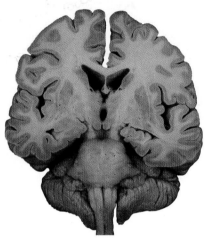

1 __Coronal Plane__
Anterior view

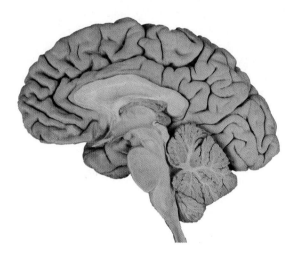

2 __Midsagittal__
Medial view

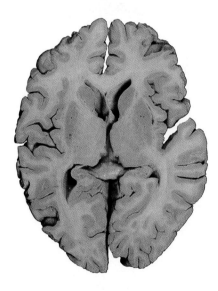

3 __transverse plane__
Superior view

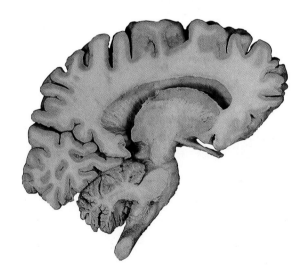

4 __Sagittal__
Medial view

Figure 2.2 Sections Through a Human Brain.

| 1 | coronal | 2 | midsagittal | 4 | sagittal | 3 | transverse |

Directional Terms

When we use everyday language we often use terms like "front" and "back," and "on top of" or "on the bottom of," to give directions. While this is perfectly appropriate in everyday language, it can create a great deal of confusion when referring to directions in the human body. For instance, when we say "on top of," we need to know what part of the body "on top of" refers to, and different people might use the term in different ways or in reference to different structures. Furthermore, they may be thinking "on top of" relative to different body positions. This becomes problematic in a medical setting because it leads to confusion, which has the potential to create severe consequences for a patient. Thus, we use an agreed-upon set of directional terms in anatomy and medicine to be as specific as possible when describing directions, but also to ensure we are all speaking the same language. **Table 2.2** lists these directional terms and gives definitions of each of them. In exercise 2.2 you will practice using these directional terms.

Table 2.2	Directional Terms
Directional Term	**Definition**
Anterior (ventral)	Toward the front of the body (the belly side)
Posterior (dorsal)	Toward the back of the body (the back side)
Superior	Above; closer to the head
Inferior	Below; closer to the feet
Cranial (cephalic)	At the head end of the body
Caudal	At the tail end of the body
Medial	Toward the midline of the body
Lateral	Away from the midline of the body
Superficial	Toward the surface of the body; on the outside
Deep	Beneath the surface of the body; on the inside
Proximal	Near; closer to the attachment point of a limb to the trunk
Distal	Far; farther from the attachment point of a limb to the trunk

EXERCISE 2.2

USING DIRECTIONAL TERMS

Figure 2.3 shows a posterior view of a human. Three locations marked on the body are denoted with the numbers 1–3. In the spaces below, describe the locations of markings 1–3 as specifically as possible using correct anatomic terminology. When you are finished, compare your answers to those of other students in your class to see how similar your answers are.

Location 1

thoracic — mammary, pectoral & sternal

Location 2

antebrachial (forearm)

Location 3

femoral — (thigh)

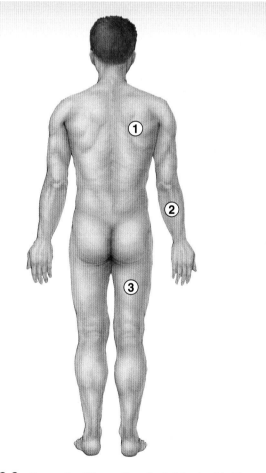

Figure 2.3 Posterior View of an Individual with Three Reference Locations (1–3) Marked.

☐ antebrachial   ☐ femoral   ☐ thoracic

INTEGRATE

LEARNING STRATEGY

The directional terms *superior* and *inferior* are used when describing one structure with respect to another structure in the trunk of the body. The directional terms *proximal* and *distal* are used when describing the position of one structure with respect to another structure on the limbs. Thus, it is more appropriate to say the elbow is located *proximal* to the wrist, rather than to say it is superior to the wrist. On the other hand, it is quite appropriate to say the thorax is located *superior* to the abdomen.

Regional Terms

Just as with directional terms, we have common, everyday terms that we use to describe regions of the body, like *arm* or *back*. The correct anatomic terms to describe regions of the body are basically synonyms for these terms, and are closer to the Latin or Greek derivative of the term. **Table 2.3** correlates some commonly used regional terms in anatomy with words we use to describe the same region using everyday language. Your main textbook has a much more inclusive table of regional terms, as well as a figure describing these regions, which you should use as references when performing exercise 2.3.

Table 2.3 Selected Regional Terms

Regional Term	Description
Axillary	Armpit
Brachial	Arm
Buccal	Cheek
Carpal	Wrist
Cephalic	Head
Cervical	Neck
Digital	Fingers or toes
Femoral	Thigh
Gluteal	Buttock
Inguinal	Groin
Lumbar	Lower back
Mental	Chin
Orbital	Eye
Plantar	Sole of the foot
Umbilical	Navel
Vertebral	Spinal column

Body Cavities and Membranes

Many organs within the body are compartmentalized and separated from each other by a *body cavity*. Compartmentalizing the organs this way allows the separate organs to perform their functions without interfering with the functioning of other organs. For example, the pumping action of the heart does not interfere with the expansion and contraction of the lungs because each organ is enclosed in its own cavity. In addition, the encasement of organs within separate cavities helps to prevent the spread of infection from one region of the body to another.

INTEGRATE

CONCEPT CONNECTION

Serous (*serosus*, watery fluid) **membranes** cover organs within the body cavities. These membranes are composed of two layers: visceral and parietal. The visceral (*viscus*, internal organ) layer membrane lies adjacent to the organ, whereas the parietal (*paries*, wall) layer membrane attaches to the wall of the body cavity. The cells that compose the serous membrane produce a slippery substance called **serous fluid** into the space between the two layers of membrane to add further protection. As organs such as the heart and lungs expand and contract, serous fluid reduces friction between the organs and surrounding structures. You will examine the tissues that compose serous membranes in chapter 5, namely epithelial and connective tissues. You will also examine the specific organ coverings and their structure in the "gross anatomy" section of subsequent chapters. Note that serous membranes surrounding each organ generally have a unique name related to that organ. For example, the *pericardium* surrounds the heart and *pleurae* surrounds the lungs.

INTEGRATE

CLINICAL VIEW
Inflammation of Serous Membranes

Serous membranes are membranes that line body cavities, such as the pericardial cavity. Inflammation of serous membranes can be a serious health risk. Clinical terms related to inflammation of serous membranes reference the specific serous membrane affected. For example, *pericarditis* (*peri-*, around + *cardio*, heart + *itis*, inflammation) is inflammation of the pericardium surrounding the heart; *pleurisy* is inflammation of the pleurae surrounding the lungs; *peritonitis* is inflammation of the peritoneum surrounding the abdominopelvic cavity. One consequence of inflammation of a serous membrane is swelling, which can impede function of surrounding organs within the body cavity. For example, pericarditis may interfere with the heart's ability to pump blood. Pleurisy may prevent adequate gas exchange. Treatment typically involves removal of the excess fluid and administration of an anti-inflammatory drug. In addition, antibiotics may be given for bacterial infections. If left untreated, the condition can be fatal.

EXERCISE 2.3

USING REGIONAL TERMS

1. Identify the body regions listed in **figure 2.4** on your own body.
2. Label figure 2.4 with the appropriate regional terms.

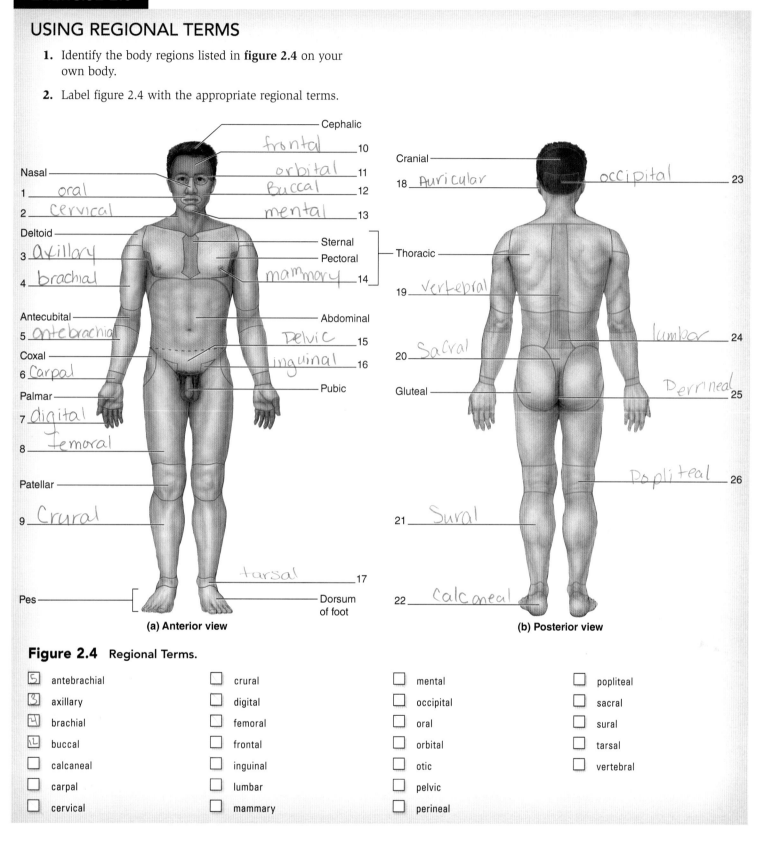

Figure 2.4 Regional Terms.

- [5] antebrachial
- [3] axillary
- [4] brachial
- [12] buccal
- [ ] calcaneal
- [ ] carpal
- [ ] cervical
- [ ] crural
- [ ] digital
- [ ] femoral
- [ ] frontal
- [ ] inguinal
- [ ] lumbar
- [ ] mammary
- [ ] mental
- [ ] occipital
- [ ] oral
- [ ] orbital
- [ ] otic
- [ ] pelvic
- [ ] perineal
- [ ] popliteal
- [ ] sacral
- [ ] sural
- [ ] tarsal
- [ ] vertebral

EXERCISE 2.4

UNDERSTANDING BODY CAVITIES

1. Observe a human torso model or a human cadaver.
2. Using your textbook as a guide, identify the body cavities listed in **figure 2.5** on the torso model or human cadaver. Then label figure 2.5 with the appropriate terms. (Answers may be used more than once.)

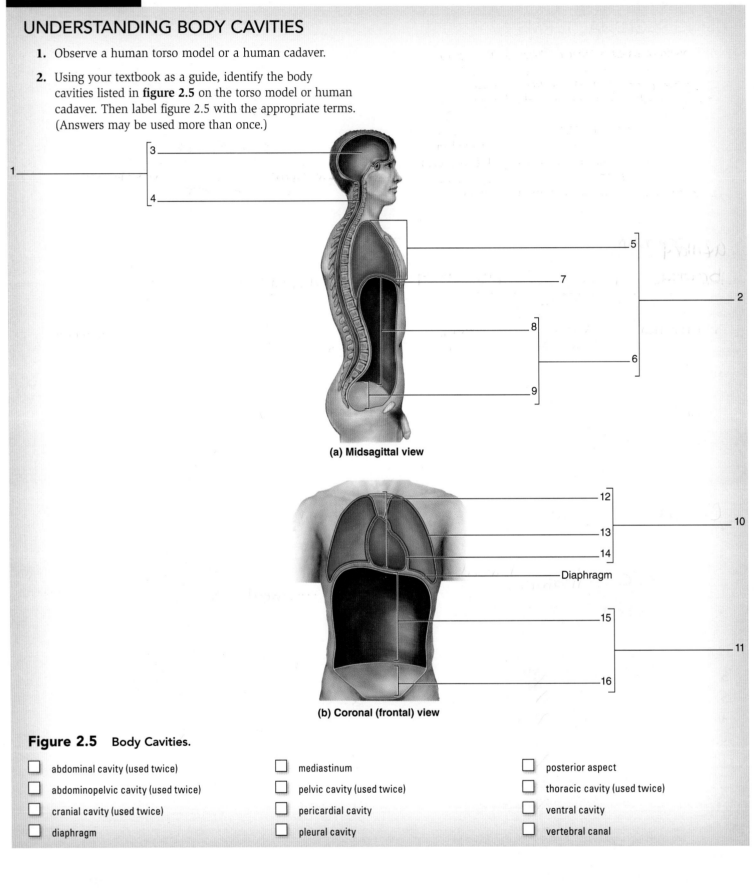

(a) Midsagittal view

(b) Coronal (frontal) view

Figure 2.5 Body Cavities.

- ☐ abdominal cavity (used twice)
- ☐ abdominopelvic cavity (used twice)
- ☐ cranial cavity (used twice)
- ☐ diaphragm
- ☐ mediastinum
- ☐ pelvic cavity (used twice)
- ☐ pericardial cavity
- ☐ pleural cavity
- ☐ posterior aspect
- ☐ thoracic cavity (used twice)
- ☐ ventral cavity
- ☐ vertebral canal

Abdominopelvic Regions and Quadrants

To describe the locations of organs in the abdominopelvic cavity, we need a set of terms that allows us to specify where within the cavity an organ or tissue is located. There are two approaches used to do this. The first approach is to divide the abdominopelvic cavity into **quadrants** (*quad-*, four). This is done by passing one imaginary line vertically through the *umbilicus* (belly button) and another horizontally through the umbilicus. The four resulting quadrants are the right upper, left upper, right lower, and left lower **(figure 2.6a)**. Because this approach is simple, it is the approach used most often in a clinical setting. The second approach is to divide the abdominopelvic cavity into **regions**. This is done by drawing one vertical line to the left of the umbilicus and another to the right of the umbilicus (at the *midclavicular line*, a vertical line that passes through the midpoint of the clavicle), then drawing one horizontal line superior to the umbilicus and another inferior to the umbilicus. The result is a grid similar to a tic-tac-toe layout with nine regions formed **(figure 2.6b)**. This approach allows us to be more specific about describing the locations of organs and tissues within the abdominopelvic cavity. The resulting regions are the umbilical, epigastric (*epi-*, above, + *gastēr*, belly), hypogastric (*hypo-*, below, + *gastēr*, belly), right and left hypochondriac (*hypo-*, below, + *chondro-*, cartilage, as in the cartilages that attach ribs to sternum), right and left lumbar (*lumbus*, a loin), and right and left iliac (*ilium*, flank). Exercise 2.5 is designed to introduce you to the locations of major organs within the abdominopelvic cavity and to help you become comfortable with the terms used to describe the quadrants and regions.

EXERCISE 2.5

LOCATING MAJOR BODY ORGANS USING ABDOMINOPELVIC REGION AND QUADRANT TERMINOLOGY

1. Using figure 2.6 as a guide, identify the following structures on a human torso model or on a human cadaver:

   ☐ left kidney   ☐ spleen
   ☐ liver   ☐ stomach
   ☐ pancreas   ☐ urinary bladder
   ☐ small intestine

2. Based on your observations of the cadaver or human torso model, complete the chart on p. 38 by indicating the quadrant(s) or region(s) in which each organ is found.

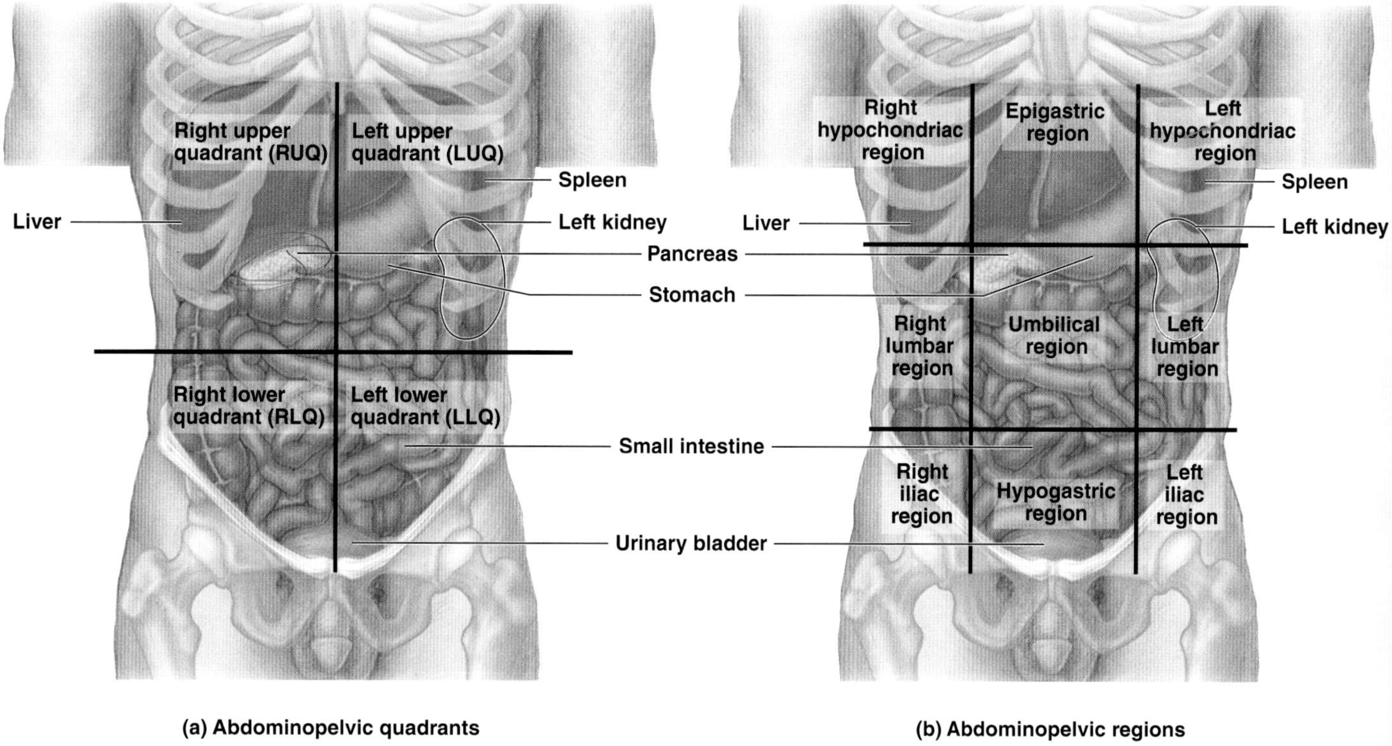

(a) Abdominopelvic quadrants

(b) Abdominopelvic regions

Figure 2.6 The Abdominopelvic Cavity. The abdominopelvic cavity can be subdivided into (a) four abdominopelvic quadrants, or (b) nine abdominopelvic regions.

(continued on next page)

(continued from previous page)

Organ	Quadrant(s)	Region(s)
Left kidney	LUQ	left hypochondriac region, left lumbar region
Liver	RUQ	right hypochondriac region
Pancreas	RUQ	Umbilical region
Small intestine	RLQ, LLQ	umbilical + hypogastric
Spleen	LUQ	Left hypochondriac
Stomach	LUQ	umbilical, Left hypochondriac + Epigastric
Urinary bladder	RLQ, LLQ	Hypogastric

CHAPTER 3

The Microscope

OUTLINE AND LEARNING OBJECTIVES

Histology 45

The Compound Microscope 45

Caring for the Compound Microscope 45

EXERCISE 3.1: PARTS OF A COMPOUND MICROSCOPE 46
- ❶ Identify the parts of a compound microscope
- ❷ Calculate total magnification of the microscope using each objective lens

Focus and Working Distance 48

EXERCISE 3.2: VIEWING A SLIDE OF THE LETTER e 48
- ❸ Define working distance, and estimate working distance of the microscope

Diameter of the Field of View 49

EXERCISE 3.3: MEASURING THE DIAMETER OF THE FIELD OF VIEW 50
- ❹ Explain how the diameter of the field of view changes as total magnification increases
- ❺ Given the diameter of the field of view at one total magnification, calculate the diameter of the field of view for a different total magnification

EXERCISE 3.4: ESTIMATING THE SIZE OF A SPECIMEN 50
- ❻ Estimate the size of a specimen using known values for the diameter of the field of view

Depth of Field 51

EXERCISE 3.5: DETERMINING DEPTH OF FIELD 51
- ❼ Explain how depth of field changes as total magnification increases

Gross Anatomy 52

The Dissecting Microscope 52

EXERCISE 3.6: PARTS OF A DISSECTING MICROSCOPE 53
- ❽ Identify the parts of a dissecting microscope
- ❾ Describe the types of specimens you would use a dissecting microscope to view

INTRODUCTION

The study of anatomy involves observation of both gross (large) and microscopic (small) structures. To get the most out of your experience observing microscopic structures, it is imperative that you know how to use a microscope properly. Although you have probably used a microscope before, pay very careful attention to the instructions in this chapter. Even though many students have been taught the proper use and care of the microscope, most continue to use poor technique. This, in turn, causes great frustration and is a major impediment to the learning process. Treat this laboratory exercise as an opportunity to refine and improve upon your technique.

In this laboratory exercise you will learn the proper use and care of the compound microscope, which is used to view slides of cells and tissues. You will also learn how depth of field and diameter of the field of view change as total magnification increases. After performing these exercises you will be prepared for the observations of histology (tissues) that you will be making in later chapters. In addition, you will learn the proper use and care of a dissecting microscope (also called a stereomicroscope), which is used to view larger specimens such as hairs, muscle fibers that have been teased apart.

Chapter 3: The Microscope

PRE-LABORATORY WORKSHEET

Name: _____
Date: _____ Section: _____

Also available at www.connect.mcgraw-hill.com

1. Match the definition in column A with the microscope part in column B.

   Column A

   ____ 1. the light source located in the base of the microscope
   ____ 2. a device on the eyepiece that is used to focus the eyepiece
   ____ 3. everything that is visible when looking through the eyepiece
   ____ 4. a measure of how much of the total thickness of the specimen is in focus
   ____ 5. the part of the microscope that connects the head to the base
   ____ 6. a part of the condenser that controls the amount of light that passes through
   ____ 7. knobs that are used to move the slide on the stage
   ____ 8. the platform that a slide is placed upon
   ____ 9. a device that connects the objective lenses to the head of the microscope
   ____ 10. lenses that are attached to the nosepiece
   ____ 11. lenses that are located within the eyepieces
   ____ 12. a switch that is used to turn the power on or off
   ____ 13. calculated by multiplying the ocular lens power by the objective lens power
   ____ 14. the distance between the mechanical stage and the tip of the objective lens
   ____ 15. a knob that is used to regulate the brightness of the light
   ____ 16. provides attachment and support for the objective lenses
   ____ 17. a knob that moves the mechanical stage up and down in small increments
   ____ 18. a tube located on the head of the microscope that a person looks into
   ____ 19. a knob that moves the mechanical stage up and down in large increments
   ____ 20. a device that is used to focus the light coming from the illuminator
   ____ 21. the bottom support of the microscope

   Column B

   a. arm
   b. base
   c. coarse adjustment knob
   d. condenser
   e. depth of field
   f. diopter adjustment ring
   g. eyepiece
   h. field of view
   i. fine adjustment knob
   j. head
   k. illuminator
   l. iris diaphragm
   m. mechanical stage
   n. mechanical stage controls
   o. nosepiece
   p. objective lens
   q. ocular lens
   r. power switch
   s. total magnification
   t. voltage regulator
   u. working distance

2. In the space below, draw or describe in words the proper procedure for holding and transporting a microscope.

3. Explain how to calculate the total magnification of a microscope. _____

4. As total magnification power increases, depth of field _____.

5. As total magnification power increases, diameter of the field of view _____.

Histology

The Compound Microscope

A compound microscope is used to view structures that are not visible to the naked eye. Most compound microscopes can magnify images anywhere from 40 to 1000 times (40× to 1000×) normal size. The total magnification of a compound microscope is achieved through a combination of objective and ocular lenses (see **table 3.1** for definitions). The term "compound" comes from the fact that total magnification is "compounded" by the use of more than one lens.

Caring for the Compound Microscope

Microscopes are very expensive instruments. You should always use great care when you transport, use, and store them.

- *Workspace:* Keep your workspace clear of all materials except for the microscope, your laboratory manual, lens paper, and microscope slides. Remove or restrict any loose articles of clothing or jewelry so they do not interfere with your work. Long necklaces that dangle in front of your neck can damage the microscope. If you have long hair, tie it back to keep it out of the way.

- *Transport: Always* use both hands when carrying the microscope. Place one hand on the base of the microscope and the other on the arm and use care to always move deliberately when you are carrying the microscope **(figure 3.1)**.

- *Lens care:* Lenses are some of the most expensive and delicate parts of the microscope and should be treated with great care. Special lens paper is the only thing that should ever be used to clean dirty lenses. This paper is designed so it will not scratch the lenses. *Never* use facial tissues, paper towels, articles of clothing, or anything else to clean the lenses because they may scratch the lenses. A special cleaning solution can be used with the lens paper, although lens paper generally does a fine job of cleaning when used alone. *Never* use saliva, water, or other fluids to moisten the lens paper.

- *Storage:* When you are finished using the microscope, move the stage to its lowest position. Make sure that no slides are left on the stage, and then rotate the nosepiece so the lowest-power objective lens is over the stage. Wrap the power cord around the base of the microscope and replace the dust cover.

- *Handling slides:* Always hold a slide by the edges so you do not leave smudges and fingerprints, which will interfere with your ability to view the slide clearly. If the slide is dirty, clean it using

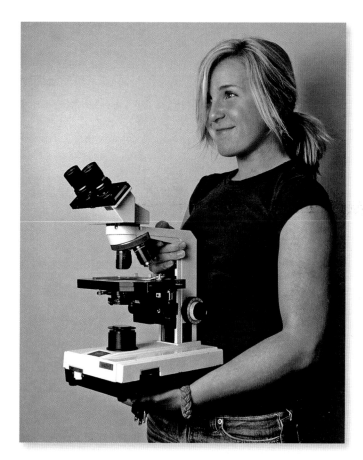

Figure 3.1 Proper Technique for Carrying a Microscope.

lens paper before placing it on the microscope stage. If a slide becomes cracked or broken, notify your instructor so that it can be disposed of properly. Broken slides should be placed in a special broken glass container, never in the garbage can.

Other things to be aware of:

- When removing the microscope cover, be careful not to pull the ocular lens off with the cover.

- When using the coarse adjustment knob, be careful not to allow the objective lens to smash into the slide, especially on high power.

EXERCISE 3.1

PARTS OF A COMPOUND MICROSCOPE

1. Obtain a compound microscope.

2. Using **figure 3.2** and table 3.1 as guides, identify all the parts listed in table 3.1 on your compound microscope.

3. Record the magnifications of the lenses on your microscope in the spaces below.

   Magnification of ocular lenses: _____

   Magnification of objective lenses:

   Scanning _____   Low _____   High _____

4. To determine the total magnification of an image, simply multiply the magnification of the ocular lens with that of the objective lens. For example, if the magnification of the ocular lens is 10× and the magnification of the objective lens is 4×, then the total magnification is 40×. In the spaces below, calculate the total magnification of your microscope for each of these objective lenses: scanning, low-power, and high-power.

   Total magnification (ocular × objective):

   Scanning _____   Low _____   High _____

5. Use the information from steps 3 and 4 to answer question 3 of the Post-Laboratory Worksheet (p. 54).

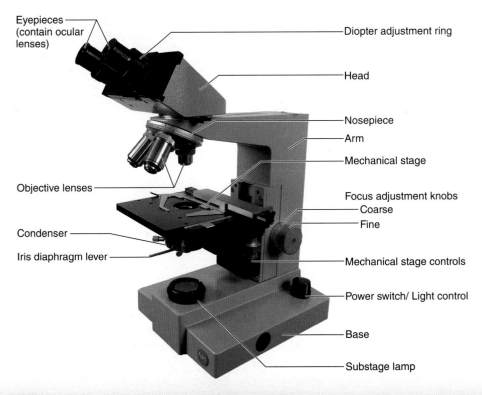

Figure 3.2 Parts of a Compound Microscope.

Table 3.1	Parts of the Microscope
Term	**Description and Usage**
Arm	Connects the head to the base. When transporting the microscope, always grasp the arm with one hand.
Base	The bottom part of the microscope. It supports the entire microscope and encases much of the electrical wiring. When transporting the microscope, one hand should always be placed under the base for support.
Condenser	A device used to focus the light coming from the illuminator/substage lamp. It generally works best in the position closest to the mechanical stage.
Coarse adjustment knob	Moves the mechanical stage up and down in relatively large increments. It is used to position a specimen into your field of view when you first begin to scan a slide at low power. It should never be used when the microscope is on the higher powers.
Depth of field	How much of the total thickness of the specimen is in focus
Diopter adjustment ring	Changes the focus of the eyepiece. Used to make adjustments when you have better vision in one of your eyes and you are not using corrective lenses. To use this device, begin by closing the eye that looks through the ocular lens containing the diopter adjustment ring. Looking through the *other* ocular lens, bring the sample into focus. Then, open the eye that was closed and bring the sample into focus for that eye using the diopter adjustment ring.
Eyepieces	The parts of the microscope that you look into. Always use both eyepieces. The eyepieces are movable so you can adjust them to the width of your own eyes.
Field of view	Everything visible when looking through the eyepieces
Fine adjustment knob	Moves the mechanical stage up and down in small increments. Once a specimen is brought into view using the coarse adjustment knob, the fine adjustment is used to bring the specimen into focus.
Head	The part of the microscope that provides attachment and support for the objective and ocular lenses. It also serves as the support for the eyepieces.
Substage lamp (illuminator)	The light source located in the base of the microscope. When it is turned on, light passes through the specimen on the stage, and through the lenses of the microscope, and it ultimately hits your eye, allowing you to see the specimen.
Iris diaphragm	A part of the condenser that can be opened or closed to control the amount of light passing through the condenser. More light (iris open) decreases contrast between structures. Less light (iris closed) increases contrast.
Light control (voltage regulator)	Typically, a rotating or sliding knob that alters the voltage going to the substage lamp to regulate the brightness of the light
Mechanical stage	Platform that holds the specimen. It contains clips to hold a slide in place and knobs for positioning the slide on the stage.
Mechanical stage controls	Knobs used to position the slide on the stage
Nosepiece	Device that connects the objective lenses to the head of the microscope, and rotates to change objective lenses
Objective lenses	A typical microscope has three objective lenses, though some have four. Each objective lens has a label that tells how much it magnifies the image. For instance, a $4\times$ objective lens magnifies the specimen four times normal size. The lowest-power objective lens ($4\times$) is also called the "scanning" objective.
Ocular lenses	Lenses located within the eyepieces, which generally magnify the specimen ten times normal size
Power switch	Usually located somewhere on the base of the microscope; used to turn the power on or off
Working distance	The distance between the mechanical stage (and the slide on it) and the tip of the objective lens

Focus and Working Distance

The purpose of exercise 3.2 is to familiarize you with the microscope and help you to obtain an appreciation for the relationship between what you see when you look directly at a slide and what you see when you look at a slide through the microscope lenses. If you experience difficulties and cannot see anything through the microscope or cannot focus the lenses, refer to **table 3.2,** "Troubleshooting the Compound Microscope."

Table 3.2	Troubleshooting the Compound Microscope
Problem	**Solution**
"No light is coming from my illuminator."	Make sure the microscope is plugged into a working power outlet
	Check the power switch and make sure it is turned on
	Check the light control/voltage regulator and make sure it isn't turned all the way down
	If the first 3 steps don't solve the problem, see your instructor. The bulb may have burned out
"I can't find anything on my slide."	Go back to scanning power. Lower the stage as far as it will go. Look at the slide on the stage and position it so that the specimen is illuminated. Look through the eyepiece and, using the coarse adjustment knob, slowly bring the stage up until the specimen comes into view.
"I can't use both eyes to view the slide." or "I have to close one eye to view the slide."	Move your head back from the eyepieces slightly. If you are too close, it is more difficult to see a single image.
	Move the eyepieces (closer together or farther apart) until you see a single image through the microscope
	If the first 2 steps don't work, get a classmate to help you measure the distance between your pupils using a ruler. Then use the ruler to move the eyepieces apart that same distance.
"I see a dark crescent in the view."	Make sure the objective lens is clicked into place
	Adjust the condenser
"I can't get the specimen in clear focus."	Make sure the slide is not upside down on the mechanical stage

EXERCISE 3.2

VIEWING A SLIDE OF THE LETTER e

EXERCISE 3.2A Focusing the Microscope

1. Obtain a compound microscope and a slide of the letter *e*.

2. *Always begin your observation with the lowest possible total magnification (scanning objective in place).* Make sure the mechanical stage is in its lowest position, closest to the base of the microscope, then place the slide on the stage and turn on the illuminator. Adjust the light control/voltage regulator so it is somewhere near the middle or low end of its range.

3. Once you are sure that light is coming out of the illuminator, position the microscope stage so the letter *e* on the slide is over the opening in the stage where light comes through. Do this by looking directly at the microscope stage (not through the eyepiece[s]). This ensures that your specimen will be in the field of view, or at least very close to it, when you look through the lens.

4. Check to see if the scanning objective is in place over the stage, and look into the eyepiece(s). Using the coarse adjustment knob, slowly move the stage up until the specimen comes into view, and then use the fine adjustment knob to bring the specimen into focus. Next, play with the adjustments of the iris diaphragm and light control/voltage regulator to see how each affects the clarity and contrast of your image.

5. In the circle below, draw what you see within the field of view, and record the total magnification.

Total magnification = _____ ×

6. How does the image you see through the microscope differ from what you see when you look directly at the slide?

_____

_____

_____

_____

7. Without changing the position of the slide or the stage, rotate the nosepiece so the low-power objective is in place. Because you had the slide in focus with the scanning objective, you should need to use only the fine adjustment knob to bring the specimen into focus with the low-power objective. Each time you increase the magnification, you should need to use only the fine adjustment knob to focus the specimen, as long as you had the specimen in clear focus at the lower magnification.

8. Repeat step 7 using the high-power objective.

EXERCISE 3.2B Working Distance

The **working distance** is the distance between the mechanical stage (and the slide upon it) and the tip of the objective lens (**figure 3.3**). The shorter the working distance, the more likely it is that the objective lens will run into the slide or the stage. Because of this, only the fine adjustment knob should be used when viewing specimens using high-power objective lenses.

1. Begin with the microscope stage at its lowest position. Place the slide of the letter *e* on the stage and bring it into focus using the scanning objective.

2. Using a millimeter ruler, measure the working distance: the distance between the top of the slide on the stage and the bottom of the objective lens. Record the distance here: _____ mm.

3. Change to the low-power objective lens and repeat the process in step 2. Record the distance here: _____ mm.

4. Change to the high-power objective and repeat the process in step 2. Record the distance here: _____ mm.

5. Use your results in answering question 3 of the Post-Laboratory Worksheet (p. 54).

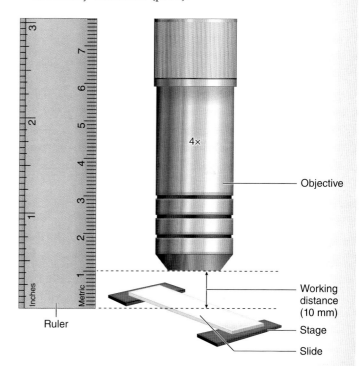

Figure 3.3 How to Measure Working Distance.

Diameter of the Field of View

The **field of view** is everything that is visible when you look through the eyepiece. You have probably noticed already that as magnification increases, the field of view decreases. In this activity you will determine the diameter of the field of view at the various magnifications of your microscope. The diameter of the field of view is simply the distance (usually given in millimeters [mm], and sometimes in micrometers [μm]) across the widest part of the field.

EXERCISE 3.3

MEASURING THE DIAMETER OF THE FIELD OF VIEW

1. Obtain a compound microscope and a stage micrometer or clear ruler with mm increments.

2. Begin with the scanning objective in place. Place the stage micrometer slide on the stage and position it so that you can see the markings within the field of view when you look through the eyepiece. Then line up the first marking with the left-hand side of the field of view at the widest part. If you are using a ruler, line up one of the mm lines with the left-hand side of the field of view at the widest part.

3. Count the number of markings you see across the widest part of the field of view and record the number: _____ mm. This is the diameter of the field of view at scanning power.

INTEGRATE

LEARNING STRATEGY

A *millimeter* (mm) is 10^{-3} meters. A *micrometer* (μm) is 10^{-6} meters. At higher magnifications, you should record the size of structures in micrometers instead of millimeters. To convert 0.2 mm to micrometers, simply move the decimal point three positions to the right. Thus, a distance of 0.2 mm is a distance of 200 μm.

Practice: If the diameter of the field of view at a magnification of 400× is 0.48 mm, what is the diameter in micrometers? _____ μm

4. Repeat steps 1 and 2 with the low-power objective in place, and record the diameter of the field of view at low power: _____ mm.

5. It is often difficult to use the method from steps 1 and 2 to determine the diameter of the field of view for the high-power objective. However, you can calculate the diameter based on the measurements you took for the scanning or low-power objectives using the following formula (LP = low-power objective, HP = high-power objective):

   diameter at LP × total magnification at LP =
   diameter at HP × total magnification at HP

Example:

- Diameter of the field of view using low-power objective: __4.8__ mm
- Total magnification using low-power objective: __40__ ×
- Diameter of field of view using high-power objective: __unknown__ (this is what you will calculate)
- Total magnification using high-power objective: __400__ ×
- Calculation: 4.8 mm · 40× = unknown · 400×
  192 = unknown · 400×
  192/400 = 0.48 mm (round off the number if necessary)

The diameter of the field of view at a magnification of 400× is 0.48 mm.

6. Use the information recorded in this exercise to answer question 3 of the Post-Laboratory Worksheet (p. 54).

EXERCISE 3.4

ESTIMATING THE SIZE OF A SPECIMEN

Now that you know the diameter of the field of view for each of the objective lenses of your microscope, you can determine the approximate size of a specimen by comparing the specimen's size to the diameter of the field of view.

1. **Figure 3.4** shows a specimen (outlined in red) at various magnifications. In the space below each image, record the approximate diameter of the specimen. Assume that the diameter of the field of view at 200× total magnification is 0.6 mm. Calculate the diameter of each field of view before estimating the length of the object.

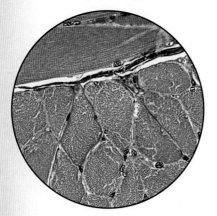 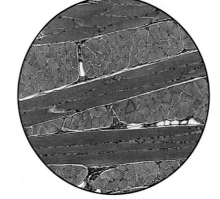 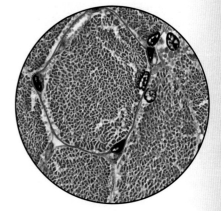

(a) Total magnification = __200×__
    Diameter of field = __0.6 mm__
    Diameter of object = _____

(b) Total magnification = __40×__
    Diameter of field = _____
    Diameter of object = _____

(c) Total magnification = __500×__
    Diameter of field = _____
    Diameter of object = _____

Figure 3.4 Estimating Specimen Size. The circle indicates the field of view. The specimen to be measured is outlined in red.

2. Is it possible to measure or estimate the length of the object in figure 3.4c? _____ If not, why not? _____ If yes, describe how to do it. _____

Depth of Field

The **depth of field** is how much of the total thickness of a specimen is in focus at each magnification. As total magnification increases, the depth of field decreases and a smaller portion of the specimen will be in focus. **Figure 3.5** demonstrates what happens to the depth of field as total magnification changes.

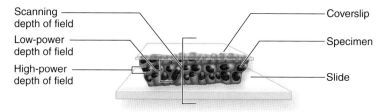

Figure 3.5 Depth of Field. Depth of field narrows as total magnification increases.

Finishing Up

When you are finished, follow proper cleanup procedures.

1. Remove the slide from the microscope stage and put it back where it belongs.
2. Turn the power switch to the "off" position, unplug the power cord, and wrap the cord neatly around the base of the microscope.
3. Rotate the nosepiece so the scanning power objective clicks into place.
4. Lower the microscope stage to its lowest position.
5. Put the dust cover back on the microscope.
6. Return the microscope to its proper storage location.

EXERCISE 3.5

DETERMINING DEPTH OF FIELD

1. Obtain a compound microscope and a slide of three crossed, colored threads.

2. Place the slide on the microscope stage and observe the slide with the scanning objective in place. Once you have the three crossed threads in the field of view, use the fine adjustment knob to focus up and down and determine which of the threads is on top, which is in the middle, and which is on the bottom. Then record your answers:

   Color of top thread: _____

   Color of middle thread: _____

   Color of bottom thread: _____

3. Now observe the slide with the low-power objective in place. Do you have to move the fine adjustment knob more or less than you did with the scanning objective in place to see all three threads? _____

4. Finally, observe the slide with the high-power objective in place. At this magnification, each individual thread will take up almost the entire diameter of the field of view. Is the entire thickness of an individual thread in focus at this magnification? _____. What does this tell you about the depth of field at this magnification?

   _____

INTEGRATE

CONCEPT CONNECTION

In each chapter of this laboratory manual, you will examine histological slides of cells, tissues, and organs. These slides will be used to determine both structure and function of the viewed specimens. For example, you will identify surface modifications on epithelial tissue, including cilia, which aid in movement, and microvilli, which aid in absorption. You will view specialized structures such as goblet cells that are responsible for mucus production in some epithelial tissues. With practice, viewing prepared histological slides is an excellent mechanism for truly learning rather than simply memorizing structures. Look for consistent structural patterns between tissue types and relate these patterns to the function of the cells, tissues, or organs. For example, you will find cilia associated with the epithelium of the respiratory tract, but you will also find cilia associated with the epithelium that lines the uterine tube. In both locations, cilia aid in movement of a substance along the epithelium (as in, responsible for clearing of the airways or in the moving of a fertilized ovum (egg) toward the uterus for implantation). As you view each histological slide, make every attempt to make connections between similar structures, their associated functions, and the body systems in which they are located.

Gross Anatomy

The Dissecting Microscope

A dissecting microscope (also called a stereomicroscope) is used to view structures that are larger than those viewed with a compound microscope, but that are often too small to see with the naked eye. The term "dissecting microscope" comes from the fact that these microscopes are useful for observing small dissections. The total magnification of a dissecting microscope is lower than that of a compound microscope, and the image is not inverted as it is with a compound microscope. If you place the letter *e* on a dissecting microscope stage and view it, you will see the letter *e* right side up, not upside down (try this if you need convincing). Recall that a compound microscope is used to view tissues that have been prepared on microscope slides. In contrast, objects that are viewed with a dissecting microscope need not be prepared in this way. Instead, the objects are simply placed on the stage of the microscope and illuminated with a light from above (and sometimes below). You generally use a dissecting microscope when you want to view something small, like a human hair close up to see its gross structure in more detail. If you want to see the actual cells and tissues that compose a human hair then you will need to use a compound microscope. Like a compound microscope, a dissecting microscope has both ocular and objective lenses, but the objective lens(es) are much lower power, and there are two of them—one for each eye. Thus, each of your eyes looks through the microscope independently of the other, which allows you to have a "stereo" or three-dimensional view of the specimen. In many cases, the objective lenses are also of varying powers, with typical magnification ranges of 1× to 3×.

As you work through this exercise, note that many of the parts of a dissecting microscope have the same names as those of a compound microscope. **Figure 3.6** demonstrates an example of a dissecting microscope. This microscope uses an external illuminator that inserts into the back of the base of the microscope (not visible in the figure). The mirror below the stage then reflects the light through the glass on the stage to illuminate a specimen placed upon the stage. Although it is often useful to illuminate specimens from below, many times you may wish to illuminate the specimen from above. Many dissecting microscopes have both above- and below-stage illuminators for this very purpose. The microscope shown in figure 3.6 has only a below-stage illuminator.

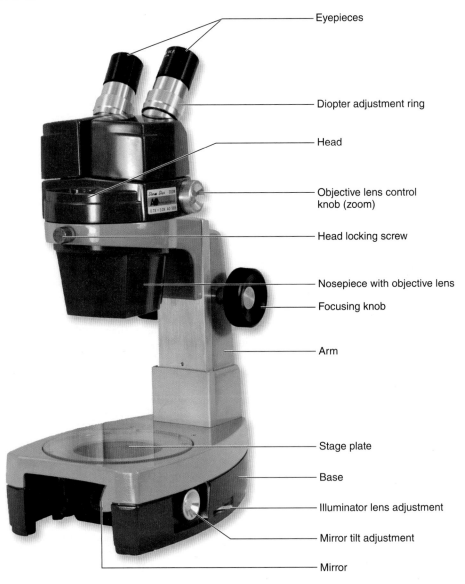

Figure 3.6 Parts of a Dissecting Microscope.

EXERCISE 3.6

PARTS OF A DISSECTING MICROSCOPE

1. Obtain a dissecting microscope.

2. Using figure 3.6 and table 3.1 as guides, identify the following structures on the dissecting microscope.

   - ☐ arm
   - ☐ base
   - ☐ diopter adjustment ring
   - ☐ eyepieces
   - ☐ focusing knob
   - ☐ head
   - ☐ head locking screw
   - ☐ illuminator lens adjustment
   - ☐ mirror
   - ☐ mirror tilt adjustment
   - ☐ nosepiece with objective lens
   - ☐ objective lens control knob
   - ☐ stage plate

3. Record the magnifications of the lenses on your microscope in the spaces below.

   Magnification of ocular lenses: _____

   Magnification of objective lenses:

   Lowest _____   Highest (if applicable) _____

4. What is the total magnification of your microscope? If there is more than one objective lens, list the lowest and highest total magnifications possible.

   Total magnification:

   Lowest _____   Highest (if applicable) _____

5. Place a pen or pencil on the microscope stage and play with the controls (focus, zoom, illuminator, etc.) while you observe the object through the ocular lenses. Place a piece of paper with text written on it on the microscope stage and view that through the ocular lenses. Is there any change in the orientation of the letters as seen through the microscope as compared to viewing the letters with your naked eye? _____ If so, describe what change(s) you see. _____

   _____

   _____

   _____

   _____

   _____

   _____

Chapter 3: The Microscope

Name: _____
Date: _____ Section: _____

POST-LABORATORY WORKSHEET

The ❶ corresponds to the Learning Objective(s) listed in the chapter opener outline.

Do You Know the Basics?

Exercise 3.1: Parts of a Compound Microscope

1. Label the parts of the compound microscope. ❶

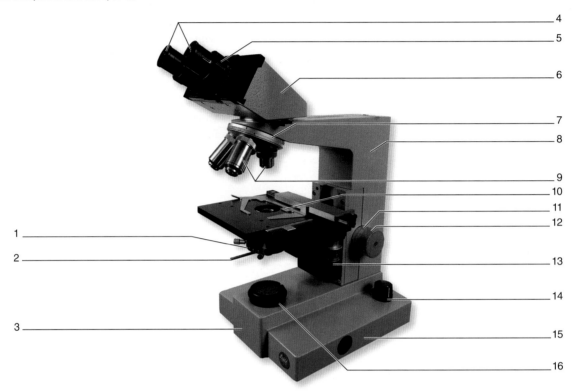

2. Describe the following about caring for a microscope: ❶

   a. Transporting _____

   b. Positioning on lab table_____

   c. Cleaning the microscope lenses_____

   d. Storing the microscope_____

Exercise 3.2: Viewing a Slide of the Letter e

3. What is the total magnification of a microscope set up with an ocular lens magnification of 10× and an objective lens magnification of 43×?_____ ❷

Exercise 3.3: Measuring the Diameter of the Field of View

4. Complete the following table with the numbers you calculated in exercises 3.1, 3.2, and 3.3. ❸❹❺

Power	Ocular Magnification	Objective Magnification	Total Magnification	Diameter of the Field of View	Working Distance
Scanning					
Low					
High					

Exercise 3.4: Estimating the Size of a Specimen

5. Refer back to your calculations on the sample "specimens" in figure 3.4. Enter information from your calculations in the spaces below. Pay attention to the units you are asked to use, because they are not all the same. ❻

A.

Total magnification = __200×__

Diameter of field = __0.6 mm__

Length of object = _____ mm

B.

Total magnification = __40×__

Diameter of field = _____ mm

Length of object = _____ mm

C.

Total magnification = __500×__

Diameter of field = _____ μm

Length of object = _____ μm

Exercise 3.5: Determining Depth of Field

6. Record your answers to the questions about colored threads in the spaces below. ❼

 Color of top thread: _____

 Color of middle thread: _____

 Color of bottom thread: _____

7. Explain why proper microscope technique requires that you always begin viewing a slide with the scanning objective first before moving to higher power objectives. Use the concept of *depth of field* in your explanation. ❼

 _____

 _____

 _____

 _____

 _____

Exercise 3.6: Parts of a Dissecting Microscope

8. Label the parts of the dissecting microscope in the figure below. ❽

 1 _____
 2 _____
 3 _____
 4 _____
 5 _____
 6 _____
 7 _____
 8 _____
 9 _____
 10 _____
 11 _____
 12 _____
 13 _____

9. List three types of specimens you might use a dissecting microscope to view: ❾

 a. _____

 b. _____

 c. _____

Chapter Three The Microscope

Can You Apply What You've Learned?
Exercise 3.1: Parts of a Compound Microscope

10. What microscope structures are used to control the amount of light illuminating the specimen?

11. What happened to the light intensity when you switched from low to high power?

12. What adjustment will you typically have to make to the light after changing from the low-power to the high-power objective?

13. a. How does working distance change as total magnification increases?

    b. What are the practical consequences of this change in working distance?

14. If you can see four cells within the field of view at its maximum diameter at a total magnification of 200×, how many cells will you be able to see at a total magnification of 500×?

PART II | ORGANIZATION OF THE HUMAN BODY

CHAPTER 4

Cell Structure and Membrane Transport

INTRODUCTION

The cell is the basic unit of life. Organisms can be unicellular or multicellular, but they must be composed of cells to be considered living entities. Human beings are, of course, multicellular organisms. Our bodies are composed of **tissues:** groups of similar cells and associated extracellular materials that function together as a unit. The study of tissues is called **histology** [*histos*, web (tissue), + *logos*, study]. To understand the study of histology, you must first become comfortable identifying cells and cellular organelles under the microscope. Most cells are easily seen with a light microscope, but most cellular organelles are too small to be seen without the use of a more powerful electron microscope. The purpose of this laboratory session is to introduce you to the microscopic appearance of cells, give you a brief review of cellular anatomy and the stages of mitosis, and help you become more confident in your use of the compound microscope.

Most of the slides that you will view in the anatomy laboratory have been stained with hematoxylin and eosin, and are labeled "H and E." This stain makes the **cytoplasm** of the cell appear pink in color and makes visible the outline of the cell where the **plasma membrane (cell membrane)** is located. The **nucleus** of the cell appears dark purple in color, and the **nucleolus** often appears as a dark spot within the nucleus. Though these structures have just been described using references to color, you should not generally ➡

OUTLINE AND LEARNING OBJECTIVES

Histology 61

Structure and Function of a Typical Animal Cell 61

EXERCISE 4.1: OBSERVING CELLULAR ANATOMY 63
1. Prepare a wet mount of human cheek cells
2. Identify nucleus, nucleolus, cytoplasm, and plasma membrane in a human cheek cell
3. Explain the importance of using an isotonic solution in preparing a wet mount
4. Associate a function with each cellular organelle

Mitosis 65

EXERCISE 4.2: OBSERVING MITOSIS IN A WHITEFISH EMBRYO 66
5. Identify the four stages of mitosis and interphase in a slide of a whitefish embryo/blastula
6. Describe the events that occur during each stage of mitosis/blastula

Gross Anatomy 66

Models of a Typical Animal Cell and Stages of Mitosis 66

EXERCISE 4.3: OBSERVING CLASSROOM MODELS OF CELLULAR ANATOMY AND MITOSIS 67
7. Identify cellular organelles and the four stages of mitosis and interphase on classroom models

Physiology 67

Mechanisms of Membrane Transport 67

EXERCISE 4.4: DIFFUSION (WET LAB) 67
8. Observe the effects of temperature, solvent viscosity, solute molecular weight, and membrane permeability on rates of diffusion

EXERCISE 4.5: OSMOSIS (WET LAB) 77
9. Observe osmosis across an artificial membrane
10. Observe the effect selective permeability of an artificial membrane has on diffusion

EXERCISE 4.6: FILTRATION (WET LAB) 79
11. Observe the process of filtration
12. Describe the association between fluid (hydrostatic) pressure and filtration rate

EXERCISE 4.7: **Ph.I.L.S.** LESSON 1: OSMOSIS AND DIFFUSION: VARYING EXTRACELLULAR CONCENTRATION 81
13. Identify independent and dependent variables in a virtual laboratory experiment
14. Measure the color of blood solutions using a virtual spectrophotometer
15. Explore the concept of osmosis using virtual blood samples with varying solute concentrations (hypertonic, isotonic, and hypotonic)
16. Determine the concentration of sodium chloride solution that is isotonic to red blood cells

MODULE 2: CELLS & CHEMISTRY

learn to recognize cell structures based on color alone. Instead, you should learn to recognize cells and cellular organelles based on *shape*. If you do this, you will have an easier time when observing slides stained with stains other than H and E, which can produce different colors.

Most animal cells are transparent, so when tissue samples are prepared for use in the anatomy and physiology laboratory, the slides are stained so cellular details will be visible when viewed under a microscope. Different parts of a cell attract the biological stains to different degrees, which makes some parts of the cell appear darker in color, and others appear lighter in color, or even transparent. The nucleus of the cell has a high attraction for most biological stains, so it is often the most recognizable part of a cell.

The exercises in the "histology" and "gross anatomy" sections of this chapter are designed to assist you with identification of cellular organelles that are visible using a light microscope, and with identification of the stages of mitosis in a whitefish embryo. The exercises in the "physiology" section of this chapter are mainly "wet lab" activities that demonstrate the processes of diffusion, osmosis, and filtration.

Physiologically, one of the most important components of a cell is the plasma membrane, which is **selectively permeable.** The selective permeability of the plasma membrane allows the cell to control both its internal and external environment through the processes of membrane transport. Mechanisms of membrane transport include simple diffusion, facilitated diffusion, osmosis, active transport and **vesicular transport.** In this laboratory session you will perform experiments that will allow you to observe the effect of factors such as temperature on the rate of diffusion, the process of osmosis across an artificial membrane such as dialysis tubing, and the process of simple filtration. You will also perform a simulated (Ph.I.L.S.) activity in which you will observe the effect of placing erythrocytes (red blood cells) in solutions that mimic changes in extracullular concentration.

Chapter 4: Cell Structure and Membrane Transport

Name: _____
Date: _____ Section: _____

PRE-LABORATORY WORKSHEET

Also available at www.connect.mcgraw-hill.com

1. Match the cell structures in column A with their functions in column B.

   Column A
   ___ 1. centrioles
   ___ 2. chromatin
   ___ 3. cytoplasm
   ___ 4. cytoskeleton
   ___ 5. rough endoplasmic reticulum (RER)
   ___ 6. Golgi apparatus
   ___ 7. lysosomes
   ___ 8. mitochondria
   ___ 9. nucleolus
   ___ 10. nucleus
   ___ 11. peroxisomes
   ___ 12. plasma membrane
   ___ 13. ribosomes

   Column B
   a. provides a selectively permeable barrier between the intracellular and extracellular environments of the cell
   b. includes cellular organelles and cytosol; cytosol contains enzymes that mediate many cytosolic reactions, such as glycolysis and fermentation
   c. contains the cell's genetic material (DNA)
   d. synthesizes rRNA and assembles ribosomes in the nucleus
   e. genetic material within the nucleus; consists of uncoiled chromosomes and associated proteins
   f. synthesizes new proteins destined for the plasma membrane, for lysosomes, or for secretion from the cell
   g. sites of protein synthesis; may be bound to the ER ("bound") or found within the cytoplasm ("free")
   h. a stack of flattened membranes that are the site where proteins from the ER are modified, packaged, and sorted for delivery to other organelles or to the plasma membrane of the cell
   i. membrane-enclosed sacs that contain digestive enzymes and function in the breakdown of intracellular structures
   j. membrane-enclosed sacs that contain catalase and other oxidative enzymes
   k. often referred to as the "powerhouse" of the cell, these organelles are the site of cellular respiration
   l. paired organelles composed of microtubules that are used to organize the spindle microtubules that attach to chromosomes during mitosis
   m. composed of protein filaments called microtubules, intermediate filaments, and microfilaments; provides the main structural support for the cell

2. Define *mitosis*: _____

3. List the four stages of mitosis (in order):

   a. _____

   b. _____

   c. _____

   d. _____

4. The stage of the cell cycle when cells are *not* undergoing mitosis is _____. During this phase, individual chromosomes are/are not (circle one) visible within the nucleus of the cell.

5. Define the following terms:

   a. *solute*

   _____

   _____

Chapter Four *Cell Structure and Membrane Transport* 59

b. *solvent*

c. *diffusion*

d. *hypotonic*

e. *hypertonic*

f. *isotonic*

g. *osmosis*

6. In the space below, list three factors that affect the rate of diffusion of a substance:

   a. _____

   b. _____

   c. _____

7. What do you expect to happen to a red blood cell when you place it in a hypertonic solution?

Histology

Structure and Function of a Typical Animal Cell

Table 4.1 lists the parts of a typical animal cell and gives descriptions of their functions and microscopic features. The parts of a typical animal cell that are most readily visible under a light microscope are the nucleus, the nucleolus, and the boundary of the cell, which is where the plasma membrane is located. Thus, as you observe animal cells under the microscope, your focus will be on finding those structures in particular. Once you have learned to recognize what parts of an animal cell are most typically visible under the light microscope, you will be ready to observe the different cell types that are presented in future laboratory exercises.

Table 4.1 Parts of a Typical Animal Cell

Organelle / Structure	Function	Microscopic Features	Word Origin	Appearance
Centrioles	Paired organelles that are used to organize the spindle microtubules that attach to chromosomes during mitosis. The area next to the nucleus that contains the centrioles is called the *centrosome*.	Visible only when a cell is actively undergoing nuclear division (mitosis)	*kentron*, center	
Chromatin	Genetic material within the nucleus; consists of uncoiled chromosomes and associated proteins	Most of the coloration seen in the nucleus, with exception of the nucleolus, consists of chromatin	*chroma*, color	
Cytoplasm	Includes cellular organelles and cytosol; cytosol contains enzymes that mediate cytosolic reactions such as glycolysis and fermentation	Clear and homogeneous in appearance; may contain granular substances such as glycogen in certain cells (e.g., hepatocytes)	*kytos*, a hollow (cell), + *plasma*, something formed	
Cytoskeleton	Provides the main structural support for the cell and is composed of microtubules, intermediate filaments, and microfilaments	Not generally visible under the light microscope	*kytos*, a hollow (cell), + *skeletos*, dried	
Endoplasmic Reticulum (ER)	Site of lipid synthesis and detoxification of drugs and alcohol (smooth ER). Additionally, rough ER synthesizes proteins destined for the cell membrane, for lysosomes, or for secretion.	Not generally visible under the light microscope. In neurons, the rough ER stains very dark and is called chromatophilic substance (Nissl bodies).	*endon*, within, + *plasma*, something formed, + *rete*, a net	

(continued on next page)

Table 4.1 Parts of a Typical Animal Cell (continued)

Organelle / Structure	Function	Microscopic Features	Word Origin	Appearance
Golgi Apparatus	A stack of flattened membranes that receive proteins from the rough ER and then modify, package, and sort them for delivery to other organelles or to the plasma membrane of the cell	Not generally visible under a light microscope	*Golgi*, Camillo, Italian histologist and Nobel laureate, 1843–1926	
Lysosomes	Membrane-enclosed sacs that contain digestive enzymes; function in the breakdown of intracellular debris	Not generally visible under a light microscope	*lysis*, a loosening, + *soma*, body	
Mitochondria	Often referred to as the "powerhouse" of the cell. These organelles are the site of cellular respiration, the metabolic pathway that utilizes oxygen in the breakdown of food molecules to produce ATP	Not generally visible under a light microscope	*mitos*, thread, + *chondros*, granule	
Nucleolus	Synthesizes rRNA and assembles ribosomes in the nucleus	Recognized as a small, dark, circular structure within the nucleus	*nucleus*, a little nut	
Nucleus	Contains the cell's genetic material (DNA)	The most noticeable feature of a cell; typically stains very dark	*nucleus*, a little nut	
Peroxisomes	Membrane-enclosed sacs that contain catalase and other oxidative enzymes. The enzymes break down lipids and toxic substances by first converting them into hydrogen peroxide and then breaking down the hydrogen peroxide into water and oxygen.	Not generally visible under a light microscope	*peroxi*, relating to hydrogen peroxide, + *soma*, body	
Plasma Membrane	Provides a selectively permeable barrier between the intracellular and extracellular environments of the cell	Visible only using an electron microscope. However, the outer border of the cell, where the cell membrane is located, is often visible under the light microscope.	*plasma*, something formed, + *membrane*, a membrane	
Ribosomes	Sites of protein synthesis: may be bound to the ER ("fixed") or found within the cytoplasm ("free")	Not generally visible under a light microscope	*ribose*, the sugar in RNA, + *soma*, body	

EXERCISE 4.1

OBSERVING CELLULAR ANATOMY

Preparing a Wet Mount of Human Cheek Cells

For this exercise you will be taking a sample of cells from the inside of your cheek and preparing a wet mount. A wet mount is a procedure that involves placing a tissue sample in a wet medium onto a microscope slide. The "wet" medium is typically an isotonic saline solution. Why is it important for the solution to be isotonic? _____
_____

The cells on the inside of your cheek are squamous cells, which are flattened cells. The inside of your cheek is lined with multiple layers of these cells. Therefore, you may gently scrape off a few cells without causing damage to the entire epithelium (lining) of the inside of your mouth. You will learn more about these cells in the context of epithelial tissues in chapter 5.

Once you have obtained these cells, you will place them on a microscope slide, apply a stain (methylene blue), and then cover them with a coverslip. The stain is necessary to visualize the cells because normal cells are nearly transparent. The stain is basophilic (base-loving) and is attracted to eosinophilic (acid-loving) components of the cell. The most eosinophilic part of the cell is the nucleus, which contains the nucleic acids DNA and RNA. Thus, the nucleus of the cell stains more intensely than other parts of the cell.

1. Obtain the following:
   a. compound microscope
   b. microscope slide and coverslip
   c. toothpick or wood applicator stick
   d. methylene blue solution with eyedropper
   e. fine tissue paper or KimWipes®

2. Place the microscope slide on a piece of white paper on your lab bench. The piece of white paper will make your observations easier.

3. Place a small drop of normal saline on the microscope slide **(figure 4.1a)**. Using the toothpick, very *gently* scrape the inside of your cheek. All you need to do is run the toothpick gently along the inside of your mouth and it will pick up a few cells. This should not be painful, and most definitely should not draw blood!

4. Next, place the tip of the toothpick in the drop of saline on your slide (figure 4.1b) and roll it around gently so the cells detach from the toothpick and drop into the drop of saline.

5. Obtain a vial of methylene blue solution. Place a single drop of methylene blue on the drop of saline containing cheek cells (figure 4.1c).

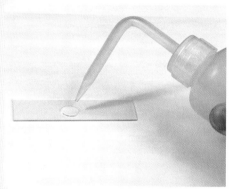

(a) Place a drop of normal saline on the slide.

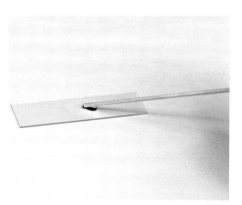

(b) After collecting cheek cells, gently roll the tip of the toothpick in the drop of saline.

(c) Place a drop of methylene blue on the slide.

(d) Slowly lower a cover slip over the drop of liquid containing cheek cells.

(e) Make sure there are no air bubbles between the cover slip and the slide. The slide is now ready to view with the microscope.

Figure 4.1 Preparing a Wet Mount of Human Cheek Cells.

(continued on next page)

(continued from previous page)

6. Obtain a coverslip and place it on the edge of the liquid on the microscope slide as shown in figure 4.1d. Carefully and slowly lower the coverslip onto the drop of liquid. Your goal is to place the coverslip over the drop of saline without introducing air bubbles. If you simply drop the coverslip on the slide you will end up with large air bubbles, which interfere with your ability to see the cells on the slide. If you carefully and slowly lower the coverslip, starting on one side and lowering it down at an angle (figure 4.1d), most air bubbles will be pushed out of the way as the coverslip is placed. Obtain a piece of tissue paper or a KimWipe® and use it to dab any excess liquid on the sides of the coverslip (if necessary). A successful wet mount of cheek cells should resemble that shown in figure 4.1e.

7. Observe the slide with your naked eye. Can you see anything on the slide? In particular, can you see any cheek cells? _____

8. Arrange the objective lens on your microscope so it is set to use the scanning objective. Place the slide containing the cheek cells on the microscope stage and bring the tissue sample into focus using the scanning objective. Next, change to a higher power and bring the tissue sample into focus once again. The cells you see should somewhat resemble those in **figure 4.2**, although they will generally be isolated from each other rather than being in a sheet as in figure 4.2.

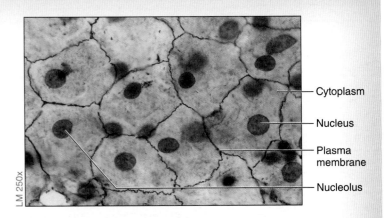

Figure 4.2 Human Cheek Cells.

9. Scan the slide until you see cheek cells in your field of view. In the space on the next page sketch the cheek cells as seen through the microscope. Label the following on your sketch:

☐ cytoplasm
☐ nucleolus
☐ nucleus
☐ plasma membrane

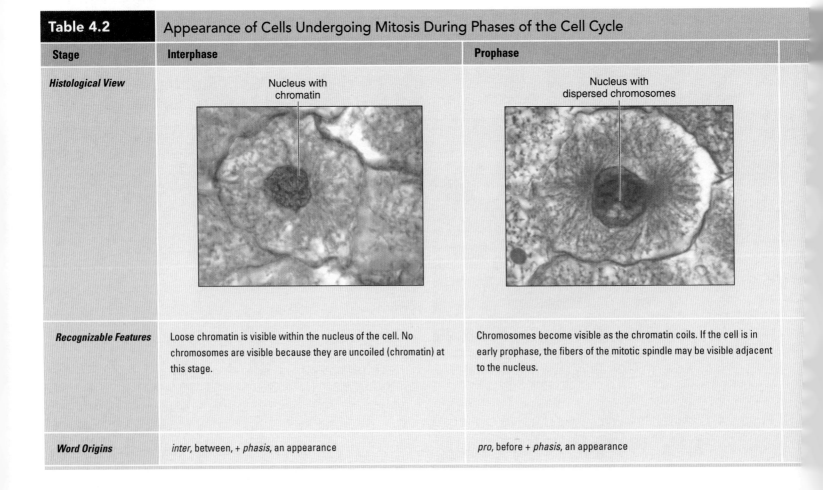

Table 4.2	Appearance of Cells Undergoing Mitosis During Phases of the Cell Cycle	
Stage	**Interphase**	**Prophase**
Histological View	Nucleus with chromatin	Nucleus with dispersed chromosomes
Recognizable Features	Loose chromatin is visible within the nucleus of the cell. No chromosomes are visible because they are uncoiled (chromatin) at this stage.	Chromosomes become visible as the chromatin coils. If the cell is in early prophase, the fibers of the mitotic spindle may be visible adjacent to the nucleus.
Word Origins	*inter*, between, + *phasis*, an appearance	*pro*, before + *phasis*, an appearance

_____ ×

Troubleshooting:

If you see cellular debris on your slide instead of intact cells, it is likely that you mistakenly used distilled water instead of saline when making your slide. Distilled water is a hypotonic solution. When cells are placed in a hypotonic solution, they will lyse, and you will see only cellular debris on the slide.

10. *Optional Activity:* **AP|R** **2: Cells & Chemistry**—Examine the "Generalized cell" dissection and test yourself on cell structures in the Quiz area.

Mitosis

The cell cycle describes the events that occur during the process of forming a new cell. As a cell passes through the stages of the cell cycle two identical daughter cells are formed from one original parent cell. The cell cycle is divided into two main phases: interphase and mitotic phase. Interphase is the time in which the genetic material is uncoiled as chromatin. Mitotic phase (*mitos*, thread) includes the processes by which cells reproduce and includes mitosis (nuclear division) and cytokinesis (*cyto-*, cell, + *kinesis*, movement), which is division of the cytoplasm. Note that casual usage of the term "mitosis" usually implies that the cytoplasm and cellular organelles have also divided. Some cells undergo mitosis without undergoing cytokinesis. The resulting cell thus becomes a *multinuclear* (*multus*, much, + *nuclear*, nucleus) cell. **Table 4.2** describes the microscopic appearance of cells in interphase and each of the four stages of mitosis. In this laboratory exercise your goal is to locate cells in each of the stages of mitosis by observing whitefish embryos (blastulas). Whitefish embryos are very small and are rapidly developing, which makes them ideal specimens for observations of cells undergoing mitosis.

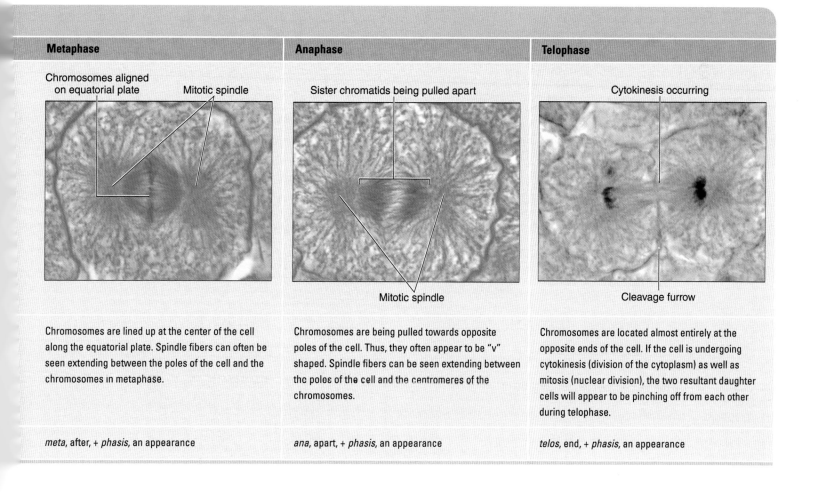

Metaphase	**Anaphase**	**Telophase**
Chromosomes are lined up at the center of the cell along the equatorial plate. Spindle fibers can often be seen extending between the poles of the cell and the chromosomes in metaphase.	Chromosomes are being pulled towards opposite poles of the cell. Thus, they often appear to be "v" shaped. Spindle fibers can be seen extending between the poles of the cell and the centromeres of the chromosomes.	Chromosomes are located almost entirely at the opposite ends of the cell. If the cell is undergoing cytokinesis (division of the cytoplasm) as well as mitosis (nuclear division), the two resultant daughter cells will appear to be pinching off from each other during telophase.
meta, after, + *phasis*, an appearance	*ana*, apart, + *phasis*, an appearance	*telos*, end, + *phasis*, an appearance

Chapter Four Cell Structure and Membrane Transport

EXERCISE 4.2

OBSERVING MITOSIS IN A WHITEFISH EMBRYO

1. Obtain a compound microscope and a prepared slide of a whitefish embryo (blastula) or other slide of cells undergoing mitosis.
2. Using table 4.2 as a guide, scan the slide and locate cells in the following four stages of mitosis and interphase. Once you locate a cell in a particular phase, switch to a higher-power objective to see the cell more clearly.
   - ☐ anaphase
   - ☐ interphase
   - ☐ metaphase
   - ☐ prophase
   - ☐ telophase
3. Sketch cells in each of these phases (as viewed under the microscope) in the spaces provided for you in the post-laboratory worksheet (question #4) on p. 85.

Gross Anatomy

Models of a Typical Animal Cell and Stages of Mitosis

In exercise 4.3 you will observe classroom models demonstrating a typical animal cell and the stages of mitosis. **Figure 4.3** is a photograph of a classroom model of a typical animal cell, which has the organelles labeled for your reference.

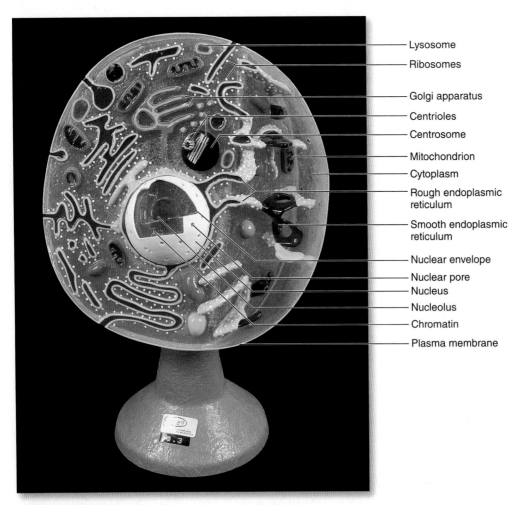

Figure 4.3 Classroom Model of a Typical Animal Cell.

EXERCISE 4.3

OBSERVING CLASSROOM MODELS OF CELLULAR ANATOMY AND MITOSIS

1. Obtain classroom models demonstrating a typical animal cell and the stages of mitosis.

2. Using figure 4.3, table 4.1, and your textbook as guides, identify the following structures on a classroom model of a typical animal cell.

   ☐ centrioles   ☐ nucleus
   ☐ chromatin   ☐ peroxisome
   ☐ cytoplasm   ☐ plasma membrane
   ☐ Golgi apparatus   ☐ nuclear pore
   ☐ lysosome   ☐ ribosomes
   ☐ mitochondria   ☐ rough ER
   ☐ nuclear envelope   ☐ smooth ER
   ☐ nucleolus

3. Using figure 4.3, table 4.2, and your textbook as guides, identify all of the stages of mitosis on a classroom model of cells undergoing mitosis.

   ☐ anaphase   ☐ metaphase   ☐ telophase
   ☐ interphase   ☐ prophase

Physiology

Mechanisms of Membrane Transport

Before you begin, use your textbook as a guide to familiarize yourself with the following concepts: kinetic energy, active transport, passive transport, molecular weight, simple diffusion, osmosis, filtration, hypertonic, hypotonic, isotonic, and membrane impermeable solute.

EXERCISE 4.4

DIFFUSION (WET LAB)

Diffusion (*diffundo*, to pour in different directions) refers to the net movement of solute particles from an area of high concentration to an area of low concentration. Diffusion occurs whenever a concentration gradient exists. If we consider a solution, the substance dissolved in the solution is the **solute,** and the substance in which the solute is dissolved is the **solvent.** When we work with solutions in the anatomy and physiology laboratory, the solvent will typically be water.

The **second law of thermodynamics** (also called the **law of entropy**) is the law governing the natural behavior of molecules, including molecules in solution. This law states that, over time, all molecules move at random toward a state of increasing disorder. **Entropy** (*entropia*, a turning toward) is a measure of the disorder of a system. Thus, for a system to become more disorderly, no energy input is required (as the saying goes, "entropy happens"). Similarly, energy must be put into a system for the system to become more orderly. How does the law of entropy apply to diffusion? Consider the situation in **figure 4.4**. The purple dots represent solute particles in solution. On the left of figure 4.4*a* the solution is more concentrated, and on the right of figure 4.4*a* the solution is more dilute. If you observe each particle in relation to the other solute particles, you will notice that where the solution is more concentrated, the particles are packed close together, and thus are more orderly. Where the solution is more dilute, the solute particles are farther away from each other and thus are more disorderly. If we follow the movement, or **flux** (*fluxus,* to flow), of the solute particles over time, we should expect the random

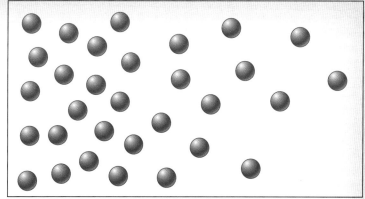

(a)

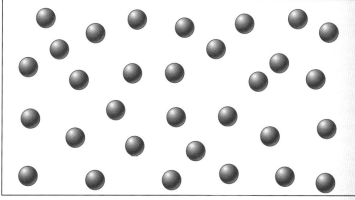

(b)

Figure 4.4 Diffusion. (a) Initially concentrated molecules.
(b) Molecule distribution after diffusion takes place.

(continued on next page)

(continued from previous page)

movement of solute particles in solution to result in a situation where the particles become as disorderly as possible. This means each particle will be as far away from another particle as is possible. Figure 4.4b shows what happens if we allow diffusion to take place.

When two solutions of differing concentrations are separated from each other by a semipermeable membrane that is permeable to the solute in question, solute particles will move at random both with and against the concentration gradient because of their inherent kinetic energy (kinetic energy is energy of *motion*). Thus, there will be two one-way fluxes: A → B and B → A **(figure 4.5)**. However, the *net* movement of particles (the difference between the two one-way fluxes: A → B and B → A) will be from an area of high concentration to an area

INTEGRATE

CONCEPT CONNECTION

The movement of substances into and out of the cell is an important concept that provides the foundation for many physiological processes. For example, passive and active transport are the basic processes that make our neurons and muscle cells "excitable." These cells, as with all cells, contain a plasma membrane that is selectively permeable, allowing water and small, mainly lipid-soluble substances to pass freely into and out of the cell by passive transport. Protein pumps embedded within the plasma membrane actively pump ions (active transport) such as Na^+ and K^+, maintaining the appropriate concentrations of solutes in the intracellular fluid (ICF) and extracellular fluid (ECF). When protein channels open, these ions can flow along their concentration gradients from an area of high concentration to an area of low concentration (passive transport). In doing so, they also carry positive charges across the membrane. The movement of charges across the membrane is responsible for generating the electrical signals called action potentials. This process will be described in much more detail in chapter 14. You will continue to see examples of passive and active transport throughout this course.

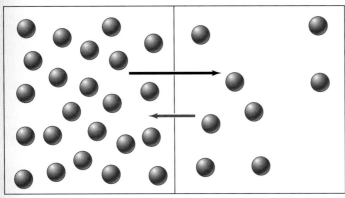
Container A   Container B
(a)

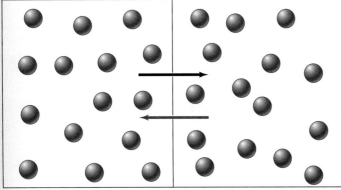
Container A   Container B
(b)

Figure 4.5 Diffusion Through a Semipermeable Membrane. (a) Diffusion of solute from A → B is greater than the diffusion of solute from B → A. Thus, the net movement is from A → B. (b) Once equilibrium has been reached, the movement from A → B is equal in magnitude but opposite in direction of the movement from B → A. Thus, there is no net movement at equilibrium, and the concentration of solute in both containers is the same.

of low concentration. Once the concentration of solute is the same on both sides of the membrane, the solution is said to be at *equilibrium*. It is important to note that although equilibrium has been reached at that point, this does not mean that solute particles have stopped moving across the membrane. On the contrary, particles continue to move at random and will still cross the membrane. However, the rate at which particles move from container A to container B will be equal in magnitude but opposite in direction to the rate at which particles move from container B to container A. Thus, *no net movement* will occur. This is how the point of equilibrium is defined.

Some factors that affect the rate at which a substance diffuses include:

1. viscosity of the solvent
2. temperature
3. molecular weight of the substance
4. permeability of the membrane

In the following exercises you will observe the effect of each of these factors on rates of diffusion. To demonstrate the effect of the first three factors, you will observe the diffusion of two different dyes, potassium permanganate (molecular weight 158) and methylene blue (molecular weight 374), in petri dishes containing water or agar. To demonstrate the effect of membrane permeability on diffusion, you will observe diffusion of iodine and/or starch across an artificial membrane (dialysis tubing).

EXERCISE 4.4A Effect of Solvent Viscosity on Rate of Diffusion

In this exercise, you will observe the movement of a solute (potassium permanganate) in two solvents (water and agar). Agar is a gel-like substance that is made from algae. Agar is 98% water, so molecules will diffuse through it. However, agar is more viscous (thicker) than water. In this experiment you will place a crystal of potassium permanganate in a petri dish containing agar and another crystal of potassium permanganate in a petri dish containing water. You will then measure the rate of diffusion in each dish. Before you begin, state a hypothesis regarding the effect that viscosity has on the rate of diffusion.

Hypothesis: _____

_____

_____

Pour a small amount of potassium permanganate into a weigh boat or other small container. Obtain a single moderate-sized crystal using tissue forceps.

Figure 4.6 Obtaining Potassium Permanganate Crystals.

1. Obtain the following:
   a. letter-size (8.5" × 11") piece of white paper
   b. petri dish containing distilled water
   c. petri dish containing agar
   d. fine tissue forceps
   e. potassium permanganate crystals
   f. one large (~30 cm) and 2 small (~15 cm) metric rulers
   g. a stopwatch or other device for recording the time

2. Obtain a small weigh boat containing potassium permanganate crystals **(figure 4.6)**. You will only need two moderate-sized crystals for this experiment, one for the agar dish and one for the water dish, so only obtain a small spoonful of crystals. Be aware! Potassium permanganate is a dye. Wear gloves when handling the crystals and take care not to spill the crystals on your lab table, books, and clothing because they will permanently dye those items. If you spill any crystals, first sweep them up with a dry brush or cloth. Try not to use a wet towel until you have cleaned up most of the crystals because the water will create a liquid dye that can spread more easily (and thus stain more items) than the powdered form.

3. Place the rulers and the petri dishes on the white paper as shown in **figure 4.7**. The large ruler will be used as a shim to keep the petri dishes level, whereas the small rulers will be used to measure rates of diffusion. Place the large ruler on the left side of the paper and orient it vertically. Place the small rulers at right angles to the large ruler, approximately 8 cm apart from each other, and with the "zero" point approximately at the center of the petri dish.

4. Label the petri dishes "A – distilled water" and "B – agar" on the paper in the space next to each dish (see figure 4.7).

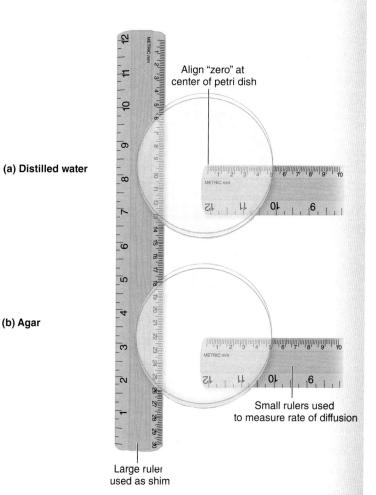

Figure 4.7 Setup of Petri Dishes and Rulers for Exercise 4.4A.

(continued on next page)

(continued from previous page)

5. You will first observe diffusion of dye crystals in **water** (**figure 4.8**). This process happens fairly quickly, so be prepared before you begin. One lab partner will place the crystal and record the distance diffused, while another will start the stopwatch and record time intervals. To begin: Carefully place a single potassium permanganate crystal in the center of the petri dish containing water (dish A) at the "0" mark on the ruler. The moment the crystal is placed in the water, start the stopwatch and observe diffusion of the dye. Record the radius of the dye spot in 30-second intervals for a total of at least 5 minutes (see **figure 4.9** for a description of how to measure diffusion distance). Record your data in **table 4.3**. In the space below, note your observations:

_____

_____

_____

_____

_____

6. You will next observe diffusion of dye crystals in **agar**. This process happens fairly slowly, so you need not be as "at the ready" as you were for the water diffusion exercise. To begin, carefully place a single potassium permanganate crystal in the center of the petri dish containing agar (dish B) at the "0" mark on the ruler. The moment the crystal is placed on the agar, start the stopwatch and observe diffusion of the dye. Record the radius of the dye spot in 15-minute intervals for a total of 2 hours. Record your data in table 4.3.

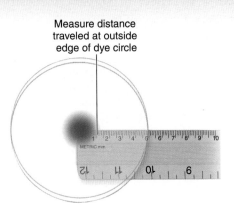

Figure 4.9 Measuring the Diffusion Distance.

7. In the space below, make a note regarding your conclusions of this experiment (i.e., did it support or refute your hypothesis?). Your notes here need not be extensive, just enough for you to refer back to when completing the post-laboratory worksheet.

_____

_____

_____

_____

_____

_____

8. Use the data from table 4.3 to complete the appropriate questions in the post-laboratory worksheet.

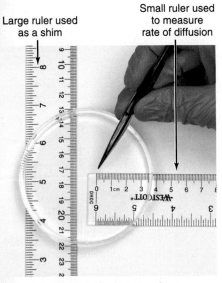

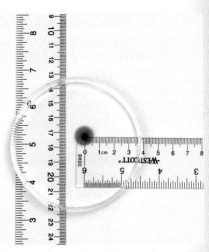

(a) Carefully place a single potassium permanganate crystal at the center of the petri dish at the "0" mark on the ruler.

(b) Start the timer as soon as the crystal is placed into the water.

(c) Measure the radius of the dye "spot" at 30-second intervals for 5 minutes. Record your data in table 4.3.

Figure 4.8 Observing Diffusion of Potassium Permanganate Crystals in Water.

Table 4.3	Effect of Solvent Viscosity on Rate of Diffusion		
Water		Agar	
Time	Distance Diffused (mm)	Time	Distance Diffused (mm)
30 sec		15 min	
1 min		30 min	
1.5 min		45 min	
2 min		1 hour	
2.5 min		1 hour 15 min	
3.0 min		1 hour 30 min	
3.5 min		1 hour 45 min	
4.0 min		2 hours	
4.5 min			
5.0 min			

EXERCISE 4.4B Effect of Temperature on Rate of Diffusion

In this exercise you will repeat the method of observing diffusion of potassium permanganate crystals in water from exercise 4.4A, but this time you will compare the rates of diffusion in water of different temperatures. Recall that molecules move at random in solution because of their inherent kinetic energy. When molecules are heated, the amount of kinetic energy increases and the molecules move faster. Likewise, when molecules are cooled, their kinetic energy decreases and they move more slowly. In this exercise you will place potassium permanganate crystals in petri dishes containing water of differing temperatures and will measure the rate of diffusion of each solute over time. Before you begin, state a hypothesis regarding the effect that temperature has on the rate of diffusion.

Hypothesis: _____
_____
_____

1. Obtain the following:

   a. letter-size (8.5" × 11") piece of white paper
   b. three empty petri dishes
   c. water from one of three flasks: (1) ice water, (2) boiling water, and (3) room temperature water
   d. fine tissue forceps
   e. potassium permanganate crystals
   f. one large (~30 cm) and 3 small (~15 cm) metric rulers
   g. a stopwatch or other device for recording the time

2. Place the rulers and the petri dishes on the white paper as you did in exercise 4.4A, only this time, set up three petri dishes instead of two. Use the large ruler as a shim to keep the petri dishes level, and the small rulers to measure rates of diffusion. Place the large ruler on the left side of the paper and orient it vertically. Place the small rulers at right angles to the large ruler, approximately 8 cm from each other, and with the "zero" point approximately at the center of the petri dish.

3. Label the petri dishes "A – ice water," "B – room-temperature water," and "C – boiling water" on the paper in the space next to each dish.

4. Obtain a small vial of potassium permanganate crystals (figure 4.8). You will need three moderate-sized crystals for this experiment, one for each dish.

5. You will first observe diffusion of dye crystals in ice water. You will need ice water. Before obtaining the ice water, measure the temperature of the water and record it here: _____. Pour some of the ice water into petri dish "A." As you saw in exercise 4.4A, the diffusion of potassium permanganate happens fairly quickly in water, so be prepared before you begin this exercise. One lab partner will place the crystal and record the distance diffused, while another will start

(continued on next page)

(continued from previous page)

the stopwatch and record time intervals. To begin: Place a single potassium permanganate crystal in the center of the petri dish containing ice water (dish A) at the "0" mark on the ruler. The moment the crystal is dropped in the water, start the stopwatch and observe diffusion of the dye. Record the radius of the dye in 30-second intervals for a total of 5 minutes. Record your data in **table 4.4**.

6. In the space below, make a note regarding your conclusions of this experiment (i.e., did it support or refute your hypothesis?). Your notes here need not be extensive, just enough for you to refer back to when completing the post-laboratory worksheet.

   _____

   _____

   _____

   _____

7. Next you will observe diffusion of dye crystals in room-temperature water. You will need room-temperature water. Before obtaining the water, measure the temperature of the water and record it here: _____. Pour some of the room-temperature water into petri dish "B." Place a single potassium permanganate crystal in the center of the petri dish containing room-temperature water (dish B) at the "0" mark on the ruler. The moment the crystal is dropped in the water, start the stopwatch and observe diffusion of the dye. Record the radius of the dye in 30-second intervals for a total of 5 minutes. Record your data in table 4.4. In the space below, note your observations:

   _____

   _____

   _____

   _____

   _____

   _____

8. Finally, you will observe diffusion of dye crystals in **boiling water.** You will need boiling water. Before obtaining the hot water, measure the temperature of the water and record it here: _____. Pour some of the boiling water into petri dish "C." Place a single potassium permanganate crystal in the center of the petri dish containing boiling water (dish C) at the "0" mark on the ruler. The moment the crystal is dropped in the water, start the stopwatch and observe diffusion of the dye. Record the radius of the dye in 30-second intervals for a total of 5 minutes. Record your data in table 4.4.

Table 4.4	Effect of Temperature on Rate of Diffusion		
	Ice Water Temp: _____	Room-Temp. Water Temp: _____	Hot/boiling Water Temp: _____
Time	Distance Diffused (mm)	Distance Diffused (mm)	Distance Diffused (mm)
30 sec			
1 min			
1.5 min			
2 min			
2.5 min			
3.0 min			
3.5 min			
4.0 min			
4.5 min			
5.0 min			

9. In the space below, make a note regarding your conclusions of this experiment (i.e., did it support or refute your hypothesis?). Your notes here need not be extensive, just enough for you to refer back to when completing the post-laboratory worksheet.

_____
_____
_____
_____
_____
_____

10. Use the data from table 4.4 to complete the appropriate questions in the post-laboratory worksheet.

EXERCISE 4.4C Effect of Solute Molecular Weight on Rate of Diffusion

The **molecular weight** of a substance is the mass of a single molecule of the substance. Molecular weight is calculated by taking the sum of the atomic weights of the consituent atoms that make up the molecule. In this exercise you will place drops of solutions composed of molecules of differing molecular weights in a petri dish containing agar and will measure the rate of diffusion of each solute over time. The two solutes you will observe are potassium permanganate (molecular weight 158) and methylene blue (molecular weight 374). Before you begin, state a hypothesis regarding the effect that molecular weight has on the rate of diffusion.

Hypothesis: _____
_____
_____
_____

INTEGRATE

CONCEPT CONNECTION

Calculating the molecular weight of a substance requires you first to know the molecular formula for the substance. Once you have the molecular formula, you simply multiply the atomic mass of each element in the formula by the number of atoms of that element and add them all up. For example, the molecular weight of table salt, with molecular formula NaCl, is calculated as shown below:

Atomic mass of Na = 23; Contribution of Na to molecular weight of NaCl = 23 × 1 (only 1 atom of Na in NaCl)

Atomic mass of Cl = 35; Contribution of Cl to molecular weight of NaCl = 35 × 1 (only 1 atom of Cl in NaCl)

Molecular weight of NaCl = 23(1) + 35(1) = 58

You have already been told the molecular weights of the two substances used in the exercises in this chapter (potassium permanganate and methylene blue), but let's see if you can calculate them on your own. In the space below, perform these calculations. Be sure to show your work so that if you obtain the wrong answer you can ask your instructor for assistance. The molecular formula for potassium permanganate is $KMnO_4$. The molecular formula for methylene blue is $C_{16}H_{18}ClN_3S$. Use your textbook or a periodic table of the elements as a guide for locating the atomic mass of each element.

(continued on next page)

(continued from previous page)

1. Obtain the following:
   a. petri dish filled with agar
   b. drinking straw or medicine dropper
   c. dropper vials containing 0.1M solutions of potassium permanganate and methylene blue
   d. metric ruler

2. In this exercise you will not need to place the metric ruler underneath the petri dish because it is difficult to read through the agar. Thus, you also will not need another metric ruler as a shim. Instead, you will measure diffusion distance by holding the ruler above the diffusion circle (without letting it touch the agar).

3. Obtain a drinking straw or medicine dropper and make a well in the agar dish as shown in **figure 4.10**. If you are using a drinking straw, press the end of the straw into the agar (figure 4.10a) and then lift it out at a slight angle so as to take agar with the straw as it is removed, leaving a well in the agar (figure 4.10b). If you are using a medicine dropper, compress the bulb before you push it into the agar, then release the bulb to suction the agar into the dropper, leaving a well in the agar. Form two wells of equal size in the agar, approximately two inches apart.

4. Obtain dropper bottles of 0.1M potassium permanganate and 0.1M methylene blue. Using a medicine dropper, carefully fill the first well with a drop of potassium permanganate as shown in figure 4.10c. Take care not to allow the dye to spill over the edges of the well. Next, using another medicine dropper, carefully fill the second well with a drop of methylene blue (figure 4.10d). Record the time: _____

5. Measure the radius of the dye diffusion every 15 minutes for 1.5–2.0 hours. Record your data in **table 4.5**.

6. In the space below, make a note regarding your conclusions of this experiment (i.e., did it support or refute your hypothesis?). Your notes here need not be extensive, just enough for you to refer back to when completing the post-laboratory worksheet.

   _____
   _____
   _____
   _____
   _____

7. Use the data from table 4.5 to complete the appropriate questions in the post-laboratory worksheet.

(a) Press the end of a straw or medicine dropper into the agar.

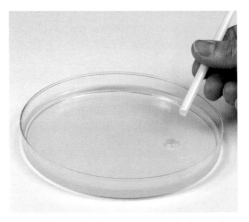

(b) Carefully withdraw the straw so as to take the agar with it to form a well in the agar.

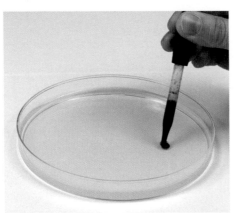

(c) Fill the well with dye, taking care not to let it spill over the edge of the well onto the surface of the agar.

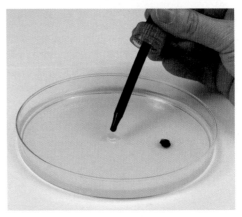

(d) Form a second well at least two inches away from the first well, and fill it with the other dye.

Figure 4.10 Observing the Effect of Solvent Molecular Weight on Diffusion Rates.

Table 4.5	Effect of Solute Molecular Weight on Rate of Diffusion	
Solute	Methylene Blue	Potassium Permanganate
Molecular Weight	374	158
Time	Distance Diffused (mm)	Distance Diffused (mm)
15 min		
30 min		
45 min		
1 hour		
1 hour 15 min		
1 hour 30 min		
1 hour 45 min		
2 hours		

EXERCISE 4.4D Effect of Membrane Permeability on Rate of Diffusion

Recall that the plasma membrane of cells is a selectively permeable membrane. This means it allows only certain molecules to pass through it. The permeability of a cell's plasma membrane is imparted by characteristics of the lipid bilayer and various proteins inserted into the bilayer. In this exercise you will demonstrate the effect of a selectively permeable membrane on diffusion. To do this, you will use an artificial membrane, dialysis tubing. You will place a starch solution into the tubing, seal the tubing, and then place it into a beaker containing water and iodine. Thus, the starch and iodine solutions will be separated by a selectively permeable membrane. Your task is to determine if the membrane is permeable to starch, iodine, or both.

How will you know if any of these substances pass through the dialysis tubing? You will use *indicator solutions* that allow you to determine if a substance is present in a solution. In this exercise, the first indicator substance you will use is **iodine**, which is yellow-brown in color when placed in water. When iodine is added to a solution of uncooked starch, a chemical reaction occurs, which produces a purple/black substance. What about testing for sugar? For this, you can use either **glucose test strips** or **Benedict's solution**. If glucose is present in a solution, a glucose test strip dipped into the solution will change color. If you are using Benedict's solution to test for glucose, you will add Benedict's solution to the sample, heat it in a water bath for several minutes, and look for it to change color.

Before you begin, state a hypothesis regarding diffusion of substances into and out of the dialysis bag.

Hypothesis (What substance(s) will diffuse into/out of the dialysis bag? How will you know?): _____

_____

_____

1. Obtain the following from your instructor:
   a. piece of 1″ diameter dialysis tubing about 8″ long
   b. ~150 mL of 10% uncooked starch solution
   c. 500 mL glass beaker filled with ~300 mL distilled water
   d. dropper bottle of Lugol's solution (iodine potassium iodide)
   e. funnel

2. Soak the dialysis tubing in warm water for at least one minute to soften it up.

(continued on next page)

(continued from previous page)

(a) After soaking the dialysis tubing in water to soften it up, tie off one end.

(b) Using a funnel, pour starch solution into the dialysis tubing. Next, tie off or clamp the open end.

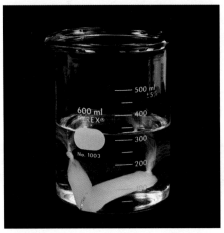

(c) After rinsing the tubing to ensure no starch is on the outside of the tubing, place it in a beaker containing water.

(d) Add a few drops of iodine solution to the beaker. Let it sit for 30 minutes to 1 hour to allow diffusion to take place.

Figure 4.11 Exercise Demonstrating Effect of Membrane Permeability on Diffusion.

3. Open the dialysis tubing by gently rubbing it in between your fingers. Tie off one end of the dialysis tubing or close it with a dialysis tubing clamp **(figure 4.11a)**.

4. Using a funnel, pour starch solution into the dialysis bag. Leave enough room so you can tie off or clamp the open end when you are finished filling it (figure 4.11b).

5. Rinse the dialysis bag with water to ensure no starch remains on the outside of the tubing.

6. Place the dialysis bag into a beaker containing approximately 300 mL of distilled water (figure 4.11c).

7. Add 3–4 drops of Lugol's solution (iodine potassium iodide) to the beaker containing the dialysis bag and water (figure 4.11d). Note any observations you make upon adding the iodine to the water here:

8. Allow the beaker to sit while you continue to work on other exercises in this chapter. Observe the setup again after 30 minutes to one hour have gone by, and note any observations here:

9. Use your observations to answer the appropriate questions in the post-laboratory worksheet.

EXERCISE 4.5

OSMOSIS (WET LAB)

Osmosis (*osmos,* a thrusting) is a special case of diffusion. That is, osmosis refers to the net movement of **water** from an area of high concentration of water (low solute concentration) to an area of low concentration of water (high solute concentration). Osmosis occurs when you have solutions with different concentrations of solutes that are separated by a selectively permeable membrane through which water can pass, but solutes cannot pass. Such solutes are referred to as "membrane impermeable" or "nonpenetrating" solutes. If the solutes cannot pass, then water will move to obtain equilibrium. In this exercise, you will be using either an animal membrane or dialysis tubing that is selectively permeable. The dialysis tubing does not allow large molecules such as sugar molecules to pass through (it is impermeable to sugar). You will place a sugar solution (molasses, a hypertonic solution) in a thistle tube **(figure 4.12)**, cover the opening with dialysis tubing, and place it in a beaker of distilled water (a hypotonic solution). If water moves out of the tube into the beaker, the level of sugar solution in the thistle tube will decrease. If water moves into the tube from the beaker, the level of sugar solution in the thistle tube will increase. This exercise requires two people working together, so partner up with another student in the lab to do the exercise. Before you begin, state a hypothesis regarding the effect of membrane selective permeability on diffusion and osmosis.

Hypothesis: _____

_____

_____

_____

1. Obtain the following:
    a. stand and clamp
    b. molasses
    c. 3" diameter disc of animal membrane (alternative: 2" long piece of 1.5" wide dialysis tubing)
    d. 100 mL glass beaker
    e. distilled water
    f. thistle tube
    g. stopwatch or other timing device
    h. wax pencil

2. Soak the animal membrane or dialysis tubing in warm water for a few minutes to soften it up. If using dialysis tubing, loosen and open it by rubbing it between your thumb and index finger. Cut along one side of the tubing to open it and create a flat sheet of tubing. Place the tubing back into the water to soak until you are ready to use it.

3. Fill the glass beaker about halfway with distilled water.

4. Set up the thistle tube, beaker, stand, and clamp as shown in figure 4.12. Next, turn the thistle tube so the funnel side is up. Have one lab partner place a finger over the bottom end of the tube to form a seal.

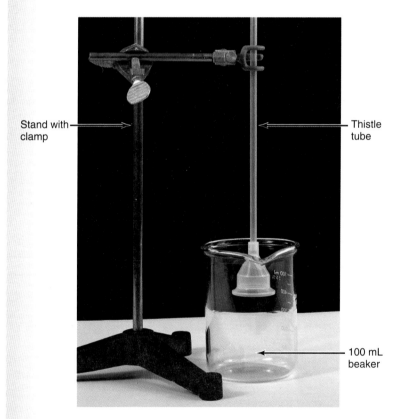

Figure 4.12 Setup for Osmosis Experiment with Thistle Tube. Make sure you have the stand, clamp, and beaker set up the way you want them before you proceed to fill the thistle tube.

(continued on next page)

(continued from previous page)

5. With one person holding the thistle tube with his/her finger covering the bottom end of the tube, have the other person pour molasses into the funnel part of the thistle tube until it begins to leave the funnel part and enter the tube **(figure 4.13a)**. Try to fill the tube ~1/2" because the molasses will run back into the funnel part of the thistle tube when you turn it back over.

6. Obtain the animal membrane or piece of dialysis tubing from the beaker of water and place it over the top of the thistle tube. Secure it with a rubber band (figure 4.13b).

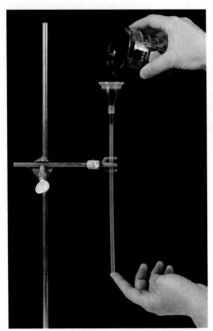

(a) Keeping a finger over the bottom of the tube, pour molasses into the funnel until it fills the bulb completely and runs into the thin part of the thistle tube.

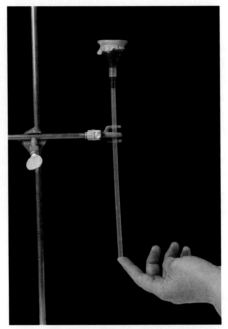

(b) Still keeping a finger on the bottom of the tube, wrap the membrane over the top of the funnel and secure it with a rubber band.

(c) Turn the tube over and affix it to the stand clamp. Mark the starting level of the molasses with a wax pencil.

(d) Lower the thistle tube into the water. Record the level of the molasses every 15 minutes for at least 2 hours.

Figure 4.13 Osmosis Exercise Using an Animal Membrane and Thistle Tube.

7. Carefully turn the tube over so the funnel part is facing the beaker and table. Let the molasses settle into the funnel for a couple of minutes, then mark the starting level with a wax pencil (figure 4.13c).

8. Lower the thistle tube into the glass beaker containing water so it is suspended in the water as shown in figure 4.13d. Tighten the clamp on the tube so it will remain suspended. Record the start time:

   _____

9. After lowering the thistle tube into the beaker of water, start a timer. Record the change in meniscus level in **table 4.6** at 15-minute intervals for 30 minutes to several hours.

10. In the space below, make a note regarding your conclusions of this experiment (i.e., did it support or refute your hypothesis?). Your notes here need not be extensive, just enough for you to refer back to when completing the post-laboratory worksheet.

   _____
   _____
   _____
   _____
   _____
   _____

Table 4.6	Osmosis
Time (min)	Height of Molasses Column (mm)
0	0
15	
30	
45	
60	
75	
90	
105	
120	
135	
150	
165	
180	

EXERCISE 4.6

FILTRATION (WET LAB)

Filtration (*filtro*, to strain through) is a process you are likely familiar with if you have ever brewed coffee using a filtration brewing system. It is a process by which solutes are separated from a solvent when passed through a filter. The size of pores in the filter will prevent solute particles that are larger than the size of the pores from passing through, while admitting any solute particles that are smaller than the size of the pores to pass. At times, the charge on a filter can also prevent substances from passing through. For example, if the substance is small and negatively charged and the filter is also negatively charged, even if the substance is small enough to fit through the pores in the filter, it will likely be prevented from passing because of the repellent forces of the negative charges. Notably, if enough fluid pressure, or **hydrostatic pressure** (*hydro*, water + *statikos*, causing to stand), is exerted on the fluid, it can overcome the influence of like charges repelling each other. In the case of making coffee using a filtration system, you place coffee grounds in a filter and run hot water over the grounds. The extracts of the coffee, such as caffeine, readily pass through the filter, but the coffee grounds do not pass because they are too large. The resulting coffee that ends up in your mug is a **filtrate**.

Although there are multiple factors that can affect filtration, in this exercise we will look at only two of those factors:

1. *Particle size*—particles smaller than the size of the pores in the filter will pass through. Particles larger than the size of the pores in the filter will remain in the filter.

2. *Hydrostatic pressure*—a higher column of fluid exerts more pressure and thus increases filtration rate.

In this exercise you will make a solution containing substances of varying sizes. You will pour this solution into a funnel containing a paper filter. Then you will measure the rate at which fluid passes through the filter and fills up a graduated cylinder. You will compare the rate of filtration when the filter is completely full to the rate of filtration when the filter is only 50% full. The solution will contain the following:

   a. distilled water

   b. 10% uncooked starch solution

   c. 10% dark corn syrup solution

   d. charcoal or ground black pepper

(continued on next page)

(continued from previous page)

Before you begin, state a hypothesis regarding the effect of particle size and hydrostatic pressure on filtration rate. Will there be a difference in filtration rate when the filter is 100% vs. 50% full?

Hypothesis: _____

_____

_____

_____

1. Obtain the following:

    a. a ring stand and ring clamp to hold a glass funnel

    b. one piece of filter paper (18.5 cm diameter)

    c. glass funnel (10 cm diameter)

    d. 100 mL glass beaker

    e. 100 mL graduated cylinder

    f. uncooked dry starch

    g. dark corn syrup

    h. charcoal or black pepper

    i. stopwatch or other timing device

    j. Lugol's iodine (to test for starch)

    k. glucose test strips (to test for glucose)

Figure 4.14 Setup for Filtration Exercise.

2. Set up the ring stand with the funnel and graduated cylinder as shown in **figure 4.14**. To prepare the filter paper, first fold it in half. Then fold it in half once again and open it up into a cone that will fit into the glass funnel. Place the filter in the funnel.

3. Mix the following solution:

    a. 100 mL distilled water

    b. 1 teaspoon charcoal or black pepper

    c. 1 Tablespoon dark corn syrup

    d. 1 Tablespoon dry uncooked corn starch

4. One lab partner will be in charge of timing, the other will be in charge of pouring the solution into the funnel and recording the times in **table 4.7**. Before you pour the solution into the funnel, give it a quick stir so that solids are not settled at the bottom of the beaker. Then, quickly pour the solution into the filter-lined funnel.

Table 4.7	Filtration Rate
Volume (mL)	Time (sec)
5	
10	
15	
20	
25	
30	
35	
40	
45	
50	
55	
60	
65	
70	
75	
80	
85	
90	

The liquid should come up to just below the lip of the filter paper. Start the timer. Note the time it takes to fill the cylinder in increments of 5 mL, and record these times in table 4.7. NOTE THE TIME WHEN THE COLUMN OF FLUID IN THE FUNNEL IS AT 50 mL. When that happens, record the time here: _____. This will allow you to compare filtration rates.

5. Allow filtration to continue until the level of filtrate in the cylinder reaches 80 or 90 mL. When filtration is complete, remove the funnel containing the filter. Do you see anything that was retained by the filter? If so, record it here:

_____

_____

_____

_____

_____

_____

_____

6. Next, perform the following tests on the filtrate in the graduated cylinder:

   a. Observe: Are there **charcoal** or black pepper flecks present in the filtrate? _____

   b. Dip a glucose test strip into the filtrate and look for a color change. Is **glucose** present in the filtrate? _____

   c. Place a couple of drops of Lugol's iodine into the filtrate and look for a color change. Is there **starch** in the filtrate? _____

7. In the space below, make a note regarding your conclusions of this experiment (i.e., did it support or refute your hypothesis?). Your notes here need not be extensive, just enough for you to refer back to when completing the post-laboratory worksheet.

_____

_____

_____

_____

_____

_____

EXERCISE 4.7

Ph.I.L.S. LESSON 1: OSMOSIS AND DIFFUSION: VARYING EXTRACELLULAR CONCENTRATION

The purpose of this laboratory exercise is to demonstrate the principle of osmosis and the effect of osmosis on human cells. You will observe the effect of placing an erythrocyte (red blood cell) into several solutions of differing concentrations of sodium chloride [NaCl]. Placing erythrocytes in the various concentrations of NaCl allows you to observe the effects on the integrity of the erythrocyte plasma membrane due to the movement of water by osmosis. After performing this exercise and analyzing your data, you will be able to reasonably predict the physiological range of NaCl that keeps an erythrocyte membrane intact. You will also be able to reasonably predict what would happen to an erythrocyte membrane if the NaCl in the extracellular fluid were to become too high (hypertonic) or too low (hypotonic).

Before you begin, familiarize yourself with the following concepts (use your textbook as a guide):

- Normal intracellular and extracellular concentrations of sodium (Na^+) for a typical cell
- The mechanisms of transport used to move Na^+ into and out of a typical cell
- Differences between hypotonic, hypertonic, and isotonic solutions

1. Open Ph.I.L.S. Lesson 1: Osmosis and Diffusion: Varying Extracellular Concentration.

2. Read the objectives and introduction for the lab exercise. Contained in the objectives and introduction are several hyperlinks to videos that will help you to understand the experimental process. Once you have read through the pre-laboratory introduction, complete the pre-laboratory quiz.

(continued on next page)

(continued from previous page)

3. After completing the pre-laboratory quiz, read the explanation of the experiment in the "wet lab" section.

4. The laboratory exercise will open when you click "Continue" after you have read the "wet laboratory" section of the exercise **(figure 4.15)**.

5. Click the power switch on the front of the spectrophotometer to turn it on.

6. Using the arrow controls, set the wavelength at 510 nm. (This is the wavelength at which cell membranes absorb light if they are intact.)

7. Click and drag the green pipette to the stock solution of blood. Click the up arrow to draw 1 mL of stock blood solution into the pipette.

8. Drag the pipette to an empty tube and release the blood into the tube by pressing the down arrow. Note: If you are experiencing difficulty, click on the bottom of the pipette tube to get it to move.

9. Repeat steps 7 and 8 until all of the test tubes are filled with blood.

10. To calibrate the spectrophotometer, adjust the transmittance value to "0" using the arrows above the "zero" buttons.

11. Open the lid of the "holder" of the spectrophotometer (if unsure, click on the word "spectrophotometer" in the instructions at the bottom of the screen) and click and drag the "blank" (tube on the far left of the rack with "0" on it) into the spectrophotometer. The blank contains no salt solution and serves as a control for the remaining salt solutions. Close the lid and set the transmittance at 100 using the arrows at "Calibrate."

12. Click the lid of the spectrophotometer and then the tube (top of tube visible in the spectrophotometer) to remove it. The tube will pop straight up out of the spectrophotometer. Then click the tube and drag it back to the rack.

13. The journal will open automatically and display the transmittance value in a table and in graphic form.

14. After viewing, close the journal by clicking on the X in the upper right corner of the graph.

15. Click and drag the next tube to the spectrophotometer and close the lid. (A transmittance reading is taken when the lid is closed.) Open the lid. The journal will open automatically and you can view the data.

16. Click on the test tube to remove it. (The tube will come up out of the spectrophotometer and the journal will close automatically.) Then drag it back to the rack.

17. Repeat steps 15 and 16 until all tubes have been tested. Record the data in **table 4.8** below.

18. With the graph still on the screen, transfer the data points to figure 4.16 on the next page. If you happen to accidentally close the program, you can regraph the data using the data your recorded in table 4.8 below.

19. **Notes about Transmittance:** More light is transmitted in samples where the erythrocytes have taken on water and ruptured open (hemolysis). (Erythrocytes are destroyed and no longer block the light as effectively, so more light is transmitted through the sample.) Less light is transmitted in samples where the erythrocytes have lost water and shriveled (crenated). (Erythrocytes are losing water and becoming more "compact," thus

Table 4.8	Ph.I.L.S. Lesson 1: Varying Extracellular Concentration
[NaCl] (mM)	**Transmittance (%)**
0	100
50	
100	
120	
140	
160	
180	
200	
220	
240	

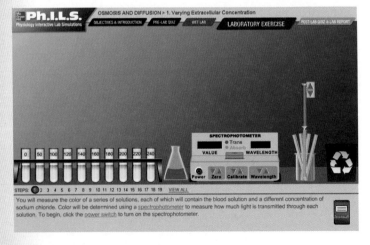

Figure 4.15 Opening Screen for the Ph.I.L.S. Exercise "Osmosis and Diffusion: Varying Extracellular Concentration."

they block the light more effectively and less light is transmitted through the sample.)

> Swelling cells: high transmittance
>
> Shrinking cells: low transmittance

20. Complete the post-laboratory quiz (click open post-laboratory quiz and laboratory report) by answering the ten questions on the computer screen.

21. Read the conclusion on the computer screen.

22. Make note of your observations in the space below:

_____

_____

_____

_____

_____

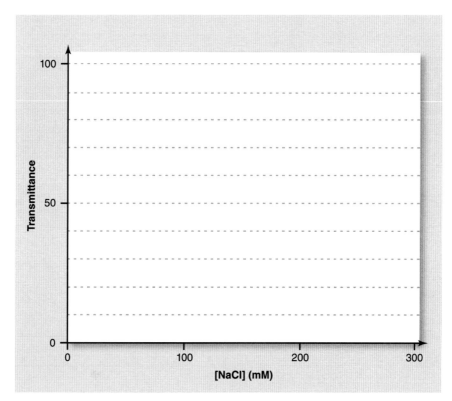

Figure 4.16 Graph of Transmittance vs. salt concentration from Ph.I.L.S. Exercise "Osmosis and Diffusion: Varying Extracellular Concentration."

INTEGRATE

CLINICAL VIEW
Hemodialysis

Patients with reduced kidney function have an inability to remove wastes and excess fluid from the blood. The result is an accumulation of toxic substances in the blood, an increased volume of interstitial fluid (which results in edema, or swelling), and reduced urinary output. When kidney function becomes critically low, these patients must undergo a procedure called **hemodialysis** (*hemo-*, blood + *dialysis*, a separation), which utilizes the same properties of membrane transport discussed in this chapter. In hemodialysis, the patient's arterial blood is passed through a series of selectively permeable tubes.

These tubes consist of a membrane that allows substances such as metabolic wastes and electrolytes to pass through, but prevents the passage of larger substances such as plasma proteins. This mimics the filtration membrane that is normally functional in the kidneys. A dialysis fluid (called dialysate) surrounds the tubes. This fluid matches the composition of the plasma (fluid) portion of blood with the exception of metabolic wastes. As a result, metabolic wastes flow along their individual concentration gradients from the blood to the surrounding dialysate. Equally important, there is no net movement (and thus, loss) of electrolytes or larger proteins from the blood into the dialysate. The filtered blood is returned through the patient's vein. This process must be repeated weekly, depending on the extent of kidney damage.

Chapter 4: Cell Structure and Membrane Transport

Name: _____
Date: _____ Section: _____

POST-LABORATORY WORKSHEET

The ❶ corresponds to the Learning Objective(s) listed in the chapter opener outline.

Do You Know the Basics?

Exercise 4.1: Observing Cellular Anatomy

1. Label the structures in this diagram of a typical cell: ❷

   1 _____
   2 _____
   3 _____
   4 _____
   5 _____
   6 _____
   7 _____
   8 _____
   9 _____
   10 _____
   11 _____
   12 _____
   13 _____

2. Explain the importance of using an isotonic solution in preparing a wet mount. ❸

   _____
   _____
   _____

3. Briefly describe the function of each of the following cellular organelles: ❹

   centrioles   _____
   chromatin   _____
   cytoplasm   _____
   cytoskeleton   _____
   endoplasmic reticulum   _____
   Golgi apparatus   _____
   lysosomes   _____
   mitochondria   _____
   nucleolus   _____
   nucleus   _____
   peroxisomes   _____
   plasma membrane   _____
   ribosomes   _____

Exercise 4.2: Observing Mitosis in a Whitefish Embryo

4. In the spaces below, draw a picture of a whitefish embryo in each of the four stages of mitosis and interphase. To the right of your drawing write a brief description of the major cellular events that occur in each stage. ❺

STAGE: Interphase

EVENTS: _____

_____

_____

_____

STAGE: Prophase

EVENTS: _____

_____

_____

_____

STAGE: Metaphase

EVENTS: _____

_____

_____

_____

STAGE: Anaphase

EVENTS: _____

_____

_____

_____

STAGE: Telophase

EVENTS: _____

_____

_____

_____

Exercise 4.4A: Diffusion: Effect of Solvent Viscosity on Rate of Diffusion

5. In the space below, graph your results from table 4.3: Effect of **Solvent Viscosity** on Rate of Diffusion. You should have two data lines, one for each solvent (agar and water). Do not forget to label your axes with the appropriate units.

6. Based on your results, what do you conclude about the effect of **solvent viscosity** on rate of diffusion?

_____

_____

_____

_____

7. Briefly explain the reason *why* **solvent viscosity** has an effect on rate of diffusion. That is, explain the mechanism by which an increase or decrease in **solvent viscosity** affects the rate of diffusion of a given substance.

_____

_____

_____

_____

_____

Exercise 4.4B: Diffusion: Effect of Temperature on Rate of Diffusion

8. In the space below, graph your results from table 4.4: Effect of **Temperature** on Rate of Diffusion. You should have three data lines, one for each temperature. Do not forget to label your axes with the appropriate units. ❽

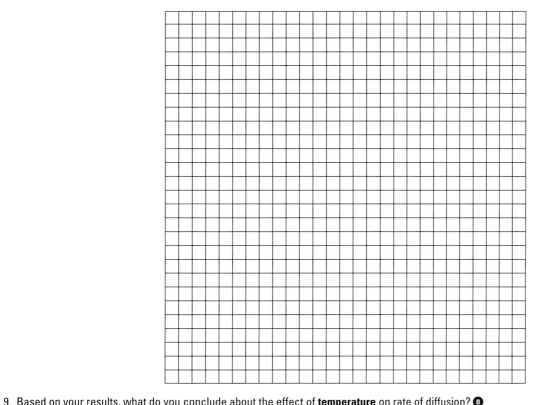

9. Based on your results, what do you conclude about the effect of **temperature** on rate of diffusion? ❽

_____
_____
_____
_____
_____
_____
_____

10. Briefly explain the reason *why* **temperature** has an effect on rate of diffusion. That is, explain the mechanism by which an increase or decrease in **temperature** affects the rate of diffusion of a given substance. ❽

_____
_____
_____
_____
_____
_____
_____
_____

Chapter Four *Cell Structure and Membrane Transport*

Exercise 4.4C: Diffusion: Effect of Solute Molecular Weight on Rate of Diffusion

11. In the space below, graph your results from table 4.5: Effect of **Solute Molecular Weight** on Rate of Diffusion. Do not forget to label your axes with the appropriate units. ❽

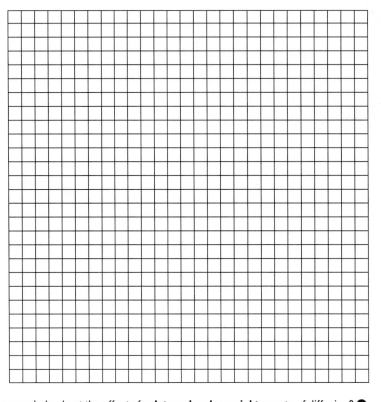

12. Based on your results, what do you conclude about the effect of **solute molecular weight** on rate of diffusion? ❽

_____
_____
_____
_____
_____
_____
_____

13. Briefly explain the reason *why* **solute molecular weight** has an effect on rate of diffusion. That is, explain the mechanism by which an increase or decrease in **solute molecular weight** affects the rate of diffusion of a given substance. ❽

_____
_____
_____
_____
_____
_____
_____

Exercise 4.4D: Diffusion: Effect of Membrane Permeability on Rate of Diffusion

14. Was there a change in the color of the water in the beaker when you added iodine? _____ Based on your answer to this question, did starch diffuse out of the dialysis tubing into the water in the beaker? ❽ _____

15. Was there a change in the color of the starch solution in the dialysis bag after allowing it to sit in the iodine solution for a while? _____ Based on your answer to this question, did iodine diffuse into the dialysis tubing from the water in the beaker? ❽ _____

16. Based on the results of this exercise, summarize the permeability of the dialysis tubing with respect to starch and iodine. ❽

_____
_____
_____
_____

Exercise 4.5: Osmosis [Wet Lab]

17. If the concentration of membrane-impermeable solutes in the intracellular fluid of a cell is higher than that in the extracellular fluid, which of the following will occur? ❾

    a. There will be a net movement of water INTO the cell.

    b. There will be a net movement of water OUT OF the cell.

    c. There will be a net movement of solutes INTO the cell.

    d. There will be a net movement of solutes OUT OF the cell.

    e. There will be a net movement of solutes OUT OF the cell and water INTO the cell.

    f. There will be a net movement of solutes INTO the cell and water OUT OF the cell.

Exercise 4.6: Filtration [Wet Lab]

18. In the space below, graph your results from table 4.7: **Filtration Rate.** Do not forget to label your axes with the appropriate units. After you have graphed your results, put a star (*) on the graph to indicate the time at which the graduated cylinder reached 50 mL. Is there a difference in these two slopes? ⓫

19. The slope (X/Y) of the line on your graph is an estimate of the **filtration rate** (mL/sec). Observe the slope of the line before and after the point at which the graduated cylinder reached 50 mL. Is there a difference in these two slopes? ⓫ ⓬

_____

20. The height of the column of fluid in the funnel affects the hydrostatic pressure of the fluid being filtered. Based on your results, what do you conclude about the effect of **hydrostatic pressure** (fluid height) on filtration rate? ⓬

_____
_____
_____
_____

Exercise 4.7: Ph.I.L.S. Lesson 1: Osmosis and Diffusion: Varying Extracellular Concentration

21. a. Define *independent variable*: _____

    b. Define *dependent variable*: _____

22. a. In the virtual experiment, describe why the same volume of blood is placed in each vial. _____

    b. Identify the control in this experiment. _____

23. Predict what would happen if the spectrophotometer were set to measure the incorrect wavelength. _____

24. Using the graph you created in figure 4.16, explain the relationship between transmittance and [NaCl]. In your explanation, be sure to note the status of the erythrocytes at the various concentrations of salt solution.

    _____

Can You Apply What You've Learned?

25. One major function of liver cells (hepatocytes) is to detoxify alcohol. Based on this function, what organelle(s) do you predict hepatocytes would contain large numbers of? Why?

    _____

26. Certain cells within the pancreas function in the synthesis and secretion of the hormone insulin, which is a protein. Based on this function, what organelle(s) do you predict these cells would contain large numbers of? Why?

    _____

27. Your patient enters the emergency department with considerable blood loss from a blunt force trauma. You need to increase her bodily fluids to maintain blood pressure and systemic perfusion. You quickly administer an intravenous solution of NaCl. To avoid causing more harm, the tonicity of the I.V.

    solution should be _____.

28. While on vacation your brother runs out of contact solution. In an attempt to improvise, he uses tap water to hydrate his contact. Immediately upon application of the contact to his eye, he starts complaining of pain in his eye. The pain is bad enough that he removes the contact and is afraid he will not be able to wear them for the rest of the vacation. Explain why tap water caused your brother such significant discomfort.

29. You are in the laboratory observing a fresh sample of erythrocytes. In an attempt to make them easier to see, you decide to dilute the sample. Following dilution you make another slide, but you are unable to find any cells. What happened to your cells? (Hint: You may have grabbed the wrong solution to make the dilute sample.)

Can You Synthesize What You've Learned?

30. Chemotherapy treatments are given to cancer patients in an attempt to halt or slow the growth of a tumor, which is composed of rapidly-dividing cells. Certain chemotherapy drugs exert their actions by interfering with mitosis. For example, some drugs act to prevent microtubules from lengthening or shortening. Microtubules are protein filaments that attach to chromosomes and centrioles, forming the mitotic spindle, which moves the chromosomes during mitosis. Based on this role of microtubules during mitosis, with which stage(s) of mitosis would these drugs most likely interfere?

31. What is the function of rough endoplasmic reticulum? Why do you think neurons might contain so much rough endoplasmic reticulum? (Hint: Neurons are cells that need to be able to transport numerous ions into and out of the cell.)

32. Exercise 4.4D (Effect of Membrane Permeability on Rate of Diffusion) made you curious, and you now want to know if dialysis tubing is permeable to glucose. In the space below, describe an experiment that you might perform to determine if the dialysis tubing is permeable to glucose.

CHAPTER 5

Histology

OUTLINE AND LEARNING OBJECTIVES

Histology Slides 94

Histology 96

Epithelial Tissue 96

EXERCISE 5.1: IDENTIFICATION AND CLASSIFICATION OF EPITHELIAL TISSUE 98
1. Identify the different types of epithelial tissues and their structures when viewed through a microscope
2. Fully classify an epithelial tissue. This includes classifying the tissue by cell shape, number of layers, and surface modifications
3. Identify the following specialized cells and surface modifications: goblet cells, keratinization, cilia, and microvilli
4. Identify apical and basal surfaces of epithelial tissues
5. Associate basic epithelial structures with functions

Connective Tissue 104

EXERCISE 5.2: IDENTIFICATION OF EMBRYONIC CONNECTIVE TISSUE 105
6. Identify mesenchyme

EXERCISE 5.3: IDENTIFICATION AND CLASSIFICATION OF CONNECTIVE TISSUE PROPER 106
7. Identify collagen and elastic fibers
8. Identify fibroblasts and adipocytes
9. Identify the types of connective tissue proper: areolar, adipose, reticular, dense regular, dense irregular, and elastic

EXERCISE 5.4: IDENTIFICATION AND CLASSIFICATION OF SUPPORTING CONNECTIVE TISSUE 111
10. Recognize the histological features unique to cartilage and bone
11. Compare and contrast the structure of the three types of cartilage
12. List locations in the body where each type of cartilage is found
13. Identify osteons, osteoblasts, and osteocytes in a slide of compact bone

EXERCISE 5.5: IDENTIFICATION AND CLASSIFICATION OF FLUID CONNECTIVE TISSUE 114
14. Describe the properties of fluid connective tissue that characterize it as connective tissue
15. Identify the cells and extracellular matrix of fluid connective tissue

Muscle Tissue 114

EXERCISE 5.6: IDENTIFICATION AND CLASSIFICATION OF MUSCLE TISSUE 116
16. Compare and contrast the three types of muscle tissue
17. List locations in the body where each type of muscle tissue is found

Nervous Tissue 117

EXERCISE 5.7: IDENTIFICATION AND CLASSIFICATION OF NERVOUS TISSUE 118
18. Identify neurons and glial cells in a slide of nervous tissue
19. Describe the structural and functional differences between neurons and glial cells

INTRODUCTION

In this laboratory session you will begin your practical study of **histology**, tissue biology (*histo*, tissue, + *logos*, the study of). Tissues consist of multiple cells that function together as a unit. An understanding of histology is important in the health sciences because many times the first manifestation of disease is seen at the tissue level of organization. For example, when a patient presents to his or her doctor with a tumor (*tumere*, to swell), one of the primary methods for determining the type of tumor and whether it is cancerous is to do a biopsy (take a tissue sample) and look at the tissue through a microscope. Thus, understanding normal histology is important for understanding histopathology (*histo*, tissue + *pathos*, disease).

This chapter introduces you to the key features of the four basic tissue types—epithelial, connective, muscle, and nervous. The text and figures in this chapter provide you with examples of where each type of tissue is located in the body.

In this laboratory session you should expect to spend nearly all of your time looking at tissues through the microscope. At first you may think you are simply looking at a lot of slides with "pink and purple stuff" on them, and you will have difficulty ➡

MODULE 3: TISSUES

identifying particular tissue types and structures. The exercises in this chapter address each of the four basic tissue types. You may work through the exercises in any order, or in the order your instructor chooses. However, be sure to complete all the tasks involving a particular tissue type before moving on to another type of tissue.

Histology Slides

As you get yourself oriented when looking at a histology slide through the microscope, one of the most important things to keep in mind is the issue of the *size of the tissue sample* you are viewing. **Figure 5.1** shows a life-size prepared slide that is similar to the slides you will find in the laboratory. Let us consider briefly how this slide was prepared. The process of making a histology slide involves five general steps:

1. Obtain a tissue sample.
2. Prepare the tissue sample for slicing.
3. Cut thin slices of the tissue using a special knife called a microtome.
4. Transfer the tissue slices to a microscope slide.
5. Stain the slide.

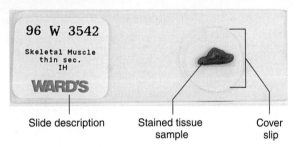

Figure 5.1 Life-Size Histology Slide.

After a tissue sample has been obtained, it must be made rigid so that it will be easy to slice. This is done either by freezing the tissue or by embedding the tissue in a block of paraffin wax. The next step in preparing the histology slide involves slicing the frozen sample (or wax block) into very thin slices (on the order of micrometers—1 μm is 10^{-6} meters) using a special knife called a **microtome** (*micro*, small, + *tome* or *temmein*, to cut). The slices are cut so thin that often only a single layer of cells is contained in the slice. Once the slices are made, they are transferred to a microscope slide, which is then covered with a cover slip. Finally, the samples are stained to make intracellular and extracellular structures visible. The most common method of staining, hematoxylin and eosin (H and E), makes most structures appear pink or purple and makes the nucleus of the cell, in particular, easily visible.

Observe the life-size histology slide shown in figure 5.1. Measure the dimensions of the slide. What are they? _____ by _____ cm. These measurements will give you an idea of the size limits of what can fit on a microscope slide.

Always ask yourself the "size" question when you first view a slide. Then, when you see a slide labeled "small intestine" (for example), you will expect to see only a *portion* of the wall of the small intestine, not a cross section of the entire organ—because that would not fit on the slide. This practice will become more natural for you as you gain experience viewing slides.

Chapter 5: Histology

Name: Carly Guerra
Date: 9-24   Section: A

PRE-LABORATORY WORKSHEET

Also available at www.connect.mcgraw-hill.com

1. List the four basic tissue types.
    a. Connective
    b. Nerve
    c. Muscle
    d. Epithelium

2. Bone is classified as a(n) _____ tissue.

3. Which two of the four basic tissue types are excitable?
    a. _____
    b. _____

4. The basic tissue type that contains an extensive extracellular matrix (ECM) is _____.

5. List the three types of muscle tissue.
    a. _____
    b. _____
    c. _____

6. The only basic tissue type that exhibits polarity (has both apical and basal surfaces) is _____.

7. The basic tissue type that contains a basement membrane is _____ tissue.

8. The two specialized tissue types that are characterized as supporting connective tissues are _____ and _____.

9. Fluid connective tissue includes both _____ and _____.

10. The only type of connective tissue that is avascular (lacking blood vessels) is _____.

Chapter Five *Histology*   95

Histology

Epithelial Tissue

Epithelial tissues cover the body surfaces, line the body cavities, and form the majority of glands. As such, they will have a free edge (see Learning Strategy on p. 97). They are characteristically highly **cellular** (mostly composed of cells, with little extracellular material) and **avascular** (no blood supply). Epithelial cells exhibit **polarity;** they have a distinct *basal* (bottom) and *apical* (top) surface. On their basal surface, they have a specialized extracellular structure called a **basement membrane,** which anchors the epithelium to the underlying tissues. The characteristics used to classify epithelial tissue include: (a) the shape of the cells on the *apical* surface of the epithelium **(table 5.1),** (b) the number of layers of cells (simple, stratified, or pseudostratified), and (c) presence of any surface modifications **(table 5.2).**

Figure 5.2 is a flowchart for classification of epithelial tissues that you can use when you are given an unknown slide of epithelial tissue. It will help you decide how to classify the epithelial tissue.

Table 5.1 Epithelial Cell Shapes

Cell Shape	Squamous	Cuboidal	Columnar	Transitional
Micrograph	Squamous cells / Lumen (LM 130x)	Cuboidal cells / Lumen (LM 165x)	Columnar cells / Lumen (LM 200x)	Transitional cell / Lumen (LM 180x)
Description	Cells are flattened and have irregular borders	Cells are as tall as they are wide	Cells are taller than they are wide	Cells change shape depending on the stress on the epithelial tissue. The cells change between a cuboidal shape and a more flattened, squamous shape.
Generalized Functions	If the epithelium is only one cell layer thick, it provides a very thin barrier for *diffusion*. If the epithelium is several layers thick, the cells specialize in *protection* (as in epidermal cells of the skin).	The shape of the cell allows more room for cellular organelles (e.g., mitochondria, endoplasmic reticulum). These cells generally function in *secretion* and/or *absorption*.	The large size of the cell allows even more room for cellular organelles (e.g., mitochondria, endoplasmic reticulum, etc.). These cells generally function in *secretion* and/or *absorption*.	The fact that these cells change shape means that they are good at *resisting stretch* without being torn apart from each other. They are only found lining structures of the urinary tract (ureters, urinary bladder, and portions of urethra).
Identifying Characteristics	In cross section, the nucleus is the most visible structure. It will be very flattened. In a surface view of the epithelium, the cell borders will be irregular in shape.	Generally cuboidal cells are identified by their very round, plump nucleus, and by equal amounts of cytoplasm in the spaces between the nucleus and the plasma membrane on all sides	The nuclei of columnar cells can be either oval or round in shape, and they generally line up in a row. If the nuclei are round, you will see more cytoplasm between the nucleus and the plasma membrane on the apical side of the nucleus than on the other three sides.	Transitional cells are located on the apical surface of the epithelium. However, they appear much more rounded or dome-shaped than typical cuboidal cells, and they are sometimes binucleate.

Table 5.2 Cell Surface Modifications and Specialized Cells of Epithelial Tissues

Surface Modification	Cilia	Goblet Cells	Keratinization	Microvilli
Micrograph	(LM 200x) Lumen; Cilia	(LM 100x) Goblet cell; Columnar epithelial cell; Mucin within goblet cell; Lumen; Goblet cell nucleus; Location of basement membrane	(LM 40x) Lumen; Keratinization	(LM 165x) Brush border of microvilli; Lumen
Description and Function	Small hairlike structures that extend from the apical surface of epithelial cells. Cilia actively move to *propel substances along the apical surface of an epithelial sheet*. Cilia move substances in only one direction.	Goblet cells are named for their shape. They are rounded near the apical surface and they narrow toward their basal surface. Goblet cells contain many small mucin granules and function in the *production of mucus*. The mucus is used to *assist in transport* of substances along an epithelial sheet, to *provide a protective barrier* along the apical surface of the epithelium, or to provide *lubrication*.	Stratified squamous epithelial cells of the skin contain keratin (an intermediate filament). Bundles of keratin fill up entire cells and bind to desmosomes, which firmly anchor the dead squamous epithelial cells together. The layers of cells appear to be a single homogeneous unit. Keratin imparts *strength* and *protection* to dead skin epithelial cells.	Microvilli are extremely small extensions of the plasma membrane of the apical surface of cells. Microvilli *increase the surface area* of the cell to enhance the process of *absorption*.
Identifying Characteristics	When cilia are present, and the slide is viewed at sufficient magnification, you can see what appear to be individual "hairs" on the apical surface of the epithelial cells	Goblet cells are named for their shape. They are rounded near the apical surface and they narrow toward their basal surface. The shape is similar to the shape of a goblet wine glass. The mucin inside the cells does not typically take up biological stains, so the cells often appear white or "empty." If the slide is stained specifically for mucin, then the goblet cells will appear dark.	Keratinization is recognized as a homogeneous, acellular-looking portion of a stratified squamous epithelium	Individual microvilli can be seen only when the specimen is viewed with an electron microscope. Thus, you will *not* be able to see individual microvilli with a light microscope. Instead, the apical surface of the epithelium will appear to be "fuzzy." For this reason, epithelia containing microvilli are often said to have a "brush border."

INTEGRATE

LEARNING STRATEGY

When you are viewing a slide for the purpose of identifying an epithelial tissue, remember that epithelial tissues always form linings and coverings of organs. The slides you will be observing typically contain more tissues than just epithelial tissues. To locate the epithelial tissue, first look for any white space, or "empty" space, on the slide. This space will typically be the outside of an organ or the lumen (inside) of the organ. The tissue that lies directly adjacent to the empty space will usually be an epithelial tissue.

INTEGRATE

LEARNING STRATEGY

A **simple** epithelium is only one cell layer thick. A **stratified** epithelium is two or more cell layers thick. Be aware that cell shape can vary in stratified epithelium. To avoid confusion, always identify the shapes of cells on the apical surface when classifying stratified epithelium.

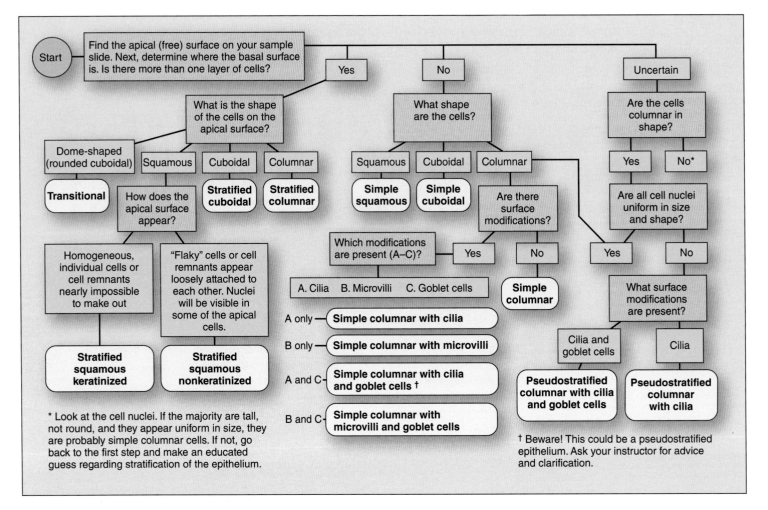

Figure 5.2 Flowchart for Classifying Epithelial Tissues.

EXERCISE 5.1

IDENTIFICATION AND CLASSIFICATION OF EPITHELIAL TISSUE

EXERCISE 5.1A Simple Squamous Epithelium

1. Obtain a slide of a small vein in cross section **(figure 5.3)**.

2. Place the slide on the microscope stage and bring the tissue sample into focus on low power.

3. Look for any "empty" space on the slide. The empty space on the slide will be either the inside of the vessel (the **lumen** of the vessel) or the outside edge of the tissue sample.

4. Locate the lumen of the vessel and move the microscope stage so the lumen is at the center of the field of view. The lumen of all blood vessels and lymph vessels, and the four chambers of the heart, is lined with a simple squamous epithelium called *endothelium*. This type of epithelium is always somewhat difficult to see because it is extremely thin.

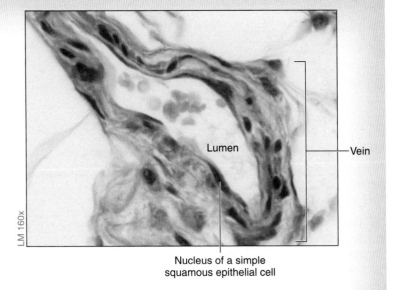

Figure 5.3 Simple Squamous Epithelium. Cross section through a small vein lined with epithelial cells, which are simple squamous epithelial cells.

5. Once you have identified the lumen of the vessel and know where the epithelial tissue is located (lining the lumen), change to high power. Look for the flattened nuclei of the squamous epithelial cells that line the lumen. It is unlikely that you will be able to see much, if any, of the cellular cytoplasm because the cells are extremely thin. Because simple squamous epithelium is extremely thin, it functions in diffusion—a process that occurs over very short distances (distances of approximately 10 μm). Diffusion (see chapter 4) is the movement of particles from an area of high concentration to an area of low concentration, and is one mechanism by which substances are transported in the body.

6. Using figure 5.3 and tables 5.1 and 5.2 as guides, identify the following structures on the slide:

    ☐ lumen of vein

    ☐ nucleus of squamous epithelial cell

7. In the following space, sketch simple squamous epithelium as seen through the microscope. Be sure to identify all the structures listed in step 6 on your drawing.

 400 ×

INTEGRATE

CONCEPT CONNECTION

Simple squamous epithelium lining the cardiovascular system is called **endothelium**. Simple squamous epithelium lining body cavities is called **mesothelium**. Thus, these terms (endothelium and mesothelium) indicate not only the type of epithelium (simple squamous), but also the location of the epithelium.

EXERCISE 5.1B Simple Cuboidal Epithelium

1. Obtain a slide of the kidney **(figure 5.4)**.

2. Place the slide on the microscope stage and bring the tissue sample into focus on low power.

3. Locate the lumen of a tubule in cross section (figure 5.4), and then identify the cells that lie next to the lumen. These cells should have plump, round nuclei and approximately equal amounts of cytoplasm surrounding each nucleus. These are cuboidal epithelial cells, which line the kidney tubules and function in secretion and absorption of substances across the epithelium (table 5.1).

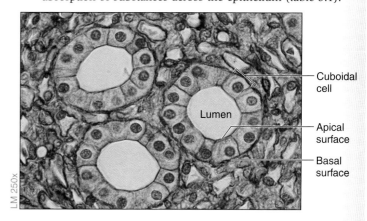

Figure 5.4 Simple Cuboidal Epithelium. Cross section of three kidney tubules demonstrating simple cuboidal epithelium.

4. Using figure 5.4 and table 5.1 as guides, identify the following structures on the slide:

    ☐ apical surface   ☐ cuboidal cell

    ☐ basal surface   ☐ lumen of tubule

5. In the following space, sketch simple cuboidal epithelium as seen through the microscope. Be sure to identify all the structures listed in step 4 on your drawing.

400 ×

(continued on next page)

(continued from previous page)

EXERCISE 5.1C Simple Columnar Epithelium (nonciliated)

1. Obtain a slide of the small intestine (**figure 5.5**).

2. Place the slide on the microscope stage and bring the tissue sample into focus on low power.

3. This slide will show only a part of the wall of the intestine, so you will need to find some empty space on the slide first and then look for epithelium next to that empty space. Once you have the epithelium in the center of the field of view, switch to high power. Look for epithelial cells with oval, elongated nuclei that have most of their cytoplasm on the apical side of the nucleus. Columnar cells are taller than they are wide, and their nuclei generally appear to be lined up in a row. The nuclei can be either elongated or round in shape.

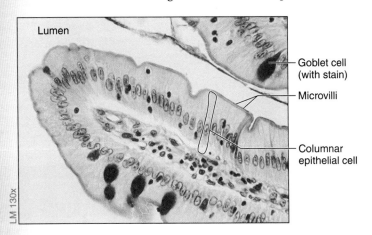

Figure 5.5 Simple Columnar Epithelium with Microvilli. Simple columnar epithelium with microvilli lining the lumen of the small intestine.

4. This epithelium also demonstrates **goblet cells,** which secrete mucin, and **microvilli,** which increase the surface area of the epithelial cells for absorption (table 5.2).

5. Using figure 5.5 and tables 5.1 and 5.2 as guides, identify the following structures on the slide:

   ☐ columnar epithelial cell   ☐ lumen of intestine
   ☐ goblet cell   ☐ microvilli

6. In the following space, sketch simple columnar epithelium as seen through the microscope. Be sure to identify all the structures listed in step 5 on your drawing.

_400_ ×

7. Are any surface modifications present in the slide of the small intestine? If so, describe them: _____
_____
_____

EXERCISE 5.1D Simple Columnar Epithelium (ciliated)

1. Obtain a slide of a uterine tube (**figure 5.6**).

2. Place the slide on the microscope stage and bring the tissue sample into focus on low power.

3. Look for a tubular structure cut in cross section (figure 5.6), and identify its lumen. Look for columnar epithelial cells next to the lumen. Once you have the epithelium in the field of view, switch to high power. When cilia are present, you should be able to make out the individual "hair"-like cilia on the apical surface of the epithelial cells.

4. Using figure 5.6 and tables 5.1 and 5.2 as guides, identify the following structures on the slide:

   ☐ cilia
   ☐ lumen of uterine tube
   ☐ simple columnar epithelial cell

5. In the following space, sketch simple columnar epithelium with cilia as seen through the microscope. Be sure to identify all the structures listed in step 4 in your drawing.

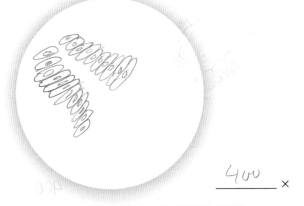

_400_ ×

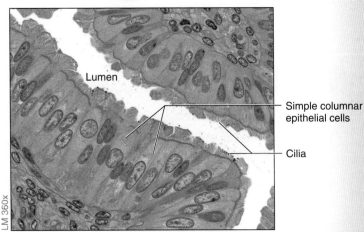

Figure 5.6 Simple Columnar Epithelium with Cilia. Simple columnar ciliated epithelium lining the lumen of the uterine tube.

EXERCISE 5.1E Stratified Squamous Epithelium (nonkeratinized)

1. Obtain a slide of the trachea and esophagus (figure 5.7a,b).

2. Place the slide on the microscope stage and bring the tissue sample into focus on low power.

3. The slide shown contains two different organs in cross section, the trachea and the esophagus. The epithelium you are looking for lines the esophagus and consists of multiple layers of cells.

4. Locate the lumen of the esophagus (figure 5.7a), then identify stratified squamous nonkeratinized epithelium lining the lumen of the esophagus. The cells on the apical surface of this epithelium will appear flattened. This nonkeratinized stratified squamous epithelium is sometimes referred to as stratified squamous "moist" epithelium. It lines surfaces within the body that experience friction and abrasion, but where water loss is not a problem (for example, lining the oral cavity, esophagus, and vagina).

5. Using figure 5.7b and tables 5.1 and 5.2 as guides, identify the following structures on the slide.

   ☐ lumen of esophagus
   ☐ squamous epithelial cells
   ☐ stratified squamous nonkeratinized epithelium

6. In the following space, sketch stratified squamous nonkeratinized epithelium as seen through the microscope. Be sure to identify the structures listed in step 5 on your drawing.

400 ×

7. Keep the slide on the microscope stage and proceed to exercise 5.1F (next page), where you will focus on the epithelium lining the trachea.

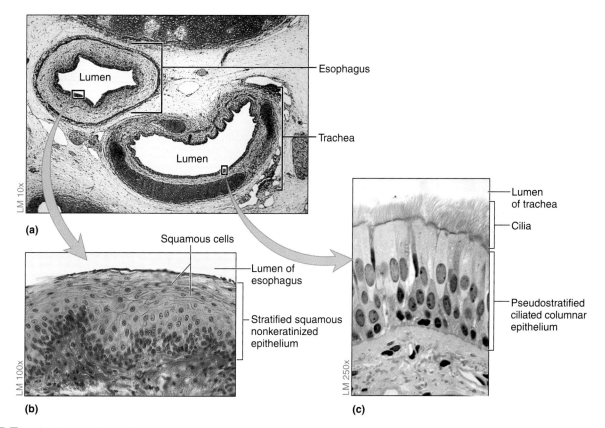

Figure 5.7 Trachea and Esophagus. (a) Cross section through the trachea and esophagus. (b) Stratified squamous nonkeratinized epithelium lining the esophagus. (c) Pseudostratified ciliated columnar epithelium lining the trachea.

(continued on next page)

(continued from previous page)

EXERCISE 5.1F Pseudostratified Columnar Epithelium

1. Obtain a slide of the trachea and esophagus (figure 5.7a).

2. Place the slide on the microscope stage and bring the tissue sample into focus on low power.

3. This slide will contain two organs in cross section, the trachea and the esophagus. The epithelium you are looking for lines the trachea and contains ciliated cells.

4. Locate the lumen of the trachea (figure 5.7a), then identify pseudostratified columnar epithelium (figure 5.7c) lining the trachea. Once you have the epithelium in the field of view switch to high power. This epithelium is characterized by the presence of columnar cells that are not all the same height. All cells contact the basement membrane, but not all reach the apical surface. Because not all cells reach the apical surface of the epithelium, the nuclei will *not* be nicely lined up in rows as with simple columnar epithelium. Instead, the nuclei will appear to be layered. Hence, the name of this epithelium: **pseudostratified** (*pseudo-*, false). Most, but not all, pseudostratified epithelia also contain cilia and goblet cells.

5. Using figure 5.7c and tables 5.1 and 5.2 as guides, identify the following structures on the slide:

    ☐ cilia

    ☐ columnar epithelial cell

    ☐ lumen of trachea

6. In the following space, sketch pseudostratified columnar epithelium as seen through the microscope. Be sure to label the structures listed in step 5 in your drawing.

_400_ ×

EXERCISE 5.1G Stratified Cuboidal or Stratified Columnar Epithelium

1. Obtain a slide of a merocrine sweat gland (**figure 5.8**).

2. Place the slide on the microscope stage and bring the tissue sample into focus on low power.

3. Locate the lumen of a duct of the sweat gland (figure 5.8). Identify stratified cuboidal epithelium next to the lumen. Once you have the epithelium in the field of view switch to high power. Stratified cuboidal epithelium is found lining the ducts of merocrine sweat glands, which are located in the dermis of the skin. Stratified cuboidal epithelium is generally only two cell layers thick. How many layers do you see on the slide? Your laboratory may have other slides available that demonstrate stratified *columnar* epithelium lining the ducts of other exocrine glands. As with stratified cuboidal epithelium, stratified columnar epithelium is also rarely more than two cell layers thick.

4. Using figure 5.8 and tables 5.1 and 5.2 as guides, identify the following structures on the slide:

    ☐ basement membrane

    ☐ cuboidal epithelial cell

    ☐ lumen of the duct

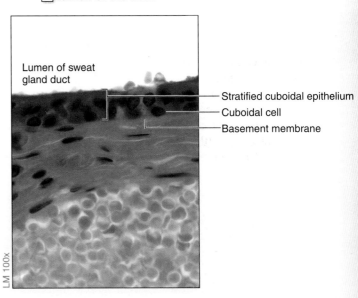

Figure 5.8 **Stratified Cuboidal Epithelium.** Epithelium lining the duct of a merocrine sweat gland.

5. In the following space, sketch stratified cuboidal (or stratified columnar) epithelium as seen through the microscope. Be sure to label the structures listed in step 4 on your drawing.

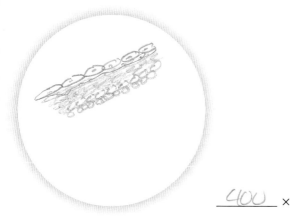

_400_ ×

EXERCISE 5.1H Transitional Epithelium

1. Obtain a slide of the urinary bladder (**figure 5.9**).

2. Place the slide on the microscope stage and bring the tissue sample into focus on low power.

3. This slide will show only a part of the wall of the bladder, so you will need to locate the empty space first and then look for epithelium next to that empty space. Locate the transitional epithelium. Once you have the epithelium in the field of view, switch to high power. Transitional epithelium is stratified, and it can look very similar to stratified squamous epithelium. However, the cells on the apical surface will appear cuboidal in shape, they will be much more rounded or dome-shaped than typical cuboidal cells, and they may contain more than one nucleus per cell (these cells are sometimes referred to as "dome" or "umbrella" cells).

4. Observe the cells on the basal surface of the transitional epithelium. These cells are usually columnar in shape, in contrast to the cells on the basal surface of a stratified squamous epithelium, which tend to be more cuboidal in shape. Transitional epithelium is found lining the urinary bladder and other urine-draining structures. Its structure allows the epithelium to stretch easily to accommodate the passage or storage of urine without causing the epithelial cells to tear apart.

5. Using figure 5.9 and tables 5.1 and 5.2 as guides, identify the following structures on the slide:

    ☐ dome-shaped epithelial cells
    ☐ lumen of urinary bladder
    ☐ transitional epithelium

6. In the following space, sketch transitional epithelium as seen through the microscope. Be sure to label the structures listed in step 5 in your drawing.

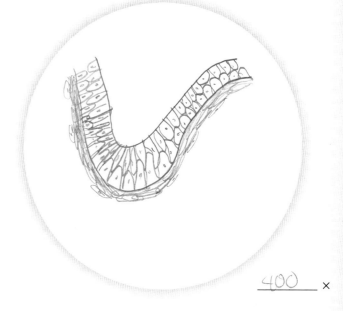

_400_ ×

7. *Optional Activity:* **AP|R** 3: Tissues—Watch the "Epithelial Tissue Overview" animation.

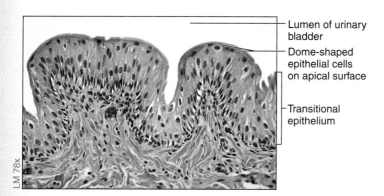

Figure 5.9 **Transitional Epithelium.** Epithelium lining the urinary bladder.

Labels: Lumen of urinary bladder; Dome-shaped epithelial cells on apical surface; Transitional epithelium

INTEGRATE

LEARNING STRATEGY

Students often confuse the terms *basement membrane* and *basal surface*. To clarify, the **basement membrane** is a connective tissue structure that lies at the **basal surface** of the epithelium. The two terms are not synonymous. That is, the basement membrane is a *structure;* the basal surface is a *location.* Also note that the basement membrane is an extracellular structure on which the epithelium rests. The function of the basement membrane is to anchor the epithelium to the underlying connective tissue, provide physical support for the epithelium, and act as a barrier to regulate the passage of large molecules between the epithelium and the underlying tissues.

INTEGRATE

CONCEPT CONNECTION

The presence of epithelial modifications such as cilia, microvilli, and goblet cells in epithelial tissues reveals the overall function of the epithelium in question. In the exercises in this chapter, you are viewing representative examples of each type of epithelial tissue. These examples will be revisited in the histology sections of subsequent chapters, where they will be covered in more detail. Rather than memorizing the locations of these tissues at this time, focus on identifying how the modifcation relates to tissue function.

Cilia aid in *movement/transport,* microvilli aid in *absorption,* and goblet cells aid in *lubrication.* For example, consider the epithelial tissue lining the upper respiratory structures: pseudostratified ciliated columnar epitheum. This epithelium contains goblet cells, which produce mucus, and cilia, which trap and transport mucus and debris along the epithelial surface. Together, these modifications form a "mucus escalator" that prevents foreign objects and pathogens from entering lower respiratory structures. Damage to this epithelium, specifically the cilia, is detrimental, as it renders the lower respiratory tract more prone to infection.

Connective Tissue

All connective tissues share three basic components: cells, protein fibers, and ground substance. The different types of protein fibers include collagen, elastic, and reticular (see **table 5.3**). Connective tissues are derived from an embryonic tissue called **mesenchyme (figure 5.10).** There are three broad categories of mature connective tissues: connective tissue proper, supporting connective tissues, and fluid connective tissues. **Connective tissue proper** is the "glue" that holds things together (such as tendons and ligaments) and the "stuffing" that fills in spaces (such as the fat that fills in spaces between muscles). The category of connective tissue proper includes both loose connective tissue (areolar, adipose, and reticular) and dense connective tissue (dense regular, dense irregular, and elastic) as described in **table 5.4. Supporting connective tissues** (cartilage and bone) are specialized connective tissues that provide support and protection for the body. **Fluid connective tissues** (blood and lymph) are specialized connective tissues that function to transport substances throughout the body. The details of bone and blood are covered in chapters 7 and 20 of this laboratory manual.

Connective Tissue Proper

Connective tissue proper is a kind of "grab bag" category that contains all of the unspecialized connective tissues (that is, any connective tissue other than supporting or fluid connective tissues). These tissues are used either to hold things together (as with tendons and ligaments) or to fill up space (as with adipose tissue). Loose connective tissues have a loose association of cells and fibers, whereas dense connective tissues have cells and fibers that are densely packed together, which makes them much tougher than loose connective tissues. Table 5.4 summarizes the characteristics of the different types of connective tissue proper. **Figure 5.11** is a flowchart for classification of connective tissue proper that you can use when you are given an unknown slide of connective tissue. It will help you to classify the connective tissue.

Table 5.3	Connective Tissue Fibers		
Fiber Type	**Collagen**	**Elastic**	**Reticular**
Micrograph	Fibroblast nucleus, Collagen fibers (LM 200x)	Elastic fibers, Fibroblast nucleus (LM 200x)	Reticular fibers (LM 250x)
Identifying Characteristics	Thicker than elastic or reticular fibers and usually somewhat *pale* in color. They stain either *pink or blue*, depending on the stain used.	Appear either as fine, thin *black* fibers (when a silver stain is used) or as thin, *wavy* fibers (when silver stain is not used)	Composed of a fine, thin type of collagen. The fibers can be seen only when a silver stain is applied, and they appear as an *irregular network* of thin *black* fibers (*reticular, network*). Reticular fibers are much shorter than elastic fibers.
Functions	Collagen fibers are good at *resisting tensile (stretching) forces.* They are good at resisting stretch in only one direction, along the long axis of the fiber.	Elastic fibers have the *ability to stretch and recoil.* They will often stretch to greater than 150% of their resting length without damage. Too much stretch will break the fiber.	Reticular fibers *form a delicate inner supporting framework* for highly cellular organs such as the liver, spleen, and lymph nodes

EXERCISE 5.2

IDENTIFICATION OF EMBRYONIC CONNECTIVE TISSUE

1. Obtain a slide of mesenchyme (figure 5.10).
2. Place the slide on the microscope stage and bring the tissue sample into focus on low power, then switch to high power.
3. Notice that there are no visible fibers within the extracellular matrix (mature fibers do not yet exist within mesenchyme). The **mesenchymal cells** are recognized by their large oval nuclei. Mesenchyme has the ability to differentiate into any of the mature connective tissue cell types. That is, mesenchymal cells can differentiate into any of the adult connective tissue cell types (e.g., fibroblasts, chondroblasts, osteoblasts).
4. Using figure 5.10 as a guide, identify the following structures on the slide:
   - ☐ ground substance
   - ☐ mesenchymal cell
5. In the following space, sketch mesenchyme as seen through the microscope. Be sure to label the structures listed in step 4 in your drawing.

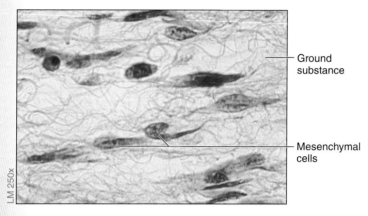

Figure 5.10 Mesenchyme. Mesenchyme is embryonic connective tissue.

_____ ×

Table 5.4	Connective Tissue Proper	
Classification	**Description and Function**	**Location(s)**
Loose Connective Tissue		
Areolar (Figure 5.12)	Consists of a loose arrangement of collagen and elastic fibers with numerous fibroblasts. It loosely anchors structures to each other or fills in spaces between organs.	Located in the superficial fascia below the skin, which anchors skin to underlying muscle. It is also found surrounding many organs
Adipose (Figure 5.13)	Characterized by adipocytes, which appear to be large, "empty" cells because the process of preparing the tissue removes all lipid within the cells. Collagen fibers located between the adipocytes hold the tissue together. Adipose tissue functions in insulation, protection, and energy storage.	Subcutaneous (under the skin). It also surrounds organs such as the kidneys, where it provides protection and fills in potential spaces such as within the axilla and popliteal fossa
Reticular (Figure 5.14)	Composed of reticular fibers, which form an inner supporting framework for highly cellular organs such as the liver	The inner stroma of organs such as the spleen, liver, and lymph nodes (*stroma*, bed)
Dense Connective Tissue		
Dense regular (Figure 5.15)	Composed of regular bands of collagen fibers all oriented in the same direction. The flattened nuclei of fibroblasts can be seen between bundles of collagen fibers. Dense regular connective tissue is good at resisting tensile forces in one direction only.	Tendons (connect muscle to bone) and ligaments (connect bone to bone)
Dense irregular (Figure 5.16)	Composed of bundles of collagen fibers arranged in many directions. Fibroblast nuclei can be seen between bundles of collagen fibers. Some nuclei appear round (if cut in cross section) and some appear flattened (if cut in longitudinal section). Many of the collagen fibers appear wavy because they are not all cut along the same plane. This tissue is tough and resists tensile forces applied in multiple directions.	Organ capsules, dermis of the skin, periosteum (outer covering of bone), perichondrium (outer covering of cartilage)
Elastic (Figure 5.17)	Consists of both collagen and elastic fibers all oriented in the same direction. Collagen fibers are thick and typically stain light pink or purple. Elastic fibers appear thin and black (if stained) or wavy. Fibroblast nuclei can be seen between the densely packed fibers. Elastic connective tissue is extensible and allows structures to stretch and recoil back to their original shape.	Walls of large arteries such as the aorta and some ligaments

Chapter Five Histology

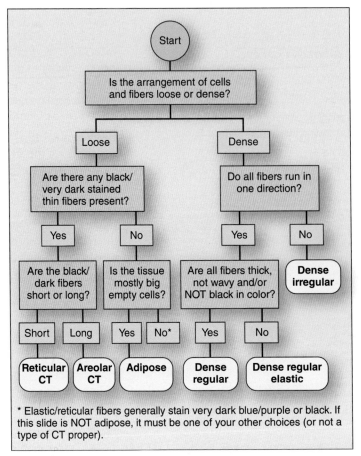

Figure 5.11 Flowchart for Classifying Connective Tissue Proper.

EXERCISE 5.3

IDENTIFICATION AND CLASSIFICATION OF CONNECTIVE TISSUE PROPER

EXERCISE 5.3A Areolar Connective Tissue

1. Obtain a slide of areolar connective tissue (**figure 5.12**).

2. Place the slide on the microscope stage and bring the tissue sample into focus on low power. Then change to high power.

3. You will see several small cells scattered throughout the slide. Most of these cells are **fibroblasts,** which secrete the elastic and collagen fibers (table 5.4). Collagen fibers will generally be light pink in color and somewhat thick. Elastic fibers will generally be long and thin, and will stain very dark or black.

4. Using figure 5.12 and tables 5.3 and 5.4 as guides, identify the following structures on the slide:

   ☐ collagen fibers   ☐ fibroblasts
   ☐ elastic fibers

5. In the following space, sketch areolar connective tissue as seen through the microscope. Be sure to label the structures listed in step 4 on your drawing.

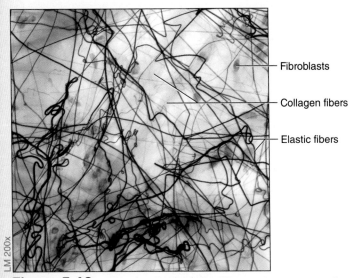

Figure 5.12 **Areolar Connective Tissue.** Areolar connective tissue contains predominantly fibroblasts, collagen fibers, and elastic fibers.

_____ ×

EXERCISE 5.3B Adipose Connective Tissue

1. Obtain a slide of adipose connective tissue **(figure 5.13)**.

2. Place the slide on the microscope stage and bring the tissue sample into focus on low power. Then change to high power.

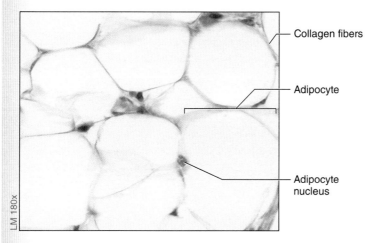

Figure 5.13 Adipose Connective Tissue. Adipose connective tissue is characterized by large adipocytes held together with a loose arrangement of collagen fibers.

3. Using figure 5.13 and tables 5.3 and 5.4 as guides, identify the following structures on the slide:

   ☐ adipocyte   ☐ collagen fibers
   ☐ adipocyte nucleus

4. In the following space, sketch adipose connective tissue as seen through the microscope. Be sure to label all the structures listed in step 3 in your drawing.

_____ ×

EXERCISE 5.3C Reticular Connective Tissue

1. Obtain a slide of reticular connective tissue **(figure 5.14)**.

2. Place the slide on the microscope stage and bring the tissue sample into focus on low power. Then change to high power.

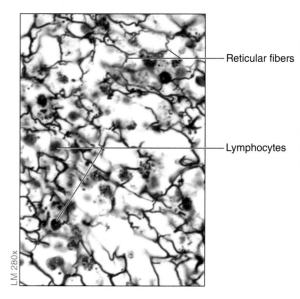

Figure 5.14 Reticular Connective Tissue. Reticular connective tissue consists of reticular fibers (a fine, thin form of collagen) that are visible only when a special stain is used, which makes them appear black.

3. Using figure 5.14 and tables 5.3 and 5.4 as guides, identify the following on the slide:

   ☐ lymphocytes   ☐ reticular fibers

4. In the following space, sketch reticular connective tissue as seen through the microscope. Be sure to label all the structures listed in step 3 in your drawing.

_____ ×

(continued on next page)

(continued from previous page)

INTEGRATE

LEARNING STRATEGY

Though a variety of stains are used for visualization of cells and tissues, certain tissues are visible only when special stains are used. In connective tissues, elastic and reticular fibers can be seen only when a special stain is used that stains the fibers black. Thus, if you see black fibers in a slide, you know you are looking at either elastic or reticular fibers. If the black fibers you see are very long and thin or wavy, they are elastic fibers. If they are short and form a network (*rete*, network), they are reticular fibers.

EXERCISE 5.3D Dense Regular Connective Tissue

1. Obtain a slide of a tendon or ligament **(figure 5.15)**.

2. Place the slide on the microscope stage and bring the tissue sample into focus on low power. Then change to high power.

3. Using figure 5.15 and tables 5.3 and 5.4 as guides, identify the following structures on the slide:

   ☐ collagen fibers   ☐ fibroblast nuclei

4. In the following space, sketch dense regular connective tissue as seen through the microscope. Be sure to label the structures listed in step 3 in your drawing.

_____ ×

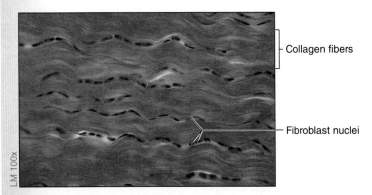

Figure 5.15 Dense Regular Connective Tissue. Dense regular connective tissue consists of bundles of collagen fibers all oriented in the same direction, with fibroblasts located between the bundles of collagen fibers.

INTEGRATE

LEARNING STRATEGY

Sometimes it helps to relate what you see through the microscope to something you already know. For example, the regular arrangement of collagen fibers in dense regular connective tissue often resembles uncooked lasagna noodles stacked upon each other, whereas the irregular arrangement of collagen and elastic fibers in areolar connective tissue often resembles a piece of abstract art.

EXERCISE 5.3E Dense Irregular Connective Tissue

1. Obtain a slide of skin **(figure 5.16)**. Skin consists of two major layers, an outer epidermis, composed of stratified squamous epithelial tissue, and an inner dermis, composed of connective tissue (areolar and dense irregular).

(a)

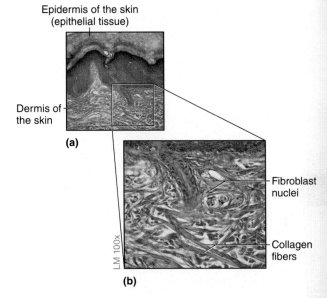

(b)

Figure 5.16 Dense Irregular Connective Tissue. Dense irregular connective tissue is located in the dermis of the skin (a). Dense irregular connective tissue consists of bundles of collagen fibers oriented in many different directions, with fibroblasts located between the bundles of collagen fibers (b).

2. Place the slide on the microscope stage and bring the tissue sample into focus using the scanning objective. Identify the epithelial tissue in the epidermis (figure 5.16a), then move the stage so the lens focuses on the underlying connective tissue in the dermis (figure 5.16b). Switch to a higher-power objective.

3. Using figure 5.16 and tables 5.3 and 5.4 as guides, identify the following structures on the slide:

   ☐ collagen fibers   ☐ fibroblast nucleus

4. In the following space, sketch dense irregular connective tissue as seen through the microscope. Be sure to label the structures listed in step 3 in your drawing.

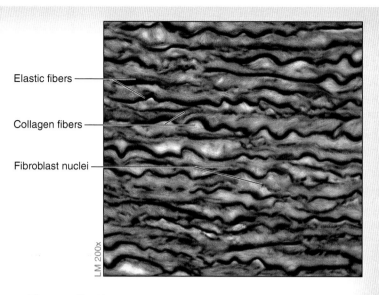

_____ ×

Figure 5.17 Elastic Connective Tissue. The wall of the aorta consists of dense regular elastic connective tissue, which contains both collagen and elastic fibers (which stain black). All the fibers are oriented in the same direction. In this slide, the collagen fibers are dark pink and the elastic fibers are black and wavy. Some fibroblast nuclei are visible in between the bundles of fibers.

4. In the following space, sketch elastic connective tissue as seen through the microscope. Be sure to label the structures listed in step 3 in your drawing.

EXERCISE 5.3F Elastic Connective Tissue

1. Obtain a slide of the aorta or an elastic artery **(figure 5.17)**.

2. Place the slide on the microscope stage and bring the tissue sample into focus on low power. Then switch to high power.

3. Using figure 5.17 and tables 5.3 and 5.4 as guides, identify the following structures on the slide:

   ☐ collagen fibers   ☐ fibroblasts
   ☐ elastic fibers

_____ ×

Supporting Connective Tissue

Cartilage

Cartilage is a specialized connective tissue whose function is to provide strong, yet flexible, support. Cartilage is unique as a connective tissue in that it is avascular. A dense irregular connective tissue covering, called the perichondrium, surrounds all types of cartilage except fibrocartilage. The innermost part of the perichondrium contains immature cartilage cells called chondroblasts. The function of the chondroblasts is to secrete the fibers and ground substance that compose the extracellular matrix of cartilage. As a chondroblast secretes extracellular matrix, it eventually becomes completely surrounded by the matrix; at this point it is considered a mature cell, a chondrocyte. The space in the matrix where a chondrocyte sits is a lacuna (*lacus*, a lake). The function of a chondrocyte is to maintain the matrix that has already been formed. All types of cartilage contain chondrocytes in lacunae and have a ground substance that consists largely of glycosaminoglycans (GAGs). The three types of cartilage differ mainly in the type and arrangement of fibers within the matrix. **Table 5.5** summarizes the characteristics of the different types of cartilage.

Bone

Bone (table 5.5) is a specialized connective tissue whose function is to provide strong support. It protects vital organs and provides strong attachment points for skeletal muscles. Similar to cartilage, bone is surrounded by a dense irregular connective tissue covering, called the periosteum. The innermost part of the periosteum contains precursor bone cells called osteoprogenitor cells, which develop into immature bone cells called osteoblasts. Osteoblasts secrete the extracellular matrix (fibers and ground substance) of bone. When they become completely enveloped by the bony matrix, the osteoblasts become mature osteocytes.

There are two types of bone tissue: compact bone (dense) and spongy bone (cancellous). The structural and functional unit of compact bone is an osteon, which consists of concentric layers of bony matrix (lamellae). Along the lamellae are lacunae that contain osteocytes. The details of the two types of bone tissue will be covered in chapter 7.

Figure 5.18 is a flowchart for classification of supporting connective tissues that you can use when you are given an unknown slide of connective tissue. It will help you decide how to classify the connective tissue.

Table 5.5	Supporting Connective Tissue: Cartilage and Bone		
Tissue Type	**Description**	**Functions and Locations**	**Identifying Characteristics**
Hyaline cartilage	Chondrocytes are located within lacunae. Contains a perichondrium of dense irregular connective tissue. Extracellular matrix consists of diffuse collagen fibers spread throughout a semirigid ground substance, which is composed mainly of glycoproteins and water.	Hyaline cartilage provides *strong, semiflexible support* for structures such as the nasal septum, costal cartilages, articular cartilages, larynx, and tracheal "C" rings	Hyaline cartilage is recognized by the chondrocytes in lacunae and the fact that no fibers are visible in the extracellular matrix
Fibrocartilage	Very similar to hyaline cartilage except there is no perichondrium and the collagen fibers form thick visible bundles	The organization and density of the collagen fibers make this cartilage particularly effective at *resisting compressive forces*. It is found in areas where compressive forces are high, such as intervertebral discs, the pubic symphysis, and the menisci of the knee joint.	Bands of fibers are easily visible. In addition, chondrocytes will appear to be lined up in rows because the thick bands of collagen fibers force them into this configuration.
Elastic cartilage	Very similar to hyaline cartilage in all aspects. However, the addition of elastic fibers to the matrix makes the cartilage much more flexible than hyaline cartilage.	The addition of elastic fibers within the cartilage provides *flexible support*. Locations include the epiglottis, the lining of the auditory tube and the external ear.	Elastic fibers are stained black in most preparations. The elastic fibers are generally higher in concentration near the lacunae.
Compact bone	Composed of osteons (Haversian systems), which are concentric layers of bony matrix (lamellae) surrounding a central canal. Lamellae have lacunae located along them, with osteocytes inside the lacunae. Canaliculi connect adjacent lacunae, and perforating canals run perpendicular to the central canals.	Provides strong, rigid support. It is thickest in the diaphysis of long bones, but is also found as a thin layer forming the peripheral component of all bone.	Multiple osteons packed tightly together. No marrow spaces.

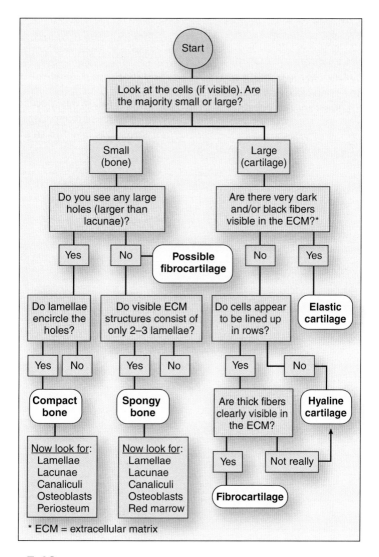

Figure 5.18 Flowchart for Classifying Supporting Connective Tissues.

EXERCISE 5.4

IDENTIFICATION AND CLASSIFICATION OF SUPPORTING CONNECTIVE TISSUE

EXERCISE 5.4A Hyaline Cartilage

1. Obtain a slide of the trachea and esophagus (see exercise 5.1F, p. 102).

2. Place the slide on the microscope stage and bring the tissue sample into focus on low power.

3. Find the lumen of the trachea and identify the epithelial tissue that lines the trachea (see exercise 5.1F, p. 102).

4. Move the microscope stage so that the tissue deep to the epithelium of the trachea is in the center of the field of view. Here you will find a plate of hyaline cartilage **(figure 5.19)**.

5. Observe the slide on the lowest magnification possible. Identify the perichondrium, which surrounds the cartilage plate.

6. Look for small nuclei on the inner surface of the perichondrium. These are the nuclei of chondroblasts.

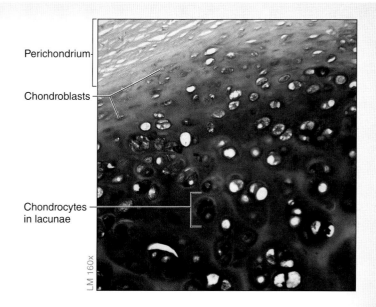

Figure 5.19 Hyaline Cartilage. Hyaline cartilage contains prominent lacunae with chondrocytes. No fibers are visible in the extracellular matrix. At the top of this photograph you can see the perichondrium (light pink) with small, flattened chondroblasts on its inner surface.

(continued on next page)

Chapter Five Histology

(continued from previous page)

7. Next, identify the chondrocytes located within lacunae. In hyaline cartilage you will not be able to identify any fibers in the matrix because the fibers are spread very diffusely throughout the matrix. Instead, the matrix will appear uniform and smooth.

8. Using figure 5.19 and table 5.5 as guides, identify the following structures on the slide:

   ☐ chondroblasts   ☐ lacunae

   ☐ chondrocytes   ☐ perichondrium

9. In the following space, sketch hyaline cartilage as seen through the microscope. Be sure to label all the structures listed in step 8 in your drawing.

_____ ×

EXERCISE 5.4B Fibrocartilage

1. Obtain a slide of an intervertebral disc **(figure 5.20)**.

2. Place the slide on the microscope stage and bring the tissue sample into focus on low power. Then change to high power.

3. Using figure 5.20 and table 5.5 as guides, identify the following structures on the slide:

   ☐ chondrocytes   ☐ bundle of collagen fibers

4. In the following space, sketch fibrocartilage as seen through the microscope. Be sure to label all the structures listed in step 3 in your drawing.

_____ ×

EXERCISE 5.4C Elastic Cartilage

1. Obtain a slide of elastic cartilage **(figure 5.21)**.

2. Place the slide on the microscope stage and bring the tissue sample into focus on low power. Then change to high power.

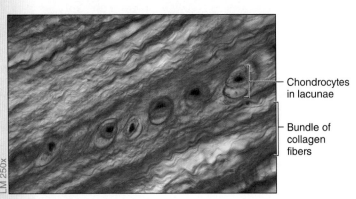

Figure 5.20 Fibrocartilage. Fibrocartilage contains visible bundles of collagen fibers within its matrix. The chondrocytes (in their lacunae) often appear to line up in rows.

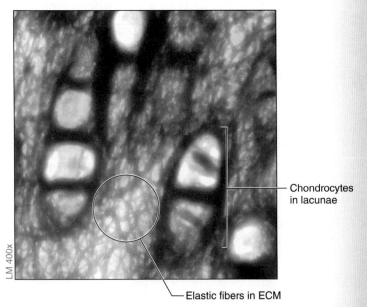

Figure 5.21 Elastic Cartilage. Elastic cartilage contains chondrocytes in lacunae and visible elastic fibers (which stain black) in its extracellular matrix.

3. Using figure 5.21 and table 5.5 as guides, identify the following structures on the slide:

   ☐ chondroblasts   ☐ lacunae
   ☐ chondrocytes   ☐ perichondrium
   ☐ elastic fibers

4. In the following space, sketch elastic cartilage as seen through the microscope. Be sure to label all the structures listed in step 3 in your drawing.

_____ ×

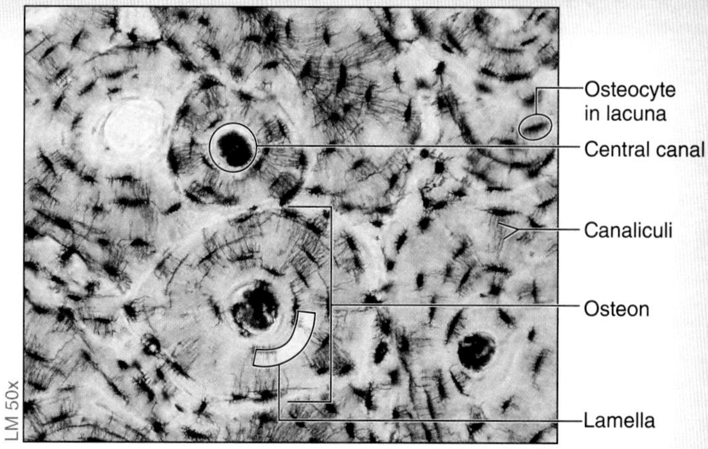

Figure 5.22 Compact Bone. Compact bone is composed of multiple osteons. Each osteon is characterized by a central canal surrounded by concentric layers of bony matrix called lamellae.

Labels: Osteocyte in lacuna, Central canal, Canaliculi, Osteon, Lamella. LM 50x

4. In the following space, sketch compact bone as seen through the microscope. Be sure to label all the structures listed in step 3 in your drawing.

_____ ×

EXERCISE 5.4D Bone

1. Obtain a slide of compact bone **(figure 5.22)**.

2. Place the slide on the microscope stage and bring the tissue sample into focus on low power. Then change to high power.

3. Using figure 5.22 and table 5.5 as guides, identify the following structures on the slide:

   ☐ canaliculus   ☐ lamella
   ☐ central canal   ☐ osteocyte
   ☐ lacuna   ☐ osteon

Fluid Connective Tissue

Fluid connective tissue (blood and lymph) are specialized in that the extracellular matrix of these tissues consists of a liquid ground substance and soluble fibers that become insoluble only in response to tissue injury. These tissues will be covered in detail in chapters 20 and 23. In this chapter we will consider only the basic characteristics of blood as a connective tissue. The cell types in blood are **erythrocytes** (red blood cells), **leukocytes** (white blood cells), and **platelets** (thrombocytes). Platelets are not actually cells. Instead they are cytoplasmic fragments of cells called **megakaryocytes,** which are found in the bone marrow. The extracellular matrix of blood consists of blood **plasma.** The ground substance is liquid, composed mainly of water and a number of dissolved substances. In addition, blood contains soluble proteins, some of which are called clotting proteins, such as **fibrinogen.** Fibrinogen becomes insoluble **fibrin** and forms fibers in response to tissue injury.

EXERCISE 5.5

IDENTIFICATION AND CLASSIFICATION OF FLUID CONNECTIVE TISSUE

1. Obtain a slide of a blood smear (**figure 5.23**).

2. Place the slide on the microscope stage and bring the tissue sample into focus on low power. Then bring the sample into focus using the oil immersion lens.

3. Using figure 5.23 as a guide, identify the following structures on the slide:
   - ☐ erythrocytes
   - ☐ leukocytes
   - ☐ platelets (thrombocytes)

4. In the following space, sketch blood as seen through the microscope. Be sure to label all the structures listed in step 3 in your drawing.

_____ ×

5. *Optional Activity:* **AP|R** **3: Tissues**—Watch the "Connective Tissue Overview" animation.

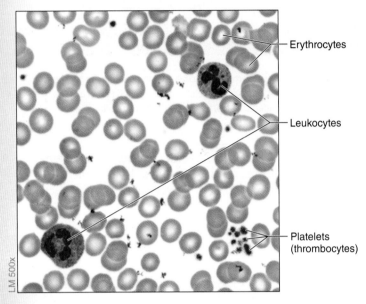

Figure 5.23 Blood. Blood is a fluid connective tissue containing erythrocytes (red blood cells), leukocytes (white blood cells), thrombocytes (platelets), and an extracellular matrix called plasma.

INTEGRATE

CLINICAL VIEW
Carcinomas and Sarcomas

Clinically, a tumor derived from epithelial tissues is called a *carcinoma* (*karkinos*, crab + -*oma*, tumor), and a tumor derived from connective tissues is called a *sarcoma* (*sarc-*, flesh + -*oma*, tumor). Carcinomas are considered noninvasive when they do not penetrate the basement membrane that lies between the epithelial and connective tissue layers. Once rapidly dividing cells penetrate the basement membrane, the cancer is considered invasive. While epithelial tissue is avascular, the underlying connective tissue is not. Thus, invading cancer cells can easily metastasize (*meta-*, change + -*ize*, an action) to other locations of the body through blood and lymphatic vessels. Sarcomas, which arise from connective tissues, pose a similar risk. However, they tend to grow and metastasize much more readily than carcinomas because of the highly vascular nature of connective tissues.

Muscle Tissue

Muscle tissue is both excitable and contractile. Excitable tissues are able to generate and propagate electrical signals called action potentials. As a contractile tissue, muscle has the ability to actively shorten and produce force. There are three types of muscle tissue: skeletal muscle, cardiac muscle, and smooth (visceral) muscle. The three types of muscle tissue are distinguished by the presence or absence of visible striations, the shape of the cells, and the number and location of nuclei. Skeletal muscle is found in the voluntary muscles that move the skeleton and the facial skin. Cardiac muscle is found in the heart, and smooth muscle is found in the walls of soft viscera, such as the blood vessels, stomach, urinary bladder, intestines, and uterus. **Table 5.6** compares the three types of muscle tissue, and the flowchart in **figure 5.24** explains steps you can use to identify muscle tissues.

Table 5.6	Muscle Tissue		
Type of Muscle	**Description**	**Generalized Functions**	**Identifying Characteristics**
Skeletal	Elongate, cylindrical cells with multiple nuclei. Nuclei are peripherally located. Tissue appears striated (light and dark bands along the length of the cell).	Provides voluntary movement	Length of cells (extremely long), striations, multiple peripheral nuclei
Cardiac	Short, branched cells with 1 to 2 nuclei. Nuclei are centrally located. Dark bands (intercalated discs) are seen where two cells join. Tissue appears striated (light and dark bands along the length of each cell).	Responsible for pumping action of the heart	Branched, uninucleate cells, striations, intercalated discs
Smooth	Elongate, spindle-shaped cells (fatter in the center, narrowing at the ends) with single, "cigar-shaped" or "spiral" nuclei. Nuclei are centrally located. No striations are apparent.	Creates movement within viscera such as intestines, bladder, uterus, and stomach. Moves blood through blood vessels.	Spindle-shaped cells, no striations, cigar-shaped nuclei that are centrally located

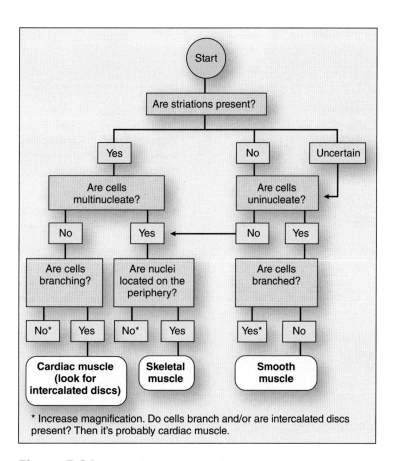

Figure 5.24 Flowchart for Classifying Muscle Tissues.

EXERCISE 5.6

IDENTIFICATION AND CLASSIFICATION OF MUSCLE TISSUE

EXERCISE 5.6A Skeletal Muscle Tissue

1. Obtain a slide of skeletal muscle (**figure 5.25**).

2. Place the slide on the microscope stage and bring the tissue sample into focus on low power. Then change to high power.

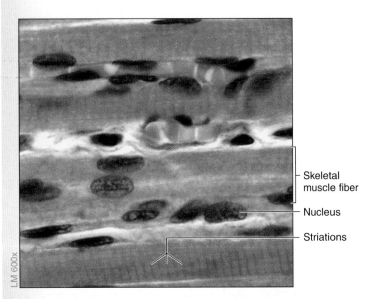

Figure 5.25 Skeletal Muscle. Skeletal muscle fibers are elongated and striated, and contain multiple peripherally located nuclei.

3. Using figure 5.25 and table 5.6 as guides, identify the following structures on the slide:

   ☐ nucleus   ☐ striations
   ☐ skeletal muscle fiber

4. In the following space, sketch skeletal muscle as seen through the microscope. Be sure to label all the structures listed in step 3 in your drawing.

EXERCISE 5.6B Cardiac Muscle Tissue

1. Obtain a slide of cardiac muscle (**figure 5.26**).

2. Place the slide on the microscope stage and bring the tissue sample into focus on low power. Then change to high power.

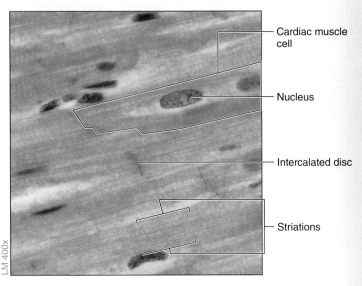

Figure 5.26 Cardiac Muscle. Cardiac muscle cells are short and branched, striated, and generally contain only one centrally located nucleus.

3. Using figure 5.26 and table 5.6 as guides, identify the following structures on the slide:

   ☐ cardiac muscle cell   ☐ nucleus
   ☐ intercalated disc   ☐ striations

4. In the following space, sketch cardiac muscle as seen through the microscope. Be sure to label all the structures listed in step 3 in your drawing.

EXERCISE 5.6C Smooth Muscle Tissue

1. Obtain a slide of smooth muscle **(figure 5.27)**.

2. Place the slide on the microscope stage and bring the tissue sample into focus on low power. Then change to high power.

3. Using figure 5.27 and table 5.6 as guides, identify the following structures on the slide:

   ☐ smooth muscle cell   ☐ nucleus

4. In the following space, sketch smooth muscle as seen through the microscope. Be sure to label all the structures listed in step 3 in your drawing.

 ×

5. *Optional Activity:* **AP|R** **3: Tissues**—Watch the "Muscle Tissue Overview" animation.

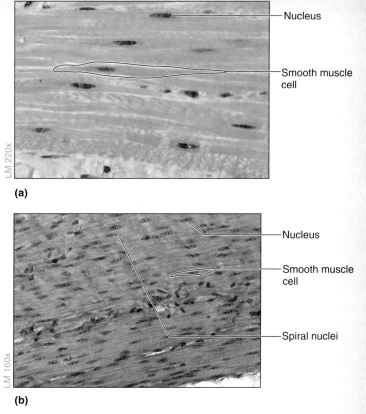

Figure 5.27 Smooth Muscle. Smooth muscle fibers are short and spindle-shaped, not striated, and contain only one centrally located nucleus (a). In figure 5.27b you can see some nuclei that have taken on a "spiral" shape. This happens when the muscle fibers contract. As the fibers shorten, the nuclei start to coil up, or "spiral."

Nervous Tissue

Nervous tissue is characterized by its *excitability*: the ability to generate and propagate electrical signals called action potentials. Nervous tissue is composed of two basic cell types **(table 5.7)**. **Neurons** are excitable cells that send and receive electrical signals. They have a limited ability to divide and multiply in the adult brain. **Glial cells** are supporting cells that support and protect neurons. Glial cells maintain the ability to divide and multiply in the adult brain. Glial cells constitute over 60% of the cells found in neural tissue.

Table 5.7	Nervous Tissue	
Cell Type	**Description**	**Generalized Functions**
Neurons	Though varied in shape, most neurons appear to have numerous branches coming off the cell body (soma). Neurons contain large amounts of rough endoplasmic reticulum (ER). The ribosomes and rough ER stain very dark and are collectively called chromatophilic substance.	These cells are responsible for *generating and transmitting information via electrical impulses* within the nervous system. Thus, they are "excitable" cells.
Glial cells	Even more varied in shape than neurons, glial cells are generally much smaller than neurons with fewer (if any) branching processes	These cells are the *general supporting cells* of the nervous system. They protect, nourish, and support the excitable cells, the neurons.

EXERCISE 5.7

IDENTIFICATION AND CLASSIFICATION OF NERVOUS TISSUE

1. Obtain a slide of nervous tissue (**figure 5.28**).

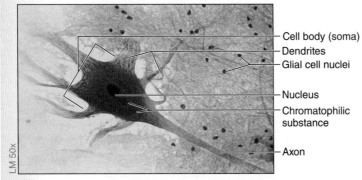
(a)

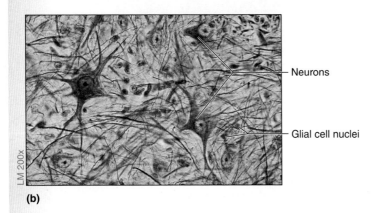
(b)

Figure 5.28 Nervous Tissue. (a) Neurons are very large cells with a prominent nucleus and nucleolus, and dark-staining endoplasmic reticulum (chromatophilic substance). (b) Glial cells are much smaller than neurons, and are also more abundant.

2. Place the slide on the microscope stage and bring the tissue sample into focus on low power. Then change to high power.

3. Using figure 5.28 and table 5.7 as guides, identify the following structures on the slide:

   ☐ axon of neuron   ☐ dendrite of neuron
   ☐ cell body of neuron   ☐ nucleus of glial cell
   ☐ chromatophilic   ☐ nucleus of neuron
      substance

4. In the following space, sketch nervous tissue as seen through the microscope. Be sure to label all the structures listed in step 3 in your drawing.

_____ ×

5. *Optional Activity:* **AP|R** **3: Tissues**—Watch the "Nervous Tissue Overview" animation.

Chapter 5: Histology

Name:_____
Date:_____ Section:_____

POST-LABORATORY WORKSHEET

The ❶ corresponds to the Learning Objective(s) listed in the chapter opener outline.

Do You Know the Basics?

Exercise 5.1: Identification and Classification of Epithelial Tissue

1. The epithelial type that protects against abrasion is _____. ❶

2. Fully classify the epithelial tissue shown in this micrograph: ❷

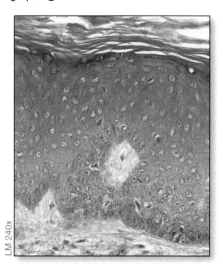

_____
_____

3. Fully classify the epithelial tissue shown in this micrograph: ❷

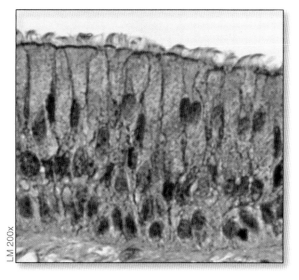

_____
_____

4. Endothelium is a _____ epithelium that lines the walls of blood vessels and the heart. ❷

5. List and give the function of two cell surface modifications and specialized cells found in epithelial tissues. ❸

Surface Modifications/Specialized Cells	*Function*
a. _____	_____
b. _____	_____
c. _____	_____
d. _____	_____

Chapter Five *Histology*

6. If you are observing a cross section of a tube that is lined with epithelial tissue (e.g., the gut tube), the epithelial surface that faces the lumen of the tube is the _____ surface. ❹

Exercise 5.2: Identification of Embryonic Connective Tissue

7. Embryonic connective tissue is called _____. ❻

Exercise 5.3: Identification and Classification of Connective Tissue Proper

8. The extracellular matrix of connective tissues is composed of _____ and _____. ❼

9. For each category of connective tissue listed in the following table, write in the major cell types, the fiber types, and the characteristics of the ground substance of the tissue. Refer to your textbook if you need help. ❼–⓮

	Connective Tissue Proper	Cartilage	Bone	Fluid Connective Tissue
Cells				
Fibers				
Ground Substance				

Exercise 5.4: Identification and Classification of Supporting Connective Tissue

10. The three types of cartilage are differentiated from one another by the type and arrangement of _____. ❿

11. The two types of bone tissue are _____ and _____. ⓭

Exercise 5.5: Identification and Classification of Fluid Connective Tissue

12. The two types of fluid connective tissue are _____ and _____. ⓮

13. The cells in fluid connective tissue are _____. The fibers in fluid connective tissue are _____. ⓯

14. In the following table, compare and contrast the characteristics of the various types of muscle tissue. ⓰

Characteristic	Skeletal Muscle	Cardiac Muscle	Smooth Muscle
Location and Number of Nuclei			
Cell Shape			
Presence or Absence of Striations			

15. The only kind of muscle tissue that does NOT have striations is _____. ⓰

16. List a location in the body where you might find smooth muscle tissue. _____ ⓱

Exercise 5.7: Identification and Classification of Nervous Tissue

17. The two main cell types found in nervous tissue are _____ and _____. ⓲

Can You Apply What You've Learned?

18. Compare and contrast epithelial and connective tissues with respect to the following:

Characteristic	Epithelial Tissues	Connective Tissues
Cell Number and Arrangement		
Polarity		
Extracellular Matrix		
Vascularity		

19. Identify each of the following connective tissues by its description.

    _____ Tightly packed collagen fibers aligned parallel to one another (tissue that withstands stress in one plane)
    _____ Meshwork of reticular fibers
    _____ Collagen fibers are not visible; ground substance appears clear and glassy (chondrocytes in lacunae)
    _____ Loose array of collagenous, reticular, and some elastic fibers; (abundant distribution of blood vessels; ground substance is abundant and viscous)
    _____ Collagen fibers arranged in concentric rings (hard ground substance); (osteocytes in lacunae)
    _____ Bundles and clumps of collagen fibers extending in all directions; (forms sheaths)
    _____ Clear ground substance; biconcave pink cells and larger cells with dark-staining nucleus
    _____ Densely packed elastic fibers
    _____ Densely packed elastic fibers; chondrocytes in lacunae
    _____ Irregular bundles of very light wavy collagen fibers densely interwoven; fibers barely noticeable; chondrocytes in lacunae
    _____ Composed primarily of adipose cells

20. What does the presence of cilia indicate about the function of an epithelial tissue?

    _____
    _____

21. Place a ✓ by the statements about connective tissue that are accurate.

    ____ Forms deep layers of skin (dermis)

    ____ Forms glands (e.g., salivary, pancreas)

    ____ Forms a structural framework for organs

    ____ Forms bones, cartilage, ligaments, and tendons

    ____ Highly vascular (with some exceptions)

    ____ Conducts neural impulses

    ____ Contains bundles of nerve fibers and muscle fibers

    ____ Includes blood and components of blood vessel walls

22. Fill in the blank with the name of the connective tissue that is described in each of the items below.

    a. Makes up the cartilage discs between the vertebrae and between the pubic bones.

    _____

    b. Transports materials within blood vessels.

    _____

    c. Inner supporting framework of spleen, liver, and lymph nodes.

    _____

d. Tough tissue that connects bone to bone and muscle to bone. It can be pulled along its length and usually does not tear or stretch.

e. Cartilage that retains its original shape after being deformed, such as in the cartilage of the ear.

f. Allows the aorta (a large blood vessel) to stretch with each pulse of blood pumped into it from the heart and then return to its original size.

g. Forms the "C" shaped cartilage rings of the trachea.

h. Helps keep you warm and stores excess fuel (energy) molecules.

i. Forms the ends of long bones, the larynx, the costal cartilages of the ribs, and the embryonic skeleton.

23. Explain the characteristics that make blood a connective tissue. (Hints: What embryonic connective tissue is blood derived from? What is the composition of the extracellular matrix? What is the cellular component? What are the fibers?)

24. Write the name of the specific tissue type below each photo.

(a) _____

(b) _____

(c) _____

(d) _____

(e) _____

(f) _____

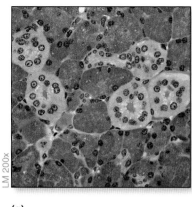
(g) _____

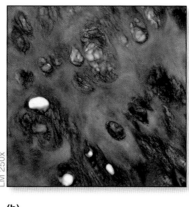
(h) _____

25. A friend of yours was recently diagnosed with a tumor called a glioma. What basic tissue type is the tumor derived from? _____

26. The biceps brachii, deltoid, and pectoralis muscles are composed of what type of muscle tissue? _____

Can You Synthesize What You've Learned?

27. For each of the examples given below, do you think the structure is likely to fit in its entirety on a microscope slide? (Circle yes or no for each.)

    a. A cross section of a pinky finger   Yes   No
    b. A cross section of the large intestine   Yes   No
    c. A cross section of a small blood vessel   Yes   No
    d. A cross section of the humerus (your arm bone)   Yes   No
    e. A cross section of a ureter (~0.5 cm in diameter)   Yes   No

28. You are interested in developing a tissue that can resist stretch in multiple directions, yet also remains very flexible. Given what you know about the properties of human tissues, describe how you might design such a tissue.

    _____
    _____
    _____
    _____
    _____

29. Cartilage is unique as a connective tissue in that it is avascular. What limitations do you think avascularity imposes on cartilage as a tissue?

    _____
    _____
    _____
    _____
    _____

30. Two of the surface modifications of epithelial cells, microvilli and cilia, are commonly confused with each other. In the space below, describe how you would teach a fellow student to distinguish between these two surface modifications.

    _____
    _____
    _____
    _____
    _____

31. Most tumors arise in cells that are constantly undergoing cell division, or mitosis (for example, skin cells). A patient has recently been diagnosed with a brain tumor. What type of cell do you think the tumor most likely arose from—a neuron or a glial cell?

32. You have a patient who has recently been diagnosed with an osteosarcoma. In the space below, explain what cells this tumor most likely arose from and why.

PART III | SUPPORT AND BODY MOVEMENT

CHAPTER 6

Integument

OUTLINE AND LEARNING OBJECTIVES

Histology 127

The Epidermis 127

EXERCISE 6.1: LAYERS OF THE EPIDERMIS 128
1. Describe the layers of the epidermis of thick skin, from deep to superficial
2. Compare and contrast thick skin and thin skin
3. Identify regions on the body where thick and thin skin are found

EXERCISE 6.2: PIGMENTED SKIN 129
4. Identify keratinocytes containing melanin granules in pigmented skin

The Dermis 130

EXERCISE 6.3: LAYERS OF THE DERMIS 130
5. Differentiate between the papillary and reticular layers of the dermis

EXERCISE 6.4: MEROCRINE (ECCRINE) SWEAT GLANDS AND SENSORY RECEPTORS 131
6. Identify merocrine sweat glands, tactile (Meissner) corpuscles, and lamellated (Pacinian) corpuscles
7. Describe the function of merocrine sweat glands, tactile corpuscles, and lamellated corpuscles

EXERCISE 6.5: THE SCALP—HAIR FOLLICLES AND SEBACEOUS GLANDS 133
8. Describe the parts of a hair follicle
9. Describe the location, structure, and function of sebaceous glands

EXERCISE 6.6: AXILLARY SKIN—APOCRINE SWEAT GLANDS 135
10. Identify apocrine sweat glands
11. Describe how apocrine sweat glands differ from merocrine sweat glands in both location and function

EXERCISE 6.7: STRUCTURE OF A NAIL 136
12. Describe the structure of a nail

Gross Anatomy 137

Integument Model 137

EXERCISE 6.8: OBSERVING CLASSROOM MODELS OF INTEGUMENT 137
13. Identify skin layers and accessory structures on a model of the integument

 **MODULE 4: INTEGUMENTARY SYSTEM**

INTRODUCTION

In this laboratory session you will encounter the body's largest organ system, the **integument** (*integumentum*, a covering). We often mistakenly refer to the integument as the "skin." However, the two terms do not refer to the same thing. The **skin** is the outer covering of the body and is composed of an **epidermis** (*epi*, upon, + *derma*, skin) and **dermis** (*derma*, skin). The integument is an organ system that includes the skin *plus* all of the accessory structures of skin, such as sensory organs, glands, hair, and nails. The integument is a truly remarkable system, and we typically take its functions for granted—unless, of course, we happen to injure large portions of it. Damage to large areas of the integument (as might happen to a burn victim) can easily be life-threatening.

In this laboratory session, the materials at your disposal will include histological slides of the skin and its accessory structures, and models of integument. While it is important that you learn to identify the structures of the integument through the microscope and on the models. However, do not forget to use your own body as an example as you work through this laboratory exercise. For instance, while you view slides of thin skin and thick skin, be sure to also find locations on your own body that contain thin skin and thick skin. See and feel the differences between the two for yourself, and try to correlate what you see and feel on your own body with what you see through the microscope. Finally, be sure to compare the appearance and texture of your own skin and hair with that of another student in your lab who has skin or hair of a different color or texture than your own.

125

Chapter 6: Integument

PRE-LABORATORY WORKSHEET

1. List four functions of the integument.
   a. _Protection_
   b. _Keeps from Drying out_
   c. _____
   d. _____

2. What type of epithelium composes the epidermis of the skin? _Stratified Squamous_

3. What type of connective tissue constitutes most of the dermis of the skin? _irregular connective tissue_

4. List the five layers of the epidermis of thick skin, starting with the layer closest to the basal surface.
   a. _Stratum basale_
   b. _Stratum Spinosum_
   c. _Stratum granulosum_
   d. _Stratum lucidum_
   e. _Stratum Corneum_

5. List the two layers of the dermis.
   a. _Papillary_
   b. _dermal_

6. List three accessory structures located within the dermis.
   a. _Sebaceous glands_
   b. _Sweat gland_
   c. _tactile receptor_

7. The main protein that composes a hair is
   a. elastin   b. fibrin   **c. keratin**   d. collagen

8. The subcutaneous layer of the skin (hypodermis) is mainly composed of _Connective tissue_ tissue.

9. **True** or false: Thick skin is located only on the palms of the hands and soles of the feet.

10. **True** or false: Sweat glands are composed of epithelial tissue.

Histology

The Epidermis

The **skin** is composed of an **epidermis** (*epi*, upon, + *derma*, skin) and **dermis** (*derma*, skin). The epidermis is the outermost layer of the skin and consists of keratinized stratified squamous epithelial tissue. Thus there are no blood vessels in the epidermal layers. The main cell type composing the epithelial tissue in the skin is a **keratinocyte,** so named because of its role in synthesizing the protein **keratin.** Keratin is an insoluble protein that imparts strength to the skin and makes it almost completely waterproof. Other cell types include melanocytes that produce melanin pigment, and tactile (Merkel) cells that when compressed release chemicals to stimulate nerve endings. The epidermis of *thin skin* (for example, see figure 6.3b), which is located on most of the body's surface, has four distinct layers, whereas the epidermis of *thick skin* (see figure 6.1), which is located only on the palms of the hands and soles of the feet, has five distinct layers. **Table 6.1** summarizes the layers of the epidermis and their characteristics.

INTEGRATE

LEARNING STRATEGY

The terms *thick skin* and *thin skin* refer only to the thickness of the epidermal layer. Technically, the thickest skin on the body is located on the back. However, that thickness takes into account not only the thickness of the epidermis, but that of the dermis as well. The thick skin on the palms of the hands and the soles of the feet is actually rather thin when considering the total thickness of the epidermis and the dermis combined. To understand this a little better, compare the total thickness of the skin on your back (by feel) to the total thickness of the skin on the middle arch of the sole of your foot or palm of your hand.

Table 6.1 Layers of the Epidermis

Epidermal Layer	Description	Word Origin
Stratum Basale	The deepest layer is only one cell thick. At higher magnification, the cells appear large, with round nuclei, and you may be able to see cells undergoing cell division (*mitosis*). In addition, melanin granules are most concentrated in this layer of the skin. You will be able to view melanin granules well only in a sample of pigmented skin.	*stratum*, layer, + *basalis*, situated near the base
Stratum Spinosum	Named for the spiny appearance of the keratinocytes when viewed at high magnification. This layer is composed of several layers of keratinocytes that appear "prickly" or "spiny." The "spines" are artifacts that are formed during tissue preparation. During this process the cytoplasm of the cells shrinks, while the cytoskeletal elements and desmosomes remain intact keeping neighboring cells attached.	*stratum*, layer, + *spina*, a thorn
Stratum Granulosum	Consists of three to five layers of cells that appear granular and darker in color than those of the underlying stratum spinosum. The graininess of the cells in this layer makes it quite distinct. Keratinocytes begin to die within the stratum granulosum as organelles and nuclei degenerate and the cells accumulate bundles of keratin.	*stratum*, layer, + *granulum*, a small grain
Stratum Lucidum	Present only in thick (glabrous) skin, which is located on the palm of the hand and sole of the foot. Consists of 2 to 3 layers of dead cells containing the translucent protein eleidin, which is an intermediate product in keratin formation. This clear layer sometimes stains darker than the remainder of the stratum corneum, depending on the dye used.	*stratum*, layer, + *lucidus*, clear
Stratum Corneum	Consists of multilayers (20 to 30) dead, scaly, interlocking keratinocytes. On the outer (apical) surface of the stratum corneum you may be able to view layers of dead cells in the process of **desquamation** (*squamosus*, scaly, fr. *squama*, scale).	*stratum*, layer, + *cornu*, horn, hoof

INTEGRATE

LEARNING STRATEGY

Follow these steps to help you identify the epidermal layers through the microscope:

1. **Determine if the layer is more superficial or deep.** Remember the stratum corneum is the most superficial layer (next to the free surface), whereas the stratum basale is the deepest epidermal layer.
2. **Examine the shape of the cells.** The stratum corneum and lucidum contain squamous cells, the stratum spinosum contains polygonal cells, and the stratum basale contains cuboidal to low columnar cells.
3. **Determine if the keratinocytes have a nucleus or not.** Living keratinocytes (as found in the stratum granulosum, spinosum, and basale) have prominent nuclei. Dead keratinocytes (as found in the stratum lucidum and corneum) no longer have a nucleus.
4. **Count the number of cells in the layer.** The stratum corneum contains 20 to 30 layers of cells. The stratum lucidum, granulosum, and spinosum contain two to five layers of cells. The stratum basale contains only one layer of cells.
5. **Determine if the cytoplasm of the cells contain visible granules.** The only layer with visible (and generally dark) cytoplasmic granules is the stratum granulosum.

EXERCISE 6.1

LAYERS OF THE EPIDERMIS

1. Obtain a histology slide of **thick skin** (from the palm of the hand or sole of the foot) **(figure 6.1).** Thick skin is also called *glabrous* skin, meaning it lacks hair (*glaber*, smooth). It does, however, contain all five epidermal layers, so it is an ideal slide for demonstrating characteristics of the epidermis.

2. Place the slide on the microscope stage and scan the slide at low power. Look for any empty or clear space present in the slide, and try to find the apical surface of the skin epithelium lying next to the clear space.

3. Once you have located the skin epithelium, switch to a higher power. Scan the slide until the basal surface of the epithelium is in the center of the field of view. Here you will see the junction between the epidermis and the dermis. Notice that the basal layer is thrown into folds that extend into the dermis. These folds are called **epidermal ridges.** Epidermal ridges are thickened layers of the epidermis that are located between extensions of the papillary layer of the dermis called **dermal papillae** (*papilla*, nipple). The increased area of adhesion between the epidermis and the dermis in the areas of epidermal ridges and dermal papillae helps prevent the epidermis and dermis from coming apart in areas where the skin experiences large frictional forces. Epidermal ridges are larger in the tips of the finger and are used to form fingerprints. Observe the fingerprints on your own hand.

4. Using figure 6.1 and table 6.1 as guides, identify the following structures on the slide of thick skin:

   ☑ stratum basale ☐ stratum spinosum
   ☐ stratum granulosum ☑ stratum lucidum
   ☑ stratum corneum ☐ epidermal ridges

5. In the space below, sketch the layers of the epidermis of thick skin as seen through the microscope. Be sure to label all of the structures listed in step 4 in your drawing.

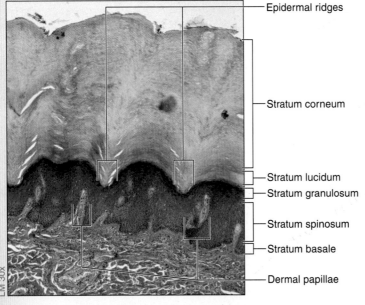
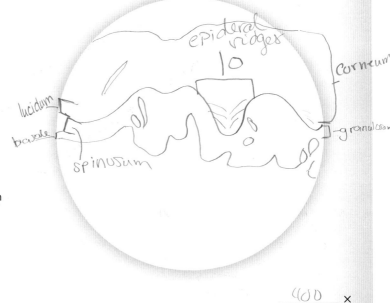

Figure 6.1 Thick Skin. Epidermal layers in thick skin (found on the palms of the hands and soles of the feet).

EXERCISE 6.2

PIGMENTED SKIN

The color of a person's skin is partially due to the pigment **melanin** (*melas*, black). Melanin is produced by cells called **melanocytes,** which are found along the base of the stratum basale of the epidermis. Melanocytes have long dendritic (*dendrites*, relating to a tree) processes that extend between keratinocytes and transfer the melanin granules they produce to the adjacent keratinocytes. The melanin then accumulates on the apical side of the nuclei of the keratinocytes, protecting them from harmful ultraviolet light that can damage their DNA. A single melanocyte produces melanin for a number of adjacent keratinocytes (the melanocyte plus the cells served by it are referred to as an *epidermal-melanin unit*). Some areas of the body have a higher density of melanocytes than others. For example, the skin of the areola of the breast has more melanocytes than the skin on the rest of the breast. However, the *total number of melanocytes in light- and dark-skinned individuals is the same.* Differences in skin color have to do with the *amount of melanin* produced by melanocytes. The melanocytes of dark-skinned individuals simply produce greater amounts of melanin than those of light-skinned individuals.

1. Obtain a slide of pigmented skin and place it on the microscope stage. Observe at low power to locate the epidermis, and then change to a higher-power lens.

2. Focus on the stratum spinosum of the pigmented skin. You should be able to notice distinct black/brown melanin granules within the keratinocytes **(figure 6.2)**. Can you identify any melanocytes in the slide?_____ Melanocytes are located in the stratum basale and have small, round nuclei and pale-staining cytoplasm. Please note that melanocytes are not easily identified on the standard slides found in the laboratory.

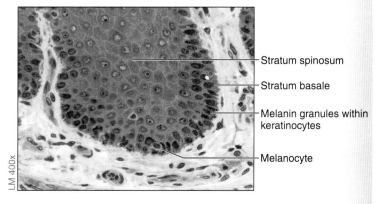

Figure 6.2 Pigmented Skin. Pigmented skin containing melanin granules (brown) within keratinocytes, particularly in the stratum basale of the epidermis. Melanocytes themselves are distinguished by a halo of clear cytoplasm that surrounds the nucleus.

INTEGRATE

CLINICAL VIEW
Melanoma

Melanoma (*melano-*, black + *-oma*, tumor) is an aggressive and sometimes deadly skin cancer that involves uncontrolled proliferation of melanocytes in the stratum basale layer of the epidermis. Melanoma usually presents as a mole on the surface of the skin that exhibits the following "ABCD" characteristics: **asymmetry (A), irregular borders (B), varied colors (C),** and a **diameter of greater than 6 mm (D) (figure 6.3a).** Doctors test the suspicious lesion by taking a biopsy of the affected tissue to confirm the diagnosis (figure 6.3b). Once it has been identified as melanoma, doctors classify the severity of the case by identifying the "stage" of cancer. Tumors that remain in the outer portions of the epidermis are classified as stage I. These are the most simple types of melanomas to treat. Stage II occurs when the tumor invades the underlying dermis, yet there are no signs of spread to local lymph nodes. Because the dermis contains both lymph vessels and blood vessels, there is the risk of metastasis, or spreading of the cancer cells from one part of the body to another. When the cancer spreads to nearby lymph nodes or tissues, it is classified as stage III. Once there is evidence of metastasis to distant organs, it is classified as stage IV. Treatment may involve removal of the tumor and affected lymph nodes, as well as radiation and chemotherapy to attack rapidly dividing cells. Unfortunately, these treatments also affect healthy cells that divide rapidly, such as epithelial cells of the skin and digestive tract. The result may be a loss of hair, increased skin irritations, and severe nausea and vomiting.

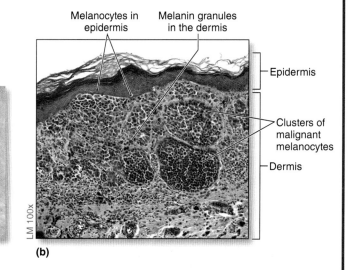

Figure 6.3 Malignant Melanoma. (a) This mole demonstrates the A,B,C,Ds of melanoma: A = asymmetry, B = irregular borders, C = multicolored, D = diameter >6 mm. (b) Histopathology of a malignant melanoma. Note the large clusters of melanocytes in the dermis, and the brownish melanin pigment granules scattered throughout.

The Dermis

The **dermis** of the skin is more complex than the epidermis because it contains numerous skin appendages such as hair follicles, glands, blood vessels, and nerves. The dermis consists of two layers: an outer **papillary layer,** and an inner **reticular layer (figure 6.4).** The papillary layer is the part of the dermis that contains "nipplelike" extensions that project toward the epidermis (the dermal papillae). The papillary layer is generally quite thin and composed of areolar connective tissue. The reticular layer is named for its "networked" appearance (*rete*, a net), not because it contains reticular fibers. In fact, the major fiber type found in this layer is the collagen that composes the dense irregular connective tissue. These collagen fibers are interwoven into a meshwork that surrounds structures of the dermis such as hair follicles, sweat glands, sebaceous glands, blood vessels, and nerves. The slide descriptions in this section of the laboratory manual will guide you through both general observation of the structure and function of the dermis and the specific structures and functions of the various skin appendages found within the dermis.

INTEGRATE

LEARNING STRATEGY

The next time you see a piece of leather, look at it carefully and think about your own skin. The very tough tissue that composes most of the leather is dense irregular connective tissue (collagen fibers) from the dermis of the animal's skin.

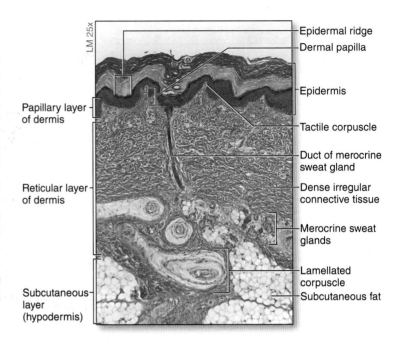

Figure 6.4 **Dermis.** Thick skin demonstrating several skin appendages, such as tactile corpuscles, lamellated corpuscles, and merocrine sweat glands, in the dermis.

EXERCISE 6.3

LAYERS OF THE DERMIS

1. Obtain a slide of thick skin or thin skin and place it on the microscope stage.

2. Scan the slide at low power to identify the skin tissue, and then bring the dermal layer of the skin into the center of the field of view.

3. Switch to high power and distinguish the papillary dermis from the reticular dermis **(figure 6.5).** Locate a **tactile (Meissner) corpuscle** within a dermal papilla. Tactile corpuscles are sensory receptors for fine touch (see table 6.3).

4. Using figures 6.4 and 6.5 as guides, identify the following structures of the dermis on the slide:

   - dense irregular connective tissue
   - dermal papillae
   - papillary dermis
   - reticular dermis
   - tactile (Meissner) corpuscle

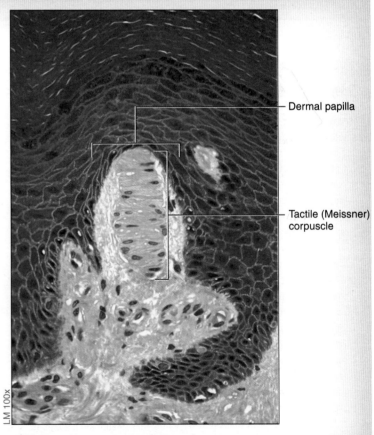

Figure 6.5 **Tactile Corpuscle.** High-magnification view of a dermal papilla containing a tactile corpuscle.

5. In the space to the right, sketch the dermis as seen through the microscope. Be sure to label all the structures listed in step 4 in your drawing.

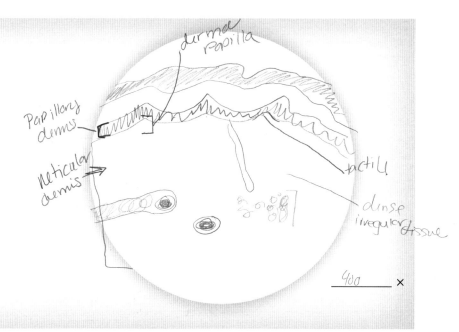

EXERCISE 6.4

MEROCRINE (ECCRINE) SWEAT GLANDS AND SENSORY RECEPTORS

1. Obtain a slide of thick skin or thin skin and place it on the microscope stage. Begin by observing the epidermal and dermal layers of the skin. As you view the current slide, focus your attention on the numerous skin appendages found within the dermis.

2. Locate the coiled, tubular glands found deep within the reticular dermis (these glands open to the skin surface, and you might see a duct traveling to the surface of the skin). These are **merocrine (eccrine) sweat glands** (*meros*, to share, + *krino*, to separate) (figure 6.4), which are located in skin covering nearly the entire body (as opposed to apocrine sweat glands, which are located only in the axillary region, pubic region, and areola). **Table 6.2** lists the types of glands located in the dermis and describes their locations and functions.

3. Focus your attention on sensory receptors found in the dermis. **Table 6.3** summarizes the characteristics of

Table 6.2	Glands in the Dermis				
Gland	**Location**	**Description**	**Mode of Secretion**	**Function**	**Word Origin**
Apocrine Sweat Glands	Axilla, areola of the breast, pubic, and anal regions	Coiled tubular glands located next to hair follicles. Ducts open into the hair follicle.	*Merocrine*—exocytosis of vesicles containing product into the duct of the gland	Produce a thick, slightly oily sweat that may have pheromone-like properties	*apo-*, away from, + *krino*, to separate or secrete
Merocrine (eccrine) Sweat Glands	Most of the surface of the body	Coiled tubular glands whose main secretory portions are found deep within the reticular layer of the dermis. Ducts open to the surface of the skin.	*Merocrine*—exocytosis of vesicles containing product into the duct of the gland	Produce the thin, watery sweat that cools the body	*meros*, share, + *krino*, to separate or secrete
Sebaceous Glands	Wherever hair follicles are found. Particularly abundant on the scalp.	Glands located next to hair follicles. Ducts commonly open into the hair follicle.	*Holocrine*—disintegrated whole cells filled with product are discharged into the duct of the gland	Produce sebum, an oily substance that lubricates the skin surface, keeps it from drying out, and inhibits the growth of bacteria	*sebaceous*, relating to sebum; oily; *holos*, whole, + *krino*, to separate or secrete

(continued on next page)

(continued from previous page)

Table 6.3		Sensory Receptors in the Dermis	
Sensory Receptor	**Location**	**Structure/Appearance**	**Function**
Free Nerve Ending	At epidermal/dermal junction	Dendritic endings of sensory neurons. There is no special structure at the end of the nerve (hence "free" nerve ending).	Light touch, temperature, pain, and pressure
Lamellated (Pacinian) Corpuscle	Deep in the dermis and hypodermis	Dendritic endings of sensory neurons ensheathed with an inner core of neurolemmocytes and outer concentric layers of connective tissue. Resemble an onion.	Sensation of deep pressure and high frequency vibration
Tactile (Merkel) Cell	At epidermal/dermal junction	Flattened dendritic endings of sensory neurons that end adjacent to specialized tactile (Merkel) cells. Round cell located in the stratum basale of the epidermis. It associates with a sensory nerve ending in the dermis (a tactile disc).	Sensation of fine touch, textures, and shapes
Tactile (Meissner) Corpuscle	In dermal papillae	Highly intertwined dendritic endings enclosed by modified neurolemmocytes and connective tissue. Oval-shaped structure with cells that appear almost layered on top of each other.	Sensation of fine, light touch and texture

sensory receptors in the dermis. Many of the receptors are difficult to identify histologically, so you should also refer to the gross anatomy section of this manual and identify them on classroom models of integument. Observe the many dermal papillae found in the slide. Center one or more papillae in the field of view and increase the magnification.

4. Using figure 6.5 as a guide, try to locate the tiny sensory receptors called **tactile (Meissner) corpuscles** (*tactus*, to touch, + *corpus*, body) within the papillae. These sensory receptors, appropriately located near the surface of the skin, are responsible for sensing fine touch.

5. After identifying the tactile corpuscles, change back to a lower magnification and scan the lower reticular dermis and subcutaneous regions of the skin. Look for a large onion-shaped organ. This is a sensory receptor called a **lamellated (Pacinian) corpuscle** (*lamina*, plate, + *corpus*, body) (figure 6.4). Such sensory organs, located deep within the dermis, are responsible for sensing deep pressure applied to the skin. The other main sensory receptors found within the dermis, the **tactile (Merkel) cells** and **free nerve endings**, are not easily identifiable through the light microscope so you must rely on classroom models for their identification.

6. Finally, see if you can identify cross sections through the numerous small blood vessels located in the dermis.

7. In the space below, sketch the skin appendages you were able to locate.

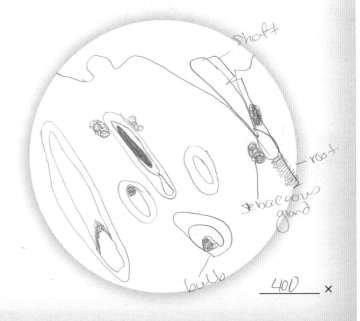

INTEGRATE

CONCEPT CONNECTION

The skin is an organ that is composed of both epithelial and connective tissues. Recall that epithelial tissues cover surfaces and are avascular, whereas connective tissue provides support and varies in degrees of vascularity. The epidermis of the skin, the outermost layer, is composed of keratinized stratified squamous epithelial tissue. The dermis of the skin is composed of two types of connective tissue. The papillary layer contains areolar connective tissue, whereas the reticular layer contains dense irregular connective tissue. It is critical to relate structure to function. That is, the epidermis contains *stratified* epithelium; therefore, it provides *protection* for underlying structures. The dermis contains largely dense irregular connective tissue, which is composed of large quantities of collagen. Therefore, the dermis provides support and resistance to stress in multiple directions. These qualities allow the skin to aid in temperature regulation, prevention of excessive water loss, and protection against abrasion and foreign pathogens.

EXERCISE 6.5

THE SCALP—HAIR FOLLICLES AND SEBACEOUS GLANDS

1. Obtain a slide of the **scalp** and place it on the microscope stage. Begin by observing the epidermal and dermal layers of the scalp, just as you did with previous slides. The scalp epithelium is thin, but the dermis is thick and contains numerous **hair follicles** (**figure 6.6** and **table 6.4**).

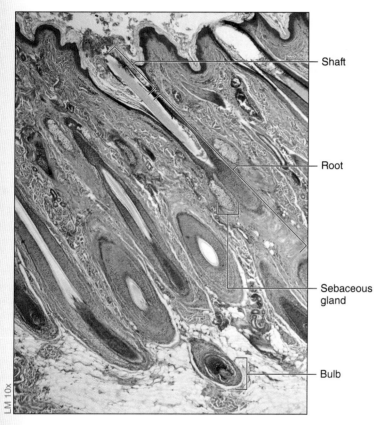

Figure 6.6 Scalp. Skin of the scalp, demonstrating several hair follicles.

2. Scan the slide until you locate a hair follicle that is sliced longitudinally, so you can see the entire hair, from the base of the hair follicle to where it exits the skin. Notice that the color of the cells lining the hair follicle is similar to the color of the cells within the epidermis. This is because hair follicles are derivatives of the epidermis and they develop in the direction of the stratum basale. There are three distinct regions to a hair: (1) the **shaft,** which is the portion of the hair that exits the skin surface; (2) the **root,** which is the portion of the hair within the skin itself; and (3) the **bulb,** which is the swelled base of the hair.

3. Observe the bulb of the hair at higher magnification (**figure 6.7**). Notice the **papilla,** a cone-shaped structure in the middle of the base of the follicle. The papilla is part of the dermis, and is separated from the hair follicle by the basement membrane of the hair follicle epithelium (this basement membrane continues external to the hair follicle as the "glassy membrane"). The papilla contains sensory nerve endings and numerous blood vessels, which are important in supplying nutrients to the developing hair.

4. Return to a lower magnification and look for the oil-secreting **sebaceous glands** (figure 6.6) that connect to the hair follicles in the region of the hair roots. Sebaceous glands secrete an oily substance, **sebum,** into the hair follicle.

5. Carefully observe several hair follicles to locate the small **arrector pili** muscles (*arrector,* that which raises, + *pili,* hair) that attach to the base of the hair follicle (not visible in figure 6.6; see figure 6.11). When these smooth muscles contract, they pull at the base of the hair follicle. This causes the hair to stand up straight, rather than lie flat against the surface of the skin. Muscle contraction pulls down on the skin, while the area where the hair shaft comes out of the skin remains elevated. Thus, in humans it gives the appearance of "goose bumps" or "goose pimples."

Table 6.4	Parts of a Hair Follicle
Structure	**Description and Function**
Connective Tissue (Dermal) Root Sheath	The connective tissue of the dermis (mainly dense collagen fibers) that surrounds the entire hair follicle
Cortex	Constitutes the bulk of the hair; composed predominantly of keratin
Cuticle Layer	The outer portion of the hair itself; composed of several layers of hard plates of keratin that surround the cortex of the hair
External Root Sheath	The outer layers of the hair follicle, which are continuous with the stratum basale and stratum spinosum of the epidermis
Glassy Membrane	A specialized basement membrane located external to the external root sheath and internal to the connective tissue that surrounds the hair follicle (the connective tissue root sheath)
Internal Root Sheath	A sheath derived from epithelial tissue that lies between the external root sheath and the hair itself

(continued on next page)

(continued from previous page)

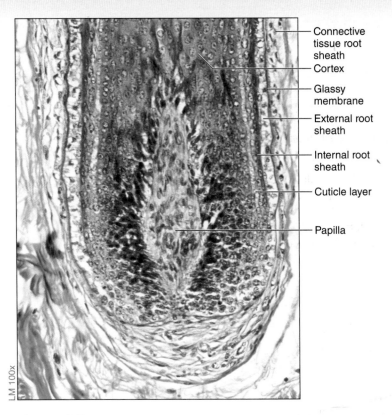

Figure 6.7 Close-up View of a Hair Bulb.

6. Using figures 6.6 and 6.7 and tables 6.2 and 6.3 as guides, identify the following structures related to the hair follicle:

   ☐ arrector pili muscle ☐ root of hair follicle
   ☐ bulb of hair follicle ☐ sebaceous gland
   ☐ hair follicle ☐ shaft of hair follicle
   ☐ papilla of hair follicle

7. In the space below, sketch a hair follicle and associated skin appendages as viewed through the microscope. Be sure to label all the structures listed in step 6 in your drawing.

8. Next, scan the slide until you find a hair follicle in cross section (**figure 6.8**).

9. Using figure 6.8 and table 6.4 as guides, identify the following structures:

   ☐ connective tissue (dermal) root sheath ☐ external root sheath
   ☐ cortex ☐ glassy membrane
   ☐ cuticle layer ☐ internal root sheath

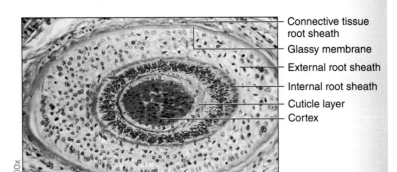

Figure 6.8 Cross Section of a Hair Follicle.

____400____ ×

EXERCISE 6.6

AXILLARY SKIN—APOCRINE SWEAT GLANDS

1. Obtain a slide of **axillary skin** and place it on the microscope stage. Locate a hair follicle on the slide. Notice that the hair follicles in this skin are oriented at a fairly steep angle with respect to the apical surface of the epithelium. This, in part, is what makes the hairs in the axillary region curly.

2. Look for glands that open into the hair follicle. These are **apocrine sweat glands** (*apo-*, away from, + *krino*, to separate or secrete) **(figure 6.9)**. Apocrine sweat glands are predominantly located in the axillary, pubic, and anal regions of the body, though they are also located in the areola of the nipple and in men's facial hair. These glands produce their secretions the same way that merocrine glands do, via exocytosis. However, apocrine glands produce a secretion that is thicker and oilier than that of the merocrine glands. Thus, although the term *apocrine* historically referred to a different mode of secretion,* these glands are still referred to as apocrine glands.

3. In the space below, sketch a hair follicle from axillary skin and its associated apocrine sweat gland.

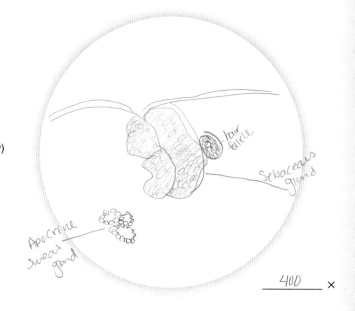

Figure 6.9 Apocrine Sweat Glands. Axillary skin demonstrating apocrine sweat glands, which open to the hair follicle.

*Traditionally, *apocrine* referred to a process by which the apical part of the cell discharged into the duct of the gland along with the secretion. However, we now know that this is not how apocrine glands produce their secretions.

EXERCISE 6.7

STRUCTURE OF A NAIL

1. Obtain a slide of a **nail**. A nail is a very specialized skin appendage that arises from epithelial tissue. **Table 6.5** lists the parts of a nail, and **figure 6.10** shows a longitudinal section of a nail.

2. Using figure 6.10 and table 6.5 as guides, identify the following structures on the slide:

   ☑ body   ☑ hidden border
   ☐ eponychium   ☑ hyponychium

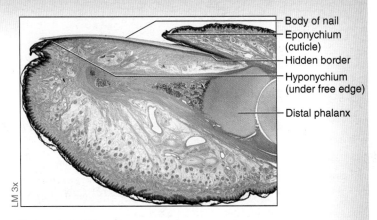

Figure 6.10 Longitudinal Section of a Nail.

3. In the space below, sketch a nail as seen through the microscope. Be sure to label all of the structures listed in step 2 in your drawing.

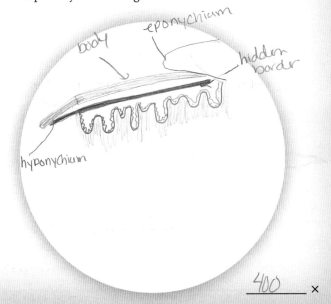

Table 6.5	Parts of a Nail	
Structure	**Description**	**Word Origins**
Body	The main portion of the nail	
Eponychium	The fold of skin at the root of the nail that folds over the body of the nail	*epi*, upon, + *onyx*, nail
Hidden Border	The portion of the nail that lies beneath the eponychium	
Hyponychium	The skin underneath the free border of the nail	*hypo*, under, + *onyx*, nail
Lunula	A white, curved area at the base of the nail	*luna*, moon

CHAPTER 7

The Skeletal System: Bone Structure and Function

OUTLINE AND LEARNING OBJECTIVES

Histology 147

Bone Tissue 147

EXERCISE 7.1: COMPACT BONE 148
1. Identify the histological features of compact bone
2. Describe the structure of an osteon

EXERCISE 7.2: SPONGY BONE 149
3. Identify the histological features of spongy bone

EXERCISE 7.3: ENDOCHONDRAL BONE DEVELOPMENT 150
4. Describe the process of endochondral bone development
5. Identify the five functional layers of the epiphyseal plate

Gross Anatomy 152

Classification of Bones 152

EXERCISE 7.4: IDENTIFYING CLASSES OF BONES BASED ON SHAPE 153
6. List and identify the structural classifications of bones of the human skeleton

Structure of a Typical Long Bone 153

EXERCISE 7.5: COMPONENTS OF A LONG BONE 154
7. Identify the components of a typical long bone

EXERCISE 7.6: COW BONE DISSECTION 155
8. Identify specific types of bone and cartilage tissues through gross observation of a fresh cow bone

Survey of the Human Skeleton 157

EXERCISE 7.7: THE HUMAN SKELETON 157
9. Identify the bones of the human skeleton

INTRODUCTION

Our skeletal system is composed of bones, as well as cartilage, ligaments, and other connective tissues that stabilize or connect the bones. Bones are organs, which are composed primarily of bone tissue that serve as the rigid framework to support the body. Bone is a highly specialized connective tissue that is both rigid and flexible. Most of us tend to view bone as static, nonliving tissue because the type of bones we commonly see are those left behind from the carcasses of animals or the preserved skeletons we see in the anatomy laboratory. In reality, bone is a dynamic, versatile tissue that is in a constant state of turnover and is one of the most metabolically active kinds of tissue within the body. Bone has a rich blood supply and an amazing ability to alter its structure in response to the changing stresses placed upon it. Indeed, prolonged absence of stress will cause bone to lose density and strength, whereas increased stress causes bone to increase in density and strength.

The focus of this chapter is to introduce you to the general micro- and macrostructure of bone, and to provide you with a brief overview of the structure and function of the human skeleton. The detailed structures and functions of the bones making up the human skeleton are covered in chapters 8 and 9.

 **MODULE 5: SKELETAL SYSTEM**

Chapter 7: The Skeletal System: Bone Structure and Function

PRE-LABORATORY WORKSHEET

1. List four functions of the skeletal system.
   a. Protection & Support
   b. Movement
   c. Hemopoiesis
   d. Storage of minerals & energy reserves

2. Name the two types of bone tissue.
   a. Connective tissue
   b. Spongy

3. Describe the following structures:
   a. *osteoprogenitor cell* _____
   b. *osteoblast* _____
   c. *osteocyte* _____
   d. *osteoclast* _____
   e. *lamella* _____
   f. *lacuna* _____
   g. *canaliculus* _____

4. Fill in the blanks:
   a. The end of a long bone is the Proximal or Distal epiphysis.
   b. The shaft of a long bone is the Diaphysis.
   c. The area where the end of a long bone meets the shaft of a long bone is the Metaphysis.

5. In an adult skeleton, Yellow bone marrow fills the medullary cavity of a long bone, and red bone marrow is located within the proximal epiphyses of long bones such as the humerus and femur.

6. The outer covering of bone is the periosteum and it is composed of irregular connective tissue.

7. List the two types of bone formation.
   a. Intermembranous ossification
   b. Endochondral

8. List and define the two divisions of the human skeleton.
   a. _____
   b. _____

Histology

In these exercises you will look at the histology of compact bone tissue, spongy bone tissue, and endochondral bone development. **Table 7.1** summarizes the characteristics of the types of bone cells you will observe.

Bone Tissue

You may observe two types of bone histology preparations in the laboratory. In the first type of preparation, **ground bone** (see figure 7.1), a hard bone sample is ground down into a section that is thin enough to be viewed under a microscope. This type of preparation destroys living cells. Thus, no osteocytes, osteoblasts, or other cells can be seen. In the second type of preparation, **decalcified bone** (see figure 7.2), the rigid, mineralized (calcified) matrix of the bone is dissolved away so the tissue is soft enough to be sectioned in the traditional manner. This type of preparation preserves living cells such as osteocytes and osteoblasts, but the fine structure of the bony matrix cannot be seen because of the removal of the rigid portion of the matrix. However, details of the bony matrix—such as central (Haversian) canals, perforating (Volkmann) canals, and canaliculi—are preserved.

Table 7.1	Types of Bone Cells		
Cell Name	**Description and Function**	**Drawing**	**Word Origins**
Osteoprogenitor Cell	A stem cell derived from mesenchyme that differentiates into an osteoblast. These cells are located in both the endosteum and in the inner layer of the periosteum.	Nuclei	(*osteon*, bone + *pro*, before + *genesis*, origin)
Osteoblast	A small, immature bone cell derived from an osteoprogenitor cell that functions to lay down new bone for bone growth, remodeling, and repair. These cells often appear lined up in rows next to a trabecula of spongy bone.	Nucleus	(*osteon*, bone + *blastos*, germ)
Osteocyte	A mature bone cell derived from an osteoblast that functions to maintain the matrix surrounding it. These cells are located within lacunae.	Nucleus	(*osteon*, bone + *kytos*, a hollow; a cell)
Osteoclast	A very large, multinucleate, phagocytic cell derived from bone marrow cells that also produce monocytes. Osteoclasts break down bone and are often found on the opposite side of a trabecula of spongy bone as the layer of osteoblasts.	Nuclei	(*osteon*, bone + *klastos*, broken)

EXERCISE 7.1

COMPACT BONE

1. Obtain a slide of **ground compact bone (figure 7.1)** and place it on the microscope stage. Bring the tissue sample into focus on low power and locate an **osteon.** Remember, you will not be able to see any living cells in this tissue sample.

2. With the osteon at the center of the field of view, increase the magnification. The **central (Haversian) canals** will be the largest holes visible in the sample. (The blood vessels and nerves that are housed within these canals can be viewed in the decalcified compact bone.) In contrast, **lacunae** will appear to be much smaller holes and there will only be one cell. (Osteocytes present in lacunae can be viewed in the slide of decalcified compact bone.) **Concentric lamellae** extend outward from the central canal in concentric rings. **Lacunae** that house osteocytes are found at the border between two adjacent lamellae. Interstitial lamellae can be seen between osteons. In comparison, circumferential lamallae can be viewed by scanning the outer border of the bone sample, which surround the entire diaphysis of the bone. Lacunae appear very dark in color, and at very high magnification they may appear to be empty because of the lack of living osteocytes in the tissue.

    Observe this slide at the highest magnification possible (without using the oil immersion lens) and notice what appear to be tiny little "cracks" or fractures that run perpendicular to the central canals. These are the tiny **canaliculi,** which in living bone contain cytoplasmic extensions of osteocytes. Within the canaliculi osteocytes connect to each other via gap junctions, which allow them to exchange substances with each other.

3. Finally, scan the slide and see if you can find any large canals that run perpendicular to the central canals. These canals are **perforating (Volkmann) canals,** which convey blood vessels from the outer periosteum into the central canals.

4. Obtain a slide of **decalcified compact bone (figure 7.2)** and place it on the microscope stage. Bring the tissue sample into focus on low power and locate an osteon. This slide allows you to view the living structures (cells, blood vessels, and nerves) within bone.

5. With the osteon at the center of the field of view, increase the magnification. The **central (Haversian) canals** will be the largest holes visible in the sample and will appear to have a lot of "junk" inside them. The "junk" consists of blood vessels and nerves. In contrast, **lacunae** will appear to be much smaller holes and there will only be one cell (an osteocyte) inside each lacuna. **Lamellae** are sometimes difficult to make out, but once you identify where the lacunae are, you will have an idea of where the lamellae are as well (lacunae are found at the border between two adjacent lamellae). As you scan toward the outer border of the bone sample, you should be able to identify **circumferential lamellae,** which surround the entire diaphysis of the bone.

6. Scan the outer edge of the bone tissue on the slide and identify the **periosteum** of the bone. The periosteum contains an outer layer of dense irregular connective tissue and an inner layer of **osteoprogenitor cells**. Can you identify this layer of osteoprogenitor cells on your slide?

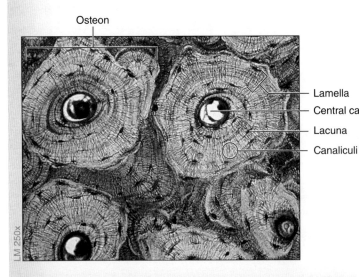

AP|R Figure 7.1 Ground Compact Bone. No perforating canals are visible in this section.

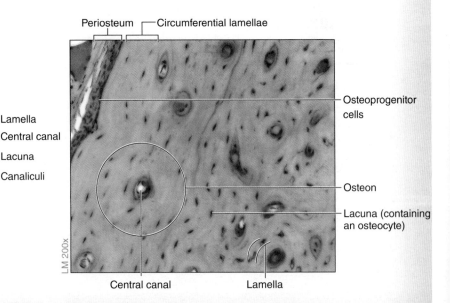

Figure 7.2 Decalcified Compact Bone.

7. In the spaces below, sketch ground compact bone and decalcified compact bone as viewed through the microscope. Using table 7.1 and figures 7.1 and 7.2 as guides, label the following on your sketches:

☐ central canal ☐ osteocyte
☐ circumferential lamellae ☐ osteon
☐ concentric lamellae ☐ osteoprogenitor cell
☐ interstitial lamellae ☐ perforating canal
☐ lacuna ☐ periosteum
☐ osteoblast

Ground compact bone *Decalcified compact bone*

_____ × _____ ×

EXERCISE 7.2

SPONGY BONE

1. Obtain a slide of decalcified **spongy (cancellous) bone** (figure 7.3) and place it on the microscope stage. The sample of spongy bone will appear most similar to the slide of decalcified compact bone. The main difference between the two is that the spongy bone does *not* contain osteons, so you will not see central canals or perforating canals.

2. Observe the slide at high magnification and look for areas where several small cells are lined up next to each other on the edge of a trabecula. These are **osteoblasts** (figure 7.3a) that actively secrete new bony matrix. Can you identify any very large, multinucleate cells on the slide? If so, these are bone-resorbing **osteoclasts** (figure 7.3b). Both osteoclasts and osteoblasts are actively involved in the process of bone remodeling.

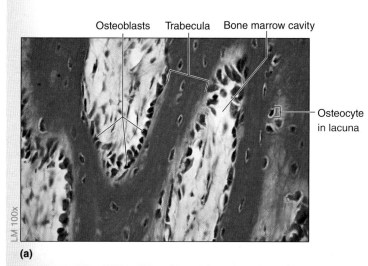

Figure 7.3 **Spongy Bone.** Decalcified sections of spongy bone showing (a) osteoblasts and (b) osteoclasts.

(continued on next page)

(continued from previous page)

3. In the space to the right, sketch spongy bone as viewed through the microscope. Using table 7.1 and figure 7.3 as guides, label the following on your sketch:

   ☐ bone marrow cavity   ☐ osteoclast
   ☐ lacuna   ☐ osteocyte
   ☐ osteoblast   ☐ trabecula

   _____ ×

EXERCISE 7.3

ENDOCHONDRAL BONE DEVELOPMENT

1. Obtain a slide of **developing long bone (figure 7.4)** and place it on the microscope stage. This slide will typically contain the developing femur of a young mammal. The femur develops using **endochondral ossification**, a process by which a hyaline cartilage model of the bone is gradually replaced by bone tissue.

2. Scan the slide at the lowest magnification and identify the parts of the bone—specifically, the epiphysis, diaphysis, and metaphysis. In the metaphysis of a developing long bone is an **epiphyseal (growth) plate**. Name the type of tissue you expect to find at this location: _____

_____

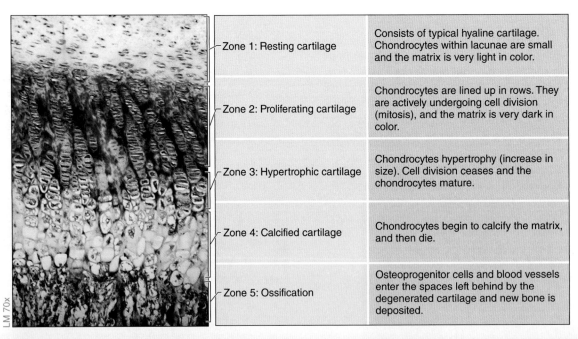

Figure 7.4 Epiphyseal Plate. The five functional layers of the epiphyseal plate.

3. Position the metaphysis at the center of the field of view and increase the magnification. Using figure 7.4 as a guide, identify the five functional layers within the epiphyseal plate.

4. In the space below, sketch the epiphyseal growth plate as viewed through the microscope. Using figure 7.4 and your textbook as guides, label the following on your sketch:

☐ zone of calcified cartilage
☐ zone of hypertrophic cartilage
☐ zone of ossification
☐ zone of proliferating cartilage
☐ zone of resting cartilage

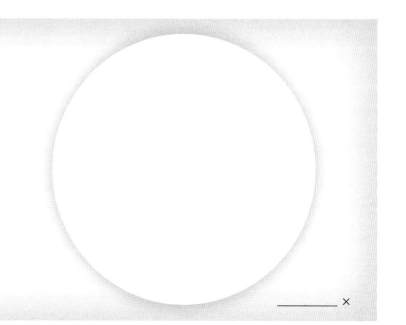

_____ ×

INTEGRATE

CONCEPT CONNECTION

Recall that bone is a supporting connective tissue that protects organs and allows for body movement by providing attachment sites for muscles. Bone matrix is comprised of both organic and inorganic components. *Osteoid,* the organic portion of bone, contains a dense collection of collagen fibers and proteoglycans that provide strength and flexibility. The inorganic portion of bone contains crystals of the mineral compound hydroxyapatite. Hydroxyapatite is formed through a reaction between calcium phosphate and calcium hydroxide. The presence of hydroxyapatite in the mineral matrix of bone allows the bone to store a large supply of the minerals calcium and phosphate. This mineralized matrix also provides rigidity and incompressiblility to bone. As bone tissue is loaded by mechanical stresses, osteoblasts are stimulated to secrete osteoid, which leads to calcification of the bone matrix. When mechanical stresses are reduced (for example, when a limb is immobilized in a cast for some time), osteoclasts are stimulated to secrete substances to break down the osteoid and hydroxyapatite. The result of the breakdown of the bony matrix is the release of calcium and phosphorus into the blood. In addition to the effect of mechanical loading and unloading on the process of bone formation and remodeling, hormones are also influential in regulating these processes. For example, when blood calcium levels drop below physiological levels, cells in the parathyroid glands release parathyroid hormone (PTH) that targets osteoclasts. This results in an increase in bone resorption. Bone resorption causes a decrease in bone density. Thus, calcium is taken from the bone and transported into the blood. In contrast, when blood calcium levels increase above physiological levels, cells in the thyroid release calcitonin. Calcitonin targets osteoblasts to increase the deposition of calcium from the blood into the bony matrix. Maintaining blood calcium at homeostatic levels is of the utmost importance to regulation of many bodily functions: It is critical for muscle contraction and activity of excitable cells. Maintenance of blood calcium levels is so important that regulatory mechanisms prioritize breaking down bone—even to the point of sacrificing the structural integrity of the bone—just to be able to maintain blood calcium levels.

Gross Anatomy

Classification of Bones

Bones of the human skeleton appear in various shapes and sizes, depending on their function. The four classes of bone as determined by shape are long, short, flat, or irregular based on the overall shape of the bone. These are summarized in **table 7.2** and an example of each is shown in **figure 7.5**.

Table 7.2	Classification of Bones Based on Shape	
Class of Bone	**Description**	**Examples**
Long	Greater in length than width	Most limb bones (e.g., humerus, femur, metacarpals)
Short	Nearly equal in length and width	Wrist and ankle bones (e.g., carpals, tarsals)
Flat	Have thin, flat surfaces; may be slightly curved	Many skull bones (e.g., frontal, parietal)
Irregular	Complex shape that does not fit into other classifications	Vertebrae, some skull bones (e.g., vertebrae, sphenoid)

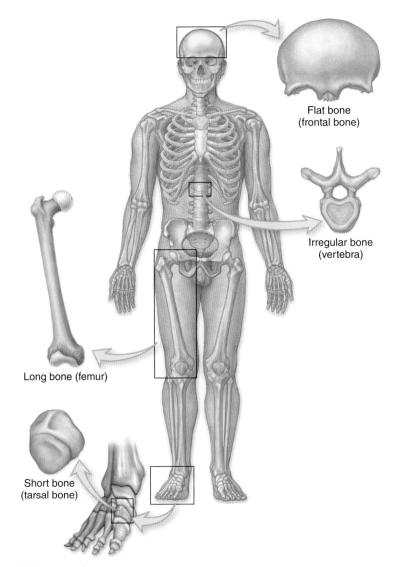

Figure 7.5 Structural Classifications of Bones.

EXERCISE 7.4

IDENTIFYING CLASSES OF BONES BASED ON SHAPE

1. Obtain a box containing disarticulated human bones. As you pull each bone out of the box, classify the bone according to its shape. Note that there is a fair bit of variability in the shape of bones within each classification.

2. **Figure 7.6** is a photograph of several disarticulated human bones. In the space next to each bone, name the category to which each bone belongs: long, short, flat, or irregular.

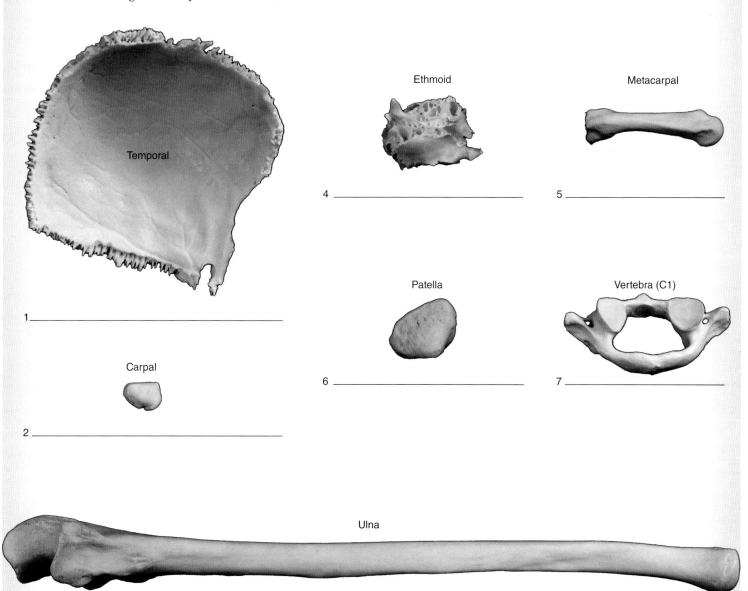

Figure 7.6 Identifying Bones Based on Structure. Identify the structural classification of each bone in this photograph: long, short, flat, or irregular.

Structure of a Typical Long Bone

A typical long bone such as the femur **(figure 7.7)** is composed of a long shaft, called the **diaphysis** (*dia*, through, + *physis*, growth); rounded ends, called **epiphyses** (*epi*, upon, + *physis*, growth); and articulation points between the two, called **metaphyses** (*meta*, between, + *physis*, growth). Within the shaft is a large cavity called the **medullary cavity,** which in the adult is filled with **yellow bone marrow** (adipose tissue). The walls of the diaphysis are composed of a thick layer of **compact bone** tissue. The epiphyses of the bone are surrounded by a thin layer of compact bone and have **articular cartilages** on the ends. **Spongy bone** tissue is found within the epiphyses of the bone. In the fetus, the marrow spaces between the trabeculae of spongy bone are composed of red bone marrow. However, in the adult they are composed mainly of yellow bone marrow because of the conversion of red marrow to yellow marrow that occurs as the skeleton matures. In the adult, red bone marrow

154 Chapter Seven *The Skeletal System: Bone Structure and Function*

is primarily limited to the proximal epiphyses of the humerus and femur, the sternum, and the iliac crest. Observing a fresh bone specimen allows you to see many of the tissues that normally associate with bones, such as periosteum, articular cartilages, muscles, tendons, ligaments, marrow, and blood vessels. In the following exercises you will first observe a fresh specimen of a long bone from a cow, and then study an articulated human skeleton to familiarize yourself with the major bones of the human body.

EXERCISE 7.5

COMPONENTS OF A LONG BONE

1. Obtain both a femur (not cut) and a femur that has been cut along its longitudinal axis.

2. Using figure 7.7 identify the following components of a long bone:

   ☐ compact bone ☐ epiphysis
   ☐ diaphysis ☐ medullary cavity
   ☐ epiphyseal line ☐ spongy bone

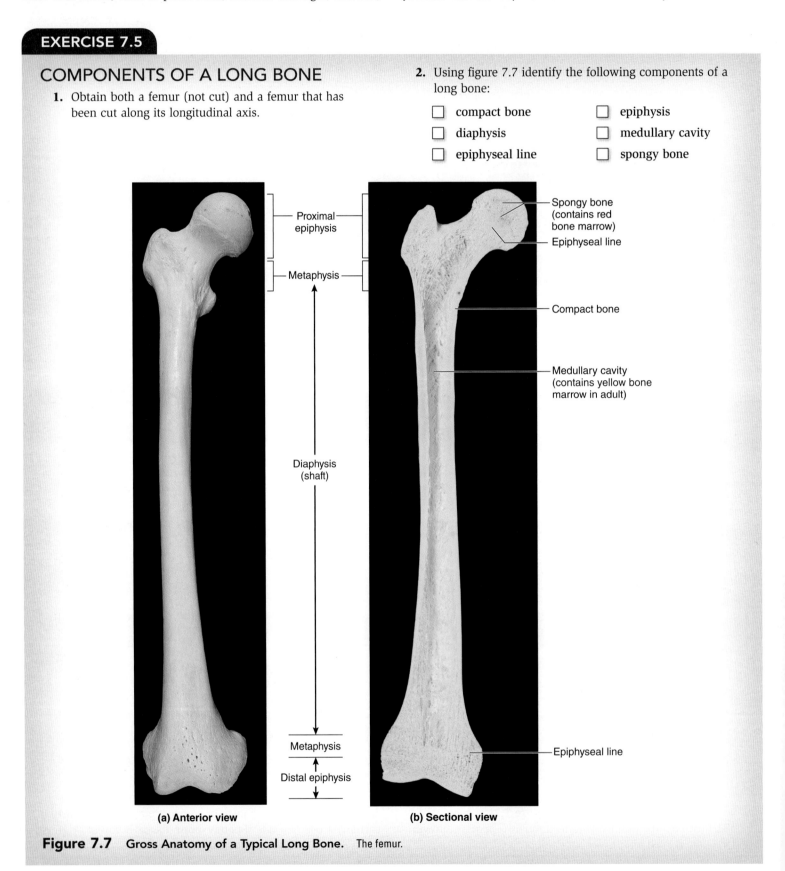

(a) Anterior view (b) Sectional view

Figure 7.7 Gross Anatomy of a Typical Long Bone. The femur.

EXERCISE 7.6

COW BONE DISSECTION

1. Obtain a dissecting pan, blunt probe, forceps, and a fresh cow bone cut in a longitudinal section. Place the bone in the dissecting pan and begin your observations. Find the large medullary cavity that is filled with yellow bone marrow **(figure 7.8)**. Notice the thick layer of compact bone that surrounds the medullary cavity.

2. If you have a dissecting microscope or magnifying glass at your disposal, use it to focus in on the compact bone tissue. Notice how "bloody" it appears. Can you see tiny dots of blood within the compact bone? These result from the rupture of tiny blood vessels in the tissue when the bone was sectioned. This observation should reaffirm for you that living bone is a highly vascular, metabolically active tissue—quite different from the appearance of preserved bones.

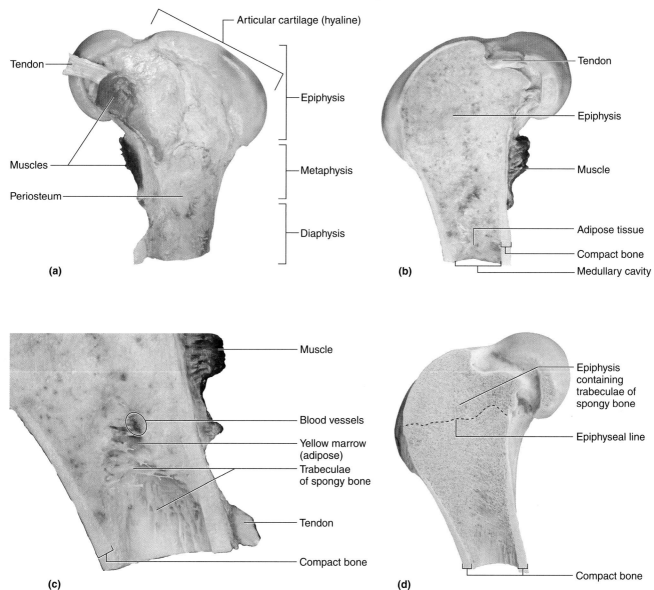

Figure 7.8 **Fresh Cow Bone.** (a) Exterior view, (b) interior view, (c) close-up of the medullary cavity. (d) The same bone, with all of the fat, muscle, tendons, cartilage, and blood vessels removed.

(continued on next page)

(continued from previous page)

3. Using the blunt probe, carefully begin to clean out the adipose tissue from the medullary cavity (figure 7.8b). While you do this, gently probe for medium- to large-size blood vessels embedded within the adipose tissue. If you find a blood vessel that travels from the outer surface of the diaphysis into the medullary cavity, this is most likely the **nutrient artery,** an artery that grows into the diaphysis of the bone during the initial stages of ossification of the bone.

4. As you continue to clean out the inside of the diaphysis of the bone and progress toward the inside of the epiphysis, you will begin to feel small, hard "strings" of bony tissue. These are the **trabeculae** (*trabs,* beam) of spongy bone, which are located within the epiphysis and lining the inside of the diaphysis. These trabeculae will make it difficult to clean out the adipose tissue in the epiphysis, but spend some time poking around in there—this will give you a better understanding of the arrangement of the trabeculae within. You may observe a portion of the epiphysis that appears much more "bloody" than the rest of the inside of the bone. This area consists of **red bone marrow**—a **hemopoietic** (*hemo-,* blood, + *poiesis,* a making) tissue that produces blood cells. Because you are observing adult cow bones, you will observe very little red marrow; most of the spaces will be filled with yellow marrow.

5. Turn your attention to the structures on the outside of the bone. Using forceps, pick away at the dense connective tissue on the outside of the diaphysis (figure 7.8). This is the periosteum, or outer covering of the bone. The periosteum acts as an attachment point for tendons and ligaments and as an anchoring point for blood vessels and nerves that enter the bone.

6. Observe the outside of the epiphysis and look for the cut portion of a tendon or ligament where it attaches to the periosteum (figure 7.8a,c). If you carefully observe either one, you will see that it consists of a regular arrangement of shiny, white fibers. These are the tough collagen fibers that give the tendon or ligament its great tensile strength.

7. Observe the shiny, white cartilage on the ends of the bone. This is the articular cartilage, which is composed of hyaline cartilage. Notice that the periosteum ends where the articular cartilage begins.

8. If you are fortunate enough to have a cow tibia as your bone sample, try to identify the C-shaped pads of fibrocartilage located on top of the articular cartilages. These are the **menisci** (s. *meniscus*) of the knee joint **(figure 7.9).** If your cow bone does not have a meniscus, find another group in your laboratory whose cow bone has a meniscus so you can observe

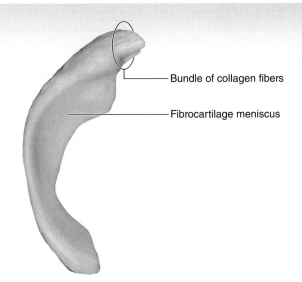

Figure 7.9 Meniscus. One of the C-shaped meniscal cartilage pads removed from the end of the tibia of a fresh cow bone.

its gross structure. Within the meniscus you will see more of the shiny, white collagen fibers that you saw in the tendons and ligaments. Recall from chapter 5 that fibrocartilage contains thick bundles of collagen fibers in its extracellular matrix. Note that the meniscus is somewhat "tied in to" the connective tissues surrounding the joint.

9. In the space below, sketch the fresh cow bone you dissected. Using figure 7.8 as a guide, label the following on your sketch:

   ☐ articular cartilage ☐ periosteum
   ☐ compact bone ☐ spongy bone
   ☐ medullary cavity ☐ tendon/ligament

10. When you are finished with your dissection, dispose of the cow bones according to your laboratory instructor's directions, and clean your dissection instruments and workspace.

Survey of the Human Skeleton

The human skeleton consists of two divisions: (1) the **axial skeleton**, which includes bones of the cranium, vertebrae, ribs, and sternum, and (2) the **appendicular skeleton**, which includes bones of the pectoral girdle, upper limb, pelvic girdle, and lower limb. The **pectoral girdle** consists of the scapula and clavicle. These bones provide support and attachment points for muscles that connect the limbs to the axial skeleton. The **pelvic girdle** consists of the bones composing the **os coxae** (*os,* bone, + *coxa,* hip): the ilium, ischium, and pubis. These bones protect contents of the pelvic cavity and provide support and attachment points for muscles that connect the lower limb to the axial skeleton. Learning the names of all the bones in the human body and their features can seem like a Herculean task. Often it is best to break a big task into many small tasks that can be tackled individually in short segments of time. As you begin the task of learning all of the bones and bone features of the human skeleton, an excellent "first task" is to simply learn the names of the bones that compose the skeleton. In this exercise you will do just that. The exercises in chapters 8 and 9 will then guide you in the task of learning the names of specific features unique to each bone of the skeleton.

EXERCISE 7.7

THE HUMAN SKELETON

1. Observe an articulated human skeleton.

2. Using your textbook as a guide, identify the bones listed in **figure 7.10** on the articulated skeleton. Then label them in figure 7.10.

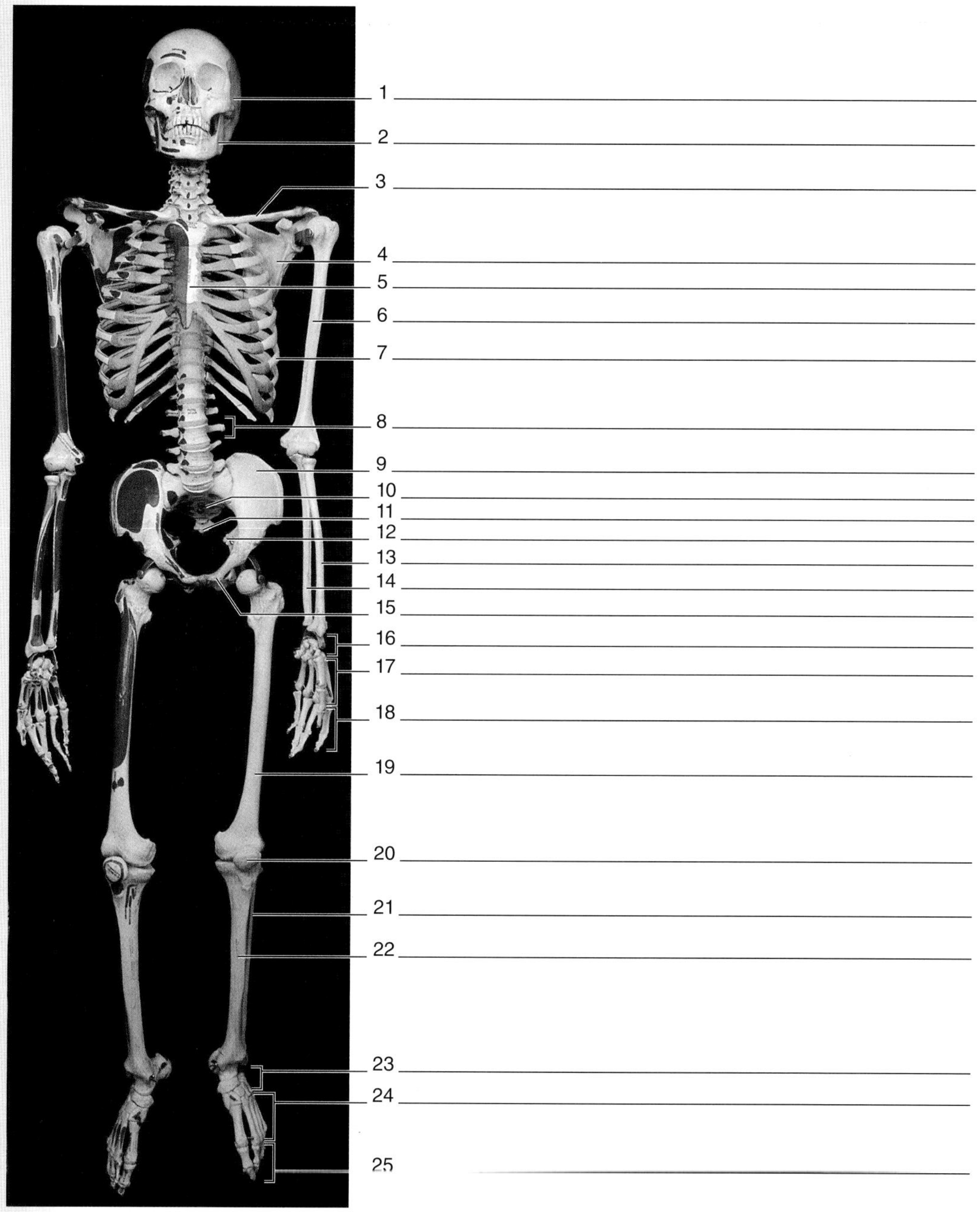

- ☐ carpals
- ☐ clavicle
- ☐ coccyx
- ☐ femur
- ☐ fibula
- ☐ humerus
- ☐ ilium
- ☐ ischium
- ☐ mandible
- ☐ metacarpals
- ☐ metatarsals
- ☐ patella
- ☐ phalanges (of the foot)
- ☐ phalanges (of the hand)
- ☐ pubis
- ☐ radius
- ☐ rib
- ☐ sacrum
- ☐ scapula
- ☐ skull
- ☐ sternum
- ☐ tarsals
- ☐ tibia
- ☐ ulna
- ☐ vertebra

1. _____
2. _____
3. _____
4. _____
5. _____
6. _____
7. _____
8. _____
9. _____
10. _____
11. _____
12. _____
13. _____
14. _____
15. _____
16. _____
17. _____
18. _____
19. _____
20. _____
21. _____
22. _____
23. _____
24. _____
25. _____

(a) Anterior view

Figure 7.10 The Human Skeleton.

(continued on next page)

(continued from previous page)

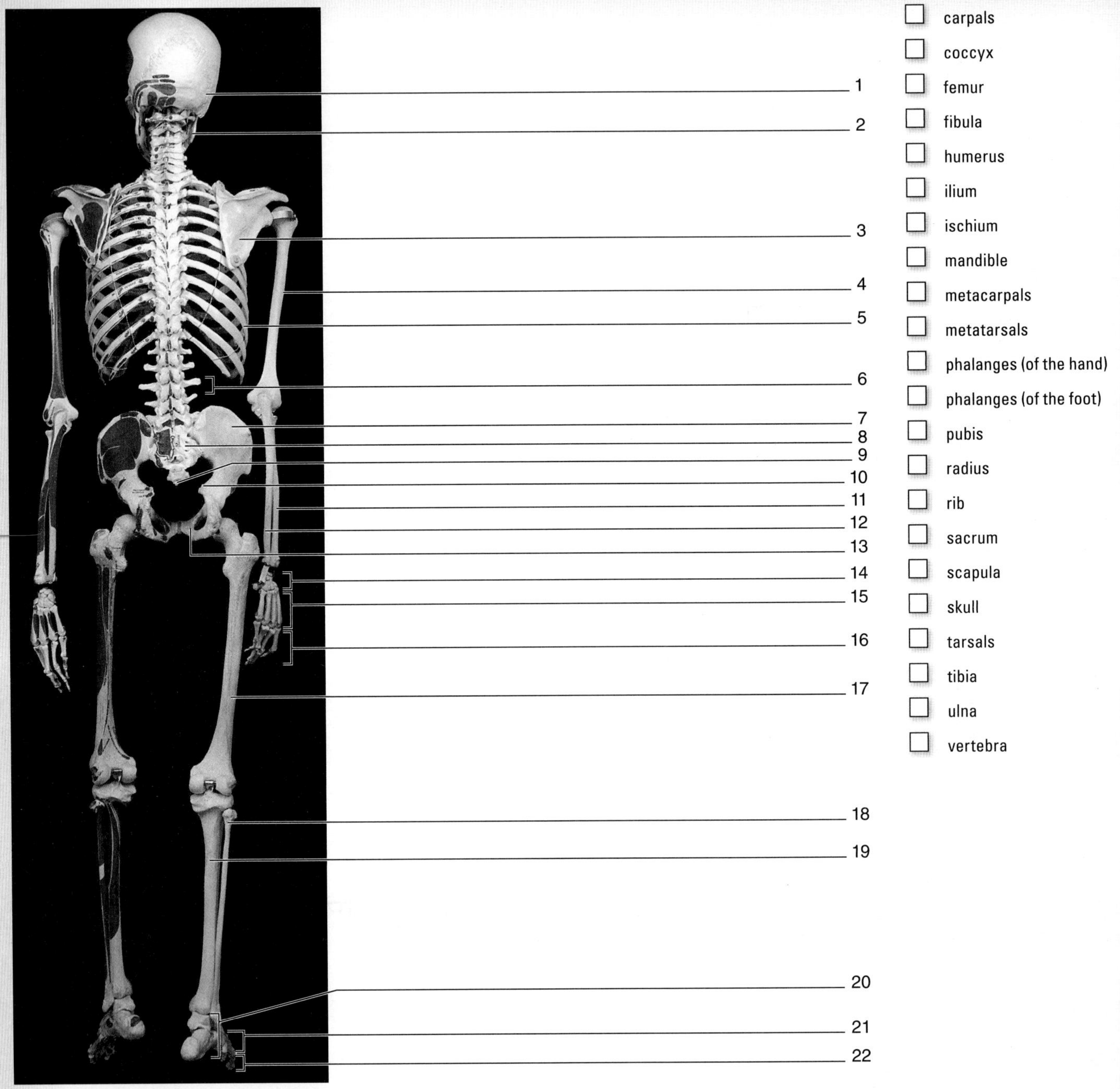

☐ carpals
☐ coccyx
☐ femur
☐ fibula
☐ humerus
☐ ilium
☐ ischium
☐ mandible
☐ metacarpals
☐ metatarsals
☐ phalanges (of the hand)
☐ phalanges (of the foot)
☐ pubis
☐ radius
☐ rib
☐ sacrum
☐ scapula
☐ skull
☐ tarsals
☐ tibia
☐ ulna
☐ vertebra

(b) Posterior view

Figure 7.10 The Human Skeleton *(continued)*.

INTEGRATE

CLINICAL VIEW
Bones and Mechanical Stress

Bone remodels itself in response to applied mechanical stresses that are placed upon it. In this instance, "stress" refers to force per unit area of tissue. For example, weight-bearing exercises "load" bone, thereby increasing stress on the tissue. The increased force directly impacts osteoblast activity. That is, there is an increase in the deposition of bone matrix and calcification with increasing force. We load bones daily when maintaining our upright posture due to the downward force of gravity. However, the bone remodeling process is disrupted when gravity is greatly reduced. For example, astronauts who spend only a few days in outer space under microgravity (reduced gravity) conditions experience significant bone density loss. Research suggests that microgravity inhibits osteoblast activity and stimulates osteoclast activity. The net result is an increase in bone resorption without bone matrix deposition. To counteract this change, astronauts attempt to simulate the mechanical loading they normally would experience on earth using resistance training (lifting weights). Unfortunately, such exercise has not been able to reverse the reduction in new bone formation that astronauts experience.

Chapter 7: The Skeletal System: Bone Structure and Function

Name: _____
Date: _____ Section: _____

POST-LABORATORY WORKSHEET

The ❶ corresponds to the Learning Objective(s) listed in the chapter opener outline.

Do You Know the Basics?

Exercise 7.1: Compact Bone

1. Label the components of compact bone on the following diagram. ❶❷

1. _____
2. _____
3. _____
4. _____
5. _____
6. _____
7. _____
8. _____
9. _____
10. _____
11. _____
12. _____
13. _____
14. _____
15. _____

2. In the following table, list the four types of bone cells and describe the function of each. ❶

Bone Cell	Function
Osteoprogenitor Cells	Stem cell derived from mesechyme that differentiates into an osteoblast. Cells located in both endosteum & inner layer of periosteum
Osteoblast	Small immature cells derive from an osteoprogenitor cell that function to lay down new bone for growth, remodeling & repair
Osteocytes	derive from osteoblast that function to maintain the matrix surrounding. These cells are located w/in lacunae
Osteoclasts	large, multinucleor, phagocytic cell derived from bone marrow cells that also produce monocytes.

Exercise 7.2: Spongy Bone

3. Label the components of spongy bone on the following diagram. ❸

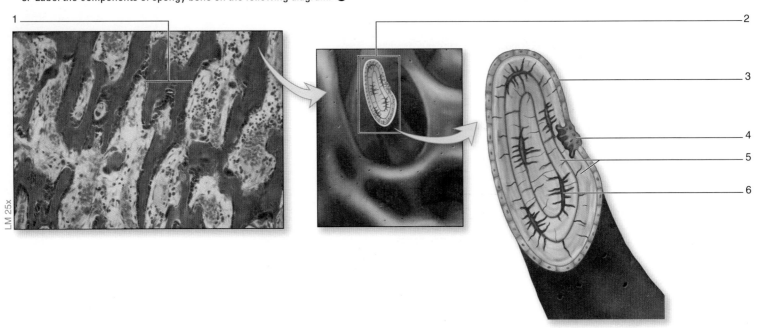

Exercise 7.3: Endochondral Bone Development

4. Which of the following bones does NOT undergo endochondral bone development (ossification)? ❹

    a. femur

    (b.) temporal

    c. radius

    d. first metacarpal

    e. tibia

5. The figure below is a light micrograph from developing endochondral bone. The dotted line divides the epiphyseal plate into two major regions, a and b. One region is **new bone tissue** and the other is **hyaline cartilage.** On the lines next to the labels (a or b), write the name of the tissue (new bone tissue or hyaline cartilage) that composes the majority of the region. ❺

    a. _hyaline cartilage_

    b. _new bone tissue_

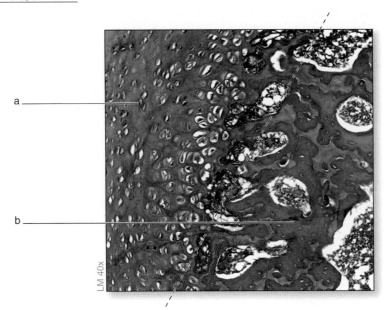

Chapter Seven *The Skeletal System: Bone Structure and Function* 161

Exercise 7.4: Identifying Classes of Bones Based on Shape

6. A vertebra is classified as a(n) ___irregular___ bone based on shape.

Exercise 7.5: Components of a Long Bone

7. Label the parts of a typical long bone on the following diagram.

1. Articular Cartilage
2. Spongy bone
3. Compact bone
4. medullary cavity
5. Yellow marrow
6. periosteum
7. epiphyseal line
8. Proximal epiphysis
9. Diaphysis
10. Distal epiphysis

Exercise 7.6: Cow Bone Dissection

8. Label the parts of a cow bone.

1. tendon
2. muscles
3. periosteum
4. Articular Cartilage (hyaline)
5. epiphysis
6. metaphsis
7. Diaphysis

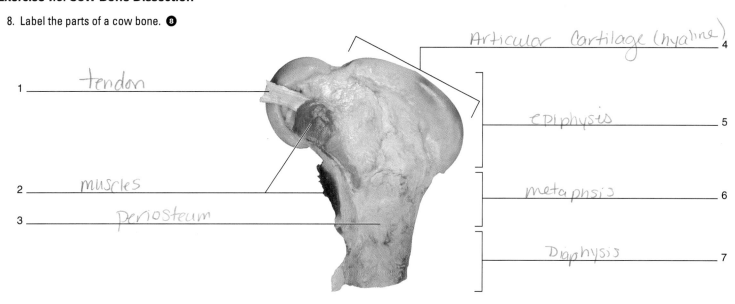

162　Chapter Seven　*The Skeletal System: Bone Structure and Function*

Exercise 7.6: The Human Skeleton

9. Label all of the major bones of the human skeleton on the following figure.

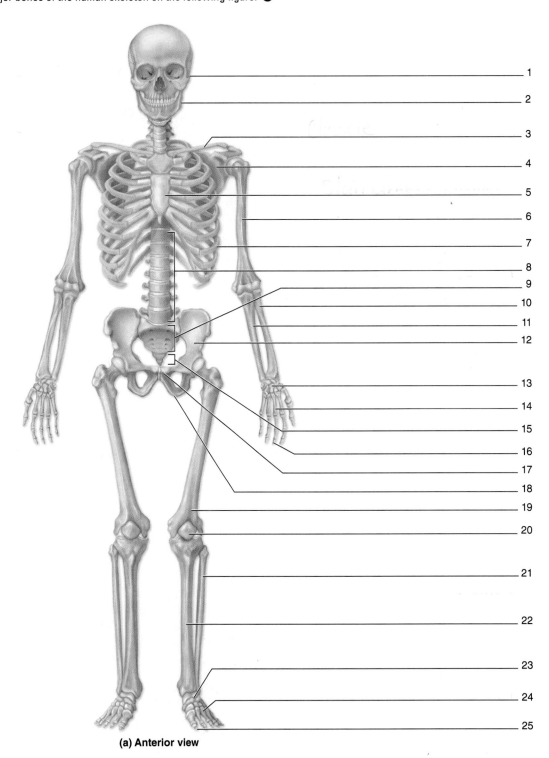

(a) Anterior view

10. Label all of the major bones of the human skeleton on the following figure.

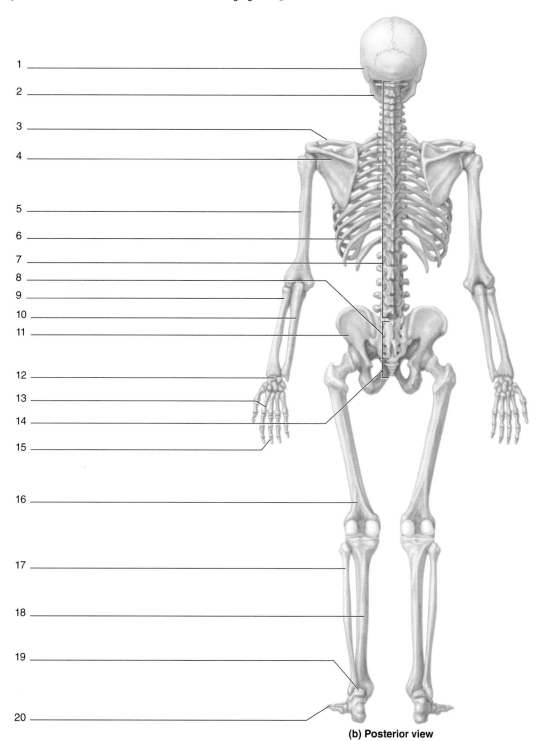

1. _____
2. _____
3. _____
4. _____
5. _____
6. _____
7. _____
8. _____
9. _____
10. _____
11. _____
12. _____
13. _____
14. _____
15. _____
16. _____
17. _____
18. _____
19. _____
20. _____

(b) Posterior view

Can You Apply What You've Learned?

11. You are a biomedical researcher who is interested in designing a drug to make bones increase their density. You have decided you can take one of two approaches: (1) design a drug that stimulates bone to be built faster, or (2) design a drug that prevents bone from breaking down. To begin this process, you need to study the cells that would likely respond to either drug 1 or drug 2. What cells are these?

a. _____Osteoblasts_____ (build bone)
b. _____Osteoclasts_____ (break down bone)

Can You Synthesize What You've Learned?

12. Compare and contrast the histologic structure of compact and spongy bone.

13. Describe the process of endochondral bone development (ossification).

14. When a bone fractures and subsequently undergoes the process of repair, what bone cells will be involved in the repair process, and how will each bone cell participate in the process?

15. What process must stop in order for the epiphyseal plate to close? (That is, which layer of the plate must stop its development first?)

16. Many people who have torn a ligament in the knee also have a torn meniscus. Why do you think it is so common for individuals who tear knee ligaments to also tear a meniscus?

CHAPTER 8

The Skeletal System: Axial Skeleton

OUTLINE AND LEARNING OBJECTIVES

Gross Anatomy 169

The Skull 169

EXERCISE 8.1: ANTERIOR VIEW OF THE SKULL 171
1. Identify the bony landmarks visible in an anterior view of the human skull
2. List the bones that form the orbit of the eye
3. List the bones that form the nasal cavity
4. Identify the features of the mandible

EXERCISE 8.2: ADDITIONAL VIEWS OF THE SKULL 176
5. Identify the bony landmarks visible in lateral, posterior, superior, and inferior views of the human skull
6. List the bones that compose the zygomatic arch

EXERCISE 8.3: SUPERIOR VIEW OF THE CRANIAL FLOOR 179
7. Distinguish the three cranial fossae
8. Identify the bony landmarks found within each cranial fossa

EXERCISE 8.4: BONES ASSOCIATED WITH THE SKULL 181
9. Identify the features of the hyoid bone

The Fetal Skull 181

EXERCISE 8.5: THE FETAL SKULL 182
10. Identify the features of the fetal skull

The Vertebral Column 183

EXERCISE 8.6: VERTEBRAL COLUMN REGIONS AND CURVATURES 184
11. Describe the normal curvatures of the vertebral column

EXERCISE 8.7: STRUCTURE OF A TYPICAL VERTEBRA 185
12. Describe the features of a typical vertebra

EXERCISE 8.8: CHARACTERISTICS OF INDIVIDUAL VERTEBRAE 186
13. Identify the features unique to cervical, thoracic, lumbar, sacral, and coccygeal vertebrae
14. Describe modifications of the atlas (C_1) and axis (C_2) that permit them to allow for extensive movements of the neck
15. Describe the modifications of thoracic vertebrae that permit articulation with the ribs
16. Compare and contrast the features of a typical vertebra with the features of the fused vertebrae that compose the sacrum

The Thoracic Cage 190

EXERCISE 8.9: THE STERNUM 191
17. Identify the features of the sternum
18. Explain the clinical significance of the sternal angle

EXERCISE 8.10: THE RIBS 192
19. Describe the parts of a typical rib
20. Describe how a typical rib articulates with a thoracic vertebra

INTRODUCTION

The skeletal system (typically composed of 206 bones in an adult) is organized into two divisions: the axial skeleton and the appendicular skeleton. The term **axial** is a derivative of the term *axis*. The axial skeleton consists of bones that form the main axis of the body—the skull, vertebrae, ribs, and sternum, which collectively form the body's core structural foundation. These bones are also critical for protection of the body's most vital organs, such as the brain, heart, and lungs.

In this chapter you will begin a task that can seem daunting to anatomy and physiology students: the process of learning a large number of structures in a short period of time. Most students find this process of trying to learn all of the processes, projections, foramina (holes), and other markings of the bones to be an enormous effort in memorization. It need not be if you approach it with the proper attitude. That is, as you view each structure, study it closely and contemplate its function. Is the process an attachment point for a ligament, tendon, or a muscle? Does the opening or hole serve as a passageway for a nerve, artery, or vein? Is the smooth surface where the bone articulates with another bone? This is an important step in ➡

MODULE 5: SKELETAL SYSTEM

helping you to integrate form and function. While completing this exercise, you should view both individual disarticulated bones and the same bones on an articulated (completed) skeleton. This will increase your understanding of how individual bones fits in with the rest of the axial skeleton.

In this laboratory session you will observe the bones that compose the axial skeleton on both an articulated human skeleton, on disarticulated bones of the human skeleton, or on bone models. You will identify the major landmarks and identifying features of each bone, and will associate the observed structures with their associated functions.

Chapter 8: The Skeletal System: Axial Skeleton

Name: _____
Date: _____ Section: _____

PRE-LABORATORY WORKSHEET

Also available at www.connect.mcgraw-hill.com

1. Match the general structure described in column A with the marking listed in column B.

   a. **Articulating Surfaces**

Column A	Column B
____ smooth, grooved, pulley-like articular process	condyle
____ small, flat, shallow, articulating surface	facet
____ large, smooth, rounded, articulating, oval structure	head
____ prominent, rounded epiphysis	trochlea

   b. **Depressions**

Column A	Column B
____ flattened or shallow depression	alveolus
____ deep pit or socket in the maxillae or mandible	fossa
____ narrow groove	sulcus

   c. **Projections for tendon and ligament attachment**

Column A	Column B
____ massive, rough projection found only on the femur	crest
____ large, rough projection	epicondyle
____ pointed, slender process	line
____ low ridge	process
____ small, round projection	ramus
____ narrow, prominent ridgelike projection	spine
____ projection adjacent to a condyle	trochanter
____ angular extension of a bone relative to the rest of the structure	tubercle
____ any marked bony prominence	tuberosity

   d. **Openings and spaces**

Column A	Column B
____ cavity or hollow space in a bone	canal/meatus
____ passageway through a bone	fissure
____ rounded passageway through a bone	foramen
____ narrow, slitlike opening through a bone	sinus

2. List the bones that compose the axial skeleton.

   _____
   _____
   _____
   _____

3. In a typical human, how many vertebrae (individual or fused) does each section of the vertebral column contain?

   Cervical _____   Thoracic _____   Lumbar _____   Sacral _____   Coccygeal _____

4. All ribs articulate with _____ vertebrae.

5. List the three parts of the sternum, from superior to inferior portion.

   a. _____

   b. _____

   c. _____

6. a. What is a paranasal sinus? _____

   b. Which bones contain paranasal sinuses? _____, _____, _____, _____

7. Which skull bone is the only one that is mobile (movable)? _____

8. The sella turcica is a feature of the _____ bone.

9. The foramen magnum (through which the spinal cord exits the cranial cavity) is a feature of the _____ bone.

10. What is a fontanel? _____

11. The suture that forms between the frontal and parietal bones is called the _____ suture.

Gross Anatomy

The Skull

The bones that make up the skull are separated into two functional categories: the **cranial bones** (frontal, parietal, temporal, occipital, sphenoid, and ethmoid) and the **facial bones** (maxilla, mandible, zygomatic, nasal, lacrimal, palatine, inferior nasal conchae, and vomer). The roof of the cranium—the **calvaria**, or skullcap—is the dome-shaped part of the skull that protects the brain. In an adult, all of the skull bones, with the exception of the mandible, are fused to each other via synarthrotic (immovable) joints called **sutures**.

Table 8.1 describes each bone of the skull individually and lists the best view(s) for observing its features. Because your laboratory experience is practical, you will view the skull from six points of reference: anterior view, lateral view, posterior view, superior view, inferior view, and superior view of the cranial floor. Your first goal will be to identify the individual bones visible in each view. Your second goal will be to identify all processes, foramina, and major features (often formed from multiple bones) that are visible in each view of the skull. As you work on the second goal, always try to relate the bony processes, fossae, and foramina to the individual bone(s) from which they are formed.

Table 8.1 The Axial Skeleton: Skull Bones and Important Bony Landmarks

Major Bone	Bone Features	Description and Related Structures of Importance	Best View
Ethmoid	Cribriform plate	Forms the roof of the nasal cavity and part of the floor of the cranial cavity	Superior view of cranial floor
	Cribriform foramina	The olfactory nerve (CN I) passes through en route to the brain	Superior view of cranial floor
	Crista galli	A projection that serves as the attachment point for the falx cerebri	Superior view of cranial floor
	Superior nasal concha	Forms superior lateral wall of nasal cavity and causes turbulent air flow	Anterior
	Middle nasal concha	Forms middle lateral wall of nasal cavity and causes turbulent air flow	Anterior
	Perpendicular plate	Forms superior part of nasal septum	Anterior
Frontal	Frontal sinus	A cavity within the frontal bone	Superior view of cranial floor
	Supraorbital foramen (notch)	A hole (that is sometimes just a notch) on the superior ridge of the orbit	Anterior
	Superciliary arch	Process that forms the brow ridges; more pronounced in males than in females	Anterior
Inferior Nasal Conchae	NA	Curved bone that forms inferior part of lateral wall of the nasal cavity and causes turbulent air flow	Anterior
Lacrimal	Lacrimal groove	Forms the medial, inferior aspect of orbit of eye. Groove connects orbital and nasal cavities.	Lateral
Mandible	Alveolar processes	Cavities in the mandible that form the tooth "sockets"	Lateral
	Angle	The portion of the mandible connecting the body to the ramus, forming a right angle	Lateral
	Body	The anterolateral portion of the mandible	Lateral
	Coronoid process	Insertion point for the temporalis muscle	Lateral
	Head	Forms a joint with the mandibular fossa of the temporal bone (temporomandibular joint)	Lateral
	Mandibular foramen	Passageway for the mandibular branch of the trigeminal nerve (CN V_3)	Lateral
	Mental foramen	Passageway for the mental artery and nerve (CN V_3)	Anterior
	Mental protuberance	An anterior projection of the mandible that forms the chin	Lateral
	Ramus	The part of the bone that forms an angle with the body of the mandible	Lateral
Maxilla	Infraorbital foramen	A passageway for the infraorbital artery and nerve	Lateral
	Incisive foramen (fossa)	Contains arteries and nerves passing from the nasal cavity into the oral cavity	Inferior
	Palatine process	Forms anterior floor and part of lateral wall of nasal cavity	Inferior

(continued on next page)

Table 8.1	The Axial Skeleton: Skull Bones and Important Bony Landmarks *(continued)*		
Major Bone	**Bone Features**	**Description and Related Structures of Importance**	**Best View**
Nasal	NA	Forms most of the bridge of the nose	Frontal
Occipital	External occipital protuberance	A large projection that can be palpated on the posterior aspect of the head; serves as a muscle attachment point	Posterior
	Foramen magnum	Large hole for passage of spinal cord	Superior view of the cranial floor
	Hypoglossal canal	The hypoglossal nerve (CN XII) travels through	Superior view of the cranial floor
	Jugular foramen	The jugular vein and a number of nerves (CN IX, X, and XI) travel through	Superior view of the cranial floor
	Occipital condyle	Smooth surface for articulation with atlas (first vertebra)	Inferior
Palatine	NA	Forms the posterior floor of nasal cavity, part of the orbit, and part of the hard palate	Inferior
Parietal	NA	L-shaped bone that forms the lateral, superior wall of the cranial cavity	Lateral
Sphenoid	Foramen ovale	The mandibular branch of the trigeminal nerve (CN V_3) travels through	Superior view of the cranial floor
	Foramen rotundum	The maxillary branch of the trigeminal nerve (CN V_2) travels through	Superior view of the cranial floor
	Foramen spinosum	Middle meningeal artery and vein and a branch of the trigeminal nerve (CN V) travel through	Superior view of the cranial floor
	Greater wing	Forms parts of the posterior orbit of the eye and the middle cranial fossa	Superior view of the cranial floor
	Inferior orbital fissure	The maxillary branch of the trigeminal nerve (CN V_2) and the infraorbital artery and vein travel through	Superior view of the cranial floor
	Lesser wing	Forms part of the anterior cranial fossa	Superior view of the cranial floor
	Optic foramen	The optic nerve (CN II) travels through	Superior view of the cranial floor
	Sella turcica	A "Turkish saddle"-shaped depression that houses the pituitary gland	Superior view of the cranial floor
	Superior orbital fissure	The oculomotor (CN III), trochlear (CN IV), trigeminal (CN V_1), and abducens (CN VI) nerves travel through	Superior view of the cranial floor
Temporal	Carotid canal	The internal carotid artery and associated nerves travel through.	Inferior
	External acoustic (auditory) meatus	The opening into the external auditory canal	Lateral
	Foramen lacerum	Largely covered by cartilage in a living human; no one structure passes completely through it	Superior view of the cranial floor
	Internal acoustic (auditory) meatus	The facial (CN VII) and vestibulocochlear nerves (CN VIII) travel through	Superior view of the cranial floor
	Mandibular fossa	The point of articulation with the head of the mandible, forming the temporomandibular joint	Lateral
	Mastoid process	Serves as an attachment point for muscles of the neck	Lateral
	Petrous part	Houses structures for hearing and equilibrium; separates middle and posterior cranial cavities	Superior
	Squamous part	Forms the inferior, posterior part of the temporal fossa	Lateral
	Styloid process	Serves as an attachment point for muscles controlling the tongue	Lateral
	Zygomatic process	A projection that articulates with the temporal process of the zygomatic bone	Lateral
Vomer	NA	Forms inferior and posterior part of nasal septum	Inferior
Zygomatic	Frontal process	Articulates with the frontal bone	Lateral
	Maxillary process	Articulates with the zygomatic process of the maxillary bone	Lateral
	Temporal process	Articulates with the zygomatic process of the temporal bone	Lateral

EXERCISE 8.1

ANTERIOR VIEW OF THE SKULL

EXERCISE 8.1A Anterior View of the Skull

1. Obtain a skull and observe it from an anterior view **(figure 8.1)**. An anterior view of the skull reveals much of the detail of the facial bones. Facial bones play a role in mastication (chewing) and in the protection and support of special sensory organs such as the eye. **Table 8.2** describes bony structures of the face and lists the bones that compose each structure.

2. Using table 8.2 and your textbook as guides, identify the structures listed in figure 8.1 on a skull or model in the laboratory. Then, label figure 8.1.

3. *Optional Activity:* **AP|R** 5: Skeletal System—Watch the "Skull" animation to see how the bones of the skull fit together.

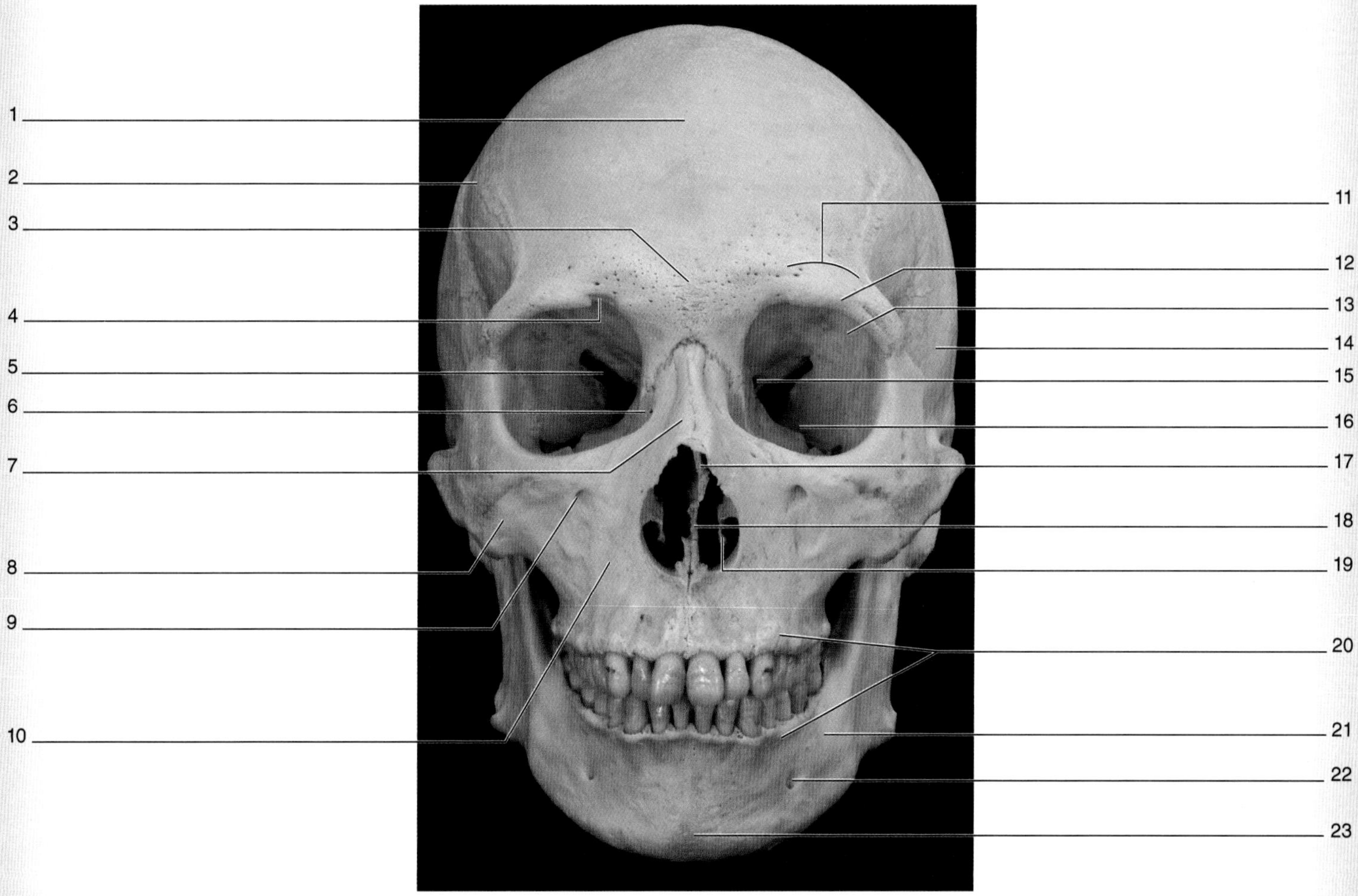

Figure 8.1 Anterior View of the Skull.

☐ alveolar processes
☐ frontal bone
☐ glabella
☐ inferior nasal concha
☐ inferior orbital fissure
☐ infraorbital foramen
☐ lacrimal bone
☐ mandible

☐ maxilla
☐ mental foramen
☐ mental protuberance
☐ nasal bone
☐ optic canal
☐ parietal bone
☐ perpendicular plate of ethmoid
☐ sphenoid bone

☐ superciliary arch
☐ superior orbital fissure
☐ supraorbital foramen (notch)
☐ supraorbital margin
☐ temporal bone
☐ vomer
☐ zygomatic bone

(continued on next page)

(continued from previous page)

Table 8.2 — The Axial Skeleton: Anterior View of the Skull

Facial Structure	Major Bone	Bone Feature	Description and Related Structures of Importance
Forehead	Frontal	Squamous part	Remnant of a fetal joint between the two parts of the frontal bone
		Coronal suture	Suture between frontal and parietal bones
		Metopic suture	Suture between the two parts of the frontal bone; only named in the adult if the suture persists
		Squamous suture	Suture between frontal and temporal bones
		Glabella	Prominent bony ridge located immediately superior to the nasal bone
		Superciliary arch	The "brow" ridges, which are located superior to the supraorbital margin
Orbit	Frontal	Supraorbital margin	Bony support and protection of the superior border of the orbit
		Supraorbital foramen	Supraorbital artery and nerve travel through. Sometimes the supraorbital foramen is just a notch
	Sphenoid	Optic canal	Optic nerve (CN II) travels through.
		Superior orbital fissure	Oculomotor (CN III), trochlear (CN IV), trigeminal (CN V_1), and abducens (CN VI) nerves travel through
		Inferior orbital fissure	Maxillary branch of trigeminal nerve (CN V_2) and infraorbital artery and vein travel through
	Ethmoid		Forms medial wall and part of the posterior wall of the orbit
	Lacrimal	Lacrimal fossa	Drains tears from the surface of the eye into the nasal cavity
	Maxilla		Forms medial and inferior walls of the orbit
		Infraorbital foramen	Infraorbital nerve (a branch of CN V) and artery travel through
	Zygomatic		Forms lateral border and wall of the orbit
Nose	Nasal		Forms most of the bridge and the anterior portion of the bony skeleton of the nose
	Maxilla	Frontal processes	Form lateral aspect of the bony skeleton of the nose
	Frontal	Nasal spine	Forms superior aspect of the bony skeleton of the nose
Nasal Septum	Ethmoid	Perpendicular plate	Forms superior portion of the nasal septum
	Vomer		Forms the posterior-inferior portion of the nasal septum
Nasal Cavity	Ethmoid	Cribriform foramina	Holes that olfactory nerves (CN I) travel through to get to the CNS
		Cribriform plate	Forms roof of nasal cavity
		Superior and middle conchae	Curved bony structures that form superior part of lateral wall and cause turbulent air flow
	Inferior nasal concha		Curved bone that forms inferior part of lateral wall and causes turbulent air flow
	Maxilla	Palatine process	Forms anterior floor and part of lateral wall of the nasal cavity
	Palatine		Forms posterior floor of nasal cavity
Oral Cavity (Buccal)	Palatine		Forms posterior roof of oral cavity
	Maxilla	Palatine process	Forms anterior roof of oral cavity
		Incisive foramen	Contains arteries and nerves passing from the nasal cavity into the oral cavity
		Alveolar processes	Form joints with the teeth
	Mandible	Alveolar processes	Form joints with the teeth
		Body	Forms anterior portion of lower border of oral cavity
		Ramus	Forms lateral portion of lower border of oral cavity
Chin	Mandible	Body	The anterolateral portion of the mandible
		Angle	The portion of the mandible connecting the body to the ramus, forming a right angle
		Mental foramen	Mental artery and nerve (CN V_3) travel through
		Ramus	Part of the bone that forms an angle with the body of the mandible
		Alveolar processes	Form joints with the teeth
		Mental protuberance	Anterior projection of mandible, forming the anterior projection of the chin

EXERCISE 8.1B The Orbit

1. Observe the **orbit (figure 8.2)** on a skull or model.

2. The orbit is the bony casing that supports and protects the eyeball. Parts of the frontal, zygomatic, maxillary, ethmoid, and lacrimal bones form the anterior border of the orbit. The ethmoid and lacrimal bones form most of the medial wall of the orbit, and the sphenoid bone forms most of the posterior wall. Identify the walls and borders of the orbit on your specimen.

3. Using table 8.2 and your textbook as guides, identify the features of the orbit listed in figure 8.2 on a skull. Then, label figure 8.2.

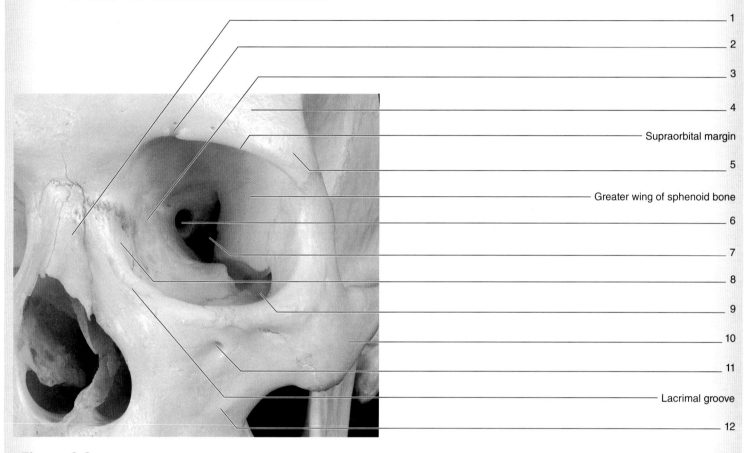

Figure 8.2 The Orbit. Anterior view.

☐ ethmoid bone
☐ frontal bone
☐ inferior orbital fissure
☐ infraorbital foramen
☐ lacrimal bone
☐ maxilla
☐ nasal bone
☐ optic foramen
☐ superior orbital fissure
☐ supraorbital foramen (notch)
☐ zygomatic bone
☐ zygomatic process of frontal bone

EXERCISE 8.1C The Nasal Cavity

1. Observe the **nasal cavity (figure 8.3)** on a skull or model.

2. The nasal cavity is a large, complex cavity that is separated into two halves by a **nasal septum** (*saeptum*, a partition). The ethmoid bone forms parts of the roof, septum, and lateral walls. The **cribriform** (*cribrum*, a sieve, + *forma*, a form) **plate** of the ethmoid bone forms most of the roof. The **palatine processes of the maxillary bones** and the **palatine bones** form the floor of the nasal cavity (and the roof of the oral cavity). The **nasal bones** form most of the bridge of the nose. Finally, the bony portion of the nasal septum is formed from the **perpendicular plate of the ethmoid bone** and the **vomer**.

3. Using table 8.2 and your textbook as guides, identify the features of the nasal cavity listed in figure 8.3 on a skull. Then, label figure 8.3.

(continued on next page)

174 Chapter Eight *The Skeletal System: Axial Skeleton*

(continued from previous page)

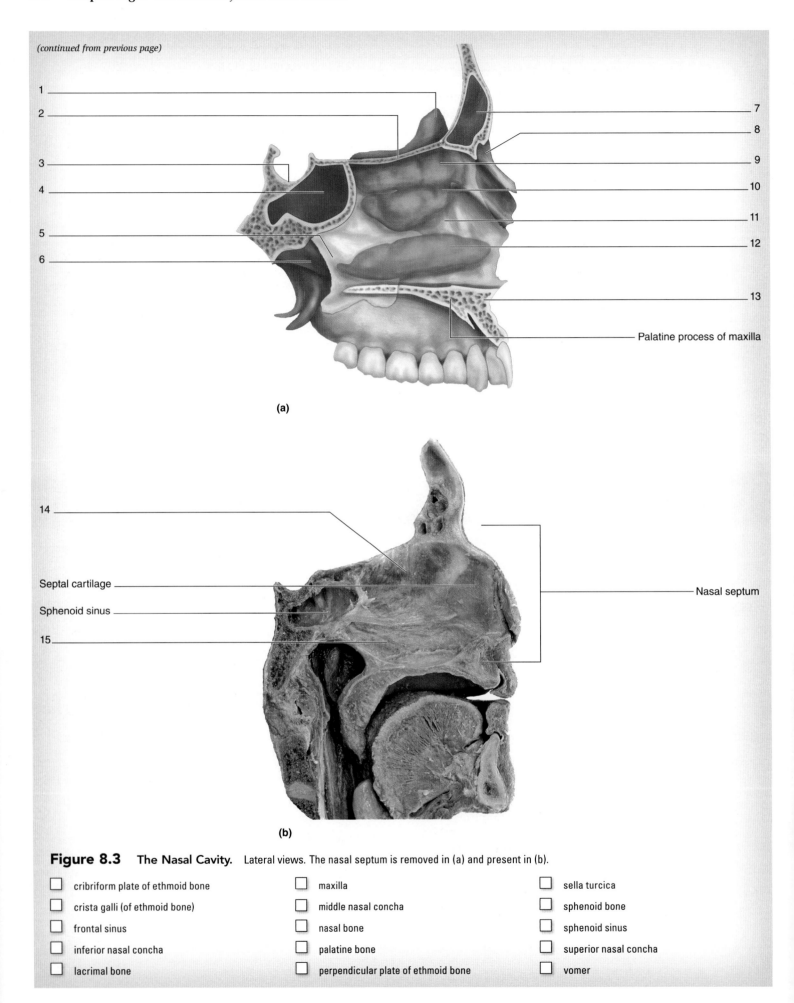

Figure 8.3 **The Nasal Cavity.** Lateral views. The nasal septum is removed in (a) and present in (b).

☐ cribriform plate of ethmoid bone
☐ crista galli (of ethmoid bone)
☐ frontal sinus
☐ inferior nasal concha
☐ lacrimal bone
☐ maxilla
☐ middle nasal concha
☐ nasal bone
☐ palatine bone
☐ perpendicular plate of ethmoid bone
☐ sella turcica
☐ sphenoid bone
☐ sphenoid sinus
☐ superior nasal concha
☐ vomer

INTEGRATE

CONCEPT CONNECTION

The nasal cavity plays an important role in respiration and olfaction (the sense of smell). The role of the nasal cavity in respiration is to moisten, filter, and warm the air before it enters lower respiratory structures. The role of the nasal cavity in olfaction is to house the olfactory epithelium, which is responsible for detecting odorants in the air. Olfactory receptor cells are embedded within this epithelium, and their nerve processes project through the foramina of the cribriform plate of the ethmoid bone to reach the olfactory nerve. This pathway relays sensory input regarding the sense of smell to the brain. As odorants enter the nasal cavity, they dissolve in mucus in the nasal cavity, then bind to receptors on olfactory "hairs," thereby exciting these sensory neurons. Nerve signals, or action potentials, are then transmitted along the olfactory nerves to the olfactory bulb. From there, the signals are transmitted to the brain for the conscious perception of smell. This process will be revisited in chapter 18 in the discussion of special senses.

EXERCISE 8.1D The Mandible

1. Obtain an isolated mandible or observe the mandible on an articulated skeleton or complete skull.

2. The **mandible** (figure 8.4) is unique among skull bones, because it is the only bone that is movable. It shares an articulation with the temporal bone, forming the **temporomandibular joint**. Here, the **head of the mandible** articulates with the mandibular fossa of the temporal bone. The **mandibular condyle** is the rounded projection on the head of the mandible that actually forms the joint with the temporal bone. Place your fingers just anterior to your ears and then open and close your mouth to feel the movement of the joint formed between the mandible and temporal bone (the temporomandibular joint).

3. The mandible also has a prominent **coronoid process** (*corona*, crown, + *eidos*, form, resemblance) that serves as the insertion point for the temporalis muscle.

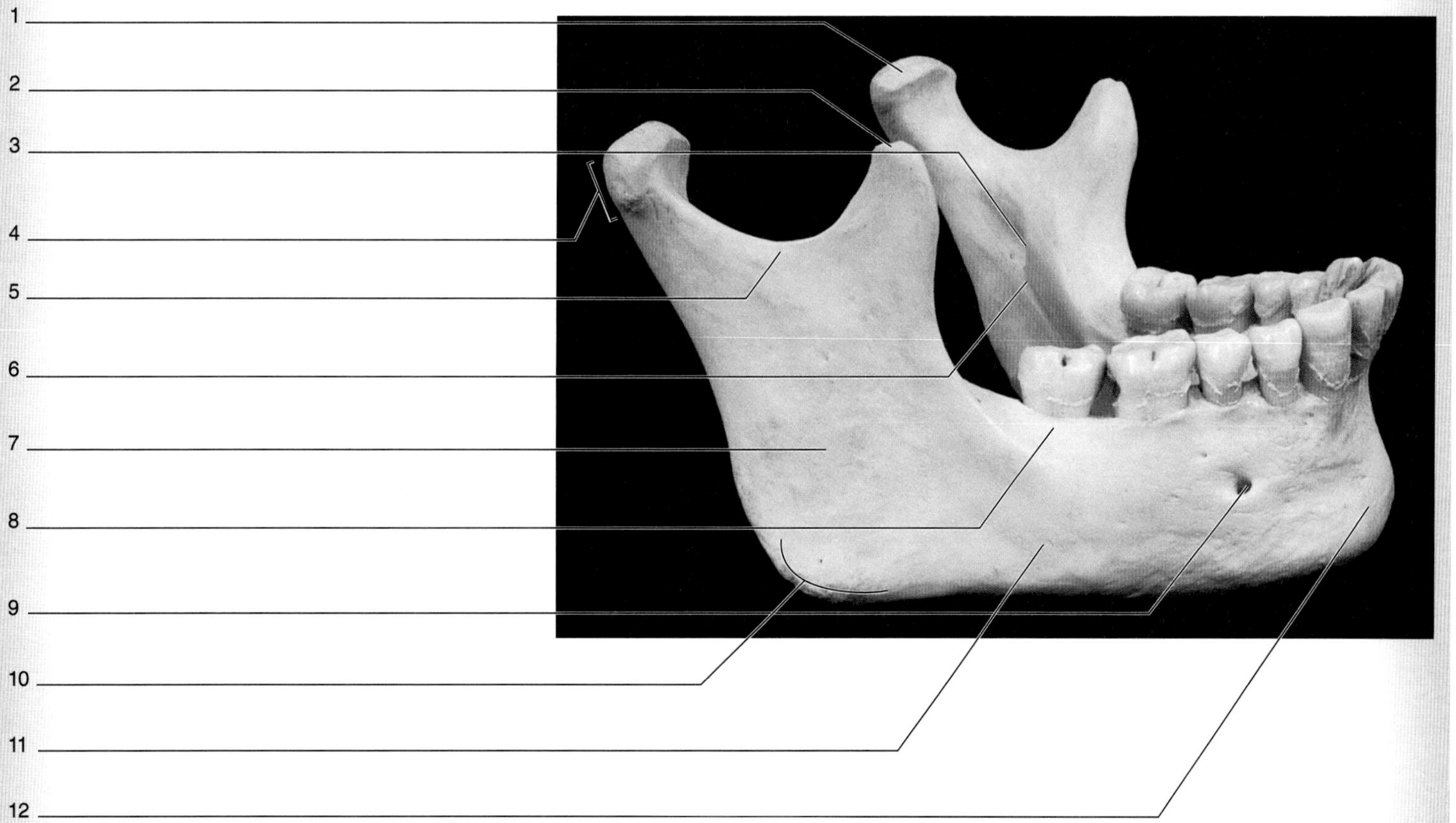

1 _____
2 _____
3 _____
4 _____
5 _____
6 _____
7 _____
8 _____
9 _____
10 _____
11 _____
12 _____

Figure 8.4 The Mandible. Lateral view.

- ☐ alveolar process
- ☐ angle
- ☐ body
- ☐ condylar process
- ☐ coronoid process
- ☐ head of mandible
- ☐ mandibular foramen
- ☐ mandibular notch
- ☐ mental foramen
- ☐ mental protuberance
- ☐ mylohyoid line
- ☐ ramus

(continued on next page)

176 Chapter Eight The Skeletal System: Axial Skeleton

(continued from previous page)

4. The mandible contains two paired prominent foramina. The first, found on the inner (medial) surface of the ramus of the mandible, is the **mandibular foramen**. A continuation of the mandibular branch of the trigeminal nerve (CN V$_3$) passes through this foramen into the interior of the bone. It then sends branches to the **alveolar processes** of the mandible to innervate the roots of the teeth in a living individual. When you are having dental work done on the teeth of your lower jaw, the dentist will direct a needle containing anaesthetic at the mucosa surrounding this foramen in order to bathe the nerve branches that travel into the mandible. Finally, the **mental foramen** (*mental*, chin) is located just superior to the lower border of the mandible at about the midpoint of the body of the mandible. The mental artery and nerve travel through the mental foramen in a living individual.

5. Using table 8.2 and your textbook as guides, identify the structures listed in figure 8.4 on a mandible. Then, label figure 8.4.

EXERCISE 8.2

ADDITIONAL VIEWS OF THE SKULL

EXERCISE 8.2A Lateral View of the Skull

1. Obtain a skull and observe it from a lateral view (**figure 8.5**).

2. The most notable feature in a lateral view of the skull is the **zygomatic arch,** which is the bony structure that forms the superior part of a person's cheek. It is formed from the zygomatic process of the temporal bone, and the temporal process of the zygomatic bone.

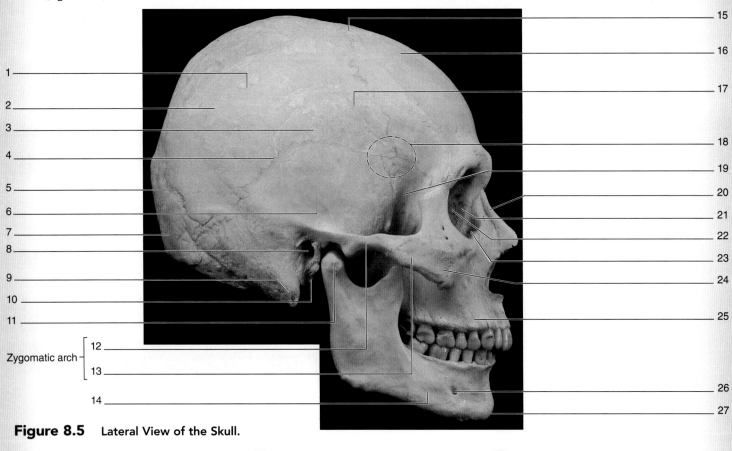

Figure 8.5 Lateral View of the Skull.

- ☐ body of mandible
- ☐ coronal suture
- ☐ ethmoid bone
- ☐ external acoustic meatus
- ☐ frontal bone (used twice)
- ☐ greater wing of sphenoid bone
- ☐ head of mandible
- ☐ inferior temporal line
- ☐ lacrimal bone
- ☐ lacrimal groove
- ☐ lambdoid suture
- ☐ mastoid process
- ☐ maxilla
- ☐ mental foramen
- ☐ mental protuberance
- ☐ nasal bone
- ☐ occipital bone
- ☐ parietal bone (used twice)
- ☐ parietal eminence
- ☐ pterion
- ☐ squamous part of temporal bone
- ☐ squamous suture
- ☐ styloid process
- ☐ superior temporal line
- ☐ temporal process of zygomatic bone
- ☐ zygomatic bone
- ☐ zygomatic process of temporal bone

3. Using table 8.1 and your textbook as guides, identify the structures listed in figure 8.5 on a skull. Then, label figure 8.5.

EXERCISE 8.2B Posterior View of the Skull

1. Obtain a skull and observe it from a posterior view **(figure 8.6)**.

2. The most notable feature in a posterior view of the skull is the **lambdoid suture** (*lambda*, the Greek letter λ, + *eidos*, resemblance), which is named for its resemblance to the Greek letter lambda. Can you see the resemblance? Draw the letter lambda in the space provided here: _____

3. Using table 8.1 and your textbook as guides, identify the structures listed in figure 8.6 on a skull. Then, label figure 8.6.

EXERCISE 8.2C Superior View of the Skull

1. Obtain a skull (with the skullcap intact) and observe it from a superior view **(figure 8.7)**.

2. The most notable feature in a superior view of the skull is the **sagittal suture.**

3. Using table 8.1 and your textbook as guides, identify the structures listed in figure 8.7 on a skull. Then, label figure 8.7.

1 _____
2 _____
3 _____
4 _____
5 _____
6 _____
7 _____
8 _____
9 _____
10 _____

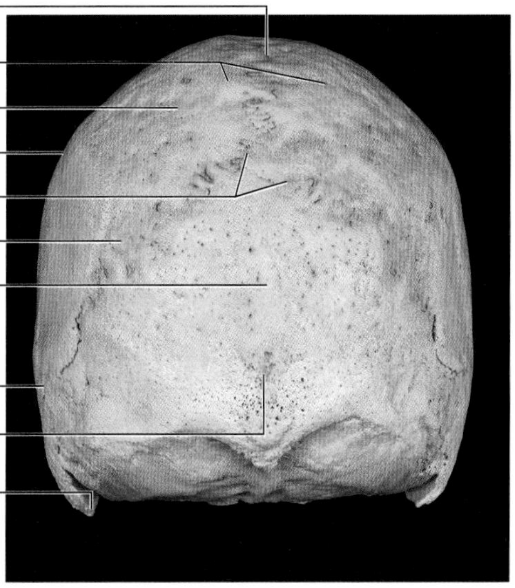

☐ external occipital protuberance
☐ lambdoid suture
☐ mastoid process
☐ occipital bone
☐ parietal bone
☐ parietal eminence
☐ parietal foramina
☐ sagittal suture
☐ sutural (Wormian) bone
☐ temporal bone

Figure 8.6 Posterior View of the Skull.

1 _____
2 _____
3 _____
4 _____
5 _____
6 _____
7 _____
8 _____

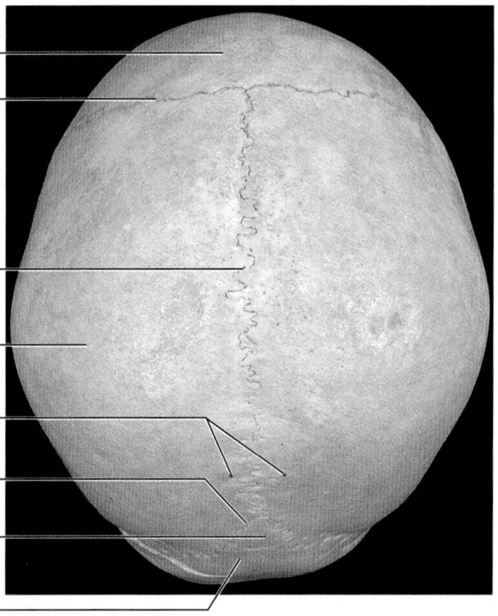

☐ coronal suture
☐ frontal bone
☐ lambdoid suture
☐ occipital bone
☐ parietal bone
☐ parietal foramina
☐ sagittal suture
☐ sutural (Wormian) bone

Figure 8.7 Superior View of the Skull.

(continued on next page)

(continued from previous page)

EXERCISE 8.2D Inferior View of the Skull

1. Turn the skull over and observe it from an inferior view **(figure 8.8)**. Many of the structures visible in this view are foramina that nerves and blood vessels pass through to get into and out of the cranial cavity.

2. Obtain a broom straw or other nonmarking pointing device (NO pens or pencils!) from your instructor. As you identify each foramen in the inferior view, pass the broom straw through the foramen and see where it comes out in the cranial floor. This activity will help you visualize the pathways that structures traveling through the foramina take to get into or out of the cranial cavity.

3. The **foramen lacerum** (*lacero*, to tear to pieces) is unique among cranial foramina. It is one of the longest canals in the skull (about a centimeter in length). However, no single structure passes completely through it from one opening to the other. Instead, several structures pass through small portions of the canal. Such structures include a number of nerves as well as the internal carotid artery. As the internal carotid artery travels superiorly from the thorax into the cranial cavity, it passes first through the **carotid canal** and then enters the superior portion of the foramen lacerum as it proceeds toward the brain.

4. Using table 8.1 and your textbook as guides, identify the structures listed in figure 8.8 on a skull. Then label figure 8.8.

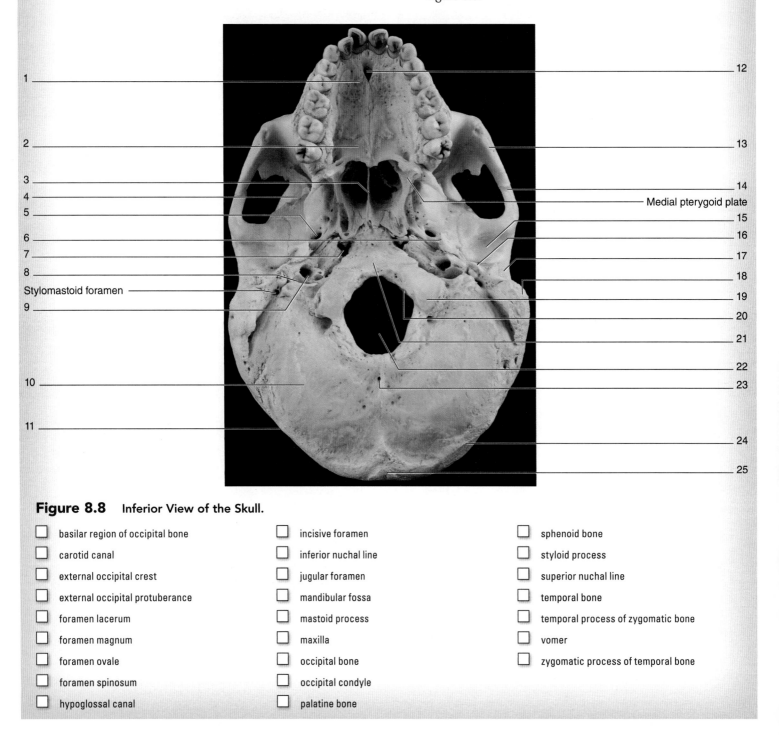

Figure 8.8 Inferior View of the Skull.

- ☐ basilar region of occipital bone
- ☐ carotid canal
- ☐ external occipital crest
- ☐ external occipital protuberance
- ☐ foramen lacerum
- ☐ foramen magnum
- ☐ foramen ovale
- ☐ foramen spinosum
- ☐ hypoglossal canal
- ☐ incisive foramen
- ☐ inferior nuchal line
- ☐ jugular foramen
- ☐ mandibular fossa
- ☐ mastoid process
- ☐ maxilla
- ☐ occipital bone
- ☐ occipital condyle
- ☐ palatine bone
- ☐ sphenoid bone
- ☐ styloid process
- ☐ superior nuchal line
- ☐ temporal bone
- ☐ temporal process of zygomatic bone
- ☐ vomer
- ☐ zygomatic process of temporal bone

Chapter Eight The Skeletal System: Axial Skeleton 179

EXERCISE 8.3

SUPERIOR VIEW OF THE CRANIAL FLOOR

1. Obtain a skull and remove the superior part of the cranium so the cranial floor is visible **(figure 8.9)**.

2. Notice how the floor of the cranium is separated into three fossae: anterior, middle, and posterior. The **lesser wing of the sphenoid bone** forms the border between anterior and middle cranial fossae. The **petrous part of the temporal bone** (*petrosus*, a rock) forms the "rocky" border between the middle and posterior cranial fossae.

3. Using colored pencils, color and label the anterior, middle, and posterior cranial fossae in figure 8.9. As you color in the fossae, pay attention to the structures that form the natural divisions between these fossae.

4. Observe the **sella turcica** (*sella*, saddle, + *turcica*, Turkish) in the central portion of the sphenoid bone. This structure gets its name from its resemblance to a Turkish saddle, which contains large, prominent horns **(figure 8.10)**. The anterior part of the sella turcica contains a slight projection called the **tuberculum sellae** (*tuber*, a knob). Posterior to that is the **hypophyseal fossa** (*hypophysis*, an undergrowth), which houses the pituitary gland in a living human. The pituitary gland is a small pea-shaped endocrine gland that connects to the brain via a small stalk called the infundibulum (*infundibula*, a funnel). The larger projection in the posterior part of the sella is the **dorsum sellae**, which connects laterally to the two **posterior clinoid processes** (*klino*, to slope).

Figure 8.10 Photograph of a Turkish Saddle.

5. The temporal bone has two major portions: a lateral **squamous part,** which forms part of the lateral wall of the cranium, and a thick **petrous part,** which forms the border between the middle and posterior cranial fossae (although it is considered part of the middle cranial fossa).

6. Locate the petrous part of the temporal bone. Notice how it forms a sort of "rocky" ridge within the cranial floor. This portion of the temporal bone is very large and bulky because it contains the structures for hearing (the cochlea) and equilibrium/balance (the semicircular canals). These structures cannot be seen from the surface of the bone. However, if you were to cut open the petrous part of the temporal bone, you would find them inside.

7. Find the internal and external acoustic (auditory) meatuses. The **internal acoustic (auditory) meatus** is the opening into the **internal auditory canal,** which is a passageway for the nerves that carry sensory information from the cochlea and semicircular canals to the brain. The **external acoustic (auditory) meatus** is the opening into the **external auditory canal,** which is a passageway through which sound waves travel to reach the tympanic membrane (eardrum).

8. Using table 8.1 and your textbook as guides, identify the structures listed in **figure 8.11** on a superior view of the cranial floor. Then label figure 8.11.

INTEGRATE

LEARNING STRATEGY

It is useful to focus your observations on one fossa at a time as you identify bony structures within the cranial floor. This is a natural way to divide the features of the cranial floor into manageable pieces of material. If you use this approach, when you take a laboratory practical exam and are asked to identify one of the many foramina in the cranial floor, you will be able to narrow down your choices if you first identify the cranial fossa where the foramen is located.

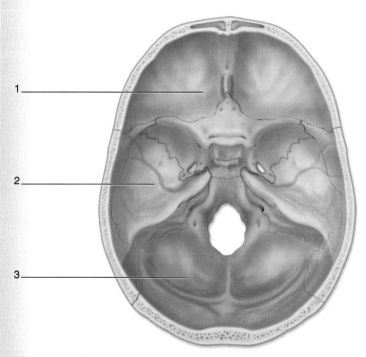

1. _____
2. _____
3. _____

Figure 8.9 Cranial Fossae.
☐ anterior cranial fossa   ☐ posterior cranial fossa
☐ middle cranial fossa

(continued on next page)

(continued from previous page)

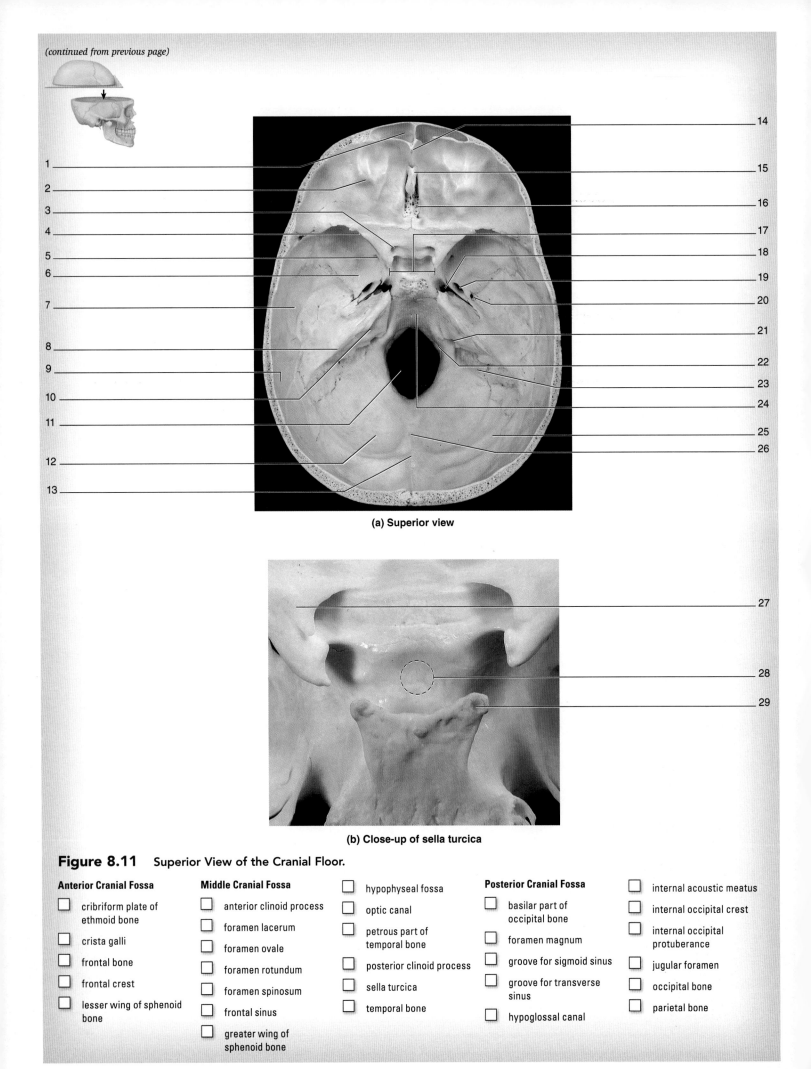

(a) Superior view

(b) Close-up of sella turcica

Figure 8.11 Superior View of the Cranial Floor.

Anterior Cranial Fossa
- ☐ cribriform plate of ethmoid bone
- ☐ crista galli
- ☐ frontal bone
- ☐ frontal crest
- ☐ lesser wing of sphenoid bone

Middle Cranial Fossa
- ☐ anterior clinoid process
- ☐ foramen lacerum
- ☐ foramen ovale
- ☐ foramen rotundum
- ☐ foramen spinosum
- ☐ frontal sinus
- ☐ greater wing of sphenoid bone

- ☐ hypophyseal fossa
- ☐ optic canal
- ☐ petrous part of temporal bone
- ☐ posterior clinoid process
- ☐ sella turcica
- ☐ temporal bone

Posterior Cranial Fossa
- ☐ basilar part of occipital bone
- ☐ foramen magnum
- ☐ groove for sigmoid sinus
- ☐ groove for transverse sinus
- ☐ hypoglossal canal

- ☐ internal acoustic meatus
- ☐ internal occipital crest
- ☐ internal occipital protuberance
- ☐ jugular foramen
- ☐ occipital bone
- ☐ parietal bone

EXERCISE 8.4

BONES ASSOCIATED WITH THE SKULL

The hyoid bone and the auditory ossicles are bones of the axial skeleton that are associated with the skull, but they are not part of the skull proper. The auditory ossicles are part of the hearing apparatus and will be covered in chapter 18. This exercise will concentrate on the features of the hyoid bone.

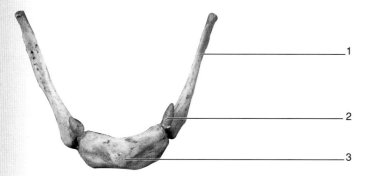

Figure 8.12 **The Hyoid Bone.** The hyoid bone is not in direct contact with any other bone of the skeleton.

☐ body   ☐ greater cornu   ☐ lesser cornu

1. Observe the hyoid bone (**figure 8.12**) on an articulated skeleton.

2. The **hyoid bone** (*hyoeidēs*, shaped like the Greek letter upsilon, υ) is the only bone in the body with no direct articulation with another bone. Muscles that move the tongue and pharynx attach to the hyoid.

3. Palpate your own hyoid by placing your thumb and index finger just medial to the angle of your mandible on either side and then moving them from side to side. Can you feel the rigid structure that you are moving?

4. Observe the articulated skeleton in your laboratory and look at the placement of the hyoid bone with respect to the mandible.

5. Using your textbook as a guide, identify the structures listed in figure 8.12 on a hyoid bone. Then label figure 8.12.

The Fetal Skull

Like all bones in the body, the skull bones of the fetus are still developing. Recall that they develop via intramembranous ossification, which involves replacing a connective tissue membrane with bone tissue. Thus, when a fetus is born, the sutures between skull bones have not yet formed, which allows the head to distort as it moves through the birth canal. The spaces between the plates of bone in the developing skull still consist of connective tissue membranes, which are largest in places where more than two bones come together. These membranes are **fontanels** (*fontaine*, a small fountain), and can be felt as "soft spots" on a baby's head. The fontanelles will close between the ages of 2 and 3.

Chapter Eight The Skeletal System: Axial Skeleton

EXERCISE 8.5

THE FETAL SKULL

1. Observe a fetal skull or a model of a fetal skull (**figure 8.13**).
2. Based on your observations of the adult skull, locate the major cranial bones on the fetal skull (e.g., frontal, parietal, occipital).
3. Using your textbook as a guide, identify the structures listed in figure 8.13 on a fetal skull or model of a fetal skull. Then label them in figure 8.13.

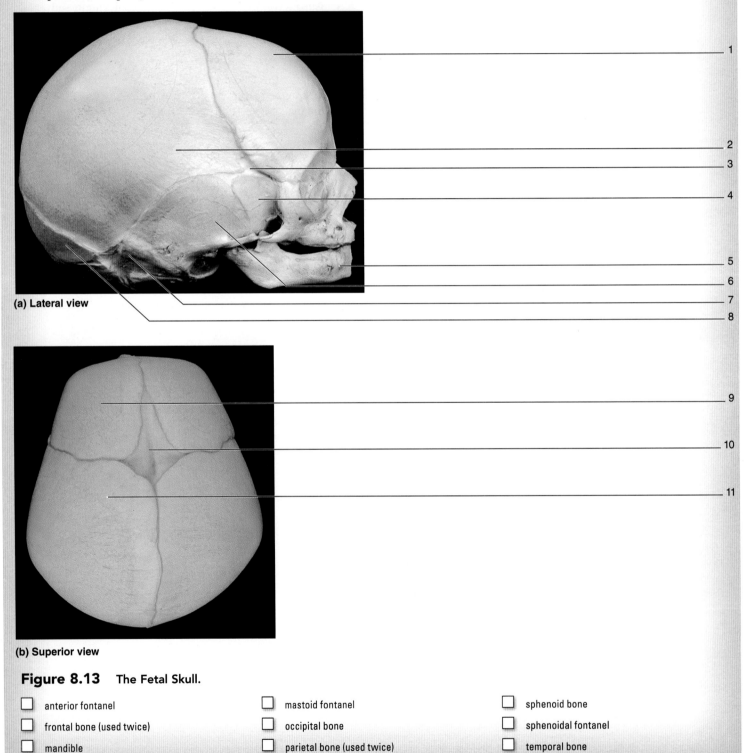

(a) Lateral view

(b) Superior view

Figure 8.13   The Fetal Skull.

☐ anterior fontanel
☐ frontal bone (used twice)
☐ mandible
☐ mastoid fontanel
☐ occipital bone
☐ parietal bone (used twice)
☐ sphenoid bone
☐ sphenoidal fontanel
☐ temporal bone

The Vertebral Column

The vertebral column lies at the core of the human skeleton. It quite literally is the "backbone" that anchors nearly every major component of the skeletal support system. The vertebral column is divided into five major regions: **cervical, thoracic, lumbar, sacral,** and **coccygeal.** The vertebrae themselves change size and shape rather drastically from the cervical region to the coccygeal region. These changes reflect the different weight-bearing demands placed on the vertebrae in each region. **Cervical vertebrae** are small and light because they are not supporting a lot of weight (relatively speaking), and they are specialized to allow a lot of movement of the neck, particularly rotation. **Thoracic vertebrae** are specialized to provide articulation points for the ribs. **Lumbar vertebrae** are very large, bulky vertebrae that are specialized for supporting the weight of the entire vertebral column and body structures above them. They do not allow much movement, but instead are designed to keep the vertebral column stable. The **sacrum,** which consists of fused vertebrae, is specialized to provide a stable anchoring point for the bones of the pelvic girdle. Finally, the **coccyx** consists of 3–5 small vertebrae, which have fused together during development. It serves as an attachment point for several ligaments and for muscles of the pelvic floor. **Table 8.3** summarizes the characteristics of each type of vertebra.

Table 8.3 The Axial Skeleton—Vertebral Column

Vertebrae	Number of Vertebrae	Bone Features	Description and Related Structures of Importance
Typical Vertebra	32	Lamina	Connects transverse process to spinous process on either side of each vertebra
		Pedicle	Connects body to transverse process
		Transverse processes	Processes that are directed laterally (one on each side)
		Spinous process	A process that is directed posteriorly
		Inferior articular process	Contains a facet that forms a joint with the superior articular process of the vertebra immediately inferior
		Superior articular process	Contains a facet that forms a joint with the inferior articular process of the vertebra immediately superior
		Vertebral foramen	Hole within each vertebra; spinal cord extends through "stacked" vertebral foramina
		Body	The largest part of the vertebra. Intervertebral discs are found between bodies of adjacent vertebrae.
		Intervertebral foramina	Formed when two vertebra come together; passageway for exit of spinal nerves
Cervical (C)	7	Body	Small body, oval/kidney bean shape
		Spinous process	Horizontal, bifid (forked) spine of cervical vertebrae 3–6
		Vertebral foramen	Large (especially with respect to size of the body), slight oval shape
		Transverse processes	Each contains a transverse foramina
		Transverse foramen	Contain the vertebral artery
Atlas (C_1)		Body	Has no body; the body has become the dens (odontoid process) of the axis
		Arch	Contains the articular surface for the dens of the axis and the posterior tubercle (no spinous process)
Axis (C_2)		Body	Has odontoid process (dens), which is the fused body of C_1.
Vertebra Prominens (C_7)		Spinous processes	Very large and blunt, not bifid, not covered by ligamentum nuchae. Therefore, is the first spinous process easily felt under the skin.
Thoracic (T)	12	Body	Heart-shaped, contains demifacets for articulation of the head of a rib
		Spinous process	Points inferiorly
		Vertebral foramen	Relatively small; circular in shape; houses the spinal cord
		Transverse processes	Contain facets for articulation with the tubercle of a rib
		Costal facets	Located on the lateral surface of the body and transverse processes; these form joints with the ribs
Lumbar (L)	5	Body	Very large, heavy
		Spinous process	Short and blunt, square shaped, horizontal; location for the spinal cord
		Vertebral foramen	Small (especially with respect to size of body), round; houses the spinal cord
		Transverse processes	Short and tapered at the ends
Sacrum (S)	5 (fused)	Anterior sacral foramina	Passageway for exit of anterior (ventral) rami of sacral spinal nerves
		Posterior sacral foramina	Passageway for exit of posterior (dorsal) rami of sacral spinal nerves
		Median sacral crest	Represents fused spinous processes of sacral vertebrae (S_1–S_4)
		Auricular processes	Earlike (*auris,* ear) processes that articulate with the iliac bones
		Superior articular processes	Contains a facet to form a joint with the inferior articular processes of L_5
		Sacral hiatus	The opening at the inferior end of the sacral canal; formed by unfused laminae of S_5
		Sacral promontory	The anteriosuperior border of the body of S_1
Coccyx (Co)	3 to 5 (fused)	Cornu (horns)	Small projections that point superiorly (part of Co_1)

EXERCISE 8.6

VERTEBRAL COLUMN REGIONS AND CURVATURES

1. Observe the vertebral column of an articulated skeleton (**figure 8.14**).

2. Using colored pencils, color and label the regions of the vertebral column in figure 8.14. Use a different color for each region. As you shade in each region, count the number of vertebrae that make up the region and write that number in the appropriate space in figure 8.14.

3. As the vertebral column develops, it forms several curvatures because of the stresses placed on it. The first curvatures to develop during the fetal period are **primary curvatures.** These form in the thoracic and sacral regions due to growth of the viscera. The second curvatures, which develop after birth, are **secondary curvatures.** These form in the cervical and lumbar regions. The cervical curvature forms when an infant begins to lift his head and the lumbar curvature forms when an infant begins to stand on his feet.

4. Locate all of the curvatures of the vertebral column on an articulated skeleton, and label the curvatures on figure 8.14.

☐ cervical curvature
☐ cervical vertebrae
☐ coccygeal vertebrae
☐ lumbar curvature
☐ lumbar vertebrae
☐ sacral curvature
☐ sacrum
☐ thoracic curvature
☐ thoracic vertebrae

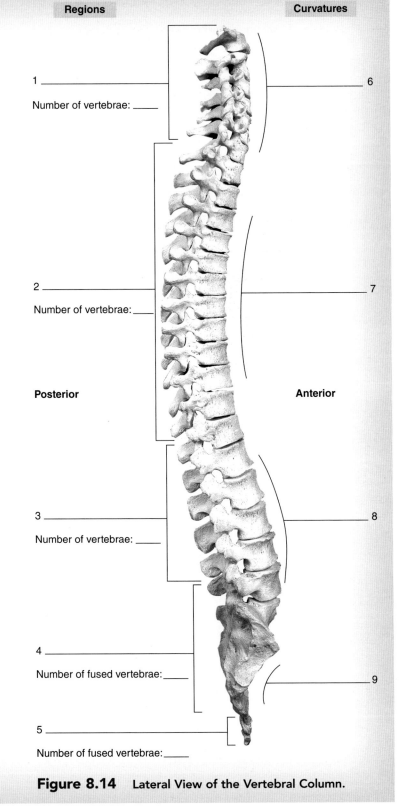

Regions

1 _____
Number of vertebrae: ___

2 _____
Number of vertebrae: ___

3 _____
Number of vertebrae: ___

4 _____
Number of fused vertebrae: ___

5 _____
Number of fused vertebrae: ___

Posterior   Anterior

Curvatures

6 _____

7 _____

8 _____

9 _____

Figure 8.14 Lateral View of the Vertebral Column.

EXERCISE 8.7

STRUCTURE OF A TYPICAL VERTEBRA

1. Obtain a **thoracic vertebra (figure 8.15)** from your instructor or bone box—this will be your example of a "typical" vertebra. It is helpful to begin your study of the vertebral column by taking a "typical" vertebra and identifying its component parts. Then as you begin observing normal variability among vertebrae you will have an easier time finding the appropriate structures.

2. Looking at the vertebra from the superior view, notice the large **body.** The body is generally the largest part of the vertebra, and connections between adjacent vertebral bodies (with intervertebral discs in between) provide the main support of the vertebral column. Just posterior to the body is the large foramen called the **vertebral (spinal) foramen.** The spinal cord extends through the "stacked" vertebral foramina, also called the vertebral canal. The vertebral foramen itself is formed by a structure that is collectively referred to as the **vertebral arch.** The vertebral arch is composed of two sets of processes and the structures that connect them. We will return to a description of this arch after you have identified the major processes of the vertebra.

3. Observe the vertebral processes that project posteriorly and laterally from the vertebral arch. The largest vertebral processes are the **spinous processes,** which are directed posteriorly, and the **transverse processes,** which are directed laterally.

4. Return your attention to the vertebral arch. Notice the bony connections between the vertebral body and the transverse processes. These structures are called **pedicles** (L. *pediculus*, dim. of *pes*, foot). The word *pedicle* comes from a word meaning "foot." If you use your imagination a bit, can you imagine how the vertebral arch stands upon the body on its "feet"? Now notice the bony connections between the transverse processes and the spinous process. These structures are called **laminae** (*lamina*, layer).

5. Next, turn the vertebra so that you are observing it from an anterior view. Notice that there are two prominent structures that project superiorly from the vertebral arch and two that project inferiorly. The projections are respectively called the **superior articular processes** and **inferior articular processes.** Note that on each process there is a smooth, flat surface. These surfaces are called **facets** (*facette*, face). The term *facet* literally means "a little face." (This is the same term used to describe the surfaces on a diamond.) Each vertebra contains upon its superior and inferior processes a pair of **superior articular facets** and **inferior articular facets.** These facets are the surfaces that form the joints between vertebrae, as described in the next step (6).

6. Pick up a second vertebra that articulates (forms a joint) with the vertebra in your hand. As you put the two together, observe how the superior facets and inferior facets articulate with each other to form a joint. These joints are much more mobile than the intervertebral joints (the joints between the vertebral bodies), and they are the sites where most of the movement is allowed by the vertebral column.

7. Now that you have the two vertebrae articulated with each other, look at them from a lateral view. Notice the foramen that forms between the pedicles of adjacent vertebrae. This is the **intervertebral foramen.** This foramen is the location where spinal nerves (nerves that come off of the spinal cord) exit the vertebral canal to travel to their destinations throughout the body.

8. Using table 8.3 and your textbook as guides, identify the structures listed in figure 8.15 on a typical vertebra. Then label them in figure 8.15.

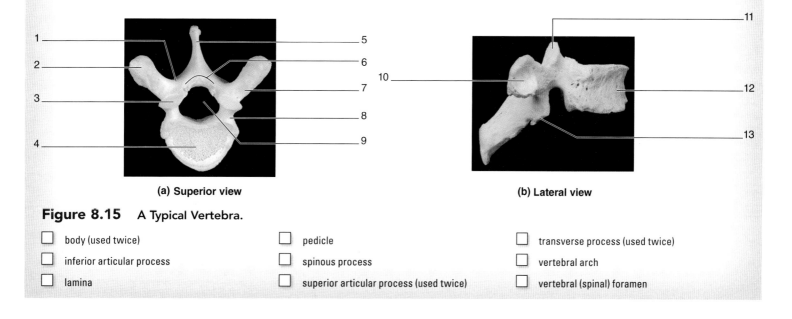

(a) Superior view   (b) Lateral view

Figure 8.15 A Typical Vertebra.

- ☐ body (used twice)
- ☐ inferior articular process
- ☐ lamina
- ☐ pedicle
- ☐ spinous process
- ☐ superior articular process (used twice)
- ☐ transverse process (used twice)
- ☐ vertebral arch
- ☐ vertebral (spinal) foramen

EXERCISE 8.8

CHARACTERISTICS OF INDIVIDUAL VERTEBRAE

INTEGRATE

LEARNING STRATEGY

As you observe individual vertebrae, always try to keep function in mind. Ask yourself questions such as these: Why does this vertebra have such a large/small body? Why does this vertebra have such a large/small vertebral canal? How does this vertebra "fit" with other aspects of the skeletal system? Asking yourself these questions will help you identify each type of vertebra correctly. Finally, when you are observing an isolated vertebra, always be sure to identify the same vertebra on an articulated skeleton so you can develop an appreciation for how the complete vertebral column is assembled.

EXERCISE 8.8A Typical Cervical Vertebrae

1. Obtain a cervical vertebra (*cervix*, neck) **(figure 8.16)**. As you identify the features of each of the individual **cervical vertebrae,** think about how modifications of the cervical vertebrae allow for a great deal of movement in the neck region of the vertebral column.

2. Observe the vertebra from a superior view. Typical cervical vertebrae have a small, oval body and a large triangular vertebral foramen. The spinous process of some of them is forked, or **bifid** (*bifidus*, cleft in two parts).

3. Observe cervical vertebrae on an articulated skeleton. Notice how the fork on one vertebra fits over the top of the spinous process of the vertebra below it.

4. Observe the transverse processes on a cervical vertebra. Notice that it has a hole, or **transverse foramen,** in it. This foramen protects an artery, the **vertebral artery,** as it travels from the thorax to the cranial cavity to supply the brain with blood.

5. Using table 8.3 and your textbook as guides, identify the structures listed in figure 8.16 on a cervical vertebra. Then label them in figure 8.16.

The Atlas (C_1) and Axis (C_2)

The Greek Titan, Atlas, held up the heavens on his shoulders. The first cervical vertebra is named the **atlas** because it holds up the head in much the same way. The second cervical vertebra is called the **axis** because it forms an axis of rotation for the first cervical vertebra to rotate about. Both of these vertebrae are specialized to allow for extensive flexion, extension, and rotational movements of the neck. As you observe the special modifications of the atlas and axis, try to visualize how these modifications allow extensive movement of the head and neck.

EXERCISE 8.8B The Atlas (C_1)

1. Obtain an **atlas** (C_1) and an axis (C_2).

2. Notice that the atlas is missing a body **(figure 8.17)**. During development, the tissue that would normally become the body of the atlas fuses with the body of the axis, (and separates from the atlas) forming the *dens* (odontoid process) of the axis. This modification allows the atlas to rotate around the axis.

3. Instead of laminae and pedicles, the atlas has an **anterior arch** and **posterior arch**. Notice the **articular facet for the dens** on the inner surface of the anterior arch. Also note that instead of a spinous process there is a smaller **posterior tubercle** on the posterior arch.

4. Observe the **superior articular facets** of the atlas. These facets are oriented horizontally in the atlas, rather than vertically as with the other vertebrae. These facets articulate with the **occipital condyles.** Observe the occipital bone and atlas on an articulated skeleton to see how these structures fit together to form the **atlanto-occipital joint.** This joint allows flexion and extension movements of the neck—as when we nod our heads to indicate "yes."

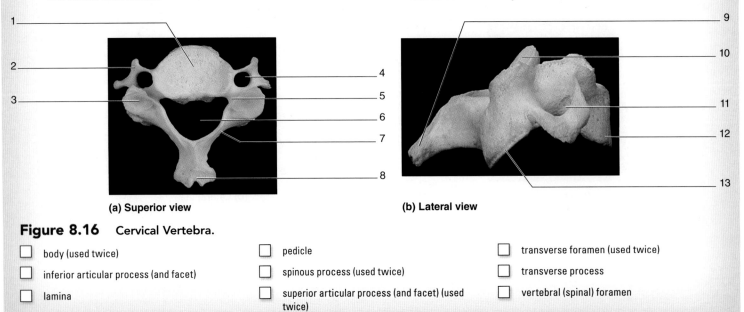

(a) Superior view (b) Lateral view

Figure 8.16 Cervical Vertebra.

- ☐ body (used twice)
- ☐ inferior articular process (and facet)
- ☐ lamina
- ☐ pedicle
- ☐ spinous process (used twice)
- ☐ superior articular process (and facet) (used twice)
- ☐ transverse foramen (used twice)
- ☐ transverse process
- ☐ vertebral (spinal) foramen

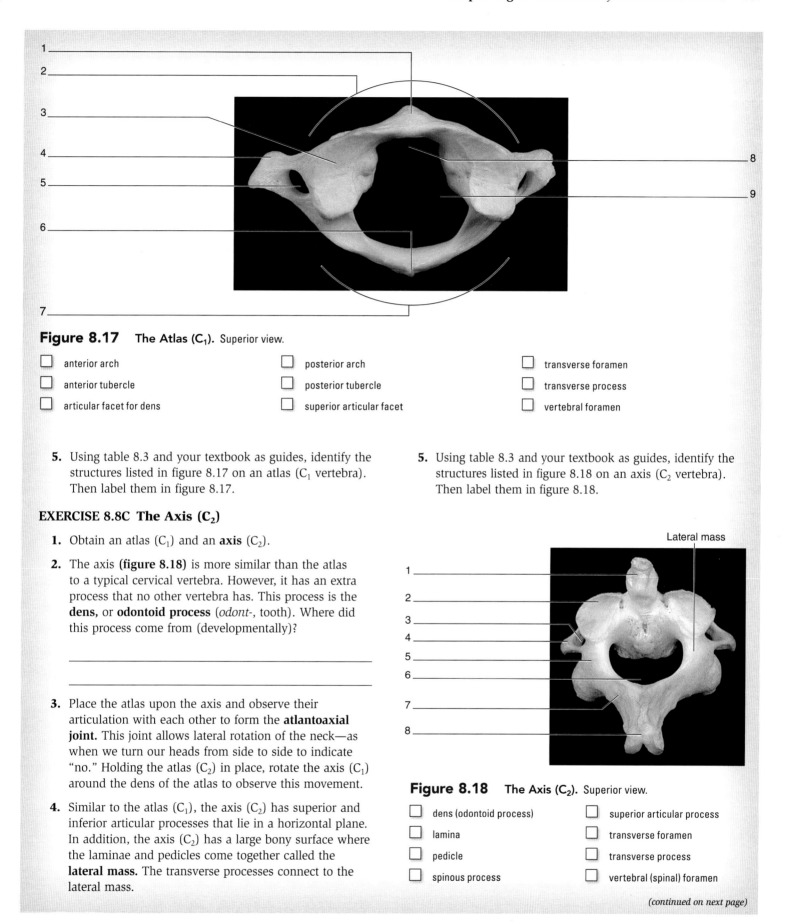

Figure 8.17 The Atlas (C₁). Superior view.

☐ anterior arch ☐ posterior arch ☐ transverse foramen
☐ anterior tubercle ☐ posterior tubercle ☐ transverse process
☐ articular facet for dens ☐ superior articular facet ☐ vertebral foramen

5. Using table 8.3 and your textbook as guides, identify the structures listed in figure 8.17 on an atlas (C₁ vertebra). Then label them in figure 8.17.

EXERCISE 8.8C The Axis (C₂)

1. Obtain an atlas (C₁) and an **axis** (C₂).

2. The axis **(figure 8.18)** is more similar than the atlas to a typical cervical vertebra. However, it has an extra process that no other vertebra has. This process is the **dens,** or **odontoid process** (*odont-*, tooth). Where did this process come from (developmentally)?

_____

_____

3. Place the atlas upon the axis and observe their articulation with each other to form the **atlantoaxial joint.** This joint allows lateral rotation of the neck—as when we turn our heads from side to side to indicate "no." Holding the atlas (C₂) in place, rotate the axis (C₁) around the dens of the atlas to observe this movement.

4. Similar to the atlas (C₁), the axis (C₂) has superior and inferior articular processes that lie in a horizontal plane. In addition, the axis (C₂) has a large bony surface where the laminae and pedicles come together called the **lateral mass.** The transverse processes connect to the lateral mass.

5. Using table 8.3 and your textbook as guides, identify the structures listed in figure 8.18 on an axis (C₂ vertebra). Then label them in figure 8.18.

Figure 8.18 The Axis (C₂). Superior view.

☐ dens (odontoid process) ☐ superior articular process
☐ lamina ☐ transverse foramen
☐ pedicle ☐ transverse process
☐ spinous process ☐ vertebral (spinal) foramen

(continued on next page)

188 Chapter Eight *The Skeletal System: Axial Skeleton*

(continued from previous page)

EXERCISE 8.8D Thoracic Vertebrae

1. Obtain a **thoracic vertebra (figure 8.19)**. Thoracic vertebrae are the only vertebrae that articulate with the ribs. Thus, these vertebrae have special articular surfaces (facets) in locations where the ribs and vertebrae meet and form joints.

2. Observe the thoracic vertebra from a superior view. Thoracic vertebrae typically have a heart-shaped body (medium in size), a round vertebral foramen, a spinous process that projects inferiorly, and superior and inferior articular processes with surfaces that lie in the frontal plane.

3. Look at the relationship between the ribs and vertebrae on an articulated skeleton. Notice that the **tubercle** of a rib articulates with the transverse process of a thoracic vertebra. Notice also that the head of the rib articulates at the junction between two vertebral bodies. Thus, it articulates with the **inferior costal facet** of the vertebra superior to it and the **superior costal facet** of the vertebra inferior to it.

4. Using table 8.3 and your textbook as guides, identify the structures listed in figure 8.19 on a thoracic vertebra. Then label them in figure 8.19.

EXERCISE 8.8E Lumbar Vertebrae

1. Obtain a **lumbar vertebra (figure 8.20)**. Lumbar vertebrae have very large, round or oval bodies, small vertebral foramina, and a short and blunt spinous process that projects posteriorly. The superior and inferior articular processes have facets that face medial and lateral, respectively.

2. Using table 8.3 and your textbook as guides, identify the structures listed in figure 8.20 on a lumbar vertebra. Then label them in figure 8.20.

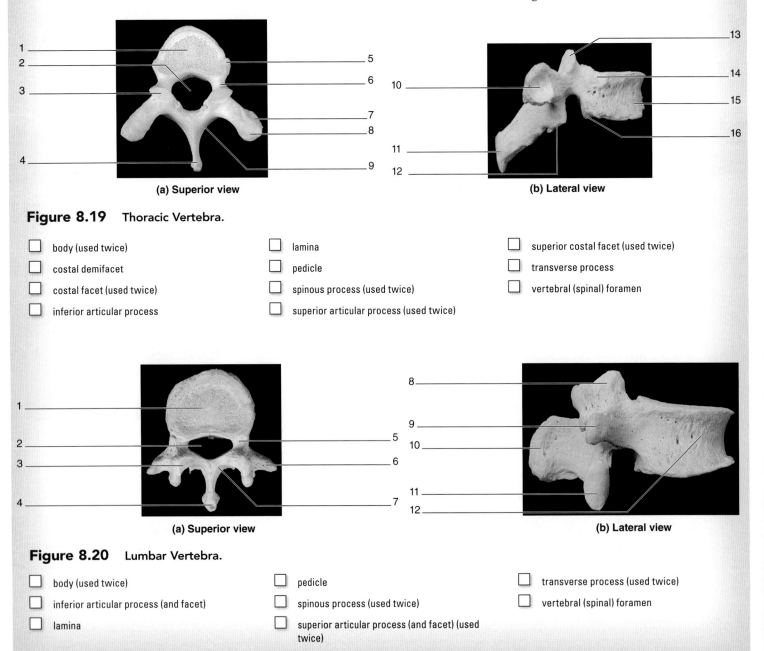

Figure 8.19 Thoracic Vertebra.

- ☐ body (used twice)
- ☐ costal demifacet
- ☐ costal facet (used twice)
- ☐ inferior articular process
- ☐ lamina
- ☐ pedicle
- ☐ spinous process (used twice)
- ☐ superior articular process (used twice)
- ☐ superior costal facet (used twice)
- ☐ transverse process
- ☐ vertebral (spinal) foramen

Figure 8.20 Lumbar Vertebra.

- ☐ body (used twice)
- ☐ inferior articular process (and facet)
- ☐ lamina
- ☐ pedicle
- ☐ spinous process (used twice)
- ☐ superior articular process (and facet) (used twice)
- ☐ transverse process (used twice)
- ☐ vertebral (spinal) foramen

EXERCISE 8.8F The Sacrum and Coccyx

1. Obtain a sacrum and coccyx **(figure 8.21),** or observe them on an articulated skeleton.

2. The **sacrum** (*sacr-*, sacred) forms by the fusion of five primitive vertebrae that subsequently form a single bony structure. As you identify the features of the sacrum, one of your goals is to recognize the parts of a typical vertebra within the sacrum.

3. The coccyx is usually composed of three to five small bones. The vertebrae have only two prominent structures, the *cornu* (*cornu*, horn) and the *transverse processes*.

4. Using table 8.3 and your textbook as guides, identify the structures listed in figure 8.21 on a sacrum and coccyx. Then label them in figure 8.21.

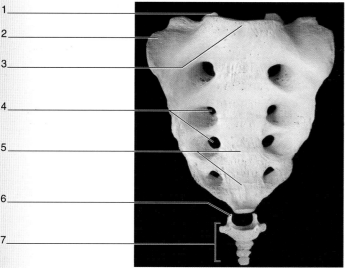
(a) Anterior view

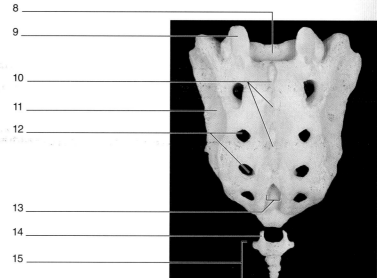
(b) Posterior view

Figure 8.21 Sacrum and Coccyx.

- [ ] ala
- [ ] anterior sacral foramina
- [ ] auricular surface
- [ ] coccygeal cornu
- [ ] coccyx
- [ ] median sacral crest
- [ ] posterior sacral foramina
- [ ] sacral canal
- [ ] sacral hiatus
- [ ] sacral promontory
- [ ] superior articular facet
- [ ] superior articular process
- [ ] transverse ridges

INTEGRATE

CLINICAL VIEW
Spina Bifida

Abnormalities in the developing skeleton can significantly impact structure and function. For example, spina bifida (*spina-*, backbone + *-bi*, two + *-fid*, to split) is a condition whereby the neural tube fails to develop properly in utero. Often, this is also associated with the failure of one or more vertebral laminae to fuse. The condition varies in severity, from cases where patients are unaware that they even have spina bifida to cases where patients suffer significant disabilities due to involvement of the spinal cord. In the most severe cases, patients present with a protruding, fluid-filled sac containing critical structures such as the spinal cord. This may occur in any region of the vertebral column, although it most commonly occurs in the lumbar or sacral regions of the vertebral column. This often results in significant motor and sensory deficits below the site of protrusion. In addition, patients typically suffer from hydrocephalus (*hydro-*, water + *-cephal*, head) due to an interruption in the flow of cerebrospinal fluid. Treatment of severe cases of spina bifida involves closure of the protruding structures and insertion of a shunt to drain excess cerebrospinal fluid. The causes of spina bifida remain unknown. However, recent research suggests that increased intake of folic acid during pregnancy reduces the incidence of such neural tube defects.

The Thoracic Cage

The **thoracic cage** consists of the **sternum, ribs,** and **thoracic vertebrae.** Its main function is to protect vital organs such as the heart and lungs. However, the bones of the thoracic cage also serve as important attachment sites for muscles involved with respiratory movements and muscles involved with movements of the back, chest, and neck. **Table 8.4** summarizes the key features of the sternum and ribs. Refer to table 8.3 for descriptions of the features of the thoracic vertebrae, which articulate with the ribs.

Table 8.4	The Axial Skeleton: Sternum and Ribs	
Bone	**Bone Features**	**Description and Related Structures of Importance**
STERNUM		
Manubrium	Clavicular notch	Point of articulation with the clavicle
	Suprasternal notch	Depression at the superior border
	Notch for rib 1	Location of articulation with costal cartilage of rib 1
	Sternal angle	Joint between the manubrium and the body; point of articulation with the costal cartilage of rib 2
Body	Notches for ribs 2–7	Point of articulation for the costal cartilages of ribs 2–7. The notch for rib 2 is a partial notch.
	Xiphisternal joint	Joint between the body and the xiphoid; point of articulation with the superior part of the costal cartilage of rib 7
Xiphoid	Partial notch for rib 7	Point of articulation for the inferior part of the costal cartilage of rib 7
RIBS		
Typical Rib	Head	The part of a rib that articulates with the bodies of the thoracic vertebrae
	Superior articular facet	A facet on the head of a rib that articulates with the inferior costal facet on the body of the vertebra that lies one level above it (i.e., superior articular facet of rib 6 with T_5)
	Inferior articular facet	A facet on the head of a rib that articulates with the superior costal facet on the body of the numerically equivalent thoracic vertebra (i.e., inferior articular facet of rib 6 to T_6)
	Shaft	The main part (body) of a rib, which begins at the angle of the rib and projects anteriorly
	Neck	A narrow region where the head meets the tubercle of the rib
	Tubercle	A projection at the junction between the shaft and neck; contains a facet for articulation with the transverse process of a thoracic vertebra
	Angle	The location where the rib curves anteriorly
	Costal groove	A groove on the inferior, deep border of the shaft; contains the intercostal artery, vein, and nerve
	Cup	The point of articulation for a costal cartilage
First Rib	Scalene tubercle	Attachment for the anterior scalene muscle
	Groove for subclavian artery	A depression indicating the location where the subclavian artery passes out of the thoracic cavity
	Groove for subclavian vein	A depression indicating the location where the subclavian vein passes into the thoracic cavity
	Articular facet	A singular facet on the head of the rib (a typical rib has two facets)
Second Rib	All markings of a typical rib	Unique features of rib 2 are a rough tuberosity and a shallow costal groove
11th and 12th Ribs	Articular facet	A singular facet on the head of the rib (a typical rib has two facets)
	Tubercle	Absent
	Neck	Absent

EXERCISE 8.9

THE STERNUM

1. Observe the thoracic cage on an articulated skeleton and locate the sternum **(figure 8.22)**. The sternum has three sections: the **manubrium,** the **body,** and the **xiphoid process** (*xiphos*, sword). The depression on the superior part of the manubrium is the **suprasternal notch.**

2. Palpate the sternal notch on your own body. Keeping your fingers on your manubrium, move your fingers inferiorly until you feel a rough ridge. This is the **sternal angle.** The sternal angle is located where the manubrium meets the body of the sternum. It is an important clinical landmark, because this is where the second rib articulates with the sternum.

3. Using table 8.4 and your textbook as guides, identify the structures listed in figure 8.22 on a sternum. Then label them in figure 8.22.

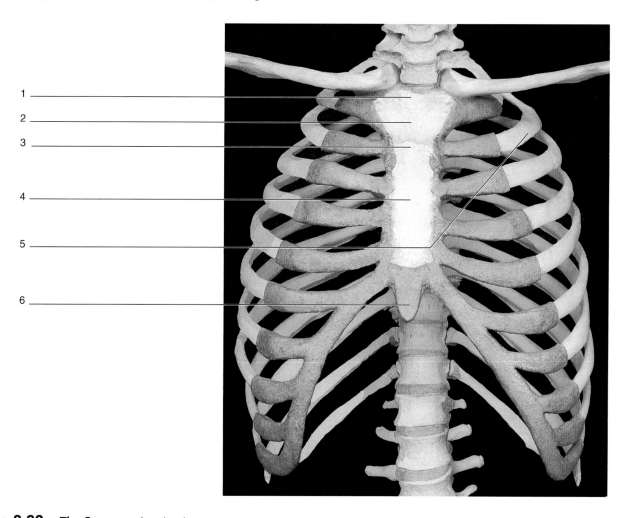

Figure 8.22 **The Sternum.** Anterior view.

☐ body ☐ second rib ☐ suprasternal notch
☐ manubrium ☐ sternal angle ☐ xiphoid process

EXERCISE 8.10

THE RIBS

EXERCISE 8.10 The Ribs

1. There are twelve pairs of **ribs**, one pair for each thoracic vertebra. Ribs 1–7 are called **true ribs** because each true rib connects individually to the sternum by separate cartlaginous extensions called costal cartilages. Ribs 8–12 are called **false ribs** because their costal cartilages do not attach directly to the sternum. The two pairs of false ribs (ribs 11 and 12) are called **floating ribs** because they have no connection to the sternum. Obtain a typical rib (any rib other than ribs 1, 2, 11, or 12) **(figure 8.23)**.

2. As you observe the features of a typical rib, pay particular attention to the surfaces of the rib that articulate with the thoracic vertebrae.

3. Observe the articulations between the ribs and the thoracic vertebrae on an articulated skeleton, and review the unique features of thoracic vertebrae that allow them to form articulations with the ribs.

4. Using table 8.4 and your textbook as guides, identify the structures listed in figure 8.23 on a typical rib. Then label them in figure 8.23.

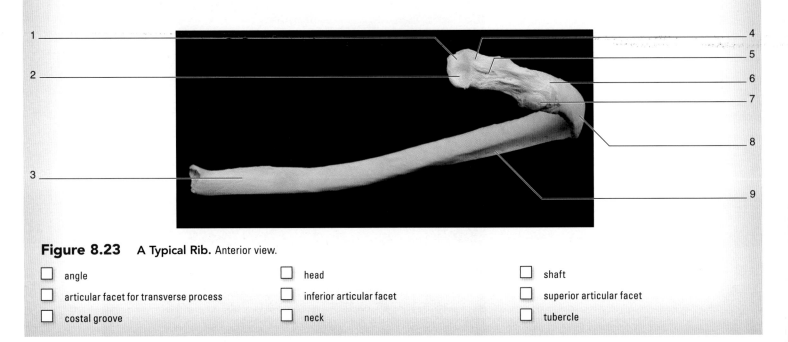

Figure 8.23 A Typical Rib. Anterior view.

- ☐ angle
- ☐ articular facet for transverse process
- ☐ costal groove
- ☐ head
- ☐ inferior articular facet
- ☐ neck
- ☐ shaft
- ☐ superior articular facet
- ☐ tubercle

Chapter 8: The Skeletal System: Axial Skeleton

Name: _____
Date: _____ Section: _____

POST-LABORATORY WORKSHEET

The ❶ corresponds to the Learning Objective(s) listed in the chapter opener outline.

Do You Know the Basics?

Exercise 8.1: Anterior View of the Skull

1. The bony structure that forms the cheek is called the ❶ _____.

2. List all of the major bones that form the orbit of the eye. ❷
   _____
   _____
   _____

3. Name the two bones that form the nasal septum. ❸
   _____

4. The _____ process of the mandible articulates with the temporal fossa. ❹

Exercise 8.2: Additional Views of the Skull

5. a. What does the term *petrous* mean? ❺ _____

   b. Why is this term used to describe the petrous part of the temporal bone?
   _____
   _____

   c. What structures are found deep within the petrous part of the temporal bone?
   _____
   _____

6. The zygomatic arch is composed of the _____ and the _____ . ❻

Exercise 8.3: Superior View of the Cranial Floor

7. Describe the three cranial fossae. Include a description of the major bony structures used to separate the fossae from each other. ❼
   _____
   _____
   _____
   _____
   _____
   _____
   _____

8. a. To what does the term *sella* turcica refer? ❽
   _____
   _____

   b. What endocrine gland is found within the sella turcica in a living human?
   _____
   _____
   _____

9. a. What does the term *lacero* mean? ⑧

   b. Why do you think the term *lacero* is used to describe the foramen lacerum?

   c. What is unique about the foramen lacerum (as compared to other cranial foramina)?

Exercise 8.4: Bones Associated with the Skull

10. The only bone of the axial skeleton that does not articulate with any other bone is the _____. ⑨

Exercise 8.5: The Fetal Skull

11. a. What is a fontanel? ⑩

    b. By what age do most of the fontanels completely close? _____

Exercise 8.6: Vertebral Column Regions and Curvatures

12. The primary curvatures of the vertebral column are the _____ and _____ curvatures. The secondary curvatures of the vertebral column are the _____ and _____ curvatures. ⑪

Exercise 8.7: Structure of a Typical Vertebra

13. The vertebral arch is composed of two main parts, the _____, which connect transverse and spinous processes, and the _____, which connect transverse processes to the body of the vertebra. ⑫

Exercise 8.8: Characteristics of Individual Vertebrae

14. a. What foramen is present in cervical vertebrae, but is not present in other vertebrae? ⑬

    b. What structure runs through this foramen in a living human?

15. How do the superior and inferior articular processes of the atlas differ from the same processes on a "typical" vertebra? ⑭

16. a. What is the dens (odontoid process)? ⑭

    b. From what structure did the dens arise developmentally?

17. What are the two locations on a thoracic vertebra where the ribs articulate? ⑮

    a. _____
    b. _____

18. The anterior and posterior sacral foramina are the equivalent of the _____ foramina in other regions of the vertebral column. ⑯

Exercise 8.9: The Sternum

19. a. The three sets of bones that compose the thoracic cage are _____, _____, and _____.

    b. The sternal angle is the point of articulation between the _____ and the _____ of the sternum.

20. What is the clinical significance of the sternal angle?

_____

_____

Exercise 8.10: The Ribs

21. What structures run in the costal groove?

_____

_____

Can You Apply What You've Learned?

22. What is a functional consequence of the shape (and arrangement) of the superior and inferior articular processes of the lumbar vertebrae? (Hint: Put two of them together and see what movement is, or is not, allowed.) _____

_____

23. The optic nerve extends from the eye toward the brain by traveling through the optic foramen. The optic foramen is a hole in what bone? _____

_____

24. Match the structure described in column A with the bone it is associated with listed in column B (some answers are used more than once).

    Column A
    ___ 1. mastoid process and styloid process
    ___ 2. external auditory canal
    ___ 3. foramen magnum
    ___ 4. skull bone that articulates with atlas (C_1)
    ___ 5. sella turcica, greater/lesser wings, pterygoid plates
    ___ 6. crista galli, perpendicular plate, and cribriform plate
    ___ 7. most superior skull bone
    ___ 8. major bone forming anterior cranial fossa
    ___ 9. major bone forming posterior cranial fossa
    ___ 10. the only bone of the body that does not articulate with another bone

    Column B
    a. ethmoid
    b. frontal
    c. hyoid
    d. occipital
    e. parietal
    f. sphenoid
    g. temporal

25. Since there are 12 ribs (in both the male and female) and each rib articulates with a thoracic vertebra, there are _____ thoracic vertebrae. These vertebrae are designated _____ to _____. Superior to the thoracic vertebrae there are _____ cervical vertebrae, designated _____ to _____. Inferior to the thoracic vertebrae there are _____ lumbar vertebrae, designated _____ to _____. Inferior to the lumbar vertebrae is the sacrum formed by the fusion of _____ vertebrae. The coccyx is the most inferior of the vertebrae formed by the fusion of _____ vertebrae. The total number of vertebrae in the vertebral column (after fusion) is _____.

Can You Synthesize What You've Learned?

26. Explain the difference between the vertebral foramen and the intervertebral foramen.

_____

_____

196 Chapter Eight *The Skeletal System: Axial Skeleton*

27. Examine the skeleton and the different types of vertebrae to answer these questions.
    a. In the photos below, circle the structures that differentiate cervical, thoracic, and lumbar vertebrae on the photo of each vertebra. Then describe the feature(s) in the numbered space(s) below each figure.

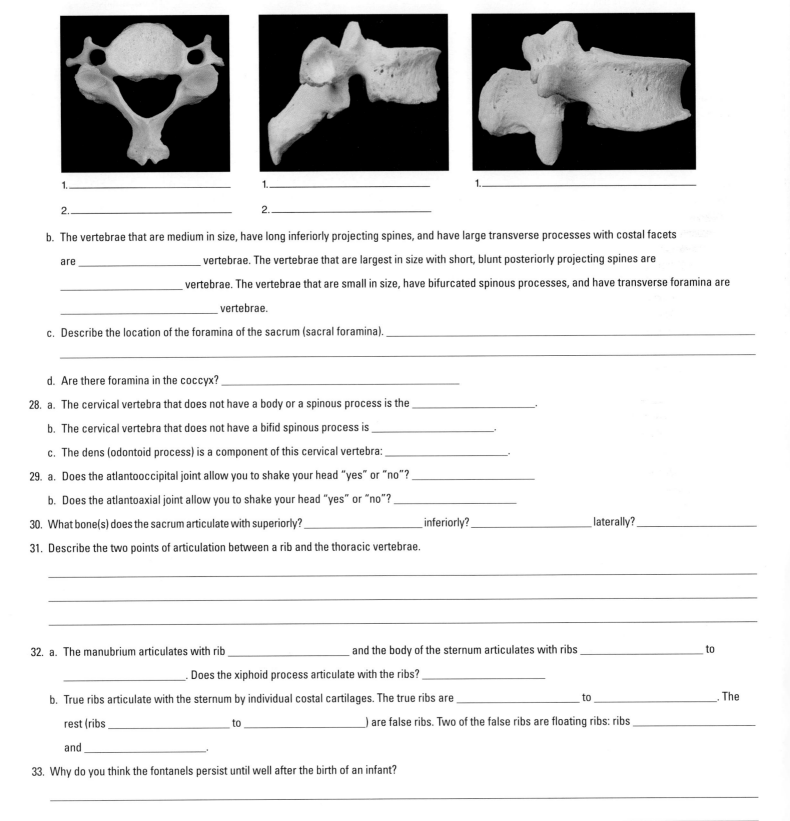

1. _____   1. _____   1. _____
2. _____   2. _____

   b. The vertebrae that are medium in size, have long inferiorly projecting spines, and have large transverse processes with costal facets are _____ vertebrae. The vertebrae that are largest in size with short, blunt posteriorly projecting spines are _____ vertebrae. The vertebrae that are small in size, have bifurcated spinous processes, and have transverse foramina are _____ vertebrae.

   c. Describe the location of the foramina of the sacrum (sacral foramina). _____
   _____

   d. Are there foramina in the coccyx? _____

28. a. The cervical vertebra that does not have a body or a spinous process is the _____.
    b. The cervical vertebra that does not have a bifid spinous process is _____.
    c. The dens (odontoid process) is a component of this cervical vertebra: _____.

29. a. Does the atlantooccipital joint allow you to shake your head "yes" or "no"? _____
    b. Does the atlantoaxial joint allow you to shake your head "yes" or "no"? _____

30. What bone(s) does the sacrum articulate with superiorly? _____ inferiorly? _____ laterally? _____

31. Describe the two points of articulation between a rib and the thoracic vertebrae.
    _____
    _____
    _____

32. a. The manubrium articulates with rib _____ and the body of the sternum articulates with ribs _____ to _____. Does the xiphoid process articulate with the ribs? _____

    b. True ribs articulate with the sternum by individual costal cartilages. The true ribs are _____ to _____. The rest (ribs _____ to _____) are false ribs. Two of the false ribs are floating ribs: ribs _____ and _____.

33. Why do you think the fontanels persist until well after the birth of an infant?
    _____
    _____
    _____
    _____

34. How might the bifid spinous processes of cervical vertebrae affect anterior-posterior movement in the cervical region of the vertebral column?

35. How do the superior and inferior articular processes of the atlas differ from the same processes on a "typical" vertebra. How does this difference contribute to the special movement allowed at the atlanto-occipital and atlantoaxial joints?

36. Notice that when you put two lumbar vertebrae together, little to no lateral rotation is allowed because of the shape of the articulating bones. Why do you think they interact in this manner?

37. What part of the vertebral column is removed when a laminectomy is performed?

38. Needles inserted into the thoracic cavity must always be placed along the superior border of a rib so as not to injure important structures. What important structures could be damaged by insertion of a needle too close to the inferior border of a rib? (Hint: Refer to table 8.4.)

The Skeletal System: Appendicular Skeleton

CHAPTER 9

INTRODUCTION

The term **appendicular** comes from the term *appendage*. A dictionary might define an appendage as *something that is added or attached to an item that is larger or more important, as an adjunct*. While our appendages may be described as simply "added" or "adjunct" structures, most of us would consider our arms and legs to be essential. In chapter 8 you looked at the bones that constitute the main structural support of the body: the bones of the axial skeleton. In this chapter you will study the bones composing the **pectoral girdle** and the **pelvic girdle,** which act as attachment points for muscles and ligaments that anchor the upper limbs and lower limbs to the axial skeleton, respectively. In addition, you will study the bones of the **upper limb** and the **lower limb,** which act as attachment points for muscles and ligaments that allow movement and dexterity of the limbs. The more successfully you accomplish the task of learning bony features in this chapter, the better prepared you will be to understand muscle actions when you get to chapter 12. Bony features such as tuberosities, trochanters, and tubercles exist because of the pulling action of ➡

OUTLINE AND LEARNING OBJECTIVES

Gross Anatomy 202

The Pectoral Girdle 202

EXERCISE 9.1: BONES OF THE PECTORAL GIRDLE 203
1. Identify the bony landmarks of the pectoral girdle
2. List the two bones that compose the pectoral girdle
3. Identify the joints of the pectoral girdle

The Upper Limb 205

EXERCISE 9.2: BONES OF THE UPPER LIMB 207
4. Identify the landmarks of upper limb bones, and relate them to muscular attachments
5. Associate the shapes of carpal bones with their respective names

EXERCISE 9.3: SURFACE ANATOMY REVIEW—PECTORAL GIRDLE AND UPPER LIMB 213
6. Palpate the prominent bony structures of the pectoral girdle and upper limb

The Pelvic Girdle 214

EXERCISE 9.4: BONES OF THE PELVIC GIRDLE 215
7. Identify the bony landmarks of the pelvic girdle
8. List the three bones that compose the pelvic girdle

The Lower Limb 218

EXERCISE 9.5: BONES OF THE LOWER LIMB 220
9. Identify the landmarks of lower limb bones, and relate them to muscular attachments
10. Associate the shapes of tarsal bones with their respective names
11. Compare and contrast the structure and function of tarsal and carpal bones

EXERCISE 9.6: SURFACE ANATOMY REVIEW—PELVIC GIRDLE AND LOWER LIMB 227
12. Palpate the prominent bony structures of the pelvic girdle and lower limb

MODULE 5: SKELETAL SYSTEM

muscles that attach to them and stress them during development. For example, without a sternocleidomastoid muscle pulling on the mastoid process of the temporal bone, a mastoid process would not exist! As you observe the features of the bones, remember that all bony features tell a story about the development and history of the entire musculoskeletal system.

As you observe the different bones and their bony features, refer to the tables in this chapter for derivatives of the names of the features. If you understand the word origins, it will be easier for you to connect a structure with its name. For example, the *conoid tubercle* of the clavicle gets its name from its conical shape (*konoeides*, cone-shaped). The *coracoid process* of the scapula is named for its resemblance to a crow's beak (*karakodes*, like a crow's beak). The two names look very similar and can be easily confused if you do not pay close attention to their meanings.

In this laboratory session, you will study the bones that compose the appendicular skeleton on an articulated human skeleton, on disarticulated bones of the human skeleton, or on bone models. You will identify the major landmarks and identifying features of each bone, and will associate the observed structures with their functions.

Chapter 9: The Skeletal System: Appendicular Skeleton

Name: _____

Date: _____ Section: _____

PRE-LABORATORY WORKSHEET

Also available at www.connect.mcgraw-hill.com

1. List the two bones that compose the pectoral girdle.

    a. _____

    b. _____

2. List the three bones that make up the pelvic girdle.

    a. _____

    b. _____

    c. _____

3. In the anatomic position, the radius lies <u>medial/lateral</u> to the ulna. (Circle the correct answer.)

4. In the anatomic position, the tibia lies <u>medial/lateral</u> to the fibula. (Circle the correct answer.)

5. The carpal bones are located in the _____.

6. Tarsal bones are located in the _____.

7. Match the description given in column A with the appropriate bone listed in column B.

    Column A

    ____ 1. a bone that has two large tubercles on its proximal end

    ____ 2. a bone that has two large trochanters on its proximal end

    ____ 3. a bone that has an olecranon process

    ____ 4. a bone found in the wrist

    ____ 5. a sesamoid bone found in the knee

    ____ 6. the largest bone in the leg

    Column B

    a. tibia
    b. ulna
    c. humerus
    d. femur
    e. patella
    f. carpal

8. In what region of the body is the calcaneus located? _____

9. A bone that has both an acromial and a coracoid process is the _____.

10. In what region of the body is the pisiform bone located? _____

Gross Anatomy

The Pectoral Girdle

The **pectoral girdle** (*girdle*, a belt) consists of the paired clavicles and scapulae. The function of these bones is to act as the bony support for muscles that attach the upper limb to the axial skeleton and to attach each upper limb to the axial skeleton by one bony joint, the sternoclavicular joint. **Table 9.1** lists the bones of the pectoral girdle and describes their key features.

Table 9.1	Appendicular Skeleton: Pectoral Girdle		
Bone	**Bony Landmark**	**Description**	**Word Origin**
Clavicle *clavicula, a small key*	Acromial end	The lateral end of the bone, which is flattened horizontally	*akron*, tip, + *-omos*, shoulder
	Conoid tubercle	A small "cone-shaped" tubercle on the lateral, inferior end of the bone	*konoeides*, cone-shaped
	Costal tuberosity	A rough impression on the inferior surface of the sternal end of the bone that serves as the attachment point for the costoclavicular ligament	*costa*, rib
	Sternal end	The medial end of the bone, which is triangular in shape	*sternon*, chest
Scapula *scapula, the shoulder blade*	Acromion	The large process at the lateral tip of the scapular spine, which projects laterally and slightly anteriorly; articulates with acromial end of clavicle	*akron*, tip, + *-omos*, shoulder
	Coracoid process	The smaller of the two major scapular processes, which projects anteriorly; attachment site for muscles	*korakodes*, like a crow's beak
	Glenoid fossa	A shallow depression that forms the articulation between the scapula and the humerus	*glenoeides*, resembling a socket
	Inferior angle	The angle between the medial and lateral borders	*inferior*, lower
	Infraglenoid tubercle	A rough projection at the inferior border of the glenoid fossa; attachment point for the long head of the triceps brachii muscle	*infra*, below, + *glenoeides*, resembling a socket
	Infraspinous fossa	A large depression inferior to the scapular spine; origin for the infraspinatus and teres minor muscles	*infra*, below, + *spina*, spine
	Lateral (axillary) border	The border of the scapula that has the glenoid fossa on its superior part	*axilla*, armpit
	Medial (vertebral) border	The longest border of the scapula; contains very few notable features	*medialis*, middle
	Spine	A long "spiny" process on the posterior surface; attachment point for trapezius and deltoid muscles	*spina*, spine
	Subscapular fossa	A large depression on the anterior surface of the bone; origin of the subscapularis muscle	*sub*, under, + *spina*, spine
	Superior angle	The angle between the superior and medial borders	*superus*, above
	Superior border	The border from which the coracoid and acromial processes project	*superus*, above
	Supraglenoid tubercle	A rough projection at the superior border of the glenoid fossa; attachment point for the long head of the biceps brachii muscle	*supra*, on the upper side, + *glenoeides*, resembling a socket
	Suprascapular notch	A small, deep notch just medial to the coracoid process; the suprascapular nerve, artery, and vein pass through the notch	*supra*, on the upper side, + *scapula*, shoulder blade
	Supraspinous fossa	A large depression superior to the scapular spine; origin of the supraspinatus muscle	*supra*, on the upper side, + *spina*, spine

Chapter Nine The Skeletal System: Appendicular Skeleton   203

EXERCISE 9.1

BONES OF THE PECTORAL GIRDLE

EXERCISE 9.1A The Clavicle

1. Obtain a clavicle or observe the clavicle on an articulated skeleton **(figure 9.1)**.

2. The **clavicle** forms part of the only articulation between the axial skeleton and the upper limb, at the **sternoclavicular joint.** The clavicle has very few muscular attachments, compared to other bones. Rather than acting as a rigid attachment point for muscles, the clavicle functions more like a strut that pushes the shoulders laterally and keeps them from collapsing anteriorly toward the sternum. The superficial location of the clavicle allows most parts of the bone to be easily palpated.

3. Palpate your own clavicle. As you do this, pay attention to the two ends of the clavicle and name the bones that articulate with each end of the clavicle.

4. Using table 9.1 and your textbook as guides, identify the structures listed in figure 9.1 on the clavicle. Then label them in figure 9.1. (Answers may be used more than once).

5. Pick up the clavicle and hold it in front of you. Notice its S shape. How can you tell if it is a right clavicle or a left clavicle?

_____

_____

EXERCISE 9.1B The Scapula

1. Obtain a scapula or observe the scapula on an articulated skeleton **(figure 9.2)**.

2. The **scapula** (*scapula*, shoulder blade) is a large, irregular bone that is not directly attached to the axial skeleton. Recall that the only location where the bones of the pectoral girdle attach to the axial skeleton is where the sternal end of the clavicle articulates with the manubrium of the sternum at the sternoclavicular joint.

3. Using table 9.1 and your textbook as guides, identify the structures listed in figure 9.2 on the scapula. Then label them in figure 9.2. (Answers may be used more than once.)

4. How can you tell whether the bone you are holding is a right scapula or a left scapula?

_____

_____

_____

5. Pick up the scapula and hold it in front of you. Find the glenoid fossa. With what structure does this fossa articulate (articulate = form a joint)?

_____

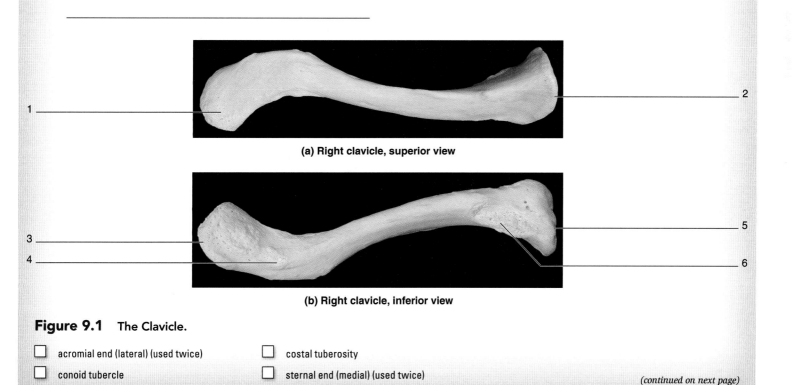

(a) Right clavicle, superior view

(b) Right clavicle, inferior view

Figure 9.1 The Clavicle.

☐ acromial end (lateral) (used twice)
☐ conoid tubercle
☐ costal tuberosity
☐ sternal end (medial) (used twice)

(continued on next page)

(continued from previous page)

1. _____
2. _____
3. _____
4. _____
5. _____

6. _____
7. _____
8. _____
9. _____
10. _____
11. _____

(a) Anterior view

12. _____
13. _____
14. _____
15. _____
16. _____

17. _____
18. _____
19. _____
20. _____
21. _____

(b) Lateral view

22. _____
23. _____
24. _____
25. _____
26. _____
27. _____
28. _____

29. _____
30. _____
31. _____
32. _____

(c) Posterior view

Figure 9.2 The Right Scapula.

- ☐ acromion (used three times)
- ☐ coracoid process (used three times)
- ☐ glenoid cavity (used three times)
- ☐ inferior angle (used three times)
- ☐ infraglenoid tubercle
- ☐ infraspinous fossa
- ☐ lateral (axillary) border (used three times)
- ☐ medial (vertebral) border (used twice)
- ☐ spine (used three times)
- ☐ subscapular fossa (used twice)
- ☐ superior angle (used twice)
- ☐ superior border (used twice)
- ☐ supraglenoid tubercle
- ☐ suprascapular notch (used twice)
- ☐ supraspinous fossa

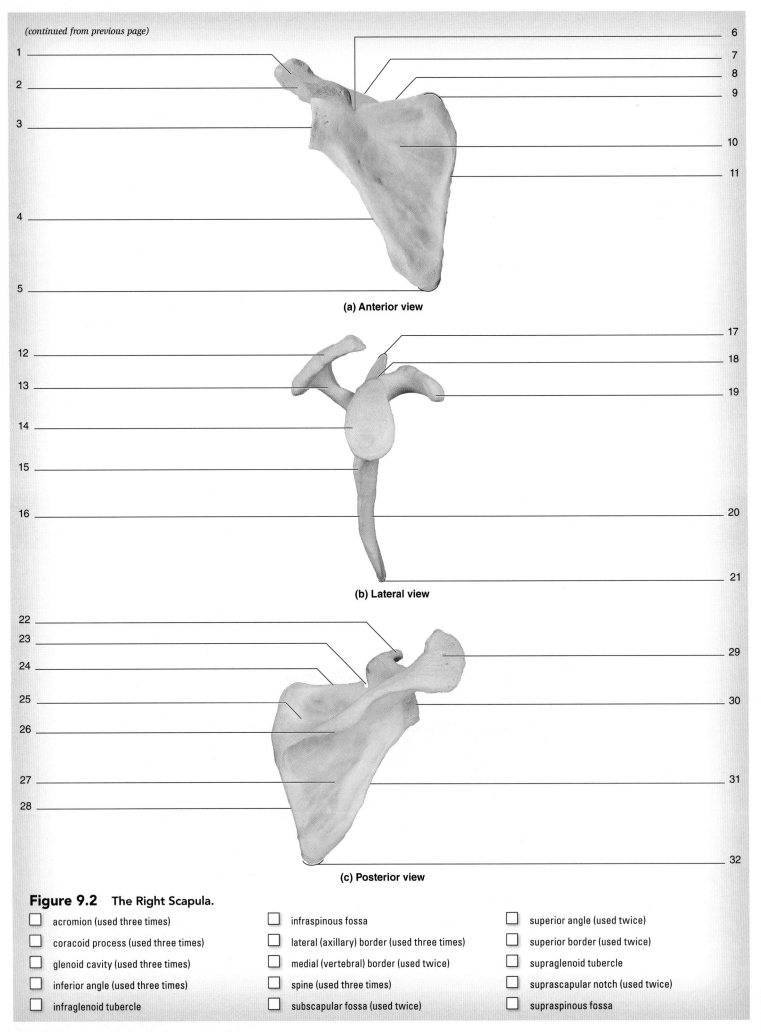

204

6. Now that you have observed both the clavicle and the scapula, obtain a right clavicle and right scapula and articulate them with each other. In the box to the right, sketch the relationship between the clavicle and the scapula.

7. What is the name of the joint formed between the clavicle and the scapula? _____

8. Palpate this joint on your own upper limb or your lab partner's upper limb. To do this, begin by palpating your clavicle. Then "walk" your fingers laterally until you reach the tip of your shoulder, where you will feel the acromion of the scapula. If you raise and lower your upper limb while keeping your fingers on the bones, you will be able to feel the joint between the clavicle and scapula.

The Upper Limb

The upper limb consists of the humerus, radius, ulna, carpals, metacarpals, and phalanges. Nearly all of the projections on these bones serve as attachment points for the muscles that move the upper limb. **Table 9.2** lists the bones of the upper limb and describes their key features.

Table 9.2 The Appendicular Skeleton: Upper Limb

Bone	Bony Landmark	Description	Word Origin
Humerus *humerus, shoulder*	Anatomical neck	The narrow part between the head and the tubercles; location of the former epiphyseal (growth) plate	NA
	Capitulum	The rounded surface (condyle) that articulates with the radius	*caput*, head
	Coronoid fossa	A depression for the coronoid process of the ulna	*corona*, crown, + *eidos*, resemblance, + *fossa*, trench
	Deltoid tuberosity	A rough projection on the proximal diaphysis	*delta*, triangle, + *eidos*, resemblance
	Greater tubercle	A large lateral projection on the proximal epiphysis	*tuber*, a knob
	Head	A rounded projection on the medial side of the proximal epiphysis; a facet that forms half of the glenohumeral joint	*head*, the rounded extremity of a bone
	Intertubercular sulcus (groove)	A groove between the greater and lesser tubercles; sometimes referred to as the bicipital groove	*inter-*, between, + *tuber*, knob
	Lateral epicondyle	A rough ridge proximal to the capitulum	*epi-*, above, + *kondylos*, knuckle
	Lesser tubercle	A small, rough projection on the anterior side of the proximal epiphysis	*tuber*, a knob
	Medial epicondyle	A rough ridge proximal to the trochlea	*epi-*, above, + *kondylos*, knuckle
	Olecranon fossa	A depression for the olecranon process of the ulna	*olecranon*, the head of the elbow, + *fossa*, trench
	Radial fossa	A depression for the head of the radius	*radius*, spoke of a wheel, + *fossa*, trench
	Supracondylar ridges	Sloped ridges (medial and lateral) proximal to the epicondyles	*supra-*, on the upper side, + *kondylos*, knuckle
	Surgical neck	The location where the head meets the shaft; the most common site of fracture	NA
	Trochlea	The rounded surface (condyle) that articulates with the ulna	*trochileia*, a pulley

(continued on next page)

Table 9.2 The Appendicular Skeleton: Upper Limb (continued)

Bone	Bony Landmark	Description	Word Origin
Ulna ulna, *elbow*	Coronoid process	A projection on the anterior surface of the ulna that articulates with the humerus	*corona*, crown, + *eidos*, resemblance
	Olecranon	A large projection on the proximal ulna, which forms the point of the elbow and serves as an attachment point for the triceps brachii muscle	*olecranon*, the head of the elbow
	Radial notch	A depression on the lateral, proximal surface of the ulna that articulates with the head of the radius	*radius*, spoke of a wheel
	Styloid process	A pointed process on the distal ulna that forms the medial aspect of the wrist	*stylos*, pillar, + *eidos*, resemblance
	Trochlear notch	A ridge on the middle of the anterior surface of the proximal epiphysis of the ulna that separates the two depressions that articulate with the condyles of the humerus	*trochileia*, a pulley
	Tuberosity of ulna	A process on the anterior surface of the proximal ulna that serves as an attachment point for the brachialis muscle	*ulna*, elbow
Radius radius, *spoke of a wheel*	Head	The disc-shaped proximal end of the radius	NA
	Neck	Location where the head of the radius meets the shaft of the bone	NA
	Radial tuberosity	A large projection on the medial surface distal to the proximal epiphysis of the bone that serves as an attachment point for the biceps brachii muscle	*radio-*, ray
	Styloid process of the radius	A small, pointed projection on the distal radius	*stylos*, pillar, + *eidos*, resemblance
Carpals carpus, *wrist* **Proximal row**	Scaphoid	A large, "boat-shaped" bone that articulates with the radius	*skaphe*, boat, + *eidos*, resemblance
	Lunate	A "moon-shaped" bone that articulates with the radius	*luna*, moon
	Triquetrum	A triangular bone on the medial, proximal aspect of the wrist; articulates with the ulna	*triquetrus*, three-cornered
	Pisiform	A "pea-shaped" bone on the medial, palmar surface of the wrist	*pisum*, pea, + *forma*, appearance
Distal row	Trapezium	A "table-shaped" bone that lies at the base of the first metacarpal (base of the thumb)	*trapezion*, a table
	Trapezoid	A "table-shaped" bone that lies at the base of the second metacarpal	*trapezion*, a table, + *eidos*, resemblance
	Capitate	A "head-shaped" bone that lies in the center of the wrist, at the base of the third metacarpal	*caput*, head
	Hamate	A "hook-shaped" bone that lies at the base of the fifth metacarpal	*hamus*, a hook
Metacarpals meta-, *after*, + carpus, *wrist*	Base	The proximal epiphysis of the metacarpal	NA
	Body	The diaphysis of the metacarpal	NA
	Head	The distal epiphysis of the metacarpal	NA
Phalanges	NA	Bones of the fingers and thumb	*phalanx*, line of soldiers
II through V	Proximal	The phalanx closest to the palm of the hand	*proximus*, nearest
	Middle	The middle phalanx	NA
	Distal	The small, cone-shaped distal bone of the digits	*distalis*, away
Pollex (I) pollex, *thumb*	Proximal	The phalanx closest to the palm of the hand	*proximus*, nearest
	Distal	The small, cone-shaped distal bone of the thumb	*distalis*, away

EXERCISE 9.2

BONES OF THE UPPER LIMB

EXERCISE 9.2A The Humerus

1. Obtain a **humerus** or observe the humerus on an articulated skeleton **(figure 9.3)**.

2. Using table 9.2 and your textbook as guides, identify the structures listed in figure 9.3 on the humerus, and then label them in figure 9.3. (Answers may be used more than once.)

3. How can you tell whether the bone you are holding is a right humerus or a left humerus?

   _____

   _____

   _____

4. How can you determine which is the proximal end of the humerus?

   _____

   _____

   _____

5. How can you distinguish the anterior surface of the humerus from the posterior surface?

   _____

   _____

   _____

6. Obtain a scapula and a humerus and articulate them with each other. In the box below, sketch the relationship between the scapula and the humerus.

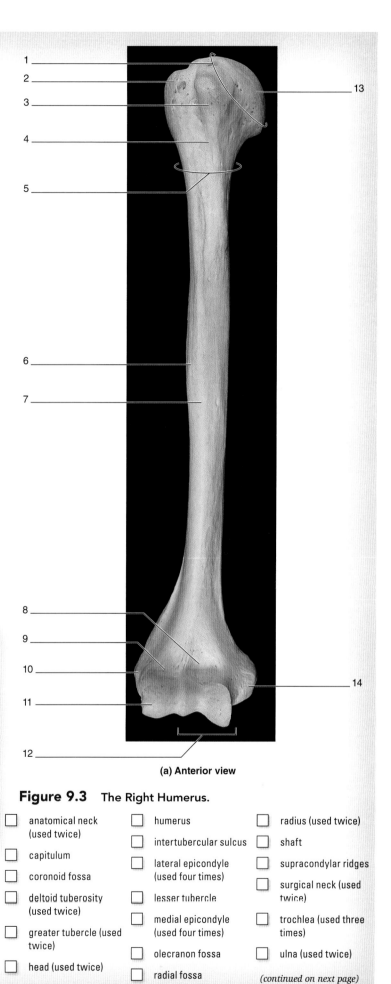

(a) Anterior view

Figure 9.3 The Right Humerus.

- ☐ anatomical neck (used twice)
- ☐ capitulum
- ☐ coronoid fossa
- ☐ deltoid tuberosity (used twice)
- ☐ greater tubercle (used twice)
- ☐ head (used twice)
- ☐ humerus
- ☐ intertubercular sulcus
- ☐ lateral epicondyle (used four times)
- ☐ lesser tubercle
- ☐ medial epicondyle (used four times)
- ☐ olecranon fossa
- ☐ radial fossa
- ☐ radius (used twice)
- ☐ shaft
- ☐ supracondylar ridges
- ☐ surgical neck (used twice)
- ☐ trochlea (used three times)
- ☐ ulna (used twice)

(continued on next page)

Chapter Nine *The Skeletal System: Appendicular Skeleton* 207

208 Chapter Nine The Skeletal System: Appendicular Skeleton

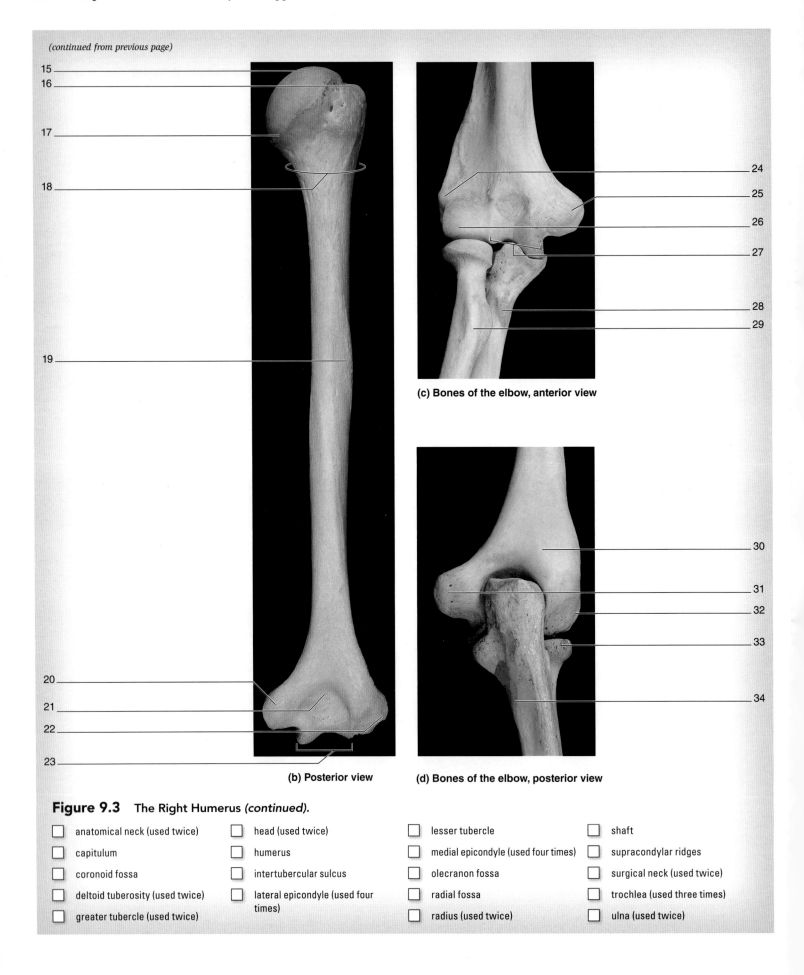

(b) Posterior view

(c) Bones of the elbow, anterior view

(d) Bones of the elbow, posterior view

Figure 9.3 The Right Humerus *(continued)*.

- [ ] anatomical neck (used twice)
- [ ] capitulum
- [ ] coronoid fossa
- [ ] deltoid tuberosity (used twice)
- [ ] greater tubercle (used twice)
- [ ] head (used twice)
- [ ] humerus
- [ ] intertubercular sulcus
- [ ] lateral epicondyle (used four times)
- [ ] lesser tubercle
- [ ] medial epicondyle (used four times)
- [ ] olecranon fossa
- [ ] radial fossa
- [ ] radius (used twice)
- [ ] shaft
- [ ] supracondylar ridges
- [ ] surgical neck (used twice)
- [ ] trochlea (used three times)
- [ ] ulna (used twice)

EXERCISE 9.2B The Radius

1. Obtain a **radius** or observe the radius on an articulated skeleton **(figure 9.4)**.

2. Using table 9.2 and your textbook as guides, identify the structures listed in figure 9.4 on the radius, and then label them in figure 9.4.

3. How can you tell whether the bone you are holding is a right radius or a left radius?

   _____
   _____
   _____

4. How can you determine which end of the radius is the proximal end?

   _____
   _____
   _____

5. How can you distinguish the anterior surface of the radius from the posterior surface?

   _____
   _____
   _____

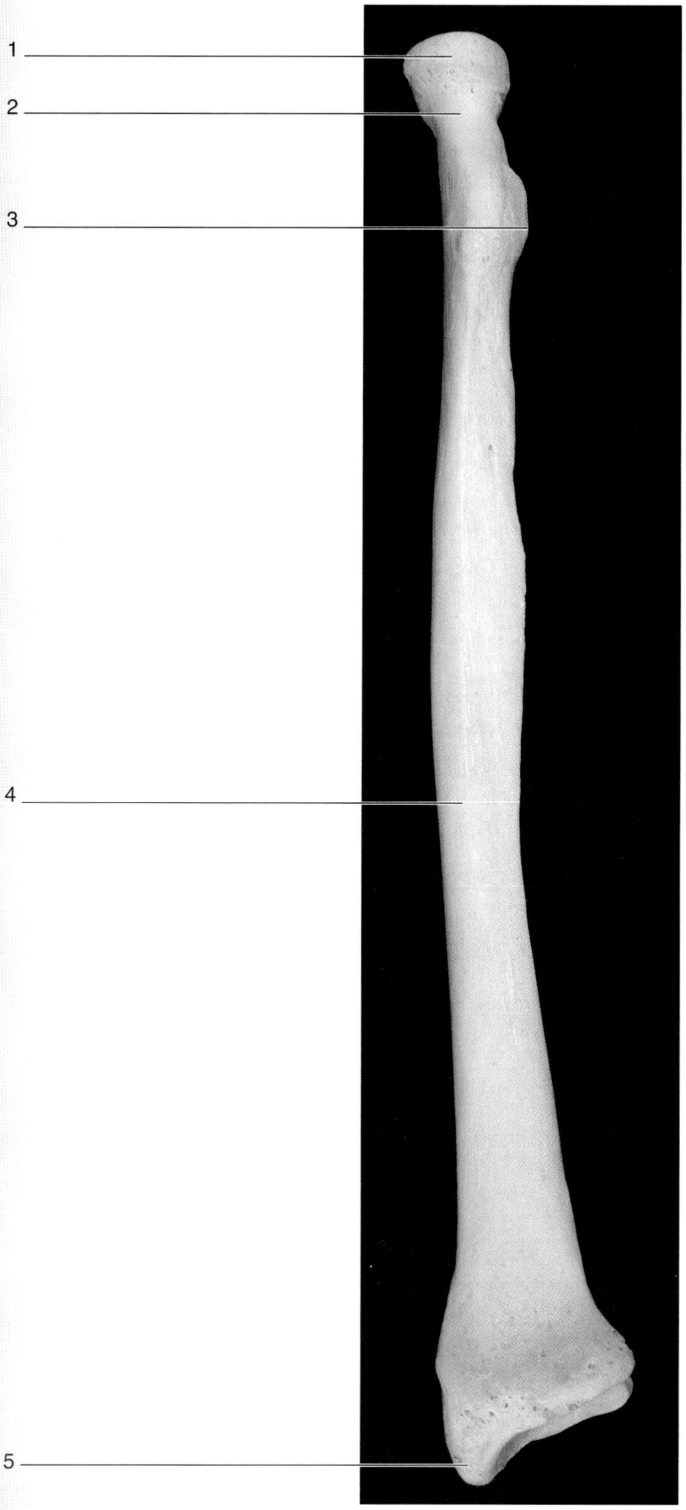

1 _____
2 _____
3 _____
4 _____
5 _____
6 _____

(a) Right radius, anterior view

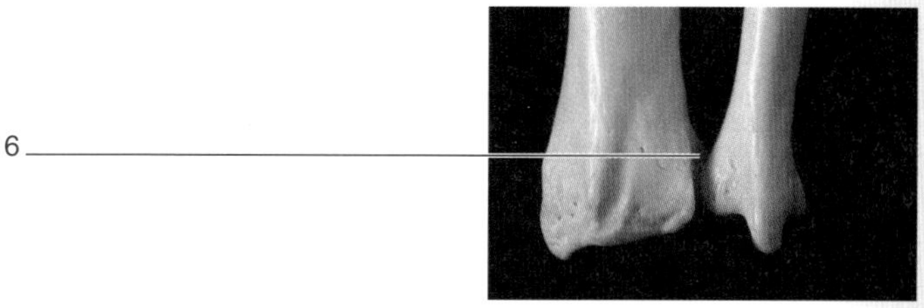

(b) Distal radius

☐ head   ☐ shaft
☐ neck   ☐ styloid process of radius
☐ radial tuberosity   ☐ ulnar notch

Figure 9.4 The Radius.

(continued on next page)

(continued from previous page)

EXERCISE 9.2C The Ulna

1. Obtain an **ulna** or observe the ulna on an articulated skeleton **(figure 9.5)**.

2. Using table 9.2 and your textbook as guides, identify the structures listed in figure 9.5 on the ulna, and then label them in figure 9.5.

3. How can you tell whether the bone you are holding is a right ulna or a left ulna?

___

4. How can you determine which end of the ulna is the proximal end?

___

5. How can you distinguish the anterior surface of the ulna from the posterior surface?

___

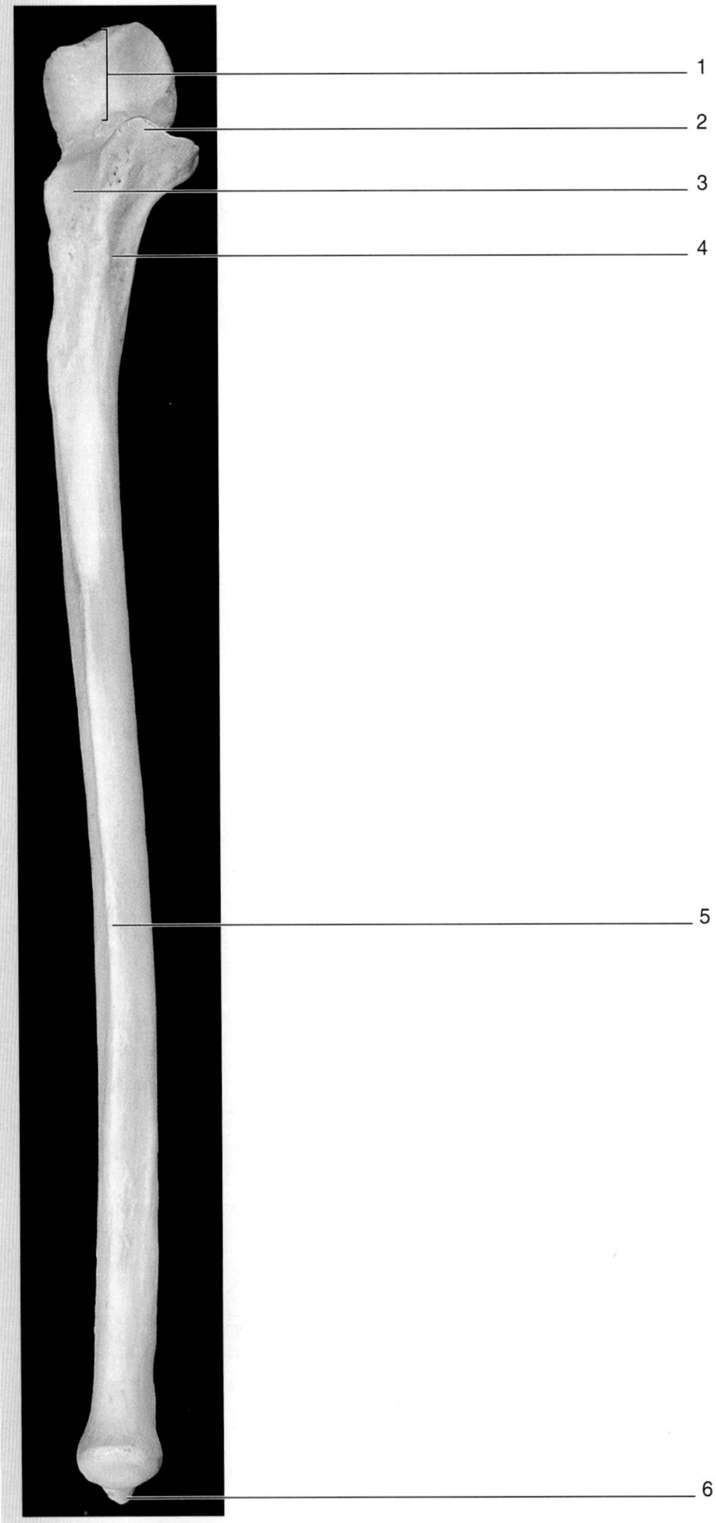

☐ coronoid process (used twice)   ☐ styloid process of ulna
☐ olecranon   ☐ trochlear notch
☐ radial notch   ☐ tuberosity of ulna
☐ shaft of ulna

(a) Right ulna, anterior view

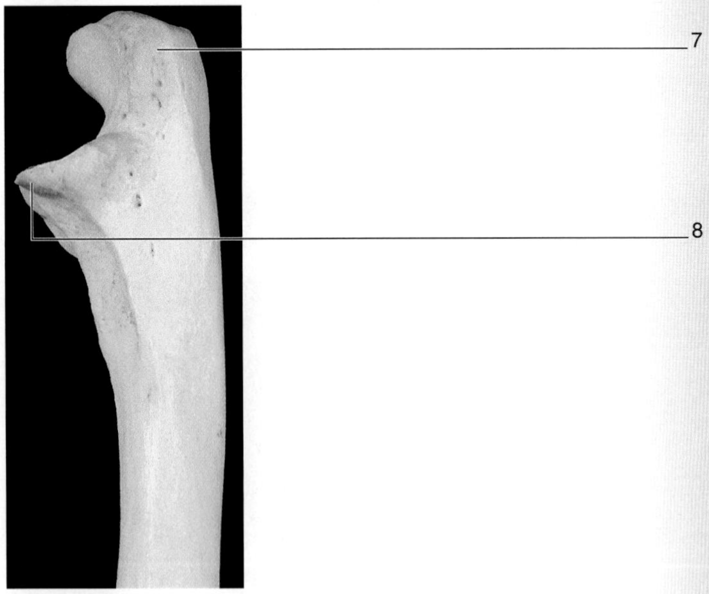

(b) Right ulna, medial view

Figure 9.5 The Ulna.

6. Now that you have observed the humerus, radius, and ulna, obtain these bones from the right side of the body and articulate them with each other. In the box below, sketch the relationship between the right humerus, radius, and ulna from an anterior view.

Proximal row

Distal row

EXERCISE 9.2D The Carpals

1. Obtain articulated bones of the wrist (carpals), or observe the **carpal** bones on an articulated skeleton (**figure 9.6**). The term *carpus* means "wrist." Thus, the carpal bones are the bones of the wrist.

2. The eight carpal bones are arranged in two rows of four bones each. The proximal row, which is adjacent to the radius and ulna, contains the scaphoid, lunate, triquetrum, and pisiform. The bones of the distal, which are adjacent to the metacarpals include the trapezium, trapezoid, capitate, and hamate. Use the spaces to the right to draw the carpal bones of the proximal and distal rows as they would be positioned in an anterior view. Then label the bones in each row.

3. Using word origins to associate the shape of each carpal bone with its name will help you identify the bones individually. Refering to table 9.2 as a guide, complete the following chart to help you remember the carpal bones.

Carpal Bone	Word Origin	Bone Shape/Appearance
Scaphoid		
Lunate		
Triquetrum		
Pisiform		
Trapezium		
Trapezoid		
Capitate		
Hamate		

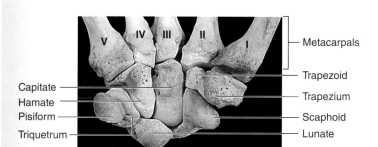

(a) Right wrist and hand, anterior view

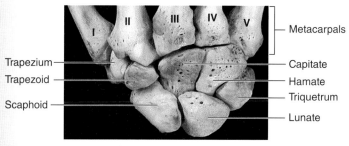

(b) Right wrist and hand, posterior view

Figure 9.6 The Carpals.

(continued on next page)

212 Chapter Nine The Skeletal System: Appendicular Skeleton

(continued from previous page)

4. *Optional Activity:* **AP|R** 5: Skeletal System—Visit the Quiz area for focused drill and practice on the bones of the appendicular skeleton.

EXERCISE 9.2E The Metacarpals and Phalanges

1. Obtain articulated bones of the hand (metacarpals and phalanges), or observe bones of the hand on an articulated skeleton **(figure 9.7)**.

2. **Metacarpals:** The prefix *meta-* means "after." Thus, the metacarpal bones are the bones that come after the carpus, or wrist, and they are located in the palm of the hand. There are five metacarpal bones, numbered I through V. The first metacarpal is on the lateral surface of the hand and forms the base of the thumb, or pollex. Palpate the palm of your hand to feel the metacarpal bones.

3. **Phalanges:** The term *phalanx* means "a line of soldiers." The next time you are typing on a keyboard writing a term paper or some other assignment, think about how your little "soldiers" are marching along doing great work for you.

4. The **pollex** is the thumb. How do the phalanges of the pollex (metacarpal I) differ from the phalanges of digits II–V?

   _____

   _____

   _____

5. Using table 9.2 and your textbook as guides, identify the metacarpals and phalanges on an articulated skeleton. Then label them in figure 9.7.

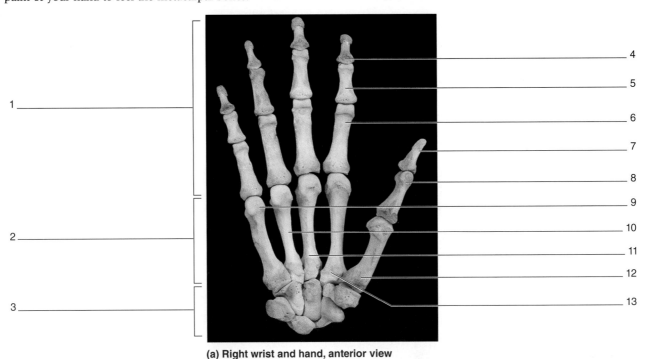

(a) Right wrist and hand, anterior view

Figure 9.7 The Metacarpals and Phalanges.

☐ carpals
☐ distal phalanx
☐ distal phalanx of pollex
☐ metacarpals
☐ metacarpal I
☐ metacarpal II
☐ metacarpal III
☐ metacarpal IV
☐ metacarpal V
☐ middle phalanx
☐ phalanges
☐ proximal phalanx
☐ proximal phalanx of pollex

INTEGRATE

LEARNING STRATEGY

Here is a mnemonic that may help you remember the names of the carpal bones, and the order in which they are found: **S**o **L**ong **T**op **P**art, **H**ere **C**omes **T**he **T**humb.
So Long Top Part = Scaphoid, Lunate, Triquetrum, Pisiform
Here Comes The Thumb = Hamate, Capitate, Trapezium, Trapezoid
    Proximal row first, moving from lateral to medial (anatomic position).
Distal row second, moving from lateral to medial (anatomic position).

EXERCISE 9.3

SURFACE ANATOMY REVIEW—PECTORAL GIRDLE AND UPPER LIMB

1. Palpate the manubrium and suprasternal (jugular) notch on yourself. If you move your fingers just lateral from the sternal notch, you will palpate the joint between the manubrium and the proximal end of the clavicle: the **sternoclavicular joint.** Recall that the only bony attachment between the pectoral girdle and the axial skeleton is at the sternoclavicular joint.

2. Palpate along the **clavicle** and make note of the curvatures of the clavicle as you move your fingers from medial to lateral. At the tip of the shoulder, you will feel the joint between the lateral aspect of the clavicle and the **acromial process** of the scapula: the **acromioclavicular joint.**

3. Continue to palpate along the acromial process as it curves posteriorly and becomes the **spine of the scapula.**

4. Palpate the inferior, lateral border of the deltoid muscle. Where the deltoid attaches to the humerus, at the **deltoid tuberosity,** you can often feel part of the diaphysis of the humerus because there is very little muscle between the bone and the skin at that point.

5. Moving distally to your elbow, palpate the large **olecranon** of the ulna. This is the bony process that rests on a table when you lean on your elbows.

6. Just proximal from the olecranon on the medial aspect of the elbow, palpate the **medial epicondyle of the humerus.** If you place your thumb in the hollow between the olecranon of the ulna and the medial epicondyle of the humerus, you may be able to feel the cablelike **ulnar nerve.** This nerve is what causes the pain or tingly sensations that you feel when you hit your "funny bone."

7. Palpate the olecranon once again. Continue to palpate along the ulna distally until you come to the wrist joint. The bump you feel on the medial aspect of your wrist is the **styloid process of the ulna.** If you palpate the corresponding location on the lateral aspect of the wrist, you will feel the **styloid process of the radius.**

8. Finally, palpate the small metacarpal and phalangeal bones of the hand (see figure 9.7). As you do this, review the names of the bones.

9. Using your textbook as a guide, label the surface anatomy structures in **figure 9.8.**

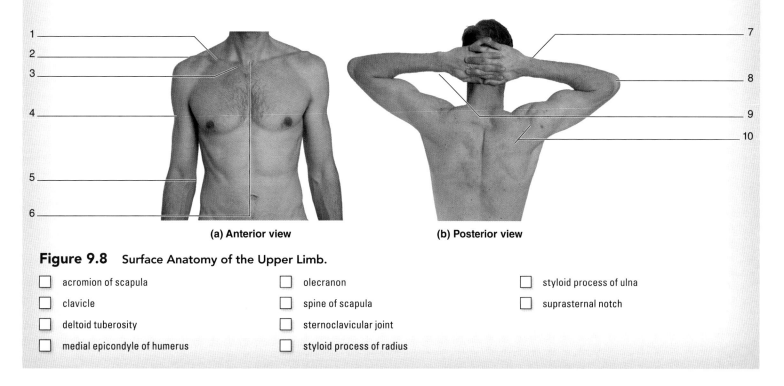

Figure 9.8 Surface Anatomy of the Upper Limb.

- ☐ acromion of scapula
- ☐ clavicle
- ☐ deltoid tuberosity
- ☐ medial epicondyle of humerus
- ☐ olecranon
- ☐ spine of scapula
- ☐ sternoclavicular joint
- ☐ styloid process of radius
- ☐ styloid process of ulna
- ☐ suprasternal notch

Chapter Nine The Skeletal System: Appendicular Skeleton

The Pelvic Girdle

The **pelvic girdle** consists of the paired **ilium, ischium,** and **pubis** bones. Together, the three bones compose the *os coxae* (*os*, bone, + *coxa*, hip). **Table 9.3** lists the bones composing the os coxae and describes their key features. A complete pelvis is formed when right and left os coxae come together and articulate with the sacrum. Unlike the bones of the pectoral girdle, the bones of the pelvic girdle fuse together during development to become one solid structure.

In the following exercises you will first consider the features of the os coxae as a whole. You will then consider the features of the individual bones that compose the os coxae. Finally, you will observe the structural differences between the male and female pelvic girdles to discover the functional differences between them.

Table 9.3 The Appendicular Skeleton: Pelvic Girdle

Bone	Bony Landmark	Description	Word Origin
Os Coxae *os*, bone, + *coxa*, hip	Acetabulum	A bony socket that forms the articulation with the head of the femur	*acetabula*, a shallow cup
	Linea terminalis (pelvic brim)	An oblique ridge on the inner surface of the ilium and pubic bones that separates the true pelvis (below) from the false pelvis (above); consists of the pubic crest, pectineal line, and arcuate line	*linea*, line, + *terminalis*, ending
	Lunate surface	The half-moon-shaped (curved) smooth surface on the superior border of the acetabulum, which articulates with the head of the femur	*luna*, moon
	Obturator foramen	A large, oval hole in the inferior part of the os coxae located anterior and medial	*obturo*, to occlude
Ilium *ilium, flank*	Ala	Concave, curved anterior region of the ilium	*ala*, wing
	Anterior gluteal line	A rough line running obliquely on the lateral surface of the ilium from the iliac crest to the greater sciatic notch; attachment site for gluteal muscles	*gloutos*, buttock
	Anterior inferior iliac spine	A process inferior to the anterior superior iliac spine; origin of rectus femoris muscle	*spina*, a spine
	Anterior superior iliac spine	A projection at the anteriormost part of the iliac crest; lateral attachment point for the inguinal ligament; origin for sartorius muscle	*spina*, a spine
	Arcuate line	An oblique line between the ilium and ischium that composes the iliac part of the linea terminalis (pelvic brim) of the bony pelvis	*arcuatus*, bowed
	Auricular surface	An "earlike" rough surface on the medial aspect of the ilium; the point of articulation with the sacrum	*auris*, ear
	Greater sciatic notch	A deep notch on the posterior surface of the ilium inferior to the posterior iliac spines; sciatic nerve descends adjacent to this notch	*sciaticus*, the hip joint
	Iliac crest	The superior border of the ilium, beginning at the sacrum and ending on the lateral aspect of the hip	*crista*, a ridge
	Iliac fossa	A large fossa on the anteromedial (internal) surface of the ilium inferior to the iliac crest	*ilium*, flank, + *fossa*, a trench
	Iliac tuberosity	A large projection on the posterior, superior aspect of the ilium	*ilium*, flank, + *tuber*, a knob
	Inferior gluteal line	A rough line running transversely on the lateral surface of the ilium just superior to the acetabulum; attachment site for gluteal muscles	*gloutos*, buttock
	Posterior gluteal line	A rough line running vertically on the lateral surface of the ilium from the iliac crest to the posterior rim of the greater sciatic notch; attachment site for gluteal muscles	*gloutos*, buttock
	Posterior inferior iliac spine	A small projection on the posterior inferior point of the ilium	*spina*, a spine
	Posterior superior iliac spine	A projection on the posterior superior point of the ilium	*spina*, a spine
Ischium *ischion, hip*	Body of ischium	Bulky bone superior to the ramus of ischium	*ischion*, hip, + *spina*, a spine
	Ischial spine	A small, sharp spine on the posterior aspect of the ischium	*ischion*, hip
	Lesser sciatic notch	A notch located immediately inferior to the ischial spine on the posterior surface of the ilium	*sciaticus*, the hip joint
	Ramus of ischium	The inferior part of the ischium that connects to the pubis anteriorly and forms the inferior part of the obturator foramen	*ramus*, branch
	Ischial tuberosity	A large, rough projection on the posterior, inferior surface of the ischium; attachment point for hamstring muscles	*ischion*, hip, + *tuber*, a knob

Table 9.3	The Appendicular Skeleton: Pelvic Girdle *(continued)*		
Bone	**Bony Landmark**	**Description**	**Word Origin**
Pubis pubis, *pubic bone*	Inferior pubic ramus	The inferior part of the pubis that joins with the ischium	*ramus*, branch
	Pectineal line	Rough ridge on the medial surface of the superior ramus of the pubis	*pectineal*, relating to the pubis
	Pubic crest	A ridge on the lateral part of the superior ramus of the pubis	*crista*, crest
	Pubic tubercle	A projection composing the anteriormost point of the bone; medial attachment point for the inguinal ligament	*tuber*, a knob
	Superior pubic ramus	The superior part of the pubis that joins with the ilium	*ramus*, branch
	Symphysial surface	Site of articulation for pubic bones at pubic symphysis	*symphysis*, a growing together

EXERCISE 9.4

BONES OF THE PELVIC GIRDLE

EXERCISE 9.4A The Os Coxae

1. Obtain an **os coxae** or observe the os coxae on an articulated skeleton **(figure 9.9)**.

2. Using table 9.3 and your textbook as guides, identify the structures listed in figure 9.9 on the os coxae. Then label them in figure 9.9. (Some answers may be used more than once.)

3. How can you tell whether the bone you are holding is a right os coxae or a left os coxae?

   _____

   _____

   _____

4. How can you determine which end of the os coxae is the superior end?

   _____

   _____

   _____

5. How can you distinguish the anterior surface of the os coxae from the posterior surface?

   _____

   _____

   _____

6. With which part of the vertebral column does the os coxae articulate?

   _____

EXERCISE 9.4B Male and Female Pelves

1. Obtain a male pelvis and a female pelvis and lay them next to each other on your workspace with the anterior surfaces facing toward you. **Figure 9.10** demonstrates the features of male and female pelves.

2. There are numerous features that help distinguish a **male pelvis** from a **female pelvis.** For example, a female pelvis is generally wider and more flared; it has a broader subpubic angle; a smaller, triangular obturator foramen; a wide, shallow greater sciatic notch; and ischial spines that rarely project into the pelvic outlet. All of these are adaptations that allow for childbirth. However, no method of sexing a pelvis is completely foolproof. At the very least, you can make a determination that a pelvis is more malelike than femalelike, or vice versa. As you observe male and female pelves, develop your own method of differentiating the male pelvis from the female pelvis. If you have difficulty, consult your textbook for assistance.

INTEGRATE

LEARNING STRATEGY

One method of estimating female versus male subpubic angles is to compare them to the angles formed between the digits on your hand when you spread them apart. For example, the angle between your thumb and index finger when you spread them apart approximates the wide subpubic angle of a female pelvis, whereas the angle formed between your index and middle fingers when you spread them apart approximates the narrower subpubic angle of a male pelvis.

(continued on next page)

(continued from previous page)

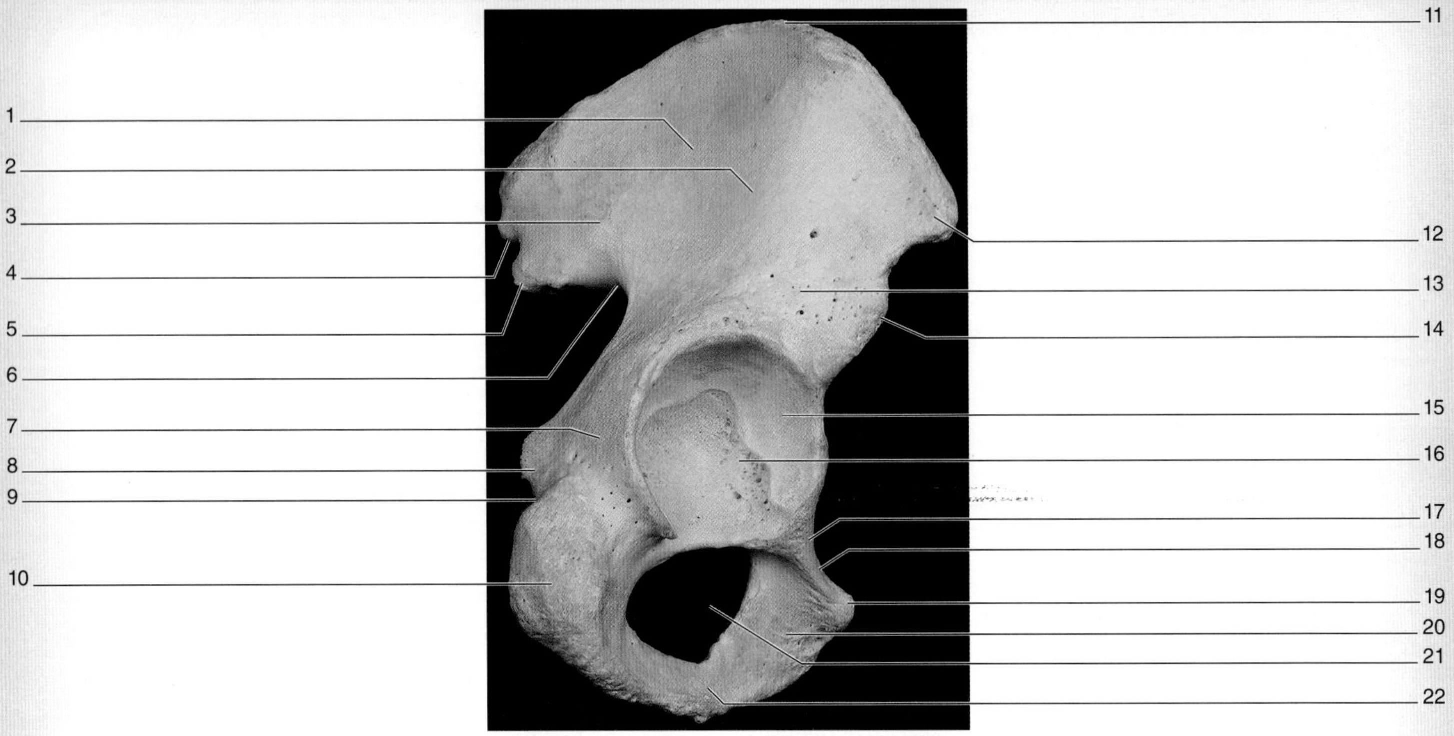

1 _____
2 _____
3 _____
4 _____
5 _____
6 _____
7 _____
8 _____
9 _____
10 _____

11 _____
12 _____
13 _____
14 _____
15 _____
16 _____
17 _____
18 _____
19 _____
20 _____
21 _____
22 _____

(a) Lateral view

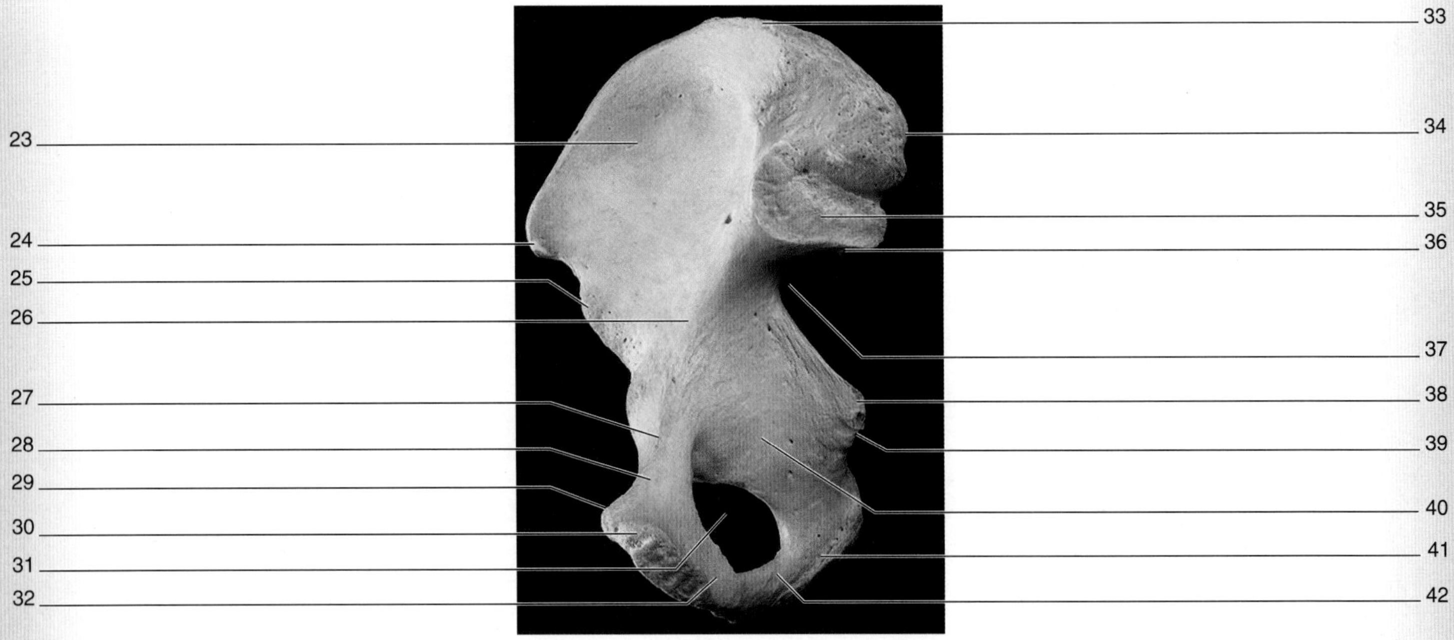

23 _____
24 _____
25 _____
26 _____
27 _____
28 _____
29 _____
30 _____
31 _____
32 _____

33 _____
34 _____
35 _____
36 _____
37 _____
38 _____
39 _____
40 _____
41 _____
42 _____

(b) Medial view

Figure 9.9 The Right Os Coxae.

- ☐ acetabulum
- ☐ ala
- ☐ anterior gluteal line
- ☐ anterior inferior iliac spine (used twice)
- ☐ anterior superior iliac spine (used twice)
- ☐ arcuate line
- ☐ auricular surface
- ☐ body of ischium (used twice)
- ☐ greater sciatic notch (used twice)
- ☐ iliac crest (used twice)
- ☐ iliac fossa
- ☐ inferior gluteal line
- ☐ inferior pubic ramus (used twice)
- ☐ ischial spine (used twice)
- ☐ ischial tuberosity (used twice)
- ☐ lesser sciatic notch (used twice)
- ☐ lunate surface
- ☐ obturator foramen (used twice)
- ☐ pectineal line
- ☐ posterior gluteal line
- ☐ posterior inferior iliac spine (used twice)
- ☐ posterior superior iliac spine (used twice)
- ☐ pubic crest
- ☐ pubic tubercle (used twice)
- ☐ ramus of ischium (used twice)
- ☐ superior pubic ramus (used twice)
- ☐ symphysial surface of pubic bone

3. In the spaces below, sketch the male and female pelves. Then list the features you used to distinguish the male pelvis from the female pelvis. Use the photos in figure 9.10 as a guide if you do not have samples of both a male and a female pelvis in your laboratory.

Male Pelvis

Female Pelvis

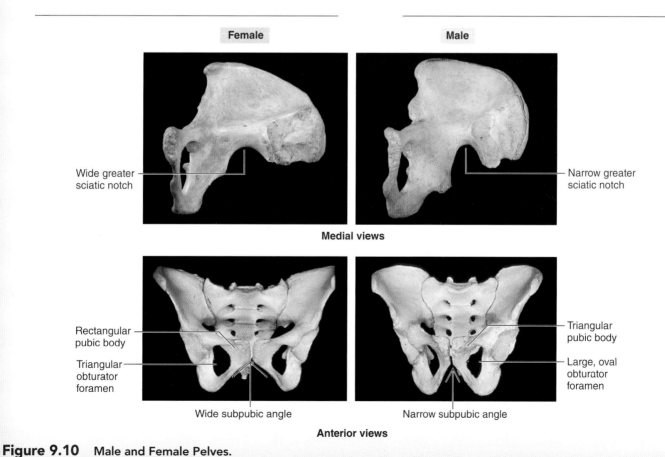

Figure 9.10 Male and Female Pelves.

INTEGRATE

CLINICAL VIEW
Pregnancy and Childbirth

Figure 9.10 demonstrates several of the features that distinguish the female pelvis from the male pelvis that make childbirth possible. Specifically, females generally have a wider, flared pelvis and greater subpubic angle. These features allow a baby's head to pass through the pelvic outlet during childbirth. Females also usually have ischial spines that remain clear of the pelvic outlet, so as not to impede the passage of the baby through the pelvic outlet. In addition to these anatomical differences between female and male pelves there are hormones that play an important role in creating a "roomy" birth canal. Specifically, the hormone *relaxin*, which is primarily secreted by the ovaries during pregnancy, promotes remodeling of connective tissue. Relaxin increases the flexibility of the pubic symphysis, the slightly movable fibrocartilaginous joint that connects the right and left pubic bones. Overall, these factors increase separation of the pubic bones in the latest stages of pregnancy. In rare cases, the pubic bones separate more than the physiological limit (10 mm), a condition called *diastasis symphysis pubis*. Factors that predispose a woman to this condition are the baby's position during birth, a rapid delivery, or excessive relaxation of the symphysis pubis joint. Symptoms of diastasis symphysis pubis include pain and decreased hip mobility. Treatment involves bracing the pelvic region.

The Lower Limb

The lower limb consists of the femur, patella, tibia, fibula, tarsals, metatarsals, and phalanges. Nearly all of the projections on these bones are attachment points for the muscles that move the limb. Table 9.4 lists the bones of the lower limb and describes their key features.

Table 9.4 The Appendicular Skeleton: Lower Limb

Bone	Bony Landmark	Description	Word Origin
Femur *femur, thigh*	Adductor tubercle	A small projection proximal or superior to a medial condyle	*tuber*, a knob
	Fovea	A circular depression within the head of the femur	*fovea*, a dimple
	Gluteal tuberosity	A projection on the proximal aspect of the linea aspera; gluteus maximus muscle attaches to it	*gloutos*, buttock, + *tuber*, a knob
	Greater trochanter	A very large projection on the lateral surface of the proximal epiphysis	*trochanter*, a runner
	Head	A very large, ball-shaped structure on the proximal end of the femur; articulates with acetabulum of os coxae	NA
	Intercondylar fossa	A depression on the distal end of the femur between the two condyles; serves as an attachment point for the cruciate ligaments of the knee	*inter-*, between, + *kondylos*, knuckle
	Intertrochanteric crest	A large ridge that runs between the greater and lesser trochanters on the posterior surface	*inter-*, between, + *trochanter*, a runner
	Intertrochanteric line	A shallow ridge that runs between the greater and lesser trochanters on the anterior surface	*inter-*, between, + *trochanter*, a runner
	Lateral condyle	A large, rounded surface that articulates with the lateral condyle of the tibia	*kondylos*, knuckle
	Lateral epicondyle	A rough surface superior to the lateral condyle	*epi*, above, + *kondylos*, knuckle
	Lateral supracondylar line	Distal, lateral branch of linea aspera	*supra*, above
	Lesser trochanter	A large projection on the medial surface of the proximal epiphysis; attachment site of many thigh muscles	*trochanter*, a runner
	Linea aspera	A "rough line" that runs along the posterior surface of the diaphysis	*linea*, line, + *aspera*, rough
	Medial condyle	A rounded surface that articulates with the medial condyle of the tibia	*kondylos*, knuckle
	Medial epicondyle	A rough surface superior to the medial condyle	*epi*, above, + *kondylos*, knuckle
	Medial supracondylar line	Medial, lateral branch of linea aspera	*supra*, above
	Neck	The narrow portion where the head meets the shaft of the bone; this is the part of the bone that is fractured in a "broken hip"	NA
	Patellar surface	A smooth depression on the anterior surface of the distal epiphysis; location where the patella articulates with the femur	(patella) *patina*, a shallow disk
	Pectineal line	A line on the posterior, superior aspect of the femur that serves as an attachment point for the pectineus muscle	*pectineal*, ridged or comblike
	Popliteal surface	A triangular region on the posterior aspect of the distal femur; delineated by the lateral and medial supracondylar lines	*popliteal*, the back of the knee
	Shaft	The diaphysis of the bone	NA

Table 9.4	The Appendicular Skeleton: Lower Limb (continued)		
Bone	**Bony Landmark**	**Description**	**Word Origin**
Tibia tibia, *the large shin bone*	Anterior border	A ridge on the anterior surface extending distally from the tibial tuberosity; commonly referred to as the "shin"	NA
	Fibular articular facet	Articulates with head of fibula	NA
	Intercondylar eminence	A prominent projection between the two condyles on the proximal epiphysis	*eminentia*, a raised area on a bone
	Lateral condyle	A large, flat surface on the lateral aspect of the proximal epiphysis; the point of articulation with the lateral condyle of the femur	*kondylos*, knuckle
	Medial condyle	A large, flat surface on the medial aspect of the proximal epiphysis; point of articulation with the medial condyle of the femur	*kondylos*, knuckle
	Medial malleolus	A projection on the medial surface of the distal epiphysis	*malleus*, hammer
	Shaft	The diaphysis of the bone	NA
	Tibial tuberosity	A projection on the anterior surface of the proximal epiphysis; attachment point for the patellar ligament (quadriceps femoris muscle attachment)	*tibia*, shin bone
Fibula fibula, *a clasp or buckle*	Head	The rounded proximal end of the bone	NA
	Lateral malleolus	A projection on the lateral surface of the distal epiphysis	*malleus*, hammer
	Neck	The narrow portion where the head meets the diaphysis	NA
	Shaft	The diaphysis of the bone	NA
Tarsals tarsus, *a flat surface*	Calcaneus	Bone that forms the heel of the foot; attachment point for the calcaneal tendon	*calcaneus*, the heel
	Cuboid	A cube-shaped bone located at the base of the fourth metatarsal	*kybos*, cube, + *eidos*, resemblance
	Intermediate cuneiform	A wedge-shaped bone located at the base of the second metatarsal	*cuneus*, wedge, + *forma*, shape
	Lateral cuneiform	A wedge-shaped bone located at the base of the third metatarsal	*cuneus*, wedge, + *forma*, shape
	Medial cuneiform	A wedge-shaped bone located at the base of the first metatarsal	*cuneus*, wedge, + *forma*, shape
	Navicular	A bone shaped like a boat ("ship") located just anterior to the talus	*navis*, ship
	Talus	The major weight-bearing bone of the ankle; articulates with the tibia and fibula	*talus*, ankle
Metatarsals meta-, *after*, + tarsus, *a flat surface*	Base	The proximal epiphysis of the bone	NA
	Head	The distal epiphysis of the bone	NA
	Shaft	The diaphysis of the bone	NA
Phalanges phalanx, *line of soldiers*			
II through V	Proximal	The phalanx closest to the metatarsal bones	*proximus*, nearest
	Middle	The middle phalanx	NA
	Distal	The small, cone-shaped distal bone of the digits	*distalis*, away
Hallux hallux, *the big toe*	Proximal	The phalanx closest to the sole of the foot	*proximus*, nearest
	Distal	The distal phalanx	*distalis*, away

Chapter Nine The Skeletal System: Appendicular Skeleton

EXERCISE 9.5

BONES OF THE LOWER LIMB

EXERCISE 9.5A The Femur

1. Obtain a **femur** or observe the femur on an articulated skeleton **(figure 9.11)**.

2. Using table 9.4 and your textbook as guides, identify the structures listed in figure 9.11 on the femur. Then label them in figure 9.11.

3. How can you tell whether the bone you are holding is a right femur or a left femur?

   _____

   _____

   _____

4. How can you determine which end of the femur is the proximal end?

   _____

   _____

   _____

5. How can you distinguish the anterior surface of the femur from the posterior surface?

   _____

   _____

   _____

6. Obtain a right os coxae and a right femur and articulate them with each other. In the box below, sketch the relationship between the bones of the os coxae and the femur.

7. What is the name of the joint formed between the bones of the os coxae and the femur?

   _____

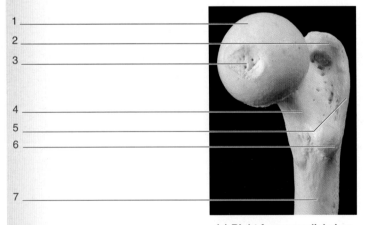

(a) Right femur, medial view

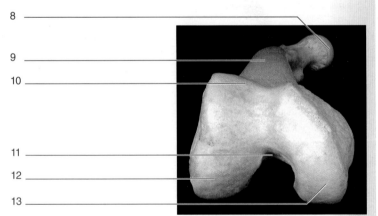

(b) Right femur, inferior view

Figure 9.11   The Right Femur.

☐ adductor tubercle (used twice)
☐ fovea (used twice)
☐ gluteal tuberosity
☐ greater trochanter (used three times)
☐ head (used four times)
☐ intercondylar fossa (used twice)
☐ intertrochanteric crest (used three times)
☐ intertrochanteric line
☐ lateral condyle (used twice)
☐ lateral epicondyle (used three times)
☐ lateral supracondylar line
☐ lesser trochanter (used three times)

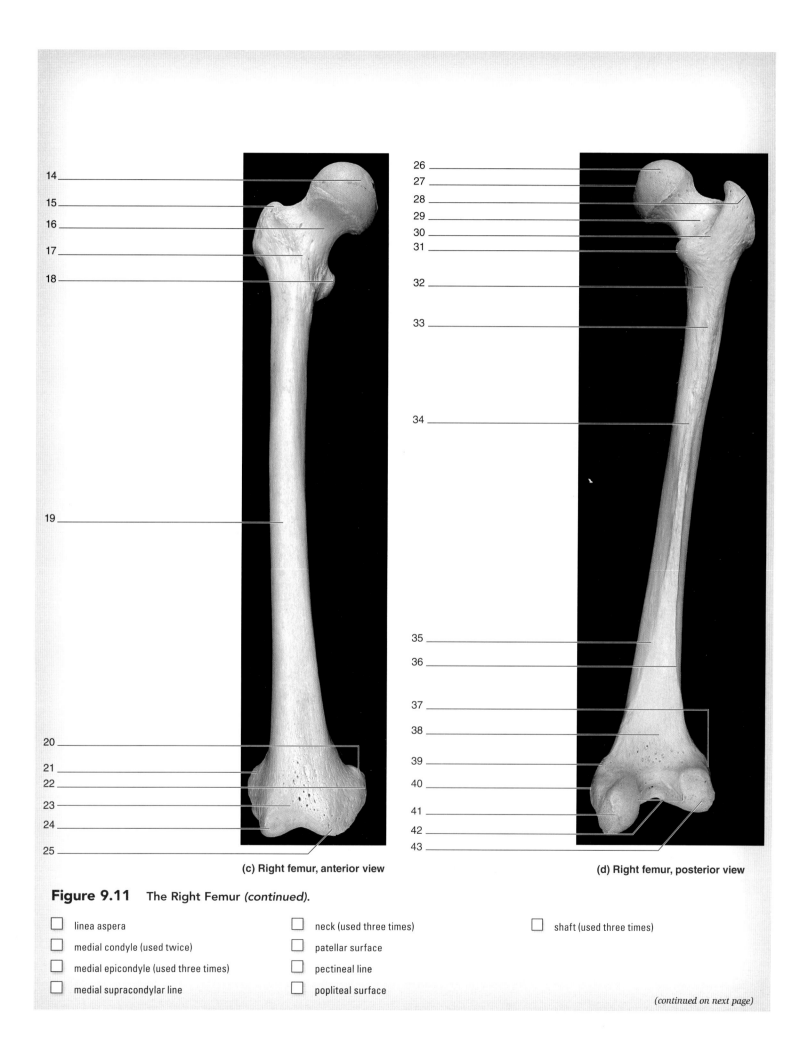

(c) Right femur, anterior view

(d) Right femur, posterior view

Figure 9.11 The Right Femur *(continued)*.

- ☐ linea aspera
- ☐ medial condyle (used twice)
- ☐ medial epicondyle (used three times)
- ☐ medial supracondylar line
- ☐ neck (used three times)
- ☐ patellar surface
- ☐ pectineal line
- ☐ popliteal surface
- ☐ shaft (used three times)

(continued on next page)

(continued from previous page)

EXERCISE 9.5B The Tibia

1. Obtain a **tibia** or observe the tibia on an articulated skeleton. **(figure 9.12).**

2. Using table 9.4 and your textbook as guides, identify the structures listed in figure 9.12 on the tibia. Then label them in figure 9.12. (Answers may be used more than once).

3. How can you tell whether the bone you are holding is a right tibia or a left tibia?

   _____

   _____

   _____

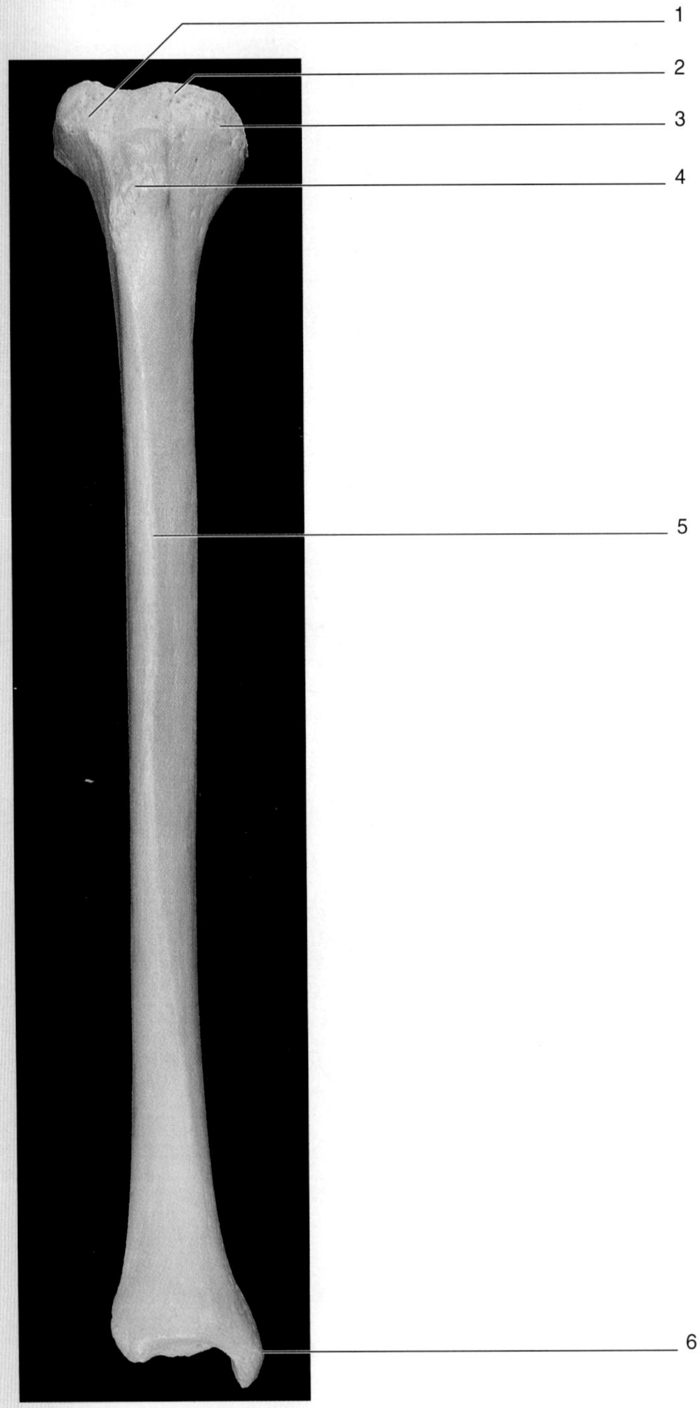

(a) Right tibia, anterior view

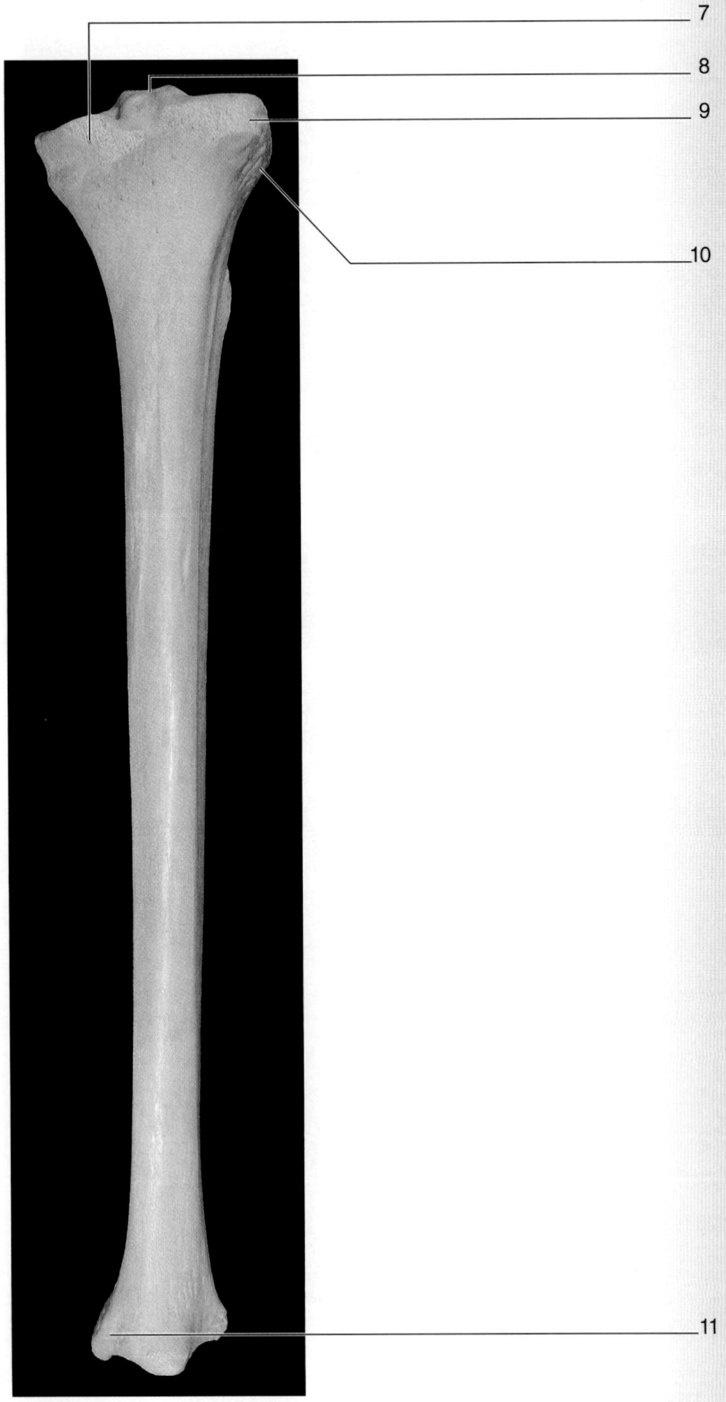

(b) Right tibia, posterior view

Figure 9.12   The Tibia.

- ☐ anterior border
- ☐ fibular articular facet
- ☐ intercondylar eminence (used twice)
- ☐ lateral condyle (used twice)
- ☐ medial condyle (used twice)
- ☐ medial malleolus (used twice)
- ☐ tibial tuberosity

4. How can you determine which end of the tibia is the proximal end?

   _____

   _____

   _____

5. How can you distinguish the anterior surface of the tibia from the posterior surface?

   _____

   _____

   _____

EXERCISE 9.5C The Fibula

1. Obtain a **fibula** or observe the fibula on an articulated skeleton (**figure 9.13**).

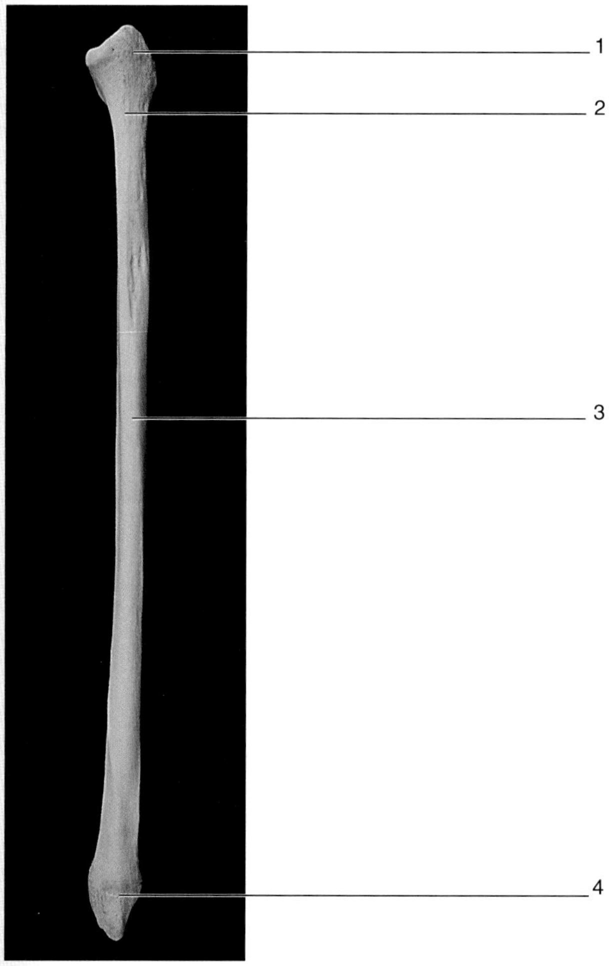

Figure 9.13 The Right Fibula. Lateral view.

- ☐ head
- ☐ lateral malleolus
- ☐ neck
- ☐ shaft

2. Using table 9.4 and your textbook as guides, identify the structures listed in figure 9.13 on the fibula. Then label them in figure 9.13.

3. How can you tell whether the bone you are holding is a right fibula or a left fibula?

   _____

   _____

   _____

4. How can you determine which end of the fibula is the proximal end?

   _____

   _____

   _____

   _____

5. How can you distinguish the anterior surface of the fibula from the posterior surface?

   _____

   _____

   _____

6. Now that you have observed the femur, tibia, and fibula, obtain these bones (all from the same side of the body) and articulate them with each other. In the box below, sketch the relationship between the right femur, tibia, and fibula from an anterior view. Include in your sketch the location of the patella.

(continued on next page)

(continued from previous page)

7. Does the fibula form part of the knee joint?

   _____

8. Does the fibula form part of the ankle joint?

   _____

EXERCISE 9.5D The Tarsals

1. Obtain bones of the ankle (tarsals) or observe the **tarsal** bones on an articulated skeleton (**figures 9.14** and **9.15**).

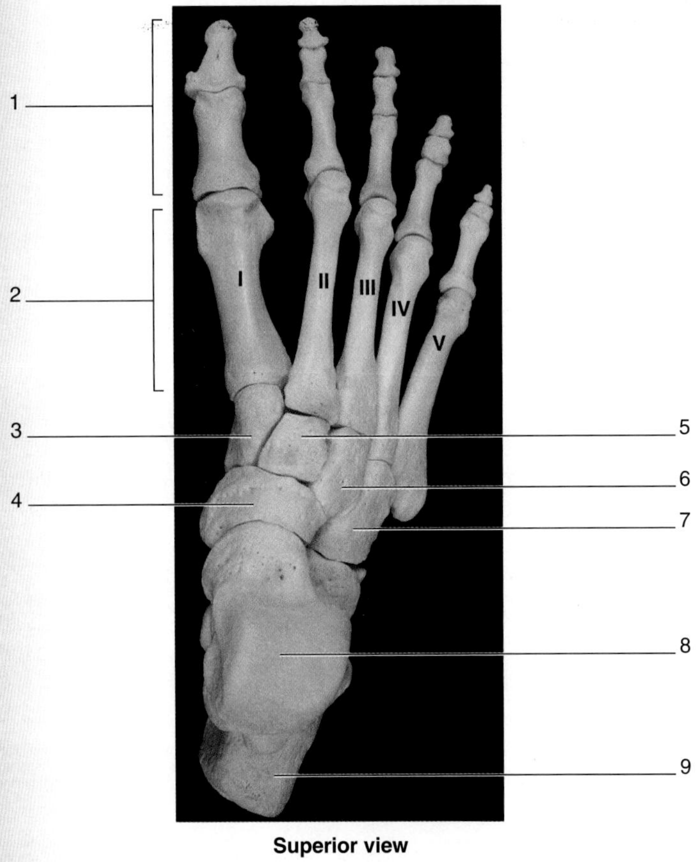

Figure 9.14 The Tarsals.

☐ calcaneus ☐ lateral cuneiform ☐ navicular
☐ cuboid ☐ medial cuneiform ☐ phalanges
☐ intermediate cuneiform ☐ metatarsals ☐ talus

Superior view

2. The term *tarsal* means "flat surface," as in the sole of the foot. The tarsal bones are much larger than their counterparts in the wrist (the carpal bones). The largest differences in structure are seen in the talus and calcaneus, two bones that form a major part of the weight-bearing ankle joint.

3. Using table 9.4 and your textbook as guides, identify the tarsal bones on an articulated skeleton. Then label them in figure 9.14.

4. The seven tarsal bones are arranged in two rows. The proximal row includes the talus, calcaneus, and navicular. The bones of the distal row—the medial cuneiform, intermediate cuneiform, lateral cuneiform, and cuboid—articulate with the metatarsal bones. Use the spaces below and on the following page to diagram the tarsal bones of the proximal and distal rows as they would be positioned in a superior view. Then label the bones in each row.

Proximal Row

Distal Row

5. The names of the tarsal bones provide good clues for identifying each bone. Using table 9.4 as a guide, complete the following chart to help you remember the shapes of the tarsal bones.

Tarsal Bone	Word Origin	Bone Shape/Appearance
Talus		
Calcaneus		
Navicular		
Medial Cuneiform		
Intermediate Cuneiform		
Lateral Cuneiform		
Cuboid		

6. Now that you have observed the tibia, fibula, calcaneus, and talus, obtain these bones (all from the same side of the body) and articulate them with each other. In the box below, sketch the relationship between the tibia, fibula, calcaneus, and talus.

7. Which bones or parts of bones compose the ankle joint?

_____

_____

_____

(continued on next page)

(continued from previous page)

EXERCISE 9.5E The Metatarsals and Phalanges

1. Obtain bones of the foot (metatarsals and phalanges) or observe the bones of the foot on an articulated skeleton (figures 9.14 and 9.15).

2. **Metatarsals:** The term *meta-* means "after." Thus, the metatarsal bones are the bones that come after the tarsus. These bones are found in the sole of your foot. Palpate the sole of your foot and feel for the metatarsal bones. There are five metatarsal bones, numbered I through V. The first metatarsal is on the medial surface of the foot and articulates distally with the base of the big toe, or *hallux*.

3. The **hallux** is the big toe. How do its phalangeal bones differ from the phalanges of digits II–V?

   _____

   _____

   _____

4. Using table 9.4 and your textbook as guides, identify the metatarsal bones and phalanges on an articulated skeleton. Then label them in figure 9.15.

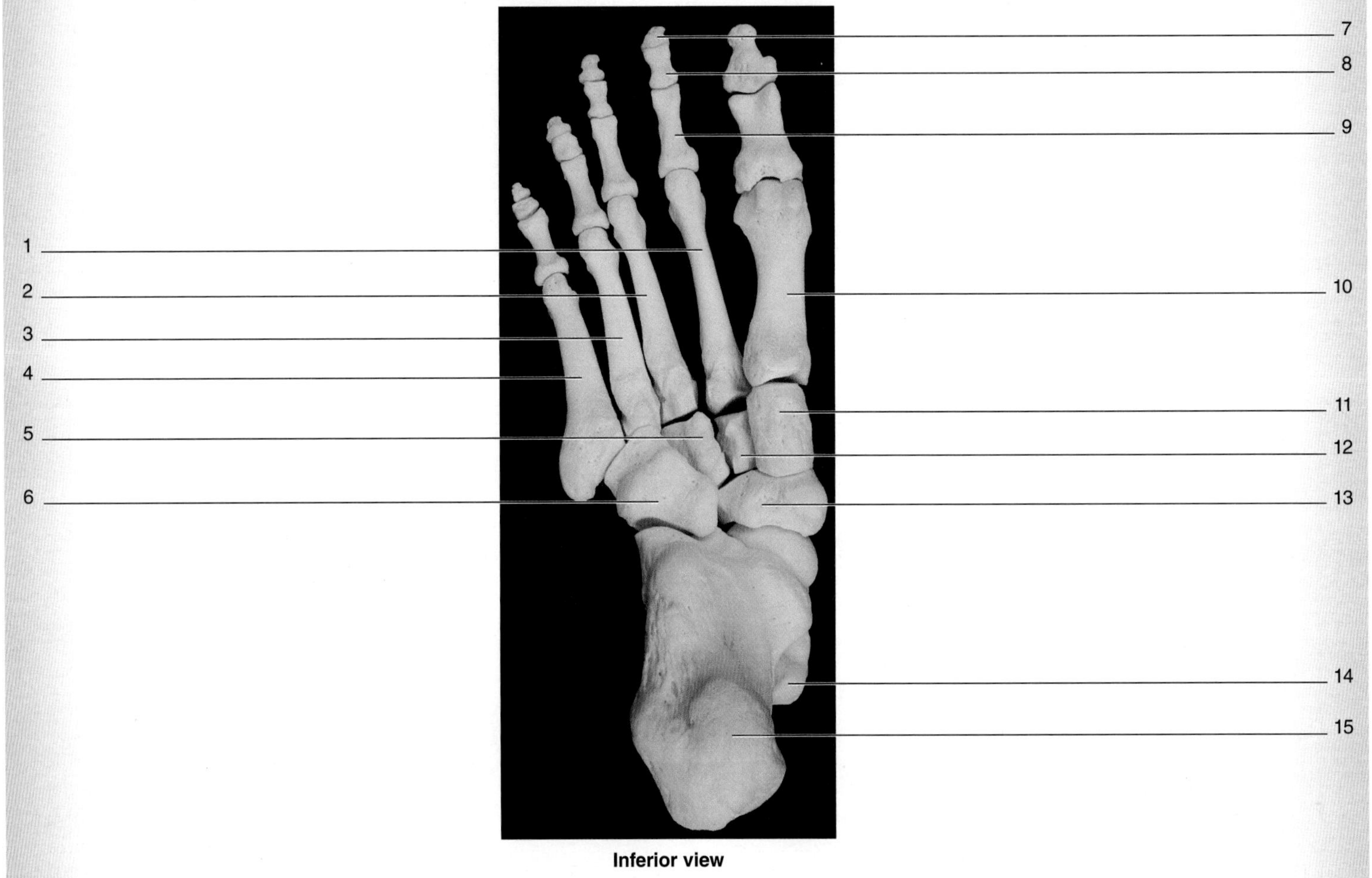

Inferior view

Figure 9.15 The Metatarsals and Phalanges.

Metatarsals
- ☐ metatarsal I
- ☐ metatarsal II
- ☐ metatarsal III
- ☐ metatarsal IV
- ☐ metatarsal V

Phalanges
- ☐ proximal phalanx
- ☐ middle phalanx
- ☐ distal phalanx

Tarsals
- ☐ calcaneus
- ☐ cuboid
- ☐ intermediate cuneiform
- ☐ lateral cuneiform
- ☐ medial cuneiform
- ☐ navicular
- ☐ talus

EXERCISE 9.6

SURFACE ANATOMY REVIEW—PELVIC GIRDLE AND LOWER LIMB

1. Place your hands on your hips. The bony ridge you feel under your skin is the **iliac crest.** The iliac crest is located at vertebral level L_1/L_2. Palpate anteriorly along the iliac crest until you come to the large bony projection on the anterior surface, the **anterior superior iliac spine.**

2. Palpate laterally along the iliac crest until your fingers are on the lateral aspect of the hip. Palpate the soft tissue of the gluteus medius muscle just inferior to the lateral portion of the iliac crest.

3. As you continue to palpate inferiorly, you will soon feel another bony projection. This is the **greater trochanter of the femur.** Depending upon the amount of fat and muscle in your gluteal region, you may also be able to palpate the **ischial tuberosities** deep within the posterior region of the buttocks.

4. Palpate the **patella** on the anterior aspect of the knee.

5. Place your thumb and pinky finger (fifth phalanx) on the medial and lateral aspects of the knee. Here you will feel the large **medial** and **lateral epicondyles of the femur.**

6. Palpate the lateral epicondyle of the femur. Now move your fingers distally until you feel the knoblike **head of the fibula.** As you continue to palpate distally, you will be unable to feel the shaft of the fibula because of the fibularis muscles that overlie it.

7. As you palpate distally along the fibula toward the ankle joint, you will eventually feel the **lateral malleolus** of the fibula in the ankle joint.

8. Palpate the patella once again. Distal to the patella on the anterior surface of the leg, you will feel the **tibial tuberosity.** Continue to palpate distally along the tibia. Notice that you can feel the entire shaft of the bone because it is only covered with skin and a little bit of fat. The subcutaneous part of the tibia here is commonly referred to as the shin.

9. As you palpate the most distal end of the tibia, you will feel the **medial malleolus** of the tibia in the ankle joint.

10. Palpate the large **calcaneus** in your heel. The talus and other tarsal bones are more difficult to palpate. However, if you wiggle your toes, you can see and palpate the **metatarsal bones** and the **phalanges.** As you palpate these bones, review their names. Recall that the first metatarsal is the metatarsal that lines up with the base of the big toe, or hallux, and the numbering continues, II through V, as you move toward the lateral aspect of the foot.

11. Using your textbook and atlas as guides, label the surface anatomy structures in **figure 9.16.** (Answers may be used more than once).

(continued on next page)

INTEGRATE

CONCEPT CONNECTION

Red bone marrow is the site of hemopoiesis, or blood cell formation. In young children, red bone marrow is found within most bones. However, by adulthood most red bone marrow is replaced by yellow bone marrow, which is mainly adipose tissue. In an adult, red bone marrow is limited to the ribs, sternum, vertebrae, pelvis, flat bones of the skull, and proximal epiphyses of the femur and humerus. Red bone marrow contains stem cells known as hemocytoblasts, which differentiate into *erythrocytes* (red blood cells), *leukocytes* (white blood cells), and *megakaryocytes*, which produce platelets. Chemical "factors," including hormones released in the blood, determine the ultimate fate of these stem cells. Bone marrow donation involves harvesting red bone marrow from either the iliac crest or the sternum (locations where red bone marrow is abundant and relatively easy to access). As with blood transfusions, donor and recipient bone marrow must "match" to prevent rejection. Properties of erythrocytes and leukocytes will be explored in chapters 20 (blood) and 23 (lymphatic and immune systems), respectively.

228 Chapter Nine *The Skeletal System: Appendicular Skeleton*

(continued from previous page)

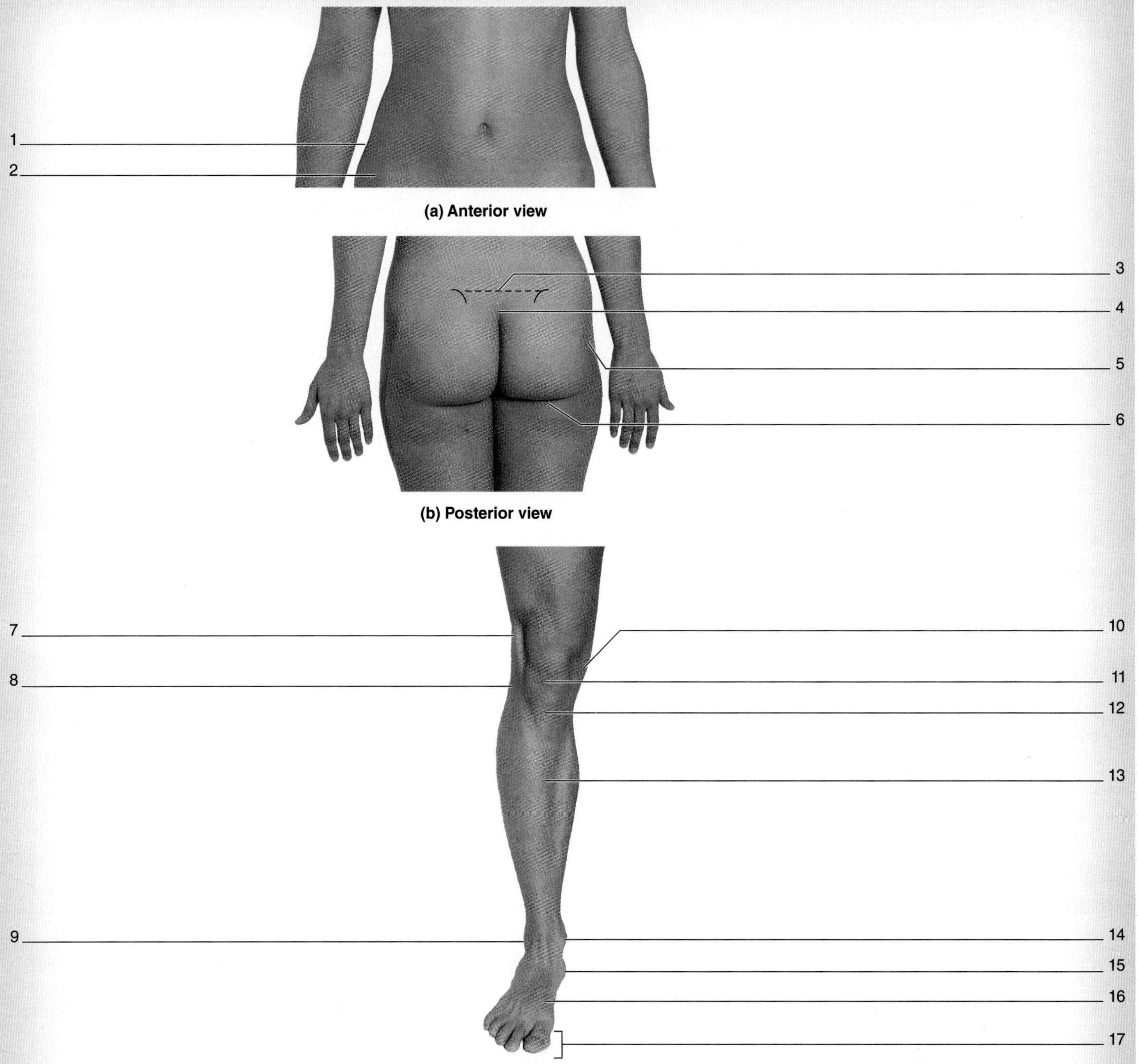

1 _____
2 _____

(a) Anterior view

3 _____
4 _____
5 _____
6 _____

(b) Posterior view

7 _____
8 _____
9 _____
10 _____
11 _____
12 _____
13 _____
14 _____
15 _____
16 _____
17 _____

(c) Anterior view

Figure 9.16 Surface Anatomy of the Lower Limb.

- ☐ anterior superior iliac spine
- ☐ calcaneus
- ☐ coccyx
- ☐ greater trochanter of femur
- ☐ head of fibula
- ☐ iliac crest
- ☐ ischial tuberosity
- ☐ lateral epicondyle of femur
- ☐ lateral malleolus
- ☐ medial epicondyle of femur
- ☐ medial malleolus
- ☐ metatarsals
- ☐ patella
- ☐ phalanges
- ☐ sacrum
- ☐ shaft of tibia
- ☐ tibial tuberosity

Chapter 9: The Skeletal System: Appendicular Skeleton

Name: _____
Date: _____ Section: _____

POST-LABORATORY WORKSHEET

The ❶ corresponds to the Learning Objective(s) listed in the chapter opener outline.

Do You Know the Basics?

Exercise 9.1: Bones of the Pectoral Girdle

1. The head of the humerus articulates with the _____ of the scapula. ❶

2. The pectoral girdle consists of the _____ and _____. ❷

3. The only point of articulation between the pectoral girdle and the axial skeleton is the _____. ❷

4. The joint between the clavicle and scapula is the _____ joint. The joint between the scapula and the humerus is the _____ joint. The joint between the humerus and the ulna is the _____ joint. The joint between the ulna, radius, and carpal bones is the _____ joint. ❸

Exercise 9.2: Bones of the Upper Limb

5. Match the description in column A with the bone or marking listed in column B. ❹

Column A

____ 1. bone of the arm
____ 2. bone of the forearm that aligns with the thumb
____ 3. bone of the forearm that aligns with digit V (pinky finger)
____ 4. middle bone of index finger
____ 5. part of the scapula that articulates with the humerus
____ 6. part of the scapula that articulates with the clavicle
____ 7. attachment site on the scapula for muscles of the arm
____ 8. bone that articulates with both sternum and scapula
____ 9. first bone of palm of hand
____ 10. long bony ridge on the posterior aspect of the scapula
____ 11. bones of the wrist
____ 12. middle bone of ring finger

Column B

a. acromion
b. carpals
c. clavicle
d. corocoid process
e. glenoid cavity
f. humerus
g. metacarpal I
h. middle phalanx II
i. radius
j. spine of scapula
k. middle phalanx IV
l. ulna

6. The anatomic name for the thumb is the _____. ❹

7. The carpal bone that is shaped like a half moon is the _____. The carpal bone that is shaped like a "table" and is located at the base of the *first* metacarpal (thumb) is the _____ , whereas the carpal bone that is shaped like a "table" and is located at the base of the *second* metacarpal (index finger) is the _____. The smallest carpal bone, which is shaped like a "pea," is the _____. The carpal bone that is shaped like a "hook" and is located at the base of the *fifth* metacarpal (pinky) is the _____. ❺

Exercise 9.3: Surface Anatomy Review: Pectoral Girdle and Upper Limb

8. Which process of the scapula can be palpated at the tip of the shoulder? _____ ❻

9. The tip of the elbow is formed by the _____ process. The radial process that can be palpated at the wrist is the _____ process. The process that can be palpated on the medial aspect of the elbow is the _____ process, which is part of the _____ bone. ❻

Chapter Nine The Skeletal System: Appendicular Skeleton

Exercise 9.4: Bones of the Pelvic Girdle

10. Match the structure in column A with the bone or bone feature listed in column B.

    Column A
    - ____ three bones fuse to form this bone of the pelvis
    - ____ specific bone that your weight rests on when you sit
    - ____ most anterior bone of the os coxae
    - ____ "hip" bone
    - ____ hole in the os coxae
    - ____ structure the femur articulates with at the hip

    Column B
    a. acetabulum
    b. ilium
    c. ischium
    d. obturator foramen
    e. os coxae
    f. pubic

11. The pelvic girdle consists of the _____, _____, and _____.

Exercise 9.5: Bones of the Lower Limb

12. The joint between the os coxae and the femur is the _____ joint. The joint between the femur and the tibia is the _____ joint. The joint between the tibia, fibula, and talus is the _____ joint.

13. The anatomic name for the big toe is the _____.

14. The tarsal bone that forms the heel is the _____. The tarsal bone that is shaped like a "ship" and is located anterior to the talus is the _____. The major weight-bearing tarsal bone, which articulates with the tibia and fibula to form the ankle joint, is the _____. The tarsal bones that are shaped like a "wedge" and are located at the base of the first, second, and third metatarsals are the _____.

15. The carpal and tarsal bones form the wrist and ankle joint, respectively. Is there a difference in number of tarsals as compared to carpals? _____ Do you see any resemblances between any of the carpal and tarsal bones? _____ Question 21 will provide you an opportunity to note any similarities and differences.

Exercise 9.6: Surface Anatomy Review: Pelvic Girdle and Lower Limb

16. When you place your hands on your hips and palpate the ridge of bone located there, you are palpating the _____. The _____ is the bone palpated on the anterior part of the knee. Just superior to the knee, the processes that can be palpated on the lateral and medial aspects of the knee are the _____ and _____, respectively. To palpate the calcaneus, you palpate on the _____. The _____ is a process that is palpated on the anterior leg just distal to the knee joint. The medial malleolus, which can be palpated on the medial aspect of the ankle, is a part of the _____, whereas the lateral malleolus, which can be palpated on the lateral aspect of the ankle, is a part of the _____.

Can You Apply What You've Learned?

17. What are the structural and functional differences between the anatomical and surgical necks of the humerus?

    _____
    _____
    _____

18. Construct a mnemonic to assist you with your recall of the carpal bones. Write your mnemonic here:

    _____
    _____
    _____

19. Compare and contrast both the appearance and location of the carpal and tarsal bones.

    _____
    _____
    _____

20. Construct a mnemonic device to assist you with your recall of the tarsal bones. Write your mnemonic here:

_____

_____

_____

21. List three of the features that help you differentiate a male pelvis from a female pelvis.

   Male Pelvis

   a. _____

   b. _____

   c. _____

   Female Pelvis

   a. _____

   b. _____

   c. _____

22. What bone or bony process of the upper limb serves the same function as the patella in the lower limb? (Hint: The patella's function is to act as a lever. It forces the tendons of the muscles on the anterior surface of the thigh farther away from the center of rotation of the joint in order to give the muscles a greater mechanical advantage—greater leverage.) _____

23. Explain why one end of the clavicle is called the sternal end, and the other is called the acromial end.

_____

_____

_____

Can You Synthesize What You've Learned?

24. What bony landmark of the upper limb is used as a reference point for locating the ulnar nerve at the elbow?

_____

25. Which of the two necks of the humerus do you think is more likely to fracture in an accident?

_____

_____

_____

26. Observe the relationship between the carpal bones and the distal portion of the radius and ulna on an articulated skeleton. Which of the carpal bones do you think is most likely to fracture when someone falls on an outstretched hand?

_____

_____

_____

27. Why do you think it is functionally important that the bones of the os coxae fuse together rather than remain independent bones?

_____

_____

_____

28. A fracture to which of the leg bones (tibia or fibula) would result in the greatest loss of function of the lower limb? Why?

_____

_____

_____

CHAPTER 10

Articulations

OUTLINE AND LEARNING OBJECTIVES

Gross Anatomy 236

Fibrous Joints 236

EXERCISE 10.1: FIBROUS JOINTS 236
1. Describe the structure of fibrous joints
2. Classify the three types of fibrous joints by structure and function (movement)
3. Give examples of places in the body where each type of fibrous joint is located

Cartilaginous Joints 237

EXERCISE 10.2: CARTILAGINOUS JOINTS 238
4. Describe the structure of cartilaginous joints
5. Classify the two types of cartilaginous joints by structure and function (movement)
6. Give examples of places in the body where each type of cartilaginous joint is located

Synovial Joints 239

EXERCISE 10.3: GENERAL STRUCTURE OF A SYNOVIAL JOINT 239
7. Identify the components of a synovial joint
8. Describe the functions of each component of a synovial joint

EXERCISE 10.4: STRUCTURAL CLASSIFICATIONS OF SYNOVIAL JOINTS 240
9. Describe the structural classifications of synovial joints
10. Give examples of places in the body where synovial joints are located

EXERCISE 10.5: PRACTICING SYNOVIAL JOINT MOVEMENTS 241
11. Demonstrate the movements that occur at synovial joints

EXERCISE 10.6: THE KNEE JOINT 243
12. Identify and describe the structures that compose the knee joint

INTRODUCTION

If someone asks whether you have ever injured a joint, you will probably think of your ankle, knee, hip, wrist, elbow, and shoulder joints. However, you may not have considered the fact that the sutures between your skull bones are also joints, as are the connections between your ribs and sternum. A joint, or **articulation** (*articulatio*, a forming of vines), is formed wherever one bone comes together with another bone. The study of joints is called **arthrology** (*arthron*, a joint, + *logos*, the study of). Joints are classified both by extent of movement (synarthrotic, amphiarthrotic, or diarthrotic) and by structure (fibrous, cartilaginous, and synovial). Your main textbook defines these classifications and describes the joint movements (e.g., flexion, extension) that occur at diarthrotic synovial joints. Before engaging in the laboratory activities in this chapter, review these joint definitions and the descriptions of joint movements from your textbook so you will have a firm grip on the terminology, which will prepare you for the laboratory. For your reference, **table 10.1** summarizes the *functional* (movement) classifications of joints. The *structural* classifications of joints are presented in the exercises in this chapter. As you work through the structural classifications of joints, also practice using the functional terms from table 10.1 to describe the movement allowed at each joint.

 **MODULE 5: SKELETAL SYSTEM**

In this laboratory session you will begin by briefly investigating the nature of fibrous and cartilaginous joints. The remainder of the laboratory session will be focused on an in-depth look at the structure and function of a representative synovial joint: the knee joint. All of the exercises in this chapter can be performed either on cadaver specimens or on joint models, so the text is written simply to guide you through the process of finding the structures associated with each of the joints using whatever materials are available in the classroom. Under each heading you will find a brief description of the joint(s) in question and instructions regarding what you are to observe in the laboratory. Use the tables and figures provided in this manual, which contain detailed information about the joints you will be observing, to guide you in your observations.

Table 10.1	Functional (Movement) Classification of Joints		
Type of Joint	**Description**	**Examples**	**Word Origin**
Amphiarthrosis	A slightly mobile joint	Syndesmosis, synchondrosis, and symphysis	*amphi*, on both sides, + *arthron*, a joint
Diarthrosis	A freely mobile joint	All synovial joints	*di-*, two, + *arthron*, a joint
Synarthrosis	An immobile joint	Skull suture and tooth gomphosis	*syn*, together, + *arthron-*, a joint

Chapter 10: Articulations

Name: _____
Date: _____ Section: _____

PRE-LABORATORY WORKSHEET

Also available at www.connect.mcgraw-hill.com

|ANATOMY & PHYSIOLOGY

1. Define the following terms that relate to the classification of joints based on movement:

   a. *synarthrotic* _____

   _____

   b. *amphiarthrotic* _____

   _____

   c. *diarthrotic* _____

   _____

2. List the three types of fibrous joints. For each type, give an example of a location in the body where that type of joint is found.

   Type of Fibrous Joint   **Location**

   a. _____   a. _____

   b. _____   b. _____

   c. _____   c. _____

3. List the two types of cartilaginous joints. For each type, give an example of a location in the body where it is found.

   Type of Cartilaginous Joint   **Location**

   a. _____   a. _____

   b. _____   b. _____

4. Describe how synovial joints differ from fibrous and cartilaginous joints.

   _____

   _____

   _____

   _____

5. All synovial joints are classified as _____ when classified based on the extent of movement.

6. a. Explain the relationship between the amount of movement allowed at a joint and the stability of the joint.

   _____

   _____

   _____

   _____

   b. Is this relationship direct or inverse? _____

Gross Anatomy

Fibrous Joints

Fibrous joints are characterized by having some amount of fibrous connective tissue connecting neighboring bones. Fibrous joints are classified as either synarthrotic (immobile) or amphiarthrotic (slightly mobile), based on the amount of movement allowed. Structurally, they are classified as sutures, syndesmoses, or gomphoses **(table 10.2)**.

Table 10.2	Classification of Fibrous Joints		
Fibrous Joints	**Structure and Description**	**Examples**	**Word Origin**
Suture	Found exclusively between skull bones; consists of a small amount of connective tissue (the sutural ligament) holding the bone surfaces together	Lambdoid suture, sagittal suture	*sutura*, a seam
Syndesmosis	Consists of large surfaces of bones that are anchored together by a connective tissue membrane called an interosseous membrane	Distal radioulnar joint; distal tibiofibular joint	*syn-*, together, + *desmos*, a band
Gomphosis	Consists of a cone-shaped peg fitting into a socket and anchored by the periodontal membrane	Teeth articulating with alveolar processes of mandible or maxilla	*gomphos*, nail, + *-osis*, condition

EXERCISE 10.1

FIBROUS JOINTS

1. *Suture*—Observe a skull on an articulated skeleton or by itself. Observe the numerous **sutures** (*sutura*, a seam) between the cranial bones. Notice how tightly the bones fit together. In a preserved skeleton, the sutural joints may be somewhat loose because the membranous connective tissue that normally holds the bones together was destroyed when the skeleton was prepared. However, in a living adult the skull bones are held tightly together by membranous connective tissue that lies in the spaces between the bones, and also by the interlocking shapes of the articulating bones. Recall that the flat bones of the skull form by **intramembranous ossification.** The fibrous connective tissue within the sutures is a remnant of the original membrane that served as the structural framework for the developing bones. How would you classify a sutural joint based on the amount of *movement* allowed?

   _____

2. *Syndesmosis*—Observe the radius and ulna on an articulated skeleton. Even though the articulated skeleton no longer contains a **syndesmosis** (*syn-*, together, + *desmos*, a band) joint (it was destroyed during preparation of the skeleton), you can still observe the movement at the distal radioulnar joint. Take hold of the radius and ulna, and then move the bones to mimic the actions of supination and pronation of the forearm (table 10.6). Notice how the radius pivots around the ulna during this motion. In a living human, these two bones are anchored to each other by an **interosseous membrane (figure 10.1)**, which composes the syndesmosis between these two bones. This membrane also contributes to the connective tissue that separates the forearm into anterior and posterior compartments. How would you classify a syndesmosis based on the normal amount of *movement* allowed?

   _____

3. *Gomphosis*—Observe the maxilla and mandible on a skull or on an articulated skeleton. The joints between the teeth and their sockets are **gomphosis** joints (*gomphos*, nail, + *-osis*, condition). The teeth might be very loose (or absent) in the skeleton because the **periodontal membrane** that normally holds them tightly in place was destroyed when the skeleton was prepared. If the skull has some empty tooth sockets (alveolar processes), observe the shape of the inside of the socket that the cone-shaped root of the tooth fits into. How would you classify a gomphosis based on the amount of normal *movement* allowed?

   _____

4. Using table 10.2 and your textbook as guides, label the types of fibrous joints in figure 10.1.

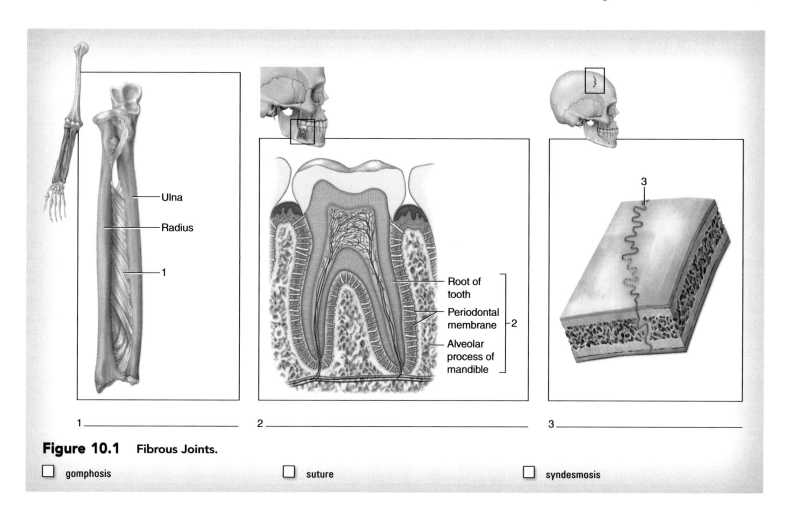

Figure 10.1 Fibrous Joints.

☐ gomphosis ☐ suture ☐ syndesmosis

Cartilaginous Joints

Cartilaginous joints are characterized by having bone connected to bone with only cartilage between the bones. The joints are classified based on the type of cartilage found in the joint: either **hyaline cartilage** or **fibrocartilage.** Cartilaginous joints are classified as synarthroses or amphiarthroses, based on extent of movement. Structurally, they are classified as synchondroses and symphyses **(table 10.3)**.

Table 10.3	Classification of Cartilaginous Joints		
Cartilaginous Joints	**Structure and Description**	**Examples**	**Word Origin**
Symphysis	Consists of bone connected to bone by fibrocartilage	Pubic symphysis; intervertebral discs	*symphysis*, a growing together
Synchondrosis	Consists of bone connected to bone by hyaline cartilage	Sternocostal joints; epiphyseal plates	*syn-*, together, + *chondrion*, cartilage

EXERCISE 10.2

CARTILAGINOUS JOINTS

1. **Synchondrosis**—A **synchondrosis** (*syn-*, together, + *chondrion*, cartilage) consists of bone connected to bone with **hyaline cartilage.** Examples are the sternocostal joints (the joints between the ribs and sternum) and the epiphyseal plates. Synchondroses should not be confused with synovial joints (or vice versa). Although synovial joints have hyaline cartilage as *part* of their structure (the articular cartilages), that cartilage is not the *only* thing found between the bones.

2. Observe the articulations between the ribs and sternum on an articulated skeleton. On the articulated skeleton, the "cartilage" between the ribs and sternum is some sort of replacement material such as plastic or rubber. In a living human, this would be hyaline cartilage. However, observing the skeleton will give you an idea of the structure of a synchondrosis.

3. How would you classify the sternocostal joints based on the amount of *movement* allowed? _____

4. **Symphysis**—A **symphysis** (*symphysis*, a growing together) joint consists of bone connected to bone with **fibrocartilage.** Examples are the pubic symphysis and the intervertebral discs. The intervertebral discs have a more complex structure than the pubic symphysis. They consist of an outer band of fibrocartilage, the **anulus fibrosus,** which surrounds a gel-like interior, the **nucleus pulposus.** A "ruptured," "herniated," or "slipped" disc occurs when the annulus fibrosis tears and the nucleus pulposus leaks out. The leaked nucleus pulposus can compress nerve fibers and cause neurological problems such as pain and numbness.

5. Observe the **pubic symphysis** and the **intervertebral discs** on the articulated skeleton. On the articulated skeleton, the "cartilage" is some sort of replacement material. In a living human this would be fibrocartilage.

6. How would you classify a symphysis joint based on the amount of *movement* allowed? _____

7. Using table 10.3 and your textbook as guides, label the types of cartilaginous joints shown in **figure 10.2**.

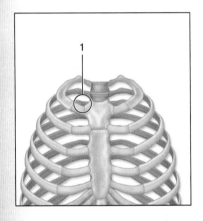
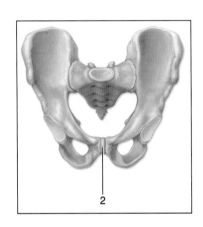
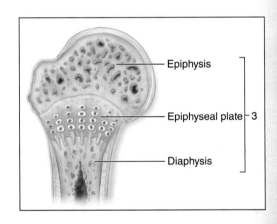

1 _____   2 _____   3 _____

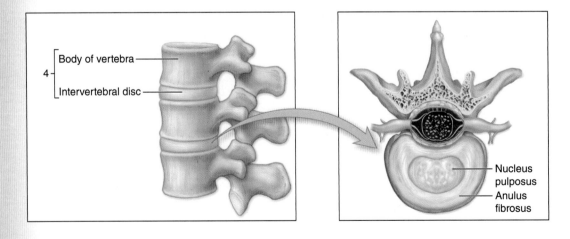

4 _____

Figure 10.2 Cartilaginous Joints.

☐ symphysis (used twice)   ☐ synchondrosis (used twice)

Synovial Joints

Synovial joints have a complex structure that includes a joint cavity filled with fluid. The term *synovial* literally means "together with egg" (*syn*, together, + *ovum*, egg). This term refers to the fluid inside the joint (the synovial fluid), which has the consistency and appearance of egg white. The joint cavity filled with synovial fluid allows the articulating bones to move easily past one another with very little friction between the bones. The joint cavity allows synovial joints to have a great deal of movement. Thus, all synovial joints are classified as diarthroses based upon the extent of movement. Structurally they are classified based upon the extent of the shapes of the articulating bones (for example: ball-and-socket). Table 10.5 summarizes the structural classifications of synovial joints.

EXERCISE 10.3

GENERAL STRUCTURE OF A SYNOVIAL JOINT

1. Observe a model of a synovial joint, preferably a model of the knee joint.

2. Using **table 10.4** and your textbook as guides, identify on the model the features of a typical synovial joint listed in **figure 10.3**. Then label them in figure 10.3.

3. *Optional Activity:* **AP|R** **5: Skeletal System**—Watch the "Synovial Joint" animation for a summary of synovial joint structure and types.

Table 10.4	Components of Synovial Joints	
Structure	**Description**	**Word Origin**
Articular Capsule	Consists of two layers: an outer fibrous capsule and an inner synovial membrane	*arthron*, a joint, + *capsa*, a box
Articular Cartilages	Hyaline cartilage found on the epiphyses of the articulating bones	*arthron*, a joint
Bursae and Tendon Sheaths (most joints)	Either small round sacs (bursae) or elongated structures that wrap around tendons (tendon sheaths); lined with synovial membrane and filled with synovial fluid; function to reduce friction between joint structures	*bursa*, a purse
Fibrous Layer of Articular Capsule	A dense irregular connective tissue that anchors the two articulating bones to each other; anchors to the periosteum of the articulating bones; thickenings of the fibrous capsule form several joint ligaments	*fibra*, fiber
Menisci (some joints)	Crescent-shaped pads of fibrocartilage found within the joint that provide cushioning between the articulating bones	*meniskos*, crescent
Synovial Cavity	A cavity within the joint that is lined by a synovial membrane and filled with synovial fluid	*syn*, together, + *ovum*, egg, + *cavus*, hollow
Synovial Fluid	A very slippery fluid consisting of hyaluronic acid and other glycoproteins; primary function is to reduce friction in the joint	*syn*, together, + *ovum*, egg, + *fluidus*, to flow
Synovial Membrane	A thin connective tissue membrane that lines all structures within the joint, including intra-articular ligaments; responsible for the formation of synovial fluid	*syn*, together, + *ovum*, egg, + *membrana*, a skin

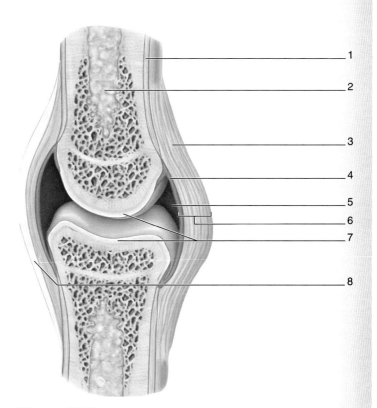

Figure 10.3 Diagram of a Representative Synovial Joint.

☐ articular capsule ☐ periosteum
☐ articular cartilage ☐ synovial (joint) cavity
☐ fibrous layer of articular capsule ☐ synovial membrane
☐ ligament ☐ yellow bone marrow

EXERCISE 10.4

STRUCTURAL CLASSIFICATIONS OF SYNOVIAL JOINTS

Synovial joints are classified using several different methods, most of which involve a description of the shape of the bones and the type of movement allowed at the joint **(table 10.5)**. For example, in a ball-and-socket joint, the "ball" (rounded) part of one bone fits into the "socket" (concave) part of another bone.

1. Observe an articulated skeleton to see examples of synovial joints.
2. Using table 10.5 and your textbook as guides, identify the joints in **figure 10.4** on the articulated skeleton. Then label each type of joint in figure 10.4.

Table 10.5	Structural Classification of Synovial Joints		
Surface Shape	**Structure and Description**	**Examples**	**Word Origin**
Ball-and-Socket	Formed when a spherical head fits into a concave socket	Hip joint; shoulder joint	*ball*, a round mass, + *soccus*, a shoe or sock
Condylar (ellipsoid)	Formed when a convex oval surface fits into an elliptical concavity	Radiocarpal joints; metacarpophalangeal joints	*kondylos*, knuckle, or *ellips*, oval, + *eidos*, form
Hinge (gynglymoid)	Formed when a convex surface fits into a concave surface and allows movement along only a single plane	Knee joint; elbow joint	*gynglymos*, a hinge joint
Pivot	Formed when a round surface fits into a ring formed by a ligament and a depression in another bone	Atlantoaxial joint	*pivot*, a post upon which something turns
Plane (gliding)	Formed when two flat surfaces come together	Intermetatarsal joints; some intercarpal joints	*planus*, flat
Saddle (sellar)	Formed when bones having both concave and convex surfaces come together such that the concave surfaces are at right angles to each other	Carpometacarpal joint of the thumb; ankle joint; calcaneocuboid joint	*sella*, saddle

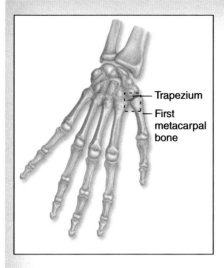
1 _____

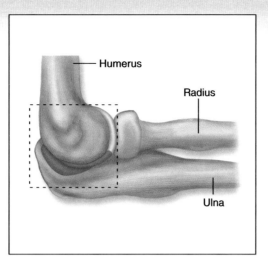
2 _____

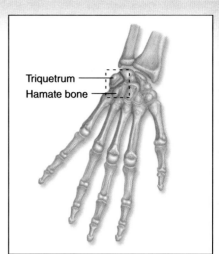
3 _____

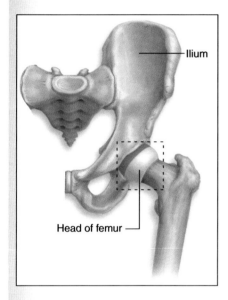
4 _____

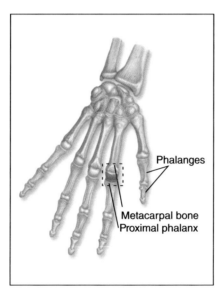
5 _____

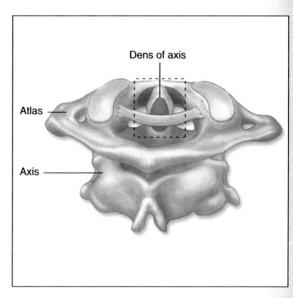
6 _____

Figure 10.4 Structural Classifications of Synovial Joints.

☐ ball-and-socket ☐ hinge ☐ plane
☐ condylar ☐ pivot ☐ saddle

EXERCISE 10.5

PRACTICING SYNOVIAL JOINT MOVEMENTS

1. All synovial joints are classified as **diarthrotic** (*di-*, two, + *arthron*, a joint) based upon movement. There are several types of movement possible with synovial joints. **Table 10.6** summarizes the types of movements possible at synovial joints.

2. Practice all of the movements listed in table 10.6 with your laboratory partner until you feel confident that you can demonstrate or describe them all. This knowledge is critically important for your success in the later chapters that cover muscle actions. Muscle actions are described in terms of joint movements (e.g., flexor digitorum = muscle that flexes the digits). Thus, if you are able to describe the movement of a particular joint, and you identify a muscle that acts about the joint to create that movement, you will most likely be able to describe the action of the muscle without having to look it up in a reference source.

3. *Optional Activity:* **AP|R** 5: Skeletal System—Review the series of joint movement animations to see examples of multiple movements at each joint.

(continued on next page)

(continued from previous page)

Table 10.6 Movements of Synovial Joints

Movement	Description	Opposite Movement	Word Origin
Abduction	Movement of a body part away from the midline	Adduction	*ab*, from, + *duco*, to lead
Adduction	Movement of a body part toward the midline	Abduction	*ad-*, toward, + *duco*, to lead
Circumduction	Movement of the distal part of an extremity in a circle	NA	*circum-*, around, + *duco*, to lead
Depression	Movement of a body part inferiorly	Elevation	*depressio*, to press down
Dorsiflexion	Movement of the ankle such that the foot moves toward the dorsum (back)	Plantar flexion	*dorsum*, the back, + *flexus*, to bend
Elevation	Movement of a body part superiorly	Depression	*e-levo-atus*, to lift up
Eversion	Movement of the ankle such that the plantar surface of the foot faces laterally	Inversion	*e-*, out, + *versus*, to turn
Extension	An increase in joint angle	Flexion	*extensio*, a stretching out
Flexion	A decrease in joint angle	Extension	*flexus*, to bend
Inversion	Movement of the ankle such that the plantar surface of the foot faces medially	Eversion	*in-*, inside, + *versus*, to turn
Opposition	Placement of the thumb (pollex) such that it crosses the palm of the hand and can touch all of the remaining digits	Reposition	*op -*, against, + *positio*, a placing
Plantar Flexion	Movement of the ankle such that the foot moves toward the plantar surface	Dorsiflexion	*plantaris*, the sole of the foot, + *flexus*, to bend
Pronation	Movement of the palm from anterior to posterior	Supination	*pronatus*, to bend forward
Protraction	Movement of a body part anteriorly (especially mandible and scapula)	Retraction	*pro-*, before + *tractio*, to draw
Reposition	Movement of the thumb to anatomic position after opposition	Opposition	*re-*, backward, + *positio*, a placing
Retraction	Movement of a body part posteriorly (especially mandible and scapula)	Protraction	*re-*, backward, + *tractio*, to draw
Rotation	Movement of a body part around its axis	NA	*rotatio*, to rotate
Supination	Movement of the palm from posterior to anterior	Pronation	*supinatus*, to bend backward

EXERCISE 10.6

THE KNEE JOINT

In this exercise we will observe the structure of the knee joint as an example of a synovial joint. The knee joint is complex, as are most synovial joints, and it contains several modifications, such as bursae, tendon sheaths, and menisci. The knee joint also contains several strong ligaments, which help to stabilize the joint. **Table 10.7** summarizes the structures composing the knee joint.

Many of us have some peripheral knowledge of the knee joint, having known individuals who have suffered a ruptured ACL, torn meniscus, or other knee injury, even if we have no idea what an ACL is. The ACL (anterior cruciate ligament) is one of two **cruciate ligaments** (*cruciatus*, resembling a cross) found in the knee joint, and a ruptured ACL is a knee injury that is common in football players, downhill skiers, and others involved in contact sports.

1. Observe a model of the knee joint or the knee joint of a cadaver.

2. The knee joint is actually two joints: the **tibiofemoral joint,** which is classified as a synovial hinge joint that acts in flexion and extension (though it allows for some rotational movement as well), and the **patellofemoral joint,** which is classified as a planar (gliding) joint. The bony structure of the tibiofemoral joint includes the medial and lateral femoral condyles, which sit on top of the medial and lateral tibial condyles. The bony structure of the patellofemoral joint includes the patellar surface of the femur, which consists of the smooth anterior surface between the femoral condyles, and the medial and lateral facets of the patella.

3. Using table 10.7 and your textbook as guides, identify the structures listed in **figure 10.5** on the model of the knee joint. Then label them in figure 10.5. (Answers may be used more than once.)

Table 10.7 Structures of the Knee Joint

	Description	Word Origin
Ligaments		
Anterior Cruciate Ligament	Connects the anterior intercondylar eminence of the tibia to the medial surface of the lateral condyle of the femur	ante-, in front of, + *cruciatus*, shaped like a cross
Fibular (lateral) Collateral Ligament	Connects the lateral epicondyle of the femur to the head of the fibula	co, together, + *latus*, side
Patellar Ligament	Connects the patella to the tibial tuberosity	*patina*, a shallow disk
Posterior Cruciate Ligament	Connects the posterior intercondylar eminence of the tibia to the anterior part of the lateral surface of the medial condyle of the femur	post-, behind, + *cruciatus*, shaped like a cross
Tibial (medial) Collateral Ligament	Connects the medial epicondyle of the femur to the medial surface of the tibia; its deep surface is anchored to the medial meniscus	co, together, + *latus*, side
Menisci		
Lateral Meniscus	A crescent-shaped pad of fibrocartilage located between the lateral condyle of the femur and the lateral condyle of the tibia	*meniskos*, crescent
Medial Meniscus	A crescent-shaped pad of fibrocartilage located between the medial condyle of the femur and the medial condyle of the tibia	*meniskos*, crescent
Bursae		
Infrapatellar Bursa	Located between the proximal tibia and the patellar ligament	infra, below, + *patina*, a shallow disc, + *bursa*, a purse
Prepatellar Bursa	Located between the patella and the overlying skin	pre-, before, + *patina*, a shallow disc, + *bursa*, a purse
Suprapatellar Bursa	Located between the distal femur and the quadriceps femoris tendon; communicates with the synovial cavity of the knee joint	supra, on the upper side, + *patina*, a shallow disc, + *bursa*, a purse
Tendon		
Quadriceps Femoris Tendon	Connects the quadriceps femoris muscle to the patella	quad, four + *femoris*, femur

(continued on next page)

244   Chapter Ten   Articulations

(continued from previous page)

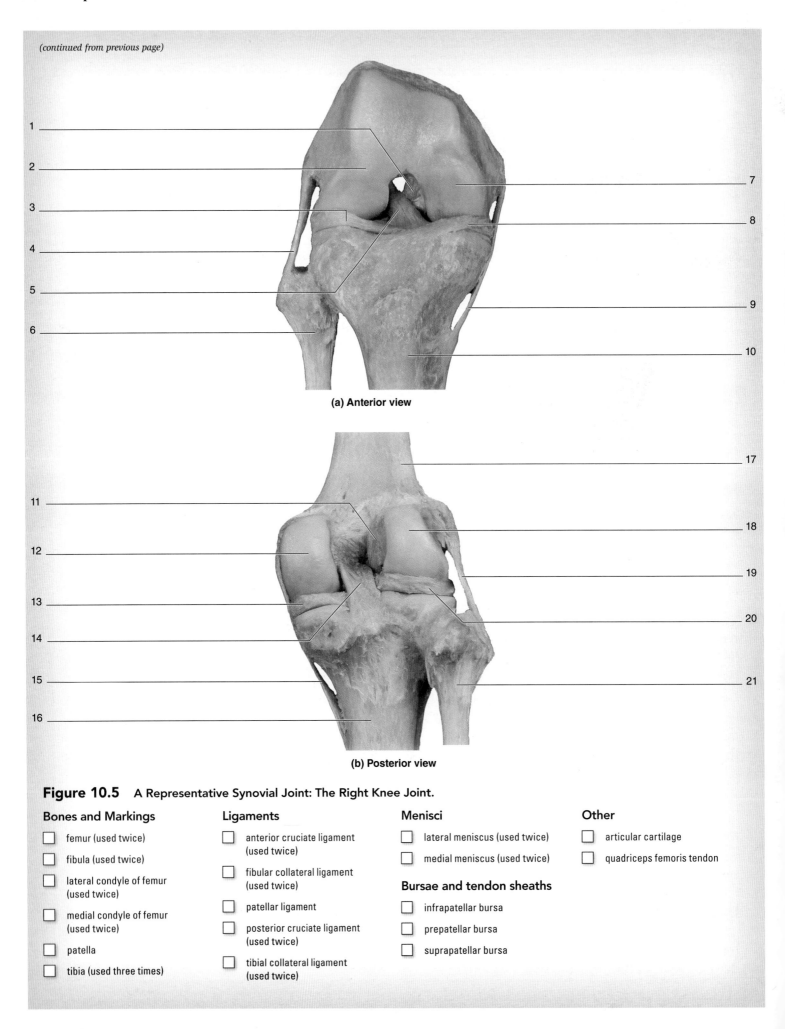

Figure 10.5   A Representative Synovial Joint: The Right Knee Joint.

Bones and Markings
- ☐ femur (used twice)
- ☐ fibula (used twice)
- ☐ lateral condyle of femur (used twice)
- ☐ medial condyle of femur (used twice)
- ☐ patella
- ☐ tibia (used three times)

Ligaments
- ☐ anterior cruciate ligament (used twice)
- ☐ fibular collateral ligament (used twice)
- ☐ patellar ligament
- ☐ posterior cruciate ligament (used twice)
- ☐ tibial collateral ligament (used twice)

Menisci
- ☐ lateral meniscus (used twice)
- ☐ medial meniscus (used twice)

Bursae and tendon sheaths
- ☐ infrapatellar bursa
- ☐ prepatellar bursa
- ☐ suprapatellar bursa

Other
- ☐ articular cartilage
- ☐ quadriceps femoris tendon

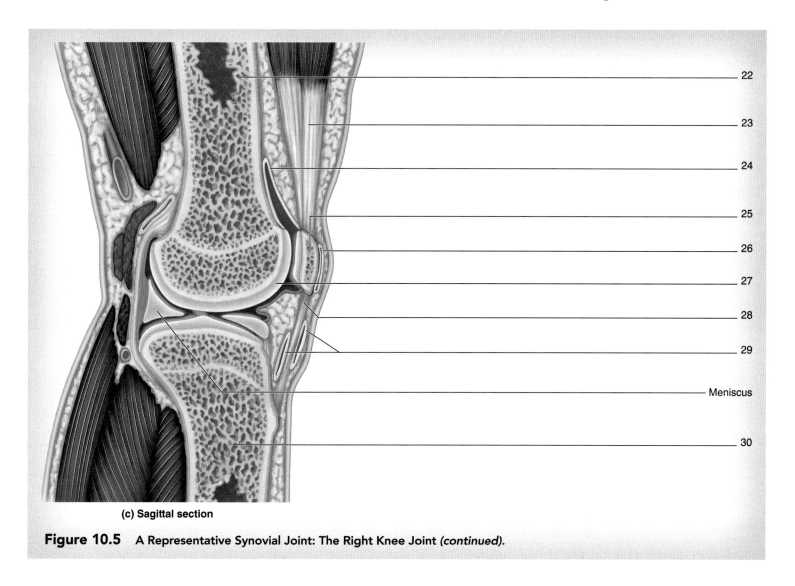

(c) Sagittal section

Figure 10.5 A Representative Synovial Joint: The Right Knee Joint *(continued)*.

Labels: 22, 23, 24, 25, 26, 27, 28, 29, Meniscus, 30

INTEGRATE

CONCEPT CONNECTION

Movement of synovial joints requires the activation of muscles that surround the joint. For example, extending the knee joint (increasing the joint angle) requires contracting the quadriceps muscles, and flexing the knee joint (decreasing the joint angle) requires contracting the hamstring muscles. These two muscle groups are antagonists of each other. That is, when extending the knee joint, quadriceps muscles act as **agonists,** whereas the hamstring muscles act as **antagonists.** The individual muscles that compose the quadriceps and hamstring muscle groups will be discussed in chapter 13. This allows the joint to move in the desired direction without interference from muscles pulling in the wrong direction. The neural control mechanism that creates the opposing actions is called *reciprocol innervation.* The action of both agonist and antagonist muscles is important for stabilization of synovial joints.

INTEGRATE

CLINICAL VIEW
Hip Replacement Surgery

What comes to mind when you think of someone's "hip"? Many people think this is the part of the body that lies superficial to the iliac crest. In fact, the term *ilium* literally means "hip." When you "put your hands on your hips," you are putting them on your iliac crests. However, when someone falls and suffers a fractured hip, it has little to do with the iliac bones at all. Instead, it concerns a fracture of another bone that composes part of the hip joint: the femur. Specifically, a fractured hip refers to a fracture of the neck of the femur.

The hip joint is the joint formed between the acetabulum of the os coxae and the head of the femur. This joint suffers wear and tear with age, and is a common site of osteoarthritis. In the elderly, particularly elderly women, the bones that compose the hip joint may become brittle over time. These individuals are at increased risk of developing a femoral or "hip" fracture if they fall. Unfortunately, hip fractures do not heal particularly well, because the fracture often disrupts the blood supply to the head of the femur. Therefore, instead of repairing the fractured bones, a surgeon might opt to completely replace the hip joint.

Hip replacement surgery involves cleaning out the acetabulum. In many cases it must be reconstructed with bone grafts to create a more efficient and complete socket that will be a better fit for the two parts of the prosthetic hip. Within the acetabulum, the surgeon places an artificial cup, which will compose the "socket" of this "ball-and-socket" joint. Next, the fractured head and neck of the femur are removed, and a prosthetic is placed on the proximal end of the femur, thus composing the new "ball" of the "ball-and-socket" joint. **Figure 10.6** shows a few different types of prosthetics. Figure 10.6*a* is an example of a very old hip prosthetic. Notice how large the ball is. Figure 10.6*b* is an example of a prosthetic with a much smaller head, and figure 10.6*c* shows a similar prosthetic that is still within the bone. Notice in figure 10.6*c* that the greater and lesser trochanters of the femur are still intact. This is necessary to maintain the connections between the bone and the gluteal muscles and other muscles that act about the hip. Such muscles are separated during the surgery so the surgeon can access the hip joint. However, they must be reconnected to the bone after the surgery.

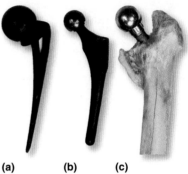

(a)   (b)   (c)

Figure 10.6 Three Different Examples of Prostheses Used for Hip Replacements.

Chapter 10: Articulations

Name: _____
Date: _____ Section: _____

POST-LABORATORY WORKSHEET

The ❶ corresponds to the Learning Objective(s) listed in the chapter opener outline.

Do You Know the Basics?

Exercise 10.1: Fibrous Joints

1. The fibrous connective tissue component that holds together a gomphosis (holds a tooth into a socket) is called the _____. ❶

2. A fibrous joint between skull bones is called a _____. ❷

3. An interosseous membrane is an example of a _____. ❷

4. Give an example of a location of the type of joint described in question #3. ❸ _____

Exercise 10.2: Cartilaginous Joints

5. A synchondrosis is composed of _____. ❹

6. A symphysis is composed of _____. ❹

7. Based on extent of movement, a synchondrosis joint is classified as _____, whereas a symphysis joint is classified as _____. ❺

8. List two locations where synchondroses are located in the body: ❻
   a. _____
   b. _____

9. List two locations where symphyses are located in the body: ❻
   a. _____
   b. _____

Exercise 10.3: General Structure of a Synovial Joint

10. What are the main components of synovial fluid? ❼
    _____

11. What type of tissue forms the synovial membrane? ❼
    _____

12. What is the function of the articular cartilages? ❽
    _____

13. What is the function of the synovial membrane? ❽
    _____

14. How are synovial joints classified based on extent of movement? ❽
    _____

248 Chapter Ten *Articulations*

Exercise 10.4: Structural Classifications of Synovial Joints

15. Match the type of synovial joint given in column A with its description in column B.

    Column A

    a. ball and socket
    b. condylar
    c. hinge
    d. plane
    e. pivot
    f. saddle

    Column B

    ____ head fits into a socket
    ____ biaxial; oval convex and concave
    ____ uniaxial; two flat surfaces
    ____ uniaxial; round surface in ring
    ____ resembles shape of a saddle
    ____ uniaxial; convex and concave surfaces

16. The atlantoaxial joint is classified as a(n) _____ joint.

17. Give an example of a synovial hinge joint.

_____

18. Give an example of a synovial ball-and-socket joint.

_____

Exercise 10.5: Practicing Synovial Joint Movements

19. For the following synovial joints, list the type(s) of movement allowed at the joint (e.g., flexion, extension, abduction, adduction, rotation, etc.).

Synovial Joint	Type of Movement
Shoulder	
Knee	
Carpals	
Wrist	
Ankle	
Atlas/axis	
Interphalangeal	
Metacarpal/phalangeal	
Hip	
Thumb	

20. Match the type of movement (gliding, angular, or rotation) in column A with its description listed in column B.

    Column A

    a. abduction
    b. adduction
    c. circumduction
    d. extension
    e. flexion
    f. gliding
    g. hyperextension
    h. lateral flexion
    i. rotation

    Column B

    ____ angle of joint decreases
    ____ movement of limb away from midline
    ____ angle of joint increases
    ____ movement of limb toward the midline
    ____ vertebral column moves in either lateral direction
    ____ extension past 180 degrees
    ____ flat surfaces slide past one another
    ____ limb pivots around its long axis
    ____ distal part of a limb moves in a circle

21. Match the type of special movement (gliding, angular, or rotation) in column A with its description in column B. ⓫

    Column A

    a. depression
    b. dorsiflexion
    c. elevation
    d. eversion
    e. inversion
    f. opposition
    g. plantar flexion
    h. pronation
    i. protraction
    j. retraction
    k. supination

    Column B

    ____ sole of foot is lateral (outward)
    ____ palm is turned posterior or inferior
    ____ movement of a body part superiorly
    ____ body part is moved anterior in the horizontal plane
    ____ sole of foot is medial (inward)
    ____ movement of a body part inferiorly
    ____ dorsum of foot moves superiorly
    ____ palm is turned superior or anterior
    ____ body part is moved posterior in horizontal plane
    ____ dorsum of foot is moved inferiorly
    ____ special movement of thumb across palm toward fingers

22. For each movement listed below, identify the opposite movement. ⓫

    a. abduction _____

    b. flexion _____

    c. depression _____

    d. dorsiflexion _____

    e. eversion _____

    f. pronation _____

    g. opposition _____

    h. protraction _____

Exercise 10.6: The Knee Joint

23. Match the appropriate knee joint structure in column A with its description in column B. ⓬

    Column A

    a. ACL and PCL
    b. meniscus
    c. fibular (lateral) collateral ligament
    d. patella
    e. patellar tendon
    f. bursae
    g. patellar ligament
    h. tibial (medial) collateral ligament

    Column B

    ____ ligament connecting femur to tibia
    ____ ligaments that cross in the middle of the knee joint
    ____ sesamoid bone located in the knee joint
    ____ ligament connecting femur to fibula
    ____ structure connecting patella to tibial tuberosity
    ____ structure connecting quadriceps muscles to patella
    ____ fibrocartilaginous structure within knee joint
    ____ synovial sacs located above, below, and anterior to the patella

Can You Apply What You've Learned?

24. Where do the articular cartilages arise developmentally?

    _____

    _____

    _____

25. Define *bursitis*.

26. How does a symphysis joint differ from a synchondrosis in terms of both structure and extent of movement?

27. Explain why the knee joint is *not* classified as a synchondrosis, even though it has hyaline cartilage.

28. Fully classify the glenohumeral joint, using both functional (movement) and structural terms.

Can You Synthesize What You've Learned?

29. Why do you think the anterior cruciate ligament is so often injured during contact sports?

30. Injuries to the tibial collateral ligament are often accompanied by a torn medial meniscus. After observing the fibrous tissues that form the joint capsule of the knee, why do you think this pattern of injury is so common?

CHAPTER 11

The Muscular System: Muscle Structure and Function

OUTLINE AND LEARNING OBJECTIVES

Histology 254

Skeletal Muscle Tissue 254

EXERCISE 11.1: HISTOLOGY OF SKELETAL MUSCLE FIBERS 256
- ① Identify skeletal muscle tissue through the microscope, and describe the features unique to skeletal muscle tissue
- ② Name the visible bands that form the striations in skeletal muscle tissue

EXERCISE 11.2: CONNECTIVE TISSUE COVERINGS OF SKELETAL MUSCLE 257
- ③ Describe the layers of connective tissue that surround skeletal muscle tissue

The Neuromuscular Junction 257

EXERCISE 11.3: THE NEUROMUSCULAR JUNCTION 259
- ④ Define motor unit and describe how the concept of a motor unit applies to neuromuscular junctions

Smooth Muscle Tissue 259

EXERCISE 11.4: SMOOTH MUSCLE TISSUE 260
- ⑤ Identify smooth muscle tissue through the microscope, and describe the features unique to smooth muscle tissue
- ⑥ Describe how two layers of smooth muscle tissue act as antagonists

Cardiac Muscle Tissue 260

EXERCISE 11.5: CARDIAC MUSCLE TISSUE 261
- ⑦ Identify cardiac muscle tissue through the microscope, and describe the features unique to cardiac muscle tissue
- ⑧ Compare and contrast the structure of skeletal, smooth, and cardiac muscle tissues

Gross Anatomy 261

Gross Anatomy of Skeletal Muscles 261

EXERCISE 11.6: NAMING SKELETAL MUSCLES 261
- ⑨ Explain some of the logic behind the naming of skeletal muscles

EXERCISE 11.7: ARCHITECTURE OF SKELETAL MUSCLES 264
- ⑩ Use anatomic terminology to describe the architecture of skeletal muscles, and describe how the architecture of a skeletal muscle is related to its action

Organization of the Human Musculoskeletal System 265

EXERCISE 11.8: MAJOR MUSCLE GROUPS AND FASCIAL COMPARTMENTS OF THE LIMBS 266
- ⑪ Describe the location and major actions of the major muscle groups of the body
- ⑫ Describe the fascial compartments of the limbs, and explain the major actions associated with each fascial compartment

Physiology 268

Force Generation of Skeletal Muscle 268

EXERCISE 11.9: MOTOR UNITS AND MUSCLE FATIGUE (HUMAN SUBJECT) 269
- ⑬ Compare and contrast the characteristics of Type I, Type IIa, and Type IIb muscle fibers
- ⑭ Describe the sequence of motor unit recruitment
- ⑮ Explain the relationship between motor unit recruitment, muscle force, and fatigue

EXERCISE 11.10: CONTRACTION OF SKELETAL MUSCLE (WET LAB) 270
- ⑯ Describe the effect of adding ATP alone, ATP + salts, or salts only on contraction of glycerinated muscle
- ⑰ Explain the relationship between ATP supply to skeletal muscle and rigor mortis

EXERCISE 11.11: Ph.I.L.S. LESSON 4: STIMULUS-DEPENDENT FORCE GENERATION 274
- ⑱ Describe the events of a muscle twitch
- ⑲ Describe the relationship between stimulus intensity and maximum isometric twitch force
- ⑳ Describe the threshold voltage, and explain what happens when suprathreshold stimuli are applied to skeletal muscle

EXERCISE 11.12: Ph.I.L.S. LESSON 5: THE LENGTH-TENSION RELATIONSHIP 276
- ㉑ Describe the relationship between muscle length and tension for a maximal isometric contraction
- ㉒ Describe what is happening at the sarcomere level when a muscle is contracting isometrically at its optimal length

EXERCISE 11.13: Ph.I.L.S. LESSON 6: PRINCIPLES OF SUMMATION AND TETANUS 278
- ㉓ Describe the relationship between stimulus frequency and muscle tension
- ㉔ Demonstrate the concepts of summation, incomplete tetanus, and complete tetanus in a virtual experiment

EXERCISE 11.14: Ph.I.L.S. LESSON 7: EMG AND TWITCH AMPLITUDE 279
- ㉕ Describe the relationship between muscle tension and EMG amplitude

MODULE 6: MUSCULAR SYSTEM

INTRODUCTION

As you are reading the text on this page, skeletal muscles connected to your eyeballs (the *extrinsic* eye muscles) are contracting to produce the very fine movements necessary for your eyes to track the words on the paper. At the same time, smooth muscles within the ciliary bodies of the eyes are contracting to alter the shape of the lens so the image on the page is focused clearly upon your retina. In comparison, contraction of cardiac muscle in your heart is creating the force necessary to propel blood through your arteries to deliver oxygen and glucose to the working tissues within your eyes, your brain, and the rest of your body. Clearly, you require the use of muscle tissue to read the words on this page. Indeed, properly functioning muscle tissue is essential for your very survival.

Muscle tissue is one of the most metabolically active tissues in the body and is the one tissue capable of creating movement of the body or body organs. All types of muscle tissue are characterized by the following properties: *excitability, contractility, elasticity,* and *extensibility.* **Excitable** tissues are able to generate and propagate special electrical signals called *action potentials* (APs). **Contractile** tissues actively shorten themselves and produce force. **Elastic** tissues return to their original shape following either contraction or stretching. **Extensible** tissues are able to be lengthened by the pull of an external force (such as an external weight or the action of an opposing muscle). **Conductive** tissues allow electrical signals such as action potentials to travel along the plasma membrane of the cell.

There are three types of muscle tissue: skeletal muscle, smooth muscle, and cardiac muscle. **Skeletal muscle** comprises the voluntary muscles that move the skin of the face and the skeleton, **smooth muscle** is found mainly in the walls of the viscera (such as the blood vessels, stomach, urinary bladder, intestines, and uterus), and **cardiac muscle** is found in the heart. These three types of muscle tissue are distinguished from each other based on location, neural control (voluntary vs. involuntary), the presence or absence of visible striations (striped appearance), the shape of the cells, and the number of nuclei per cell. In this laboratory session, you will explore each of these features to gain an appreciation for the similarities and differences between the three types of muscle tissue. You will also begin your exploration of the human musculoskeletal system, and finally, you will observe many of the physiological properties of skeletal muscle.

In this laboratory session you will observe the detailed structure of skeletal, smooth, and cardiac muscle tissues as viewed through the microscope. In addition, you will observe common morphologies of skeletal muscles and begin to develop an understanding of the logic behind the naming of skeletal muscles. You will also begin your study of the skeletal muscles of the body. The exercises in this chapter will give you an overview of the organization of the human musculoskeletal system. If you take to this introductory task and learn the information well, you will be prepared to learn the detailed names, bony attachments, and actions of the individual muscles of the body that will be introduced in chapters 12 and 13. By starting with an understanding of the overall organization of the human musculoskeletal system and the functions of specific muscle groups, the task of remembering the minutiae for each individual muscle will become much easier and will seem less challenging.

Finally, you will perform some exercises that will allow you to gain an understanding of some of the physiological properties of skeletal muscle tissue. These exercises include looking at the relationship between motor units, muscle recruitment, and fatigue using a human subject; observing the effect of ATP and salts on glycerinated skeletal muscle fibers; and observing both the relationship between stimulation intensity and force generation by skeletal muscle and the relationship between muscle length and tension produced, using physiology simulation exercises in Ph.I.L.S.

INTEGRATE

LEARNING STRATEGY

Terminology related to the action of muscle fibers can be very confusing. Recall from chapter 10 that the term *flexion* refers to any joint movement in which the angle between two bones is decreased. However, people often use the term *flex* (as in "flex your bicep for me") when they are asking you to *contract* a muscle. The term *contract* is the term scientists prefer to use to describe the action of a muscle fiber producing force. Even so, the term *contract* can also be misleading at times! Literally, to contract means to shorten. All muscle fibers have the ability to shorten when they produce force. However, skeletal muscle fibers can also produce force while remaining at a fixed length, or even while being forcibly lengthened.

Chapter 11: The Muscular System: Muscle Structure and Function

Name: _____

Date: _____ Section: _____

PRE-LABORATORY WORKSHEET

Also available at www.connect.mcgraw-hill.com

|ANATOMY & PHYSIOLOGY

1. List the three types of muscle tissue, and give a location in the body where each type of tissue is found.

   Muscle Tissue   **Location in Body**

   a. _____   a. _____

   b. _____   b. _____

   c. _____   c. _____

2. What does it mean for a tissue to be excitable?

   _____

   _____

3. What does it mean for a tissue to be contractile?

   _____

4. The structural and functional unit of skeletal muscle is called a _____.

5. Define *tendon:* _____

   _____

6. Intercalated discs are characteristic of _____ muscle tissue.

7. Cylindrical bundles of contractile proteins located inside skeletal muscle fibers are called _____.

8. Skeletal muscles are given names that reflect location, shape, attachments, or other features related to the muscles. These names are based on Latin and Greek word roots. Match the word root given in column A with its corresponding meaning in column B.

   Column A   **Column B**

   ____ 1. *brevis*   a. around

   ____ 2. *capitis*   b. at an angle

   ____ 3. *endo-*   c. belly

   ____ 4. *epi-*   d. between

   ____ 5. *gastro-*   e. head

   ____ 6. *inter-*   f. large

   ____ 7. *magnus*   g. short

   ____ 8. *oblique*   h. straight

   ____ 9. *peri-*   i. upon

   ____ 10. *rectus*   j. within

Chapter Eleven *The Muscular System: Muscle Structure and Function* **253**

Histology

In the following exercises you will observe the histology of skeletal, smooth, and cardiac muscle tissues. **Table 11.1** lists the characteristics of the three types of muscle tissue, which you can use as a reference as you work through the exercises.

Skeletal Muscle Tissue

Skeletal muscle cells are some of the largest cells in the body. Their enormous size results from the fusion of hundreds of *myoblasts* (embryonic muscle cells) during development into a single muscle fiber. A mature muscle fiber is multinucleate, containing 200 to 300 nuclei per millimeter of fiber length. The nuclei of normal skeletal muscle fibers are located directly under the *sarcolemma* (plasma membrane of a muscle fiber).

A muscle, such as the biceps brachii muscle of the arm, consists of several bundles, or **fascicles** (*fascis*, bundle), of muscle fibers (**figure 11.1a**). Each fascicle contains hundreds of long, cylindrical **muscle fibers** (figure 11.1b). Each muscle fiber contains within it a number of cylindrical bundles of contractile proteins called **myofibrils** (figure 11.1c). The myofibrils contain the **myofilaments** actin and myosin (the main contractile proteins of muscle), which are arranged into **sarcomeres** (the structural and functional unit of skeletal muscle, figure 11.1d). The regular arrangement of actin and myosin into sarcomeres gives each myofibril a striated or banded appearance, with visible *A bands* (dark bands), *I bands* (light bands), and *Z discs* (a dark line in the middle of an I band) (figure 11.1d,e).

An adult skeletal muscle fiber typically contains about 2000 myofibrils per cell. Because intermediate filaments within the muscle fiber anchor and align the Z discs of adjacent myofibrils, the entire muscle fiber takes on the same regular striated appearance of A bands and I bands found in the myofibrils when viewed through the light microscope.

When viewed through the light microscope, the A band appears dark in color. The dark color is due to the presence of thick filaments composed of **myosin**. Because the filaments are relatively thick, they absorb light, which is what causes the A band to appear dark. On the other hand, the I band contains thin filaments composed primarily of **actin**. Because the filaments are so thin, light passes easily through them, causing the I band to appear light. Finally, there are anchoring proteins found at the **Z discs,** which are dense like the A bands and cause light to be absorbed in that location. Thus, Z discs appear dark and are visible in the middle of each I band when the slide is viewed at high power.

Figure 11.1 demonstrates the relationships between the gross structure of a skeletal muscle and the microstructure of a skeletal muscle fiber. Though you will not be able to see myofibrils or myofilaments when you observe the muscle tissue through the microscope, you should be able to visualize A bands, I bands, and Z discs, and relate these bands to the corresponding arrangement of myofilaments into sarcomeres. That is, you should be able to relate what you see through the microscope to the structures shown in the drawing of a sarcomere in figure 11.1.

Table 11.1	Muscle Tissues		
Type of Muscle	**Description**	**Functions**	**Identifying Characteristics**
Skeletal	Elongate, cylindrical cells with multiple nuclei that are peripherally located. Tissue appears striated (alternating light and dark bands along the length of the cell).	Produces voluntary movement of the skin and the skeleton	Length of cells (extremely long); striations; multiple, peripheral nuclei
Smooth	Elongate, spindle-shaped cells (fatter in the center, narrowing at the tips) with single, "cigar-shaped" nuclei that are centrally located. No striations.	Produces movement within visceral organs such as intestines, bladder, uterus, and stomach (e.g., propels food through the gastrointestinal tract)	Spindle-shape of the cells; lack of striations; cigar-shaped or spiral nuclei
Cardiac	Short, branched cells with a single nucleus that is centrally located. Dark lines (intercalated discs) are seen where two cells bind; tissue appears striated (light and dark bands along the length of each cell).	Performs the contractile work of the heart. Responsible for the pumping action of the heart.	Branched, uninucleate cells; striations; intercalated discs

Chapter Eleven *The Muscular System: Muscle Structure and Function* **255**

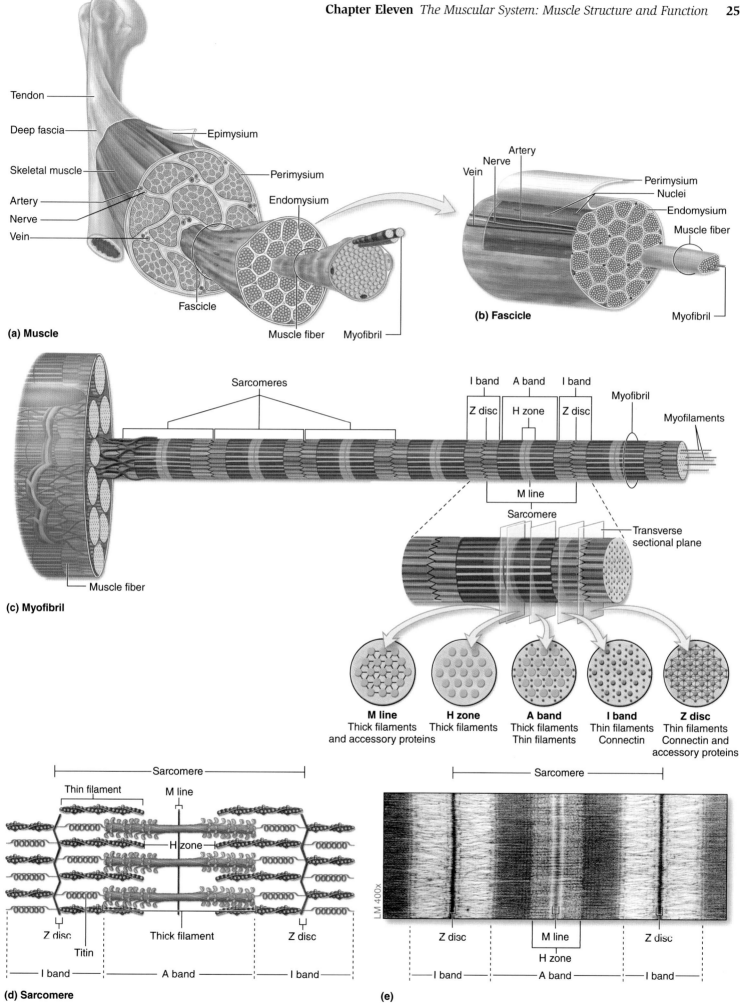

AP|R Figure 11.1 Levels of Structural Organization of Skeletal Muscle. (a) A whole muscle, (b) a muscle fascicle, (c) a myofibril, (d) a sarcomere, (e) an electron micrograph of a sarcomere.

EXERCISE 11.1

HISTOLOGY OF SKELETAL MUSCLE FIBERS

1. Obtain a slide of skeletal muscle tissue and place it on the microscope stage.

2. Bring the tissue into focus using the scanning objective. Switch to low power, bring the tissue sample into focus once again, and then switch to high power. Many slides of skeletal muscle contain muscle fibers shown in both longitudinal section and cross section. Scan the slide and identify muscle fibers shown in both longitudinal section and cross section (**figures 11.2** and **11.3**).

3. Focus on muscle fibers shown in longitudinal section (figure 11.2) and observe them at high power. Identify an individual muscle fiber. Notice the numerous peripherally located nuclei. Most of the nuclei that you see on the slide belong to the muscle fibers. However, about 5% to 15% of the visible nuclei are those of **satellite cells,** myoblast-like cells located between the muscle fibers. Satellite cells give skeletal muscle a limited ability to repair itself after injury. Through the light microscope you will not be able to tell which nuclei belong to satellite cells.

4. Using table 11.1 and figure 11.2 as guides, identify the following structures on the slide of skeletal muscle tissue:

   ☐ A band ☐ sarcomere (if visible)
   ☐ I band ☐ Z disc (if visible)
   ☐ nucleus

5. In the space below, compose a sketch of skeletal muscle fibers as viewed through the microscope. Be sure to label the structures listed above in your drawing.

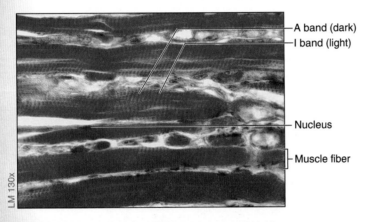

Figure 11.2 Skeletal Muscle in Longitudinal View.
Z discs and sarcomeres are not visible in this micrograph.

EXERCISE 11.2

CONNECTIVE TISSUE COVERINGS OF SKELETAL MUSCLE

Now that you have observed the structure of individual muscle fibers, it is time to direct your attention to the relationship between individual muscle fibers and whole muscles. A whole skeletal muscle, such as the biceps brachii muscle, consists of many individual muscle fibers bundled together with connective tissue (figure 11.3). Each individual muscle fiber is covered by a thin layer of connective tissue called the **endomysium** (*endo*, within, + *mys*, muscle). Several muscle cells are bundled together into **fascicles** (*fascis*, bundle) by a surrounding layer of connective tissue called the **perimysium** (*peri*, around, + *mys*, muscle). Finally, the entire skeletal muscle is surrounded by a layer of connective tissue called the **epimysium** (*epi*, upon, + *mys*, muscle). The epimysium is an extension of the **deep fascia**, which will be discussed shortly.

1. You should still have the slide of skeletal muscle on the microscope stage. View the slide at low magnification and move the stage until you can see muscle fibers in cross section. In this view you will be able to identify the connective tissues that bundle the muscle fibers together to form a whole skeletal muscle.

2. Using figure 11.3 as a guide, identify the following structures:*

   ☐ endomysium   ☐ fascicle   ☐ nucleus
   ☐ epimysium   ☐ muscle fiber ☐ perimysium

3. In the space below, compose a brief sketch of the connective tissue coverings of skeletal muscle as viewed through the microscope. Be sure to label the structures listed in step 2 in your drawing.

_____ ×

*If the slide is skeletal muscle from a small mammal such as a mouse, you should be able to identify all three layers of connective tissue. However, if it is from a larger mammal, you will most likely be able to identify only endomysium and perimysium.

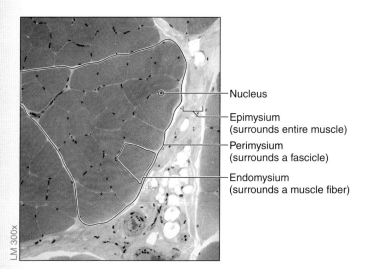

Figure 11.3 Connective Tissue Coverings of Skeletal Muscle. Muscle fibers are seen in a cross-sectional view.

The Neuromuscular Junction

Within each skeletal muscle fascicle are several individual muscle fibers. However, these fibers are generally of mixed fiber type (e.g., "fast" vs. "slow") and are innervated by several somatic motor neurons. **Figure 11.4a** shows the anatomical relationship of a motor neuron and a skeletal muscle. Each motor neuron branches out into several axonal branches that innervate muscle fibers dispersed throughout the muscle. A single motor neuron and all muscle fibers innervated by that motor neuron constitute a **motor unit.** When a nerve signal (an action potential) travels down the motor neuron to stimulate the muscle, all fibers within that motor unit contract. The point of interaction between an axonal branch with an individual skeletal muscle fiber is called a **neuromuscular junction** (figure 11.4b–d). Each neuromuscular junction is composed of an oval-shaped synaptic knob of a motor neuron, a motor end plate of a skeletal muscle fiber, and the synaptic cleft (the space between the two). In this exercise you will observe a neuromuscular junction in a histology slide that has been stained to show axonal branches of the motor neuron and synaptic knobs.

258 Chapter Eleven *The Muscular System: Muscle Structure and Function*

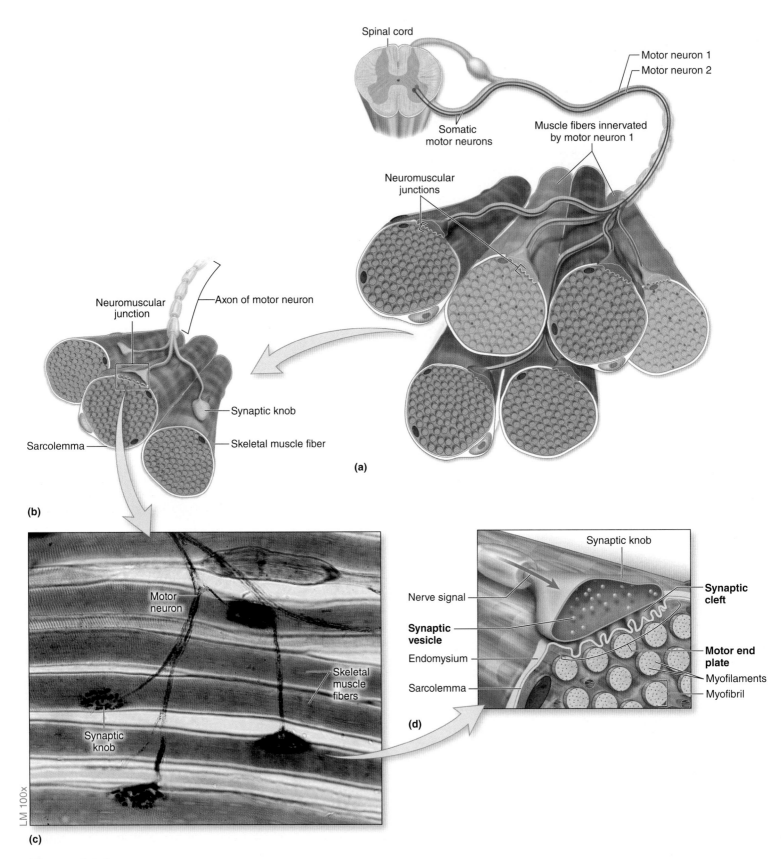

Figure 11.4 **The Neuromuscular Junction.** (a) Axons of somatic motor neurons that arise in the spinal cord extend axons to skeletal muscles. Within the skeletal muscle, each motor neuron branches out to innervate several individual muscle fibers to form a motor unit. (b) Each of the fibers within the motor unit has an area where the synaptic knob of the motor neuron contacts the muscle fiber; the neuromuscular junction. (c) Histologically, the axonal branches and synaptic knobs can be viewed as darkly stained structures associated with skeletal muscle fibers. (d) At each synaptic knob, there are synaptic vesicles that contain the neurotransmitter acetylcholine (ACh).

EXERCISE 11.3

THE NEUROMUSCULAR JUNCTION

1. Obtain a slide of a neuromuscular junction and place it on the microscope stage.

2. Bring the tissue sample into focus using the scanning objective. Switch to low power, bring the tissue sample into focus once again, and then switch to high power. Scan the slide until you locate skeletal muscle fibers. Then look for darkly staining thin fibers with "knobby" ends. The "fibers" are the **axonal branches** of a somatic motor neuron and the oval-shaped "knobs" are the neuromuscular junctions that contain the synaptic knobs of the somatic motor neuron (figure 11.4).

3. Using figure 11.4 as a guide, identify the following structures on the slide of the neuromuscular junction:

   ☐ axonal branches of the somatic motor neuron
   ☐ skeletal muscle fibers
   ☐ synaptic knob

4. In the space below, compose a brief sketch of the neuromuscular junction as viewed through the microscope. Be sure to label the structures listed in step 3.

Smooth Muscle Tissue

Spindle-shaped cells with cigar-shaped or spiral nuclei and an absence of striations is a characteristic of smooth muscle tissue. Individual cells have tapered ends, and there is only one centrally located nucleus per cell. Most of the nuclei will appear to be somewhat cigar-shaped. However, in cells that have contracted, the nuclei take on a corkscrew or spiral appearance **(figure 11.5a)**, which can be a key identifying feature.

Smooth muscle is generally found in two layers around tubular organs such as the small intestine. The most common arrangement is an *inner circular layer* and an *outer longitudinal layer* of smooth muscle cells (figure 11.5b).

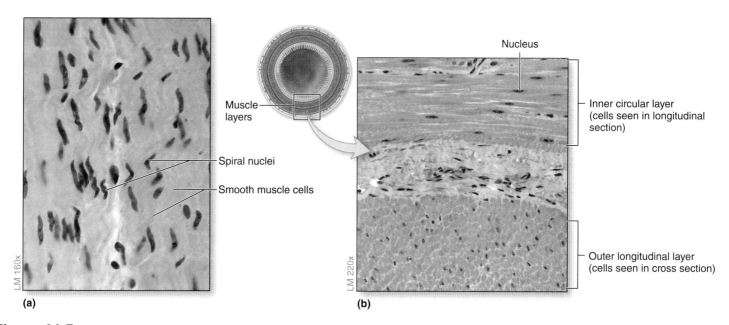

Figure 11.5 Smooth Muscle Tissue. (a) Close-up view of smooth muscle demonstrating spiral nuclei, which are visible when the muscle fibers are contracted; (b) circular and longitudinal layers of smooth muscle tissue.

EXERCISE 11.4

SMOOTH MUSCLE TISSUE

1. Obtain a slide containing smooth muscle tissue (figure 11.5) and place it on the microscope stage.

2. Bring the tissue into focus using the scanning objective. Switch to low power, bring the tissue sample into focus once again, and then switch to high power. If you are viewing a slide of the small intestine, identify smooth muscle cells that have been cut in both longitudinal section and cross section. Note that the cells do not appear to be of uniform diameter when viewed in cross section. This is because some cells are sliced through the tapered ends, while others are sectioned through the thickest part of the fiber, which contains the nucleus.

3. Using table 11.1 and figure 11.5 as guides, identify the following structures on the slide of smooth muscle tissue:

   ☐ inner circular layer
   ☐ muscle cell in cross section
   ☐ muscle cell in longitudinal section
   ☐ nucleus
   ☐ outer longitudinal layer

4. In the space below, compose a brief sketch of smooth muscle cells as viewed through the microscope. Be sure to label the structures listed in step 3 in your drawing.

_____ ×

INTEGRATE

LEARNING STRATEGY

A common error is to confuse smooth muscle tissue and dense regular connective tissue. A few key features will help you to distinguish them from each other. First, fibroblast nuclei (found in dense regular connective tissue) appear flattened and are located between fibers, whereas smooth muscle fiber nuclei are plumper and can be seen *within* the cells. In addition, there will be relatively fewer nuclei per unit area in dense regular connective tissue than in smooth muscle tissue. Finally, the appearance of spiral or corkscrew nuclei is a good clue that the tissue is smooth muscle, not dense regular connective tissue. See if you can find a spiral nucleus on the slide of smooth muscle tissue to make sure you understand what it is.

Cardiac Muscle Tissue

Branching cells with centrally located nuclei are characteristic of **cardiac muscle tissue (figure 11.6).** Like skeletal muscle fibers, cardiac muscle cells are striated. The striations appear because cardiac muscle cells also contain sarcomeres within myofibrils. Unlike skeletal muscle cells, the myofibrils in cardiac muscle cells are not anchored to each other with intermediate filaments at the Z discs. The resulting slight offset of myofibrils within the muscle fiber means that the banding pattern isn't always as clear in cardiac muscle as it is in skeletal muscle.

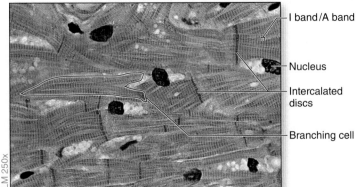

Figure 11.6 Cardiac Muscle Tissue.

EXERCISE 11.5

CARDIAC MUSCLE TISSUE

1. Obtain a slide of cardiac muscle tissue and place it on the microscope stage.

2. Bring the tissue sample into focus using the scanning objective. Switch to low power, bring the tissue sample into focus once again, and then switch to high power. Notice that the cells contain only one or two nuclei, and that the cells are short and branched. Where two cells come together, you will notice a darkly stained line. This is an **intercalated disc.** Intercalated discs contain numerous **desmosomes,** which function to hold the cells together, and **gap junctions,** which allow electrical signals to be transmitted very rapidly from one cell to the next.

3. Using table 11.1 and figure 11.6 as guides, identify the following structures on the slide of cardiac muscle tissue:

   ☐ A band   ☐ intercalated disc
   ☐ branching cells   ☐ nucleus
   ☐ I band

4. In the space below, compose a brief sketch of cardiac muscle cells as viewed through the microscope. Be sure to label the structures listed in step 3.

_____ ×

Gross Anatomy

Gross Anatomy of Skeletal Muscles

The names of skeletal muscles often seem overly complex. However, if you begin to understand the basis of their names, you will find them far less mysterious. Over time you will begin to discover that the name of a muscle often gives you a clue to its location, size, shape, action, or attachment points. The efforts you have made to learn the Latin and Greek word roots of anatomical terms thus far will become even more valuable as you work through the next three chapters.

EXERCISE 11.6

NAMING SKELETAL MUSCLES

Table 11.2 summarizes some of the common ways skeletal muscles are named and gives word origins for the muscle names. Plan to spend some time on your own mastering these word origins so you will be better prepared to handle the material to come in chapters 12 and 13. To help you in this effort, the post-laboratory worksheet has a number of questions related to the naming of skeletal muscles.

Table 11.2	Common Methods for Naming Skeletal Muscles		
Name	Meaning	Word Origin	Example
Naming Skeletal Muscles Based on Shape			
Deltoid	Triangular	*delta*, the Greek letter delta (a triangle), + *eidos*, resemblance	Deltoid
Gracilis	Slender	*gracilis*, slender	Gracilis
Lumbrical	Wormlike	*lumbricus*, earthworm	Lumbricals

(continued on next page)

(continued from previous page)

Table 11.2 Common Methods for Naming Skeletal Muscles *(continued)*

Name	Meaning	Word Origin	Example
Rectus	Straight	*rectus*, straight	Rectus abdominis
Rhomboid	Diamond-shaped	*rhombo-*, an oblique parallelogram with unequal sides, + *eidos*, resemblance	Rhomboid major
Teres	Round	*teres*, round	Teres major
Trapezius	A four-sided geometrical figure having no two sides parallel	*trapezion*, a table	Trapezius
Naming Skeletal Muscles Based on Size			
Brevis	Short	*brevis*, short	Adductor brevis
Latissimus	Broadest	*latissimus*, widest	Latissimus dorsi
Longissimus	Longest	*longissimus*, longest	Longissimus capitis
Longus	Long	*longus*, long	Adductor longus
Major	Bigger	*magnus*, great	Teres major
Minor	Smaller	*minor*, smaller	Teres minor
Naming Skeletal Muscles Based on the Number of Heads and/or Bellies			
Biceps	2 heads	*bi*, two, + *caput*, head	Biceps brachii
Digastric	2 bellies	*bi*, two, + *gastro*, belly	Digastric
Quadriceps	4 heads	*quad*, four, + *caput*, head	Quadriceps femoris
Triceps	3 heads	*tri*, three, + *caput*, head	Triceps brachii
Naming Skeletal Muscles Based on Position			
Abdominis	Abdomen	*abdomen*, the greater part of the abdominal cavity	Rectus abdominis
Anterior	On the front surface of the body	*ante-*, before, in front of	Serratus anterior
Brachii	Arm	*brachium*, arm	Biceps brachii
Dorsi	Back	*dorsum*, back	Latissimus dorsi
Femoris	Thigh	*femur*, thigh	Rectus femoris
Infraspinatus	Below the scapular spine	*infra-*, below, + *spina*, spine	Infraspinatus
Interosseous	In between bones	*inter*, between + *osseus*, bone	Interossei
Oris	Mouth	*oris*, mouth	Orbicularis oris
Pectoralis	Chest	*pectus*, chest	Pectoralis major
Posterior	On the back surface of the body	*posterus*, following	Serratus posterior
Supraspinatus	Above the scapular spine	*supra-*, on the upper side, + *spina*, spine	Supraspinatus

Table 11.2 Common Methods for Naming Skeletal Muscles (continued)

Name	Meaning	Word Origin	Example
Naming Skeletal Muscles Based on Depth			
Externus	External	*external*, on the outside	Obturator externus
Internus	Internal	*internal*, away from the surface	Obturator internus
Profundus	Deep	*pro*, before, + *fundus*, bottom	Flexor digitorum profundus
Superficialis	Superficial	*super*, above, + *facies*, face	Flexor digitorum superficialis
Naming Skeletal Muscles Based on Action			
Abductor	Moves a body part away from the midline	*ab*, from, + *ductus*, to bring toward	Abductor pollicis brevis
Adductor	Moves a body part toward the midline	*ad*, toward, + *ductus*, to bring toward	Adductor pollicis
Constrictor	Acts as a sphincter and closes an orifice	*cum*, together, + *stringo*, to draw tight	Superior pharyngeal constrictor
Depressor	Flattens or lowers a body part	*de-*, away, + *pressus*, to press	Depressor anguli oris
Dilator	Causes an orifice to open, or dilate	*dilato*, to spread out	Dilator pupillae
Extensor	Causes an increase in joint angle	*ex-*, out of, + *-tensus*, to stretch	Extensor carpi ulnaris
Flexor	Causes a decrease in joint angle	*flectus*, to bend	Flexor carpi ulnaris
Levator	Raises a body part superiorly	*levo* + *atus*, a lifter	Levator scapulae
Pronator	Turns the palm of the hand from anterior to posterior	*pronatus*, to bend forward	Pronator teres
Supinator	Turns the palm of the hand from posterior to anterior	*supino* + *atus*, to bend backward	Supinator

EXERCISE 11.7

ARCHITECTURE OF SKELETAL MUSCLES

The overall architecture of a skeletal muscle affects how the muscle functions. When you observe a whole muscle, the individual fibers and fascicles are visible, making it relatively easy to see how the fascicles are arranged within the muscle. Recall that when skeletal muscle contracts, it generally gets shorter and brings the two attachment points closer to each other. Thus, the orientation of the muscle fascicles compared to the attachment points of the muscle will directly affect the force produced by the muscle and the complexity of the muscle's actions. For instance, **pennate** architecture (*penna*, feather) allows a muscle to produce greater force per distance shortened than **parallel** architecture. In addition, muscles with more than two attachments (for example, biceps and triceps) produce more complex movements than muscles with only two attachments (one proximal attachment and one distal attachment). **Table 11.3** summarizes the common patterns of fascicle arrangement that contribute to skeletal muscle architecture.

1. Using classroom models of skeletal muscles or a prosected human cadaver (if available), observe the arrangement of fascicles in several different skeletal muscles of the body.

2. Locate the following muscles on the classroom models or the human cadaver. Then identify them on **figure 11.7**.

    - ☐ deltoid
    - ☐ extensor digitorum
    - ☐ gastrocnemius
    - ☐ orbicularis oculi
    - ☐ pectoralis major
    - ☐ rectus femoris
    - ☐ sartorius
    - ☐ trapezius
    - ☐ triceps brachii

3. Using table 11.3 and your textbook or atlas as guides, fill in the table below with the names of the muscles numbered in figure 11.7. Then, indicate the type of architecture represented by each muscle.

Muscle Number	Muscle Name	Architecture
1		
2		
3		
4		
5		
6		
7		
8		
9		

Table 11.3 Common Architectures of Skeletal Muscles

Name	Unipennate	Bipennate	Multipennate	Circular	Convergent	Parallel
Word Origin	*uni*, one, + *penna*, feather	*bi*, two, + *penna*, feather	*multi*, many, + *penna*, feather	*circum*, around	*cum*, together, + *vergo*, to incline	*para*, alongside

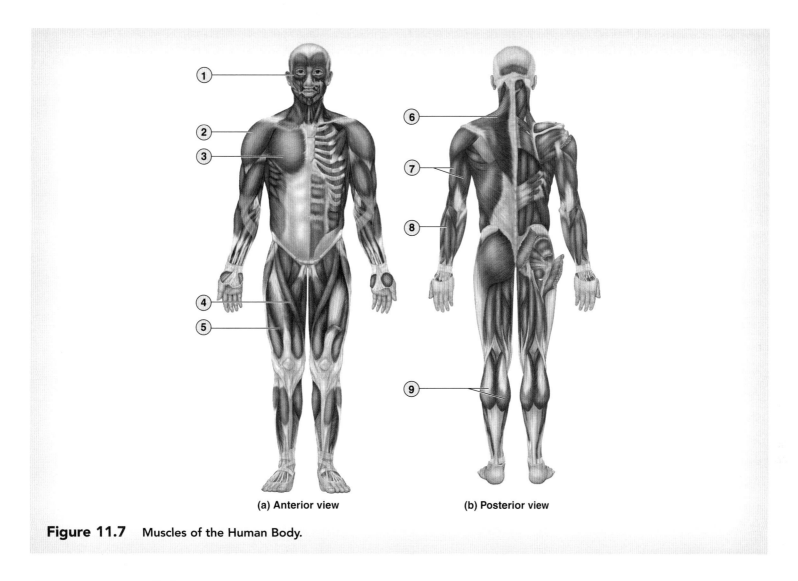

(a) Anterior view (b) Posterior view

Figure 11.7 Muscles of the Human Body.

Organization of the Human Musculoskeletal System

Table 11.4 lists the major muscle groups of the body and the common actions of each group of muscles. Obviously, each individual muscle within a group of muscles has its own specific actions. However, skeletal muscles do not act alone, so it makes sense to learn the muscles as groups and to learn the major actions of each group. Muscles contained within a group commonly act as synergists (muscles that facilitate the prime mover) and often have a common nerve and blood supply. Thus, damage to the nerve and blood supply often affects a whole group of muscles, not just an individual muscle.

In the limbs, muscles are not only *functionally* organized into groups; they are also *physically* organized into groups. Muscle groups

Table 11.4	Major Muscle Groups and Their Actions
Major Muscle Group	**General Description of Muscle Actions**
Muscles of facial expression	Produce facial expressions
Muscles of mastication	Used for chewing
Suprahyoid and infrahyoid muscles	Move the tongue and pharynx
Muscles of the neck	Rotate, flex, and extend the head
Muscles of respiration	Involved in breathing movements
Muscles of the abdomen	Flex, bend, and rotate the spine
Muscles of the back and spine	Extend, bend, and rotate the spine
Muscles of the pelvic floor	Support pelvic contents and form sphincters around structures such as the urethra and the anus
Muscles that act about the pectoral girdle	Stabilize the pectoral girdle and anchor the upper limb to the pectoral girdle
Muscles that act about the pelvic girdle	Stabilize the pelvic girdle and anchor the lower limb to the pelvic girdle

are separated from each other by extensions of the deep fascia that form **compartments (table 11.5).** Why is it important to learn this relationship? As an example, consider the muscles of the arm. In this laboratory session you will learn that the muscles located in the anterior compartment of the arm act primarily in flexion of the shoulder and elbow joints. In chapter 16 you will learn that this entire group of muscles is innervated by a single bundle of nerve fibers, the musculocutaneous nerve. Even if you forget the names of the individual muscles in the anterior compartment of the arm, you will be able to predict the loss of function that would result if the musculocutaneous nerve was damaged, based on your knowledge of compartments and muscle groups.

Table 11.5	Fascial Compartments of the Limbs and Their General Muscle Actions
Compartment	**General Description of Muscle Actions**
Compartments of the Arm	
Anterior	Flexion of shoulder and elbow
Posterior	Extension of shoulder and elbow
Compartments of the Forearm	
Anterior	Flexion of the wrist and digits (fingers)
Posterior	Extension of the wrist and digits (fingers)
Compartments of the Thigh	
Anterior	Extension of the knee, flexion of the hip
Posterior	Extension of the hip, flexion of the knee
Medial	Adduction of the thigh
Compartments of the Leg	
Anterior	Dorsiflexion and inversion of the ankle, extension of the digits (toes)
Posterior	Plantarflexion and inversion of the ankle, flexion of the digits (toes)
Lateral	Eversion of the ankle

EXERCISE 11.8

MAJOR MUSCLE GROUPS AND FASCIAL COMPARTMENTS OF THE LIMBS

1. Using tables 11.4 and 11.5 as guides, identify each of the major muscle groups and compartments of the body.

2. As you identify each muscle group or compartment, practice performing the actions listed in the tables. Try to feel the contraction of the muscles by palpating the skin overlying the muscle groups. If you have difficulty remembering joint actions, refer back to chapter 10 for descriptions.

3. **Figures 11.8** and **11.9** represent cross sections of the arm (brachium) and thigh, respectively. The details in cross-sectional views such as these are difficult to negotiate the first time through. Thus, right now just focus on the organization of the muscles into compartments. Visualize how the cross-sectional diagram relates to your own arm or thigh. As you perform the actions listed for each compartment, correlate the muscles used to perform each action with a specific compartment of the arm or thigh.

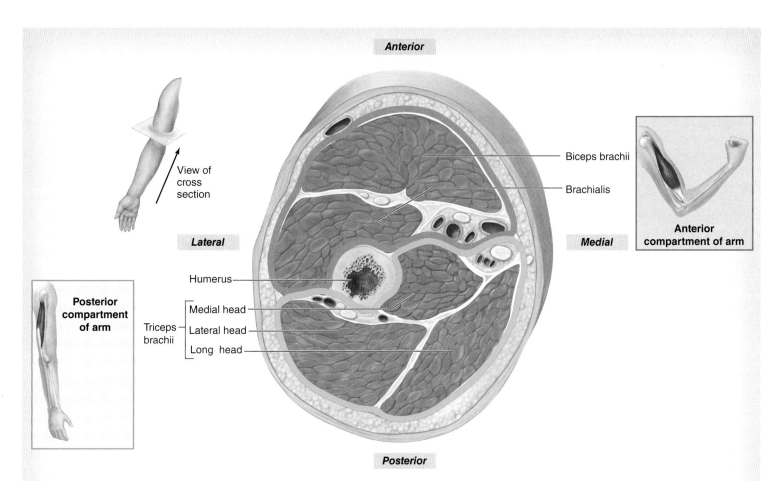

Figure 11.8 Fascial Compartments of the Right Arm (Brachium).

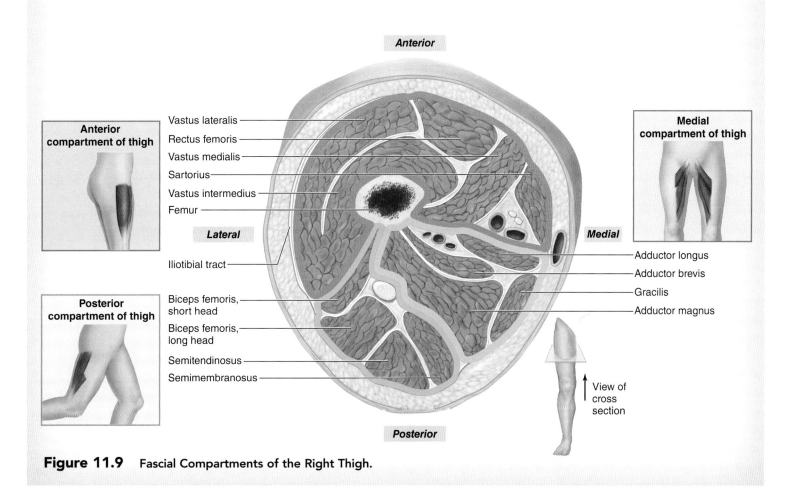

Figure 11.9 Fascial Compartments of the Right Thigh.

Physiology

Force Generation of Skeletal Muscle

When we wish to increase the total force generated by a skeletal muscle *in vivo*, we do so by an orderly recruitment of motor units. A **motor unit** consists of a single motor neuron and all the fibers that it innervates. For most activities, the pattern of motor unit recruitment occurs such that the smallest motor units, which consist of Type I fibers, are recruited first. Successive increases in muscle force are then achieved by recruitment of increasingly larger motor units. Because each motor unit consists of muscle fibers that are all of the same type, this also means that recruitment follows the pattern of Type I fibers first, then Type IIa, and finally Type IIb. This order of recruitment (I → IIa → IIb) ensures that successive addition of larger and larger motor units will cause incremental increases in total force produced by the muscle (**Figure 11.10**).

Table 11.6 summarizes many of the properties of the three main muscle fiber types. We have already talked about the relationship between fiber type and motor unit size. Now notice the relationship between fiber type and resistance to fatigue (row 3 in table 11.6). Notice that the largest motor units, which produce the greatest force, are also the first to fatigue. This is why we find it impossible to sustain a maximal contraction for a long period of time (such as when lifting a very heavy weight). On the flip side, the smallest motor units are very fatigue resistant. Thus, activities that require only tonic contraction of small motor units (such as standing still) can be maintained almost indefinitely.

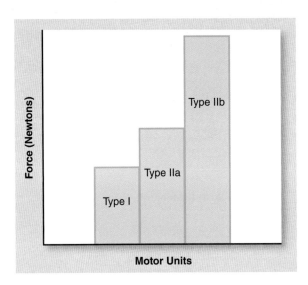

Figure 11.10 Recruitment of Motor Units. The first motor units recruited consist of Type I fibers, which produce very little force. Further recruitment brings in Type IIa fibers, which produce more force. Type IIb fibers are recruited last. Maximal force generation by a muscle requires recruitment of all fibers in the muscle.

Table 11.6	Properties of Skeletal Muscle Fiber Types		
Fiber Type	Type I (SO; Slow Oxidative)	Type IIa (FOG; Fast Oxidative-glycolic)	Type IIb (FG; Fast-glycolic)
Fiber cross-sectional area	Small	Medium	Large
Order of recruitment	First	Second	Third
Resistance to fatigue	Large	Medium	Small
Size of motor units	Small	Medium	Large
Speed of contraction	Slow	Fast	Fast
Twitch force	Small	Medium	Large
Twitch speed	Slow	Fast	Fast

EXERCISE 11.9

MOTOR UNITS AND MUSCLE FATIGUE (HUMAN SUBJECT)

In this exercise you will observe the relationship between motor unit recruitment and fatigue by successively working motor units to the point of fatigue. You will begin with a maximal contraction holding a heavy weight until the point of fatigue and will measure the time it takes to reach fatigue. You will then hold successively lighter weights until the point of fatigue and will measure that time as well.

In this experiment, what is the independent variable?

_____

What is the dependent variable? _____

Before you begin, state your hypothesis regarding the relationship between force generation and muscle fatigue.

_____

_____

Obtain the Following:
- hand barbells of the following weight increments: 1 lb, 2 lb, 5 lb, 10 lb, 25 lb
- stopwatch (or other mechanism for keeping track of time)

1. Choose a volunteer to be your subject. Your subject will need to initially guess the maximum weight that he or she will be able to hold in an outstretched arm for a period of time. Begin with the heaviest weight your subject can comfortably hold (even if only for a few seconds). To start your first trial, have the subject pick up the weight with one hand, extend the arm laterally so that the arm is positioned parallel to the floor, and hold the weight in that position for as long as he or she can.

2. As soon as your subject gets the upper limb into position with the weight, start your stopwatch.

3. Observe the subject's upper limb for signs of fatigue (shaking, trembling, and an inability to hold the weight steady). As soon as the subject can no longer hold the weight in position (i.e., upper limb falls below 90 degrees), stop the timing and have the subject lay the weight down. Record both the weight used and the time to fatigue in **table 11.7**.

4. Trial #2: Have your subject repeat the process above using a lighter weight. For example, if he started with a 25-lb weight, have him hold the 10-lb weight for trial #2. Again, record the weight and time to fatigue in table 11.7.

5. Repeat the process with successive trials until you run out of lighter weights for the subject to hold. At that time, do one last trial having the subject hold his or her arm out without holding any weight at all. For each trial, record the weight used and the time to fatigue. The weight for the last trial should be zero.

6. In the space below, note any observations and/or initial thoughts you have about the outcome of this experiment:

_____

_____

_____

Table 11.7	Results for Exercise 11.9 (Motor Units and Muscle Fatigue)	
Trial #	Weight (lb)	Time to fatigue (min)
1		
2		
3		
4		
5		

Note: It is possible you may not have five trials.

INTEGRATE

CLINICAL VIEW
Amyotrophic Lateral Sclerosis (ALS)

Amyotrophic lateral sclerosis (ALS), or Lou Gehrig's Disease, is so named for the famous baseball player, Lou Gehrig, who was diagnosed with this fatal disease at the age of 36 after suffering a dramatic decline in the quality of his game. ALS impacts somatic motor neurons, whose cell bodies are located in the posterior (ventral) horn of the spinal cord (see chapter 16). These neurons innervate skeletal muscle fibers. While the etiology of ALS is still unclear, it is known that ALS leads to the destruction of these motor neurons. As motor neurons die and retract from their respective muscle fibers, there is a loss in function in the nerve-muscle connection, or neuromuscular junction. With this loss comes an inability to excite the skeletal muscle fiber. Without excitation, there is no contraction, and if you don't use it, you lose it! That is, muscle fibers will begin to atrophy (*a-*, without + *-trophy*, nourishment) due to disuse. As you can imagine, this also impacts motor unit recruitment, as a loss of motor neurons results in atrophy of all of the fibers that the motor neurons innervate. As a compensatory mechanism, healthy motor neurons begin sprouting collateral processes to make connections with the fibers that are no longer receiving electrical signals from the central nervous system. Often, muscle fiber types will begin to "group" together, such that Type I fibers may be concentrated in one particular compartment of a muscle. This is in contrast to a healthy muscle where muscle fiber types are often distributed throughout a compartment. As the disease progresses, patients may begin to experience slurred speech, muscle fatigue, and clumsiness. Unfortunately, ALS is an aggressive disease that attacks all somatic motor neurons in the body, including those responsible for innervating skeletal muscles that aid in respiration, such as the diaphragm. Over time, often in as little as 3 to 5 years, there is so much denervation and muscle atrophy that patients are no longer able to effectively contract the diaphragm and die of respiratory failure.

EXERCISE 11.10

CONTRACTION OF SKELETAL MUSCLE (WET LAB)

In this exercise you will be observing muscle contraction using a preparation of glycerinated muscle fibers. The process of preparing these fibers (glycerination) removes ions (e.g., Na^+, K^+, Cl^-, Ca^{2+}) and ATP from the tissue. It also disrupts the troponin/tropomyosin complex in such a way that tropomyosin no longer covers the myosin binding sites on actin **(figure 11.11)**. Thus, if the fibers are given the appropriate materials (particularly ATP), actin and myosin will be able to interact, crossbridge cycling will take place, and the fibers will contract. The salt solutions that you will be adding contain two specific salts: KCl (potassium chloride) and $MgCl_2$ (magnesium chloride). Myosin has a high affinity for these salts. Ironically, magnesium interferes with contraction of muscle *in situ*, but has been observed to *increase* the strength of contraction in glycerinated muscle. In this exercise you will test to see if this is indeed true.

Before you begin, familiarize yourself with the following concepts:
- Structure of actin and myosin
- Role of calcium in inducing muscle contraction
- Events of the crossbridge cycle

Figure 11.12 is a figure of a single crossbridge cycle. Using your textbook as a guide, fill in the descriptions of the events that happen at each stage of the crossbridge cycle.

In this exercise you will observe the effect of adding ATP alone, ATP plus salts (KCl and $MgCl_2$), and salts alone (KCl and $MgCl_2$) on contraction of preserved (glycerinated) skeletal muscle. You will observe contraction of the muscle fibers using a microscope, although the contraction can also be observed grossly without the use of a microscope—you just won't be able to see the bands.

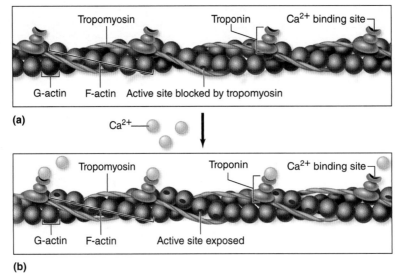

Figure 11.11 **The Role of Troponin/Tropomyosin Complex in Muscle Contraction.** (a) In resting skeletal muscle *in vivo*, the troponin/tropomyosin complex blocks the active sites on actin, thus preventing contraction. (b) When calcium binds to troponin, it causes the troponin/tropomyosin complex to move such that the active sites on the actin molecules are exposed. This allows myosin to bind to actin (forming a crossbridge) and contraction to occur. Glycerinated muscle, such as that used in exercise 11.10, has active sites on actin exposed as in part (b) of this figure, but without requiring the addition of calcium.

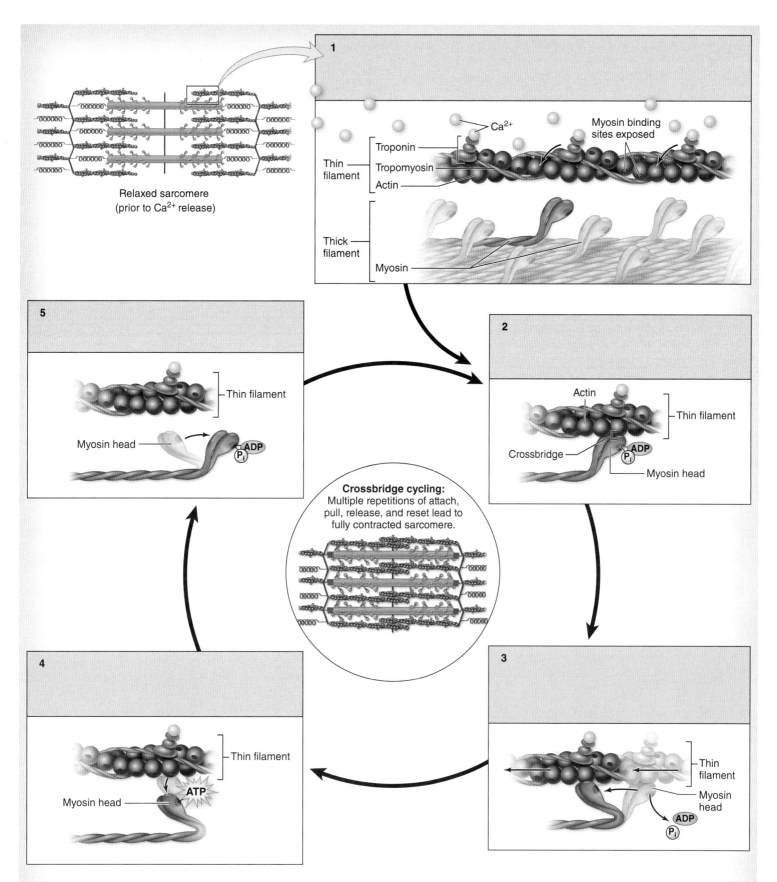

Figure 11.12 The Crossbridge Cycle. Using your textbook as a guide, fill in the blanks describing the steps of the crossbridge cycle.

(continued on next page)

(continued from previous page)

EXERCISE 11.10A Observation of Muscle Fibers

Obtain the Following:
- dissecting needles (2)
- fine-tissue forceps ("watchmaker" forceps)
- ~ 2 cm long section of muscle tissue in petri dish covered in glycerol
- solutions: (1) ATP only, (2) $KCl/MgCl_2$ only, (3) ATP + $KCl/MgCl_2$
- medicine dropper or pipette
- 5 glass microscope slides
- 5 glass microscope slide cover slips
- millimeter ruler
- compound microscope
- dissecting microscope

1. Obtain the petri dish with the muscle tissue. Using fine forceps and the dissecting needles, tease the fibers apart longitudinally until you get strands that are as small (thin) as possible, as shown in **figure 11.13**. You'll want to tease the sample apart enough to get at least 5 separate bundles of fibers because you will be using each of the bundles to perform a separate exercise in this series. Keep the petri dish covered, and in a cool spot away from any heat source, while you are performing these exercises.

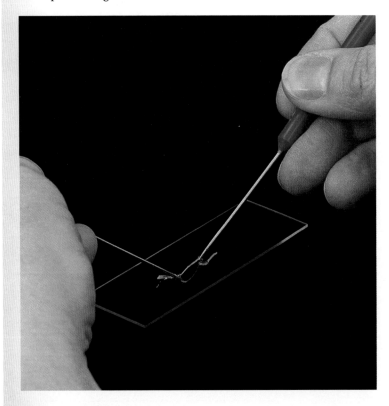

Figure 11.13 Glycerinated Muscle Fibers. Using forceps and a dissecting needle, or two dissecting needles (shown here), gently tease apart the muscle fiber bundle in an attempt to obtain the thinnest possible strands for exercise 11.10.

2. Place the compound microscope on your laboratory bench and obtain the glass slides and cover slips. Transfer one small bundle of muscle fibers from the petri dish to a glass slide. Cover the slide with a cover slip, and place it on the microscope stage.

3. Bring the tissue sample into focus using the scanning objective. Because the tissue has not been stained, it may be somewhat difficult to see. Closing the iris diaphragm on the microscope may assist you in obtaining better tissue contrast (see chapter 3, The Microscope). Next, change to a higher-power objective and bring the tissue sample into focus once again. Look for the following on the slide:

   ☐ nuclei ☐ striations
   ☐ skeletal muscle fibers

4. Observe the striations carefully and compare them to the drawing in figure 11.1. Which of the following are visible in your sample (check any and all that you can make out)?

   ☐ A band ☐ I band
   ☐ H zone ☐ Z disc

5. In the space below, sketch the muscle fibers as viewed through the microscope. This sketch is to aid you later on in these laboratory exercises.

_____ ×

EXERCISE 11.10B Observing the Effect of ATP and Salts on Contraction of Glycerinated Muscle Fibers

In this exercise you will observe the effect of adding solutions containing ATP, salts ($MgCl_2$ and KCl) only, and ATP plus salts on contraction of glycerinated muscle fibers. Refer back to the exercise introduction for assistance in formulating your hypotheses.

Before you begin, state your hypotheses regarding how each of the following influence muscle contraction:

a. Addition of ATP: _____

_____

b. Addition of salts (KCl/$MgCl_2$) only: _____

_____

c. Addition of ATP plus salts (MgCl): _____

_____

1. Obtain three microscope slides and cover slips. Label three of the slides "A," "B," and "C" to correspond to the hypotheses that you have stated above. Then obtain the solutions and three separate eyedroppers (one for each solution so you do not cross-contaminate the solutions).

Observe the effect of adding ATP

2. Obtain the slide labeled "A," a millimeter ruler, and the solution labeled "ATP Alone."

3. Obtain a bundle of fibers from the petri dish and place it on the microscope slide labeled "A."

4. Measure the starting length of the muscle fibers and record it in **table 11.8**.

5. Using a clean eyedropper, flood the muscle fiber with the ATP-only solution. Record your observations:

_____

_____

6. After approximately 1 minute, measure the ending length of the muscle fibers and record it in table 11.8.

7. Obtain another clean glass slide. Tease a few fibers from the bundle of fibers treated with ATP and place them on the clean slide. Place a cover slip over the fibers and then observe them with the compound microscope. Record your observations: _____

_____

Observe the effect of adding salts only (KCl/$MgCl_2$)

8. Obtain the slide labeled "B," a millimeter ruler, and the solution labeled "Salt Solution."

9. Obtain a bundle of fibers from the petri dish and place it on the microscope slide labeled "B."

10. Measure the starting length of the muscle fibers and record it in table 11.8.

11. Using a clean eyedropper, flood the muscle fiber with the salt solution. Record your observations: _____

_____

12. After approximately 1 minute, measure the ending length of the muscle fibers and record it in table 11.8.

Observe the effect of adding ATP + salts (KCl/MgCl)

13. Obtain the slide labeled "C," a millimeter ruler, and the solution labeled "ATP Plus Salt Solution."

14. Obtain a bundle of fibers from the petri dish and place it on the microscope slide labeled "C."

15. Measure the starting length of the muscle fibers and record it in table 11.8.

16. Using a clean eyedropper, flood the muscle fiber with the ATP plus salt solution. Record your observations:

_____

_____

17. After approximately 1 minute, measure the ending length of the muscle fibers and record in table 11.8.

Table 11.8 Results for Exercise 11.10 (Contraction of Skeletal Muscle)

Slide	Solution	Starting Length (mm)	Ending Length (mm)
A	ATP only		
B	Salts only (KCl/$MgCl_2$)		
C	ATP + Salts (MgCl)		

EXERCISE 11.11 Ph.I.L.S.

Ph.I.L.S. LESSON 4: STIMULUS-DEPENDENT FORCE GENERATION

The purpose of this exercise is to demonstrate how stimulus intensity relates to maximum isometric twitch force generated in a skeletal muscle. In the experiment, you will apply electrical stimuli over a range of voltages to a frog gastrocnemius muscle. Each stimulus mimics an action potential delivered to a muscle by a motor neuron. In this Ph.I.L.S. exercise, increases in voltage will serve as the increased stimulus. As the voltage is increased from 0 volts up to 1.6 volts, you should observe how an increase in stimulus intensity relates to increases in maximal isometric twitch force produced by the muscle. A **muscle twitch** is the response of a muscle to a single action potential. In this simulated Ph.I.L.S. experiment, the muscle is being held at a fixed length; hence, the contraction is an **isometric** contraction. In reality, muscles *in vivo* are stimulated by a series of action potentials delivered by the motor neuron (hence, they do not "twitch"). However, studying the twitch response by muscles *in vitro* (or in this case, through a simulated *in vitro* experiment) allows us to understand some of the basic physiological properties of skeletal muscle.

The lowest stimulus that generates a muscle twitch is known as the *threshold stimulus*. As stimulus intensity increases above the threshold stimulus, the maximum isometric twitch force generated by the skeletal muscle also increases. This occurs as more motor units are recruited, which results in an increase in muscle force. Subsequent recruitment of motor units will continue to increase the total muscle force until the muscle reaches a maximal contraction (i.e., all motor units have been recruited). The smallest stimulus that elicits a maximal twitch contraction in the muscle is called the *maximal stimulus*. If you continue to stimulate the muscle with voltages greater than the maximal stimulus, there will be no further increases in force generation. These stimuli are termed *supramaximal stimuli*.

Before you begin, familiarize yourself with the following concepts (use your textbook as a reference):

- Phases of an isometric muscle twitch (latent period, contraction period, relaxation period)
- Innervation of skeletal muscle fibers by motor neurons
- Motor units
- The differences between threshold, submaximal, maximal, and supramaximal stimuli

State your hypothesis regarding the influence of voltage on maximal isometric twitch force.

_____

_____

_____

_____

1. Open Lesson 4: Stimulus-Dependent Force Generation.
2. Read the objectives and introduction and take the pre-laboratory quiz.
3. After completing the pre-laboratory quiz, read through the wet lab.
4. The lab exercise will open when you have completed the wet lab (figure 11.14).
5. Turn on the power of the data acquisition unit (DAQ).
6. Connect the force transducer to the data acquisition unit (DAQ) by dragging the black plug to recording input 1.
7. Connect the stimulating electrodes to the stimulator outputs by dragging the blue plug to the negative output and the red plug to the positive output.
8. Set the stimulus voltage to 1.0 volts. This voltage will give you a baseline reading.
9. To stimulate the muscle, click the "shock" button in the control panel (a graph will appear with a red line indicating the force generated by the muscle and a blue line indicating the stimulus applied).
10. Measure the muscle force (magnitude of contraction) by positioning the crosshairs (using the mouse) at the top of the wave. Click to make a black arrow appear. If you are not in the correct location, reposition by dragging the arrow to a new position. Now position the crosshairs at the bottom of the deflection (after the wave, or 0 volts) and click. The result will display in the yellow data panel. Amplitude (amp) is the change in force of the muscle, which was recorded by the transducer when the muscle contracted.
11. Click the "journal" panel (red rectangle at bottom right of screen) to enter your value into the journal. A table with volts and amplitude (amp) and a graph will appear. Note the range in values for volts from 0 to 1.6 volts.

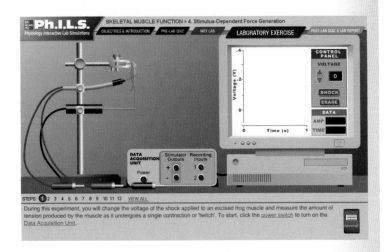

Figure 11.14 Opening Screen for Ph.I.L.S. Lesson 4: Stimulus-Dependent Force Generation.

12. Close the journal window by clicking on the "X" in the right-hand corner.

13. Adjust the voltage by one-tenth of a volt and repeat the procedure in steps 8–10 to record your data. Continue to repeat this exercise, increasing the voltage by 0.1 volts for each new trial, until you have measured the amplitude that corresponds to applied voltages of 0 to 1.6 volts. Record the data from the Ph.I.L.S. "journal" in **table 11.9** below.

Table 11.9	Results of Ph.I.L.S. Lesson 4: Stimulus-Dependent Force Generation
Volts	**Amplitude/Force**
0.0	
0.1	
0.2	
0.3	
0.4	
0.5	
0.6	
0.7	
0.8	
0.9	
1.0	
1.1	
1.2	
1.3	
1.4	
1.5	
1.6	

14. Graph data you recorded in table 11.9 on the axis below.

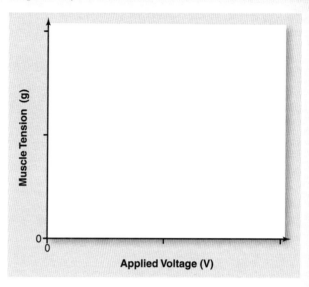

15. Make note of any pertinent observations here:

_____
_____
_____

EXERCISE 11.12 Ph.I.L.S.

Ph.I.L.S. LESSON 5: THE LENGTH-TENSION RELATIONSHIP

The sliding filament theory of muscle contraction states that muscle contraction occurs as thick and thin filaments slide past each other. This process occurs as myosin heads interact with binding sites on actin. The degree of overlap between thick and thin filaments determines how many crossbridges are formed and consequently how much force the muscle will be able to produce during an isometric contraction. The "optimal length" of a muscle is the length at which the maximum number of myosin crossbridges are allowed to interact with the binding sites on actin without interference. It just so happens that a skeletal muscle at its normal *in vivo* resting length is also at its "optimal length," and thus is able to produce the maximal amount of tension when stimulated to contract isometrically. As the length of the muscle (and hence, the sarcomeres within) gets longer or shorter than the optimal/resting length, force decreases because fewer crossbridges are formed between myosin and actin are possible.

In this experiment, you will again be using a frog gastrocnemius muscle. The muscle will be held at a fixed length and then stimulated to induce a maximum contraction (hence, the contraction will be isometric, and you will produce maximum isometric force). You will record the force produced during the trial in the "journal." You will then repeat the experiment by varying the length of the muscle (make it longer and/or shorter than the starting length), stimulating it to contract maximally once again, measuring the force generated, and recording it in the "journal." When you have completed all of the trials, you will graph your results to demonstrate the effect of muscle length on maximum isometric force production.

Before you begin, familiarize yourself with the following concepts (use your main textbook as a reference):

- Sliding filament theory of muscle contraction
- Events of the crossbridge cycle
- Optimal length of a skeletal muscle

State your hypothesis regarding the influence of muscle length on maximum isometric force.

_____

_____

_____

_____

_____

1. Open Ph.I.L.S. Lesson 5, The Length-Tension Relationship.
2. Click the "power" switch on the data acquisition unit (DAQ) to begin the experiment.
3. Connect the force transducer to the DAQ by dragging the black plug to recording input 1.
4. Connect the stimulating electrodes to the stimulator outputs by dragging the blue plug to the negative output and the red plug to the positive output.
5. Click the "zoom" button in the middle of your computer screen to get an enlarged view of the muscle preparation. The muscle is already stretched to 26 mm at the beginning of the experiment (notice the measurement on the blue ruler next to the muscle).
6. Set the voltage on the control panel by clicking the up or down arrows on the control panel (right side of your screen). Set the voltage at 1.8 volts.
7. Click the "shock" button on the control panel. This will elicit a contraction (the red line on the screen is the tension and the blue line is the stimulus).
8. Position the crosshair at the tallest portion of the tension peak, and click the mouse button. Then, position the crosshair over the flat portion of the tension recording and press the mouse button. Notice that numbers have appeared in the data window of the control panel.
9. Click the "journal" button in the lower right-hand corner to enter the data into the on-screen table. Then enter your data in **table 11.10** below. Close the journal by clicking the "X" in the upper right-hand corner of the journal window.

Table 11.10	Results from Ph.I.L.S. Lesson 5: The Length-Tension Relationship
Length (mm)	**Tension (g)**

10. Click the up arrow on the muscle clamp to increase the length of the skeletal muscle by 0.5 mm.
11. Repeat steps 7–10 until you have fully stretched the muscle.
12. Take the post-laboratory quiz and read the conclusion to complete the experiment.

13. Graph data you recorded in table 11.10 on the axis below.

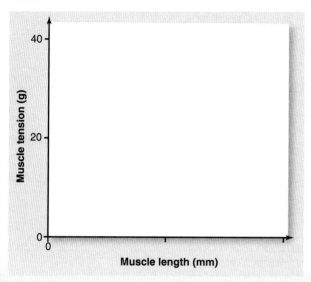

14. Make note of any pertinent observations here.

_____
_____
_____
_____

INTEGRATE

CONCEPT CONNECTION

With the study of muscle tissues comes the comparison between two important muscle types: skeletal and cardiac. The mode of activation for these two muscle types differs. Skeletal muscle is voluntary, meaning that we "think" about moving our skeletal muscles, and it happens through the processes already discussed in this chapter. In contrast, cardiac muscle is involuntary, meaning that we do not think about moving our hearts. This is probably a good thing, as it is inconvenient to "think" about contracting the heart every time we need to circulate blood through the body. Instead, cardiac muscle contracts ("beats") *intrinsically* (on its own). The rate of contraction of cardiac muscle is *extrinsically* regulated by the autonomic nervous system. The electrical conduction system of the heart will be explored further in chapter 21.

While the mode of activation for these two muscle types may differ, there are some structural similarities between skeletal and cardiac muscle. Both skeletal and cardiac muscle tissues are striated. Therefore, both contain sarcomeres that are responsible for force generation. Recall that sarcomere length determines force output. This is the foundation for the length-tension curve of muscle. There is an optimal sarcomere length for maximal force generation. Typically, the sarcomeres contained in skeletal muscles operate close to their optimal length. However, the sarcomeres in cardiac muscle "rest" at a shorter-than-optimal length for maximal force production. These sarcomeres must be lengthened in order to generate maximal force. In the heart, can you imagine what force has the ability to "stretch" cardiac muscle fibers? You probably guessed it—blood! As blood enters the heart, it fills the heart chambers (atria and ventricles), with walls composed of cardiac muscle tissue. The greater the volume of blood entering the ventricles, the greater the stretch on the cardiac muscle. When the sarcomeres are stretched to their optimal length, the cardiac muscle responds by generating increasing amounts of force. We call this *intrinsic* regulation of cardiac muscle contraction. The relationship between length and tension in cardiac muscle is also referred to as *Starling's Law of the Heart*.

EXERCISE 11.13 Ph.I.L.S.

Ph.I.L.S. LESSON 6: PRINCIPLES OF SUMMATION AND TETANUS

Action potential propagation along the sarcolemma of skeletal muscle fibers (cells) results in the release of calcium from the sarcoplasmic reticulum into the surrounding sarcoplasm (cytoplasm within muscle cell). The duration of the electrical event is rapid (1–2 ms) when compared to the duration of the mechanical contraction (100–200 ms), principally because of the time that it takes to actively pump the calcium back into the sarcoplasmic reticulum. The presence of calcium in the sarcoplasm prolongs the period of crossbridge formation. As action potentials increase in frequency, more calcium is available in the sarcoplasm; therefore, muscle tension increases.

The purpose of this laboratory exercise is to observe the resulting muscle tension as the frequency of action potentials increases. As the frequency increases, muscle twitches occur more rapidly, ultimately resulting in summation of muscle tension.

Before you begin, familiarize yourself with the following concepts (use your main textbook as a reference):

- Voltage-gated calcium channels in skeletal muscle
- Crossbridge cycling
- A single muscle twitch
- Summation of muscle twitch
- Incomplete and complete tetanus

State a hypothesis regarding the effect that action potential frequency has on calcium concentration in the sarcoplasm and resulting muscle tension.

_____

_____

_____

_____

_____

1. To begin the experiment, open Ph.I.L.S. Lesson 6: Principles of Summation and Tetanus (see **figure 11.15**).
2. Click the "power" button on the data acquisition unit (DAQ).
3. Connect the force transducer to the data acquisition unit by clicking and dragging the black plug to recording input 1.

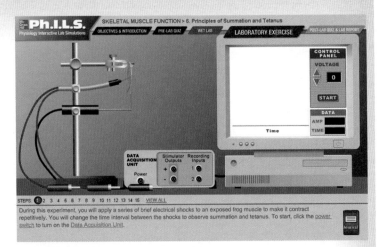

Figure 11.15 Opening Screen for Ph.I.L.S. Lesson 6: Principles of Summation and Tetanus.

4. Connect the stimulating electrodes to the stimulator outputs on the data acquisition unit by clicking and dragging the blue plug to the negative terminal and clicking and dragging the red plug to the positive terminal.
5. On the control panel, set the voltage to a value between 1.6 and 2.0 volts by clicking the up and down arrows.
6. To determine the time interval required to elicit summation, click the "start" button to begin delivering stimuli to the muscle. The default time interval is 500.
7. At a time interval of 500, you are observing single muscle twitches. To determine the interval that will elicit summation, decrease the interval between stimuli by clicking the down arrow on the control panel. Observe the changes seen in the twitches as you decrease the time interval.
8. Continue the above procedure until you observe summation. Click the stop button once summation is achieved.
9. Click the "journal" in the lower right corner of the screen. Record the time interval where you first observe summation in **table 11.11**. Click the "x" in the upper right corner of the journal window to continue the experiment.
10. Click the "start" button to begin stimulating the muscle again.

Table 11.11	Results from Ph.I.L.S. Lesson 6: Principles of Summation and Tetanus
Type of Contraction	**Time Interval of Stimulation**
Summation	
Incomplete Tetanus	
Complete Tetanus	

11. Continue to decrease the time interval by clicking the down arrow until you observe incomplete tetanus. Record the time interval where the muscle reached incomplete tetanus in table 11.11.

12. Click the "start" button to begin stimulating the muscle again.

13. Continue to decrease the time interval by clicking the "down" arrow until you observe complete tetanus. Record the time interval where the muscle reached complete tetanus in table 11.11.

14. Make note of any pertinent observations here:

_____
_____
_____

EXERCISE 11.14

Ph.I.L.S. LESSON 7: EMG AND TWITCH AMPLITUDE

The purpose of this laboratory exercise is to record the electrical activity from skeletal muscles using virtual surface electrodes, to generate an electromyogram (EMG). Pressure generated from a contracting muscle and EMG readings will be obtained in a virtual subject, and this data will be used to estimate the number of muscle fibers recruited to generate muscle tension.

Before you begin, familiarize yourself with the following concepts (use your main textbook as a reference):

- The physiological principles of the neuromuscular junction
- The anatomical and physiological properties of a motor neuron
- The depolarization and repolarization of a muscle cell
- The contractile properties of skeletal muscle
- The "all-or-none" principle of skeletal muscle contraction

State your hypothesis regarding the relationship between muscle force and EMG.

_____
_____
_____
_____
_____

1. Open the Ph.I.L.S Lesson 7: EMG and Twitch Amplitude (see **figure 11.16**).

2. Click the power switch on the data acquisition (DAQ) unit to begin the lab.

3. Connect the pressure transducer by clicking and dragging the blue plug to the recording input number 2.

4. Click and drag the hand dynamometer to the subject's right hand.

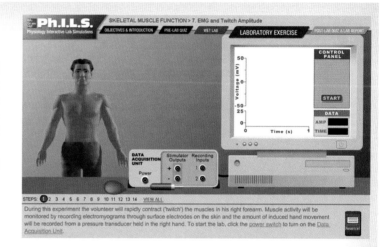

Figure 11.16 Opening Screen for Ph.I.L.S. Lesson 7: EMG and Twitch Amplitude.

5. To set up the EMG, connect the electrodes to the data acquisition unit by clicking and dragging the black plug to recording input number 1.

6. Click and drag the green electrode to a position just above your subject's right wrist. Click and drag the red electrode to the upper right forearm of your subject (just inside the elbow joint). Click and drag the black electrode to a position between the red and green electrodes on the forearm.

7. Just below the subject's right hand, a slider bar with arrows has appeared. This is used to control the amount of forearm contraction. To increase the contraction strength, click the up arrow. To decrease the contraction strength, click the down arrow.

8. To begin the experiment, adjust the grip strength by clicking the down arrow until the setting on the slider bar is set at 1.

9. Click "start" on the control panel. The EMG is the upper red tracing and is recorded in millivolts (mV). The pressure is the lower blue tracing. Move the mouse to the highest point on the blue tracing and click the mouse button to mark the spot. Move the mouse to a

(continued on next page)

(continued from previous page)

flat part of the tracing and click the mouse button again. Data now appears in the "data window" of the control panel. Click on "journal" in the lower right-hand corner of the screen. Enter the pressure data in **table 11.12**.

10. To record the EMG, move the mouse pointer to the highest peak in the EMG recording and click the mouse button. Move the mouse pointer to the lowest point on the EMG recording and again, click the mouse button. Data appears in the window on the control panel. Enter this EMG data in the table below.

11. Repeat steps 8–10 by increasing the forearm contraction from 2 to 6. Record your data for all points in table 11.12.

Table 11.12	Results from Ph.I.L.S. Lesson 7: EMG and Twitch Amplitude	
Recording (Value)	**Pressure Amplitude**	**EMG Amplitude**
1		
2		
3		
4		
5		
6		

12. Graph the data you generated on the axis below.

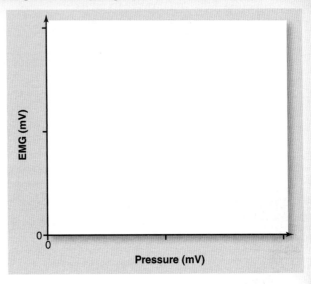

13. Make note of any pertinent observations here:

_____
_____
_____

Chapter 11: The Muscular System: Muscle Structure and Function

Name: _____
Date: _____ Section: _____

POST-LABORATORY WORKSHEET

The ❶ corresponds to the Learning Objective(s) listed in the chapter opener outline.

Do You Know the Basics?

Exercise 11.1: Histology of Skeletal Muscle Fibers

1. The nervous control of skeletal muscle is _____. ❶

2. Match the zone or band in column A with the appropriate description in column B. ❷

    Column A   *Column B*

    _____ 1. H zone   a. the dark band in skeletal muscle
    _____ 2. A band   b. edges of a sarcomere are determined by this
    _____ 3. Z disc   c. contains only thin filaments (actin)
    _____ 4. M line   d. a light region in the middle of the A band
    _____ 5. I band   e. a dark line in the middle of the A band

Exercise 11.2: Connective Tissue Coverings of Skeletal Muscle

3. Match the connective tissue covering listed in column A with the structure it surrounds in column B. ❸

    Column A   *Column B*

    _____ 1. epimysium   a. fascicle
    _____ 2. endomysium   b. skeletal muscle fiber
    _____ 3. perimysium   c. entire skeletal muscle

Exercise 11.3: The Neuromuscular Junction

4. A somatic motor neuron and all the muscle fibers innervated by that motor neuron comprises a _____
   _____. ❹

Exercise 11.4: Smooth Muscle Tissue

5. Describe the properties that are unique to smooth muscle tissue. ❺

   _____

6. In a tube-shaped structure that is surrounded by smooth muscle tissue, when the inner circular layer of fibers contracts, the diameter of the tube _____ and the length of the tube _____. When the outer longitudinal layer of muscle contracts, the diameter of the tube _____ and the length of the tube _____. ❻

Exercise 11.5: Cardiac Muscle Tissue

7. Describe the properties that are unique to cardiac muscle tissue. ❼

   _____

8. For the following, fill in the blank with the name of the muscle tissue that is described. ❽

    a. Muscle tissue of the heart _____

    b. A component of the iris that changes the size of the pupil in the eye _____

    c. Muscle tissue that attaches to the skeleton _____

    d. Changes the size of the opening (lumen) of the air passageway (bronchioles) _____

    e. Responsible for forcing the baby out through the birth canal _____

    f. Changes size of the lumen of blood vessels and helps regulate blood pressure _____

    g. Regulates movement of material from one area to another (e.g., the movement of digested food from the stomach into the small intestine is accomplished by this muscle tissue) _____

    h. Helps (through involuntary control) to expel urine from the bladder and feces from the digestive tract _____

282 Chapter Eleven The Muscular System: Muscle Structure and Function

9. Match the appropriate muscle type in column B with a characteristic listed in column A. You may use an answer choice more than once and you may list more than one type of muscle tissue for each characteristic.

   Column A
   _____ 1. branched fibers
   _____ 2. central nuclei
   _____ 3. cylindrical fibers
   _____ 4. inner circular and outer longitudinal layers
   _____ 5. intercalated discs
   _____ 6. multinucleate
   _____ 7. satellite cells
   _____ 8. spindle-shaped cells
   _____ 9. spiral nuclei
   _____ 10. striated

   Column B
   a. skeletal muscle
   b. cardiac muscle
   c. smooth muscle

10. What characteristic is a common feature of all three types of muscle?

Exercise 11.6: Naming Skeletal Muscles

11. Match the appropriate word origin in column B with its meaning listed in column A.

    Column A
    _____ 1. smaller
    _____ 2. round
    _____ 3. a lifter
    _____ 4. around
    _____ 5. abductor
    _____ 6. widest
    _____ 7. straight
    _____ 8. two-headed
    _____ 9. slender
    _____ 10. internal

    Column B
    a. per-
    b. rectus
    c. latissimus
    d. biceps
    e. minor
    f. internus
    g. levator
    h. to move away
    i. teres
    j. gracilis

Exercise 11.7: Architecture of Skeletal Muscles

12. Match the muscle architecture shown in column B with the appropriate description listed in column A.

    Column A
    _____ 1. bipennate
    _____ 2. circular
    _____ 3. convergent
    _____ 4. multipennate
    _____ 5. parallel
    _____ 6. unipennate

    Column B
    a., b., c., d., e., f.

Exercise 11.8: Major Muscle Groups and Fascial Compartments of the Limbs

13. Match the muscle actions listed in column B with the appropriate compartment and/or major muscle group listed in column A.

Column A

_____ 1. anterior compartment of the arm
_____ 2. muscles of the abdomen
_____ 3. medial compartment of the thigh
_____ 4. lateral compartment of the leg
_____ 5. muscles of mastication
_____ 6. muscles of the neck
_____ 7. posterior compartment of the forearm
_____ 8. anterior compartment of the thigh
_____ 9. muscles of facial expression
_____ 10. anterior compartment of the leg

Column B

a. dorsiflex and invert ankle
b. flex shoulder and elbow
c. adduct the thigh
d. create facial expressions
e. extend the wrist and digits
f. flex, bend, and rotate the spine
g. allow one to chew
h. evert the ankle
i. rotate, flex, and extend the head
j. extend the knee; flex the hip

14. In column B, write the letter of the structure labeled in the cross section of the arm that matches the description given in column A.

Column A

anterior (flexor) compartment of the arm
humerus
posterior (extensor) compartment of the arm

Column B

1. _____
2. _____
3. _____

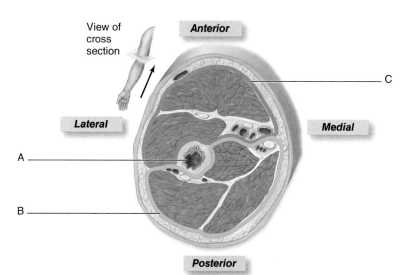

Fascial compartments of the arm (brachium)

15. In column B, write the letter of the structure labeled in the cross section of the thigh that matches the description given in column A.

    Column A
    medial compartment of the thigh
    anterior compartment of the thigh
    posterior compartment of the thigh

    Column B
    1. _____
    2. _____
    3. _____

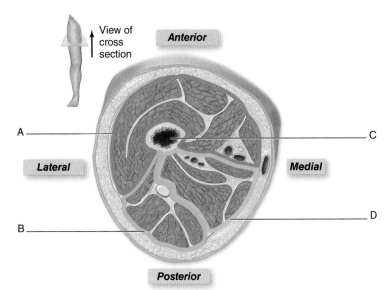

Fascial compartments of the thigh

Exercise 11.9: Motor Units and Muscle Fatigue (Human Subject)

16. Match the muscle fiber type listed in column B with the appropriate characteristic listed in column A.

    Column A
    _____ 1. largest fiber cross-sectional area
    _____ 2. slow-twitch
    _____ 3. very low fatigue resistance
    _____ 4. smallest fiber cross-sectional area
    _____ 5. highly fatigue resistant
    _____ 6. composes the smallest motor units
    _____ 7. recruited last in a muscular contraction
    _____ 8. produces the lowest amount of twitch force
    _____ 9. fast-twitch with moderate resistance to fatigue
    _____ 10. intermediate (medium) fiber cross-sectional area and motor unit size

    Column B
    a. Type I
    b. Type IIa
    c. Type IIb

17. In the spaces below, list the types of motor units that are recruited first, second, and third, as increased force is demanded of skeletal muscle.

    a. _____
    b. _____
    c. _____

18. If a muscular contraction requires the recruitment of Type IIb muscle fibers, the force produced will be _____,
    the speed of contraction will be _____, and the contraction _____will/will not_____ be able to be sustained for a significant length of time.

Chapter Eleven *The Muscular System: Muscle Structure and Function* **285**

Exercise 11.10: Contraction of Skeletal Muscle (Wet Lab)

19. In the spaces below, describe the results of adding each of the following solutions to the glycerinated muscle preparation: ⓰
    a. ATP only _____
    b. Salts only _____
    c. ATP + Salts _____

20. When the supply of ATP in a muscle runs out, what is the position of the myosin crossbridges with respect to actin? _____ What effect does this have on the ability of the muscle fiber to contract? _____. Finally, what is the condition the muscle is in called? ⓱ _____

Exercise 11.11: Ph.I.L.S. Lesson 4: Stimulus-Dependent Force Generation

21. Discuss the difference between a twitch and a sustained muscle contraction (tetanus). ⓲ _____
_____
_____

22. Identify, in order, the three phases of a muscle twitch: ⓲
    a. _____
    b. _____
    c. _____

23. Describe how force generation in muscle is influenced by voltages greater than the maximal stimulus. ⓳ _____
_____
_____

24. Discuss why muscle contraction fails to occur with subthreshold stimuli. ⓳ _____
_____
_____

25. A stimulus that is not large enough to produce a muscle twitch is called a _____ stimulus. A stimulus that is large enough to produce a muscle twitch, but not large enough to produce summation of twitches, is called a _____ stimulus. A stimulus that is large enough that it no longer causes an increase in muscle force is called a _____ stimulus. ⓳

26. What was the threshold stimulus (voltage)? ⓴ _____
_____

27. What was the maximal stimulus (voltage)? ⓴ _____

Exercise 11.12: Ph.I.L.S. Lesson 5: The Length-Tension Relationship

28. What sarcomere length corresponded to the optimal muscle length in skeletal muscle? ㉑ _____
_____

29. Describe the events that occur at the level of the sarcomere to create the force at each of the numbered regions of the length-tension curve shown in the figure below. ㉒

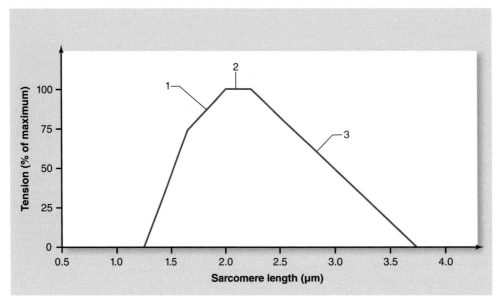

a. _____

b. _____

c. _____

Exercise 11.13: Ph.I.L.S. Lesson 6: Principles of Summation and Tetanus

30. Describe the underlying mechanisms for incomplete and complete tetanus in a skeletal muscle fiber. ㉓ _____

Exercise 11.14: Ph.I.L.S. Lesson 7: EMG and Twitch Amplitude

31. Describe the relationship between EMG amplitude and muscle tension. ㉕ _____

Can You Apply What You've Learned?

32. Place the three types of muscle tissues in order of size (smallest to largest). _____

33. The proximal joint of the front limb of a dog contains two compartments of muscles, an anterior and a posterior compartment. If the muscles in the anterior compartment are all flexors, what is the common action of all the muscles in the posterior compartment? _____

34. You have discovered a new muscle in the body that extends along the tibia. The muscle is a long muscle that contains four heads. Based on your knowledge of word origins, suggest a logical name for this muscle. _____

35. The graph below represents a single isometric muscle twitch. First, name each of the labeled phases (1, 2, and 3). Then describe the physiological events that occur during each of the labeled phases 1, 2, and 3.

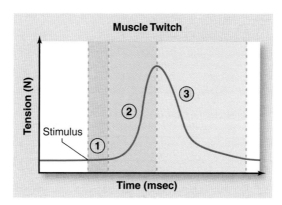

1. Phase _____

   Description of events:

   _____
   _____
   _____

2. Phase _____

   Description of events:

   _____
   _____
   _____

3. Phase _____

   Description of events:

   _____
   _____
   _____

36. Describe the events that must take place for a weight lifter to generate enough tension to lift a 100-lb load.

   _____
   _____
   _____

Can You Synthesize What You've Learned?

37. Tetrodotoxin, also known as "zombie powder," is synthesized by the puffer fish as a defense mechanism. Tetrodotoxin's mechanism of action is to block voltage-gated sodium (Na+) channels in motor neurons, thus preventing them from transmitting action potentials. What effect, then, do you predict that tetrodotoxin will have on the generation of force by a muscle?

38. Using your knowledge of the crossbridge cycle, explain why muscles remain in a contracted state (rigor) after death, even though there is no longer a supply of ATP.

CHAPTER 12

The Muscular System: Axial Muscles

INTRODUCTION

Imagine a time when you slept on a very uncomfortable mattress and woke up the next morning to find the muscles in your back and neck incredibly stiff and sore. For the next few days, every action seemed to cause pain. The classic comment on such an experience is, "I discovered muscles I never knew I had!" It seems that the only time we really contemplate the muscles of our back and neck is when they are injured. At such a time it is impossible *not* to be consciously aware of these muscles—they constantly hurt. This example is a reminder of the importance of the muscles of the back, neck, and thorax for creating the most basic motions of our bodies.

In this laboratory session, you will identify, name, and explore the structure and function of muscles that move the axial skeleton including the muscles of facial expressions, chewing, eye movements, swallowing, the muscles of the neck, and vertebral column. You will also study the muscles responsible for such vital bodily functions as breathing and coughing. Before you begin your studies, find out from your instructor which muscles you will be responsible for ➡

OUTLINE AND LEARNING OBJECTIVES

Gross Anatomy 292

Muscles of the Head and Neck 292

EXERCISE 12.1: MUSCLES OF FACIAL EXPRESSION 292
1. Identify the muscles of facial expression, and describe their actions and innervation

EXERCISE 12.2: MUSCLES OF MASTICATION 295
2. Identify the muscles of mastication, and describe their actions and innervation

EXERCISE 12.3: EXTRINSIC EYE MUSCLES 296
3. Identify the six extrinsic eye muscles, and describe the actions performed by each muscle

EXERCISE 12.4: MUSCLES THAT MOVE THE TONGUE 297
4. Explain the difference between extrinsic and intrinsic tongue muscles
5. Identify the intrinsic tongue muscles, and describe their actions

EXERCISE 12.5: MUSCLES OF THE PHARYNX 298
6. Identify muscles of the pharynx, and explain their role in swallowing

EXERCISE 12.6: MUSCLES OF THE NECK 299
7. Identify the major muscles that act about the neck, and describe their actions
8. Describe how the sternocleidomastoid and splenius capitis muscles can act as both synergists and antagonists to each other

Muscles of the Vertebral Column 303

EXERCISE 12.7: MUSCLES OF THE VERTEBRAL COLUMN 303
9. Describe the major muscle groups that stabilize and move the vertebral column

Muscles of Respiration 307

EXERCISE 12.8: MUSCLES OF RESPIRATION 308
10. Correlate specific respiratory movements (e.g., expiration, coughing) with the muscles responsible for causing the movements
11. List the three major structures that pass through the diaphragm, and explain the significance of the location through which each passes

Muscles of the Abdominal Wall 310

EXERCISE 12.9: MUSCLES OF THE ABDOMINAL WALL 310
12. Describe the layered arrangement of the abdominal musculature
13. Explain how the aponeuroses of the external obliques, internal obliques, and transverse abdominis form the rectus sheath

EXERCISE 12.10: THE RECTUS SHEATH, INGUINAL LIGAMENT, AND INGUINAL CANAL 312
14. Locate the inguinal canal and inguinal ligament on a cadaver or classroom model
15. Describe the structures that pass through the inguinal canal in both males and females

 **MODULE 6: MUSCULAR SYSTEM**

learning. Then, highlight or star those muscles on the tables throughout this chapter so you may focus only on the muscles you are required to know for your course. In addition, before you begin the more detailed study of the axial musculature in this chapter, make sure you have successfully completed the Gross Anatomy section in chapter 11. Those exercises introduced you to the major muscle groups and the common actions of those muscle groups (table 11.4).

Chapter 12: The Muscular System: Axial Muscles

PRE-LABORATORY WORKSHEET

Name: _____
Date: _____ Section: _____

Also available at www.connect.mcgraw-hill.com

1. Briefly state the difference between a muscle of *facial expression* and a muscle of *mastication*.

   _____

   _____

   _____

2. Define the following terms:

   *aponeurosis:*_____

   *inguinal:*_____

   *extension:*_____

   *zygomatic:*_____

   *extrinsic:*_____

3. The largest and most important respiratory muscle is a dome-shaped muscle that separates the thoracic cavity from the abdominal cavity. This muscle is called the _____.

4. List the four muscles that compose the anterior abdominal wall.

   a. _____

   b. _____

   c. _____

   d. _____

5. The three attachment points for the sternocleidomastoid muscle are

   a. _____

   b. _____

   c. _____

6. List the three groups of muscles that form the *erector spinae*.

   a. _____

   b. _____

   c. _____

Chapter Twelve *The Muscular System: Axial Muscles*   291

Chapter Twelve: The Muscular System: Axial Muscles

INTEGRATE

LEARNING STRATEGY

Follow the procedure below as you study each major group of muscles. This will help you build your knowledge and understanding in logical steps. Once you have mastered the information from step 1, you will be ready to move on to subsequent steps. You will find that each successive step requires a more detailed level of understanding. However, you will also find yourself prepared to deal with these increasing levels of knowledge if you follow the steps.

Stepwise Approach to Learning Muscles of the Body

1. Describe the general location of the muscle group.
2. Describe the general actions that all muscles of the group have in common.
3. List the names of all muscles belonging to that group (or just the ones your instructor requires of you).
4. Identify the muscles on a model or a cadaver (this can be done concurrently with step 3).
5. Learn the outliers—the muscles that DO NOT share the common actions of the group.
6. Learn specific points of attachment and actions of the muscles your instructor wants you to know.

Follow these suggestions: Study one muscle group at a time, take a lot of breaks, and study using frequent, short time intervals. You will be amazed at how well you are able to absorb the knowledge, and you might be pleasantly surprised at your results.

Gross Anatomy

Muscles of the Head and Neck

Muscles of the head and neck include the muscles of the face (muscles of facial expression and muscles of mastication), extrinsic eye muscles, muscles that move the tongue, and muscles that move the neck.

EXERCISE 12.1

MUSCLES OF FACIAL EXPRESSION

1. Observe muscles of the face on a human cadaver or a classroom model.

2. Muscles of the face are separated into two groups based on function and innervation: muscles of facial expression and muscles of mastication (chewing). **Muscles of facial expression (table 12.1)** allow you to express emotions such as fright, delight, confusion, and surprise. These muscles are unique in that they have distal attachments on skin instead of bone. Thus, when the muscles contract, they pull on the skin. This movement is easily seen on the surface of the face as a facial expression. Over time, the pulling of these muscles on the face causes a characteristic wrinkling of the skin. For this reason, it's a good idea to practice using your "smiling" muscles much more often than using your "frowning" muscles—you'll look happier in old age! All of the muscles of facial expression are innervated by the **facial nerves,** which are the paired seventh cranial nerves (CN VII). A person with damage to a facial nerve will be unable to demonstrate facial expressions on the affected side.

3. Look into a mirror (you may have to perform this part of the exercise at home if there isn't a mirror in your laboratory classroom) and practice the following facial expressions. As you perform each expression, name the muscles used to create the expression (use table 12.1 as a guide).

   - ☐ anger or doubt
   - ☐ happiness (smiling or laughter)
   - ☐ kissing (close mouth, purse cheeks, close eyes)
   - ☐ sadness (frowning)
   - ☐ surprise or delight

4. Using table 12.1 and your textbook as guides, identify all of the muscles of facial expression listed in **figure 12.1** on the cadaver or model of the face. Then label them in figure 12.1. (Answers may be used more than once.)

5. *Optional Activity:* **AP|R** 6: Muscular System—Watch the muscle action animations to review the actions of many of the muscles mentioned in chapters 12 and 13. You can also try the action, origin, and insertion questions found in the quiz area for challenging drill and practice.

Table 12.1 Muscles of Facial Expression*

Muscle	Origin	Insertion	Action	Innervation	Word Origin
Buccinator	Mandible, molar region of mandible and maxilla	Orbicularis oris (corners of the lips)	Presses cheek against molar teeth, as in chewing, whistling, or playing a wind instrument	Facial (CN VII)	*bucca*, cheek
Corrugator Supercilii	Superciliary arch	Skin of eyebrow	Creates vertical wrinkles in medial forehead, as in frowning	Facial (CN VII)	*corrugo*, to wrinkle, + *superus*, above, + *cilium*, eyelid

Table 12.1	Muscles of Facial Expression* (continued)				
Muscle	**Origin**	**Insertion**	**Action**	**Innervation**	**Word Origin**
Depressor Anguli Oris	Mandible (anterolateral surface of the body)	Muscles and skin in the lower lip near the angle of the mouth	Pulls corners of the mouth inferior, as in frowning	Facial (CN VII)	*depressus*, to press down, + *angulus*, angle, + *oris*, mouth
Depressor Labii Inferioris	Mandible (between the midline and the mental foramen)	Oribicularis oris and skin of the lower lip	Depresses the lower lip, as in expressions of doubt and sadness	Facial (CN VII)	*depressus*, to press down, + *labia*, lip, + *inferior*, lower
*Epicranius (Occipitofrontalis)***	Epicranial aponeurosis	Skin of the forehead (frontalis); superior nuchal line (occipitalis)	Elevates the eyebrows and causes horizontal wrinkles in the forehead, as in expressions of surprise or delight	Facial (CN VII)	*occiput*, the back of the head, + *frontalis*, in front
Levator Anguli Oris	Maxilla (lateral portion)	Skin at the superior corner of the mouth	Elevates the corners of the mouth and pulls them laterally, as in smiling	Facial (CN VII)	*levatus*, to lift, + *labia*, lip, + *superus*, above
Levator Labii Superioris	Maxilla (inferior to infraorbital foramen)	Orbicularis oris and skin of the upper lip	Elevates the upper lip, as in expressions of sadness or seriousness	Facial (CN VII)	*levatus*, to lift, + *anguli*, angle, + *oris*, mouth
Mentalis	Mandible (incisive fossa)	Skin of the chin	Wrinkles the skin of the chin and elevates and protrudes the lower lip, as in expressions of doubt	Facial (CN VII)	*mentum*, the chin
Nasalis	Maxilla	Alar cartilages of the nose	Flares the nostrils, widens the anterior nasal aperture	Facial (CN VII)	*nasus*, nose
Orbicularis Oculi	Skin around the margin of the orbit of the eye	Skin surrounding the eyelids	Closes the eyelids as in blinking	Facial (CN VII)	*orbiculus*, a small disk, + *oculus*, eye
Orbicularis Oris	Deep surface of skin of maxilla and mandible	Mucous membrane of the lips	Purses and protrudes the lips, closes the mouth	Facial (CN VII)	*orbiculus*, a small disk, + *oris*, mouth
Platysma	Fascia superficial to the deltoid and pectoralis major muscles at ribs 1 and 2	Mandible (lower border) and skin of the cheek	Stretches the skin of the anterior neck, depresses the lower lip, as in expressions of fright	Facial (CN VII)	*platys*, flat, broad
Procerus	Nasal bones and nasal cartilages	Aponeurosis at the bridge of the nose and the skin of the forehead	Depresses the eyebrows and elevates the nose producing wrinkles in the skin of the nose, as in frowning and squinting the eyes	Facial (CN VII)	*procerus*, long or stretched out
Risorius	Fascia overlying the masseter muscles	Orbicularis oris and skin of the corner of the mouth	Pulls the corners of the mouth laterally, as in expressions of laughter or smiling	Facial (CN VII)	*risus*, to laugh
Zygomaticus (Major and Minor)	Zygomatic bone	Skin and muscle at corner of the mouth	Pulls the corners of the mouth posteriorly and superiorly, as in smiling	Facial (CN VII)	*zygon*, yoke

*All muscles in this table are innervated by the facial nerve (CN VII).
**The epicranius consists of the epicranial aponeurosis and the occipitofrontalis muscle, which has two bellies: frontal belly of occipitofrontalis and occipital belly of occipitofrontalis.

(continued on next page)

294 Chapter Twelve The Muscular System: Axial Muscles

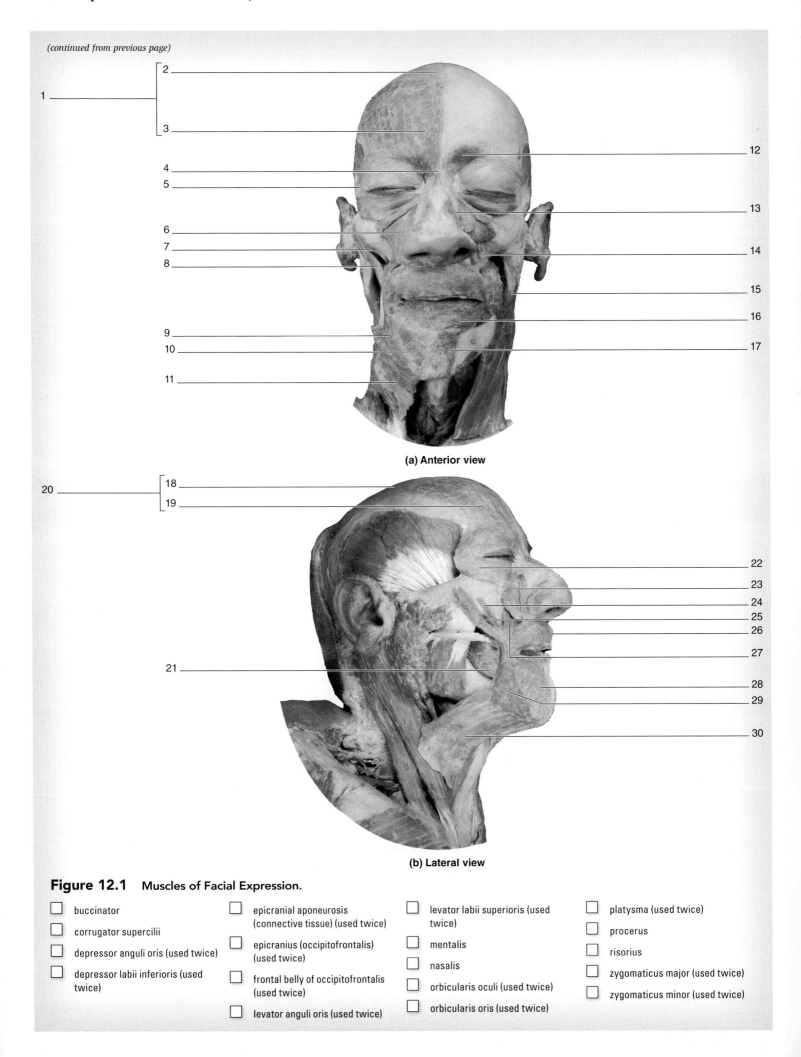

Figure 12.1 Muscles of Facial Expression.

☐ buccinator
☐ corrugator supercilii
☐ depressor anguli oris (used twice)
☐ depressor labii inferioris (used twice)
☐ epicranial aponeurosis (connective tissue) (used twice)
☐ epicranius (occipitofrontalis) (used twice)
☐ frontal belly of occipitofrontalis (used twice)
☐ levator anguli oris (used twice)
☐ levator labii superioris (used twice)
☐ mentalis
☐ nasalis
☐ orbicularis oculi (used twice)
☐ orbicularis oris (used twice)
☐ platysma (used twice)
☐ procerus
☐ risorius
☐ zygomaticus major (used twice)
☐ zygomaticus minor (used twice)

EXERCISE 12.2

MUSCLES OF MASTICATION

1. Observe muscles of the face on a human cadaver or a classroom model.

2. **Muscles of mastication** (*masticate*, to chew) are used in chewing **(table 12.2)**. These muscles attach to the only mobile bone of the skull, the mandible. The muscles of mastication are innervated by branches of the **trigeminal nerves,** which are the paired fifth cranial nerves (CN V). Damage to the trigeminal nerve causes an inability to chew on the affected side.

3. The two most powerful muscles of mastication are the **masseter** and **temporalis** muscles. Place your fingers over the angle and ramus of the mandible (just below your cheek) and close your jaw forcefully (elevate the mandible) to feel contraction of the masseter. Then repeat the process, only this time place your fingers over your temples to feel the contraction of the temporalis muscle. Muscles that depress the mandible (open the mouth) are the infrahyoid muscles, which are covered in exercise 12.6.

4. Using table 12.2 and your textbook as guides, identify all of the muscles of mastication listed in **figure 12.2** on a cadaver or the model of the face. Then label them in figure 12.2. (Answers may be used more than once.)

Table 12.2 Muscles of Mastication*

Muscle	Origin	Insertion	Action	Innervation	Word Origin
Lateral Pterygoid**	Sphenoid (greater wing and lateral pterygoid plate)	Mandible (neck)	Protracts the mandible and depresses the chin; produces a side-to-side motion of the mandible	Trigeminal (CN V)	*pteryx*, wing, + *eidos*, resemblance
Masseter	Zygomatic arch	Mandible (lateral surface of the ramus)	Elevates and protracts the mandible	Trigeminal (CN V)	*masetér*, chewer
Medial Pterygoid**	Sphenoid (lateral pterygoid plate) and maxilla	Mandible (medial surface of the ramus and neck)	Elevates and protracts the mandible; produces a side-to-side motion of the mandible	Trigeminal (CN V)	*pteryx*, wing, + *eidos*, resemblance
Temporalis	Temporal line	Mandible (coronoid process)	Elevates and retracts the mandible	Trigeminal (CN V)	*tempus*, temple

*All muscles in this table are innervated by the trigeminal nerve (CN V).
**Lateral and medial pterygoids, when acting alone (alternating one side at a time), produce a side-to-side grinding motion.

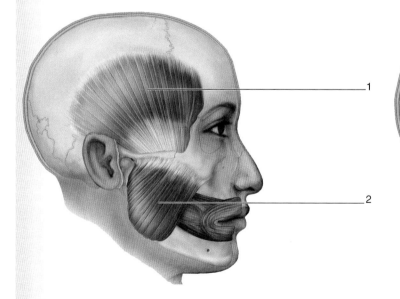

(a) Superficial lateral view   (b) Deep lateral view

Figure 12.2 Muscles of Mastication.

☐ lateral pterygoid   ☐ masseter   ☐ medial pterygoid   ☐ temporalis (used twice)

EXERCISE 12.3

EXTRINSIC EYE MUSCLES

1. Observe a model of the eye with extrinsic eye muscles.

2. The **extrinsic**, or extraocular (*extra-*, outside of, + *oculus*, eye), muscles of the eye **(table 12.3)** allow us to move our eyes up, down, side to side, and at an angle. These muscles originate on bone and insert onto the sclera of the eye. The **sclera** is the tough white connective tissue covering of the eyeball. They are named based on location, shape, or function, so they are quite easy to identify and remember. Our coverage of them in this exercise will be brief. However, we will return to these muscles when we cover the cranial nerves and their functions in chapter 16. If you understand the functions of these muscles well now, you will find it much easier to understand the signs and symptoms of cranial nerve disorders when you get to the exercises in chapter 16.

3. Ask your laboratory partner to look in different directions and observe his or her eye movements. As his or her eyes move, name the muscles (in *both* eyes because they will be different!) used to create the movement (use table 12.3 as a guide).

4. Using table 12.3 and your textbook as guides, identify the **extrinsic eye muscles** listed in **figure 12.3** on the model of the eye. Then label them in figure 12.3.

Table 12.3	Extrinsic Muscles of the Eye				
Muscle	**Origin**	**Insertion**	**Action**	**Innervation**	**Word Origin**
Inferior Oblique	Maxilla (anterior portion of orbit)	Sclera on the anterior, lateral surface of the eyeball, deep to the lateral rectus muscle	Elevates, abducts, and laterally rotates the eyeball	Oculomotor (CN III)	*inferior*, lower, + *obliquus*, slanting
Inferior Rectus	Sphenoid (tendinous ring around optic canal)	Sclera on the anterior, inferior surface of the eyeball	Depresses, adducts, and medially rotates the eyeball	Oculomotor (CN III)	*inferior*, lower, + *rectus*, straight
Lateral Rectus	Sphenoid (tendinous ring around optic canal)	Sclera on the anterior, lateral surface of the eyeball	Abducts the eyeball	Abducens (CN VI)	*lateralis*, lateral, + *rectus*, straight
Medial Rectus	Sphenoid (tendinous ring around optic canal)	Sclera on the anterior, medial surface of the eyeball	Adducts the eyeball	Oculomotor (CN III)	*medialis*, middle, + *rectus*, straight
Superior Oblique	Sphenoid (tendinous ring around optic canal)	Sclera on the posterior, superiolateral surface of the eyeball just deep to the belly of the superior rectus muscle	Depresses, abducts, and laterally rotates the eyeball	Trochlear (CN IV)	*superus*, above, + *obliquus*, slanting
Superior Rectus	Sphenoid (tendinous ring around optic canal)	Sclera on the anterior, superior surface of the eyeball	Elevates, adducts, and medially rotates the eyeball	Oculomotor (CN III)	*superus*, above, + *rectus*, straight
Levator Palpebrae Superioris*	Sphenoid (lesser wing anterior and superior to the optic canal)	Skin of the superior eyelid	Elevates the upper eyelid	Oculomotor (CN III)	*levatus*, to lift, + *palpebra*, eyelid, + *superus*, above

*This muscle, while associated with the eye, does not attach to, or move, the eyeball itself.

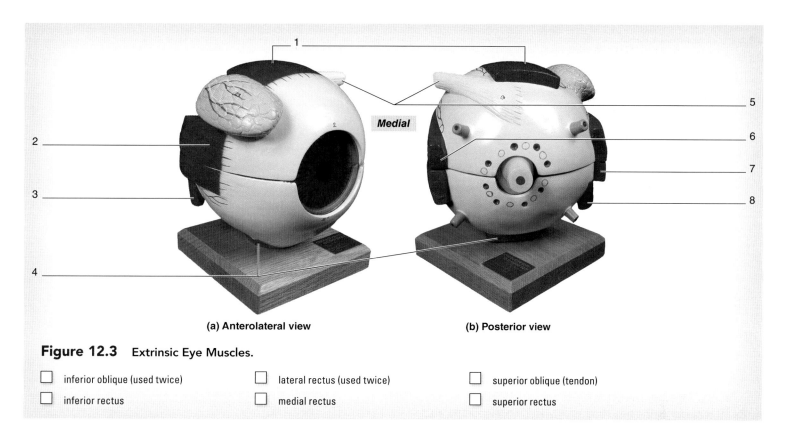

Figure 12.3 Extrinsic Eye Muscles.

☐ inferior oblique (used twice)   ☐ lateral rectus (used twice)   ☐ superior oblique (tendon)
☐ inferior rectus   ☐ medial rectus   ☐ superior rectus

EXERCISE 12.4

MUSCLES THAT MOVE THE TONGUE

1. Observe a model of a head that has been cut in a midsagittal plane so the muscles of the tongue are visible.

2. The tongue has both intrinsic and extrinsic muscles (**table 12.4**). The *intrinsic* muscles, within the tongue, change the shape of the tongue. The *extrinsic* muscles, which connect the tongue to bony structures of the head and neck, cause fine movements of the tongue necessary (e.g., to form speech and manipulate food).

3. Stick your tongue out, and manipulate it in varies ways (e.g., curl it, flip it). As you perform these actions, decide if the muscles used are intrinsic or extrinsic muscles of the tongue.

4. Using table 12.4 and your textbook as guides, identify the **muscles that move the tongue** listed in **figure 12.4** on a cadaver or on a model of the head and neck. (Note: These are all extrinsic muscles of the tongue.) Then label them in figure 12.4.

Table 12.4	Muscles That Move the Tongue				
Muscle	**Origin**	**Insertion**	**Action**	**Innervation**	**Word Origin**
Genioglossus	Mandible (mental spine)	Hyoid bone (body) and inferior portion of the tongue	Depresses and protrudes the tongue	Hypoglossal (CN XII)	*geneion*, chin, + *glossa*, tongue
Hyoglossus	Hyoid bone (body and greater horn)	Inferior and lateral aspects of the tongue	Depresses and retracts the tongue	Hypoglossal (CN XII)	*Hyo-*, hyoid bone, + *glossa*, tongue
Palatoglossus	Soft palate (palatine aponeurosis)	Inferiolateral aspect of the tongue	Depresses the soft palate and elevates the posterior aspect of the tongue	Vagus (CN X)	*palatum*, palate, + *glossa*, tongue
Styloglossus	Styloid process of the temporal bone	Inferior and lateral aspects of the tongue	Retracts and elevates the tongue during swallowing	Hypoglossal (CN XII)	*stylo*, styloid process, + *glossa*, tongue

(continued on next page)

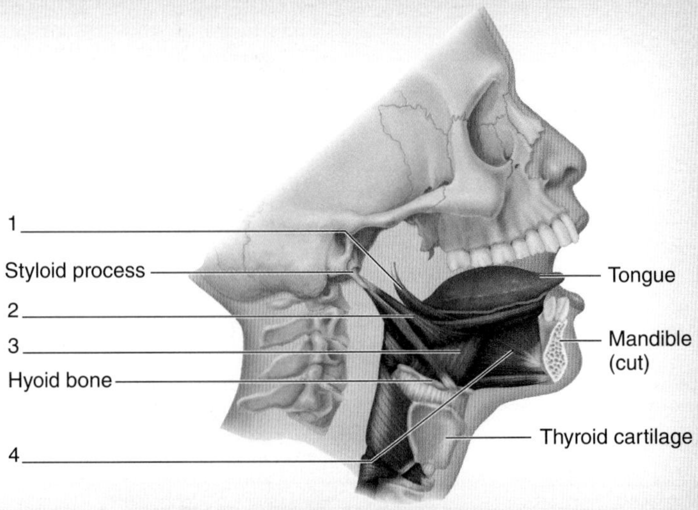

Figure 12.4 Muscles That Move the Tongue.

☐ genioglossus ☐ palatoglossus
☐ hyoglossus   ☐ styloglossus

Right lateral view

EXERCISE 12.5

MUSCLES OF THE PHARYNX

1. Observe a prosected cadaver, a model of the larynx, or a model of the head and neck. (The models should also show the muscles of the pharynx.)

2. The pharyngeal muscles (**table 12.5**) are used during the swallowing process. To get a feel for how these muscles function, swallow some saliva or a drink of water and pay attention to the role these muscles play during swallowing. Notice that as you swallow,

Table 12.5 Muscles of the Pharynx

Muscle	Origin	Insertion	Action	Innervation	Word Origin
Laryngeal Elevators					
Levator Veli Palatini	Temporal bone (petrous portion) and cartilage of the auditory tube	Soft palate (palatine aponeurosis)	Elevates the soft palate, as in swallowing and yawning	Vagus (CN X)	*levatus*, to lift, + *velum*, veil, + *palatum*, palate
Tensori Veli Palatini	Sphenoid bone, region around auditory tube	Soft palate	Tenses soft palate and opens auditory tube when swallowing or yawning	Trigeminal N (CN V)	*tensus*, to stretch, + *velum*, veil, + *palatine*, palate
Palate Muscles					
Palatopharyngeus	Soft palate (palatine aponeurosis)	Lateral wall of the pharynx	Elevates the larynx and pharynx	Vagus (CN X)	*palatum*, palate, + *pharyngo-*, pharynx
Stylopharyngeus	Styloid process of temporal bone	Larynx (thyroid cartilage)	Elevates the larynx and pharynx	Glossopharyngeal (CN IX)	*stylo-*, styloid process, + *pharyngo-*, pharynx
Tensor Veli Palatini	Sphenoid bone (pterygoid process)	Soft palate (palatine aponeurosis)	Elevates the soft palate	Trigeminal (CN V)	*tensus*, to stretch, + *velum*, veil, + *palatum*, palate
Pharyngeal Constrictors					
Inferior Constrictor	Larynx (thyroid and cricoid cartilages)	Posterior median raphe	Constricts the pharynx	Vagus (CN X)	*inferior*, lower, + *constringo*, to draw together
Middle Constrictor	Hyoid bone	Posterior median raphe	Constricts the pharynx	Vagus (CN X)	*middle*, middle, + *constringo*, to draw together
Superior Constrictor	Sphenoid bone (pterygoid process)	Posterior median raphe	Constricts the pharynx	Vagus (CN X)	*superus*, above, + *constringo*, to draw together

your larynx moves superiorly by suprahyoid muscles (included in exercise 12.6). (Your larynx is best felt by palpating your thyroid cartilage, or "Adam's apple." It is located in the upper portion of anterior neck.) The muscles you are focusing on here are the ones you feel at the back of your throat at the very end of the swallowing process. The pharyngeal constrictors move swallowed materials (e.g., food, saliva) into the esophagus.

3. Using table 12.5 as a guide, identify the following **muscles of the pharynx** listed in **figure 12.5** on a cadaver, on a model of the head and neck, or on a model of the larynx. Then label them in figure 12.5.

☐ inferior constrictor
☐ levator veli palatini
☐ middle constrictor
☐ stylopharyngeus
☐ superior constrictor
☐ tensor veli palatini

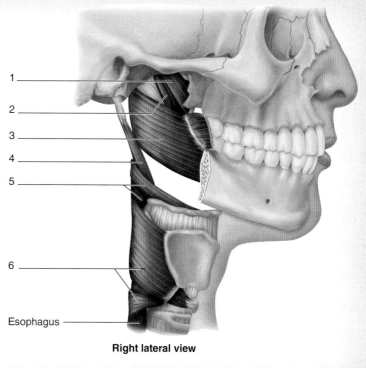

Right lateral view

Figure 12.5 Muscles of the Pharynx.

EXERCISE 12.6

MUSCLES OF THE NECK

1. Observe a prosected cadaver or a classroom model demonstrating muscles of the neck.

2. The most prominent muscle in the anterior neck is the **sternocleidomastoid** muscle. The attachment of this muscle can be easily palpated through the skin. To do this, place your fingers just behind your ears to palpate the **mastoid process**, the insertion of the sternocleidomastoid. Next, palpate your sternum, just lateral to the sternal notch as you rotate your neck from right to left, and feel for the tendons of the **sternal head** of the sternocleidomastoid. Finally, palpate just superior to the medial third of the clavicle while laterally flexing your neck to see if you can feel contraction of the **clavicular head** of the sternocleidomastoid.

INTEGRATE

CONCEPT CONNECTION

An **agonist** (*agon*, a contest), or **prime mover**, is a muscle used to create a given action about a joint. An **antagonist** (*anti*, against, + *agon*, a contest) is a muscle whose action opposes the action of the agonist. In contrast, a synergist (*syn*, together) is a muscle whose action assists the agonist. The sternocleidomastoid and splenius capitis are muscles that provide a good example of how muscles can act as either synergists or antagonists, depending upon the movement required. These muscles can act as synergists for the action of lateral rotation of the neck, whereas they act as antagonists for the actions of flexion and extension of the neck. Palpate the sternal head of the sternocleidomastoid on your right side, and then laterally rotate your head to the right. Do you feel tension in the muscle? _____ While palpating the muscle, laterally rotate your head to the left. Do you feel tension in the muscle? _____ Based on your observations, in which direction does the right sternocleidomastoid rotate the neck? _____

The most superficial muscle on the posterior side of the neck is the trapezius, a muscle that can move the neck, but is more of a prime mover of the scapula (see table 13.1). If you palpate the posterior side of your neck as you rotate your neck, you can feel the trapezius muscle and a smaller muscle located deep to the trapezius, the splenius capitis, which is a prime mover of the neck (see table 12.6 for reference to its actions). Place your right hand over the posterior aspect of the right side of your neck and rotate your head to the right. Keeping your hand in the same location, rotate your head to the left. In which direction of rotation do you feel more tension in the muscles on the back of the neck? _____ Based on your observations, in which direction does the right splenius capitis muscle rotate the neck? _____

In summary: To rotate your neck to the right, you use the sternal head of the sternocleidomastoid on the _____ side of the neck, and the splenius capitis muscle on the _____ side of the neck.

(continued on next page)

(continued from previous page)

Understanding the locations of the attachments, bellies, and borders of the two heads of this muscle is important because the sternocleidomastoid is a major clinical landmark in the neck.

3. Using **table 12.6** as a guide, identify the muscles of the neck listed in **figure 12.6** on a cadaver or a model of the head and neck. Then label them in figure 12.6. (Some muscles may be used more than once.)

Table 12.6	Muscles of the Neck				
Muscle	**Origin**	**Insertion**	**Action**	**Innervation**	**Word Origin**
Muscles That Move the Neck					
Longissimus Capitis	T_1–T_4 (transverse processes) and C_4–C_7 (articular processes)	Mastoid process of temporal bone	Bilateral: extends the neck Unilateral: laterally rotates the neck to the same side	Cervical and thoracic spinal nerves	*longissimus*, longest, + *caput*, head
Semispinalis Capitis	T_1–T_5 (spinous processes), C_4–C_7 (articular processes)	Occipital bone (between superior and inferior nuchal lines)	Bilateral: extends the neck Unilateral: laterally flexes the neck to the same side	Dorsal rami of cervical spinal nerves	*semi*, half, + *spina*, spine, + *caput*, head
Splenius Capitis	Ligamentum nuchae and T_1–T_6 (spinous processes)	Superior nuchal line of occipital bone (lateral aspect) and mastoid process of temporal bone	Bilateral: extends the neck Unilateral: laterally rotates and laterally flexes the neck to the same side	Dorsal rami of spinal nerves	*splenion*, a bandage, + *caput*, head
Sternocleidomastoid	Sternal head: manubrium of the sternum Clavicular head: clavicle (medial third)	Mastoid process of temporal bone	Bilateral: flexes the neck Unilateral: laterally rotates the neck to the opposite side	Accessory (CN XI)	*sterno-*, sternum, + *cleido-*, clavicle, + *mastoid*, resembling a breast
Suprahyoid Muscles					
Digastric	Mastoid process (digastric groove on medial aspect)	Mandible (lower border near the midline)	When the mandible is fixed, it elevates the hyoid (posterior belly). When the hyoid is fixed, it depresses the mandible (anterior belly).	Posterior belly: facial (CN VII) Anterior belly: mandibular branch of the trigeminal (CN V)	*di-*, two, + *gastro-*, belly
Geniohyoid	Mandible (mental spine)	Hyoid bone	Elevates the hyoid	Hypoglossal (CN XII)	*geneion*, chin, + *hyoides*, shaped like the letter U
Mylohyoid	Mandible (mylohyoid line)	Hyoid bone	Elevates the floor of the mouth and tongue. When the hyoid is fixed it depresses the mandible.	Mandibular branch of the trigeminal (CN V)	*myle*, a mill, + *hyoides*, shaped like the letter U
Stylohyoid	Styloid process of temporal bone	Hyoid bone	Elevates the hyoid	Facial (CN VII)	*stylos*, pillar, + *hyoides*, shaped like the letter U
Infrahyoid Muscles					
Omohyoid	Scapula (between the superior angle and the scapular notch)	Hyoid bone	Depresses the hyoid	Cervical spinal nerves C1–C3 through ansa cervicalis	*omos*, shoulder, + *hyoides*, shaped like the letter U

Table 12.6 — Muscles of the Neck (continued)

Muscle	Origin	Insertion	Action	Innervation	Word Origin
Sternohyoid	Manubrium of sternum (posterior surface) and first costal cartilage	Hyoid bone	Depresses the hyoid	Cervical spinal nerves C1–C3 through ansa cervicalis	*sternon*, chest, + *hyoides*, shaped like the letter U
Sternothyroid	Manubrium of sternum (posterior surface) and first costal cartilage	Thyroid cartilage	Depresses the larynx	Cervical spinal nerves C1–C3 through ansa cervicalis	*sternon*, chest, + *thyroid*, shaped like a shield
Thyrohyoid	Thyroid cartilage	Hyoid bone	Moves hyoid toward the larynx	First cervical spinal nerve C1 via hypoglossal (CN XII)	*thyro-*, thyroid, + *hyoides*, shaped like the letter U

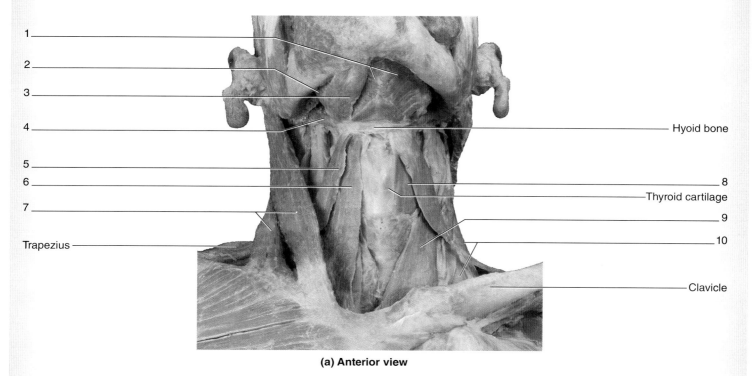

(a) Anterior view

Figure 12.6 Muscles of the Head and Neck.

- ☐ digastric (anterior belly) (used twice)
- ☐ digastric (posterior belly) (used twice)
- ☐ mylohyoid (used twice)
- ☐ omohyoid (inferior belly) (used twice)
- ☐ omohyoid (superior belly) (used twice)
- ☐ scalenes (used twice)
- ☐ splenius capitis (used three times)
- ☐ sternocleidomastoid (used three times)
- ☐ sternohyoid (used twice)
- ☐ sternothyroid (used twice)
- ☐ stylohyoid
- ☐ thyrohyoid

INTEGRATE

LEARNING STRATEGY

When studying muscles, or any other anatomical structure for that matter, remember that directional terms like "anterior/posterior" or "medial/lateral" are used in the naming of muscles only when necessary. That is, they are generally used only when there are two or more similar muscles with the same name. For example: If there is a "major" (e.g., zygomaticus major), then there must also be a "minor" (e.g., zyogmaticus minor); if there is a "superior" (e.g., superior oblique), there must also be an "inferior" (e.g., inferior oblique). This may seem simple, but it is not always obvious to the beginning student of anatomy.

(continued on next page)

(continued from previous page)

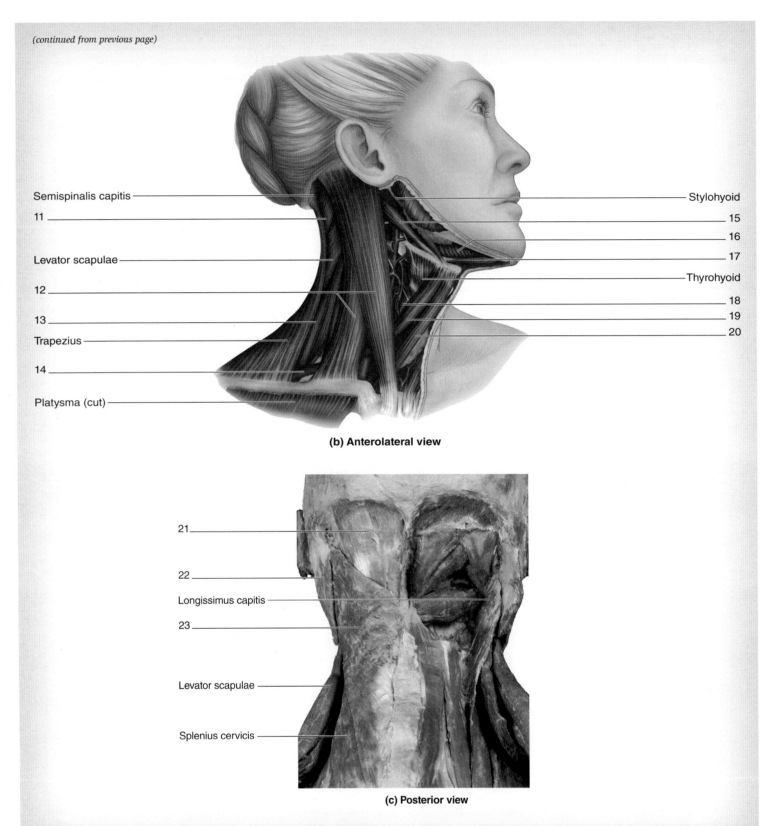

(b) Anterolateral view

(c) Posterior view

Figure 12.6 Muscles of the Head and Neck *(continued)*.

INTEGRATE

CLINICAL VIEW
Stroke

A **stroke** occurs when one or more parts of the brain suffer(s) a lack of oxygen from either a reduction in blood flow—as can happen with a blood clot or narrowed artery (ischemic stroke)—or from rupture of a blood vessel in the brain (hemorrhagic stroke). The consequences of a stroke can be great, depending on the portion of the brain affected. In chapter 15 you will study the parts of the brain that function in planning movements, perceiving sensations, and maintaining homeostasis. As you may imagine, loss of oxygen or rupture of a blood vessel to the part of the brain responsible for regulating bodily functions such as heart rate can have very serious implications. Patients who suffer from a stroke often experience symptoms such as gross motor impairments, slurred speech, an inability to swallow, and a "droopy" face. These symptoms often indicate damage to the specific areas of the brain responsible for each modality. For example, the command for voluntary movement of skeletal muscles arises from the frontal lobe.

Therefore, damage to the frontal lobe may result in an inability to stimulate skeletal muscles in various regions of the body. The nuclei for the cranial nerves innervate skeletal muscles of the head and neck. For example, you have learned that the paired facial nerves (CN VII) innervate the muscles of the face responsible for creating facial expressions. The cell bodies of the facial nerves are located within the brainstem. If the supply of oxygen is reduced in this critical part of the brain, nerve signals will not be sent from the cranial nerve nuclei out to the skeletal muscles. This results in muscle paralysis so the patient is unable to demonstrate facial expressions on the affected side. The patient may also experience an inability to elevate the right upper lip. Likewise, damage to the nuclei for the paired trigeminal (CN V) nerves may cause a patient to experience difficulty with mastication (chewing). Damage to the hypoglossal (CN XII) nerve may make swallowing or speech difficult. Such symptoms give clinicians some clues regarding what structures or areas of the brain have been damaged by a stroke. Typically, a magnetic resonance imaging (MRI) scan is performed on the brain to confirm the exact location of damage.

Muscles of the Vertebral Column

Muscles that move the vertebral column are complex in both location and function. Your ability to identify all of the muscles listed here will depend on the degree to which the cadaver in your laboratory is dissected or the type(s) of models available to you in the laboratory. To simplify the process of learning these muscles, initially focus your attention on learning how the groups of muscles are arranged from superficial to deep, and from medial to lateral. Then turn your attention to specifically identifying the individual muscles belonging to each group.

EXERCISE 12.7

MUSCLES OF THE VERTEBRAL COLUMN

1. Observe a prosected cadaver or a classroom model of the thorax/abdomen that demonstrates muscles of the vertebral column (**table 12.7**).

2. Using table 12.7 as a guide, identify the **muscles of the vertebral column** listed in **figure 12.7** on a cadaver or on models of the thorax and abdomen. Then label them in figure 12.7.

3. The largest muscles of the back that move the vertebral column are collectively referred to as the **erector spinae** (*erector*, to make erect or straight, + *spina*, spine). The erector spinae consist of three muscle groups that form long columns along both sides of the vertebral column. The muscle groups, from medial to lateral, are the spinalis, longissimus, and iliocostalis. Deep to the erector spinae is the **transversospinal** group of muscles, so named because they attach to transverse and spinous processes of adjacent vertebrae, and several smaller muscles that create fine movements of the vertebral column: interspinales and intertransversarii. The deeper muscles can be difficult to identify on cadavers or models. In general, the **semispinalis** muscles are the most superficial of the transversospinal muscles, span 5–6 vertebrae, and are most highly developed in the cervical and upper thoracic regions of the vertebral column. The **multifidus** lie deep to the semispinalis muscles, span 3–4 vertebrae, and are most highly developed in the lumbar region of the vertebral column. Finally, the **rotatores** are the deepest muscles of the transversospinal group, span 1–2 vertebrae, and are most highly developed in the lower thoracic region of the vertebral column.

Table 12.7 Muscles of the Vertebral Column

Muscle Group	Individual Muscles	Origin	Insertion	Action	Innervation	Word Origin
Superficial Layer — Splenius Muscles						
Splenius Muscles	The splenius muscles are thick, flat muscles on the lateral and posterior aspect of the neck	Midline	Cervical vertebrae and skull	Hold the deep neck muscles in position; extend, laterally flex, and laterally rotate the neck	Dorsal rami of spinal nerves	*splenius*, bandage

(continued on next page)

(continued from previous page)

Table 12.7 Muscles of the Vertebral Column (continued)

Muscle Group	Individual Muscles	Origin	Insertion	Action	Innervation	Word Origin
REGIONS	Capitis (splenius capitis)	Ligamentum nuchae and T_1–T_6 vertebrae (spinous processes)	Superior nuchal line (lateral aspect)	Bilateral: extends the neck Unilateral: laterally rotates and laterally flexes the neck to the same side Bilateral: extends the neck	Dorsal rami of spinal nerves	*splenion*, a bandage, + *caput*, head
	Cervicis (splenius cervicis)	Nuchal ligament and C_7–T_4 (spinous processes)	C_1–C_3 vertebrae (posterior tubercles)	Unilateral: laterally rotates and laterally flexes the neck to the same side	Dorsal rami of spinal nerves	*cervix*, neck
Intermediate Layer — Erector Spinae (Sacrospinalis) Muscles						
Erector Spinae	The erector spinae muscles compose the intermediate layer of back muscles. They are arranged into groups.	Broad tendon covering the posterior iliac crest, the lumbar vertebrae, and the sacrum	Vertebrae and ribs	Bilateral: extends the vertebral column and the head/neck Unilateral: laterally bends the vertebral column	Dorsal rami of spinal nerves	*erector*, to make erect, + *spina*, spine
GROUPS	Iliocostalis (lateral group)	Broad tendon covering the posterior iliac crest, the lumbar vertebrae, and the sacrum	Ribs (angles of lower ribs) and cervical vertebrae (transverse processes)	Extends the vertebral column	Dorsal rami of spinal nerves	*ilium*, groin, + *costal*, rib
	Longissimus (intermediate group)	Broad tendon covering the posterior iliac crest, the lumbar vertebrae, and the sacrum	Ribs (between tubercles and angles), cervical and thoracic vertebrae (transverse processes), and mastoid process of temporal bone	Extends the neck and vertebral column and laterally rotates the head	Dorsal rami of spinal nerves	*longissimus*, longest
	Spinalis (medial group)	Broad tendon covering the posterior iliac crest, the lumbar vertebrae, and the sacrum	Vertebrae (spinous processes of upper thoracic), and skull	Extends the neck and vertebral column and laterally rotates the head	Dorsal rami of spinal nerves	*spina*, spine
Spinal Flexors — Quadratus Lumborum						
	Quadratus lumborum	Iliac crest and transverse processes of lower lumbar vertebrae	Rib 12, transverse processes of upper lumbar vertebrae	Abducts the trunk	Ventral rami of lumbar spinal nerves	*quadratus*, square, + *lumbus*, loin

Table 12.7	Muscles of the Vertebral Column *(continued)*					
Muscle Group	**Individual Muscles**	**Origin**	**Insertion**	**Action**	**Innervation**	**Word Origin**
Deep Layer—Transversospinalis Muscles						
Transversospinal Group	The transversospinal muscles are the deepest muscles of the back and lie between transverse and spinous processes of adjacent vertebrae	Transverse processes of inferior vertebrae	Spinous process of cervical and thoracic vertebrae 1–3 levels above the vertebra of origin and/or the posterior aspect of the occipital bone	Extends and rotates the vertebral column; stabilizes the vertebrae during local movements of the vertebral column	Dorsal rami of spinal nerves	*transverse*, across, + *spina*, spine
Multifidus	NA	T_1–T_3 vertebrae (transverse processes), C_4–C_7 (articular processes), ilium and sacrum	Spinous process of vertebra located 2–4 segments superior to vertebra of origin	Assists with local extension and rotation of the vertebral column	Dorsal rami of spinal nerves	*multus*, much, + *findo*, to cleave
Rotatores	NA	Transverse processes of all vertebrae (most developed in the thoracic region)	Vertebral arch (between lamina and transverse process) of vertebra superior to the vertebra of origin	Assists with local extension and rotation of the vertebral column	Dorsal rami of spinal nerves	*rotatus*, to rotate
Semispinalis Group	The semispinalis muscles are the deepest muscles of the back and lie between transverse and spinous processes of adjacent vertebrae	Transverse processes of inferior vertebrae	Spinous process of vertebra above and/or the posterior aspect of the occipital bone	Extends and rotates the vertebral column; stabilizes the vertebrae during local movements of the vertebral column	Dorsal rami of spinal nerves	*semis*, half, + *spina*, spine
REGIONS	Capitis (semispinalis capitis)	Inferior cervical and superior thoracic vertebrae (spinous and transverse processes)	Occipital bone (between superior and inferior nuchal lines)	Extends and rotates the vertebral column; stabilizes the vertebrae during local movements of the vertebral column	Dorsal rami of spinal nerves	*caput*, head
	Cervicis (semispinalis cervicis)	T_1–T_6 vertebrae (transverse processes)	C_2–C_3 vertebrae (spinous processes)	Extends and rotates the vertebral column; stabilizes the vertebrae during local movements of the vertebral column	Dorsal rami of spinal nerves	*cervix*, neck
	Thoracis (semispinalis thoracis)	T_6–T_{10} vertebrae (transverse processes)	C_5–T_4 vertebrae (spinous processes)	Extends and rotates the vertebral column; stabilizes the vertebrae during local movements of the vertebral column	Dorsal rami of spinal nerves	*thoracis*, thorax

(continued on next page)

(continued from previous page)

Table 12.7 Muscles of the Vertebral Column (continued)

Muscle Group	Individual Muscles	Origin	Insertion	Action	Innervation	Word Origin
Minor Deep Back Muscles						
	Interspinales	Cervical and lumbar vertebrae (superior surfaces of spinous processes)	Spinous process of the vertebra superior to the vertebra of origin (inferior surface)	Extends and rotates the vertebral column	Dorsal rami of spinal nerves	*inter*, between, + *spina*, spine
	Intertransversarii	Cervical and lumbar vertebrae (transverse processes)	Transverse processes of vertebra above or below vertebra of origin	Laterally flexes the vertebral column	Dorsal rami of spinal nerves	*inter*, between, + *transversarii*, relating to the transverse process

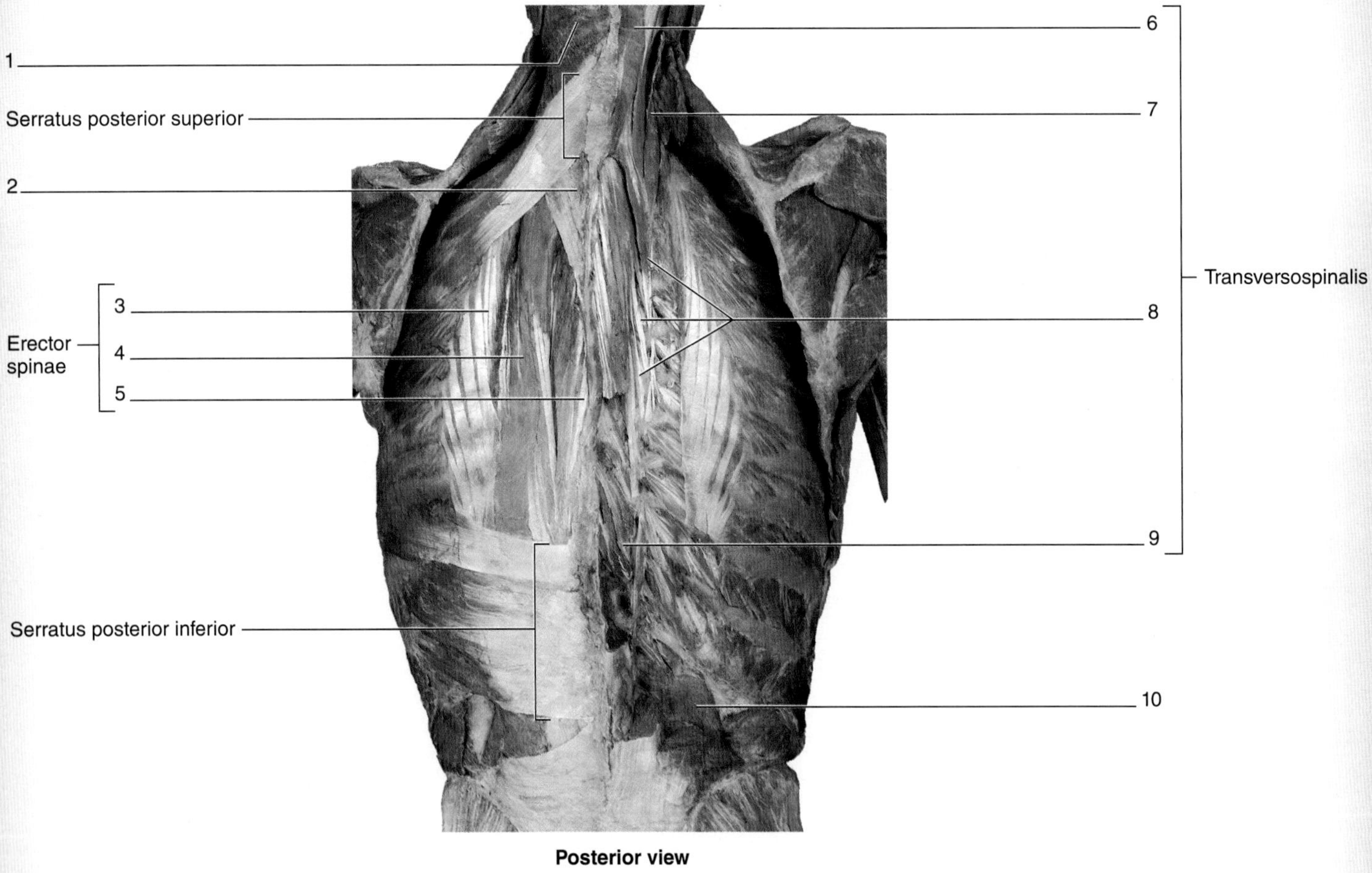

Figure 12.7 Muscles of the Vertebral Column.

Posterior view

Labels: Serratus posterior superior, Erector spinae (3, 4, 5), Serratus posterior inferior, Transversospinalis, 1, 2, 6, 7, 8, 9, 10

☐ iliocostalis
☐ longissimus
☐ multifidus
☐ quadratus lumborum
☐ semispinalis capitis
☐ semispinalis cervicis
☐ semispinalis thoracis
☐ spinalis
☐ splenius capitis
☐ splenius cervicis

Muscles of Respiration

The muscles of the thoracic cage and the diaphragm are the primary **muscles of respiration (table 12.8)**. These muscles include the diaphragm, external and internal intercostals, the transversus thoracis, and the scalenes.

Table 12.8	Muscles of Respiration				
Muscle	**Origin**	**Insertion**	**Action**	**Innervation**	**Word Origin**
Diaphragm	Inferior borders of rib 12, sternum, and the xiphoid process; costal cartilages of ribs 6–12, and lumbar vertebrae	Central tendon	Prime mover for inspiration; flattens when contracted, and increases intra-abdominal pressure and the size of the thoracic cavity	Phrenic nerves	*diaphragma*, a partition wall
External Intercostals	Inferior border of superior rib	Superior border of the rib below	Elevates the ribs	Intercostal nerves (continuations of thoracic spinal nerves)	*externus*, on the outside, + *inter*, between, + *costal*, rib
Internal Intercostals	Superior border of inferior rib	Inferior border of the rib above	Depresses the ribs	Intercostal nerves (continuations of thoracic spinal nerves)	*internus*, away from the surface, + *inter*, between, + *costal*, rib
Transversus Thoracis	Posterior surface of the lower half of the body of the sternum	Costal cartilages of ribs 2–6 (posterior surface)	Depresses the ribs	Intercostal nerves (continuations of thoracic spinal nerves)	*transversus*, crosswise, + *thoracis*, thorax
Anterior Scalene	C_3–C_6 (transverse processes)	First rib (scalene tubercle)	Elevates the first rib	Cervical plexus	*skalenos*, uneven
Middle Scalene	C_2–C_6 (transverse processes)	First rib (posterior to groove for subclavian artery)	Elevates the first rib	Cervical plexus	*skalenos*, uneven
Posterior Scalene	C_4–C_6 (transverse processes)	Second rib (lateral surface)	Elevates the second rib	Cervical and brachial plexuses	*skalenos*, uneven

EXERCISE 12.8

MUSCLES OF RESPIRATION

1. Observe a prosected cadaver or a classroom model of the thorax/abdomen that demonstrates muscles of the thoracic cage.

2. Using table 12.8 and your textbook as guides, identify the muscles of respiration listed in **figure 12.8** on a cadaver or on a model of the thorax. Then label them in figure 12.8. (Answers may be used more than once.)

3. *Intercostals*—The majority of the muscle mass of the **external intercostals** is located on the posterior and lateral thorax, extending from the vertebral column to the **midclavicular line** (a vertical line that passes through the middle of the clavicle; see **figure 12.9a**). The **external intercostal membrane** lies in place of the external intercostal muscles in the space between the midclavicular line and the sternum. Notice that the muscle fibers of the external intercostals are arranged obliquely, pointing in an inferomedial direction. The external intercostals elevate the ribs during inspiration.

4. On a cadaver you can identify the external intercostal membrane because the connective tissue fibers parallel the direction of the muscle fibers of the external intercostals. In fact, if you look closely, you will see muscle adjacent to the sternum in the intercostal spaces, but that muscle is an *internal* intercostal muscle. Internal intercostals depress the ribs during forced expiration. If you are observing classroom models, you will most likely not be able to identify this membrane.

5. In contrast to the external intercostals, the majority of the muscle mass of the **internal intercostals** is located on the anterior surface of the thorax. These muscles extend from the sternum to the **scapular line** (a vertical line that passes through the inferior angle of the scapula; **figure 12.9b**) on the posterior thorax. The **internal intercostal membrane** lies in place of the internal intercostal muscles on the posterior thorax between the scapular line and the vertebral column. The muscle fibers of the internal intercostals are arranged obliquely, pointing in an inferolateral direction, at right angles to the fibers of the external intercostals.

6. If possible, remove the breastplate on the cadaver (or on a model of the thorax) and observe its interior surface. Here you will see, lying adjacent to the inferior part of the sternum, the **transversus thoracis** muscle, which consists of several muscle bellies running obliquely. The transversus thoracis assists in depression of the ribs during forced expiration.

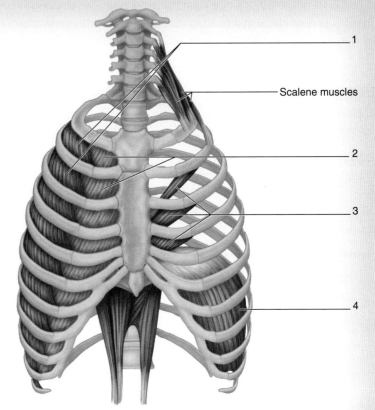

(a) Anterior view

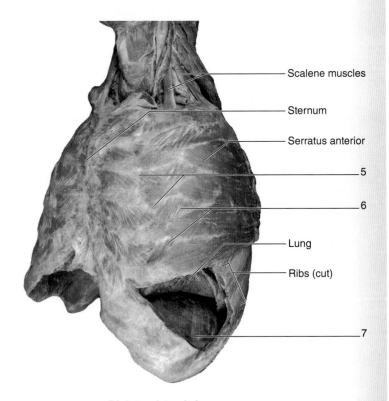

(b) Anterolateral view

Figure 12.8 Muscles of Respiration.

- ☐ diaphragm (used twice)
- ☐ external intercostals (used twice)
- ☐ internal intercostals (used twice)
- ☐ transverse thoracis

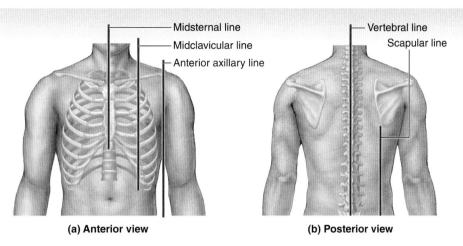

Figure 12.9 Points of Reference on the Anterior and Posterior Thorax.

7. In the space below, make a representative drawing of the intercostal muscles, noting the fiber orientation for each of them.

8. *Diaphragm*—The **diaphragm** has a broad origin along the bones that constitute the lower border of the thoracic cage. The muscle has a unique central attachment, the **central tendon of the diaphragm.** Several important structures pass through the diaphragm, including the aorta, inferior vena cava, and esophagus. The aorta and inferior vena cava have passages through the central tendon, whereas the passage for the esophagus is surrounded by the muscle fibers of the diaphragm. Very practical consequences result from this arrangement. As the diaphragm contracts during inspiration, it changes from its dome-shape to a more flattened position. The muscle fibers of the diaphragm squeeze the esophagus and act as a sphincter to prevent stomach contents from being pushed back into the esophagus. In contrast, because the aorta and inferior vena cava pass through the central tendon, they are not constricted during inspiration. Remove the breastplate from the cadaver or model and observe the diaphragm. The structures within the diaphragm are best seen from an inferior view. If possible, try to observe the diaphragm from both superior and inferior points of view.

9. Using your textbook as a guide, label the following items on **figure 12.10**.

   ☐ aortic opening (hiatus) ☐ esophageal opening (hiatus)
   ☐ caval opening (for inferior vena cava)

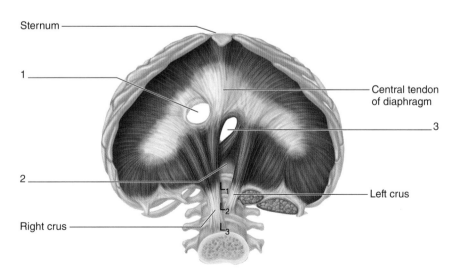

Figure 12.10 Diaphragm. Inferior View.

Muscles of the Abdominal Wall

Muscles of the abdominal wall include the external obliques, internal obliques, transversus abdominis, and rectus abdominis.

EXERCISE 12.9

MUSCLES OF THE ABDOMINAL WALL

1. Observe a prosected cadaver or a model of the thorax/abdomen that demonstrates muscles of the abdominal wall (**table 12.9** and figures 12.11–12.13).

2. The muscle fiber orientation of the abdominal muscles parallels the muscle fiber orientation of the intercostal muscles. Like the external intercostals, the **external obliques** have muscle fibers that point in an inferomedial direction. Like the internal intercostals, the **internal obliques** have muscle fibers that point in an inferolateral direction. Deep to the internal obliques, the **transversus abdominis** has muscle fibers that run in a transverse, or horizontal, direction. All three of these muscles have broad, flat tendons called **aponeuroses** (*apo-*, from, + *neuron*, sinew). The aponeuroses of these muscles begin at the midclavicular line and extend to a central, tendonlike structure called the **linea alba**. In addition, the aponeuroses of these muscles form the rectus sheath, which surrounds the fourth abdominal muscle, the **rectus abdominis**.

3. Using table 12.9 as a guide, identify the **abdominal muscles** listed in **figure 12.11** on a cadaver or on a classroom model of the abdomen. Then label them in figure 12.11.

Table 12.9 Muscles of the Abdominal Wall

Muscle	Origin	Insertion	Action	Innervation	Word Origin
External Oblique	Anterior surface of inferior 8 ribs	Linea alba and the anterior iliac crest	Flexes and rotates the trunk; compresses the abdominal viscera	Spinal nerves T8–T12, L1	*externus*, on the outside, + *obliquus*, slanting
Internal Oblique	Thoracolumbar fascia, lateral half of inguinal ligament, and iliac crest	Linea alba, iliac crest, pubic tubercle, and the inferior border of last 4 ribs; costal cartilages of ribs 8–10	Flexes and rotates the trunk; compresses the abdominal viscera	Spinal nerves T8–T12, L1	*internus*, away from the surface, + *obliquus*, slanting
Rectus Abdominis	Pubic symphysis and crest	Xiphoid process and the costal cartilages of ribs 5–7	Flexes and rotates the trunk; compresses the abdominal viscera	Spinal nerves T7–T12	*rectus*, straight, + *abdominis*, the abdomen
Transversus Abdominis	Lateral third of inguinal ligament, iliac crest, costal cartilages of inferior 6 ribs	Linea alba and pubic crest	Compresses the abdominal viscera	Spinal nerves T8–T12, L1	*transversus*, crosswise, + *obliquus*, slanting

Structures Related to Abdominal Musculature

Structure	Attachment 1	Attachment 2	Description	Innervation	Word Origin
Aponeurosis	NA	NA	A broad, flat tendon, such as those connecting abdominal muscles to the linea alba	NA	*apo-*, from, + *neuron*, sinew
Inguinal Canal	NA	NA	An oblique passage in the anterior abdominal wall located superior to the inguinal ligament; it is formed from the aponeuroses of the external and internal oblique muscles	NA	*inguen*, groin

Chapter Twelve The Muscular System: Axial Muscles

Table 12.9	Muscles of the Abdominal Wall *(continued)*				
Structure	**Attachment 1**	**Attachment 2**	**Description**	**Innervation**	**Word Origin**
Inguinal Ligament	Anterior superior iliac spine	Pubic tubercle	A structure formed from the aponeurosis of the external oblique muscle; an important anatomical landmark in the inguinal region	NA	*inguen*, groin
Linea Alba	Xiphoid process of the sternum	Pubic symphysis	Literally, the "white line"; a tendinous structure that acts as the insertion point for the oblique and transversus abdominis muscles	NA	*linea*, line, + *alba*, white
Rectus Sheath	NA	NA	A connective tissue sheath that surrounds the rectus abdominis muscle and is formed from the aponeuroses of the external oblique, internal oblique, and transversus abdominis muscles	NA	*rectus*, referring to the rectus abdominis muscle
Tendinous Intersections	NA	NA	Tendinous bands that run across a muscle. In this case, the tendinous intersections are the structures that separate the parts of the rectus abdominis muscle.	NA	*tendo-*, to stretch out, + *inscriptio*, to write on

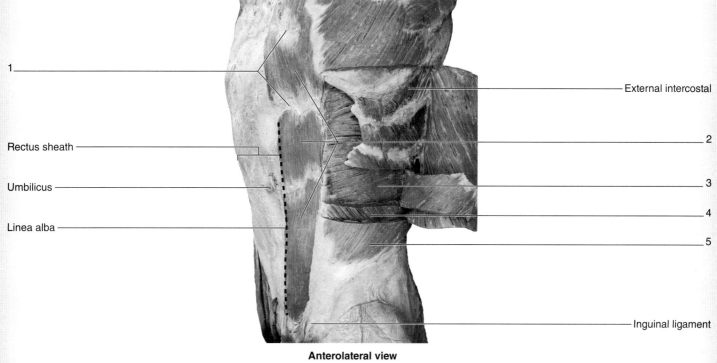

Anterolateral view

Figure 12.11 Muscles of the Abdominal Wall.

☐ external oblique ☐ rectus abdominis ☐ transversus abdominis
☐ internal oblique ☐ tendinous intersections

EXERCISE 12.10

THE RECTUS SHEATH, INGUINAL LIGAMENT, AND INGUINAL CANAL

1. Observe a prosected cadaver or a classroom model of the thorax/abdomen that demonstrates muscles of the abdominal wall. In this exercise you will focus on important structures associated with the abdominal musculature.

2. The **rectus sheath** (figure 12.12; see table 12.9) is a structure formed from the aponeuroses of the external obliques, internal obliques, and the transversus abdominis muscles (this relationship is accurate only for the sheath superior to the umbilicus, so your observations should be made in that location). The sheath has an *anterior border* formed by the aponeurosis of the external oblique and part of the aponeurosis of the internal oblique. It has a *posterior border* formed by the remaining part of the aponeurosis of the internal oblique and the aponeurosis of the transversus abdominis. The rectus abdominis muscle is located within the rectus sheath, and is divided into four sections by connective tissue partitions called **tendinous intersections**. The tendinous intersections effectively divide one very long muscle into four smaller muscles, arranged in a series. This allows the muscle as a whole to make a greater overall change in length, in addition to increasing its force of contraction.

3. The **inguinal ligament** (figure 12.13; table 12.9) is formed by the *aponeurosis of the external oblique* muscle. This ligament extends from the anterior superior iliac spine (ASIS) laterally, to the pubic tubercle medially. Instead of being a straight ligament, the ligament folds back upon itself, forming a trough. The inguinal ligament is an important landmark of the abdomen and thigh. In addition, the trough formed by the aponeurosis of the external oblique forms part of the *inguinal canal*.

4. The **inguinal canal** (figure 12.13; table 12.9) is an oblique passageway in the inferior abdominal wall. Its floor is formed by the trough of the **aponeurosis of the external oblique**. Its roof is formed by fibers of the **aponeurosis of the internal oblique**. Within this canal, structures pass from the abdomen into the subcutaneous tissues of the **perineum** (*perneon*, the area between the thighs below the pelvic diaphragm). In males, structures that pass through compose the **spermatic cord**, which consists of the testicular artery, vein, and nerve; lymphatic vessels; and the ductus deferens. In females, the **round ligament of the uterus,** a suspensory ligament, passes through. The **superficial inguinal ring** is located lateral to the pubic symphysis. It is composed of fibers of the aponeuroses of the external oblique aponeurosis, and it is the path of exit for structures that pass through the inguinal canal. On a male cadaver or on a classroom model of the abdomen, identify the spermatic cord as it passes through the superficial inguinal ring. Attempt to identify the round ligament of the uterus on a female cadaver, keeping in mind that identification of the round ligament of the uterus is often quite difficult.

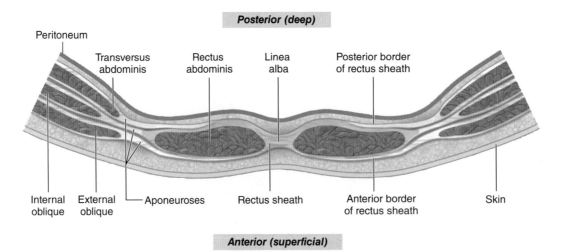

Figure 12.12 The Rectus Sheath. A transverse section through the abdominal wall demonstrates components of the rectus sheath.

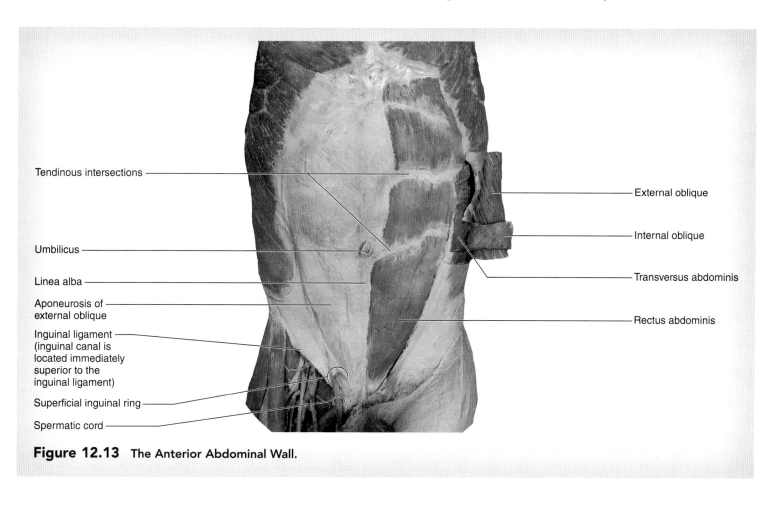

Figure 12.13 The Anterior Abdominal Wall.

Chapter 12: The Muscular System: Axial Muscles

Name: _____
Date: _____ Section: _____

POST-LABORATORY WORKSHEET

The ❶ corresponds to the Learning Objective(s) listed in the chapter opener outline.

Do You Know the Basics?

Exercise 12.1: Muscles of Facial Expression

1. Match the muscles of facial expression with the movements produced by each. ❶

   Column A
   ____ blink or close eyes
   ____ close mouth
   ____ wrinkle forehead, raise eyebrows
   ____ tense skin of neck
   ____ compress cheeks

   Column B
   a. buccinator
   b. epicranius
   c. orbicularis oculi
   d. orbicularis oris
   e. platysma

2. The muscles of facial expression are innervated by the _____ nerve. ❶

Exercise 12.2: Muscles of Mastication

3. Match the muscles of mastication with the movements produced by each. ❷

   Column A
   ____ grind food side to side (2 answers)
   ____ elevate mandible (3 answers)
   ____ depresses the chin
   ____ protract mandible, depress chin (3 answers)

   Column B
   a. temporalis
   b. medial pterygoid
   c. lateral pterygoid
   d. masseter

4. The muscles of mastication are innervated by the _____ nerve. ❷

Exercise 12.3: Extrinsic Eye Muscles

5. For each of the eye movements listed below, list the extrinsic eye muscle that creates that action (consider only movement of the right eye). ❸

   Look lateral only: _____
   Look medial only: _____
   Look up and in (medial): _____
   Look up and out (lateral): _____
   Look down and in (medial): _____
   Look down and out (lateral): _____

Exercise 12.4: Muscles That Move the Tongue

6. Explain the difference in location of the intrinsic muscles of the tongue and extrinsic muscles of the tongue. ❹

   _____
   _____
   _____

7. The genioglossus muscle is a(n) intrinsic/extrinsic (circle one) muscle of the tongue. ❺

Exercise 12.5: Muscles of the Pharynx

8. The pharyngeal muscles that narrow the diameter of the pharynx during swallowing are called pharyngeal _____. ❻

Exercise 12.6: Muscles of the Neck

9. Match the muscles listed in column A with the class of muscles each belongs to in column B. Answers may be used more than once. ❼

   Column A

   ____ mylohyoid

   ____ longissimus capitis

   ____ omohyoid

   ____ stylohyoid

   Column B

   a. infrahyoid muscle

   b. muscle that moves the neck

   c. suprahyoid muscle (2 answers)

10. Explain how the sternocleidomastoid and splenius muscles can act as either synergists or antagonists for the actions of neck flexion, extension, and lateral rotation. ❽

   _____
   _____
   _____
   _____
   _____
   _____
   _____

Exercise 12.7: Muscles of the Vertebral Column

11. The erector spinae is a collective term for posterior/anterior (circle one) groups of muscles that function to straighten the spine. ❾

12. List the three muscle groups that compose the erector spinae: ❾

    a. _____

    b. _____

    c. _____

Exercise 12.8: Muscles of Respiration

13. When the diaphragm is contracted, is it flat or dome-shaped? _____ ❿

14. There are two sets of intercostals muscle (muscles between the ribs). The _____ intercostals are superficial to the _____ intercostals. ❿

15. What is the primary muscle used for breathing? ❿

    _____
    _____

16. Which muscles assist in forced inspiration? (Hint: these muscles lift the ribs or elevate the spine.) ❿

    _____
    _____

17. Which muscles assist in forced expiration? (Hint: these muscles depress the ribs or compress the abdomen.) ❿

    _____
    _____

18. List the three major structures that pass through the diaphragm. ⓫

    a. _____

    b. _____

    c. _____

Chapter Twelve The Muscular System: Axial Muscles

Exercise 12.9: Muscles of the Abdominal Wall

19. List the muscles of the abdominal wall.

    a. _____

    b. _____

    c. _____

    d. _____

20. There are two sets of oblique abdominal muscles (muscles running at an angle). The most superficial is the _____. Deep to this is the _____.

21. Match the muscle of the abdominal wall listed in column A with the direction of the fibers listed in column B.

    Column A

    ____ external obliques

    ____ internal obliques

    ____ rectus abdominis

    ____ transverse abdominis

    Column B

    a. horizontal

    b. inferiolateral (down and away from middle)

    c. inferioromedial (down and toward middle)

    d. vertical

22. The inguinal ligament is formed from the aponeurosis of which abdominal muscle? _____

Exercise 12.10: The Rectus Sheath, Inguinal Ligament, and Inguinal Canal

23. What structures divide the rectus abdominis into four segments (forming the traditional "six-pack" of muscles? _____ The fibrous sheath that encloses the rectus abdominis is the _____. The sheath comes together at the midline to form a vertical fibrous strip called the _____ (which sometimes darkens during pregnancy).

24. What structures run through the inguinal canal?

    Males: _____

    Females: _____

Can You Apply What You've Learned?

25. If the facial nerve is severed, which of the following muscles will not be paralyzed? _____

    a. buccinator

    b. masseter

    c. frontalis

    d. zygomaticus

    e. epicranius

26. Explain the major difference in muscular movement produced by the masseter and the temporalis during mastication (chewing). _____

27. Explain how the suprahyoid and infrahyoid muscles are named. _____

28. a. Describe the action of the sternocleidomastoid if both heads are contracted simultaneously. _____

    b. Describe the action of the sternocleidomastoid if only the right clavicular head contracts. _____

29. Explain how the external obliques, internal obliques, and transversus abdominis muscles relate to the rectus sheath.

Can You Synthesize What You've Learned?

30. In a **indirect inguinal hernia**, abdominal contents, such as a loop of small intestine, pass into the inguinal canal. Why do you think such hernias are more common in males than in females?

31. In people with gastroesophageal reflux disease (GERD), acidic stomach contents enter into the lower part of the esophagus, thus damaging the lining of the esophagus. GERD is often caused by weakness in part of the diaphragm. Using your knowledge of the diaphragm and the locations where important structures pass through, hypothesize a mechanism by which a weak diaphragm could contribute to the development of GERD.

CHAPTER 13

The Muscular System: Appendicular Muscles

OUTLINE AND LEARNING OBJECTIVES

Gross Anatomy 323

Muscles That Move the Pectoral Girdle and Glenohumeral Joint 323

EXERCISE 13.1: MUSCLES THAT MOVE THE PECTORAL GIRDLE AND GLENOHUMERAL JOINT 324

1. *Identify the muscles that connect the pectoral girdle to the thorax, and describe their actions*
2. *Demonstrate the actions of the trapezius, and name muscles that act as synergists for the actions of elevation and adduction of the scapula*
3. *Describe the consequences of damage to the serratus anterior*
4. *List the four muscles that make up the rotator cuff, and explain the functional importance of these muscles*

Upper Limb Musculature 326

EXERCISE 13.2: COMPARTMENTS OF THE ARM 326

5. *Identify the muscles of the anterior and posterior compartments of the arm, and describe their locations and actions*
6. *Explain the role of the biceps brachii in the actions of forearm flexion and forearm supination*

EXERCISE 13.3: COMPARTMENTS OF THE FOREARM 329

7. *Identify the muscles of the anterior and posterior compartments of the forearm, and describe their locations and actions*
8. *Demonstrate the location of the palmaris longus on yourself*
9. *Describe the relationship between the tendons of the flexor digitorum superficialis and the flexor digitorum profundus as they attach to the phalanges*

EXERCISE 13.4: INTRINSIC MUSCLES OF THE HAND 334

10. *Identify the intrinsic muscles of the hand*
11. *Describe the locations of the thenar eminence and the hypothenar eminence*
12. *Describe the actions of the dorsal and palmar interossei*

Muscles That Move the Hip Joint 336

EXERCISE 13.5: MUSCLES THAT MOVE THE HIP 336

13. *Identify the muscles that act about the hip, and describe their actions*
14. *Explain the roles of the gluteus medius and gluteus minimus in locomotion*
15. *Explain the importance of the piriformis as a clinically relevant landmark*
16. *Describe the composition, location, and function of the iliopsoas muscle*

Lower Limb Musculature 339

EXERCISE 13.6: COMPARTMENTS OF THE THIGH 339

17. *Identify the muscles of the anterior, medial, and posterior compartments of the thigh, and describe their locations and actions*
18. *Name the two muscles of the anterior compartment of the thigh that flex the hip joint*
19. *Identify the borders of the femoral triangle, and explain the clinical significance of the femoral triangle*

EXERCISE 13.7: COMPARTMENTS OF THE LEG 344

20. *Identify the muscles of the anterior, lateral, and posterior compartments of the leg, and describe their locations and actions*
21. *Name two muscles that act as antagonists to the fibularis muscles for the function of everting the ankle*
22. *Name the muscles that compose the triceps surae*

EXERCISE 13.8: INTRINSIC MUSCLES OF THE FOOT 347

23. *Identify the intrinsic muscles of the foot, and describe their locations and actions*

 **MODULE 6: MUSCULAR SYSTEM**

INTRODUCTION

The **appendicular muscles** are the muscles that most of us tend to be most familiar with. Anyone who has gone to a gym and lifted weights, or who has admired the muscular body of a basketball player or gymnast, has at least some familiarity with these muscles. They are the muscles we use to perform active tasks such as typing, walking, running, and lifting. They are the bulging muscles we see in the arms and legs of elite athletes. When lifting a heavy weight, we can feel the muscles in our upper limbs working, so it is fairly easy to determine which muscles we are using to perform the particular action. If we cannot figure this out right away from the fatigue or pain we experience immediately in the working muscle, we most definitely will figure it out in the next two days as delayed-onset muscle soreness (DOMS) sets in.

All of these activities give us a fascination and curiosity about the muscles responsible for causing movement. As you work through the task of learning names, attachments, and actions for the appendicular muscles, try to identify the muscles on your own body. Practice using them. An excellent way to do this is to go to a gym that has weight machines. Observe the illustrations on the weight machines that demonstrate what muscle or muscles the exercise is meant to work on. Then practice the movement *without using any weights* (especially if you have never lifted weights before). As you perform the stated action of the muscle(s), feel the muscle(s) produce tension under your skin. Once you can make the connection between the muscles in your own body and the muscles you see on the cadaver, on models, or in photographs, you will begin to truly appreciate their functions.

In this laboratory session, you will identify, name, and explore the structure and function of muscles that move the appendicular skeleton on a model, photo, or cadaver. Before you begin, find out from your laboratory instructor which muscles you will be responsible for on your practical exams. By doing this first, you can place check marks in the summary tables of this chapter allowing you to focus on those muscles you are required to know for your course. In addition, before you begin the more detailed study of the appendicular musculature in this chapter, make sure you have completed the exercises related to appendicular musculature in chapter 11. Those exercises introduced you to the major muscle groups and the common actions of those groups. Refer to tables 11.4 and 11.5 to review the muscle compartments of the upper and lower limbs, and the major actions of the muscles in each compartment.

Chapter 13: The Muscular System: Appendicular Muscles

PRE-LABORATORY WORKSHEET

Also available at www.connect.mcgraw-hill.com

1. Match the compartment listed in column A with the action(s) described in column B. Actions are those performed by the majority of muscles within the listed compartment.

   Column A
   - ____ 1. anterior arm
   - ____ 2. posterior arm
   - ____ 3. anterior forearm
   - ____ 4. posterior forearm
   - ____ 5. anterior thigh
   - ____ 6. posterior thigh
   - ____ 7. medial thigh
   - ____ 8. anterior leg
   - ____ 9. posterior leg
   - ____ 10. lateral leg

   Column B
   a. adduction of the thigh
   b. dorsiflexion and inversion of the ankle, extension of the digits
   c. extension of the hip, flexion of the knee
   d. extension of shoulder and elbow
   e. extension of the knee, flexion of the hip
   f. flexion of shoulder and elbow
   g. extension of the wrist and digits
   h. eversion of the ankle
   i. flexion of the wrist and digits
   j. plantarflexion and inversion of the ankle, flexion of the digits

2. Match the bony process (or small bone) listed in column A with the appropriate bone (or appendage) listed in column B. Not all answer choices will be used, and each answer choice may be used more than once.

   Column A
   - ____ 1. greater trochanter
   - ____ 2. calcaneus
   - ____ 3. lesser tubercle
   - ____ 4. olecranon process
   - ____ 5. pisiform
   - ____ 6. cuneiform
   - ____ 7. lateral malleolus
   - ____ 8. coracoid process

   Column B
   a. carpal bone
   b. clavicle
   c. femur
   d. fibula
   e. humerus
   f. radius
   g. scapula
   h. tarsal bone
   i. tibia
   j. ulna

3. Define the following terms:

   a. *prime mover (agonist)* _____

   b. *synergist* _____

   c. *antagonist* _____

4. Describe the region of the body to which each of the following terms refers:

   a. *upper limb* _____

   b. *arm* _____

Chapter Thirteen *The Muscular System: Appendicular Muscles*

c. *forearm* _____

_____

d. *lower limb* _____

_____

e. *thigh* _____

_____

f. *leg* _____

_____

5. List the four muscles that compose the rotator cuff.

   a. _____

   b. _____

   c. _____

   d. _____

6. The hamstring muscles include:

   a. _____

   b. _____

   c. _____

7. Identify the most superficial muscle of the chest. _____

Gross Anatomy

Muscles That Move the Pectoral Girdle and Glenohumeral Joint

The **pectoral girdle** consists of the clavicle and scapula. As you learned in chapter 10, the shoulder joint (glenohumeral joint) is a highly movable joint. This is largely because there is only minimal bony attachment between the pectoral girdle and the axial skeleton (the only point of bony attachment is the sternoclavicular joint). Because of this arrangement, there must be very strong muscular attachments between the pectoral girdle and the axial skeleton. **Table 13.1** lists the characteristics of muscles whose primary action is about the pectoral girdle, and **table 13.2** lists muscles that move the glenohumeral joint.

Table 13.1	Muscles That Move the Pectoral Girdle				
Muscle	**Origin**	**Insertion**	**Action***	**Innervation**	**Word Origin**
Pectoralis Minor	Ribs 3–5 (anterior surface)	Coracoid process of scapula	Protracts and depresses the scapula	Medial pectoral nerve	*pectus*, chest, + *minor*, smaller
Serratus Anterior	Ribs 1–8 (outer surface)	Scapula (medial border)	Protracts the scapula and rotates it superiorly; most important for holding the scapula flat against the rib cage; damage causes a "winging" of the scapula	Long thoracic nerve	*serratus*, a saw, + *anterior*, the front surface
Trapezius (Upper Portion)	Occipital bone, ligamentum nuchae, spine of C_7	Clavicle (lateral third) and scapula (acromial process)	Elevates the scapula and rotates it superiorly	Accessory nerve (CN XI)	*trapeza*, a table
Trapezius (Middle Portion)	Spines of T_1–T_5	Spine of scapula	Retracts, adducts, and stabilizes the scapula	Accessory nerve (CN XI)	*trapeza*, a table
Trapezius (Lower Portion)	Spines of T_1–T_{12}	Spine of scapula	Depresses the scapula and rotates it superiorly	Accessory nerve (CN XI)	*trapeza*, a table
Levator Scapulae	Transverse processes of C_1–C_4	Scapula (superior vertebral border)	Elevates the scapula and tilts the glenoid inferiorly	Dorsal scapular nerve	*levatus*, to lift, + *scapula*, the shoulder blade
Rhomboid Minor	Spines of C_7–T_1	Scapula (medial border)	Retracts, adducts, and stabilizes the scapula; tilts the glenoid inferiorly	Dorsal scapular nerve	*rhomboid*, resembling an oblique parallelogram, + *minor*, smaller
Rhomboid Major	Spines of T_2–T_5	Scapula (medial border)	Retracts, adducts, and stabilizes the scapula; tilts the glenoid inferiorly	Dorsal scapular nerve	*rhomboid*, resembling an oblique parallelogram, + *major*, larger

*Only actions that apply to movement of the pectoral girdle are listed.

Table 13.2 Muscles That Move the Glenohumeral Joint

Muscle	Origin	Insertion	Action*	Innervation	Word Origin
Pectoralis Major	Sternum, medial clavicle, and costal cartilages 2–6	Lateral part of intertubercular groove of humerus	Flexes and adducts arm	Medial pectoral nerve	*pectus*, chest, + *major*, larger
Pectoralis Minor	Anterior surfaces of ribs 3–5	Coracoid process of scapula	Flexes, adducts, medially rotates arm	Lateral pectoral nerve	*pectus*, chest, + *minor*, smaller
Coracobrachialis	Coracoid process of scapula	Midhumerus (medial surface)	Flexes and adducts arm	Musculocutaneous nerve	*coraco*, referring to the coracoid process, + *brachium*, arm
Biceps Brachii (Long Head)	Supraglenoid tubercle	Radial tuberosity	Flexes arm (weak)	Musculocutaneous nerve	*bi*, two, + *caput*, head, + *brachium*, arm
Biceps Brachii (Short Head)	Coracoid process of scapula	Radial tuberosity	Flexes arm (weak)	Musculocutaneous nerve	*bi*, two, + *caput*, head, + *brachium*, arm
Deltoid (Anterior Fibers)	Clavicle (lateral third)	Humerus (deltoid tuberosity)	Flexes, adducts, medially rotates arm	Axillary nerve	*deltoid*, resembling the Greek letter delta (a triangle)
Deltoid (Middle Fibers)	Acromion, clavicle (lateral part), and scapula (lateral spine)	Humerus (deltoid tuberosity)	Abducts arm	Axillary nerve	*deltoid*, resembling the Greek letter delta (a triangle)
Deltoid (Posterior Fibers)	Spine of scapula	Humerus (deltoid tuberosity)	Extends, adducts, laterally rotates arm	Axillary nerve	*deltoid*, resembling the Greek letter delta (a triangle)
Supraspinatus (RC)	Supraspinous fossa	Humerus (greater tubercle)	Stabilizes arm, assists in abduction	Suprascapular nerve	*supra*, above, + *spina*, referring to the spine of the scapula
Latissimus Dorsi	Spinous process of T_6–L_5, iliac crest, and ribs 10–12	Humerus (intertubercular groove)	Extends, adducts, medially rotates arm	Thoracodorsal nerve	*latissimus*, widest, + *dorsum*, the back
Teres Major	Posterior surface of scapula at inferior angle	Crest of lesser tubercle on anterior humerus	Extends, adducts, medially rotates arm	Lower subscapular nerve	*teres*, round, + *major*, larger
Triceps Brachii (Long Head)	Infraglenoid tubercle of scapula	Olecranon process of ulna	Extends, adducts arm	Radial nerve	*tri*, three, + *caput*, head, + *brachium*, arm
Infraspinatus (RC)	Infraspinous fossa of scapula	Greater tubercle of humerus	Adducts and laterally rotates arm	Suprascapular nerve	*infra*, below, + *spina*, referring to the spine of the scapula
Teres Minor (RC)	Inferior angle of scapula	Greater tubercle of humerus	Adducts and laterally rotates arm	Axillary nerve	*teres*, round, + *minor*, smaller
Subscapularis (RC)	Subscapular fossa	Lesser tubercle of humerus	Medially rotates arm	Upper and lower scapular nerves	*sub*, under, + *scapula*, the shoulder blade

*Only actions that apply to movement of the glenohumeral joint are listed.
RC = a rotator cuff muscle, which is important in stabilization of glenohumeral joint, in addition to functions above.

EXERCISE 13.1

MUSCLES THAT MOVE THE PECTORAL GIRDLE AND GLENOHUMERAL JOINT

1. Observe a prosected human cadaver or a classroom model demonstrating muscles of the thorax and upper limb.
2. Using tables 13.1 and 13.2 and your textbook as guides, identify the **muscles that move the pectoral girdle and glenohumeral joint** listed in **figure 13.1** on the cadaver or on a classroom model. Then label them in figure 13.1.
3. *Trapezius:* The largest muscle that moves the pectoral girdle is the superficial **trapezius** muscle, which acts to anchor the scapula to the entire superior two-thirds of the vertebral column and to the back of the head. As noted in table 13.1, this muscle performs multiple actions on the scapula, depending upon which part of the muscle contracts at a given time. Deep to the trapezius are smaller, more numerous muscles that act as **synergists** of the trapezius muscle. These include the

Chapter Thirteen The Muscular System: Appendicular Muscles 325

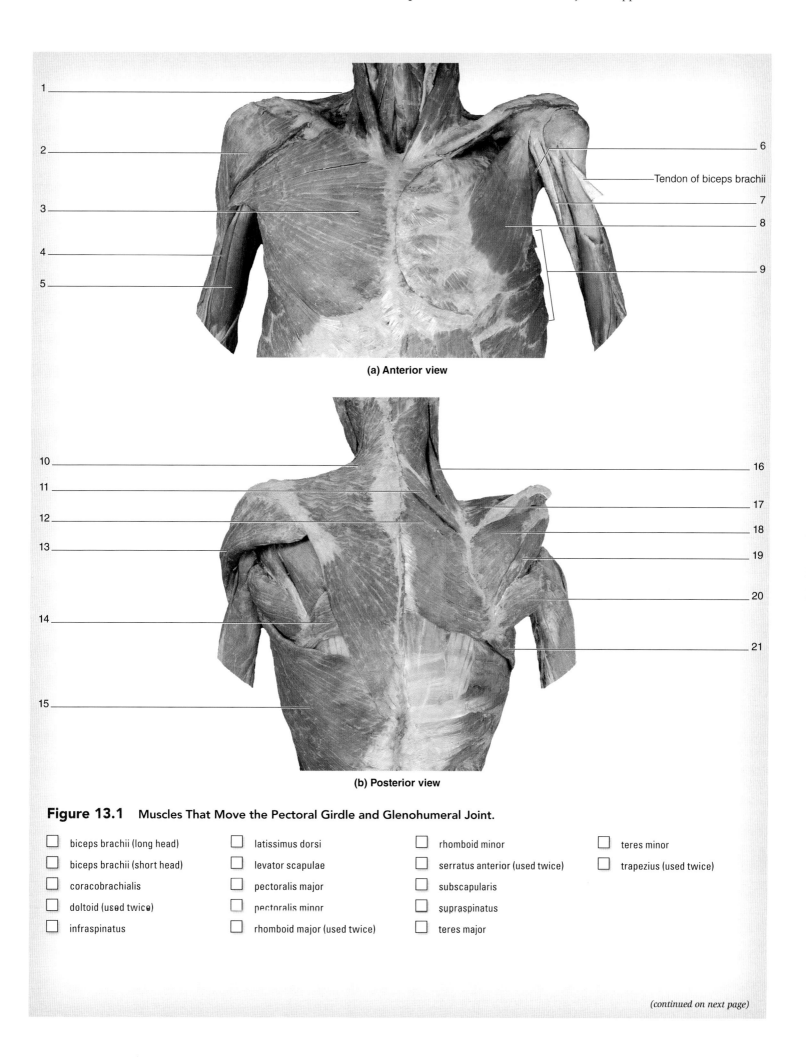

(a) Anterior view

(b) Posterior view

Figure 13.1 Muscles That Move the Pectoral Girdle and Glenohumeral Joint.

- [ ] biceps brachii (long head)
- [ ] biceps brachii (short head)
- [ ] coracobrachialis
- [ ] deltoid (used twice)
- [ ] infraspinatus
- [ ] latissimus dorsi
- [ ] levator scapulae
- [ ] pectoralis major
- [ ] pectoralis minor
- [ ] rhomboid major (used twice)
- [ ] rhomboid minor
- [ ] serratus anterior (used twice)
- [ ] subscapularis
- [ ] supraspinatus
- [ ] teres major
- [ ] teres minor
- [ ] trapezius (used twice)

(continued on next page)

326 Chapter Thirteen *The Muscular System: Appendicular Muscles*

(continued from previous page)

  rhomboids (major and minor) and the **levator scapulae.** As you identify these muscles on the cadaver or models, think about the actions they share with the trapezius. Practice performing the actions listed in table 13.1 and make sure that they make sense to you in light of the location and fiber orientation of the muscles themselves.

4. *Serratus Anterior:* One of the most important muscles for stabilization of the scapula is the **serratus anterior.** This muscle's main action is to prevent the scapula from pulling away from the rib cage ("winging"). To test the action of the serratus anterior muscle, stand and face a wall. With your arms held straight out horizontally in front of you, place your palms flat against the wall in front of you and then push against the wall. If the serratus anterior is working properly, your scapula will remain flat against your rib cage. If not, your scapula will "wing" and its medial border will be pushed away from your rib cage.

5. *Pectoralis Minor:* The **pectoralis minor** is a small muscle that can have multiple actions, depending upon what other muscles of the thorax and shoulder are contracting at the same time. If the scapula is fixed, the pectoralis minor assists in elevation of the rib cage (as in inspiration). If the scapula is not fixed, the pectoralis minor acts to pull the scapula anteriorly.

6. *Rotator Cuff Muscles:* The glenohumeral (shoulder) joint is the most mobile joint in the body, but this mobility comes at the expense of stability. Because of this, the muscles that connect the upper limb to the pectoral girdle perform two critical functions: stabilization of the glenohumeral joint, and movement of the glenohumeral joint. The **supraspinatus, infraspinatus, subscapularis,** and **teres minor** compose a musculotendinous cuff, the **rotator cuff,** whose main function is to stabilize the glenohumeral joint. More than any other factor, the strength of these muscles determines how stable the joint is. Weaknesses in these muscles contribute to many musculoskeletal problems in the glenohumeral joint. Rotator cuff injuries are common in baseball players and in older adults who fall on an outstretched arm. The term *rotator cuff* comes from the fact that these muscles act to medially or laterally rotate the humerus, in addition to stabilizing the glenohumeral joint.

7. In the space below, make a representative drawing of the three rotator cuff muscles that are seen in a posterior view.

INTEGRATE

LEARNING STRATEGY

The following mnemonic can help you remember the names of the rotator cuff muscles. "A baseball pitcher who tears his rotator cuff **SITS** out the season." (**SITS: S** = supraspinatus, **I** = infraspinatus, **T** = teres minor, **S** = subscapularis.) To remember it is the teres **minor** (not the teres major) that forms part of the rotator cuff, you can say that this injured pitcher will then be relegated to the **minor** leagues.

Upper Limb Musculature

The upper limb consists of the **arm** and the **forearm.** Each is divided into anterior and posterior compartments. The **anterior compartment** is collectively referred to as a **flexor compartment** because its muscles act to flex the arm, elbow, or wrist and fingers. The **posterior compartment** is collectively referred to as an **extensor compartment** because its muscles act to extend the arm, elbow, or wrist and fingers. Understanding the compartmental nature of the muscles not only is important for simplifying the task of learning the muscles within each compartment—it also helps when you learn the distribution of peripheral nerves. In most cases, one nerve innervates one compartment. If you understand the common function of the muscles in an entire compartment, you will easily be able to determine the deficits an individual will suffer when a particular nerve is damaged.

EXERCISE 13.2

COMPARTMENTS OF THE ARM

EXERCISE 13.2A Anterior Compartment of the Arm

1. Observe a prosected human cadaver or a classroom model demonstrating muscles of the upper limb.

2. Using **table 13.3** and your textbook as guides, identify the **muscles of the anterior compartment of the arm** listed in **figure 13.2** on the cadaver or on a classroom model. Then label them in figure 13.2. As you identify these muscles, make note of the following:

   Common Actions: Muscles in this compartment flex the arm or forearm.
   Exceptions: The coracobrachialis muscle acts only about the glenohumeral joint.
   The brachialis muscle acts only about the elbow joint.

Chapter Thirteen The Muscular System: Appendicular Muscles

Table 13.3 Anterior (Flexor) Compartment of the Arm

Muscle	Origin	Insertion	Action	Innervation	Word Origin
Biceps Brachii (Long Head)	Supraglenoid tubercle of scapula	Radial tuberosity	Supinates forearm and flexes forearm (weak)	Musculocutaneous nerve	*bi*, two, + *caput*, head, + *brachium*, arm
Biceps Brachii (Short Head)	Coracoid process of scapula	Radial tuberosity	Supinates forearm and flexes forearm (weak)	Musculocutaneous nerve	*bi*, two, + *caput*, head, + *brachium*, arm
Coracobrachialis	Coracoid process of scapula	Midhumerus (medial surface)	Flexes arm and adducts (weak)	Musculocutaneous nerve	*coraco*, referring to the coracoid process, + *brachium*, arm
Brachialis	Humerus (lower half of anterior surface)	Ulna (coronoid process and ulnar tuberosity)	Flexes forearm	Musculocutaneous nerve	*brachium*, arm

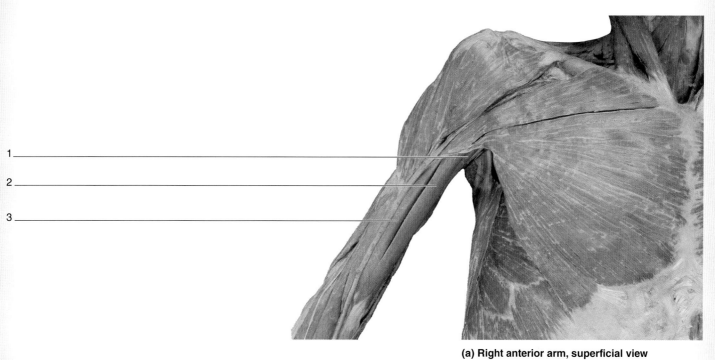

(a) Right anterior arm, superficial view

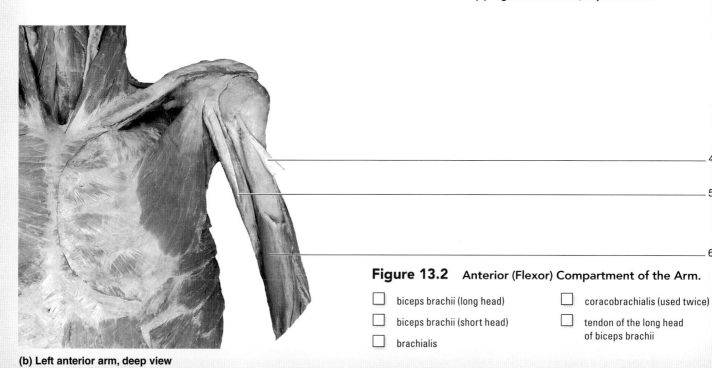

(b) Left anterior arm, deep view

Figure 13.2 Anterior (Flexor) Compartment of the Arm.

☐ biceps brachii (long head)
☐ biceps brachii (short head)
☐ brachialis
☐ coracobrachialis (used twice)
☐ tendon of the long head of biceps brachii

(continued on next page)

(continued from previous page)

3. *Brachialis and Biceps Brachii:* A common misconception is that the biceps brachii muscle is the most powerful flexor of the forearm (elbow). It is not. You can test this by performing the following activity. With your right wrist pronated, place the palm of your left hand over the belly of the biceps brachii on your right arm. Flex your elbow and note the degree to which the biceps brachii muscle produces tension. Next, supinate your wrist and perform the flexion action again. Did the biceps brachii produce more tension when the wrist was pronated or supinated? _____ You should have noticed a large difference in the amount of tension produced under the two conditions. In fact, the **prime mover,** or **agonist,** for the action of forearm flexion is the **brachialis** muscle, not the biceps brachii muscle. The biceps brachii is a **synergist** for this action. The biceps brachii itself is a very powerful **supinator** of the wrist and forearm. Thus, when we perform actions that require both flexion of the elbow and supination of the wrist and forearm, this is when we use our biceps brachii to its full capacity. Think about this the next time you are trying to twist off a stuck bottle lid.

4. Note the following when identifying the biceps brachii muscle on models or cadavers. The **long head of the biceps brachii** disappears into the intertubercular groove of the humerus on its way to its attachment on the supraglenoid tubercle, only a small portion of its tendon is visible. On the other hand, the entire tendon of the **short head of the biceps brachii** is visible attaching to the coracoid process of the scapula. Thus, on models and cadavers, the short head of the biceps brachii will generally appear to be longer.

EXERCISE 13.2B Posterior Compartment of the Arm

1. Observe a prosected human cadaver or a classroom model demonstrating muscles of the upper limb.

2. The posterior compartment of the arm consists of one muscle, the triceps brachii, with three heads (long, lateral, and medial).

3. Using **table 13.4** and your textbook as guides, identify the **muscles of the posterior compartment of the arm** listed in **figure 13.3** on the cadaver or on a classroom model. Then label them in figure 13.3. As you identify these muscles, make note of the following:

Common Actions: Extend the forearm (elbow).

Exception: The long head also extends the arm.

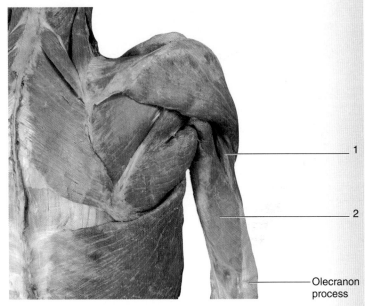

Right posterior arm, superficial view

Figure 13.3 Posterior (Extensor) Compartment of the Arm.

☐ lateral head of triceps brachii
☐ long head of triceps brachii

INTEGRATE

LEARNING STRATEGY

The *long* head of the *triceps brachii* attaches to the *infraglenoid tubercle* (of the scapula), while the *long* head of the *biceps brachii* attaches to the *supraglenoid tubercle* (of the scapula). Also, the medial head of the triceps brachii can be visualized only if the lateral head is bisected. It is a deep muscle, and it attaches directly to the humerus.

Table 13.4	Posterior (Extensor) Compartment of the Arm				
Muscle	**Origin**	**Insertion**	**Action**	**Innervation**	**Word Origin**
Triceps brachii					*tri*, three, + *caput*, head, + *brachium*, arm
Long Head	Infraglenoid tubercle of scapula	Olecranon of ulna	Extends forearm, extends arm	Radial nerve	
Lateral Head	Proximal half of posterior humerus	Olecranon of ulna	Extends forearm	Radial nerve	
Medial Head	Distal half of posterior humerus	Olecranon of ulna	Extends forearm	Radial nerve	

EXERCISE 13.3

COMPARTMENTS OF THE FOREARM

EXERCISE 13.3A Anterior Compartment of the Forearm

1. Observe a prosected human cadaver or a classroom model demonstrating muscles of the upper limb.

2. The muscles of the **anterior compartment of the forearm** cause flexion of the wrist and the digits (fingers). They are arranged into two layers—a superficial group and a deep group. In most cases, the name of the muscle tells exactly what the muscle does or where the muscle is located. Thus, although the names of the muscles seem long, they can be very useful if you understand the meanings of the names.

3. Using **table 13.5** and your textbook as guides, identify the muscles of the anterior compartment of the forearm listed in **figure 13.4** on the cadaver or on a classroom model. Then label them in figure 13.4.

4. *Identification of Superficial Muscles and Tendons:* Place your left palm on the medial epicondyle of your right humerus **(figure 13.5)**. In this position, the order of the muscles on your right forearm, from lateral to medial, is

   Index finger—pronator teres (PT)
   Middle finger—flexor carpi radialis (FCR)
   Ring finger—palmaris longus (PL)
   Pinky finger—flexor carpi ulnaris (FCU)

   As you perform this exercise, flex your wrist and digits to identify the tendons, from lateral to medial, of the flexor carpi radialis, palmaris longus (if present; see below), and flexor carpi ulnaris.

5. *Palmaris Longus:* Approximately 15% to 20% of humans do not have a palmaris longus muscle. The palmaris longus passes *superficial* to the **flexor retinaculum**

Table 13.5 Anterior (Flexor) Compartment of the Forearm

Muscle	Origin	Insertion	Action	Innervation	Word Origin
Superficial Group					
Pronator Teres	Humerus (medial epicondyle)	Radius (lateral shaft)	Pronates forearm and flexes elbow (weak)	Median nerve	*pronatus*, to bend forward, + *teres*, round
Flexor Carpi Radialis	Humerus (medial epicondyle)	Metacarpals II and III (base)	Flexes and abducts wrist	Median nerve	*flex*, to bend, + *carpus*, wrist, + *radialis*, radius
Palmaris Longus	Humerus (medial epicondyle)	Palmar aponeurosis	Flexes wrist	Median nerve	*palmaris*, the palm, + *longus*, long
Flexor Digitorum Superficialis	Humerus (medial epicondyle)	Middle phalanx of digits 2–5	Flexes the metacarpophalangeal and proximal interphalangeal joints of digits 2–5	Median nerve	*flex*, to bend, + *digit*, finger, + *superficialis*, surface
Flexor Carpi Ulnaris	Humerus (medial epicondyle)	Pisiform, hamate, and metacarpal V (base)	Flexes and adducts wrist	Ulnar nerve	*flex*, to bend, + *carpus*, wrist, + *ulnaris*, ulna
Deep Group					
Flexor Pollicis Longus	Radius (anterior surface) and interosseous membrane	Distal phalanx of the thumb	Flexes the distal phalanx of the thumb	Median nerve	*flexus*, to bend, + *pollex*, the thumb, + *longus*, long
Pronator Quadratus	Ulna (distal anterior shaft)	Radius (distal, anterior shaft)	Pronates the forearm	Median nerve	*pronatus*, to bend forward, + *quadratus*, square
Flexor Digitorum Profundus	Ulna (anteromedial surface) and interosseous membrane	Distal phalanx of digits 2–5	Flexes the distal phalanx of digits 2–5	Ulnar and median nerves	*flexus*, to bend, + *digitorum*, finger, + *profundus*, deep
Supinator	Humerus (lateral epicondyle) ulna (distal to radial notch)	Radius (anterolateral surface distal to radial tuberosity)	Supinates arm	Radial nerve	*supinator*, supinate

(continued on next page)

(continued from previous page)

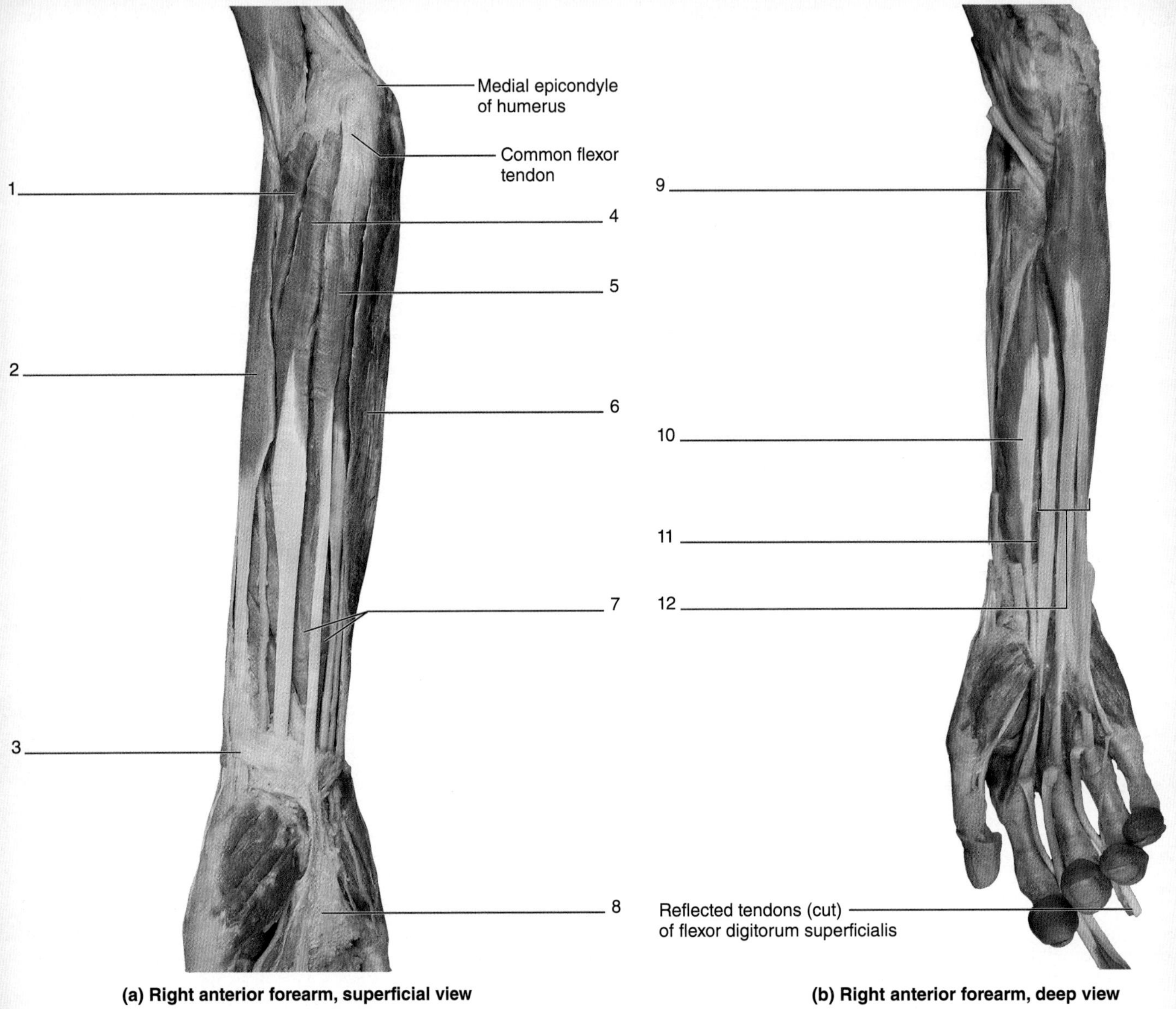

Medial epicondyle of humerus
Common flexor tendon
1
2
3
4
5
6
7
8
(a) Right anterior forearm, superficial view
9
10
11
12
Reflected tendons (cut) of flexor digitorum superficialis
(b) Right anterior forearm, deep view

Figure 13.4 Anterior (Flexor) Compartment of the Forearm.

- ☐ brachioradialis
- ☐ flexor carpi radialis
- ☐ flexor carpi ulnaris
- ☐ flexor digitorum profundus
- ☐ flexor digitorum superficialis
- ☐ flexor pollicis longus
- ☐ flexor retinaculum
- ☐ palmar aponeurosis
- ☐ palmaris longus
- ☐ pronator quadratus
- ☐ pronator teres
- ☐ supinator

(Left hand covers medial epicondyle)

Figure 13.5 Locating Superficial Anterior Forearm Muscles. Locate the superficial anterior forearm muscles by placing the left hand on the right forearm and noting the position of the digits, which correspond to the locations of the superficial forearm muscles.

(a band of connective tissue that wraps around the wrist) and inserts onto the **palmar aponeurosis.** Perform the following exercise to determine whether you have a palmaris longus muscle. With your forearm supinated, touch the tips of your first and fifth digits (thumb and pinky finger) together and flex the wrist *slightly*. If palmaris longus is present, you will see its tendon passing longitudinally in the middle of your wrist. If no tendon is visible, the muscle is probably absent.

6. *Flexor Digitorum:* There are two muscles that flex the digits, the **flexor digitorum superficialis** (superficial) and the **flexor digitorum profundus** (deep) **(figure 13.6).** These muscles, as their names suggest, flex the fingers. They do so by attaching to the phalanges. These muscles take somewhat unique routes to get to their respective attachments on the phalanges. The flexor digitorum profundus tendon attaches to the **distal phalanx.** Thus, it must pass through the tendon of the more superficial muscle (the tendon of the flexor digitorum superficialis) as it travels to the distal phalanx. Observe a cadaver or a classroom model of the hand, and locate the middle and distal phalanges of one digit. The tendon of flexor digitorum superficialis inserts onto the **middle phalanx** of the digit. Before it reaches its insertion point, it has a slit through which the tendon of flexor digitorum profundus passes. Thus, though both

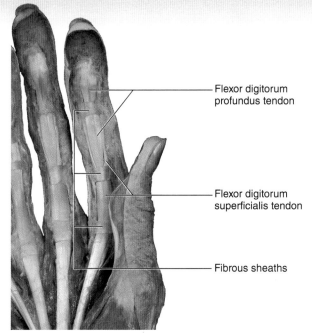

Right hand, palmar view

Figure 13.6 Flexor Tendons within the Hand. The tendon of the flexor digitorum profundus pierces the tendon of the flexor digitorum superficialis as it travels to its attachment point on the distal phalanx.

muscles flex the digits, they do so at different joints. Perform an experiment to test the function of these two muscles. With your right hand, pull the tip of the third digit (middle finger) of your left hand posteriorly. This stretches the flexor digitorum profundus muscle beyond the point at which it can contract. Now try to flex the other digits of your hand. Notice that you can flex the joint between your proximal and middle phalanges, but you cannot flex the joint between your middle and distal phalanges. Once you release your hold on the third digit, you will be able to flex all the interphalangeal joints together.

7. *Optional Activity:* **AP|R 6: Muscular System**—Visit the quiz area for drill and practice on muscles of the upper limb.

(continued on next page)

(continued from previous page)

EXERCISE 13.3B Posterior Compartment of the Forearm

1. Observe a prosected human cadaver or a classroom model demonstrating muscles of the upper limb.

2. Using **table 13.6** and your textbook as guides, identify the **muscles of the posterior compartment of the forearm** listed in **figure 13.7** on the cadaver or on a classroom model. Then label them in figure 13.7.

Table 13.6 Posterior (Extensor) Compartment of the Forearm

Muscle	Origin	Insertion	Action	Innervation	Word Origin
Brachioradialis	Humerus (lateral supracondylar ridge)	Radius (styloid process)	Flexes the forearm	Radial nerve	*brachium*, arm, + *radialis*, radius
Extensor Carpi Radialis Longus	Humerus (lateral supracondylar ridge)	Metacarpal II (base)	Extends and abducts the wrist	Radial nerve	*extendo*, to stretch out, + *carpus*, wrist, + *radialis*, radius, + *longus*, long
Extensor Carpi Radialis Brevis	Humerus (lateral epicondyle)	Metacarpal III (base)	Extends and abducts the wrist	Radial nerve	*extendo*, to stretch out, + *carpus*, wrist, + *radialis*, radius, + *brevis*, short
Extensor Digitorum	Humerus (lateral epicondyle)	Distal and middle phalanges of digits 2–5	Extends digits 2–5 at the metacarpophalangeal joint, extends the wrist	Radial nerve	*extendo*, to stretch out, + *digit*, finger
Extensor Digiti Minimi	Humerus (lateral epicondyle)	Proximal phalanx of digit 5 (little finger)	Extends the little finger at the metacarpophalangeal and interphalangeal joints	Radial nerve	*extendo*, to stretch out, + *digit*, finger, + *minimi*, smallest
Extensor Carpi Ulnaris	Humerus (lateral epicondyle) and proximal ulna	Metacarpal V (base)	Extends and adducts the wrist	Radial nerve	*extendo*, to stretch out, + *carpus*, wrist, + *ulnaris*, ulna
Supinator	Humerus (lateral epicondyle) and proximal ulna	Radius (proximal third)	Supinates the forearm	Radial nerve	*supinatus*, to move backward
Abductor Pollicis Longus	Ulna and radius (posterior surface) and interosseous membrane	Metacarpal I (base)	Abducts the thumb	Radial nerve	*abduct*, to move away from the midline, + *pollex*, thumb, + *longus*, long
Extensor Pollicis Brevis	Radius (posterior surface) and interosseous membrane	Proximal phalanx of the thumb	Extends the proximal phalanx of the thumb at the metacarpophalangeal joint	Radial nerve	*extendo*, to stretch out, + *pollex*, thumb, + *brevis*, short
Extensor Pollicis Longus	Ulna (middle third, posterior surface) and interosseous membrane	Distal phalanx of the thumb	Extends the distal phalanx of the thumb at the metacarpophalangeal and interphalangeal joints	Radial nerve	*extendo*, to stretch out, + *pollex*, thumb, + *longus*, long
Extensor Indicis	Ulna (posterior surface) and interosseous membrane	Tendon of extensor digitorum of digit 2 (index finger)	Extends the index finger	Radial nerve	*extendo*, to stretch out, + *indicis*, the forefinger

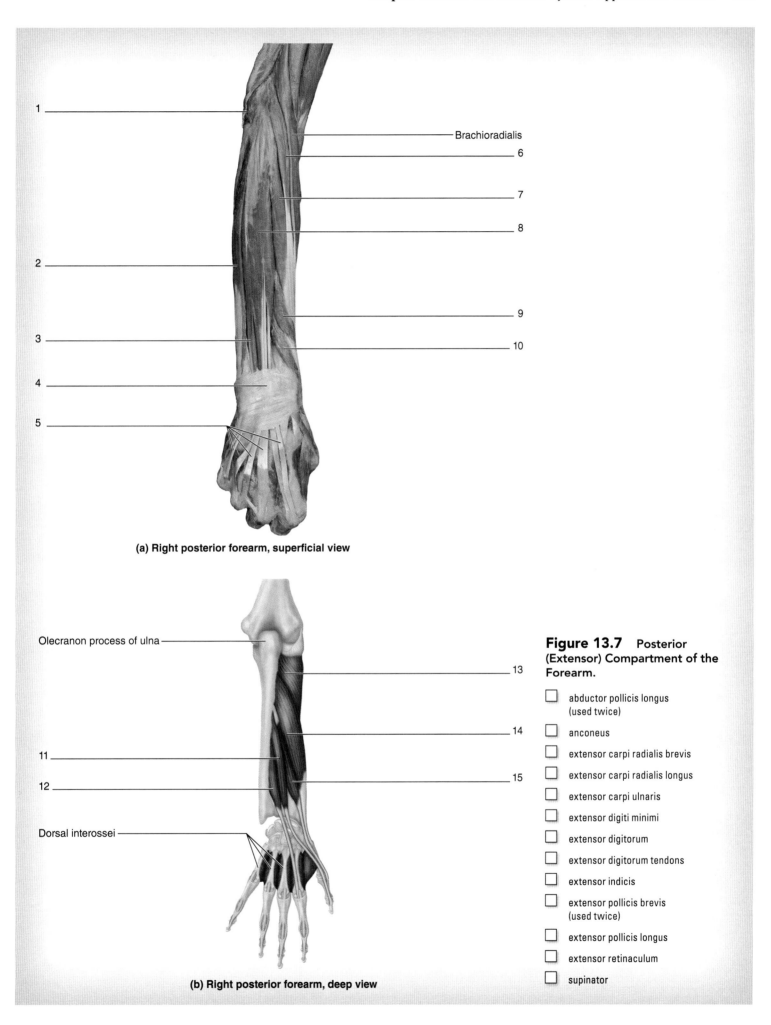

(a) Right posterior forearm, superficial view

(b) Right posterior forearm, deep view

Figure 13.7 Posterior (Extensor) Compartment of the Forearm.

- ☐ abductor pollicis longus (used twice)
- ☐ anconeus
- ☐ extensor carpi radialis brevis
- ☐ extensor carpi radialis longus
- ☐ extensor carpi ulnaris
- ☐ extensor digiti minimi
- ☐ extensor digitorum
- ☐ extensor digitorum tendons
- ☐ extensor indicis
- ☐ extensor pollicis brevis (used twice)
- ☐ extensor pollicis longus
- ☐ extensor retinaculum
- ☐ supinator

EXERCISE 13.4

INTRINSIC MUSCLES OF THE HAND

1. Observe a prosected human cadaver or a model demonstrating muscles of the hand.
2. The **intrinsic muscles of the hand** consist of the thenar and hypothenar groups of muscles and midpalmar group. The **thenar** group of muscles forms the large pad at the base of the thumb, the **thenar eminence.** Muscles in this group create special movements of the pollex (thumb), the most important of which is **opposition.** The **hypothenar** group of muscles forms a pad at the base of the fifth digit, the **hypothenar eminence.** Muscles in this group create special movements of the fifth digit (little finger). Observe both the thenar and hypothenar eminences on one of your hands. The **midpalmar** group is a group of muscles that lies between the thenar and hypothenar muscle groups. The midpalmar group consists of the lumbricals, interossei,

Table 13.7 Intrinsic Muscles of the Hand

Muscle	Origin	Insertion	Action	Innervation	Word Origin
Thenar Group					*thenar*, the palm of the hand
Abductor Pollicis Brevis	Flexor retinaculum, scaphoid and trapezium (tubercles)	Base of the proximal phalange of the thumb (lateral side)	Abducts the thumb	Recurrent branch of the median nerve	*abduct*, to move away from the midline, + *pollex*, thumb, + *brevis*, short
Flexor Pollicis Brevis	Flexor retinaculum and trapezium (tubercles)	Base of the proximal phalange of the thumb (lateral side)	Flexes the thumb	Recurrent branch of the median nerve	*flexus*, to bend, + *pollex*, thumb, + *brevis*, short
Opponens Pollicis	Flexor retinaculum and trapezium (tubercles)	Metacarpal I (lateral side)	Assists in opposition and medial rotation of the thumb	Recurrent branch of the median nerve	*oppono*, to place against, + *pollex*, thumb
Hypothenar Group					*hypo-*, under, + *thenar*, the palm of the hand
Abductor Digiti Minimi	Pisiform bone and tendon of flexor carpi ulnaris	Base of the proximal phalanx of digit 5 (medial side)	Abducts digit 5	Ulnar nerve	*abduct*, to move away from the median plane, + *digitus*, a finger, + *minimi*, smallest
Flexor Digiti Minimi Brevis	Hamate (hook) and flexor retinaculum	Base of the proximal phalanx of digit 5 (medial side)	Flexes the proximal phalanx of digit 5	Ulnar nerve	*flex*, to bend, + *digitus*, a finger, + *minimi*, smallest
Opponens Digiti Minimi	Hamate (hook) and flexor retinaculum	Metacarpal V (medial border)	Medially rotates and opposes digit 5 toward the thumb	Ulnar nerve	*oppono*, to place against + *digitus*, a finger, + *minimi*, smallest
Midpalmar Group					
Adductor Pollicis	Capitate bone, metacarpals II–III	Medial side of proximal phalanx of thumb	Adducts thumb	Ulnar nerve	*adduct*, to move toward the midline, + *pollex*, thumb
Dorsal Interossei	Metacarpals II–IV (medial and lateral surface)	Tubercle of the proximal phalanx and dorsal aponeurosis	Abducts digits 2–4	Ulnar nerve	*dorsal*, the back, + *inter-*, between, + *os*, bone
Palmar Interossei	Metacarpals (medial and lateral surface)	Tubercle of the proximal phalanx and the dorsal aponeurosis	Adducts digits 2–5	Ulnar nerve	*palmar*, the palm of the hand, + *inter-*, between, + *os*, bone
Lumbricals	Flexor digitorum profundus tendon	Lateral side of the dorsal expansion of digits 2–5	Flexes metacarpal-phalangeal joint	Ulnar and median nerves	*lumbricus*, earthworm

INTEGRATE

LEARNING STRATEGY

The **D**orsal interossei **AB**duct the digits (mnemonic is **DAB**), whereas the **P**almar interossei **AD**duct the digits (mnemonic is **PAD**).

and adductor pollicis. The **lumbricals** and **interossei** muscles are small muscles, located very deep in the hand, that create small, intricate movements of the fingers.

3. Using **table 13.7** and your textbook as guides, identify the intrinsic muscles of the hand listed in **figure 13.8** on the cadaver or on a classroom model. Then label them in figure 13.8.

4. *The Interossei:* The two groups of interossei muscles act to abduct and adduct the digits. There are two groups of interossei: dorsal and palmar. The *dorsal interossei* abduct the digits, and the *palmar interossei* adduct the digits. To feel the belly of the first dorsal interosseous muscle, forcibly press (adduct) your thumb and index finger together. Then palpate between metacarpals I and II using your other hand to feel the first dorsal interosseous muscle. Try this out on yourself now.

5. *Lumbricals:* The lumbricals are relatively easy to identify because they appear slim and "wormlike," as their name implies. There is a lumbrical muscle located between every metacarpal bone. The lumbricals flex the metacarpophalangeal joints and extend the interphalangeal joints.

6. In the space below, make a representative drawing of the intrinsic muscles of the hand.

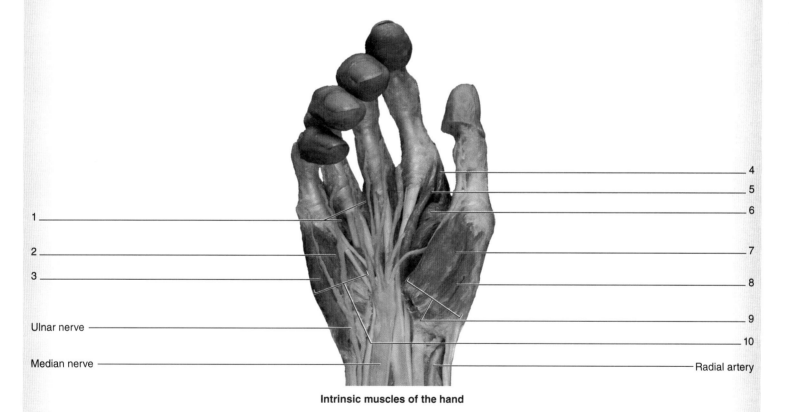

Intrinsic muscles of the hand

Figure 13.8 Intrinsic Muscles of the Hand.

- ☐ abductor digiti minimi
- ☐ abductor pollicis brevis
- ☐ adductor pollicis
- ☐ first dorsal interosseous
- ☐ flexor digiti minimi brevis
- ☐ flexor pollicis brevis
- ☐ hypothenar group
- ☐ lateral lumbricals
- ☐ medial lumbrical
- ☐ thenar group

Muscles That Move the Hip Joint

The **muscles that move the hip** include the gluteal muscles, the deep lateral rotators, and the iliopsoas muscle. The gluteal muscles and the iliopsoas are very large and powerful muscles. We use them for standing, walking, running, cycling, and just about every other locomotor activity. Failure of these muscles to work properly leads to gait disturbances.

EXERCISE 13.5

MUSCLES THAT MOVE THE HIP

1. Observe a prosected human cadaver or a model demonstrating muscles of the hip.
2. Using **table 13.8** and your textbook as guides, identify the muscles that move the hip listed in **figure 13.9** on the cadaver or on a classroom model. Then label them in figure 13.9.
3. *Gluteal Muscles:* The gluteus maximus extends the hip and *laterally* rotates the thigh. The gluteus medius and minimus muscles abduct the hip and *medially* rotate the thigh.
4. Consider the action of **hip abduction**. It may not seem that we perform this action often, except perhaps in a kickboxing class. In fact, we rarely use the gluteus

Table 13.8 Muscles That Act about the Hip

Muscle	Origin	Insertion	Action	Innervation	Word Origin
Gluteus Maximus	Dorsal ilium, sacrum, and coccyx	Gluteal tuberosity of the femur, iliotibial tract	Extends the thigh, assists in lateral rotation	Inferior gluteal nerve (L5–S2)	*gloutos*, buttock, + *maximus*, greatest
Gluteus Medius	Ilium (between the anterior and posterior gluteal lines)	Greater trochanter of femur	Abducts and medially rotates the thigh	Superior gluteal nerve (L4–S1)	*gloutos*, buttock, + *medius*, middle
Gluteus Minimus	Ilium (between the anterior and inferior gluteal lines)	Greater trochanter of femur	Abducts and medially rotates the thigh	Superior gluteal nerve (L4–S1)	*gloutos*, buttock, + *minimus*, smallest
Tensor Fascia Latae	Iliac crest (anterior aspect), anterior superior iliac spine (ASIS)	Iliotibial tract	Abducts the thigh	Superior gluteal nerve (L4–S1)	*tensus*, to stretch, + *fascia*, a band, + *latus*, side
Superior Gemellus	Ischial spine	Greater trochanter of the femur (medial surface)	Laterally rotates the thigh	Nerve to obturator internus (L5 and S1)	*geminus*, twin, + *superus*, above
Inferior Gemellus	Ischial tuberosity	Greater trochanter of femur (medial surface)	Laterally rotates the thigh	Nerve to quadratus femoris (L5 and S1)	*geminus*, twin, + *inferior*, lower
Piriformis	Anterior sacrum, sacrotuberous ligament	Greater trochanter of femur	Laterally rotates the thigh	Nerve to piriformis (S1–S2)	*pirum*, pear, + *forma*, form
Obturator Internus	Obturator membrane (posterior surface)	Margins of obturator foramen	Laterally rotates the thigh	Nerve to obturator internus (L5 and S1)	*obturo*, to occlude, + *internus*, away from the surface
Quadratus Femoris	Ischial tuberosity (lateral border)	Intertrochanteric crest of femur	Laterally rotates the thigh	Nerve to quadratus femoris (L5 and S1)	*quad*, four (sided), + *femur*, thigh
Iliopsoas					
Psoas Major	T_{12}–L_5 (bodies and transverse processes)	Femur (lesser trochanter)	Flexes the hip joint	Lumbar plexus (L2–L3)	*psoa*, the muscles of the loins
Iliacus	Iliac bone (iliac fossa)	Femur (lesser trochanter)	Flexes the hip joint	Femoral nerve (L2–L3)	*ilium*, groin

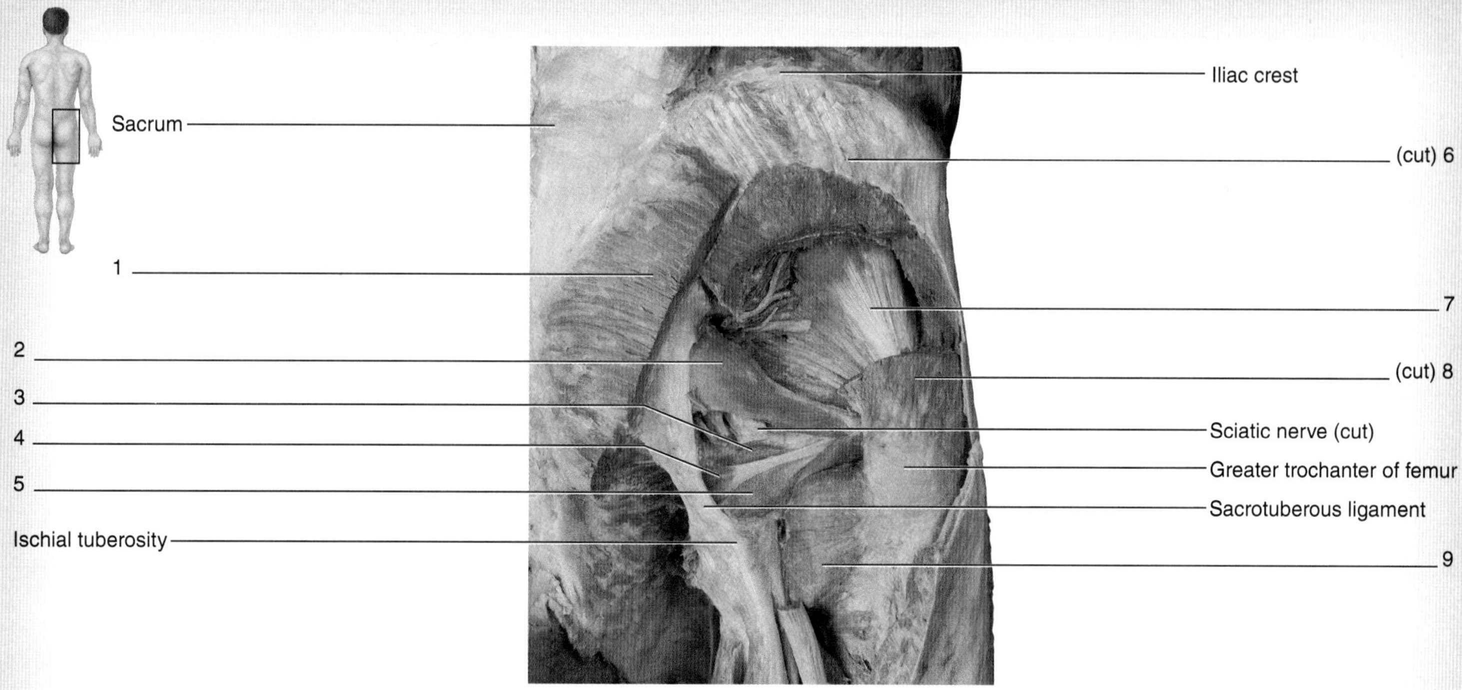

Chapter Thirteen *The Muscular System: Appendicular Muscles* **337**

Sacrum

1
2
3
4
5
Ischial tuberosity

Iliac crest
(cut) 6
7
(cut) 8
Sciatic nerve (cut)
Greater trochanter of femur
Sacrotuberous ligament
9

(a) Right thigh and hip, deep posterior view

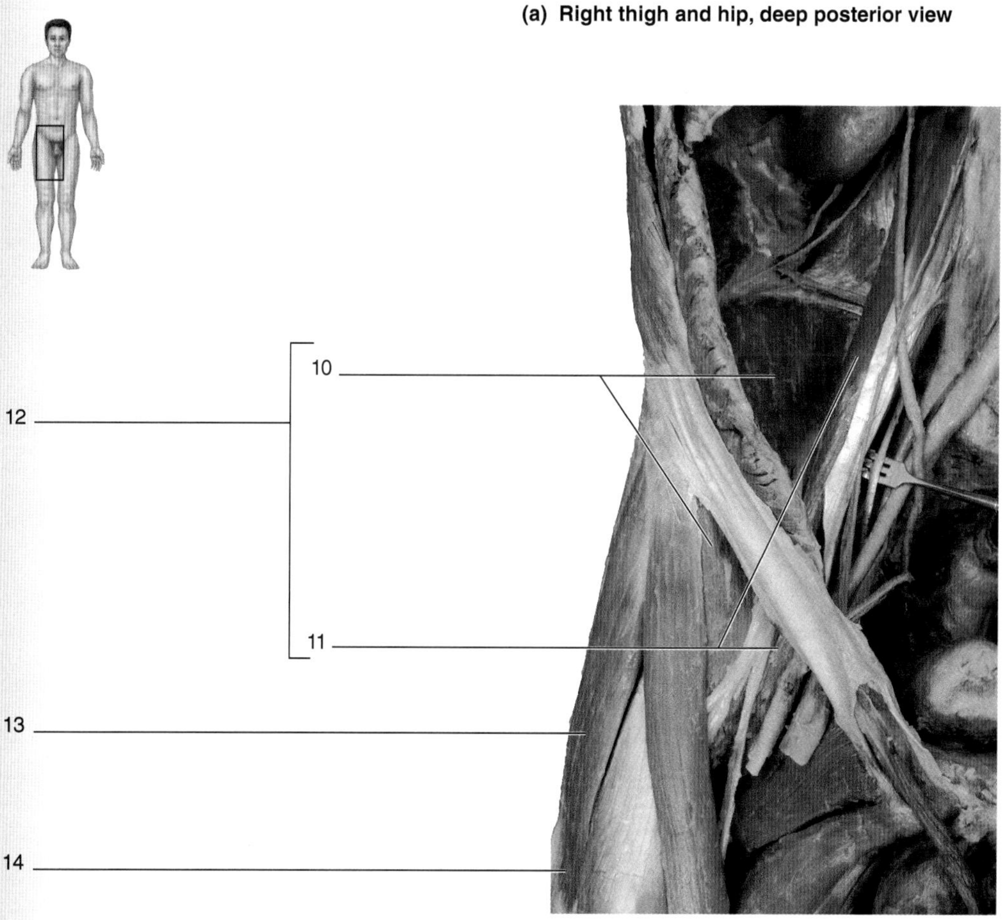

12
10
11
13
14

(b) Right thigh and hip, deep anterior view

Figure 13.9 Muscles That Move the Pelvic Girdle.

- [ ] gluteus maximus
- [ ] gluteus medius (used twice)
- [ ] gluteus minimus
- [ ] iliacus
- [ ] iliopsoas
- [ ] iliotibial tract
- [ ] inferior gemellus
- [ ] obturator internus
- [ ] piriformis
- [ ] psoas
- [ ] quadratus femoris
- [ ] superior gemellus
- [ ] tensor fascia latae

(continued on next page)

(continued from previous page)

medius and minimus muscles for extreme abduction of the hip, as in kicking the lower limb out to the side. Instead, when we attempt to abduct the hip against resistance, as when standing on one leg, these muscles prevent the hip from *adducting*, which would cause the hip on the side of the stance (standing) leg to protrude laterally **(figure 13.10).** To test the function of the gluteus medius and minimus muscles, first take a standing position. Next, place the palm of your right hand flat against your right lateral hip (superficial to the location of the gluteus medius and minimus muscles). Now flex the knee on your left lower limb bringing your left foot off the ground so you are balancing on your right limb only. Do you feel tension in the muscles deep to your palm? If you are holding your hip vertical, you will. If you then relax the gluteus medius and minimus muscles, you will find that your hip protrudes laterally and the hip tilts to the unsupported side. The gluteus medius and minimus hold the hip in the vertical position every time the limb is in its stance position. By holding the hip in such a position, they allow the foot on the swing limb to move forward without scraping against the ground. It may seem like a minor action to you, but it can be quite problematic for an individual whose muscles are not functioning properly due to nerve injury or other disease.

5. *Iliotibial Tract:* The iliotibial tract (also called the IT band) is a thickening of the deep fascia of the thigh, which extends from the anterior iliac crest to the lateral tibia. Two muscles insert onto it: the **gluteus maximus** and the **tensor fascia latae.** When these muscles become stiff from overuse, they put excessive tension on the IT band. Most commonly this causes pain in the knee joint because the tight IT band compresses structures in the lateral knee and the underlying vastus lateralis muscle.

6. *Lateral Rotators:* Several small muscles extend between the margins of the obturator foramen of the pelvis and insert onto the greater trochanter of the femur. They are called lateral rotators because lateral rotation of the thigh is their primary action. However, they are also very important in stabilizing the hip joint, in much the same way that the rotator cuff muscles stabilize the shoulder joint. A very important lateral rotator is the **piriformis** muscle because it is a major landmark in the gluteal region. For example, the sciatic nerve exits inferior to this muscle, and the superior and inferior gluteal arteries and nerves are named for their passages above and below the piriformis, respectively.

7. Observe a prosected cadaver or a model in which the gluteus maximus muscle has been reflected or cut away. First identify the gluteus medius and minimus muscles. Inferior to the gluteus medius is the pear-shaped piriformis muscle. Note the tough **sacrotuberous ligament** (figure 13.9a) that overlies the medial attachment of the piriformis, and the large **sciatic nerve,** which exits inferior to the piriformis. Finally, locate the following lateral rotators, from superior to inferior:

   ☐ superior gemellus
   ☐ obturator internus (in this location, you will see only its tendon)
   ☐ inferior gemellus
   ☐ quadratus femoris

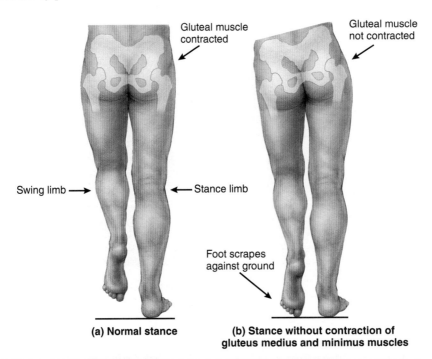

(a) Normal stance (b) Stance without contraction of gluteus medius and minimus muscles

Figure 13.10 Actions of the Gluteal Muscles during Locomotion. The gluteus medius and minimus support the hip on the right side of the body when the right leg is the stance (supporting) leg. This allows the foot on the swing leg to clear the ground.

8. *Iliopsoas:* The **iliacus** and **psoas major** are considered together as the **iliopsoas** muscle. They have separate origins, but they come together and insert together on the **lesser trochanter** of the femur. These muscles are difficult to identify on the cadaver and models if you look at the anterior thigh to find them, because only a small portion of the insertion is visible there. The bellies of these muscles are located deep within the pelvis on the posterior abdominopelvic wall. Thus, when identifying them on the cadaver or on models, be sure to view the inside of the abdominopelvic cavity where the muscles originate. The longer of the two muscles is the psoas major. It is an important landmark in the pelvis because structures such as the ureters course superficial to it. In the anterior thigh, the iliopsoas muscle forms the floor of the *femoral triangle* (see figure 13.11). You must look deep to the structures in the femoral triangle (the femoral nerve, artery, and vein) to see the iliopsoas muscle here.

INTEGRATE

CLINICAL VIEW
Piriformis Syndrome

The piriformis muscle is a "pear-shaped" muscle that lies in close proximity to important structures within the gluteal region, such as the sciatic nerve, and the gluteal arteries and nerves. Piriformis syndrome is a painful condition that results from inflammation or overuse of the piriformis muscle. The incidence of piriformis syndrome is relatively common in athletes such as runners and cyclists, who may develop an imbalance in the strength of the piriformis muscle as compared to the gluteal muscles. Specifically, the syndrome occurs when the piriformis muscle (which laterally rotates the thigh) is stronger than the gluteus medius and gluteus minimus muscles (which are responsible for medial rotation of the thigh). As the piriformis muscle becomes inflamed or experiences spasms, it may also compress the underlying sciatic nerve, resulting in sciatica. **Sciatica** is a tingling, painful, or even numbing sensation that travels down the path of the sciatic nerve. Patients complain of shooting pain that runs from the gluteal region down the lateral aspect of the thigh, and toward the leg. Often the pain may be exacerbated when the body is held in certain positions, such as prolonged sitting or standing. The symptoms of piriformis syndrome can be reduced with the administration of anti-inflammatory drugs and through stretching exercises.

Lower Limb Musculature

The lower limb is composed of the **thigh** (from hip to knee), the **leg** (from knee to ankle), and the **foot**. The thigh and the leg each consist of three compartments. As with the upper limb, each compartment consists of muscles that perform similar actions, and each compartment is served by one peripheral nerve branch.

EXERCISE 13.6

COMPARTMENTS OF THE THIGH

EXERCISE 13.6A Anterior Compartment of the Thigh

1. Observe a prosected human cadaver or a classroom model demonstrating muscles of the thigh.

2. Using **table 13.9** and your textbook as guides, identify the **muscles of the anterior compartment of the thigh** listed in **figure 13.11** on the cadaver or on a classroom model. Then label them in figure 13.11. As you identify these muscles, make note of the following:

   Common Actions: Extend the knee.
   Exceptions: The rectus femoris and sartorius muscles flex the hip joint. Sartorius also flexes the knee joint.

3. *Sartorius:* The word *sartorius* literally means "tailor." If you remember this and think about the typical cross-legged position tailors take to mend something by hand, it will be easy to remember the actions of the sartorius muscle. Try this out for yourself. Sit in a chair and cross your right leg over your left so that your right ankle is resting on your left knee. Next, note the positions of your right hip and knee joints. The hip is flexed and laterally rotated, and the knee is also flexed. These are the actions of the sartorius muscle.

4. *Femoral Triangle:* The **femoral triangle** is a triangular space in the upper, anterior thigh. Its borders are the **sartorius,** the **adductor longus,** and the **inguinal ligament.** It is clinically relevant because of the structures located within the space, which include the femoral nerve, artery, and vein. The only structures superficial to these are fat and skin, thus it is a location where the vascular system can be accessed with relative ease.

(continued on next page)

(continued from previous page)

Table 13.9 Anterior Compartment of the Thigh

Muscle	Origin	Insertion	Action	Innervation	Word Origin
Sartorius	ASIS (anterior superior iliac spine)	Medial tibia	Crosses legs: flexes the hip and knee, abducts and laterally rotates the thigh	Femoral nerve (L2–L4)	*sartor*, a tailor
Quadriceps Femoris Group					
Rectus Femoris	Anterior inferior iliac spine	Tibial tuberosity	Extends the knee, flexes the hip	Femoral nerve (L2–L4)	*rectus*, straight, + *femur*, thigh
Vastus Lateralis	Femur (medial and posterior)	Tibial tuberosity	Extends the knee	Femoral nerve (L2–L4)	*vastus*, great, + *lateralis*, to the side
Vastus Intermedius	Femur (medial and lateral)	Tibial tuberosity	Extends the knee	Femoral nerve (L2–L4)	*vastus*, great, + *intermedius*, in between
Vastus Medialis	Femur (inferior and posterior)	Tibial tuberosity	Extends the knee	Femoral nerve (L2–L4)	*vastus*, great, + *medialis*, medial

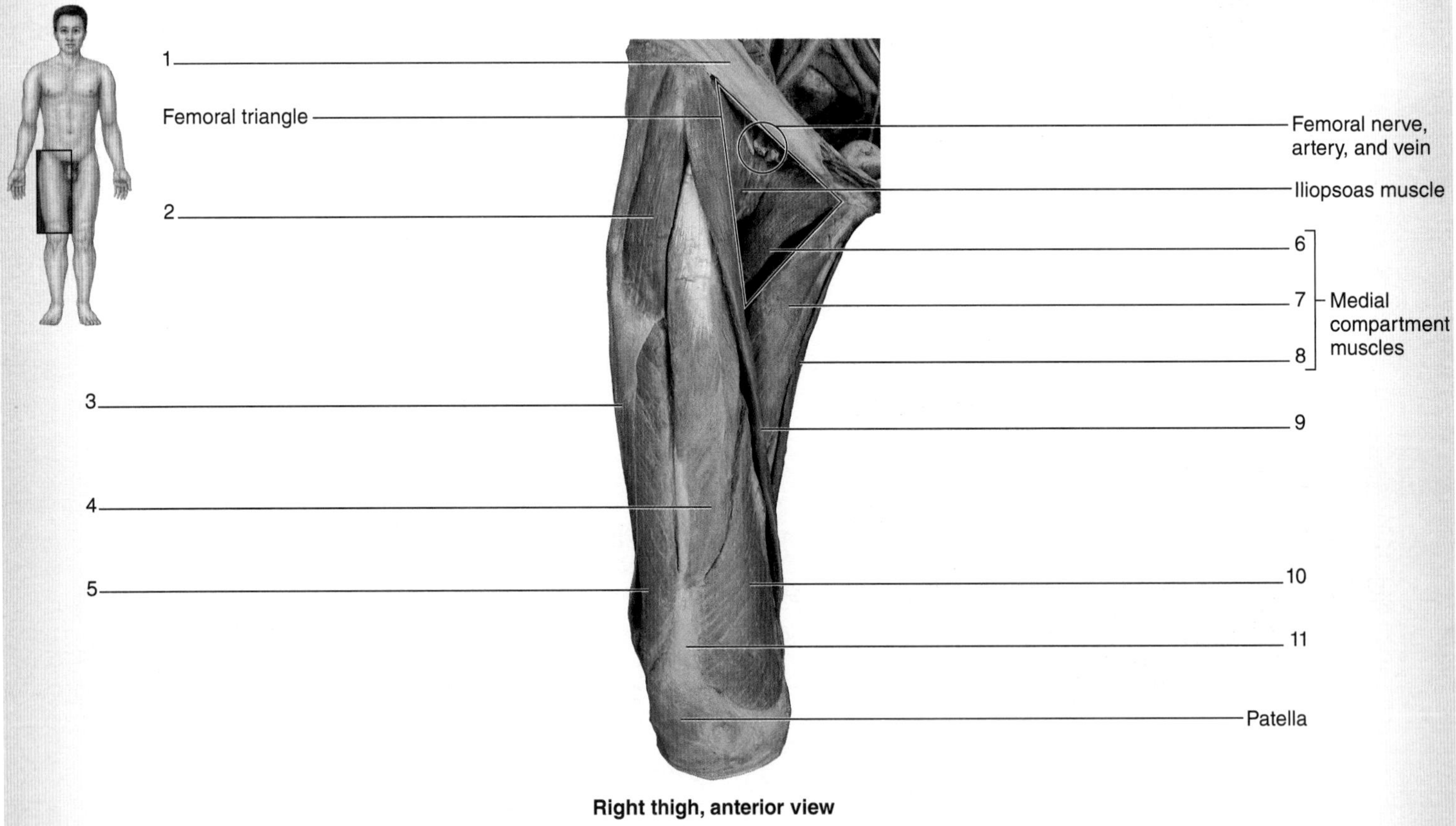

Right thigh, anterior view

Figure 13.11 Anterior Compartment of the Thigh.

- ☐ adductor longus
- ☐ gracilis
- ☐ iliotibial tract
- ☐ inguinal ligament
- ☐ pectineus
- ☐ quadriceps tendon
- ☐ rectus femoris
- ☐ sartorius
- ☐ tensor fasciae latae
- ☐ vastus lateralis
- ☐ vastus medialis

5. In the space below, make a representative drawing of the femoral triangle. Be sure to draw and label the structures that form its boundaries, and draw and label the contents found within the triangle.

EXERCISE 13.6B Medial Compartment of the Thigh

1. Observe a prosected human cadaver or a classroom model demonstrating muscles of the thigh.

2. Using **table 13.10** and your textbook as guides, identify the **muscles of the medial compartment of the thigh** listed in **figure 13.12** on the cadaver or on a classroom model. Then label them in figure 13.12. As you identify these muscles, make note of the following:

Common Action: Adduct the thigh.

Exceptions: Although also an adductor of the thigh, the gracilis muscle does not attach to the femur (it attaches to the tibia). Thus, the gracilis is also the only muscle of the group to cross and flex the knee joint.

INTEGRATE

LEARNING STRATEGY

The femoral triangle contains some very important structures. The mnemonic for remembering the contents of the femoral triangle is **NAVEL**, which lists the contents of the femoral triangle from lateral to medial:

N—femoral **N**erve

A—femoral **A**rtery

V—femoral **V**ein

E—**E**mpty space

L—**L**ymphatic vessel

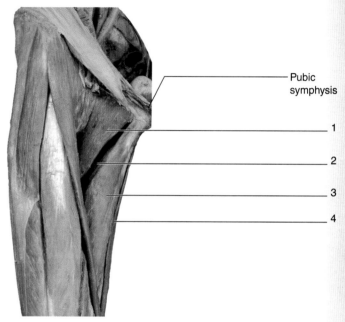

Right thigh, anterior view

Figure 13.12 Medial Compartment of the Thigh.
The adductor magnus is not visible in this photo.

☐ adductor brevis   ☐ gracilis
☐ adductor longus   ☐ pectineus

Table 13.10	Medial Compartment of the Thigh				
Muscle	**Origin**	**Insertion**	**Action**	**Innervation**	**Word Origin**
Pectineus	Pubis	Femur (pectineal line)	Adducts and laterally rotates the thigh	Obturator or femoral nerve (L2–L4)	*pectineal*, a ridged or comblike structure
Adductor Longus	Pubis	Femur (linea aspera)	Adducts and laterally rotates the thigh	Obturator nerve (L2–L4)	*adduct*, to bring toward the median plane, + *longus*, long
Adductor Brevis	Pubis	Femur (linea aspera)	Adducts and laterally rotates the thigh	Obturator nerve (L2–L3)	*adduct*, to bring toward the median plane, + *brevis*, short
Adductor Magnus	Pubis	Femur (linea aspera)	Adducts and laterally rotates the thigh	Obturator nerve (L2–L4)	*adduct*, to bring toward the median plane, + *magnus*, large
Gracilis	Pubis	Tibia (medial condyle)	Adducts the thigh and flexes the leg	Obturator nerve (L2–L4)	*gracilis*, slender

(continued on next page)

(continued from previous page)

3. *Gracilis:* In addition to being the outlier in terms of its attachments, this muscle is a weak adductor of the thigh—so weak, in fact, that it can be removed without a patient's experiencing great loss in function. When a patient needs a muscle graft, a common muscle used for this purpose is the **gracilis.** The superficial location and limited function of this muscle make it a good candidate for grafting procedures.

4. In the space below, make a representative drawing of the muscles of the medial compartment of the thigh.

5. *Optional Activity:* **AP|R** 6: **Muscular System**—Visit the quiz area for drill and practice on muscles of the lower limb.

INTEGRATE

LEARNING STRATEGY

- All the muscles in the medial compartment of the thigh attach to the pubic bone and, except for the gracilis, to the linea aspera of the femur.

- Think of these muscles as "the short one" (adductor brevis), "the long one" (adductor longus), "the really big one" (adductor magnus), and "the graceful one" (gracilis) to make them easier to identify relative to each other.

- As you view the anterior thigh, you will see these muscles medial to the sartorius. From superior to inferior, the order of muscles is pectineus, adductor longus, and adductor magnus. The adductor brevis is located deep to the pectineus and adductor longus muscles, so you will have to move them aside to see it clearly.

EXERCISE 13.6C Posterior Compartment of the Thigh

1. Observe a prosected human cadaver or a classroom model demonstrating muscles of the thigh.

2. The **posterior compartment of the thigh** contains only three muscles: the semimembranosus, the semitendinosus, and the biceps femoris. These muscles are collectively referred to as the **hamstring** muscles. They received this name because of their association with the analogous muscles in pigs. The meat we call ham comes from these posterior thigh muscles of a pig. In a

Table 13.11	Posterior Compartment of the Thigh				
Muscle	**Origin**	**Insertion**	**Action**	**Innervation**	**Word Origin**
Biceps Femoris					*bi*, two, + *caput*, head, + *femur*, thigh
Long Head	Ischial tuberosity	Head of fibula	Flexes the leg, extends the hip	Tibial nerve (L4–S1)	
Short Head	Linea aspera of femur	Head of fibula	Flexes the leg	Fibular nerve (L5–S1)	
Semimembranosus	Ischial tuberosity	Medial tibia	Flexes the leg, extends the hip	Tibial nerve (L4–S1)	*semi*, half, + *membrana*, a membrane
Semitendinosus	Ischial tuberosity	Medial tibia	Flexes the leg, extends the hip	Tibial nerve (L4–S1)	*semi*, half, + *tendinosus*, tendon

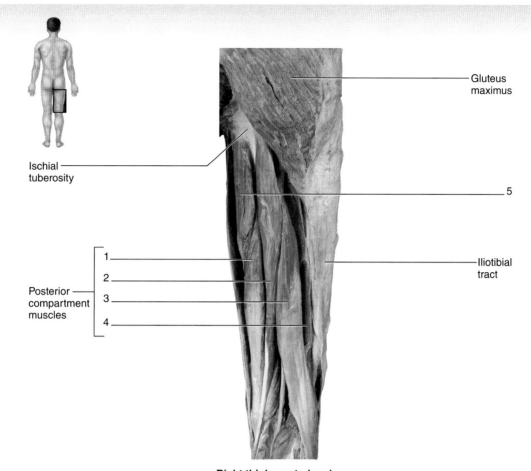

Right thigh, posterior view

Figure 13.13 Posterior Compartment of the Thigh.

☐ adductor magnus ☐ biceps femoris (short head) ☐ semitendinosus
☐ biceps femoris (long head) ☐ semimembranosus

slaughterhouse, pig thighs are hung by the long tendons of these muscles, hence the term *hamstrung*. In fact, when we say that we feel "hamstrung," we say that we feel (figuratively) crippled. If our hamstring muscles were severed, we literally would be crippled.

3. Using **table 13.11** and your textbook as guides, identify the **muscles of the posterior compartment of the thigh** listed in **figure 13.13** on the cadaver or on a classroom model. Then label them in figure 13.13. As you identify these muscles, make note of the following:

Common Actions: Extend the hip and flex the leg.

Exceptions: The short head of the biceps femoris does not cross the hip joint. Thus, it does not extend the hip.

4. *Semitendinosus and Semimembranosus:* These two muscles are named for the appearance of their connective tissue components. The **semitendinosus** is the more superficial of the two muscles. At its *distal* end it forms a long, slender tendon; hence the name ending "tendinosus." The **semimembranosus** muscle is the deeper of the two muscles. At its *proximal* end, it forms a broad, flat tendon that looks like a membrane; hence the name ending "membranosus."

EXERCISE 13.7

COMPARTMENTS OF THE LEG

EXERCISE 13.7A Anterior Compartment of the Leg

1. Observe a prosected human cadaver or a classroom model demonstrating muscles of the leg.

2. Using **table 13.12** and your textbook as guides, identify the **muscles of the anterior compartment of the leg** listed in **figure 13.14** on the cadaver or on a classroom model. Then label them in figure 13.14. As you identify these muscles, make note of the following:

   Common Actions: Dorsiflex the ankle and extend the digits.

   Exception: The tibialis anterior does not act on the digits. In addition, it inverts the ankle as it dorsiflexes.

EXERCISE 13.7B Lateral Compartment of the Leg

1. Observe a prosected human cadaver or a classroom model demonstrating muscles of the leg.

2. Using **table 13.13** and your textbook as guides, identify the **muscles of the lateral compartment of the leg** listed in **figure 13.15** on the cadaver or on classroom models. Then label them in figure 13.15.

3. *Fibularis Longus and Brevis:* The fibularis muscles were formerly referred to as peroneus muscles. Thus, you should be aware that the terms relate to the same muscles. The **fibularis (peroneus) longus** has a very long tendon that lies superficial to the belly of the **fibularis (peroneus) brevis** muscle. The tendons of both muscles pass immediately posterior to the lateral malleolus en route to their attachments on the plantar surface of the foot. Their major action is to evert the ankle.

INTEGRATE

LEARNING STRATEGY

pEroneus (fibularis) muscles always Evert, whereas tIbialis muscles always Invert.

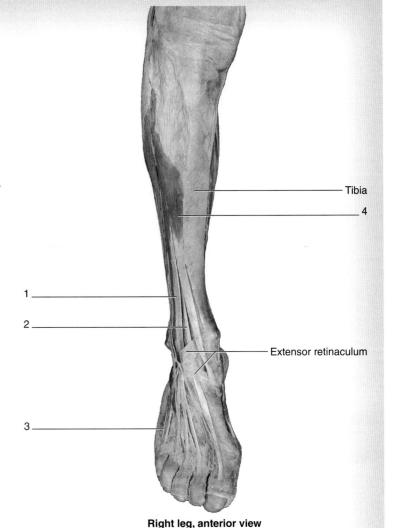

Right leg, anterior view

Figure 13.14 Anterior Compartment of the Leg.

☐ extensor digitorum longus ☐ fibularis tertius tendon
☐ extensor hallucis longus ☐ tibialis anterior

Table 13.12	Anterior Compartment of the Leg				
Muscle	**Origin**	**Insertion**	**Action**	**Innervation**	**Word Origin**
Tibialis Anterior	Tibia (lateral condyle and upper two-thirds)	Middle cuneiform (inferior surface) and metatarsal I	Dorsiflexes and inverts the foot	Deep fibular nerve	*tibia*, the shinbone, + *anterior*, the front surface
Extensor Hallucis Longus	Fibula (anteromedial surface) and interosseous membrane	Distal phalanx of big toe	Extends the great toe and dorsiflexes the ankle	Deep fibular nerve	*extendo*, to stretch out, + *hallux*, the great toe, + *longus*, long
Extensor Digitorum Longus	Tibia (lateral condyle and proximal three-fourths)	2nd and 3rd phalanges of digits 2–5	Extends digits 2–5 and dorsiflexes the ankle	Deep fibular nerve	*extendo*, to stretch out, + *digit*, a toe, + *longus*, long
Fibularis Tertius	Fibula (anterior, distal surface) and interosseous membrane	Base of metatarsal V	Dorsiflexes and weakly everts foot	Deep fibular nerve	*fibularis*, fibula, + *tertius*, third

Table 13.13 Lateral Compartment of the Leg

Muscle	Origin	Insertion	Action	Innervation	Word Origin
Fibularis Longus	Head of fibula	Metatarsal I and middle cuneiform	Everts the foot and plantar flexes the ankle	Superficial fibular nerve	*fibularis*, relating to the fibula, + *longus*, long
Fibularis Brevis	Distal fibula	Metatarsal V	Everts the foot and plantar flexes the ankle	Superficial fibular nerve	*fibularis*, relating to the fibula, + *brevis*, short

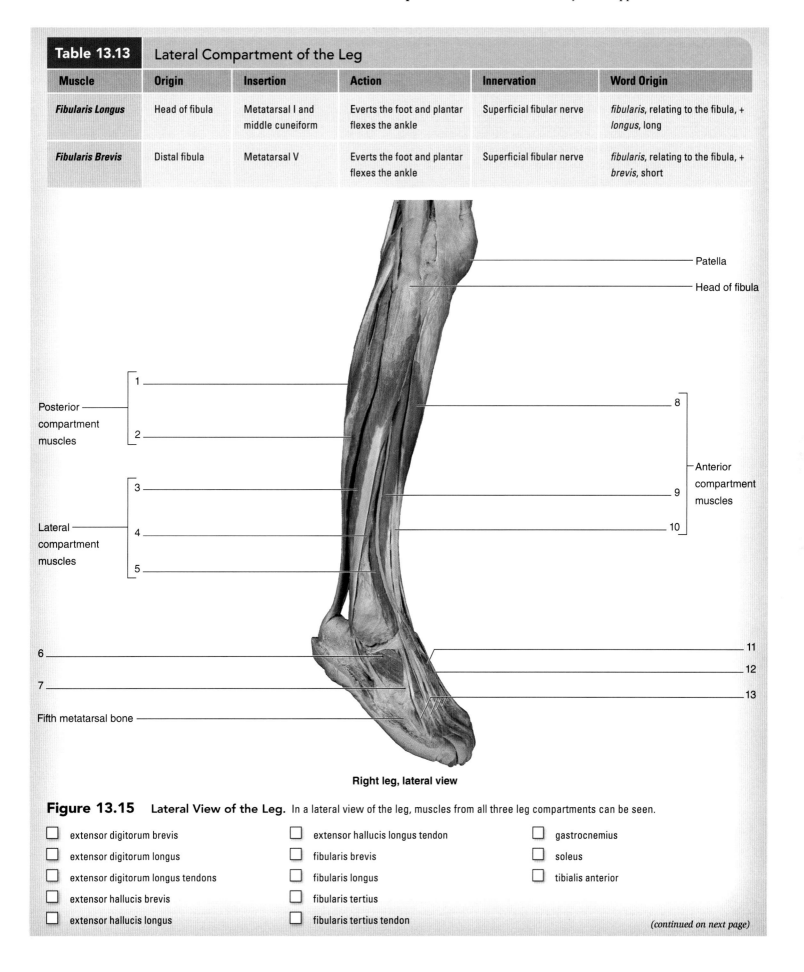

Right leg, lateral view

Figure 13.15 Lateral View of the Leg. In a lateral view of the leg, muscles from all three leg compartments can be seen.

- ☐ extensor digitorum brevis
- ☐ extensor digitorum longus
- ☐ extensor digitorum longus tendons
- ☐ extensor hallucis brevis
- ☐ extensor hallucis longus
- ☐ extensor hallucis longus tendon
- ☐ fibularis brevis
- ☐ fibularis longus
- ☐ fibularis tertius
- ☐ fibularis tertius tendon
- ☐ gastrocnemius
- ☐ soleus
- ☐ tibialis anterior

(continued on next page)

346 Chapter Thirteen *The Muscular System: Appendicular Muscles*

(continued from previous page)

EXERCISE 13.7C Posterior Compartment of the Leg

1. Observe a prosected human cadaver or a classroom model demonstrating muscles of the leg.

2. Using **table 13.14** and your textbook as guides, identify the **muscles of the posterior compartment of the leg** listed in **figure 13.16** on the cadaver or on a classroom model. Then label them in figure 13.16.

3. *Triceps Surae:* The **triceps surae** includes the **soleus** and the two-headed **gastrocnemius** muscle. The tendons of these muscles come together to insert onto the calcaneus via the calcaneal (Achilles) tendon. The most superficial of the muscles is the gastrocnemius, though the lateral edges of the soleus can be seen as they poke out beneath the inferior portion of the gastrocnemius. The **plantaris** is a small muscle that can be seen only by reflecting or removing the gastrocnemius and soleus muscles. The plantaris has a very small belly, located just below the popliteal fossa in the back of the knee. It has a very long, flat tendon that often resembles ribbon used to decorate presents. The muscle itself is a very weak flexor of the leg. If you are viewing a cadaver and cannot find the plantaris, this may be because it isn't there. The plantaris is absent in approximately 6% to 8% of humans, and it is more commonly absent in the left leg than in the right leg. Interestingly, because its actions are minimal and its tendon is long, the tendon is often removed and used for tendon grafts.

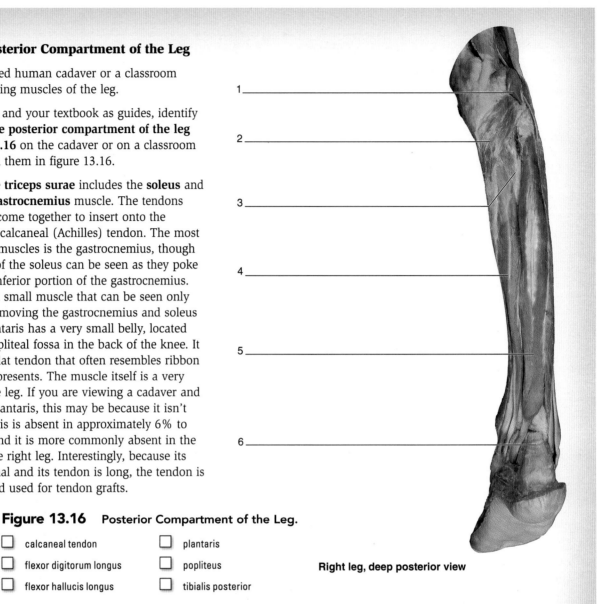

Figure 13.16 Posterior Compartment of the Leg.

- ☐ calcaneal tendon
- ☐ plantaris
- ☐ flexor digitorum longus
- ☐ popliteus
- ☐ flexor hallucis longus
- ☐ tibialis posterior

Right leg, deep posterior view

Table 13.14	Posterior Compartment of the Leg				
Muscle	**Origin**	**Insertion**	**Action**	**Innervation**	**Word Origin**
Gastrocnemius	Femoral condyles	Calcaneus	Plantar flexes the ankle and flexes the leg (weak)	Tibial nerve (L4–S1)	*gaster*, belly, + *kneme*, leg
Soleus	Proximal tibia, fibula	Calcaneus	Plantar flexes the ankle	Tibial nerve (L4–S1)	*solea*, a sandal, sole of the foot
Plantaris	Femur (supracondylar ridge)	Calcaneus	Plantar flexes the ankle and flexes the leg (weak)	Tibial nerve (L4–S1)	*plantar*, relating to the sole of the foot
Popliteus	Lateral condyle of femur	Posterior, proximal surface of tibia	Flexes leg, unlocks the knee joint	Tibial nerve	*poplit*, back of the knee
Tibialis Posterior	Proximal tibia and fibula	Medial cuneiform and navicular	Inverts the foot and plantar flexes the ankle	Tibial nerve (L5–S1)	*tibia*, the shinbone, + *posterior*, the back surface
Flexor Digitorum Longus	Posterior tibia	Distal phalanx of digits 2–5	Flexes digits 2–5	Tibial nerve (L5–S1)	*flexus*, to bend, + *digit*, a toe, + *longus*, long
Flexor Hallucis Longus	Inferior two-thirds of fibula	Distal phalanx of the great toe	Flexes the great toe	Tibial nerve (L5–S1)	*flexus*, to bend, + *hallux*, the great toe, + *longus*, long

EXERCISE 13.8

INTRINSIC MUSCLES OF THE FOOT

1. Observe a prosected human cadaver or a classroom model demonstrating muscles of the foot.

2. The intrinsic muscles of the plantar surface of the foot are arranged in four layers, named layer 1 to layer 4 from superficial to deep. The foot also contains two neurovascular planes (a neurovascular plane is a region where the nerves and blood vessels that serve the region are located). The first is located between muscle layers 1 and 2, and the second is located between layers 3 and 4.

3. Using **table 13.15** and your textbook as guides, identify the **intrinsic muscles of the foot** listed in **figure 13.17** on the cadaver or on a classroom model. Then label them in figure 13.17.

4. *Adductor Hallucis:* The **adductor hallucis** muscle is located in the third plantar muscle layer. It is a good landmark for the identification of other muscles in the foot. In addition, its two heads (oblique and transverse) appear like the number 7 when viewed together. To identify this muscle and others in the deeper layers of the foot, the tough **plantar aponeurosis** must first be cut and reflected.

5. *Interossei:* Like in the hand, there are two groups of interosseous muscles in the foot, though they are called dorsal and *plantar*, instead of dorsal and *palmar*. The functions of the interossei in the foot are the same as in the hand: The dorsal interossei abduct the digits, whereas the plantar interossei adduct the digits.

INTEGRATE

LEARNING STRATEGY

Conveniently, the mnemonic for remembering the functions of the interossei in the foot is the same as in the hand: **DAB** and **PAD**. The **D**orsal interossei **AB**duct the digits **(DAB)**, whereas the **P**lantar interossei **AD**duct the digits **(PAD)**.

Table 13.15 Intrinsic Muscles of the Foot

Muscle	Origin	Insertion	Action	Innervation	Word Origin
Dorsal Musculature					
Extensor Digitorum Brevis	Calcaneus and extensor retinaculum (inferior surface)	Digits 2–4 (dorsal surfaces)	Extends the proximal phalanx of digits 2–4	Deep fibular nerve	*dorsal*, the back, + *inter-*, between, + *os*, bone
Extensor Hallucis Brevis	Calcaneus and extensor retinaculum (inferior surface)	Dorsal surface of the great toe	Extends the proximal phalanx of the great toe	Deep fibular nerve	*palmar*, the palm of the hand (foot), + *inter-*, between, + *os*, bone
Plantar Musculature—Layer 1					
Abductor Digiti Minimi	Calcaneus and plantar aponeurosis	Base of the proximal phalanx of digit 5 (lateral surface)	Abducts and flexes digit 5 (the little toe)	Lateral plantar nerve	*abduct*, to move away from the median plane, + *digitus*, toe, + *minimi*, smallest
Abductor Hallucis	Calcaneus and plantar aponeurosis	Base of the proximal phalanx of digit 1 (medial side)	Abducts the great toe	Medial plantar nerve	*abduct*, to move away from the median plane, + *hallux*, the great toe
Flexor Digitorum Brevis	Calcaneus and plantar aponeurosis	Middle phalanx of digits 2–5	Flexes digits 2–5	Medial plantar nerve	*flex*, to bend, + *digitus*, toe, + *brevis*, short
Plantar Musculature—Layer 2					
Lumbricals	Tendons of the flexor digitorum longus	Tendons of the extensor digitorum longus	Flexes the proximal phalanges and extends the distal phalanges of digits 2–5	Medial and lateral plantar nerve	*lumbricus*, earthworm
Quadratus Plantae	Calcaneus (medial surface and lateral margin)	Tendon of the flexor digitorum longus (lateral side)	Flexes digits 2–5	Lateral plantar nerve	*quadratus*, having four sides, + *plantae*, sole of the foot

(continued on next page)

(continued from previous page)

Table 13.15 — Intrinsic Muscles of the Foot *(continued)*

Muscle	Origin	Insertion	Action	Innervation	Word Origin
Plantar Musculature—Layer 3					
Adductor Hallucis (Oblique Head)	Base of metatarsals II–V	Base of the proximal phalanx of the great toe	Adducts the great toe and helps maintain the transverse arch of the foot	Lateral plantar nerve	*adduct*, to move toward the median plane, + *hallux*, the great toe
Adductor Hallucis (Transverse Head)	Capsules of the metatarsophalangeal joints III–V	Base of the proximal phalanx of the great toe	Adducts the great toe and helps maintain the transverse arch of the foot	Lateral plantar nerve	*adduct*, to move toward the median plane, + *hallux*, the great toe
Flexor Digiti Minimi Brevis	Base of metatarsal V	Proximal phalanx of digit 5	Flexes the proximal phalanx of digit 5 (the little toe)	Lateral plantar nerve	*flex*, to bend, + *digitus*, a toe, + *minimi*, smallest, + *brevis*, short
Flexor Hallucis Brevis	Cuboid and lateral cuneiform (plantar surfaces)	Proximal phalanx of the great toe	Flexes the proximal phalanx of the great toe	Medial plantar nerve	*flex*, to bend, + *hallux*, the great toe, + *brevis*, short
Plantar Musculature—Layer 4					
Dorsal Interossei	Metatarsals I–V (adjacent surfaces)	Proximal phalanges of digits 2–4 (sides)	Abducts digits 2–4 and flexes the metatarsophalangeal joints	Lateral plantar nerve	*dorsal*, the back, + *inter-*, between, + *os*, bone
Plantar Interossei	Bases of metatarsals III–V	Bases of proximal phalanges of digits 3–5 (medial side)	Adducts digits 2–4 and flexes the metatarsophalangeal joints	Lateral plantar nerve	*plantae*, the sole of the foot, + *inter-*, between, + *os*, bone

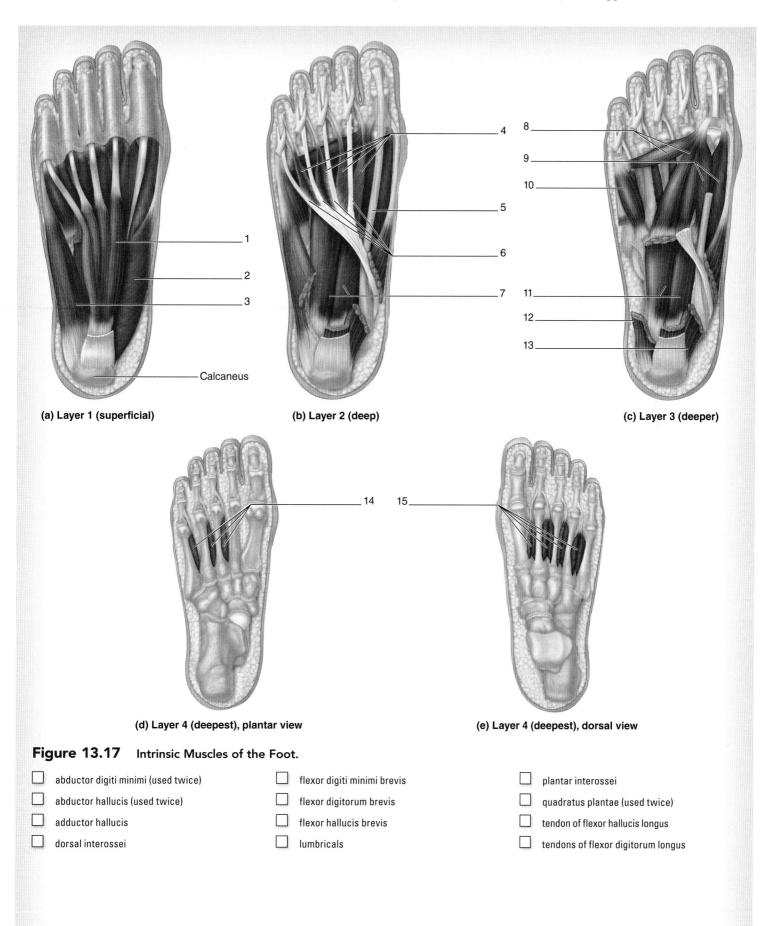

Figure 13.17 Intrinsic Muscles of the Foot.

- [ ] abductor digiti minimi (used twice)
- [ ] abductor hallucis (used twice)
- [ ] adductor hallucis
- [ ] dorsal interossei
- [ ] flexor digiti minimi brevis
- [ ] flexor digitorum brevis
- [ ] flexor hallucis brevis
- [ ] lumbricals
- [ ] plantar interossei
- [ ] quadratus plantae (used twice)
- [ ] tendon of flexor hallucis longus
- [ ] tendons of flexor digitorum longus

> **INTEGRATE**
>
> ## CONCEPT CONNECTION
>
> Sensory receptors embedded in muscles and tendons sense precise information regarding muscle length and tension. These specialized receptors, called *proprioceptors* (*proprio-,* one's own), relay this information to the central nervous system, enabling you to perceive the position of your body in space. For example, when you close your eyes, you generally still have some awareness of the position of your limbs. Likewise, when you walk into a dark room, you are still able to navigate through the space, in part due to proprioception. Two important types of proprioceptors are muscle spindles and Golgi tendon organs, which detect muscle length and tension, respectively. Muscle spindles lie parallel to muscle fibers so that they experience the same changes in length as the muscle fibers. Golgi tendon organs are embedded within the tendons of muscles. As muscle force builds during a muscle contraction, the sensory receptor becomes compressed. Both muscle spindles and Golgi tendon organs also play a role in reflexes, or preprogrammed responses to stimuli. For example, the stretch reflex involves the muscle spindle, and it is often tested clinically with a hammer tap to the patellar ligament (see chapter 17). The tap causes a rapid stretch of the quadriceps muscle group, which also stretches the muscle spindles. The result is a contraction of the entire quadriceps group, thereby causing knee extension, which counteracts the stretch that was applied. It is a protective mechanism to prevent too much force on a muscle from causing injury. Proprioceptors are distributed in all of your skeletal muscles. However, the number of muscle spindles and Golgi tendon organs varies from one muscle to another.

Chapter 13: The Muscular System: Appendicular Muscles

Name:_____
Date:_____ Section:_____

POST-LABORATORY WORKSHEET

The ❶ corresponds to the Learning Objective(s) listed in the chapter opener outline.

Do You Know the Basics?

Exercise 13.1: Muscles That Move the Pectoral Girdle and Glenohumeral Joint

1. Match the muscles listed in column A with the movement produced by each listed in column B. ❶

    Column A

    a. deltoid
    b. latissimus dorsi
    c. pectoralis major
    d. teres major

    Column B

    ____ prime mover of arm extension; adducts and medially rotates arm (big muscle)
    ____ prime mover of arm flexion; adducts and medially rotates arm (big muscle)
    ____ extends, adducts, and medially rotates arm
    ____ flexes and medially rotates arm; prime mover of arm abduction; extends and laterally rotates arm

2. A synergist of the trapezius for the action of *adduction* of the scapula is _____. A synergist of the trapezius for the action of *elevation* of the scapula is _____. ❷

3. Damage to the serratus anterior muscle is tested by having a patient push against a wall with an outstretched upper limb. If there is weakness or paralysis of the muscle, the scapula will (circle one) ❸

    a. adduct excessively.
    b. abduct excessively.
    c. "wing" (medial border moves posterior).
    d. depress excessively.

4. List the four muscles that compose the rotator cuff: ❹

    a. _____
    b. _____
    c. _____
    d. _____

Exercise 13.2: Compartments of the Arm

5. The arm is divided into a(n) _____ compartment, which consists largely of forearm flexors, and a(n) _____ compartment, which consists largely of forearm extensors. ❺

6. The biceps brachii is a prime mover for the action of _____. The biceps brachii is a synergist with the brachialis muscle for the action of _____. ❻

Exercise 13.3: Compartments of the Forearm

7. Match the muscles listed in column A with the forearm compartment they belong to in column B. There will be more than one answer for each compartment. ❼

    Column A

    a. extensor indicis
    b. palmaris longus
    c. flexor carpi ulnaris
    d. supinator
    e. pronator quadratus

    Column B

    ____ anterior compartment
    ____ posterior compartment

352 Chapter Thirteen *The Muscular System: Appendicular Muscles*

8. Describe how to demonstrate the presence of the palmaris longus muscle on yourself. _____
_____
_____

9. The tendons of the flexor digitorum superficialis muscle attach to the proximal/middle/distal (circle one) phalanges, whereas the tendons of the flexor digitorum profundus attach to the proximal/middle/distal (circle one) phalanges.

Exercise 13.4: Intrinsic Muscles of the Hand

10. Mark an "X" next to each muscle below that is an intrinsic muscle of the hand.

    _____ a. dorsal interossei

    _____ b. lumbricals

    _____ c. flexor digitorum brevis

    _____ d. abductor pollicis brevis

    _____ e. abductor digiti minimi

11. The large pad at the base of the thumb is called the _____, and it contains muscles collectively referred to as _____ muscles. The large pad at the base of the pinky finger (digit V) is called the _____, and it contains muscles collectively referred to as _____ muscles.

12. The dorsal interossei _____ the digits, whereas the palmar interossei _____ the digits.

Exercise 13.5: Muscles That Move the Hip

13. The _____ muscle is a major *extensor* of the hip. Its action is antagonized by the _____ muscle, a major *flexor* of the hip.

14. During normal locomotion, two muscles on the stance limb, which abduct the hip, must function correctly for the foot of the swing limb to clear the ground. List these two muscles in the spaces below.

    a. _____

    b. _____

15. A deep muscle of the posterior hip that is an important clinical landmark due to its association with the gluteal arteries and nerves, and the sciatic nerve, is the _____ muscle.

16. The iliopsoas muscle is composed of two muscles with a common insertion. The two muscles that compose the iliopsoas are the _____ and the _____. The common insertion point for this muscle is the _____.

Exercise 13.6: Compartments of the Thigh

17. Match the muscle listed in column A with the appropriate compartment of the thigh in column B.

    Column A

    _____ 1. rectus femoris

    _____ 2. biceps femoris

    _____ 3. adductor magnus

    _____ 4. sartorius

    _____ 5. vastus lateralis

    _____ 6. semitendinosus

    _____ 7. gracilis

    Column B

    a. anterior compartment

    b. medial compartment

    c. posterior compartment

18. The only two muscles of the anterior compartment of the thigh that cross the hip joint are the _____ and the _____. The action of these two muscles about the hip is to _____ the hip.

19. List the three structures that form the borders of the femoral triangle:

    a. _____

    b. _____

    c. _____

Exercise 13.7: Compartments of the Leg

20. Match the muscle listed in column A with the appropriate compartment of the leg in column B. ⓴

    Column A

    _____ 1. soleus

    _____ 2. extensor digitorum longus

    _____ 3. flexor hallucis longus

    _____ 4. tibialis anterior

    _____ 5. gastrocnemius

    _____ 6. fibularis (peroneus) brevis

    _____ 7. fibularis (peroneus) longus

    Column B

    a. anterior compartment

    b. lateral compartment

    c. posterior compartment

21. The fibularis longus and brevis evert the foot at the ankle joint. Name two muscles whose primary action is to oppose the action of the fibularis muscles (i.e., that invert the foot at the ankle joint). ㉑

    a. _____

    b. _____

22. The three muscles that compose the triceps surae are ㉒

    a. _____

    b. _____

    c. _____

Exercise 13.8: Intrinsic Muscles of the Foot

23. Mark an "X" next to each muscle below that is an intrinsic muscle of the foot. ㉓

    ____ a. soleus

    ____ b. flexor digitorum brevis

    ____ c. abductor hallucis

    ____ d. tibialis anterior

    ____ e. gastrocnemius

    ____ f. quadratus femoris

    ____ g. dorsal interossei

    ____ h. flexor carpi ulnaris

Can You Apply What You've Learned?

24. Name three muscles that attach to the coracoid process of the scapula.

    a. _____

    b. _____

    c. _____

25. One action of the gluteal muscles (gluteus maximus, medius, and minimus) is to rotate the hip.

    a. Which gluteal muscle(s) laterally rotate the hip? _____

    b. Which gluteal muscle(s) medially rotate the hip? _____

26. Which muscle of the medial compartment of the thigh does NOT insert onto the linea aspera of the femur? _____

27. Which muscle of the quadriceps muscle group is the only muscle to cross the hip joint? _____ What is the *origin* (proximal attachment) of this muscle? _____. What is the *insertion* (distal attachment)? _____

Chapter Thirteen — The Muscular System: Appendicular Muscles

28. Describe the difference between the patellar *tendon* and the patellar *ligament*, and describe the location of each. (Hint: Think about the specific definitions of *tendon* and *ligament*.)

29. How can you distinguish between the semitendinosus muscle and the semimembranosus muscle?

30. The tendons of which three of the rotator cuff muscles hold the humerus into the glenoid cavity on the posterior surface? _____, _____, _____. The tendon of the rotator cuff muscle that holds the humerus into the glenoid cavity on the anterior surface is the tendon of the _____ muscle.

31. What two muscles insert onto the iliotibial tract (IT band)?

    a. _____   b. _____

32. The gluteus maximus muscle is a major extensor of the hip. What muscle acts as the major antagonist to the gluteus maximus muscle? (Hint: The muscle you name would be a prime mover for hip *flexion*.) _____

33. Describe the involvement of the biceps in the following movements:

    a. Flexion of the forearm:

    b. Supination of the forearm/wrist:

Can You Synthesize What You've Learned?

34. The palmaris longus muscle is absent in approximately 20% of the population. Describe a simple test that can be used to demonstrate presence or absence of this muscle in an individual.

35. A patient is told he has a "torn hamstring." In the space below, list the possible muscles that could have been torn in this case. Then, describe the actions that would demonstrate weakness in the muscle as a result of the tear.

36. Compartment syndrome is a disorder in which inflammation within the compartment of a limb creates pressure such that blood flow is reduced, nerves become damaged, and the muscles start to become nonfunctional (and eventually necrotic, if the situation is not treated). A patient is experiencing compartment syndrome in the anterior compartment of the leg, which has weakened the muscles in this compartment. As a result, what actions would this patient have difficulty performing?

PART IV | COMMUNICATION AND CONTROL

CHAPTER 14

Nervous Tissues

OUTLINE AND LEARNING OBJECTIVES

Histology 358

EXERCISE 14.1: GRAY AND WHITE MATTER 358
1. Identify gray matter and white matter in a slide of nervous tissue
2. Describe the parts of a typical neuron

Neurons 359

EXERCISE 14.2: GENERAL MULTIPOLAR NEURONS—ANTERIOR HORN CELLS 360
3. Identify anterior horn cells in a slide of the cross section of the spinal cord

EXERCISE 14.3: CEREBRUM—PYRAMIDAL CELLS 360
4. Identify pyramidal cells and white matter (nerve tracts) in slides of the cerebral cortex

EXERCISE 14.4: CEREBELLUM—PURKINJE CELLS 361
5. Identify Purkinje cells in a slide of the cerebellum

Glial Cells 362

EXERCISE 14.5: ASTROCYTES 362
6. Identify astrocytes in a histology slide of the cerebellum stained with the silver impregnation method

EXERCISE 14.6: EPENDYMAL CELLS 363
7. Identify ependymal cells in a histology slide of the cerebrum or choroid plexus

EXERCISE 14.7: NEUROLEMMOCYTES (SCHWANN CELLS) 364
8. Identify neurolemmocytes and neurofibril nodes in a histology slide of a longitudinal section of a peripheral nerve

EXERCISE 14.8: SATELLITE CELLS 364
9. Identify a posterior root ganglion (peripheral nerve ganglion) and its associated unipolar neurons and satellite cells

Peripheral Nerves 365

EXERCISE 14.9: COVERINGS OF A PERIPHERAL NERVE 366
10. Locate epineurium, perineurium, and endoneurium in a cross-sectional slide of a peripheral nerve

Physiology 366

Resting Membrane Potential 366

EXERCISE 14.10: Ph.I.L.S. LESSON 8: RESTING POTENTIAL AND EXTERNAL [K⁺] 367
11. Describe the normal intracellular and extracellular concentrations of potassium in a cell and the effect this has on resting membrane potential

EXERCISE 14.11: Ph.I.L.S. LESSON 9: RESTING POTENTIAL AND EXTERNAL [Na⁺] 369
12. Describe the normal intracellular and extracellular concentrations of sodium in a cell and describe the effect this has on resting membrane potential

Action Potential Propagation 371

EXERCISE 14.12: Ph.I.L.S. LESSON 10: THE COMPOUND ACTION POTENTIAL 372
13. Demonstrate threshold by increasing membrane potential to evoke compound action potentials
14. Demonstrate recruitment by relating external applied voltage to compound action potential amplitude

EXERCISE 14.13: Ph.I.L.S. LESSON 11: CONDUCTION VELOCITY AND TEMPERATURE 374
15. Explain the consequence of changing temperature on the conduction velocity of action potentials

EXERCISE 14.14: Ph.I.L.S. LESSON 12: REFRACTORY PERIODS 375
16. Associate absolute and relative refractory periods in excitable cells with specific ionic events that take place on the membrane of the cell

MODULE 7: NERVOUS SYSTEM

INTRODUCTION

Many of the structures that make up the nervous system are not particularly impressive in the absence of any knowledge of what they do, particularly when compared with very dynamic-looking organs such as the heart. Imagine how unimpressed early anatomists must have been with the brain. All they saw was a very dense, oatmeal-colored mass of tissue with no known function. Their observations of brain tissue could hardly have been as fascinating or romantic as their observations of the heart. At first, the most inspiring names they could find for brain tissues were terms like "gray matter" and "white matter." Today we know that the nervous system is much more fascinating than we could ever have imagined, and that scientists are barely beginning to grasp its complexity. Most information about the complexity of the nervous system has been gained through dynamic studies of the brain using methods such as magnetic resonance imaging (MRI), as opposed to static observations of gross structures.

As a student of anatomy and physiology beginning to learn about the gross structures of the nervous system, you will have to trust what others tell you about the *functions* of those structures because these functions cannot be observed at the gross level. However, when you observe nervous tissue through the microscope, it becomes much more interesting and you will begin to understand some of the relationships between structure and function. In particular, as you view various types of neurons, try to associate the parts of the neuron (e.g., soma, axon, dendrite) with what you are learning in the lecture about the functions of the nerve cell membrane in those areas and its ability to generate graded or action potentials.

In this laboratory session you will observe the histological structure of nervous tissues. As you proceed to subsequent chapters that cover the gross structures of the nervous system, it will be important for you to relate what you observe microscopically with what you observe grossly. In addition, you will perform some simulated experiments on nervous tissues using the Ph.I.L.S. program to begin to investigate the functional properties of the neuron cell membrane (neurolemma).

Nervous tissue is characterized by its **excitability.** Recall that excitability is the ability to generate and propagate electrical signals called action potentials. Nervous tissue is composed of two basic cell types. Neurons are the excitable cells. They have a limited ability to divide and multiply in the adult. Neuroglia (glial cells) are supporting cells whose functions are to support and protect neurons. Glial cells retain the ability to divide and multiply, and they constitute more than half of the matter found in neural tissue. In this laboratory session you will observe the histological structure of both types of cells.

Chapter 14: Nervous Tissues

PRE-LABORATORY WORKSHEET

Name:_____
Date:_____ Section:_____

Also available at www.connect.mcgraw-hill.com

1. For each name, write the letter for the structure as labeled in the diagram below.

   ____ axon   ____ neurofibril node (node of Ranvier)
   ____ axon hillock   ____ nucleolus
   ____ dendrite   ____ nucleus
   ____ myelin sheath/neurolemmocyte (Schwann cell)   ____ cell body (soma)
   ____ chromatophilic substance   ____ synaptic knobs

2. Match the cell type in column B with its description in column A.

   Column A

   ____ 1. cells that myelinate axons in the central nervous system
   ____ 2. cells that help reinforce the blood-brain barrier
   ____ 3. glial cells found in peripheral nerve ganglia
   ____ 4. cells that myelinate axons in the peripheral nervous system
   ____ 5. cells that engage in phagocytosis in response to tissue injury
   ____ 6. cells that line the ventricles of the brain

   Column B

   a. astrocytes
   b. ependymal cells
   c. microglia
   d. oligodendrocytes
   e. satellite cells
   f. neurolemmocytes (Schwann cells)

3. Match the connective tissue structure in column B with its description in column A.

   Column A

   ____ 1. surrounds a fascicle of axons
   ____ 2. surrounds an individual axon
   ____ 3. surrounds the entire nerve

   Column B

   a. endoneurium
   b. epineurium
   c. perineurium

Chapter Fourteen *Nervous Tissues*   357

Histology

At the gross level, nervous tissue is classified as either white matter or gray matter. **White matter** consists mainly of myelinated axons. Bundles of white matter in the central nervous system are **tracts,** and bundles of white matter in the peripheral nervous system are **nerves.** White matter is white because of the presence of myelin, a fatty substance produced by glial cells that is used to insulate axons to increase the speed of conducting action potentials along the axon. **Gray matter** consists mainly of neuron cell bodies, dendrites, and unmyelinated axons. Collections of gray matter within the central nervous system are **nuclei,** and collections of gray matter within the peripheral nervous system are **ganglia.** Gray matter is gray because of an absence of myelin. In this laboratory session you will begin by observing the histological appearance of white matter and gray matter and then will proceed to observe specific cell types found within nervous tissues.

EXERCISE 14.1

GRAY AND WHITE MATTER

1. Obtain a slide of a cross section of the spinal cord.

2. Observe the slide with your naked eye before placing it on the microscope stage. Notice that the spinal cord consists of an inner, butterfly-shaped core of **gray matter** surrounded by an outer region of **white matter (figure 14.1).** What neuronal structures do you expect to find within the gray matter? _____

   What neuronal structures to you expect to find within the white matter? _____

3. In the space below, sketch your microscopic observations. Label the following in your drawing:

   ☐ gray matter   ☐ white matter

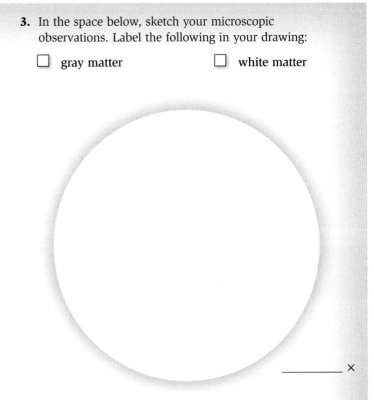

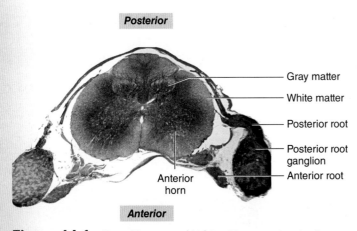

Figure 14.1 Gray Matter and White Matter. A stained, histological cross section through the spinal cord showing gray matter (pink) and white matter (purple).

Neurons

Neurons are typically the largest cells seen in any slide of nervous tissue. They have very large, light-colored nuclei with prominent nucleoli. The **cell body** (soma) of the neuron contains a large amount of rough endoplasmic reticulum that stains darkly and is called **chromatophilic substance** (Nissl bodies). Generally **dendrites** are not easily seen, even at high magnification. On the other hand, one can often identify a single large **axon** leaving the cell body. There will be an absence of chromatophilic substance at the location where the axon leaves the cell body, an area of the neuron called the **axon hillock**. Table 14.1 summarizes the parts of a typical neuron and their functions.

Neurons are classified based on the number of cellular processes that extend from the cell body. **Multipolar** neurons (table 14.2) have many dendrites and a single axon. **Bipolar** neurons have one dendrite and one axon. Finally, **unipolar** neurons have a single process (*uni*, one).

Table 14.1	Parts of a Neuron	
Part of the Neuron	**Description and Function**	**Word Origin**
Axon	A single large process of a neuron that conducts action potentials away from the cell body of a neuron	*axon*, axis
Axon Hillock	A light-staining region where the axon leaves the cell body of a neuron; devoid of chromatophilic substance; the location where action potentials are generated by a neuron	*hillock*, a small elevation
Cell Body (Soma)	The part of the neuron that contains the nucleus and cellular organelles	*soma*, body
Chromatophilic (Nissl) Substance	Dark-staining material found within the soma of a neuron, but absent in the axon hillock region; rough endoplasmic reticulum that functions in the production of membrane-associated proteins	*chroma*, color, + *phileo*, to love
Dendrite	Multiple branching processes of a neuron that transmits graded potentials to the cell body	*dendron*, tree
Neurofibril Node (Node of Ranvier)	A bare region on a myelinated axon where there is an absence of myelin and where action potentials are propagated	*neuron*, nerve, + *fibrilla*, fiber; *Louis Ranvier*, French pathologist
Synaptic Knobs	Swellings on the ends of an axon that form synapses with either another neuron or an effector organ (i.e., glands and muscle fibers)	*syn-*, together, + *hapto*, to clasp, + *knob*, a protuberance
Telodendria (axon terminals)	Branches at the end of an axon, with each process containing a synaptic knob at its end	*telos*, end, + *dendron*, tree

Table 14.2	Multipolar Neurons in the Cerebrum and Cerebellum	
Neuron Type	**Location**	**Features**
Anterior Horn Cell	Anterior horn of the spinal cord	Very large neuron with prominent chromatophilic substance. Cell body is irregularly shaped with multiple dendrites extending from the soma
Pyramidal Cell	Cerebral cortex	Cell body has a triangular shape and multiple dendrites extend from the cell body
Purkinje Cell	Cerebellum	Cell bodies appear "basketlike" with a rounded area facing the granular layer of the cerebellum, and a tuft of dendrites extending into the molecular layer of the cerebellum

EXERCISE 14.2

GENERAL MULTIPOLAR NEURONS—ANTERIOR HORN CELLS

1. Obtain a slide of the *spinal cord in cross section*.
2. Place the slide on the microscope stage and bring the tissue sample into focus on low power **(figure 14.2a)**.

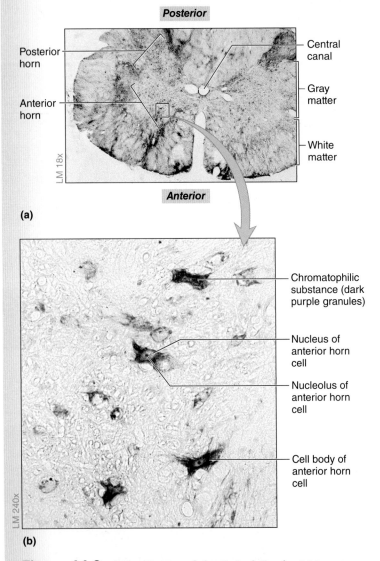

(a)
(b)

Figure 14.2 Gray Matter of the Spinal Cord. (a) Cross section of spinal cord. (b) Close up showing large motor neurons (anterior horn cells).

Again, note the inner core of gray matter surrounded by an outer region of white matter. Move the microscope stage so the inner gray matter is in the center of the field of view and then switch to high power (figure 14.2b). Look for very large, multipolar neurons found in the anterior (ventral) horn of the gray matter. These cells are called **anterior horn cells,** based on their location. They are large somatic motor neurons whose axons exit the spinal cord and travel through peripheral nerves to skeletal muscles in the body. Because their axons are so long, the cell bodies of these neurons are quite large and easy to see.

3. Using tables 14.1 and 14.2 and figure 14.2 as guides, identify the following structures on the spinal cord slide:

   - ☐ cell body of anterior horn cell
   - ☐ chromatophilic substance
   - ☐ gray matter
   - ☐ nucleolus of anterior horn cell
   - ☐ nucleus of anterior horn cell
   - ☐ white matter

4. In the space below, sketch anterior horn cells as seen through the microscope.

EXERCISE 14.3

CEREBRUM—PYRAMIDAL CELLS

1. Obtain a slide of the cerebrum that has been stained with Nissl stain. Nissl stain colors the **rough endoplasmic reticulum** (the chromatophilic substance) of the neurons an intense blue color.
2. Place the slide on the microscope stage and bring the tissue sample into focus on low power. Identify areas of gray matter and white matter on the slide **(figure 14.3)**.
3. Bring an area of gray matter to the center of the field of view and switch to high power. Note the very large cells located within the gray matter. These cells are **neurons.** Note the very large nuclei of the *neurons*, which are surrounded by many cells with much smaller nuclei, the *glial cells*. Locate large neurons whose cell bodies appear triangular in shape. These cells are special neurons of the cerebrum called **pyramidal cells** (figure 14.3, table 14.2).

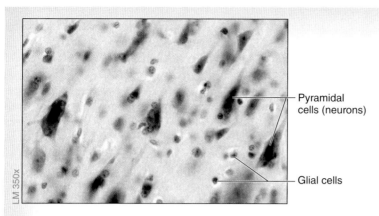

Figure 14.3 Pyramidal Cells in the Cerebral Cortex. The pyramidal cells (neurons) are the large triangular cells. The smaller nuclei visible in the slide are mainly those of glial cells.

4. In the space below, sketch what you observed through the microscope. Label the following in your drawing:

☐ glial cells   ☐ pyramidal cells

_____ ×

EXERCISE 14.4

CEREBELLUM—PURKINJE CELLS

1. Obtain a slide of the **cerebellum**. Place it on the microscope stage and bring the tissue sample into focus on low power **(figure 14.4a)**.

2. Note the folds of tissue with an outer region, the *molecular layer*, and an inner region, the *granular layer* (figure 14.4a). Deep to the folds is the **white matter** of the cerebellum. Move the microscope stage to bring the junction between the molecular and granular layers to the center of the field of view. Then switch to high power (figure 14.4b). Identify the very large cells located at this juncture. These cells are **Purkinje cells** (table 14.2), also known as basket cells, which are large multipolar neurons of the cerebellum (the layer where they are located is the *Purkinje cell layer*).

3. In the space below, sketch what you observed at high power. Label the following in your drawing:

☐ granular layer   ☐ Purkinje cells
☐ molecular layer   ☐ white matter

_____ ×

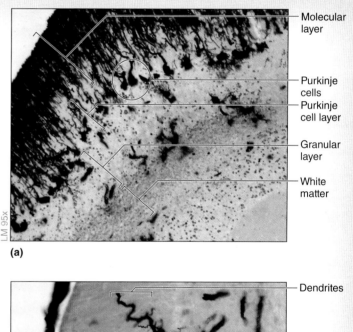

(a)

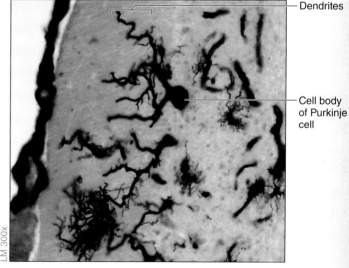

(b)

Figure 14.4 Cerebellum—Purkinje Cells. The cerebellum, demonstrating Purkinje cells between the granular and molecular layers. (a) Low power. (b) High power.

Glial Cells

Glial cells, or **neuroglia** (literally, "nerve glue"), are much more abundant than neurons in the nervous system, and they retain the capacity for cell division. Thus, most brain tumors arise from glial cells and are referred to as *gliomas*. The two most abundant glial cells in the central nervous system are **astrocytes** and **oligodendrocytes.** Because of their large size, astrocytes and oligodendrocytes are collectively referred to as *macroglia*. The central nervous system also contains resident macrophages called **microglia.** As their name implies, microglia are very tiny cells. Their name is a bit of a misnomer, however, because these cells become very large, phagocytic cells when tissue injury occurs. You will not see microglial cells in the laboratory. Special epithelial cells found lining the fluid-filled ventricles of the brain are **ependymal cells.**

The peripheral nervous system contains only one type of glial cell, but it is named differently—and functions differently—depending upon its location. These glial cells have the same embryonic origin, but not the same function, which is why they are given a different name. Thus, we typically refer to the peripheral nervous system as having two glial cell types: *neurolemmocytes (Schwann cells)* and *satellite cells.* **Table 14.3** summarizes the characteristics of each type of glial cell.

Table 14.3 Glial Cells

Cell Name	Location	Description	Function(s)	Word Origin
Astrocytes	CNS	Star-shaped cells that are very abundant in the central nervous system	General supporting cells in the CNS. Transfer nutrients to neurons from the blood. Reinforce the blood-brain barrier. Maintain the extracellular environment around neurons.	*astron*, star, + *kytos*, cell
Oligodendrocytes	CNS	Cells with several long processes that wrap around axons in the CNS	Myelinate axons in the central nervous system. Each oligodendrocyte can myelinate multiple axons. Myelination allows for faster nerve signal conduction.	*oligos*, few, + *dendro-*, like a tree, + *kytos*, cell
Microglia	CNS	Small cells with oval nuclei and multiple branching processes	Microglia are derived from blood monocytes, and they are the resident macrophages in the CNS. They are normally very small (hence the name), but transform into very large, phagocytic cells (macrophages) when tissues are injured.	*mikros*, small, + *glia*, glue
Ependymal Cells	CNS	Cuboidal to columnar-shaped cells with microvilli and cilia on their apical surfaces	Ependymal cells are epithelial cells that line the ventricles (fluid-filled spaces) of the brain and the central canal of the spinal cord. Unlike other epithelia, they DO NOT have a basement membrane. They play a role in the production and circulation of CSF.	*ependyma*, an upper garment
Neurolemmocytes (Schwann Cells)	PNS	Large cells that wrap their plasma membrane around peripheral nerve axons	Myelinate axons in the peripheral nervous system. Each neurolemmocyte can myelinate only part of one axon; typically takes many neurolemmocytes to myelinate an entire axon. Myelination allows for faster nerve signal conduction.	*neuron*, nerve, + *lemma*, husk, + *kytos*, cell; Theodor Schwann, German histologist and physiologist
Satellite Cells	PNS	Small glial cells found surrounding the cell bodies within a ganglion (e.g., posterior root ganglion)	Satellite cells sit right outside the cell bodies of somatic sensory neurons, hence the appearance of "satellites" around those neurons. They provide general support for the neurons and are analogous in function to astrocytes in the CNS.	*satelles*, attendant

Glial Cells of the Central Nervous System

EXERCISE 14.5

ASTROCYTES

1. Obtain a slide of the cerebrum or cerebellum that has been stained with silver stain. Silver stain makes the general supporting cells of the central nervous system, the **astrocytes,** stain very dark so they are visible through the microscope (**figure 14.5**).

2. Place the slide on the microscope stage and bring the tissue sample into focus on low power. Then switch to high power and bring the tissue sample into focus once again.

3. The two most prominent cell types you will see are the large neurons (can you identify pyramidal cells or Purkinje cells?) and the smaller astrocytes. Astrocytes, as their name implies, are shaped like stars (*astron-*, star). They have multiple long cellular processes that wrap themselves around neurons and around blood vessels in the central nervous system. These processes allow them to perform one of their main functions, which is to transport nutrients from the blood to the neurons.

Chapter Fourteen *Nervous Tissues*   363

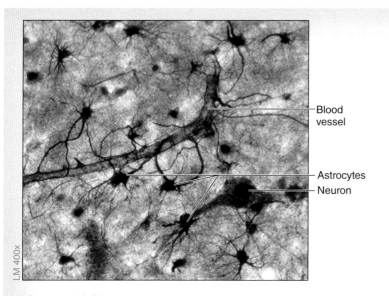

Figure 14.5 Astrocytes. Two astrocytes in the center of the slide have processes that can be seen touching a blood vessel.

4. In the space below, sketch an astrocyte as seen through the microscope.

_____ ×

EXERCISE 14.6

EPENDYMAL CELLS

1. Obtain a slide of the brain and place it on the microscope stage. Bring the tissue sample into focus on low power. Then switch to high power and bring the tissue sample into focus once again. Look for the "empty" spaces on the slide, because those spaces will likely be the ventricular spaces that are lined with ependymal cells.

2. **Ependymal cells (figure 14.6)** are cuboidal to columnar-shaped epithelial cells that line the ventricles of the brain and the central canal of the spinal cord.

They also form the outer layer of the choroid plexus. They have both *cilia* and *microvilli* on their apical surfaces, but they are unique as epithelia in that they do not have a basement membrane.

3. Using figure 14.6 as a guide, locate ependymal cells on the slide.

4. In the space below, sketch ependymal cells as seen through the microscope. Label the following in your drawing:

   ☐ ependymal cells   ☐ ventricular space

Figure 14.6 Ependymal Cells. Cuboidal ependymal cells lining a ventricular surface of the brain.

_____ ×

Glial Cells of the Peripheral Nervous System

EXERCISE 14.7

NEUROLEMMOCYTES (SCHWANN CELLS)

1. Obtain a slide of a longitudinal section of a peripheral nerve and place it on the microscope stage (figure 14.7). A nerve is a bundle of myelinated axons, so you should expect to observe axons, myelin sheaths, and connective tissue on this slide.

2. Bring the tissue sample into focus on low power and note the wavy appearance of the axons. The axons appear wavy because the nerve is not stretched. When a nerve is stretched, the axons also stretch out. The fact that peripheral nerves have elasticity is important because it allows the nerves to stretch during movement without being damaged. This elasticity is imparted to the nerves by the connective tissue coverings that surround the axons.

3. Observe the slide on high power. Note the very dark lines, which are the axons. Each axon is surrounded by light material, which is the **myelin sheath**. Note the elongated nuclei seen throughout the slide. These nuclei belong to **neurolemmocytes (Schwann cells)**, the glial cells responsible for myelinating axons in the peripheral nervous system. Each individual neurolemmocyte myelinates only a portion of a single axon and a single axon typically has hundreds of neurolemmocytes along its length. The bare areas of axon found between myelin sheaths are called **neurofibril nodes**, or **nodes of Ranvier**.

4. Scan the slide to see if you can locate a neurofibril node.

5. In the space below, sketch a peripheral nerve as seen through the microscope. Label the following in your drawing:

   ☐ axon ☐ neurofibril node
   ☐ myelin sheath ☐ neurolemmocyte nucleus

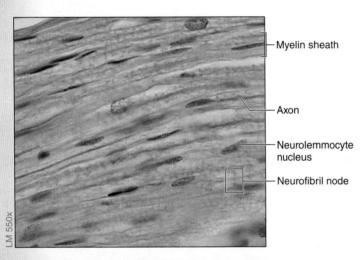

Figure 14.7 Neurofibril Nodes. Longitudinal section of a nerve demonstrating neurofibril nodes.

EXERCISE 14.8

SATELLITE CELLS

1. Obtain a slide of a spinal ganglion (the slide may be labeled dorsal root ganglion, posterior root ganglion, or peripheral nerve ganglion) and place it on the microscope stage.

2. Bring the tissue sample into focus on low power and locate both the nerve *root* and the *ganglion* (figure 14.8a).

3. Recall that the peripheral nervous system is mainly composed of nerves, which are bundles of myelinated axons. Occasionally collections of neuron cell bodies are found along these nerves. A collection of neuron cell bodies in the peripheral nervous system is called a **ganglion** (*ganglion*, swelling). The name comes from the fact that the area where the neuron cell bodies aggregate appears as a "swelling" on the cordlike nerve. The posterior roots of the spinal cord contain swellings called **posterior (dorsal) root ganglia** (peripheral nerve ganglia). These structures contain the neuron cell bodies of somatic sensory neurons, which are classified as unipolar neurons. Unipolar neurons have only a single process extending from the cell body of the neuron. The absence of multiple dendrites coming right off the cell bodies of the neurons allows the glial cells (which are also found in the ganglion) to lie directly adjacent to the cell bodies of the neurons. They look like small "satellites" surrounding the neuron cell body, and are called **satellite cells**. Satellite cells are general supporting cells for neurons within the posterior root ganglia.

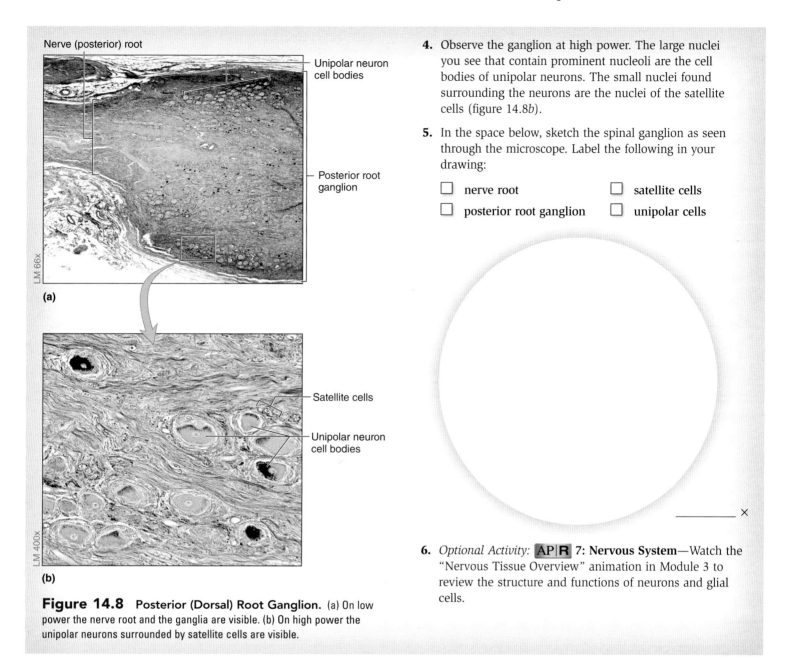

Figure 14.8 Posterior (Dorsal) Root Ganglion. (a) On low power the nerve root and the ganglia are visible. (b) On high power the unipolar neurons surrounded by satellite cells are visible.

4. Observe the ganglion at high power. The large nuclei you see that contain prominent nucleoli are the cell bodies of unipolar neurons. The small nuclei found surrounding the neurons are the nuclei of the satellite cells (figure 14.8b).

5. In the space below, sketch the spinal ganglion as seen through the microscope. Label the following in your drawing:

   ☐ nerve root   ☐ satellite cells
   ☐ posterior root ganglion   ☐ unipolar cells

6. *Optional Activity:* **AP|R** 7: **Nervous System**—Watch the "Nervous Tissue Overview" animation in Module 3 to review the structure and functions of neurons and glial cells.

Peripheral Nerves

Both cranial nerves (nerves extending from the brain) and spinal nerves (nerves extending from the spinal cord) are collections of myelinated axons that are bundled together with connective tissue. The bundling pattern is exactly like the bundling pattern seen with skeletal muscle, and the prefixes used for the names of the connective tissue coverings are also the same. An entire peripheral nerve is bundled by connective tissue called **epineurium**. Within an entire nerve, the axons are grouped into bundles called **fascicles**. Each fascicle is surrounded by a connective tissue covering called the **perineurium**. Within each fascicle are individual axons. Each individual axon is surrounded by a connective tissue covering called the **endoneurium**. Between the fascicles are blood vessels that supply nutrients to the nerve.

366 Chapter Fourteen *Nervous Tissues*

EXERCISE 14.9

COVERINGS OF A PERIPHERAL NERVE

1. Obtain a slide of a cross section of a peripheral nerve and place it on the microscope stage.

2. Bring the tissue sample into focus on low power. Using **figure 14.9** as a guide, identify epineurium, fascicles, and perineurium.

3. Observe the slide at high power so you can see the individual *axons* in cross section. Each axon is surrounded by a myelin sheath, which is then surrounded by a connective tissue covering of endoneurium. In between the fascicles, locate the many small blood vessels.

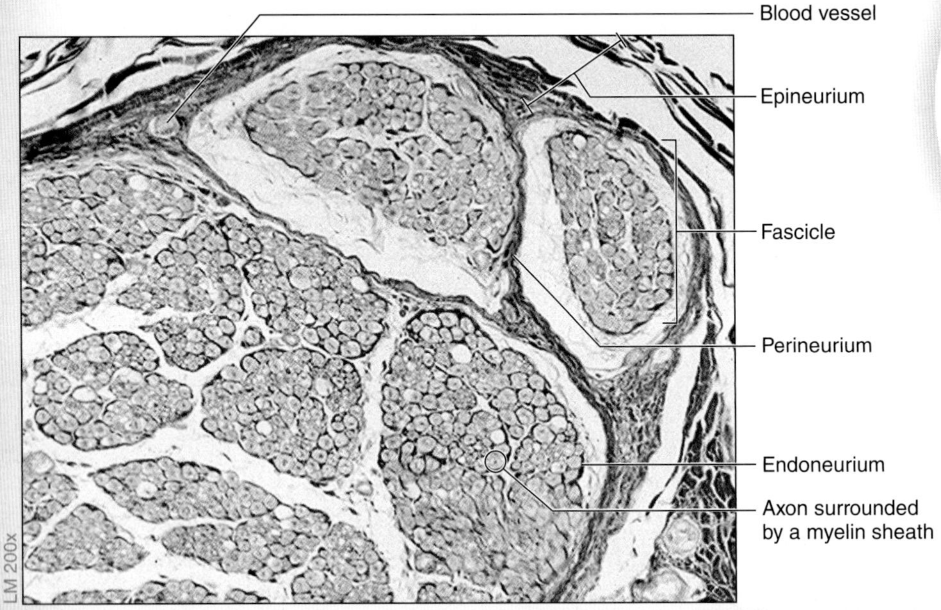

4. In the space below, sketch a cross section of a peripheral nerve as seen through the microscope. Label the following in your drawing:

   ☐ axon
   ☐ blood vessel
   ☐ endoneurium
   ☐ epineurium
   ☐ fascicle
   ☐ myelin sheath
   ☐ perineurium

Figure 14.9 Cross Section through a Peripheral Nerve.

Physiology

Resting Membrane Potential

The plasma membrane of an excitable cell is composed of the same biomolecular structures discussed in chapter 4. That is, the selectively permeable membrane is a phospholipid bilayer, oriented such that the hydrophilic (*hydro-*, water + *-philos*, fond) heads are pointed toward the intracellular and extracellular fluid and the hydrophobic (*hydro-* water + *-phobos*, fear) tails are pointed toward one another within the interior of the membrane. Proteins, including transport pumps, receptors, and ion channels, provide mechanisms to move ions into and out of the cell.

The **resting membrane potential** is very negative, meaning that the inside of the membrane is negative compared to the outside of the membrane. The establishing and maintaining of the resting membrane potential is dependent on K^+ leak channels, Na^+ leak channels, and Na^+/K^+ pumps. The typical resting membrane potential for neurons is −70 millivolts (mV). However, rather than simply memorizing a value of the resting membrane potential, know that this number may fluctuate between −40 mV and −90 mV as the membrane's permeability to particular ions changes. Most significantly, the changes in the concentration of extracellular Na^+ or K^+ have the ability to alter the resting membrane potential. The membrane potential deviates from resting when gated channels specific for ions (e.g., Na^+, K^+) are opened. In the following laboratory exercises you will investigate the changes in resting membrane potential that result from changing the concentration of these ions in the extracellular fluid.

EXERCISE 14.10

Ph.I.L.S. LESSON 8: RESTING POTENTIAL AND EXTERNAL [K⁺]

In this exercise, you will observe the changes in resting membrane potential of crayfish muscle fibers when they are placed in solutions of increasing concentrations of KCl. This simulates the effect of altering the [K⁺] in normal extracellular fluid. This exercise—along with exercise 11 (Resting Potential and External [Na⁺])—emphasizes the contribution of K⁺ and Na⁺ to development of the resting membrane potential (E_m). In addition, you will begin to understand why it is so important physiologically for the body to maintain concentration gradients of K⁺ and Na⁺ between intracellular fluid (ICF) and extracellular fluid (ECF) within narrow limits. Large variations in the concentrations of either of these two ions can have drastic consequences for the functioning of excitable cells.

While the study cells examined in this simulation are crayfish muscle fibers, the concepts apply to any excitable cells, including neurons. In fact, early studies that were performed to understand the function of excitable cells were done on squid axons because they are so large that microelectrodes could be placed in them in much the same way that the simulated microelectrodes are placed in crayfish muscle in this exercise.

Before you begin, familiarize yourself with the following concepts (use your textbook as a reference):

- Function of the Na⁺/K⁺ ATPase
- Normal concentrations of Na⁺ and K⁺ in the ICF and ECF
- Resting membrane potential
- Permeability of the cell membrane to Na⁺ and K⁺ ions at rest
- Relative contributions of Na⁺ and K⁺ to development of the resting membrane potential
- Depolarization, hyperpolarization, and repolarization

Before you begin, state your hypothesis regarding the influence of extracellular [K⁺] on resting membrane potential.

_____

_____

_____

1. Open Ph.I.L.S. Lesson 8: Resting Potential and External [K⁺].
2. Read the objectives and introduction, and take the pre-laboratory quiz.
3. After completing the pre-laboratory quiz, read through the wet lab.
4. The lab exercise will open when you have completed the wet lab.
5. To start the experiment, click the "power" switch on the data acquisition unit and the electrometer (figure 14.10).
6. Click and drag the "blue plug" to input 1 on the data acquisition unit to connect the electrometer.
7. To record the output of the electrometer, click the "start" button on the control panel. A red line appears on the virtual monitor. This red line will represent the voltage measured from the crayfish preparation.
8. Zero the electrometer by clicking the "zero" button near the read-out on the electrometer.
9. The microelectrode is a thin blue line above the crayfish preparation. To insert the electrode into the sample, click the "down control" on the micromanipulator (large black wheel on the bottom left of the screen). Watch the red line as the electrode is inserted into the crayfish muscle fiber. When you see a change in the voltage register on the electrometer, release the down control.
10. The electrode has penetrated the cell membrane and is now measuring the membrane potential. Record that value here: E_m (normal saline) = _____ Then click the "journal" to enter the data into the Ph.I.L.S. journal. You have now completed the control part of the experiment. Subsequent recordings will be your experimental recordings where you vary the [K⁺] in the ECF.
11. To begin the experiment, raise the micromanipulator so the electrode is no longer in the muscle fiber by clicking the "up control" arrow.

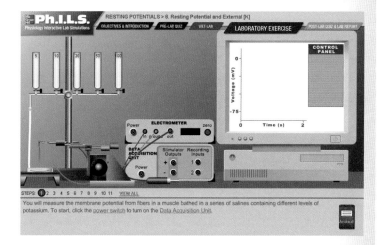

Figure 14.10 Setup of the Wet Lab for Ph.I.L.S. Exercise 8: Resting Potential and External [K⁺]. An isolated muscle fiber has been placed in the apparatus and is hooked up to both stimulating and recording electrodes. Test tubes above the setup are arranged so that you will be able to bathe the muscle in solutions containing varying concentrations of K⁺.

(continued on next page)

(continued from previous page)

12. Perfuse the sample with 5 mEq/L K⁺ solution by clicking the "tap" below the appropriately labeled test tube.

13. Reposition the microelectrode by clicking the "lateral controls" to the left, right, anterior, or posterior. You simply need to find an intact portion of membrane to make your new recording.

14. Click the "zero" on the electrometer.

15. Insert the microelectrode into the crayfish muscle preparation by clicking the "down control" on the micromanipulator. Remember to watch for the red line to move, indicating the microelectrode is recording a change in voltage across the membrane. When the red line changes, release the down control and record the membrane potential in the Ph.I.L.S. journal, and in **table 14.4** below.

16. Repeat step 15 until you have five measurements from different portions of the membrane in the 5 mEq/L K⁺ solution. Record your data in table 14.4.

17. Repeat steps 9 through 16 with the 10, 20, 50, and 100 mEq/L K⁺ solutions.

18. After you have completed all of your measurements, press the "CALC" button in the journal to perform the logarithmic calculations for the data that you will need to graph. Enter the calculated values in **table 14.5**, and then plot your data on the graph below.

19. Make note of any pertinent observations here.

_____
_____
_____

Table 14.4	Raw Data for Ph.I.L.S. Lesson 8: Resting Potential and External [K⁺]				
Tube	5 mEq/L K⁺ E_m	10 mEq/L K⁺ E_m	20 mEq/L K⁺ E_m	50 mEq/L K⁺ E_m	100 mEq/L K⁺ E_m
1					
2					
3					
4					
5					

Table 14.5	Log Data for Ph.I.L.S. Lesson 8: Resting Potential and External [K⁺]	
[K⁺]	Av. E_m	#
5		
10		
20		
50		
100		

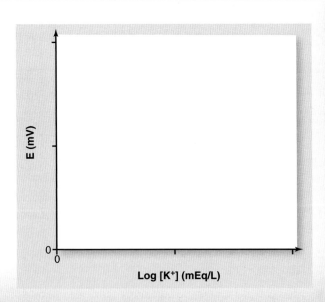

EXERCISE 14.11 Ph.I.L.S.

Ph.I.L.S. LESSON 9: RESTING POTENTIAL AND EXTERNAL [Na⁺]

This exercise is a logical extension of exercise 14.10. In that exercise, you observed the changes in resting membrane potential of crayfish muscle fibers when they were placed in solutions of increasing concentrations of KCl, which simulated the effect of altering the [K⁺] in normal extracellular fluid. In this exercise, you will simulate the effect of altering the [Na⁺] in normal extracellular fluid by placing the crayfish muscle fibers into solutions of increasing concentrations of NaCl. Because of the enormous contribution of K⁺ and Na⁺ in forming the resting membrane potential (RMP) of a cell, this exercise, as well as exercise 14.10, are important first steps for you to take to understand how changes in the concentrations of these ions in the ECF alter membrane potential. This will help you understand why it is so important physiologically for the body to maintain concentrations of K⁺ and Na⁺ within such narrow limits in intracellular fluid (ICF) and extracellular fluid (ECF). Subsequent exercises (exercises 14.12 to 14.14) will allow you to investigate the role these two ions play in causing *changes* in membrane potential: the mechanism neurons and other excitable cells use to create signals such as graded potentials and action potentials. As with exercise 14.10, the study tissue in this simulation consists of crayfish muscle fibers, which are excitable cells.

Before you begin, familiarize yourself with the following concepts (use your textbook as a reference):

- Function of the Na⁺/K⁺ ATPase
- Normal concentrations of Na⁺ and K⁺ in the ICF and ECF
- Resting membrane potential
- Permeability of the cell membrane to Na⁺ and K⁺ ions at rest
- Relative contributions of Na⁺ and K⁺ to development of the resting membrane potential
- Depolarization, hyperpolarization, and repolarization

Before you begin, state your hypothesis regarding the effect of extracellular [Na⁺] on resting membrane potential.

_____

_____

_____

1. Open Ph.I.L.S. Lesson 9: Resting Potential and External [Na⁺].

2. Read the objectives and introduction, and take the pre-laboratory quiz.

3. After completing the pre-laboratory quiz, read through the wet lab.

4. The lab exercise will open when you have completed the wet lab.

5. To start the experiment, click the "power" switch on the data acquisition unit and the electrometer.

6. Click the "power" switch on the electrometer and connect the electrometer to the data acquisition unit by clicking and dragging the "blue plug" to recording input 1.

7. To record the output of the electrometer, click the "start" button on the control panel. A red line appears on the virtual monitor. This red line will represent the voltage measured from the crayfish preparation.

8. Zero the electrometer by clicking the "zero" button near the read-out on the electrometer.

9. The microelectrode is a thin blue line above the crayfish preparation. To insert the electrode into the sample, click the "down control" on the micromanipulator (large black wheel on the bottom left of the screen). Watch the red line as the electrode is inserted into the crayfish muscle fiber. When you see a change in the voltage register on the electrometer, release the down control.

10. The electrode has penetrated the plasma membrane and is now measuring the membrane potential. Record that value here: E_m (normal saline) = _____ Then click the "journal" to enter the data into the Ph.I.L.S. journal. You have now completed the control part of the experiment. Subsequent recordings will be your experimental recordings when you vary the [Na⁺] in the ECF.

11. To begin the experiment, raise the micromanipulator so the electrode is no longer in the muscle fiber by clicking the "up control" arrow.

12. Perfuse the sample with 200 mM NaCl solution by clicking the "tap" below the appropriately labeled test tube.

13. Reposition the microelectrode by clicking the "lateral controls" to the left, right, anterior, or posterior. You simply need to find an intact portion of membrane to make your new recording.

(continued on next page)

(continued from previous page)

Table 14.6	Raw Data for Ph.I.L.S. Lesson 9: Resting Potential and External [Na⁺]				
Tube	10 mM NaCl E_m	20 mM NaCl E_m	50 mM NaCl E_m	100 mM NaCl E_m	200 mM NaCl E_m
1					
2					
3					
4					
5					

14. Click the "zero" on the electrometer.

15. Insert the microelectrode into the crayfish muscle preparation by clicking the "down control" on the micromanipulator. Remember to watch for the red line to move, indicating the microelectrode is recording a change in voltage across the membrane. When the red line changes, release the down control and record the membrane potential in the Ph.I.L.S. journal, and in **table 14.6**.

16. Repeat step 15 until you have five measurements from different portions of the membrane in the 200 mM NaCl solution. Then plot your data on the graph on the right.

17. Repeat steps 9 through 16 with the 100, 50, 20 and 10 mM NaCl solutions.

18. Complete the post-lab worksheet for this exercise.

19. Make note of any pertinent observations here.

_____
_____
_____

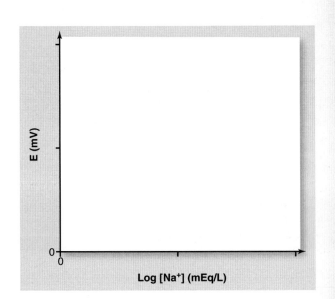

INTEGRATE

LEARNING STRATEGY

In discussions of energy (see chapter 4, p. 67) we discussed two main forms of energy: **potential energy,** which is stored energy (e.g., water held behind a hydroelectric dam) and **kinetic energy,** which is energy of movement (e.g., water released from a hydroelectric dam). As more potential energy is stored, more energy is available to convert to kinetic energy, which can then be used to do work. In the example of a hydroelectric dam, when the water is released from the dam it moves turbines within the dam. Moving the turbines creates electrical energy, which can then be used to provide power to people's homes.

In an excitable cell such as a neuron, the separation of positive and negative charges by the selectively permeable plasma membrane represents a form of potential energy similar to holding water behind a dam. The amount of energy that is *stored* across the membrane is the **membrane *potential*.** The unit of measure for electrical potential energy is a **volt.** Thus, we speak of a resting membrane potential, or **membrane voltage.** As another example, a 9-volt battery stores 9 volts of electrical potential energy. To make that energy useful, we must hook the battery up to wires, which will allow the electrical charges to flow toward each other (e.g., positive moves toward negative and vice versa). This is akin to releasing water from a dam to get energy out of it. In an excitable cell, opening ion channels allows stored charges to move (e.g., Na⁺ flowing into the cell or K⁺ flowing out of the cell). This is a form of *electrical kinetic energy*, and is called a **current** (*currere,* running).

Action Potential Propagation

Although the plasma membranes of all nerve and muscle cells have a negative resting membrane potential (RMP), not all parts of the membrane have the ability to form the same types of electrical signals. The two main types of electric signals that these cells create are graded potentials and action potentials. **Graded potentials** are small, transient changes in membrane potential that become weaker with time and distance along the plasma membrane. **Action potentials** are very large, specific changes in membrane potential that maintain their intensity with time and distance along the plasma membrane. Instead, action potentials are propagated along the membrane until they reach the end of the membrane or a part of the membrane that is not excitable. Any cell that has a membrane capable of generating action potentials is called an **excitable cell.** The two types of excitable cells in the body are nerve and muscle cells.

Depolarization of an excitable membrane beyond a certain threshold level (which is usually around −55 mV for a neuron) will result in generation of an **action potential** on the membrane. The action potential involves a specific, sequential opening and closing of fast, voltage-gated sodium channels and then slow voltage-gated potassium channels. The voltage-gated sodium channels quickly open in response to membrane depolarization and then rapidly close. This rapid closure inactivates the voltage-gated sodium channels and prevents the depolarization of the membrane. In this state, the inactivated sodium channels cannot be opened (activated) again until they reset to the resting state. This resetting process takes time. As a consequence, after an action potential has been generated at the membrane, there is a period of time when the membrane cannot produce another action potential, no matter how great the stimulus that is applied. This period is called the **absolute refractory period.**

The initial membrane depolarization that occurs with an action potential activates voltage-gated potassium channels in addition to the sodium channels. These are slow ion channels that do not open until after the peak of the action potential has been reached. When opened, these channels allow potassium ions to diffuse out of the cell, causing repolarization of the membrane. The driving force for potassium ions to leave the cell results in potassium leaving the cell, which causes the membrane potential to become more negative than −70 mV, or **hyperpolarize** before returning to resting membrane potential once again **(figure 14.11).** If another stimulus is applied to the membrane during this hyperpolarization period, the membrane potential is essentially starting at a potential that is more negative than normal (e.g., starting at −80 mV instead of the usual −70 mV). Thus, the stimulus must be much greater than normal for the membrane potential to reach threshold. Because an action potential can be generated, but it requires a larger stimulus to do so, this time period is called the **relative refractory period.**

In many nerve cells, action potentials must travel long distances along the axons to reach the target organs. For example, motor neurons that send signals to the muscles in the big toe have cell bodies located in the lumbar region of the spinal cord. The axons of those motor neurons start in the spinal cord and travel along the lower limb nerves for several meters before reaching ("innervating") the muscles in the foot. For action potential propagation to be effective over long distances, **myelination** of the axon by glial cells is required. Myelin forms an insulating sheath around the axon, which allows the action potential to travel much faster than it would along a "bare" axon. It does so by preventing leakage of ions. The speed at which an action potential is propagated along an axon is the **conduction velocity.** When comparing "bare" to myelinated axons, the conduction velocity achieved by the myelinated axon can be up to 200 times faster than that of the "bare" axon—a huge difference!

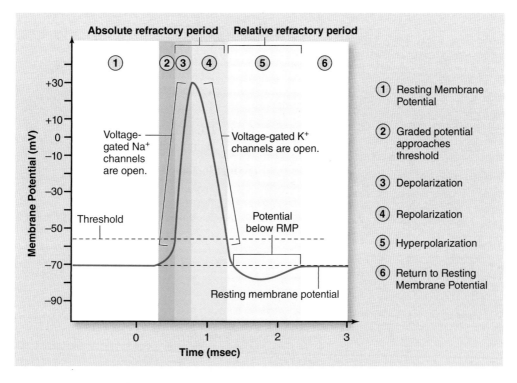

Figure 14.11 Absolute and Relative Refractory Periods and the Ionic Events Associated with Each.

EXERCISE 14.12 Ph.I.L.S.

Ph.I.L.S. LESSON 10: THE COMPOUND ACTION POTENTIAL

The structure of a nerve is such that there are multiple axons contained within a bundle of connective tissue. An external stimulus (voltage) that allows the membrane of an axon to reach threshold will initiate an action potential. Increasing the stimulus intensity will recruit more neurons, thereby producing compound action potentials. Compound action potentials can be recorded at the postsynaptic membrane. The purpose of this exercise is to vary the intensity of external stimuli on an excised frog sciatic nerve to demonstrate both threshold and recruitment.

Before you begin, familiarize yourself with the following concepts (use your main textbook as a reference):

- The role of sodium (Na^+) and potassium (K^+) in generating and maintaining membrane potential
- How the concept of threshold applies to the generation of an action potential
- The functionality of voltage-gated channels
- The definition of depolarization and hyperpolarization in terms of membrane potential

State a hypothesis regarding the effect of stimulus intensity on threshold and recruitment.

_____

_____

_____

1. Open Ph.I.L.S. Lesson 10: Action Potentials: The Compound Action Potential **(figure 14.12)**.

2. Read the objectives and introduction, take the pre-laboratory quiz, and read through the wet lab. The laboratory exercise will open when you have completed the wet lab.

3. To start the experiment, click the "power switch" to turn on the data acquisition unit (DAQ).

4. Click and drag the "recording cables" to the matching colored post on the nerve chamber.

5. Click and hold the "tap" until all saline is drained from the chamber.

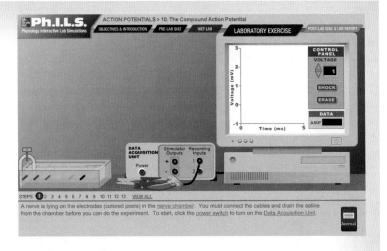

Figure 14.12 Setup of the Wet Lab for Ph.I.L.S. Lesson 10: The Compound Action Potential. An isolated nerve has been placed in the apparatus and is hooked up to both stimulating and recording electrodes.

6. The default setting for the voltage is 1.0 volts. Click the "shock" button to apply a stimulus to the nerve.

7. Measure the voltage of the compound action potential by positioning the crosshairs (using the mouse) at the top of the wave on the graph. Click on the wave and a black arrow will appear. If it is not in the correct location, reposition the arrow by clicking and dragging it to a new position. Now position the crosshairs at the bottom of the deflection (0 volts) and click. The result will display in the (yellow) data panel on the data acquisition unit.

8. Click the "journal" panel (red rectangle at the bottom right of screen) to enter your value into the journal. A table with volts and amplitude (amp) and a graph will appear. Note the range in values for volts from 0 to 1.6 volts.

9. Close the journal window by clicking on the "X" in the right-hand corner.

10. To complete the table and produce the graph, repeat steps 8 to 10 for the entire range of voltage (0–1.6 mV). Record your data in **table 14.7**.

Table 14.7	Membrane Potentials with Applied External Stimuli
Volts	Amplitude
0.0	
0.1	
0.2	
0.3	
0.4	
0.5	
0.6	
0.7	
0.8	
0.9	
1.0	
1.1	
1.2	
1.3	
1.4	
1.5	
1.6	

11. After finishing the laboratory exercise, print the line tracing by clicking on the "P" in the bottom left monitor screen.

12. Complete the post-laboratory quiz (click open post-laboratory quiz and laboratory report) by answering the ten questions on the computer screen.

13. Construct a graph of a compound action potential in the space below. Be sure to indicate where voltage-gated sodium channels and voltage-gated potassium channels are opened and closed.

14. Make note of any pertinent observations here.

_____
_____
_____

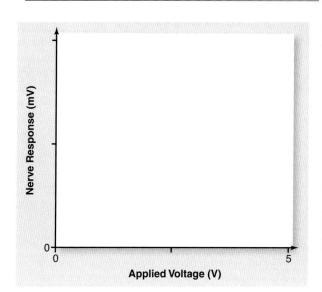

INTEGRATE

CONCEPT CONNECTION

The membrane potential of an excitable cell is regulated by active and passive **membrane transport** mechanisms. Active pumps move Na^+ out of the cell and K^+ into the cell. The movements of Na^+ and K^+ in this case are caused by active transport because the pump is moving the ions *against* their respective concentration gradients in a ratio of 3 Na^+ out of the cell for every 2 K^+ into the cell. This leads to a separation of charge across the plasma membrane due to the membrane's selective permeability. The movement of Na^+ or K^+ ions into or out of the cell is driven by passive transport mechanisms. At rest, there are leak channels for both Na^+ and K^+, although there are many more leak channels for K^+ than for Na^+. Therefore, the cell is more permeable to K^+ than Na^+ at rest. These ions pass freely through their respective leak channels along their individual electrochemical gradients (e.g., K^+ moves out of the cell, from an area of high concentration to an area of low concentration). There is also an opposing electrical gradient due to the negatively charged phosphate and proteins on the inside of the cell attempting to attract the K^+ ions back into the cell. Eventually, there is no net movement of K^+, as the electrical and chemical gradients are equal in magnitude and opposite in direction. In addition to leak channels, ions may pass through either chemically gated or voltage-gated channels. Opening many gates simultaneously creates a large permeability change. Again, passive transport mechanisms allow for these ions to move along their respective electrochemical gradients through these channels. Rather than memorizing the movement of ions into or out of the plasma membrane, simply refer to the governing principles of membrane transport for guidance.

EXERCISE 14.13 Ph.I.L.S.

Ph.I.L.S. LESSON 11: CONDUCTION VELOCITY AND TEMPERATURE

In this experiment, you will measure the conduction velocity of action potentials as they propagated along myelinated axons in the frog sciatic nerve. The nerve will be excised from the frog hind limb and placed in a bath containing an isotonic (Ringer's) solution. Electrical shocks will be applied to one end of the excised nerve, which will generate action potentials in the nerve. The currents associated with the action potentials that are generated will be recorded at the other end of the nerve. Subsequently, the stimulating and recording electrodes will be moved closer to each other, another electrical shock will be applied, and the currents will be measured once again. The conduction velocity of the nerve will then be calculated by comparing the stimulus-response times for the two lengths of nerve. This technique will then be used to study the effect of decreasing temperature on the conduction velocity of action potentials within the nerve.

Before you begin, familiarize yourself with the following concepts (use your textbook as a reference):

- Graded potentials
- Action potentials and excitable cells
- The effect of temperature on membrane transport
- The definition of depolarization and hyperpolarization in terms of membrane potential

Before you begin, state your hypothesis regarding the effect of temperature on action potential propagation.

_____
_____
_____

1. Open Ph.I.L.S. Lesson 11: Conduction Velocity and Temperature.

2. When you observe the opening screen **(figure 14.13)** you will see the sciatic nerve has already been removed from the frog and placed in the nerve chamber. To begin, click "power" on the data acquisition unit.

3. Activate the temperature probe by clicking the "power" button.

4. Click and drag the "red clip" to the red post on the nerve chamber. Then, click and drag the "blue clip" to the blue post, and the "green clip" to the green post (these clips are connected to the recording cables and ground cables).

5. Two additional clips, orange and black, will now appear. These are the stimulating cables. Click and drag the "orange clip" to the orange post and the "black clip" to the black post.

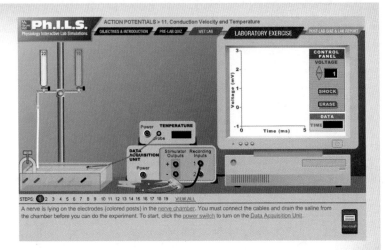

Figure 14.13 Setup of the Wet Lab for Ph.I.L.S. Lesson 11: Conduction Velocity and Temperature. An isolated nerve has been placed in the apparatus and is hooked up to both stimulating and recording electrodes.

6. Before the stimulus can be applied, the saline must be drained from the nerve chamber to prevent the current from passing through the saline instead of the neuron. To drain the saline from the nerve chamber, click the "tap" and hold it until all the saline has been removed.

7. The initial shock or stimulus voltage is set at the default of 1.0 volt. Click the "shock" button to apply a single stimulus to the nerve. The response from the nerve is displayed on the control panel screen (the red line trace). This response is a compound action potential (CAP).

8. Next, you will change the distance the stimulus will travel. Disconnect the black and orange clips by clicking on each of them. Then, reconnect the orange clip to the gray post and connect the black clip to the orange post. This moves the stimulating and recording cables 10 mm closer to each other.

9. Click the "shock" button to produce a compound action potential (at 1.0 volt) to the nerve.

10. There will now be two responses recorded on the display screen. The one on the right is the original CAP, and the one on the left is the CAP that was recorded after you moved the stimulating and recording cables closer (by 10 mm). You will now measure the difference in time between the two trials. Drag the mouse over the tracings and you will see two crosshairs appear. First click the highest point on the CAP on the left, then click the highest point on the CAP on the right. Press the "journal" icon to have the program calculate the conduction velocity (conduction velocity = distance/time). Record your data in **table 14.8.**

11. Moisten the nerve with 22°C Ringer's solution by clicking and holding the tap. Then repeat steps 5 through 10.

12. Moisten the nerve with 10°C Ringer's solution by clicking and holding the tap. Then repeat steps 5 through 10.

13. Make note of any pertinent observations here.

_____

_____

_____

Table 14.8 Raw Data for Ph.I.L.S. Lesson 11: Conduction Velocity and Temperature			
Temperature	Distance (mm)	Time (ms)	Conduction Velocity (Distance/Time)
22°C			
22°C			
10°C			
10°C			

EXERCISE 14.14

Ph.I.L.S. LESSON 12: REFRACTORY PERIODS

In this experiment, you will observe the response of myelinated axons to successive stimuli in a frog sciatic nerve. The nerve will be excised from the frog hind limb and placed in a bath containing an isotonic (Ringer's) solution. Electrical shocks will be applied to one part of the excised nerve, which will generate an action potential in the nerve. Subsequently, you will apply a series of two stimuli to the same region of the neuron, and will record the amplitude of the second action potential. You will be looking for changes in the peak amplitude of the second action potential to observe both absolute and relative refractory periods in the nerve.

Before you begin, familiarize yourself with the following concepts (use your textbook as a reference):

- The difference between graded potentials and action potentials
- The meaning of a threshold potential as it applies to excitable cells
- The function of voltage-gated Na$^+$ and K$^+$ channels in the generation of an action potential
- Absolute refractory period of a nerve
- Relative refractory period of a nerve

Before you begin, state your hypothesis regarding the influence of refractory period on action potential propagation:

_____

_____

_____

1. Open Ph.I.L.S. Lesson 12: Refractory Periods.

2. When you observe the opening screen (similar to that in figure 14.13) you will see that the sciatic nerve has already been removed from the frog and placed in the nerve chamber. To begin, click "power" on the data acquisition unit.

3. Click and drag the "red clip" to the red post on the nerve chamber and release the mouse button to connect the clip. Repeat the procedure for the blue and green clips.

4. Two additional clips, orange and black, will now appear. These are the stimulating cables. Click and drag the "orange clip" to the orange post and the "black clip" to the black post.

5. Before the stimulus can be applied, the saline must be drained from the nerve chamber to prevent the current from passing through the saline instead of the neurons. To drain the saline from the nerve chamber, click the "tap" and hold it until all the saline has been removed.

6. The voltage is set at the default of 0.7 V. Click the "shock" button on the control panel to see the nerve response. The recording will include a graded potential (larger peak) and a stimulus artifact (smaller peak).

7. To determine the minimal stimulus required to elicit a compound action potential, increase the voltage by clicking the "up arrow" on the control panel until the voltage reaches 0.8 V. Click the "shock" button and observe the change in the amplitude of the recording.

8. Repeat step 7 until you determine the *minimum* stimulus required to elicit a maximal peak height of the compound action potential. Record the amplitude of this peak on the line at the top of **table 14.9.**

9. To determine the refractory period of the nerve, you will be delivering two stimuli (shocks) at different time intervals. To begin, set the time interval to 1 millisecond by clicking the "down arrow."

10. Click "shock" to deliver the two stimuli with a 1 millisecond interval between.

(continued on next page)

Table 14.9 — Raw Data for Ph.I.L.S. Lesson 12: Refractory Periods

Maximal Peak Amplitude: _____

Interval	Amplitude
1.0	
1.5	
2.0	
2.5	
3.0	
3.5	
4.0	
4.5	
5.0	
5.5	
6.0	
6.5	
7.0	
7.5	
8.0	
8.5	
9.0	

11. Two compound action potentials will be produced. Drag the cursor over the top of the tallest part of the first peak and click the mouse button to drop a marker at the peak. Drag the cursor over the tallest part of the second peak and click the mouse button to drop a marker at that peak as well. The difference in amplitude between the two peaks will be displayed in the data window. Click the "journal" button in the lower right-hand corner to enter the data in the journal. Then enter the data into table 14.9.

12. Click the "up arrow" in the time interval box to set the interval between stimuli to 1.5 milliseconds. Click "shock" and repeat the procedure in step 11 to measure and record the difference in peak amplitudes.

13. Repeat step 12 until you reach the maximum time interval of 9 milliseconds.

14. Complete the post-laboratory worksheet for this exercise.

15. Graph the data from table 14.8 on the figure below.

16. Make note of any pertinent observations here.

_____
_____
_____

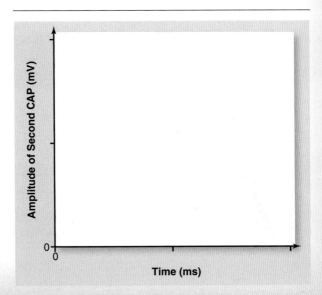

INTEGRATE

CLINICAL VIEW
Potassium Channel Blockers

Demyelinating diseases, such as multiple sclerosis, result in the loss of myelin that surrounds axons. Myelin insulates axons, thereby preventing the leakage of ions. Losing myelin leads to slower nerve conduction velocities, as more positive charge (K^+) can "leak" out of the cell along the length of the axon. *Dalfampridine*, a drug that blocks voltage-gated potassium channels, is now FDA approved for patients suffering from multiple sclerosis (MS). The drug has been shown to increase nerve conduction velocities due to its mechanism of prolonging the action potential. That is, blocking voltage-gated potassium channels blocks the rapid repolarization phase of the action potential, thereby prolonging its duration. This blockage keeps the cell depolarized for longer, thereby increasing the conduction velocity. MS patients report an improvement in sensory and motor function while taking dalfampridine. Incidentally, potassium channel blockers are also used to treat arrhythmias in the heart. In this application, the drug targets the rapid repolarization phase of cardiac muscle cells rather than neurons. However, the result is the same: The excitable cell remains depolarized for longer. In cardiac muscle cells, prolonging the repolarization phase of the action potential allows for a slower heart rate, therefore allowing the heart more time to fill with blood. Regulation of heart rate and cardiac output will be explored further in chapter 21.

Chapter 14: Nervous Tissues

Name:_____
Date:_____ Section:_____

POST-LABORATORY WORKSHEET

The ❶ corresponds to the Learning Objective(s) listed in the chapter opener outline.

Do You Know the Basics?

Exercise 14.1: Gray and White Matter

1. a. A collection of neuron cell bodies in the CNS is called a(n) _____. ❶

   b. A collection of neuron cell bodies in the PNS is called a(n) _____.

   c. A bundle of myelinated axons in the CNS is called a(n) _____.

   d. A bundle of myelinated axons in the PNS is called a(n) _____.

2. Which of the following parts of a neuron contain excitable membranes? Check all that apply. ❷

   ☐ dendrites   ☐ soma   ☐ axon hillock   ☐ axon

3. The branches at the end of an axon are called _____, and they contain swellings called _____. ❷

4. a. What is chromatophilic substance? ❷

   _____

   b. Where is chromatophilic substance found?

   _____

   c. The part of a neuron distinguished by its notable *absence* of chromatophilic substance is called the _____.

Exercise 14.2: General Multipolar Neurons—Anterior Horn Cells

5. A multipolar neuron has many _____ and a single _____. The large multipolar neurons located in the anterior horn of the spinal cord are called _____. ❸

Exercise 14.3: Cerebrum—Pyramidal Cells

6. a. Large multipolar neurons located in the cerebral cortex are called _____. ❹

   b. Briefly describe the appearance of these cells.

   _____
   _____

Exercise 14.4: Cerebellum—Purkinje Cells

7. a. Large multipolar neurons located in the cerebellum are called _____. ❺

   b. Briefly describe the appearance of these cells.

   _____
   _____

Exercise 14.5: Astrocytes

8. List three functions of astrocytes. ❻

   a. _____

   b. _____

   c. _____

Exercise 14.6: Ependymal Cells

9. Ependymal cells are unique as epithelia in that they do not have a _____. ❼

Exercise 14.7: Neurolemmocytes (Schwann Cells)

10. Glial cells in the CNS derived from circulating monocytes are _____. ❽

11. The counterpart of a neurolemmocyte (Schwann cell) in the CNS is a(n) _____. ❽

378 Chapter Fourteen *Nervous Tissues*

12. a. What is a neurofibril node (node of Ranvier)? ⑧

   b. What is the function of a neurofibril node?

Exercise 14.8: Satellite Cells

13. Satellite cells are glial cells that surround the cell bodies of _____ neurons. These are located within a structure called

   the _____, which lies just outside and lateral to the spinal cord. ⑨

Exercise 14.9: Coverings of a Peripheral Nerve

14. A fascicle of axons is surrounded by a connective tissue sheath called the _____. ⑩

Exercise 14.10: Ph.I.L.S. Lesson 8: Resting Potential and External [K⁺]

15. In a typical excitable cell, the concentration of K⁺ in the ECF is (circle one) <u>greater than/less than</u> the concentration of K⁺ in the ICF. ⑪

16. Explain why you had to move the electrode to a different section of the membrane for each of the measurements. ⑪ _____

17. Using your knowledge of electrochemical gradients, describe the mechanism that explains why the resting membrane potential changes as the concentration of potassium in the ECF increased. ⑪ _____

Exercise 14.11: Ph.I.L.S. Lesson 9: Resting Potential and External [Na⁺]

18. In a typical excitable cell, the concentration of Na⁺ in the ECF is (circle one) <u>greater than/less than</u> the concentration of Na⁺ in the ICF. ⑫

19. Describe whether sodium or potassium has a greater effect on resting membrane potential. ⑪ ⑫ _____

20. Using your knowledge of electrochemical gradients, describe the mechanism that explains why resting membrane potential changed as the concentration of sodium in the ECF increased. ⑫ _____

Exercise 14.12: Ph.I.L.S. Lesson 10: The Compound Action Potential

21. Define compound action potential. ⑬ ⑭ _____

22. Explain the relationship between increasing voltage stimulation and the recruitment of nerve fibers. ⑬ ⑭ _____

Exercise 14.13: Ph.I.L.S. Lesson 11: Conduction Velocity and Temperature

23. Which of the following type of signals has the ability to travel a great distance without losing its strength? ⑬ (circle one) <u>action potential/graded potential</u>

24. One way to increase conduction velocity of action potentials along an axon (aside from changing the temperature) is to _____ the axon. ⑬

25. How did the distance between the stimulating electrodes affect conduction velocity in the frog sciatic nerve? ⑬ _____

26. Describe how conduction velocity was impacted when temperature was lowered from 22°C to 10°C. In your explanation, refer to the effect of temperature on membrane transport mechanisms. ⑬ _____

27. What do you predict would happen to conduction velocity if the temperature were raised to 37°C? ⑬ _____
_____

Exercise 14.14: Ph.I.L.S. Lesson 12: Refractory Periods

28. Explain the roles of Na⁺ and K⁺ in action potential generation. _____ ⑭
_____
_____

29. The _____ refractory period is associated with voltage-gated potassium channels. It occurs during the _____ period of the action potential. During this refractory period, another action potential (circle one) <u>can/cannot</u> be generated, no matter how strong the stimulus. ⑭

30. The _____ refractory period is associated with voltage-gated sodium channels. It occurs during the _____ period of the action potential. During this refractory period, another action potential (circle one) <u>can/cannot</u> be generated, no matter how strong the stimulus. ⑭

31. Describe what happened to the amplitude of the two peaks when the time interval between action potentials increased. ⑭ _____
_____

Can You Apply What You've Learned?

32. Describe the structure, function, and specific location of satellite cells.
_____
_____
_____

33. Describe the structure and function of microglia.
_____
_____
_____

34. Give the location where each of the neuron types listed below is found:
    a. Purkinje cell _____
    b. anterior horn cell _____
    c. pyramidal cell _____

35. Compare and contrast the structure, location, and function of neurolemmocytes and oligodendrocytes:
_____
_____
_____
_____
_____

36. Is it more common for brain tumors to arise from neurons or from glial cells? _____
    Explain your answer:
_____
_____
_____

Can You Synthesize What You've Learned?

37. A patient with an *astrocytoma* (*-oma,* tumor) has a tumor that is derived from _____. This tumor would be located in what part of the nervous system? _____

Chapter Fourteen Nervous Tissues

38. Multiple sclerosis (MS) is a disease in which the immune system attacks myelin sheaths. Often a patient suffering from MS will experience a bout of illness, which involves muscle weakness and sensory alterations, followed by a period of recovery, in which normal function partially or completely returns. Using your knowledge of glial cells, explain the biological basis for the partial or complete recovery of nerve function in an MS patient.

39. Which type of neural tissue is more involved with the integration of information (rather than the transmission or conduction of information)—white matter or gray matter? Explain.

40. When a peripheral nerve is compressed, blood vessels that run between fascicles of axons are also compressed. What might be a consequence of this?

41. In the United States, some states have legalized capital punishment (the death penalty). If this is done via a "lethal injection," the substance that is injected is a very high dose of potassium chloride (KCl), which will stop the heart from beating. This is due to the effect of the change in extracellular $[K^+]$ on the resting membrane potential of the cardiac muscle cells, which are excitable cells. Explain how a high dose of KCl will stop the beating of the heart.

42. Clinically, many medications are administered intravenously. These medications are often dissolved in a solution of NaCl. Explain why medications can safely be delivered in a NaCl solution, but would be deadly if they were administered in a KCl solution.

43. Novocaine is used by dentists as an anaesthetic so that a tooth can be extracted or a cavity filled. Novocaine works to block action potentials along sensory neurons typically responsible for the perception of pain. Propose a possible mechanism for this drug.

44. An ice fisherman has fallen through the ice and is under water for 15 minutes. He is rescued and is slowly warmed back up to normal body temperature. Upon recovery, he is found to have little to no brain damage. Using what you learned in this exercise, explain how the cold water may have helped protect his brain function.

45. If a muscle had a very long refractory period, what consequence do you think this would have on the ability of the muscle to achieve a tetanic contraction?

CHAPTER 15

The Brain and Cranial Nerves

OUTLINE AND LEARNING OBJECTIVES

Gross Anatomy 384

The Meninges 384

EXERCISE 15.1: CRANIAL MENINGES 384
1. Identify dura mater, arachnoid mater, pia mater, cranial dural septa, and dural venous sinuses
2. Describe the pathway fluid takes as it flows through the dural venous sinuses
3. Trace cerebrospinal fluid (CSF) from where it is formed within the ventricles until it enters a dural venous sinus

Ventricles of the Brain 388

EXERCISE 15.2: BRAIN VENTRICLES 389
4. Identify the components of the brain ventricular system on a model of the brain ventricles
5. Correlate each ventricular space with the secondary brain vesicle from which it developed
6. Trace the flow of cerebrospinal fluid (CSF) from its site of production in the choroid plexus to its site of drainage in the arachnoid villi

The Human Brain 390

EXERCISE 15.3: SUPERIOR VIEW OF THE HUMAN BRAIN 394
7. Identify structures visible in a superior view of the human brain and describe their functions

EXERCISE 15.4: LATERAL VIEW OF THE HUMAN BRAIN 395
8. Identify structures visible in a lateral view of the human brain and describe their functions

EXERCISE 15.5: INFERIOR VIEW OF THE HUMAN BRAIN 396
9. Identify structures visible in an inferior view of the human brain and describe their functions

EXERCISE 15.6: MIDSAGITTAL VIEW OF THE HUMAN BRAIN 397
10. Identify structures visible in a midsagittal view of the human brain and describe their functions

Cranial Nerves 398

EXERCISE 15.7: IDENTIFICATION OF CRANIAL NERVES ON A BRAIN OR BRAINSTEM MODEL 398
11. List the names, numbers, and general functions (motor, sensory, or mixed) of each of the cranial nerves
12. Identify the cranial nerves on a human brain or model of a human brain

The Sheep Brain 401

EXERCISE 15.8: SHEEP BRAIN DISSECTION 401
13. Associate the structures of a sheep brain with those of a human brain, noting structures that differ greatly from each other in size and function

Physiology 408

Testing Cranial Nerve Functions 408

EXERCISE 15.9: TESTING SPECIFIC FUNCTIONS OF THE CRANIAL NERVES 410
14. Perform common tests of cranial nerve function

MODULE 7: NERVOUS SYSTEM

INTRODUCTION

Recall a time when you were unable to figure out a word problem in your math class. At the time you or someone else may have said, "Come on, use your gray matter!"—meaning "Use your brain! Think!" Having completed the exercises in chapter 14, you now know that your brain consists not only of gray matter (neuron dendrites and cell bodies, but also of white matter (fiber tracts).

To help you make sense of the structures you will view in this laboratory session, the tables in this chapter list the names of the structures along with brief summaries of associated functions and information about the derivation of the names of the structures. Keep in mind that brain structures were named well before their functions were known. Most of the names do not relate to function; instead, they derive from the general appearance of the structure. For instance, "mammillary bodies" have little or nothing to do with the female reproductive system, but to early anatomists they appeared to be "little breasts" on the inferior surface of the brain. The area known as the thalamus apparently looked like a ➡

"bed" or "bedroom." As you will see, the thalamus has very little, if anything, to do with the bedroom or sleep.

As you learn the structures of the brain, try to imagine how and why each structure was given its name. This will make it easier for you to recall the names of structures. In addition, pay attention to the function(s) of each structure you identify. Knowledge of function is even more important than knowledge of structure, because when an area of the brain is damaged, you will expect to see deficiencies directly related to the functions of that area of the brain. The same holds true for the cranial nerves, which transmit information between the brain and structures of the head, neck, and thoracic and abdominal viscera.

In this laboratory session you will first examine the protective coverings of the brain: the meninges. You will then observe the ventricular system of the human brain and relate the development of ventricular structures to the development of the brain. These exercises will be followed with the study of preserved human brains or models of the human brain, and dissection of a representative mammalian brain: the sheep brain.

You will then begin your study of the *cranial nerves*, which are nerves that arise from the brain (as contrasted with *spinal nerves*, which arise from the spinal cord). The cranial nerve exercises in this laboratory session are organized so that if you complete them in the proper sequence, your knowledge of the cranial nerves will become more detailed as you proceed. The first exercises will help you establish a base knowledge about the locations and names of the nerves. Subsequent exercises will build on that base knowledge and help you learn the detailed functions, locations, and common disorders of the cranial nerves.

Chapter 15: The Brain and Cranial Nerves

PRE-LABORATORY WORKSHEET

Name: _____
Date: _____ Section: _____

Also available at www.connect.mcgraw-hill.com

1. a. Define *gyrus*:

   _____

   _____

   b. Define *sulcus*:

   _____

   _____

2. The three meninges that cover the brain and spinal cord (from outermost to innermost) are the

   a. _____

   b. _____

   c. _____

3. The ventricles of the brain that are located within the cerebral hemispheres are the _____.

4. Dural venous sinuses carry _____.

5. The _____ separates the frontal lobe from the parietal lobe of the brain.

6. The _____ lobe of the brain is involved with the sense of vision.

7. The _____ separates the cerebral hemispheres of the brain from the cerebellar hemispheres of the brain.

8. The three parts of the brainstem are the _____, _____, and _____.

9. A cranial nerve is a nerve that originates from the _____.

10. Somatic motor neurons innervate _____ muscle tissue. Visceral motor neurons innervate _____ or _____ muscle tissue, or _____.

11. An example of a special sensory organ is _____.

Gross Anatomy

The Meninges

The **meninges** are connective tissue coverings of the brain and spinal cord and consist of the dura mater, arachnoid mater, and pia mater. These coverings perform many functions and their structure varies slightly depending on whether they are covering the brain or the spinal cord. In **exercise 15.1** we will consider meningeal structures as they apply to the brain, and in chapter 16 we will consider meningeal structures as they apply to the spinal cord.

EXERCISE 15.1

CRANIAL MENINGES

1. Obtain a model of the dura mater of the brain or a cadaveric specimen of the head with intact dural structures.

2. Observe the dura mater **(figure 15.1)**. The **dura mater** (*dura*, hard, + *mater*, mother) is composed of two layers: The outer **periosteal layer** (which is simply the periosteum that lines the internal portion of the cranial bones), and the inner **meningeal layer.** The meningeal layer folds inward to form cranial dural septae and dural venous sinuses (figure 15.1b). **Cranial dural septae** are partitions that stabilize and support the brain. **Dural venous sinuses** are modified veins that transport venous blood and cerebrospinal fluid (CSF) from the brain to the internal jugular vein. The largest dural venous sinus is the **superior sagittal sinus**. Eventually all sinuses drain into the **sigmoid** (*sigmoid*, S-shaped) **sinus,** which drains into the **internal jugular vein.** Table 15.1 includes a list of dural venous sinuses and describes their general locations.

3. If you have a cadaveric specimen available, locate the superior sagittal sinus and note the small granular structures located in and around the sinus. These structures are called **arachnoid villi (granulations)** and serve as sites for absorption of cerebrospinal fluid from the subarachnoid space into the dural venous sinuses.

4. Obtain a human brain with the arachnoid mater and pia mater intact. If you do not have one available, this activity will be completed when you perform the sheep brain dissection. On the human brain, note the thin, transparent covering that lies over the surface of the brain and does not follow the sulci (shallow grooves). This is the **arachnoid mater** (*arachnoid*, shaped like a spiderweb). Deep to the arachnoid mater is the finest, thinnest meningeal layer, the **pia mater** (*pia*, soft or tender). The pia mater is in direct contact with the neural tissue of the brain and follows all of the contours (gyri and sulci) on its surface. The space between the arachnoid mater and the pia mater is the **subarachnoid space.** What fluid is normally found in the subarachnoid space? _____

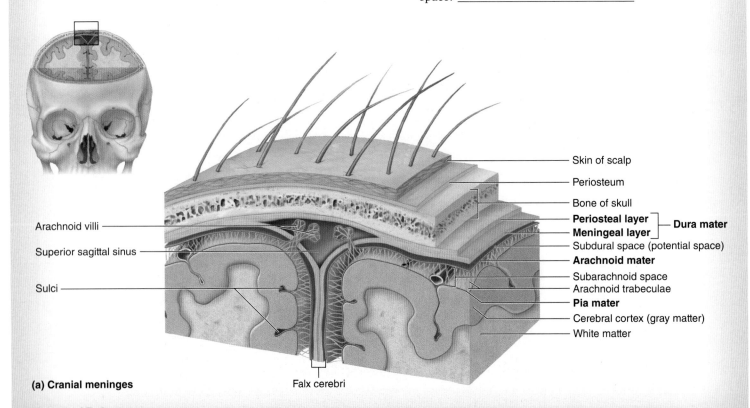

(a) **Cranial meninges**

Figure 15.1 The Meninges. (a) Coronal section through the superior sagittal sinus. (b) Dural venous sinuses and cranial dural septa.

Table 15.1 Meninges, Dural Septa, and Dural Venous Sinuses

Structure	Description and Function	Word Origin
Meningeal Layer		
Dura Mater	Very tough, durable membrane composed of dense irregular connective tissue that protects CNS structures within the cranial cavity and vertebral canal; composed of two layers	*durus*, hard, + *mater*, mother
Periosteal Layer	Outer layer of dura mater that composes the inner periosteum of the cranial bones and anchors the dura mater tightly to the cranial bones (not present within the vertebral canal). In most places within the cranial cavity it is anchored to, or continuous with, the meningeal layer.	*periosteal*, relating to the periosteum, + *dura*, relating to the dura mater
Meningeal Layer	Inner layer of dura mater that forms cranial dural septa and dural venous sinuses within the cranial cavity	*meninx*, membrane, + *dura*, relating to the dura mater
Arachnoid Mater	Thin, loose connective tissue membrane that lies adjacent to the dura mater; contains numerous weblike extensions called arachnoid trabeculae, which are composed of collagen and elastic fibers that anchor it to the pia mater; forms a space (the subarachnoid space) for cerebrospinal fluid (CSF) to circulate around the brain and spinal cord	*arachno*, spider cobweb, + *eidos*, resemblance, + *mater*, mother
Pia Mater	A thin, highly vascular, areolar connective tissue membrane in direct contact with the brain and spinal cord. It follows all the surface contours (gyri and sulci) of the brain and is generally inseparable from brain tissue.	*pia*, soft or tender, + *mater*, mother
Cranial Dural Septa	Folds of the meningeal layer that extend inward to form partitions within the cranial cavity that stabilize and support the brain	*durus*, hard, + *septa*, fold
Diaphragma Sellae	Forms a "roof" over the sella turcica of the sphenoid bone and contains a hole for the passage of the infundibulum	*diaphragm*, diaphragm, + *sella*, saddle

(continued on next page)

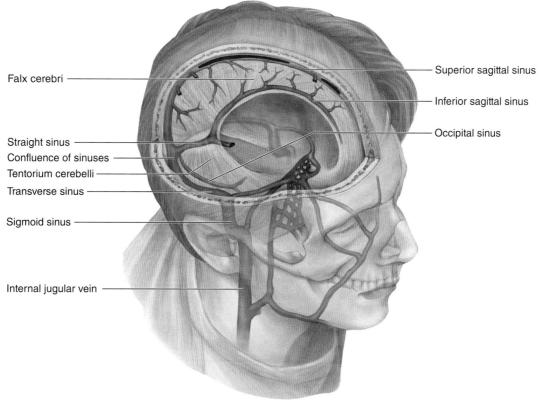

(b) Dural venous sinuses

(continued from previous page)

Table 15.1	Meninges, Dural Septa, and Dural Venous Sinuses *(continued)*	
Structure	**Description and Function**	**Word Origin**
Falx Cerebelli	Located between the two cerebellar hemispheres	*falx*, sickle, + *cerebelli*, relating to the cerebellum
Falx Cerebri	Located between the two cerebral hemispheres, within the longitudinal fissure of the brain; anchored anteriorly to the crista galli of the ethmoid bone	*falx*, sickle, + *cerebelli*, relating to the cerebrum
Tentorium Cerebelli	Drapes across the cerebellar hemispheres horizontally within the transverse fissure of the brain between the cerebellum and the cerebrum	*tentorium*, a tent, + *cerebelli*, relating to the cerebellum
Dural Venous Sinuses	Spaces formed between layers of meningeal and periosteal layers of the dura mater within the cranial cavity; transport CSF and venous blood from the subarachnoid space and the brain to the internal jugular vein	*durus*, hard, + *vena*, a blood vessel, + *sinus*, a channel
Inferior Sagittal Sinus	Located within the inferior part of the falx cerebri. Receives CSF and venous blood from the subarachnoid space and the brain (respectively) and drains into the straight sinus.	*inferior*, below, + *sagitta*, an arrow (relating to the sagittal plane), + *sinus*, a channel
Occipital Sinus	Smallest sinus within the cranial cavity; located in the margin of the tentorium cerebelli; drains into the confluence of sinuses	*sinus*, a channel + *occipital*, inner surface of occipital bone
Sigmoid Sinuses	Located in the posterior cranial fossa just posterior to the petrous part of the temporal bone and extending into the jugular foramen. Transports CSF and venous blood from the transverse sinuses to the internal jugular vein.	*sinus*, a channel, + *sigmoid*, shaped like an S
Confluence of Sinuses	Located within the posterior cranial cavity deep to the external occipital protuberance. Transmit CSF and venous blood from the superior sagittal sinus and the straight sinus to the transverse sinuses.	*confluens*, to flow together, + *sinus*, a channel
Straight Sinus	Located at the junction between the falx cerebri, falx cerebelli, and tentorium cerebelli. Transmit CSF and venous blood from the inferior sagittal sinus to the confluence of sinuses.	*sinus*, a channel, + straight
Superior Sagittal Sinus	Largest sinus within the cranial cavity; located within the superior portion of the falx cerebri. Transports CSF and venous blood from the subarachnoid space and the brain (respectively) to the confluence of sinuses.	*superior*, above, + *sagitta*, an arrow (relating to the sagittal plane), + *sinus*, a channel
Transverse Sinuses	Located posterior to the tentorium cerebelli. Runs from the confluence of sinuses, along the posterior aspect of the occipital bone, to the posterior cranial fossa just posterior to the petrous part of the temporal bone. Transports CSF and venous blood from the confluence of sinuses to the sigmoid sinuses.	*transversus*, across, + *sinus*, a cavity
Meningeal Spaces		
Epidural Space	Space between the dura mater and the walls of the vertebral canal (there is no epidural space within the cranial cavity). Clinically relevant for administration of anesthetic (an "epidural"). Anesthetic within this space numbs the roots of the nerves that exit the spinal cord. Thus, areas located below the spinal cord level where anesthetic is administered become completely devoid of sensation.	*epi*, above, + *dura*, relating to the dura mater
Subarachnoid Space	Space between the arachnoid mater and the pia mater where CSF flows as it circulates around the brain and spinal cord. Blood vessels are located within this space.	*sub*, under, + *arachnoid*, relating to the arachnoid mater
Subdural Space	Potential space between the dura mater and the arachnoid mater. In a healthy individual, this space does not exist, but traumatic injury may cause bleeding into the subdural space (subdural hematoma).	*sub*, under, + *dural*, relating to the dura mater

Chapter Fifteen *The Brain and Cranial Nerves* **387**

5. Using table 15.1 and figure 15.1 as guides, identify the **meningeal structures** listed in **figure 15.2** on a model or cadaveric specimen. Then label them in figure 15.2. (Answer choices may be used more than once.)

6. *Optional Activity:* **AP|R** **7: Nervous System**—Watch the "Meninges" and "Dural Sinus Blood Flow" animations to reinforce your understanding of these structures and their relationships.

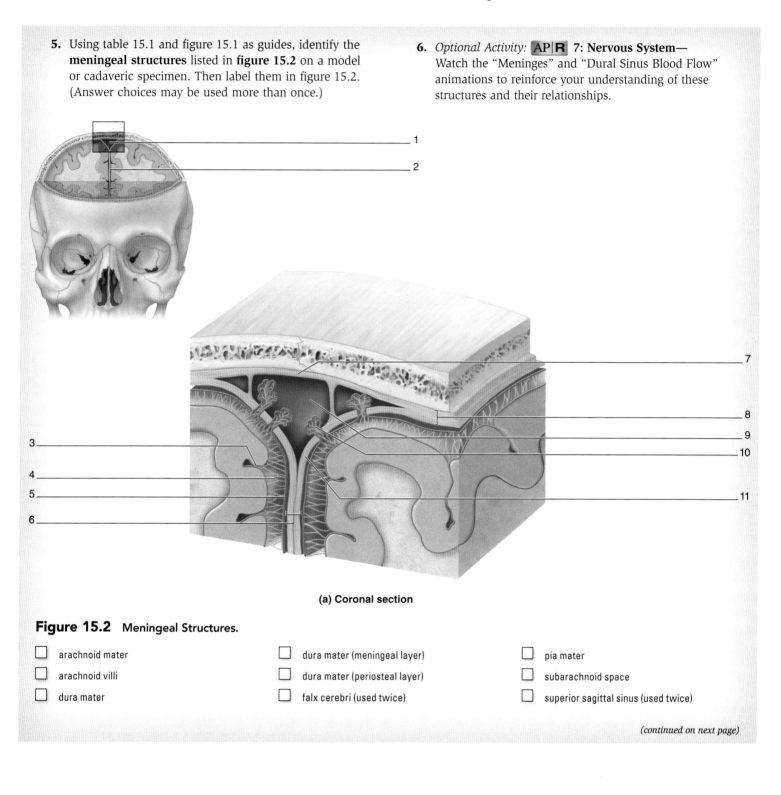

(a) Coronal section

Figure 15.2 Meningeal Structures.

☐ arachnoid mater
☐ arachnoid villi
☐ dura mater

☐ dura mater (meningeal layer)
☐ dura mater (periosteal layer)
☐ falx cerebri (used twice)

☐ pia mater
☐ subarachnoid space
☐ superior sagittal sinus (used twice)

(continued on next page)

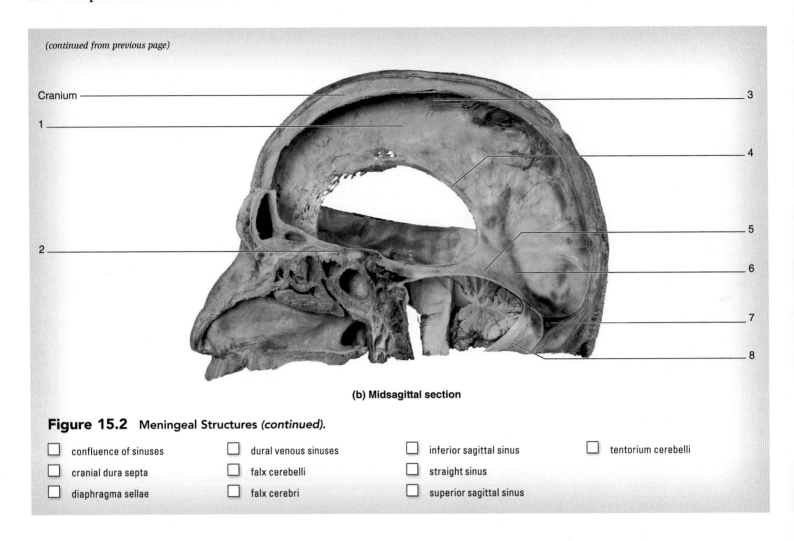

(b) Midsagittal section

Figure 15.2 Meningeal Structures *(continued).*

☐ confluence of sinuses ☐ dural venous sinuses ☐ inferior sagittal sinus ☐ tentorium cerebelli
☐ cranial dura septa ☐ falx cerebelli ☐ straight sinus
☐ diaphragma sellae ☐ falx cerebri ☐ superior sagittal sinus

Ventricles of the Brain

The **ventricles** are fluid-filled spaces within the central nervous system that are complex in shape. In exercise 15.2 you will view a *cast* of the ventricles, which is produced by filling the ventricular spaces with plastic, allowing the plastic to harden, and then removing the brain tissue so only the cast is left. A cast allows you to visualize the three-dimensional structure of the ventricles without the brain literally "getting in the way." If your laboratory does not have casts of the brain ventricles, this exercise can be performed using table 15.2 and figures in your textbook.

Recall that the central nervous system initially develops as a neural tube. As it grows, the neural tube begins to change size and shape. The cephalic end develops into the brain, while the rest develops into the spinal cord. Both the brain and the spinal cord contain fluid-filled spaces inside. The pattern of growth of the neural tissue surrounding the neural tube changes the size and shape of the fluid-filled spaces within. Thus, because the spinal cord remains mostly a tubular structure as it grows, the fluid-filled space inside, the **central canal**, remains tubular. On the other hand, because the cephalic (brain) end of the neural tube undergoes extensive folding as it grows, the fluid-filled spaces within develop into irregular shapes. These shapes tell a story about how the parts of the brain developed.

The cephalic end of the neural tube first develops into three **primary vesicles** (prosencephalon, mesencephalon, and rhombencephalon) and then into five **secondary vesicles** (telencephalon, diencephalon, mesencephalon, metencephalon, and myelencephalon). **Figure 15.3** lists the secondary vesicles of

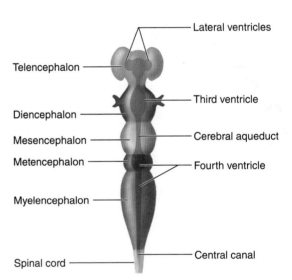

Figure 15.3 Secondary Brain Vesicles and Associated Ventricular Structures of the Brain.

the brain and the parts of the ventricular system that develop from each of them. For instance, the telencephalon undergoes extensive growth as it develops into the cerebral hemispheres. This growth results in the horseshoe-shaped structure of the **lateral ventricles** in the adult brain. On the other hand, the mesencephalon does not undergo such extensive growth as it develops into the midbrain. Hence the fluid-filled space inside, the **cerebral aqueduct,** remains tubular in shape within the adult brain.

EXERCISE 15.2

BRAIN VENTRICLES

1. Obtain a cast of **the ventricles of the brain (figure 15.4)**.

2. Identify the ventricles of the brain listed in **table 15.2** on the cast of the ventricles. As you identify each of the ventricles, relate each ventricle to the **secondary brain vesicle** from which it developed (table 15.2 and figure 15.3).

3. Using table 15.2 and your textbook as guides, identify the ventricular structures listed in figure 15.4 on a **cast of the brain ventricles.** Then label them in figure 15.4. (Answers may be used more than once.)

4. *Optional Activity:* **AP|R 7: Nervous System**—Watch the "Brain Ventricles" and "CSF Flow" animations to solidify your understanding of how cerebrospinal fluid flows through the brain ventricles.

INTEGRATE

LEARNING STRATEGY

As you observe whole brains or brain models in the laboratory, always begin by locating the ventricular spaces and associating each ventricular space with a secondary brain vesicle (table 15.2). Next, identify the adult brain structures that surround the ventricular spaces. Finally, correlate each adult brain structure to the secondary brain vesicle from which it formed. For example, the lateral ventricles (ventricular space), which are part of the telencephalon (secondary brain vesicle), are surrounded by the cerebral hemispheres (adult brain structures). Thus, the cerebral hemispheres (brain structure) are derived from the telencephalon (secondary brain vesicle).

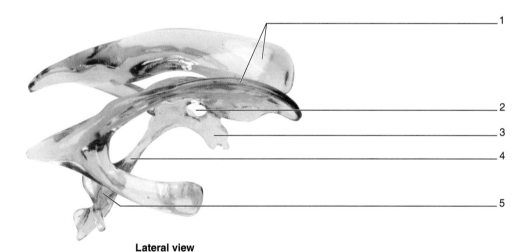

Lateral view

Figure 15.4 Cast of the Ventricles of the Brain.

☐ cerebral aqueduct ☐ lateral ventricles
☐ fourth ventricle ☐ third ventricle
☐ interventricular foramen

(continued on next page)

(continued from previous page)

Table 15.2	Ventricles of the Brain		
Structure	**Location and Description**	**Secondary Brain Vesicle Derivative**	**Word Origin**
Lateral Ventricles	Horseshoe-shaped ventricles within the cerebral hemispheres containing anterior and posterior horns whose shape follows the developmental shape of the cerebral hemispheres	Telencephalon	*latus*, to the side, + *ventriculus*, belly
Third Ventricle	Narrow, quadrilateral-shaped ventricle located in the midsagittal plane inferior to the corpus callosum and medial to the thalamic nuclei; surrounded by structures of the diencephalon	Diencephalon	*ventriculus*, belly
Cerebral Aqueduct	Narrow channel that lies in the midbrain between the cerebral peduncles and the tectal plate (corpora quadrigemina)	Mesencephalon	*aquaeductus*, a canal
Fourth Ventricle	Diamond-shaped ventricle located anterior to the cerebellum and posterior to the pons	Metencephalon	*ventriculus*, belly

The Human Brain

In the next series of exercises you will identify structures that are visible in four views of the human brain: superior, lateral, inferior, and midsagittal. **Table 15.3** lists the main brain structures, the views in which each is visible, and a description of its function.

As you identify parts of the brain, follow these steps to make things easier:

- Name the structures in a logical order, such as the order in which the structures appear from anterior to posterior. If you concentrate on learning the structures in an orderly fashion, you will be more able to recall the information later on.
- Do not think of brain regions as isolated structures. The structures are easier to identify within the context of their surroundings. If you do not believe this, think about it the next time you are traveling to your anatomy & physiology class—would you know how to get to the classroom if all of the buildings on campus (with the exception of the one to which you are traveling) were suddenly moved around?
- Always associate a function with each structure you identify. Use table 15.3 as a reference.

INTEGRATE

LEARNING STRATEGY

Because structure/function relationships are not easily visualized when it comes to the brain, consider making flashcards using table 15.3 as a reference. List the brain structure on one side of the card and its function on the other side. Once you have mastered the names and locations of the listed structures, quiz your lab partners on the structure and function of each item.

Table 15.3	Brain Structures Visible in Superficial Views of Whole or Sagittally-Sectioned Brains			
Brain Structure	**Description**	**Function(s)**	**Word Origin**	**Views Where Visible**
Central Sulcus	A deep groove that runs along the coronal plane	Separates the frontal lobe from the parietal lobe	*central*, in the center, + *sulcus*, a furrow	Superior, lateral, and midsagittal
Cerebellum	The second largest part of the brain, after the cerebral hemispheres	Regulation of muscle tone (a low-level muscle contraction), coordination of motor activity, and maintenance of balance and equilibrium	*cerebellum*, little brain	Lateral, inferior, and midsagittal
Cerebral Aqueduct	The part of the ventricular system of the brain that lies in the midbrain	Carries cerebrospinal fluid from the third ventricle to the fourth ventricle	*aquaeductus*, a canal	Midsagittal
Cerebral Peduncles	Fiber tracts located between the midbrain and the pons	The fibers connect the forebrain (cerebral hemispheres and diencephalon) to the hindbrain (medulla oblongata, pons, and cerebellum)	*cerebrum*, brain, + *pedunculus*, a little foot	Inferior, midsagittal
Cingulate Gyrus	A gyrus located just superior to the corpus callosum	This area of the brain is not well understood. It is predominantly motor and may play a role in the limbic system (such as controlling motor functions with a strong emotional component).	*cingo*, to surround, + *gyros*, circle	Midsagittal
Tectal Plate (corpora quadrigemina)	Consists of four twin bodies, the superior and inferior colliculi	Control center for visual and auditory reflexes	*corpus*, body, + *quad*, four, + *geminus*, twin	Midsagittal
Corpus Callosum	A fiber tract located superior to the lateral ventricles	Contains axons that connect the lobes of the two cerebral hemispheres to each other (except for the temporal lobes)	*corpus*, body, + *callosus*, thick-skinned	Midsagittal
Epithalamus	A small projection extending posteriorly from the superior portion of the third ventricle	Contains the pineal body (pineal gland) along with other structures	*epi*, above, + *thalamos*, a bed or bedroom	Midsagittal
Fornix	An arching fiber tract located inferior to the septum pellucidum	Connects limbic system structures to each other	*fornix*, arch	Midsagittal
Fourth Ventricle	Ventricle of the brain that lies between the cerebellum and brainstem and contains choroid plexus	CSF from the cerebral aqueduct flows into this ventricle before moving on to the central canal of the spinal cord and exiting into the subarachnoid space through the median and lateral apertures	*ventriculus*, belly	Midsagittal
Frontal Lobe	Lies deep to the frontal bone	Controls conscious movement of skeletal muscle; contains Broca's area, which controls motor speech. Controls conjugate eye movement (the ability to move the eyes together). Higher level functions include judgment and foresight (the ability to think before acting).	*frontal*, in the front, + *lobos*, lobe	Superior, lateral, inferior, and midsagittal
Hypothalamus	Located deep to the walls of the inferior part of the third ventricle	Regulates body temperature, metabolism (hunger/thirst), sleep, sex, and emotional control (limbic system functions). It is also a "master" endocrine gland, controlling hormone secretion from the pituitary gland.	*thalamos*, a bed or bedroom	Midsagittal
Inferior Colliculus	A pair of oval projections that make up the inferior part of the tectal plate (corpora quadrigemina)	Controls *auditory reflexes*, such as the sudden turning of the head toward the source of a very loud sound	*inferior*, lower, + *colliculus*, a mound or hill	Midsagittal
Infundibulum	A funnel-shaped inferior extension of the brain located immediately posterior to the optic chiasm	Consists of tracts that connect the hypothalamus to the posterior pituitary (pars nervosa)	*infundibulum*, a funnel	Inferior and midsagittal
Intermediate Mass (Interthalamic Adhesion)	A fiber tract that crosses the third ventricle. In a midsagittal section of the brain you will see the cut end of this structure in the middle of the third ventricle.	A fiber tract that connects the two thalamic nuclei to each other. It is absent in about 20% of human brains.	*intermediate*, in the middle, + *mass*, a mass	Midsagittal

(continued on next page)

Table 15.3	Brain Structures Visible in Superficial Views of Whole or Sagitally-Sectioned Brains *(continued)*			
Brain Structure	**Description**	**Function(s)**	**Word Origin**	**Views Where Visible**
Lateral Sulcus	A deep, horizontal groove between the frontal/parietal lobes and the temporal lobe	Separates the frontal and parietal lobes from the temporal lobe	*latus*, the side, + *fissure*, a deep furrow	Lateral
Longitudinal Fissure	A deep fissure between the two cerebral hemispheres	Separates the two cerebral hemispheres; the falx cerebri occupies this fissure in a living human	*longus*, long, + *fissure*, a deep furrow	Superior and inferior
Mammillary Bodies	Two small bumplike ("breast-shaped") structures of the hypothalamus located immediately posterior to the infundibulum	Involved in short-term memory processing; part of the limbic system (the emotional brain). Also involved with suckling and chewing reflexes.	*mammillary*, shaped like a breast	Inferior and midsagittal
Medulla Oblongata	The most inferior aspect of the brainstem, forming the transition zone between the brain and spinal cord	Contains the centers for regulation of respiration and cardiac function, and contain nuclei of the *reticular activating system*, which is a group of nuclei that are important in regulating wakefulness and selective attention	*medius*, middle, + *oblongus*, rather long	Lateral, inferior, and midsagittal
Occipital Lobe	Lies deep to the occipital bone	Primary visual area (the first area of the cerebral cortex where visual information synapses, after the thalamus). Visual association area (the ability to interpret visual information).	*occiput*, the back of the head, + *lobos*, lobe	Superior, lateral, and midsagittal
Olfactory Bulbs	Swellings connected to the anterior end of the olfactory tracts that lie on the inferior surface of the frontal lobes of the brain lateral to the longitudinal fissure	Location where cranial nerve I (CN I), the olfactory nerves, first synapse after passing through the cribriform plate of the ethmoid bone	*olfactus*, to smell, + *bulbus*, a globular structure	Inferior
Olfactory Tracts	Nerve fibers that extend from the olfactory bulbs posteriorly to the junction where the frontal lobes meet the optic chiasm	Carry the axons of neurons from the olfactory bulbs toward structures in other areas of the brain involved with olfaction	*olfactus*, to smell, + *tractus*, a drawing out	Inferior
Optic Chiasm	The X-shaped structure formed where the two optic nerves join, with most fibers crossing to the opposite side; located just anterior to the infundibulum	Location where fibers from both optic nerves cross over and travel in the optic tract on the opposite side. Not all fibers from the optic nerves cross over.	*optikos*, relating to the eye or vision, + *chiasma*, two crossing lines	Inferior and midsagittal
Parietal Lobe	Lies deep to the parietal bone	Receives sensory input from the skin and proprioceptors. Higher level functions include logical reasoning (math, problem solving).	*parietal*, a wall, + *lobos*, lobe	Superior, lateral, and midsagittal
Parieto-occipital Sulcus	Small groove that runs along the coronal plane	Separates the parietal lobe from the occipital lobe	*parieto-occipital*, between the parietal and occipital lobes, + *sulcus*, a furrow	Superior, lateral, and midsagittal
Pineal Body (Gland)	Small gland found within the epithalamus. You will not be able to establish the difference between the epithalamus and the pineal gland on gross observation alone.	Secretes the hormone melatonin from its precursor molecule, serotonin, in response to *decreased* light levels. Melatonin has an effect on circadian rhythms. May also play a role in establishing the onset of puberty.	*pineal*, shaped like a pinecone	Midsagittal
Pons	Appears as a large mass just superior to the medulla oblongata	A "bridge" of nerve tracts that connect the cerebral hemispheres to the cerebellar hemispheres. Contains centers for control of respiration.	*pons*, bridge	Lateral, inferior, and midsagittal

Table 15.3 Brain Structures Visible in Superficial Views of Whole or Sagitally-Sectioned Brains (continued)

Brain Structure	Description	Function(s)	Word Origin	Views Where Visible
Postcentral Gyrus	Fold of brain tissue located immediately posterior to the central sulcus	Primary somatic sensory area of the brain. Sensory information that comes in from the body travels to this area of the cerebral cortex.	*post*, after, + *central*, relating to the central sulcus, + *gyros*, circle	Superior, lateral, and midsagittal
Precentral Gyrus	Fold of brain tissue located immediately anterior to the central sulcus	Primary somatic motor area of the brain. Neurons from this gyrus are somatic motor neurons that initiate motor signals to control voluntary muscle activity.	*pre*, before, + *central*, relating to the central sulcus, + *gyros*, circle	Superior, lateral, and midsagittal
Septum Pellucidum	A thin membrane located between the corpus callosum (above) and fornix (below)	Contains neurons and glial cells, and forms a thin connection between the corpus callosum above and the fornix below; also forms a thin wall between the two anterior horns of the lateral ventricles	*saeptum*, a partition, + *pellucidus*, allowing the passage of light	Midsagittal
Superior Colliculus	A pair of rounded projections that make up superior part of the corpora quadrigemina (tectal plate)	Controls visual reflexes, such as the sudden turning of the head toward the source of a flashing light	*superus*, above, + *colliculus*, a mound or hill	Midsagittal
Temporal Lobe	Lies deep to the temporal bone	Primary auditory and auditory association area of the brain; conscious perception of smell	*tempus*, time	Lateral
Thalamus	Paired nuclei located deep to the lateral walls of the third ventricle. A pin pierced through the lateral wall of the third ventricle adjacent to the intermediate mass will pass into the thalamic nuclei.	Primary relay center for all sensory information coming into the brain (except olfaction)	*thalamos*, a bed or bedroom	Midsagittal
Third ventricle	The part of the ventricular system of the brain that lies within the diencephalon	CSF from the lateral ventricles flows into this ventricle before moving on to the cerebral aqueduct	*ventriculus*, belly	Midsagittal
Transverse Fissure	A deep fissure between the cerebrum and the cerebellum	Separates the cerebral hemispheres from the cerebellar hemispheres. The tentorium cerebelli lies in this fissure	*transversus*, across, + *fissure*, a deep furrow	Lateral

INTEGRATE

CONCEPT CONNECTION

The hypothalamus is the homeostatic center of the brain, as it regulates such fundamental processes in the body as metabolism (thirst/hunger), body temperature, sleep, and sex. Although the hypothalamus is composed of nervous tissue, it is also considered an endocrine organ, as it plays a critical role in the release of hormones. Its connection with the pituitary gland, which lies inferior to the hypothalamus, is commonly called the hypothalamo-hypophyseal axis. The pituitary gland is composed of two lobes, posterior and anterior. The posterior pituitary, or neurohypophysis, is formed from an inferior growth of nervous tissue from the hypothalamus during development. Neurosecretory cells with their cell bodies in the hypothalamus produce hormones. The axons of these cells extend inferiorly through the infundibulum to the posterior pituitary. Upon stimulation, these neurosecretory cells release the hormone from their axon terminals in the posterior pituitary via exocytosis, much like acetylcholine is released from somatic motor neurons at the neuromuscular junction. The hormones subsequently enter the blood and are transported throughout the body. Two hormones are released from the posterior pituitary: antidiuretic hormone (ADH) and oxytocin.

Hormones released by the anterior pituitary, or adenohypophysis, do so in response to hormones released by the hypothalamus. In this case, neurosecretory cells in the hypothalamus release hormones through axon terminals in the median eminence. These hormones enter a portal vein and are transported to the anterior pituitary gland. There, the hypothalamic hormones stimulate cells in the anterior pituitary to release their hormones. As was the case with the posterior pituitary, these hormones then enter the systemic circulation and are carried to their distant targets. The specific hormones released and the cells responsible for hormone production will be covered in detail in chapter 19.

EXERCISE 15.3

SUPERIOR VIEW OF THE HUMAN BRAIN

1. Obtain a human brain or models of a human brain.
2. Observe the superior surface of the brain **(figure 15.5)**. The most prominent feature in this view is the **longitudinal fissure,** which separates the two cerebral hemispheres from each other. A major sulcus you must identify is the **central sulcus**. Identification of the central sulcus is difficult, though not impossible, on a real human brain. The following are two features to look for:

   - The pre- and postcentral gyri should become continuous with each other on the lateral aspect of the central sulcus just above the lateral sulcus. This means the central sulcus will not enter the lateral sulcus.
   - The central sulcus will dip down into the longitudinal fissure.

3. Once you have identified the central sulcus, you will be able to identify the associated lobes and gyri located nearby.
4. Using table 15.3 and your textbook as guides, identify the structures listed in figure 15.5 on the superior view of the brain. Then label them in figure 15.5.

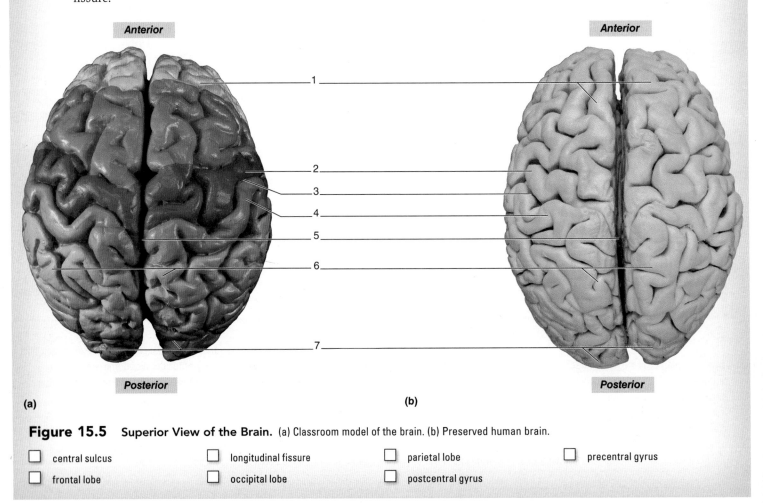

Figure 15.5 Superior View of the Brain. (a) Classroom model of the brain. (b) Preserved human brain.

- ☐ central sulcus
- ☐ frontal lobe
- ☐ longitudinal fissure
- ☐ occipital lobe
- ☐ parietal lobe
- ☐ postcentral gyrus
- ☐ precentral gyrus

EXERCISE 15.4

LATERAL VIEW OF THE HUMAN BRAIN

1. Obtain a human brain or models of a human brain.

2. Observe the lateral surface of the brain **(figure 15.6)**. As with the superior view, you should be able to see the **central sulcus** by identifying the location where the pre- and postcentral gyri become continuous with each other just above the **lateral sulcus.**

3. Using table 15.3 and your textbook as guides, identify the structures listed in figure 15.6 on the lateral view of the brain. Then label them in figure 15.6.

4. *Optional Activity:* **AP|R** 7: **Nervous System**—Watch the "Divisions of Brain" animation for an overview of the regions of the brain and their general functions.

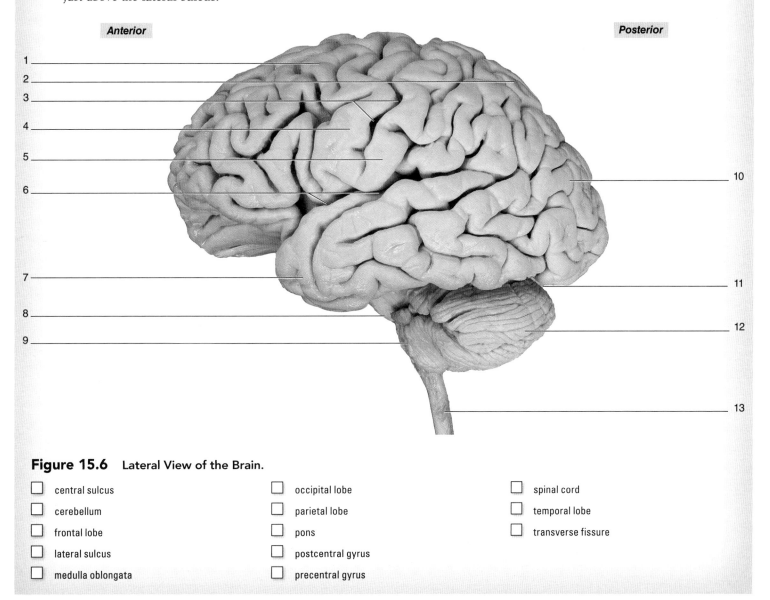

Figure 15.6 Lateral View of the Brain.

- ☐ central sulcus
- ☐ cerebellum
- ☐ frontal lobe
- ☐ lateral sulcus
- ☐ medulla oblongata
- ☐ occipital lobe
- ☐ parietal lobe
- ☐ pons
- ☐ postcentral gyrus
- ☐ precentral gyrus
- ☐ spinal cord
- ☐ temporal lobe
- ☐ transverse fissure

EXERCISE 15.5

INFERIOR VIEW OF THE HUMAN BRAIN

1. Obtain a human brain or models of a human brain.

2. Observe the inferior surface of the brain **(figure 15.7)**. This view is considerably more complicated than the superior or lateral views because of the cranial nerves that arise from the diencephalon and hindbrain. Do not concern yourself with the cranial nerves at this time. You will come back to them later in this chapter.

   One of the more problematic structures to identify in this view on a real brain is the infundibulum. When a brain is removed from the cranium, the pituitary gland almost always gets left behind within the sella turcica of the sphenoid bone because of a tough dural septa called the **diaphragma sellae,** which lies between the pituitary gland and the rest of the brain. The only structure left connected to the brain is the stalk of tissue that connects the pituitary gland to the hypothalamus, which is the **infundibulum,** or **pituitary stalk.** If you are observing a model of the brain, the pituitary gland may be shown intact. If not, the infundibulum can be identified as a small strand of tissue that is located directly posterior to the **optic chiasm** and directly anterior to the **mammillary bodies.**

3. Using table 15.3 and your textbook as guides, identify the structures listed in figure 15.7 on the inferior view of the brain. Then label them in figure 15.7.

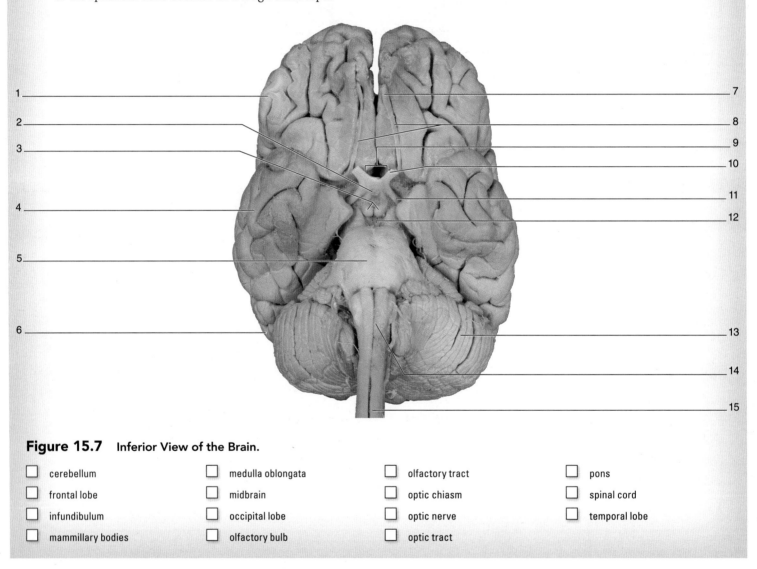

Figure 15.7 Inferior View of the Brain.

- ☐ cerebellum
- ☐ frontal lobe
- ☐ infundibulum
- ☐ mammillary bodies
- ☐ medulla oblongata
- ☐ midbrain
- ☐ occipital lobe
- ☐ olfactory bulb
- ☐ olfactory tract
- ☐ optic chiasm
- ☐ optic nerve
- ☐ optic tract
- ☐ pons
- ☐ spinal cord
- ☐ temporal lobe

EXERCISE 15.6

MIDSAGITTAL VIEW OF THE HUMAN BRAIN

1. Obtain a human brain or brain model that has been sectioned along the midsagittal plane and observe its medial surface **(figure 15.8)**.

2. In the very center of your view, you will see the **third ventricle.** The third ventricle is the central depressed area that appears to have a cut nerve in the center. The "cut nerve" isn't really a nerve, but it is similar. It is a fiber tract called the **interthalamic adhesion** (or **intermediate mass**), which connects the two thalamic nuclei to each other. Use the interthalamic adhesion, the thalamus, and the third ventricle as reference points for identification of other structures in this view. Many of the structures that are located around the third ventricle of the brain belong to a system called the **limbic system** (*limbus*, border), so named because structures of the limbic system are located at the border of the third ventricle and the brainstem. The limbic system is referred to as the "emotional brain" because its structures play a role in our emotions.

3. Using table 15.3 and your textbook as guides, identify the structures listed in figure 15.8 on the midsagittal view of the brain. Then label them in figure 15.8.

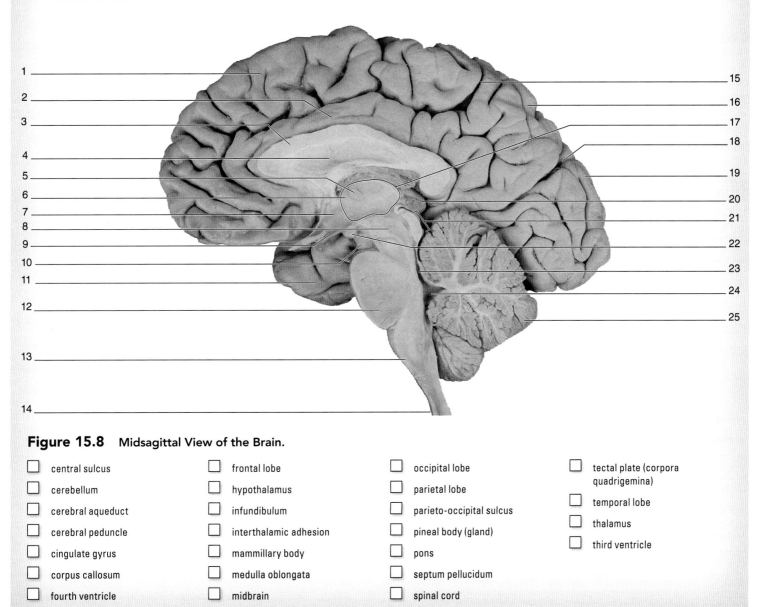

Figure 15.8 Midsagittal View of the Brain.

- ☐ central sulcus
- ☐ cerebellum
- ☐ cerebral aqueduct
- ☐ cerebral peduncle
- ☐ cingulate gyrus
- ☐ corpus callosum
- ☐ fourth ventricle
- ☐ frontal lobe
- ☐ hypothalamus
- ☐ infundibulum
- ☐ interthalamic adhesion
- ☐ mammillary body
- ☐ medulla oblongata
- ☐ midbrain
- ☐ occipital lobe
- ☐ parietal lobe
- ☐ parieto-occipital sulcus
- ☐ pineal body (gland)
- ☐ pons
- ☐ septum pellucidum
- ☐ spinal cord
- ☐ tectal plate (corpora quadrigemina)
- ☐ temporal lobe
- ☐ thalamus
- ☐ third ventricle

INTEGRATE

CLINICAL VIEW
Epilepsy

Epilepsy is a neurological condition characterized by excessive patterns of synchronous electrical activity of the brain. The resulting **seizure** may vary in severity depending upon the strength, duration, and location of synchronous electrical activity. Seizures may impact a local area or a large portion of the brain. An electroencephalogram (EEG), which records the electrical activity of the brain with surface electrodes placed on the skull, is used clinically to diagnose epilepsy. This technique is similar to recording electrical activity of skeletal muscles with an electromyogram (EMG), or recording electrical activity of cardiac muscle with an electrocardiogram (ECG). Possible symptoms that occur during a seizure include confusion, sensitivity to light and sounds, slurred speech, and a glazed stare. Seizures that impact large portions of the brain may be classified as *grand mal* (*grand*, big + *malus*, bad) seizures. In a grand mal seizure, patients lose consciousness and experience whole-body convulsions. Once patients regain consciousness, they often have no recollection of the event. Treatment depends upon the severity of the seizures. Drugs that influence the excitability of neurons may be prescribed. In patients where the electrical activity is localized to one portion of the brain and there is no response to drug therapy, doctors may implant stimulators that can apply a "resetting" current to nerves or brain tissue. This is similar to a pacemaker that resets the electrical activity in the heart. In the most severe cases, doctors may surgically remove affected brain tissue.

Cranial Nerves

An inferior view of the brain allows you to see all of the cranial nerves at the location where they arise from the brain. Cranial nerves are numbered, starting from the anterior (rostral) part of the brain and moving posterior (caudal), using Roman numerals I through XII. **Figure 15.9** shows the inferior surface of the brain and the cranial nerves. The olfactory bulbs (where CN I synapses) and the optic nerves (CN II) are very large, easily identifiable structures located on the inferior surface of the frontal lobes of the brain. The remainder of the cranial nerves (CN III through CN XII) are generally smaller and are located closer together in the region of the midbrain, pons, and medulla, thus making their identification a little more challenging. In **exercise 15.7** you will identify the cranial nerves on a brain or on a model of the brain.

EXERCISE 15.7

IDENTIFICATION OF CRANIAL NERVES ON A BRAIN OR BRAINSTEM MODEL

1. Obtain a human brain or a model of a human brain or brainstem.
2. Turn the brain over and observe its inferior surface (figure 15.9). Note all of the small nerves exiting the brain from various locations. These are the cranial nerves **(table 15.4)**.
3. Using table 15.4 and figure 15.9 as guides, identify the twelve cranial nerves on the inferior surface of the brain.
4. As you continue observing the inferior surface of the brain, notice where each nerve exits the brain or brainstem. Knowledge of the general area where a specific nerve emerges can be helpful in identifying the nerves, even if by process of elimination. Complete the following chart for cranial nerves III through XII to organize your thoughts. Use figure 15.9 as a guide.

Table 15.4	Names of Cranial Nerves		
Nerve Number	**Name**	**Foramina of Exit**	**Word Origin**
I	Olfactory	Olfactory foramina in the cribiform plate of the ethmoid bone	*olfacio*, to smell
II	Optic	Optic canal	*optikos*, relating to the eye or vision
III	Oculomotor	Superior orbital fissure	*oculo-*, the eye, + *motorius*, moving
IV	Trochlear	Superior orbital fissure	*trochileia*, a pulley
V	Trigeminal	Superior orbital fissure (V_1—ophthalmic) Foramen rotundum (V_2—maxillary) Foramen ovale (V_3—mandibular)	*tri-*, three, + *geminus*, twins
VI	Abducens	Superior orbital fissure	*abductio-*, to move away from the median plane

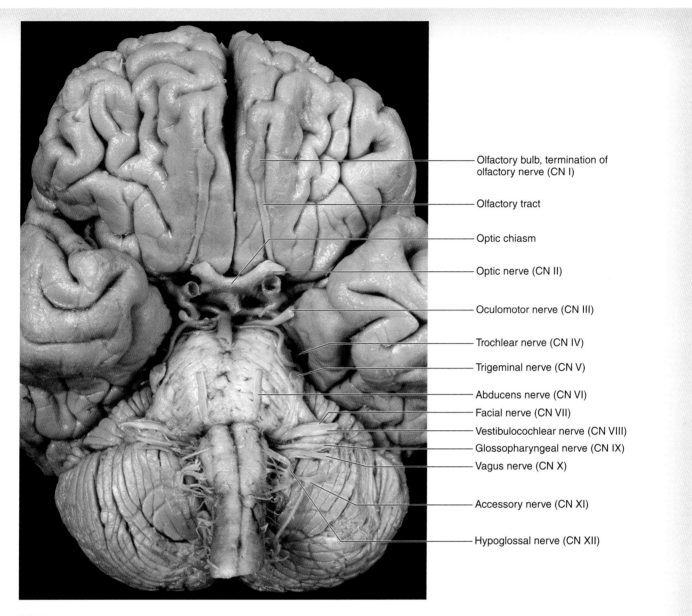

Figure 15.9 Cranial Nerves on the Inferior Surface of the Brain.

Table 15.4	Names of Cranial Nerves *(continued)*		
Nerve Number	**Name**	**Foramina of Exit**	**Word Origin**
VII	Facial	Internal acoustic meatus (exits via the stylomastoid foramen)	*facialis*, relating to the face
VIII	Vestibulocochlear	Internal acoustic meatus	*vestibulum*, entrance, + *cochlea*, snail shell
IX	Glossopharyngeal	Jugular foramen	*glossus*, tongue, + *pharyngeus*, pharynx
X	Vagus	Jugular foramen	*vagus*, wanderer
XI	Accessory	Jugular foramen (accessory division) Foramen magnum (spinal division)	*spina*, spine, + *accessory*, an extra structure
XII	Hypoglossal	Hypoglossal canal	*hypo*, beneath, + *glossus*, tongue

(continued on next page)

(continued from previous page)

Point of Exit	Cranial Nerve
Midbrain	
Pons	
Medulla Oblongata	

5. **Mnemonic devices,** simple phrases or plays on words that aid in memory recall, make remembering long strings of anatomical names easier. One popular technique is to associate a word with the first letter of the corresponding word you want to remember. A mnemonic that will help you recall the names of the cranial nerves in numerical order is

   Oh **O**nce **O**ne **T**akes **T**he **A**natomy **F**inal, **V**ery **G**ood **V**acations **A**re **H**eavenly.

   Mnemonics work best when they mean something to you. Develop your own mnemonic for remembering the cranial nerves. _____

   _____

   _____

   _____

6. *Optional Activity:* **AP|R** **7: Nervous System**—Visit the quiz area to test yourself on cranial nerve location, composition, and related foramina.

INTEGRATE

LEARNING STRATEGY

If you are observing a real brain instead of a model, you might not see the trochlear nerve (CN IV). It is very small, and often detaches from the brain when the brain is removed from the skull. If this is the case, make sure you identify the trochlear nerve on a model of the human brain. This nerve is unique because it is the only cranial nerve that originates from the posterior part of the brainstem, rather than the anterior or medial part.

The Sheep Brain

In the following exercises you will identify many of the same structures you identified on a human brain on a sheep brain. Sheep brains share many similarities with human brains, and are readily available for laboratory studies. The experience of dissecting a real brain (as opposed to observing a plastic model of a brain) will give you an appreciation for the true appearance and texture of the brain and its associated structures. You may notice that the brain tissue itself is relatively delicate. Even so, the tissue is much more solid than it would be if you were observing a fresh brain because it has been fixed with chemicals. Living brain tissues are *extremely* delicate, and have the consistency of firm gelatin. In these dissection exercises you will identify both brain structures *and* cranial nerves on the sheep brain. Later in this chapter you will observe the structure, function, and identification of cranial nerves on a human brain.

EXERCISE 15.8

SHEEP BRAIN DISSECTION

EXERCISE 15.8A Dura Mater

1. Obtain a sheep brain, dissecting tray, and dissecting tools (forceps, scissors, a scalpel, and gloves). Place the sheep brain in the dissecting tray and take turns with your laboratory partner(s) observing its gross structure. This exercise requires a sheep brain with dura mater intact. If the dura mater is missing, proceed to the section "Inferior View of the Sheep Brain" (p. 403). Otherwise begin here.

2. Using your hands or blunt forceps, feel the toughness of the dura mater. Notice how it surrounds the entire brain. What type of tissue is the dura mater composed of?

3. Observe the dura mater on the superior surface of the brain **(figure 15.10b)**, and locate the following structures:

   ☐ confluence of sinuses   ☐ transverse sinuses
   ☐ superior sagittal sinus

4. Rotate the brain so it is resting on its superior surface and you are viewing its inferior surface (figure 15.10a).

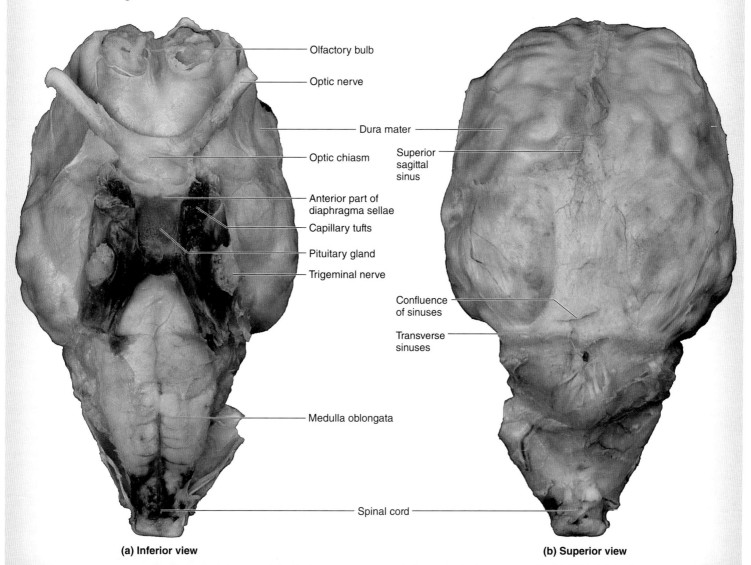

(a) Inferior view   (b) Superior view

Figure 15.10 Anatomical Landmarks of the Sheep Brain.

(continued on next page)

402 Chapter Fifteen *The Brain and Cranial Nerves*

(continued from previous page)

Notice the relatively large **pituitary gland** projecting inferior to the dura mater. Your goal in this part of the dissection will be to cut away the dura mater without disconnecting the pituitary from the rest of the brain. Notice the capillary tufts found just posterior and lateral to the pituitary gland. Just lateral to these capillaries on both sides you will see the large **trigeminal nerves** (CN V). Using a blunt probe, feel the dura mater surrounding the base of the pituitary gland. This dural membrane is the **diaphragma sellae,** which lies between the pituitary and the rest of the brain (except where the pituitary stalk exits the sella turcica of the sphenoid bone).

5. To free the connections between the dura and the rest of the brain without breaking off the pituitary gland, you will first cut around the trigeminal nerves and capillary tufts. **Figure 15.11a** shows where to make the initial incision. Cut *around* (lateral to) the optic chiasm, diaphragma sellae, pituitary gland, and trigeminal nerves to make a complete circle, which will free the dura mater from its attachments.

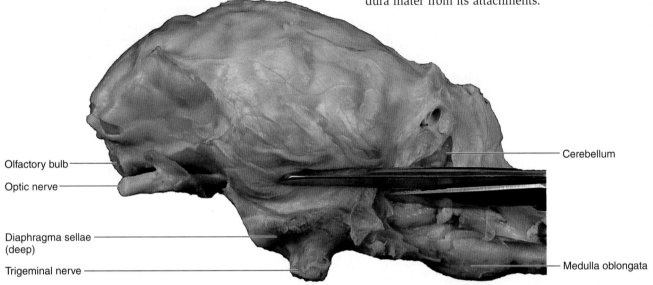

(a) Lateral view

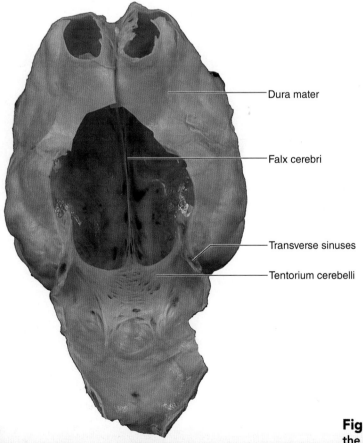

(b) Inferior view

Figure 15.11 Incisions to Remove the Dura Mater from the Sheep Brain. (a) Initial incision in the dura mater. (b) Appearance of the dura mater after it has been cut away from the brain.

6. Next, make an anterior cut in the dura mater along the midsagittal plane between the olfactory bulbs and olfactory tracts (figure 15.11*b*). Once you have freed the dura mater from its connections to the diaphragma sellae, gently pull the dura away from the brain. Pull in a posterior, superior direction so the falx cerebri and tentorium cerebelli slip out of their respective fissures without damaging the delicate brain tissues. As you pull the dura mater away from the brain, gently tease away any remaining connections. Figure 15.11*b* shows what the dura mater should look like after you have cut it away and removed it from the brain.

7. Once you have completely freed the dura from the brain, observe it closely and compare the dural septa and sinuses in the sheep brain to those you identified in the human brain. What dural septa is missing in a sheep brain that is present in a human brain?

_____

EXERCISE 15.8B Inferior View of the Sheep Brain

1. Obtain a sheep brain without the dura mater intact, or use the brain from which you just removed the dura mater. Place it in the dissecting pan on its superior surface so you are looking at its inferior surface **(figure 15.12)**.

2. When sheep brains are collected by a commercial vendor for use in the laboratory, the dura mater is first separated from the cranial bones. In such specimens, most of the dura has been dissected away from the cranium and the only part remaining is the diaphragma sellae, a membrane between the sella turcica and the rest of the brain (see figure 15.11*a*). Surrounding the diaphragm sellae are some capillary tufts and large cranial nerves, the trigeminal nerves (CN V).

3. Using figure 15.12 as a guide, identify the following on the sheep brain you are dissecting:

   ☐ capillary tufts   ☐ optic chiasm
   ☐ diaphragma sellae   ☐ pituitary gland
   ☐ olfactory bulb   ☐ trigeminal nerves
   ☐ olfactory tract   (CN V)

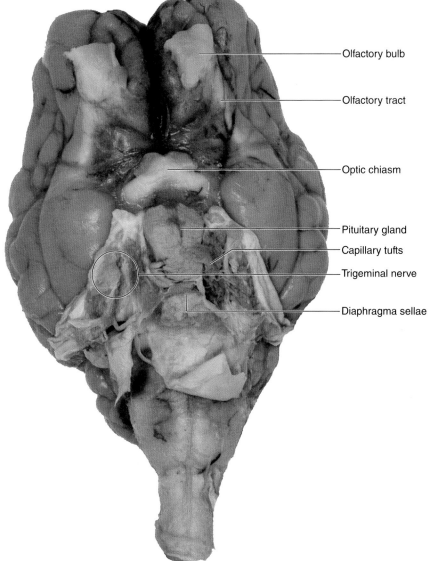

Figure 15.12 **Inferior View of the Sheep Brain.** The dura mater surrounding the pituitary gland has not yet been removed.

(continued on next page)

404 Chapter Fifteen *The Brain and Cranial Nerves*

(continued from previous page)

4. Next you will dissect the diaphragma sellae and the capillary tufts away from the pituitary gland without damaging the cranial nerves, without detaching the pituitary from the infundibulum, and without detaching the trigeminal nerves from the brain. You will need to dissect carefully, because it is very easy to accidentally detach these structures from the brain if you pull too hard on the diaphragma sellae while you are trying to remove it.

5. Gently lift the dura mater posterior to the pituitary gland to see the small nerves that enter the dura mater on its deep surface **(figure 15.13).**

6. Using scissors or a scalpel, detach the nerves where they enter the dura mater. Cut the nerves where they attach to the dura (not where they attach to the brain!) and then cut the dura and bony material away, removing as much of it as possible while keeping the pituitary intact. You will need to be careful as you do this, because the connection between the pituitary and the rest of the brain is delicate.

7. Using **figure 15.14a** as a guide, identify the following structures in the inferior view of the sheep brain you are dissecting:

   ☐ cerebellum
   ☐ cerebral peduncle
   ☐ frontal lobe
   ☐ longitudinal fissure
   ☐ medulla oblongata
   ☐ olfactory bulb
   ☐ olfactory tract
   ☐ optic chiasm
   ☐ optic nerve (CN II)
   ☐ pituitary gland
   ☐ pons
   ☐ spinal cord
   ☐ temporal lobe
   ☐ transverse fissure

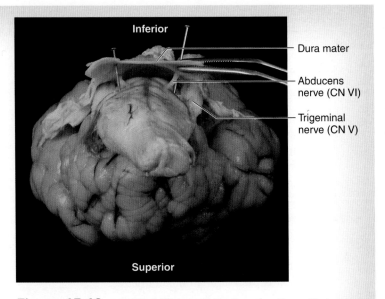

Figure 15.13 Cranial Nerves Entering the Dura Mater. The abducens and trigeminal nerves can be seen exiting the inferior surface of the sheep brain and piercing the dura mater.

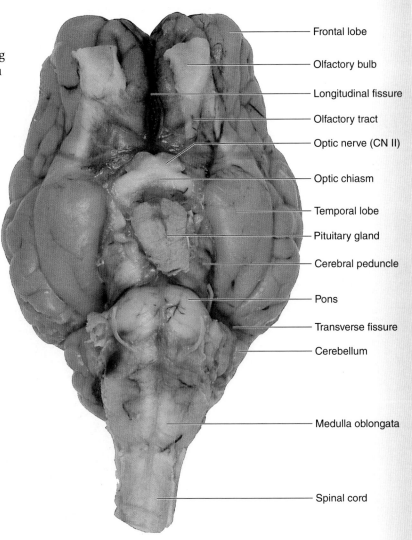

(a) Inferior view, pituitary gland intact; dura matter removed

Figure 15.14 Inferior Views of the Sheep Brain.

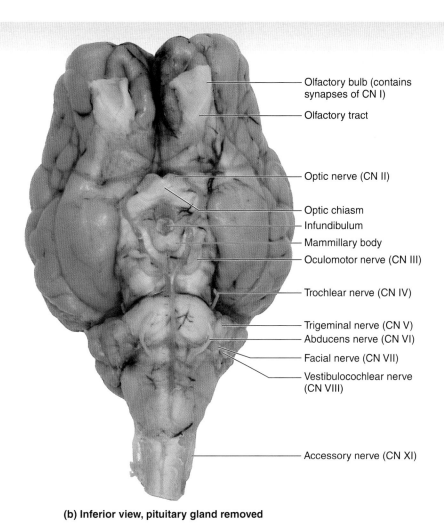

(b) Inferior view, pituitary gland removed

Figure 15.14 Inferior Views of the Sheep Brain *(continued)*.

8. Gently lift the pituitary to observe the **mammillary body** (figure 15.14b). Note that the sheep brain has only a single mammillary body, whereas the human brain has two.

9. Finally, attempt to identify the cranial nerves listed in figure 15.14b. Identification of cranial nerves IX (glossopharyngeal nerve) through XII (hypoglossal nerve) may not be possible because these nerves are very small and may have been damaged or torn off the brain as the dura mater was removed from the brain, or as the brain was removed from the cranium.

(continued on next page)

(continued from previous page)

EXERCISE 15.8C Superior View of the Sheep Brain

1. Place the brain in the dissecting tray with the inferior side facing down (away from you) **(figure 15.15a)**. Note the thin, transparent **arachnoid mater** that covers the entire surface of the brain without dipping into the **sulci** (grooves) between the **gyri** (folds) of the brain. Note the numerous **blood vessels** that lie between the arachnoid mater and the **pia mater**. The space occupied by the blood vessels is also a space where cerebrospinal fluid flows in the living animal. What is the name of this space?

   _____

2. Using figure 15.15a as a guide, identify the following structures in the superior view of the sheep brain you are dissecting:

   ☐ arachnoid mater   ☐ longitudinal fissure
   ☐ blood vessels   ☐ spinal cord
   ☐ cerebellum   ☐ sulcus
   ☐ cerebrum   ☐ transverse fissure
   ☐ gyrus

3. Pick up the brain and gently pull the cerebellum away from the cerebrum so you can see into the transverse fissure. Using figure 15.15b as a guide, identify the following structures:

   ☐ cerebellum   ☐ pineal gland
   ☐ cerebrum   ☐ superior colliculus
   ☐ inferior colliculus

EXERCISE 15.8D Midsagittal and Coronal Sections of the Sheep Brain

1. Some of you will perform a midsagittal section of the sheep brain; others will perform a coronal section. Ask your instructor which section to make before you begin cutting. Although you will perform only one of these dissections, be sure to observe brains that have been sectioned along both planes.

2. *Midsagittal Section:* Place the sheep brain in your dissecting tray with its superior surface facing up (facing you). Using a scalpel, cut the brain in half along the midsagittal plane. Start your cut on the anterior end of the brain by placing the scalpel blade within the longitudinal fissure. What is the first structure the scalpel blade will cut through?

   _____

3. Once you have cut the brain in half, observe its medial surface. Using **figure 15.16** as a guide, identify the following structures on the sheep brain you are dissecting:

   ☐ central canal (of spinal cord)   ☐ medulla oblongata
                                       ☐ optic chiasm
   ☐ cerebellum   ☐ pineal gland
   ☐ cerebral aqueduct   ☐ pituitary gland
   ☐ cerebral peduncle   ☐ pons
   ☐ cerebrum   ☐ spinal cord
   ☐ corpus callosum   ☐ superior colliculus
   ☐ fornix   ☐ thalamus
   ☐ fourth ventricle
   ☐ mammillary body

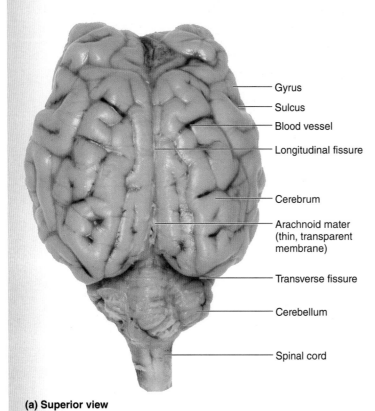

(a) Superior view
— Gyrus
— Sulcus
— Blood vessel
— Longitudinal fissure
— Cerebrum
— Arachnoid mater (thin, transparent membrane)
— Transverse fissure
— Cerebellum
— Spinal cord

(b) Posterior view
— Cerebrum
— Pineal gland
— Superior colliculus
— Inferior colliculus
— Cerebellum

Figure 15.15 Superior and Posterior Views of the Sheep Brain. (a) Superior view of the sheep brain. (b) Posterior view; cerebral hemispheres are pulled away from the cerebellum to reveal deeper structures.

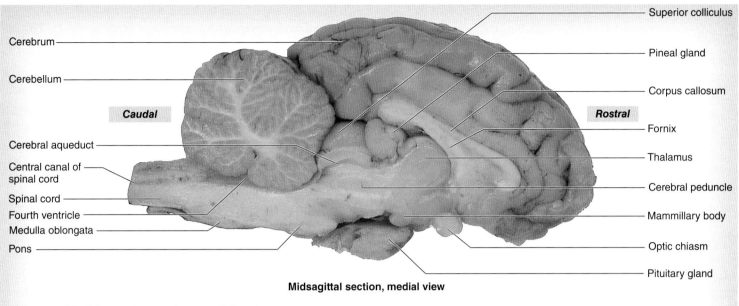

Figure 15.16 Midsagittal View of the Sheep Brain.

4. *Coronal Section:* Place the sheep brain in your dissecting tray with the superior surface down (facing away from you), and identify the pituitary gland (or the pituitary stalk, if the pituitary gland has been removed). Using a scalpel, cut the brain in half along a coronal plane that travels through the pituitary gland and continues toward the cerebral hemispheres.

5. Once you have cut the brain in half, observe the cut surface. Using **figure 15.17** as a guide, identify the following structures on the sheep brain you are dissecting:

☐ cerebral cortex   ☐ internal capsule
☐ cerebral peduncle   ☐ lateral ventricle
☐ corona radiata   ☐ longitudinal fissure
☐ corpus callosum   ☐ pons
☐ fornix   ☐ thalamus
☐ hypothalamus   ☐ third ventricle

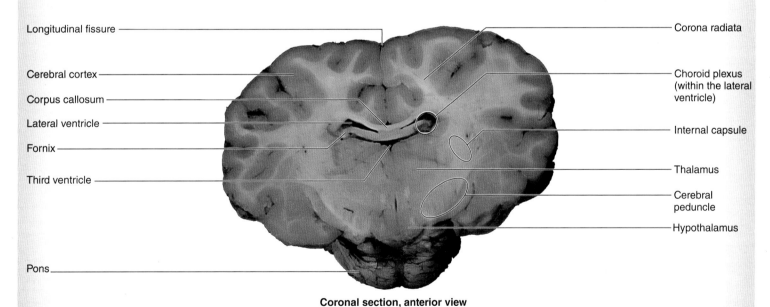

Figure 15.17 Coronal Section through the Sheep Brain.

(continued on next page)

(continued from previous page)

6. The **corpus callosum, fornix, internal capsule, corona radiata,** and **cerebral peduncles** are all fiber tracts. Recall from chapter 14 that a fiber tract is a bundle of myelinated axons. What is the function of these fiber tracts? _____

   _____

7. Using forceps, open the lateral ventricles a bit and see if you can identify **choroid plexus** in the wall of the ventricle. Upon gross observation, the choroid plexus kind of looks like "junk" inside the ventricle, but it is really a tuft of capillaries covered by ependymal cells.

   What two structures make up the choroid plexus?

   _____

   What is the function of the choroid plexus?

   _____

   _____

8. When you have finished your dissection, collect all the organic material from your dissecting pan and dispose of it in the proper containers (ask your laboratory instructor what these are). Dispose of the scalpel blades in the sharps container, and throw used paper towels and gloves into the garbage. Clean your dissecting tools and dissecting pan and return them to the proper storage area, and disinfect your laboratory workstation.

Physiology

Testing Cranial Nerve Functions

Cranial nerves carry different types of information from the brain to or from the target organ(s) for each nerve. **Sensory** (afferent) information travels from the target organ to the brain, whereas **motor** (efferent) information travels from the brain to the target organ.

The types of information carried by cranial nerves are classified in **table 15.5** on the right.

Table 15.6 lists the type of information carried by each cranial nerve and lists the specific functions of each cranial nerve.

INTEGRATE

LEARNING STRATEGY

A mnemonic device to remember if a cranial nerve is sensory (S), motor (M) or both (B) you can use the mnemonic device: **S**ome **S**ay **M**arry **M**oney **B**ut **M**y **B**rother **S**ays **B**ig **B**rains **M**atter **M**ore.

Table 15.5	Modalities within Cranial Nerves
Somatic Motor	Motor output to skeletal muscle
Somatic Sensory	Sensation from cutaneous touch, pain, temperature, position, and pressure receptors
Special Sensory	Somatic sensation from special sensory organs such as the eye and ear
Visceral Motor	Motor output to cardiac muscle, smooth muscle, and glands
Visceral Sensory	Sensation from sensory receptors within visceral organs (e.g., stretch receptors in bladder wall)

Table 15.6 Detailed Functions and Divisions of the Cranial Nerves

Nerve Number	Name	Type of Nerve	Functions
I	Olfactory	Special sensory	Sensory nerve of smell
II	Optic	Special sensory	Sensory nerve of vision
III	Oculomotor	Somatic motor	Motor to levator palpebrae superioris (raises the upper eyelid) and all extrinsic eye muscles except for the superior oblique and the lateral rectus
		Visceral motor (parasympathetic)	Motor to the ciliary body and iris (pupillary sphincter muscles) of the eye
IV	Trochlear	Somatic motor	Motor to superior oblique muscle of the eye
V	Trigeminal	Mixed	Sensation from the cornea, scalp, forehead, face, and teeth; motor to the muscles of mastication (chewing)
V_1	Ophthalmic branch	Somatic sensory	Sensory from the cornea, scalp, and forehead
V_2	Maxillary branch	Somatic sensory	Sensory from the face, cheeks, and maxillary (upper) teeth
V_3	Mandibular branch	Somatic motor	Motor to the muscles of mastication
		Somatic sensory	Sensory from the chin, mandibular (lower) teeth, and tongue
VI	Abducens	Somatic motor	Motor to the lateral rectus muscle of the eye
VII	Facial	Mixed	Sensory for taste from the anterior two-thirds of the tongue; motor to facial muscles, lacrimal glands, and salivary gland
		Somatic motor	Motor to the muscles of facial expression
		Special sensory	Sensory for taste from the anterior two-thirds of the tongue
		Visceral motor (parasympathetic)	Motor to the lacrimal glands, and the submandibular and sublingual salivary glands
VIII	Vestibulocochlear	Special sensory	Sensory nerve of hearing and balance
IX	Glossopharyngeal	Mixed	Sensory for taste from the posterior one-third of the tongue, sensation from the ear and pharynx, motor to the stylopharyngeus muscle and the parotid salivary gland
		Somatic motor	Motor to the stylopharyneus muscle
		Somatic sensory	Cutaneous sensation from the external ear; general sensation from posterior one-third of tongue
		Special sensory	Sensory for taste from the posterior one-third of the tongue
		Visceral motor (parasympathetic)	Motor to the parotid salivary gland
		Visceral sensory	Sensory from the carotid sinus and carotid body, mastoid air cells, pharynx, middle ear, and tympanic cavity
X	Vagus	Mixed	Motor to the pharynx and thoracic and abdominal viscera, sensation from the pharynx and ear, and sense of taste from pharynx and epiglottis (portion of larynx)
		Somatic motor	Motor to muscles of the pharynx, larynx, and palate (except stylopharyngeus and tensor veli palatini)
		Somatic sensory	Sensory from the external acoustic meatus, tympanic membrane, dura mater of the posterior cranial fossa, and auricle of the ear
		Special sensory	Sensory fibers for taste from the palate and epiglottis
		Visceral motor (parasympathetic)	Motor to glands of the pharynx and larynx, and smooth muscle of the heart, lungs, and abdominal viscera
		Visceral sensory	Sensory from the pharynx, larynx, bronchi, aorta, and abdominal viscera
XI	Accessory	Somatic motor	Motor to the trapezius and sternocleidomastoid muscles
XII	Hypoglossal	Somatic motor	Motor to the intrinsic and extrinsic muscles of the tongue

EXERCISE 15.9

TESTING SPECIFIC FUNCTIONS OF THE CRANIAL NERVES

Neurological tests similar to some of those described in this exercise are performed by physicians when testing for damage to one or more cranial nerves. These tests can indicate if a cranial nerve is damaged. However, they are not infallible. For instance, an inability to hear could indicate damage to the vestibulocochlear nerve. However, the damage could also reside in the auditory cortex of the brain. Each of the tests described here was chosen because it is both easy and quick to perform. **Table 15.7** lists the common disorders of the cranial nerves along with potential signs and symptoms, and causes of the disorders.

Table 15.7 Common Disorders of the Cranial Nerves

Nerve Number	Name	Signs and Symptoms of Damage	Potential Cause of the Disorder
I	Olfactory	Inability to smell (anosmia)	A fracture of the cribriform plate of the ethmoid can damage the olfactory nerves
II	Optic	Blindness on the affected side (hemianopia)	Intracranial tumor or stroke that damages the nerve or tract prevents visual information from reaching the brain
III	Oculomotor	Pupil dilation (mydriasis). Eye deviates down and out (strabismus) from muscle paralysis resulting in double vision (diplopia). Eyelid droops (ptosis).	Increased intracranial pressure is a common cause of compression of the nerve. The parasympathetic fibers that innervate the pupillary sphincter muscle are located on the surface of the nerve, so pupil dilation is often the first sign of nerve damage or increased intracranial pressure. Nerve damage results in paralysis of all extraocular muscles except the superior oblique and lateral rectus muscles. Paralysis of the levator palpebrae superioris muscle causes ptosis.
IV	Trochlear	Difficulty turning the eye inferior and lateral, which leads to double vision (diplopia)	Nerve damage results in paralysis of the superior oblique muscle
V	Trigeminal	Trigeminal neuralgia (tic douloureux), a sudden, intense pain along the course of one of the divisions of the nerve	Pressure on the nerve from the artery that courses alongside it stimulates sensory fibers within the nerve. Pain is often triggered by touching structures inside the mouth.
VI	Abducens	Eye deviates medially (adducts), causing double vision (diplopia)	Any disorder that increases intracranial pressure (for example, a stroke) can cause this nerve to be crushed against the clivus (sloped portion) of the sphenoid bone
VII	Facial	Bell palsy—paralysis of the muscles of facial expression on the side of the face with the affected nerve. Loss of taste sensation on the anterior two-thirds of the tongue (ageusia). Decreased salivation (hypoptyalism).	A viral infection that causes inflammation of the facial nerve is the most likely source. This problem often resolves itself within a couple of months.
VIII	Vestibulocochlear	Loss of balance and equilibrium, nausea, vomiting, and dizziness or inability to hear (anacusis)	Acoustic neuroma—a tumor originating in neurolemmocytes within the internal acoustic meatus—causes compression of the nerve
IX	Glossopharyngeal	Difficulty swallowing (dysphagia). Loss of taste sensation on the posterior one-third of tongue (ageusia). Decreased salivation (hypoptyalism).	Nerve damage interrupts the sensory component of the swallowing reflex
X	Vagus	Difficulty swallowing (dysphagia) or hoarseness (dysphonia)	Nerve damage interrupts the motor component of the swallowing reflex. Hoarseness results from paralysis of the muscles of the larynx.
XI	Accessory	Difficulty elevating the scapula or rotating the head	Nerve damage results in paralysis of the sternocleidomastoid and/or trapezius muscles
XII	Hypoglossal	When sticking out the tongue, it moves in the direction of the damaged nerve	Compression of nerve from increased intracranial pressure

EXERCISE 15.9A Olfactory (CN I)

The olfactory nerves **(figure 15.18)** are unique as cranial nerves in that there are more than two of them (the rest of the cranial nerves are paired—a right and left for each), and they are constantly being replaced. The nerves lie within the nasal epithelium, and their axons project through the **olfactory foramina** within the cribriform plate of the ethmoid bone (table 15.4). The olfactory neurons then synapse with neurons within the **olfactory bulbs** and the signals are sent to the brain via the **olfactory tracts.**

1. Obtain vials of peppermint, lemon, and vanilla oils.
2. Have your laboratory partner close his eyes. Open the vial of peppermint oil and pass it just under your laboratory partner's nose, then ask him to breathe in and identify the smell.
3. Repeat this process with the vials of lemon and vanilla oils. Allow some time between applications of the different oils. Damage to the olfactory nerves results in an inability to identify odors. Excessive smoking or inflammation of the nasal mucosa as a result of a viral infection can inhibit the sense of smell, and the sense of smell also declines with age.

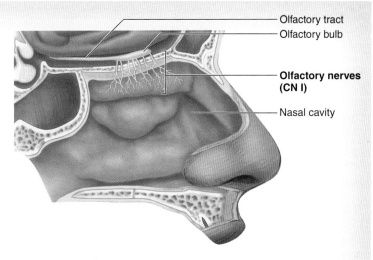

Figure 15.18 Location of the Olfactory Nerves (CN I).

EXERCISE 15.9B Optic (CN II)

Each of the optic nerves **(figure 15.19)** begins as axons of ganglion cells located within the **retina** of the eye. Different parts of the retina correspond to different portions of the visual field. The axons of these nerves exit the eye at the optic disc, or "blind spot," of the eye, at which point they become the **optic nerve,** which then travels posteriorly toward the diencephalon. Anterior to the infundibulum, most of the fibers cross over at the prominent **optic chiasm** (*chiasma,* a crossing of two lines) to the other side of the brain. The fibers then continue to travel posteriorly to reach the visual cortex in the occipital lobe of the brain. Figure 15.19 demonstrates the pattern of flow of visual information from the retina to the brain. When damage to the retina or optic nerve is suspected, visual field tests are performed to discover the location of the damage. These tests of visual function are beyond the scope of this course and will not be performed in this laboratory session.

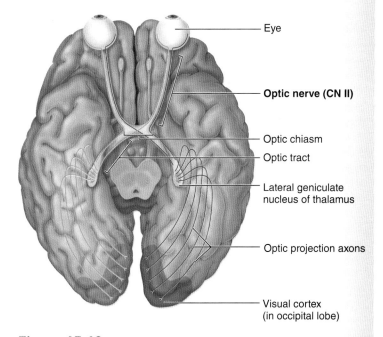

Figure 15.19 The Optic Nerve (CN II).

(continued on next page)

(continued from previous page)

EXERCISE 15.9C Oculomotor (CN III)

The oculomotor nerves **(figure 15.20)** send motor fibers to the majority of the extraocular eye muscles as well as to the muscles that control pupil diameter (tables 15.6 and 15.7).

1. Obtain a small flashlight. Look into one of your laboratory partner's eyes and observe the size of the pupil. While looking into her eye, gently shine the light into the eye (if it is a bright light, just bring it near the eye so that more light enters the eye—the goal here is not to blind your laboratory partner with the light).

   Did you observe a change in pupil diameter?_____

   If so, what happened?_____

2. Repeat the above activity, but this time observe the pupil of the other eye.

   Did you observe a change in pupil diameter?_____

   If so, what happened?_____

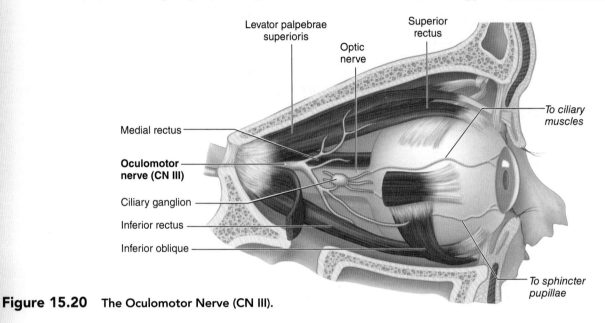

Figure 15.20 The Oculomotor Nerve (CN III).

EXERCISE 15.9D Trochlear (CN IV)

The trochlear nerve **(figure 15.21)** controls only one extraocular eye muscle—the superior oblique (tables 15.6 and 15.7). Ask your laboratory partner to look down and out (inferior and lateral). Weakness or an inability to perform this action indicates a weak superior oblique muscle or damage to the trochlear nerve.

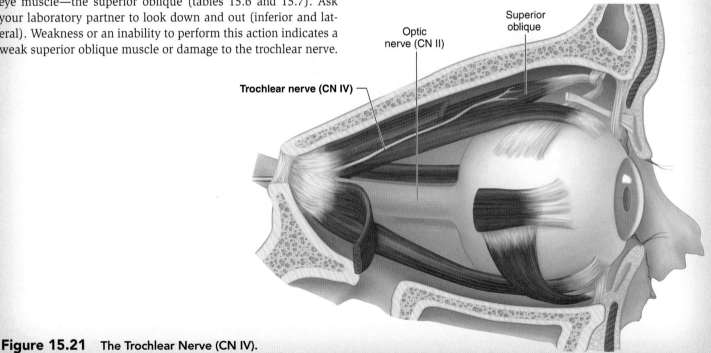

Figure 15.21 The Trochlear Nerve (CN IV).

EXERCISE 15.9E Trigeminal (CN V)

The trigeminal nerve **(figure 15.22)** is the largest and most complex of the cranial nerves. It has three branches: the ophthalmic, maxillary, and mandibular branches, which are named V_1, V_2, and V_3, respectively. It is the predominant nerve carrying sensory information from the face, but it also carries motor output to the muscles of mastication.

1. Obtain a feather and a cotton ball.
2. Have your laboratory partner close his eyes, and then proceed to gently touch his face with the feather in the sensory distribution areas of the trigeminal nerve shown in figure 15.22b. An inability to feel this sensation in one or more locations indicates damage to a branch of the trigeminal nerve.
3. Have your laboratory partner keep his eyes open and look up and away from you. *Very gently and lightly* touch a few strands of the fibers from the cotton ball to his cornea. This is a test for the *corneal reflex*, whose sensory component is carried by a branch of the trigeminal nerve. Touching the cornea with the cotton should cause him to blink. An absent corneal reflex indicates damage to the ophthalmic branch (V_1) of the trigeminal nerve (contact lens wearers may also have a diminished or absent corneal reflex).

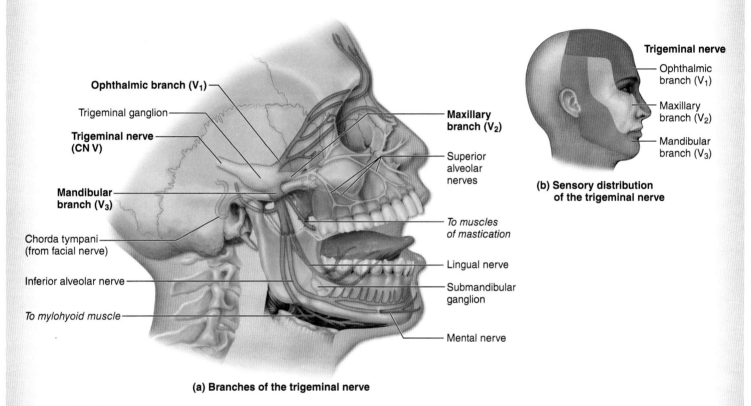

Figure 15.22 The Trigeminal Nerve (CN V).

EXERCISE 15.9F Abducens (CN VI)

The abducens nerve **(figure 15.23)** controls only one extraocular eye muscle—the lateral rectus (tables 15.6 and 15.7). Ask your laboratory partner to look laterally to the right. Weakness or an inability to do so indicates a weak lateral rectus muscle in the right eye or damage to the right abducens nerve.

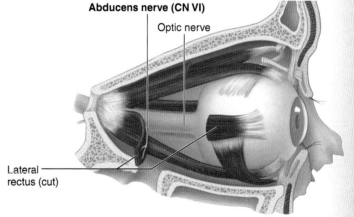

Figure 15.23 The Abducens Nerve (CN VI).

(continued on next page)

(continued from previous page)

EXERCISE 15.9G Facial (CN VII)

The facial nerve **(figure 15.24)** has several functions. Two major functions are to carry motor output to the muscles of facial expression and to carry sensory information to the brain from taste buds on the anterior two-thirds of the tongue **(figure 15.25)**.

1. Obtain vials of salt and sugar, and a cup of drinking water.

2. Ask your laboratory partner to close her eyes, open her mouth, and stick out her tongue. Place a few grains of salt on her tongue, and see if she can positively identify the taste as salty.

3. Have your laboratory partner take a drink of water to refresh her taste buds before performing the next test. Once again, ask her to close her eyes, open her mouth, and stick out her tongue. Place a few grains of sugar on the tip of her tongue and see if she can positively identify the taste as sweet. An inability to identify these tastes may indicate damage to the facial nerve.

4. Ask your laboratory partner to demonstrate facial expressions, such as surprise, happiness, sadness, and confusion. An inability to express these emotions facially may indicate paralysis of the muscles of facial expression, a common consequence of damage to the facial nerve.

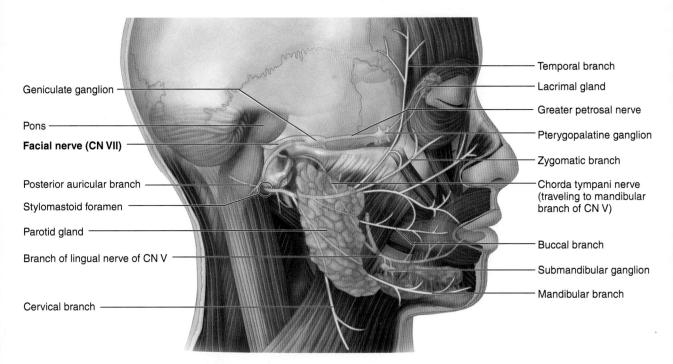

Figure 15.24 The Facial Nerve (CN VII).

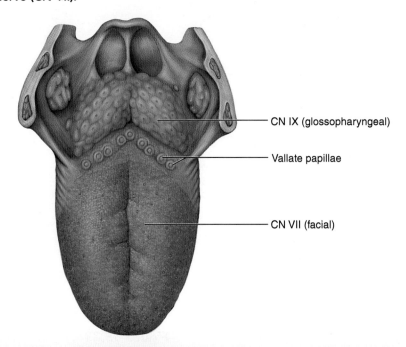

Figure 15.25 Innervation of the Taste Buds of the Tongue.

EXERCISE 15.9H Vestibulocochlear (CN VIII)

The vestibulocochlear nerve **(figure 15.26)** is two nerves: the **vestibular nerve,** which transmits nerve signals from the vestibule and semicircular canals to the brain regarding balance and equilibrium and the **cochlear nerve,** which transmits nerve signals from the cochlea to the brain regarding sound. Both nerves enter the petrous part of the temporal bone through the **internal auditory canal.** Figure 15.26 demonstrates the special sensory structures within this bone that are targets of the vestibulocochlear nerve.

1. Obtain a tuning fork. Ask your laboratory partner to close his eyes.
2. Holding the tuning fork by its base, gently strike it on the table, and then hold it near your laboratory partner's ear and ask him if he hears anything. An inability to hear the vibrations caused by the tuning fork can indicate damage to the vestibulocochlear nerve.

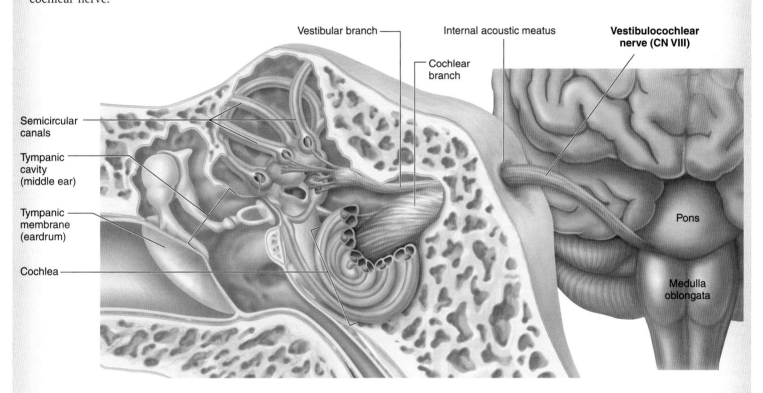

Figure 15.26 The Vestibulocochlear Nerve (CN VIII).

(continued on next page)

(continued from previous page)

EXERCISE 15.9I Glossopharyngeal (CN IX) and Vagus (CN X)

The glossopharyngeal and vagus nerves (**figures 15.27** and **15.28**, tables 15.6 and 15.7) both carry information to and from the soft palate, pharynx, and larynx to control reflexes such as the coughing and gagging reflexes. In addition, the glossopharyngeal nerve carries sensory information from the taste buds on the posterior one-third of the tongue (see figure 15.25). The vagus nerve is unique as a cranial nerve because it innervates many structures within the thoracic and abdominal cavities. It is the predominant pathway for parasympathetic information to travel from the brain to the visceral organs of the body. Though tests for the functioning of the glossopharyngeal and vagus nerves are not easy to perform, you will try to observe at least some of the functions of these nerves with this exercise.

1. Ask your laboratory partner to open her mouth and say "Ah" while you observe her soft palate and uvula.

2. Unilateral drooping of the soft palate or deviation of the uvula to one side may indicate damage to either the glossopharyngeal or the vagus nerves. The glossopharyngeal nerve carries sensory information from the pharynx to the brain, while the vagus nerve carries motor information back out to the muscles that raise the palate and that are used in swallowing.

3. Another test for the functioning of these two nerves is to test for the gag reflex. This will not be attempted in the lab because inexperienced testing of this reflex could cause your laboratory partner to choke or, at the very least, become very uncomfortable.

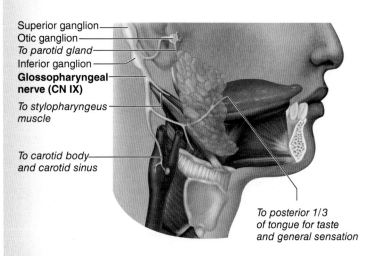

Figure 15.27 The Glossopharyngeal Nerve (CN IX).

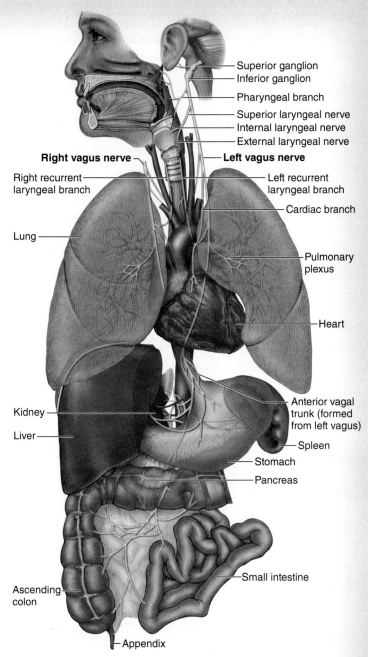

Figure 15.28 The Vagus Nerve (CN X).

EXERCISE 15.9J Accessory (CN XI)

The accessory nerve **(figure 15.29)** carries motor information to the trapezius and sternocleidomastoid muscles (tables 15.5 and 15.6).

1. Ask your laboratory partner to elevate her scapula ("shrug" her shoulders) to test the function of the trapezius muscle.

2. Next, ask her to rotate her head first to the right and then to the left to test the function of the sternocleidomastoid muscle. Damage to the accessory nerve would cause both of these actions to be weak or impossible due to paralysis of the muscles.

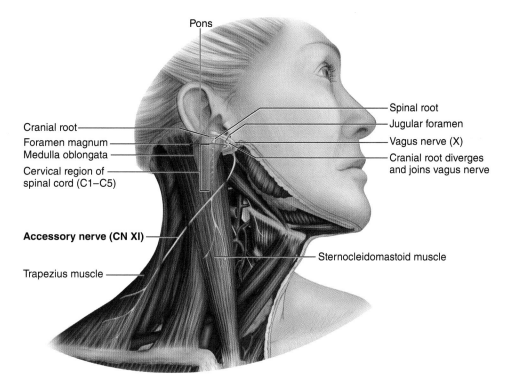

Figure 15.29 The Accessory Nerve (CN XI).

EXERCISE 15.9K Hypoglossal (CN XII)

The hypoglossal nerve **(figure 15.30)** innervates intrinsic and extrinsic muscles of the tongue (tables 15.6 and 15.7). Ask your laboratory partner to stick out his tongue. He should be able to stick out his tongue without his tongue deviating to one side. If the hypoglossal nerve is damaged, the tongue will deviate to the side of the damaged nerve.

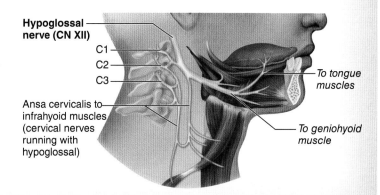

Figure 15.30 The Hypoglossal Nerve (CN XII).

Chapter 15: The Brain and Cranial Nerves

Name:_____
Date:_____ Section:_____

POST-LABORATORY WORKSHEET

The ❶ corresponds to the Learning Objective(s) listed in the chapter opener outline.

Do You Know the Basics?

Exercise 15.1: Cranial Meninges

1. Dura mater forms several major septa that dip into the major fissures of the brain. List the proper dural septum that lies in each of the described fissures below. ❶

    a. between the right and left hemispheres in the longitudinal fissure _____

    b. between the cerebrum and cerebellum _____

2. The dura mater is composed of _____ connective tissue. ❶

3. CSF and venous blood flowing out of the superior sagittal sinus directly flows into the _____, and then on to the _____ sinuses. ❷

4. All dural venous sinuses eventually drain into the _____ vein. ❸

Exercise 15.2: Brain Ventricles

5. Match the description in column A with the appropriate structure listed in column B. ❹

    Column A

    ____ the two most superior ventricles

    ____ a hole between the lateral ventricle and the third ventricle

    ____ the ventricle surrounded by the diencephalon

    ____ a channel located between the third and fourth ventricles

    ____ the ventricle located between the brainstem and cerebellum

    ____ capillaries that aid in production of CSF

    ____ a membrane that separates the two lateral ventricles

    Column B

    a. cerebral aqueduct
    b. choroid plexus
    c. fourth ventricle
    d. interventricular foramen
    e. lateral ventricle
    f. septum pellucidum
    g. third ventricle

6. Match the ventricular space listed in column A with the secondary brain vesicle from which it developed in column B. ❺

    Column A

    ____ lateral ventricles

    ____ third ventricle

    ____ cerebral aqueduct

    ____ fourth ventricle

    Column B

    a. diencephalon
    b. mesencephalon
    c. myelencephalon
    d. telencephalon

7. Cerebrospinal fluid circulates between the _____ mater and the _____ mater. ❻

Exercise 15.3: Superior View of the Human Brain

8. Match the description in column A with the appropriate structure listed in column B.

   Column A

   ____ a groove between the right and left cerebral hemispheres

   ____ a groove between the temporal and parietal lobes

   ____ a groove between the frontal lobe and the remaining lobes

   ____ ridges of brain tissue

   ____ shallow grooves on the surface of the brain

   ____ a groove that separates the cerebrum from the cerebellum

   ____ a fiber tract that connects the right and left cerebral hemispheres

   ____ the most anterior lobe of the brain

   ____ the most superior lobe(s) of the brain

   ____ the most lateral lobe(s) of the brain

   Column B

   a. central sulcus
   b. corpus callosum
   c. frontal lobe
   d. gyri
   e. lateral sulcus
   f. longitudinal fissure
   g. parietal lobe
   h. sulci
   i. temporal lobe
   j. transverse fissure

Exercise 15.4: Lateral View of the Human Brain

9. List the four major lobes that are visible on the surface of the cerebrum. (The insula is a lobe deep to the temporal lobe.)

   a. _____

   b. _____

   c. _____

   d. _____

Exercise 15.5: Inferior View of the Human Brain

10. List the three parts of the brainstem.

    a. _____

    b. _____

    c. _____

Exercise 15.6: Midsagittal View of the Human Brain

11. You can realize that a structure belongs to the diencephalon because "_____" is part of its name.

12. The _____ is a fiber tract that connects the left cerebral hemisphere to the right cerebral hemisphere.

13. Brain structure such as the mammillary bodies, which are involved with the suckling reflex and emotions, are part of the _____ system.

Exercise 15.7: Identification of Cranial Nerves on a Brain or Brainstem Model

14. Complete the table for the cranial nerves that extend from the three regions of the brainstem

Region of Brainstem	Cranial Nerves Extending from This Region of Brainstem
Medulla Oblongata	
Pons	
Midbrain	

15. Write the names of the numbered structures shown on the illustration of the inferior brain in the spaces provided.

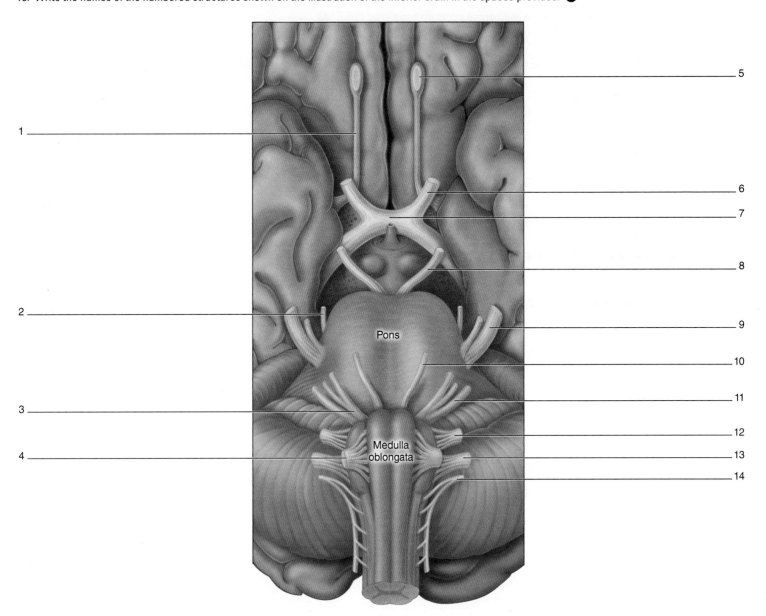

1. _____
2. _____
3. _____
4. _____
5. _____
6. _____
7. _____
8. _____
9. _____
10. _____
11. _____
12. _____
13. _____
14. _____

Exercise 15.8: Sheep Brain Dissection

16. Label the following figure of the inferior view of a sheep brain.

17. Label the following figure of a midsagittal section of a sheep brain.

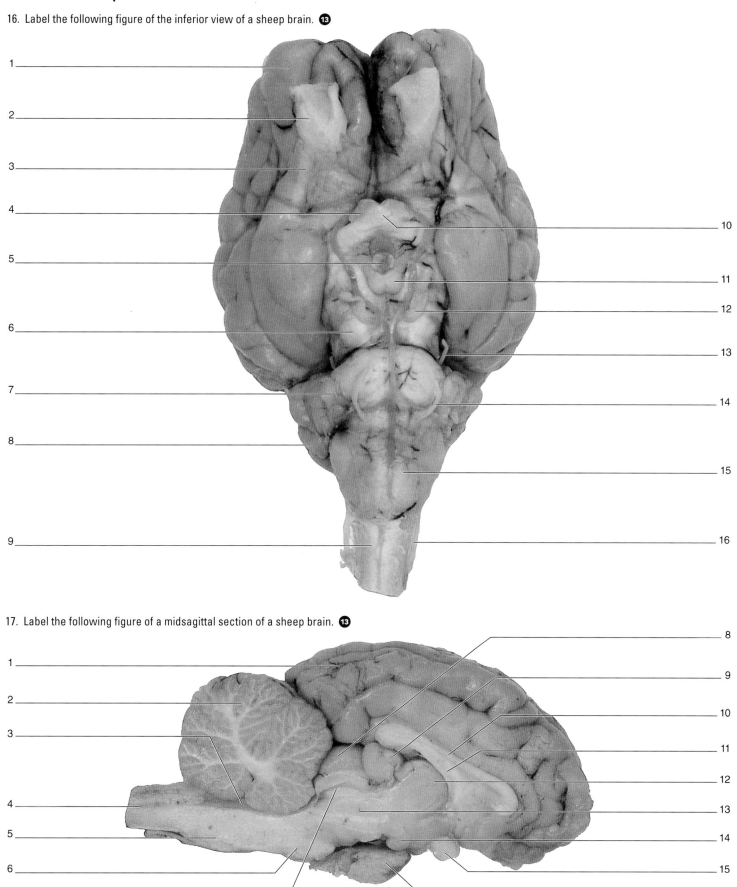

Chapter Fifteen *The Brain and Cranial Nerves* 421

422   Chapter Fifteen *The Brain and Cranial Nerves*

18. Label the following figure of a coronal section of a sheep brain.

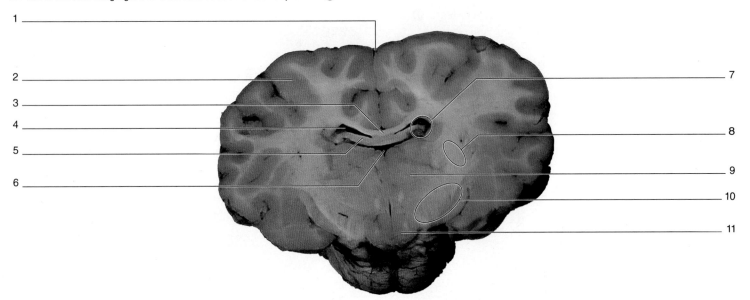

1 _____
2 _____
3 _____
4 _____
5 _____
6 _____
7 _____
8 _____
9 _____
10 _____
11 _____

Exercise 15.9: Testing Specific Functions of the Cranial Nerves

19. Match the disorder listed in column A with the cranial nerve associated with that disorder listed in column B.

Column A

____ 1. tongue deviates to one side when it is stuck out of the mouth
____ 2. soft palate droops on one side (2 answers)
____ 3. weakness in elevation of the scapula
____ 4. inability to smell
____ 5. pupillary reflexes are absent
____ 6. inability to taste bitter (sensed by posterior taste buds of the tongue)
____ 7. inability to pucker the lips
____ 8. blindness
____ 9. inability to laterally rotate the eye
____ 10. corneal reflex is absent
____ 11. difficulty turning the eye inferior and lateral
____ 12. inability to maintain balance and equilibrium

Column B

a. olfactory
b. optic
c. oculomotor
d. trochlear
e. trigeminal
f. abducens
g. facial
h. vestibulocochlear
i. glossopharyngeal
j. vagus
k. accessory (spinal accessory)
l. hypoglossal

20. Three of the twelve cranial nerves carry somatic motor fibers to extrinsic muscles of the eye. In the table below, list the nerves that carry somatic motor fibers to the extrinsic eye muscles, and then name the muscle(s) innervated by each nerve.

Cranial Nerve	Muscles Innervated

21. Several cranial nerves innervate structures of the tongue. In the table below, list the nerves that innervate structures of the tongue, and then list the functions of each nerve.

Cranial Nerve	Tongue Structures Innervated by Nerve

22. Four of the twelve cranial nerves carry parasympathetic motor output. In the table below, list the nerves that carry parasympathetic motor output, and then give the parasympathetic function(s) of each nerve.

Cranial Nerve	Parasympathetic Motor Function(s) of the Nerve

Can You Apply What You've Learned?

23. Match the following brain structures with the main region of the brain they are associated with by placing checks in the appropriate columns.

Area of the Brain	Brainstem	Cerebellum	Cerebrum	Diencephalon
Arbor Vitae				
Cerebellar Cortex				
Cerebral Cortex				
Corpus Callosum				
Fourth Ventricle				
Hypothalamus				
Intermediate Mass				
Lateral Ventricles				
Medulla Oblongata				
Midbrain				
Pineal Gland				
Pons				
Tectal Plate				
Thalamus				
Third Ventricle				
Vermis				

24. Describe the structural relationship between the olfactory bulbs, optic nerves, optic chiasm, optic tract, hypothalamus, infundibulum, pituitary gland, and mammillary bodies.

_____

_____

_____

_____

25. You should have noticed that the tectal plate (corpora quadrigemina) is much larger in the sheep brain, relative to total brain size, than in the human brain. Using this information, answer the following questions:

   a. What is the function of the superior colliculus?

   _____

   _____

   b. What is the function of the inferior colliculus?

   _____

   _____

   c. What does the difference in size between the superior and inferior colliculi tell you about the influence this region of the brain has on the overall functioning of a sheep versus a human? (That is, compare how much influence this area of the brain has on control over body functions.)

   _____

   _____

   _____

26. Compare and contrast structures of the human brain and the sheep brain by filling in the following table. That is, write down information about the relative size of the structure compared to the size of the entire brain. Then, based on function, explain why the structure might be more important for survival of the human or the sheep.

Brain Structure	Human Brain	Sheep Brain
Frontal Lobe		
Inferior Colliculi		
Mammillary Bodies		
Medulla Oblongata		
Olfactory Bulbs		
Pineal Body (Gland)		
Superior Colliculi		

27. What would be the effect of severing the corpus callosum?

_____

_____

_____

Can You Synthesize What You've Learned?

28. If the passage of fluid is blocked at the confluence of sinuses, into which sinuses will fluid back up?

29. An **acoustic neuroma** is a tumor that arises from neurolemmocytes (Schwann cells) surrounding the vestibular portion of the vestibulocochlear nerve (CN VIII). The tumor is benign, but generally grows within the confined space of the petrous part of the temporal bone, thus compressing the nerve and creating problems with balance and hearing loss. What nerve other than CN VIII would you expect to be affected by this tumor (due to its close proximity)?

30. When a light is shined into a patient's right eye, an examiner expects to see a change in pupil diameter in both eyes. The response, called the *consensual light reflex,* is used to test the function of two cranial nerves. The reflex involves one cranial nerve sending the afferent (sensory) signal toward the brain, and another cranial nerve sending the efferent (motor) signal out to the pupil.

   a. Which cranial nerve carries the afferent (sensory) signal to the brain?

   b. Which cranial nerve carries the efferent (motor) signal to the brain?

CHAPTER 16

The Spinal Cord and Spinal Nerves

OUTLINE AND LEARNING OBJECTIVES

Histology 430

Spinal Cord Organization 430

EXERCISE 16.1: HISTOLOGICAL CROSS SECTIONS OF THE SPINAL CORD 430
1. Describe the organization of gray matter and white matter in the spinal cord
2. Identify spinal cord structures as seen in cross section when viewed through the microscope
3. Identify the cervical, thoracic, lumbar, and sacral parts of the spinal cord in histological cross section

Gross Anatomy 433

The Spinal Cord 433

EXERCISE 16.2: GROSS ANATOMY OF THE SPINAL CORD 434
4. Compare and contrast the arrangement of the meninges covering the spinal cord and the brain
5. Identify the gross anatomical structures of the spinal cord on classroom models

Peripheral Nerves 435

EXERCISE 16.3: THE CERVICAL PLEXUS 436
6. Identify the phrenic nerve and describe its function

EXERCISE 16.4: THE BRACHIAL PLEXUS 437
7. Describe the organizational scheme of the brachial plexus
8. Identify the major nerves of the brachial plexus
9. Identify the nerves that innervate each compartment of the upper limb

EXERCISE 16.5: THE LUMBOSACRAL PLEXUS 440
10. Identify the major nerves of the lumbosacral plexus
11. Identify the nerves that innervate each compartment of the lower limb

INTRODUCTION

When we hear that someone has fractured a vertebra in an accident, we probably think the worst: "They are going to be paralyzed!" Though paralysis is common when the spinal cord is severed, the amount of paralysis and subsequent loss of function is highly dependent upon what part of the spinal cord is injured.

We know that the spinal cord is important for transmitting nerve signals between the body and the brain. An understanding of where nerve signals enter and exit the spinal cord is important for understanding the degree of paralysis that might result from a trauma. For instance, the most common vertebral fractures occur in the lower lumbar region (L_3–L_5). Did you know that a fracture here cannot sever the spinal cord and rarely results in paralysis? At the end of this laboratory exercise you will understand why. In contrast, a fracture high in the vertebral column, such as between the atlas (C_1) and axis (C_2), is commonly fatal. Not because of paralysis or loss of sensation from the limbs, but because the nerve that controls ➡

MODULE 7: NERVOUS SYSTEM

the diaphragm (the phrenic nerve) can no longer stimulate the muscle to contract, and breathing ceases.

Injury to a spinal nerve or the peripheral nerves that branch from them are most often the result of peripheral nerve compression or irritation. Some familiar examples are *carpal tunnel syndrome,* which causes pain, weakness, or numbness in the hand, and *sciatica,* a condition in which pain or numbness occurs in the lower back and may radiate to buttock, posterior thigh, leg, and foot. Understanding these clinical conditions requires an understanding of the organization of the spinal cord and spinal nerves.

In this laboratory session we will investigate the structure and function of the spinal cord and spinal nerves, with a focus on major nerves of the body and the structures they serve.

You may need to look at more than one model to be able to see all of the anatomic structures of the spinal cord and spinal nerves. For instance, to see all the parts of the brachial plexus, you may first need to observe a model of the head, neck, and thorax and then view a model of the upper limb. As far as the spinal cord is concerned, understanding the cross-sectional anatomy of the spinal cord is critical for understanding the links between the central and peripheral nervous systems.

Chapter 16: The Spinal Cord and Spinal Nerves

Name: _____
Date: _____ Section: _____

PRE-LABORATORY WORKSHEET

Also available at www.connect.mcgraw-hill.com

1. Which of the three meningeal layers forms the dural venous sinuses?
   _____

2. Cerebrospinal fluid flows between

   a. the dura mater and the arachnoid mater.

   b. the dura mater and the pia mater.

   c. the arachnoid mater and the pia mater.

   d. the cranial bones and the dura mater.

3. The spinal cord lies within the _____ canal of the vertebral column.

4. True or False: A nerve is a bundle of myelinated axons. _____

5. True or False: Both cranial and spinal nerves are part of the peripheral nervous system. _____

6. Which of the following sections of the spinal cord does not form a nerve plexus?

   a. cervical

   b. thoracic

   c. lumbar

   d. sacral

7. The muscles in the anterior compartment of the arm share what common action(s)?
   _____

8. The muscles in the posterior compartment of the thigh share what common action(s)?
   _____

9. The muscles in the anterior compartment of the leg share what common action(s)?
   _____

10. Which plexus (cervical, brachial, lumbar, or sacral) is composed of rami, trunks, divisions, and cords? _____

11. Define *tract*: _____

12. Define *ganglion*: _____

13. Somatic sensory neurons are classified as _____ neurons (based on neuron structures).

14. Somatic motor neurons are classified as _____ neurons (based on neuron structures).

Chapter Sixteen *The Spinal Cord and Spinal Nerves* 429

Histology

Spinal Cord Organization

Recall that the brain consists of an outer cortex of gray matter and an inner core of white matter (with some internal nuclei of gray matter as well). The spinal cord is organized just the opposite, with an outer cortex of white matter surrounding an inner core of gray matter. The gray matter of the spinal cord is organized into horns, so named because of their appearance. Posterior horns contain axons from sensory neurons that enter the spinal cord through the posterior roots, and also contain the cell bodies of interneurons (association neurons). **Anterior horns** contain cell bodies of somatic motor neurons, whose axons exit the spinal cord through the anterior roots and extend to skeletal muscle. **Lateral horns,** located only in the thoracic region and first two lumbar segments of the spinal cord, contain the cell bodies of visceral motor neurons, with axons that exit the spinal cord through the anterior roots and extend to visceral effectors (smooth muscle, cardiac muscle, and glands). The **gray commissure** is a horizontal bar of gray matter that surrounds the narrow central canal.

The white matter on each side of the spinal cord is partitioned into three **funiculi** (*funis,* cord) and are identified based on their anatomic position: posterior funiculus, lateral funiculus, and anterior funiculus. Each funiculus is composed of myelinated axons that transmit nerve signals through the spinal cord between the brain and body. The anterior funiculi are interconnected by the **white commissure.**

The spinal cord varies in diameter in different regions. It is the largest in the cervical region (called the **cervical enlargement**) and lumbar region (called the **lumbar enlargement**). These enlargements exist because of the increased number of neuron cell bodies in these regions that make the additional connections to the upper and lower limbs, respectively. In addition, there is a difference in the relative percentage of gray matter versus white matter in the different parts of the spinal cord. In general, there is more white matter at the cranial end of the spinal cord and more gray matter at the caudal end of the spinal cord. As you observe histological cross sections of the spinal cord, make note of the similarities and differences between the different parts of the spinal cord and try to associate these with functions. Also make note of similarities and differences between the organization of gray matter and white matter in the spinal cord as compared to the brain.

EXERCISE 16.1

HISTOLOGICAL CROSS SECTIONS OF THE SPINAL CORD

1. Obtain slides containing cross sections of the spinal cord (**figure 16.1**).

2. As you observe each slide, first observe it with the naked eye and attempt to identify the different parts of the spinal cord (cervical, thoracic, lumbar, and sacral) based on cross-sectional area and relative amounts of gray vs. white matter. **Table 16.1** compares the general characteristics of tissue sections from each spinal cord part, and **table 16.2** summarizes the features you will observe in a typical spinal cord cross section.

3. Place the slide on the microscope stage and bring the tissue sample into focus on low power.

4. Using figure 16.1, tables 16.1 and 16.2 and your textbook as guides, identify the following structures on the histology slide of the spinal cord.

   - ☐ anterior funiculus
   - ☐ anterior horn
   - ☐ anterior median fissure
   - ☐ anterior root
   - ☐ central canal
   - ☐ gray commissure
   - ☐ gray matter
   - ☐ lateral funiculus
   - ☐ lateral horn
   - ☐ nerve roots of cauda equina
   - ☐ posterior funiciculus
   - ☐ posterior horn
   - ☐ posterior median sulcus
   - ☐ posterior root
   - ☐ posterior root ganglion
   - ☐ white commissure
   - ☐ white matter

5. In the spaces below, draw a simple cross section of each spinal cord region, making note of the major differences between the regions. Be sure to indicate the magnification.

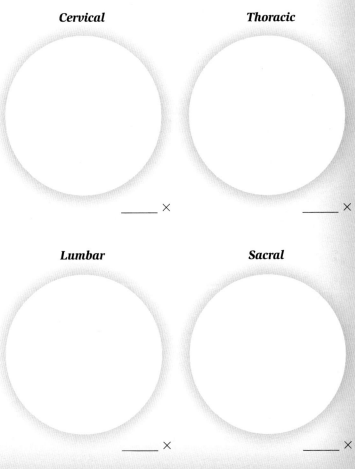

Cervical   *Thoracic*

_____ ×   _____ ×

Lumbar   *Sacral*

_____ ×   _____ ×

Chapter Sixteen The Spinal Cord and Spinal Nerves 431

INTEGRATE

LEARNING STRATEGY

One way to correctly distinguish anterior from posterior horns is to remember that *posterior* horns reach all the way to the *back*. That is, in the posterior horns the gray matter extends all the way to the edge of the spinal cord, whereas in the anterior horns the gray matter does not extend to the edge of the spinal cord.

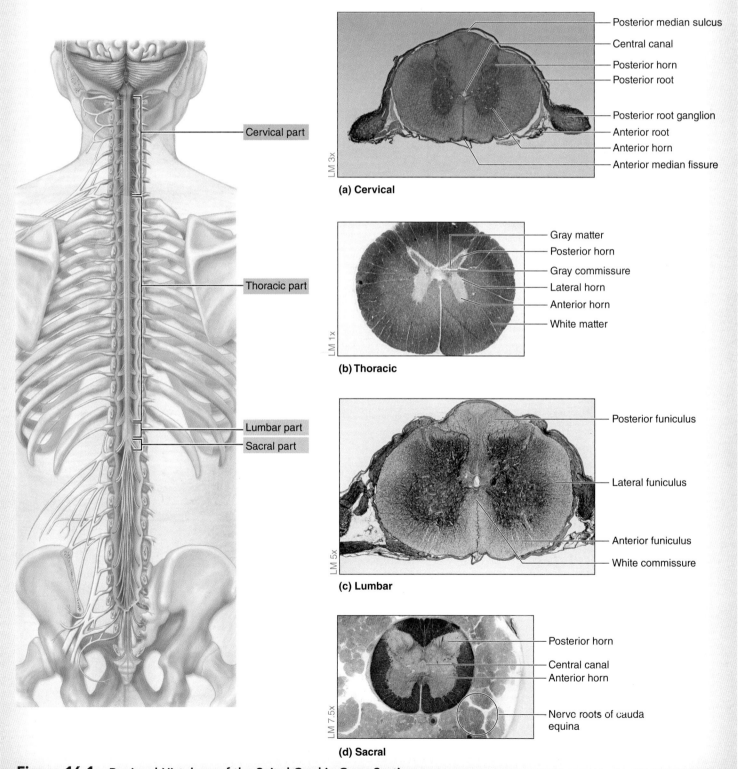

Figure 16.1 Regional Histology of the Spinal Cord in Cross Section.

(continued on next page)

INTEGRATE

CLINICAL VIEW
Lumbar Puncture versus Epidural Anesthesia

A **lumbar puncture** (spinal tap) involves inserting a needle between two vertebrae (L_3–L_4) in the region of the cauda equina of the spinal cord (below where the spinal cord typically ends in an adult at L_1 vertebra). The needle pierces the dura mater and arachnoid mater within the vertebral canal to enter the subarachnoid space, thus allowing for the withdrawal of cerebrospinal fluid (CSF). In contrast, an **epidural** (a procedure to administer an anesthetic to women during childbirth) does not penetrate any menigeal layers within the vertebral canal. Medications are injected into the epidural space, which contains mostly blood vessels and fat. Of note, when a lumbar puncture is performed on a young child or infant, the needle must be placed lower than vertebral level L_3–L_4 because the spinal cord ends at a lower level in these individuals.

Table 16.1 Regional Characteristics of the Spinal Cord

Region of the Spinal Cord	Relative Size	Predominant Tissue
Cervical	Relatively large cross-sectional area due to an abundance of nerves entering and exiting from the cervical and brachial plexuses	White matter
Thoracic	Relatively small cross-sectional area. A distinguishing feature is the presence of lateral horns, which contain the cell bodies of sympathetic motor neurons.	White matter
Lumbar	The largest cross-sectional area, due to an abundance of nerves entering and exiting from the lumbar and sacral plexuses. The anterior horns are large.	Gray matter
Sacral	The smallest cross-sectional area. Cross sections of numerous nerve roots surrounding the spinal cord are usually visible if the section goes through both the spinal cord and the roots of the cauda equina, which run adjacent to the spinal cord.	Gray matter

Table 16.2 Histology of the Spinal Cord in Cross Section

Structure	Description	Word Origin
Anterior Funiculus	Tracts of white matter that occupy the anterior region of the spinal cord between the anterior gray horn and the anterior median fissure	*anterior*, the front, + *funis*, cord
Anterior Horns	Gray matter on the anterolateral part of the spinal cord that primarily houses the cell bodies of somatic motor neurons (neurons that extend to skeletal muscle)	*anterior*, the front, + *horn*, resembling a horn in shape
Anterior Median Fissure	Deep groove on the anterior surface of the spinal cord	*anterior*, the front, + *median*, the middle, + *fissure*, a deep furrow
Central Canal	Small channel within the gray commissure of the spinal cord that is continuous with the ventricles of the brain; appears as a hole in a microscopic slide of the spinal cord	*central*, in the center, + *canalis*, a duct or channel
Gray Commissure	A horizontal bar of gray matter that surrounds the central canal	*commissura*, a seam
Gray Matter	Butterfly-shaped inner region of the spinal cord. It contains cell bodies of motor neurons or interneurons, depending upon the location.	literally, a substance that appears gray in color
Lateral Funiculus	Tracts of white matter that occupy each lateral side of the spinal cord	*latus*, to the side, + *funis*, cord
Lateral Horns	Small lateral extensions of gray matter found in the T_1–L_2 region of the spinal cord, and containing cell bodies of visceral motor neurons (neurons that extend to smooth muscle, cardiac muscle, and glands)	*latus*, to the side, + *horn*, resembling a horn in shape
Posterior Funiculus	Tracts of white matter that occupy the posterior region of the spinal cord between the posterior gray horn and the posterior median sulcus	*posterior*, the back, + *funis*, cord

Table 16.2	Histology of the Spinal Cord in Cross Section *(continued)*	
Structure	**Description**	**Word Origin**
Posterior Horns	Gray matter on the posterolateral part of the spinal cord that primarily houses the axons of sensory neurons and the cell bodies of interneurons (neurons contained completely within the central nervous system)	*posterior*, the back, + *horn*, resembling a horn in shape
Posterior Median Sulcus	Shallow groove located on the posterior surface of the spinal cord	*posterior*, the back, + *median*, the middle, + *sulcus*, a furrow
White Commissure	White matter that interconnects the anterior funiculi	*commissura*, a seam
White Matter	Outer portion of the spinal cord consisting of bundles of myelinated axons organized into tracts	literally, a substance that appears white in color

Gross Anatomy

The Spinal Cord

The spinal cord, part of the central nervous system, is surrounded by the same three meninges that surround the brain: the dura mater, arachnoid mater, and pia mater (figures 16.2 and 16.3). The dura mater surrounding the spinal cord consists of only a single tissue layer, whereas the dura mater surrounding the brain consists of two layers (periosteal layer and meningeal layer). In addition, there are extensions of pia mater for two structures that help stabilize the spinal cord within the vertebral canal: the denticulate ligaments and filum terminale. **Denticulate ligaments** are extensions of the pia mater that anchor the spinal cord to the arachnoid mater and dura mater at regular intervals all along the length of the spinal cord. The filum terminale is an extension of the pia mater at the caudal end of the spinal cord that anchors it inferiorly to the coccyx. **Table 16.3** lists some of the key features of the spinal cord that you will identify upon gross observation, and describes their functions.

Table 16.3	Gross Anatomy of the Spinal Cord	
Structure	**Description**	**Word Origin**
Anterior (Motor) Root*	A nerve root extending from the ventrolateral surface of the spinal cord that contains axons of somatic motor neurons and visceral motor neurons	*anterior*, the front, + *root*, the beginning part
Cauda Equina	A collection of anterior and posterior roots that extend inferiorly from the lumbar and sacral parts of the spinal cord and lie within the vertebral canal. It is named for the fact that the bundle of nerve roots resembles a horse's tail.	*cauda*, tail, + *equinus*, horse
Conus Medullaris	The cone-shaped distal tip of the spinal cord	*konos*, cone, + *medius*, middle
Denticulate Ligaments	Extensions of the pia mater located between anterior and posterior roots. These "ligaments" anchor the spinal cord to the arachnoid and dura mater at intervals between the locations where anterior and posterior roots pierce the dura mater.	*denticulus*, a small tooth, + *ligamentum*, a bandage
Filum Terminale	An extension of pia mater beyond the distal end of the spinal cord. It begins at the conus medullaris and attaches to the distal end of the dura mater at the coccyx.	*filum*, thread, + *terminatio*, ending
Posterior (Sensory) Root*	A nerve root extending from the posterolateral surface of the spinal cord that contains axons of sensory neurons	*posterior*, the back, + *root*, the beginning part
Posterior Root Ganglion*	A swelling on the posterior root that contains cell bodies of sensory neurons and satellite cells (glial cells)	*posterior*, the back, + *root*, the beginning part, + *ganglion*, a swelling
Rootlets (Radicular Fila)	Small branches of nerve fibers coming off the spinal cord that come together to form the anterior and posterior roots	*radicula*, a spinal nerve root, + *filum*, thread
Spinal Nerves	Nerves that form when anterior and posterior roots converge. A spinal nerve exits the vertebral canal through an intervertebral foramina, which forms between adjacent vertebrae.	*spinal*, relating to the spinal cord, + *nevus*, a white, cordlike structure

*The anterior root is also known as the ventral root. The posterior root is also known as the dorsal root.

Chapter Sixteen The Spinal Cord and Spinal Nerves

EXERCISE 16.2

GROSS ANATOMY OF THE SPINAL CORD

1. Observe a whole spinal cord from a cadaver (meninges intact) or a model of the spinal cord.

2. Using table 16.3 and your textbook as guides, identify the structures listed in **figure 16.2** on a human spinal cord or a model of the spinal cord. Then label them in figure 16.2.

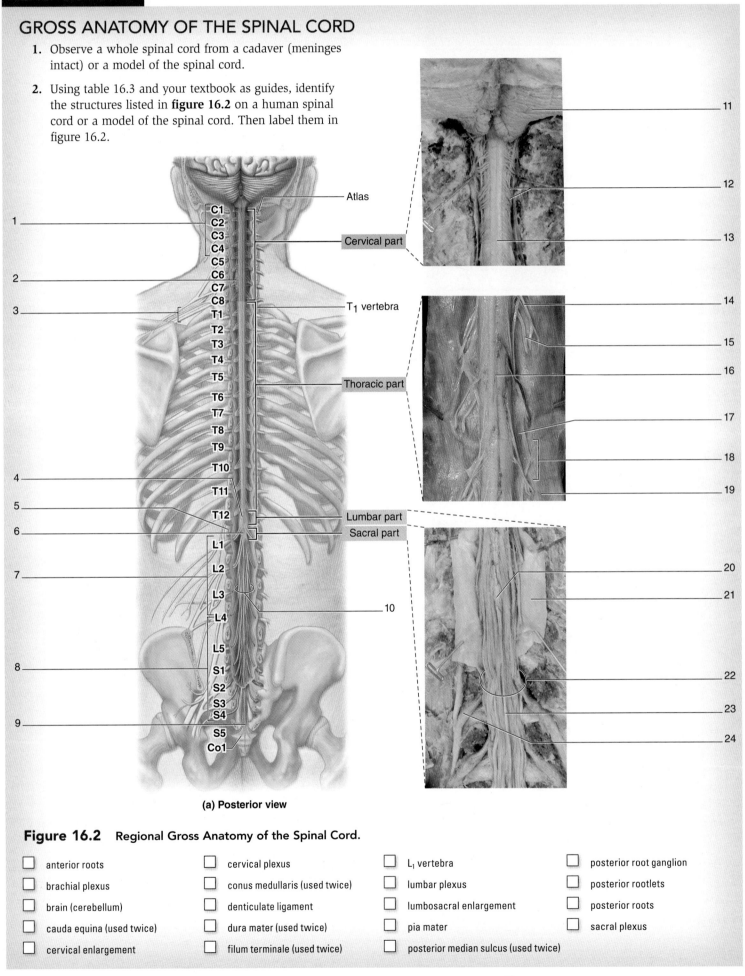

(a) Posterior view

Figure 16.2 Regional Gross Anatomy of the Spinal Cord.

- ☐ anterior roots
- ☐ brachial plexus
- ☐ brain (cerebellum)
- ☐ cauda equina (used twice)
- ☐ cervical enlargement
- ☐ cervical plexus
- ☐ conus medullaris (used twice)
- ☐ denticulate ligament
- ☐ dura mater (used twice)
- ☐ filum terminale (used twice)
- ☐ L₁ vertebra
- ☐ lumbar plexus
- ☐ lumbosacral enlargement
- ☐ pia mater
- ☐ posterior median sulcus (used twice)
- ☐ posterior root ganglion
- ☐ posterior rootlets
- ☐ posterior roots
- ☐ sacral plexus

3. Observe a model of a cross section of the spinal cord.

4. Using tables 16.2 and 16.3 and your textbook as guides, identify the structures listed in **figure 16.3** on the spinal cord model. Then label them in figure 16.3.

5. Describe key features on the anterior and posterior surface of the spinal cord (as seen in cross section) that will enable you to distinguish the anterior surface from the posterior surface on laboratory practical exams.

_____
_____
_____
_____
_____
_____

6. *Optional Activity:* **AP|R** **7: Nervous System**—Watch the "Typical Spinal Nerve" animation to help you visualize how spinal nerves relate to the spinal cord.

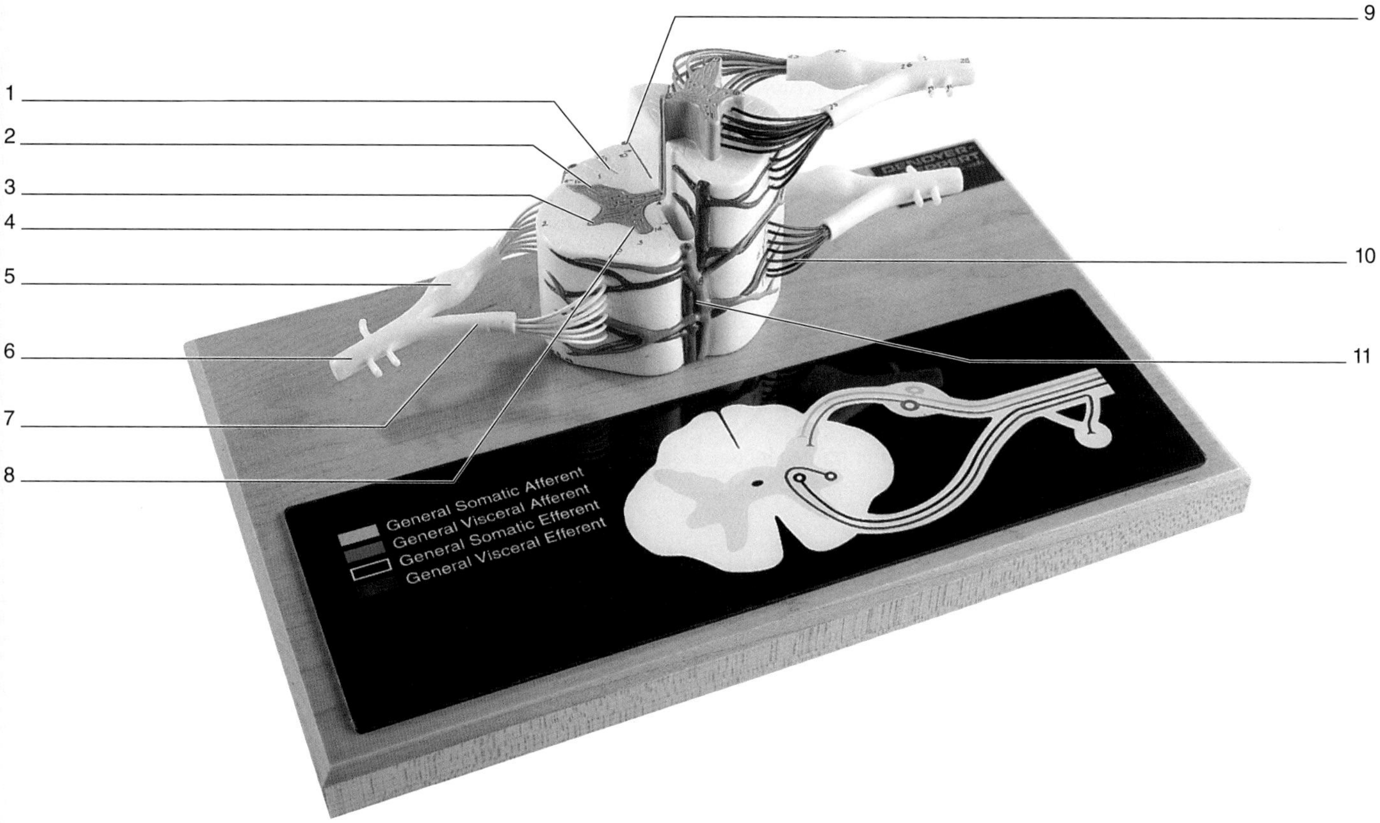

Figure 16.3 Model of a Spinal Cord in Cross Section.

☐ anterior horn ☐ anterior rootlets ☐ posterior horn ☐ posterior rootlets
☐ anterior median fissure ☐ lateral horn ☐ posterior median sulcus ☐ spinal nerve
☐ anterior root ☐ posterior funiculus ☐ posterior root ganglion

Peripheral Nerves

The peripheral nervous system consists of cranial nerves and spinal nerves. The structure and function of the cranial nerves was covered in chapter 15. In this laboratory exercise we will investigate the structure and function of the **spinal nerves.** After spinal nerves exit the vertebral canal through intervertebral foramina, they immediately branch into posterior and anterior **rami.** Throughout the length of the vertebral column, the **posterior rami** innervate both skin and muscles of the back that move the vertebral column (this excludes muscles located on the back that move the pectoral girdle and upper limb). In the thoracic region of the vertebral column, the **anterior rami** become **intercostal nerves** that supply skin, bone, and muscle of the thoracic cage. In the cervical, lumbar, and sacral regions of the vertebral column, the anterior rami form complex networks called **plexuses** (*plexus*, a braid), which then branch to form peripheral nerves that innervate skin, muscle, and bones of the limbs. There are four major nerve plexuses: cervical, brachial, lumbar, and sacral. In this laboratory exercise we will investigate each of these plexuses and explore the major peripheral nerves that arise from each plexus. Because the lumbar and sacral plexuses share some anterior rami, we will consider them together as the **lumbosacral plexus.**

EXERCISE 16.3

THE CERVICAL PLEXUS

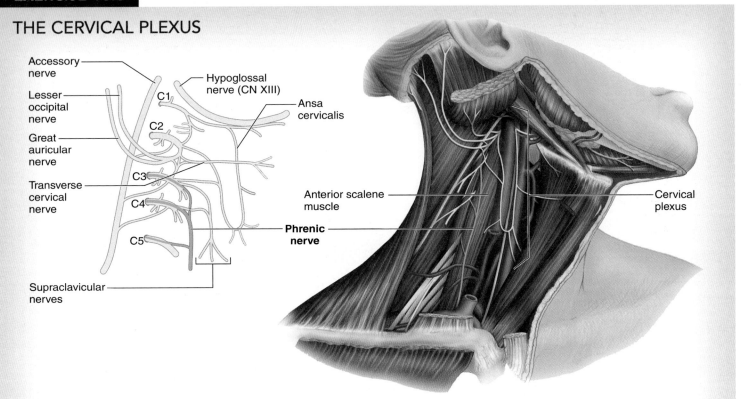

Figure 16.4 The Cervical Plexus and Phrenic Nerve in the Posterior Triangle of the Neck.

1. Observe a prosected human cadaver or a model of the head, neck, and thorax demonstrating nerves of the cervical plexus.

2. The cervical plexus arises from the anterior rami of spinal nerves C1–C4. Most of the nerves extending from the cervical plexus transmits nerve signals from skin of the neck and portions of the head and shoulders, and motor information to anterior neck muscles. Perhaps the single most important nerve branching from the cervical plexus is the **phrenic nerve,** which innervates the diaphragm (*phrenic*, relating to the diaphragm or mind—historically the diaphragm was thought to be the seat of the mind). In the neck, the phrenic nerves can be seen on the inferior part of the anterior scalene muscle, but they are difficult to identify in this location **(figure 16.4).** They are most easily identified within the thoracic cavity **(figure 16.5).** Here they travel within the mediastinum, between the pleural and pericardial cavities. If the heart or lungs have been removed from the cadaver, the phrenic nerves will be easy to identify (if they are still intact). If the heart and lungs remain (or if you are looking at a model of the thorax), you will need to look at the location where the pleural and pericardial cavities meet to locate the phrenic nerves.

3. Using your textbook as a guide, identify the structures listed in figure 16.5 on a human cadaver or a human torso model. Then label them in figure 16.5.

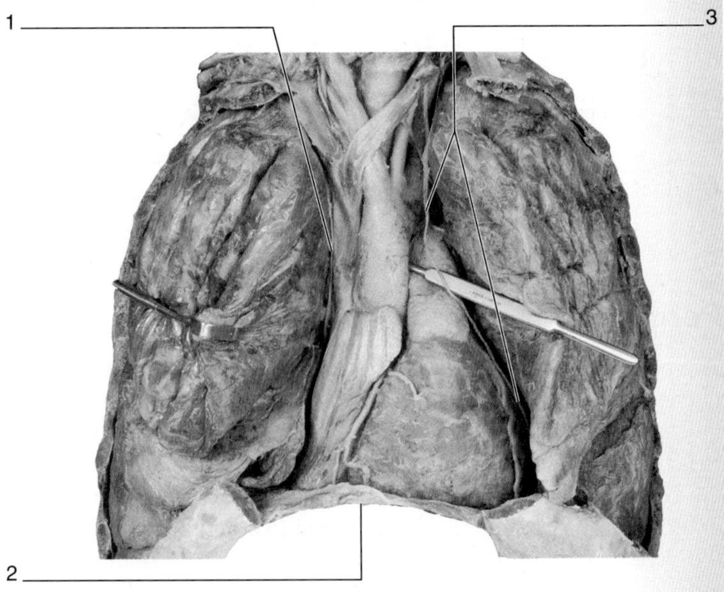

Figure 16.5 The Phrenic Nerves in the Thoracic Cavity.

☐ diaphragm ☐ right phrenic nerve
☐ left phrenic nerve

INTEGRATE

LEARNING STRATEGY

The phrenic nerve arises from the anterior rami of spinal nerves C3–C5, with the majority of fibers coming from C4. A handy way to remember this is "C3, C4, and C5 keep the body alive." If the phrenic nerves are unable to stimulate the diaphragm, breathing will cease and death will follow.

EXERCISE 16.4

THE BRACHIAL PLEXUS

1. Observe a prosected human cadaver or a model of the axilla and upper limb demonstrating nerves of the brachial plexus.

2. The **brachial plexus** arises from the anterior rami of spinal nerves C5–T1. The overall organization of the brachial plexus is complex, but follows an organized pattern. We will investigate the branching pattern of the brachial plexus in detail in this laboratory exercise. However, please note that the branching pattern of the brachial plexus is unique to the brachial plexus. Other plexuses (cervical, lumbar, and sacral) have their own branching patterns, which we will not explore in detail.

3. **Table 16.4** describes the organization of the brachial plexus. In this laboratory session, you will focus on identification of the **trunks, cords,** and **branches** of the brachial plexus. There are three **trunks,** which pass between the anterior and middle scalene muscles of the neck. The **superior trunk** forms from the anterior rami of C5 and C6, the **middle trunk** is a continuation of the anterior ramus of C7, and the **inferior trunk** forms from the anterior rami of C8 and T1. Each trunk divides into two **divisions,** anterior and posterior. All posterior divisions come together to form the **posterior cord,** and the anterior divisions come together to form either the **medial cord** or **lateral cord.** The cords are named for their location relative to the axillary artery. The best way to identify the terminal branches that arise from the medial and lateral cords of the brachial plexus (the musculocutaneous, median, and ulnar nerves) is to first locate the medial and lateral cords around the axillary artery and then use your fingers to spread apart the terminal branches that arise from the cords. Notice that the connections between the cords and branches appear to form a letter 'M' **(figure 16.6).**

4. *The Posterior Cord:* There are only two major nerves **(table 16.5)** that arise from the posterior cord: the **axillary nerve,** which remains in the axillary region to innervate the deltoid and teres minor muscles; and the **radial nerve,** a large nerve that continues to travel posteriorly along the arm and forearm and innervate skin and muscle along the way (thus, a nerve from the *posterior* cord innervates all structures in the *posterior* compartments of both the arm and the forearm including the triceps brachii).

5. *The Medial and Lateral Cords:* The medial and lateral cords form three main terminal nerves: the musculocutaneous, median, and ulnar nerves. The **musculocutaneous nerve** forms from the lateral cord and innervates muscles of the anterior compartment of the arm (biceps brachii, coracobrachialis, and brachialis). It is most easily identified where it pierces through the coracobrachialis muscle. The **ulnar nerve**

INTEGRATE

LEARNING STRATEGY

The segments of the brachial plexus, from proximal to distal, are **R**ami, **T**runks, **D**ivisions, **C**ords, and **B**ranches. A mnemonic for remembering this is, "**R**eally **T**ired, **D**rink **C**offee—**B**lack."

Table 16.4	Organization of the Brachial Plexus		
Structure	**Description**	**Number**	**Names**
Rami	These are the anterior rami of cervical spinal nerves. The rami combine to form trunks as they pass between the anterior and medial scalene muscles of the neck.	5	C5, C6, C7, C8, T1
Trunks	Located between the anterior and middle scalene muscles of the neck	3	Superior, middle, inferior
Divisions	Each trunk divides into an anterior and posterior division. The posterior divisions of each trunk come together to form the posterior cord. The anterior divisions of each trunk come together to form the medial and lateral cords.	6	Anterior and posterior division (for each trunk)
Cords	Located in the axilla and named for their location relative to the axillary artery	3	Medial, lateral, and posterior
Branches (Terminal Nerves)	Each terminal nerve innervates a compartment of the arm or forearm, with the exception of the median and ulnar nerves, which both innervate the anterior compartment of the forearm	5	Axillary, radial, musculocutaneous, median, ulnar

(continued on next page)

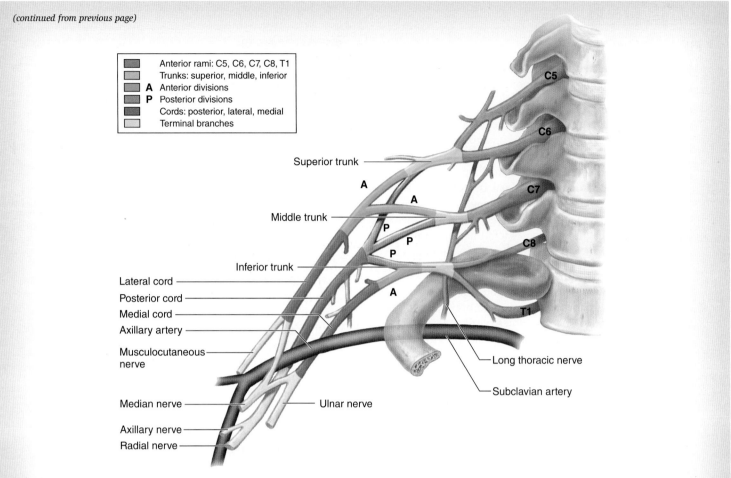

Figure 16.6 Organizational Scheme of the Brachial Plexus.

(continued from previous page)

forms from the medial cord and innervates muscles on the ulnar surface of the forearm along with most intrinsic muscles of the hand. It is most easily identified where it passes superficially behind the medial epicondyle of the humerus. In this location it is vulnerable to injury. When it is struck, as when you bang your elbow against a counter, you feel a tingly sensation along the course of the nerve. For this reason, this region of the elbow supplied by the ulnar nerve is commonly referred to as the "funny bone." The **median nerve** forms from branches of both medial and lateral cords. The median nerve is most easily identified in the distal, anterior compartment of the wrist, where it passes through the **carpal tunnel** into the hand. If your laboratory has a cadaver, see if you can pass a blunt probe through the carpal tunnel alongside the median nerve. This will allow you to better understand why the median nerve is so easily irritated by repetitive motions of the wrist. Its passage through the carpal tunnel is very narrow. As the flexor muscles of the forearm contract, their tendons rub against the median nerve, causing inflammation that results in **carpal tunnel syndrome**.

7. Using tables 16.4 and 16.5 and your textbook as guides, identify the structures of the brachial plexus listed in **figure 16.7** on a human cadaver or on models of the upper limb. Then label them in figure 16.7.

INTEGRATE

CLINICAL VIEW
Additional Nerves of the Brachial Plexus

The **long thoracic** nerve arises from the anterior rami of C5–C7 and innervates the **serratus anterior muscle.** Unlike most nerves, which lie deep to the muscles they innervate, the long thoracic nerve lies superficial to the muscle, which makes it particularly susceptible to injury. Injury can result in paralysis of the serratus anterior muscle. Remembering the discussion of the serratus anterior muscle in chapter 13, describe how to diagnose a paralyzed or nonfunctional serratus anterior muscle.

Table 16.5	Major Nerves of the Brachial Plexus			
Nerve	**Description**	**Motor Innervation**	**Sensory Innervation**	**Word Origin**
Axillary (C5–C6)	Arises from the posterior cord and runs around the surgical neck of the humerus en route to the shoulder. It can be easily injured when the surgical neck of the humerus is fractured.	Deltoid and teres minor	Skin on the dorsolateral aspect of the shoulder	*axilla*, the armpit
Long Thoracic (C5–C7)	Arises from the roots of C5, C6, and C7 and continues superficially to the serratus anterior muscle, which predisposes it to injury	Serratus anterior	None	*longus*, long, + *thoracis*, relating to the thorax
Median (C5–T1)	Arises from both the medial and lateral cords. Extends along the anteriomedial region of the arm, passes deep to the bicipital aponeurosis in the cubital fossa, then enters the forearm. After passing through the forearm, it courses deep to the flexor retinaculum through the carpal tunnel and into the hand. The tight passage through the carpal tunnel predisposes the nerve to injury (carpal tunnel syndrome).	Anterior compartment of the forearm except for the flexor carpi ulnaris and ulnar half of the flexor digitorum profundus	Skin of the lateral three and one-half digits on the palmar surface of the hand (and the distal aspect of the same digits on the dorsal surface), skin of the medial palm proximal to its innervation of the digit	*median*, in the middle
Musculocutaneous (C5–C7)	Arises from both the medial and lateral cords of the brachial plexus. It pierces through the coracobrachialis muscle and then extends between the brachialis and biceps brachii muscles.	Anterior compartment of the arm; biceps brachii, brachialis, and coracobrachialis	Skin overlying the lateral surface of the forearm	*muscus*, a mouse (muscle), + *cutaneous*, relating to the skin
Radial (C5–T1)	Arises from the posterior cord and extends deep within the posterior compartment of the arm adjacent to the humerus	Posterior (extensor) compartments of the arm and forearm	Skin overlying the posterior compartments of the arm and forearm and the dorsum of the hand (except the tips of the fingers and the entire fifth digit)	*radialis*, relating to the radius
Ulnar (C8–T1)	Arises from the medial cord, and continues along the medial aspect of the arm. It travels superficially immediately posterior to the medial epicondyle of the humerus (where it is easily injured).	Anterior forearm muscles on the medial side including the flexor carpi ulnaris and ulnar half of flexor digitorum profundus, as well as most intrinsic hand muscles	Skin of the medial one and one-half digits on the palmar surface of the hand, skin of the lateral surface of the hand on both palmar and dorsal surfaces	*ulnar*, relating to the ulna

(continued on next page)

440 Chapter Sixteen *The Spinal Cord and Spinal Nerves*

(continued from previous page)

8. In the space to the right, make a simple line drawing of the brachial plexus and label the structures using the list in figure 16.7.

INTEGRATE

LEARNING STRATEGY

The *cord* of the brachial plexus is *medial*, as in located medial to the axillary artery; the *nerve* is *median*, as in located in the middle (of the branches that form from medial and lateral cords).

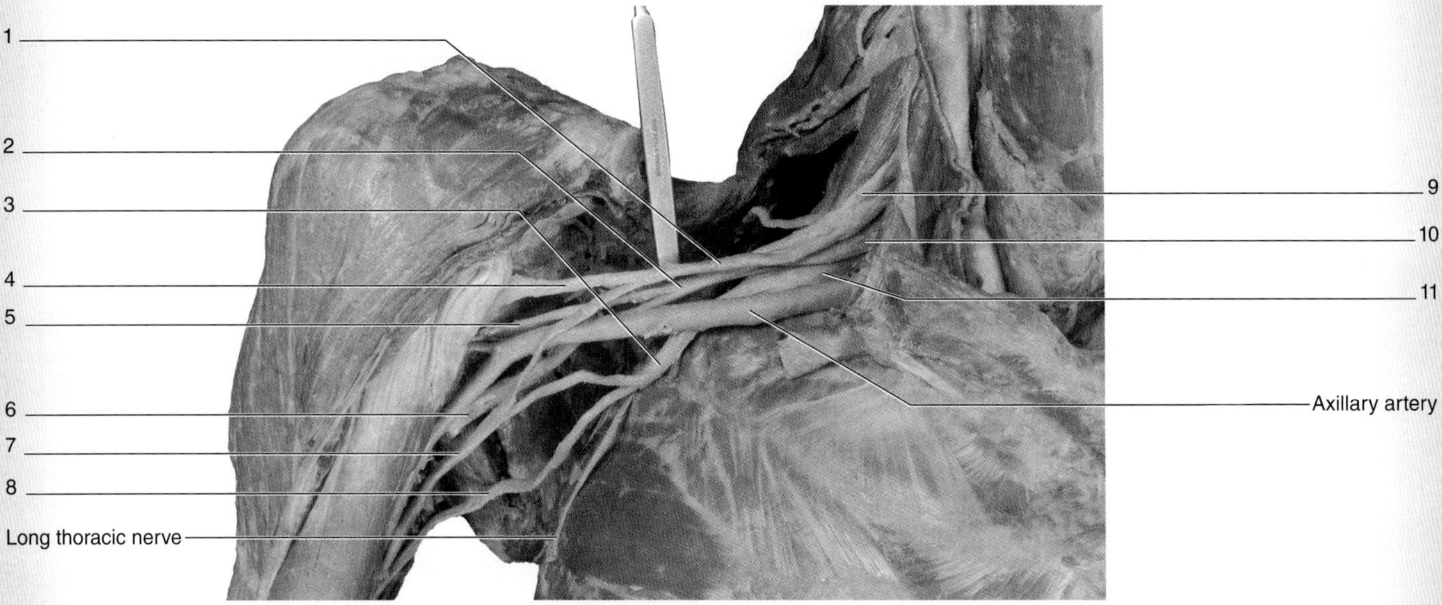

Right axilla, anterior view

Figure 16.7 Major Nerves of the Brachial Plexus.

- ☐ axillary nerve
- ☐ inferior trunk
- ☐ lateral cord
- ☐ medial cord
- ☐ median nerve
- ☐ middle trunk
- ☐ musculocutaneous nerve
- ☐ posterior cord
- ☐ radial nerve
- ☐ superior trunk
- ☐ ulnar nerve

EXERCISE 16.5

THE LUMBOSACRAL PLEXUS

1. Observe a prosected human cadaver or a model of the abdomen and lower limb demonstrating nerves of the lumbosacral plexus.

2. The lumbosacral plexus is really two plexuses, the **lumbar plexus** (arising from the anterior rami of L1–L4) and the **sacral plexus** (arising from the anterior rami of L4–S4). The branching pattern of the nerves is unique, as with the other plexuses in the body. However, we will not concern ourselves with the branching pattern in this laboratory exercise. We will focus on the major nerves that arise from the lumbosacral plexus **(table 16.6)**.

Table 16.6 Major Nerves of the Lumbosacral Plexus

Nerve	Description	Motor Innervation	Sensory Innervation	Word Origin
Common Fibular (S) (L4–S2)	Begins at the bifurcation of the sciatic nerve proximal to the popliteal fossa, passes superficially near the head of the fibula, wraps around the neck of the fibula, and then divides into superfcial and deep branches	Biceps femoris muscle, short head	Skin on the proximal posterolateral surface of the leg	*fibular*, relating to the fibula
Deep Fibular (S) (L4–S1)	Arises from the common fibular nerve at the neck of the fibula, passes through the extensor digitorum longus muscle into the anterior compartment of the leg	Anterior compartment of the leg, dorsal musculature of the foot	Skin on the first interdigital cleft of the foot	*deep*, situated at a deeper level than a corresponding structure, + *fibular*, relating to the fibula
Femoral (L) (L2–L4)	Runs along the lateral border of the psoas major muscle, travels under the inguinal ligament into the femoral triangle to innervate the anterior compartment of the thigh	Anterior compartment of the thigh	Skin of the anterior thigh and leg	*femoral*, relating to the femur or thigh
Inferior Gluteal (S) (L5–S2)	Exits the pelvis through the greater sciatic foramen inferior to the piriformis	Gluteus maximus	NA	*inferior*, lower, + *gloutos*, the buttock
Obturator (L) (L2–L4)	Runs along the medial border of the psoas major muscle, travels through the obturator foramen into the medial compartment of the thigh	Medial compartment of the thigh	Skin on the medial surface of the thigh	*obturatus*, to occlude or stop up
Pudendal (S) (S2–S4)	Exits the pelvis through the greater sciatic foramen inferior to the piriformis muscle, then travels through the lesser sciatic foramen to enter the perineum	Perineal muscles, external anal sphincter, and external urethral sphincter	External genitalia in both males and females	*pudendal*, that which is shameful
Sciatic (S) (L4–S3)	Exits the pelvis through the greater sciatic foramen inferior to the piriformis muscle, then enters the posterior compartment of the thigh through a groove between the ischial tuberosity and greater trochanter of the femur	Posterior compartment of the thigh (the tibial division of the sciatic is responsible for innervation of all posterior compartment muscles except for the short head of biceps femoris)	NA	*sciaticus*, the hip joint
Superficial Fibular (S) (L5–S2)	Arises from the common fibular nerve at the neck of the fibula and descends within the lateral compartment of the leg	Lateral compartment of the leg	Skin on the distal, lateral surface of the leg and the dorsal surface of the foot	*superficialis*, the surface, + *fibular*, relating to the fibula
Superior Gluteal (S) (L4–S1)	Exits the pelvis through the greater sciatic foramen superior to the piriformis	Gluteus medius, gluteus minimus, and tensor fascia lata muscles	NA	*superus*, above, + *gloutos*, the buttock
Tibial (S) (L4–S3)	Begins at the bifurcation of the sciatic nerve proximal to the popliteal fossa, runs along the tibialis posterior muscle, and then branches into two plantar nerves at the ankle	Posterior compartment of the leg, plantar musculature of the foot	NA	*tibial*, relating to the tibia

S = Sacral; L = Lumbar

(continued on next page)

(continued from previous page)

3. *Nerves Within the Gluteal Region:* The **gluteal nerves** (superior and inferior) are named for their exit location relative to the piriformis muscle of the deep buttock **(figure 16.8)**. The **superior gluteal nerve** arises superior to the piriformis, while the **inferior gluteal nerve** arises inferior to the piriformis. A small, but very important, nerve that also arises in this region is the **pudendal nerve** (*pudendal,* that which is shameful).

4. Using table 16.6 and your textbook as guides, identify the nerves and muscles in the gluteal region listed in figure 16.8 on a human cadaver or a model of the gluteal region. Then label them in figure 16.8.

INTEGRATE

LEARNING STRATEGY

If you remember that the word *pudendal* means "that which is shameful," it will be easy to remember the structures it innervates. It innervates all of our "shameful" structures—namely, the external genitalia. In addition, its fibers control the contraction of the external sphincters surrounding the anus and urethra. Wouldn't it be "shameful" if this nerve were damaged? A very minimal anesthetic given during childbirth is a **pudendal nerve block.** It blocks sensation from the birth canal (vagina), but the mother still receives sensations from the contracting uterus.

5. The **sciatic nerve** deserves special attention. The sciatic nerve arises inferior to the piriformis muscle and travels into the posterior compartment of the thigh. Along the way, it passes close to the ischial tuberosity. If you sit on a hard surface for a long period of time, the areas served by the nerve can lose sensation due to a blockage of blood supply to the fibers within the nerve. Once pressure is relieved and blood flow is restored, the sensations return as a painful "pins and needles" sensation. The next time this happens to you, note that the pins-and-needles sensation occurs on the posterior thigh and in the entire leg. The anterior and medial thigh are spared, because they receive innervation from the femoral and obturator nerves, respectively. The sciatic nerve is the *largest nerve in the body.* It is very large because it is actually two nerves, the **common fibular nerve** and the **tibial nerve.** These two nerves are bundled together in a common connective tissue sheath. The section where the two are bundled together is what we refer to as the sciatic nerve proper. Most commonly the two nerves separate from each other proximal to the popliteal fossa of the knee. However, it is not uncommon for them to separate much higher up in the posterior thigh. It is the **tibial division** of the sciatic nerve that is actually responsible for innervating the posterior compartment of the thigh. Because the nerve often remains bundled, we usually say it is the sciatic nerve that innervates the posterior compartment of the thigh.

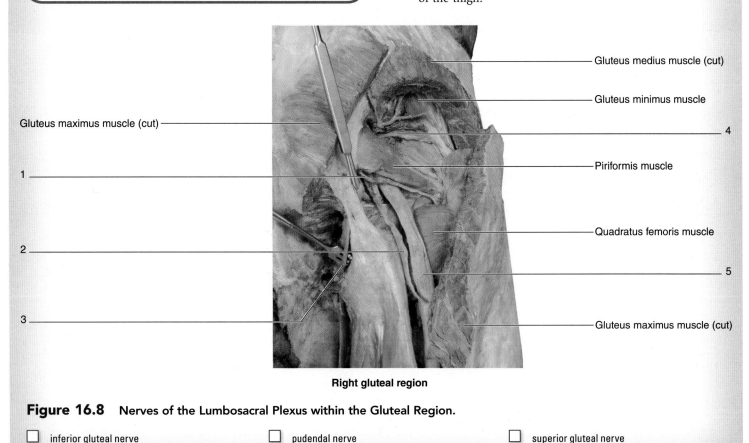

Right gluteal region

Figure 16.8 Nerves of the Lumbosacral Plexus within the Gluteal Region.

- ☐ inferior gluteal nerve
- ☐ pudendal nerve
- ☐ superior gluteal nerve
- ☐ posterior femoral cutaneous nerve
- ☐ sciatic nerve

6. *Nerves of the Lower Limb:* It will be easiest to learn the nerves of the lower limb if you can relate each nerve to a single compartment. Here is where your knowledge of limb compartments from chapter 13 becomes important. Recall that there are three compartments of the thigh (anterior, posterior, and medial) and three compartments of the leg (anterior, posterior, and lateral). Each compartment, for the most part, receives innervation from a single nerve of the lumbosacral plexus. If you first associate one nerve with one compartment, you will then be prepared to learn the "exceptions" to the rules. The general rules are as follows:

Compartment	Nerve
Anterior thigh	Femoral nerve
Posterior thigh	Sciatic nerve (tibial division)
Medial thigh	Obturator nerve
Anterior leg	Deep fibular nerve
Posterior leg	Tibial nerve
Lateral leg	Superficial fibular nerve

7. Using table 16.6 and your textbook as guides, identify the nerves and muscles listed in **figure 16.9** in the lower limb of a human cadaver or a model of the lower limb. Then label them in figure 16.9.

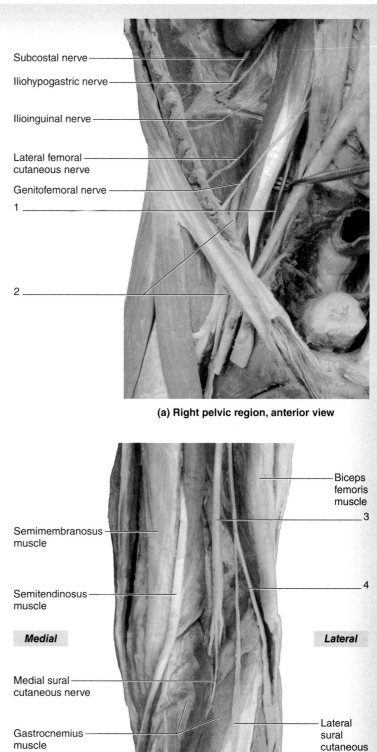

Figure 16.9 Nerves of the Lumbosacral Plexus. (a) Pelvis and anterior thigh. (b) Posterior view of lower limb showing popliteal region.

☐ common fibular nerve
☐ femoral nerve
☐ obturator nerve
☐ tibial nerve

INTEGRATE

CONCEPT CONNECTION

Spinal nerves, like cranial nerves, are part of the peripheral nervous system. All sensory information coming into the central nervous system "travels" within peripheral nerves. For example, when lamellated (Pacinian) corpuscles embedded in the dermis of the skin detect a sensation of deep pressure, they initiate action potentials, which propagate along a sensory neuron within a spinal nerve to enter the posterior horn of the spinal cord. From there, sensory information ascends to the brain so that the brain can *perceive* (rather than just "sense") the deep pressure stimulus. In fact, it is the brain that interprets exactly where, and precisely how much, pressure was applied. Although processing by the brain is necessary to provide this conscious awareness of sensory input, the brain is by no means required for "processing" *all* sensory and motor information. For example, spinal reflexes, such as the patellar tap reflex (see chapter 17), are neural circuits that require only the functioning of spinal nerves and the spinal cord. The overall process for the flow of information in reflexes and more complex sensory modalities is essentially the same: (1) peripheral nerves transmit sensory information to the central nervous system, (2) the central nervous system integrates this sensory information, and (3) peripheral nerves carry motor information away from the central nervous system to the appropriate effector organs.

Chapter 16: The Spinal Cord and Spinal Nerves

Name:_____
Date:_____ Section:_____

POST-LABORATORY WORKSHEET

The ❶ corresponds to the Learning Objective(s) listed in the chapter opener outline.

Do You Know the Basics?

Exercise 16.1: Histological Cross Sections of the Spinal Cord

1. The outer cortex of the spinal cord consists of _____ matter, whereas the inner medulla of the spinal cord consists of _____ matter. ❶

2. a. Which region(s) of the spinal cord contain lateral horns? ❷
   _____

   b. What structures are located within the lateral horns?
   _____
   _____

3. The cell bodies of _____ neurons are located in the posterior root ganglion. The posterior gray horn contains the cell bodies of _____. The anterior gray horn contains the cell bodies of the _____ neurons. ❷

4. There are two enlargements of the spinal cord, each of which corresponds with large numbers of nerve fibers extending from the spinal cord to innervate the limbs. The enlargement associated with the upper limb is the _____ and that associated with the lower limb is the _____. ❸

Exercise 16.2: Gross Anatomy of the Spinal Cord

5. Compare and contrast the features of the brain and spinal cord by filling in the following table. ❹

Structure	Unique Features within the Brain	Unique Features within the Spinal Cord
Dura mater		
Arachnoid mater		
Pia mater		
Central space (for example, ventricles)		
Location of gray matter		
Location of white matter		

6. The inferior end of the spinal cord lies at what vertebral level in the adult? ❺ _____

7. The groove that extends the length of the spinal cord on its anterior surface is the _____. The groove that extends the length of the spinal cord on its posterior surface is the _____. ❺

8. The structure that consists of dorsal and ventral roots within the lumbar region of the vertebral canal and the sacral canal (extending from the inferior end of the spinal cord) is the _____. ❺

Exercise 16.3: The Cervical Plexus

9. The cervical plexus is formed from the anterior rami of these spinal nerves: _____ ❻

10. Arguably the most important nerve arising from the cervical plexus is the _____ nerve, which arises from the anterior rami of these spinal nerves: _____. ❻

446 Chapter Sixteen *The Spinal Cord and Spinal Nerves*

Exercise 16.4: The Brachial Plexus

11. List the levels of organization of the brachial plexus in order, beginning with the anterior rami and ending with the terminal branches.

    a. Roots (anterior rami)

    b. _____

    c. _____

    d. _____

    e. Branches

12. Using colored pencils, color in the segments of the brachial plexus in the illustration below as per the colors in the key. Then label the nerves indicated by the leader lines.

    Anterior rami: C5, C6, C7, C8, T1
    Trunks: superior, middle, inferior
    Anterior divisions
    Posterior divisions
    Cords: posterior, lateral, medial
    Terminal branches

13. Match the region or compartment of the upper limb listed in column A with the nerve that innervates the majority of muscles in that region/compartment listed in column B. (Answer choices may be used more than once.)

    Column A

    ____ 1. shoulder (e.g., deltoid muscle)

    ____ 2. most of the anterior forearm (lateral aspect)

    ____ 3. anterior forearm (medial aspect and intrinsic muscles of the hand)

    ____ 4. anterior arm (e.g., biceps brachii)

    ____ 5. posterior arm (e.g., triceps brachii muscle)

    ____ 6. posterior forearm

    Column B

    a. ulnar n.

    b. musculocutaneous n.

    c. axillary n.

    d. radial n.

    e. median n.

Exercise 16.5: The Lumbosacral Plexus

14. The lumbosacral plexus contains the largest nerve in the human body, which is the _____ nerve. This nerve is really two nerves, the _____ nerve and the _____ nerve.

15. Match the region or compartment of the lower limb listed in column A with the nerve that innervates the majority of muscles in that region/compartment listed in column B. ⑪

   Column A
   ___ 1. medial thigh
   ___ 2. anterior leg
   ___ 3. posterior thigh
   ___ 4. lateral leg
   ___ 5. anterior thigh
   ___ 6. posterior leg

   Column B
   a. femoral n.
   b. tibial n.
   c. superficial fibular n.
   d. deep fibular n.
   e. sciatic n.
   f. obturator n.

Can You Apply What You've Learned?

16. Fill in the table below with features of the main nerve plexuses.

Plexus	Formed from Anterior Rami of These Spinal Nerves	Major Nerves Formed from the Plexus
Cervical		
Brachial		
Lumbosacral		

17. Explain, in your own words, why the cervical region of the spinal cord has more white matter than the other regions of the spinal cord.

    _____
    _____
    _____
    _____
    _____
    _____

18. In your own words, describe the organization (branching scheme) of the brachial plexus.

    _____
    _____
    _____
    _____

19. Fill in the following table with the nerve that innervates each compartment or listed structure.

Location and/or Muscles	Nerve
Anterior compartment of the arm	
Anterior compartment of the forearm	
Anterior compartment of the leg	
Anterior compartment of the thigh	
Deltoid and teres minor muscles	
Lateral compartment of the leg	
Medial compartment of the thigh	
Medial forearm and most intrinsic hand muscles	
Posterior compartment of the arm	
Posterior compartment of the forearm	
Posterior compartment of the leg	
Posterior compartment of the thigh	

Can You Synthesize What You've Learned?

20. If a posterior *root* is severed, what loss of function will result?

21. As relates to spinal cord injuries, what is an advantage of having the phrenic nerve arise from the cervical plexus instead of arising from the thoracic part of the spinal cord (even though the throacic spinal cord is closer in physical location to the diaphragm)?

22. A patient's spinal cord is severed between the first and second cervical vertebrae. This is a life-threatening situation. Given what you know about the peripheral nerves that exit the spinal cord in this region, explain why the individual will die without immediate medical intervention. (Hint: What spinal nerve exits at this location, and how does it relate to the major nerve plexuses of the body?)

CHAPTER 17

The Somatic and Autonomic Nervous Systems and Human Reflex Physiology

OUTLINE AND LEARNING OBJECTIVES

Gross Anatomy 452

Somatic Nervous System 452

EXERCISE 17.1: IDENTIFYING COMPONENTS OF A REFLEX ON A CLASSROOM MODEL 452
1. Compare and contrast the somatic and autonomic nervous systems
2. Identify the components of a somatic reflex on a classroom model or diagram

Autonomic Nervous System 453

EXERCISE 17.2: PARASYMPATHETIC DIVISION 453
3. Identify the major anatomical components of the parasympathetic division of the ANS
4. Identify the four cranial nerves that carry parasympathetic information, their associated ganglia, and their target organs

EXERCISE 17.3: SYMPATHETIC DIVISION 455
5. Identify the major anatomical components of the sympathetic division of the ANS
6. Trace the four major sympathetic outflow pathways

Physiology 459

Somatic Reflexes 459

EXERCISE 17.4: PATELLAR REFLEX 460
7. Describe the anatomical components of the patellar reflex
8. Determine the spinal cord levels that are tested using the patellar reflex test

EXERCISE 17.5: PLANTAR REFLEX 460
9. Describe the structural components of the plantar reflex
10. Determine the spinal cord levels that are tested using the plantar reflex test

EXERCISE 17.6: CORNEAL REFLEX 461
11. Describe the structural components of the corneal reflex, including a description of which cranial nerves carry the afferent and efferent signals
12. Describe the protective function of the corneal reflex

Autonomic Reflexes 461

EXERCISE 17.7: PUPILLARY REFLEXES 462
13. Describe the structural components of the pupillary reflex, including a description of which cranial nerves carry the afferent and efferent signals
14. Describe the protective function of the pupillary reflex

INTRODUCTION

Think about a time when you did something "reflexively." If someone asked you why you did it, you might have replied something like, "It was a reflex. I didn't think about it." This response is more correct than you may realize, for in a reflex reaction there is no "thinking" about the response. Specifically, this means that the afferent (sensory) and efferent (motor) signals do not pass through the "thinking" part of the nervous system, such as the frontal lobe of the brain (the premotor cortex). This makes the response extremely fast, which is important because most reflexes in the human body have a protective function.

Reflexes are preprogrammed responses that our nervous system uses to detect and respond to certain incoming sensory input. A classic example is the reflex that causes you to immediately pull your hand away from a hot stove.

As you recall this experience, do you remember how you likely noticed the pain associated with the injured skin only AFTER you pulled away from the hot stove? This is because the movement of the injured body part away from the stimulis is part of a reflex. In this reflex, the flow of information is only to the spinal cord, ➡

MODULE 7: NERVOUS SYSTEM

making it very fast. Why do these reflexes exist? Consider this: If you had to wait until the pain signal from the hot stove "registered" in your brain, the injury to your tissues would likely be greater than if you had pulled your hand away immediately.

There are generally five structures involved in a reflex: a **receptor,** a **sensory neuron,** the **control center,** a **motor neuron,** and an **effector** organ. **Table 17.1** summarizes the functions of each of these components. Reflexes can be classified based on the location of the control center or the type of effector involved. The classification system based on the *region* of the central nervous system where the control center is located divides reflexes into **spinal reflexes,** in which the control center is the spinal cord, and **cranial reflexes,** in which the control center is the brain. The classification system based on the *type of effector* involved divides reflexes into **somatic reflexes,** in which the effector is skeletal muscle, and **autonomic reflexes,** in which the effector is cardiac muscle, smooth muscle, or a gland.

The human **nervous system** is divided into two major functional components: the somatic nervous system and the autonomic nervous system **(figure 17.1).** The **somatic autonomic nervous (SNS)** can be thought of as the "voluntary" division—it involves sensations that are usually conscious. This would include sensory input from the five special senses (touch, taste, hearing, vision, and smell) and proprioceptors within our joints, tendons, and muscles. In addition, motor output to skeletal (voluntary) muscle tissue is also considered part of our somatic nervous system. In contrast, the **autonomic nervous system (ANS)** can be thought of as the "involuntary" division—it involves sensations that we are usually unaware of (such as blood pressure or oxygen levels in the blood) and motor output to cardiac muscle, smooth muscle, and glands. Both divisions of the nervous system contain reflexes.

In this laboratory session you will begin by reviewing the gross anatomical structure of a simple reflex arc, and the anatomical organization of the parasympathetic and sympathetic divisions of the autonomic nervous system (ANS). Exercises that are designed to test the function of specific somatic reflexes and autonomic reflexes will follow. Such tests are important in a clinical situation because they allow a clinician to test the region of the nervous system involved in the reflex for signs of dysfunction.

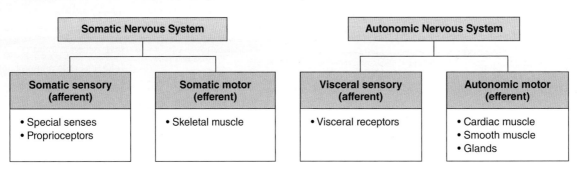

Figure 17.1 Functional Organization of the Nervous System.

Chapter 17: The Somatic and Autonomic Nervous Systems and Human Reflex Physiology

Name:_____
Date:_____ Section:_____

PRE-LABORATORY WORKSHEET

Also available at www.connect.mcgraw-hill.com

1. Match the following components of the reflex arc listed in column A with their correct definition listed in column B.

   Column A
   _____ effector
   _____ interneuron
   _____ motor neuron
   _____ receptor
   _____ sensory neuron

   Column B
   a. neuron that sends impulses to the CNS
   b. neuron that sends impulses away from CNS
   c. neuron within the CNS
   d. detects the stimulus
   e. carries out the response

2. Match the types of reflexes listed in column A with their components (control center and effector) listed in column B.

   Column A
   _____ spinal somatic
   _____ spinal autonomic
   _____ cranial somatic
   _____ cranial autonomic

   Column B
   a. brain and skeletal muscle
   b. brain and cardiac muscle, smooth muscle, or glands
   c. spinal cord and skeletal muscle
   d. s spinal cord and cardiac muscle or glands

3. Which three of the reflexes will you perform during this lab (circle the three that are applicable): spinal somatic, spinal autonomic, cranial somatic, or cranial autonomic?

4. The cranial nerve that carries the sense of vision from the retina of the eye is _____.

5. The cranial nerve that carries somatic sensation from the cornea of the eye is _____.

6. The muscle whose contraction causes the eye to close shut is the _____.

7. The cranial nerve that innervates the muscle indicated in question 6 is _____.

8. The patellar tendon is a common tendon for which muscle group? _____.

9. Explain the difference between the patellar tendon and the patellar ligament:
   _____
   _____
   _____

10. The _____ division of the autonomic nervous system (ANS) controls "rest and digest" activities, whereas the _____ division of the ANS controls the "fight or flight" response.

11. Activation of the sympathetic division of the ANS will cause the pupils of the eye to (circle one) <u>dilate/constrict.</u>

Chapter Seventeen: The Somatic and Autonomic Nervous Systems and Human Reflex Physiology

Gross Anatomy

Somatic Nervous System

Recall that the human nervous system is divided into two major parts: the **somatic nervous system (SNS)** and the **autonomic nervous system (ANS)**. In this exercise you will observe the components of a typical *somatic* reflex on a classroom model.

EXERCISE 17.1

IDENTIFYING COMPONENTS OF A REFLEX ON A CLASSROOM MODEL

1. Observe a classroom model or a diagram of a reflex arc.

2. Using **table 17.1** and your textbook as guides, identify the structures listed in **figure 17.2** on a classroom model or a diagram of the reflex arc. Then label figure 17.2.

Table 17.1	Components of a Reflex	
Component	**Description and Function**	**Example(s)**
Receptor	Sensory organ designed to detect a specific stimulus	Tactile corpuscle in skin
Sensory Neuron	Unipolar neuron that carries information from a receptor to the control center	Somatic sensory neuron
Control Center	Area within the brain or spinal cord that controls a particular reflex response	Superior colliculus in brain
Motor Neuron	Multipolar neuron that carries information from the control center to an effector	Somatic motor neuron
Effector Organ	Muscles or glands that carry out the response to a stimulus	Skeletal, smooth, or cardiac muscle; glands

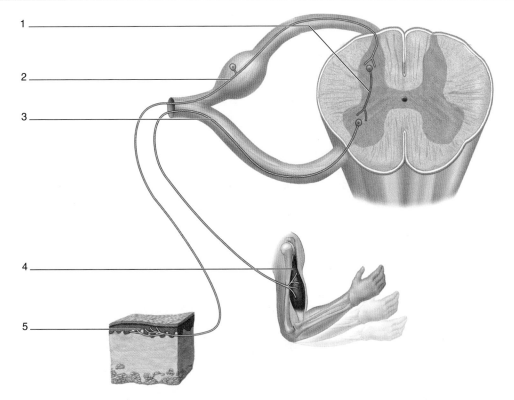

Figure 17.2 Components of a Reflex. Label this figure using the terms below.

☐ control center   ☐ effector   ☐ motor neuron   ☐ receptor   ☐ sensory neuron

Autonomic Nervous System

Recall the autonomic nervous system regulates cardiac muscle, smooth muscle, and glands. Because of the involuntary nature of the ANS, most visceral effectors are dually innervated by the ANS. This means that effector organs (e.g., heart, gastrointestinal tract) receive innervation from two different divisions of the ANS. These two divisions are the **parasympathetic** (craniosacral), which functions to control "rest and digest" functions, and **sympathetic** (thoracolumbar) division, which is activated in emergency situations and during the "fight-or-flight" response **(figure 17.3)**. Generally, one division will increase the organ's activity and the other will decrease the organ's activity. For example, the cells that control heart rate are innervated by both parasympathetic and sympathetic nerve fibers. When the parasympathetic division is activated, heart rate decreases; whereas when the sympathetic division is stimulated, heart rate increases. In this exercise you will observe the structure of these two divisions by labeling line drawings of the two divisions (using your textbook as a guide). In the "Physiology" section of this chapter you will explore the function of these two divisions of the ANS.

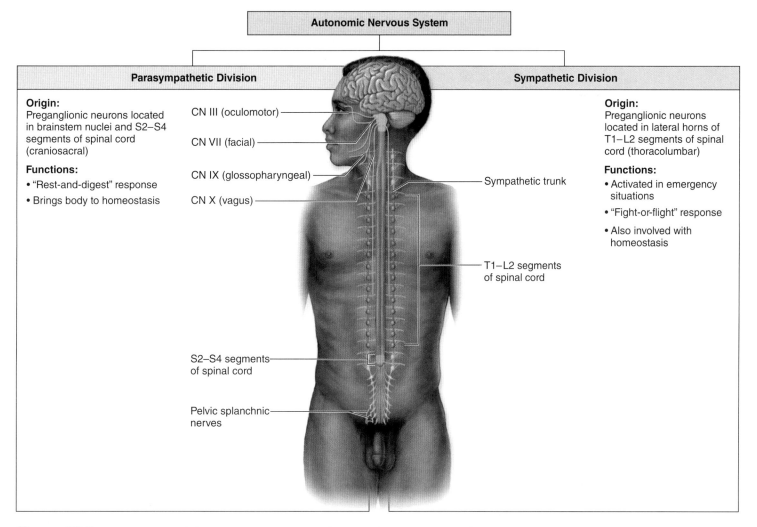

Figure 17.3 Comparison of the Parasympathetic and Sympathetic Divisions of the ANS.

EXERCISE 17.2

PARASYMPATHETIC DIVISION

1. Observe **figure 17.4,** which is an overview of the parasympathetic pathways. Notice that the term "parasympathetic" refers to the fact that the nerve fibers of this division originate from the CNS adjacent to where the nerve fibers of the sympathetic division originate (*para*, alongside). The parasympathetic division, then, arises from portions of the CNS that are superior and inferior to the components of the sympathetic division. Specifically, the parasympathetic division arises from the brain (from fibers contained within cranial nerves III, VII, IX, and X) and the sacral region of the spinal cord. This is where the alternate term for this division, *craniosacral*, arises. Figure 17.4 depicts several key structures of the parasympathetic division of the ANS. Using your textbook as a guide, identify the structures listed in figure 17.4.

(continued on next page)

454 Chapter Seventeen The Somatic and Autonomic Nervous Systems and Human Reflex Physiology

(continued from previous page)

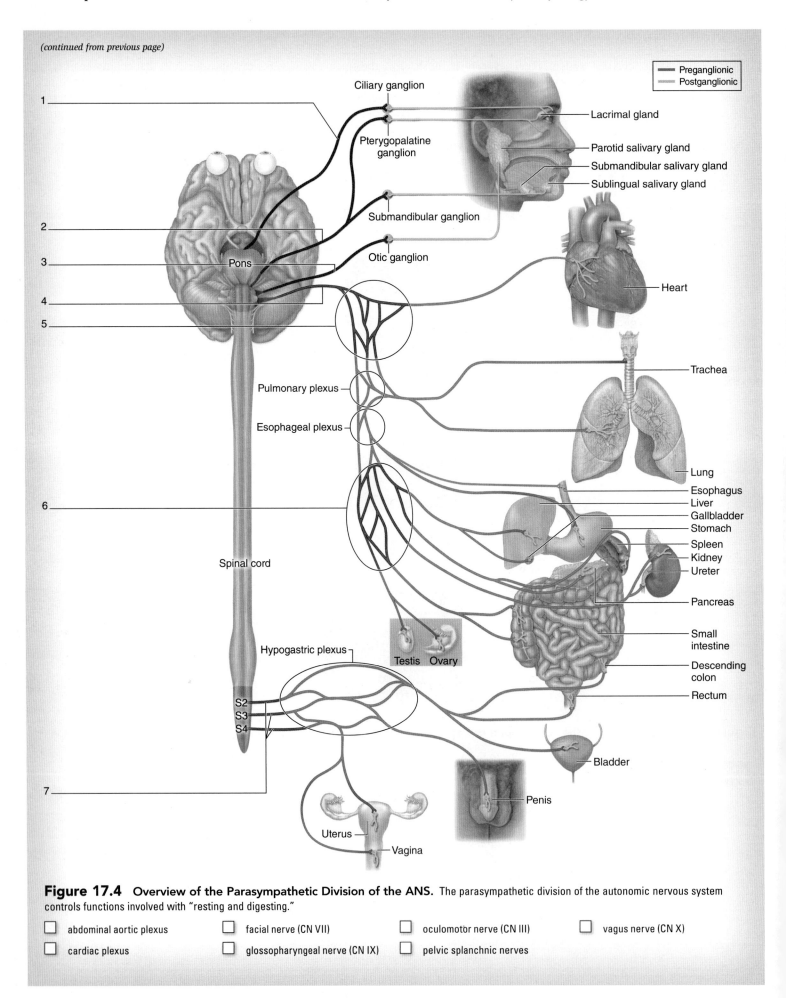

Figure 17.4 Overview of the Parasympathetic Division of the ANS. The parasympathetic division of the autonomic nervous system controls functions involved with "resting and digesting."

- ☐ abdominal aortic plexus
- ☐ cardiac plexus
- ☐ facial nerve (CN VII)
- ☐ glossopharyngeal nerve (CN IX)
- ☐ oculomotor nerve (CN III)
- ☐ pelvic splanchnic nerves
- ☐ vagus nerve (CN X)

INTEGRATE

LEARNING STRATEGY

The "big picture" way of remembering where parasympathetic innervation of visceral organs arises is to learn a couple of basic rules:

1. Specialized structures within the head (such as salivary glands) receive parasympathetic innervation through cranial nerves III, VII, IX, and X.
2. Nearly all thoracic and abdominal viscera receive parasympathetic innervation via the vagus nerve (CN X). Recall from chapter 16 that vagus means "wanderer" (because it wanders throughout the body). The vagus nerve is the only cranial nerve to innervate structures below the head.
3. Viscera within the pelvis are innervated by fibers that arise from the sacral spinal cord (e.g., structures of both the reproductive and urinary systems).

Clinically, this means that someone who has been paralyzed from the neck down will still have parasympathetic innervation of the majority of the vital viscera (e.g., heart, small intestine). However, the individual will lose function of involuntary sphincters such as those that control the flow of urine from the urinary bladder.

EXERCISE 17.3

SYMPATHETIC DIVISION

1. The sympathetic division of the ANS arises from the thoracolumbar regions of the spinal cord. Unfortunately for students trying to learn this material, the pathways for sympathetic fibers are more complicated than those for parasympathetic fibers. One reason for this is that in order for the sympathetic response to have widespread and immediate effects, it must activate organs throughout the entire body in unison to initiate the "fight-or-flight" response. The biggest difference you will notice between the sympathetic division and the parasympathetic division is the location of the ganglia. Most parasympathetic ganglia are located within the target organs themselves **(table 17.2)**. In contrast, sympathetic ganglia are located in either **sympathetic trunk ganglia** (paravertebral ganglia) or **prevertebral ganglia**. In addition, there are three separate pathways for preganglionic fibers to reach these ganglia from the spinal cord, with additional nerve fibers that extend to the **adrenal medulla**. In this exercise you will trace these pathways. As you trace each pathway, consider the organs that are regulated.

(continued on next page)

(continued from previous page)

2. Use table 17.2 as a guide and think about the "fight-or-flight" response initiated by the sympathetic division. Then, for each of the following organs, first identify the pathway for nerve signals to reach the organ, then name the specific effect created by the nervous stimulation.

   _____

   _____

   Heart (both rate and strength of contraction):

   Pathway: _____

   Response: _____

   Small intestine:

   Pathway: _____

   Response: _____

   Blood vessels within skeletal muscle:

   Pathway: _____

   Response: _____

   Adrenal medulla:

   Pathway: _____

   Response: _____

3. **Figure 17.5** depicts several key structures of the sympathetic nervous system. Using your textbook as a guide, identify the structures listed in the figure. Then label figure 17.5.

4. **Figure 17.6** contains four parts, each representing one outflow pattern for the sympathetic nervous system. Using two colored pens or pencils and your textbook as a guide, draw the pathway for preganglionic neurons in one color and postganglionic neurons in the other.

Table 17.2	Comparison of Parasympathetic and Sympathetic Divisions	
Feature	**Parasympathetic Division**	**Sympathetic Division**
Divergence of Axons	Few (1 axon innervates < 4 ganglionic cell bodies)	Extensive (1 axon innervates > 20 ganglionic cell bodies)
Function	Conserves energy and replenishes energy stores; maintains homeostasis; "rest-and-digest" division	Prepares body to cope with emergencies and intensive muscle activity; "fight-or-flight" division
Length of Postganglionic Axon	Short	Long
Length of Preganglionic Axon	Long	Short
Location of Ganglia	Terminal ganglia located close to the target organ; intramural ganglia located within wall of target organ	Sympathetic trunk (paravertebral) ganglia located on either side of vertebral column; prevertebral (collateral) ganglia located anterior to vertebral column and descending aorta
Location of Ganglionic Neuron Cell Bodies	Terminal or intramural ganglion	Sympathetic trunk ganglion (paravertebral) or prevertebral ganglion
Location of Preganglionic Neuron Cell Bodies	Brainstem and lateral gray matter in S2–S4 segments of spinal cord	Lateral horns in T1–L2 segments of spinal cord
Rami Communicantes	None	White rami attach to T1–L2 spinal nerves; gray rami attach to all spinal nerves

Chapter Seventeen The Somatic and Autonomic Nervous Systems and Human Reflex Physiology   457

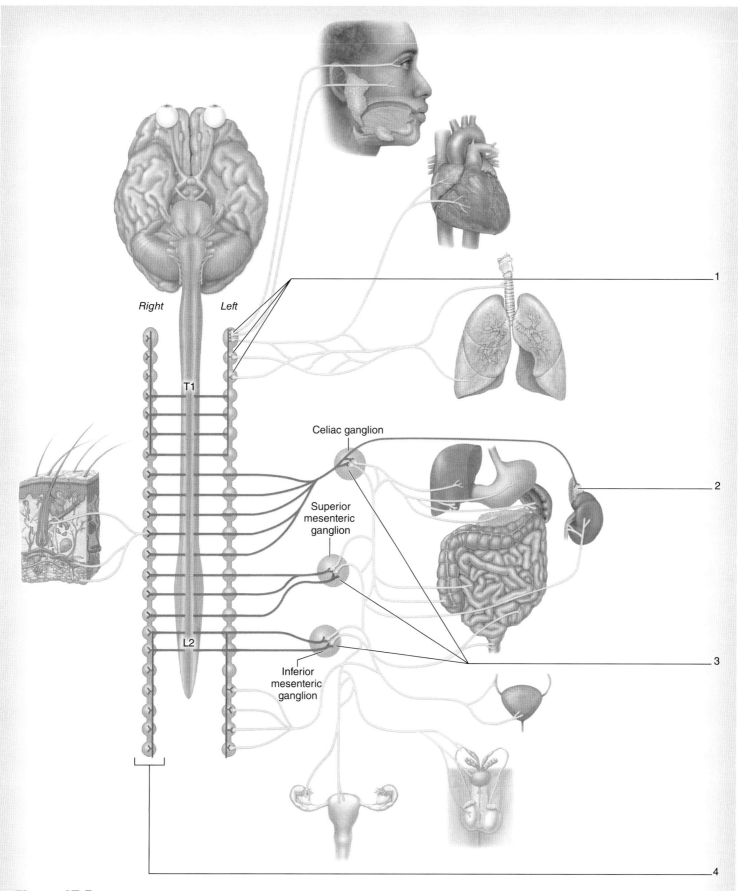

Figure 17.5 Overview of the Sympathetic Division of the ANS. The sympathetic division of the autonomic nervous system controls the "fight-or-flight" response. Label the figure using the terms below.

- [ ] adrenal medulla
- [ ] prevertebral ganglia
- [ ] sympathetic trunk
- [ ] sympathetic trunk ganglia (paravertebral)

(continued on next page)

(continued from previous page)

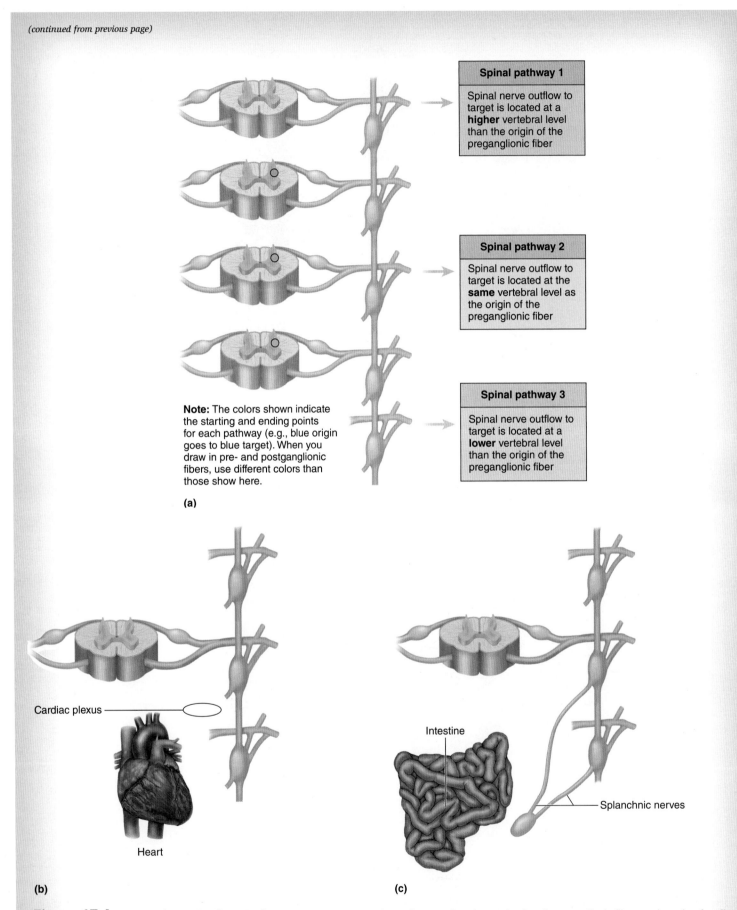

Figure 17.6 Sympathetic Outflow Pathways. Using two colored pencils or markers (one color for all preganglionic fibers and another for all postganglionic fibers), draw in the following sympathetic outflow pathways: (a) Three basic spinal nerve pathways. (b) Postganglionic sympathetic nerve pathway. (c) Splanchnic nerve pathway. (d) Adrenal medulla pathway.

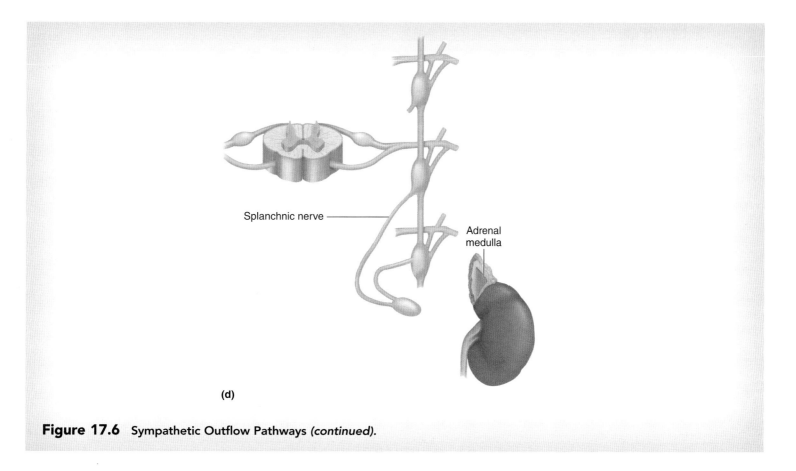

(d)

Figure 17.6 Sympathetic Outflow Pathways *(continued).*

INTEGRATE

CONCEPT CONNECTION

Recall that somatic motor neurons as well as ANS motor neurons (sympathetic division) are located in the gray matter of the spinal cord. Specifically, somatic motor neurons are located in the anterior (ventral) horn of the spinal cord gray matter, whereas ANS motor neurons are located in the lateral horn of the spinal cord gray matter. Whether projecting to voluntary muscles (somatic) or involuntary muscles and/or glands (autonomic), these motor neurons have axons that exit the spinal cord via the same route. Since both somatic and autonomic neurons are providing motor output, through which root (anterior or posterior) will the fibers travel out of the spinal cord? _____

Physiology

Somatic Reflexes

Somatic reflexes result in the contraction of skeletal muscle (remember: Somatic is associated with *voluntary* movement). The somatic spinal reflexes you will test here are stretch reflexes. A **stretch reflex** is initiated when muscle spindles (specialized stretch receptors within skeletal muscle) are stretched. Nerve signals are transmitted along sensory neurons from the muscle spindle to the spinal cord where they synapse with motor neurons in the gray matter of the spinal cord. Motor neurons subsequently transmit nerve signals along axons to stimulate the skeletal muscle to contract. Stretch reflexes are important in regulating subconscious muscle movements, including those involved with balance and posture. Abnormal responses may be an indication of certain diseases (e.g., syphilis, late-stage alcoholism, diabetic neuropathy) or damage to the peripheral nerves or spinal cord.

In the following exercises you will observe two common somatic spinal reflexes: the **patellar reflex** and the **plantar reflex.** When performing the reflex tests, think about the anatomical structures that are involved. For example, if you were to test a reflex involving the triceps brachii, you might remember that the nerve that innervates the triceps brachii is the radial nerve, which is a branch of the brachial plexus. Recall from chapter 16 that the brachial plexus arises from spinal nerves C5–T1. Thus, testing this reflex is a way of testing the function of the spinal cord levels between C5 and T1. In an actual clinical situation, a physician is trained to identify spinal cord levels even more precisely than this. However, just by going through the thought process, you will begin to understand the concept of how to assess the function of certain regions of the CNS using reflex tests.

460 Chapter Seventeen *The Somatic and Autonomic Nervous Systems and Human Reflex Physiology*

EXERCISE 17.4

PATELLAR REFLEX

1. Obtain a rubber percussion hammer and choose a student to be the subject.

2. Have the subject sit on the lab table or a chair with her knees bent, legs completely relaxed, and feet not touching the floor (this is why a lab table is a better location than a chair).

3. Gently strike the subject's patellar ligament with the *blunt* side of the percussion hammer **(figure 17.7)** and note your observations:

   _____

   _____

4. This exercise tests the function of the components of the stretch reflex, namely muscle spindles located within the muscles that are stretched when the patellar ligament is tapped. Recall from chapter 13 that the patellar ligament is an extension of the patellar tendon. The patellar tendon is associated with what group of muscles?_____

   What somatic nerve innervates this group of muscles?

   _____

   From what somatic nerve plexus does this nerve extend?

   _____

   What spinal nerves contribute to this plexus? _____

   _____

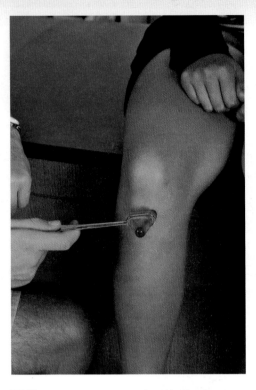

Figure 17.7 Testing the Patellar Reflex. The subject's leg should be completely relaxed and feet should not touch the floor. Using the blunt end of a percussion hammer, gently strike the patellar ligament of the subject and note what happens.

EXERCISE 17.5

PLANTAR REFLEX

1. Obtain a rubber percussion hammer and choose a student to be the subject.

2. Have the subject remove his shoe and sock from one foot and lie supine on the lab table.

3. Move the metal end of the rubber hammer firmly over the lateral aspect of the sole of the foot from the heel to the base of the great toe **(figure 17.8)** and note your observations:

   _____

   _____

4. This exercise tests the function of cutaneous receptors. Stimulation of cutaneous receptors on the plantar surface of the foot evokes a spinal reflex that, in turn, stimulates the flexor muscles of the leg (e.g., flexor digitorum longus). What somatic nerve innervates this group of muscles (Hint: it extends down the posterior leg)? _____

   _____

   From what somatic nerve plexus does this nerve extend?

   _____

   What spinal nerves contribute to this plexus? _____

   _____

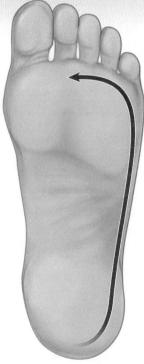

Figure 17.8 Testing the Plantar Reflex. Have the subject lie down on the lab table. Use the metal end of the percussion hammer so that you can press hard on the foot. Press the bottom of the foot, following the arc denoted by the arrow in this figure. Note the reaction of the subject.

INTEGRATE

CLINICAL VIEW
Babinski Reflex

In infants, the normal response to the plantar reflex test (figure 17.8) is opposite the normal response of adults because the axons within this nerve are not yet completely myelinated. That is, the infant response to the plantar reflex test involves the toes spreading apart (abduction) and the foot dorsiflexing. This reflex is called the *Babinski reflex* (toe abduction and foot dorsiflexion denotes a "positive" Babinski sign). If an adult responds to the plantar reflex test with a positive Babinski sign, it indicates an abnormal response. Specifically, this abnormality suggests damage to the corticospinal pathway (or pyramidal neurons) within the central nervous system, as this motor pathway is responsible for modulating the spinal reflex. The cause of such damage must be explored further by the physician.

EXERCISE 17.6

CORNEAL REFLEX

The corneal reflex is an example of a protective reflex. The **corneal reflex** involves touch receptors located within the cornea that detect pressure on the cornea. Sensory input (afferent signal) is relayed along the *trigeminal nerve* (CN V) to nuclei within the brain (control center). Motor output (efferent signal) is relayed from the brain along the *facial nerve* (CN VII) to skeletal muscles that surround the eyes—the orbicularis oculi muscles (effectors). When the cornea is touched, the contraction of the orbicularis oculi muscles causes the eyes to blink if both nerves are functioning properly.

1. Obtain a cotton ball or piece of soft tissue paper. Choose a study subject.

2. Lightly touch the cotton/tissue paper to the corner of the subject's cornea **(figure 17.9)** and note the response:

_____

_____

Figure 17.9 Testing the Corneal Reflex. Have the subject hold both of his or her eyes open. *Very carefully* touch the cornea of the subject's right eye with a piece of cotton or a soft tissue, and note the subject's reaction.

Autonomic Reflexes

Autonomic reflexes, like somatic reflexes, involve both sensory input and motor output. However, because these reflexes are *autonomic*, the motor output is to smooth muscle, cardiac muscle, or glands. Thus, the reflex typically occurs below the level of conscious awareness. Autonomic reflexes can be either spinal or cranial. The reflexes we will observe in this section are examples of cranial reflexes. **Cranial reflexes** involve transmission of nerve signals along cranial nerves. They are important in clinical tests to determine the function of specific cranial nerves or regions of the brain where the appropriate cranial nerves arise.

EXERCISE 17.7

PUPILLARY REFLEXES

The pupillary reflex is another example of a protective reflex. The **pupillary reflex** involves light receptors located within the retina, which detect light entering the eye. Sensory input (afferent signal) is relayed along the *optic nerve* (CN II) to nuclei within the brain (control center). Motor output (efferent signal) is then relayed from the brain along the *oculomotor nerve* (CN III) to the pupillary sphincter muscle within the iris of the eye (effector). When a bright light is shone into the eye, the normal pupillary response is constriction of the pupil. This is a protective function because too much light entering the eye can damage the retina. Testing for the pupillary reflex is important when brain trauma or other brain damage (e.g., a stroke) is suspected. For example, increased intracranial pressure can crush the cranial nerves that transmit either afferent or efferent signals. If this happens, the pupils will remain dilated when a light is shone into the eye. Altering the size of the pupil is controlled by both divisions of the autonomic nervous system. Activation of the sympathetic division of the ANS results in dilation of the pupil, so as to increase visual acuity. Conversely, activation of the parasympathetic division of the ANS results in the constriction of the pupil.

1. Obtain a flashlight and a small ruler, and choose a study subject.

2. Measure the size of the subject's pupil using the ruler. DO NOT touch the ruler to the subject's eyes. Simply hold it up in front of each eye and get as close as you can to a precise measurement. Record the initial pupil diameters below:

    Right pupil diameter: Left pupil diameter:

    _____ mm   _____ mm

3. Shielding one eye (by holding a hand vertically between the subject's eyes), shine a flashlight into the other eye **(figure 17.10)**.

4. Measure the resulting pupil diameters below:

    Right pupil diameter: Left pupil diameter:

    _____ mm   _____ mm

Figure 17.10 Testing the Pupillary Reflex. Have the subject shield one eye by holding his or her hand in front of the nose as shown. Shine a light in one eye and note the change in pupil diameter in *both* eyes.

Chapter 17: The Somatic and Autonomic Nervous Systems and Human Reflex Physiology

Name: _____
Date: _____ Section: _____

POST-LABORATORY WORKSHEET

The ❶ corresponds to the Learning Objective(s) listed in the chapter opener outline.

Do You Know the Basics?

Exercise 17.1: Identifying Components of a Reflex on a Classroom Model

1. Somatic motor neurons innervate _____ muscle. Autonomic motor neurons innervate _____ and _____ muscle and _____. ❶

2. Somatic reflexes can be either _____, if the information travels through the brain, or _____, if the information travels through the spinal cord. ❷

Exercise 17.2: Parasympathetic Division

3. The parasympathetic division of the ANS is also called the _____ division because of its anatomical location. ❸

4. List the four cranial nerves that carry parasympathetic information: ❹
   a. _____
   b. _____
   c. _____
   d. _____

5. The parasympathetic nerves that innervate structures within the pelvic cavity arise from the _____. ❹

Exercise 17.3: Sympathetic Division

6. The sympathetic division of the ANS is also called the _____ division because of its anatomical location. ❺

7. Sympathetic ganglia that connect to spinal nerves via white and grey communicating rami are the _____ ganglia. ❺

8. The major sympathetic ganglia that lie on top of the unpaired abdominal blood vessels (celiac trunk, superior mesenteric artery, inferior mesenteric artery) are also known as _____ ganglia because of their anatomic location relative to the vertebral column. ❺

9. List the four major sympathetic outflow pathways. ❻
   a. _____
   b. _____
   c. _____
   d. _____

Exercise 17.4: Patellar Reflex

10. The sensory organs involved in the patellar reflex are called _____ and they are located within this group of muscles. _____. ❼

11. The spinal nerves that are tested using the patellar reflex test are _____ through _____. ❽

Exercise 17.5: Plantar Reflex

12. The effector group of muscles tested in the plantar reflex are the _____ muscles of the leg. ❾

13. The spinal nerves that are tested using the plantar reflex test include _____ through _____. ❿

Exercise 17.6: Corneal Reflex

14. The _____ nerve carries afferent input of the corneal reflex _____, whereas the _____ nerve carries efferent output of the corneal reflex. ⓫

15. The function of the corneal reflex is _____. ⓬

464 Chapter Seventeen *The Somatic and Autonomic Nervous Systems and Human Reflex Physiology*

Exercise 17.7: Pupillary Reflexes

16. The _____ nerve carries afferent input of the corneal reflex, whereas the _____ nerve carries efferent output of the corneal reflex. ⓭

17. The function of the pupillary reflex is _____. ⓮

Can You Apply What You've Learned?

18. Match each of the effectors listed in column A with the cranial nerve that innervates the effector listed in column B.

    Column A
    - _____ lacrimal gland
    - _____ heart
    - _____ bronchioles
    - _____ salivary glands (under tongue)
    - _____ stomach and other digestive organs
    - _____ kidneys
    - _____ salivary gland (parotid)
    - _____ ciliary muscles and iris of eye

    Column B
    a. facial
    b. glossopharyngeal
    c. oculomotor
    d. vagus

19. In the *spinal nerve* outflow route for the sympathetic division of the ANS, the preganglionic and postganglionic fibers synapse in the _____ ganglion (either at the same level or at different levels) and extend through the _____ nerve to these effectors: _____

20. In the *sympathetic nerve* outflow route for the sympathetic division of the ANS, the preganglionic and postganglionic fibers synapse in the _____ ganglion (either at the same level or at different levels) and extend through the _____ nerve to these effectors: _____

21. In the *splanchnic nerve* outflow route for the sympathetic division of the ANS, the preganglionic fiber extends through the _____ nerve to the _____ ganglion, where it synapses with the postganglionic fiber. The postganglionic fiber extends to these effectors: _____

22. Only one nerve fiber (a preganglionic fiber) is required in the outflow route for the sympathetic division of the ANS to the _____ _____. When stimulated, this organ releases _____ and _____ into the bloodstream.

Can You Synthesize What You've Learned?

23. Patients experiencing hypertension (high blood pressure) may be prescribed drugs called beta-blockers. These drugs bind to receptors on cardiac muscle cells and block the action of the sympathetic nerves to the heart. Describe how administration of a beta-blocker will influence heart rate and strength of contraction.

24. A visit to the eye doctor often involves "dilation" of the pupils. Doctors will place drops of a drug called *tropicamide* in each eye and wait a period of time for the drops to take effect. The drug binds to receptors on the pupillary sphincter muscles to block action of the parasympathetic nervous system in these muscles. Describe why application of these eye drops leads to increased sensitivity to light.

25. Susan, a 50-year-old female, is experiencing motor and sensory deficits in her limbs. Specifically, she exhibits hyperactive stretch reflexes and bilateral positive Babinski signs. Describe each of these results, including how each can be tested in a clinical setting. Finally, discuss possible locations of the injury.

26. John, a 28-year-old male, enters the emergency room after experiencing a severe blow to the head while playing football. Doctors test both corneal and pupillary reflexes and both are abnormal. Describe each of these results, including how each can be tested in a clinical setting. Finally, discuss possible locations of the injury.

CHAPTER 18

General and Special Senses

OUTLINE AND LEARNING OBJECTIVES

Histology 468

General Senses 468

EXERCISE 18.1: TACTILE (MEISSNER) CORPUSCLES 468
1. Identify dermal papillae and tactile corpuscles when viewed with a microscope
2. Describe the structure and function of tactile corpuscles

EXERCISE 18.2: LAMELLATED (PACINIAN) CORPUSCLES 469
3. Identify the reticular layer of the dermis and lamellated corpuscles when viewed with a microscope
4. Describe the structure and function of lamellated corpuscles

Special Senses 470

EXERCISE 18.3: GUSTATION (TASTE) 470
5. Identify different types of tongue papillae, and identify taste buds when viewed with the microscope
6. Describe the structure and function of taste buds

EXERCISE 18.4: OLFACTION (SMELL) 472
7. Identify olfactory epithelium when viewed with a microscope
8. Identify the general location and describe the function of basal cells, olfactory receptor cells, and supporting cells within olfactory epithelium

EXERCISE 18.5: VISION (THE RETINA) 474
9. Identify the retina when viewed with a microscope
10. Identify the rod and cone layer, bipolar cell layer, and ganglion cell layer in a histology slide of the retina
11. Describe the functional importance of the fovea centralis and the optic disc, and describe how the histology of the retina is modified in these areas to match those functions

EXERCISE 18.6: HEARING 476
12. Identify the cochlea when viewed with a microscope
13. Identify the scala vestibuli, scala media, and scala tympani of the cochlea when viewed with a microscope
14. Identify the histological structure of the spiral organ when viewed with a microscope
15. Describe how cells within the spiral organ function to transmit the special sense of hearing to the brain

Gross Anatomy 479

General Senses 479

EXERCISE 18.7: SENSORY RECEPTORS IN THE SKIN 479
16. Identify the different types of sensory receptors on a classroom model of the skin

17. Describe the general location and function of tactile discs, tactile corpuscles, lamellated corpuscles, and free nerve endings

Special Senses 481

EXERCISE 18.8: GROSS ANATOMY OF THE EYE 481
18. Identify structures of the eye on a classroom model of the eye

EXERCISE 18.9: COW EYE DISSECTION 485
19. Identify internal and external eye structures on a fresh cow eye
20. Describe the ways in which a cow eye is both similar to and different from a human eye

EXERCISE 18.10: GROSS ANATOMY OF THE EAR 487
21. Identify structures of the ear on a classroom model of the ear
22. Describe the gross structure, location, and functions of the external, middle, and inner ear

Physiology 491

General Senses 491

EXERCISE 18.11: TWO-POINT DISCRIMINATION 491
23. Describe the relationship between receptive field and the density of sensory receptors in the skin
24. Explain what areas of the body have the greatest density of sensory receptors in the skin

EXERCISE 18.12: TACTILE LOCALIZATION 492
25. Explain why it is more difficult to locate the exact source of sensory input when a stimulus is applied to the tips of the fingers as compared to the back of the forearm

EXERCISE 18.13: GENERAL SENSORY RECEPTOR TESTS: ADAPTATION 493
26. Explain the process of sensory adaptation

Special Senses 494

EXERCISE 18.14: GUSTATORY TESTS 494
27. List the five basic taste sensations and describe the type of substance that is responsible for causing each sensation

EXERCISE 18.15: OLFACTORY TESTS 495
28. Describe the effect that reduced olfactory sensation has on the ability to taste

EXERCISE 18.16: VISION TESTS 497
29. Describe how to test for visual acuity and how to report the results of the test
30. Explain the relationship between flexibility of the lens and near-point accommodation
31. Describe how to test for a person's blind spot, and explain how the test works

EXERCISE 18.17: HEARING AND EQUILIBRIUM TESTS 499
32. Explain the use of the Rinne and Weber tests, and what abnormal findings for each of the tests may indicate
33. Explain the purpose of the Romberg and Barany tests, and what abnormal findings for each of the tests may indicate

 MODULE 7: NERVOUS SYSTEM

INTRODUCTION

Your ability to read this chapter and perform the described laboratory activities is dependent upon the proper functioning of your sensory systems. Your special sense of vision is used to read the words typed on the page. In the laboratory classroom, you rely on the special sense of hearing to respond to the instructions delivered to you by your instructor and to discuss the material with your classmates. Your special sense of equilibrium maintains your body's three-dimensional position in space so you don't fall over as you move about the classroom. In addition, sensory receptors in your skin constantly perceive stimuli from your environment relaying touch, pressure, and temperature sensations, thus allowing you to orient yourself and protect your body from harm. Indeed, if you are reading this chapter introduction at home, you may be enjoying a cup of coffee or other food or drink as you read. Your enjoyment of such things is highly dependent upon the special senses of olfaction (smell) and gustation (taste). In fact, our enjoyment of things like music, warm baths, food, drink, and a good book are all completely dependent upon the variety of sensory receptors to detect stimuli from our environment. Of course, it is our brain that actually *interprets* such sensory input. However, without the ability to detect the appropriate input, there would be nothing for our brains to interpret. Through the laboratory exercises within this chapter we will explore the intricate structure and function of many of the amazing, beautiful, and incredibly *functional* sensory receptors and organs found throughout our bodies.

In this laboratory session you will explore the histology of a select few somatic sensory receptors and special sensory organs of the body and the gross structure of two special sensory organs (the eye and ear) through observations of classroom models and by dissection of a cow eye. In the remaining lab exercises, you will perform specific tests that involve the function of general senses and special senses.

Chapter 18: General and Special Senses

Name: _____
Date: _____ Section: _____

PRE-LABORATORY WORKSHEET

Also available at www.connect.mcgraw-hill.com

1. Explain the difference between *general senses* and *special senses*.

   _____
   _____
   _____
   _____
   _____
   _____

2. List the five special senses.

   a. _____

   b. _____

   c. _____

   d. _____

   e. _____

3. Of the five special senses you listed in question 2, which senses have receptor organs located within the petrous part of the temporal bone?

   a. _____

   b. _____

4. Tactile corpuscles are located within the _____ layer of the dermis.

5. Lamellated corpuscles are located within the _____ layer of the dermis.

6. The optic nerve carries the special sense of _____.

7. The olfactory nerves extend through the _____ of the ethmoid bone.

8. The vestibulocochlear nerve carries the special senses of _____ and _____.

9. Taste receptors on the posterior third of the tongue are innervated by the _____ nerve, while taste receptors on the anterior two-thirds of the tongue are innervated by the _____ nerve.

Chapter Eighteen *General and Special Senses* 467

Histology

General Senses

Sensory receptors can be classified based on receptor distribution in the body. The two distribution types are the general senses and the special senses. Receptors of the **general senses** are dendritic endings of sensory neurons that are located throughout the body. They include both somatic sensory receptors of the skin, joints, muscles, and tendons, which detect light touch temperature, pain, and pressure and visceral sensory receptors of internal organs, which respond, for example, to changes in stretch of an organ wall and carbon dioxide levels. In comparison, receptors of the **special senses** are complex structures located only in the head, and are associated with taste, smell, vision, hearing, and equilibrium.

Here, we will focus on general **somatic** senses in the skin (table 18.1).

Table 18.1	Sensory Receptors in Thick Skin			
Receptor	**Location**	**Structure**	**Senses**	**Word Origin**
Free Nerve Ending	In the dermis, ending in glands and hair follicles, with some extending into the epidermis	An unmodified, unencapsulated nerve ending	Sustained touch, pain, temperature, itching, and hair movement	*free*, referring to the fact that there are no connective tissue coverings
Lamellated (Pacinian) Corpuscle	Deep within the reticular layer of the dermis	Concentric layers of connective tissue surrounding a nerve ending at the core	Deep pressure and high-frequency vibration	*lamina*, plate, + *corpus*, body
Tactile (Meissner) Corpuscle	Within dermal papillae, especially in lips, palms of the hand, eyelids, nipples, and external genitalia	An oval structure consisting of modified neurolemmocytes encapsulating a nerve ending	Fine, discriminitive touch to determine textures and shapes; light touch	*tactus*, to touch, + *corpus*, body
Tactile Disc (Merkel Disc)	At the junction between the dermis and the epidermis	An association between a modified keratinocyte in the epidermis, called a tactile cell, and a specialized nerve ending in the dermis, called a tactile disc	Fine, light touch	*tactus*, to touch

EXERCISE 18.1

TACTILE (MEISSNER) CORPUSCLES

1. Obtain a histology slide of thick skin and place it on the microscope stage. Bring the tissue sample into focus using the scanning objective.

2. Observe the slide on low power and review the layers of the skin (**figures** 6.1 and **18.1**).

   Epidermis: stratum basale, stratum spinosum, stratum granulosum, stratum lucidum, and stratum corneum
   Dermis: papillary layer and reticular layer

3. Move the microscope stage so the junction between the dermis and epidermis is in the center of the field of view, and locate the **dermal papillae** (figure 18.1). Next, move the stage so a single papilla is in the center of the field of view, then change to high power.

4. Sensory receptors called **tactile corpuscles** are located within the dermal papillae of thick skin. These receptors are oval in shape with a surrounding capsule of connective tissue (figure 18.1 and table 18.1), and they function in sensing light touch. Look for tactile corpuscles within the dermal papilla. If you don't see one, scan the slide for other papillae that may have tactile corpuscles within.

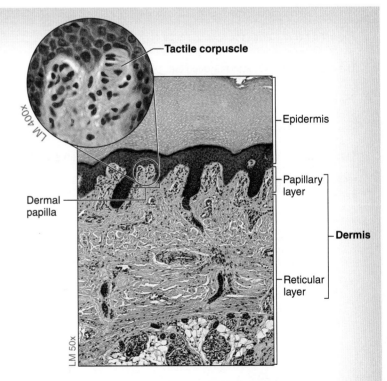

Figure 18.1 Thick Skin and Tactile (Meissner) Corpuscles. Tactile corpuscles are found within the dermal papillae of the skin and are sensory receptors for fine touch.

5. In the space below, make a sketch of a tactile corpuscle within a dermal papilla as seen through the microscope.

_____ ×

EXERCISE 18.2

LAMELLATED (PACINIAN) CORPUSCLES

1. Obtain a histology slide of thick skin and place it on the microscope stage.

2. Observe the slide on low power and identify the dermis and epidermis (figure 18.1).

3. Move the microscope stage so the deepest part of the reticular layer of the dermis is in the center of the field of view. Within this portion of the dermis you will see cross sections through numerous sweat glands and blood vessels. You will also see cross sections of **lamellated corpuscles.** Lamellated corpuscles resemble onions in cross section because they are composed of dendritic endings of sensory neurons ensheathed with an inner core of neurolemmocytes and outer concentric layers of connective tissue (**figure 18.2** and table 18.1).

4. Locate a lamellated corpuscle and then move the microscope stage so the lamellated corpuscle is in the center of the field of view. Change to a higher power to view its structure in more detail. Lamellated corpuscles function in the sensation of deep pressure. When enough pressure is applied to the surface of the skin, the layers of connective tissue surrounding the central sensory receptor are compressed, and initiating nerve signals to the brain.

5. In the space below, make a sketch of a lamellated corpuscle as seen through the microscope, making note of its location in the skin.

Figure 18.2 Lamellated Corpuscles. Lamellated corpuscles are located deep within the reticular layer of the dermis and are sensory receptors for deep pressure.

_____ ×

Special Senses

Special senses are specialized organs within the head that respond specifically to the modalities of olfaction, taste, vision, hearing, and equilibrium. In the next set of laboratory exercises, we will explore the structure and function of these special sensory organs.

EXERCISE 18.3

GUSTATION (TASTE)

Gustation (taste) is one of the most pleasurable sensations humans can experience. However, to have a complete sense of gustation, our olfactory sense must also be functioning. In fact, without the ability to smell, the ability to taste suffers tremendously. Perhaps one reason gustation is so pleasurable is that the gustatory and olfactory pathways relay sensory input to the limbic system, our emotional brain.

Table 18.2 Tongue Papillae

Type of Papilla	Location	Description	Function	Word Origin
Filiform	Anterior two-thirds of the tongue	The most numerous of the papillae. They are conical in shape and contain sensory nerve endings.	General sensation; assist in detecting texture and manipulating of food	*filum*, thread, + *forma*, shape
Foliate	Sides of the tongue	Large papillae with deep crypts that house taste buds in their walls in infants and children; not well developed in humans	Taste sensation	*foliate*, resembling a leaf
Fungiform	Anterior part of the tongue; most numerous at the tip	Less numerous than filiform papillae. They are shaped like mushrooms and contain some taste buds.	Taste sensation	*fungus*, a mushroom, + *forma*, shape
Vallate (Circumvallate)	Posterior part of the tongue along the sulcus terminalis (a V-shaped structure on the posterior tongue)	The largest papillae. They contain deep crypts that house taste buds in their lateral walls.	Taste sensation	*circum-*, around, + *vallum*, wall

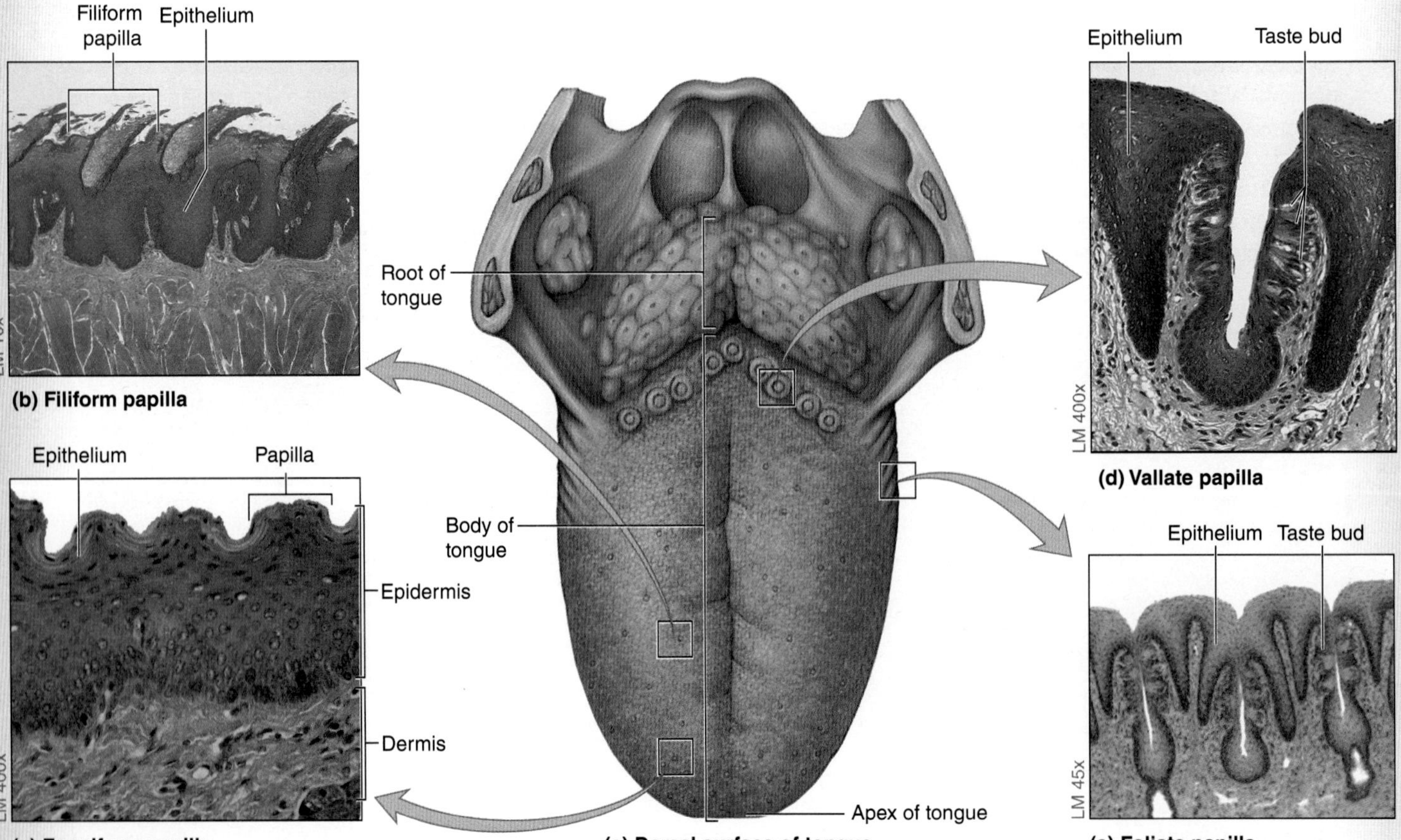

Figure 18.3 Taste Buds. Gustation (taste) requires taste buds, which are associated with tongue papillae. (a) Dorsal surface of tongue, (b) Filiform papilla, (c) Fungiform papilla, (d) Vallate papilla, (e) Foliate papilla.

There are four types of papillae located on the tongue: filiform, fungiform, vallate (circumvallate), and foliate. The detailed structure, function, and locations of the tongue papillae are shown in **figure 18.3** and summarized in **table 18.2**. The large papillae are the **foliate** and **vallate (circumvallate)** papillae. The vallate papillae house more than half of our taste buds in the walls of the crypts that surround each papilla. The sensory receptor specialized for gustatory sensation is a **taste bud** (gustatory bud) (**table 18.3** and **figure 18.4**). Taste buds are particularly concentrated on the papillae of the tongue, but are also located throughout the oral cavity and pharynx. In this exercise we will explore the types of tongue papillae and the location and function of taste buds associated with the papillae.

1. Obtain a histology slide of the tongue or a histology slide demonstrating mammalian *vallate papillae* (figure 18.3d). Scan the slide at low power and identify the papillae on the surface of the tongue. Move the microscope stage so one or two papillae are in the center of the field of view. Then increase the magnification, first to medium and then to high power. Locate a taste bud at the edge of the papilla (figure 18.4).

2. Using figures 18.3 and 18.4 and tables 18.2 and 18.3 as guides, identify the following structures (Note: you may not be able to identify all of the types of papillae on a single slide):

   ☐ basal cells
   ☐ filiform papillae
   ☐ foliate papillae
   ☐ fungiform papillae
   ☐ gustatory cells
   ☐ supporting cells
   ☐ taste pores
   ☐ vallate (circumvallate) papillae

Table 18.3	Cells Associated with Taste Buds	
Structure	**Description**	**Function**
Basal Cells	These are small stem cells found at the base of the taste bud	Precursor cells to the supporting cells and gustatory cells
Gustatory Cells	These cells have round nuclei and are very light in color. They are epithelial cells which are composed of microvilli twisted around each other, and a taste hair on the apical surface.	Detect chemicals dissolved in solution; transmit nerve signals for taste sensation to the CNS
Supporting Cells (Sustentacular Cells)	Located between the gustatory cells, these cells have more oval-shaped nuclei and stain darkly	Support the gustatory cells by producing a glycoprotein; may also function in taste sensation

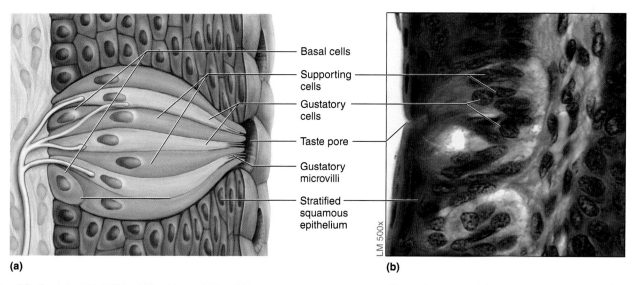

Figure 18.4 Detailed Structure of a Taste Bud. (a) Illustration. (b) Photomicrograph.

(continued on next page)

472 Chapter Eighteen *General and Special Senses*

3. Locate a taste bud at the edge of the papilla in the crevice (crypt) between two papillae (figure 18.4).

4. In the space to the right, sketch the histology of a vallate papilla as seen through the microscope, noting the location of the taste buds.

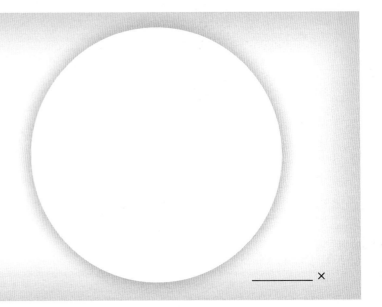

_____ ×

EXERCISE 18.4

OLFACTION (SMELL)

Olfaction is the sense of smell. It is an important sensory modality, not just for smell, but also for gustation (taste). Olfactory sensation is detected by special sensory cells found within the epithelium lining the roof of the nasal cavity—**olfactory epithelium (figure 18.5a and table 18.4)**. The **olfactory receptor cells** are neurons, but they are unique because they are continuously replaced. The olfactory receptor cells compose the **olfactory nerves (CN I)**. The axons of the olfactory nerves travel through the **cribriform foramina** within the cribriform plate of the ethmoid bone to synapse with neurons within the **olfactory bulbs,** which subsequently transmit nerve signals to the brain via the **olfactory tract**.

1. Obtain a histology slide of olfactory epithelium (figure 18.5b) and place it on the microscope stage.

2. Bring the tissue sample into focus on low power. Increase magnification to medium power, and then to high power so you clearly can see the cells composing the olfactory epithelium. It will be difficult to distinguish between the three major cell types of the olfactory epithelium: basal cells, olfactory receptor cells, and supporting cells. In general, the nuclei closest to the basement membrane of the epithelium are the nuclei of *basal cells* (the cells appear triangular in shape), the nuclei in the middle of the epithelium are the nuclei of *olfactory receptor cells*, and the nuclei

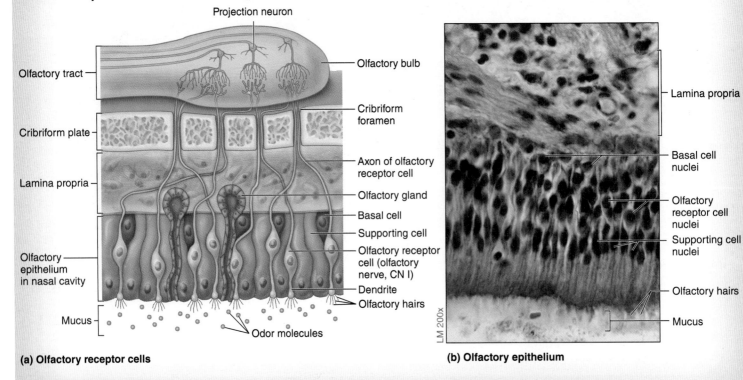

(a) **Olfactory receptor cells**

(b) **Olfactory epithelium**

Figure 18.5 Olfactory Epithelium. (a) Cell types within the olfactory epithelium. (b) Histology of the olfactory epithelium.

Table 18.4 Olfactory Epithelium

Structure	Description	Function	Word Origin
Basal Cells	Cuboidal (triangular) cells whose nuclei are located near the basal surface of the olfactory epithelium	Thought to be the precursor cells to the olfactory receptor cells and supporting cells	*basalis*, situated near the base
Cribriform Plate of the Ethmoid Bone	The superior portion of the ethmoid bone containing olfactory foramina	Forms the roof of the nasal cavity	*cribriform*, shaped like a sieve, + *ethmoid*, resembling a sieve
Cribriform Foramina	Numerous small holes in the cribriform plate of the ethmoid bone	Allow for the passage of axons of the olfactory receptor cells from the olfactory epithelium to the olfactory bulb, where they will synapse with other neurons	*olfactus*, to smell, + *foramen*, to pierce
Olfactory Bulb	A white swelling on the superior and anterior region of the cribriform plate of the ethmoid bone	The location where axons from olfactory receptor cells synapse with other neurons that will transmit the nerve signals to the brain via the olfactory tract	*olfactus*, to smell, + *bulb*, a globular structure
Olfactory Hairs (Cilia)	Surface modifications of olfactory receptor cells that contain microtubules; specialized cilia in that they are not motile	Apical ends are the site of interaction between dissolved odoriferous substances and the olfactory receptor cells	*olfactus*, to smell
Olfactory Receptor Cells	Columnar epithelial cells whose large nuclei are located in the middle of the olfactory epithelium (between apical and basal surfaces) and have cilia on their apical surfaces	Bipolar neurons that function as sensory receptors for olfaction (smell)	*olfactus*, to smell, + *recipio*, to receive, + *cella*, a chamber
Olfactory Tract	A white extension of the olfactory bulb that lies inferior to the frontal lobe of the brain	Consists of myelinated axons of neurons that originate in the olfactory bulb and carry olfactory information to the brain	*olfactus*, to smell, + *tractus*, a drawing out
Supporting (Sustentacular) Cells	Columnar epithelial cells with nuclei located near the apical surface of the olfactory epithelium	Surround and support the specialized olfactory receptor cells	*supporto*, to carry, + *cella*, a chamber

closest to the apical surface of the epithelium are the nuclei of *supporting cells*.

3. Using figure 18.5 and table 18.4 as guides, identify the following structures on the histology slide of olfactory epithelium:

   ☐ basal cells
   ☐ ofactory hairs (cilia)
   ☐ olfactory receptor cells
   ☐ supporting cells

4. In the space below, make a sketch of the olfactory epithelium as seen through the microscope, labeling all of the structures listed in step 3.

EXERCISE 18.5

VISION (THE RETINA)

The **retina (figure 18.6)** of the eye is called the **neural tunic**, because this layer is composed of neural tissue. The retina develops as a direct outgrowth of the brain. Thus, the retina is the only part of the brain visible to us without surgical intervention (though you do need an ophthalmoscope). Axons from neurons within the retina travel to the brain through the **optic nerve (CN II)**. The retina is responsible for translating information that comes into the eye as light rays into electrical signals (action potentials) that the brain can "read." This information is relayed from ganglion cells in the retina that extend into the optic nerves and travel to the **thalamus,** then on to the **occipital lobe** of the brain. Here, the image is truly processed and "visualized." The retina is a very complex yet beautifully organized structure. In this laboratory exercise we will explore the structure and function of the retina by observing the cells that are visible histologically. In the gross anatomy section of this laboratory exercise, you will place the retina in context of other structures of the eye.

1. Obtain a histology slide of the retina and place it on the microscope stage.

2. Bring the tissue sample into focus on low power, then move the microscope stage so the retina (figure 18.6) is in the center of the field of view. Switch to medium power and bring the tissue sample into focus once again.

3. Using **table 18.5** and figure 18.6 as guides, identify the following structures on the slide of the retina (you may need to change to high power to view all of the structures or to see them in greater detail):

   ☐ bipolar cell layer   ☐ pigmented layer
   ☐ choroid   ☐ rod and cone layer
   ☐ ganglion cell layer   ☐ sclera

4. With the medium-power objective in place, scan the slide and locate the **fovea centralis (figure 18.7a)**. The fovea centralis is a thinner than normal area of the retina. The fovea centralis is devoid of bipolar and ganglion cell layers and has the highest concentration of cones of the entire retina. When light focuses on the fovea, visual acuity is at its highest. When we focus our gaze on something, we move our eyes so the light entering the eye hits the fovea and the image becomes most clear.

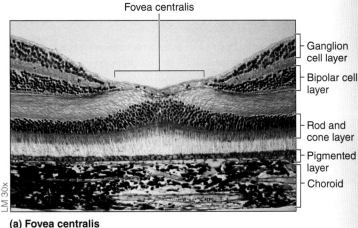

(a) Fovea centralis

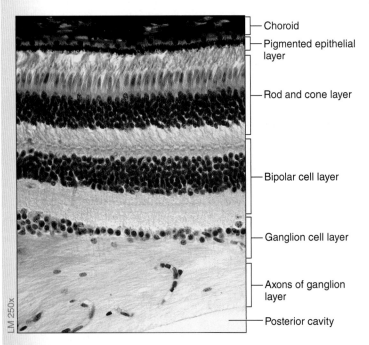

Figure 18.6 Histology of the Retina.

(b) Optic disc

Figure 18.7 Specialized Areas of the Neural Tunic of the Eye. (a) The fovea is the area of the retina where visual acuity is the highest. Ganglion and bipolar cell layers are absent, and there is an abundance of cones in the rod and cone layer. (b) The optic disc is the area of the retina where the axons of ganglion cells exit the retina to become the optic nerve. There are no rods or cones in this area, which is why the optic disc is also referred to as the "blind spot" of the retina.

Table 18.5	The Retina		
Structure	**Description**	**Function**	**Word Origin**
Bipolar Cell Layer	Middle cell layer of the retina, composed of cells with intermediate-sized nuclei; as the name suggests, these neurons are bipolar neurons	Receives signals from rods and cones and transmits electrical signals to ganglion cells	*bipolar*, relating to bipolar neurons
Choroid	The vascular and pigmented layer of the eye located between the retina (internal) and the sclera (external). It is recognized histologically by numerous blood vessels and by dark staining characteristics of the melanin.	Blood vessels of the choroid supply nutrients to the tissues of the retina and sclera, and the pigment absorbs excess light waves	*choroideus*, like a membrane
Cones	Photoreceptor cells with a light-transducing portion located in the outermost layer of the retina and with nuclei located in the layer just internal to that. You will not be able to distinguish rods from cones using a light microscope.	Photoreceptor cell specializing in color vision	*conus*, shaped like a cone
Fovea Centralis	Histologically, an area of the retina devoid of bipolar and ganglion cell layers. Its photoreceptor layer is composed exclusively of cones.	Area of highest visual acuity in the eye. When you focus on something, you move your eyes so the light entering the eye is focused on the fovea.	*fovea*, a pit, + *centralis*, in the center
Ganglion Cell Layer	Innermost layer of the retina, composed of cells with very large nuclei	Receives information from bipolar cells and sends that information to the brain	*ganglion*, a swelling or knot
Retina	Referred to as the "neural tunic" of the eye; consists of numerous layers of neurons involved in phototransduction	Phototransduction: transduction of light waves that enter the eye into nerve signals (action potentials) that can be interpreted by the brain	*rete*, a net
Rod and Cone Layer	Outermost layer of the retina (closest to the choroid and sclera), containing the light-transducing portions of photoreceptor cells (rods and cones). The layer immediately internal to this layer contains the nuclei of rods and cones, the smallest and most numerous nuclei of the retina.	Layer of the retina where light waves are initially transduced into neuronal action potentials; the cells in this layer synapse with neurons in the bipolar cell layer	NA
Rods	Photoreceptor cells with a light-transducing portion located in the outermost layer of the retina and with nuclei located in the layer just internal to that; not distinguishable from cones using a light microscope	Photoreceptor cells specializing in black-and-white vision; very sensitive, most useful when light is dim	*rod*, shaped like a rod
Sclera	Dense irregular connective tissue that surrounds the entire eye except for its anterior aspect where the cornea is located; histologically, the most external layer, composed of collagen fibers and fibroblasts	Protects the eye, serves as an attachment point for extraocular eye muscles, and helps maintain the round shape of the eye	*skleros*, hard

5. Scan the slide and locate the **optic disc,** the location where the optic nerve leaves the eye (figure 18.7*b*). The **optic disc** (blind spot) is easily identifiable because all retinal layers are absent at this location. Notice that the optic disc and optic nerve are approximately the same color as the cells in the ganglion cell layer of the retina. This should help you understand that the ganglion cell layer, the optic disc, and the optic nerve all contain parts of the same cells—namely, the axons of ganglion cells. The axons of the ganglion cell layer leave the eye at the optic disc. Once they enter the optic nerve, the axons are myelinated.

(continued on next page)

476 Chapter Eighteen *General and Special Senses*

(continued from previous page)

6. In the space below, make a sketch of the retina as seen through the microscope. Then label the following layers on your sketch: rod and cone layer, bipolar cell layer, and ganglion cell layer.

_____ ×

7. *Optional Activity:* **AP|R** **7: Nervous System**—Watch the "Vision" animation to learn the sequence of events involved in vision and the functions of cells in the retina.

EXERCISE 18.6

HEARING

Hearing is a function of the **cochlea.** This special sensory organ is located within the petrous part of the temporal bone. We will explore the location, gross structure, and function of this organ in the gross anatomy section of this chapter. In this section we will focus on the histological features of the highly specialized epithelium that lines the cochlea, the **spiral organ** (Organ of Corti), which will provide us with insight into how this organ performs its function: transformation of sound waves into nerve signals that can be interpreted by the brain.

1. The **cochlea (figure 18.8)** is the organ responsible for transducing fluid vibrations received at the oval window into electrical signals that are sent to the thalamus and then on to the temporal lobe of the brain, where they are interpreted.

2. Within the cochlea, the **spiral organ** rests upon the **basilar membrane,** within the scala media (cochlear duct).

3. Obtain a slide of the *cochlea* (**figure 18.9**) and place it on the microscope stage. Bring the tissue sample into focus on low power and then increase the magnification. Move the microscope stage until a single cross section through the cochlea is in the center of the field of view.

4. Using figures 18.8 and 18.9 and **table 18.6** as guides, identify the three chambers within the cochlea and the membranes that separate the chambers from each other:

    ☐ basilar membrane   ☐ scala vestibuli
    ☐ scala media   ☐ vestibular membrane
       (cochlear duct)
    ☐ scala tympani

5. Once you have identified the *scala media* (cochlear duct), move the microscope stage so the scala media is in the center of the field of view. Increase the magnification to high power and focus in on the *spiral organ* (organ of Corti) (figure 18.9).

6. Using table 18.6 and figure 18.9 as guides, identify the following structures on the slide of the cochlea:

    ☐ basilar membrane   ☐ scala vestibuli
    ☐ cochlear nerve   ☐ spiral ganglion
    ☐ endolymph   ☐ supporting cells
    ☐ perilymph   ☐ tectorial membrane
    ☐ scala media   ☐ vestibular membrane
    ☐ scala tympani

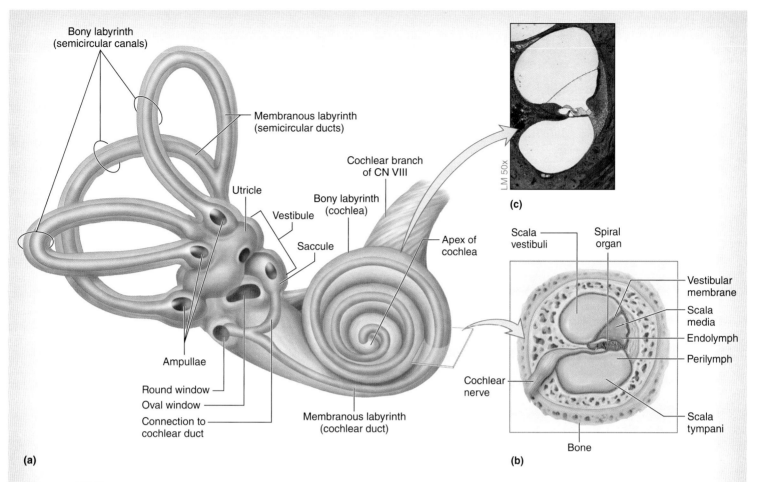

Figure 18.8 The Cochlea. (a) The semicircular canals and cochlea are part of the inner ear. (b) The cochlea houses the spiral organ, which contains specialized cells that translate sound waves into sensory impulses. (c) Light micrograph demonstrating a cross section through the cochlea.

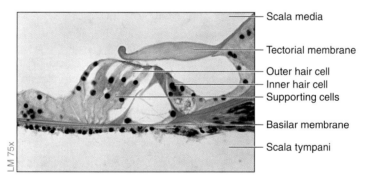

Figure 18.9 Histology of the Spiral Organ.

Table 18.6	The Cochlea		
Structure	**Description**	**Function**	**Word Origin**
Basilar Membrane	Forms the floor of the scala media and supports the hair cells of the spiral organ	Vibrates in response to fluid vibrations of the perilymph in the scala tympani	*basalis*, situated near the base, + *membrana*, a membrane
Endolymph	Fluid within the scala media (cochlear duct) that has a composition similar to that of intracellular fluid (high in potassium)	Nourishes the epithelial cells of the spiral organ	*endon*, within, + *lympha*, a clear fluid

(continued on next page)

(continued from previous page)

Table 18.6	The Cochlea *(continued)*		
Structure	**Description**	**Function**	**Word Origin**
Hair Cells (Sensory Cells)	Highly specialized epithelial cells of the spiral organ that contain stereocilia, which are embedded in the tectorial membrane	Bending of the stereocilia generates action potentials in the hair cells; axons of sensory cells project into the spiral ganglion and synapse with neurons that relay the signals to the brain	*sensus*, to sense
Helicotrema	A connection between the scala vestibuli and scala tympani at the end of the cochlea; shaped like a half moon	Allows perilymph from the scala tympani to communicate with perilymph from the scala vestibuli	*helix*, a spiral, + *trema*, a hole
Perilymph	Fluid contained within the scala vestibuli and scala tympani that has a composition similar to that of extracellular fluid (high in sodium)	Transmits pressure waves through the scala tympani and scala vestibuli; vibrations of the stapes at the oval window cause the pressure waves, and they are dampened when they reach the round window	*peri-*, around, + *lympha*, a clear fluid
Scala Media (Cochlear Duct)	Middle chamber of the cochlea, containing the spiral organ and filled with endolymph	Contains the special sensory organ of sound, the spiral organ	*scala*, a stairway, + *medialis*, middle
Scala Tympani	Chamber inferior to the scala media; filled with perilymph	Transmits pressure waves of the perilymph from the helicotrema to the round window	*scala*, a stairway, + *tympani*, relating to the tympanic membrane
Scala Vestibuli	Chamber superior to the scala media; filled with perilymph	Transmits pressure waves of the perilymph from the oval window to the helicotrema	*scala*, a stairway, + *vestibulum*, entrance court
Spiral Ganglion	Group of neuron cell bodies located on the cochlear part of the cochlear nerve	Contains the cell bodies of bipolar neurons, which receive input from sensory cells of the spiral organ and then relay those signals to the brain	*spiralis*, a coil, + *ganglion*, a swelling or knot
Spiral Organ (Organ of Corti)	Composed of specialized epithelium that rests on the basilar membrane; the epithelium is composed of sensory hair cells and supporting cells	Special inner and outer sensory hair cells have stereocilia embedded in the tectorial membrane; when the basilar membrane vibrates, the stereocilia bend and the sensory hair cells send action potentials to the brain	*spiralis*, a coil, + *organum*, a tool, instrument
Tectorial Membrane	Gelatinous membrane in which the cilia of hair cells of the spiral organ are embedded	Does not vibrate itself, but when the basilar membrane vibrates, the cilia of the hair cells embedded in the tectorial membrane bend, causing the hair cells to generate action potentials	*tectus*, to cover, + *membrana*, a membrane
Vestibular Membrane	Thin membrane between the scala vestibuli and the scala media	Forms a partition separating endolymph within the scala media from perilymph within the scala vestibuli	*vestibulum*, relating to the vestibule, + *membrana*, a membrane
Vestibulocochlear Nerve (CN VIII)	Cranial nerve arising from axons of sensory cells of the spiral organ and from the vestibular apparatus	Transmits nerve signals from the cochlea and the vestibular apparatus to the brain	*audio*, to hear

7. In the space below, make a sketch of a cross section of the cochlea as seen through the microscope, and label the locations of the structures listed in step 6.

_____ ×

Gross Anatomy

General Senses

The definitions of general senses and special senses are given at the beginning of the histology section of this chapter. If you are starting your laboratory observations with the gross anatomy exercises, read the introduction to general senses on p. 468 before you proceed.

Most sensory receptors responsible for general sensation (things such as touch, pain, pressure, temperature) are located in the skin. Thus, in exercise 18.7 you will observe a classroom model of the skin, with special emphasis on sensory receptors located within the skin.

EXERCISE 18.7

SENSORY RECEPTORS IN THE SKIN

1. Observe a classroom model of thick skin **(figure 18.10)**. Some somatic sensory receptors in the skin, such as tactile menisci and free nerve endings, are too small to view under the microscope, so you will identify them on models instead.

2. Locate a tactile disc. **Tactile discs** (Merkel discs) are located at the dermal/epidermal junction of thick skin, although their location is not restricted to dermal papillae, unlike tactile corpuscles. The function of tactile discs is the sensation of light touch.

3. Find some free nerve endings on the skin model. **Free nerve endings** are just that—nerve endings with no specialized cells surrounding them. The ends of these neurons are located near the epidermal/dermal junction, with many of them extending into the epidermis. They also have endings in hair follicles and glands. Free nerve endings function in the sensation of sustained touch, temperature, itching, and pain. Imagine a time when you had a blister that ripped open. A blister forms when the dermis separates from the epidermis, and fluid accumulates between the layers. When the epidermis is removed from a blister, it is very painful! Why?

(continued on next page)

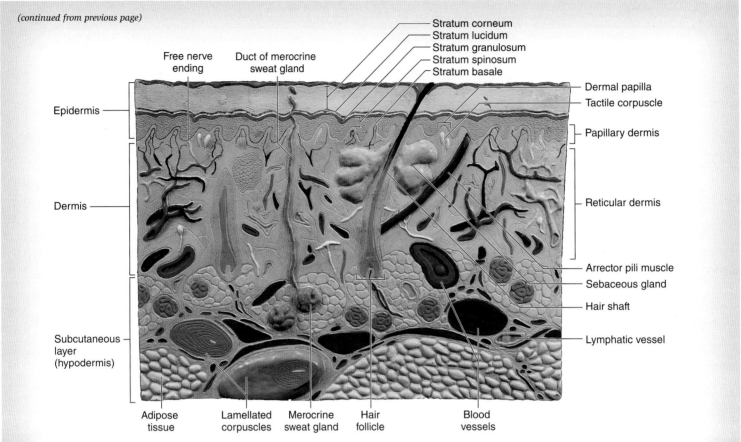

Figure 18.10 Skin. Model of the skin demonstrating sensory receptors such as free nerve endings, and lamellated corpuscles. This model is misleading in that it demonstrates hair follicles *and* thick skin (epidermis). Thick skin does not contain hair follicles.

The free nerve endings in the dermis are now exposed to the environment, and this causes them to generate action potentials. This is why large superficial wounds ("scrapes") on the skin are much more painful than deep wounds. We perceive a deep wound as a more severe wound (as it typically is), but are often confused that it causes us less pain than a more superficial wound. The greater the surface area exposed, the greater the number of nerve endings stimulated. Hence, the common exclamation, "It's only a scrape, but it hurts like crazy!"

4. Using figure 18.10, table 18.1, and your textbook as guides, identify the following on a model of the skin:

   ☐ dermal papillae
   ☐ free nerve endings
   ☐ lamellated corpuscle
   ☐ papillary dermis
   ☐ reticular dermis
   ☐ stratum basale
   ☐ stratum corneum
   ☐ stratum granulosum
   ☐ stratum lucidum
   ☐ stratum spinosum
   ☐ tactile corpuscle
   ☐ tactile disc

5. In the space below, make a sketch of thick skin and label the locations of the sensory receptors listed in table 18.1.

Chapter Eighteen *General and Special Senses*

Special Senses

The definitions of general senses and special senses are given at the beginning of the histology section of this chapter. If you are starting your laboratory observations with the gross anatomy exercises, read the introduction to special senses on p. 470 before you proceed. The gross anatomy exercises in this section focus on two special sensory organs: the eye and the ear.

EXERCISE 18.8

GROSS ANATOMY OF THE EYE

In exercise 18.9 we will explore the anatomy of the eye, using a cow eye as a model organ. Though there is some variance in structure between a cow eye and a human eye, cow eyes are much easier to obtain and they are larger, which greatly facilitates making internal observations of the eye. In this laboratory exercise you will identify all structures listed in **tables 18.7** and **18.8** on a classroom model or on yourself, because not all structures associated with the eye are visible in the dissected cow eye or in the histology slide of the retina.

Table 18.7	External Structures of the Eye		
Structure	**Description**	**Function**	**Word Origin**
Cornea	Transparent tissue on the anterior surface of the eye consisting of an external layer of stratified squamous epithelium, a middle layer of regularly arranged collagen fibers, and an inner layer of endothelium	The primary structure used to refract (bend) light waves in the eye	*corneus*, horny
Extrinsic Muscles	Muscles that originate from a common tendinous ring and insert on the sclera of the eye	Responsible for movement of eye within the orbit	*extra*, outside of, + *oculus*, the eye
Fibrous Tunic	Tough outer connective tissue covering of the eye, composed of the sclera and cornea	Included elsewhere in this table, for sclera and cornea	*fibro-*, fiber + *tunic*, a coat
Lacrimal Caruncle	Small, fleshy mound of tissue at the medial aspect of the eye containing ciliary glands (modified sweat glands) and tarsal glands	Ciliary glands that form secretions to lubricate eyelashes	*caruncula*, a small fleshy mass
Lacrimal Gland	Almond-shaped serous gland located in the superior and lateral aspect of the orbit	Secretes tears	*lacrima*, a tear
Lacrimal Puncta	Two small openings in the lacrimal caruncle; the tiny holes on the "bump" at the inferomedial aspect of the eye	Opening for the drainage of lacrimal fluid into the lacrimal sac	*lacrima*, a tear, + *punctum*, a prick or point
Lacrimal Sac	A swelling at the superior part of the nasolacrimal duct, medial to the lacrimal bone, lateral to the nasal bone, and deep to the maxilla	Receives lacrimal fluid from the lacrimal canals and transports it to the nasolacrimal duct	*lacrima*, a tear
Nasolacrimal Duct	A duct that runs from the lacrimal sac into the nasal cavity	Conducts lacrimal fluid from the lacrimal sac into the nasal cavity	*nasal*, relating to the nose, + *lacrima*, a tear, + *ductus*, to lead
Optic Nerve	CN II; a large nerve exiting the posteromedial region of the eye that exits the orbit through the optic foramen; consists of myelinated axons of ganglion cells	Transmits nerve signals from the eye to the brain	*optikos*, relating to the eye or vision
Orbital Fat Pad	A thick capsule of adipose connective tissue that fills in all the spaces between the eye, extrinsic eye muscles, nerves, and orbit	Cushions the eye and helps support and hold it in place	*orbital*, relating to the orbit of the eye
Sclera	Dense irregular connective tissue that surrounds the entire eye except for the eye's anterior surface where the cornea is located	Protects the eye, serves as an attachment point for extrinsic eye muscles, and helps maintain the round shape of the eye	*skleros*, hard

(continued on next page)

(continued from previous page)

Table 18.8	Internal Structures of the Eye		
Structure	**Description**	**Function**	**Word Origin**
Anterior Cavity	The space anterior to the lens and posterior to the cornea. It is subdivided by the iris into the anterior and posterior chambers, which are described elsewhere in this table.	Filled with aqueous humor	*anterior*, the front surface, + *cavus*, hollow
Anterior Chamber	The space between the cornea and the iris/pupil	Filled with aqueous humor, which is described elsewhere in this table	*anterior*, the front surface, + *camera*, an enclosed space
Aqueous Humor	Watery fluid, similar in composition to cerebrospinal fluid. It is secreted by the ciliary processes and circulates within the anterior and posterior chambers of the eye.	Provides nourishment to the avascular lens and cornea	*aqueous*, watery, + *humor*, fluid
Choroid Layer	The pigmented, vascular layer located between the retina and the sclera	Blood vessels of the choroid supply nutrients to the tissues of the retina and sclera; pigment in the choroid absorbs light after it passes through the retina	*choroideus*, like a membrane
Ciliary Body	The thickened extension of the vascular tunic, located between the choroid and the iris; composed of both the ciliary process and the ciliary muscle	Produces aqueous humor; contraction of the ciliary muscle within the ciliary body alters the shape of the lens	*cilium*, eyelid, + *bodig*, a thing or substance
Ciliary Muscle	Smooth muscle found within the ciliary body that is composed of both circular and radial fibers	Contraction of this muscle relaxes the suspensory ligaments that attach it to the lens, which increases the lens curvature to accommodate for near vision	*cilium*, eyelid
Fovea Centralis	The depression ("central pit") in the macula lutea that contains only cones and lacks blood vessels	The area of highest visual acuity in the eye	*fovea*, a pit, + *centralis*, in the center
Iris	The colored portion of the eye, which makes up the anterior portion of the vascular tunic; the dilator pupillae and sphincter pupillae muscles are located within the iris	Controls the amount of light entering the eye. Contraction of the radially-arranged dilator pupillae muscle (under sympathetic stimulation) increases pupil diameter, whereas contraction of the circularly-arranged sphincter pupillae muscle (under parasympathetic stimulation) decreases the diameter of the pupil.	*iris*, rainbow
Lens	A transparent, biconvex structure composed of highly specialized, modified epithelium	Bends light waves (refraction) so that they hit the retina optimally for clear vision	*lens*, a lentil

Table 18.8	Internal Structures of the Eye (continued)		
Structure	**Description**	**Function**	**Word Origin**
Macula Lutea	A "yellow spot" on the retina located medial to the optic disc on the posterior wall of the eye, which contains the fovea centralis within it.	Contains the fovea centralis	*macula*, a spot, + *luteus*, yellow
Optic Disc ("Blind Spot")	An area of the retina where there is an absence of photoreceptors because it is where the axons of ganglion cells exit the eye to become the optic nerve	The location where axons of ganglion cells exit the eye	*optikos*, the eye, + *discus*, disc
Ora Serrata	Anteriormost portion of the retina, which appears serrated (hence the name)	Demarcates the division of the visual retina from the nonvisual retina	*ora*, an edge, + *serratus*, a saw
Posterior Chamber	The space between the iris/pupil and the lens and ciliary body	Filled with aqueous humor, which is described elsewhere in this table	*posterior*, the back surface, + *camera*, an enclosed space
Pupil	The space (opening) in the center of the iris	The size of the pupil (which is controlled by the smooth muscle within the iris) determines the amount of light entering the eye	*pupilla*, pupil
Retina	Also called the neural tunic of the eye; it is the inner layer of the eye composed of a pigmented layer, rods, cones, bipolar cells, and ganglion cells	Transduces light that enters the eye as light waves into nerve signals (action potentials) that are interpreted by the brain	*rete*, a net
Suspensory Ligaments	Ligaments that extend between the ciliary muscles and the lens	Attaches the lens to the ciliary muscles so that contraction and/or relaxation of ciliary muscles can alter the shape of the lens	*suspensio*, to hang up, + *ligamentum*, a bandage
Tapetum Lucidum	Metallic-appearing, opalescent inner layer of the sclera; present in many animals (e.g., the cow eye) but not in humans	Scatters light waves within the eye; allows for better vision in dim limited light (humans do not have this layer)	*tapeta*, a carpet, + *lucidus*, clear
Vascular Tunic	Middle layer of the wall of the eye; consists of the choroid, the ciliary body, and the iris	Provides nourishment to structures within the eye	*vasculum*, a small vessel
Posterior Cavity (Vitreous Chamber)	A space posterior to the lens and anterior to the retina	Occupied by the vitreous humor, which is described elsewhere in this table	*vitreus*, glassy, + *camera*, an enclosed space
Vitreous Humor	Clear, gelatinous mass within the vitreous chamber (posterior cavity)	Helps maintain the round shape of the eye and is critical in keeping the retina against the wall of the eye	*vitreus*, glassy, + *humor*, fluid

(continued on next page)

484 Chapter Eighteen *General and Special Senses*

(continued from previous page)

1. Observe a classroom model of the eye **(figure 18.11)**.

2. Using figure 18.11, tables 18.7 and 18.8, and your textbook as guides, identify the following structures on a model of the eye:

   - ☐ anterior cavity
   - ☐ anterior chamber
   - ☐ choroid
   - ☐ ciliary body
   - ☐ cornea
   - ☐ fovea centralis
   - ☐ inferior oblique muscle
   - ☐ inferior rectus muscle
   - ☐ iris
   - ☐ lacrimal gland
   - ☐ lateral rectus muscle
   - ☐ lens
   - ☐ medial rectus muscle
   - ☐ optic disc
   - ☐ optic nerve
   - ☐ ora serrata
   - ☐ posterior cavity (vitreous humor)
   - ☐ posterior chamber
   - ☐ pupil
   - ☐ retina
   - ☐ sclera
   - ☐ superior oblique muscle
   - ☐ superior rectus muscle
   - ☐ suspensory ligament
   - ☐ vascular tunic

Note: You should have identified the extrinsic eye muscles in chapter 12. If necessary, refer to table 12.3 on page 296 for guidance on the structure and function of these muscles.

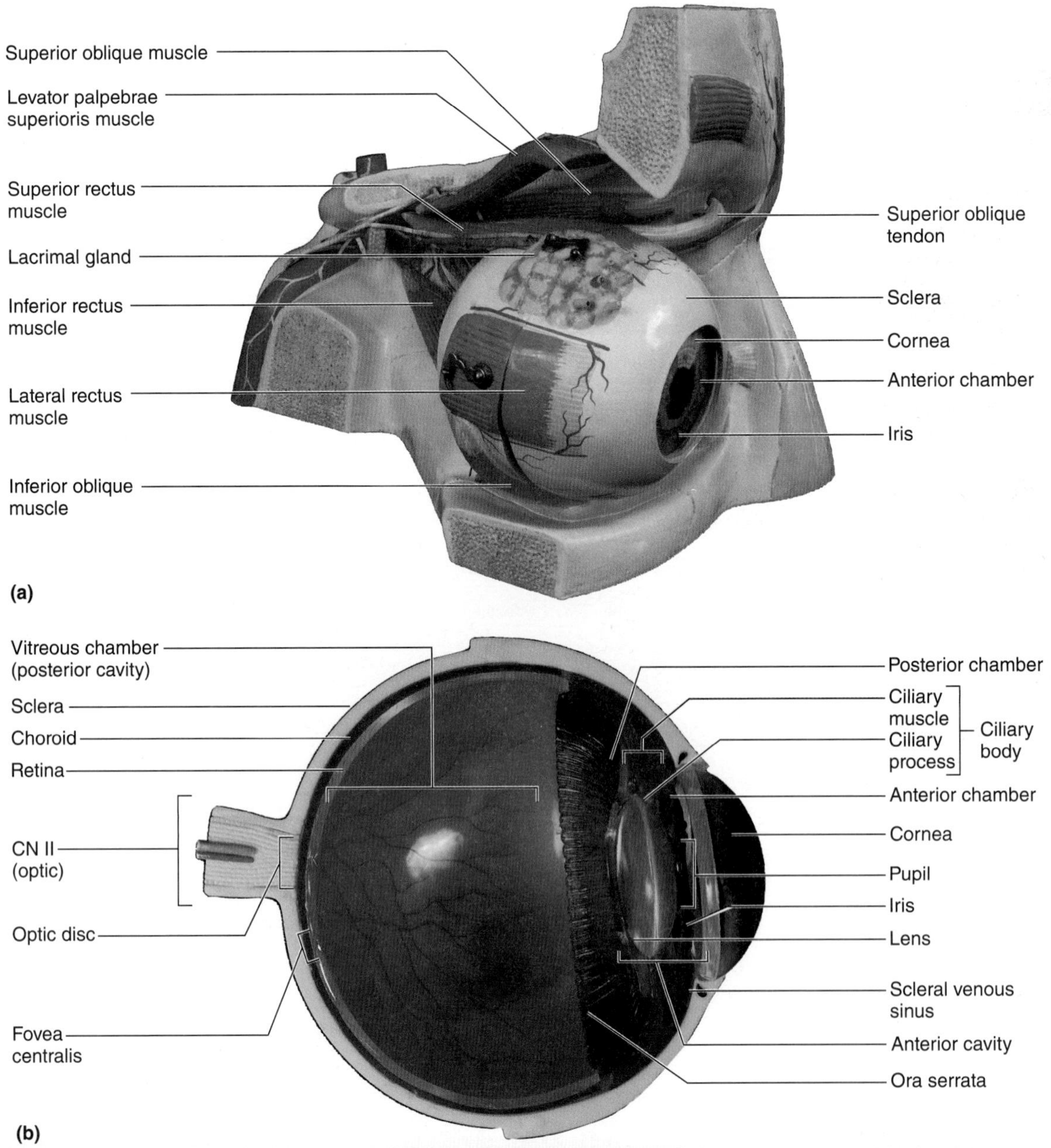

Figure 18.11 Classroom Model of the Eye. (a) Anterior view. (b) Internal view.

Chapter Eighteen *General and Special Senses* 485

3. Obtain a mirror and observe the externally visible structures of your eye. If you have no mirror, perform this observation on your lab partner. Using table 18.8 and your textbook as guides, identify all of the structures listed in **figure 18.12** on your eye (or your lab partner's eye), and then label them in figure 18.12.

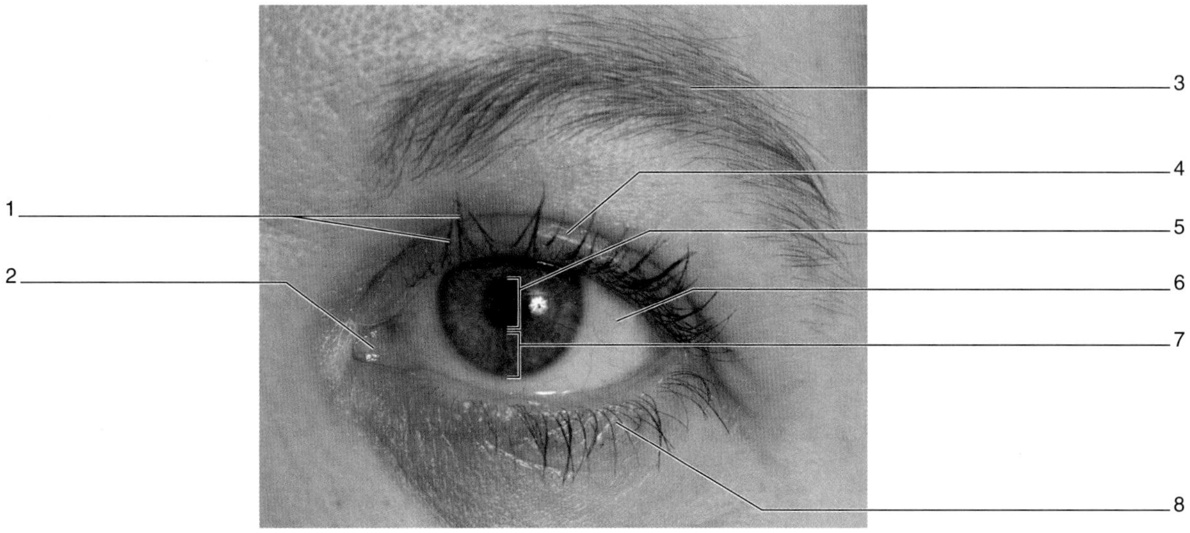

Figure 18.12 External Eye Structures.

☐ eyebrow   ☐ iris   ☐ sclera
☐ eyelashes ☐ medial canthus ☐ superior eyelid
☐ inferior eyelid ☐ pupil

EXERCISE 18.9

COW EYE DISSECTION

This exercise is designed to be done using fresh cow eyes, although preserved cow or pig eyes may be substituted if necessary. If you are dissecting a preserved cow or pig eye, the tissues will be tougher and the cornea will be opaque instead of transparent.

1. Obtain a dissecting pan, dissecting tools, and a fresh cow eye. Observe the gross structure of the eye before making any cuts. Using tables 18.7 and 18.8, and **figure 18.13a** as guides, identify the following structures:

   ☐ cornea   ☐ orbital fat pad
   ☐ extrinsic eye muscles ☐ sclera
   ☐ optic nerve

2. Using scissors and forceps (the tissue will be slippery!), remove the orbital fat pad and extraocular muscles, leaving the optic nerve intact (figure 18.13b). Once you have removed these, you will be able to view the entire eye and establish your orientation. Notice the toughness of the outer covering or **sclera** of the eye. This is the layer of tissue you will need to cut through in order to see structures within the eye.

3. Using scissors and forceps, cut the eye open by making a coronal incision through the sclera that completely encircles the eye. Once you do this, you will notice that a jellylike fluid oozes out of the posterior cavity of the eye. This fluid is the **vitreous humor** which fills the posterior cavity (**figure 18.14b**), and whose functions

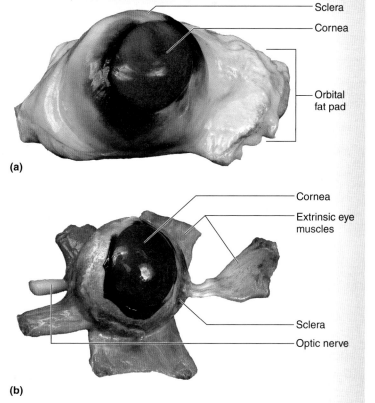

Figure 18.13 Fresh Cow Eye. (a) Before dissection with orbital fat pad intact. (b) After dissecting away the orbital fat pad to expose the extraocular muscles and optic nerve.

(continued on next page)

(continued from previous page)

include holding the **retina** against the posterolateral walls of the eye. In the cow eye, much of the **choroid,** which contains black pigment, becomes mixed into the vitreous humor once the eye is cut open. Thus, you may find that you have a gooey, black mess on your hands almost immediately. Before things get too "mixed up," identify structures in the posterior half of the dissected eye. Look for a yellowish, thin membrane that is connected to the posterior wall of the eye at only one spot (figure 18.14*a*). This is the retina. It contains neurons responsible for detecting visual stimuli and initiating nerve signals to the brain (histology of the retina was covered in detail in exercise 18.5). The retina is very delicate and easily falls away from the posterior wall of the eye when the vitreous humor is not present to hold it in place.

4. Find the location where the retina attaches to the posterior wall of the eye (it will "pucker" in this area). If you have difficulty finding this spot, first find the optic nerve on the outside of the eye and then look inside the eye for the location where the optic nerve leaves the eye. This spot within the eye is the **optic disc** (blind spot). It is called a "blind" spot because it is devoid of photoreceptors. This is where axons from the ganglion cell layer of the retina (see exercise 18.5) leave the eye and travel through the brain via the optic nerve (CN II).

5. Observe the inner walls of the posterior half of the eye. Notice the very colorful, iridescent **tapetum lucidum** (figure 18.14*a*). This structure is not present in humans, but it is present in animals that must be able see well in dim light, such as cows. The tapetum lucidum reflects light. Thus, when it is dark outside and very little light is entering the eye, the tapetum lucidum causes light waves to bounce around within the eye and increases the frequency with which light rays stimulate the retina. This makes things more visible, but the image does not become sharper. It is the reflection of light waves from the tapetum lucidum that causes a cat's eyes (or those of other animals) to "glow" when a light shines on them at night. In humans, the inside of the eye is completely coated with a black choroid, which absorbs excess light. This makes it more difficult for us to see things in the dark. On the plus side, the images that we do see are sharper.

6. Now focus your attention on the anterior portion of the eye (figure 18.14*b*). Again, it will be somewhat difficult to see many structures because the choroid covers everything, making the structures very dark. Notice the semitransparent **lens,** which is suspended in place by a ring of black-colored tissue. This tissue is the **ciliary body,** whose function is to suspend the lens. The cavity anterior to the lens and posterior to the cornea is the **anterior cavity** of the eye. The anterior cavity is further subdivided by the iris into an **anterior chamber** (between the cornea and the iris) and a **posterior chamber** (between the iris and the lens). In a living organism, the anterior cavity is filled with a clear, watery fluid called **aqueous humor.** Try to find the fine, delicate structures composing the **suspensory ligament,** which extends between the ciliary body and the lens.

7. Carefully remove the lens from the eye. You will notice that although it is not completely transparent, you can still see through it. Place it on a piece of paper containing text. As shown in **figure 18.15,** place the lens over a letter or two of text and make note of the change in appearance of the text, if any, as seen through the lens:

   _____

   _____

   _____

8. Identify the following structures on the interior of the dissected cow eye (use tables 18.7 and 18.8, and figures 18.13 through 18.15 as guides):

   ☐ anterior cavity   ☐ posterior cavity
   ☐ choroid   ☐ retina
   ☐ ciliary body   ☐ suspensory ligament
   ☐ lens   ☐ tapetum lucidum
   ☐ optic disc   ☐ vitreous humor

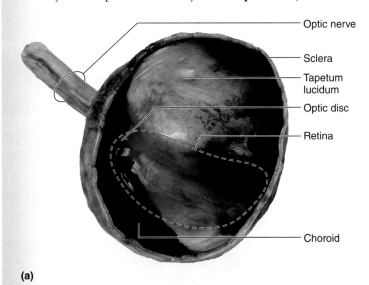

(a)

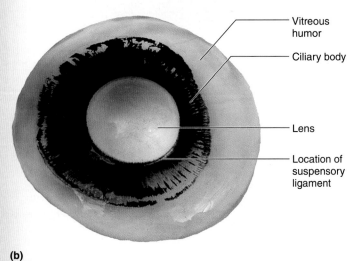

(b)

Figure 18.14 **Coronal Sections of Cow Eye.** (a) Posterior part demonstrating the retina, optic disc, and tapetum lucidum. (b) Anterior part demonstrating the choroid, lens, and iris.

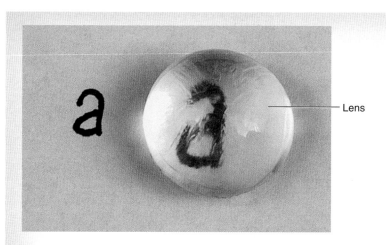

Figure 18.15 Lens. After removing the lens from the cow eye, place it over some text to see how it changes the image of the text.

9. When you have finished your dissection, clean up your workspace: Dispose of the cow eye debris in the organic waste receptacle. Dispose of used scalpel blades in the sharps container. Dispose of used paper towels and other paper waste in the wastebasket. Rinse off your dissecting tray and dissection instruments, and lay them out to dry. Finally, wipe down your laboratory workstation with disinfectant so it is clean for the next person who comes into the laboratory.

EXERCISE 18.10

GROSS ANATOMY OF THE EAR

The ear is responsible for two important special sensory modalities: equilibrium (balance) and hearing. It houses the organs for both within the *petrous part of the temporal bone*. These tiny sensory organs are nearly impossible to identify on a cadaver, thus, our exploration of the gross anatomy of the ear will be accomplished using models of the ear.

1. Obtain a model of the ear **(figure 18.16)**. On the model, first distinguish between the external-ear, middle-ear, and inner-ear cavities, and the structures that link the cavities to each other **(table 18.9)**. The **tympanic membrane** is the link between the external-ear and the middle-ear cavities, while the **oval window** is the link between the middle- and inner-ear cavities.

2. Using figure 18.16 and table 18.9 as guides, identify the following structures on the model:

   ☐ auditory tube
   ☐ auricle
   ☐ cochlea
   ☐ endolymph
   ☐ external acoustic meatus
   ☐ incus
   ☐ malleus
   ☐ ossicles
   ☐ oval window
   ☐ perilymph
   ☐ round window
   ☐ saccule
   ☐ semicircular canals
   ☐ stapedius muscle
   ☐ stapes
   ☐ tensor tympani muscle
   ☐ tympanic membrane
   ☐ utricle

3. In the space below, make a sketch of the ear and label the locations of the structures listed in step 2.

(continued on next page)

(continued from previous page)

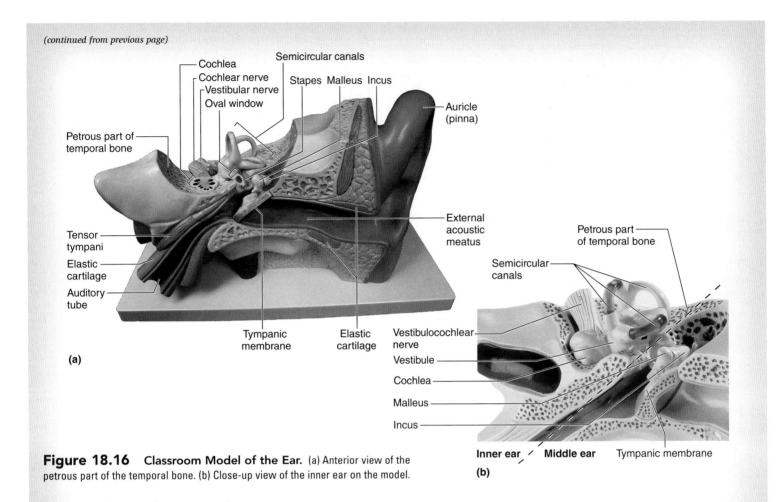

Figure 18.16 Classroom Model of the Ear. (a) Anterior view of the petrous part of the temporal bone. (b) Close-up view of the inner ear on the model.

Table 18.9	Structures of the External, Middle, and Internal Ear		
Structure	**Description**	**Function**	**Word Origin**
External Ear			
Auricle (Pinna)	External ear, composed of an elastic cartilage skeleton that is covered with skin	Collects sound waves from the environment and funnels them into the external auditory meatus	*pinna*, a wing
External Acoustic Meatus	Canal leading from the auricle of the ear to the tympanic membrane	Transmits sound waves that arrive at the auricle to the tympanic membrane, where they cause it to vibrate	*externa*, outside, + *acoustic*, relating to sound, + *meatus*, a passage
Tympanic Membrane	Drumlike, tight, thin membrane that separates the external ear cavity from the middle-ear cavity	Vibrates in response to sound waves that strike it as they reach the end of the external acoustic meatus; these vibrations cause vibrations in the ossicles of the middle-ear cavity (malleus, incus, and stapes)	*tympanon*, drum, + *membrana*, a membrane
Middle Ear	Air-filled cavity between the external ear and inner ear	Contains the ear ossicles	
Auditory Ossicles	Three tiny bones (malleus, incus, and stapes) found within the middle ear	Transmits movements caused by pressure vibrations from the tympanic membrane to the oval window of the cochlea, causing fluid pressure waves in the perilymph of the scala vestibuli	*ossiculum*, a bone
Incus	Tiny bone located within the middle-ear cavity that is shaped like an anvil	Transmits movements caused by pressure vibrations from the malleus to the stapes, thus participating in the transmission and amplification of the vibrations of the tympanic membrane	*incus*, an anvil

Table 18.9	Structures of the External, Middle, and Internal Ear *(continued)*		
Structure	**Description**	**Function**	**Word Origin**
Middle Ear *(continued)*			
Malleus	Tiny bone located within the middle-ear cavity that is shaped like a hammer	Transmits movements caused by pressure vibrations from the tympanic membrane to the incus, thus participating in the transmission and amplification of the vibrations that arrive as sound waves at the tympanic membrane	*malleus*, a hammer
Stapes	Tiny bone located within the middle-ear cavity that is shaped like a stirrup	Transmits movements caused by pressure vibrations from the incus to the oval window of the cochlea, causing fluid pressure waves in the perilymph of the scala vestibuli	*stapes*, a stirrup
Auditory Tube (Pharyngotympanic or Eustachian tube)	Tube lined with elastic cartilage that connects the middle-ear cavity to the nasopharynx	Opening of this channel allows air to enter or leave the middle-ear cavity such that the pressure in the middle ear equilibrates with the environmental pressure; this allows the tympanic membrane to vibrate freely	*audio*, to hear, + *tubus*, a canal
Oval Window	Membrane-covered opening into the scala vestibuli that is covered by the foot of the stapes	Vibrations of the stapes at the oval window causes fluid pressure waves in the perilymph of the scala vestibuli	*oval*, egg-shaped, + *window*, an opening
Stapedius Muscle	Small muscle connecting the neck of the stapes to the temporal bone	Contraction of this muscle acts to dampen vibrations of the stapes as a protective measure against excessive movement at the oval window from very loud noises	*stapedius*, relating to the stapes
Tensor Tympani Muscle	Small muscle connecting the handle of the malleus to the cartilage of the auditory tube	Contraction of this muscle pulls the malleus medially and tenses the tympanic membrane as a protective measure against excessive vibration from very loud noises	*tensus*, to stretch, + *tympani*, relating to the tympanic membrane
Inner Ear	Fluid-filled space located within the petrous part of the temporal bone that contains the cochlea, vestibule, and semicircular canals	Holds the organs responsible for the sensation of hearing (cochlea) and balance and equilibrium (vestibule and semicircular canals)	
Cochlea	Spiral-shaped organ found within the inner ear	Contains the spiral organ and associated structures that are involved in the special sense of hearing	*cochlea*, a snail shell
Saccule	Smallest membranous sac in the vestibule; connects with the cochlear duct	Contains receptors that sense linear vertical acceleration	*saccus*, a sac
Semicircular Canals	Three ringlike canals that are oriented at right angles to each other	Detect angular dynamic equilibrium	*semicircular*, shaped like a half circle, + *canalis*, a duct or channel
Spiral Organ (Organ of Corti)	Organ composed of specialized epithelium that is found within the scala media (cochlear duct) of the cochlea	Special sensory organ for hearing	*spiralis*, a coil, + *organon*, a tool or instrument
Utricle	The largest membranous sac in the vestibule	Contains receptors that sense linear horizontal acceleration	*uter*, leather bag
Vestibule	Located between the cochlea and the semicircular canals; contains the saccule and utricle	Detects both static equilibrium and linear, dynamic equilibrium	*vestibulum*, entrance court
Vestibulocochlear Nerve	CN VIII; travels through the internal acoustic meatus	Cranial nerve transmitting nerve signals associated with balance, equilibrium, and hearing to the brain	*vestibulo-*, referring to the vestibule, + *cochlea*, referring to the cochlea

(continued on next page)

(continued from previous page)

4. After you have identified all of the gross structures of the ear, review the sequence of events required for the transmission of sound waves from the environment to the cochlea **(figure 18.17)**. As you do this, name all of the structures involved in the sequence. The sequence is as follows:

   ① Sound waves are "funneled" into the **external acoustic meatus** by the contours of the outer ear **(auricle)** and cause vibrations of the **tympanic membrane.** The **auditory tube** ensures that air pressure in the middle ear is the same as air pressure in the environment so the tympanic membrane can vibrate freely. It is lined with elastic cartilage and remains collapsed unless there is a large difference in pressure between the environment and the middle-ear cavity. When a difference in pressure exists, the auditory tube opens briefly and air moves to equalize the pressure in the middle ear with the pressure in the environment.

   ② Vibrations of the tympanic membrane cause the **auditory ossicles** (malleus, incus, and stapes) to vibrate. Excessive vibrations (from a loud noise) cause a reflexive contraction of the **tensor tympani** and **stapedius** muscles to dampen the vibrations of the ear ossicles and help protect the delicate cells of the inner ear.

   ③ Vibration of the foot of the stapes against the **oval window** cause vibrations of the **perilymph** within the scala vestibuli.

   ④ Vestibular membrane moves cause pressure waves in the endolymph within the scala media (cochlear duct). This displaces the basilar membrane (in different regions depending on the frequency). Hair cells of the spiral organ bend, initiating nerve signals that are transmitted along the cochlear division of the vestibulocochlear nerve (CN VIII) to the brain.

   ⑤ Pressure waves are absorbed by the round window.

5. *Optional Activity:* **APR 7: Nervous System**—Watch the "Hearing" animation to review the sequence of events involved in hearing.

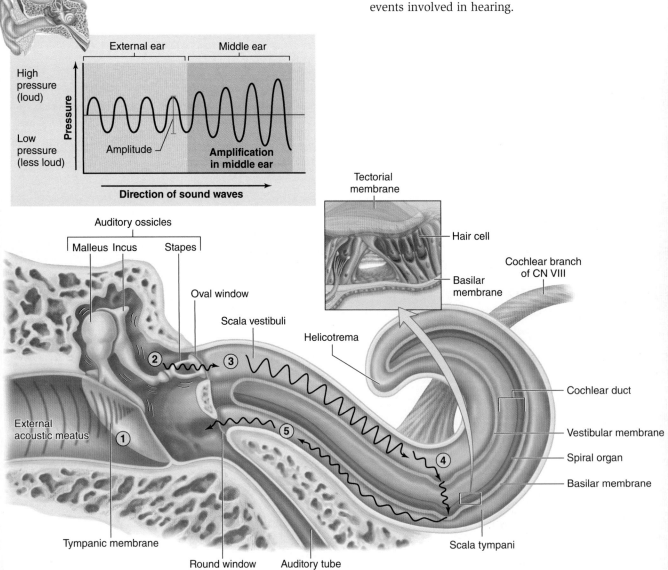

Figure 18.17 Soundwave Pathways through the Ear. Sound waves enter the external ear, are conducted through the ossicles of the middle ear, and then are detected by a specific region of the spinal organ in the inner ear.

Physiology

General Senses

In the following laboratory activities you will explore some of the functional properties of sensory receptors such as the tactile (Meissner) and lamellated (Pacinian) corpuscles, observed in the histology sections of this manual. You will begin by performing a variety of exercises that test some of the functional properties of tactile (touch) receptors. General sensory perception is complicated because several sensations share the same pathways as they travel to the brain. This convergence of pathways is the mechanism behind referred pain and "burning"

cold. Due to time limitations in the laboratory, the general sensation physiology exercises in this chapter will be those that demonstrate the **size of receptive fields** in different areas of the body, the association between size of receptive fields and **density of sensory receptors**, and the process of sensory **adaptation.**

In the remaining physiology exercises, you will explore some of the functional properties of the sensory receptors involved with the special senses (gustation, olfaction, vision, hearing, and equilibrium).

EXERCISE 18.11

TWO-POINT DISCRIMINATION

In this exercise you will compare the ability of an individual to discriminate between tactile sensations that are applied to two different locations on the skin at the same time (two-point discrimination). To do this, you will press a semi-sharp object into the skin. Which of the general sensory receptors is responsible for responding to this mode of sensation (assume the test will not cause pain)?

_____

The two-point discrimination test is an easy way to determine the size of the **receptive field,** which is associated with the relative density of sensory receptors in the skin. You will perform the test on skin covering different regions of your body and will then compare the two-point discrimination results from these different regions to evaluate the difference in receptive field size between different regions of the body.

Before you begin, state a hypothesis regarding the difference (if any) in the density of sensory receptors on the back of the hand as compared to the palm of the hand:

_____

_____

1. Obtain a two-point discrimination tool, or a metric ruler and a set of calipers **(figure 18.18).**

2. Choose a lab partner to be your subject. You will first test for two-point discrimination on skin of the subject's cheek. Start with the calipers or two-point discrimination tool all the way closed. Next, gently touch the subject's cheek with the calipers. The subject will feel only one point being touched.

3. Remove the calipers, open them one segment, then touch the subject's cheek again. Continue this process until the subject reports feeling two distinct points being touched instead of one. Now remove the calipers while holding the calipers to ensure they don't move. Place the tip of the calipers on a metric ruler and measure the distance. If you are using a two-point discrimination tool, simply take the final measurement from the tool. Record your results in **table 18.10.**

4. Complete steps 2 and 3 for each of the regions listed in table 18.10. Then record your results in table 18.10.

Table 18.10	Results of the Two-Point Discrimination Test
Body Region	**Two-Point Discrimination Distance (mm)**
Face (cheek)	
Face (lips)	
Posterior Neck	
Forearm (anterior)	
Forearm (posterior)	
Hand (posterior/back surface)	
Hand (anterior/palmar surface)	
Fingertip	
Leg	

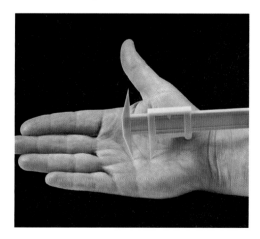

Figure 18.18 Two-Point Discrimination Test. Testing two-point discrimination on the palm of the hand using calipers.

EXERCISE 18.12

TACTILE LOCALIZATION

In this exercise you will test the ability of the brain to detect the precise location on the body that has been touched by a stimulus. This is called **tactile localization**. As with the two-point discrimination test, we might expect that the brain is better able to localize a stimulus on some areas of the body as compared to others, and that this ability is related to the size of the receptive field. When you touch the subject, the subject's brain will not necessarily know exactly what point was touched. Instead, the brain will respond to the stimulus by sensing that a larger area surrounding the touch point was stimulated. This larger area is a *receptive field*.

Before you begin, state a hypothesis regarding the difference (if any) in the receptive field of sensory receptors within the skin on the back of the hand as compared to the skin on the palm of the hand.

_____

_____

1. Obtain two or three markers of different colors.
2. Choose a person to be your subject. Have your subject close his or her eyes.
3. Touch the palm of the subject's hand with a washable marker (**figure 18.19**).
4. Hand the subject a marker of a different color than the one you used. Then ask the subject (still keeping his or her eyes closed) to try to touch the exact point that you touched, using his or her marker.
5. Measure the distance between the two points in millimeters (mm) and record it in **table 18.11**.
6. Repeat the process two more times. As the examiner, you will always be touching the same point, but your subject will not necessarily touch that same point. Calculate the average distance for the three trials and record it in table 18.11.
7. Complete steps 2–6 for each of the regions listed in table 18.11. Then record your results in table 18.11.

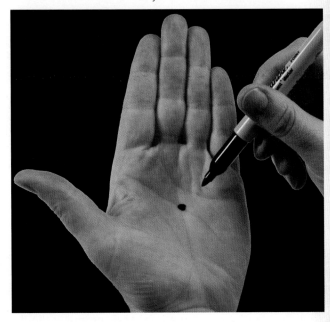

Figure 18.19 Tactile Localization Test. Testing tactile localization on the palm of the hand.

Table 18.11	Results of the Tactile Localization Test			
Body Region	**Test 1 Distance (mm)**	**Test 2 Distance (mm)**	**Test 3 Distance (mm)**	**AVERAGE Distance (mm)**
Palm of Hand				
Back of Hand				
Fingertip				
Anterior Forearm				
Back of Neck				
Anterior Leg				

EXERCISE 18.13

GENERAL SENSORY RECEPTOR TESTS: ADAPTATION

In this exercise you will test the ability of the brain to detect a constantly applied stimulus over time. A general property of our nervous system is that our brain tends to place emphasis and focus on incoming stimuli that are *changing* as opposed to stimuli that remain constant. Thus, when you initially put your clothing, glasses, jewelry, or other items on in the morning, you feel the clothing slide across your skin, the glasses rest upon your ears and nose, and your watch rest on your wrist. Soon afterwards, if these items don't move considerably, then you stop noticing them. Have you ever found yourself looking for your glasses only to discover they are right there on your head? This phenomenon is called **sensory adaptation.** Here, sensory receptors adapt so that the initial sensation (pressure, etc.) becomes a new "set point" for the brain. The brain subsequently ignores the constant signals coming in from that stimulus. This allows the brain to respond only when there is a *change* in the stimulus.

1. Obtain five pennies.
2. Choose a person to be your subject. Have your subject close his or her eyes.
3. Place a coin on the anterior surface of the subject's forearm approximately 2 cm proximal from the wrist **(figure 18.20a).**
4. Ask the subject to tell you when he or she can no longer feel the coin on his or her arm. Record the time elapsed in **table 18.12.**
5. Repeat the test, placing coins at the locations on the forearm listed in table 18.12. Record the time elapsed for each test in table 18.12.
6. Finally, perform the test one more time on the spot you used in step 3, only this time perform the test using five pennies stacked on top of each other (figure 18.20b).

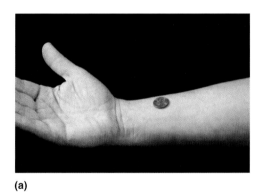 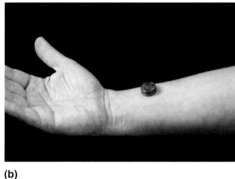

(a) (b)

Figure 18.20 Sensory Adaptation Test. Testing sensory adaptation on the anterior surface of the forearm with (a) one object and (b) multiple objects.

Table 18.12	Results of the Adaptation Test	
Coin Placement on Anterior Forearm	Number of Coins	Time Elapsed When Subject No Longer Perceives the Coins (Seconds)
2 cm proximal from wrist	1	
Midway between wrist and elbow	1	
1 cm distal from elbow	1	
2 cm proximal from wrist	5	
Midway between wrist and elbow	5	
1 cm distal from elbow	5	

Special Senses

Gustation (Taste)

There are approximately 4000 gustatory receptors ("taste buds") in the body, many of which are associated with the papillae of the tongue that you observed in the histology section of this chapter. In addition to this location of receptors on the tongue, there are also taste receptors located on the soft palate, pharynx, and epiglottis. Each receptor is specialized to respond to one of five primary taste sensations: salty, sweet, sour, bitter, and umami ("meaty"). **Table 18.13** summarizes the types of substances that elicit each of the five basic taste sensations. In this laboratory exercise you will test the specificity and location of the different types of taste sensed by these taste receptors.

Table 18.13	The Five Basic Taste Sensations	
Taste Sensation	**Substances Stimulating the Sensation**	**Examples**
Bitter	Alkaloids, which are plant-derived, nitrogen-containing basic compounds	Quinine, nicotine, caffeine, unsweetened cocoa
Salt	Metal ions	Table salt (NaCl)
Sour	Acids, which are substances in foods that release hydrogen ions (H^+)	Lemon juice, vinegar
Sweet	Natural sugars or artificial sweeteners	Corn syrup, Splenda®, granulated sugar
Umami	Amino acids such as glutamate and aspartate	Meats, cheeses, tomatoes, anchovies

EXERCISE 18.14

GUSTATORY TESTS

Before you begin, state a hypothesis regarding the region(s) of the oral cavity and pharynx that you think may have the largest number of receptors for each of the taste sensations.

_____

_____

1. Obtain seven cotton swabs and a small sample of each of the following:

   ☐ monosodium glutamate solution (umami)
   ☐ salt solution (salt)
   ☐ sugar solution (sweet)
   ☐ tonic water (bitter)
   ☐ vinegar (sour)
   ☐ water (for rinsing mouth)

2. One student will be the tester and one student will be the subject. As the tester, you must not tell the subject which solutions you are testing. Dip a cotton swab in one of the solutions and apply it to the roof of the subject's mouth (the palate), the inside of his or her cheeks, and the surface of his or her tongue.

3. Place a check mark in **table 18.14** next to all of the regions where the subject is able to taste the solution.

4. Discard the swab in a biohazard waste container.

5. Have the subject drink some plain water to rinse out his or her mouth.

6. Dip a clean swab into a different solution and repeat step 2. Continue to do this until the subject has tasted all of the solutions.

7. Use two cotton swabs. Dip one in the salt solution and one in the monosodium glutamate solution. Swab the subject's tongue with both solutions simultaneously and ask the subject to describe the taste. Record the subject's description of the taste here:

_____

_____

_____

Table 18.14	Results of the Gustatory Tests		

Place a check in the boxes where the subject was able to detect each taste.

	Palate	Cheeks	Tongue
Bitter			
Salty			
Sour			
Sweet			
Umami			

Olfaction (Smell)

As you observed in exercise 18.4, the sense of olfaction involves sensory receptors that are embedded in a special epithelium that lines the roof of the nasal cavity, the **olfactory epithelium.** This epithelium also contains goblet cells, which produce mucin that when hydrated forms mucus. As odorants enter the nasal cavity, they dissolve in the mucus. The olfactory receptor cells bind the odorant molecules and are stimulated. Nerve signals are initiated in the brain that are interpreted as the sense of olfaction (smell). As you have likely experienced before, the sense of smell is highly associated with the sense of gustation (taste). Recall a time when you had a "stuffy nose." The blockage of the sense of smell may have greatly affected your appetite because of its effect on gustation. In this exercise you will test the effect of olfaction on the perception of gustatory (taste) sensation.

EXERCISE 18.15

OLFACTORY TESTS

EXERCISE 18.15A Effect of Olfaction on Taste I

1. Obtain six cotton swabs, a blindfold, and vials containing samples of each of the following:

   ☐ almond extract   ☐ peppermint oil
   ☐ clove oil   ☐ vanilla
   ☐ lemon extract   ☐ wintergreen oil

2. One student will be the tester and one student will be the subject. As the tester, you must not tell the subject which substances you are testing. Have the subject close his or her eyes (or use the blindfold if that is easier) and pinch his or her nose closed.

3. Dip a cotton swab in one of the solutions and apply it to the subject's tongue. Ask the subject if he or she can taste the substance. Place a check in the appropriate box in **table 18.15** if the subject was able to taste the odorant with the nose closed.

4. After a few seconds, have the subject release his or her nose and breathe in. Ask him or her to identify the odorant again. Place a check in the appropriate box in table 18.15 if the subject was able to taste the odorant with the nose open.

5. Briefly note your conclusions in the space below:

   _____
   _____
   _____

Table 18.15 Results of the Tests for the Effect of Olfaction on Taste I

Place a check in the boxes where the subject was able to detect each taste.

Odorant	Nose Closed	Nose Open
Almond		
Clove		
Lemon		
Peppermint		
Vanilla		
Wintergreen		

(continued on next page)

(continued from previous page)

EXERCISE 18.15B Effect of Olfaction on Taste II

1. Obtain the following:
   - ☐ apple slices
   - ☐ blindfold
   - ☐ potato slices

2. One student will be the tester and one student will be the subject. As the tester, you must not tell the subject which substances you are testing. Have the subject close his or her eyes (or use the blindfold if that is easier). Then have the subject pinch his or her nose closed and stick out his or her tongue.

3. Apply either the apple or the potato slice to the subject's tongue. Ask the subject if the substance can be tasted. Place a check in the appropriate box in **table 18.16** if the subject was able to correctly identify the taste of the substance.

4. After a few seconds, have the subject once again close his or her eyes and stick out his or her tongue. Place the apple on the tongue and the potato under the nose (or vice versa). Place a check in the appropriate box in table 18.16 if the subject was able to correctly identify the substance.

5. Briefly note your conclusion in the space below:

_____

_____

_____

EXERCISE 18.15C Olfactory Adaptation

1. Obtain six cotton swabs, a blindfold, a timer or other mechanism for keeping time, and vials containing samples of each of the following:
   - ☐ almond extract
   - ☐ clove oil
   - ☐ lemon extract
   - ☐ peppermint oil
   - ☐ vanilla
   - ☐ wintergreen oil

2. One student will be the tester and one student will be the subject. As the tester, you must not tell the subject which odors you are testing. Have the subject close his or her eyes (or use the blindfold if that is easier).

3. Dip a cotton swab in one of the solutions and hold the swab 1–2 inches away from the subject's nose. Start the timer as soon as you place the swab next to the subject's nose. Ask the subject to identify the odor and to tell you when he or she can no longer smell the odor. The odor will "go away" with time because of adaptation of the olfactory receptor cells to the odor. Record the time taken for adaptation to occur in **table 18.17**.

4. Repeat step 3 for all of the remaining odorants.

5. Briefly note your conclusion in the space below:

_____

_____

_____

Table 18.16 Results of Tests for the Effect of Olfaction on Taste Sensation

Subject pinches nose closed and tester places item on subject's tongue.

Correct Identification of Taste?	Yes	No
Apple		
Potato		

Tester places one item on subject's tongue and the other under subject's nose (subject's nose is open).

Item on Tongue	Item under Nose	Correct Taste Perceived?
Apple	Potato	
Potato	Apple	

Table 18.17 Results of Olfactory Adaptation Tests

Record the time for adaptation for each of the following odorants.

Odorant	Time for Adaptation (Seconds)
Almond	
Clove	
Lemon	
Peppermint	
Vanilla	
Wintergreen	

Vision

Recall from exercise 18.5 that the special sense of vision involves stimulation of photoreceptors housed in the neural tunic of the eye, called the **retina**. Specifically, light changes the conformation of light-sensitive proteins in the photoreceptors of the retina: the rods and cones. **Rods** are photoreceptors that specialize in black-and-white vision, whereas **cones** are photoreceptors that specialize in color vision. When light stimulates these photoreceptors, nerve signals are transmitted along the optic nerve to the thalamus and then to the occipital lobe of the cerebrum, where the information is perceived as a visual image. Incidentally, there are no rods or cones where the optic nerve exits the eye, making that location a visual *blind spot*. Refer to your textbook as a guide as you review the specific pathways involved in processing visual information.

EXERCISE 18.16

VISION TESTS

The vast majority of us have first-hand knowledge of the special sense of vision. A trip to the eye doctor involves testing your **visual acuity**. An ophthalmologist assesses **vision** with a Snellen eye chart (**figure 18.21**), and from there determines if corrective lenses are necessary. You may have also experienced that as you age, you are forced to hold things you are reading farther away from your eyes in order to focus on an image or the words on a page. This **near-point accommodation** decreases dramatically with age due to the decreased elasticity in the lens of the eye. The result is an inability to change the shape of the lens, as is required when focusing on near objects. In the following exercises, you will test such aspects of vision as acuity, near-point accommodation, and the visual blind spot.

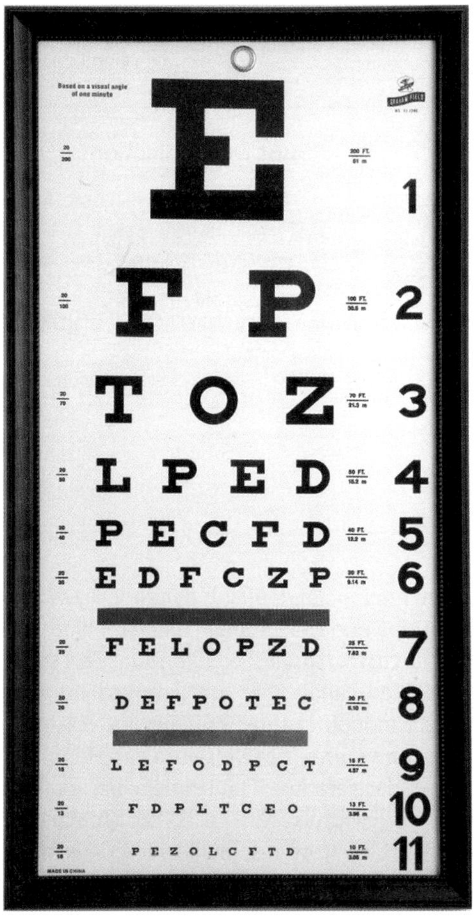

EXERCISE 18.16A Visual Acuity

1. Obtain a Snellen eye chart to test visual acuity. Hang the eye chart at eye level on a wall. Be sure that the wall is well illuminated.

2. Have the subject stand 20 feet from the eye chart, while a partner remains near the eye chart to validate the subject's responses.

3. Have the subject cover his or her left eye and read the lowest possible line of letters with his or her right eye. Record the visual acuity in **table 18.18**.

4. Repeat step 3 with the subject's left eye. Note that if the subject wears corrective lenses, this test can be conducted with and/or without glasses. A ratio of 1 (e.g., 20/20) indicates normal vision. A ratio of greater than 1 (20/15) indicates greater visual acuity; a ratio of less than 1 (20/30) indicates lesser visual acuity.

Figure 18.21 Snellen Eye Chart. Testing visual acuity using the Snellen eye chart.

Table 18.18	Results of Visual Acuity Tests
\multicolumn{2}{l}{Record visual acuity for the right and left eye, with and without corrective lenses.}	
Eye	**Visual Acuity**
Right Eye	
Right Eye (with corrective lenses)	
Left Eye	
Left Eye (with corrective lenses)	

(continued on next page)

(continued from previous page)

EXERCISE 18.16B Near-Point Accommodation

1. Obtain a pen or pencil and hold the object at arm's length in front of you. Cover your left eye and slowly move the object toward your right eye until the image of the object is no longer clear (i.e., is blurry or appears as two objects). Have your lab partner measure the distance from your right eye to the object. Record the near-point accommodation for the right eye in **table 18.19**.

2. Repeat step 1 by covering the right eye and moving the object toward the left eye. Record near-point accommodation for the left eye in table 18.19. Note that if the subject wears corrective lenses, this test can be conducted with and/or without glasses.

Table 18.19	Results of Near-Point Accommodation Tests

Record near-point accommodation for the right and left eye, with and/or without corrective lenses.

Eye	Near-Point Accommodation (cm)
Right Eye	
Right Eye (with corrective lenses)	
Left Eye	
Left Eye (with corrective lenses)	

EXERCISE 18.16C Blind Spot Determination

1. Hold **figure 18.22** an arm's length away from your face, approximately 46 cm (18 inches) away. Close your left eye and focus on the "X" with your right eye.

2. Move the figure toward your face while continuing to focus on the "X." Stop when the black dot no longer appears in your field of view. Have your lab partner measure the distance between your eye and the image. Record the distance between the image and the right eye in **table 18.20**.

3. Flip your lab book upside-down so the dot appears on the left of the image in figure 18.22. Repeat steps 1–2 by closing your right eye and focusing on the "X" with your left eye. Record the distance between the image and your left eye in table 18.20.

Figure 18.22 Blind Spot Determination.

Table 18.20	Results for Blind Spot Determination Test

Record the blind spot distance for the right and for the left eye.

Eye	Distance (cm)
Right Eye	
Left Eye	

Hearing and Equilibrium

The inner ear is composed of the cochlea, vestibule, and semicircular canals. The **cochlea** is the organ of hearing (see p. 476).

The vestibule and semicircular canals are also contained within the inner ear. The **vestibule** monitors *static equilibrium*, when the body is not in motion, and *linear dynamic equilibrium*, when the body is accelerating or decelerating in a linear plane. The **semicircular canals** detect *angular dynamic equilibrium*. The semicircular canals are three fluid-filled canals that are arranged orthogonally (at right angles) to detect motion in three dimensions. For example, the vestibule detects when you are traveling in a car along a straight road or accelerating and decelerating. The semicircular canals detect angular motion when you sharply turn a corner. Interestingly, hair cells, the same type of receptors responsible for detecting hearing in the cochlea, also detect static and dynamic equilibrium in the vestibule and semicircular canals.

Stimulation of the receptors within the vestibule and semicircular canals initiates nerve signals that are transmitted along the vestibular branch of the vestibulocochlear nerve (CN VIII) to the thalamus and cerebral cortex, as well as to the nuclei for the three cranial nerves that control the movement of the eye: oculomotor (CN III), trochlear (CN IV), and abducens (CN VI). Refer to your textbook as a guide for reviewing the specific neural pathways involved in hearing and equilibrium.

INTEGRATE

CLINICAL VIEW
Cochlear Implant

A cochlear implant is a medical device that attempts to mimic the vibrations experienced by hair cells of the cochlea, thereby performing the same amplification of sound performed typically by the cochlea. The implanted structure then transmits this signal by stimulating the vestibulocochlear nerve, which in turn allows the signal to reach the cerebral cortex, thereby allowing someone to perceive sound.

EXERCISE 18.17

HEARING AND EQUILIBRIUM TESTS

Health care providers may perform tests for proper vestibular function if they suspect a problem. For example, a patient that presents with motor deficits may undergo a Romberg test, a clinically relevant equilibrium test. In a Romberg test, the patient stands still with feet together and eyes closed while the tester watches for any swaying motion. In the following laboratory exercises, you will perform tests of both hearing and equilibrium. Since the sensory receptors for both of these senses are housed within the inner ear, and sensory information for both is transmitted along the vestibulocochlear nerve, these tests are contained within one laboratory exercise. Specifically, you will test for both different types of **deafness** (absence of sound perception) and for vestibular dysfunction.

Distinguishing between **nerve deafness** and **conduction deafness** requires important clinical tests, namely, the Rinne and Weber tests. The **Rinne test** evaluates a person's ability to perceive sound from vibrations in the air as compared to vibration applied directly to the temporal bone. The **Weber test** evaluates a person's ability to perceive sound in both ears equally, which is a specific test for nerve conduction deafness. The **Barany test** demonstrates the reflex between the movement of the fluid in the semicircular canals and the extrinsic eye muscles.

Equilibrium tests are complicated because many anatomical structures, such as the vestibular apparatus, eyes, proprioceptors, cerebellum and the cranial nerves are involved. In the following laboratory exercises, you will perform a Romberg test. A **Romberg test** is administered during a neurological exam when a patient exhibits motor or sensory deficits. It is also performed when a person is suspected of drunk driving. You will also perform an exercise that demonstrates the influence of vestibular input on eye movement.

EXERCISE 18.17A Hearing Test: Rinne

1. Obtain a tuning fork and choose an individual to be the subject. When you tap the tuning fork on a hard surface, it will begin to vibrate at a certain wavelength and produce sound.

2. Tap the tuning fork on a hard surface and hold it near the subject's ear **(figure 18.23a)**. Ask the subject to tell you about the sound and make a note of the subject's response here: _____

   _____

3. Gently tap the tuning fork on a hard surface and place the vibrating instrument on the subject's mastoid process (figure 18.23b). Again, ask the subject about the sound and about any differences in perceived sound. Make a note of the subject's response here:

   _____

   _____

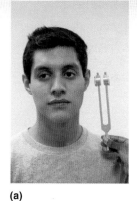 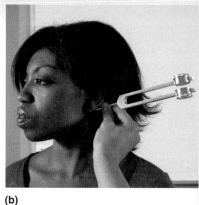

(a)   (b)

Figure 18.23 Rinne Hearing Test. (a) Strike the tuning fork and place the device near the ear. (b) Place the vibrating tuning fork on the mastoid process.

EXERCISE 18.17B Hearing Test: Weber

1. Obtain a tuning fork and choose an individual to be the subject.

2. Tap the tuning fork on a hard surface. Place the vibrating instrument on the center of the subject's forehead **(figure 18.24)**. Ask the subject to compare his or her perception of sound in the right versus left ear, and make a note of the subject's response here:

   _____

   _____

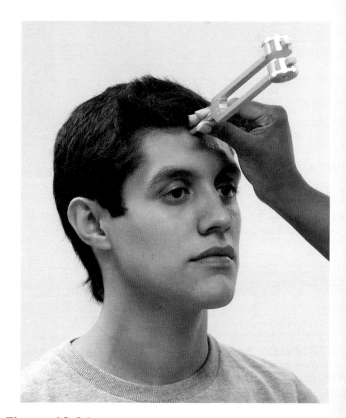

Figure 18.24 Weber Hearing Test. Place the vibrating tuning fork on the center of the forehead.

(continued on next page)

(continued from previous page)

EXERCISE 18.17C Equilibrium Test: Romberg

1. Choose an individual to be the subject. Ask the subject to stand for one minute with both feet together and both hands by his side **(figure 18.25a)**. For ease, have the subject stand in front of a surface where you can either see the subject's shadow or mark the subject's location (i.e., a whiteboard). Note any exaggerated swaying movements to the left or right here:

   _____
   _____
   _____

2. Have the subject repeat step 1 with his eyes closed. Compare the magnitude of sway in the eyes open versus eyes closed condition. Make a note of it here:

   _____
   _____
   _____

3. Have the subject rotate 90 degrees, such that his shoulder is adjacent to the whiteboard (figure 18.25b). Be sure that you can visualize the position easily with a shadow or outline of the initial position. Look for any forward-backward swaying motion when the subject's eyes are open. Make a note of it here:

   _____
   _____
   _____

4. Have the subject close his eyes and repeat step 3. Look for any differences in forward-backward sway in the eyes open versus eyes closed condition. Make a note of any differences in sway here:

   _____
   _____
   _____

(a)

(b)

Figure 18.25 Romberg Test. The subject stands with feet together and hands by his side, with his (a) back toward the whiteboard and (b) shoulders adjacent to the whiteboard. Stand close to the subject so you may catch him if he loses his balance.

INTEGRATE

CONCEPT CONNECTION

A Romberg test does more than simply test the signals of the vestibular system. Rather, it evaluates the integration of vestibular information, visual information, and proprioception. We rely on all three signals to maintain an upright posture. When one or more of these senses is compromised, our ability to balance (particularly when impaired) is compromised. A Romberg test involves intentionally removing one signal, the visual input, and then observing any disruption in balance. Disruptions in balance are observed when there is an increase in sway. If a person exhibits a positive Romberg sign (increased sway with eyes closed), it indicates that there may be deficits with proprioception. Specifically, there is likely impairment of the dorsal column motor pathway, which transmits proprioceptive information from the sensory receptors to the central nervous system. The Romberg test demonstrates to us that vestibular input is not sufficient by itself to allow a person to maintain an upright posture.

CHAPTER 19

The Endocrine System

OUTLINE AND LEARNING OBJECTIVES

Histology 512

EXERCISE 19.1: THE HYPOTHALAMUS AND PITUITARY GLAND 512

1. Differentiate between the anterior and posterior lobes of the pituitary gland on a histology slide of the pituitary gland
2. Identify the cell types present within the anterior lobe of the pituitary
3. Describe the relationship between neurons within the paraventricular and supraoptic nuclei of the hypothalamus and the secretion of hormones by the posterior lobe of the pituitary

EXERCISE 19.2: THE PINEAL GLAND 515

4. Identify the pineal gland on a histology slide
5. Describe the structure and functions of pinealocytes and pineal sand
6. Describe the functional importance of pineal sand to radiologists

EXERCISE 19.3: THE THYROID AND PARATHYROID GLANDS 516

7. Identify thyroid follicles, parafollicular cells, and the parathyroid glands on a histology slide of the thyroid gland
8. List the hormones produced by the thyroid follicles, parafollicular cells, and chief cells and describe their functions
9. Briefly describe the function of thyroid hormone, calcitonin, and parathyroid hormone and describe their functions

EXERCISE 19.4: THE ADRENAL GLANDS 517

10. Identify the zona glomerulosa, zona fasciculata, zona reticularis, and adrenal medulla on a histology slide of the adrenal gland
11. List the hormones produced by the zona glomerulosa, zona fasciculata, zona reticularis, and adrenal medulla
12. Briefly describe the functions of the hormones aldosterone, cortisol, androgens, norepinephrine, and epinephrine

EXERCISE 19.5: THE ENDOCRINE PANCREAS—PANCREATIC ISLETS (OF LANGERHANS) 520

13. Identify pancreatic islets on a histology slide of the pancreas
14. Describe the structure and function of the endocrine part of the pancreas
15. Briefly describe the functions of the hormones insulin, glucagon, and somatostatin
16. List the cell types that produce insulin, glucagon, and somatostatin

Gross Anatomy 521

Endocrine Organs 522

EXERCISE 19.6: GROSS ANATOMY OF ENDOCRINE ORGANS 522

17. Identify the classical endocrine glands on classroom models or a human cadaver

Physiology 524

Metabolism 524

EXERCISE 19.7: Ph.I.L.S. LESSON 17: THYROID GLAND AND METABOLIC RATE 524

18. Calculate the rate of oxygen consumption in a virtual experiment
19. Explain how cooling influences thyroid hormone release, and how this, in turn, affects the rate of oxygen consumption
20. Discuss the role of the thyroid gland in regulating oxygen consumption and metabolic rate

MODULE 8: ENDOCRINE SYSTEM

INTRODUCTION

How long has it been since you last ate? Even if it has been several hours since your last meal, your blood glucose and blood calcium levels remain remarkably stable, fluctuating only minor amounts around your body's normal physiological level (unless, of course, you have a disease such as diabetes). Maintenance of blood glucose and blood calcium levels are physiological imperatives, for if their levels are too high or too low, severe impairment of nervous and muscular activity will occur. However, it is not often that our blood glucose or calcium levels move drastically out of the normal range. This is because they are tightly regulated by the **endocrine system.** The hormones insulin and glucagon, produced by the pancreas, regulate blood glucose levels, and the hormones calcitonin and parathyroid hormone, produced by the thyroid and parathyroid glands, respectively, regulate blood calcium levels.

This brief description of these hormones and the variables they regulate provides only a glimpse at the functioning of the endocrine system, a system of chemical messengers that travel in the blood and act on distant target cells. The endocrine system **(figure 19.1)** consists of a number of "classical" endocrine organs, such as the pituitary gland, adrenal glands, pancreatic islets, and thyroid gland. However, many organs in the body contain cells or tissues that produce and secrete hormones. For instance, cells in the walls of the stomach secrete hormones that regulate appetite, gastric motility, and acid secretion. Cells in the testes (males) and ovaries (females) are responsible for the secretion of hormones that regulate the maturation of sperm and eggs, respectively. While the thymus is considered an endocrine organ due to its secretion of the hormone *thymosin,* it plays a larger role in the immune system, as described in chapter 23.

In this chapter, we will explore the structure and function of the classical endocrine organs. Many of the organs we will explore contain cells that secrete hormones (thus the organ has an endocrine role). However, the entire organ is not necessarily referred to as an endocrine *gland* because the organ has other roles as well. The few organs that are strictly endocrine in nature are generally quite small. Only the thyroid and adrenal glands can be viewed easily on a cadaver. Thus, our exploration of the endocrine system in the anatomy and physiology laboratory will be carried out predominantly at the microscopic level. However, in this exercise you will be asked to locate the major (large) endocrine organs on the cadaver or on classroom models. As you examine histology slides, make associations between the tissues you are viewing, the gross anatomic location where the tissue is found,

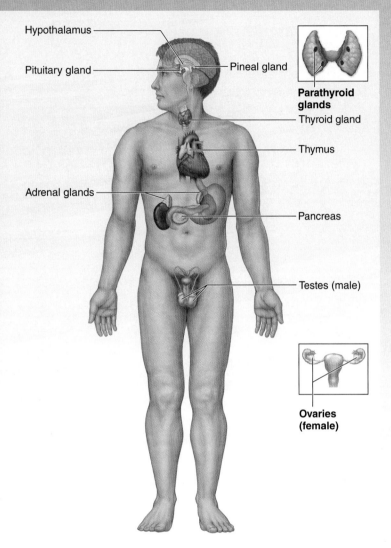

Figure 19.1 Classical Endocrine Glands of the Human Body.

and the function(s) of the hormone(s) secreted by each organ. An activity that explores the physiology of metabolism is also included in this chapter. In subsequent chapters covering the cardiovascular, lymphatic, digestive, urinary, and reproductive systems, we will explore the structure and function of endocrine cells and tissues (and associated hormones) related to each system in particular.

Chapter 19: The Endocrine System

PRE-LABORATORY WORKSHEET

Name: _____
Date: _____ Section: _____

Also available at www.connect.mcgraw-hill.com

1. Describe the basic differences between endocrine and exocrine glands.

   _____
   _____
   _____
   _____
   _____
   _____
   _____

2. Describe the embryonic origins of the anterior and posterior lobes of the pituitary gland.

   _____
   _____
   _____
   _____
   _____
   _____
   _____
   _____

3. Where are parafollicular C cells located? _____

4. What endocrine gland consists of a cortex and medulla, with each part embryonically derived from different tissues? _____

5. What endocrine gland consists of follicles formed by simple cuboidal epithelium? _____

6. What structure secretes hormones that regulate the secretion of hormones by the anterior pituitary gland? _____

7. What endocrine gland secretes the hormone melatonin? _____

8. The _____ are four small endocrine glands on the posterior part of the thyroid gland that secrete a hormone that regulates blood calcium levels.

9. Name the major endocrine gland that secretes sex steroid hormones in the male and in the female.

   a. male _____

   b. female _____

Chapter Nineteen *The Endocrine System* 511

Histology

In the following exercises, the microscopic anatomy of the various endocrine glands is described in detail. The tables in these exercises list the glands, the hormones they produce, and the functions of each of the listed hormones. Your focus in the following exercises will be to differentiate the various endocrine glands from each other when viewing them under the microscope.

Endocrine Glands

EXERCISE 19.1

THE HYPOTHALAMUS AND PITUITARY GLAND

The **pituitary** gland, or hypophysis (*hypophysis*, an undergrowth), is a remarkable organ. This small organ plays a significant role in endocrine regulation of the body. The pituitary gland is only about the size and shape of a pea, but it secretes hormones that regulate the growth and development of nearly every other organ in the body. The secretion of pituitary hormones is tightly controlled by cells within the **hypothalamus,** a part of the brain whose structure and function was discussed in chapter 15. Your textbook covers the structure of, function of, and relationships between the hypothalamus and pituitary gland in detail. Because it is not possible to visualize the details of these relationships in the laboratory, we will focus our studies here on the structure and function of the pituitary gland alone. **Table 19.1** summarizes the cells and structures that compose the pituitary gland and lists the hormones secreted by each cell type, along with the releasing or inhibiting hormones secreted by the hypothalamus that influence the secretion of hormones by the pituitary gland.

1. Obtain a histology slide of the pituitary gland **(figure 19.2).** Before placing it on the microscope stage, observe the slide with your naked eye. Notice there is a distinctive difference in color between the two parts, or lobes, of the pituitary gland. The darker area is the **anterior pituitary** (also called the anterior lobe, or **adenohypophysis** [*adeno-*, a gland, + *hypophysis*, pituitary]), whereas the lighter area is the **posterior pituitary** (also called the posterior lobe, or **neurohypophysis** [*neuro-*, relating to nervous tissue, + *hypophysis*, pituitary]).

2. Place the slide on the microscope stage and bring the tissue sample into focus on low power. Once again, identify the anterior and posterior lobes (figure 19.2a). Recall that the anterior pituitary is derived embryologically from an outpocketing of the roof of the mouth and consists of epithelial tissue. Thus, the cells have an appearance that is characteristic of glandular epithelial tissue (figure 19.2b). The posterior pituitary is derived embryologically from a downgrowth of the diencephalon of the brain and consists of nervous tissue. Thus, the cells have an appearance that is characteristic of nervous tissue (figure 19.2b).

3. *Anterior Pituitary*—Using figures 19.2 and 19.3 as guides, identify the anterior pituitary gland. Then move the microscope stage so the anterior pituitary is in the center of the field of view.

4. Increase the magnification so you can see the glandular nature of the cells **(figure 19.3).** There are three cell types within the anterior pituitary: acidophils, basophils, and chromophobes. The first two kinds of cells are named for their "love" (*-phil,* to love) of acidic or basic dyes. *Acidophils* attract acidic dyes, and appear red. *Basophils* attract basic dyes and appear blue. *Chromophobes* (*chroma,* color, + *phobos,* fear) are cells that attract neither acidic nor basic dyes, and are thought to be cells that have released their hormone(s). You will not be required to identify the various cell types within the anterior lobe, nor will you be required to know which cells produce which hormones. However, the cell types and hormones produced by each cell form the basis for a couple of handy mnemonic devices to remember the names of the hormones produced by the anterior pituitary (provided in step 5), which you may find helpful.

5. The mnemonics for remembering which hormones are produced by which cells of the anterior pituitary are as follows:

    Mnemonic for acidophils: **GPA** (as in Grade Point Average: If you do not remember this information, it could be harmful to your GPA.)

    G = growth hormone (GH)
    P = prolactin (PRL)
    A = acidophil

    Mnemonic for basophils: **B-FLAT** (as in the musical note: If you remember this information, it will be beneficial to your GPA and you will be happily singing to the tune of B-flat!)

    B = basophil
    F = follicle-stimulating hormone (FSH)
    L = luteinizing hormone (LH)
    A = adrenocorticotropic hormone (ACTH)
    T = thyroid-stimulating hormone (TSH)

6. *Posterior Pituitary*—Using figure 19.2 as a guide, identify the posterior pituitary. Then move the microscope stage so the posterior pituitary is in the center of the field of view.

Chapter Nineteen The Endocrine System

Table 19.1 Histology of the Pituitary Gland

Pituitary Gland	Cells	Description	Hormones Produced	Action of Pituitary Hormone(s)	Hypothalamic Releasing or Inhibiting Hormone	Word Origin
Anterior Pituitary (Adenohypophysis)	Acidophils	Appear red in color due to their attraction for acidic stains	GH, PRL	NA	NA	*acidus*, sour (relating to acidic dyes), + *phil*, to love
			Growth hormone (GH)	Stimulates the liver and other tissues to produce IGF-1 (insulin-like growth factor-1), which promotes bone and muscle growth	Growth-hormone-releasing hormone (GHRH) stimulates release, whereas growth-hormone-inhibiting hormone (GHIH, somatostatin) inhibits release	*grōthr*, growth, + *hormon*, to set in motion
			Prolactin (PRL)	Stimulates the mammary glands to develop and produce milk	Prolactin-inhibiting hormone (PIH) inhibits release	*pro-*, before, + *lac*, milk
	Basophils	Appear blue in color due to their attraction for basic stains	FSH, LH, ACTH, TSH	See below	See below	*baso-*, basic (relating to basic dyes), + *phil*, to love
			Follicle-stimulating hormone (FSH)	Stimulates the growth and maturation of ovarian follicles (females); stimulates spermatogenesis (males)	Gonadotropin-releasing hormone (GnRH) stimulates release, whereas the hormone inhibin inhibits its release	*folliculus*, a small sac (referring to the ovarian follicles)
			Luteinizing hormone (LH)	Induces ovulation, stimulates the production of estrogen and progesterone by cells of the corpus luteum (females); stimulates interstitial cells to produce testosterone (males)	Gonadotropin-releasing hormone (GnRH) stimulates release	*luteus*, yellow (referring to the corpus luteum of the female ovary)
			Adrenocorticotropic hormone (ACTH)	Stimulates the growth, development, and secretion of steroid hormones by the adrenal cortex (e.g., cortisol)	Corticotropin-releasing hormone (CRH) stimulates release	*adrenocortico*, referring to the adrenal cortex, + *trophe*, nourishment
			Thyroid-stimulating hormone (TSH)	Stimulates the secretion of thyroid hormones by the thyroid gland	Thyrotropin-releasing hormone (TRH) stimulates release	*thyroid*, shaped like a shield
	Chromophobes	Appear very light in color due to a lack of staining	Thought to be devoid of hormone, hence the lack of staining properties	NA	NA	*chroma*, color, + *phobos*, fear

Pituitary Gland	Cells	Description	Hypothalamic Hormones Stored and Released	Action of Pituitary Hormone(s)	Hypothalamic Nucleus Containing Neuron Cell Bodies	Word Origin
Posterior Pituitary (Neurohypophysis)	Axon terminals	Axon terminals store hormone that was produced in the cell bodies of the neurons within the hypothalamus	Oxytocin	Stimulates uterine contractions and milk ejection by mammary glands	Paraventricular nucleus and supraoptic nucleus	*axon*, axis, + *terminus*, the limit; *para-*, next to, + *ventricular*, relating to the third ventricle of the brain, + *nucleus*, a collection of neuron cell bodies; *oxytokos*, swift birth
	Axon terminals	Axon terminals store hormone that was produced in the cell bodies of the neurons within the hypothalamus	Antidiuretic hormone (ADH, vasopressin)	Increases water retention by the kidneys (increases blood volume and blood pressure); vasoconstriction	Paraventricular nucleus and supraoptic nucleus	*axon*, axis, + *terminus*, the limit; *supra-*, above, + *optic*, relating to the optic tract, + *nucleus*, a collection of neuron cell bodies; *anti*, against, + *diuresis*, excretion of urine
	Pituicytes	Derived from glial cells; have processes that surround axon terminals of the hormone-secreting neurons	NA	NA	NA	*pituita*, a phlegm (relating to the pituitary gland), + *-cyte*, cell

(continued on next page)

(continued from previous page)

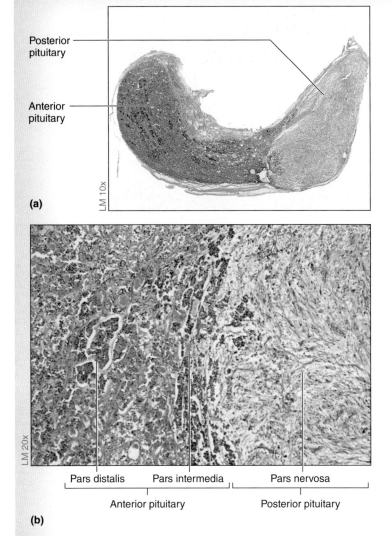

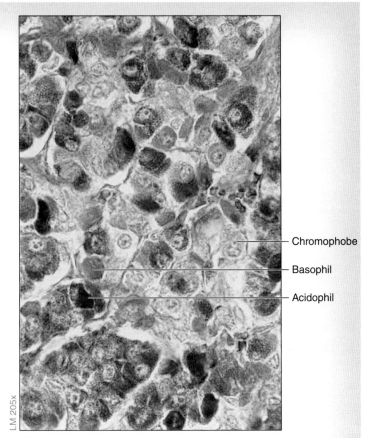

Figure 19.3 Anterior Pituitary. The three cell types in the anterior pituitary are basophils, which stain blue, acidophils, which stain red, and chromophobes, which don't take up biological stains.

Figure 19.2 Pituitary Gland. (a) Low-magnification view of both parts of the pituitary. (b) Medium-magnification view of the anterior and posterior pituitary.

7. Increase the magnification so you can see the cells clearly **(figure 19.4)**. The majority of the nuclei visible within the slide are the nuclei of **pituicytes,** which are derived from glial cells. Pituicytes surround the axon terminals of neurons whose cell bodies are located in the paraventricular and supraoptic nuclei of the hypothalamus. These neurons secrete the hormones oxytocin and antidiuretic hormone (ADH, vasopressin).

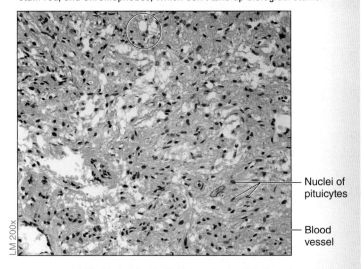

Figure 19.4 Posterior Pituitary. The majority of the nuclei seen in this micrograph are of pituicytes, which are glial cells.

8. Using figures 19.2 through 19.4, and table 19.1 as guides, identify the following structures on the slide of the pituitary gland.

- ☐ acidophils
- ☐ anterior pituitary
- ☐ basophils
- ☐ chromophobes
- ☐ pituicytes
- ☐ posterior pituitary

9. In the space below, sketch the pituitary gland as seen through the microscope, labeling the structures listed in step 8.

_____ ×

EXERCISE 19.2

THE PINEAL GLAND

The **pineal gland** (*pineus*, relating to a pine, shaped like a pinecone), also called the **pineal body,** is a small region in the epithalamus of the brain. Its primary cells, called **pinealocytes (figure 19.5),** secrete the hormone **melatonin.** Melatonin's effects in humans are unsubstantiated, but in other organisms melatonin is responsible for regulation of circadian rhythms. In humans, it may have a role in determining the onset of puberty. Pinealocytes are innervated by neurons from the sympathetic nervous system, and their secretion of hormone is affected by the amount of light received by the individual, which is relayed to the pineal through these neurons. Melatonin secretion increases when light levels are low (at night) and decreases when light levels are high (during the day). Pinealocytes appear in groups of cells within the pineal gland.

They are surrounded by glial cells, whose function is similar to that of astrocytes in other parts of the brain. Clinically, one of the most important features of the pineal gland is the presence of calcium concretions, termed **"pineal sand"** (corpora arenacea). These concretions are easily visible in radiographs of the head, and provide radiologists with a landmark that is consistent and easy to identify. The number of concretions in the pineal gland increases with age.

1. Obtain a histology slide of the pineal gland. Place the slide on the microscope stage and bring the tissue sample into focus on low power.

2. Using figure 19.5 as a guide, identify the following structures on the slide:

   - ☐ pinealocytes
   - ☐ pineal sand

3. In the space below, sketch the pineal gland as seen through the microscope at medium or high magnification, labeling pinealocytes and pineal sand.

Figure 19.5 The Pineal Gland. Histology of the pineal gland.

_____ ×

EXERCISE 19.3

THE THYROID AND PARATHYROID GLANDS

The **thyroid gland** (figure 19.6; table 19.2) is a butterfly-shaped gland located anterior to the trachea and inferior to the thyroid cartilage of the larynx. It consists of two main lobes connected to each other anteriorly by a narrow **isthmus** (*isthmus*, neck). The functional units of the thyroid gland are **thyroid follicles,** which are lined with a simple cuboidal epithelium. Inside each follicle is a mass of **colloid,** consisting largely of precursors to the thyroid hormones, called **thyroglobulins**, which are the precursor molecules for the formation of the **thyroid hormones** (T3 and T4, hormones released following stimulation of the thyroid gland). The spaces between the follicles contain another cell type, called **parafollicular cells.** These cells secrete the hormone **calcitonin.**

Embedded within the posterior portion of the thyroid gland are a series of small glands (usually four) called **parathyroid glands.** These glands consist of two cell types: **chief (principal) cells,** which are smaller, more abundant cells with relatively clear cytoplasm that produce **parathyroid hormone** (PTH, parathormone), and **oxyphil cells,** which are larger, less abundant cells with granular pink cytoplasm, and whose function is unknown.

1. Obtain a histology slide of the thyroid and parathyroid glands and place it on the microscope stage. Bring the tissue sample into focus on low power and then change to high power.

2. Using figure 19.6 and table 19.2 as guides, identify the following structures on the slide:

   Thyroid gland
   ☐ colloid
   ☐ follicular cells
   ☐ parafollicular cells
   ☐ thyroid follicles

   Parathyroid gland
   ☐ chief (principal) cells
   ☐ oxyphil cells

3. In the space below, sketch the thyroid and parathyroid glands as seen through the microscope, labeling all of the structures listed above.

_____ x

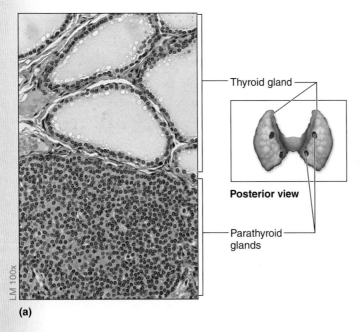

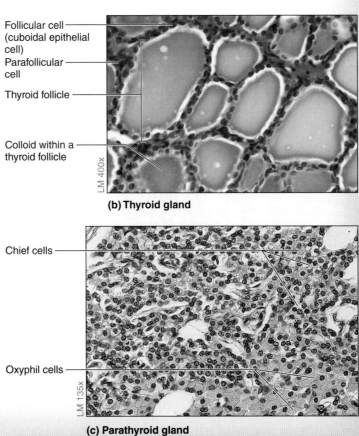

Figure 19.6 Thyroid and Parathyroid Glands. (a) The thyroid gland consists of large thyroid follicles and the parathyroid gland consists of tightly packed cells. (b) Thyroid follicles, which produce the thyroid hormones, and parafollicular cells, which produce calcitonin. (c) High-magnification view of the parathyroid gland demonstrating chief cells and oxyphil cells.

Table 19.2 Histology of the Thyroid and Parathyroid Glands

Cell Types	Description	Hormones Produced	Action of Hormone	Mechanism of Action	Word Origin
Thyroid Gland	A shield- or butterfly-shaped gland located anterior to the trachea and inferior to the thyroid cartilage. Consists of two lobes connected by a narrow isthmus anteriorly.				*thyroid*, shaped like an oblong shield
Follicular Cells	Simple cuboidal epithelial cells that form the thyroid follicles; they have very dark nuclei	Thyroid hormones (T3 and T4/Thyroxine)	Increase basal metabolic rate (BMR); important in early development of the central nervous system	Stimulates or inhibits transcription of certain genes in target cells	*folliculus*, a small sac
Parafollicular Cells	Lighter-staining cells found in the interstitial spaces between the thyroid follicles; larger than the follicular cells, with nuclei that have an appearance similar to a clock face	Calcitonin	Decrease blood calcium levels	Inhibits the action of osteoclasts, increases urinary excretion of calcium	*para*, next to, + *folliculus*, a small sac
Parathyroid Glands	4–6 small glands located on the posterior surface of the thyroid gland				*para*, next to, + *thyroid*, shaped like an oblong shield
Chief (Principal) Cells	Relatively small cells that contain a centrally located, round nucleus with one or more nucleoli	Parathyroid hormone (parathormone)	Increase blood calcium levels	Indirectly increases the action of osteoclasts, decreases urinary excretion of calcium, and stimulates synthesis of calcitriol, which increases dietary absorption of calcium	*Principal*, the predominant cell type of a gland
Oxyphil Cells	Larger than chief cells and more reddish in color	Unknown	NA	NA	*oxys*, sour acid, + *-phil*, to love

EXERCISE 19.4

THE ADRENAL GLANDS

The adrenal (suprarenal) glands are located directly superior to each kidney (*ad*, to, + *ren*, kidney). They are similar to the pituitary gland in that they are composed of two regions (**figure 19.7**), each with a separate embryological origin. The outer region, the **adrenal cortex,** is derived from mesoderm and has the appearance of typical glandular epithelium. The inner region, the **adrenal medulla,** is derived from modified postganglionic sympathetic neurons (which are derived from neural crest cells) and has the appearance of nervous tissue. The cells of the adrenal cortex synthesize steroid hormones (specifically, **corticosteroids,** *cortico,* relating to the adrenal cortex, + *steroid*, steroid hormone such as cortisol), whereas the cells of the adrenal medulla synthesize **catecholamine** hormones (that is, hormones derived from the amino acid tyrosine and that contain a catechol ring; includes epinephrine and norepinephrine). The entire gland is surrounded by a dense irregular connective tissue **capsule,** which protects the gland and helps anchor it to the superior border of the kidney.

The adrenal cortex has three recognizable zones, and each zone has cells that predominantly secrete one category of corticosteroid hormones. Characteristics of the zones, and descriptions of the hormones secreted by cells within each zone, are summarized in **table 19.3**.

1. Obtain a histology slide of the adrenal gland (figure 19.7) and place it on the microscope stage. Bring the tissue sample into focus on low power and identify the two major regions, the cortex and the medulla (figure 19.7b).

2. Move the stage so the *adrenal cortex* (figure 19.7b) is in the center of the field of view. Then change to high power.

3. Using figure 19.7 and table 19.3 as guides, identify the following zones of the adrenal cortex from outermost to innermost: zona glomerulosa, zona fasciculata, and zona reticularis.

(continued on next page)

518 Chapter Nineteen *The Endocrine System*

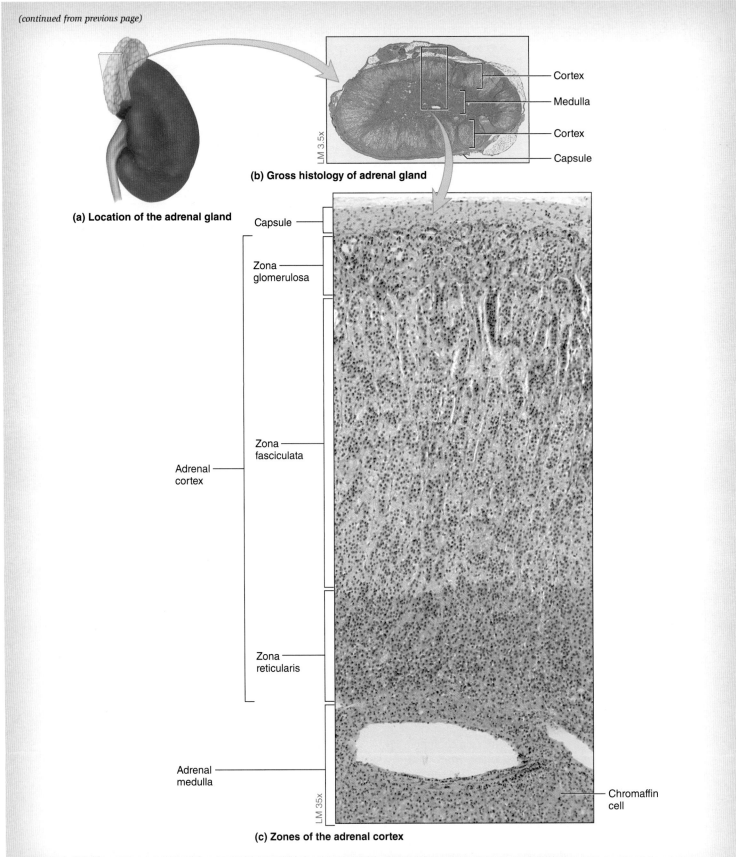

Figure 19.7 Adrenal Glands. (a) The adrenal glands are located superior to each kidney. (b) Low-magnification micrograph of the adrenal glands. (c) High-magnification micrograph demonstrating the three layers of the adrenal cortex and part of the adrenal medulla.

Table 19.3 Histology of the Adrenal Glands

Adrenal Gland Region	Zone and/or Cells	Description	Hormones Produced	Action of Hormone(s)	Word Origin
Adrenal Cortex	Zona glomerulosa	The outermost region of the adrenal cortex, containing "balls" of cells, located immediately deep to the capsule of the adrenal gland	Mineralocorticoids: aldosterone	Increases sodium and water retention by the kidneys (thus increasing blood volume and blood pressure); vasoconstriction	*zona*, zone, + *glomus*, a ball of yarn
	Zona fasciculata	The middle and largest region of the adrenal cortex, containing long cords, or "bundles," of cells	Glucocorticoids: cortisol	Increases glucose synthesis through protein breakdown and gluconeogenesis (the production of new glucose from amino acids), and lipolysis of adipose tissue	*zona*, zone, + *fasciculus*, a bundle
	Zona reticularis	The innermost region of the adrenal cortex, containing a "network" of cells, located between the zona fasciculata and the cells of the adrenal medulla	Glucocorticoids and gonadocorticoids: androgens	Androgens are similar in structure and function to the male sex steroid hormone testosterone; in females they are responsible for sex drive, in males they have little function because they are secreted in very low amounts compared to testosterone secretion by the testes	*zona*, zone, + *rete*, a net
Adrenal Medulla	Chromaffin cells	Large, spherical cells that have a yellowish-brown tint when specially stained due to their reaction with chrome salts	Catecholamines: epinephrine and norepinephrine	Epinephrine, secreted in the largest quantity (80–90% of all hormone secretion by the adrenal medulla), exhibit numerous effects including increases heart rate and contractility (force of heart muscle contraction); norepinephrine is a powerful vasoconstrictor	*chroma*, color, + *affinis*, affinity, attraction for

4. Change back to the low-power objective, move the microscope stage so the *adrenal medulla* is in the center of the field of view, and then change back to high power. The nuclei you see within the adrenal medulla are nuclei of *chromaffin cells*, which are modified postganglionic sympathetic neurons. What hormone(s) do these cells secrete? _____

5. Using figure 19.7 and table 19.3 as guides, identify the following structures on the slide of the adrenal glands:
   - [ ] adrenal cortex
   - [ ] adrenal medulla
   - [ ] capsule
   - [ ] chromaffin cells
   - [ ] zona fasciculata
   - [ ] zona glomerulosa
   - [ ] zona reticularis

6. In the space below, sketch the adrenal gland as seen through the microscope, labeling all of the structures listed in step 5.

7. *Optional Activity:* **AP|R** 8: Endocrine System—Review the histology slides of the adrenal (suprarenal) gland, as well as the pituitary gland, thyroid gland, and endocrine pancreas.

EXERCISE 19.5

THE ENDOCRINE PANCREAS—PANCREATIC ISLETS (OF LANGERHANS)

The **pancreas** is largely an exocrine gland. **Exocrine** glands produce substances that are secreted into ducts. The majority of the cells within the exocrine pancreas produce digestive enzymes and bicarbonate, which are secreted into ducts that empty into the small intestine. The clusters of exocrine cells are **pancreatic acini** (singular: acinus). Interspersed among the acini are small islands of cells that have an endocrine function. **Endocrine** glands secrete hormones into the blood. The endocrine part of the pancreas consists of the **pancreatic islets** (islets of Langerhans). These islets contain hormone-secreting cells and a rich supply of blood capillaries. Four distinct cell types compose the islets, although they cannot be distinguished in a normal histological preparation. Thus, when you observe the slide of the pancreas, your goal will be to simply identify the pancreatic islets, not the specific cell types within. Nonetheless, you should know which cell type secretes each hormone. The cell types and hormones secreted by each cell type are listed in **table 19.4**.

1. Obtain a histology slide of the pancreas, place it on the microscope stage, and bring the tissue sample into focus on low power. The majority of the cells you will see, which will look somewhat like cuboidal epithelial cells, are the exocrine cells of the pancreas (**figure 19.8**). As you scan the slide you will also see some small clusters of cells, the **pancreatic islets,** which (typically) stain lighter in color than the exocrine cells.

2. Locate a pancreatic islet and move the microscope stage so the islet is in the center of the field of view.

3. Increase the magnification so you can see the islet in greater detail. Using figure 19.8 and table 19.4 as guides, identify the following on the slide:

   ☐ pancreatic acini   ☐ secretory ducts
   ☐ pancreatic islets

Table 19.4 Histology of the Pancreatic Islets (of Langerhans)

Cell Types	Description	Hormones Produced	Action of Hormone	Mechanism of Action	Word Origin
Alpha Cells	Compose about 30% of islet cells; located on the periphery of the islet	Glucagon	Increases blood glucose levels	Stimulates glycogenolysis (breakdown of glycogen) and gluconeogenesis (formation of new glucose from amino acids in the liver, and lipolysis in adipose connective tissue)	*alpha,* the first letter of the Greek alphabet, + *glucose,* sugar, + *ago,* to lead
Beta Cells	Compose about 65% of islet cells; located in the center of the islet	Insulin	Decreases blood glucose levels	Stimulates glucose, amino acid, and potassium uptake by cells, stimulates glycogenesis (formation of glycogen from glucose) in the liver, and increases lipolysis in adipose connective tissue	*beta,* the second letter of the Greek alphabet, + *insula,* an island
Delta Cells	Compose about 4% of islet cells; located on the periphery of the islet	Somatostatin	Inhibits the release of glucagon and insulin by alpha and beta cells	Inhibits the release of glucagon and insulin when nutrient levels in the bloodstream are high	*delta,* the 4th letter of the Greek alphabet, + *soma,* body, + *stasis,* standing still
F Cells	All other (rare) cell types in the pancreatic islets (~1%) are grouped together and given this name	Pancreatic polypeptide	Inhibits somatostatin release from delta cells	Regulates secretion of somatostatin from delta cells	NA

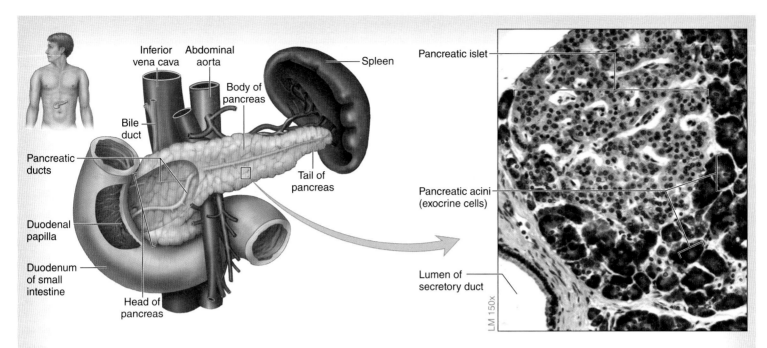

Figure 19.8 Endocrine Portion of Pancreas. The endocrine part of the pancreas consists of the pancreatic islets, which are the lighter-colored regions in this micrograph. The darker cells surrounding the islet are acinar cells, which compose the exocrine part of the pancreas.

4. In the space below, sketch a pancreatic islet and the exocrine cells that surround it as seen through the microscope. Be sure to label the structures listed in step 3.

Gross Anatomy

Some endocrine glands are difficult to identify on the cadaver. However, most classical glands and associated structures can be identified on the cadaver or on classroom models, so these will be the focus of this section.

Endocrine Organs

EXERCISE 19.6

GROSS ANATOMY OF ENDOCRINE ORGANS

1. Observe a human cadaver or classroom models of the brain, thorax, abdomen, and skull.

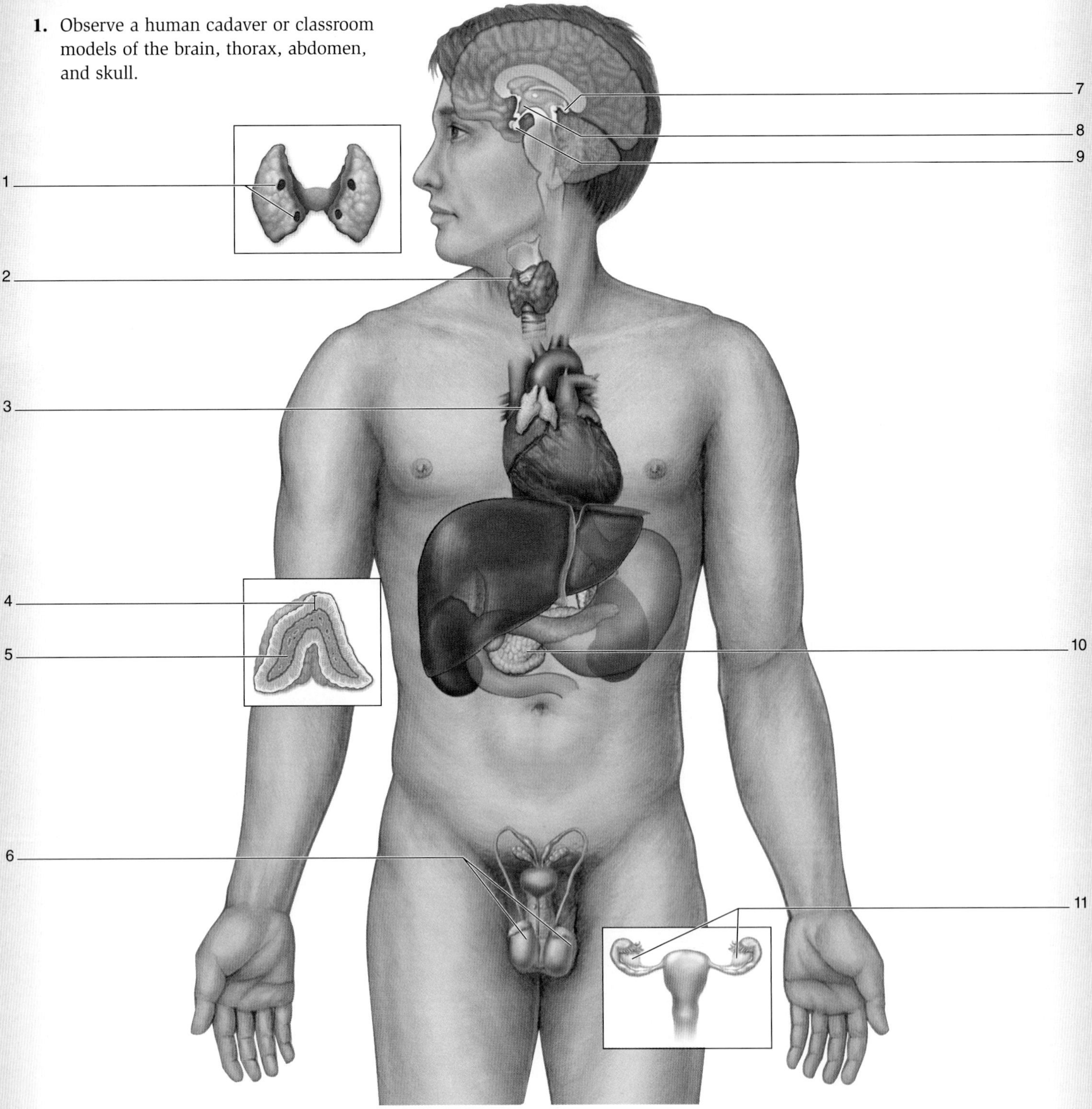

(a) Major endocrine glands

Figure 19.9 Labeling Major Endocrine Glands of the Body. (a) Major endocrine glands of the body. (b) Line art of sagittal section of the brain.

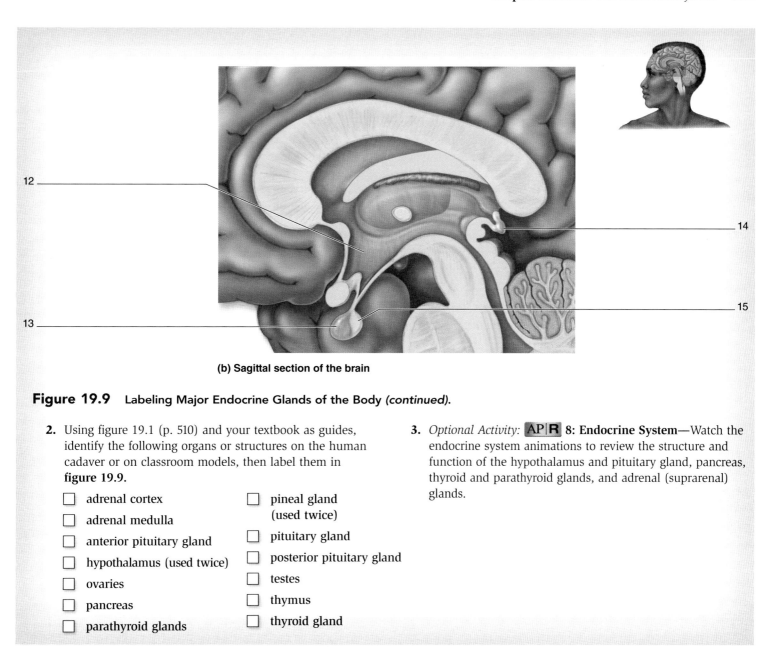

(b) Sagittal section of the brain

Figure 19.9 Labeling Major Endocrine Glands of the Body *(continued).*

2. Using figure 19.1 (p. 510) and your textbook as guides, identify the following organs or structures on the human cadaver or on classroom models, then label them in **figure 19.9**.

   ☐ adrenal cortex
   ☐ adrenal medulla
   ☐ anterior pituitary gland
   ☐ hypothalamus (used twice)
   ☐ ovaries
   ☐ pancreas
   ☐ parathyroid glands
   ☐ pineal gland (used twice)
   ☐ pituitary gland
   ☐ posterior pituitary gland
   ☐ testes
   ☐ thymus
   ☐ thyroid gland

3. *Optional Activity:* **AP|R** 8: Endocrine System—Watch the endocrine system animations to review the structure and function of the hypothalamus and pituitary gland, pancreas, thyroid and parathyroid glands, and adrenal (suprarenal) glands.

INTEGRATE

CLINICAL VIEW
Corticosteroids

Excessive levels of cortisol may suppress the immune system by inhibiting inflammation and white blood cell activation and proliferation, and they impair repair of connective tissues. *Corticosteroids* are a class of drugs that are often prescribed for individuals who are suffering from widespread inflammation. For example, patients diagnosed with the autoimmune disease rheumatoid arthritis (RA) experience chronic inflammation in their synovial joints (see exercise 10.3). Since individuals suffering from this disease have abnormal activation and proliferation of immune cells, corticosteroids may be beneficial in limiting the progression of the disease. However, administration of the drug has some serious side effects, such as causing an increase in blood glucose levels and promoting the breakdown of tissue proteins over time. For these reasons, corticosteroids must be utilized with caution, particularly in elderly patients.

INTEGRATE

CONCEPT CONNECTION

Among the various proteins present in the plasma membranes of all cells are transporters that are responsible for movement of substances into and out of the cells. These transporters exhibit *specificity*, which means that they will assist only a particular substance across the plasma membrane. The movement of substances from an area of high concentration to an area of low concentration with the assistance of such transporters is called **facilitated diffusion** (see chapter 4). As cells utilize glucose as a substrate for cellular respiration, these transporters continue to assist the movement of glucose from the extracellular environment to the intracellular environment.

Physiology

Metabolism

Metabolism is regulated by **thyroid hormone (TH)**. The release of TH is controlled indirectly through the hypothalamus. The hypothalamus detects a decrease in blood levels of TH (and other stimuli), which triggers the release of **thyrotropin releasing-hormone (TRH)** into the hypothalamo-hypophyseal portal vein. TRH triggers cells in the anterior pituitary to release **thyroid-stimulating hormone (TSH).** TSH is transported in the general circulation to target cells within the **thyroid gland.** TSH stimulates the **follicular cells** of the thyroid gland to release thyroid hormone in its two forms: **triiodothyronine (T_3)** and **tetraiodothyronine (T_4, thyroxine).** T_3 is the more active hormone. Most T_4 is converted to T_3 when it reaches its target cells. For the purposes of this exercise, T_3 and T_4 will be collectively referred to as thyroid hormone (TH).

Thyroid hormone exerts its effects by binding to intracellular receptors and stimulating the transcription of genes that code for certain proteins. For example, TH stimulates the transcription of genes that code for the Na^+/K^+ pump in excitable cells, which causes these cells to increase production of these pumps. To meet the demands of increased numbers of these pumps, the cell increases its rate of cellular respiration to synthesize the additional ATP that is required for this active transport process. TH acting elsewhere in the body stimulates the increase of heart and respiratory rates. This increases the transport of oxygenated blood to active cells throughout the body. The overall effect of TH is to **increase metabolic rate, increase oxygen consumption,** and **increase body temperature.** TH also decreases the level of circulating amino acids in the blood by promoting their uptake into cells for protein synthesis. In addition, thyroid hormone stimulates **lipolysis** and **glycogenolysis,** thereby increasing both fatty acid and blood glucose levels in the blood.

EXERCISE 19.7 Ph.I.L.S.

Ph.I.L.S. LESSON 17: THYROID GLAND AND METABOLIC RATE

The purpose of this laboratory exercise is to observe the effects of TH on body temperature and metabolic rate. Metabolic rates of two female white mice will be measured based on values of oxygen consumption rates. This simulation will calculate the least square linear regression, which determines the *dependent* (y) and *independent* (x) variables, plots the data points, and determines the line of best fit for the data points. We can use the equation that generates the line ($y = mx + b$) to predict the value of the dependent variable if we know the value of independent variable.

The two mice in the simulation have eaten either (1) normal mouse "chow," or (2) mouse "chow" containing 0.15% propylthiouracil (PTU). The production of TH is significantly reduced by the compound PTU. Finally, the rate at which each mouse consumes oxygen will be measured at different temperatures in order to determine the effect of cooling on metabolic rate.

Before you begin, familiarize yourself with the following concepts (use your textbook as a reference):

- negative feedback regulation
- the hypothalamus and its anatomical and physiological relationship with the anterior pituitary gland
- follicular tissue of the thyroid gland
- the cellular effects of thyroid hormone (TH)
- the relationship between oxygen consumption and metabolic rate

Before you begin, state your hypothesis regarding the influence of intake of mouse chow with TPU on the levels of TH, oxygen consumption, and metabolic rate in the mouse.

_____

_____

_____

1. Open the Ph.I.L.S Lesson 17: Endocrine Function: Thyroid Gland and Metabolic Rate.

2. Read the objectives and introduction and take the pre-laboratory quiz.

3. After completing the pre-laboratory quiz, read through the wet laboratory. Be sure to click on the hot-linked words to open and view the videos that are indicated in red.

4. The lab exercise will open when you click "Continue" after completing the wet laboratory (**figure 19.10**).

5. Click the power switch to turn on the scale and then click the power switch to turn on the thermostat.

6. Click "Tare" to set the scale to zero.

7. To weigh a mouse, click on one of the mice and drag it to the scale.

8. After weighing the mouse, place it in the chamber by clicking and dragging it to the chamber.

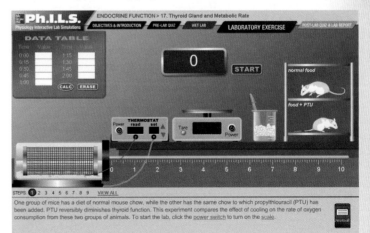

Figure 19.10 Opening Screen for the Laboratory Exercise on Endocrine Function: Thyroid Gland and Metabolic Rate.

9. Remove the pipette from the beaker, which contains bubbles, and place the tip of the pipette to the calibration tube (at the 10).

10. Measure the initial position of the soap bubbles (if at the end of the tube it will be 10 mm) and record the position at 0:00 time in the table in the program.

11. After clicking the "Start" button, measure the position of the soap bubble within the calibration tube at 15-second intervals (you will hear a beep) and record it in the data table. (The timer will begin when you hit "Start.") You may find it more manageable to click the "Pause" button after each 15-second interval, measure, record, and then click "Start" to resume the experiment.

12. When finished, click "Pause."

13. Click "Calc" to see the graph (linear regression) in the program. The "Journal" will open and show you a graph of the data that you have collected up to this point in the laboratory experiment. The journal also includes a table. Notice that the table (with temperatures ranging from 8°C to 24°C) will have the first data point entered for the average oxygen consumed per minute (total amount of oxygen divided by 2 minutes). Enter this data in **table 19.5** on this page. No points will appear on the graph until all data points are entered.

14. To complete the process again with the same animal, repeat steps 9–14 but decrease the temperature 2 degrees (by clicking on the down arrow on the thermostat) prior to beginning the steps until data is recorded for all temperatures. Be sure to record the data in table 19.5 (as indicated in step 13) every time you finish with one trial.

15. After returning the first animal to the cage, repeat steps 5–14 for the second animal.

Table 19.5 Data for Ph.I.L.S. Lesson 17: Thyroid Gland and Metabolic Rate

Temp (°C)	Normal (24.2 g)	+PTU (g)
24	1.44	
22		
20		
18		
16		
14		
12		
10		
8		

16. Plot your data on the grid below.

17. Complete the post-laboratory quiz by answering the ten questions on the computer screen.

18. Make note of any pertinent observations here:
_____
_____
_____

Linear Regression of O$_2$ Consumption

(Graph: O$_2$/m vs Temperature °C)

Chapter 19: The Endocrine System

POST-LABORATORY WORKSHEET

The ❶ corresponds to the Learning Objective(s) listed in the chapter opener outline.

Do You Know the Basics?

Exercise 19.1: The Hypothalamus and Pituitary Gland

1. Discuss the defining histological features that distinguish the anterior from the posterior lobe of the pituitary gland. ❶ _____

2. List the hormones produced by acidophils and basophils of the anterior pituitary. ❷

3. What are *chromophobes*? ❷

4. What structures are located within the paraventricular and supraoptic nuclei of the hypothalamus? ❸

5. List the two hormones released in the posterior pituitary. ❸

Exercise 19.2: The Pineal Gland

6. Describe features that allow for identification of the pineal gland on a histological slide. ❹

7. The primary cells of the pineal gland, _____, are responsible for secreting the hormone _____. ❺

8. Describe why pineal sand is important to radiologists. ❻ _____

Exercise 19.3: The Thyroid and Parathyroid Glands

9. Describe the structure and function of the thyroid gland. ❽ _____

10. List the two hormones produced by the thyroid gland and identify the cell types that secrete each hormone. ❽ _____

11. Identify the major hormone produced by the parathyroid gland, identify the cell type responsible for its secretion, and describe the hormone's action. ❾

Exercise 19.4: The Adrenal Glands

12. Describe how the zones of the adrenal cortex can be distinguished on a histological slide. ❿

13. List the hormones secreted by each region of the adrenal gland. ⓫

    zona glomerulosa _____

    zona fasciculata _____

    zona reticularis _____

    medulla _____

526   Chapter Nineteen   *The Endocrine System*

14. For each hormone listed in question 13, describe its physiological function. 🔵 _____
_____
_____

Exercise 19.5: The Endocrine Pancreas—Pancreatic Islets (of Langerhans)

15. Describe defining histological features of the pancreatic islets. 🔵
_____
_____

16. List the hormones secreted by each cell type of the pancreatic islets (of Langerhans). 🔵

    alpha cells _____ delta cells _____

    beta cells _____ F cells _____

17. For each hormone listed in question 16, describe its physiological function. 🔵
_____
_____

Exercise 19.6: Gross Anatomy of Endocrine Organs

18. List the endocrine glands that can clearly be identified on human cadavers or classroom models of the brain, thorax, abdomen, and skull. 🔵
_____
_____

Exercise 19.7: Ph.I.L.S. Lesson 17: Thyroid Gland and Metabolic Rate

19. Describe how linear regression can be used to calculate the rate of oxygen consumption in the mouse. 🔵
_____
_____

20. Identify the region of the brain that serves as the body's thermostat. 🔵 _____
_____

21. What hormones are released when body temperature drops below a physiological set point? How will these hormones influence metabolic rate? 🔵
_____
_____

22. Discuss the relationship between oxygen consumption and metabolic rate. 🔵 _____
_____
_____

Can You Apply What You've Learned?

23. Identify each of the following glands by writing in the name of the gland under the photo.

(a) _____ (b) _____ (c) _____ (d) _____

24. Based on your understanding of the location of the pituitary gland with respect to other brain structures, explain why individuals with pituitary gland tumors often also experience visual disorders.

25. Predict the oxygen consumption in an individual with hypothyroidism. Then compare the rate of oxygen consumption in an individual with hypothyroidism with that of an individual with hyperthyroidism.

26. Body builders have been known to inject insulin to increase muscle mass. Explain this reasoning and discuss any risks that may be posed by this practice.

27. Given the physiological effects of cortisol release, identify some health risks that may come from long-term use of corticosteroids.

Can You Synthesize What You've Learned?

28. Patients suffering from hyperparathyroidism are also at a greater risk for bone fracture. What do you predict blood calcium levels in a patient suffering from this condition would look like? Explain why these patients often experience a decrease in bone density.

29. Athletes may be interested in calculating their basal metabolic rate (BMR) to determine the number of calories their body utilizes at rest. To do this, use a calorimeter, which measures oxygen consumption to measure BMR. The calorimeter calculates metabolic rate based on the rate of oxygen consumption. Describe how this machine likely accomplishes this task.

PART V | MAINTENANCE AND REGULATION

CHAPTER 20

The Cardiovascular System: Blood

OUTLINE AND LEARNING OBJECTIVES

Histology 533

EXERCISE 20.1: IDENTIFICATION OF FORMED ELEMENTS ON A PREPARED BLOOD SMEAR 533

1. Identify the following formed elements on a human blood smear slide: erythrocytes, platelets, eosinophils, basophils, neutrophils, monocytes, and lymphocytes
2. Describe the functions of the formed elements of the blood

EXERCISE 20.2: IDENTIFICATION OF MEGAKARYOCYTES ON A BONE MARROW SLIDE 536

3. Identify megakaryocytes on a slide of red bone marrow
4. Explain the process of platelet formation

Gross Anatomy 537

EXERCISE 20.3: IDENTIFICATION OF FORMED ELEMENTS OF THE BLOOD ON CLASSROOM MODELS OR CHARTS 537

5. Identify the following formed elements on classroom models or charts of blood cells: erythrocytes, platelets, eosinophils, basophils, neutrophils, monocytes, and lymphocytes

Physiology 538

Blood Diagnostic Tests 538

EXERCISE 20.4: DETERMINATION OF WHITE BLOOD CELL COUNTS 539

6. Count the number of white and red blood cells on a human blood smear slide
7. Compare your blood cell counts to the expected numbers if the sample is "normal"

EXERCISE 20.5: DETERMINATION OF HEMATOCRIT 540

8. Measure the packed cell volume in a centrifuged sample of blood
9. Explain the normal range for hematocrit and the significance of hematocrits that fall out of the normal range

EXERCISE 20.6: DETERMINATION OF HEMOGLOBIN CONTENT 541

10. Measure the concentration of hemoglobin in the blood using a hemoglobinometer

EXERCISE 20.7: DETERMINATION OF COAGULATION TIME 542

11. Measure the time required for blood to coagulate

EXERCISE 20.8: DETERMINATION OF BLOOD TYPE 544

12. Apply serum containing anti-A and anti-B antibodies to blood samples to observe agglutination, and determine blood type (A, B, AB, or O)
13. Explain the mechanism behind using this method to "type" blood

EXERCISE 20.9: DETERMINATION OF BLOOD CHOLESTEROL 545

14. Measure the concentration of cholesterol in blood plasma

MODULE 9: CARDIOVASCULAR SYSTEM

INTRODUCTION

Blood is a bodily fluid that has fascinated humans for centuries. Its rich red color adds to both its beauty and its mystery. For many of us, the sight of blood is quite disturbing because it usually indicates that something is very wrong! It's no wonder that for centuries humans have been fascinated by this most colorful bodily fluid. Early Greek physicians believed that blood was one of the four bodily "humors" (*humor*, a bodily fluid). The other bodily fluids were phlegm, black bile, and yellow bile (table 20.1). Each of these humors was associated with an element—air, water, earth, or fire—and had particular characteristics: hot, cold, wet, or dry. Before you put any confidence in this idea, know that modern science recognizes over ninety-two naturally occurring elements, and earth, air, fire, and water are not among them! Greek physicians believed that a balance of the four humors was required for optimal health of both the body and the mind, and that disease was the result of an imbalance between the bodily fluids. It was this concept of disease that made bloodletting a popular treatment for disease. By draining some blood from a patient, the physician supposedly was putting the patient's humors back into balance. In those days, studying anatomy or medicine wasn't nearly as complicated as it is today. On the other hand, we can all be thankful that we didn't have to be patients back then. Ironically, to this day we continue to refer to this early concept of medicine, even though most of us are largely unaware we are doing so. We refer to an infant who cries all the time as being "colicky." We refer to someone who is sad as feeling "melancholy." The next time you are feeling sad, consider that if you had lived in the year 400 B.C., your doctor would have diagnosed you as having an excess of "black bile." Then have a good laugh and be thankful that you are living in the twenty-first century. If that isn't enough to give you a good laugh, consider what early Greeks thought to be the source of phlegm: the brain. That thought may be enough to transform your mood from melancholy to "sanguine," or lively and optimistic. Are you feeling lively and optimistic yet? Then you are ready to tackle the current "sanguine" topic: blood!

In this chapter we will explore the normal constituents of human blood. Knowledge of the normal appearance and abundance of the various types of blood cells provides a basis for recognizing when cells or amounts have become abnormal, and what that means. **Tables 20.2** and **20.3** summarize the characteristics of the formed elements of blood, which are the structures that are the focus of this laboratory exercise. You will identify the formed elements of blood and also explore the functional relevance of each of these components, learning how these cell types play a critical role in both the cardiovascular and immune systems. You will also perform some common blood tests such as determination of hemoglobin content of blood, determination of blood type, and of blood cholesterol levels. Considerably more detail on the structure and function of blood is covered in your main textbook for this course. Thus, it is important to make sure that you have read and understood this information prior to performing the laboratory exercises.

Table 20.1 The Four Greek Humors

Bodily Fluid (Humor)	Element	Characteristics	Source	Mood	Mood Characteristics	Word Origin
Black Bile	Earth	Dry and cold	Spleen	Melancholy	Sad, depressed	*melas*, black, + *chole*, bile
Blood	Air	Hot and wet	Heart	Sanguine	Lively and optimistic	*sanguis*, blood
Phlegm	Water	Cold and wet	Brain	Phlegmatic	Calm and unexcitable	*phlegma*, inflammation
Yellow Bile	Fire	Hot and dry	Liver	Choleric	Irritable	*chole*, bile

Table 20.2 Characteristics of the Formed Elements of Blood

Formed Element	Description	Function	Size (diameter)	Percentage of Formed Elements	Word Origin
Erythrocytes (Red Blood Cells, RBCs)	Very uniform in size and shape. Shaped like biconcave discs, lack a nucleus, and are orange to red in color due to their attraction for eosinophilic stains.	Transport oxygen and carbon dioxide, and participate in regulation of the pH of blood	~7.5 μm	99%	*erythros*, red, + *kytos*, a hollow (cell)
Leukocytes (White Blood Cells, WBCs)	Generally blue to purple in color due to the basophilic staining properties of their nuclei. A summary of the characteristics of the various leukocytes is provided in table 20.3.	Protect the body from pathogens, fight infections, and remove dead or damaged cells from the body	7–21 μm	<1%	*leukos*, white, + *kytos*, a hollow (cell)
Platelets	Small, purple-colored structures often mistaken for "junk" on the slide. Platelets are cell fragments of megakaryocytes and are tiny compared to erythrocytes and leukocytes.	Play a role in hemostasis (blood clotting). When activated, the surfaces become spiny instead of smooth and they become very "sticky," which helps to form the platelet plug that helps stop bleeding from the blood vessel wall.	~2 μm	1%	*platys*, flat

Table 20.3 Leukocyte Characteristics

Leukocyte	Description	Function	Percentage of Leukocytes	Diameter Relative to That of an RBC	Nuclear Shape	Color of Cytoplasmic Granules	Common Conditions Causing an Increase in Abundance	Word Origin
Neutrophils	Have very light, lavender-colored cytoplasmic granules; the multi-lobed nucleus can be seen easily.	Important in fighting bacterial infections. They migrate out of the blood toward the site of infection, where they phagocytize bacteria and damaged tissues.	50–70%	~1.3x	Multi-lobed (~5 lobes)	Lavender (pale blue/purple)	Bacterial infection	*neutro-*, neutral, + *philos*, to love
Eosinophils	Have orange to reddish cytoplasmic granules that are fairly light in color; the bluish-colored, bi-lobed nucleus can be seen easily	Fight parasitic infections and mediate (neutralize) the effects of histamines. Phagocytize antigen–antibody complexes and allergens.	1–4%	~1.3x	Bi-lobed	Orange or red (eosinophilic)	Parasitic infections and allergic reactions	*acidus*, sour, referring to acidic dyes such as eosin, + *philos*, to love
Basophils	Very rare and therefore difficult to locate on a blood smear. Cytoplasmic granules stain very dark blue, so the nucleus is generally not visible, though it is usually bi-lobed.	Release histamine and heparin, and are involved in the inflammatory response	0.5–1%	~1.3x	Bi-lobed	Blue or purple (basophilic)	Tissue injury and allergic reactions	*baso-*, referring to basic dyes, + *philos*, to love
Lymphocytes	Nearly the same size as an erythrocyte. Nuclei are very large and blue/purple in color, and are surrounded by a halo of pale blue cytoplasm.	Responsible for the specific immune response to infection. Each type of lymphocyte has a specific function in fighting pathogens.	20–40%	~1–2x	Large and round, nearly fills the entire cell	NA	Viral infections, autoimmune diseases	*lympho-*, referring to the lymphatic system, + *kytos*, a hollow (cell)
Monocytes	Recognized by their very large size, at least twice the size of an erythrocyte. Have a large, blue/purple nucleus that often has a small indentation in it, making it "horseshoe-shaped," which helps distinguish it from a lymphocyte.	Monocytes are circulating cells. When they migrate out of the bloodstream, they become large, phagocytic cells called macrophages. They have little to no function in circulating blood.	2–8%	~2–3x	Large and horseshoe-shaped	NA	Bacterial infection	*monos*, single, + *kytos*, a hollow (cell)

Chapter 20: The Cardiovascular System: Blood

Name: _____

Date: _____ **Section:** _____

PRE-LABORATORY WORKSHEET

Also available at www.connect.mcgraw-hill.com

1. List the three components of blood that separate from each other when a blood sample is centrifuged (see figure 20.5 for reference).

   a. _____

   b. _____

   c. _____

2. Explain the basis for the categorization and naming of the two major categories of white blood cells (granulocytes and agranulocytes).

   _____
   _____
   _____
   _____

3. Explain the basis for the categorization and naming of the three types of granulocytes (eosinophils, basophils, and neutrophils).

   _____
   _____
   _____
   _____

4. Define *hemopoiesis*:

   _____
   _____
   _____

5. Megakaryocytes give rise to formed elements called _____.

6. A measurement of the percentage of blood volume composed of red blood cells (erythrocytes) is the _____.

7. The red-pigmented protein found in red blood cells that binds oxygen and carbon dioxide for transport in the blood is: _____.

8. The formed element of the blood that is responsible for initiating coagulation is: _____.

Histology

CAUTION:
The exercises in this laboratory may involve the use of human, animal, or artificial blood. Blood samples, whether human or animal, are considered biohazardous wastes and should be handled with care. Be sure to take precautionary measures while handling blood samples by wearing nitrile gloves and safety goggles. Handle sharp objects with caution. Biohazardous wastes and sharps must be disposed of in the proper containers. Please review procedures for handling sharps and biohazardous wastes in chapter 1 of this manual and consult with your laboratory instructor for further instructions on laboratory policies and procedures for handling and disposing of these substances.

EXERCISE 20.1

IDENTIFICATION OF FORMED ELEMENTS ON A PREPARED BLOOD SMEAR

EXERCISE 20.1A Making a Human Blood Smear

In exercise 20.1A either you will use a prepared slide of human blood, or you will be asked to make your own slide using your own blood. To make a blood smear using your own blood, use the following procedure.

Obtain the Following:
- four glass slides
- lancet
- cotton balls
- alcohol prep pads
- vial of Wright's stain

1. Clean the slides with soap and water and let them dry. It takes two slides to make a blood smear, and you may have to try several times before you get a good preparation. This is why you will most likely need to clean at least four slides.

2. Using the alcohol prep pad, clean the tip of the middle or ring finger on the hand you use the least (e.g., if you are right-handed, use the tip of the finger on your left hand).

3. Place the tip of the lancet on the clean finger, and pull the trigger so it pierces the skin. If you are using a lancet without a trigger **(figure 20.1)**, poke the skin with a quick, strong jab.

4. Squeeze the finger until a small drop of blood appears on the surface. Blot away the first drop of blood using the cotton ball and then squeeze the finger again until another drop appears. Gently touch the drop of blood to the top surface of the microscope slide about 1.5–2 cm from the edge of the slide **(figure 20.2, step 1)**. You do not need a lot of blood to make the smear, so try not to squeeze the finger too hard.

5. Orient the slide so the end with the drop of blood on it is closest to you. Then place the edge of a second microscope slide on the drop of blood and hold it at a 45-degree angle to the slide with the drop of blood on it (figure 20.2, step 2).

6. Quickly push the second slide over the first slide (push away from you) to smear the blood (figure 20.2, step 3). You are aiming for a very thin, transparent layer of blood on the slide. If the layer is too thick, you will have difficulty seeing the cells because they will be too closely packed together. If you don't end up with an even smear, or a good spread, start over with two new slides. It might take a couple of tries before you are able to get a good, even smear. Discard any slides that made contact with blood in a bowl containing a bleach solution or sharps container. Allow the slide with the blood smear on it to air dry.

7. Place a couple of drops of Wright's stain on the blood smear slide so you end up with enough stain on the slide to cover the smear, but not so much that it overflows the slide. Keep track of the number of drops you use (number of drops = _____).

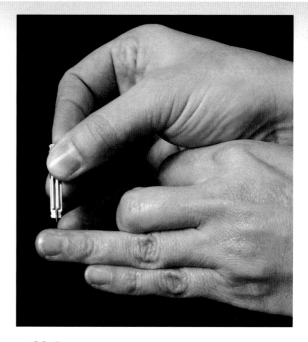

Figure 20.1 Lancing Finger for Blood Sample.

① Place a small drop of blood approximately 2 cm from the edge of the slide.

② Place the edge of a second slide on the drop of blood and hold it at a 45° angle to the first slide.

③ Push the second slide away from you to smear the blood on the first slide.

Figure 20.2 Preparation of a Blood Smear.

(continued on next page)

(continued from previous page)

8. Let the slide stand for 3–4 minutes, until the smear begins to take on a blue/green color.

9. Add to the slide a volume of distilled water equal to the volume of Wright's stain you used. Generally this means you will use the same number of drops of distilled water as you used of Wright's stain. Move the slide around and blow on it a little bit to mix the distilled water and stain on the slide.

10. Let the slide stand for 5–10 minutes. During this time, blow on the slide occasionally to keep the distilled water and stain mixed.

11. Rinse the slide with distilled water for 2–3 minutes, and then tilt the slide to allow the excess water to drip off of the slide. Allow the slide to air dry. Once the slide is dry, it will be ready to examine.

EXERCISE 20.1B Viewing Formed Elements on a Prepared Blood Smear

1. Obtain a prepared slide of human blood or use the slide you prepared (see exercise 20.1A). Place the slide on the microscope stage and observe at low power. You will not be able to see much at this magnification, but, as always, it is important to progress from low to high power so you will continue to keep the sample in focus before progressing to the oil immersion objective.

2. Change from low to medium, then to high power, making sure to bring the slide into focus at each power. After you have the slide in focus on high power, obtain a vial of immersion oil. Rotate the nosepiece (the part of the microscope that holds the objective lenses) so the high-power and oil immersion objectives lie on either side of the slide (**figure 20.3**). Place a drop of immersion oil over the center of the slide (where you see the light coming through the slide), and then carefully rotate the nosepiece to bring the immersion objective into place over the slide. Bring the slide into focus.

3. Using tables 20.2 and 20.3 and **figure 20.4** as guides, identify the following formed elements on the prepared blood slide. Note that some elements will be much easier to locate than others, due to their relative abundance in whole blood.

   ☐ basophils   ☐ monocytes
   ☐ eosinophils   ☐ neutrophils
   ☐ erythrocytes   ☐ platelets
   ☐ lymphocytes

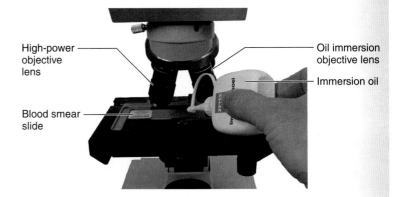

Figure 20.3 Orienting the Objective Lenses for Placement of a Drop of Immersion Oil on the Prepared Blood Slide.

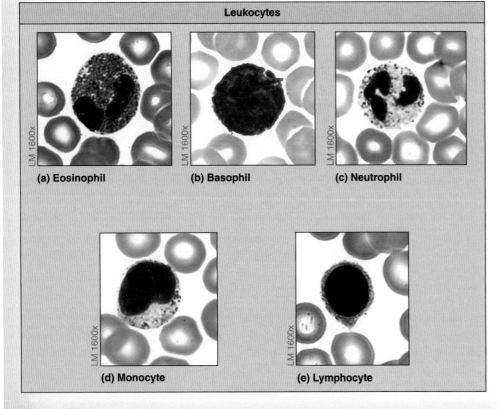

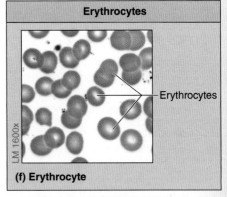

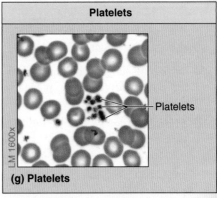

Figure 20.4 Formed Elements in the Blood.

4. After you locate each formed element, make a drawing of it in the appropriate circle below. Use colored pencils to shade each structure the way it appears under the microscope. In addition, make note of the approximate diameter of each cell, using the diameter of an erythrocyte (7.5 μm) as a reference.

Erythrocyte

_____ ×

Diameter: _____ μm

Platelet(s)

_____ ×

Diameter: _____ μm

Eosinophil

_____ ×

Diameter: _____ μm

Basophil

_____ ×

Diameter: _____ μm

Neutrophil

_____ ×

Diameter: _____ μm

INTEGRATE

LEARNING STRATEGY

Although lymphocytes usually have circular nuclei and monocytes usually have indented, or "horseshoe-shaped," nuclei, sometimes monocytes can also have what look like circular nuclei. In these cases, monocytes can appear very similar to lymphocytes. To prevent yourself from misidentifying a monocyte as a lymphocyte, always consider the size of the cell in addition to the shape of the nucleus. A monocyte is going to be 2–3 times larger than a red blood cell, whereas a lymphocyte will be much closer to the same size as a red blood cell.

(continued on next page)

(continued from previous page)

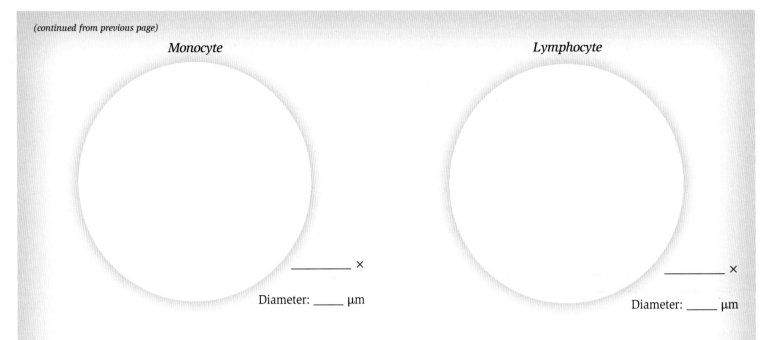

Monocyte

Diameter: _____ μm

Lymphocyte

Diameter: _____ μm

5. When you have finished viewing the slide, clean the microscope and slide thoroughly to remove all traces of oil from the instruments. To do this, rotate the nosepiece so the high-power and oil immersion lenses are on either side of the slide once again (figure 20.3). Use *lens paper* (do not use anything else or it may damage the lenses!) and lens cleaning solution to carefully and thoroughly clean the oil from both the oil immersion objective and the blood smear slide. Blood smear slides prepared using your own blood should be disposed of by placing them in a bleach solution for disinfection.

EXERCISE 20.2

IDENTIFICATION OF MEGAKARYOCYTES ON A BONE MARROW SLIDE

All the formed elements of the blood are produced in the red bone marrow through the process of **hemopoiesis** (*hemo-*, blood, + *poiesis*, a making). Identifying the precursor cells of circulating blood cells is a complicated endeavor, which is generally undertaken only in upper-level histology courses. In this laboratory exercise you will not seek to identify all of the precursor cells for each lineage of blood cells. Instead, you will only locate the precursor cells of platelets (*megakaryocytes*) and investigate their structure and function. However, as you engage in the process of identifying megakaryocytes, try to also obtain an appreciation for the general appearance of the precursor cells of erythrocytes and leukocytes that you see on the slide.

1. Obtain a prepared slide of red bone marrow and place it on the microscope stage. Bring the slide into focus on high power. Using **figure 20.5** as a guide, identify the following on the slide:

    ☐ bone tissue   ☐ platelets
    ☐ megakaryocytes   ☐ red bone marrow

2. **Megakaryocytes** (*megas*, big, + *karyo*, kernel [referring to the nucleus] + *kytos*, a hollow, cell) are extremely large cells with enormous nuclei. They are easily identifiable on prepared slides of bone marrow. Megakaryocytes remain within the bone marrow. However, their products, **platelets,** are continuously delivered to the bloodstream.

Megakaryocytes produce long extensions called proplatelets. While still attached to the megakaryocyte, these proplatelets extend through the blood vessel wall (between the endothelial cells) in the bone marrow. The force from the blood flow "slices" these proplatelets into the fragments we know as platelets.

3. In the space below, sketch a megakaryocyte as seen through the microscope. Be sure to indicate the relative size of the megakaryocyte as compared to the size of the developing blood cells that you observed on the bone marrow slide.

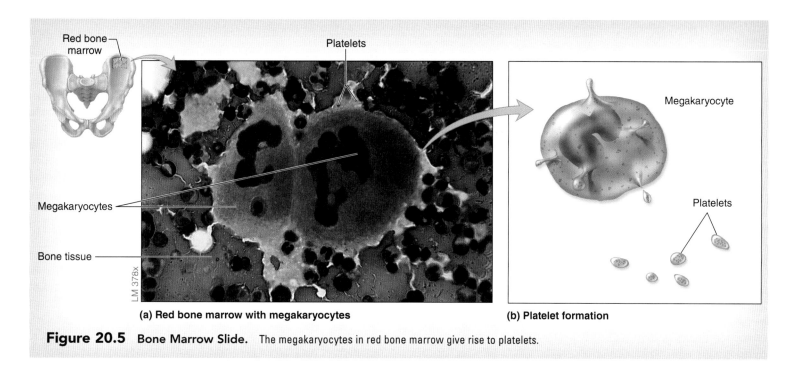

Figure 20.5 **Bone Marrow Slide.** The megakaryocytes in red bone marrow give rise to platelets.

INTEGRATE

CLINICAL VIEW
Bone Marrow Transplants

Leukemia is a disease characterized by excessive proliferation of abnormal white blood cells (WBCs). Patients who have experienced a progression of the disease may opt to undergo a bone marrow transplant. White blood cells are produced in a process called leukopoiesis, which occurs in the bone marrow. A bone marrow transplant procedure involves the destruction of the leukemia patient's existing bone marrow, typically through radiation or chemotherapy. New hematopoeitic stem cells are harvested from a donor and injected into the patient's blood. Recall that adult bone marrow that is still undergoing hemopoeisis is located in the proximal epiphysis of the humerus and femur, sternum, and iliac crest. Typically, bone marrow is harvested from the donor's iliac crest because it is relatively easy to access and causes the fewest complications for the donor.

After the donor's cells are transplanted into the recipient, they will normally migrate to the recipient's spleen and the process of leukopoiesis will begin anew. Care must be taken to ensure that the donor's bone marrow is a good match and will not elicit an immune response in the patient. As patients undergo this procedure, their white blood cells become depleted, making them very susceptible to infection. For this reason, patients must limit their exposure to pathogens until their white blood cell count has risen to safe levels once again.

Gross Anatomy

EXERCISE 20.3

IDENTIFICATION OF FORMED ELEMENTS OF THE BLOOD ON CLASSROOM MODELS OR CHARTS

1. Observe classroom models or charts demonstrating blood cells, and identify each of the following blood cell types on the models or charts.

   ☐ basophils ☐ monocytes
   ☐ eosinophils ☐ neutrophils
   ☐ erythrocytes ☐ platelets
   ☐ lymphocytes

2. *Optional Activity:* **AP|R** **9: Cardiovascular System—** Watch the "Hemopoiesis" animation to review the formation and characteristics of each of the formed elements.

Physiology

Blood Diagnostic Tests

Blood is an amazing fluid with a beautiful red color due to the presence of hemoglobin within red blood cells (RBCs). **Hemoglobin** is a protein in RBCs that is essential for transporting oxygen throughout the body. However, blood transports much more than just oxygen. Blood also transports nutrients, waste products, carbon dioxide, hormones, and heat throughout the body. Proper circulation of blood to the tissues is essential for their survival. In medicine, a sample of a patient's blood can yield important clues as to what diseases may be affecting the individual. A blood test involves first collecting a blood sample, typically performed by a *phlebotomist* (*phleps*, vein, + *tome-*, to cut), and placing the sample in a test tube (**figure 20.6,** step 1). The test tube is then placed in a centrifuge, which spins the sample at high speed to separate the component parts (figure 20.6, step 2). Because the formed elements like WBCs and RBCs are heavy, they fall to the bottom of the test tube, while the blood plasma remains floating on the top (figure 20.6, step 3). WBCs and platelets form a thin layer called a *buffy* coat, which sits on top of the layer of RBCs. The iron present in the hemoglobin in RBCs makes RBCs heavier than WBCs, which explains why RBCs lie in the layer below WBCs in the test tube.

Additional tests may be performed on a sample of blood. For example, a **white blood cell (WBC) count** performed on a sample of the patient's blood can determine whether or not the patient has a bacterial infection. An excess of neutrophils will indicate that the patient has such an infection, whereas a normal abundance of neutrophils might be a clue that an infection is viral rather than bacterial in nature. An abnormally low number of WBCs (leukopenia) may be indicative of a patient's inability to fight infection. In addition, certain pathological conditions may lead to the uncontrolled proliferation of abnormal WBCs, as in leukemia (see Clinical View: Bone Marrow Transplant, p. 573).

The number of red blood cells in the blood (RBC count) is also clinically relevant. A low RBC count is characterized by a reduced oxygen-carrying capacity (or anemia). The RBC count can be approximated in a blood sample by measuring the **hematocrit,** or the packed cell volume (PCV). Because RBCs make up the majority of the formed elements in the blood, the PCV is positively correlated with the number of red blood cells.

A test to determine the time required for blood to clot, or **coagulation time,** may be used assess platelet number and function. When a blood vessel is damaged, platelets are the formed elements involved in **hemostasis** (*hemo-*, blood, + *stasis*, stability), the process which stops blood flow. Hemostasis involves three phases: vascular spasm, platelet plug formation, and coagulation. Both the formation of a platelet plug and the reactions required for coagulation involve platelets. **Platelet plug** formation occurs when a blood vessel is damaged and collagen fibers beneath the endothelium are exposed. Platelets stick to the exposed collagen, undergo morphologic changes, and secrete chemicals that attract additional platelets and promote both coagulation and vessel repair. **Coagulation** involves a cascade of reactions that lead to the formation of a blood clot, called a **thrombus.** In addition to platelets, these reactions require substances such as calcium, clotting factors, and vitamin K. The result is a meshwork of insoluble protein, fibrin, which traps substances contained within the blood. When the inside of a vessel is damaged, the formation of a thrombus typically takes 3 to 6 minutes. Both the formation of the platelet plug and coagulation are regulated by positive feedback mechanisms in order to minimize blood loss.

Another clinically relevant blood test that may be performed is determination of a person's blood type (A, B, AB, or O). **Blood type determination** requires some understanding of an immune response; when RBCs clump together (agglutination), it is indicative of an immune response called agglutination. Agglutination occurs when antibodies circulating in the blood bind to antigens on the surface of RBCs. As concerns blood types, people with blood type A have an antibody to type B blood. People with blood type B have an antibody to type A blood. People with blood type AB have neither A nor B antibodies. People with blood type O have both type A and type B antibodies. Another common surface antigen on erythrocyte membranes determines the Rh blood type. The Rh blood type is determined by the presence or absence of the Rh surface antigen, often called either Rh factor, or surface antigen D. When the Rh factor is present, the individual is said to be Rh positive (Rh+). When the Rh factor is absent, the individual is said to be Rh negative (Rh–). Individuals with type O negative (O–) blood are typically referred to as *universal donors* because their blood can be donated to individuals of any blood type without causing a transfusion reac-

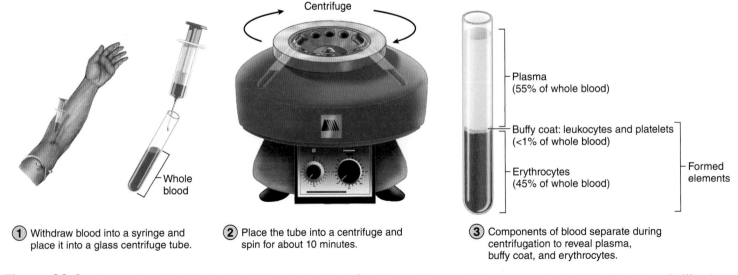

① Withdraw blood into a syringe and place it into a glass centrifuge tube.

② Place the tube into a centrifuge and spin for about 10 minutes.

③ Components of blood separate during centrifugation to reveal plasma, buffy coat, and erythrocytes.

Figure 20.6 Separation of a Whole Blood Sample by Centrifugation. Using a centrifuge to separate whole blood into plasma (55%) and formed elements (45%) is the first step in determining the composition of whole blood.

tion in the recipient. Individuals with AB positive (AB+) blood are referred to as *universal recipients* because they produce no antibodies that would cause a transfusion reaction.

A test to determine **total blood cholesterol,** including the blood values for low density lipoproteins (LDLs) and high density lipoproteins (HDLs), may be important in evaluating an individual's risk for cardiovascular disease. Excess LDLs not used by peripheral tissues may be deposited in the walls of blood vessels, and are therefore considered to be "bad cholesterol." Conversely, HDLs *remove* lipids from the walls of blood vessels and transport the lipids to the liver for waste removal. Therefore, HDLs are considered to be "good cholesterol." Clinically, elevated levels of total blood cholesterol, increased numbers of LDLs, and decreased numbers of HDLs have all been implicated as risk factors for cardiovascular disease. Total blood cholesterol exceeding 200 mg/dl is considered "high." A ratio of HDL to LDL that exceeds 0.3 is considered healthy.

In the following laboratory exercises, you will perform clinically relevant blood tests, including a differential WBC count, hematocrit, hemoglobin concentration, coagulation time, blood typing, and cholesterol testing.

INTEGRATE

CONCEPT CONNECTION

The formation of antibodies involves adaptive immunity (or acquired immunity), a process that exhibits the characteristics of *specificity* and *memory*. When a B- or T-lymphocyte recognizes an antigen on the surface of a cell, the lymphocyte becomes activated through a series of complex steps, which are outlined in your textbook. Once activated, lymphocytes proliferate, forming daughter cells (clones). B-lymphocyte proliferation yields plasma cells, which produce antibodies. Each subsequent exposure to a specific antigen causes a dramatic increase in the number of antibodies produced because of the increased number of clones and plasma cells circulating in the blood. Once antibodies are formed, they bind to antigens to form antigen–antibody complexes. While antibodies do not directly kill the invading pathogens, antigen–antibody complexes assist in the removal of pathogens from the body by making the pathogens become more susceptible to phagocytosis.

EXERCISE 20.4

DETERMINATION OF WHITE BLOOD CELL COUNTS

1. Obtain a slide of a human blood smear or prepare one of your own. See exercise 20.1A for preparing a human blood smear using your own blood.

2. Place the human blood smear slide on the microscope stage. Using the scanning objective (see exercise 3.1), move the slide on the stage so the highest density of blood cells appears in the field of view.

3. Rotate the coarse adjustment knob until WBCs are visible and in focus within the field of view. Because of the stain, WBCs will appear as small *purple* dots in a sea of RBCs.

4. To properly view blood cells, you will need to use the oil immersion (highest-power) objective of the microscope. See exercise 20.1B for instructions on how to view specimens using an oil immersion objective. Rotate the nosepiece such that the highest-power or oil immersion objective is directed toward the blood smear slide.

5. Turn the fine-adjustment knob until the WBCs come into focus.

6. Gently move the slide in a back-and-forth motion and count white blood cells as they cross the field of view (see **figure 20.7**). Be careful to count each cell only once.

7. Record the number of each type of WBC you observe in **table 20.4.**

8. Calculate the total number of WBCs observed (by adding up the numbers of each type of WBC counted) and record this in table 20.4.

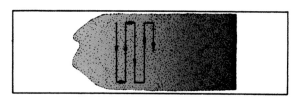

Figure 20.7 Method for Performing Differential WBC Count. Scan the slide in the illustration above.

Table 20.4	Differential White Blood Cell Data Table	
White Blood Cell Type	**Number (#) Observed**	**Percent (%) Observed**
Basophils		
Eosinophils		
Lymphocytes		
Monocytes		
Neutrophils		
Total Number WBC		

(continued on next page)

(continued from previous page)

9. Calculate the percentage for each type of WBC and record this in table 20.4. For example, the following formula demonstrates how to calculate the percentage of neutrophils observed: Percent (%) = [# Neutrophils Observed / Total # WBCs Observed] × 100.

10. Compare your calculated percentages for each type of WBC with the normal values provided in table 20.3.

11. Make note of any pertinent observations here:

_____
_____
_____

EXERCISE 20.5

DETERMINATION OF HEMATOCRIT

Obtain the Following:
- heparinized capillary tube
- Seal-ease™
- alcohol swabs
- lancet

1. Choose a subject. Before you begin, make sure you have read and are following all instructions for safe handling of human blood (see box on p. 533). Use an alcohol swab to cleanse the tip of the subject's middle or ring finger and prick the finger with a lancet. Wipe away the first drop of blood with the alcohol swab.

2. Press the colored end of the heparinized capillary tube to the drop of blood. Be sure to hold the tube tilted slightly up, to allow the blood to fill the tube by capillary action. Be careful not to allow the capillary tube to drop below a horizontal plane, because that increases the potential for bubbles to form in the blood within the tube.

3. Fill the capillary tube with blood approximately three-fourths of the length of the tube **(figure 20.8a)**.

4. Press the blood-containing end of the capillary tube into the Seal-ease™ to seal the tube (figure 20.8a).

5. Place the capillary tube in a groove of the microhematocrit centrifuge, such that the sealed end of the tube is adjacent to the outer rubber gasket of the centrifuge. Be sure to balance the centrifuge by placing samples opposite the tube (figure 20.8b).

6. Close the centrifuge cover and run the centrifuge for 4 to 5 minutes.

7. Once the centrifuge has completed the cycle, remove the capillary tube.

8. The total volume of blood can be measured with either a microhematocrit reader or a metric ruler. If using a ruler, measure the entire length of blood within the capillary tube in millimeters (figure 20.8c). This will include all three layers described above. Record the total blood length in **table 20.5**.

9. Measure the RBC layer alone in millimeters. Record the RBC length in table 20.5.

10. Compute the percentage of RBCs by dividing the length of the RBC layer by the length of the total blood, and then multiplying by 100. Record your results in table 20.5.

11. Compare your calculated hematocrits (percentage of RBCs in whole blood) with expected values (males: 40–54%; females: 37–47%).

12. Make note of any pertinent observations here:

_____
_____
_____

Table 20.5	Hematocrit Data Table
Total blood length (mm)	
RBC length (mm)	
Hematocrit (%)	

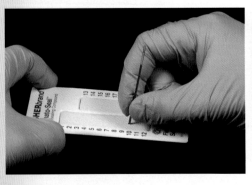
(a)

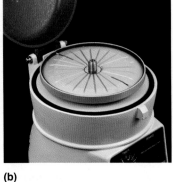
(b)

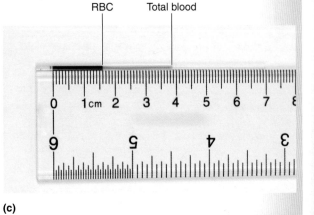
(c)

Figure 20.8 Determination of Hematocrit. (a) Fill the capillary tube with blood and plug one end with Seal-ease™. (b) Place the capillary tube in the microhematocrit centrifuge with the sealed end extending to the outside. (c) Measure the lengths of RBCs and total blood within the capillary tube.

EXERCISE 20.6

DETERMINATION OF HEMOGLOBIN CONTENT

Obtain the Following:
- alcohol swab
- hemoglobinometer
- hemolysis applicator
- lancet
- lens paper

1. Choose a subject. Before you begin, make sure you have read and are following all instructions for safe handling of human blood (see box on p. 533).

2. Disassemble the hemoglobinometer by removing the blood chamber and separating the two plates of glass from the metal clips. Note that there is a larger piece of glass that contains a U-shaped area with surrounding depressions and a smaller piece of glass that is flat on both sides.

3. Use lens paper to clean the glass plates. Ensure that no streaks appear on the glass plates, as this can introduce artifacts when measuring blood hemoglobin levels.

4. Clean the tip of the middle finger or ring finger with an alcohol swab, place the tip of the lancet on the clean finger, and pull the trigger so it pierces the skin (figure 20.1). Wipe away the first drop of blood with the alcohol swab.

5. Place a drop of blood onto the U-shaped area of the larger piece of glass **(figure 20.9a)**.

6. Use the hemolysis applicator to stir the blood such that the RBCs rupture (lyse) and the blood appears clear (~45 seconds) (figure 20.9b).

7. Place the smaller, flat piece of glass onto the U-shaped area as a cover slip.

8. Slide both pieces of glass into the metal clips of the blood chamber, and replace the blood chamber into

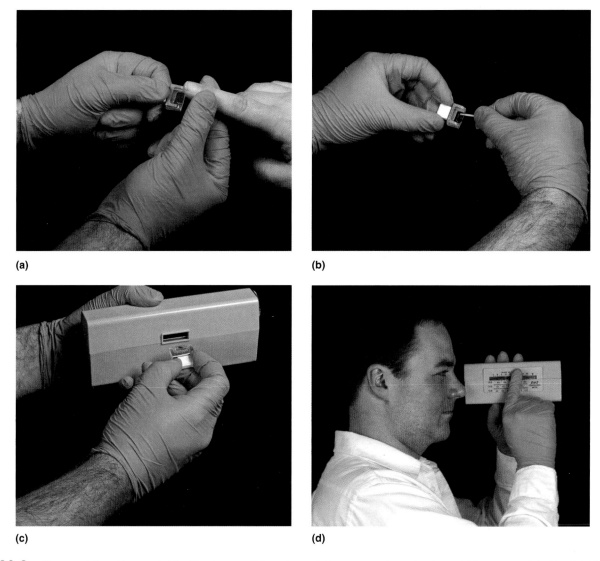

Figure 20.9 Determining Hemoglobin Content. Using a hemoglobinometer to determine hemoglobin content of the blood. (a) Gently place a drop of blood on the U-shaped part of the glass in the blood chamber. (b) Use a hemolysis applicator stick to stir the blood. (c) Slide the blood chamber into the slot on the side of the hemoglobinometer. (d) Use your thumb to turn on the light while using your index finger to move the slide until the colors seen through the window have matching intensity. Read the hemoglobin content from the side of the hemoglobinometer.

(continued on next page)

(continued from previous page)

the side slot of the hemoglobinometer (figure 20.9c). Be sure that the blood chamber is secure within the hemoglobinometer.

9. Locate the light switch on the underside of the hemoglobinometer. Using your left hand, hold the hemoglobinometer while resting your left thumb on the switch. Look into the eyepiece and turn the light on (figure 20.9d).

10. Notice there are two green areas in your field of view. Use your right hand to adjust the slide on the right side of the hemoglobinometer until the two halves of the green field match in color.

11. Record the resulting blood hemoglobin level: _____ g/dl

12. To clean, remove the blood chamber and place the glass plates and clip into a 10% bleach solution.

Tallquist Method

As an alternative, the reasonably accurate and inexpensive *Tallquist method* may be used to estimate blood hemoglobin level. If performing this activity, use a **Tallquist test kit**, which contains absorbent paper and an associated color scale. Repeat step 5 above and place the drop of blood on the absorbent paper. Allow the blood to dry (ensuring that the blood has not dried so much so that it appears brown in color) and match the blood to the color scale provided in the kit.

EXERCISE 20.7

DETERMINATION OF COAGULATION TIME

Obtain the Following:
- heparinized capillary tube
- alcohol swab
- lancet
- triangular file

1. Choose a subject. Before you begin, make sure you have read and are following all instructions for safe handling of human blood (see box on p. 533). Use an alcohol swab to cleanse the tip of the subject's middle or ring finger and prick the finger with a lancet. Wipe away the first drop of blood with the alcohol swab.

2. Load the heparinized capillary tube with blood (exercise 20.5, step 2).

3. Fill the capillary tube with blood approximately three-fourths of the length of the tube.

4. Lay the capillary tube flat on a paper towel and record the initial time: _____

5. Use the triangular file to weaken a small area of the capillary tube just adjacent to its end **(figure 20.10a)**. Once the tube is filed slightly, use both hands to hold the tube on either side of the filed section. Break the tube away from you, ensuring that both ends remain close so that you can observe a fibrin clot (if any). If the clot has not yet formed, the tube will break cleanly

INTEGRATE

CLINICAL VIEW
Normal Ranges for Laboratory Blood Tests

Coagulation Time:	5–8 min.
Hematocrit:	M: 39–49%, F: 35–45%
Hemoglobin:	M: 13.5–17.5 g/dL, F: 12.0–16.0 g/dL
Total Cholesterol = HDL + LDL (HDL = high density lipoprotein, LDL = low density lipoprotein)	
LDL Cholesterol:	Recommended: <130 mg/dL, moderate risk: 130–159 mg/dL, high risk: >160 mg/dL
HDL Cholesterol:	M: >29 mg/dL, F: >35 mg/dL

Reference: Goldman, L. and A.I. Schafer, *Goldman's Cecil Medicine 24e*, Elsevier: New York, 2012.
HDL = "good" cholesterol
LDL = "bad" cholesterol

(figure 20.10*b*). If the clot has formed, fibrin will extend from each end (figure 20.10*c*).

6. Repeat step 5 at 1-minute intervals until the fibrin clot is observed. Record the time required for coagulation: _____ minutes

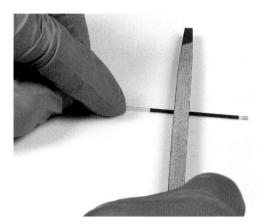
(a)

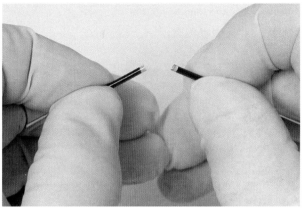
(b)

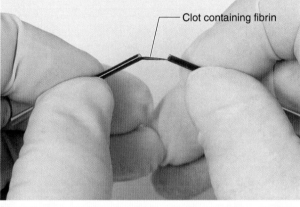
(c)

Figure 20.10 Determining Coagulation Time.
(a) Use a triangular file to score the slide. (b) Break the slide to see if coagulation has occurred (no coagulation has yet occurred in this photo). (c) Appearance of fibrin within a clot after coagulation has occurred.

EXERCISE 20.8

DETERMINATION OF BLOOD TYPE

Obtain the Following:
- blood sample (human, animal, or artificial)
- two clean microscope slides
- antibody sera (anti-A, anti-B, anti-Rh)
- droppers
- toothpicks
- wax pencil
- alcohol swabs and lancets (if using human blood)

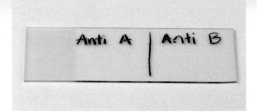
(a)

1. Draw a vertical line down the center of one microscope slide with a wax marking pencil, dividing the slide into equal halves. Label the left side "anti-A" and the right side "anti-B" **(figure 20.11a)**.

2. Place one drop of blood on the left side of the dividing line and place one drop of blood on the right side of the dividing line. If using human blood, use an alcohol swab to cleanse the finger and prick the finger with a lancet (figure 20.1).

3. Place one drop of blood in the center of the second slide.

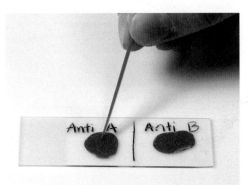
(b)

4. Place one drop of anti-A serum on the drop of blood on the left side of the dividing line on the first slide. Place one drop of anti-B serum on the drop of blood on the right side of the first slide. Place one drop of anti-Rh serum on the drop of blood on the second slide.

5. Mix each blood/anti-serum sample with a clean toothpick (figure 20.11b). Be sure to use a separate toothpick for each mixture. Transferring the toothpick to multiple mixtures will contaminate the samples. Note that an Rh typing box may be required if using human blood because thorough mixing of the blood and anti-Rh serum at a slightly elevated temperature is required.

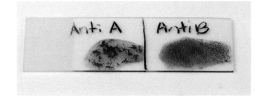
(c)

Figure 20.11 Blood Typing. (a) Prepare the slide as demonstrated here. (b) Mix anti-sera and blood sample with toothpick. (c) Observe the slides after 2–3 minutes, looking for agglutination reactions.

6. After 2 minutes, observe the blood samples. Agglutination in any of the samples indicates a positive result (figure 20.11c).

7. Place the microscope slides in a bowl containing a 10% bleach solution, and dispose of the lancets in a sharps container. Consult with your laboratory instructor for proper disposal of all contaminated materials.

8. Make note of any pertinent observations here:

_____
_____
_____

INTEGRATE

CLINICAL VIEW
Blood Transfusions

Blood transfusions involve the transfer of blood from one person, the **donor**, to another person, the **recipient**. It is critical that the blood from both donor and recipient are compatible, or the consequences of the transfusion may be lethal. To determine whether or not a blood recipient can safely receive a donor's blood, it must be determined if the recipient will "recognize" the blood as foreign or not. That is, it must be determined if the donor's blood has RBCs with proteins that are not matched to other proteins on the recipient's RBCs. For example, a recipient with A⁺ blood receives A⁻ blood from a donor. Will the recipient reject the blood? No, because the RBCs of the recipient have both the A protein and the Rh protein. The donor's blood contains only A protein. These are compatible. What if a recipient with A⁻ blood receives A⁺ blood from the donor? Will the recipient reject the blood? Yes, because the RBCs of the recipient have only the A protein, whereas the donor's blood contains both A protein and Rh protein. Since the recipient does not have Rh protein, the recipient "sees" the Rh protein as a "foreign" protein and rejects the blood. Rejection involves an agglutination reaction, as observed in exercise 20.8.

EXERCISE 20.9

DETERMINATION OF BLOOD CHOLESTEROL

Obtain the Following:
- alcohol swab
- cholesterol test card
- color scale
- lancet

1. Choose a subject. Before you begin, make sure you have read and are following all instructions for safe handling of human blood (see box on p. 533). Use an alcohol swab to cleanse the tip of the subject's middle or ring finger and prick the finger with a lancet (figure 20.1). Wipe away the first drop of blood with the alcohol swab.

2. Place a drop of the blood onto the cholesterol test card and allow the blood to dry.

3. Wait 3 minutes and remove and dispose of the sample strip contained within the cholesterol test card.

4. Compare the remaining blood test spot with the color scale provided with the cholesterol test card. Record the blood cholesterol level: _____ mg/dl. Note: normally cholesterol tests are done after the subject has fasted for several hours. Therefore, a high blood cholesterol reading taken after a meal may not be completely accurate.

Chapter 20: The Cardiovascular System: Blood

Name:_____
Date:_____ Section:_____

POST-LABORATORY WORKSHEET

The ❶ corresponds to the Learning Objective(s) listed in the chapter opener outline.

Do You Know the Basics?

Exercise 20.1: Identification of Formed Elements on a Prepared Blood Smear

1. The diameter of an erythrocyte is _____. ❶

2. Label the formed elements of blood shown in the following figure. ❶

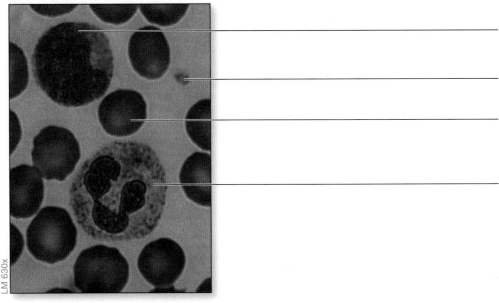

1 _____
2 _____
3 _____
4 _____

3. The rarest type of leukocyte is _____. ❶

4. The most abundant type of leukocyte is _____. ❶

5. Lymphocytes are responsible for initiating the _____ immune response to infection. ❷

6. The two substances secreted by human basophils are _____ and _____. ❷

7. A blood cell that is a precursor to tissue macrophages is a(n) ❷ _____.

Exercise 20.2: Identification of Megakaryocytes on a Bone Marrow Slide

8. Small cellular fragments that are involved in the process of hemostasis are _____. ❸

9. Platelets are derived from _____, which reside in the bone marrow. ❹

Exercise 20.3: Identification of Formed Elements of the Blood on Classroom Models or Charts

10. An agranulocyte with a large, spherical nucleus surrounded by pale blue cytoplasm is a(n) _____. ❺

11. A leukocyte that contains reddish-colored granules and a bi-lobed nucleus is a(n) _____. ❺

Exercise 20.4: Determination of White Blood Cell Counts

12. Formed elements that compose less than 1% of whole blood and are involved in defense against disease are _____. ❻

13. A significant increase in the number of circulating neutrophils may indicate a(n) _____ infection. ❼

14. A significant increase in the number of circulating eosinophils may indicate a(n) _____ infection. ❼

Exercise 20.5: Determination of Hematocrit

15. In determining hematocrit, if the length of RBCs in the capillary tube is 45 millimeters and the length of total blood is 95, what is the hematocrit? ⑧ _____

16. What is normal range for hematocrit? ⑨ Male _____ Female _____

17. What is the purpose of determining hematocrit? ⑨ _____

Exercise 20.6: Determination of Hemoglobin Content

18. Describe why blood containing decreased levels of hemoglobin appears lighter in color than blood containing normal levels of hemoglobin. ⑩

_____
_____

19. Describe the physiological consequences of decreased hemoglobin in the blood. ⑩

_____
_____

Exercise 20.7: Determination of Coagulation Time

20. What is the normal time required for blood to coagulate? ⑪ _____

21. Propose two different abnormalities of blood that, if present, might cause coagulation time to increase (take longer than usual). ⑪

_____
_____
_____

Exercise 20.8: Determination of Blood Type

22. If a person has blood type A⁺, what proteins are located in the red blood cell's membrane? ⑫ _____

_____

23. In determining blood type, if an agglutination reaction only occurs with antibody B, what is the blood type? ⑬ _____

Exercise 20.9: Determination of Blood Cholesterol

24. Describe the basic function of the following components of total blood cholesterol. ⑭

High-Density Lipoproteins (HDLs): _____

Low-Density Lipoproteins (LDLs): _____

Can You Apply What You've Learned?

25. What are some symptoms might you expect an individual suffering from anemia to exhibit? _____

_____

26. Since leukocytes and erythrocytes are produced from the same precursor cells, predict the change in white blood cell count that might occur in a patient who has an elevated red blood cell count (polycythemia). _____

27. Predict the change in red blood cell count in a patient suffering from leukemia. _____

28. Erythropoietin (EPO), a hormone produced by the kidneys, is released into the blood when blood oxygen levels are low. EPO stimulates the production of red blood cells. Predict the resulting EPO level in the blood and red blood cell count of a person who has spent considerable time living at high altitudes, where atmospheric oxygen levels are reduced. _____

_____

29. A patient has been injured and is in need of a blood transfusion. When she arrives, doctors determine that she has type A⁺ blood. What antigens are present on her red blood cells? _____ What antibodies are present in her blood?

_____ What blood type(s) can she receive in her blood transfusion?

_____

Can You Synthesize What You've Learned?

For each of the cases below, determine which values are out of the expected range for a healthy patient and propose a diagnosis. Consult table 20.3 for the relative percentages of leukocytes in the blood. Refer to **table 20.6** for reference values for the remaining blood variables.

Table 20.6	Blood Panel Reference Values
WBC	5000–10,000 cells/mm^3
RBC	Males: ~5.4 million cells/mm^3 Females: ~4.8 million cells/mm^3
Hemoglobin (Hb)	Males: 13–18 g/dl Females: 12–16 g/dl
Hematocrit	Males: 45–52% Females: 37–48%
Platelets	150,000–400,000 cells/mm^3

30. Bonnie, a 50-year-old female, complains of severe fatigue, weakness, and dizziness. Her complete blood count (CBC) is shown below. Discuss which of her variables are out of the expected range and propose a diagnosis.

WBC	1000 cells/mm^3
RBC	2 million cells/mm^3
Hb	7.5 g/dl
Hematocrit	25%
Neutrophils	31%
Eosinophils	6%
Basophils	2%
Lymphocytes	53%
Monocytes	8%
Platelets	50,000 cells/mm^3

    ___

31. Jack, a 25-year-old male, complains of severe lower abdominal pain, vomiting, and nausea. Below is his complete blood count (CBC). Discuss which of his variables are out of the expected range and propose a diagnosis.

WBC	20,000 cells/mm^3
RBC	50 million cells/mm^3
Hb	15 g/dl
Hematocrit	50%
Neutrophils	76%
Eosinophils	2%
Basophils	0%
Lymphocytes	18%
Monocytes	4%
Platelets	300,000 cells/mm^3

    ___

32. Elizabeth is 20 weeks pregnant with her second child. Elizabeth's blood type is AB negative (AB⁻). Her physician is concerned about the possibility of a reaction between Elizabeth's blood and the baby's blood because of the blood type of the baby's father. The physician prescribes RhoGAM for Elizabeth. RhoGAM contains immunoglobulins that prevent Elizabeth from producing Rh antibodies.

   a. What does this suggest about the father's blood type?

   b. What does the physician fear may be a complication with this pregnancy in the event that RhoGAM is not administered?

CHAPTER 21

The Cardiovascular System: The Heart

OUTLINE AND LEARNING OBJECTIVES

Histology 554

EXERCISE 21.1: CARDIAC MUSCLE 554
1. Describe how cardiac muscle differs structurally and functionally from skeletal muscle

EXERCISE 21.2: LAYERS OF THE HEART WALL 555
2. Name the layers of the heart wall, and describe the structures that make up each layer

Gross Anatomy 556

EXERCISE 21.3: THE PERICARDIAL CAVITY 556
3. Name and describe the layers of the pericardial sac

EXERCISE 21.4: GROSS ANATOMY OF THE HUMAN HEART 557
4. Explain how the atria and ventricles of the heart differ in their wall structure
5. Describe the specialized structures found within the right atrium of the heart, including pectinate muscle, the coronary sinus, and the fossa ovalis
6. Differentiate between pectinate and papillary muscles
7. Describe the structure and function of the atrioventricular valves, including chordae tendineae, valve cusps, and papillary muscles
8. Compare and contrast the structure and function of the right atrioventricular valves and the left atrioventricular (mitral) valves
9. Compare and contrast the structure and function of the right and left ventricles
10. Describe the structure and function of the pulmonary and aortic semilunar valves
11. Trace the flow of blood from the right atrium, through the heart and lungs, to the aorta
12. Locate the fossa ovalis and the ligamentum arteriosum, and name the fetal structures of which they are remnants

EXERCISE 21.5: THE CORONARY CIRCULATION 562
13. Identify the right and left coronary arteries, their major tributaries, and the areas of the heart served by each vessel
14. Describe the location and function of the coronary sinus

EXERCISE 21.6: SUPERFICIAL STRUCTURES OF THE SHEEP HEART 566
15. Identify superficial structures on the sheep heart, and explain the function of each
16. Observe the visceral and parietal pericardium on the sheep heart
17. Identify the great vessels on the sheep heart
18. Trace the flow of blood through the sheep heart

EXERCISE 21.7: CORONAL SECTION OF THE SHEEP HEART 567
19. Make a coronal section through the sheep heart, and measure the thicknesses of the right and left ventricular walls
20. Explain the functional consequences of the difference in wall thickness between right and left ventricles

21. Identify internal structures of the sheep heart as seen in a coronal section

EXERCISE 21.8: TRANSVERSE SECTION OF THE SHEEP HEART 568
22. Make a transverse section through the sheep heart, and measure the thicknesses of the right and left ventricular walls
23. Explain the functional consequences of the difference in wall thickness between right and left ventricles

Physiology 569

Electrical Conduction within the Heart 569

EXERCISE 21.9: Ph.I.L.S. LESSON 19: REFRACTORY PERIOD OF THE HEART 571
24. Describe the mechanism for the refractory period in cardiac muscle
25. Compare the refractory period in skeletal and cardiac muscle tissue

EXERCISE 21.10: Ph.I.L.S. LESSON 25: ELECTRICAL AXIS OF THE HEART 573
26. Identify the component parts of an electrocardiogram (ECG) from a simulated ECG experiment and relate these to the conduction system of the heart
27. Measure the amplitude of the QRS complex from two orientations
28. Calculate the electrical axis of the heart

EXERCISE 21.11: ELECTROCARDIOGRAPHY USING STANDARD ECG APPARATUS 574
29. Determine the optimal limb lead and surface electrode placement required to measure an ECG
30. Record P, QRS, and T waves for baseline and exercise conditions using standard limb leads and surface electrodes

EXERCISE 21.12: Ph.I.L.S. LESSON 26: ECG AND HEART BLOCK 575
31. Observe changes in the ECG that are associated with varying degrees of heart conduction block as observed in simulated patients
32. Determine the clinical relevance of abnormal ECG waveforms

EXERCISE 21.13: Ph.I.L.S. LESSON 27: ABNORMAL ECGS 577
33. Compare normal and abnormal ECG traces
34. Describe the physiological basis for abnormal ECG traces in three virtual subjects

Cardiac Cycle and Heart Sounds 578

EXERCISE 21.14: Ph.I.L.S. LESSON 23: THE MEANING OF HEART SOUNDS 578
35. Relate various heart sounds to the components of an ECG

EXERCISE 21.15: AUSCULTATION OF HEART SOUNDS 579
36. Describe the role of atrioventricular and semilunar valves in ensuring one-way blood flow through the heart
37. Determine the optimal placement of a stethoscope for auscultation of heart sounds
38. Determine the clinical relevance of abnormal heart sounds

EXERCISE 21.16: Ph.I.L.S. LESSON 20: STARLING'S LAW OF THE HEART 579
39. Describe length-tension relationships in cardiac muscle

MODULE 9: CARDIOVASCULAR SYSTEM

INTRODUCTION

The heart is an amazing organ that has been viewed with awe for ages. For many centuries physicians thought the heart, not the brain, was the control center and spiritual/emotional center of the body. Perhaps this is because the structural and functional relationships in the heart are relatively straightforward and easy to see, whereas such relationships are nearly impossible to discover through gross observation of the brain. As a student of anatomy and physiology who has studied the brain, you know that our emotions come not "from the heart" but from the brain. Even though scientists and laypeople alike recognize that the brain, not the heart, controls the functioning of the rest of the body, we also recognize that the heart *is* essential for the survival of all organs and tissues in the body: If the heart fails to pump blood, and that failure results in a lack of flow of oxygenated blood to the tissues, the tissues will die.

Have you ever considered how incredible it is that your heart continues to beat, day and night, day after day, year after year, without stopping? It is an enormous job. Failure of this organ is often times fatal. Perhaps it is no surprise that heart disease is the number one cause of death for Americans (it accounts for approximately 32% of all deaths, while cancer, at number two, accounts for 23% of all deaths). A thorough understanding of the structure and function of the heart is critical for every one of us, whether or not we intend to go into the health science field. If your future career involves health care, you will be dealing with individuals suffering from heart disease on a daily basis. Even if your future career does not involve health care, how you take care of it now may very well determine the length and overall quality of your life.

In this laboratory session you will review the structure of cardiac muscle and identify heart structures on a preserved human heart or a model of the heart, and through dissection of a sheep heart. As you work through these exercises, be aware that most textbook figures of the heart are drawn to make identification of the chambers and vessels very straightforward. When you observe a real heart, you may find identification of the structures to be more challenging because the chambers do not lie directly superior, inferior, or lateral to each other as they are often depicted in textbook drawings. In fact, most of the heart structures labeled "right" (such as the right atrium and right ventricle) lie not only on the right side of the heart, but also on the *anterior* surface of the heart. Likewise, most of the structures labeled "left" (such as the left atrium and left ventricle) lie not only on the left side of the heart, but also on the *posterior* surface of the heart.

In the later exercises within this chapter you will measure the electrical activity of the heart using electrocardiography (ECG), listen to heart sounds and relate these sounds to the events of the cardiac cycle, and explore the length-tension relationship in cardiac muscle.

INTEGRATE

CONCEPT CONNECTION

As discussed in exercise 21.1, there are important distinctions made between cardiac muscle tissue and skeletal muscle tissue. Cardiac muscle, like skeletal muscle, contains striations. The striations are due to the presence of sarcomeres, the fundamental contractile unit of cardiac and skeletal muscle. Sarcomeres are composed of overlapping thick filaments composed of myosin protein and thin filaments composed of actin, troponin, and tropomyosin. The mechanism of muscle contraction in cardiac muscle is similar to that of skeletal muscle. Calcium binds to the regulatory protein troponin, causing a conformational change in troponin-tropomysin complex that exposes the myosin binding sites on actin. This allows myosin heads to bind to actin, forming crossbridges. The formation of crossbridges in both cardiac and skeletal muscle is what allows the muscle to produce force and generate movement.

EXERCISE 21.2

LAYERS OF THE HEART WALL

1. Obtain a slide demonstrating the atria of the heart. Place it on the microscope stage and observe on low power. The wall of the heart is composed of three layers, the endocardium, myocardium, and epicardium. The **endocardium** consists of the simple squamous epithelium that lines the heart, called **endothelium,** plus an underlying layer of connective tissue **(figure 21.2)**. The **myocardium** is composed of cardiac muscle and is by far the thickest layer of the heart wall. The **epicardium** is the same tissue that composes the visceral layer of the pericardial sac—a serous membrane referred to as the visceral layer of serous pericardium. What to call this layer of tissue depends on the context. When referring to it as part of the heart wall, the appropriate term is *epicardium*. When referring to it as part of the pericardial sac, the appropriate term is *visceral layer of serous pericardium*. You may see cross sections of the coronary vessels deep to the epicardium on the slide as well.

2. Using figure 21.2 and **table 21.2** as a guide, identify the following on the slide of the atrium:

   ☐ endocardium   ☐ myocardium
   ☐ epicardium

3. In the space below, sketch the layers of the heart wall as seen through the microscope. Be sure to include and label all of the structures listed in step 2 in your drawing.

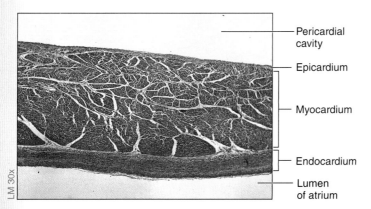

Figure 21.2 Histology of the Heart Wall. This slide demonstrates the wall layers of the atrium of the heart. In the atrium, the endocardium is thick, consisting of endothelium (a simple squamous epithelium) plus an underlying layer of areolar connective tissue. The myocardium is the thickest layer, consisting of cardiac muscle tissue. The epicardium (visceral layer of epicardium) is the thinnest layer, composed of simple squamous epithelium resting on a layer of areolar connective tissue.

Table 21.2	Layers of the Heart Wall		
Wall Layer	**Tissue**	**Description**	**Word Origin**
Endocardium	Endothelial cells plus some connective tissue and smooth muscle	This layer is relatively thick in the atria, and thin in the ventricles	*endon*, within, + *kardia*, heart
Myocardium	Composed of layers of cardiac muscle in addition to elastic fibers and loose connective tissue between the layers	This is the thickest layer of the heart wall	*mys*, muscle, + *kardia*, heart
Epicardium	Composed of a serous membrane with an underlying layer of elastic fibers and adipose connective tissue	This is the visceral pericardium. It is relatively thicker in the ventricles than in the atria.	*epi-*, upon, + *kardia*, heart

Gross Anatomy

EXERCISE 21.3

THE PERICARDIAL CAVITY

1. Observe a model of a human thorax or the thoracic cavity of a human cadaver (**figure 21.3**).

2. The heart is located within the **thoracic cavity,** a cavity that also houses the lungs, trachea, esophagus, and a variety of nerves and blood vessels. Within the thoracic cavity, the heart is located within a space called the **mediastinum.**

3. Using your textbook as a guide, identify the structures listed in figure 21.3 on a classroom model of the thorax or on a human cadaver. Then label them in figure 21.3.

4. *Optional Activity:* AP|R **9: Cardiovascular System—** Explore the "Thorax" dissections to view the heart in the thoracic cavity and appreciate the surrounding structures.

5. Within the inferior mediastinum, the heart is encased within a space called the **pericardial cavity.** The outermost covering of the pericardial cavity is composed of a double-layered **pericardial sac.** The outer portion of the pericardial sac is the **fibrous pericardium** composed of a tough, dense irregular tissue. This layer is attached inferiorly to the diaphragm and superiorly to the aorta and pulmonary trunk. The inner portion of the sac is the **parietal layer of serous pericardium.** Tightly adhered to the heart is the **visceral layer of serous pericardium** (also called the epicardium). The pericardial cavity is the fluid-filled space between the two serous membranes. Observe the pericardial cavity within the thorax and name the structures to which the pericardial sac is anchored on each of the following surfaces:

   Superior: _____

   Inferior: _____

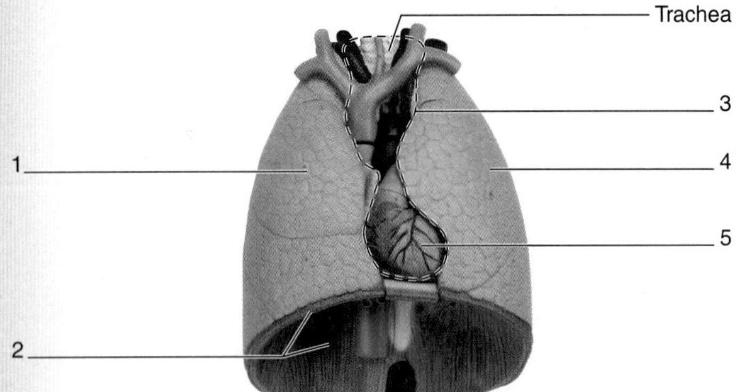

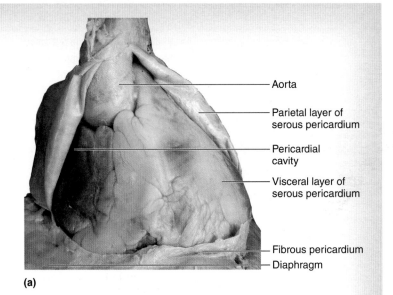

Figure 21.3 Location of the Heart within the Thoracic Cavity. Within the thoracic cavity the heart is located within the mediastinum, the space between the two pleural cavities, which house the lungs.

☐ diaphragm   ☐ mediastinum
☐ heart   ☐ right lung/pleural cavity
☐ left lung/pleural cavity

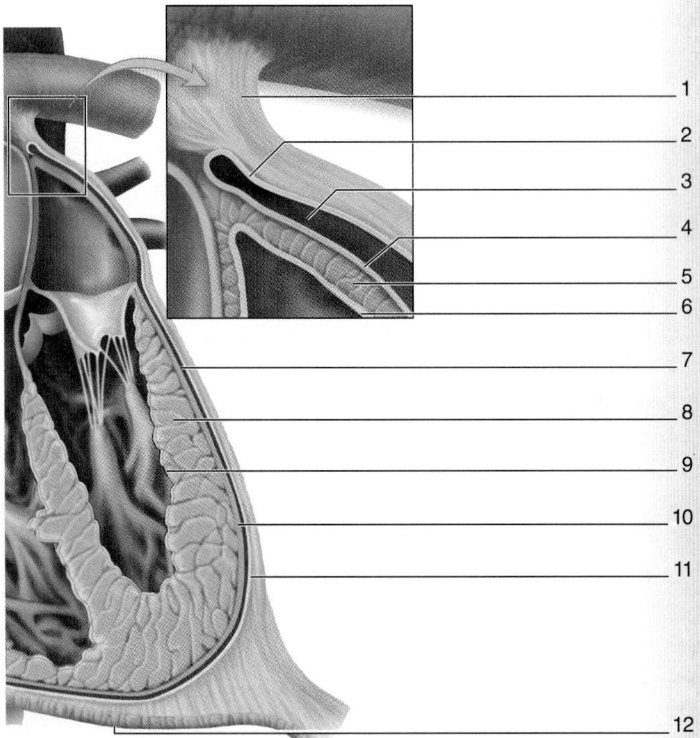

Figure 21.4 The Pericardial Sac. (a) Location of the heart within the pericardial sac of a cadaver. The parietal pericardium has been cut and reflected to reveal the heart, which is covered by visceral pericardium. (b) Layers and tissues composing the pericardial sac.

☐ diaphragm
☐ endocardium (used twice)
☐ fibrous pericardium (used twice)
☐ myocardium (used twice)
☐ parietal layer of serous pericardium
☐ pericardial cavity (used twice)
☐ visceral layer of serous pericardium (epicardium) (used twice)

6. Remove the pericardial sac from the heart (**figure 21.4a**). Upon removal of the pericardial sac, the visceral layer of serous pericardium will remain attached to the heart. Note that both the parietal layer of serous pericardium and the visceral layer of serous pericardium are composed of simple squamous epithelium called **mesothelium.** These membranes produce small amounts of **pericardial fluid,** which lubricate the surface between the heart and the pericardial sac.

7. Using figure 21.4a as a guide, identify the structures listed in figure 21.4b on a model of the thorax or on a human cadaver. Then label them in figure 21.4b. (Answers may be used more than once.)

EXERCISE 21.4

GROSS ANATOMY OF THE HUMAN HEART

Obtain a preserved human heart from a cadaver or a classroom model of the heart. If you are using a preserved heart, place it in a dissecting pan and keep it moist while you make your observations. In addition, *use only a blunt probe* to point out structures on the heart so as not to damage the heart.

1. Note the size and shape of the heart. A normal heart is about the size and shape of a human fist. Based on this information, is the heart you are observing of normal size? _____ If the heart appears to be enlarged, make note of that observation, as it can be (though it is not necessarily) indicative of heart disease. One of the first things you may notice is that the heart you are holding in your hands looks very little like the drawings in your textbook. It is a twisted organ, so identification of chambers can be challenging at first. Begin by identifying the **apex** (the pointed, inferior portion of the heart) and the **base** (the superior point where the great vessels enter and leave). In most instances, the term *base* refers to the bottom of an organ or tissue. Is this generalization true with the heart? _____
The heart has a relatively flat inferior surface, the **diaphragmatic surface,** which is the surface of the heart that lies on the superior surface of the diaphragm. Once you have identified these structures, place the heart in your right hand with the diaphragmatic surface in the palm of your hand, the apex directed toward your thumb and wrist, and the base directed toward the space between the tip of the thumb and the second digit (**figure 21.5**). You should now be viewing the anterior surface of the heart in a position that very closely resembles the heart's orientation within the thorax. If you are unsure if you are viewing the anterior surface, check with your instructor before you proceed.

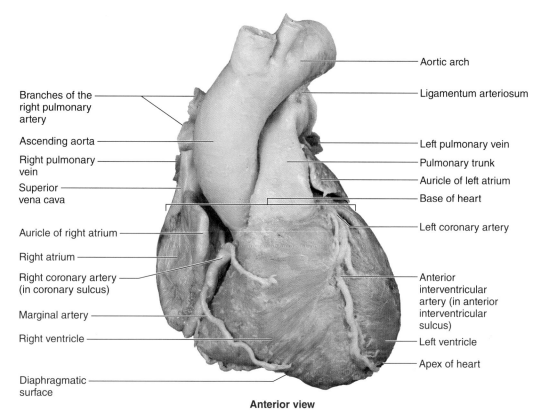

Anterior view

Figure 21.5 Orientation of the Heart in the Right Hand of the Observer. The heart shown here is oriented with the anterior surface facing the observer, which is how it should look when it is in the observer's right hand with the apex pointed toward the thumb and wrist, and the great vessels directed toward the space between the tip of the thumb and the second digit.

(continued on next page)

(continued from previous page)

2. Identify the **right and left ventricles** (figure 21.5). Because you are viewing the anterior surface of the heart, the left ventricle will be on *your* right, and the right ventricle will be on *your* left. Notice that the mass of the left ventricle fills up nearly the entire palm of your hand because it has a much thicker myocardium than the right ventricle. Identify the **right atrium** superior and lateral to the right ventricle. The left atrium is not visible in this view. In the following steps you will first identify the layers of the heart wall, and then proceed to identify the heart chambers and major heart structures within each chamber.

3. *Layers of the Heart Wall:* Using table 21.2 and figure 21.4 as guides, identify the following components of the heart wall on a heart model or on a preserved human heart.

   ☐ endocardium   ☐ myocardium
   ☐ epicardium

4. *The Right Atrium:* Holding the heart in your right hand once again, with the anterior surface directed toward you, identify the **superior and inferior vena cavae** entering the right atrium (figure 21.5). Within the thorax, these vessels run vertically and meet at the right atrium. Next, look inside the right atrium. Notice the thin strands of **pectinate muscle** in the wall of the right atrium **(figure 21.6)**. Pectinate muscle is found only in the wall of the right atrium (although some pectinate muscle is also found within the walls of both *auricles*, which are extensions of the atria), so its presence provides a good indicator that you are observing the right atrium and not the left atrium, if you are unsure. Locate the shallow depression covered with a thin membrane in the interatrial septum. This is the **fossa ovalis,** a remnant of a fetal shunt between the right and left atria called the **foramen ovale.** Is the opening completely closed off in your specimen? If not, what might some of the consequences be (this condition is called a *patent* foramen ovale [*pateo,* to lie

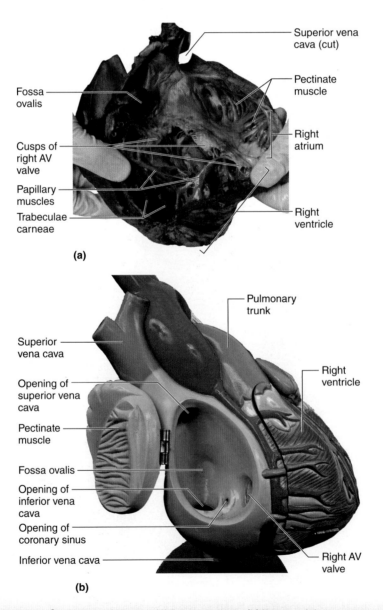

Figure 21.6 **The Right Atrium of a Human Heart.** (a) Gross specimen. (b) Heart model.

open])? _____ Just inferior to the fossa ovalis, look for the small opening of the **coronary sinus,** a vein that drains nearly all deoxygenated blood from the heart wall. Finally, observe the **right atrioventricular (AV) valve** and count the cusps. How many cusps are there? _____ Based on that information, is the right AV valve a tricuspid or bicuspid valve? _____

6. *The Right Ventricle:* Place a blunt probe in the right atrium, and direct it into the right ventricle **(figure 21.7).** If the right ventricle is not already cut open, ask your instructor to cut it open to identify the internal structures. The most prominent features in the walls of the right ventricles are the folds of cardiac muscle in the walls of the ventricle called **trabeculae carneae** (*trabs,* a beam, + *carneus,* fleshy), and the nipplelike **papillary muscles** (*papilla,* a nipple) that attach to the cusps of the right AV valve by stringlike structures called **chordae tendineae** (*chorda,* cord, + *tendo,* to stretch out). When the ventricles contract, the papillary muscles have already begun to contract and pull down on the cusps of the AV valve. Because blood is being pushed up against the inferior portion of the valve cusps as the ventricles contract, the action of the papillary muscles pulling down on the valve cusps keeps the valve closed. This prevents blood from flowing back into the right atrium, and as a result the blood is forced out through the **pulmonary trunk.**

7. Using figure 21.7 as a guide, identify the following structures in the right ventricle of a human heart or heart model:

   ☐ chordae tendineae   ☐ papillary muscles
   ☐ cusps of the right AV valve ☐ trabeculae carneae

8. Place the tip of a blunt probe in the right ventricle and pass it out through the **pulmonary trunk.** To enter the pulmonary trunk, the probe will have to pass through the **pulmonary semilunar valve.** If the vessel has been cut open, you will be able to see that the cusps of the semilunar valve are shaped like "half moons"—hence their name. How many cusps are there? _____ Where will the blood travel after it leaves the pulmonary trunk? _____

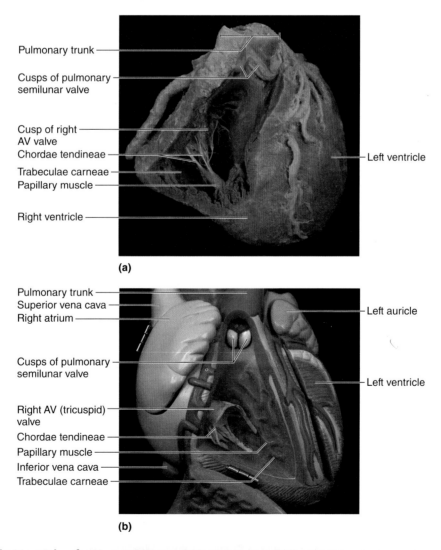

Figure 21.7 **The Right Ventricle of a Human Heart.** (a) Gross specimen. (b) Heart model.

(continued on next page)

(continued from previous page)

9. In the space below, make a simple drawing of the right atrium and right ventricle and draw arrows to indicate the flow of blood from the right atrium to the pulmonary trunk. Be sure to include and label all of the structures listed below in your drawing.

   ☐ chordae tendineae
   ☐ fossa ovalis
   ☐ inferior vena cava
   ☐ opening of coronary sinus
   ☐ papillary muscles
   ☐ pectinate muscle
   ☐ pulmonary semilunar valve
   ☐ pulmonary trunk
   ☐ right atrium
   ☐ right AV valve
   ☐ right ventricle
   ☐ superior vena cava
   ☐ trabeculae carneae

10. *The Left Atrium:* Rotate the heart until you are viewing its posterior surface **(figure 21.8).** Note the four **pulmonary veins,** which collectively drain blood into the **left atrium.** Look inside the left atrium. Notice that the wall of the left atrium is thin and smooth and does *not* have pectinate muscles (although the wall of the left *auricle* does). The left atrium is little more than an expansion of the tissue where the four pulmonary veins come together. Thus, its walls are not always easy to identify. Now that you have identified both the right atrium and left atrium, place your index finger in the right atrium and your thumb in the left atrium and again find the **fossa ovalis,** which lies in the **interatrial septum.** What is the name of the fetal structure of which the fossa ovalis is a remnant?_____ Next, observe the **left atrioventricular (AV) valve** and count the cusps. How many cusps are there?_____ Based on that information, is the left AV valve a tricuspid or bicuspid valve?_____ What is another name for this valve?_____

11. *The Left Ventricle:* Place a blunt probe in the left atrium and pass it into the **left ventricle (figure 21.9).** If the left ventricle is not already cut open, ask your instructor to cut it open so you can identify the structures within. The left ventricle contains all the same structures as the right ventricle, and the **left atrioventricular (AV) valve** functions the same way that the right AV valve functions. The biggest difference between the two chambers is the thickness of the ventricular walls.

12. Using figure 21.9 as a guide, identify the following structures in the left ventricle:

    ☐ chordae tendineae
    ☐ left AV valve
    ☐ papillary muscles
    ☐ trabeculae carneae

13. Note the difference in thickness between the myocardium in the wall of the left ventricle and that of the right ventricle. What accounts for this difference? _____
    _____
    _____

    Does the chamber size (volume) of the left ventricle appear to be greater than that of the right ventricle?_____ Do the two chambers pump the same amount of blood? _____

14. Place a probe in the left ventricle and pass it out through the **aorta.** To enter the aorta, the probe will have to pass through the **aortic semilunar valve.** If the vessel has been cut open, you will be able to see that the cusps of the semilunar valve are shaped like "half moons"—hence their name. How many cusps are there?_____ Where will the blood travel after it leaves the aorta?_____

15. Observe the outside of the arch of the aorta where it passes just superior to the pulmonary trunk. Look for a small ligament attaching the pulmonary trunk to the aorta (see figure 21.5 and figure 21.9b). This is the **ligamentum arteriosum.** Of what fetal structure is the ligamentum arteriosum a remnant?_____ This fetal structure shunts blood from the _____ to the _____, thereby shunting blood away from the _____.

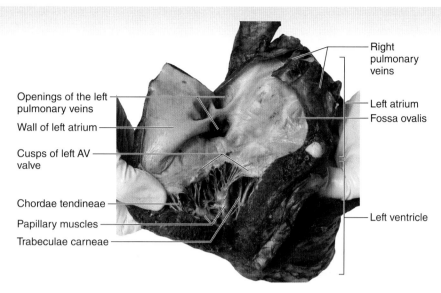

Figure 21.8 **The Left Atrium of a Human Heart.** Posterior view of a human heart with left atrium and left ventricle cut open.

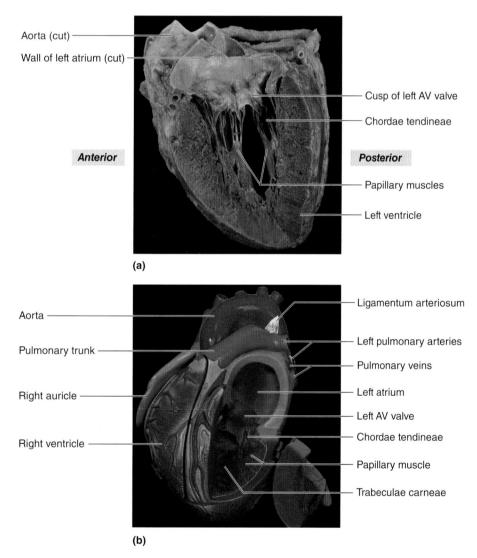

Figure 21.9 **The Left Ventricle of a Human Heart.** (a) Gross specimen. (b) Classroom model.

(continued on next page)

(continued from previous page)

16. In the space to the right, make a simple drawing of the left atrium and left ventricle and draw arrows to indicate the flow of blood from the lungs to the aorta. Be sure to include and label all of the structures listed below in your drawing.

    ☐ aorta ☐ left AV valve
    ☐ aortic semilunar valve ☐ left ventricle
    ☐ chordae tendineae ☐ papillary muscles
    ☐ fossa ovalis ☐ pulmonary veins
    ☐ left atrium ☐ trabeculae carneae

17. *Optional activity:* **APR 9: Cardiovascular System**—Watch the "Heart" animation to gain a 3D fly-through perspective of the internal heart.

EXERCISE 21.5

THE CORONARY CIRCULATION

1. Obtain a preserved human heart or a classroom model of the heart.

2. The **coronary circulation** is the circulation to the heart wall itself. Adequate blood flow through the coronary arteries is absolutely essential for the functioning of the heart. If any coronary vessel becomes blocked due to disease or other processes, the area served by the vessel may become **ischemic,** meaning it lacks blood flow (*ischio*, to keep back, + *chymos*, juice). Prolonged ischemia to heart muscle leads to **hypoxia** (*hypo-*, too little, + *oxia*, oxygen). When cardiac muscle lacks an oxygen supply for more than a few minutes, the tissue dies, or becomes **necrotic** (*nekrosis*, death). The area of dead tissue is referred to as a **myocardial infarction** (*myocardial*, referring to the myocardium, + *in-farcio*, to stuff into), otherwise known as a "heart attack" or "MI." If a myocardial infarction results in vast tissue destruction, the heart may no longer be effective as a pump, which can lead to the death of the individual. If the individual survives the attack, the body will replace the damaged tissue with scar tissue. Scar tissue is mainly composed of a dense collection of collagen fibers. **Figure 21.10** demonstrates a human heart with evidence of a healed myocardial infarction. If you are observing a human heart, does your specimen show any evidence of (scarring from) past myocardial infarctions (scar tissue is generally clear to whitish in appearance and is much tougher than muscle tissue)?

    _____

3. The blood supply to the heart arises from two main coronary vessels, the **right and left coronary arteries** (**table 21.3**). The openings into these arteries arise immediately superior to the cusps of the aortic semilunar valve. **Figure 21.11** shows the relationship between the cusps of the aortic semilunar valve and the openings to the right and left coronary arteries. Observe the aortic semilunar valve and identify the openings into the right and left coronary arteries (figure 21.11). When the left ventricle contracts and pushes blood into the aorta, the cusps of the aortic semilunar valve fold over the openings to the coronary arteries. What consequence does this have in terms of blood flow to the heart during ventricular contraction (systole)?

_____
_____
_____

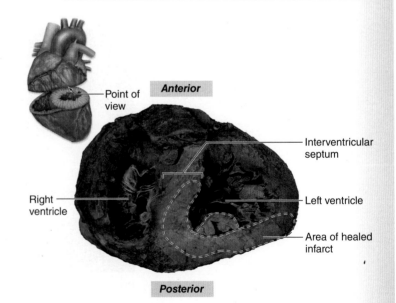

Figure 21.10 Inferior View of a Transverse Section through a Human Heart. Scar tissue, which is evidence of a previous myocardial infarction, can be seen in the interventricular septum and posterior wall of the left ventricle.

Table 21.3 Arterial Supply to the Heart

Vessel	Description	Areas Served	Word Origin
Anterior Interventricular Artery	Located in the anterior interventricular sulcus (groove). Physicians typically refer to this as the "LAD" (left anterior descending).	Anterior parts of the right and left ventricles	*inter*, between, + *ventricular*, referring to the ventricles of the heart (from *ventriculus*, belly)
Circumflex Artery	Located in the coronary sulcus between the left atrium and left ventricle	Left atrium and left ventricle (lateral part)	*circum*, around, + *flexus*, to bend
Left Coronary Artery	Located posterior to the pulmonary artery; branches into the anterior interventricular and circumflex arteries just after it emerges from behind the pulmonary artery	Anterior interventricular and circumflex arteries	*corona*, a crown
Marginal Artery	Branches off the right coronary artery at the right margin of the heart and is located on the lateral part of the right ventricle	Lateral part of the right ventricle	*margo*, border or edge
Posterior Interventricular Artery	Continuation of the right coronary artery located in the posterior interventricular sulcus	Posterior parts of the right and left ventricles	*inter*, between, + *ventricular*, referring to the ventricles of the heart (from *ventriculus*, belly)
Right Coronary Artery	Located in the coronary sulcus between the right atrium and the right ventricle. Branches include the SA nodal artery, which supplies blood to the sinoatrial node within the right atrium.	Right atrium and marginal and posterior interventricular arteries	*corona*, a crown

When the ventricles relax, the cusps of the semilunar valves fall shut as they fill with blood. Thus, they are no longer covering the openings to the coronary arteries. What consequence does this have in terms of blood flow to the heart during ventricular relaxation (diastole)?

_____
_____
_____

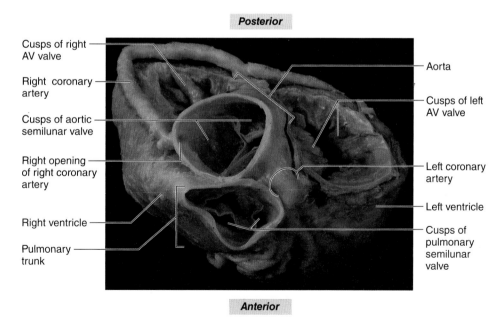

Figure 21.11 The Coronary Circulation. Superior view of the human heart with the atria removed. All four valves of the heart can be seen in this view. The openings to the right and left coronary arteries are located behind the cusps of the aortic semilunar valve. In this photo you can see where the right and left coronary arteries come off of the aorta, and you can see the cusps of both the pulmonary semilunar valve within the pulmonary trunk and the aortic semilunar valve within the aorta.

(continued on next page)

(continued from previous page)

Table 21.4	Venous Drainage of the Heart		
Vessel	**Description**	**Areas Drained**	**Word Origin**
Coronary Sinus	Located within the coronary sinus on the posterior surface of the heart. The coronary sinus is the largest vein of the heart, and its opening can be located within the right atrium of the heart just inferior to the fossa ovalis.	Entire heart; all veins of the heart drain into the coronary sinus, with the exception of a few small veins of the right ventricle, which drain directly into the right atrium	*corona*, a crown, + *sinus*, cavity
Great Cardiac Vein	Located in the anterior interventricular sulcus next to the anterior interventricular artery	Anterior parts of the right and left ventricles	*cardiacus*, heart, + *vena*, vein
Middle Cardiac Vein	Located in the posterior interventricular sulcus next to the posterior interventricular artery	Posterior parts of the right and left ventricles	*cardiacus*, heart, + *vena*, vein
Small Cardiac Vein	Located on the lateral part of the right ventricle, near the marginal artery	Lateral part of the right ventricle	*cardiacus*, heart, + *vena*, vein

4. *Cardiac Veins:* For the most part, venous drainage from the heart wall parallels the arterial supply. However, all venous blood draining the heart wall (with one exception; see table 21.4) eventually drains into one large vessel, the **coronary sinus.** The coronary sinus is located on the posterior surface of the heart within the coronary sulcus. What chamber does the coronary sinus empty into?

_____

5. Using **tables** 21.3 and **21.4** and your textbook as guides, identify the vessels shown in **figure 21.12** on the cadaver heart or on the classroom model of the heart. Then label them in figure 21.12.

6. *Optional Activity:* **APR 9: Cardiovascular System**— Study the "Heart" dissections to review the vasculature and features of the heart, then use the Quiz feature to test yourself on these structures.

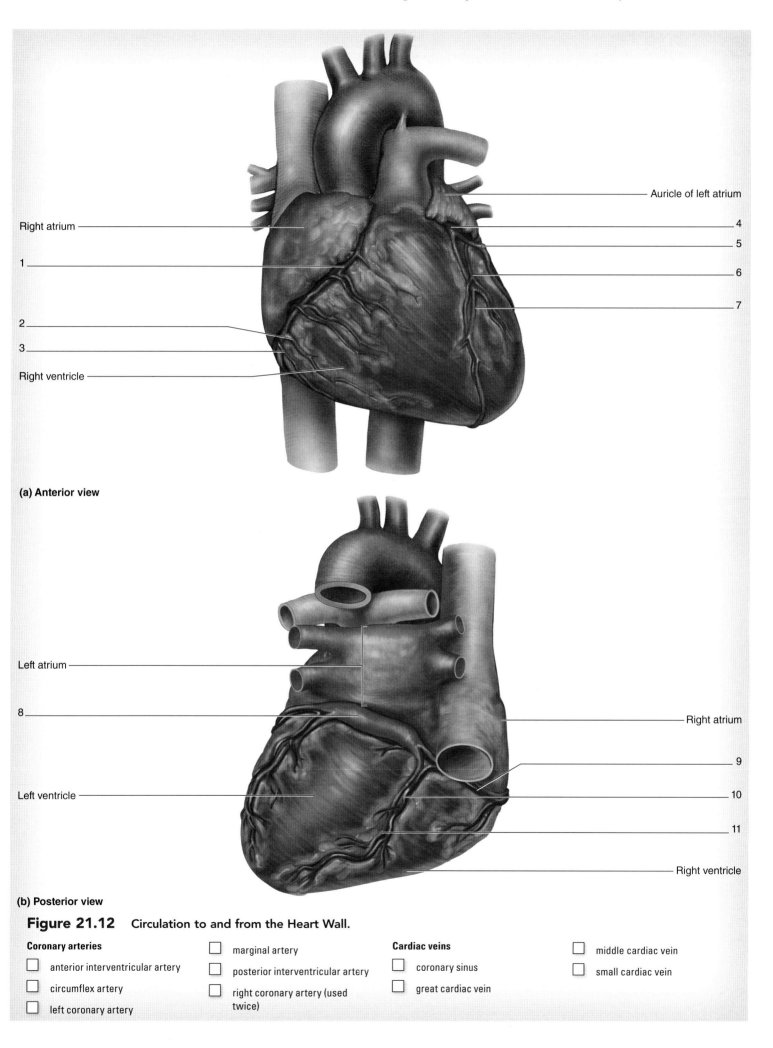

Figure 21.12 Circulation to and from the Heart Wall.

Coronary arteries
- [ ] anterior interventricular artery
- [ ] circumflex artery
- [ ] left coronary artery
- [ ] marginal artery
- [ ] posterior interventricular artery
- [ ] right coronary artery (used twice)

Cardiac veins
- [ ] coronary sinus
- [ ] great cardiac vein
- [ ] middle cardiac vein
- [ ] small cardiac vein

566 Chapter Twenty-One The Cardiovascular System: The Heart

EXERCISE 21.6

SUPERFICIAL STRUCTURES OF THE SHEEP HEART

1. Obtain a dissecting pan, dissecting instruments, gloves, and a preserved sheep heart. Rinse the heart with water to remove any dried blood or other debris, and place it in the dissecting pan to begin your superficial observations.

2. **Figure 21.13** demonstrates superficial structures of the sheep heart from both anterior and posterior views. Begin your observations by distinguishing the anterior surface from the posterior surface. One way to know you are observing the anterior surface of the heart is that you will see the fairly distinctive ruffled borders of both the right and left **auricles**. The auricles are extensions of the right and left **atria**.

3. Observe the surface of the heart closely to locate the **visceral pericardium** (epicardium). Then observe the outer surfaces of the great vessels of the heart to see if you can find any remnants of the **parietal pericardium** where it attached to these vessels. Note the large amount of fatty tissue deep to the epicardium. One of the functions of this fatty tissue is to help cushion the heart within the pericardial cavity.

4. Using figure 21.13 as a guide, identify the following superficial features on the sheep heart:

   ☐ anterior interventricular sulcus
   ☐ apex
   ☐ coronary sulcus
   ☐ left atrium
   ☐ left auricle
   ☐ left ventricle
   ☐ posterior interventricular sulcus
   ☐ right atrium
   ☐ right auricle
   ☐ right ventricle

5. Next you will identify the great vessels: the aorta, pulmonary trunk, and venae cavae. These vessels are often cut very close to their attachments to the heart, which can make identification difficult. To make the task easier, carefully remove as much of the epicardial fat as you can from the superior aspect of the heart (leave the fat in place on the ventricles for now).

6. Once you have cleaned away as much fat as possible, proceed with identification of the great vessels of the heart. Begin by viewing the anterior surface of the heart (figure 21.13a). The two most prominent vessels coming off the heart are the pulmonary trunk and the aorta. Both vessels have thick, tough walls,

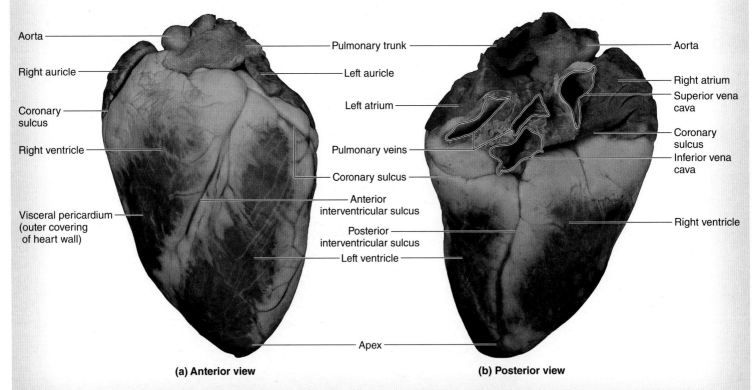

Figure 21.13 Superficial Structures of the Sheep Heart. (a) Anterior view. (b) Posterior view.

which helps with identification. The pulmonary trunk is located the most anteriorly and points to the right (from your point of view). The aorta is directly posterior to the pulmonary trunk and points to the left (from your point of view).

7. Turn the heart over to view the posterior surface (figure 21.13b). Here you can find the pulmonary veins, which will be more to the left (from your point of view), and the superior and inferior venae cavae, which will be more to the right (from your point of view). Before moving on, make sure you have identified all of the following vessels:

☐ aorta ☐ pulmonary veins
☐ inferior vena cava ☐ superior vena cava
☐ pulmonary trunk

8. To verify that you have identified the great vessels correctly, use a blunt probe or your fingers (or both) to see where each vessel comes from, or leads to, in the heart. Place the tip of the probe into the lumen of one of the vessels and see where it goes. If the vessel is large enough, put your index finger into the lumen of the vessel so you can *feel* where it goes. Now answer the following questions by giving the name of the heart chamber the probe will go into when placed into each vessel: A probe in the pulmonary trunk will lead into the _____; a probe in the aorta will lead into the _____; a probe in the pulmonary veins will lead into the _____; and a probe in the superior or inferior vena cava will lead into the _____.

In the next two exercises you will cut the entire sheep heart in half to compare wall thicknesses of the right and left ventricles. Approximately two-thirds of the dissection groups in your laboratory should make coronal sections of the heart, and the remaining third will make transverse sections. The different sections yield different views of the chambers and the structures within each chamber. All students should observe hearts that have been sectioned both ways. Ask your laboratory instructor which type of section your group should make before you begin.

EXERCISE 21.7

CORONAL SECTION OF THE SHEEP HEART

1. Obtain a scalpel or a knife with a 6-inch blade and a plastic ruler with millimeter increments. Either the scalpel or the knife will work for this next task, although the knife will make a cleaner cut. Turn the heart upside down so the base is on the dissecting pan, the apex is pointed toward you, and the anterior surface of the heart is facing your body. Make a coronal section through the entire heart to separate it into anterior and posterior portions (figure 21.14).

2. Once you have completed the cut, identify the following structures within the ventricles:

☐ chordae tendineae ☐ trabeculae carneae
☐ papillary muscles

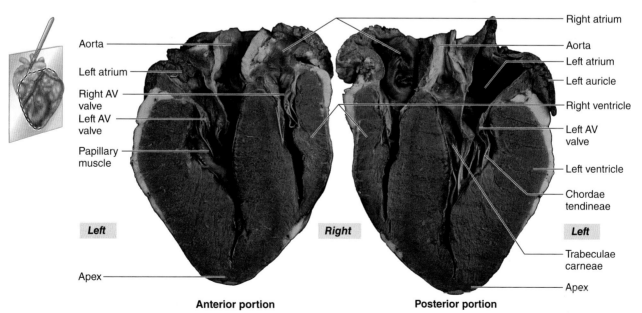

Figure 21.14 Coronal Section of the Sheep Heart.

(continued on next page)

568 Chapter Twenty-One *The Cardiovascular System: The Heart*

(continued from previous page)

3. Using a small ruler, measure the thickness of the right ventricular wall approximately 1 cm below the valve ring (where the right atrioventricular valve is located) and record it here: _____ cm. Next, measure the thickness of the left ventricular wall approximately 1 cm below the valve ring (where the left atrioventricular valve is located) and record it here: _____ cm. Approximately how much thicker is the wall of the left ventricle than the wall of the right ventricle? _____ What is the *functional* consequence of this difference? _____

_____

_____

4. Do the chambers of the right and left ventricles appear to differ in *volume* (that is, the amount of blood each could hold)? _____ Explain the consequences of having each chamber pump a different volume of blood. _____

_____

_____

_____

5. In the space below, sketch the coronal view of the ventricles. Be sure to include and label all the structures listed below in your drawing.

☐ aorta ☐ left AV valve
☐ apex ☐ papillary muscle
☐ chordae tendineae ☐ right atrium
☐ left atrium ☐ right AV valve
☐ left auricle ☐ trabeculae carneae

6. When you have finished your observations, discard the scalpel blade in the sharps container, clean your dissection instruments with soap and water and let them air dry, and put the sheep heart back into the container it came in—or dispose of it according to your instructor's directions.

EXERCISE 21.8

TRANSVERSE SECTION OF THE SHEEP HEART

1. Obtain an intact sheep heart, a scalpel or a knife, and a plastic ruler with millimeter increments. Either the scalpel or the knife will work for this next task, although the knife will make a cleaner cut. To make this cut, slice the heart transversely approximately 1 cm inferior to the coronary sulcus **(figure 21.15)** so that you will have separated the entire heart into superior and inferior portions. Observe the cut ends of the heart. Using a small ruler, measure the thickness of the right ventricular wall and record it here: _____ cm. Next, measure the thickness of the left ventricular wall and record it here: _____ cm. Approximately how much thicker is the wall of the left ventricle than the wall of the right ventricle? _____ What is the *functional* consequence of this difference?

_____

_____

_____

2. Do the chambers of the right and left ventricles appear to differ in *volume* (that is, the amount of blood each could hold)? _____

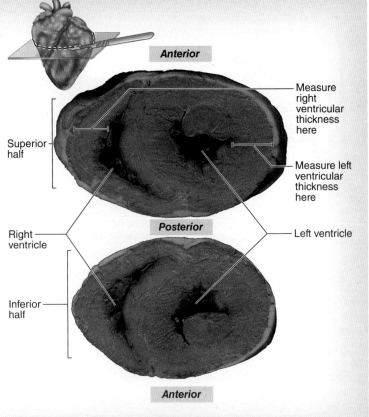

Figure 21.15 Transverse Section of the Sheep Heart.

Explain the consequences of having each chamber pump a different volume of blood.

_____

_____

_____

3. In the space below, sketch the transverse view of the ventricles. Be sure to include and label the right and left ventricles in your drawing:

4. When you have finished your observations, discard the scalpel blade in the sharps container, clean your dissection instruments with soap and water and let them air dry, and put the sheep heart back into the container it came in—or dispose of it according to your instructor's directions.

Physiology

Electrical Conduction within the Heart

Contraction of cardiac muscle results from a series of electrical potential changes (depolarization) that travel rapidly through the heart. As with skeletal muscle cells, depolarization of cardiac muscle cells is what stimulates contraction. Cardiac muscle is unique in that the heart beats *intrinsically*. That is, cardiac muscle will continue to contract even when the nerves supplying the heart have been severed. The **intrinsic conduction system** ensures that cardiac muscle within the heart depolarizes in a sequential manner (atria to ventricles) so the heart can beat as a coordinated unit. The components of the conduction system of the heart are outlined in **table 21.5.**

The sinoatrial (SA) node, the atrioventricular (AV) node, the atrioventricular (AV) bundle, the right and left bundles, and the Purkinje fibers are all capable of stimulating the heart to contract. However, because the SA node has the highest intrinsic rate, it controls the other parts of the conduction system. For this reason the SA node is called the heart's "pacemaker."

Stimulation of the heart to contract is initiated at the SA node. As these cells spontaneously depolarize, an action potential is initiated and then transmitted through the atria via gap junctions until it reaches the AV node. The action potential is delayed briefly at the AV node and then transmitted sequentially to the **AV bundle** (*bundle of His*), the **right** and **left bundle** branches, the Purkinje fibers, and then via gap junctions through the walls of the ventricles **(figure 21.16)**. A region of electrically isolated connective tissue separates the atria from the ventricles. Thus, the only mechanism for depolarization of the atria to reach the ventricles is via the AV node and AV bundle. Consequently, the atria are stimulated to contract prior to contraction of the ventricles. Note that it is depolarization that triggers the initiation of muscle contraction—first in the atria, and then in the ventricles.

Damage to any part of the conduction system of the heart may result in an irregular heart rate and other irregularities in cardiac function. In the event that the SA node is unable to set the pace of the heart, the AV node or even the Purkinje fibers may take over responsibility of setting the heart rate. While cells in other parts of the conduction system are able to depolarize spontaneously, their paces are much slower than the pace of the SA node; therefore, the heart rate decreases. Over time, if the AV node also becomes

Table 21.5	Components of the Conduction System of the Heart
Conduction System Component	**Location**
Sinoatrial (SA) Node	Within the right atrium, inferior to the superior vena cava opening
Atrioventricular (AV) Node	Base of the right atrium
AV Bundle (Bundle of His)	Interventricular septum
Right and Left Bundles	Interventricular septum
Purkinje Fibers	Ventricular walls

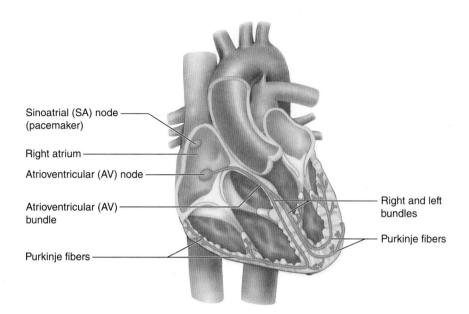

Figure 21.16 Conduction System of the Heart.

nonfunctional, the heart may be unable to deliver an adequate amount of blood to the body. An artificial pacemaker may be implanted to regulate a patient's heart rate.

The electrical changes associated with stimulation of the atria and ventricles can be detected on the body's surface using a diagnostic tool called an **electrocardiogram (ECG)**. In an ECG electrodes (called "leads") placed on the surface of the body in locations where it is easiest to pick up electrical signals from the heart. In a clinical setting, 12 leads are used to measure an ECG (thus it is called a "12-lead ECG"). In this laboratory exercise, we will use only three electrodes (left wrist, right wrist, and left ankle) to record the electrical activity associated with the heart. This arrangement roughly places the heart at the center of an equilateral triangle (*Einthoven's triangle*), where each side is the same length. The electrodes on each wrist and the left leg represent the vertices of the triangle. Voltages are measured between the electrodes to record the ECG. Lead I corresponds to the voltage difference between the left arm and the right arm; lead II corresponds to the voltage difference between the right arm and the left leg; lead III corresponds to the voltage difference between the left arm and the left leg **(figure 21.17)**. **Einthoven's Law** states that the sum of two leads results in the magnitude of the third lead. Therefore, the third lead can be calculated as the sum of two measured leads.

A normal ECG measures an overall change in voltage across the atria and ventricles and consists of three deflection waves. Each deflection wave corresponds to a distinct electrical event within the heart **(figure 21.18)**. The **P wave** corresponds to atrial depolarization, the **QRS complex** corresponds to ventricular depolarization, and the **T wave** corresponds to ventricular repolarization. Atrial repolarization does occur; however, it is masked by the large deflection caused by ventricular depolarization that occurs at approximately the same time. Figure 21.18 also depicts the P-Q segment, S-T segment, P-R interval, and Q-T interval. The P-Q segment corresponds to the atrial plateau at the sarcolemma when the cardiac muscle cells within the atria are contracting. The S-T segment corresponds to the ventricular plateau when the cardiac muscle cells within the ventricles are contracting. These segments can be measured directly and are physiologically relevant, as we will explore in the following exercises. The **P-R interval** is the time that it takes for the depolarization to spread from the SA node through the AV node, which typically ranges from 0.12 to 0.20 seconds. The **Q-T interval** is the time required for the ventricles to depolarize and repolarize, which typically ranges from 0.2 to 0.4 seconds. An ECG can be used to detect heart rates that are faster (*tachycardia*: > 100 beats/min) or slower (*bradychardia*: < 60 beats/min) than normal; the influence of drugs on cardiac function; the existence of ectopic pacemakers (ectopic = outside of the normal conduction system). Thus, the ECG is a valuable diagnostic tool.

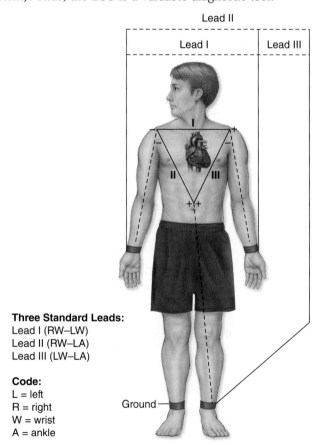

Figure 21.17 Standard Limb Leads for ECG.

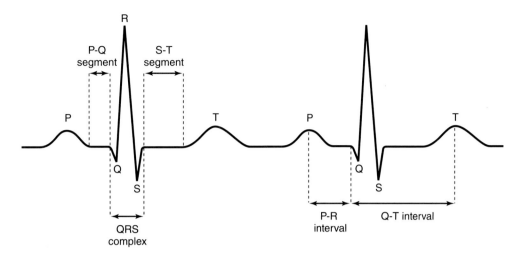

Figure 21.18 Normal Components of an ECG.

EXERCISE 21.9

Ph.I.L.S. LESSON 19: REFRACTORY PERIOD OF THE HEART

Cardiac muscle tissue must contract in a coordinated way to effectively pump the blood to the lungs and systemic tissues. Between each contraction, cardiac muscle must relax to allow time for the ventricles to fill with blood. Cardiac muscle, like skeletal muscle, has a period when it is insensitive to an electrical stimulus, known as the **refractory period.** The refractory period within cardiac muscle tissue is extended in comparison to skeletal muscle tissue. This extended refractory period results when cardiac muscle cells remain in the depolarized state because both voltage-gated K^+ channels and voltage-gated Ca^{2+} channels are open. This extended depolarized state is called the **plateau.** Because repolarization does not immediately follow depolarization, the mechanical events of cardiac muscle contraction and relaxation have time to occur prior to restimulation of cardiac muscle cells. (This prevents a sustained contraction [tetanus] from occuring.)

The purpose of this laboratory exercise is to measure the duration of the refractory period in cardiac muscle tissue. You will observe evoked contractions by delivering brief electrical shocks to exposed frog cardiac muscle tissue. Shocks will be delivered at various times during the cardiac cycle to observe when evoked contractions are and are not possible.

Before you begin, familiarize yourself with the following concepts (use your main textbook as a reference):

- Voltage-gated channels of cardiac cells
- Cardiac action potentials
- Refractory period

1. Open Ph.I.L.S Lesson 19: Refractory Period of the Heart (see **figure 21.19**).

2. To begin the experiment, click the "power" button of the data acquisition unit (DAQ).

3. Click and drag the blue plug to recording input 1 on the DAQ.

4. Click and drag the black plug to the stimulator output " + " position on the DAQ.

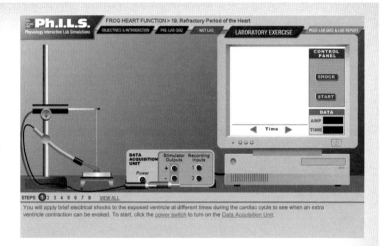

Figure 21.19 Opening Screen for the Laboratory Exercise on Refractory Period of the Heart.

5. Click the "start" button on the control panel of the virtual monitor. A red trace should appear, which represents ventricular contractions. The blue trace indicates when a shock has been applied to the tissue. Click the orange "UP" arrow on the clamp (left side of the screen) to increase the heart response on the screen.

6. Click the "shock" button to shock the heart approximately every two seconds. Click the "stop" button to cease shocking. To further increase the amplitude of the ventricular contractions, click the up arrow on the force transducer.

7. Once the recording has started, you will see a normal contractile response. Pay special attention after you shock the heart because changes in the red tracing will indicate the effect of the stimuli.

8. Make a note of any pertinent observations here:

_____

_____

_____

INTEGRATE

CLINICAL VIEW
ECG

Clinicians use ECGs to diagnose abnormalities with the conduction system of the heart. While this laboratory exercise involves using a three-lead system, a more sophisticated 12-lead system is typically used in a clinic or hospital setting. Trained specialists evaluate the ECG trace, looking for the key components described in this exercise. To illustrate the possible abnormalities that can be detected, a series of abnormal ECG traces are depicted below **(figure 21.20)**. The conditions illustrated include the following: atrial fibrillation (figure 21.20a), ventricular fibrillation (figure 21.20b), and premature ventricular contraction (figure 21.20c). **Fibrillation** involves chaotic depolarization of cardiac muscle fibers, which results in uncoordinated contraction of either the atria (atrial fibrillation) or the ventricles (ventricular fibrillation). A patient suffering from atrial fibrillation will exhibit many QRS complexes, but no clear P waves, because continuous and chaotic depolarization of the atria engages the conduction system of the heart. Observe the tracing in figure 21.20a. Can you see QRS complexes? (yes/no) Is each QRS complex followed by a T wave? (yes/no) Do all of the QRS complexes look the same? (yes/no) Look for regularly spaced P waves between the QRS complexes. Do you see them? (yes/no) In tracing a example, the P waves are very irregular, indicating **atrial fibrillation**. Given this information, why do you think the QRS complexes also look a little bit irregular in spacing and amplitude? _____

Next, observe the tracing in figure 21.20b. Can you see QRS complexes? (yes/no) Is each QRS complex followed by a T wave? (yes/no) Next, look for P waves. Do you see any? (yes/no)

Tracing b is a case of **ventricular fibrillation.** In ventricular fibrillation there are no clear P, QRS, or T waves due to aberrant electrical activity. While atrial fibrillation may lead to incomplete filling of the ventricles, ventricular fibrillation leads to inefficient ejection of blood into the systemic circulation. Without intervention, the latter results in certain death.

Observe the tracing in figure 21.20c. Can you see QRS complexes? (yes/no) Is each QRS complex followed by a T wave? (yes/no) Is each QRS complex preceded by a P wave? (yes/no)

Tracing c is a case of premature ventricular contraction (PVC). PVCs are characterized by QRS complexes that are not always preceded by a P wave. While not necessarily as detrimental as ventricular fibrillation, PVCs can also lead to a decreased cardiac output.

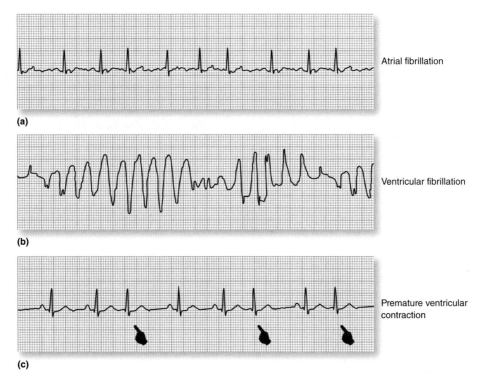

Figure 21.20 Abnormal ECG Traces. (a) Atrial fibrillation, (b) ventricular fibrillation, (c) premature ventricular contraction.

EXERCISE 21.10

Ph.I.L.S. LESSON 25: ELECTRICAL AXIS OF THE HEART

The purpose of this laboratory exercise is to measure the amplitude of the R wave of an ECG, using a virtual data acquisition unit (DAQ). Recording electrodes will be placed on the wrists of a virtual subject in lead I and III orientations. The magnitude of the QRS complex for each condition will indicate the relative distance between each electrode and the heart. This data will be used to calculate the electrical axis of the heart, which is the angle at which the heart lies in the thoracic cavity. An angle of zero would indicate the heart is situated vertically within the thoracic cavity with the atria located immediately superior to the ventricles.

Before you begin, familiarize yourself with the following concepts (use your textbook as a reference):

- Electrical activity in the heart and the ECG
- The deflection waves found in a typical ECG (P, QRS, and T)
- The function of noncontractile (i.e., connective) tissue in the heart

1. Open Ph.I.L.S. Lesson 25: Electrical Axis of the Heart.
2. Click the "power" switch on the data acquisition unit to begin the experiment.
3. Connect the ECG recording leads to the DAQ by clicking and dragging the black plug recording input 1.
4. Click and drag the red recording electrode to the left wrist of the virtual subject.
5. Click and drag the black recording electrode to the right wrist of the virtual subject.
6. Click and drag the green recording electrode to the left ankle of the virtual subject. The subject is connected and you are now ready to record an ECG.
7. Click "start" to begin the recording. You should now see the ECG being recorded in the virtual window of the control panel. Observe this for a few cycles, then click "stop."
8. Click the arrows at the bottom of the control panel screen to scroll through the record. Center one QRS complex in the middle of the screen.
9. Move the mouse pointer over the peak of the R wave and click to drop a marker at the tallest point of the peak.
10. Move the mouse pointer to the isoelectric line (baseline) between the previous cycle's T wave and the current cycle's P wave. Click the mouse to drop a second marker. The measurements appear in the data window. Click the red "journal" to enter the data in the journal, then enter it in **table 21.6** (this is lead I amplitude).
11. Switch the location of the black and green electrodes. To remove them, simply click the green electrode, then the black electrode. Now, click and drag the green electrode to the right wrist and the black electrode to the left ankle.
12. Repeat step 10 to record the amplitude and enter it in table 21.6 (this is lead III amplitude).
13. With the journal open, click on the orange "Program" button on the lower right corner of the journal and follow the instructions for calculating the axis (angle) of the heart. Using a ruler, draw a perpendicular line intersecting the lead I axis (i.e., at a right angle to the red line) at the voltage you recorded for lead II in table 21.6.
14. Using the ruler, draw a second perpendicular line intersecting the lead III axis (i.e., at a right angle to the blue line) at the voltage you recorded for lead III in table 21.6.

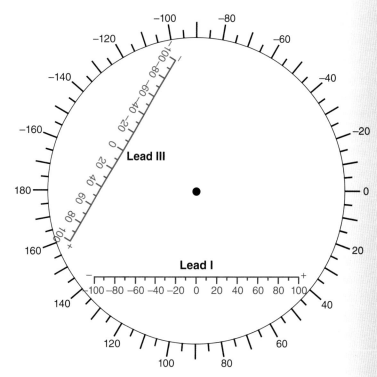

15. To calculate the angle of the heart, use the ruler to draw a line connecting the black dot in the center of the circle to the point of intersection of the two perpendicular lines you drew in steps 13 and 14. Record the electrical axis (angle) of the heart here: _____
16. Make note of any pertinent observations here:

_____

_____

Table 21.6	Limb Lead Orientation Data Table	
	Lead I	Lead III
Amplitude of R wave (mV)		

EXERCISE 21.11

ELECTROCARDIOGRAPHY USING STANDARD ECG APPARATUS

Obtain the Following:
- alcohol swabs
- electrodes
- electrode gel
- lead selector switch
- electrocardiograph (or alternative ECG recording device)

1. Choose a person to be the subject. Clean the skin on the anterior surface of the subject's forearms approximately 2–3 inches proximal to the left and right wrists, and on the skin approximately 2–3 inches superior to the medial malleolus on the left and right legs, by rubbing vigorously with an alcohol swab.

2. Apply a small quantity of electrode gel to the center of each electrode. Place an electrode firmly on the skin in each of the following locations: right arm, left arm, and left leg. Note that skin that is free of hair will make the best contact with the electrode. The right leg will serve as the ground.

3. Attach the leads of the selector switch to the electrodes.

4. Turn on the electrocardiograph and set the paper speed to 2.5 cm/second, the standard for recording an ECG.

5. Have the subject either lie supine or sit in a comfortable position.

6. Turn on the power to the ECG recording device and set the gain of the system to 1.

7. Turn the lead switch to lead I. Record the ECG for 1 minute. Label the paper "Lead I."

8. Repeat step 7 for leads II and III.

9. Remove the electrodes and clean the skin with alcohol swabs.

10. Assuming that each mm square on the recording paper corresponds to 0.04 seconds, measure from one QRS peak to the next QRS peak. Multiply this distance by 0.04 sec/mm to calculate the heart rate.

11. Measure important features of the ECG, including the durations of the QRS complex, Q-T interval and P-T interval. Record these values in **table 21.7**.

12. Have the subject exercise vigorously for at least 3 minutes. Have the patient return to the seated or supine position to record the ECG for 3 minutes using lead I.

13. Repeat steps 10 and 11 for the subject post-exercise.

14. Make note of any pertinent observations here:

_____

_____

_____

Table 21.7	ECG Component Data Table	
ECG Component	**Duration at Rest (sec)**	**Duration Post-Exercise (sec)**
QRS Complex		
Q-T Interval		
P-T Interval		

EXERCISE 21.12

Ph.I.L.S. LESSON 26: ECG AND HEART BLOCK

The purpose of this laboratory exercise is to record one normal ECG and four abnormal ECGs in subjects with varying degrees of AV block (or "heart block"). You will then explore the underlying physiological mechanisms that are responsible in each case of heart block.

Before you begin, familiarize yourself with the following concepts (use your main textbook as a reference):

- Electrical activity of the heart
- ECG
- Conduction block

1. Open Ph.I.L.S. Lesson 26: ECG and Heart Block (**figure 21.21**).

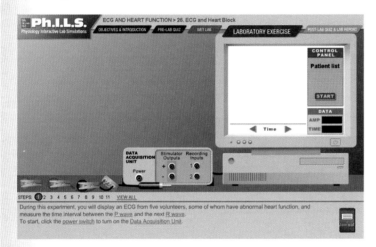

Figure 21.21 Opening Screen for the Laboratory Exercise on ECG and Heart Block.

2. Click the "power" button on the DAQ.
3. Connect the ECG electrodes to the DAQ by clicking and dragging the black plug to recording input 1.
4. Select a volunteer by moving the mouse over the patient list on the control panel. Click on one of the names.
5. Connect the ECG leads to the patient by clicking and dragging the red electrode to the left wrist, the black electrode to the right wrist, and the green electrode to the left ankle.
6. To record and measure the ECG, click the "start" button on the control panel.
7. Record at least eight complete cardiac cycles and then click the "stop" button on the control panel.
8. Click the "arrows" at the bottom of the control panel screen to scroll through the record. Center one QRS complex in the middle of the screen. Measure the P–R interval by moving the mouse to the beginning of a P wave. Click the mouse to drop a marker. Move the mouse to the peak of the R wave in the same cycle and click the mouse to drop a marker. The time interval between the P wave and the R wave is now displayed in the data window on the control panel. Enter the data in the journal by clicking the red "journal" button in the lower-right corner of the screen. Then enter the data in table 21.8 (next page).

(continued on next page)

(continued from previous page)

9. Repeat steps 5–8 for the four remaining patients and enter the data in **table 21.8**. Watch for abnormalities in the regular pattern. Measure each P-R interval and record them in table 21.8.

10. Make note of any pertinent observations here:

_____

_____

_____

Table 21.8	P-R Interval Data Table
Patient	**P-R Interval (normal P-R Interval = 0.12–20 seconds)**
Andrew	1. 2. 3. 4. 5. 6. 7. 8.
Brianna	1. 2. 3. 4. 5. 6. 7. 8.
Charlie	1. 2. 3. 4. 5. 6. 7. 8.
David	1. 2. 3. 4. 5. 6. 7. 8.
Emily	1. 2. 3. 4. 5. 6. 7. 8.

EXERCISE 21.13

Ph.I.L.S. LESSON 27: ABNORMAL ECGs

The purpose of this lab is to record normal (1) and abnormal (3) ECG traces in four virtual subjects. You will determine the physiological mechanism responsible for creating the ECG recording for each subject.

Before you begin, familiarize yourself with the following concepts (use your main textbook as a reference):

- Normal ECG recordings
- Causes of abnormal ECG recordings
- Heart block

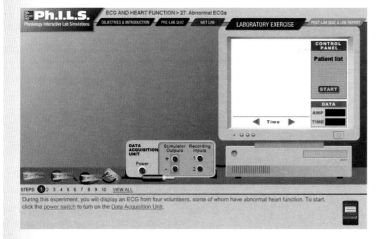

Figure 21.22 Opening Screen for the Laboratory Exercise on Abnormal ECGs.

State a hypothesis regarding the effect that damaged tissue has on the conduction system of the heart and the resulting contraction of cardiac muscle tissue. _____
_____
_____
_____

1. Open Ph.I.L.S. Lesson 27: Abnormal ECGs (**figure 21.22**).
2. Click the "power" button on the data acquisition unit (DAQ).
3. Click and drag the black plug to recording input 1.
4. Select a subject to record an ECG from by moving the pointer to the patient list on the control panel. For the first patient, select "Amy" as the subject.
5. Connect the ECG leads to the subject by dragging the red electrode to her left wrist, the black electrode to her right wrist, and the green electrode to her left ankle.
6. Click the "start" button on the control panel to begin recording the subject's ECG. Record a minimum of eight complete cycles.
7. Record observations for the first subject (Amy) in **table 21.9**.
8. Repeat steps 4 through 7 for each remaining subject.
9. Make note of any pertinent observations here:

_____
_____
_____

Table 21.9	ECG Recordings for Four Virtual Subjects	
Patient	**ECG Observations**	**ECG Pattern (sketch the pattern in the spaces below)**
Amy	P	
	QRS	
	T	
Brian	P	
	QRS	
	T	
Chris	P	
	QRS	
	T	
Deb	P	
	QRS	
	T	

Cardiac Cycle and Heart Sounds

The **cardiac cycle** refers to the events that occur in one heart beat. One complete cardiac cycle occurs when both atria and ventricles contract and then relax. The two main phases of the cardiac cycle are **diastole**, or relaxation, and **systole**, or contraction. These phases may be further subdivided as follows: (1) diastole = isovolumetric ventricular relaxation + ventricular filling, and (2) systole = isovolumetric ventricular contraction + ejection. A typical description of the cardiac cycle includes both atrial and ventricular events that occur on *the left side of the heart* only because the pressures on that side of the heart are much greater than those on the right. The volume of blood in the ventricles at the end of filling is the **end diastolic volume (EDV),** and the volume of blood remaining in the ventricles following contraction is the **end systolic volume (ESV).** The volume of blood ejected with each beat is the **stroke volume** (EDV − ESV). **Cardiac output** (CO) is simply the stroke volume (SV) multiplied by the heart rate (HR).

Blood is transported through the heart along pressure gradients, from areas of higher pressure to areas of lower pressure. **Atrioventricular (AV)** and **semilunar valves** ensure one-way flow of blood through the heart. Once pressure in the atria exceeds the pressure in the ventricles, the AV valves open and blood flows into the ventricles. Contraction of the right and left atria pushes any remaining blood into the ventricles. At this time the ventricles are relaxed and fill with blood. During ventricular systole, the pressure rises in the ventricles. When pressure in the ventricles exceeds pressure in the atria, the AV valves are forced closed. Once pressure in the ventricles exceeds pressure in the pulmonary trunk and aorta, blood pushes open the cusps of the pulmonary and aortic semilunar valves and blood is ejected from the heart into the pulmonary and systemic circuits.

The duration of a cardiac cycle varies among individuals, and also varies throughout one's lifetime. Typically, a cardiac cycle lasts 0.7–0.8 seconds, which corresponds to an average heart rate of 75–85 beats per minute. Heart sounds (*lubb dupp*) can be associated with the cardiac cycle. Turbulent blood flow that occurs as the AV valves close creates the first heart sound (*lubb*), whereas turbulent blood flow that occurs as the semilunar valves close creates the second heart sound (*dupp*). **Figure 21.23** shows the location of each of the heart valves with respect to thoracic wall surface anatomy. The yellow dots indicate the areas where each heart sound is best heard at the surface of the body. The semilunar valves are best heard in the second intercostal spaces on each side of the sternum, whereas the AV valves are best heard in the fifth intercostal spaces. Listen for the right AV valve in the fifth intercostal space, just to the right of the sternum. Listen to the left AV valve in the fifth intercostal space at the midclavicular line (see figure 12.9, p. 309). Auscultation is listening to the internal body sounds, such as the heart sounds.

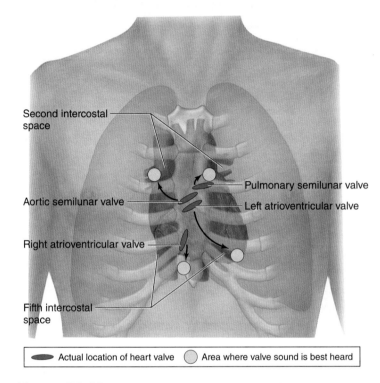

Figure 21.23 Optimal Listening Sites for Heart Sounds.

EXERCISE 21.14 Ph.I.L.S.

Ph.I.L.S. LESSON 23: THE MEANING OF HEART SOUNDS

In this laboratory exercise, a virtual data acquisition unit (DAQ) will be used to record an ECG while listening to heart sounds in a virtual subject. The mechanical events of the heart, which result in audible sounds, will be correlated with electrical events as viewed by an ECG trace.

1. Open Ph.I.L.S. Lesson 23: ECG and Heart Function: The Meaning of Heart Sounds.

2. Read the objectives and introduction, and take the pre-laboratory quiz. After completing the pre-laboratory quiz, read through the wet laboratory. View the videos that are indicated in red in the wet laboratory before beginning the experiment.

3. The laboratory exercise will open when you click "continue" after completing the wet laboratory (figure 21.24).

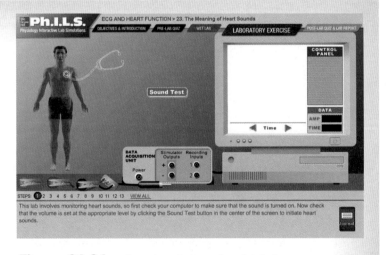

Figure 21.24 Opening Screen for the Laboratory Exercise on ECG and Heart Function: The Meaning of Heart Sounds.

4. Check that you can the hear heart sounds generated by Ph.I.L.S. on your computer by clicking the "sound test" button in the middle of the screen.

5. Click on the "practice" button in the middle of the screen to practice keeping time with the heart sounds. Once you hear the heart sounds, click on the red button to make a mark on the blue trace. Push the button when you hear the first heart sound (*lubb*) and release the button when you hear the second heart sound (*dupp*). Click "end test" once complete.

6. Click the "power" switch to turn on the DAQ.

7. To begin the recording, connect the electrodes to the DAQ by inserting the black plug into input 1 of the DAQ and attach electrodes to the subject (red electrode to the left wrist; black electrode to the right wrist; and green electrode to the left ankle). The electrodes detect the electrical activity of the heart that is conducted through the body and transmit that information to the DAQ.

8. Click "start" in the control panel. On the control panel the ECG is recorded (red line) above the heart sounds (blue line).

9. As the line tracing scrolls across the control panel screen, click the mouse on the red recording button to mark the heart sounds. Record four to six cardiac cycles. Click "stop" in the control panel.

10. Use the arrows at the bottom of the monitor screen to scroll through the trace. Determine which heart sounds correspond to the QRS complex and the T-wave. You may now print the graph of the ECG and heart sounds by clicking the "P" in the bottom left of the monitor screen.

11. Complete the post-laboratory quiz by clicking "open post-laboratory quiz and laboratory report."

12. Make a note of any pertinent observations here:

_____
_____
_____

EXERCISE 21.15

AUSCULTATION OF HEART SOUNDS

1. Obtain alcohol swabs and a stethoscope.

2. Clean the earpieces and diaphragm of the stethoscope with the alcohol swabs and allow them to dry.

3. Place the earpieces of the stethoscope into your ears (external auditory canal), then press the diaphragm of the stethoscope to your partner's skin directly over the fifth intercostal space at the midclavicular line to the left of the sternum (see figure 21.23). This space overlies the apex of the heart and is the optimal placement for auscultation of the first heart sound. Listen for a *lubb* sound.

4. Press the diaphragm to your partner's skin directly over the second intercostal space to the left of the sternum. (see figure 21.23). This is the optimal placement for auscultation of the second heart sound. Listen for a *dupp* sound.

5. Inhale and exhale slowly and deeply while you hold the diaphragm in each location. Listen for any differences in the heart sounds that occur as the subject breathes.

6. Make a note of any pertinent observations here:

_____
_____
_____

EXERCISE 21.16 Ph.I.L.S.

Ph.I.L.S. LESSON 20: STARLING'S LAW OF THE HEART

The sarcomere is the fundamental contractile unit in both skeletal muscle and in cardiac muscle. However, the optimal overlap of thin and thick filaments is different for the two types of muscle cells. In skeletal muscle maximum overlap occurs when the muscle is at rest. In comparison, maximum overlap of thin and thick filaments in cardiac muscle occurs when cardiac muscle is stretched, which is dependent on the amount of blood entering the chamber. The greater the amount of blood entering the heart chamber, the greater the stretch of the heart wall, and the greater the amount of overlap of thin and thick filaments. This provides a means of forming additional crossbridges between the thin and thick filaments and allowing cardiac muscle to contract with greater force as additional blood enters the chamber. This relationship between the degree of stretch of the heart wall and force of contraction is called **Starling's Law of the Heart.**

In this experiment, you will measure length-tension relationships in cardiac muscle tissue. Using a force transducer, you will measure force output of frog muscle tissue while administering a stretch.

Before you begin, familiarize yourself with the following concepts (use your main textbook as a reference):

- Actin and myosin
- Sliding filament theory of muscle contraction
- Length-tension relationship

State a hypothesis regarding the effect of increased sarcomere length on force output in cardiac muscle tissue. _____
_____
_____
_____

(continued on next page)

(continued from previous page)

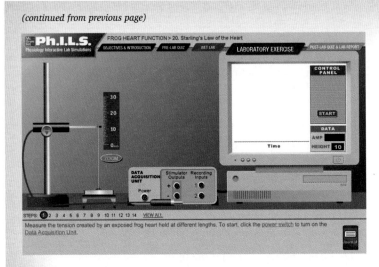

Figure 21.25 Opening Screen for the Laboratory Exercise on Starling's Law of the Heart.

1. Open Ph.I.L.S Lesson 20: Starling's Law of the Heart (**figure 21.25**).
2. To begin the experiment, click the "power" switch on the data acquisition unit (DAQ).
3. Click and drag the blue plug to recording input 1.
4. Click the "start" button on the control panel of the virtual monitor. A red trace will appear that is representative of cardiac contractions.
5. Click the "zoom" button located just below the ruler on the screen to enlarge the force transducer.

Table 21.10	Length-Tension Data
Millimeters of Passive Stretch (height of transducer)	Height of Peak (amplitude)
10	
11	
12	
13	
14	
15	
16	
17	

6. The transducer is set at 10 mm. Record a minimum of five complete cardiac cycles at this height.
7. Click the "stop" button on the virtual control panel.
8. To measure the amplitude of the contraction, move the mouse over the tracing to the top of one peak and click the mouse button.
9. Now move the mouse to the lowest point following the peak you first measured (to the next trough), and right click. Data should appear in the data window of the virtual control panel.
10. Click the red "journal" button to enter the data in the journal, and enter it in **table 21.10**. Close the journal window when you are finished.
11. Click the up arrow on the force transducer and raise it one millimeter.
12. Repeat steps 4–11 until you have stretched the heart to 17 mm. Be sure to enter the data in table 21.10, and graph your data below.

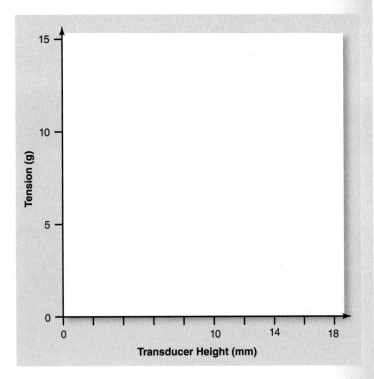

13. Make note of any pertinent observations here:

_____
_____
_____

CHAPTER 22

The Cardiovascular System: Vessels and Circulation

OUTLINE AND LEARNING OBJECTIVES

Histology 592

Blood Vessel Wall Structure 592

EXERCISE 22.1: BLOOD VESSEL WALL STRUCTURE 593

1 Name the three layers (tunics) present in the walls of all blood vessels (except capillaries)

2 Describe the basic structure of each layer of a blood vessel wall

Elastic Arteries 594

EXERCISE 22.2: ELASTIC ARTERY—THE AORTA 594

3 Describe the unique features of the tunica media of an elastic artery

4 Explain how elastic fibers in the wall of an elastic artery facilitate blood flow through the vessel

5 Define vasa vasorum

Muscular Arteries 594

EXERCISE 22.3: MUSCULAR ARTERY 595

6 Identify the internal and external elastic laminae and smooth muscle in the tunica media of a muscular artery

7 Compare and contrast elastic and muscular arteries based on structure, location, and function

Arterioles 595

EXERCISE 22.4: ARTERIOLE 595

8 Describe the role arterioles play in the regulation of blood flow

Veins 596

EXERCISE 22.5: VEIN 596

9 Identify veins through the microscope, and explain the characteristics you would use to determine that a blood vessel is a vein

10 Describe the structure and function of venous valves

Capillaries 597

EXERCISE 22.6: OBSERVING ELECTRON MICROGRAPHS OF CAPILLARIES 597

11 Describe how the endothelium differs structurally in the three types of capillaries

12 Give examples of locations in the body where each type of capillary is found

13 Describe how the three types of capillaries differ in function, and explain how the function of each type of capillary is facilitated by the capillary's structure

Gross Anatomy 600

Pulmonary Circuit 600

EXERCISE 22.7: PULMONARY CIRCUIT 600

14 Trace the flow of blood from the right ventricle of the heart through the pulmonary circulation and back to the left atrium of the heart

Systemic Circuit 601

EXERCISE 22.8: CIRCULATION TO THE HEAD AND NECK 601

15 Identify the major arteries and veins that compose the circulation to the head and neck

16 Trace the flow of blood from the left ventricle of the heart to the skin overlying the right parietal bone of the skull and back to the right atrium of the heart

EXERCISE 22.9: CIRCULATION TO THE BRAIN 603

17 Identify the major arteries and veins that compose the circulation to the brain

18 Trace the flow of blood from the left ventricle of the heart to the right parietal lobe of the brain and back to the right atrium of the heart

EXERCISE 22.10: CIRCULATION TO THE THORACIC AND ABDOMINAL WALLS 606

19 Identify the major arteries and veins that compose the circulation to the thoracic and abdominal walls

20 Trace the flow of blood from the left ventricle of the heart to the right kidney and back to the right atrium of the heart

EXERCISE 22.11: CIRCULATION TO THE ABDOMINAL CAVITY 609

21 Identify the major arteries and veins that compose the circulation to the abdominal cavity

22 Name the three major veins that compose the hepatic portal circulation

23 Describe the functional significance of the hepatic portal circulation

24 Trace the flow of blood from the left ventricle of the heart to the spleen and back to the right atrium of the heart

25 Trace the flow of blood from the left ventricle of the heart to the duodenum and back to the right atrium of the heart

26 Trace the flow of blood from the left ventricle of the heart to the sigmoid colon and back to the right atrium of the heart

EXERCISE 22.12: CIRCULATION TO THE UPPER LIMB 614

27 Identify the major arteries and veins that compose the circulation to the upper limb

28 Trace the flow of blood from the left ventricle of the heart to the anterior surface of the index finger and back along a superficial route to the right atrium of the heart

 **MODULE 9: CARDIOVASCULAR SYSTEM**

- (29) *Trace the flow of blood from the left ventricle of the heart to the capitate bone in the wrist and back along a deep route to the right atrium of the heart*

EXERCISE 22.13: CIRCULATION TO THE LOWER LIMB 618
- (30) *Identify the major arteries and veins that compose the circulation to the lower limb*
- (31) *Trace the flow of blood from the left ventricle of the heart to the dorsal surface of the big toe (hallux) and back along a superficial route to the right atrium of the heart*
- (32) *Trace the flow of blood from the left ventricle of the heart to the cuboid bone in the foot and back along a deep route to the right atrium of the heart*

Fetal Circulation 623

EXERCISE 22.14: FETAL CIRCULATION 623
- (33) *Identify the cardiovascular structures unique to the fetal circulation and describe the function of each*
- (34) *Trace the flow of blood from the left ventricle of the heart to the placenta and back to the right atrium of the heart in the fetus*
- (35) *Identify postnatal structures that are remnants of the fetal circulation*

Physiology 625

Blood Pressure and Pulse 625

EXERCISE 22.15: Ph.I.L.S. LESSON 24: ECG AND FINGER PULSE 626
- (36) *Describe pulsatile blood flow through vessels*
- (37) *Discuss the relationship between ECG and the cardiac cycle and pulse rate*

EXERCISE 22.16: BLOOD PRESSURE AND PULSE USING A STANDARD BLOOD PRESSURE CUFF 627
- (38) *Discuss how Korotkoff sounds and a sphygmomanometer are used to determine blood pressure*
- (39) *Measure pulse rate by palpating the radial artery*
- (40) *Record systolic and diastolic pressures using the auscultatory method*
- (41) *Discuss the physiological relevance of an elevated blood pressure and pulse*

EXERCISE 22.17: BIOPAC LESSON 16: BLOOD PRESSURE 628
- (42) *Measure systolic and diastolic pressures, and calculate pulse and mean arterial pressures*
- (43) *Describe changes in blood pressure associated with exercise and changes in body position*

INTRODUCTION

While proper functioning of the "heart" is necessary for adequate blood flow, our vast system of blood vessels provides necessary conduits (channels) for the blood to flow to the body's tissues. **Arteries** are blood vessels that carry blood *away* from the heart, whereas **veins** are blood vessels that carry blood *toward* the heart. Arteries carry blood toward the tissues. Arteries branch into smaller and smaller vessels, ultimately forming **arterioles** (*arteriole*, a small artery). Arterioles have the special function of controlling the flow of blood into **capillary beds.** Capillaries are the site of exchange for substances (e.g., gases, nutrients, wastes) between the blood and the tissues.

Blood flows out of capillaries into small veins called **venules,** which merge to form the larger veins that return blood to the heart. **Veins** transport blood back to the heart and serve as blood reservoirs (reservoir, a receptacle). Generally, a vein is positioned alongside each major artery and has the same name as the artery it accompanies. However, there are typically more veins draining a structure as there are arteries supplying it. In the limbs, most of the veins that do not accompany an artery are superficial veins, located just under the skin. For example, the brachium (arm) is supplied by the brachial artery, which has the brachial vein traveling next to it. The brachial artery and vein are located fairly deep within the arm, where they are protected by the musculature of the arm. In addition to the brachial vein, two superficial veins also drain blood from the arm. These are the cephalic and basilic veins. Please note that there is considerably more variation among individuals in the branching patterns and locations of veins than there is with arteries. Such variation often has clinical significance. For instance, when blood samples need to be collected from a patient, blood is commonly drawn from the median cubital vein. However, not all individuals have a median cubital vein. Lastly, when you are tracing blood flow through the venous system, remember that veins *drain* blood from an area of the body. This means that in your descriptions of blood flow through veins you should start by naming the most distal veins first, and then name the veins blood travels through as it proceeds toward the heart.

You will begin this laboratory session by investigating the histological characteristics of the different types of blood vessels (arteries, arterioles, capillaries, and veins). You will then identify the major arteries and veins of the body by locating blood vessels on a human cadaver or on classroom models of the cardiovascular system. Once you have completed the gross anatomy exercises in this chapter, you should be able to describe the pathway a drop of blood takes as it is transported from the heart to a target organ and back to the heart once again. Finally, you will explore the concept of **blood pressure,** which is the pressure that blood exerts against a blood vessel wall. You will investigate the relationships among blood pressure, blood flow, and blood vessel resistance.

Chapter 22: The Cardiovascular System: Vessels and Circulation

Name: _____
Date: _____ Section: _____

PRE-LABORATORY WORKSHEET

Also available at www.connect.mcgraw-hill.com

1. List the three layers present in all blood vessel walls (except capillaries), starting with the outermost layer.

    a. _____

    b. _____

    c. _____

2. The wall of a capillary consists of only a tunica intima. The tunica intima of a capillary (and all vessels, for that matter) is composed of _____ and a _____.

3. The three types of capillaries are (from least permeable to most permeable)

    a. _____

    b. _____

    c. _____

4. What anatomical structure is unique to veins, and prevents the backflow of blood?

    _____

5. The three arteries that branch directly off of the aortic arch are

    a. _____

    b. _____

    c. _____

6. The hepatic portal system is a system of veins that carry venous blood to the _____.

7. Which of the following is *not* a component of the hepatic portal system (circle one)?

    a. hepatic portal vein

    b. superior mesenteric vein

    c. hepatic veins

    d. inferior mesenteric vein

    e. splenic vein

8. Blood leaving the left ventricle of the heart enters the _____ circuit, while blood leaving the right ventricle of the heart enters the _____ circuit.

9. The major vein draining blood from the inferior half of the body, which empties into the right atrium of the heart, is the _____.

Chapter Twenty-Two *The Cardiovascular System: Vessels and Circulation* 591

Histology

Blood Vessel Wall Structure

All blood vessels except capillaries have three tunics (layers) forming their walls. These layers, called the tunica intima, tunica media, and tunica externa, are analogous in both structure and function to the three layers of the heart wall (endocardium, myocardium, and epicardium). The differences in structure and function between the different types of blood vessels come mainly from modifications of these three wall layers. In particular, the type of tissue that composes the tunica media greatly affects the function of the vessel. **Table 22.1** describes the general composition of the three layers of a blood vessel wall, and **table 22.2** summarizes unique features in the different types of blood vessels.

Table 22.1 Layers of a Blood Vessel Wall

Wall Layer	Location	Components	Word Origin
Tunica Intima	Innermost layer; in contact with the lumen of the vessel	Endothelium (simple squamous epithelium) and a subendothelial layer composed of areolar connective tissue	*tunic*, a coat, + *intimus*, innermost
Tunica Media	Middle layer	Varied amounts of collagen fibers, elastic fibers, and smooth muscle cells	*tunic*, a coat, + *medius*, middle
Tunica Externa	Outermost layer	Areolar connective tissue that anchors the vessel to surrounding structures	*tunic*, a coat, + *externus*, on the outside

Table 22.2 Characteristics of Wall Layers in Specific Types of Blood Vessels

Type of Vessel	Tunica Intima	Tunica Media	Tunica Externa	Diameter	Characteristics and Special Functions
Elastic Artery	Endothelium and subendothelial layer; an internal elastic lamina is present but not easily distinguished from the elastic tissue of the tunica media	Contains numerous elastic and reticular fibers; Also contains smooth muscle cells	Underdeveloped in contrast to other vessels. Contains vasa vasorum, lymphatics, and nerves.	2.5 cm–1 cm	Expansion and contraction of elastic tissues smooths out the flow of blood
Muscular Artery	Endothelium and subendothelial layer; contains a very prominent internal elastic lamina	Multiple layers of smooth muscle; numerous elastic fibers	Contains vasa vasorum, lymphatics, and nerves	1 cm–3 mm	Recoil of wall continues to propel blood through the arteries
Arteriole	Endothelium and subendothelial layer; an internal elastic lamina is present only in the largest arterioles	Contains less than six layers of smooth muscle, with no external elastic lamina	Very thin	3 mm–10 µm	Size of lumen is regulated to control the flow of blood into capillary beds
Capillary	Endothelium and a basement membrane only	NA	NA	8–10 µm	Thin wall allows for exchange between the blood and tissues
Postcapillary Venule	Endothelium and a thin subendothelial layer	Very thin with no smooth muscle cells	Very thin	10–50 µm	Drains blood from capillary beds. Site where leukocytes leave the circulation and enter the tissues via diapedesis*
Venule	Endothelium and a thin subendothelial layer	Very thin with very few smooth muscle cells	Thickest layer of the wall	50–100 µm	Venules are simply small veins, and are the counterpart to arterioles
Vein	Endothelium and subendothelial layer; infoldings form valves, which prevent the backflow of blood. Not all veins have valves.	Very thin with a small amount of smooth muscle	Thickest layer of the wall. Contains vasa vasorum.	Greater than 100 µm	Low pressure conduits; valves aid in preventing backflow of blood

*diapedesis (*dia-*, through, + *pedesis*, a leaping): the passage of leukocytes through the walls of blood vessels

EXERCISE 22.1

BLOOD VESSEL WALL STRUCTURE

1. Obtain a slide showing an artery and a vein (they may both be on the same slide, or you may have two different slides).

2. Place the slide on the microscope stage and bring the tissue sample into focus on low power. Scan the slide and look for the circular or oval cross section of a vessel. If you see more than one vessel on the slide, you will need to determine which vessel is an artery and which is a vein. In general, arteries have relatively thick walls and small lumens, whereas veins have relatively thin walls and large lumens (figure 22.1). In addition, the lumens of veins are often collapsed because of the fragile, thin nature of the blood vessel wall.

3. Once you have identified an artery and a vein, move the microscope stage so the wall of the *artery* is in the center of the field of view and increase the power on the microscope until you can see all the layers of the artery wall.

4. Using figure 22.1 and table 22.1 as guides, identify the structures listed below. Keep in mind that the innermost layer of the vessel (the tunica intima) will be incredibly thin and difficult to identify except on high power. Most likely you will see only the flattened nuclei of the endothelial cells, and very little of the rest of the cells.

    ☐ artery   ☐ tunica intima
    ☐ lumen   ☐ tunica media
    ☐ tunica externa ☐ vein

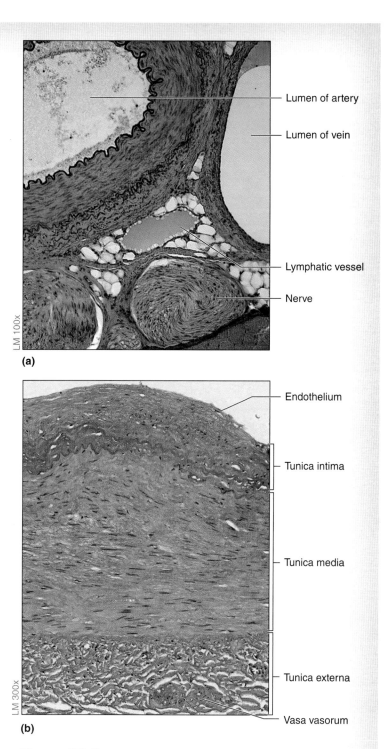

Figure 22.1 Blood Vessel Wall Structure. (a) Cross section through the center of a neurovascular bundle containing a nerve, artery, vein, and lymphatic vessel. (b) The three layers of the wall of a blood vessel: tunica intima, tunica media, and tunica externa. The tunica externa has its own blood supply, the vasa vasorum (literally, the "vessels of the vessels").

Elastic Arteries

Arteries are classified as elastic arteries, muscular arteries, and arterioles (see table 22.2). The aorta, pulmonary, brachiocephalic, common carotid, subclavian, and common iliac arteries are classified as elastic arteries (see figure 22.2, table 22.2). These arteries, located very close to the heart, have walls that are thick enough to withstand the pressure of blood that is pumped into them from the ventricles of the heart. The ventricles create enough force to move blood through these vessels, so they need very little smooth muscle in their tunica media to assist with blood flow. Instead, elastic arteries have an abundance of collagen and elastic fibers in their tunica media, which makes them both tough (collagen fibers) and expandable (elastic fibers). The ability of the vessel wall to expand and recoil as it receives blood from the ventricles, and then recoil, greatly smooths out the flow of blood through the arteries.

Large vessels such as the aorta have tiny blood vessels called *vasa vasorum* (literally, "the vessels of the vessels," see figure 22.1b) in the tunica externa. The vasa vasorum are analogous in both structure and function to the coronary arteries in the outer layer of the heart wall (epicardium). That is, the vasa vasorum supplies blood to larger vessels just as the coronary arteries supply blood to cardiac muscle tissue.

EXERCISE 22.2

ELASTIC ARTERY—THE AORTA

1. Obtain a slide of the aorta **(figure 22.2)** and place it on the microscope stage. Bring the wall of the aorta into focus on low power.

2. Using figure 22.2 and tables 22.1 and 22.2 as guides, identify the following on the slide of the aorta:
   - ☐ elastic fibers
   - ☐ lumen
   - ☐ tunica externa
   - ☐ tunica intima
   - ☐ tunica media
   - ☐ vasa vasorum

3. In the space below, sketch the wall of the aorta as seen through the microscope.

_____ ×

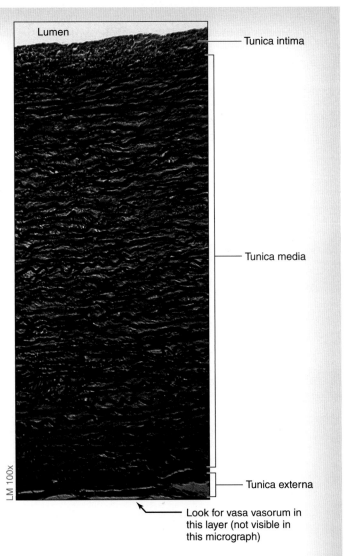

Figure 22.2 Elastic Artery. The wall of the aorta, an elastic artery, contains numerous elastic fibers (black) in the tunica media. No vasa vasorum are visible in the tunica externa in this micrograph.

Muscular Arteries

As blood moves through the elastic arteries and travels farther away from the heart, the pressure exerted by the heart is no longer great enough to keep the blood moving through the vessels. Thus, the amount of elastic tissue in the tunica media of the arteries starts to decrease and the amount of smooth muscle in the tunica media starts to increase. Contraction of the smooth muscle keeps blood moving through the arteries as the blood gets farther away from the heart. Most named vessels, including the brachial, anterior tibial, and inferior mesenteric arteries are muscular arteries. **Muscular arteries** are easily distinguished from elastic arteries by the presence of two prominent bands of elastic fibers, the **internal elastic lamina** which is the outermost layer of the tunica intima and the **external elastic lamina**, (adjacent to the tunica externa). There are several layers of smooth muscle sandwiched in between the two prominent elastic laminae (see figure 22.3, table 22.2).

EXERCISE 22.3

MUSCULAR ARTERY

1. Obtain a slide of a small, muscular artery, and place it on the microscope stage. Bring the tissue sample into focus on low power, then locate the wall of the vessel (figure 22.3).

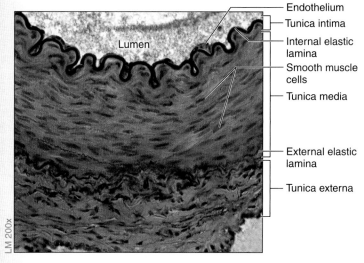

Figure 22.3 Muscular Artery. Muscular arteries contain distinct internal and external elastic laminae bordering the tunica media, which is predominantly smooth muscle.

2. Using figure 22.3 and table 22.2 as guides, identify the following on the slide of the muscular artery:

   ☐ external elastic lamina   ☐ tunica externa
   ☐ internal elastic lamina   ☐ tunica intima
   ☐ lumen   ☐ tunica media
   ☐ smooth muscle cells

3. In the space below, sketch the wall of a muscular artery as seen through the microscope.

_____ ×

Arterioles

Arterioles have a unique function in the cardiovascular system. Arterioles control the flow of blood into capillary beds (see table 22.2). As such, the most prominent feature of the wall of an arteriole is layers of circular smooth muscle in the tunica media. The smooth muscle cells of these layers are regulated to contract and relax to change the diameter of the lumen of the vessel, thus regulating the flow of blood into the capillary beds.

EXERCISE 22.4

ARTERIOLE

1. Obtain a slide of an arteriole and place it on the microscope stage. Bring the tissue sample into focus on low power. Scan the slide to locate an arteriole in cross section (figure 22.4).

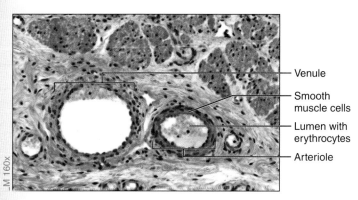

Figure 22.4 Arteriole. The tunica media of the arteriole contains circular smooth muscle that functions to alter the lumen diameter of the vessel.

2. Using figure 22.4 and tables 22.1 and 22.2 as guides, identify the following on the slide of the arteriole:

   ☐ lumen   ☐ smooth muscle

3. In the space below, sketch an arteriole as seen through the microscope.

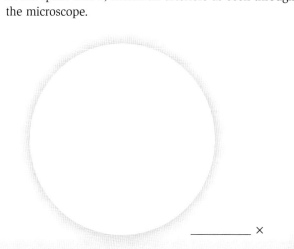

_____ ×

Veins

Veins are vessels that function to return blood to the heart at low pressure. They are characterized by having large lumens and thin walls (see figure 22.5, table 22.2). They may also contain valves, which are infoldings of the tunica intima. These valves prevent blood from flowing backward. Large veins, like large arteries, also contain vasa vasorum. **Venules** are small veins. The venules that come immediately after capillary beds, **postcapillary venules,** are the site where most white blood cells leave the circulation to enter the tissues (see table 22.2).

EXERCISE 22.5

VEIN

1. Place a slide of a large vein on the microscope stage and bring the tissue sample into focus on low power.

2. Using **figure 22.5** and table 22.2 as guides, identify the following on the slide of the large vein:

   ☐ elastic fiber
   ☐ smooth muscle cells
   ☐ tunica externa
   ☐ tunica intima
   ☐ tunica media
   ☐ valve (may not be visible on the slide)
   ☐ vasa vasorum

3. In the space below, sketch a cross-section of a large vein as seen through the microscope.

_____ ×

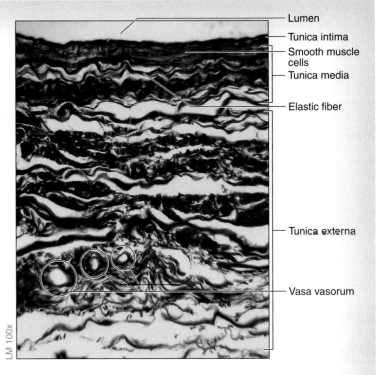

Figure 22.5 Vein. Cross section through the wall of a vein. The thickest layer of the vessel wall is the tunica externa, which contains vasa vasorum. (Spaces within the tunica externa are artifacts produced during the slide preparation.)

Capillaries

Capillaries are unique blood vessels in that they consist *only* of a tunica intima (endothelium and basement membrane). This allows for the exchange of substances between the blood and the tissues. There are several different types of capillaries in the body, which vary in their degree of permeability. For instance, capillaries in the spleen and liver, organs where a lot of blood processing takes place, allow a great deal of exchange to occur. In contrast, capillaries in the brain, an organ that must be protected from harmful substances such as toxins and viruses that might be present in the blood, allow very little exchange.

EXERCISE 22.6

OBSERVING ELECTRON MICROGRAPHS OF CAPILLARIES

1. Because capillaries are so tiny, the only way to truly appreciate their structure is to view them with an electron microscope. Obviously you do not have access to an electron microscope in your anatomy laboratory. Thus, this portion of the exercise will be performed by viewing the electron micrographs in **table 22.3**.

2. Observe the electron micrograph of a **continuous capillary** in table 22.3a. Continuous capillaries are the most common type of capillaries. They are found within muscle, the skin, the lungs, and central nervous system. They are composed of a continuous lining of simple squamous cells resting on a complete basement membrane. Identify the endothelial cells in the micrograph and then look for the areas where two endothelial cells come in contact with each other. Notice that the endothelial cells overlap each other in these areas. Where the cells come together to form a continuous lining around the lumen they do not form a complete "seal." The gaps between the endothelial cells are called **intercellular clefts.** Substances can move into and out of the blood either through endothelial cells by cellular transport processes (e.g., diffusion, pinocytosis) or between endothelial clefts by diffusion and bulk flow.

3. Observe the electron micrograph of a **fenestrated capillary** in table 22.3b. Identify endothelial cells in the micrograph. In contrast to the smooth surface of the endothelial cells of the continuous capillaries, the endothelial cells of the fenestrated capillaries appear wavy, particularly in the area adjacent to the nucleus of the endothelial cell. As with continuous capillaries, the endothelial cells of fenestrated capillaries form a continuous lining with intercellular clefts between the blood and the tissues. However, the endothelial cells themselves have **fenestrations,** which are regions where the endothelial cells are extremely thin. Fenestrations are approximately 10–100 nm and allow any substance that is smaller than the size of the fenestrations to be exchanged easily between the blood and the tissues. Thus, fenestrated capillaries allow much greater exchange than continuous capillaries, and they are found in organs where a greater amount of exchange is necessary, such as within endocrine glands, in the small intestine for absorption of nutrients, and in the kidneys for filtering blood.

4. Observe the electron micrograph of a **sinusoid** in table 22.3c. Sinusoids are located within organs that do a lot of processing of the blood, such as the liver and spleen. Sinusoids are characterized by having *discontinuous* endothelial cells, which do not overlap each other. In fact, there are open spaces between endothelial cells. The endothelial cells themselves are also fenestrated. Thus, the open spaces and fenestrations together create a minimal barrier between the blood and the tissues, which allows for maximum exchange (including the exchange of formed elements and large plasma proteins). The micrograph in table 22.3c demonstrates a sinusoidal capillary within the liver. The capillary in the top one-third of the micrograph. Inside the lumen of the capillary is a macrophage. The one large, prominent nucleus visible in the micrograph is that of a liver cell, or hepatocyte, which contains many small lipid droplets. There is another hepatocyte at the top of the micrograph as well. Observe the edges of the hepatocytes that face the lumen of the sinusoidal capillary. There you will see the thin endothelial cells that line the capillary. Now observe the area where there should be endothelium between the macrophage and the hepatocyte containing the lipid droplets and nucleus. Notice that there is no endothelial cell in between the two. Also, in the location between the macrophage and the hepatocyte above it, toward the left side of the micrograph, you can see fenestrations if you look very closely. The magnification of this micrograph is not quite high enough to see the fenestrations clearly, but they are there.

(continued on next page)

(continued from previous page)

| Table 22.3 | Characteristics of the Three Types of Capillaries |

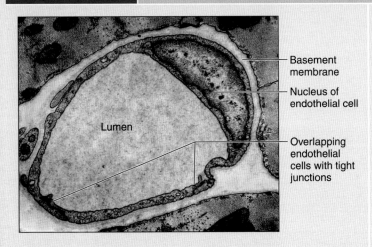
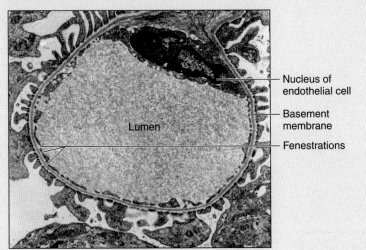

(a) Continuous Capillary		**(b) Fenestrated Capillary**	
Characteristics	Endothelial cells connected to each other by tight junctions	**Characteristics**	Same as continuous capillaries except also contain **fenestrations**
Endothelial Cells	Not fenestrated	**Endothelial Cells**	Fenestrated
Basement Membrane	Continuous	**Basement Membrane**	Continuous
Description and Function	Form a continuous lining between blood and tissues; exchange through endothelial cells and between endothelial cells via intercellular clefts	**Description and Function**	Allow for increased exchange between blood and tissues
Locations	Muscle tissue, the skin, connective tissues, exocrine glands, and central nervous system	**Locations**	Small intestine, kidneys, choroid plexus of brain, ciliary process of the eye
Word Origin	*continuus*, continued	**Word Origin**	*fenestra*, a window

5. In the spaces to the right and on the following page, sketch each of the three types of capillaries. Be sure to label the following on your drawings.

   ☐ basement membrane   ☐ fenestrations (if present)
   ☐ endothelial cells   ☐ lumen of capillary

Continuous

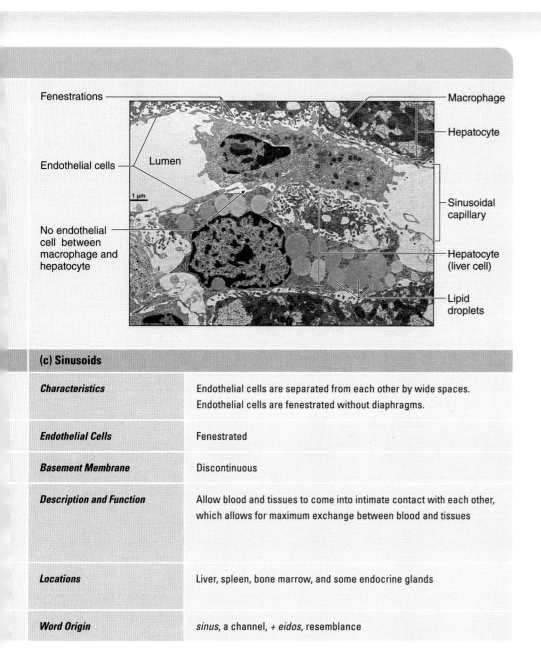

(c) Sinusoids

Characteristics	Endothelial cells are separated from each other by wide spaces. Endothelial cells are fenestrated without diaphragms.
Endothelial Cells	Fenestrated
Basement Membrane	Discontinuous
Description and Function	Allow blood and tissues to come into intimate contact with each other, which allows for maximum exchange between blood and tissues
Locations	Liver, spleen, bone marrow, and some endocrine glands
Word Origin	*sinus*, a channel, + *eidos*, resemblance

Fenestrated

Sinusoidal

Gross Anatomy

In the following laboratory exercises you will identify the major arteries and veins of the body on a prosected human cadaver or on classroom models or charts. As you identify each vessel, consider the area of the body supplied by or drained by the vessel, and ask yourself what tissue(s) or organ(s) would suffer damage if the vessel were blocked or cut.

After identifying the major vessels of each region of the body, you will be asked to trace the flow of blood from the heart to an organ within each region, and back to the heart (see Integrate: Learning Strategy box on this page). Your initial traces will be done using figures and will include only vessels that are relatively close to the target organ. However, you will also be asked to write out the complete trace in words. This means you will need to complete the trace you started in the figure by extending the list of blood vessels to include the vessels close to the heart that are not shown in the figures (for example, the aorta). When asked to do a *complete* trace, you will also need to include blood flow through the heart and the pulmonary circulation, and name all of the heart valves blood travels through on its journey. A complete trace both begins and ends in the right atrium of the heart.

INTEGRATE

LEARNING STRATEGY

The task of tracing the flow of blood through the blood vessels of the body is challenging for many students. If you find yourself having difficulty with this task, use the analogy of driving a car to your school. A drop of blood or a red blood cell represents your car, and the blood vessels represent the roads and highways. The organs represent your destination (e.g., school, the supermarket), and the heart represents your home. For example, to travel from your home to school, you must drive your car along a series of roadways that lead to the school. Each street has a name, which helps direct people driving their cars to their destinations. Similarly, as blood travels from the heart to a destination in the body (for example, the right hand), it flows along a given route. This route consists of several "streets," each with an identifying name. As you trace the flow of blood from your heart to your right hand, imagine yourself driving a car inside the blood vessels, and write down the names of the arteries you pass through as you travel to your destination. Of course, once at your destination, you must eventually return "home" to the heart. Thus, you also need to visualize the trip through the veins that you would take to get from the hand back to the heart.

Pulmonary Circuit

The **pulmonary circuit** is the system of blood vessels that carry blood from the right ventricle of the heart to the lungs and back to the left atrium of the heart. Blood leaves the right ventricle relatively deoxygenated, and returns to the left atrium highly oxygenated, having picked up oxygen in the pulmonary capillaries.

EXERCISE 22.7

PULMONARY CIRCUIT

1. Observe the thoracic cavity of a human cadaver or observe classroom models or charts demonstrating blood vessels of the heart and lungs.

2. Using your textbook as a guide, identify the structures listed in **figure 22.6** on the cadaver, models, or charts. Then label them in figure 22.6.

3. *Optional Activity:* **AP|R** **9: Cardiovascular System**—Watch the "Pulmonary and Systemic Circulation" animation to review the differences between these two circuits.

4. In the spaces to the right, trace the flow of blood from the right ventricle of the heart through the pulmonary circuit to the left atrium of the heart in words. Be sure to include heart valves.

Chapter Twenty-Two The Cardiovascular System: Vessels and Circulation

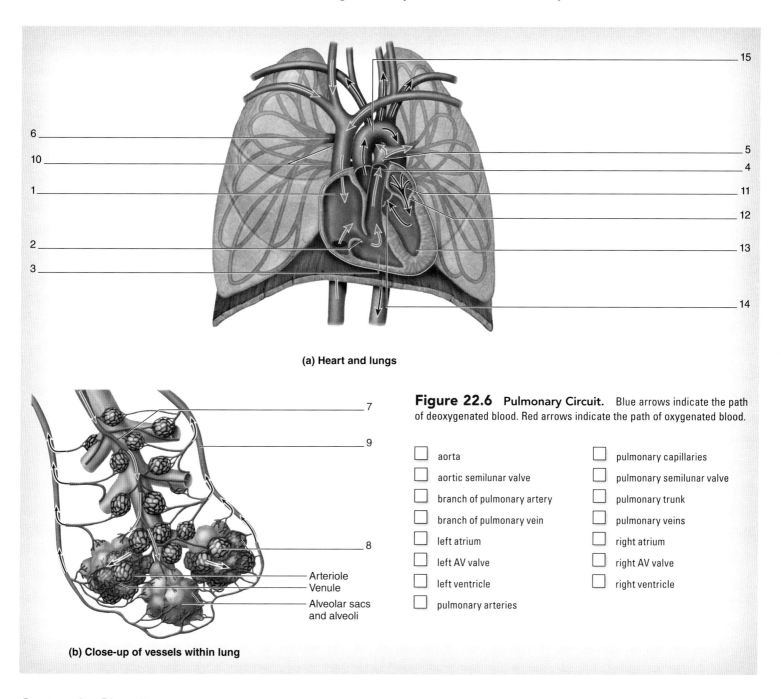

(a) Heart and lungs

(b) Close-up of vessels within lung

Figure 22.6 Pulmonary Circuit. Blue arrows indicate the path of deoxygenated blood. Red arrows indicate the path of oxygenated blood.

☐ aorta
☐ aortic semilunar valve
☐ branch of pulmonary artery
☐ branch of pulmonary vein
☐ left atrium
☐ left AV valve
☐ left ventricle
☐ pulmonary arteries
☐ pulmonary capillaries
☐ pulmonary semilunar valve
☐ pulmonary trunk
☐ pulmonary veins
☐ right atrium
☐ right AV valve
☐ right ventricle

Systemic Circuit

The **systemic circuit** is the system of blood vessels that carry blood from the left ventricle of the heart to body organs and back to the right atrium of the heart. Blood leaves the left ventricle highly oxygenated, and returns to the right atrium relatively deoxygenated.

EXERCISE 22.8

CIRCULATION TO THE HEAD AND NECK

1. Observe the head and neck regions of a prosected human cadaver or observe classroom models or charts showing blood vessels of the head and neck.

2. The major arteries carrying blood to structures of the head and neck are the external and internal carotid arteries (**figure 22.7**). The **external carotid arteries** supply most superficial structures of the head and neck. The **external jugular veins** drain most superficial areas of the scalp, face, and neck.

(continued on next page)

602 Chapter Twenty-Two *The Cardiovascular System: Vessels and Circulation*

(continued from previous page)

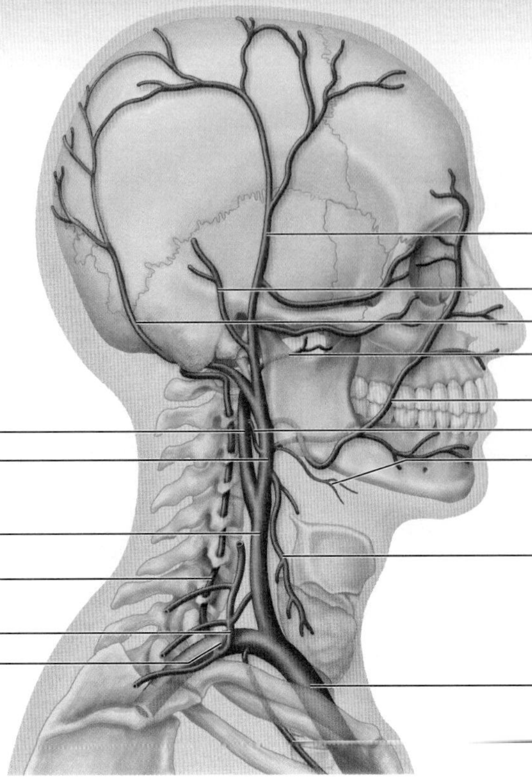

(a) Arteries, right lateral view

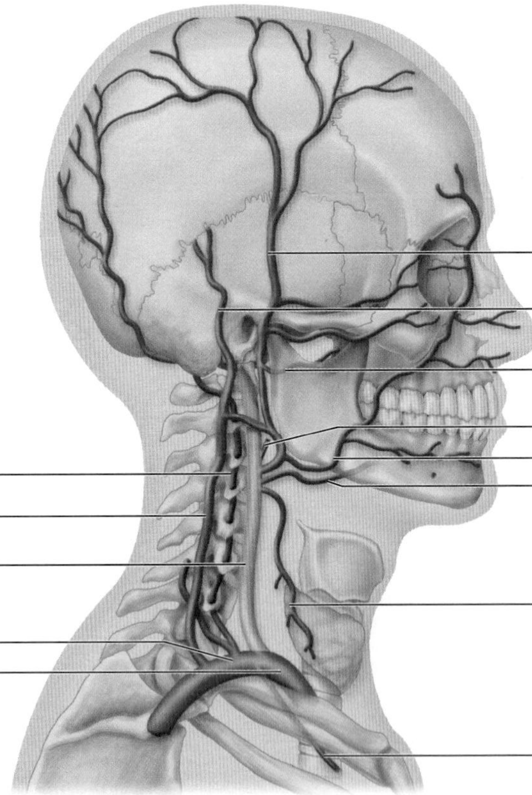

Labels on figure (a):
- 7
- Posterior auricular artery
- 8
- Maxillary artery
- 9
- Ascending pharyngeal artery
- Suprahyoid artery
- Superior thyroid artery
- 10
- Internal thoracic artery
- 1, 2, 3, 4, 5, 6

(b) Veins, right lateral view

Labels on figure (b):
- 6
- Posterior auricular vein
- Maxillary vein
- Pharyngeal vein
- 7
- Lingual vein
- 8
- Internal thoracic vein
- 1, 2, 3, 4, 5

Figure 22.7 Circulation to the Head and Neck.

(a) Arterial Supply
- ☐ brachiocephalic trunk (artery)
- ☐ common carotid artery
- ☐ external carotid artery
- ☐ facial artery
- ☐ internal carotid artery
- ☐ occipital artery
- ☐ subclavian artery
- ☐ superficial temporal artery
- ☐ thyrocervical trunk
- ☐ vertebral artery

(b) Venous Drainage
- ☐ external jugular vein
- ☐ facial vein
- ☐ internal jugular vein
- ☐ right brachiocephalic vein
- ☐ subclavian vein
- ☐ superficial temporal vein
- ☐ superior thyroid vein
- ☐ vertebral vein

Chapter Twenty-Two The Cardiovascular System: Vessels and Circulation

3. Using your textbook as a guide, identify the *arteries* listed in figure 22.7a that supply blood to the head and neck. Then label them in figure 22.7a.

4. Using your textbook as a guide, identify the *veins* listed in figure 22.7b that drain blood from the head and neck. Then label them in figure 22.7b.

5. The pathway a drop of blood takes to get from the aortic arch to the *skin overlying the anterior part of the right parietal bone of the skull* and back to the superior vena cava is shaded in **figure 22.8**. Trace this flow of blood by writing in the names of the vessels in the figure. Label the vessels in order, starting at number 1, so you are figuratively tracing the pathway and naming the vessels encountered along the way.

6. *Optional Activity:* APǀR **9: Cardiovascular System**— *Anatomy & Physiology Revealed* includes numerous dissections showing vascular supply to all body regions; review these dissections and use the Quiz feature to test yourself on each region.

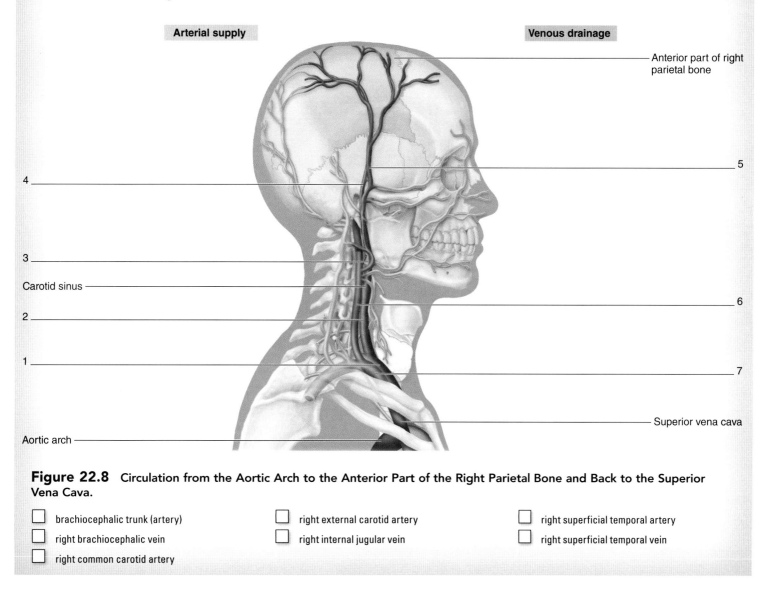

Figure 22.8 Circulation from the Aortic Arch to the Anterior Part of the Right Parietal Bone and Back to the Superior Vena Cava.

☐ brachiocephalic trunk (artery)   ☐ right external carotid artery   ☐ right superficial temporal artery
☐ right brachiocephalic vein   ☐ right internal jugular vein   ☐ right superficial temporal vein
☐ right common carotid artery

EXERCISE 22.9

CIRCULATION TO THE BRAIN

1. Observe the cranium and brain of a prosected human cadaver or observe classroom models or charts demonstrating blood vessels of the cranium and brain.

2. The major arteries carrying blood to structures of the brain are the internal carotid arteries and the vertebral arteries (see figure 22.7 and **figure 22.9**). The **internal carotid arteries** supply ~75% of the blood flow to the brain, while the **vertebral arteries** supply ~25% of the blood flow to the brain. Both pairs of vessels supply blood to the **cerebral arterial circle** (figure 22.9a). The major veins draining blood from the head, neck, and brain are the external and internal jugular veins. The **internal jugular veins** drain all blood from inside the cranial cavity plus some superficial areas of the face.

(continued on next page)

(continued from previous page)

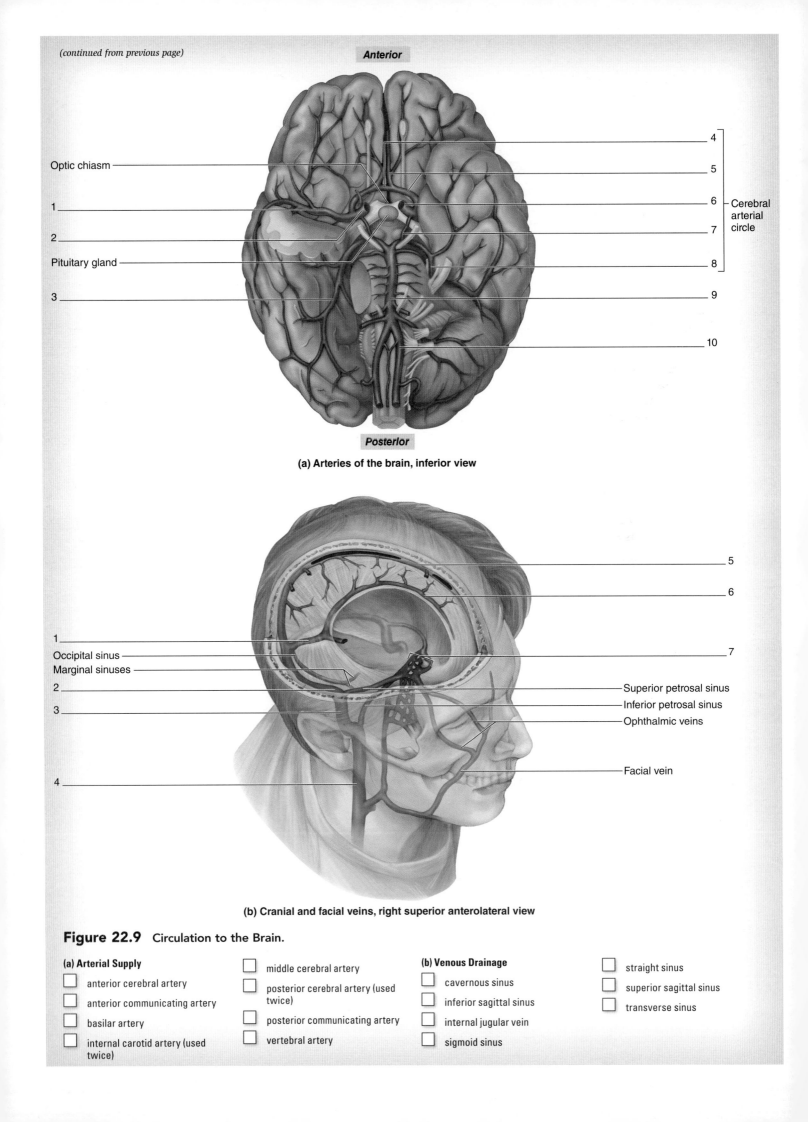

(a) Arteries of the brain, inferior view

(b) Cranial and facial veins, right superior anterolateral view

Figure 22.9 Circulation to the Brain.

(a) **Arterial Supply**
☐ anterior cerebral artery
☐ anterior communicating artery
☐ basilar artery
☐ internal carotid artery (used twice)
☐ middle cerebral artery
☐ posterior cerebral artery (used twice)
☐ posterior communicating artery
☐ vertebral artery

(b) **Venous Drainage**
☐ cavernous sinus
☐ inferior sagittal sinus
☐ internal jugular vein
☐ sigmoid sinus
☐ straight sinus
☐ superior sagittal sinus
☐ transverse sinus

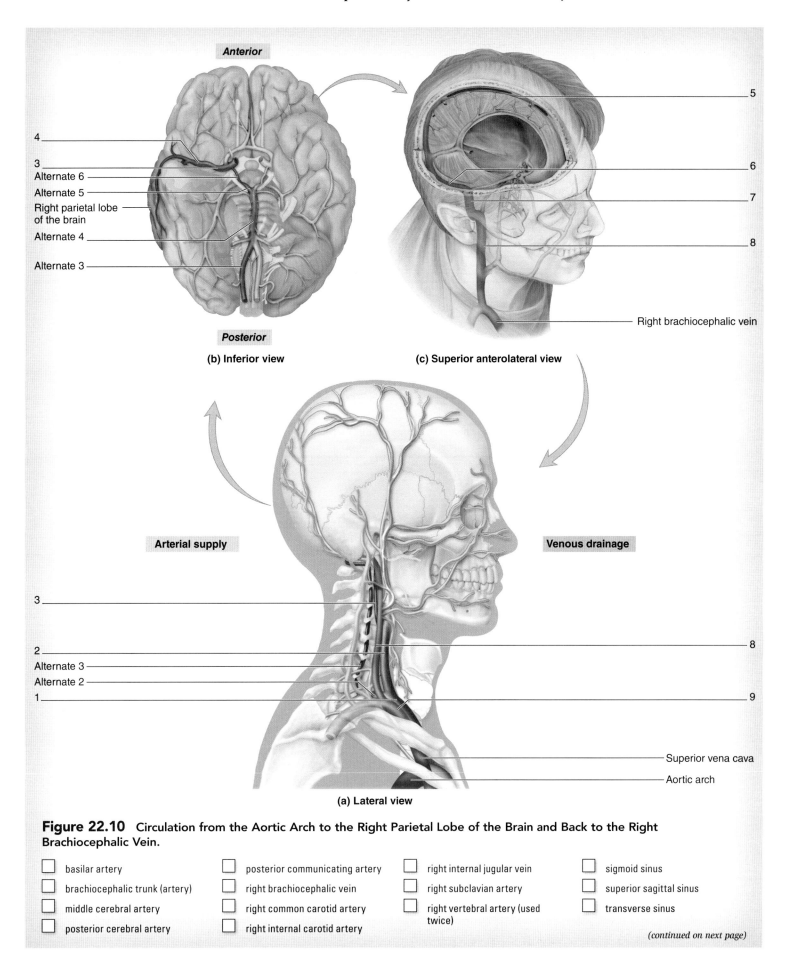

Figure 22.10 Circulation from the Aortic Arch to the Right Parietal Lobe of the Brain and Back to the Right Brachiocephalic Vein.

- basilar artery
- brachiocephalic trunk (artery)
- middle cerebral artery
- posterior cerebral artery
- posterior communicating artery
- right brachiocephalic vein
- right common carotid artery
- right internal carotid artery
- right internal jugular vein
- right subclavian artery
- right vertebral artery (used twice)
- sigmoid sinus
- superior sagittal sinus
- transverse sinus

(continued on next page)

(continued from previous page)

Blood draining from brain tissues first enters the **dural venous sinuses,** which collectively drain into the internal jugular vein (figure 22.9b).

3. Using your textbook as a guide, identify the *arteries* listed in figure 22.9a that supply blood to the brain. Then label them in figure 22.9a.

4. Using your textbook as a guide, identify the *veins* listed in figure 22.9b that drain blood from the brain. Then label them in figure 22.9b.

5. The pathway a drop of blood takes to get from the aortic arch to the *right parietal lobe of the brain* and back to the right brachiocephalic vein is shaded in **figure 22.10.** Trace this flow of blood by writing in the names of the vessels in the figure. Label the vessels in order, starting at number 1, so you are figuratively tracing the pathway and naming the vessels encountered along the way.

INTEGRATE

CLINICAL VIEW
Stroke

Maintaining adequate blood flow to organs ensures that oxygen and nutrients are delivered to the tissues. A stroke may occur when blood flow is restricted to the brain. This, in turn, restricts oxygen delivery to the tissues, or ischemia. Prolonged periods of ischemia may result in tissue death, or necrosis. As previously discussed, blood flow is influenced by changes in blood pressure and resistance. Common causes of ischemia include atherosclerotic plaques that dislodge and travel to the brain (embolism). When an embolus is present in a vessel, there is a decrease in blood vessel diameter, and an increase in resistance. Due to the inverse relationship between resistance and blood flow, this increased resistance results in a dramatic decrease in blood flow. Deficits observed following a stroke depend on the location of the ischemic event. For instance, a stroke in Broca's area may result in slurred speech, and an ischemic event in the vermis of the cerebellum may lead to motor deficits. Clinically, tests such as magnetic resonance imaging (MRI) or computed tomography (CT) are used to observe the area of infarct.

EXERCISE 22.10

CIRCULATION TO THE THORACIC AND ABDOMINAL WALLS

1. Observe the thoracic cavity on a prosected human cadaver or observe classroom models or charts demonstrating blood vessels of the thoracic and abdominal cavities.

2. Using your textbook or atlas a guide, identify the *arteries* listed in **figure 22.11a** that supply blood to the thoracic and abdominal walls. Then label them in figure 22.11a.

3. Using your textbook as a guide, identify the *veins* listed in figure 22.11b that drain blood from the thoracic and abdominal walls. Then label them in figure 22.11b.

4. Using your textbook as a guide, identify the paired and unpaired arteries that supply blood to the abdominal organs, pelvis, and perineum. Then label them in figure 22.12.

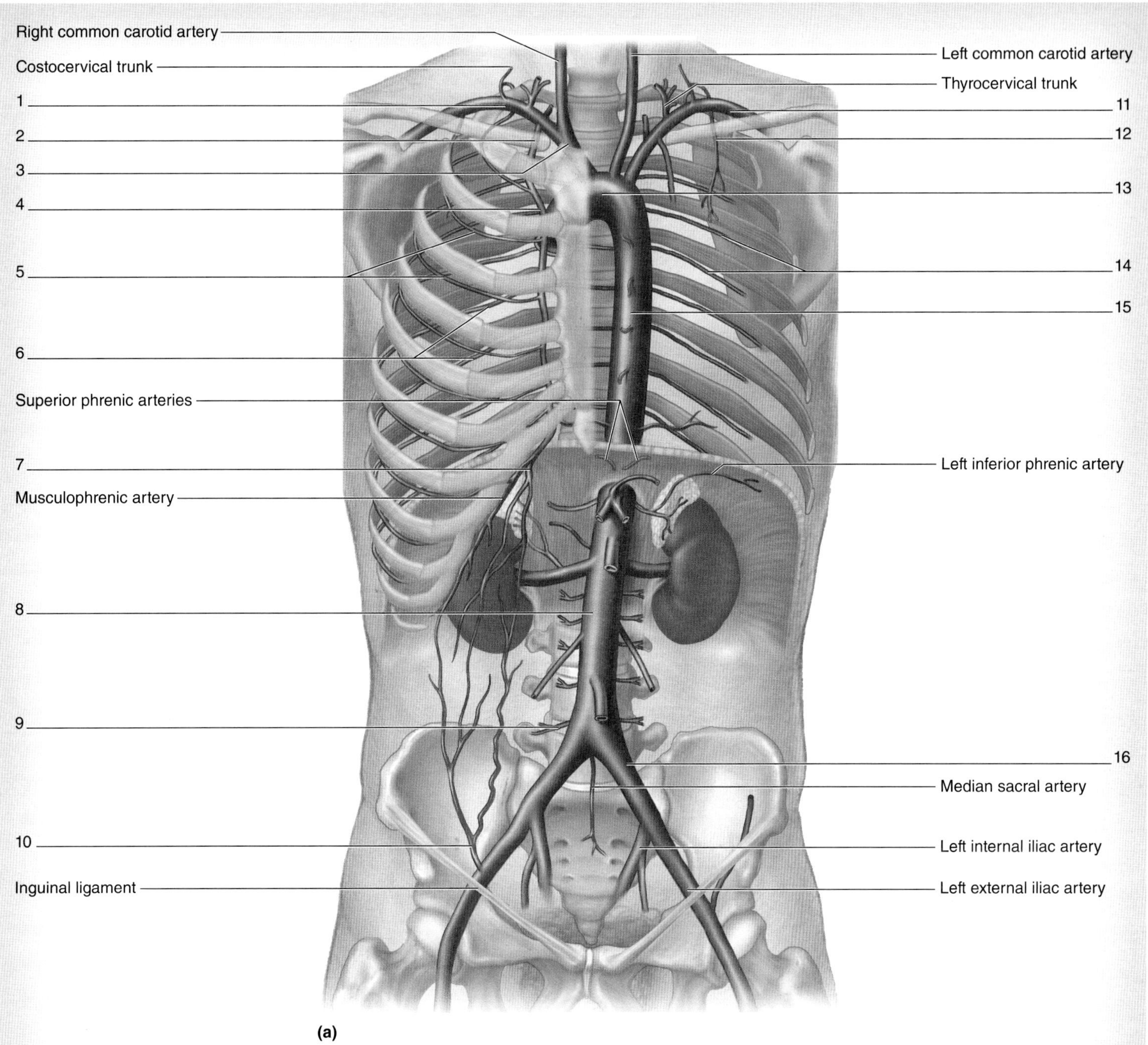

Chapter Twenty-Two *The Cardiovascular System: Vessels and Circulation* **607**

Right common carotid artery
Costocervical trunk
Left common carotid artery
Thyrocervical trunk
1 ____
2 ____
3 ____
4 ____
5 ____
6 ____
Superior phrenic arteries
7 ____
Musculophrenic artery
Left inferior phrenic artery
8 ____
9 ____
16 ____
Median sacral artery
10 ____
Left internal iliac artery
Inguinal ligament
Left external iliac artery
11 ____
12 ____
13 ____
14 ____
15 ____

(a)

Figure 22.11 Circulation to the Thoracic and Abdominal Walls.

(a) Arterial Supply
- ☐ anterior intercostal arteries
- ☐ aortic arch
- ☐ brachiocephalic trunk
- ☐ descending abdominal aorta
- ☐ descending thoracic aorta
- ☐ inferior epigastric artery
- ☐ internal thoracic artery (used twice)
- ☐ left common iliac artery
- ☐ left subclavian artery
- ☐ posterior intercostal arteries (1–2)
- ☐ posterior intercostal arteries (3–11) (used twice)
- ☐ right lumbar artery
- ☐ right subclavian artery
- ☐ superior epigastric artery

(continued on next page)

608 Chapter Twenty-Two *The Cardiovascular System: Vessels and Circulation*

(continued from previous page)

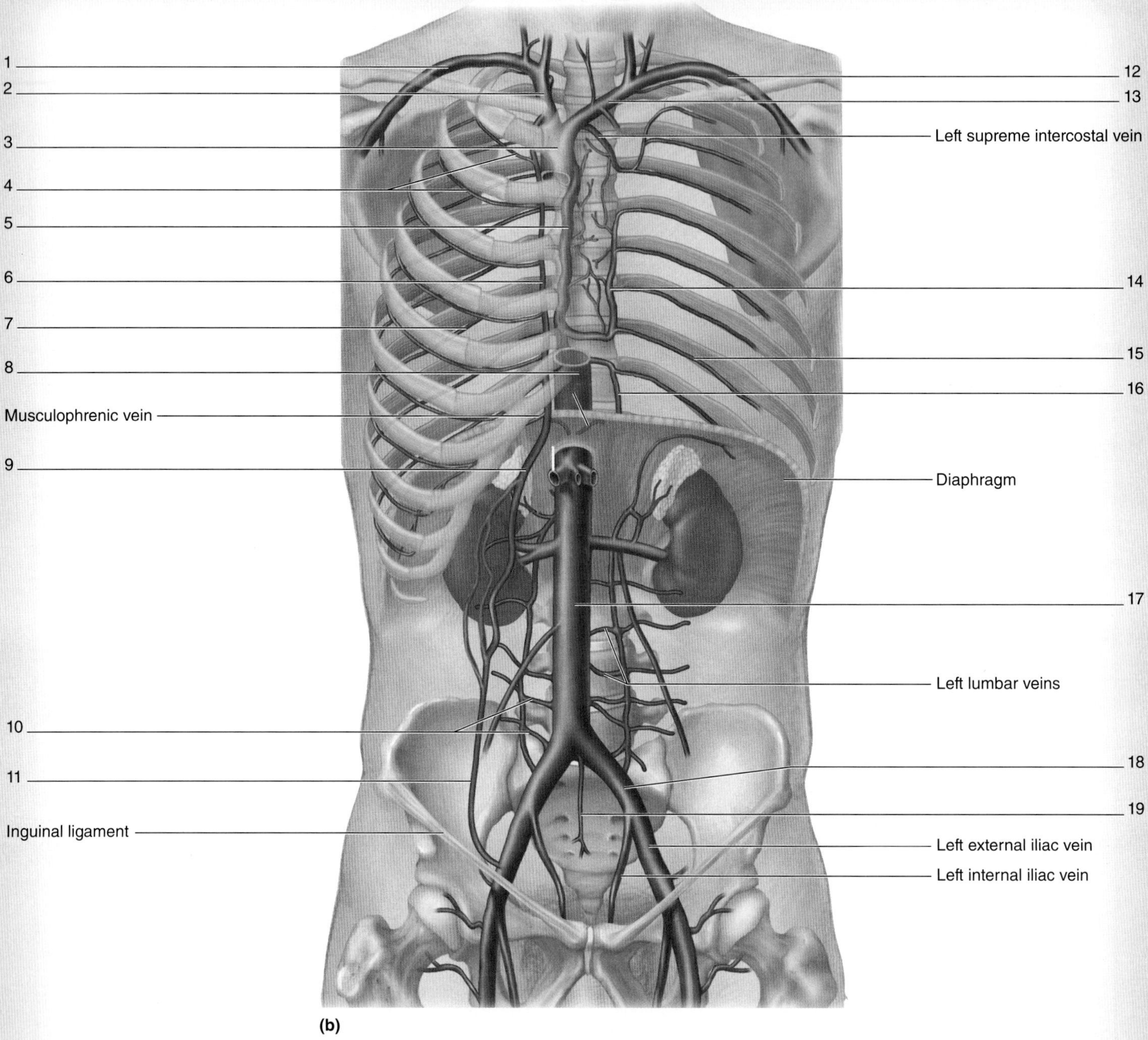

1. _____
2. _____
3. _____
4. _____
5. _____
6. _____
7. _____
8. _____
Musculophrenic vein _____
9. _____
10. _____
11. _____
Inguinal ligament _____

12. _____
13. _____
Left supreme intercostal vein
14. _____
15. _____
16. _____
Diaphragm
17. _____
Left lumbar veins
18. _____
19. _____
Left external iliac vein
Left internal iliac vein

(b)

Figure 22.11 Circulation to the Thoracic and Abdominal Walls *(continued).*

(b) Venous Drainage

- ☐ accessory hemiazygos vein
- ☐ anterior intercostal veins
- ☐ azygos vein
- ☐ hemiazygos vein
- ☐ inferior vena cava (used twice)
- ☐ internal thoracic vein
- ☐ left brachiocephalic vein
- ☐ left common iliac vein
- ☐ left posterior intercostal vein
- ☐ left subclavian vein
- ☐ median sacral vein
- ☐ right brachiocephalic vein
- ☐ right inferior epigastric vein
- ☐ right lumbar veins
- ☐ right posterior intercostal vein
- ☐ right subclavian vein
- ☐ right superior epigastric vein
- ☐ superior vena cava

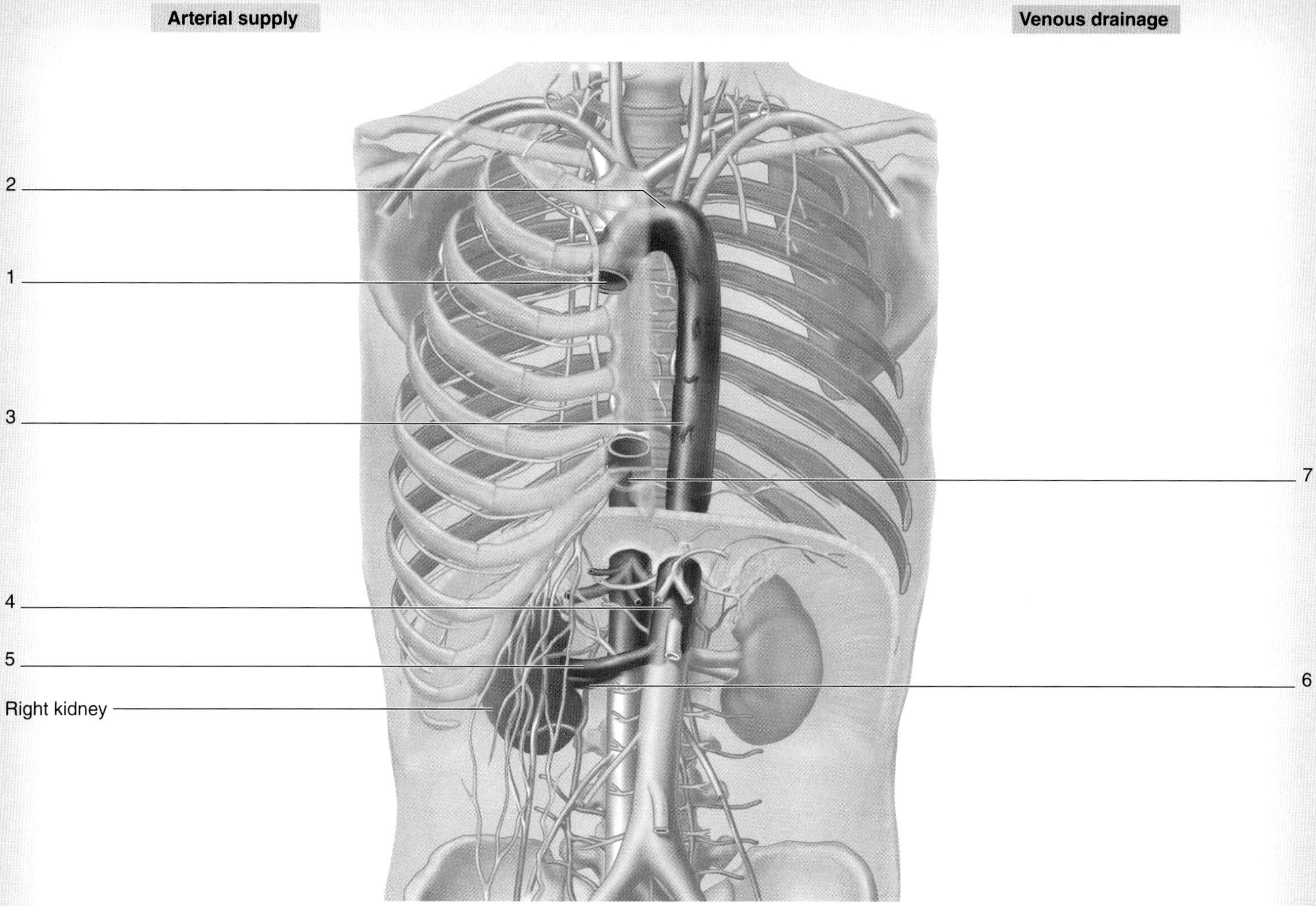

Chapter Twenty-Two *The Cardiovascular System: Vessels and Circulation* **609**

Arterial supply   **Venous drainage**

2 _____
1 _____
3 _____
                                                                                         7 _____
4 _____
5 _____   6 _____
Right kidney _____

Figure 22.12 Circulation from the Left Ventricle of the Heart to the Right Kidney and Back to the Right Atrium of the Heart.

☐ aortic arch   ☐ descending abdominal aorta   ☐ inferior vena cava   ☐ right renal vein
☐ ascending aorta   ☐ descending thoracic aorta   ☐ right renal artery

EXERCISE 22.11

CIRCULATION TO THE ABDOMINAL CAVITY

1. Observe the abdominal cavity on a prosected human cadaver and/or observe classroom models or charts demonstrating blood vessels of the abdominal cavity.

2. Using your textbook as a guide, identify the *arteries* listed in **figures 22.12** and **22.13** that supply blood to structures in the abdomen. Then label them in figures 22.12 and 22.13.

3. Venous drainage of abdominal organs is unique in that it is an example of a portal system, called the **hepatic portal system.** In this system there are two capillary beds—the first in an abdominal organ, and the second in the liver—connected to each other by a **portal vein.** An artery supplies blood to the first capillary bed, which is located in an abdominal organ such as the stomach, intestine, or spleen. Blood drains from abdominal organs into three veins: the **splenic, inferior mesenteric,** and **superior mesenteric** veins. These veins then drain into the **hepatic portal vein,** which carries blood to the second capillary bed in the liver. This blood is high in nutrient content, but also may be transporting toxins, bacteria, and other potentially dangerous substances. Because the blood flows from the abdominal organs directly to the liver, nutrients, drugs, and pathogens may be removed from the blood before the blood enters the general circulation. Thus, the liver is said to have "first pass" at the blood that drains from the

(continued on next page)

Chapter Twenty-Two *The Cardiovascular System: Vessels and Circulation*

(continued from previous page)

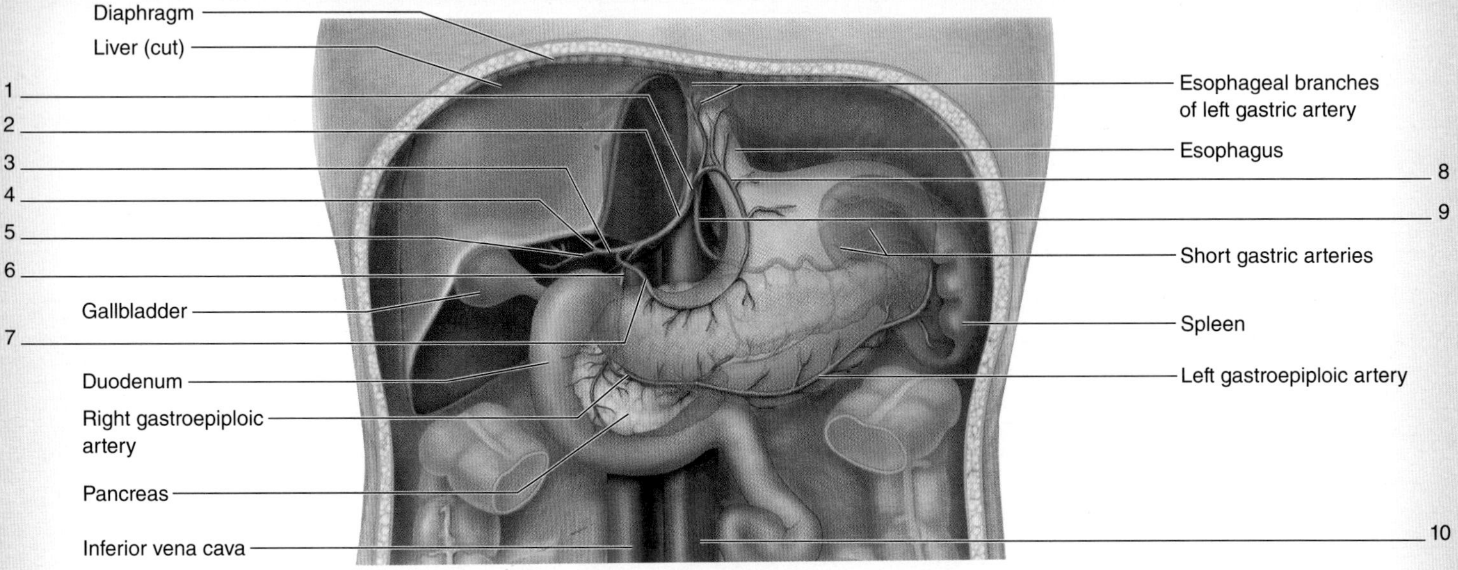

(a) Celiac trunk branches

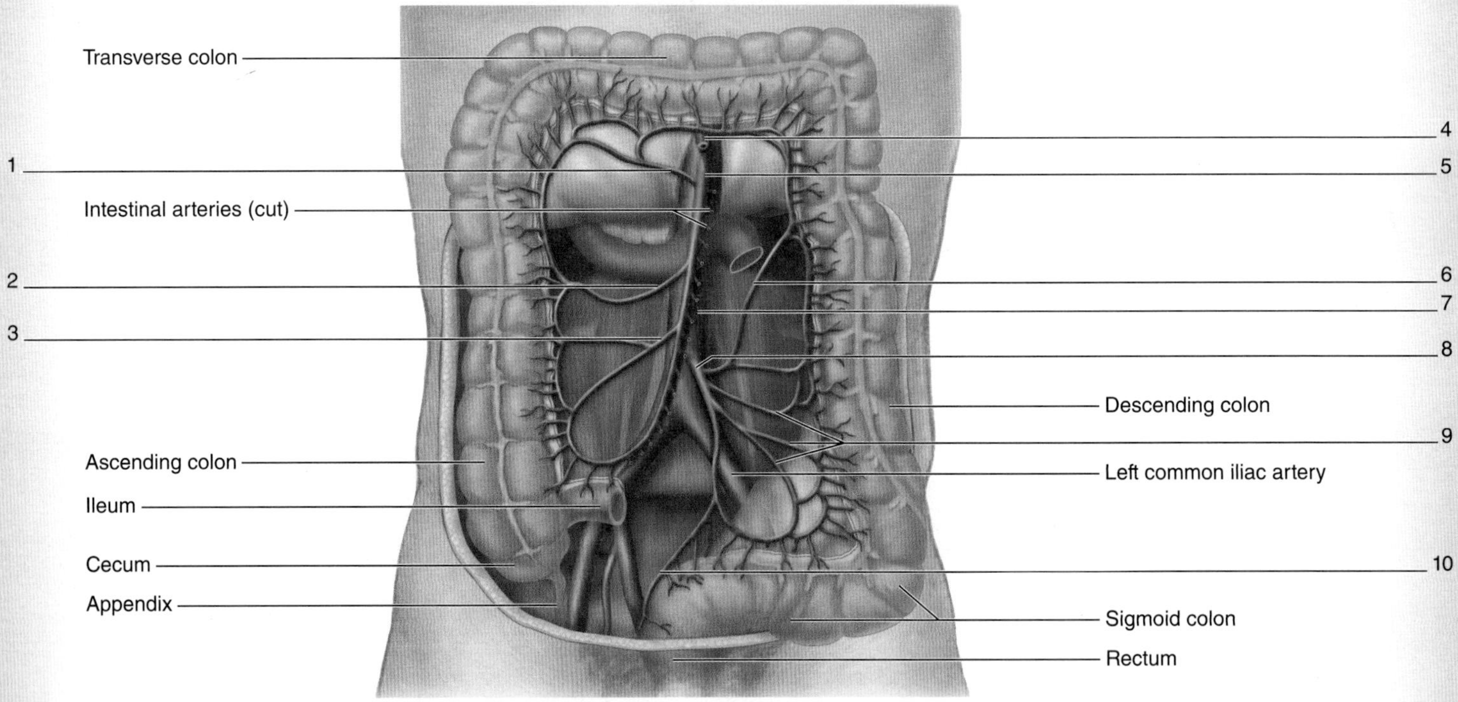

Labels shown: Diaphragm, Liver (cut), 1, 2, 3, 4, 5, 6, Gallbladder, 7, Duodenum, Right gastroepiploic artery, Pancreas, Inferior vena cava, Esophageal branches of left gastric artery, Esophagus, 8, 9, Short gastric arteries, Spleen, Left gastroepiploic artery, 10

(b) Superior and inferior mesenteric arteries

Labels shown: Transverse colon, 1, Intestinal arteries (cut), 2, 3, Ascending colon, Ileum, Cecum, Appendix, 4, 5, 6, 7, 8, Descending colon, 9, Left common iliac artery, 10, Sigmoid colon, Rectum

Figure 22.13 Arterial Supply to Abdominal Organs.

(a) Arterial Supply to the Stomach, Spleen, Pancreas, Duodenum, and Liver

- ☐ celiac trunk
- ☐ common hepatic artery
- ☐ descending abdominal aorta
- ☐ gastroduodenal artery
- ☐ hepatic artery proper
- ☐ left gastric artery
- ☐ left hepatic artery
- ☐ right gastric artery
- ☐ right hepatic artery
- ☐ splenic artery

(b) Arterial Supply to the Small and Large Intestines

- ☐ celiac trunk
- ☐ descending abdominal aorta
- ☐ ileocolic artery
- ☐ inferior mesenteric artery
- ☐ left colic artery
- ☐ middle colic artery
- ☐ right colic artery
- ☐ sigmoid arteries
- ☐ superior mesenteric artery
- ☐ superior rectal artery

abdominal organs. Finally, venous blood drains from liver capillaries into **hepatic veins,** which carry it to the inferior vena cava and back to the heart.

4. Using your textbook as a guide, identify the *veins* listed in **figure 22.14** that compose the hepatic portal system. Then label them in figure 22.14.

5. The pathway a drop of blood takes to get from the abdominal aorta to the *spleen* and back to the right atrium of the heart is shaded in **figure 22.15**. Trace this flow of blood by writing in the names of the vessels in the figure. Label the vessels in order, starting at number 1, so you are figuratively tracing the pathway and naming the vessels encountered along the way.

6. The pathway a drop of blood takes to get from the abdominal aorta to the *duodenum* and back to the right atrium of the heart is shaded in **figure 22.16**. Trace this flow of blood by writing in the names of the vessels in the figure. Label the vessels in order, starting at number 1, so you are figuratively tracing the pathway and naming the vessels encountered along the way.

7. The pathway a drop of blood takes to get from the abdominal aorta to the *sigmoid colon* and back to the right atrium of the heart is shaded in **figure 22.17**. Trace this flow of blood by writing in the names of the vessels in the figure. Label the vessels in order, starting at number 1, so you are figuratively tracing the pathway and naming the vessels encountered along the way.

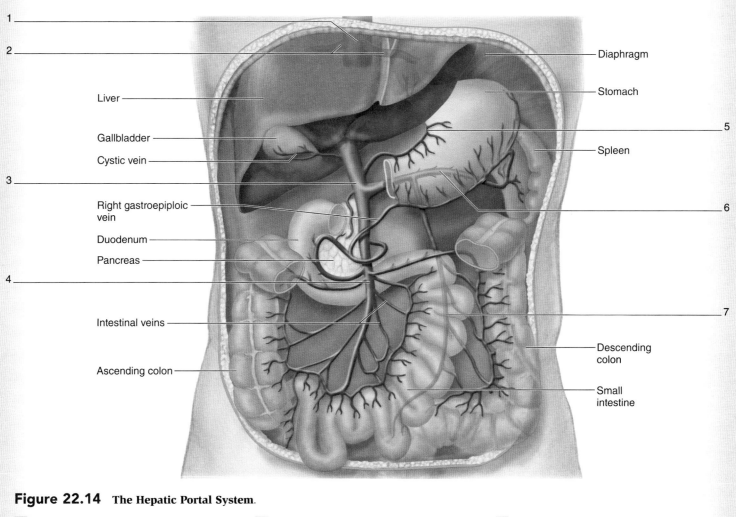

Figure 22.14 The Hepatic Portal System.

☐ gastric vein
☐ hepatic portal vein
☐ hepatic veins

☐ inferior mesenteric vein
☐ inferior vena cava
☐ splenic vein

☐ superior mesenteric vein

(continued on next page)

612 Chapter Twenty-Two *The Cardiovascular System: Vessels and Circulation*

(continued from previous page)

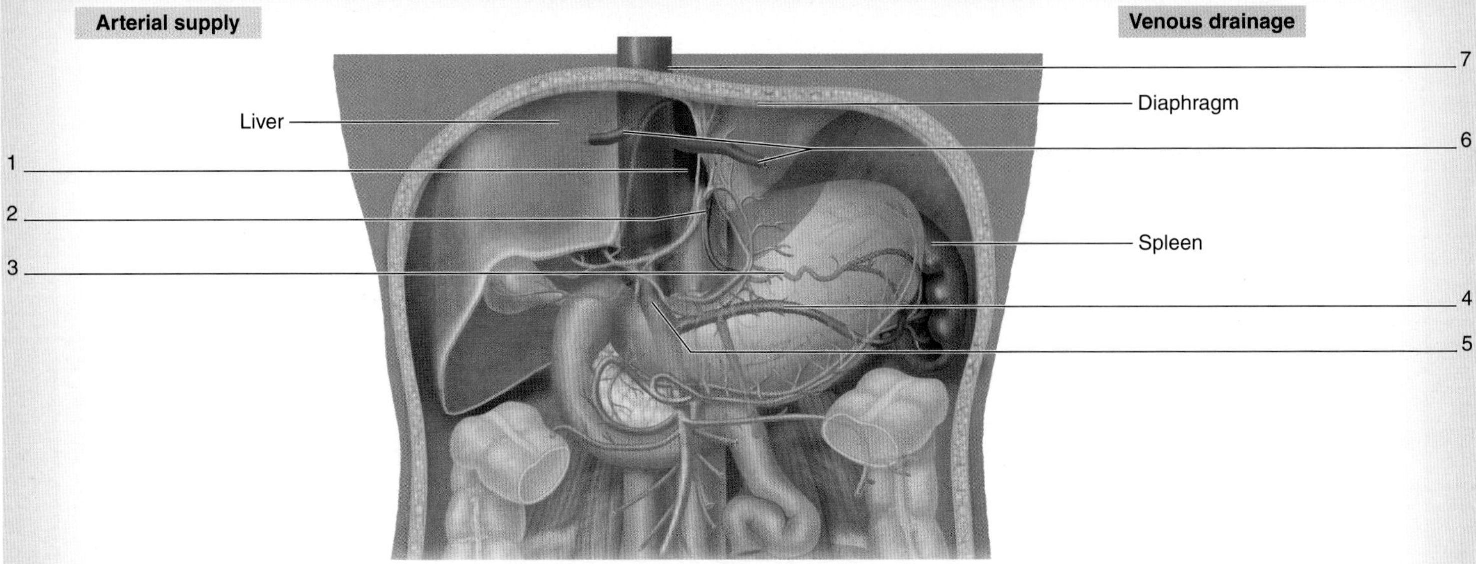

Figure 22.15 Circulation from the Abdominal Aorta to the Spleen and Back to the Right Atrium of the Heart.

- ☐ abdominal aorta
- ☐ celiac trunk
- ☐ hepatic portal vein
- ☐ hepatic veins
- ☐ inferior vena cava
- ☐ splenic artery
- ☐ splenic vein

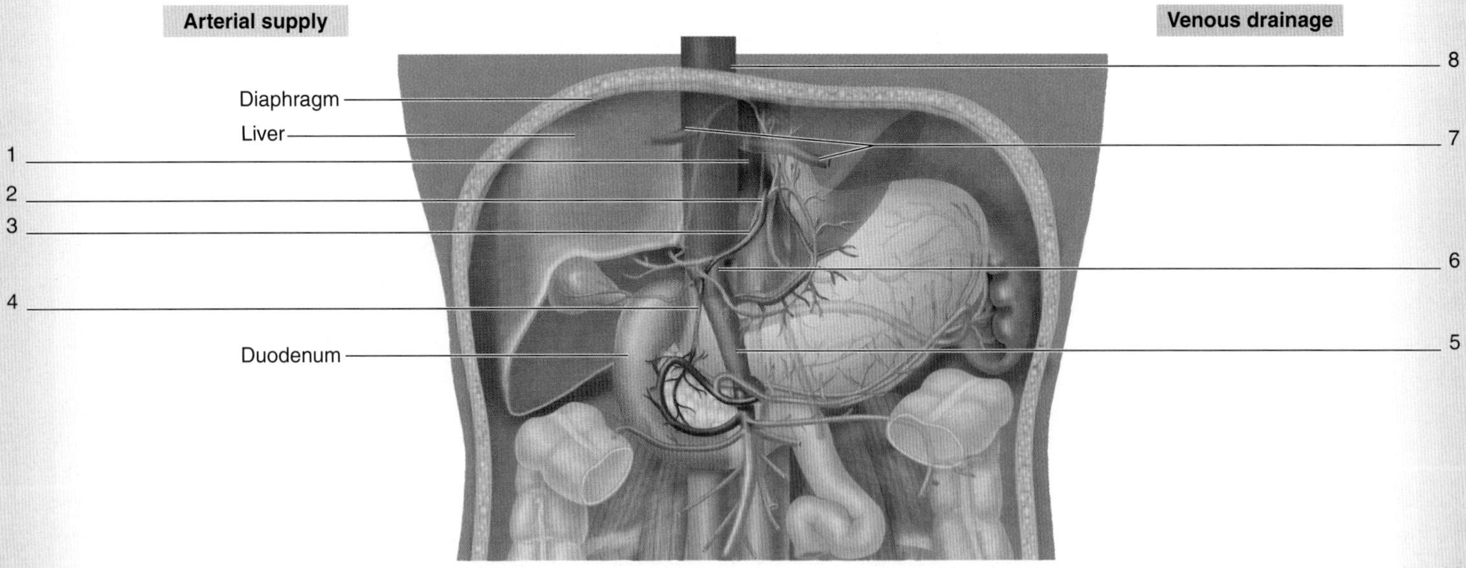

Figure 22.16 Circulation from the Abdominal Aorta to the Duodenum and Back to the Right Atrium of the Heart.

- ☐ abdominal aorta
- ☐ celiac trunk
- ☐ common hepatic artery
- ☐ gastroduodenal artery
- ☐ hepatic portal vein
- ☐ hepatic veins
- ☐ inferior vena cava
- ☐ superior mesenteric vein

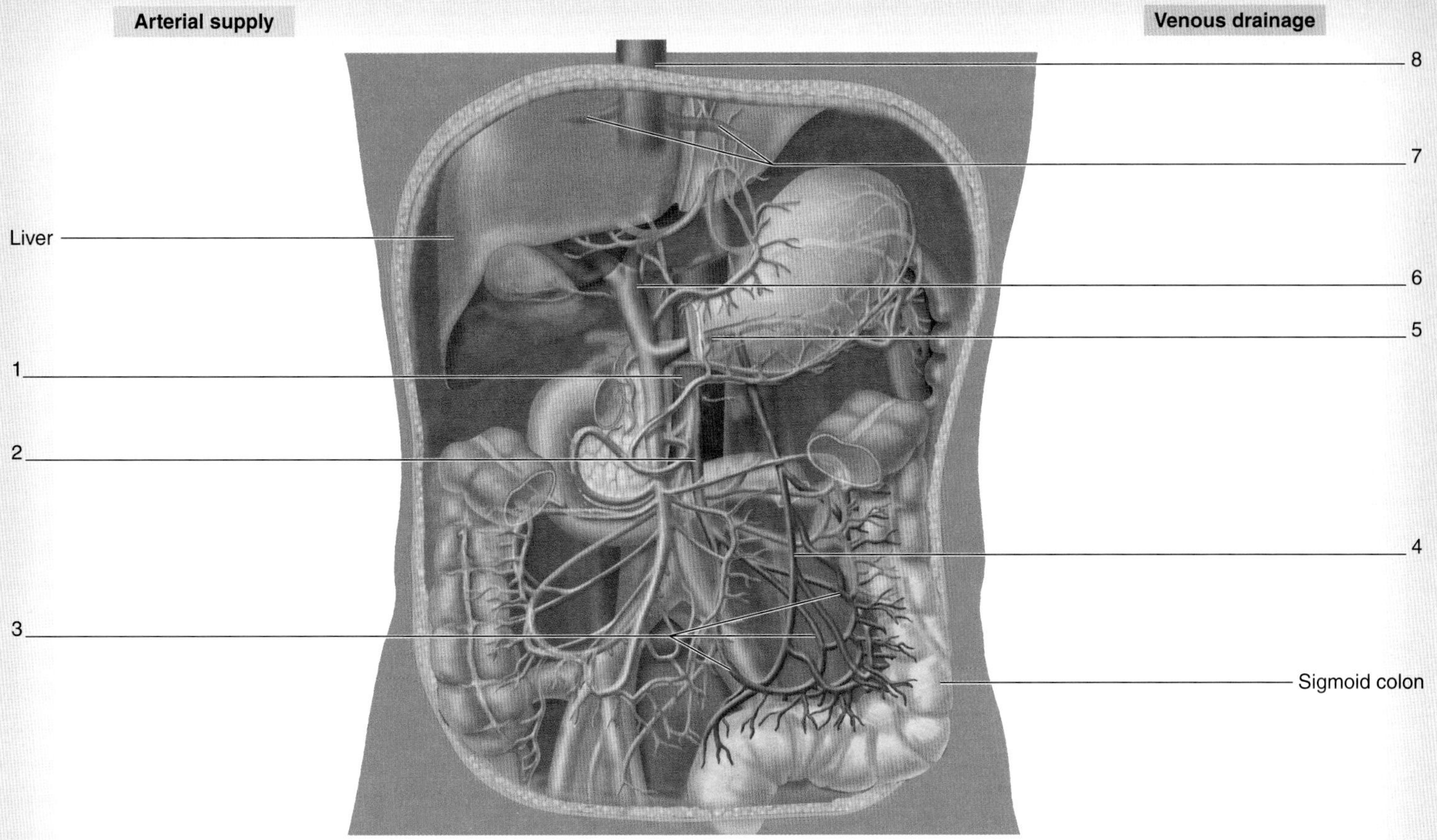

Figure 22.17 Circulation from the Abdominal Aorta to the Sigmoid Colon and Back to the Right Atrium of the Heart.

- ☐ abdominal aorta
- ☐ hepatic portal vein
- ☐ hepatic veins
- ☐ inferior mesenteric artery
- ☐ inferior mesenteric vein
- ☐ inferior vena cava
- ☐ sigmoid arteries
- ☐ splenic vein

INTEGRATE

CLINICAL VIEW
Portal-Systemic Anastomoses

Veins of the portal system are unique in that they have *no valves*. Thus, when blood backs up in this system, it can back up into systemic veins and will attempt to take an alternate route to the inferior vena cava through anastomoses (s. anastomosis). An **anastomosis** (*anastomo*, to furnish with a mouth) is a connection between two blood vessels. The major anastomoses of the portal circulation are found in veins of the esophagus and the rectum, and surrounding the umbilicus. These vessels are common sites of **varicosities** (*varix*, a dilated vein). Varicosities of the esophageal veins are extremely serious because rupture of varicose veins of the esophagus can result in an individual bleeding to death. Varicosities of the rectal veins cause **hemorrhoids** (*haimo*, blood, + *rhoia*, a flow), and varicosities of the umbilical veins form a *caput medusa* (the varicose vessels radiating out from the umbilicus resemble the snakes on Medusa's head (from Greek mythology). One cause of portal-systemic anastomoses is due to the long-term effects of alcoholism that result in liver damage leading to **cirrhosis** (*Kirrhos*, yellow, + *–osis*, condition). With cirrhosis, normal liver tissue is replaced over time with scar tissue, decreasing the size of the liver and the number of capillaries within. This creates resistance to blood flow through the liver, which causes blood to back up into the veins of the hepatic portal system.

614 Chapter Twenty-Two *The Cardiovascular System: Vessels and Circulation*

EXERCISE 22.12

CIRCULATION TO THE UPPER LIMB

1. Observe the upper limb of a prosected human cadaver or observe classroom models or charts demonstrating blood vessels of the upper limb.

2. Using your textbook as a guide, identify the *arteries* listed in **figure 22.18a** that supply blood to the upper limb. Then label them in figure 22.18a.

3. Using your textbook as a guide, identify the *veins* listed in figure 22.18b that drain blood from the upper limb. Then label them in figure 22.18b.

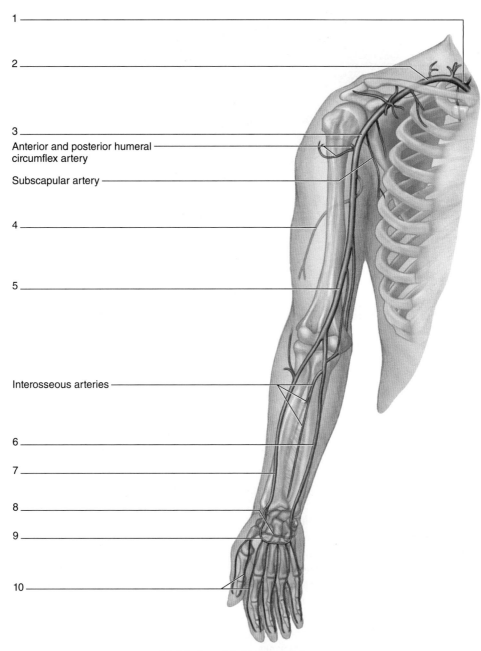

(a) Arteries of right upper limb

Figure 22.18 Circulation to the Upper Limb.

(a) Arterial Supply
- ☐ axillary artery
- ☐ brachial artery
- ☐ brachiocephalic trunk (artery)
- ☐ deep brachial artery
- ☐ deep palmar arch
- ☐ digital arteries
- ☐ radial artery
- ☐ subclavian artery
- ☐ superficial palmar arch
- ☐ ulnar artery

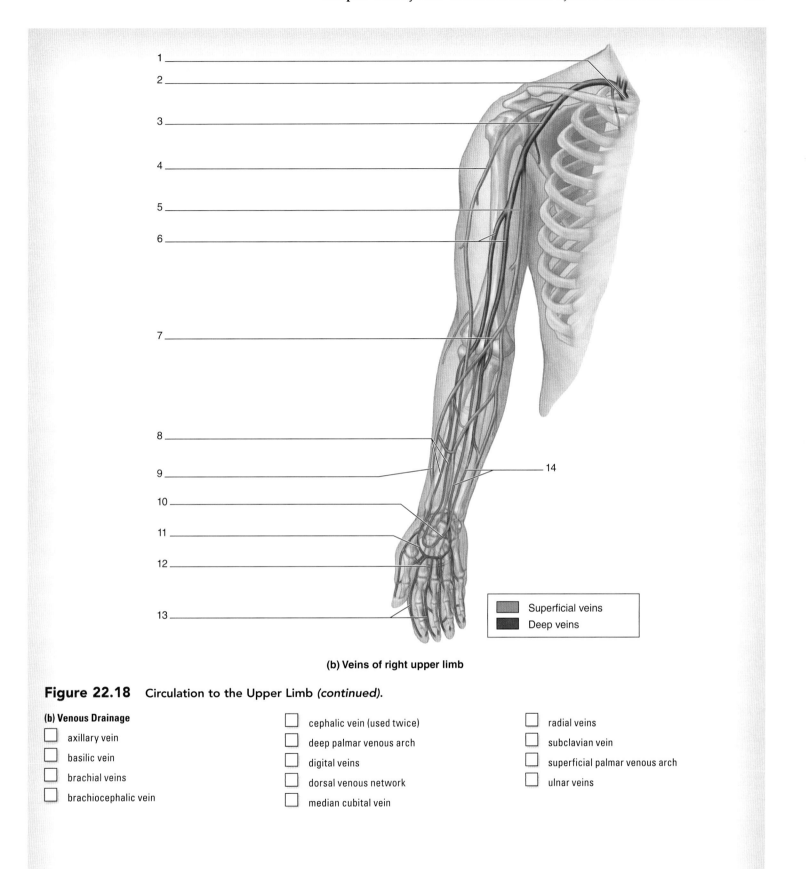

(b) Veins of right upper limb

Figure 22.18 Circulation to the Upper Limb *(continued)*.

(b) Venous Drainage
- [ ] axillary vein
- [ ] basilic vein
- [ ] brachial veins
- [ ] brachiocephalic vein
- [ ] cephalic vein (used twice)
- [ ] deep palmar venous arch
- [ ] digital veins
- [ ] dorsal venous network
- [ ] median cubital vein
- [ ] radial veins
- [ ] subclavian vein
- [ ] superficial palmar venous arch
- [ ] ulnar veins

(continued on next page)

616 Chapter Twenty-Two *The Cardiovascular System: Vessels and Circulation*

(continued from previous page)

4. *Superficial Trace*: The pathway a drop of blood takes to get from the aortic arch to the *anterior surface of the index finger* and back along a superficial route to the superior vena cava is shaded in **figure 22.19**. Trace this flow of blood by writing in the names of the vessels in the figure. Label the vessels in order, starting at number 1, so you are figuratively tracing the pathway and naming the vessels encountered along the way.

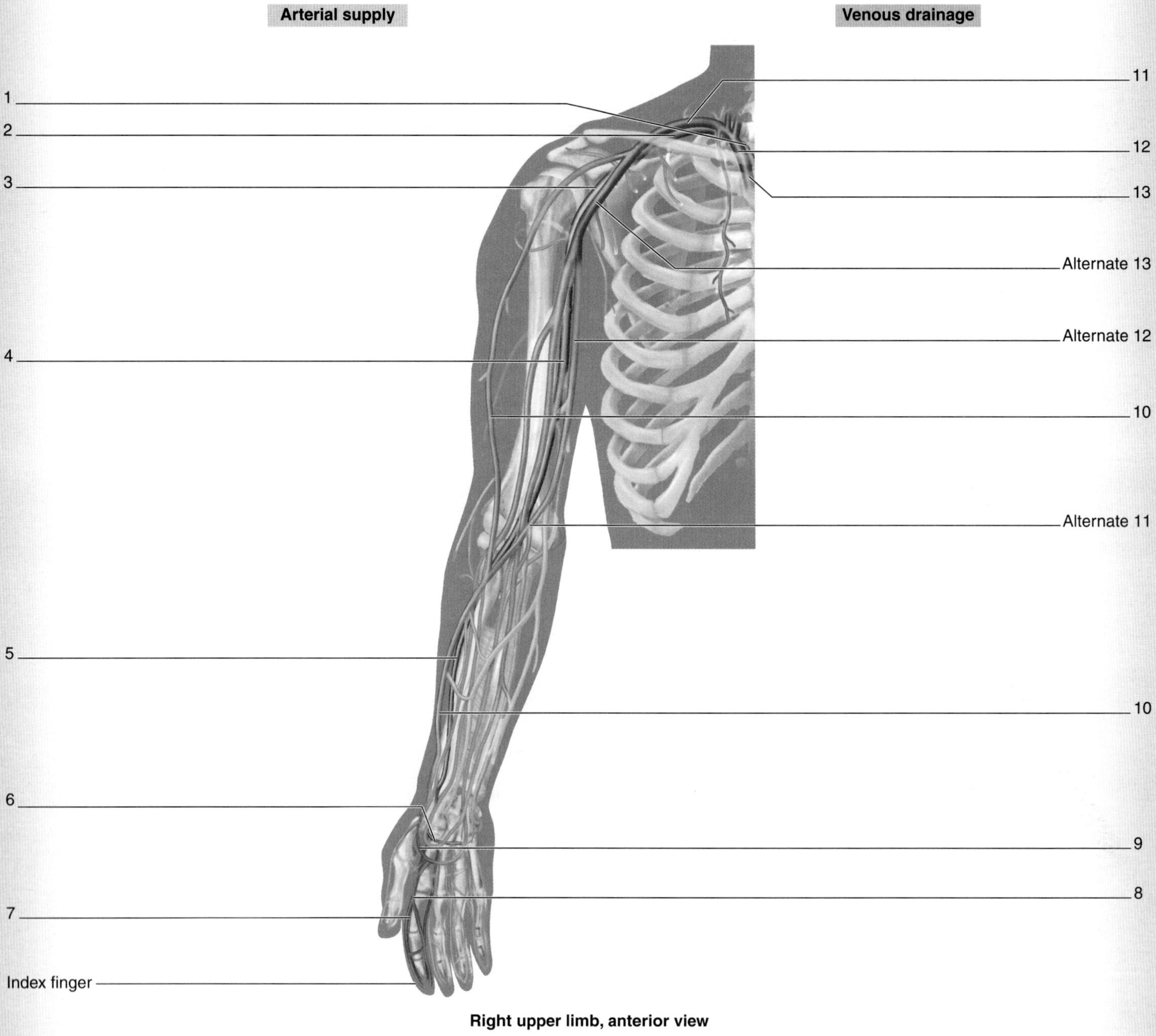

Arterial supply

1. _____
2. _____
3. _____
4. _____
5. _____
6. _____
7. _____

Index finger

Venous drainage

11. _____
12. _____
13. _____
Alternate 13 _____
Alternate 12 _____
10. _____
Alternate 11 _____
10. _____
9. _____
8. _____

Right upper limb, anterior view

Figure 22.19 Circulation from the Aortic Arch to the Anterior Surface of the Index Finger and Back along a Superficial Route to the Superior Vena Cava.

☐ axillary artery
☐ axillary vein
☐ basilic vein
☐ brachial artery
☐ brachiocephalic trunk (artery)
☐ brachiocephalic vein
☐ cephalic vein (used twice)
☐ digital artery
☐ digital vein
☐ median cubital vein
☐ radial artery
☐ subclavian artery
☐ subclavian vein
☐ superficial palmar arch
☐ superficial palmar venous arch
☐ superior vena cava

Chapter Twenty-Two *The Cardiovascular System: Vessels and Circulation* **617**

5. *Deep Trace:* The pathway a drop of blood takes to get from the aortic arch to the *capitate bone in the wrist* and back along a deep route to the superior vena cava is shaded in **figure 22.20**. Trace this flow of blood by writing in the names of the vessels in the figure. Label the vessels in order, starting at number 1, so you are figuratively tracing the pathway and naming the vessels encountered along the way.

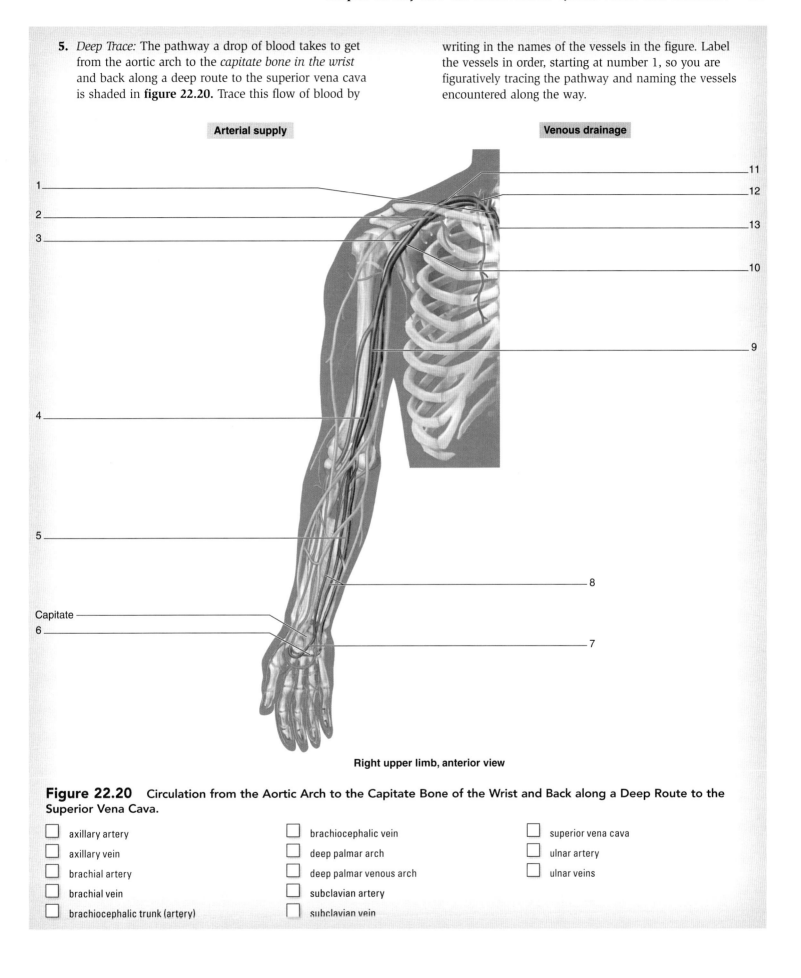

Figure 22.20 Circulation from the Aortic Arch to the Capitate Bone of the Wrist and Back along a Deep Route to the Superior Vena Cava.

- [ ] axillary artery
- [ ] axillary vein
- [ ] brachial artery
- [ ] brachial vein
- [ ] brachiocephalic trunk (artery)
- [ ] brachiocephalic vein
- [ ] deep palmar arch
- [ ] deep palmar venous arch
- [ ] subclavian artery
- [ ] subclavian vein
- [ ] superior vena cava
- [ ] ulnar artery
- [ ] ulnar veins

EXERCISE 22.13

CIRCULATION TO THE LOWER LIMB

1. Observe the lower limb of a prosected human cadaver or observe classroom models or charts demonstrating blood vessels of the lower limb.

2. Using your textbook as a guide, identify the *arteries* listed in **figure 22.21a** that supply blood to the lower limb. Then label them in figure 22.21a.

3. Using your textbook as a guide, identify the *veins* listed in figure 22.21b that drain blood from the lower limb. Then label them in figure 22.21b.

4. *Superficial Trace:* The pathway a drop of blood takes to get from the abdominal aorta to the *dorsal surface of the big toe (hallux)* and back along a superficial route to the inferior vena cava is shaded in **figure 22.22**. Trace this flow of blood by writing in the names of the vessels in the figure. Label the vessels in order, starting at number 1, so you are figuratively tracing the pathway and naming the vessels encountered along the way.

5. *Deep Trace:* The pathway a drop of blood takes to get from the abdominal aorta to the *cuboid bone in the foot* and back along a deep route to the inferior vena cava is shaded in **figure 22.23**. Trace this flow of blood by writing in the names of the vessels in the figure. Label the vessels in order, starting at number 1, so you are figuratively tracing the pathway and naming the vessels encountered along the way.

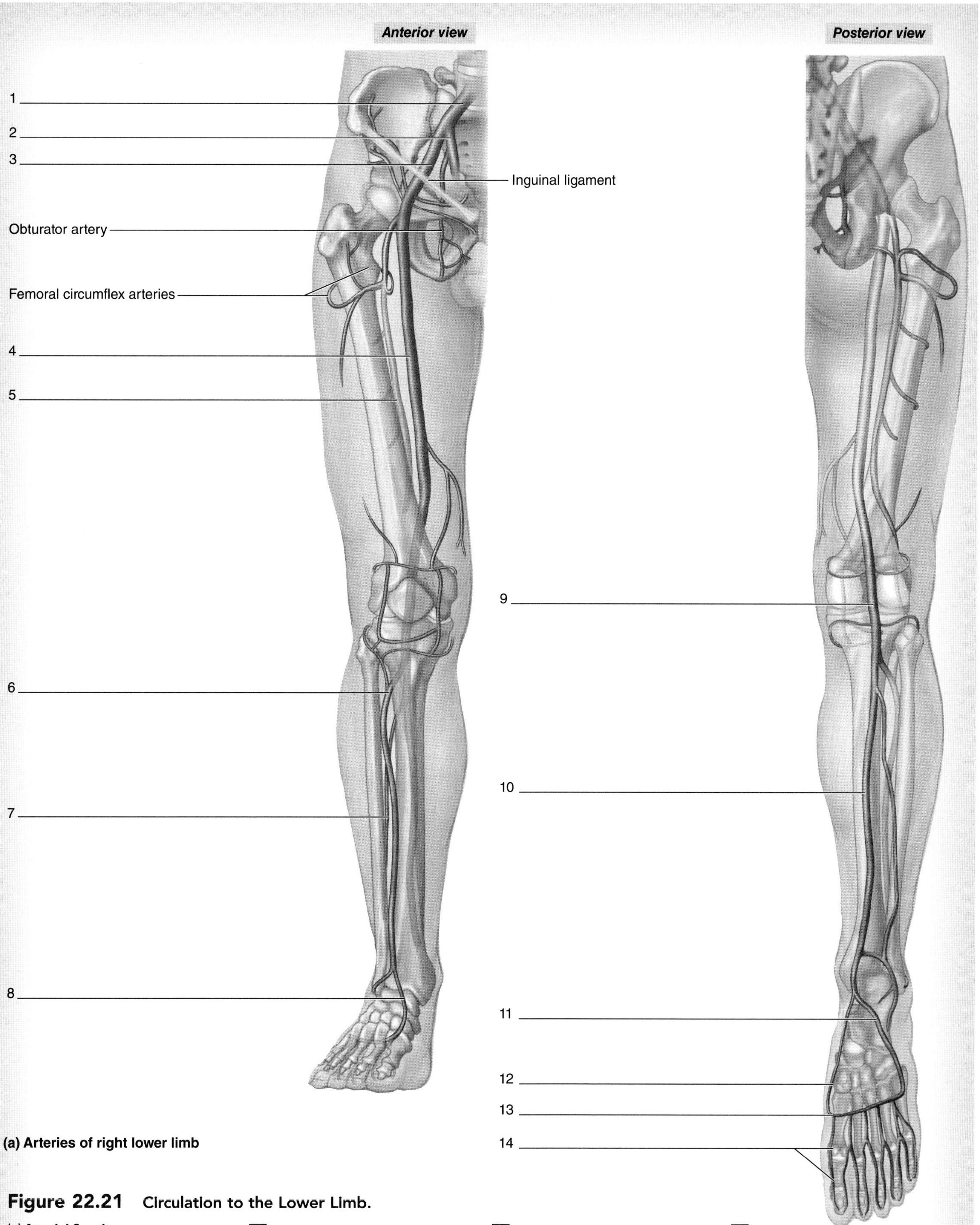

Chapter Twenty-Two The Cardiovascular System: Vessels and Circulation   619

Anterior view

1 _____
2 _____
3 _____

Obturator artery

Femoral circumflex arteries

4 _____
5 _____

— Inguinal ligament

6 _____

7 _____

8 _____

(a) Arteries of right lower limb

Posterior view

9 _____

10 _____

11 _____
12 _____
13 _____
14 _____

Figure 22.21 Circulation to the Lower Limb.

(a) Arterial Supply

☐ anterior tibial artery
☐ common iliac artery
☐ deep femoral artery
☐ digital arteries
☐ dorsalis pedis artery
☐ external iliac artery
☐ femoral artery
☐ fibular artery
☐ internal iliac artery
☐ lateral plantar artery
☐ medial plantar artery
☐ plantar arterial arch
☐ popliteal artery
☐ posterior tibial artery

(continued on next page)

620 Chapter Twenty-Two The Cardiovascular System: Vessels and Circulation

(continued from previous page)

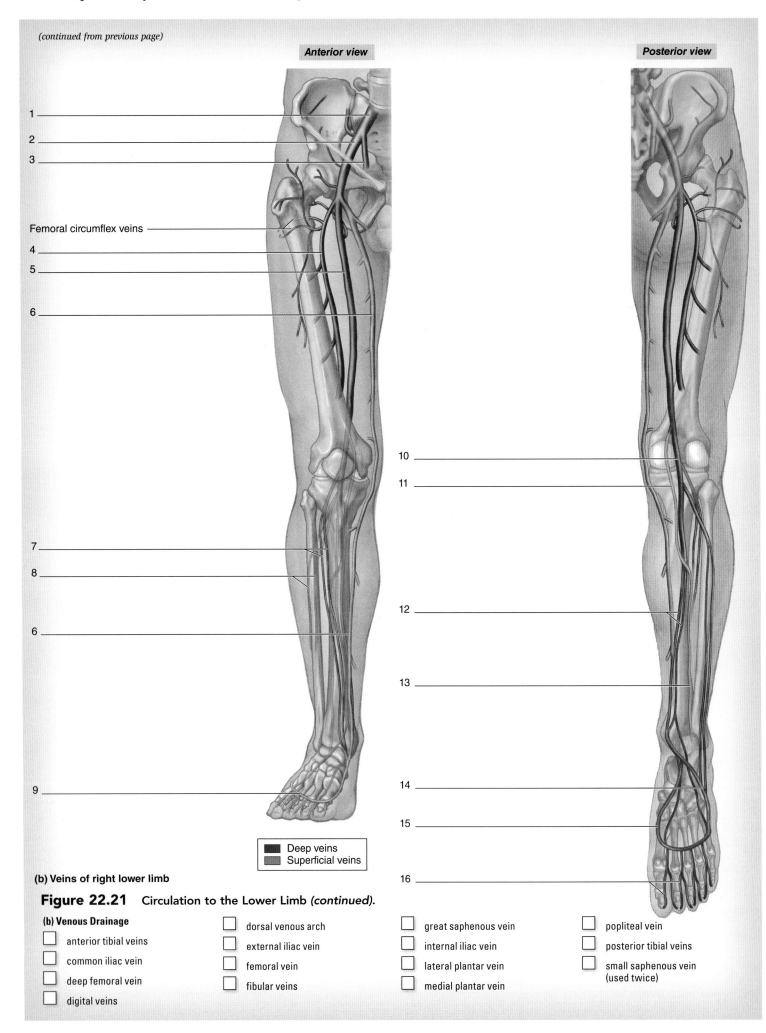

(b) Veins of right lower limb

Figure 22.21 Circulation to the Lower Limb *(continued)*.

(b) Venous Drainage

- ☐ anterior tibial veins
- ☐ common iliac vein
- ☐ deep femoral vein
- ☐ digital veins
- ☐ dorsal venous arch
- ☐ external iliac vein
- ☐ femoral vein
- ☐ fibular veins
- ☐ great saphenous vein
- ☐ internal iliac vein
- ☐ lateral plantar vein
- ☐ medial plantar vein
- ☐ popliteal vein
- ☐ posterior tibial veins
- ☐ small saphenous vein (used twice)

Chapter Twenty-Two The Cardiovascular System: Vessels and Circulation **621**

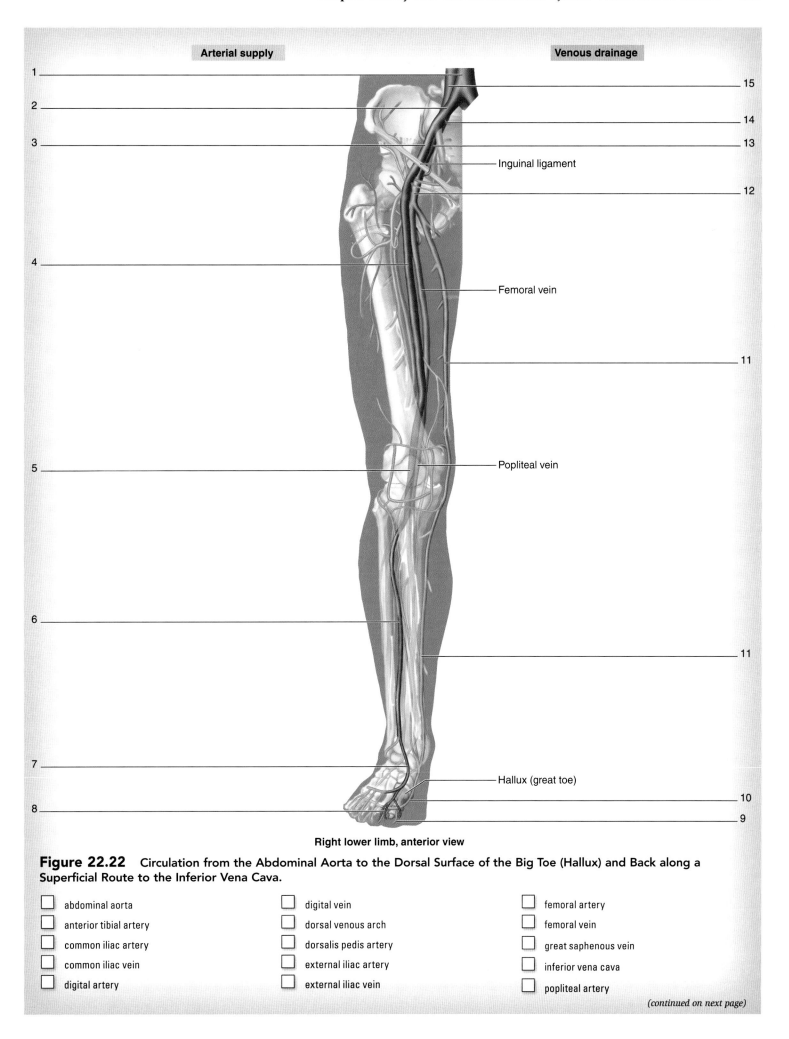

Figure 22.22 Circulation from the Abdominal Aorta to the Dorsal Surface of the Big Toe (Hallux) and Back along a Superficial Route to the Inferior Vena Cava.

- abdominal aorta
- anterior tibial artery
- common iliac artery
- common iliac vein
- digital artery
- digital vein
- dorsal venous arch
- dorsalis pedis artery
- external iliac artery
- external iliac vein
- femoral artery
- femoral vein
- great saphenous vein
- inferior vena cava
- popliteal artery

(continued on next page)

622 Chapter Twenty-Two *The Cardiovascular System: Vessels and Circulation*

(continued from previous page)

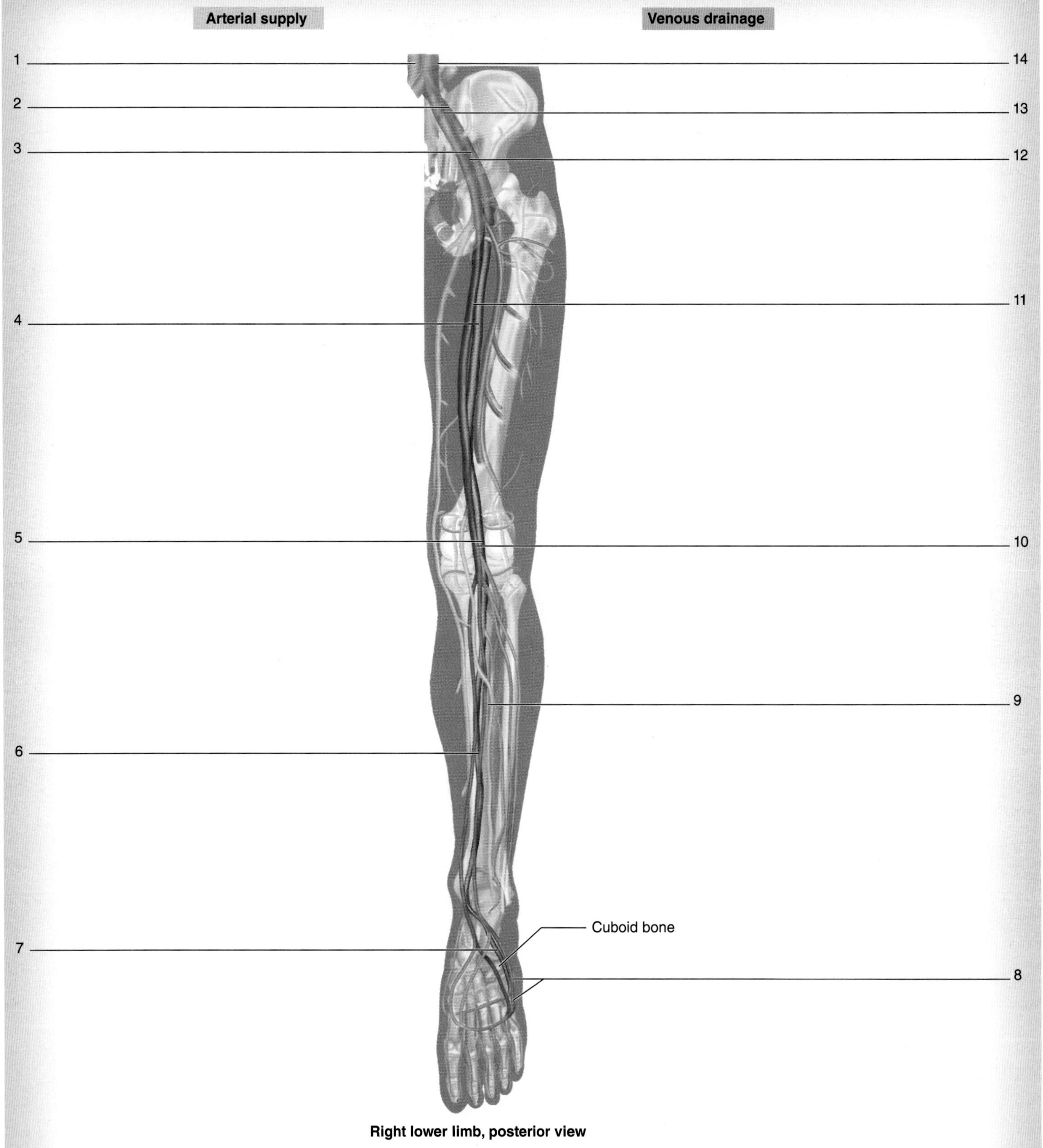

Arterial supply **Venous drainage**

1 _____   14 _____
2 _____   13 _____
3 _____   12 _____
4 _____   11 _____
5 _____   10 _____
6 _____   9 _____
7 _____   8 _____

— Cuboid bone

Right lower limb, posterior view

Figure 22.23 Circulation from the Abdominal Aorta to the Cuboid Bone in the Foot and Back along a Deep Route to the Inferior Vena Cava.

- ☐ abdominal aorta
- ☐ common iliac artery
- ☐ common iliac vein
- ☐ external iliac artery
- ☐ external iliac vein
- ☐ femoral artery
- ☐ femoral vein
- ☐ inferior vena cava
- ☐ lateral plantar artery
- ☐ lateral plantar veins
- ☐ popliteal artery
- ☐ popliteal vein
- ☐ posterior tibial artery
- ☐ posterior tibial vein

Fetal Circulation

In the fetus the lungs are nonfunctional and need only a small amount of blood to support the developing lung tissue. This blood must be oxygenated blood coming from the fetal respiratory organ: the placenta. Once the fetus is born, the circulation must change and the lungs replace the placenta as the respiratory organs. In addition, the blood returning from the placenta via the umbilical vein bypasses the liver through the ductus venosus. Thus, there are a number of **shunts** present in the fetal circulation that direct blood away from the lungs, to and from the placenta, and away from the liver. These shunts must close at birth to establish the normal postnatal circulatory pathways. In the following exercise you will identify the unique cardiovascular structures of the fetal circulation, trace the flow of blood through the fetal circulation, and identify the postnatal structures that are remnants of the fetal circulation.

EXERCISE 22.14

FETAL CIRCULATION

1. Using your textbook as a guide, label the fetal circulatory system structures listed in **figure 22.24**.

2. In the space below, trace (in words) the flow of blood from the left ventricle of the fetal heart to the placenta and back to the right atrium of the heart.

   _____

   _____

   _____

   _____

   _____

   _____

   _____

   _____

   _____

3. In **table 22.4,** write in the names of the postnatal structures that are remnants of the fetal circulation, and describe each structure's function in the fetus.

Table 22.4	Fetal Cardiovascular Structures and Associated Postnatal Structures	
Fetal Cardiovascular Structure	**Postnatal Structure**	**Function of Fetal Cardiovascular Structure**
Ductus Arteriosus		
Ductus Venosus		
Foramen Ovale		
Umbilical Arteries		
Umbilical Vein		

(continued on next page)

624 Chapter Twenty-Two *The Cardiovascular System: Vessels and Circulation*

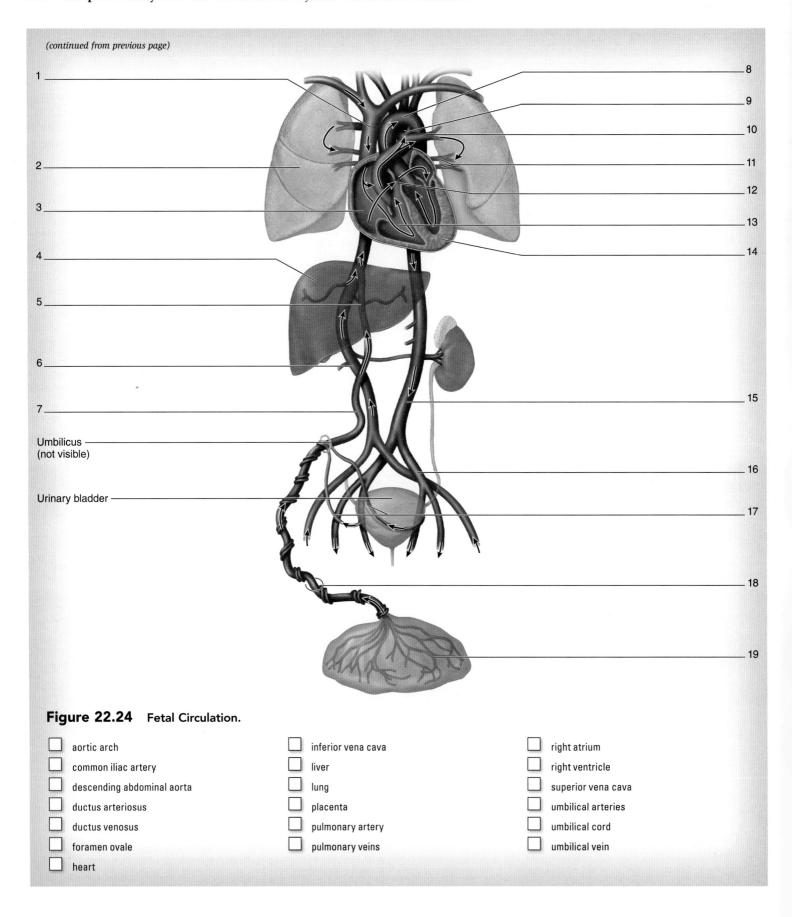

Figure 22.24 Fetal Circulation.

- [ ] aortic arch
- [ ] common iliac artery
- [ ] descending abdominal aorta
- [ ] ductus arteriosus
- [ ] ductus venosus
- [ ] foramen ovale
- [ ] heart
- [ ] inferior vena cava
- [ ] liver
- [ ] lung
- [ ] placenta
- [ ] pulmonary artery
- [ ] pulmonary veins
- [ ] right atrium
- [ ] right ventricle
- [ ] superior vena cava
- [ ] umbilical arteries
- [ ] umbilical cord
- [ ] umbilical vein

Physiology

Blood Pressure and Pulse

The blood vessels you observed in the histology and gross anatomy exercises in this chapter are the conduits that provide the passageways for blood to flow through the body. As the blood flows through these vessels, it exerts force on the vessel walls. This force (per unit area) is termed **blood pressure.** Blood flow in the systemic circulation is driven by blood pressure. Specifically, there must be a pressure gradient to drive the flow from the heart through the vessels. Contraction of the ventricles of the heart generates this pressure gradient. It follows then, that the greater the blood pressure gradient, the greater the blood flow through the vessels. Significant drops in blood pressure may lead to insufficient oxygen and nutrient delivery to the tissues. Without sufficient oxygen and nutrients, tissue death (necrosis) may occur.

Blood pressure within blood vessels is pulsatile (not smooth); that is, blood pressure waxes and wanes with each heartbeat (cardiac cycle). Pressure is highest as the ventricles contract **(systole)** and lowest when the ventricles relax **(diastole).** In a clinical setting, we refer to these two pressures (maximum and minimum) as the **systolic** and **diastolic** pressures. A clinically "normal" or "safe" blood pressure is 110/70 (110 "over" 70), where 110 corresponds to the systolic pressure and 70 corresponds to the diastolic pressure. A clinically relevant pressure is the **pulse pressure,** or the difference between the systolic blood pressure and diastolic blood pressure. The expansion and recoil of an artery associated with the pulse pressure can be palpated in arteries that are close to the body's surface—locations called **pulse points.** This is referred to simply as "taking a pulse." Counting the number of "pulses" in an artery per minute is an indirect measure of heart rate. Another clinically significant measurement is the **mean arterial blood pressure (MAP),** which is the average pressure exerted on the arterial blood vessels over time. Because the ventricles spend significantly more time in relaxation than in contraction during a single cardiac cycle, the MAP is closer in magnitude to the diastolic blood pressure. Mathematically, MAP can be calculated as follows:

$$MAP = 1/3 \text{ pulse pressure} + \text{diastolic pressure}$$

Blood pressure can also be measured indirectly with the **auscultatory method** (*auscultatare*, to listen), which involves the use of a stethoscope (*stetho-*, chest + *skop*, to look at) and **sphygmomanometer** (*sphygmo-*, pulsation + *manometer*, sparse measure). The sounds heard are referred to as **Korotkoff sounds.** To measure blood pressure using the auscultatory method, you place an inflatable cuff around the arm and inflate it until the pressure in the cuff exceeds systolic blood pressure in the brachial artery, which causes arterial blood flow in the artery to cease. Air is then released from the cuff through a valve, causing the pressure in the cuff to drop. When pressure in the the cuff falls below systolic blood pressure, blood momentarily pushes open the artery and flows through. This creates a sharp tapping sound, the first Korotkoff sound. The pressure at which the first Korotkoff sound is heard is an approximate measure of systolic blood pressure. As pressure continues to drop in the cuff, blood pressure is able to keep the brachial artery open for longer periods of time with each subsequent cardiac cycle. Each time the artery closes (during diastole), the blood flow becomes turbulent (*turbulentus*, restless), and creates a noise that can be heard through the stethoscope. Sounds are audible until the pressure in the cuff drops below the diastolic blood pressure, at which time the blood flow is smooth (no detectable turbulence). The pressure at which there is a cessation (stopping) of the sounds is an approximate measure of diastolic blood pressure.

INTEGRATE

CONCEPT CONNECTION

Blood flow is directly proportional to *changes* in pressure and inversely proportional to the resistance of the vessels.* Thus, as blood pressure increases, blood flow increases and as blood pressure decreases, blood flow decreases. In contrast, as resistance increases, blood flow decreases and as resistance decreases, blood flow increases. The following equation shows this relationship: $Q = \Delta P/R$, where Q is flow, ΔP is a change in pressure, and R is the resistance. Resistance is a property of both the blood and the blood vessels and is influenced by three main factors: viscosity of the blood, length of the vessel, and diameter of the vessel. **Viscosity** is a measure of the "thickness" of the blood. Molasses, for instance, has a much higher viscosity than water. In blood, the number of formed elements, particularly red blood cells, is the major factor that affects blood viscosity. For instance, a patient with an elevated hematocrit would have an increase in blood viscosity, thereby increasing the resistance in the vessels. In addition, increasing the length of the vessels, as in the case of obesity, increases resistance.

While viscosity and length influence overall resistance, they are factors that do not change on a minute-to-minute basis. The most dramatic way to alter the resistance of a vessel is to alter the diameter of the lumen of the vessel. Specifically, decreasing the diameter of lumen of the vessel dramatically increases the resistance. Imagine water flowing through a garden hose (small diameter) versus a fire hose (large diameter). You can imagine that the flow of water out of the fire hose is much greater than that out of the garden hose. Based on the mathematical relationship relating flow to resistance, when there are dramatic changes in resistance, there will be dramatic changes in blood flow (when blood pressure is maintained). Therefore, it is no surprise that the fastest way to change blood flow to an organ is to change the diameter of the lumen of the vessel that supplies that organ with blood. This is controlled at the level of arterioles. Recall the smooth muscle in the tunica media of these vessels. When that muscle contracts, it closes down the arteriole, dramatically increasing resistance and decreasing blood flow.

We can also consider this relationship when looking at the blood vasculature system as a whole. When systemic blood pressure drops (for example, in an accident victim who loses a great deal of blood), the sympathetic nervous system stimulates widespread constriction of blood vessels all over the body. This causes an increase in total peripheral resistance (TPR), which helps to increase blood pressure.

While the relationship between flow, pressure, and resistance may seem specific to the cardiovascular system, the same principles can be applied to any situation in which fluid flows through a tube. As you will see in chapter 24, air flow through the upper respiratory tract can be described using the same mathematical formula. In the respiratory system, you will see that air flows due to changes in pressure (ΔP), and that changes in resistance greatly influence air flow.

*In a direct relationship, when variable A increases, variable B also increases. In an inverse relationship, when variable A increases, variable B decreases.

EXERCISE 22.15 Ph.I.L.S.

Ph.I.L.S. LESSON 24: ECG AND FINGER PULSE

In this laboratory exercise, you will monitor blood flow and pulse in a virtual subject, and you will relate the subject's heart sounds and ECG to the events of the cardiac cycle.

Before you begin, familiarize yourself with the following concepts (use your main textbook as a reference):

- the cardiac cycle
- the systemic circuit of blood flow
- the pulsatile flow of blood
- arterial distensibility

1. Open Ph.I.L.S. Lesson 24: ECG and Finger Pulse (**figure 22.25**).

2. Click to the "test sound" button in the middle of the window to check if the sound is working. You should hear heart sounds; if you do not, please contact your instructor or your campus internet/computer person for help.

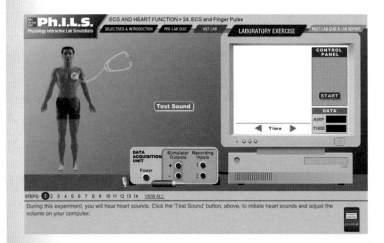

Figure 22.25 Opening Screen for Ph.I.L.S. Lesson 24: ECG and Finger Pulse.

3. Click the "power" switch on the data acquisition unit to begin the experiment.

4. Connect the finger pulse unit to the recording input by dragging the blue plug to input 2 on the data acquisition unit.

5. Click and drag the finger pulse unit to the fingers of the subject's left hand.

6. Now, connect the ECG electrodes by clicking and dragging the black plug to recording input 1 on the data acquisition unit.

7. Connect the ECG recording electrodes to the subject by clicking and dragging the red electrode to the subject's left wrist, the black electrode to his right wrist, and the green electrode to his left ankle.

8. To record and measure ECG, pulse, and heart sounds, click the "start" button on the control panel to begin recording. On the recording, the ECG is shown at the top, the pulse appears in the middle, and the heart sounds are displayed on the bottom. Click the "stop" button on the control panel after at least six heart beat cycles.

9. Click the "arrows" at the bottom of the control panel screen to scroll through your record. Center one QRS complex in the middle of the screen. Measure the time interval between an R wave and the peak of the pulse recording. Move the mouse to the top of the peak of the R wave and click the mouse to drop a marker. Then, move the mouse over the peak of the pulse wave and click to drop a second time marker. The time interval is displayed in the data window on the control panel. Enter the data below:

    Time Interval: _____

10. Make a note of any pertinent observations here:
    _____
    _____
    _____

EXERCISE 22.16

BLOOD PRESSURE AND PULSE USING A STANDARD BLOOD PRESSURE CUFF

CAUTION:
Cuff pressure should not exceed 160 mm Hg. Cuff pressures greater than 120 mm Hg should not be maintained for greater than 1 minute. Ensure that each subject is in good health and that he/she has refrained from activities or substances that may elevate heart rate and/or blood pressure (e.g., caffeine, exercise, smoking) at least 1 hour prior to testing.

Obtain the Following:
- alcohol swabs
- sphygmomanometer
- stethoscope

1. To clean the stethoscope wipe the earpieces and diaphragm with alcohol swabs.

2. Allow the subject's right arm to rest at heart level. The subject should remain as relaxed as possible throughout the exercise.

3. Locate the subject's radial artery (figure 22.18a) on the anterior surface of the right wrist.

4. Count the subject's pulse rate for one minute. Record this value in **table 22.5**.

5. Locate the right brachial artery, approximately 1.5–2 inches above the antecubital fossa (figure 22.18a). Wrap the cuff of the sphygmomanometer evenly and snugly around the subject's right arm such that the lower edge of the cuff is directly over the brachial artery, and attach Velcro® to hold in place. Ensure that the tubing and cables are not tangled or pinched.

6. Close the valve and inflate the cuff. Observe the sphygmomanometer while you are inflating the cuff and record the pressure at which the subject's pulse disappears. This corresponds to the systolic blood pressure. Deflate the cuff by opening the valve and letting air escape.

7. Firmly press the diaphragm of the stethoscope over the right brachial artery. Reinflate the cuff until you reach a pressure that is approximately 30 mm Hg greater than the systolic pressure observed in step 6.

8. Deflate the cuff at a rate of 2–3 mm Hg per second while listening for the first Korotkoff sound from the brachial artery. Make note of the pressure at which the first Korotkoff sound is heard. This is the systolic pressure. Continue to deflate the cuff while listening for the sounds to disappear. Make note of the pressure at which the sound can no longer be heard. This is the diastolic pressure. Record both the systolic blood pressure and the diastolic blood pressure in **table 22.6**.

Table 22.5 Pulse and Blood Pressure Readings while Sitting

	Sitting			
	Trial 1		Trial 2	
	Right Arm	Left Arm	Right Arm	Left Arm
Pulse Rate				
Systolic Blood Pressure				
Diastolic Blood Pressure				

Table 22.6 Pulse and Blood Pressure Readings while Lying Down, Standing, and after Vigorous Exercise

	Lying Down		Standing		Exercise	
	Right Arm	Left Arm	Right Arm	Left Arm	Right Arm	Left Arm
Pulse Rate						
Systolic Blood Pressure						
Diastolic Blood Pressure						

(continued on next page)

(continued from previous page)

9. Deflate the cuff completely.
10. Repeat this process in each arm (pulse rate: steps 4–5, blood pressure: steps 9–11). Allow the subject to rest for 2–3 minutes between recordings.
11. Next you will repeat this process for the following three conditions: 1. after allowing the subject to lie down for 3–5 minutes; 2. after allowing the subject to stand for 3–5 minutes; 3. after having the subject perform vigorous exercise for 3–5 minutes. Record one trial using the left arm and the second using the right arm (or vice versa) (pulse rate: steps 4–5; blood pressure: steps 9–11). Record your results in table 22.6.
12. Make a note of any pertinent observations here:
_____
_____
_____

EXERCISE 22.17

BIOPAC LESSON 16: BLOOD PRESSURE

CAUTION:
Cuff pressure should not exceed 160 mm Hg. Cuff pressure greater than 120 mm Hg should not be maintained for greater than 1 minute. Ensure that each subject is in good health and has refrained from activities or substances that may elevate heart rate and/or blood pressure (e.g., caffeine, exercise, smoking) at least 1 hour prior to testing.

Obtain the Following:
- surface electrodes
- BIOPAC stethoscope (SS30L)
- electrode lead set (SS2L)
- blood pressure cuff (SS19L) that fits securely around the subject's arm

1. Ensure that all the air has been expelled from the blood pressure cuff by turning the release valve counterclockwise. Turn the valve clockwise to close.
2. Prepare the BIOPAC equipment by turning the computer ON and the MP3X Data Acquisition Unit OFF. Plug in the following: BP cuff (CH 1), stethoscope (CH 2), and electrode lead set (CH 4). Turn the MP3X Data Acquisition Unit ON.
3. Clean the stethoscope earpieces and diaphragm using alcohol swabs. Clean the skin on the anterior surface of the arm 2–3 inches above the right wrist, and 2–3 inches above the medial malleolus on the right and left legs, by rubbing vigorously with an alcohol swab.
4. Apply a small quantity of electrode gel to the center of the electrodes, and place the electrode firmly on the skin in the designated locations: right arm, right leg, and left leg. Note that skin that is free of hair will make the best contact with the electrode. Attach the electrode leads to the electrodes, following the color code in **figure 22.26**.
5. Start the BIOPAC Student Lab Program. Select "L16-Bp-1" from the drop-down menu, enter the file name, and click "OK."
6. To calibrate, ensure that the cuff is not on the subject's arm. Click "calibrate" in the upper left corner of the Setup window. Inflate the cuff to 100 mm Hg and click "OK." Deflate the cuff to 40 mm Hg. Note that cuff pressure should be released at a rate of 2–3 mm Hg per second. In 20–30 seconds, the pressure should drop approximately 60 mm Hg. Click "OK." Once recording calibration data, tap the stethoscope diaphragm twice and wait for the calibration to stop. Check the calibration data to ensure that two sounds are recorded in the middle box, that the ECG is visible in the bottom box, and that the calibration data resembles **figure 22.27**. If so, continue with "Data Recording"; otherwise, "Redo Calibration."

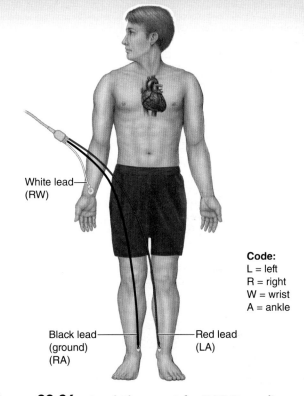

Figure 22.26 Lead Placement for ECG Recording.

7. Have the subject assume a relaxed, seated position. Allow the subject's left arm to rest at heart level and position the "artery" label over the subject's brachial artery, approximately 1.5–2 inches above the antecubital fossa. Wrap the cuff of the sphygmomanometer evenly and snugly around the subject's left arm such that the lower edge of the cuff is directly over the brachial artery,

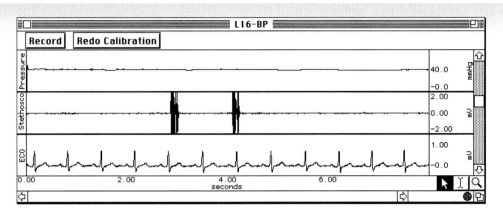

Figure 22.27 Calibration Data for ECG Recording. *Courtesy of and © BIOPAC Systems, Inc.*

and attach Velcro® to hold in place. Ensure that the tubing and cables are not tangled or pinched. Position the pressure dial indicator such that it can be easily visualized.

8. To record blood pressure and ECG data, click "record."

9. Inflate the cuff to 160 mm Hg. Click "OK." Release pressure at a rate of 2–3 mm Hg/second and insert an event marker by clicking "F9" when the Korotkoff sound for systolic pressure is heard. Continue listening and note the time ("F9") when sounds are no longer heard. Click "suspend." Fully deflate the cuff. If incorrect, click "redo."

10. Repeat step 9 for the left arm "sitting up" condition. Remove the cuff and place on the subject's right arm. Repeat step 9 two times (Trials 1 and 2) for the right arm "sitting up" condition.

11. Have the subject lie down and relax. Repeat step 9 two times (Trials 1 and 2) for the right arm "lying down" condition.

12. Disconnect the electrode lead cables from the subject by releasing the electrodes at the metal clip. Have the subject perform an exercise to elevate her heart rate (i.e., running or jumping jacks). Reattach the electrode lead cables when exercise is complete. Click "resume."

13. To perform the data analysis, click "review saved data." Note that CH 1 displays cuff pressure (mm Hg), CH 3 displays the stethoscope readings (mV), and CH 4 displays ECG (mV). View the first recording segment. To magnify the area to be measured, click on the magnifying glass icon in the lower-right corner and select the segment to expand.

14. Measurement boxes appear above the marker region in the data window. To set up, indicate the channel number and measurement type for each of three boxes: CH 1, "Value" (amplitude); CH 1, "BPM" (beats per minute); CH 1, "ΔT" (time difference).

15. Locate the "I-beam" tool to the left of the magnifying glass. Select the point on the recording segment that

Table 22.7	Systolic and Diastolic Pressures				
		Systolic Pressure (mm Hg)		Diastolic Pressure (mm Hg)	
Condition	Trial	Sound Detection	Sound Average (calculate)	Sound Detection	Sound Average (calculate)
Left Arm, Sitting	1				
	2				
Right Arm, Sitting	1				
	2				
Right Arm, Lying Down	1				
	2				
Right Arm, Exercise	1				
	2				

Courtesy of and © BIOPAC Systems, Inc. Data from BIOPAC Tables 16.2, 16.4.

(continued on next page)

(continued from previous page)

corresponds to the first event marker, or systolic blood pressure. Obtain the amplitude in the "Value" measurement box, and record this value in table 22.7. Repeat for each event marker. Calculate the averages for each condition and record in **table 22.7**.

16. Use the "I-beam" tool to select an area from one R wave to the next R wave. View the "BPM" measurement box and record the value in **table 22.8**. Repeat for two additional R waves. Calculate the average BPM for three cycles within each trial, and calculate the average across trials for each condition. Record your answers in table 22.8. Use these values, and those obtained in step 14, to calculate the mean arterial pressures and pulse pressures. Record these answers in **table 22.9**.

17. Zoom in on one ECG trace between the systolic and diastolic pressure. Use the "I-beam" tool to select the area from the peak of the R wave to the beginning of the sound detected by the stethoscope. Record the ΔT. Zoom out, locate the next recording segment, and repeat the measurement. Repeat for each recording segment and record the values in **table 22.10**.

18. Make a note of any pertinent observations here:
    _____
    _____
    _____

Table 22.8 BPM Measurements

Condition	Trial	BPM (R–R Measurements)			Average BPM (calculate)	
		Cycle 1	Cycle 2	Cycle 3	Across Cycles	Between Trials
Left Arm, Sitting	1					
	2					
Right Arm, Sitting	1					
	2					
Right Arm, Lying Down	1					
	2					
Right Arm, Exercise	1					
	2					

Courtesy of and © BIOPAC Systems, Inc. Data from BIOPAC Table 16.4.

Table 22.9 MAP and Pulse Pressure Calculations

Condition	Systolic	Diastolic	BPM	Calculations	
	Sound Average (table 22.5)	Sound Average (table 22.5)	Average BPM between Trials (table 22.6)	MAP	Pulse Pressure
Left Arm, Sitting					
Right Arm, Sitting					
Right Arm, Lying Down					
Right Arm, Exercise					

Courtesy of and © BIOPAC Systems, Inc. Data from BIOPAC Table 16.5.

Table 22.10 — Timing of Heart Sounds Relative to ECG Components

Condition	Trial	Timing of Sounds	
		Delta T	Mean (calculated)
Left Arm, Sitting	1		
	2		
Right Arm, Sitting	1		
	2		
Right Arm, Lying Down	1		
	2		
Right Arm, Exercise	1		
	2		

Courtesy of and © BIOPAC Systems, Inc. Data from BIOPAC Table 16.6.

Chapter 22: The Cardiovascular System: Vessels and Circulation

Name:_____
Date:_____ Section:_____

POST-LABORATORY WORKSHEET

The ❶ corresponds to the Learning Objective(s) listed in the chapter opener outline.

Do You Know the Basics?

Exercise 22.1: Blood Vessel Wall Structure

1. Name the three layers (tunics) present in the walls of all blood vessels (except capillaries). ❶

    a. _____

    b. _____

    c. _____

2. For each of the three layers listed above, identify the basic structure. ❷

    a. _____

    b. _____

    c. _____

Exercise 22.2: Elastic Artery—The Aorta

3. Describe defining features of an elastic artery. ❸

   _____
   _____
   _____

4. Briefly describe the role of elastic fibers within an elastic artery. ❹

   _____
   _____

5. Define *vasa vasorum* and describe its purpose. ❺

   _____
   _____

Exercise 22.3: Muscular Artery

6. Describe the defining features of a muscular artery. ❻

   _____
   _____
   _____

7. Compare and contrast elastic and muscular arteries based on structure, location, and function. ❼

   _____
   _____
   _____

Chapter Twenty-Two The Cardiovascular System: Vessels and Circulation 633

Exercise 22.4: Arteriole

8. Describe the role arterioles play in regulation of blood flow. ❽

_____
_____
_____

Exercise 22.5: Vein

9. Describe unique features of a vein. ❾

_____

10. A venous valve is an infolding of the tunica _____ of the vessel; the valve contains _____ cusps, and its function is to _____. ❿

Exercise 22.6: Observing Electron Micrographs of Capillaries

11. Describe how the endothelium differs structurally in the three types of capillaries. ⓫

_____
_____
_____

12. Fill in the chart with examples of locations in the body where each type of capillary is located and state the function of each. ⓬ ⓭

Type of Capillary	Locations and Functions
Continuous	
Fenestrated	
Sinusoidal	

Exercise 22.7: Pulmonary Circuit

13. Trace the flow of blood from the right ventricle of the heart through the pulmonary circulation and back to the left atrium of the heart. Be sure to identify heart valves through which the blood flows. ⓮

right ventricle → _____ → _____ → _____ → _____ → _____ → left atrium

634 **Chapter Twenty-Two** *The Cardiovascular System: Vessels and Circulation*

Exercises 22.8–22.13: Systemic Circuit

14. Label the diagram below with the appropriate artery names. ⑮ ⑰ ⑲ ㉑ ㉗ ㉚

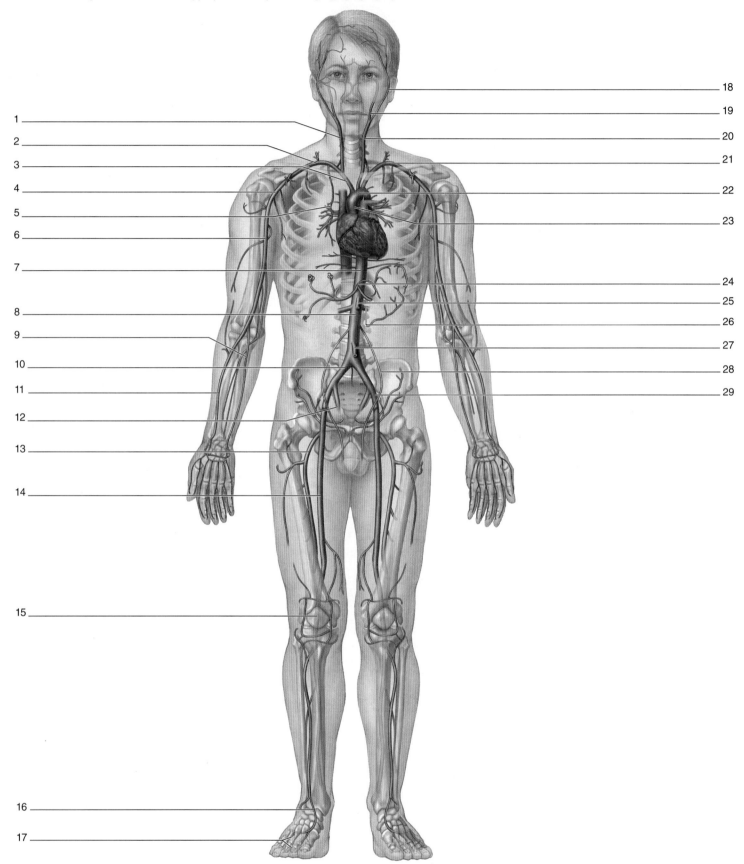

1. _____
2. _____
3. _____
4. _____
5. _____
6. _____
7. _____
8. _____
9. _____
10. _____
11. _____
12. _____
13. _____
14. _____
15. _____
16. _____
17. _____
18. _____
19. _____
20. _____
21. _____
22. _____
23. _____
24. _____
25. _____
26. _____
27. _____
28. _____
29. _____

(a) Arteries, anterior view

Chapter Twenty-Two *The Cardiovascular System: Vessels and Circulation* 635

15. Label the diagram below with the appropriate vein names.

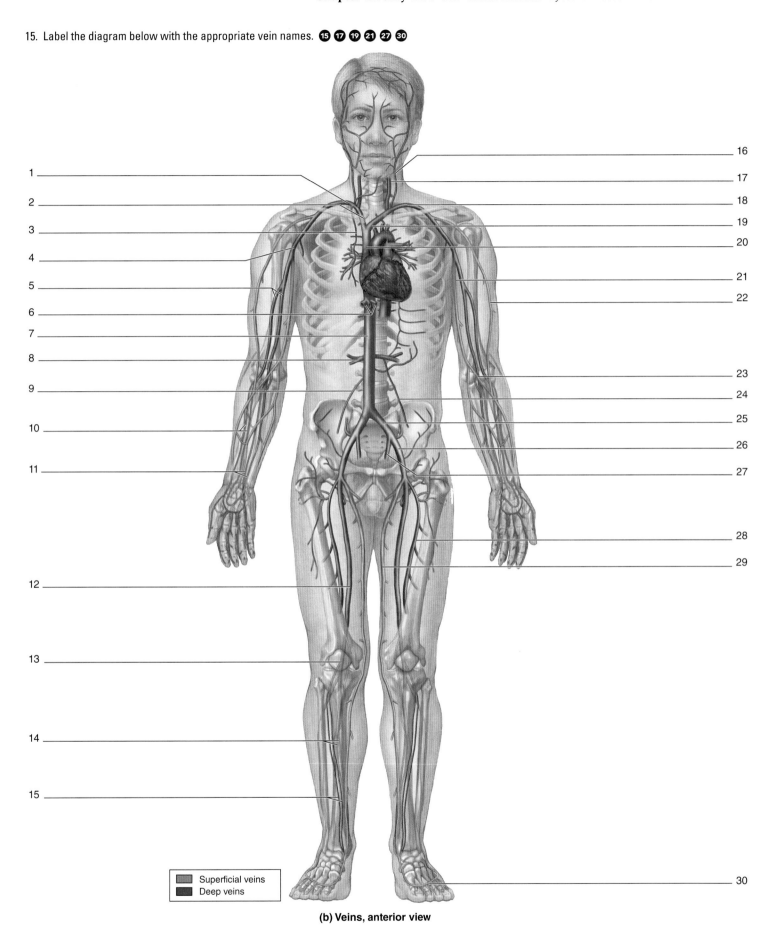

(b) Veins, anterior view

16. The vessel draining the skin overlying the right parietal bone of the skull is the _____.
17. The _____ drains blood from the brain into the right brachiocephalic vein.
18. Trace the following path of blood taken from the left ventricle of the heart, to the right kidney, and back to the right atrium.

    _____ → ascending aorta → _____ → _____ → _____ → _____ → inferior vena cava

19. Which of the following organs does *not* receive oxygenated blood from an artery that is a direct, paired branch off of the abdominal aorta?

    a. kidney

    b. spleen

    c. adrenal (suprarenal) gland

    d. gonad (testis or ovary)

20. The _____ gonadal vein empties into the renal vein on the same side of the body.
21. Name the three major veins that compose the hepatic portal circulation.

    a. _____

    b. _____

    c. _____

22. Explain the functional significance of the hepatic portal system.

    _____

    _____

    _____

23. Trace the flow of blood from the abdominal aorta to the spleen and back to the right atrium of the heart.

    abdominal aorta → _____ → _____ → _____ → _____ → _____ → inferior vena cava

24. The _____ vein drains blood from the duodenum into the hepatic portal vein.
25. The _____ vein drains blood from the sigmoid colon into the hepatic portal vein.
26. The _____ drains blood from the anterior surface of the index finger into the superficial palmar venous arch.
27. Blood draining the forearm flows from the ulnar vein into the _____.
28. Trace the path of oxygenated blood from the common iliac artery to the big toe (hallux).

    common iliac artery → _____ → _____ → popliteal artery → _____ → _____ → dorsal arterial arch → _____

29. The _____ drains blood from the superficial structures of the thigh, whereas the _____ drains blood from the deep structures of the thigh.

Exercise 22.14: Fetal Circulation

30. Circle the correct answer for each blank to complete the statement: In the fetus, the umbilical vein carry oxygenated/deoxygenated blood, while the umbilical arteries carry oxygenated/deoxygenated blood.
31. In fetal circulation, oxygenated blood flows from the placenta, to the umbilical (arteries/vein), to the right atrium of the heart. Blood returns to the placenta via the umbilical (arteries/vein).
32. Describe the function of foramen ovale in fetal circulation and identify the postnatal structure it becomes following birth.

    _____

    _____

Exercise 22.15: Ph.I.L.S. Lesson 24: ECG and Finger Pulse

33. Describe the mechanism behind pulsatile blood flow through vessels.

    _____

    _____

    _____

Chapter Twenty-Two The Cardiovascular System: Vessels and Circulation

34. Describe the blood pressure changes that occur during the cardiac cycle that allow you to detect a pulse by palpating the radial artery.

_____
_____
_____

Exercise 22.16: Blood Pressure and Pulse Using a Standard Blood Pressure Cuff

35. How may a sphygmomanometer and stethoscope be used to measure blood pressure?

_____
_____
_____

36. What properties of arteries enable pulsatile flow of blood?

_____
_____
_____

Exercise 22.17: BIOPAC Lesson 16: Blood Pressure

37. State how MAP is calculated. Then explain why MAP is closer in value to a subject's diastolic blood pressure rather than that subject's systolic blood pressure.

_____
_____
_____

38. Describe how blood pressure changes following vigorous exercise.

_____
_____

39. Trace the flow of a drop of blood from the left ventricle of the heart to the **left** ovary and back to the left ventricle of the heart by filling in the blanks below:

left ventricle → _____ → _____ → _____ → _____ → _____ → _____ → ovary →

_____ → _____ → _____ → right atrium → _____ → _____ → pulmonary semilunar valve →

_____ → _____ → capillary bed in lung → _____ → _____ → _____ → left ventricle

40. Trace the flow of a drop of blood from the left ventricle of the heart to the right styloid process of the radius (lateral wrist) and back to the left ventricle of the heart (use superficial vessels for venous return) by filling in the blanks below:

left ventricle → _____ → _____ → _____ → _____ → _____ → _____ →

_____ → capillary bed in wrist → _____ → _____ → _____ → _____ → _____ → _____ →

_____ → right atrium → _____ → _____ → _____ → _____ → _____ → capillary

bed in lung → _____ → _____ → _____ → left ventricle

Can You Apply What You've Learned?

41. Explain why large blood vessels have their own blood vessels (vasa vasorum) in the tunica externa.

_____
_____

42. A physician wishes to place a balloon catheter into the left coronary artery of his patient. A catheter is placed into the femoral artery just inferior to the inguinal ligament, and is threaded backward through the arterial system until it reaches the ascending aorta. From there, the catheter will be threaded into the left coronary artery. List all of the arteries, in order, that the catheter will pass through as it travels from the femoral artery to the left coronary artery.

   a. _____
   b. _____
   c. _____
   d. _____
   e. _____
   f. _____
   g. _____
   h. Left coronary artery

43. A physician wishes to place a central line (a catheter used to repeatedly administer drugs such as chemotherapy drugs into a patient's circulatory system) into the right atrium of the patient's heart. To do this, the physician will place the catheter into the basilic vein just superior to where it branches from the median cubital vein. The physician will then guide the catheter along the venous system until it reaches the right atrium. List, in order, the veins through which the central line passes as it travels from the basilic vein to the right atrium of the heart.

   a. _____
   b. _____
   c. _____
   d. _____
   e. Right atrium
   f. _____

44. After the central line was placed (question 43), it pinched off and failed to work when the patient held a heavy weight in her hand and the inferior movement of her clavicle compressed the central line within the subclavian vein. The physician decided to remove the central line from the basilic vein and place another central line into the internal jugular vein (IJV). The IJV provides a more direct route to the heart and avoids the problem of catheter pinch-off. List the veins, in order, through which the central line passes as it travels from the internal jugular vein to the right atrium of the heart.

   a. _____
   b. _____
   c. _____
   d. Right atrium

45. A patient suffers a stroke of the middle cerebral artery that cuts off the blood supply to the right parietal lobe of the brain. What neurological deficits would result?

   _____
   _____
   _____

46. Discuss why blood pressure might differ in the right and left arms.

   _____
   _____
   _____

47. As we age our arteries become stiff and less flexible. Discuss what might happen to the resistance and blood flow through the arteries in aging adults.

   _____
   _____
   _____

48. How would the stiffening of the arteries that occurs with age influence the time interval between the R wave (ECG) and the peak of the pulse?

Can You Synthesize What You've Learned?

49. Local factors are important for regulating blood flow to tissue capillary beds. For example, in skeletal muscle tissue that is actively contracting (as during exercise), levels of oxygen start to decrease as the oxygen is utilized, and levels of carbon dioxide increase as it is produced during cellular respiration. Keeping this in mind, answer the following questions:

    a. Do you think that increasing levels of carbon dioxide in the tissues would cause the smooth muscle forming the sphincters in arterioles to contract or relax?

    b. Why?

    c. What effect do you think increasing levels of oxygen would have on the action of arteriolar smooth muscle?

50. Specialized capillary beds are located throughout the body and are adapted for their specific physiological function. Recall from chapter 14 (Nervous Tissues) that the blood vessels that supply nervous tissues of the brain are surrounded by glial cells called astrocytes, which collectively form the blood-brain barrier. Similarly, blood vessels that form the choroid plexus within the brain ventricles are surrounded by glial cells called ependymal cells, which collectively form the blood-cerebrospinal fluid (CSF) barrier. What type of capillary (continuous, fenestrated, or sinusoids) would you expect to find in each of these areas of the brain, and why?

51. Given what you have just learned about the contents of the blood entering the liver from the hepatic portal system, discuss the type of capillaries located within the liver and state why these capillaries might be advantageous given their structure and function.

52. Preeclampsia is a condition that can occur in pregnant women that results in elevated blood pressure. While the origin of preeclampsia is unclear, there is evidence that substances that cause vasoconstriction may be released in the blood. Describe how these substances will increase blood pressure. Then discuss why this could be life threatening for both mother and child.

53. A patient enters a clinic for a routine physical examination. The nurse assigned to the patient begins to record the patient's blood pressure using a sphygmomanometer and stethoscope. As the nurse inflates the cuff to 160 mm Hg, there is an emergency down the hall that requires her attention. Rather than releasing the valve, she keeps the cuff inflated. Describe the risks to the patient that are involved when leaving the cuff inflated for greater than 1 minute.

54. Dan is diagnosed with atherosclerosis and coronary artery disease. He undergoes coronary bypass surgery to restore blood flow to his heart. Discuss why doctors might suggest that Dan limit vigorous exercise during his recovery.

CHAPTER 23

The Lymphatic System and Immunity

INTRODUCTION

The **lymphatic system** functions closely with two other body systems: the **cardiovascular system** and the **immune system**. The lymphatic system aids the cardiovascular system by returning excess fluid to the blood to maintain fluid balance, blood volume, and blood pressure. The lymphatic system facilitates the immune system in defending against foreign substances.

The terms *lymph* and *lymphatic* come from the Latin word *lympha*, which means "pure spring water." Although lymph is not exactly comparable to spring water in its composition, the fluid is relatively clear and free of suspended material because it contains no red blood cells and only limited amounts of plasma proteins. Lymphatic vessels return approximately 1–3 L of fluid to the cardiovascular system each day. When lymphatic vessels become blocked or damaged so lymph cannot flow freely through them, severe **edema** (*oidema*, a swelling) typically develops in the body part distal to the location of the obstruction.

Because most lymphatic structures are anatomically small and difficult to see in a human cadaver or on isolated organ preparations, the majority of the exercises in this chapter involve ➡

OUTLINE AND LEARNING OBJECTIVES

Histology 644

Lymphatic Vessels 644

EXERCISE 23.1: LYMPHATIC VESSELS 644
1. Identify lymphatic vessels and valves when viewed through a dissecting microscope
2. Identify lymphatic vessels and valves when viewed through a compound microscope
3. Compare and contrast the structure and function of a lymphatic vessel with the structure and function of a vein

Mucosa-Associated Lymphatic Tissue (MALT) 645

EXERCISE 23.2: TONSILS 646
4. Identify the three tonsils and their characteristic structures when viewed through a microscope
5. Distinguish pharyngeal, palatine, and lingual tonsils from each other histologically
6. Describe the structure and function of a tonsil

EXERCISE 23.3: PEYER PATCHES 647
7. Identify Peyer patches when viewed through a microscope
8. Describe the structure and function of a Peyer patch

EXERCISE 23.4: THE VERMIFORM APPENDIX 648
9. Identify the vermiform appendix and its structures when viewed through a microscope
10. Describe the structure and function of the vermiform appendix

Lymphatic Organs 648

EXERCISE 23.5: LYMPH NODES 648
11. Identify the parts of a lymph node and its structures when viewed through a microscope
12. Describe the structure and function of a lymph node
13. Trace the flow of lymph through a lymph node

EXERCISE 23.6: THE THYMUS 651
14. Identify the thymus gland and its structures when viewed through a microscope
15. Describe the structure and function of the thymus
16. Identify thymic corpuscles as viewed through the microscope, and explain their functional significance

EXERCISE 23.7: THE SPLEEN 653
17. Identify the spleen and its structures when viewed through a microscope
18. Explain the structural and functional differences between red pulp and white pulp of the spleen

Gross Anatomy 655

EXERCISE 23.8: GROSS ANATOMY OF LYMPHATIC STRUCTURES 655
19. Identify gross anatomical lymphatic structures on classroom models

 **MODULE 10: LYMPHATIC SYSTEM**

histological observations of lymphatic tissues and organs. Your histological observations will be complemented by observations of classroom models that show lymphatic organs such as lymph nodes, the spleen, and the thymus. If human cadavers are used in your laboratory, most lymphatic structures will be studied on the cadaver in conjunction with other organ systems, such as the respiratory and digestive systems, rather than within this laboratory exercise. You will also explore the functional relevance of lymphatic tissues and lymph, with a review of the immune system and the cell types involved in eliminating infectious agents.

Chapter 23: The Lymphatic System and Immunity

Name: _____

Date: _____ Section: _____

PRE-LABORATORY WORKSHEET

Also available at www.connect.mcgraw-hill.com

1. What is *lymph?*

   _____

   _____

   _____

2. List three functions of the lymphatic system:

   a. _____

   b. _____

   c. _____

3. List the three major organs of the lymphatic system:

   a. _____

   b. _____

   c. _____

4. The basic unit of lymphatic tissue is called a _____.

5. *MALT* stands for _____

   _____.

6. Lymph draining from the left side of the head and neck, the left thorax, and the entire body below the thorax, drains into a lymphatic vessel called the

   _____.

7. A lymphatic organ that filters blood is the _____.

8. A lymphatic organ located within the thoracic cavity just deep to the sternum is the _____.

Histology

Lymphatic Vessels

Lymphatic vessels are very similar in structure to veins because they carry fluid at low pressure from the tissues back toward the heart.

EXERCISE 23.1

LYMPHATIC VESSELS

1. Obtain a *dissecting* microscope and a slide demonstrating lymphatic vessels.

2. Place the slide on the microscope stage. Bring the tissue sample into focus on low power and locate a lymphatic vessel **(figure 23.1)**. Then move the microscope stage to scan the length of the vessel until you locate a valve. If necessary, change to high power and focus in on the valve to see its structure more clearly.

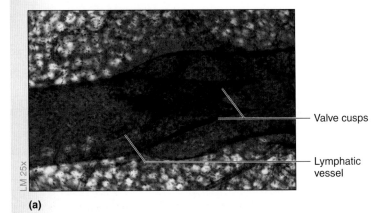

(a)

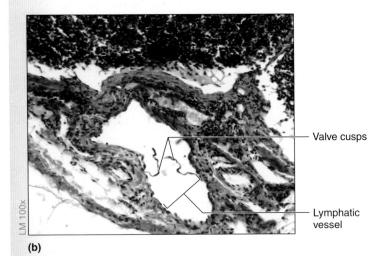

(b)

Figure 23.1 Lymphatic Vessels. (a) View of lymphatic vessels and valves as seen through a dissecting microscope. (b) Cross section through a lymphatic vessel showing a valve as seen through a compound microscope.

3. Using figure 23.1a as a guide, identify the following structures:

   ☐ lymphatic vessel   ☐ valve cusps

4. Use a compound microscope to observe a slide of lymphatic vessels.

5. Place the slide on the microscope stage. Bring the tissue sample into focus on low power and locate a cross section through a lymphatic vessel (figure 23.1b). The walls of lymphatic vessels contain the same three tunics as the walls of blood vessels, but they are not as well defined. The valves are extensions of the tunica intima of the vessel (as they are in veins). Do you see any red blood cells within the lumen of the vessel? (yes/no) Would you expect to find red blood cells within lymphatic vessels? (yes/no) Do you see any white blood cells within the lumen of the vessel? (yes/no) Would you expect to find white blood cells within lymphatic vessels? (yes/no)

6. Using figure 23.1b as a guide, identify the following structures:

   ☐ lymphatic vessel   ☐ valve cusps

7. In the spaces below, sketch the lymphatic vessel as viewed through both the dissecting and the compound microscopes.

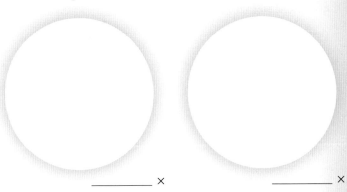

Mucosa-Associated Lymphatic Tissue (MALT)

The basic functional unit of lymphatic tissue is a **lymphatic nodule (follicle)**. A lymphatic nodule is a cluster of cells that are predominantly B-lymphocytes, with some macrophages and T-lymphocytes on its outer borders. **Figure 23.2** demonstrates a lymphatic nodule as seen through a compound microscope on high power. Notice that the central region stains lighter than the surrounding area. The central region, called the **germinal center** (*germen*, sprout), is where B-lymphocytes are most actively proliferating. Lymphatic nodules are found in many locations throughout the body. They are commonly found just deep to epithelial tissues and are even more common in locations where the epithelium changes from one type to another. Lymphatic nodules are particularly abundant in the nasal and oral cavities, and in the walls of the digestive tract. Such tissue is collectively referred to as **MALT**, or **m**ucosa-**a**ssociated **l**ymphatic **t**issue. In several regions of the respiratory and digestive tracts, the aggregations of MALT are so consistent in structure and so regular in location that they are given names. Such named aggregations of MALT include **tonsils, Peyer patches,** and the **vermiform appendix**. Table 23.1 summarizes the characteristics of MALT.

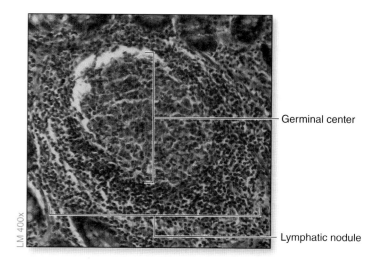

Figure 23.2 Lymphatic Nodule. The germinal center of the nodule consists of actively proliferating B-lymphocytes. The marginal zone consists of less actively proliferating B-lymphocytes as well as T-lymphocytes and macrophages.

Table 23.1	Named Aggregations of MALT (Mucosa-Associated Lymphatic Tissue)		
Named Aggregation of MALT	**Description and Location**	**Function**	**Word Origin**
Peyer Patches	Lymphatic nodules in the submucosa (deep to the epithelium) of the ileum of the small intestine	Protect against pathogens that enter the body through the intestinal mucosa	*Peyer*, Johann K. Peyer, a Swiss anatomist (1653–1712)
Tonsils	Lymphatic nodules deep to the epithelium lining the pharynx. See table 23.2 for descriptions and locations of specific tonsils.	Protect against pathogens that enter the body through the mucosa of the pharynx	*tonsilla*, a stake
Vermiform Appendix	A diverticulum (blind-ended pouch) extending from the cecum just past the ileocecal junction that contains lymphatic nodules in its walls	Protect against pathogens that enter the body through the mucosa of the cecum	*vermis*, wormlike, + *forma*, shape, + *appendix*, appendage

EXERCISE 23.2

TONSILS

1. Obtain a slide of the tonsils (lymphatic tissue) and place it on the microscope stage.
2. Tonsils are lymphatic nodules located within the walls of both the oral cavity and pharynx. They consist of lymphatic nodules interspersed between deep **crypts** (*crypt*, a pitlike depression) that open to the surface of the tonsil. The tonsils are covered superficially by the same epithelium that lines the part of the body in which they are located. There are three sets of tonsils: **pharyngeal tonsils** (called *adenoids* when they are swollen), **palatine tonsils**, and **lingual tonsils**. Table 23.2 lists characteristics of the three kinds of tonsils.

Table 23.2	Tonsils			
Tonsil	**Description and Location**	**Crypts**	**Epithelium**	**Word Origin**
Lingual	Small, paired tonsils at the base of the tongue	1–2 short crypts	Stratified squamous nonkeratinized	*lingua*, tongue
Palatine	Paired tonsils located in the fauces (the space between the mouth and pharynx) just posterior to the soft palate	10–20 deep crypts	Stratified squamous nonkeratinized	*palatum*, palate
Pharyngeal (Adenoid)	Single tonsil projecting from the roof of the nasopharynx	None	Respiratory (ciliated pseudostratified columnar)	*pharynx*, the throat, *adenos*, a gland, + *eidos*, appearance

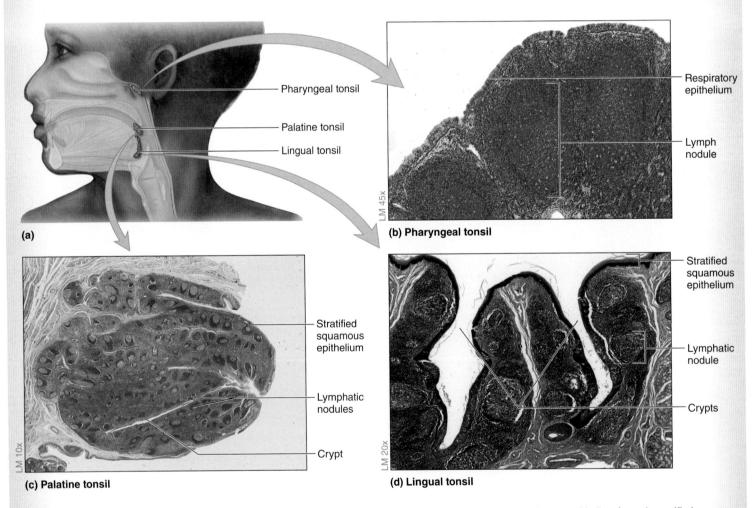

Figure 23.3 Tonsils. (a) Location of tonsils within the pharynx. (b) Pharyngeal tonsils are lined with respiratory epithelium (pseudostratified columnar with cilia and goblet cells). (c) Palatine tonsils contain numerous deep crypts and are lined with stratified squamous epithelium. (d) Lingual tonsils contain a few shallow crypts and are lined with stratified squamous epithelium.

3. Using **figure 23.3,** table 23.2, and your textbook as guides, identify the following on the microscope slide of a tonsil:

☐ crypt ☐ lymphatic nodule
☐ epithelium

4. Based on the type of epithelium covering the tonsil and the number and length of the crypts, can you determine what type of tonsil (pharyngeal, palatine, or lingual) is shown on the slide you are observing?

_____

5. In the space to the right, sketch the tonsil(s) that you observed through the microscope.

_____ ×

EXERCISE 23.3

PEYER PATCHES

1. Obtain a slide of the ileum and place it on the microscope stage.

2. Bring the tissue sample into focus on low power and scan the slide to locate the epithelial lining of the ileum, which is a simple columnar epithelium. Deep to the epithelium, look for aggregations of cells that are stained purple **(figure 23.4).** These are the aggregations of lymphocytes that compose the **Peyer patches** (aggregated lymphatic nodules of the small intestine).

3. Move the microscope stage until you have a Peyer patch at the center of the field of view, and then change to medium or high power to observe the cells within the Peyer patch.

What kind of cells are these? _____

4. Using table 23.2 and figure 23.4 as guides, identify the following structures on the slide of the ileum:

☐ epithelium ☐ Peyer patch
☐ lymphatic nodule

5. In the space below, sketch a Peyer patch as seen through the microscope.

_____ ×

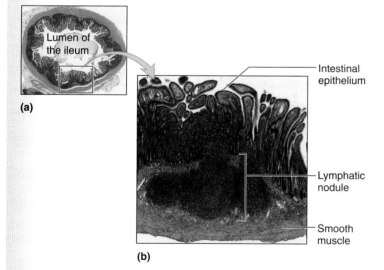

Figure 23.4 Peyer Patches. A cross section of the ileum of the small intestine is shown in (a). Deep to the intestinal epithelium and superficial to the smooth muscle that surrounds the ileum are the lymphatic nodules called Peyer patches. Peyer patches are shown in detail in (b).

EXERCISE 23.4

THE VERMIFORM APPENDIX

The **vermiform appendix** (*vermiform*, shaped like a worm) projects from the inferior region of the cecum (first part of the large intestine). It is usually about 2–4 centimeters in length and about half a centimeter in diameter. It is an evagination of the cecum and, as such, has a lumen that opens into the lumen of the cecum. Its epithelium is also a continuation of the epithelium of the cecum, and it contains numerous lymphatic nodules in its walls.

1. Obtain a slide showing a cross section of the appendix and place it on the microscope stage.

2. Bring the tissue sample into focus on low power and scan the slide until you locate the lumen of the appendix **(figure 23.5)**. Then change to medium power and bring the tissue sample into focus once again. Do you see anything within the lumen of the appendix? _____ One identifying feature of the appendix is that its lumen contains cellular debris, bacteria, and other breakdown products of food that are left over after digestion in the small intestine. What type of epithelium lines the inside of the appendix? _____

3. Observe the tissues located deep to the epithelial tissue to locate the lymphatic nodules, which are aggregates of purple-staining lymphocytes.

4. Using table 23.2 and figure 23.5 as guides, identify the following structures on the slide of the appendix:

    ☐ epithelium   ☐ lymphatic nodule
    ☐ lumen of the appendix

5. In the space below, sketch a cross section of the appendix as seen through the microscope.

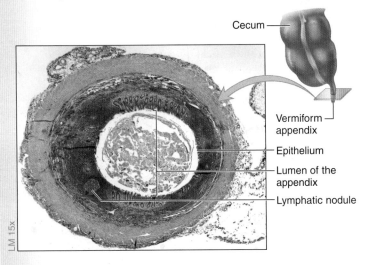

Figure 23.5 Vermiform Appendix. A cross section through the appendix, demonstrating epithelium, lymphatic nodules, and a lumen that contains breakdown products of food.

Lymphatic Organs

Lymphatic organs differ from mucosa-associated lymphatic tissue (MALT) in that they are covered by a dense connective tissue capsule and are more highly organized. Lymphatic organs include lymph nodes, the thymus, and the spleen.

EXERCISE 23.5

LYMPH NODES

Lymph nodes are small lymphatic organs located along lymphatic vessels. They filter lymph that flows through the lymphatic vessels to identify and attack lymph-borne foreign antigens.

1. Obtain a slide showing a lymph node **(figure 23.6)** and place it on the microscope stage.

2. Bring the tissue sample into focus on low power. Identify the dense irregular connective tissue **capsule,** and notice the kidney-bean shape of the lymph node (if an entire node is on the slide). A lymph node has both an outer **cortex** and an inner **medulla.** The indented region of the lymph node is called the **hilum.** This is where blood vessels enter and leave the node, and where the **efferent lymphatic vessels** drain lymph from the lymph node. **Afferent lymphatic vessels** bring lymph into the lymph node on the regions that lie opposite the hilum of the lymph node.

3. Note the dense irregular connective tissue **trabeculae** that partition the cortex into smaller regions. Each region of the cortex contains one or two **lymphatic nodules.** What types of cells are found within lymphatic nodules? _____

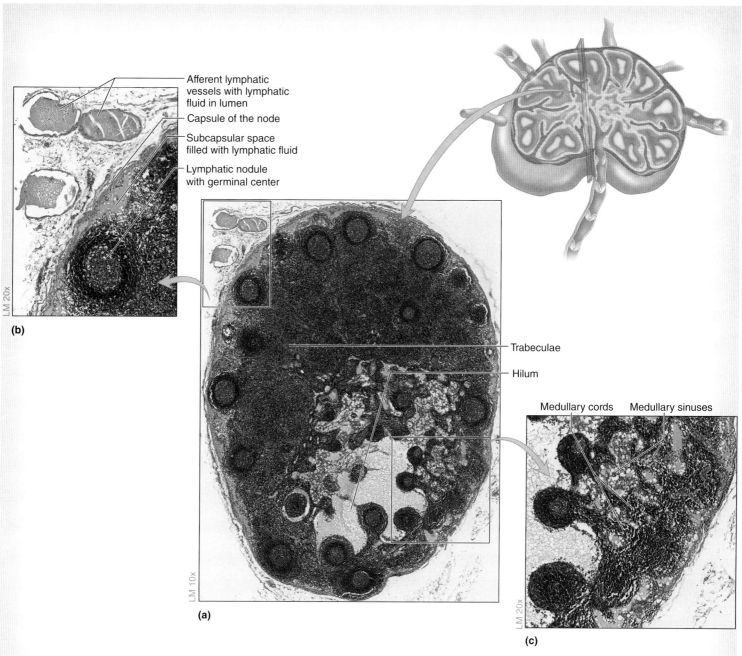

Figure 23.6 Lymph Node (a) Photomicrograph of a cross section of a lymph node that has been sectioned through the hilus. On the surface opposite the hilus, three cross sections of afferent lymphatic vessels are visible, as well as the capsule and lymphatic fluid within the subcapsular space (b). Medullary cords and spaces are visible in (c).

4. Scan the slide until the medulla is at the center of the field of view, and locate the dark-staining **medullary cords**. These support B-lymphocytes, T-lymphocytes, and macrophages. Can B- and T-lymphocytes be distinguished from each other using light microscopy?

_____

5. Identify the following structures on the tissue slide of the lymph node:

  ☐ capsule ☐ medulla
  ☐ cortex ☐ medullary cords
  ☐ hilum ☐ trabeculae
  ☐ lymphatic nodules

6. Observe the numerous spaces, called **medullary sinuses**, where lymph travels as it flows through the lymph node. **Figure 23.7** demonstrates the pathway taken by lymph as it flows through a lymph node.

7. Using figure 23.6 and **table 23.3** as guides, identify the following spaces and vessels on the slide of the lymph node:

  ☐ afferent lymphatic vessels ☐ peritrabecular spaces
  ☐ medullary sinuses ☐ subcapsular spaces

(continued on next page)

(continued from previous page)

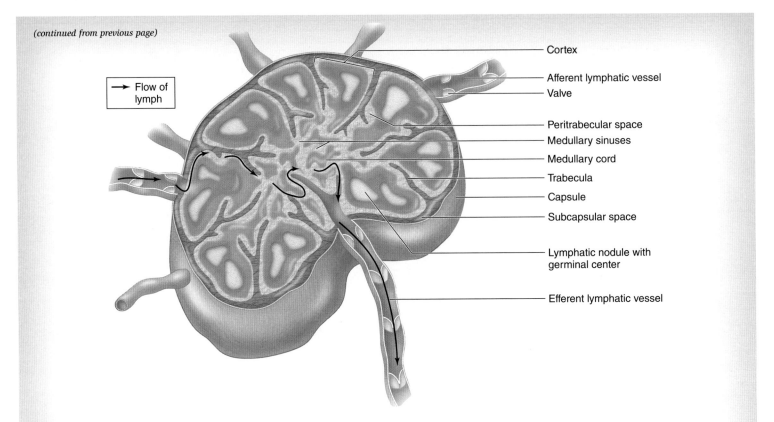

Figure 23.7 Lymph Node. Pathway that lymph takes as it flows through a lymph node.

Table 23.3	Parts of a Lymph Node	
Structure	**Description**	**Word Origin**
Afferent Lymphatic Vessel	Vessel that carries lymph into the lymph node	*affero*, to bring to
Capsule	Dense irregular connective tissue that surrounds the entire lymph node to support and protect it	*capsula*, a box
Cortex	Outer portion of the lymph node that lies deep to the capsule and contains lymphatic nodules	*cortex*, bark
Efferent Lymphatic Vessel	Vessel that carries lymph away from the lymph node; located at the hilum	*effero*, to bring out
Hilum	Indented area where efferent lymphatic vessels (and blood vessels, nerves) enter	*hilum*, a small bit
Lymphatic Nodule	Aggregate of B-lymphocytes within a germinal center, surrounded by a mantle zone, which contains T-lymphocytes, macrophages, and dendritic cells	*lympha*, clear spring water, + *nodulus*, a small knot
Medulla	Inner region of the lymph node containing medullary cords of B-lymphocytes and macrophages	*medius*, middle
Medullary Cord	Strands of B-lymphocytes, T-lymphocytes, and macrophages supported by connective tissue that are located in the medulla of a lymph node	*medius*, middle, + *chorda*, a string
Medullary Sinuses	Spaces between the medullary cords in a lymph node through which lymphatic fluid flows	*medius*, middle
Peritrabecular Space	Space located between a lymphatic nodule and a trabecula in the cortex of the lymph node	*peri-*, around, + *trabecula*, trabeculae
Subcapsular Space	Space located between a lymphatic nodule and the capsule of the lymph node	*sub-*, under, + *capsular*, capsule
Trabeculae	Invaginations of the dense irregular connective tissue capsule of a lymph node that partition the cortex into smaller compartments	*trabs*, a beam

8. In the space to the right, sketch a lymph node as seen through the microscope. Label all of the structures listed in steps 5 and 7. Then, using arrows, indicate the pathway taken by lymph as it flows through the lymph node.

_____ ×

INTEGRATE

CONCEPT CONNECTION

As discussed in chapter 5, reticular tissue is a type of connective tissue found in lymphatic organs, including the spleen and lymph nodes. Recall that reticular tissue is composed of cells and reticular fibers. Reticular fibers are a fine type of collagen that provide a loose supporting framework for the cells and aid in trapping substances. The reticular fibers help to slow the flow of lymph, allowing the contents of the lymph to come in contact with immune cells within the organ. In addition, the structure of the lymph node is such that there are many more afferent lymphatic vessels than efferent lymphatic vessels. Lymph enters the organ and meanders through the medullary sinuses, thus increasing the likelihood that a pathogen will encounter an immune cell. Once pathogens become trapped by the reticular fibers of the lymph nodes, B- and T-cells can launch an immune response. The details of the immune response are discussed in the physiology section of this chapter.

EXERCISE 23.6

THE THYMUS

The thymus is a gland located in the superior mediastinum (**figure 23.8**). It is the site of T-lymphocyte maturation, and it is also an endocrine organ, secreting the hormones **thymosin,** thymulin, and thymopoietin. The thymus is composed of two **lobes,** each separated into smaller **lobules** by connective tissue **septae** (singular: septa) **(figure 23.9).** Each lobule has a darker-staining outer cortex and a lighter-staining inner medulla. The majority of the cells within the thymus are T-lymphocytes, which are surrounded by cells of epithelial origin. Lymphocytes migrate to the thymus from the bone marrow. Once in the thymus, T-lymphocytes move from the outer cortex to the inner medulla as they complete the process of **selection.** Selection is a process by which only those T-lymphocytes that are able to recognize the MHC protein ("self" antigen) and do not attack self-antigens are permitted to leave the thymus. Selection occurs in the outer cortex, and once selection of T-lymphocytes is complete, the cells leave the thymus through venules located in the medulla. The majority of T-lymphocytes that migrate to the thymus from the bone marrow either fail to recognize the MHC protein or react to self-antigen. These cells are destroyed through a process called **apoptosis** (programmed cell death), which occurs within

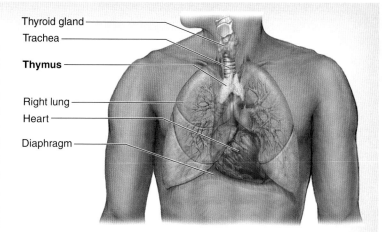

Thorax, anterior view

Figure 23.8 Location of the Thymus. The thymus is located in the superior mediastinum, deep to the sternum.

the thymic cortex. After puberty the thymus begins a gradual decline in size with age. Epithelial-derived cells in the medulla come together and form small, onionlike masses of tissue called **thymic corpuscles.** Thymic corpuscles produce chemical signals

(continued on next page)

(continued from previous page)

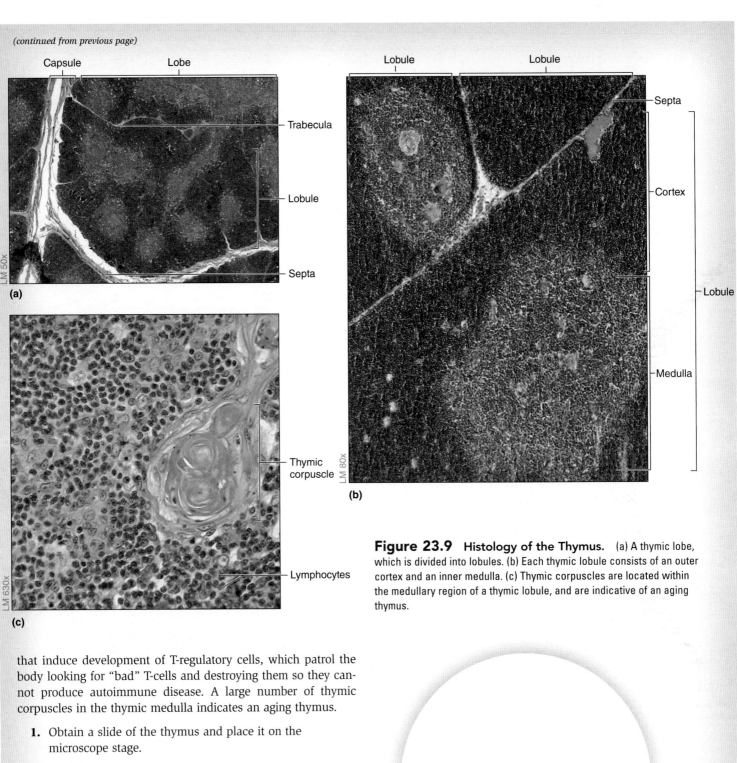

Figure 23.9 Histology of the Thymus. (a) A thymic lobe, which is divided into lobules. (b) Each thymic lobule consists of an outer cortex and an inner medulla. (c) Thymic corpuscles are located within the medullary region of a thymic lobule, and are indicative of an aging thymus.

that induce development of T-regulatory cells, which patrol the body looking for "bad" T-cells and destroying them so they cannot produce autoimmune disease. A large number of thymic corpuscles in the thymic medulla indicates an aging thymus.

1. Obtain a slide of the thymus and place it on the microscope stage.

2. Bring the tissue sample into focus at low power, then scan the slide to locate the connective tissue trabeculae that separate the gland into lobules (figure 23.9).

3. Using figure 23.9 as a guide, identify the following structures:
   - ☐ cortex
   - ☐ lobule
   - ☐ medulla
   - ☐ thymic corpuscle
   - ☐ trabeculae

4. In the space to the right, sketch the thymus as seen through the microscope. Be sure to label all the structures listed in step 3.

Chapter Twenty-Three The Lymphatic System and Immunity

EXERCISE 23.7

THE SPLEEN

The spleen is similar to a lymph node in many ways, with one major exception: The spleen filters *blood*, whereas a lymph node filters *lymph*. However, both organs filter their fluid for the purpose of identifying and eliminating foreign antigens that may be present in the fluid. The spleen receives blood through the **splenic artery,** which is a branch of the celiac trunk. Blood leaves the spleen through the **splenic vein,** which drains into the hepatic portal vein (see figures 22.13 and 22.14, pp. 610–611).

When a fresh spleen is cut, two different types of tissue are observed: Red pulp and white pulp. **Red pulp** consists of splenic sinusoids, and is red due to the large amount of blood contained within the sinusoids. It composes the majority of the splenic tissue. **White pulp** consists of aggregations of lymphocytes, and is white due to the white blood cells (leukocytes) within each mass of tissue. Because the spleen tissue sample has been stained, red and white pulp will not appear red and white in color. The red pulp will be reddish-pink in color, but the white pulp will be darker purple, with a lighter-colored central region.

1. Obtain a slide of the spleen and place it on the microscope stage. Bring the tissue sample into focus on low power and locate the dense irregular connective tissue capsule of the spleen, if present in the section on your slide **(figure 23.10).** You may also notice several connective tissue trabeculae, which are invaginations of the capsule that separate the spleen into distinct regions.

2. Notice the circular purple aggregations of cells. These aggregations constitute the white pulp of the spleen, which consists mainly of B-lymphocytes, T-lymphocytes, and some macrophages.

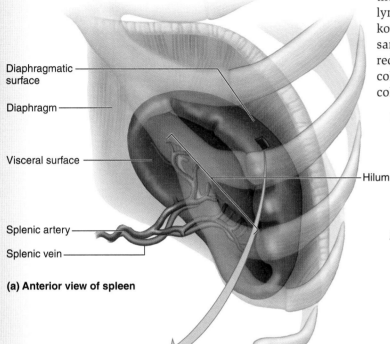
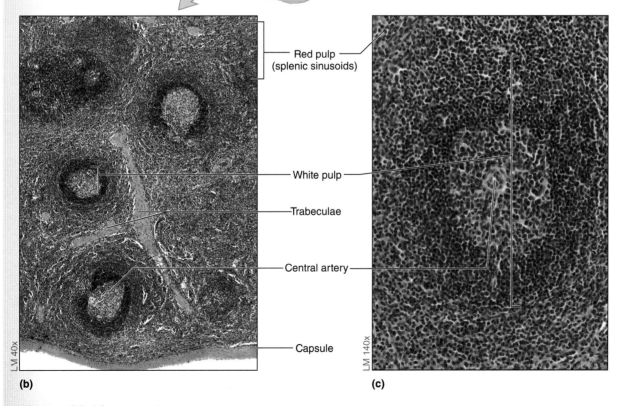

Figure 23.10 The Spleen. The spleen contains two predominant tissues: red pulp, which consists of splenic sinusoids, and white pulp, which contains lymphocytes. (a) Location of spleen. (b) Low magnification and (c) high magnification.

(continued on next page)

(continued from previous page)

3. Look for a **central artery,** which is found within each mass of white pulp. Central arteries receive blood from **trabecular arteries.** Blood coming into the spleen travels from the splenic artery to the trabecular arteries to the central arteries, and finally empties into the sinusoids or surrounding tissues of the spleen before draining into splenic veins.

4. Move the microscope stage until a mass of white pulp is in the center of the field of view and then change to high power. At this magnification you should be able to make out individual small purple-staining cells, which are lymphocytes.

5. Move the microscope stage so the red pulp of the spleen is in the center of the field of view. The red pulp of the spleen consists mainly of *splenic sinusoids* containing numerous erythrocytes. The blood-filled splenic sinusoids also contain numerous lymphocytes and macrophages. As blood travels slowly through the spleen, the lymphocytes and macrophages monitor the blood for **antigens,** which, if present, will be removed by these immune cells. In addition, the structure and function of the splenic sinusoids is designed to destroy old or abnormal erythrocytes. What is the life span of a typical erythrocyte? _____ days. (Hint: The answer is in chapter 20.)

6. Using **table 23.4** and figure 23.10 as guides, identify the following structures on the histology slide of the spleen:

   ☐ capsule   ☐ sinusoids
   ☐ central artery   ☐ trabecula
   ☐ red pulp   ☐ white pulp

7. In the space below, sketch the spleen as seen through the microscope. Be sure to label all the structures listed in step 6.

_____ ×

Table 23.4	Parts of the Spleen	
Structure	**Description**	**Word Origin**
Capsule	Dense irregular connective tissue that supports and protects the outside of the spleen; surrounds the entire spleen	*capsula*, a box
Central Arteries	Arteries located within the white pulp of the spleen	*centrum*, center, + *arteria*, the windpipe
Hilum	The indented area of the spleen where the splenic artery enters and the splenic vein exits	*hilum*, a small bit
Red Pulp	Splenic tissue consisting of splenic sinusoids lined predominantly by T-lymphocytes and macrophages	*pulpa*, flesh
Splenic Sinusoids	Very permeable capillaries, which consist of fenestrated endothelial cells and a discontinuous basement membrane	*sinus*, cavity, + *eidos*, resemblance
Trabeculae	Invaginations of the connective tissue capsule of the spleen that partition the cortex into smaller compartments	*trabs*, a beam
White Pulp	Splenic tissue consisting of lymphatic nodules that contain mainly B-lymphocytes and macrophages; each mass of white pulp contains a central artery	*pulpa*, flesh

Chapter Twenty-Three The Lymphatic System and Immunity

Gross Anatomy

EXERCISE 23.8

GROSS ANATOMY OF LYMPHATIC STRUCTURES

1. *Major Lymph Vessels of the Body:* Obtain a model of the thorax, and abdomen. Remove the heart and lungs from the thorax, and the stomach and intestines from the abdomen. Using your textbook as a guide, identify the structures listed in **figure 23.11** on the model. Then label them in figure 23.11.

2. *Mucosa-Associated Lymphatic Tissue (MALT):* Obtain a model demonstrating a midsagittal section of the head and neck, and a model of the abdomen. Using your textbook as a guide, identify the structures listed in **figure 23.12** on the model. Then label them in figure 23.12.

3. *Lymph Node:* Obtain a classroom model of a lymph node. Using table 23.3 and your textbook as guides, identify the structures listed in **figure 23.13** on the model. Then label them in figure 23.13.

4. *The Spleen:* Obtain a classroom model of the abdominal cavity, or observe the abdominal cavity of a human cadaver. Locate the spleen and identify the structures listed in **figure 23.14** on the model or the cadaver. Then label them in figure 23.14.

5. *Optional Activity:*  10: Lymphatic System—Watch the "Lymphatic System Overview" animation for a summary of the primary lymphatic structures.

INTEGRATE

LEARNING STRATEGY

Identification of the thoracic duct within the thoracic cavity of a cadaver can be challenging because of its relatively small diameter and lack of color (remember lymph does not contain erythrocytes). It travels between the esophagus and the azygos vein. Anatomists thus refer to it as the "duck between two gooses." The "duck" is the thoracic duct, and the "gooses" are the "azygoose" (azygos vein) and the "esophagoose" (esophagus).

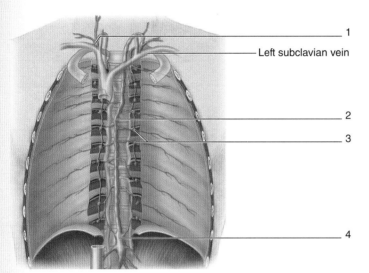

Thorax, anterior view

Figure 23.11 Major Lymph Vessels of the Body.

☐ cisterna chyli ☐ right lymphatic duct
☐ lymph nodes   ☐ thoracic duct

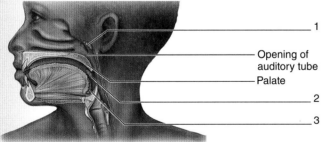

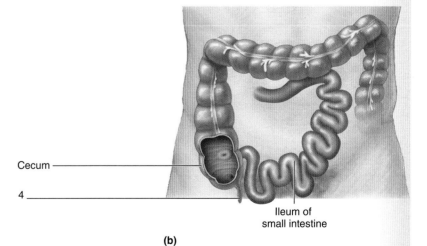

(a)   (b)

Figure 23.12 Mucosa-Associated Lymphatic Tissue (MALT).

☐ lingual tonsils   ☐ palatine tonsils   ☐ pharyngeal tonsils   ☐ vermiform appendix

(continued on next page)

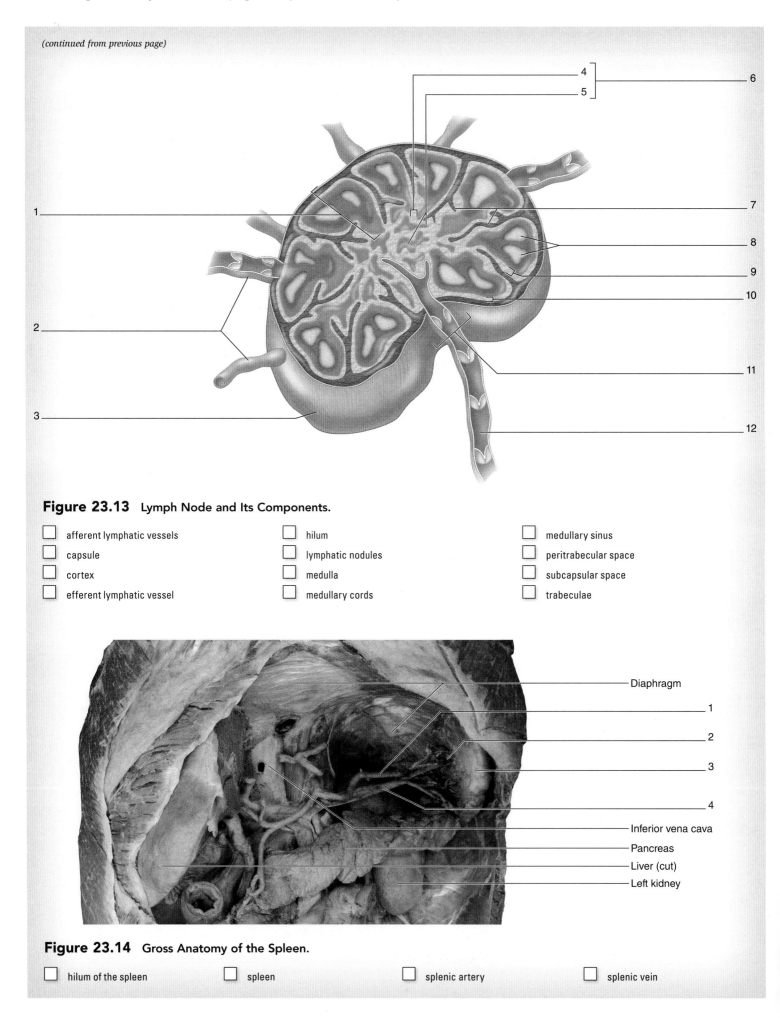

Figure 23.13 Lymph Node and Its Components.

☐ afferent lymphatic vessels
☐ capsule
☐ cortex
☐ efferent lymphatic vessel
☐ hilum
☐ lymphatic nodules
☐ medulla
☐ medullary cords
☐ medullary sinus
☐ peritrabecular space
☐ subcapsular space
☐ trabeculae

Figure 23.14 Gross Anatomy of the Spleen.

☐ hilum of the spleen
☐ spleen
☐ splenic artery
☐ splenic vein

INTEGRATE

CLINICAL VIEW
Antiserum

When a person is bitten by a venomous snake, getting medical attention as quickly as possible is essential. Without medical intervention, the venom may be lethal. Often, snake venom will target neuromuscular junctions to silence the synapses and prevent muscle contraction. This is particularly useful for the snake, for it allows the snake to quickly immobilize its prey. Unfortunately for humans, snake venom may also silence skeletal muscles that are essential for ventilation, including the diaphragm and intercostal muscles. An inability to contract these muscles means an inability to inspire, which leads to respiratory failure. Because time is limited, an **antiserum,** which contains antibodies that bind the snake venom, must be given as soon as possible.

The process of synthesizing an antiserum (often called an "antidote") involves extracting venom from a snake, diluting it, then injecting the reduced-potency venom into another animal, such as a goat or a horse. Then, the recipient animal's B-lymphocytes and T-lymphocytes identify the venom as a foreign substance and initiate an immune response against the venom. After the animal has produced the necessary antibodies, those antibodies can be extracted from the animal's blood. The extracted antibodies constitute the antiserum ("antidote") that can later be given to victims of snake bites. Having readily available antiserum against snake venom is critical for saving victims because victims do not have enough time to undergo their own immune responses—the venom is too lethal. Administration of antiserum provides a form of **passive immunity** because antibodies are given directly.

Chapter 23: The Lymphatic System and Immunity

Name: _____

Date: _____ Section: _____

POST-LABORATORY WORKSHEET

The ❶ corresponds to the Learning Objective(s) listed in the chapter opener outline.

Do You Know the Basics?

Exercise 23.1: Lymphatic Vessels

1. Describe defining features of lymphatic vessels when viewed through a dissecting microscope. ❶

   _____
   _____

2. Compare the histological structure of lymphatic vessels with that of veins. ❷ ❸

   _____
   _____

Exercise 23.2: Tonsils

3. For each of the following types of tonsils, identify characteristic structures when viewed through a microscope. ❹ ❺ ❻

   a. lingual: _____

   b. palatine: _____

   c. pharyngeal: _____

Exercise 23.3: Peyer Patches

4. Describe the histological features of Peyer patches. ❼

   _____
   _____

5. In what segment (duodenum, jejunum, ileum) of the small intestines are Peyer patches located? ❽ _____

Exercise 23.4: The Vermiform Appendix

6. Describe the appearance of the lumen of the vermiform appendix, when viewed through a microscope. ❾

   _____
   _____

7. An individual suffering from an inflamed appendix will most commonly have pain in which abdominopelvic quadrant? ❿ _____

Exercise 23.5: Lymph Nodes

8. The spaces through which lymph travels as it flows through the lymph node are _____. ⓫

9. The structures in the medulla of the lymph node that contain T-lymphocytes and macrophages are _____. ⓫

10. The vessels bringing lymph into the lymph node are _____; the vessels draining lymph from the lymph node are _____. ⓬

11. Describe the function of lymph nodes. ⓬

    _____
    _____

12. Imagine a virus that has entered a lymph node. Describe the pathway the virus would take as it traveled within the lymph node (include the types and locations of leukocytes it might encounter within the lymph node). ⓭

    _____
    _____
    _____
    _____

Exercise 23.6: The Thymus

13. The thymus contains connective tissue _____ that separate the gland into lobules.

14. What is a thymic corpuscle, and what does the presence of many thymic corpuscles tell you about the thymus? _____

_____

_____

Exercise 23.7: The Spleen

15. Describe the structure and function of splenic sinusoids.

_____

_____

16. The white pulp of the spleen is composed of _____ and functions to _____

_____

_____ .

17. The red pulp of the spleen is composed of _____ and functions to _____

_____

_____ .

Exercise 23.8: Gross Anatomy of Lymphatic Structures

18. Imagine a bacterium has entered a lymph capillary in the right arm. Describe:

    a. the pathway it would take to travel from the lymph capillary it initially enters to a lymph node in the axilla. _____

    _____

    _____

    _____

    b. the pathway it would take to get from the lymph node to the right atrium of the heart (assuming it escapes detection within the lymph nodes!).

    _____

    _____

    _____

19. Lacteals are specialized lymphatic capillaries found in the small intestine that have the special function of absorbing dietary fats. Other nutrients that are absorbed by the digestive tract are absorbed into blood capillaries, not lymphatic capillaries. Because absorbed fats are absorbed within relatively large structures (chylomicrons), they first enter lymphatic capillaries and are transported as a component of lymph within lymph vessels. Using your knowledge of the flow of both lymph and blood in the body, describe the route a dietary lipid would take to get from its location of absorption in the small intestine to its entry into the right atrium of the heart.

    _____

    _____

    _____

    _____

20. Describe the anatomic location of the three types of tonsils:

    a. lingual _____

    b. palatine _____

    c. pharyngeal _____

Chapter Twenty-Three *The Lymphatic System and Immunity*

Can You Apply What You've Learned?

21. When pharyngeal tonsils become inflamed, they are referred to as adenoids. Based on the location of the adenoids (pharyngeal tonsils), what kinds of symptoms might a patient experience due to the presence of swollen adenoids?

22. When an individual suffers from a ruptured spleen, the spleen is surgically removed so the patient does not die from internal bleeding. Using your knowledge of the circulation of the spleen, answer the following questions.

    a. Why does a ruptured spleen bleed so profusely? (Hint: What type of capillaries are found in the spleen?)

    b. Why can a patient survive without his or her spleen? (Hint: What other organ may take up the functions of the spleen?)

    c. A patient who has had his or her spleen removed will be more susceptible to certain types of infections. These would be infections that travel through the _____.

23. Discuss why lymphatic nodules are so abundant in the wall of the small intestine.

24. Describe the difference between a vaccine and an antiserum.

Can You Synthesize What You've Learned?

25. Acute appendicitis (inflammation of the vermiform appendix) requires immediate surgical removal of the appendix. In the event that the appendix bursts, its contents are released into the abdominal cavity, which can lead to widespread infection, sepsis, and death. Discuss the events that might lead to an enlarged and inflamed vermiform appendix.

26. Allergies are the result of a hypersensitivity, or enhanced immune response. Antibodies (IgE) are produced in response to an antigen, such as peanuts. These antibodies can bind directly to basophils and mast cells, which contain granules of histamine. Repeated exposure to the antigen will result in increased antibody production. These antibodies form complexes on the surface of the basophils and mast cells, thereby stressing the plasma membranes to the point of lysis. Discuss how rupturing basophils and mast cells may lead to an allergic reaction, particularly in tissues that line the respiratory tract.

CHAPTER 24

The Respiratory System

OUTLINE AND OBJECTIVES

Histology 664

Upper Respiratory Tract 664

EXERCISE 24.1: OLFACTORY MUCOSA 665

① Describe the three layers (mucosa, submucosa, adventitia) of the wall of the respiratory tract

② Identify olfactory epithelium when viewed through a microscope

③ Compare and contrast the structure and function of olfactory receptor cells, basal cells, and supporting (ciliated) cells

Lower Respiratory Tract 666

EXERCISE 24.2: THE TRACHEA 666

④ Identify the trachea and its associated structures when viewed through a microscope

⑤ Describe the epithelium that lines the trachea

⑥ Describe the function of the C-shaped tracheal cartilages and the trachealis muscle

EXERCISE 24.3: THE BRONCHI AND BRONCHIOLES 668

⑦ Identify large and small bronchi, and bronchioles when viewed through a microscope

⑧ Describe the changes that occur with epithelium, cartilage, and smooth muscle, moving from large bronchus to small bronchus, to bronchioles

Lungs 669

EXERCISE 24.4: THE LUNGS 670

⑨ Identify a slide of the lung and its associated structures when viewed through a microscope

⑩ Describe the three structures that compose the respiratory membrane

⑪ Explain the function of type I alveolar cells, type II alveolar cells, and alveolar macrophages

⑫ Identify respiratory portion structures on a histology slide of the lung

Gross Anatomy 671

Upper Respiratory Tract 671

EXERCISE 24.5: SAGITTAL SECTION OF THE HEAD AND NECK 671

⑬ Identify structures of the respiratory system in a sagittal section of the head and neck

Lower Respiratory Tract 673

EXERCISE 24.6: THE LARYNX 674

⑭ Compare and contrast vocal ligament, vocal fold, vestibular ligament, and vestibular fold, and explain how these relate to the "true vocal cords"

⑮ List the main cartilaginous structures of the larynx and identify them on classroom models of the larynx

⑯ Explain the function of the larynx in sound production

The Pleural Cavities and the Lungs 675

EXERCISE 24.7: THE PLEURAL CAVITIES 675

⑰ Describe the layers of the pleural cavities

EXERCISE 24.8: THE LUNGS 675

⑱ Identify the right and left lungs, and describe the unique features of each

⑲ Identify branches of the pulmonary arteries and veins, and bronchi in the hilum of a right and left lung

⑳ Differentiate between the left and right lung based on the number of lobes, fissures, airways, and impressions of thoracic viscera

EXERCISE 24.9: THE BRONCHIAL TREE 678

㉑ Describe the branching pattern of the bronchial tree

㉒ Associate the branches of the bronchial tree with the segments of the lungs that are served by each branch

Physiology 679

Respiratory Physiology 679

EXERCISE 24.10: MECHANICS OF VENTILATION 682

㉓ Describe Boyle's Law and describe how changes in thoracic volume result in changes in intrapulmonary pressure

㉔ Relate the concepts of pressure, flow, and resistance to the flow of air into and out of the lungs

EXERCISE 24.11: AUSCULTATION OF RESPIRATORY SOUNDS 682

㉕ Listen to the sounds made by air flowing through respiratory structures during inspiration and expiration

EXERCISE 24.12: (BIOPAC) PULMONARY FUNCTION TESTS 683

㉖ Describe the physiological relevance of tidal volume, inspiratory reserve volume, and expiratory reserve volume as measured by a spirometer

㉗ Discuss the physiological relevance of the inspiratory capacity, functional residual capacity, vital capacity, and total lung capacity. Determine how each of these measures of pulmonary function can be calculated from respiratory volumes.

MODULE 11: RESPIRATORY SYSTEM

- 28 *Discuss how spirometry may be used clinically to distinguish between obstructive and restrictive respiratory diseases*

EXERCISE 24.13: Ph.I.L.S. LESSON 31: pH & Hb-OXYGEN BINDING   686

- 29 *Relate the absorbance, measured with a virtual spectrophotometer, to the partial pressure of oxygen P_{O_2}*
- 30 *Construct an oxygen-hemoglobin dissociation curve from data collected using a virtual spectrophotometer*

- 31 *Use the constructed oxygen-hemoglobin dissociation curve to evaluate the effect of pH on hemoglobin-oxygen binding characteristics*

EXERCISE 24.14: Ph.I.L.S. LESSON 35: EXERCISE-INDUCED CHANGES   688

- 32 *Describe changes in inspiratory reserve volume, tidal volume, expiratory reserve volume, and respiratory rate during exercise*

INTRODUCTION

Breathe in deeply. What muscles did you contract to do this? As you may recall from chapter 11, the **diaphragm** and the **external intercostal muscles** are the key muscles used for **inspiration.** Contraction of these muscles increases the size of the thoracic cavity, causing the lungs to expand. As the lungs expand and the volume of the lungs increases, the pressure within the lungs becomes lower than atmospheric pressure and air flows into the lungs. This air brings a fresh supply of oxygen that is picked up by **erythrocytes** in the blood flowing through the lung capillaries. Freshly oxygenated blood will be transported back to the heart and pumped out to the body. Take another deep breath. This time focus on the process of **expiration** (breathing out). Normal expiration is a passive process, which does not require muscular effort. As soon as you stop contracting your diaphragm and external intercostal muscles, the elastic tissue of the lungs and chest cavity wall recoils resulting in a decrease in the volume of the thoracic cavity (and the lungs) increases and intrapulmonary pressure decreases. As the pressure in the lungs becomes greater than atmospheric pressure, air flows out of the lungs. This air has a higher concentration of carbon dioxide, a waste product from cellular respiration that must be removed from the body. Your study of the respiratory system will focus on the anatomical structures used in this process of conveying air into and out of the lungs. You will examine the histology of lung tissues. Their structure is ideally suited for the exchange of gases between the alveoli (air sacs) lungs and the blood.

In these laboratory exercises, you will observe the gross and histological structure of the upper respiratory tract and the lungs. You will begin by investigating the general layering pattern of the walls of the respiratory tract so that you will be able to identify specific histological structures located within each layer. This will be followed by analysis of the histology of major respiratory tract structures, including the lungs. You will observe the gross anatomy of the upper airways and the lungs, using either cadaveric specimens, fresh sheep specimens, models, or some combination of these. Finally, you will explore the processes of ventilation, gas exchange, and gas transport. You will apply these principles in the laboratory to develop an understanding of how the respiratory system plays an essential role in acid-base balance.

Chapter 24: The Respiratory System

Name: _____
Date: _____ Section: _____

PRE-LABORATORY WORKSHEET

Also available at www.connect.mcgraw-hill.com

|ANATOMY & PHYSIOLOGY

1. Describe the anatomical structure (e.g., shape, number of layers) of the *respiratory epithelium* (the epithelium found lining the nasal cavity and upper respiratory tract).

   _____
   _____
   _____

2. How many lobes compose the left lung? _____

3. Which of the following bones does *not* form a border of the nasal cavity (circle one)?

   a. vomer

   b. ethmoid

   c. mandible

   d. palatine

   e. maxilla

4. The pulmonary arteries carry _____ blood from the _____ (chamber) of the heart to the lungs, and the pulmonary veins carry _____ blood from the lungs to the _____ (chamber) of the heart. This circuit is referred to as the _____ circuit.

5. True or False: Contraction of the diaphragm decreases the volume of the thoracic cavity.

6. The respiratory membrane is composed of three structures. List these structures in the spaces below.

   a. _____

   b. _____

   c. _____

7. Which lung cells produce surfactant? _____

8. Structures that compose the respiratory portion of the respiratory tree typically contain _____.

9. A _____ bronchus leads into each lobe of a lung.

   a. main

   b. lobar

   c. tertiary

10. A _____ cavity surrounds each lung.

Chapter Twenty-Four *The Respiratory System*   **663**

Histology

Upper Respiratory Tract

The respiratory system is organized into two structural regions: An upper respiratory tract and a lower respiratory tract. The upper respiratory tract includes the nose, nasal cavity, paranasal sinuses, and pharynx. The lower respiratory tract includes the larynx, trachea, bronchi, bronchioles, alveolar ducts, and alveoli. The structures of the respiratory system are also categorized based on function. Passageways that serve primarily to transport or conduct air are part of the **conducting zone;** these structures include the passageways from the nose to the end of the terminal bronchioles. Structures that participate in gas exchange with the blood—including the respiratory bronchioles, alveolar ducts, and alveoli—are part of the **respiratory zone.**

The walls of the respiratory passageways are lined by a mucous membrane, called a mucosa. The **mucosa** (*mucosus*, mucus) consists of the epithelium resting on a basement membrane, and an underlying layer of aerolar connective tissue called the **lamina propria** (*lamina*, layer, + *proprius*, one's own). Typically, deep to the mucosa are several layers that include both the **submucosa** (*sub-*, under, + mucosa), which consists of aerolar connective tissue that contains glands, blood vessels, nerves, and the outer layer of adventitia (*adventicius,* coming from abroad), which is composed of loose connective tissue containing blood vessels and nerves. In some portions of the respiratory passageway, including the trachea and bronchi, there are partial rings or plates of hyaline cartilage, which provide support to the passageway. The cartilage is ensheathed in dense irregular tissue and is positioned between the submucosa and adventitia. **Figure 24.1** shows the layers of the tracheal wall. As you view slides of respiratory system structures, it is helpful to first identify these three layers, particularly the mucosa and submucosa, which will make it easier to identify specific cell types within each layer.

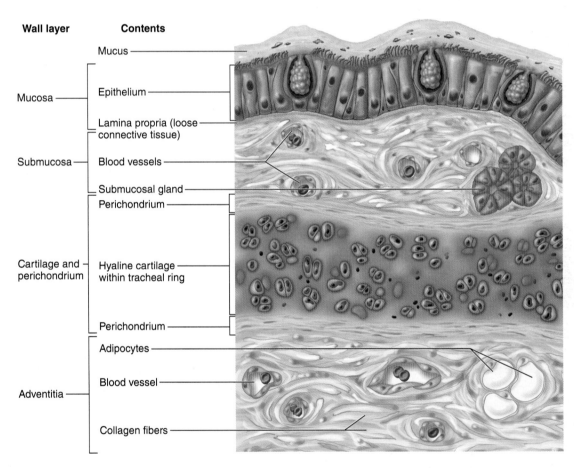

Figure 24.1 **Wall Layers of the Trachea.** The walls of all structures of the respiratory tract are composed of three layers: mucosa, submucosa, and adventitia/serosa. This drawing depicts the wall of the trachea, which has an outer covering of connective tissue referred to as an adventitia. If the outer covering were a mesothelial membrane, it would be referred to as a serosa.

EXERCISE 24.1

OLFACTORY MUCOSA

The epithelium that lines the nasal cavity is highly specialized—not only does it assist with the cleansing and warming of incoming air, but it also acts as a special sensory organ to detect odors. **Figure 24.2** demonstrates the histological appearance of the olfactory epithelium, and **table 24.1** describes the appearance and function of the various cell types and structures within the olfactory epithelium.

1. Obtain a compound microscope and a slide of olfactory epithelium. Place the slide on the microscope stage and bring the tissue sample into focus on low power.

2. Identify the **mucosa** and **epithelium.** Move the microscope stage so the epithelium is at the center of the field of view, then change to high power.

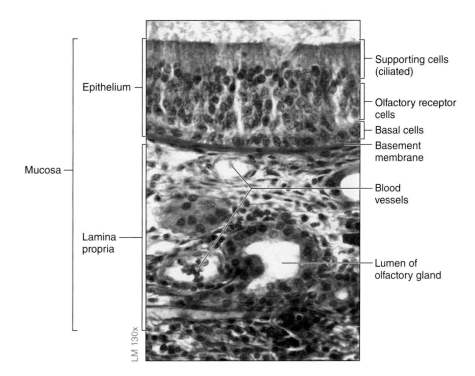

Figure 24.2 Olfactory Epithelium. Olfactory epithelium contains three cell types: basal cells, olfactory receptor cells, and supporting (ciliated) cells.

Table 24.1	Olfactory Epithelium	
Structure	**Description and Function**	**Word Origin**
Basal Cells	Small, triangular cells that lie on the basement membrane and do not reach the apical surface of the epithelium. Their *nuclei are located on the basal surface of the epithelium*. These cells are the precursor cells to the other cell types and are responsible for regeneration of the epithelium.	*basis*, bottom
Olfactory Glands	Serous-secreting glands located in the lamina propria of the olfactory epithelium that open to the surface. They secrete a serous fluid that "washes" the olfactory hairs, clearing the surface so that new odors can be sensed by the olfactory cells.	*ol-facio*, to smell
Olfactory Receptor Cells	Bipolar neurons that contain 6–8 cilia-like extensions on their apical surface called olfactory hairs. These hairs function to detect odorous substances. The cell *nuclei are located in the middle of the epithelium*. Axons of these cells travel through the cribriform foramina of the ethmoid bone to synapse with CNS neurons in the olfactory bulbs.	*ol-facio*, to smell
Supporting Cells (ciliated)	The most prominent and abundant cells. Their *nuclei are located near the apical surface of the epithelium*. They are columnar epithelial cells with cilia on their apical surfaces. They function to propel mucus and particulate matter along the apical surface of the epithelium toward the pharynx.	*cilium*, eyelash

(continued on next page)

(continued from previous page)

3. Using table 24.1 and figure 24.2 as guides, identify the following structures on the slide:

   ☐ basal cells ☐ olfactory receptor cells
   ☐ olfactory glands ☐ supporting cells (ciliated)

4. In the space to the right, sketch the olfactory epithelium as viewed through the microscope. Be sure to label all the structures listed in step 3 in your drawing.

_____×

Lower Respiratory Tract

The lower respiratory tract includes the larynx, **trachea, bronchi,** bronchioles, alveolar ducts, and alveoli. As the airways extend progressively deeper into the lungs from the trachea, into the bronchi and bronchioles, the amount of cartilage in the submucosa decreases while the amount of smooth muscle increases. In addition, several changes occur in the epithelium as it transitions from pseudostratified ciliated columnar epithelium of the trachea to the simple squamous epithelium of the alveoli within the lungs. **Table 24.2** describes the wall structure of the trachea, bronchi, and bronchioles.

EXERCISE 24.2

THE TRACHEA

1. Obtain a slide of the trachea (*tracheia*, rough artery) and place it on the microscope stage. This slide will also contain the esophagus (*oisophagos*, gullet) because the two structures are adjacent to each other *in vivo*.

2. Bring the tissue sample into focus on low power and distinguish the trachea from the esophagus. Move the microscope stage so the lumen and epithelium of the trachea are in the center of the field of view. Then switch to high power.

3. As you observe the trachea, first identify the layers of the wall of the trachea (mucosa, submucosa, and adventitia), and then identify the structures located within each layer. Table 24.2 describes the wall structure of the trachea and the structures located within each layer, and **figure 24.3** demonstrates the histological appearance of the wall of the trachea.

Table 24.2	Histology of the Trachea
Structure	**Description and Functions**
MUCOSA	Innermost layer that lines the tracheal wall; consists of respiratory epithelium resting on a basement membrane and underlying lamina propria
Epithelium	Pseudostratified ciliated columnar epithelium
Basal Cells	Small triangular cells that lie on the basal lamina and do not reach the apical surface of the epithelium. These cells are the precursor cells to the other cell types and are responsible for regeneration of the epithelium.
Ciliated Cells	The most prominent and abundant cells. These are pseudostratified ciliated columnar cells of the epithelium. They function to propel mucus and particulate matter via the "mucus escalator" along the epithelial sheet toward the pharynx.
Goblet Cells	Large round cells that appear white or clear. They secrete mucin onto the surface of the epithelium. The mucus traps particulate matter that enters the trachea while the cilia move the mucus superiorly (a "mucus escalator") toward the pharynx so that it may be swallowed.
Lamina Propria	A layer of areolar connective tissue that underlies the respiratory epithelium. In the trachea this layer can be seen in the area between the epithelial folds and the trachealis muscle or C-shaped cartilages.
SUBMUCOSA	The middle layer of the tracheal wall containing seromucous glands, blood vessels, and nerves
Submucous Glands	These produce a substance that is part watery (serous) and part viscous (mucus)

Table 24.2	Histology of the Trachea (continued)
Structure	**Description and Functions**
Tracheal Cartilages	Aspect of the trachea for expansion of the esophagus during swallowing; the trachialis muscle also alters the diameter of the trachea during coughing and sneezing. The decreased diameter of the trachea causes the air to exit more forcefully, which helps dislodge substances within the airways.
Trachealis Muscle	A layer of smooth muscle found on the posterior aspect of the trachea where the trachea lies against the esophagus. Laterally, the trachealis muscle is anchored to the ends of the cartilage C rings, and its contraction decreases the diameter of the trachea, which is important for coughing and sneezing. The decreased diameter of the trachea causes the air to exit more forcefully, which helps dislodge substances within the airways.
ADVENTITIA	Loose connective tissue on the outermost surface of the tracheal wall

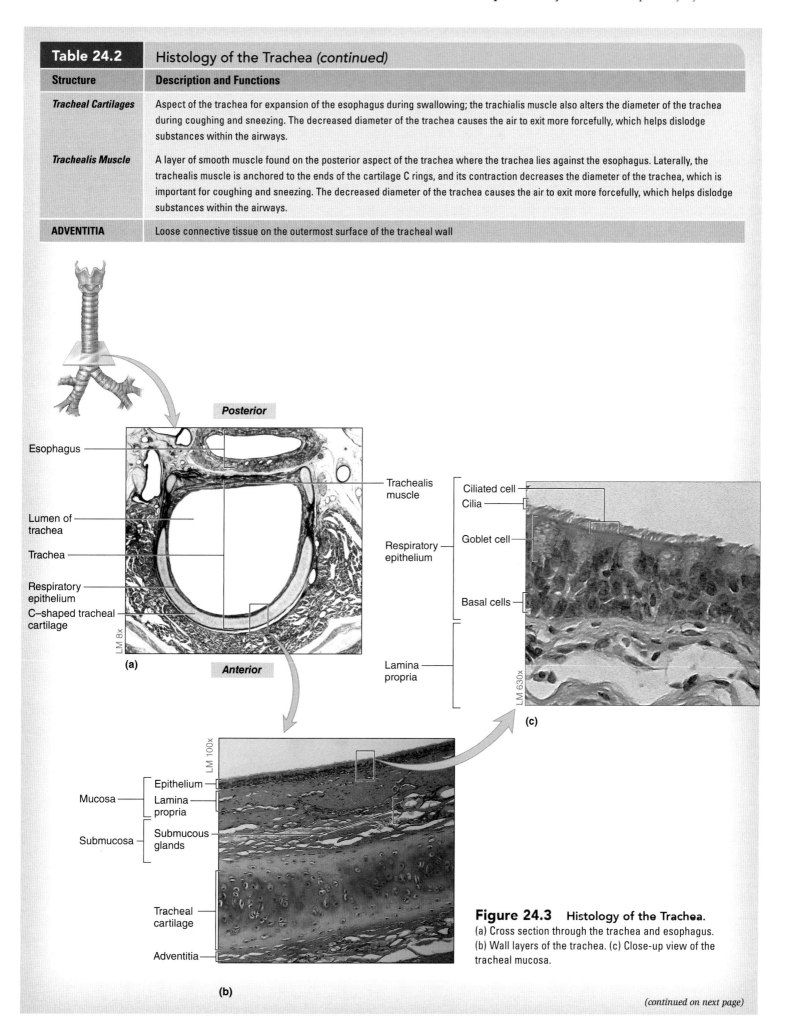

Figure 24.3 Histology of the Trachea.
(a) Cross section through the trachea and esophagus.
(b) Wall layers of the trachea. (c) Close-up view of the tracheal mucosa.

(continued on next page)

(continued from previous page)

4. Using table 24.2 and figure 24.3 as guides, identify the following structures on the slide of the trachea:

   ☐ adventitia
   ☐ basal cells
   ☐ ciliated cells
   ☐ epithelium
   ☐ goblet cells
   ☐ lamina propria
   ☐ mucosa
   ☐ submucosa
   ☐ submucous glands
   ☐ tracheal cartilages
   ☐ trachealis muscle

5. In the space below, sketch a cross section of the trachea as viewed through the microscope. Be sure to label all of the structures listed in step 4 in your drawing.

_____ ×

EXERCISE 24.3

THE BRONCHI AND BRONCHIOLES

1. Obtain a slide of the lungs and place it on the microscope stage. Bring the tissue sample into focus on low power and scan the slide to locate cross sections of **large bronchi, small bronchi,** and **bronchioles** (figure 24.4).

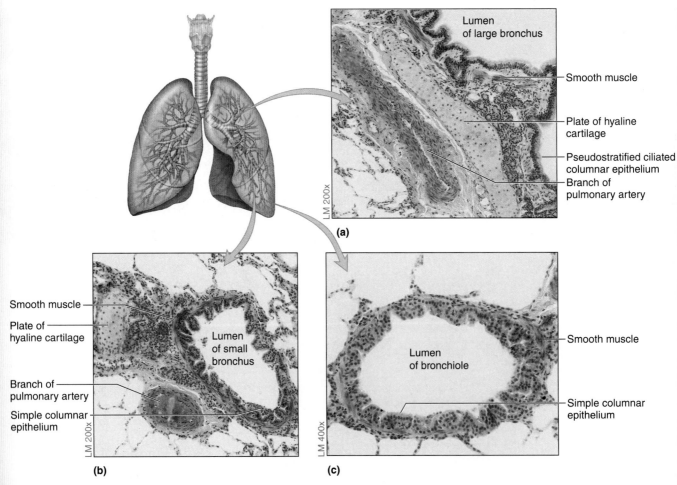

Figure 24.4 The Bronchial Tree. When making observations of the lung, you will see cross sections of portions of the bronchial tree. (a) The larger bronchi are lined with a ciliated pseudostratified columnar epithelium, and there are large plates of hyaline cartilage in the walls, along with smooth muscle. (b) The smaller bronchi are lined with a simple ciliated columnar epithelium, with smaller plates of cartilage in their walls, and relatively more smooth muscle. (c) Bronchioles (shown here) have a simple ciliated columnar epithelium, no cartilage, and smooth muscle.

2. The different characteristics of the conducting zone structures are listed in **table 24.3.** In general, as the airways travel deeper into the lung and become progressively smaller, three changes take place: (1) The epithelium transitions from pseudostratified ciliated columnar to simple ciliated columnar, and the number of cilia decrease; (2) large plates of hyaline cartilage in the walls give way to smaller and smaller pieces of cartilage with no cartilage present in bronchioles; and (3) the relative amount of smooth muscle in the airways increases.

3. Using figure 24.4 and table 24.3 as guides, identify the following structures on the slide of the lungs:

   ☐ branch of pulmonary artery
   ☐ bronchiole
   ☐ hyaline cartilage
   ☐ large bronchus
   ☐ small bronchus
   ☐ smooth muscle

4. In the space below, sketch a cross section of a bronchiole as viewed through the microscope.

_____ ×

Table 24.3	Histology of the Bronchial Tree		
Structure	**Epithelium**	**Hyaline Cartilage**	**Smooth Muscle**
Large Bronchi	Pseudostratified ciliated columnar	Large plates, which keep airway open	Encircles the lumen
Small Bronchi	Simple ciliated columnar	Small plates	Encircles the lumen
Bronchioles	Simple ciliated columnar	None	Encircles the lumen

Lungs

The lungs consist of functional units called **alveoli** (*alveus*, hollow sac), which are the sites of gas exchange between the air within the lungs and the blood. The alveoli are lined with a simple squamous epithelium, which provides the thinnest possible barrier to diffusion between the air within the alveoli and the blood within the capillaries that surround them. In a slide of the lung, you will see numerous alveoli, but you will also find cross sections of some of the smaller airways (such as respiratory bronchioles). The transition from conducting zone structures to respiratory zone structures occurs deep within the lungs, and the key feature in distinguishing them from each other is the relative thickness of the wall. Any structure with a thin wall (because it is composed of only an epithelium resting on a basement membrane) participates in gas exchange and, thus, is part of the respiratory zone.

EXERCISE 24.4

THE LUNGS

1. Obtain a slide of the lungs and place it on the microscope stage. Bring the tissue sample into focus on low power and scan the slide to locate cross sections of **bronchioles (figure 24.5)**.

2. Increase the magnification to observe the smaller airways and the air sacs (alveoli) in greater detail. **Table 24.4** describes the structure and function of the airways that compose the respiratory portion.

Table 24.5 describes the structure and function of the cell types within the alveoli. For gas exchange to occur, gas molecules must pass between the alveoli and the capillaries that surround them. The structures lining the alveoli and the capillaries collectively form the **respiratory membrane.** The respiratory membrane is composed of (1) alveolar type I cells, (2) the fused basement membrane of alveolar type I cells, and (3) the endothelial cells lining the capillaries.

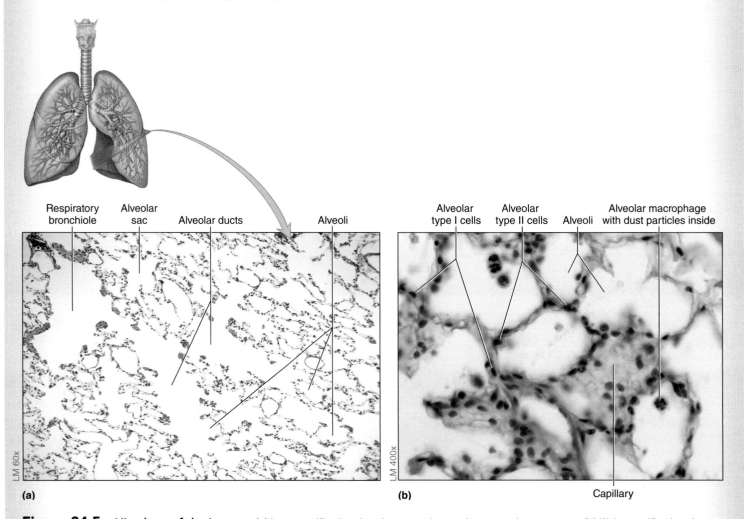

Figure 24.5 Histology of the Lungs. (a) Low-magnification view demonstrating respiratory portion structures. (b) High-magnification view showing alveolar type I cells (composed of simple squamous epithelium), alveolar type II cells (composed of simple cuboidal epithelium), and alveolar macrophages with particles of dust inside the cells.

Table 24.4	Structures Composing the Respiratory Zone	
Structure	**Description**	**Epithelium**
Respiratory Bronchioles	Thin-walled ducts with alveoli scattered along the passageway	Ciliated simple cuboidal
Alveolar Ducts	Ducts with alveoli lining the entire passageway	Simple squamous
Alveolar Sacs	Shaped like a bunch of grapes with several alveoli (the "grapes") bunched together	Simple squamous
Alveoli	A balloonlike structure that is the site of gas exchange	Simple squamous

Table 24.5	Microscopic Structures within the Lungs
Structure	**Description**
Alveolar Macrophages	Derived from monocytes. Engulf particulate matter and microorganisms.
Alveolar Type I Cell	Simple squamous epithelial cells covering 97% of the alveolar surface area. They are connected to each other with tight junctions.
Alveolar Type II Cell	Compose about 3% of alveolar surface area. Joined to type I cells by desmosomes and tight junctions. Contain a large nucleus, foamy cytoplasm, and vesicles containing **pulmonary surfactant** (a mixture of phospholipids, glycosaminoglycans, and proteins) that functions to reduce surface tension within alveoli.

3. Using tables 24.4 and 24.5, and figure 24.5 as guides, identify the following structures on the slide of the lungs:

   ☐ alveolar ducts
   ☐ alveolar macrophages
   ☐ alveolar sacs
   ☐ alveolar type I cell
   ☐ alveolar type II cell
   ☐ alveoli
   ☐ capillary
   ☐ respiratory bronchioles

4. In the space to the right, sketch the respiratory membrane. Then label the three components that make up the respiratory membrane.

5. *Optional Activity:* **AP|R** 11: Respiratory System—Watch the "Diffusion Across Respiratory Membrane" animation to visualize the microscopic structure and function of lung tissue.

Gross Anatomy

Upper Respiratory Tract

In the following exercises you will observe the gross anatomy of the respiratory structures that convey air into the lungs, the cavities that house the lungs, and the lungs themselves. Recall, however, that the process of breathing requires muscular action. You learned the details of the structure, location, and actions of the respiratory muscles in chapter 12. This is a good point in time to go back to exercise 12.9 on p. 310 and review the actions of the respiratory musculature, paying particular attention to the location and actions of the diaphragm.

Structures of the upper respiratory tract are best seen through gross observation of a sagittal section of the head and neck. This view allows you to see the gross structures within the nasal cavity that warm, moisten, and filter the air before it enters the respiratory passageways.

EXERCISE 24.5

SAGITTAL SECTION OF THE HEAD AND NECK

1. Obtain a classroom model of a sagittal section of the head and neck (or a cadaveric specimen that has been sectioned along the sagittal plane). **Figure 24.6** shows respiratory system structures that are visible in a sagittal section of the head and neck.

2. Using your textbook as a guide, identify the structures listed in **figure 24.7** on the classroom model or cadaveric specimen. Then label figure 24.7.

3. *Optional Activity:* **AP|R** 11: Respiratory System—Watch the "Respiratory System Overview" to review the structures of the upper and lower respiratory tracts.

(continued on next page)

672 Chapter Twenty-Four *The Respiratory System*

(continued from previous page)

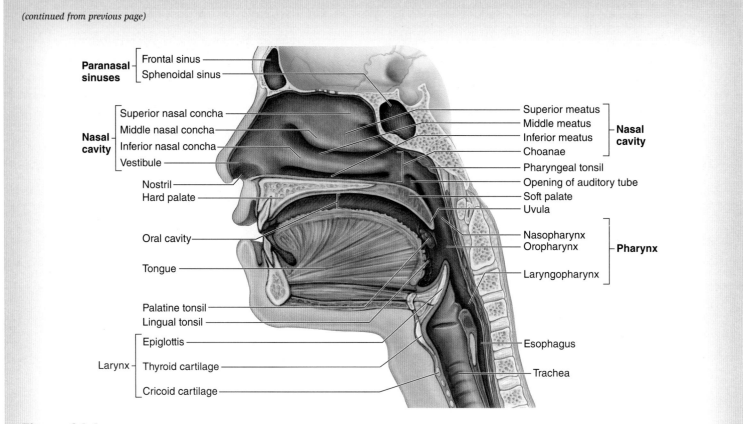

Figure 24.6 The Upper Respiratory Tract. Respiratory system structures visible in a midsagittal view of the head and neck.

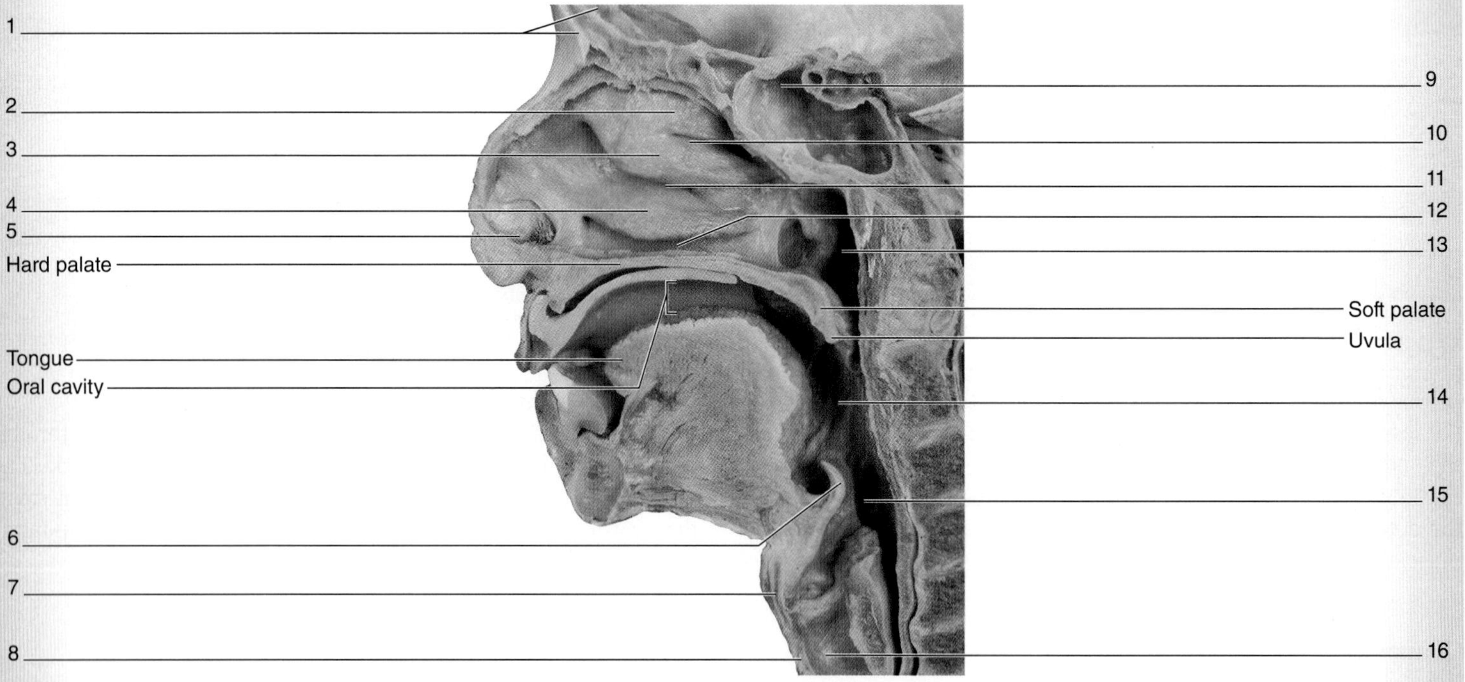

Figure 24.7 The Upper Respiratory Tract: Midsagittal View.

- ☐ cricoid cartilage
- ☐ epiglottis
- ☐ frontal sinus
- ☐ inferior meatus
- ☐ inferior nasal concha
- ☐ laryngopharynx
- ☐ lumen of larynx
- ☐ middle meatus
- ☐ middle nasal concha
- ☐ nasopharynx
- ☐ oropharynx
- ☐ sphenoidal sinus
- ☐ superior meatus
- ☐ superior nasal concha
- ☐ thyroid cartilage
- ☐ vestibule

Lower Respiratory Tract

The lower respiratory tract is composed of structures of both the conducting portion (larynx, trachea, bronchi, and bronchioles) and the respiratory portion (respiratory bronchioles, alveolar ducts, and alveoli). The **larynx** (*larynx*, organ of voice production) is much more than just an opening into the lower respiratory tract. It also houses the **vocal folds,** referred to as the "true vocal cords," which are responsible for phonation (sound production) and intrinsic muscles that control the length and tension of the vocal ligaments. The structural framework of the larynx is composed of several paired and unpaired cartilages, which act as attachment sites for the intrinsic musculature. The muscles of the larynx are all innervated by branches of the vagus nerve (CN X). Thus, lesions of the vagus nerve result in problems with phonation, such as hoarseness. **Table 24.6** describes the cartilaginous structures of the larynx, and **table 24.7** describes the noncartilaginous structures of the larynx.

Table 24.6	Cartilaginous Structures of the Larynx	
Cartilage	**Description**	**Word Origin**
Arytenoid	Small paired pyramid-shaped cartilages found on the superior, posterior aspect of the cricoid cartilage. The vocal ligaments (true vocal cords) attach to them.	*arytania*, a ladle, + *eidos*, resemblance
Corniculate	Small paired cartilages found superior to the arytenoid cartilages. The vestibular folds ("false vocal cords") attach to them.	*cornicatus*, horned
Cricoid	The second largest laryngeal cartilage, this ring-shaped cartilage serves as an attachment point for muscles	*krikos*, a ring, + *eidos*, resemblance
Cuneiform	Small paired cartilages found within the aryepiglottic fold	*cuneus*, a wedge, + *forma*, form
Epiglottis	A plate of elastic cartilage at the superior aspect of the larynx that closes over the opening to the larynx during swallowing to prevent substances from entering the larynx. It is covered by stratified squamous epithelium on its superior aspect and by respiratory epithelium on its inferior aspect.	*epi-*, above, + *glottis*, the mouth of the windpipe
Thyroid	The largest of the laryngeal cartilages, located superior to the isthmus of the thyroid gland. The vocal ligaments (true vocal cords) attach to it.	*thyreos*, an oblong shield, + *eidos*, resemblance

Table 24.7	Noncartilaginous Structures of the Larynx	
Structure	**Description and Function**	**Word Origin**
Glottis	Consists of the rima glottis plus the vocal folds	*glottis*, the mouth of the windpipe
Rima Glottidis	The space between the true vocal cords, also known as the true glottis	*rima*, a slit, + *glottis*, the mouth of the windpipe
Vestibular Folds	The "false vocal cords," which are the vestibular ligaments plus the folds of mucous membrane that cover them	*vestibulum*, a small cavity at the entrance of a canal
Vestibular Ligaments	Ligaments that stretch between the angle of the thyroid cartilage to the corniculate cartilages	*vestibulum*, a small cavity at the entrance of a canal, + *ligamentum*, a bandage
Vocal Folds ("true vocal cords")	The vocal ligaments plus the mucosa covering them. Form the "true vocal cords." Involved directly in voice production. Alterations in tension of the cords affect the pitch of the sound.	*vocalis*, pertaining to the voice, + *chorda*, cord
Vocal Ligaments	Ligaments that stretch between the thyroid and arytenoid cartilages	*vocalis*, pertaining to the voice, + *ligamentum*, a bandage

EXERCISE 24.6

THE LARYNX

1. Obtain a classroom model of the larynx **(figure 24.8)**.
2. Using tables 24.6 and 24.7, and your textbook as guides, identify the structures listed in figure 24.8 on the model of the larynx. Then label figure 24.8. (Answers may be used more than once.)
3. *Optional Activity:* **AP|R** **11: Respiratory System**—Take the "Lower Respiratory Tract" test in the Quiz section to review the larynx and other lower respiratory structures.

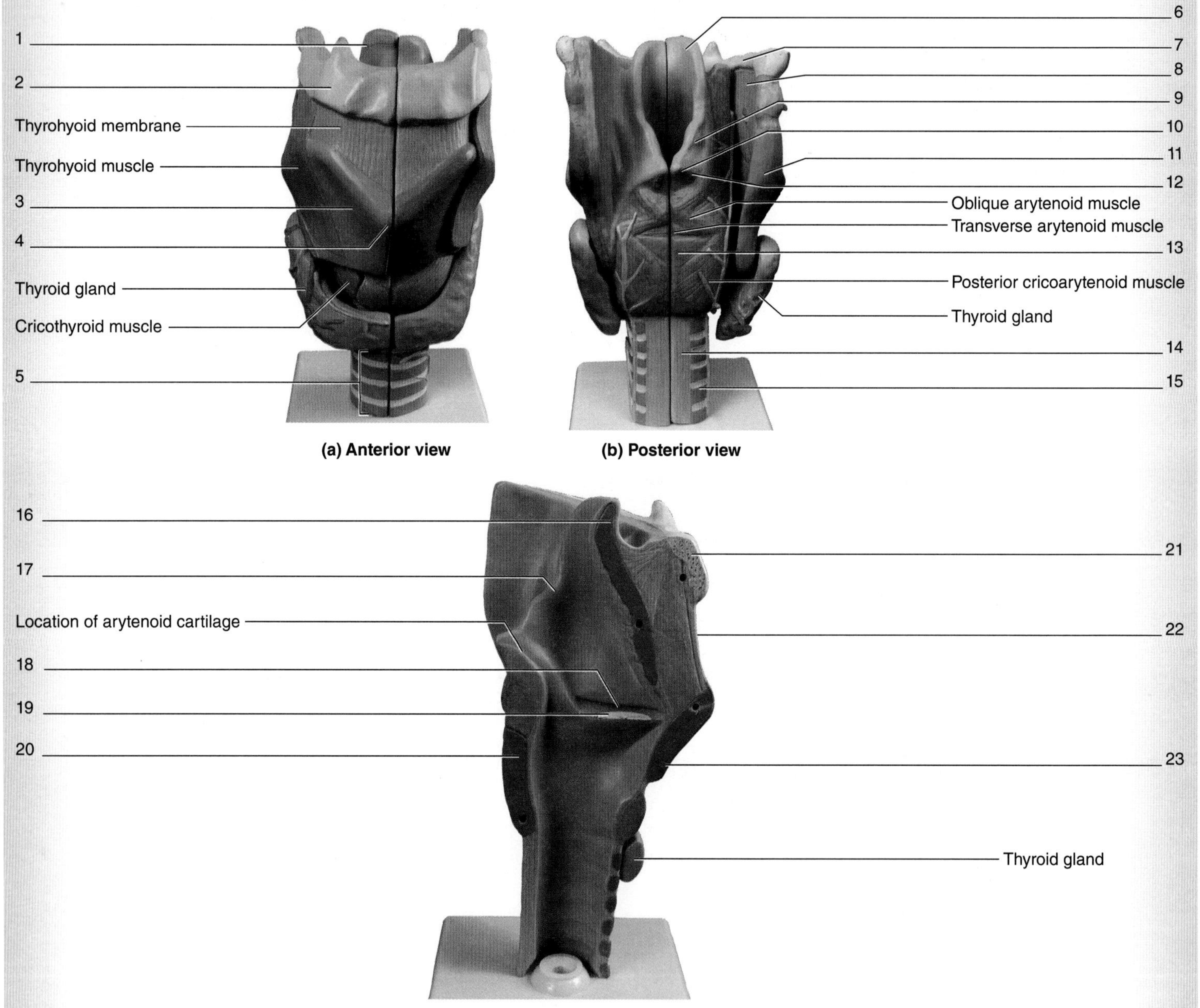

1 _____
2 _____
Thyrohyoid membrane
Thyrohyoid muscle
3 _____
4 _____
Thyroid gland
Cricothyroid muscle
5 _____

(a) Anterior view

6 _____
7 _____
8 _____
9 _____
10 _____
11 _____
12 _____
Oblique arytenoid muscle
Transverse arytenoid muscle
13 _____
Posterior cricoarytenoid muscle
Thyroid gland
14 _____
15 _____

(b) Posterior view

16 _____
17 _____
Location of arytenoid cartilage
18 _____
19 _____
20 _____
21 _____
22 _____
23 _____
Thyroid gland

(c) Midsagittal view (left side)

Figure 24.8 Classroom Model of the Larynx.

- ☐ arytenoid cartilage
- ☐ corniculate cartilage (used twice)
- ☐ cricoid cartilage (used twice)
- ☐ cuneiform cartilage
- ☐ epiglottis (used three times)
- ☐ false vocal cords
- ☐ hyoid bone (used three times)
- ☐ laryngeal prominence
- ☐ thyrohyoid membrane (used twice)
- ☐ thyroid cartilage (used three times)
- ☐ trachea
- ☐ tracheal C ring
- ☐ trachealis muscle
- ☐ true vocal cords

The Pleural Cavities and the Lungs

Within the thoracic cavity, the lungs are located within separate **pleural cavities** (*pleura*, a rib). The space within the thoracic cavity between the two lungs is the **mediastinum** (*medius*, middle). If one of the lungs collapses, the structures within the mediastinum will shift toward the side of the collapsed lung. Tightly adhered to each lung is a serous membrane called the **visceral pleura**. Adhered to the inner wall of the thoracic cavity is a serous membrane called the **parietal pleura**.

On a human cadaver the parietal pleura can often be seen as a shiny tissue attached to the innermost part of the rib cage or the superior surface of the diaphragm. Between the two serous membranes is a fluid-filled space called the **pleural cavity**. Note that the lungs are contained within the pleurae, which contain serous fluid. The serous fluid reduces friction and increases surface tension of the pleurae. The latter allows the lungs to expand and contract as the thoracic cavity changes volume, because it keeps the visceral and parietal layers "stuck" together. Incidentally, if air is introduced into this pleural space, a condition called pneumothorax (*pneumo-*, air + thorax), the lungs will not be able to overcome their natural recoil and will collapse, thereby making ventilation impossible.

EXERCISE 24.7

THE PLEURAL CAVITIES

1. Observe the thoracic cavity of a human cadaver or a model of the thorax (**figure 24.9**).

2. Using your textbook as a guide, identify the structures listed in figure 24.9 on the cadaver or model of the thorax. Then label figure 24.9.

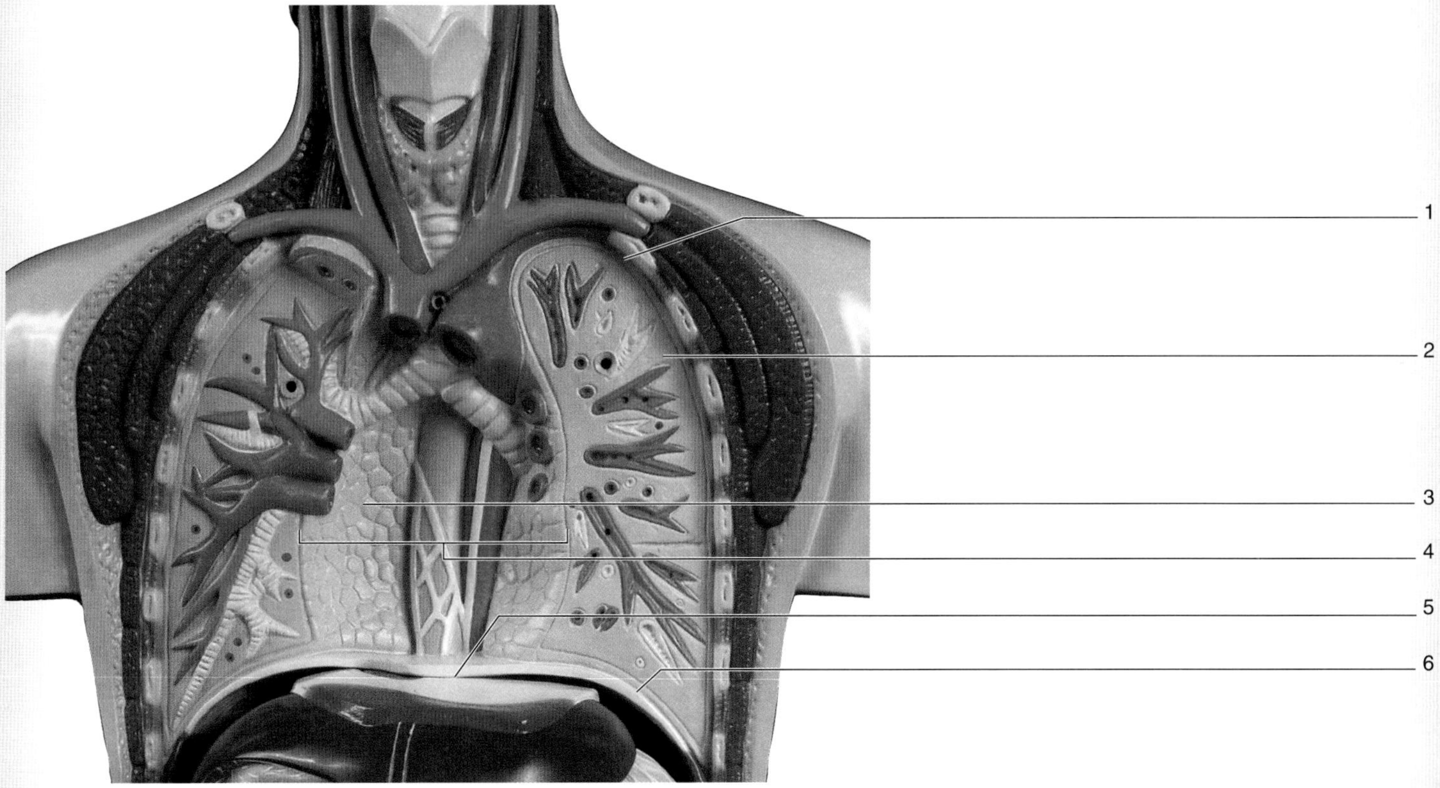

Figure 24.9 The Pleural Cavities.

- ☐ diaphragm
- ☐ left lung
- ☐ mediastinum
- ☐ parietal pleura
- ☐ right lung
- ☐ visceral pleura

EXERCISE 24.8

THE LUNGS

In this exercise you will compare and contrast the right and left lungs, and will observe the branching pattern of the respiratory tree. Although it is easy to distinguish the right and left lungs from each other based on the number of lobes (two for the left, three for the right), locating the structures that enter the hilum of the lung (pulmonary arteries, pulmonary veins, and bronchi) is more difficult. There are patterns for recognizing these structures that you will learn in this laboratory exercise. In addition, you will observe several **impressions** made in the lungs by adjacent structures. These impressions are visible in preserved human cadaver lungs and on classroom

(continued on next page)

Chapter Twenty-Four The Respiratory System

(continued from previous page)

models of the lungs, but may not be visible in fresh lungs. This is because the process of fixing the lungs with preservative also fixes the impressions of adjacent organs. The impressions are not found in fresh lungs because they have not been fixed with preservative.

EXERCISE 24.8A The Right Lung

1. Observe the lungs of a human cadaver, a fresh or preserved sheep pluck (a *pluck* contains the heart, lungs, and trachea), or a classroom model of the lungs.

2. Begin by observing the right lung **(figure 24.10)**. How many lobes does the right lung have? _____

3. Turn the lung so the hilum is visible (medial view; figure 24.10*b*). The hilum of the right lung contains branches of the pulmonary arteries and pulmonary veins, bronchi, and small bronchial arteries and veins, which represent the systemic circulation to the lungs. Which of the vessels (pulmonary arteries or pulmonary veins) do you think will have thicker walls? Recall that pulmonary arteries transport deoxygenated blood and pulmonary veins transport oxygenated blood. Explain your answer:

_____

_____

_____

4. In general, the pulmonary arteries are located on the superior aspect of the hilum of the right lung, the pulmonary veins are located on the inferior and anterior aspect of the hilum of the right lung, and the bronchi are located on the superior and posterior aspect of the hilum of the right lung. If you are viewing cadaveric lungs, these structures will be more difficult to differentiate from each other because they are not color-coded, so you will need to rely on the texture of the vessels and their locations for identification.

5. Using your textbook as a guide, identify the structures listed in figure 24.10 on the lung. Then label figure 24.10. (Some answers may be used more than once.)

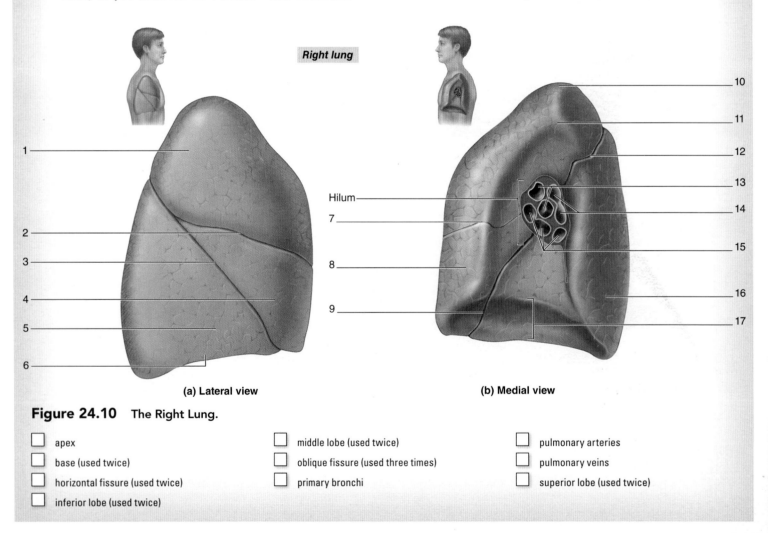

Figure 24.10 The Right Lung.

☐ apex
☐ base (used twice)
☐ horizontal fissure (used twice)
☐ inferior lobe (used twice)
☐ middle lobe (used twice)
☐ oblique fissure (used three times)
☐ primary bronchi
☐ pulmonary arteries
☐ pulmonary veins
☐ superior lobe (used twice)

EXERCISE 24.8B The Left Lung

1. Observe the lungs of a human cadaver, a fresh or preserved sheep pluck, or a classroom model of the lungs.

2. Observe the left lung (figure 24.11). How many lobes does the left lung have? _____ Is the left lung larger or smaller than the right lung? _____ Why do you think this is the case? _____
_____

3. Turn the lung so the hilum is visible (medial view; figure 24.11b). The hilum of the left lung contains branches of the pulmonary arteries and veins, bronchi, and small bronchial arteries and veins, which represent the systemic circulation to the lungs.

4. In general, the pulmonary arteries are located on the superior aspect of the hilum of the left lung, the pulmonary veins are located on the inferior and anterior aspect of the hilum of the left lung, and the bronchi are located on the superior and posterior aspect of the hilum of the left lung. If you are viewing cadaveric lungs, these structures will be more difficult to differentiate from each other because they are not color-coded, so you will need to rely on the texture of the vessels and their locations for identification. The medial view of the left lung allows you to see several impressions that accommodate nearby adjacent structures in the living human. The most prominent of these impressions is the **cardiac impression** made by the heart.

5. How will you distinguish the left lung from the right lung? _____
_____
_____

6. Using your textbook as a guide, identify the structures listed in figure 24.11 on the lung. Then label figure 24.11. (Some answers may be used more than once.)

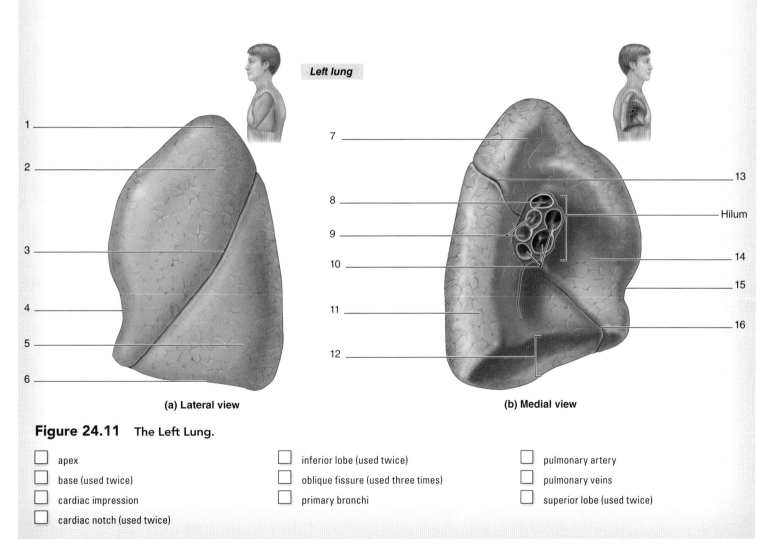

Figure 24.11 The Left Lung.

☐ apex
☐ base (used twice)
☐ cardiac impression
☐ cardiac notch (used twice)
☐ inferior lobe (used twice)
☐ oblique fissure (used three times)
☐ primary bronchi
☐ pulmonary artery
☐ pulmonary veins
☐ superior lobe (used twice)

EXERCISE 24.9

THE BRONCHIAL TREE

1. Observe the lungs of a human cadaver, a fresh or preserved sheep pluck, or a classroom model of the lungs.

2. The lungs are subdivided, from larger to smaller units, into **lobes, bronchopulmonary segments,** and **lobules.** The branching pattern of the bronchial tree follows the segmentation of the lungs, **(figure 24.12)**.

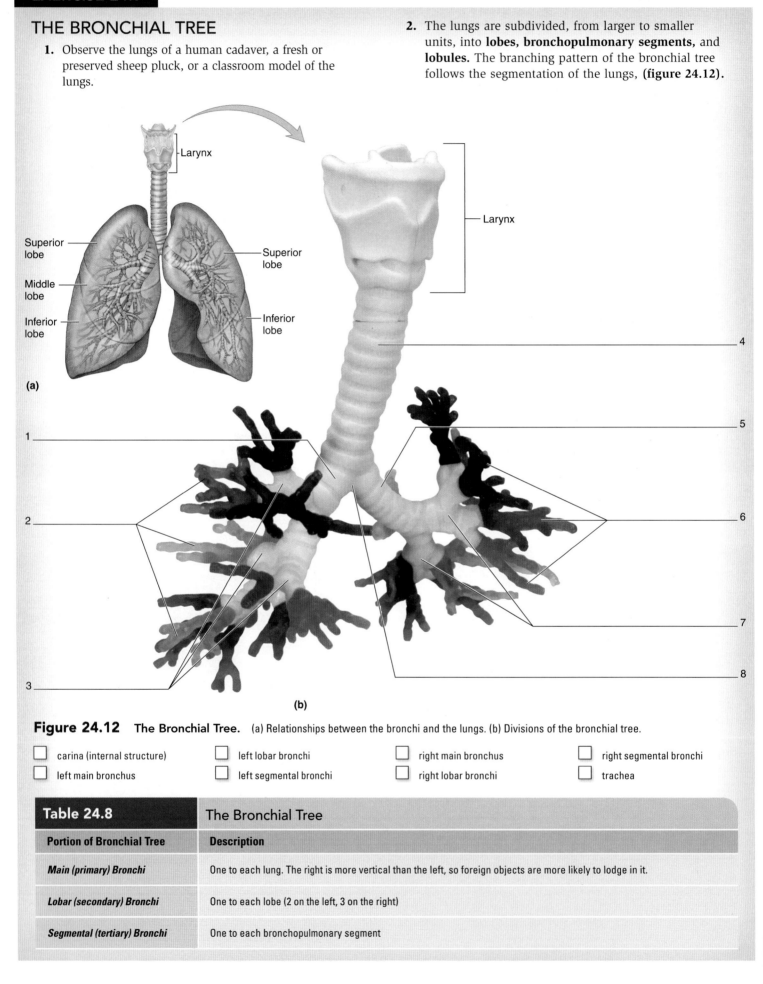

Figure 24.12 **The Bronchial Tree.** (a) Relationships between the bronchi and the lungs. (b) Divisions of the bronchial tree.

☐ carina (internal structure)　　☐ left lobar bronchi　　☐ right main bronchus　　☐ right segmental bronchi
☐ left main bronchus　　☐ left segmental bronchi　　☐ right lobar bronchi　　☐ trachea

Table 24.8	The Bronchial Tree
Portion of Bronchial Tree	**Description**
Main (primary) Bronchi	One to each lung. The right is more vertical than the left, so foreign objects are more likely to lodge in it.
Lobar (secondary) Bronchi	One to each lobe (2 on the left, 3 on the right)
Segmental (tertiary) Bronchi	One to each bronchopulmonary segment

For example, lobar (secondary) bronchi lead into lobes of the lung so there are three lobar bronchi leading into the three lobes of the right lung, and two lobar (secondary) bronchi leading into the two lobes of the left lung. **Table 24.8** describes the levels of the bronchial tree that serve each segment of the lungs. This segmental nature of the lungs and bronchial tree makes it relatively easy to remove a segment of the lung that contains a tumor (for example) without interfering with the other parts of the lung. The **carina** is the *internal* ridge between the most inferior tracheal cartilage and the start of the main bronchi. In the laboratory you will identify its location externally, where the main (primary) bronchi originate from the trachea. However, it is important to realize that the actual structure is internal. It is an important landmark for physicians performing bronchoscopy.

3. Using table 24.8 and your textbook as guides, identify the structures listed in figure 24.12 on the cadaver lungs, sheep pluck, or classroom model of the lungs. Then label figure 24.12.

Physiology

Respiratory Physiology

Breathing is critical for life and seems like a simple process to most of us. However, the mechanisms underlying much of respiratory physiology are complex. Respiration is the movement of both oxygen and carbon dioxide between the atmosphere and the body tissues, and involves four major processes: (1) pulmonary ventilation, (2) alveolar gas exchange, (3) gas transport, and (4) systemic gas exchange. **Pulmonary ventilation** is the process of air moving between the atmosphere and the alveoli. The movement of this air is dependent upon differences in pressure between the environment and the thoracic cavity/lungs. Air flow through airways is dependent on pressure differences. Changes in thoracic pressure are related to changes in thoracic volume and obey a law of physics called **Boyle's Gas Law** ($P_1V_1 = P_2V_2$). When pressure inside the lungs is less than atmospheric pressure, air flows into the lungs. In contrast, when pressure inside the lungs is greater than atmospheric pressure, air flows out of the lungs.

During **inspiration,** the diaphragm and external intercostal muscles contract and the volume of the thoracic cavity increases, and intrapulmonary pressure decreases. Air then flows from the atmosphere into the alveoli from the area of higher pressure to the area of lower pressure. During **expiration,** the diaphragm and external intercostal muscles relax and the elastic tissue of the lungs and chest cavity wall recoil, causing a decrease in the volume of the thoracic cavity and an accompanying increase in the intrapulmonary pressure. Air moves from the alveoli to the atmosphere.

Clinically, a **spirometer** can be used to record measurements that assess ventilation, such as tidal volume (TV), inspiratory reserve volume (IRV), expiratory reserve volume (ERV), and residual volume (RV). These measured volumes can then be used to calculate lung capacities including inspiratory capacity, functional residual cavity, vital capacity, and total lung capacity. Measured volumes and capacities can be compared to "normal" or "expected" values for an individual. **Table 24.9** is a quick reference guide of respiratory volumes and capacities. The table contains brief descriptions

Table 24.9 Respiratory Volumes and Capacities

Volumes				
Volume	Definition		Normal Values (Male)	Normal Values (Female)
Expiratory Reserve Volume (ERV)	The amount of air expelled from the lungs during a forced expiration, following a quiet expiration; ERV is a measure of lung and chest wall elasticity		1200 mL	700 mL
Inspiratory Reserve Volume (IRV)	The amount of air taken into the lungs during a forced inspiration, following a quiet inspiration; IRV is a measure of lung compliance		3100 mL	1900 mL
Residual Volume (RV)	Amount of air left in lungs following a forced expiration		1200 mL	1100 mL
Tidal Volume (TV)	Volume of air taken into or expelled out of lungs during a quiet breath		500 mL	500 mL
Capacities				
Capacity	Formula	Definition	Normal Values (Male)	Normal Values (Female)
Functional Residual Capacity	ERV + RV	Amount of air normally left (residual) in lungs after you expire quietly	2400 mL	1800 mL
Inspiratory Capacity	TV + IRV	Total ability to inspire	3600 mL	2400 mL
Total Lung Capacity	TV + IRV + ERV + RV	Total amount of air that can be in lungs	6000 mL	4200 mL
Vital Capacity	TV + IRV + ERV	Measure of the strength of respiration	4800 mL	3100 mL

of each measurement/calculation, any pertinent formulas for calculating lung capacities, and reference values for males and females. **Figure 24.13** illustrates the four lung volumes and four lung capacities graphically.

Air that reaches the alveoli is available for gas exchange with the blood in the pulmonary capillaries in a process called **alveolar gas exchange.** The epithelium of the alveoli, the endothelium of the alveolar capillaries, and the fused basement membranes between the two collectively form the **respiratory membrane.** For oxygen to enter the blood, or carbon dioxide to leave the blood, the gases must diffuse across this respiratory membrane. The rate at which each gas diffuses is governed by **Dalton's Law of Partial Pressures,** which states that each gas in a mixture of gases exerts an individual, or partial, pressure, in relation to its abundance in the mixture of gases. The partial pressure of each gas is calculated by multiplying the total pressure of the gas by the percentage of the individual gas. For example, if the total pressure is 760 mm Hg, and oxygen is 21% of the mixture, then the partial pressure of oxygen (P_{O_2}) = 760 mm Hg × 0.21 = 159 mm Hg. Gases will diffuse along their individual partial pressure gradients, from an area of high partial pressure to an area of low partial pressure.

Other factors that influence alveolar gas exchange include the total number of alveoli (surface area for gas exchange), the thickness of the respiratory membrane, and the solubility coefficient for the particular gas. The solubility coefficient dictates how readily a gas will dissolve in a liquid, and is governed by **Henry's Law.** Carbon dioxide has a solubility coefficient that is roughly twenty-four times that of oxygen; this means that carbon dioxide dissolves much more easily in liquid than does oxygen. Functionally, this means that carbon dioxide diffuses across the respiratory membrane much more readily than oxygen, or requires much smaller partial pressure differences to diffuse. A larger partial pressure gradient, increased surface area,

INTEGRATE

CLINICAL VIEW
Obstructive versus Restrictive Respiratory Diseases

Spirometry tests can be used clinically to distinguish between obstructive and restrictive pulmonary diseases. Obstructive diseases, such as emphysema, involve a physical barrier (obstruction) in the airway, whereas restrictive diseases, such as pulmonary fibrosis (scarring), involve an inability to physically expand the lungs (restriction). Both types of diseases decrease the amount of oxygen that can diffuse into the blood, but for very different reasons. If a patient presents with respiratory distress, doctors can measure respiratory volumes using a spirometer and calculate lung capacities to determine the disease classification. Two clinical measurements commonly used include **forced expiratory volume** in 1 second $(FEV)_1$, which corresponds to the amount of air forcefully exhaled in 1 second, and **forced vital capacity** (FVC), the amount of air forcefully exhaled after a maximal inhalation. The ratio of these two values, or $FEV_1\%$, is typically 75–80% in normal, healthy adults. Because patients with obstructive pulmonary disease have difficulty breathing out, their $FEV_1\%$ is typically reduced. Patients with an $FEV_1\%$ of less than 75% are diagnosed with an obstructive disease. In comparison, patients suffering from a restrictive disease has difficulty breathing in because of the limited ability to expand the lungs. These patients typically will have lower than normal inspiratory capacity and vital capacity. Once a patient's disease is classified as either obstructive or restrictive, further diagnostic testing can be performed and treatment plans may be devised to address the specific disease that caused the patient's respiratory distress.

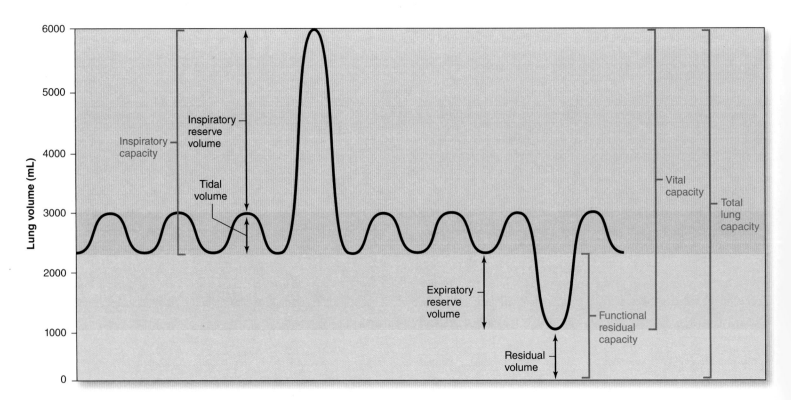

Figure 24.13 Respiratory Volumes and Capacities.

thinner respiratory membrane, and greater solubility coefficient will all increase the rate of diffusion across the respiratory membrane.

Gas exchange also occurs between systemic arterial blood and tissue cells. Here the same concepts for gas exchange apply. However, in the tissues, the membrane the gases must cross consists of the capillary endothelium (and its basement membrane) and the plasma membrane of the cell.

In addition to **exchange** of gases between the air and the blood, and the blood and the tissues, gases must also be transported within the blood. The process of moving the gases through the cardiovascular system is known as **gas transport.** Because of the low solubility of oxygen in water, 98% of the oxygen transported in the blood is bound to the heme group of hemoglobin, the major component of red blood cells (see chapter 20). The majority of carbon dioxide (~70%) is transported as bicarbonate ion. The conversion of carbon dioxide to bicarbonate ion occurs within red blood cells, which contain the enzyme *carbonic anhydrase.* Carbonic anhydrase catalyzes the reaction between carbon dioxide (CO_2) and water (H_2O) to form the weak acid carbonic acid (H_2CO_3). In solution, carbonic acid easily dissociates into hydrogen ions (H^+) and bicarbonate ions (HCO_3^-). This reaction is represented by the following equation:

$$CO_2 + H_2O \leftrightarrow H_2CO_3 \leftrightarrow H^+ + HCO_3^-$$

This reaction is important in understanding not only carbon dioxide transport as bicarbonate ion in the blood, but also the relationship between the carbon dioxide levels in the blood and the pH of the blood. As more carbon dioxide accumulates in the blood, more of it is converted into hydrogen ions. As the concentration of hydrogen ions in the blood increases, the pH decreases. Use your textbook as a guide as you perform the following exercises. These exercises will guide you in your exploration of the major respiratory processes and the relationship between pH and respiration.

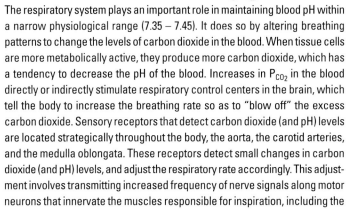

CONCEPT CONNECTION

The respiratory system plays an important role in maintaining blood pH within a narrow physiological range (7.35 – 7.45). It does so by altering breathing patterns to change the levels of carbon dioxide in the blood. When tissue cells are more metabolically active, they produce more carbon dioxide, which has a tendency to decrease the pH of the blood. Increases in P_{CO_2} in the blood directly or indirectly stimulate respiratory control centers in the brain, which tell the body to increase the breathing rate so as to "blow off" the excess carbon dioxide. Sensory receptors that detect carbon dioxide (and pH) levels are located strategically throughout the body, the aorta, the carotid arteries, and the medulla oblongata. These receptors detect small changes in carbon dioxide (and pH) levels, and adjust the respiratory rate accordingly. This adjustment involves transmitting increased frequency of nerve signals along motor neurons that innervate the muscles responsible for inspiration, including the diaphragm and external intercostal muscles. This activation will increase either the respiratory rate and/or the depth of breathing.

As respiratory rate decreases below physiological demand **(hypoventilation)**, carbon dioxide accumulates in the blood **(hypercapnia)**, leading to a drop in blood pH **(acidosis).** Conversely, as the respiratory rate increases above physiological demand **(hyperventilation)**, carbon dioxide decreases in the blood **(hypocapnia)**, leading to an increase in blood pH **(alkalosis).** Alteration of the respiratory rate, and consequently the blood pH, is a rapid process. This concept of acid-base balance will appear again in the urinary system, as the kidneys also play a role in regulating acid-base balance by secreting hydrogen ions into the filtrate for excretion in the urine. Overall, the respiratory system aids in short-term regulation of acid-base balance, and the urinary system aids in long-term regulation of acid-base balance in the blood.

EXERCISE 24.10

MECHANICS OF VENTILATION

Before you begin, state a hypothesis regarding the relationship between thoracic volume and intrapulmonary pressure in regulating the flow of air into and out of the lungs. _____

_____

_____

1. Obtain a model lung (**figure 24.14**). Note that the balloons represent the lungs, the rubber sheeting represents the diaphragm, and the glass dome represents the thoracic cavity.

2. Increase the volume of the thoracic cavity by pulling down on the "diaphragm" (rubber sheeting). Does the pressure inside the thoracic cavity increase or decrease?

   _____

   What happens to the balloons? _____

   _____

   Describe the direction of airflow. _____

   _____

3. Repeat step 2 by pushing up on the "diaphragm." Does the pressure inside the "thoracic cavity" increase or decrease? _____

   What happens to the balloons? _____

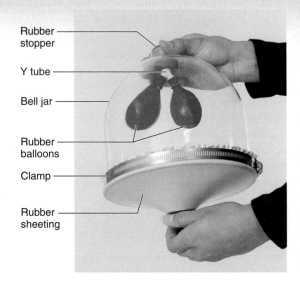

Figure 24.14 Model Lung.

   Describe the direction of airflow. _____

   _____

4. Repeat step 2 and introduce air into the "thoracic cavity" by partially releasing the rubber stopper, thereby simulating a pneumothorax. What happens to the balloons? _____

   _____

EXERCISE 24.11

AUSCULTATION OF RESPIRATORY SOUNDS

The sound of air flowing through respiratory structures can be detected by placing the diaphragm of a stethoscope on strategic locations of the chest and back, a process called auscultation (*auscultatio*, a listening). Doctors evaluating a patient with respiratory distress frequently perform this test, and they are listening for two main respiratory sounds: bronchial and vesicular. Bronchial sounds, caused by airflow through the bronchi, are often high-pitched and loud during both inspiration and expiration. Vesicular sounds, caused by air filling the alveolar sacs, are lower in pitch, quiet, and heard primarily during inspiration. When auscultating between the first and second intercostal spaces, bronchovesicular sounds of intermediate pitch can be heard during inspiration and expiration. When inflammation or infection is present in the respiratory structures, these sounds take on different characteristics. Crackles, or popping noises, particularly during inspiration, may be indicative of fluid present in the lungs. Wheezing, or high-pitched sounds particularly during expiration, may be indicative of increased resistance in the airway. Your task in this exercise is to auscultate these clinically relevant respiratory sounds.

Before you begin, state a hypothesis regarding the optimal placement of the stethoscope diaphragm that will allow for auscultation of respiratory sounds.

_____

_____

_____

1. Obtain a stethoscope and alcohol swabs, and choose a study subject. To clean the stethoscope, wipe the earpieces and diaphragm with an alcohol swab. Allow the stethoscope to dry. Insert the earpieces in your ears in preparation for auscultating respiratory sounds.

2. Locate the larynx on the subject. Press the diaphragm of the stethoscope on the skin just inferior to the subject's larynx. Auscultate the bronchial sounds as the subject breathes deeply through the mouth. While the subject is breathing, slowly move the diaphragm inferior until the sounds are no longer audible.

3. To auscultate vesicular sounds, press the diaphragm firmly on the subject's skin at the following locations: inferior to the clavicle, in the intercostal spaces, and along the medial border of the scapula. When auscultating, be sure to keep the stethoscope diaphragm in one location during both inspiration and expiration. Also remember to auscultate both right and left respiratory sounds.

EXERCISE 24.12

(BIOPAC) PULMONARY FUNCTION TESTS

In this laboratory exercise, you will measure respiratory volumes and capacities using a spirometer. Before you begin, state your hypothesis regarding how vital capacity changes with a person's size, gender, and age. _____

_____

_____

1. Obtain a BIOPAC airflow transducer (SS11LA), filter, disposable mouthpiece, noseclip, and calibration syringe.

2. Prepare BIOPAC equipment by turning the computer ON and MP3X Data Acquisition Unit OFF. Plug the airflow transducer (SS11LA) into channel 1 (CH 1). Place a disposable filter on the end of the calibration syringe and insert the entire assembly into the airflow transducer on the side labeled "inlet."

3. Start the BIOPAC Student Lab Program. Select "L12-LUNG-1" from the dropdown menu, enter the file name, and click "OK."

4. To calibrate, hold the airflow transducer/calibration syringe assembly horizontally such that the airflow transducer remains upright during the procedure (see **figure 24.15**). Pull the calibration syringe plunger out to its maximal position. Click "Calibrate." The first calibration stage runs for 8 seconds. Click "Yes" once the alert box appears. Press the syringe completely in for approximately 1 second, wait 2 seconds, and then pull the syringe completely out for 1 second. Repeat this procedure four times for a total of 5 cycles. Click "End Calibration." Compare your calibration data to **figure 24.16**. If there are differences, click "Redo Calibration." Otherwise, proceed to step 5.

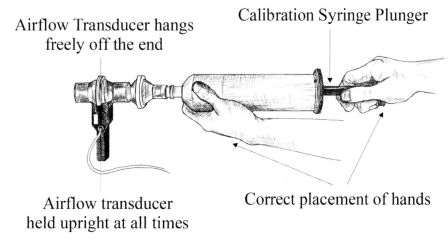

Figure 24.15 Calibration Setup and Assembly. *Courtesy of and © BIOPAC Systems, Inc.*

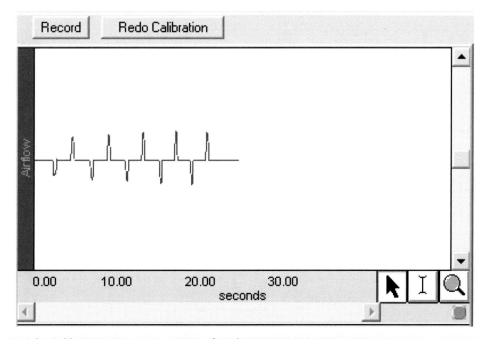

Figure 24.16 Sample Calibration Data. *Courtesy of and © BIOPAC Systems, Inc.*

(continued on next page)

684 Chapter Twenty-Four *The Respiratory System*

(continued from previous page)

5. To prepare for data collection, have the subject assume a relaxed, seated position. The subject should remain still and quiet, and the screen should not be visible to the subject during data collection. Prepare the airflow transducer by first removing the calibration syringe and then inserting a disposable mouthpiece on the side labeled "inlet." Have the subject put on a noseclip.

6. Click "Record" as the subject does the following: (1) takes five normal breaths, (2) inspires maximally once, (3) expires maximally once, (4) takes five normal breaths. Click "Stop." Review the data to ensure that positive spikes appear during inspiration and negative spikes appear during expiration. Compare your data to **figure 24.17**. If there are differences, click "Redo." Otherwise, click "Done."

7. To analyze the data, click "Review Saved Data." Note that CH 1 displays airflow and CH 2 displays volume. Review the airflow data and then turn this channel off by clicking on the channel number box and holding down the control key (PC) or option key (Mac). Note that data analysis is easier when airflow data is not visible.

8. Measurement boxes appear above the marker region in the data window. To set these up, indicate the channel

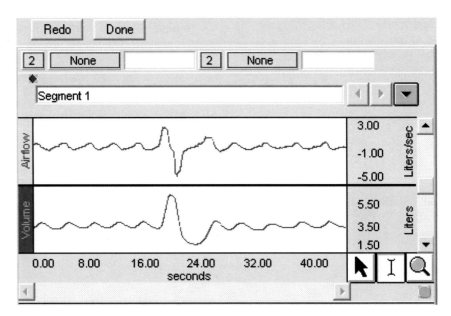

Figure 24.17 Sample Respiratory Data. *Courtesy of and © BIOPAC Systems, Inc.*

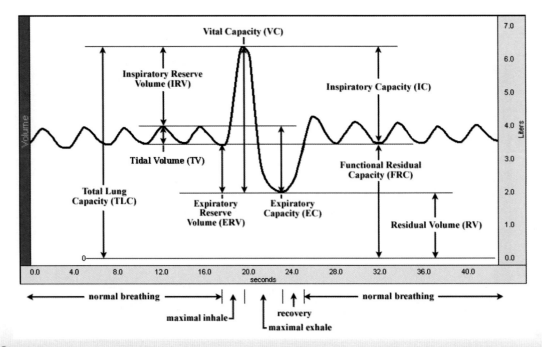

Figure 24.18 Measurement for Respiratory Volumes and Capacities. *Courtesy of and © BIOPAC Systems, Inc.*

number (CH 2) and measurement type for each of four boxes: "P-P" (max–min), "Max," "Min," "Delta" (last point–first point). To magnify a portion of data to be measured, click on the magnifying glass icon in the lower right corner and select the segment to expand.

9. Consult table 24.9 and **figure 24.18** as reference guides for the respiratory volumes and capacities. Locate the "I-beam" tool to the left of the magnifying glass. Select the point on the recording segment that corresponds to the vital capacity (VC) and record the P-P measurement in **table 24.10**.

10. To calculate the average tidal volume (TV), use the "I-beam" tool to highlight the inspiration phase of cycle 3, which should correspond to the minimum and to the maximum of the third cycle **(figure 24.19)**. Record the P-P measurement as trial "1" in table 24.10. Repeat this procedure for the expiration phase of cycle 3, and record the P-P measurement as trial "2" in table 24.10. Calculate the average for these two trials, and record the calculation in table 24.10.

11. Measure the following respiratory volumes and capacities by using the "I-beam" tool for the following values: IRV (Delta), ERV (Delta), RV (Min), IC (Delta), EC (Delta), TLC (Max). Consult figure 24.18 to review the portion of the recording that corresponds to each value. Record the data in table 24.10. Click "Exit."

12. Consult table 24.9 as a guide for the formulas to calculate each respiratory capacity. Calculate each

Table 24.10   Measured and Calculated Respiratory Volumes and Capacities

Abbreviation	Volume/Capacity	Measurement		Calculated	Reference
VC	Vital Capacity				3100 – 4800 mL
TV	Tidal Volume	(1)	(2)	AVG =	500 mL
IRV	Inspiratory Reserve Volume	Approximately 1 L			1900 – 3100 mL
ERV	Expiratory Reserve Volume				700 – 1200 mL
RV	Residual Volume				1100 – 1200 mL
IC	Inspiratory Capacity				2400 – 3600 mL
TLC	Total Lung Capacity				4200 – 6000 mL
FRC	Functional Residual Capacity				1800 – 2400 mL

Courtesy of and © BIOPAC Systems, Inc. Data from BIOPAC Table 12.2

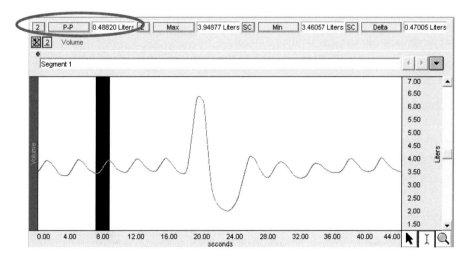

Figure 24.19   P-P Measurement. *Courtesy of and © BIOPAC Systems, Inc.*

(continued on next page)

(continued from previous page)

capacity in table 24.10 and compare your measurements with expected "normal" values. Compare measured and calculated volumes and capacities to the reference ranges provided.

13. Make note of any pertinent observations here:

_____

_____

_____

EXERCISE 24.13 Ph.I.L.S.

Ph.I.L.S. LESSON 31: pH AND Hb-OXYGEN BINDING

Hemoglobin within red blood cells binds oxygen so oxygen can be transported to the tissues. Hemoglobin's affinity for oxygen increases when the partial pressure of oxygen (P_{O_2}) is high, as in the lungs, and decreases when P_{O_2} is low, as in the tissues. The affinity of hemoglobin to bind oxygen is also influenced by blood pH. Any shift in blood pH will dramatically affect the degree to which hemoglobin binds oxygen. In this exercise you will alter the partial pressure of oxygen in a blood sample using a vacuum pump to decrease the P_{O_2} of a blood sample. Using a spectrophotometer you will then demonstrate how P_{O_2} directly affects binding of O_2 to hemoglobin. You will also demonstrate the difference in hemoglobin O_2 saturation at three different pH values (6.8, 7.4, and 8.0).

Before you begin, familiarize yourself with the following concepts (use your textbook as a reference):

- the structure of hemoglobin
- the oxygen-hemoglobin dissociation curve
- the meaning of a right shift or left shift of the oxygen-hemoglobin dissociation curve

State a hypothesis regarding the difference (if any) in the amount of oxygen that binds to hemoglobin as the partial pressure of oxygen increases and as H⁺ concentration increases.

_____

_____

_____

1. Open Ph.I.L.S. Lesson 31: pH and Hb-Oxygen Binding.

2. Read the objectives and introduction and take the pre-laboratory quiz.

3. After completing the pre-laboratory quiz, read through the wet laboratory. Be sure to click open and view the videos that are indicated in red in the wet laboratory.

4. The laboratory exercise will open when you click "Continue" after completing the wet laboratory **(figure 24.20)**.

5. Click the "power" switch to turn on the spectrophotometer.

6. To calibrate the spectrophotometer, set the wavelength at 620 nm (wavelength at which hemoglobin absorbs light if intact) and set the transmittance value at 0 using the arrows at 0.

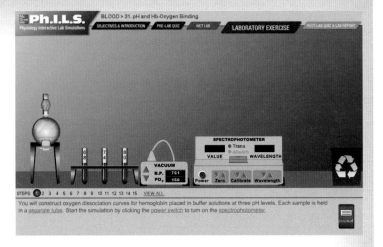

Figure 24.20 Opening Screen for Ph.I.L.S. Lesson 31: pH and Hb-Oxygen Binding.

7. Open the lid of the spectrophotometer holder (if unsure, click on the word "spectrophotometer holder" in the instructions at the bottom of the screen).

8. Click on the globe and drag it to the top of one of the tubes (it will snap in place) to form the tonometer. (To click and drag, move the mouse to position the arrow on the globe; left-click on the mouse and hold; drag the globe to the top of one of the tubes; and when positioned, release the left click.)

9. Place the tonometer into the spectrophotometer. The journal will open a table with the partial pressures of oxygen ranging from 160 mm Hg to 0 mm Hg for three different pH values—6.8, 7.4, and 8.0. Set the transmittance at 100% using the arrows at "Calibrate."

10. To record the transmittance through the sample, click the journal (red rectangle at the bottom right of screen).

11. Click on the tonometer to remove it. (The tonometer will come up out of the spectrophotometer and the journal will close automatically.) Then click and drag the tonometer back to the rack.

12. To change the partial pressure of oxygen in the sample, position the end of the vacuum tube on the top of the tonometer.

13. Click the down arrow on the vacuum once to reduce the barometric pressure (which simulates a decrease in partial pressure of oxygen).

14. Click on the vacuum tube to return it to its original position.
15. Place the tonometer in the spectrophotometer.
17. Click on "journal" to record.
18. Repeat steps 11–17 over the pressure range (160–0 mm Hg) for that sample until all values for the different partial pressures of oxygen have been recorded in the table on the computer.
19. When all values have been recorded in the table for a given sample, drag the tonometer to place the sample into the recycling can (three arrows in a circle in the bottom right of the screen) to dispose of the sample.
20. Repeat steps 7–19 with the remaining blood samples using different pH values.
21. Click on the journal to see the complete table.
22. To view the graph, click on "graph" in the lower right of the journal window.
23. Complete the post-laboratory quiz (click "open post-laboratory quiz and laboratory report") by answering the ten questions on the computer screen.
24. Complete **table 24.11** below from the table created on the screen in the journal and graph the data in the graph to the right.
25. Make note of any pertinent observations here:

_____
_____
_____

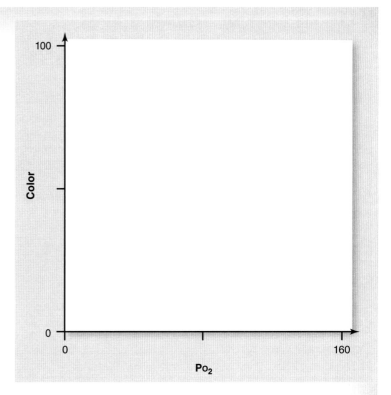

Table 24.11	Data Table for Partial Pressures of Oxygen and Blood pH		
P_{O_2}	pH 6.8	pH 7.4	pH 8.0
160			
140			
120			
100			
80			
60			
40			
20			
0			

EXERCISE 24.14 Ph.I.L.S.

Ph.I.L.S. LESSON 35: EXERCISE-INDUCED CHANGES

During exercise, the respiratory system must adapt to increased oxygen demands of working muscles. The respiratory system accomplishes this by increasing tidal volume and respiratory rate. The purpose of this laboratory exercise is to monitor changes in respiratory volumes and rates of a virtual subject before and after exercise.

Before you begin, familiarize yourself with the following concepts (use your textbook as a reference):

- inspiratory reserve volume
- expiratory reserve volume
- tidal volume
- vital capacity
- respiratory rate

State a hypothesis regarding the effect, if any, of exercise on tidal volume, inspiratory reserve volume, expiratory reserve volume, and vital capacity: _____

_____

_____

1. Open Ph.I.L.S. Lesson 35, Exercise-Induced Changes (figure 24.21).
2. Click the "power" button on the data acquisition unit to begin the experiment.
3. Click and drag the black plug to recording input 1 on the data acquisition unit.
4. Click and drag the mouthpiece to the subject's mouth.
5. Click the box marked "resting" on the control panel of the virtual monitor.

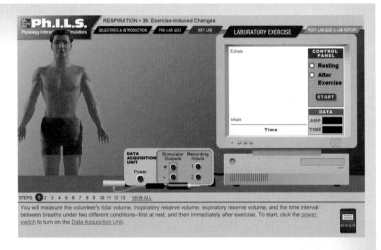

Figure 24.21 Opening Screen for Ph.I.L.S. Lesson 35: Exercise-Induced Changes.

6. To begin recording normal breathing patterns, click "start" on the control panel of the virtual monitor. A red trace will appear representing the subject's breathing pattern.
7. The apparatus will measure three tidal volumes (TV), then the subject will maximally exhale and maximally inhale. At this point the recording will stop and you can make the following measurements.
8. Measure the TV for the first breath by dragging the pointer over the first peak and clicking the mouse to drop a marker. Then move the pointer over the first trough and click the mouse. Data will appear in the window on the virtual monitor. Click the red "journal" button in the lower right of the screen. Repeat this measurement for all three breaths. Enter the TV for each of the three breaths, the mean, and the calculated breaths per minute in **table 24.12**.

Table 24.12	Tidal Volumes before and after Exercise			
Trial	TV Resting (mL)	TV Post-Exercise (mL)	Time Interval at Rest (s)	Time Interval Post-Exercise(s)
#1				
#2				
#3				
Mean				
Breaths per Minute				

9. To measure the expiratory reserve volume (ERV), move the pointer to the third trough and click the mouse to drop a marker. Then move the pointer to the lowest point on the tracing and click the mouse again to measure the distance. The data appears in the data window on the virtual monitor. Click the red "journal" button to enter it in the journal and enter the data in **table 24.13**.

Table 24.13	Lung Volumes before and after Exercise	
	Resting (mL)	Post-Exercise (mL)
IRV		
TV		
ERV		
VC		

10. To measure the inspiratory reserve volume (IRV), move the pointer to the third peak and click the mouse to drop a marker. Then move the pointer to the highest part of the trace and click the mouse to measure the height of the peak. Click the red "journal" button to enter the data in the journal, and enter the data in table 24.13.

11. To record the lung volumes following exercise, click the box marked "After Exercise" on the control panel of the virtual monitor. Repeat steps 6–10 and record the data in tables 24.12 and 24.13.

12. Make note of any pertinent observations here:

_____
_____
_____

Chapter 24: The Respiratory System

Name: _____
Date: _____ Section: _____

POST-LABORATORY WORKSHEET

The ❶ corresponds to the Learning Objective(s) listed in the chapter opener outline.

Do You Know the Basics?

Exercise 24.1: Olfactory Mucosa

1. Fully classify the epithelium that lines the nasal cavity. ❶ _____
_____

2. A fracture of the ethmoid bone, as might occur in an auto accident involving severe facial injuries, can cause an individual to lose her sense of _____. This is due to damage to the _____ nerves, which travel through the _____ of the ethmoid bone en route to the brain. ❸

Exercise 24.2: The Trachea

3. Describe the epithelium that lines the trachea. ❺
_____
_____

4. Describe the function of the C-shaped tracheal cartilages and the trachealis muscle. ❻ _____
_____

Exercise 24.3: The Bronchi and Bronchioles

5. The following questions refer to the micrograph shown below. ❼ ❽

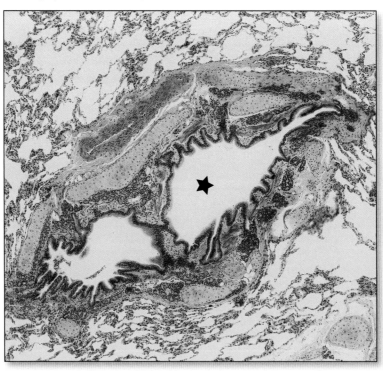

   a. What type of airway is indicated in this micrograph? (The star is in the lumen of the airway in question.)
   _____

   b. In the space below, explain your reasoning for your answer to part (a) of this question. That is, explain the histological features of the airway (e.g., epithelium, cartilage) that you used to determine your answer.
   _____
   _____
   _____

690   Chapter Twenty-Four *The Respiratory System*

Chapter Twenty-Four *The Respiratory System* 691

Exercise 24.4: The Lungs

6. Label the structures in the figure below. Then list, in order, the three structures through which a molecule of carbon dioxide must travel in order to diffuse from the blood into the alveoli.

1. _____
2. _____
3. _____
4. _____
5. _____
6. _____
7. _____
8. _____
9. _____
10. _____
11. _____

Diffusion of CO_2 Diffusion of O_2

a. _____
b. _____
c. _____

7. Hyaline membrane disease of the newborn, a disease in which the infant experiences great difficulty breathing, is characterized by a lack of pulmonary surfactant production. Which cells that produce surfactant are not yet functioning sufficiently? _____

Exercise 24.5: Sagittal Section of the Head and Neck

8. The nasal cavity contains the superior, middle, and inferior nasal _____, and the spaces between, known as the superior, middle, and inferior _____.

Exercise 24.6: The Larynx

9. Which structures are located more superior in the larynx? (circle one)

    a. the vestibular folds

    b. the vocal folds

10. Define *rima glottidis*: _____

11. Identify the three unpaired cartilages of the larynx.

    a. _____
    b. _____
    c. _____

12. Identify the three paired cartilages of the larynx.

    a. _____

    b. _____

    c. _____

13. Explain the function of the larynx in sound production.

_____

_____

Exercise 24.7: The Pleural Cavities

14. Identify the two layers of the pleura.

    a. _____

    b. _____

Exercise 24.8: The Lungs

15. The left lung is smaller than the right lung. (True/False)

16. The hilum of the right or left lung contains the following:

    a. _____

    b. _____

    c. _____

17. The right lung contains _____ lobes, whereas the left lung contains _____ lobes.

Exercise 24.9: The Bronchial Tree

18. Identify the first three branches of the bronchial tree that supply the left lung.

    a. _____

    b. _____

    c. _____

19. How many lobar bronchi lead into the right lung? _____

Exercise 24.10: Mechanics of Ventilation

20. List the four processes of respiration:

    a. _____

    b. _____

    c. _____

    d. _____

21. A(n) _____ (increase or decrease) in thoracic volume leads to an increase in intrapulmonary pressure.

22. A decrease in intrapulmonary pressure allows air to flow (in/out) of the lungs.

Exercise 24.11: Auscultation of Respiratory Sounds

23. Describe the optimal position(s) for auscultation of respiratory sounds:

    a. bronchial: _____

    b. vesicular: _____

Exercise 24.12: (BIOPAC) Pulmonary Function Tests

24. Match the terms listed in column A with the appropriate definitions given in column B.

    Column A

    _____ 1. expiratory reserve volume (ERV)
    _____ 2. functional residual capacity (FRC)
    _____ 3. inspiratory capacity (IC)
    _____ 4. inspiratory reserve volume (IRV)
    _____ 5. residual volume (RV)
    _____ 6. total lung capacity (TLC)
    _____ 7. tidal volume (TV)
    _____ 8. vital capacity (VC)

    Column B

    a. amount of air remaining in the lungs after quiet expiration
    b. measure of the strength of respiration
    c. total ability to inspire
    d. total amount of air that can be in the lungs
    e. volume of additional inspired air, above and beyond the tidal volume
    f. volume of air left in the lungs following forced expiration
    g. volume of air maximally expired above and beyond the tidal volume
    h. volume of air that enters and leaves the lungs with each normal breath

25. Describe how vital capacity changes with body size, gender, and age. _____
    _____

26. How do you predict vital capacity will change after vigorous exercise? _____
    _____

27. Spirometry can be used to distinguish _____ respiratory diseases from _____ respiratory diseases.

Exercise 24.13: Ph.I.L.S. Lesson 31: pH & Hb-Oxygen Binding

28. As the partial pressure of oxygen (P_{O_2}) increases, the percent transmittance of light _____ (increases, decreases, or stays the same).

29. There is a direct relationship between percent transmittance of light and percent saturation of hemoglobin. As the percent transmittance of light increases, the percent saturation of hemoglobin _____ (increases, decreases, or stays the same). Thus, as the partial pressure of oxygen increases, the percent saturation of hemoglobin _____ (increases, decreases, or stays the same).

30. Describe how altering the rate of respiration can influence blood pH. _____
    _____
    _____

Exercise 24:14: Ph.I.L.S. Lesson 35: Exercise-Induced Changes

31. Describe the cellular mechanism responsible for the increase in respiratory rate observed after the subject exercised. Explain whether it is an increase in oxygen use or an increase in the production of carbon dioxide that normally influences respiratory rate.
    _____
    _____
    _____
    _____

32. Describe the changes, if any, observed in tidal volume, vital capacity, inspiratory reserve volume, and expiratory reserve volume after the subject exercised.
    _____
    _____
    _____
    _____
    _____
    _____

Chapter Twenty-Four The Respiratory System

Can You Apply What You've Learned?

33. In the space below, trace the pathway a molecule of oxygen must take to travel from the nasal cavity to an alveolus in the inferior lobe of the right lung. Be sure to name all the conducting and respiratory portion structures the molecule will pass through along the way.

34. A toddler coughs when attempting to swallow a bite of hot dog, and the hot dog is directed into the respiratory tree instead of the esophagus. The piece of hot dog is most likely to become lodged in the airways leading to the _____ lung because the primary bronchus to this lung is more vertically oriented than the primary bronchus to the other lung. (Note: In 2010 the FDA listed hot dogs as a major choking hazard for children. It even suggested that hot dogs should have warning labels on them because of the large number of children who choke on them.)

35. When a person suffers from an asthma attack, there is often excessive inflammation of respiratory structures, which leads to increased mucus production and smooth muscle contraction. Assuming that asthma is an obstructive respiratory disease, predict whether a person suffering from asthma would have more difficulty inhaling or exhaling, and justify your answer.

36. Discuss abnormal respiratory sounds that might be heard in a patient suffering from asthma.

37. A patient is rushed to the hospital with a gunshot wound to the chest. Upon arrival, doctors discover that the patient is suffering from a collapsed right lung. Explain why the lung collapsed when the bullet penetrated the patient's thoracic cavity.

38. Opening a can of carbonated soda leads to the loss of "fizz" over time. Describe why this occurs, and relate this phenomenon to alveolar gas exchange.

39. When a person is hyperventilating, a home remedy is to have the person breathe deeply into and out of a paper bag. Describe how this "treatment" may be beneficial in regulating blood pH and respiratory rate.

40. Describe why a patient with a restrictive lung disease would have a decreased vital capacity.

Can You Synthesize What You've Learned?

41. The world's tallest free-standing mountain, Mount Kilimanjaro in Tanzania, Africa, stands at approximately 19,340 feet, where atmospheric pressure is roughly half that of pressure at sea level. Discuss how partial pressures of oxygen will change at the peak of the mountain as compared to sea level. Then describe how these partial pressures may affect respiratory rate and alveolar gas exchange.

42. Exposure to large quantities of carbon monoxide (CO) in a poorly ventilated space can be lethal. Unfortunately, this is a common occurrence during winter months because furnaces run the risk of emitting large quantities of CO. When inhaled, CO has a higher affinity for the heme group on hemoglobin than does oxygen. Describe how exposure to CO will influence percent oxygen saturation of hemoglobin and affect systemic gas exchange.

43. A hypersensitivity to peanuts can lead to severe inflammation of the tissues lining the respiratory tract. Often, doctors will prescribe *epinephrine* to be administered by auto injection (EpiPen®) in case of exposure. When *epinephrine* is injected intramuscularly, it binds to adrenergic receptors throughout the body (see chapter 17), including those located on the smooth muscle of respiratory structures. Based on your knowledge of the autonomic nervous and respiratory systems, predict what will happen to both resistance and airflow through the respiratory structures of a patient when *epinephrine* is administered.

The Digestive System

CHAPTER 25

OUTLINE AND LEARNING OBJECTIVES

Histology 700

Salivary Glands 700

EXERCISE 25.1: HISTOLOGY OF THE SALIVARY GLANDS 701
1. Identify mucous and serous cells in histology slides of salivary glands
2. Distinguish among the parotid, submandibular, and sublingual salivary glands when viewed through a microscope, using histological characteristics unique to each gland

The Stomach 702

EXERCISE 25.2: WALL LAYERS OF THE STOMACH 702
3. Identify the layers of the stomach wall when viewed through a microscope
4. Describe the tissues located in each layer of the stomach wall
5. Classify the epithelium that lines the stomach

EXERCISE 25.3: HISTOLOGY OF THE STOMACH 705
6. Identify gastric pits and gastric glands in histology slides of the stomach
7. Identify chief cells, mucous neck cells, parietal cells, and surface mucous cells on histology slides of the stomach

The Small Intestine 707

EXERCISE 25.4: HISTOLOGY OF THE SMALL INTESTINE 708
8. Describe the structures that increase the total surface area of the small intestine
9. Classify the epithelium that lines the small intestine
10. Distinguish the three parts of the small intestine (duodenum, jejunum, and ileum) from one another when viewed through a microscope, using histological characteristics unique to each

The Large Intestine 708

EXERCISE 25.5: HISTOLOGY OF THE LARGE INTESTINE 709
11. Identify the large intestine when viewed through a microscope
12. Classify the epithelium that lines the large intestine
13. Describe the wall layers of the large intestine

The Liver 710

EXERCISE 25.6: HISTOLOGY OF THE LIVER 710
14. Identify the liver and its associated structures when viewed through a microscope
15. Describe the structure of a hepatic lobule
16. Explain how blood flows through a hepatic lobule

The Pancreas 711

EXERCISE 25.7: HISTOLOGY OF THE PANCREAS 712
17. Identify the histological structure of the exocrine part of the pancreas when viewed through a microscope
18. Distinguish the exocrine and endocrine parts of the pancreas from each other
19. Describe the histological structure of the exocrine part of the pancreas

Gross Anatomy 712

The Oral Cavity, Pharynx, and Esophagus 712

EXERCISE 25.8: GROSS ANATOMY OF THE ORAL CAVITY, PHARYNX, AND ESOPHAGUS 712
20. Identify structures of the digestive system as seen in a sagittal section of the head and neck

The Stomach 714

EXERCISE 25.9: GROSS ANATOMY OF THE STOMACH 714
21. Identify the gross anatomic features of the stomach

The Duodenum, Liver, Gallbladder, and Pancreas 716

EXERCISE 25.10: GROSS ANATOMY OF THE DUODENUM, LIVER, GALLBLADDER, AND PANCREAS 717
22. Describe the anatomic relationships between the duodenum, liver, gallbladder, and pancreas
23. Identify the gross anatomic features of the liver
24. Trace the flow of a drop of bile from the gallbladder to the duodenum

The Jejunum and Ileum of the Small Intestine 718

EXERCISE 25.11: GROSS ANATOMY OF THE JEJUNUM AND ILEUM OF THE SMALL INTESTINE 719
25. Describe the structure of encroaching fat, arterial arcades, and vasa recta
26. Distinguish the jejunum and ileum from each other using gross anatomic characteristics

The Large Intestine 720

EXERCISE 25.12: GROSS ANATOMY OF THE LARGE INTESTINE 721
27. Identify the parts of the large intestine
28. Identify the regions of the large intestine

Physiology 722

Digestive Physiology 722

EXERCISE 25.13: DIGESTIVE ENZYMES 725
29. Identify the enzymes that chemically digest carbohydrates, proteins, fats, and nucleic acids, and specify their sites of action
30. Determine the effect of temperature on enzyme activity
31. Identify the monomer units for carbohydrates, proteins, fats, and nucleic acids, and describe how chemical digestion aids in their absorption

EXERCISE 25.14: Ph.I.L.S. LESSON 37: GLUCOSE TRANSPORT 727
32. Describe the role of the sodium/potassium ATPase pump in the transport of glucose across the wall of the small intestine

 MODULE 12: DIGESTIVE SYSTEM

INTRODUCTION

The digestive (*digero*, to force apart, dissolve) system is concerned with the breakdown of food into usable molecules needed by the body for energy, maintenance, and ultimately survival. The gastrointestinal (GI) tract, or digestive tract, consists of a long tube, which is open to the environment at both ends (the oral cavity and the anus). This means that the lining of the GI tract is open to the external environment. Thus, the interface between the lumen of the GI tract and the internal environment of the body presents a special problem. The GI tract must transport needed substances from the lumen of the gut tube into the body, while at the same time it must prevent pathogens from entering. For this reason, the subepithelial tissues of the entire GI tract are densely populated with lymphatic tissues.

The walls of the GI tract are composed of four layers that include the mucosa, submucosa, muscularis, and serosa/adventitia. In these laboratory exercises you will explore how the wall layers of the various parts of the GI tract are modified to suit the particular needs of the organ. In addition, you will explore the structure and function of organs such as the liver and pancreas—highly specialized glands that develop from the lining of the GI tract. As you observe the structures of the digestive system, you will also review the various circulatory and lymphatic structures associated with these structures.

In this laboratory session you will explore both the histological structure and the gross structure of the GI tract. The materials in the "Histology" and "Gross Anatomy" sections are organized so the order of structures studied begins at the mouth and ends at the anus. In other words, if you study the structures in the order in which they are described in this chapter, you will encounter GI tract structures in the same order that a bolus of food encounters them as it moves through the GI tract. It is not imperative that you observe the structures in this order while you are in the laboratory. This will allow you to reflect on how the different parts of the GI tract work together to accomplish digestion. In the "Physiology" section of this chapter, you will explore the processes of chemical digestion by digestive enzymes and the specific mechanisms for the absorption of glucose.

Chapter 25: The Digestive System

Name: _____
Date: _____ **Section:** _____

PRE-LABORATORY WORKSHEET

Also available at www.connect.mcgraw-hill.com

1. How many pairs of salivary glands does a human have? _____

2. What are the three segments of the small intestine?

    a. _____

    b. _____

    c. _____

3. How many lobes compose the liver? _____

4. What is the function of the gallbladder? _____

5. Into which portion of the pharynx does the oral cavity open? _____

6. List the three unpaired branches of the abdominal aorta that deliver oxygenated blood to the organs of the lower digestive tract.

    a. _____

    b. _____

    c. _____

7. Into what segment of the small intestine do the pancreatic ducts empty? _____

8. How many layers of smooth muscle are within the wall of the stomach? _____

9. A blind pouch that represents the first part of the large intestine is the _____.

10. The specific type of epithelium that lines the small and large intestines is _____
_____.

Chapter Twenty-Five *The Digestive System* 699

Histology

Salivary Glands

The **salivary glands (figure 25.1)** are accessory digestive glands composed of modified epithelial tissue that produce **saliva,** a watery secretion that helps dissolve foodstuffs and contains an enzyme, **salivary amylase,** that begins the initial digestion of carbohydrates. There are three pairs of salivary glands—parotid, submandibular, and sublingual—and all of them empty their secretions into the oral cavity through ducts. Salivary glands contain two cell types: serous cells and mucous cells. **Serous cells** produce watery secretions containing proteins, electrolytes, and the enzymes salivary amylase and lysozyme, and **mucous cells** produce mucin that when hydrated becomes mucus. Both serous fluid and mucus function as lubricants for the passage of food through the esophagus. Definitions of structures related to salivary glands are described in **table 25.1,** and details of the structure and function of each of the salivary glands are listed in **table 25.2.**

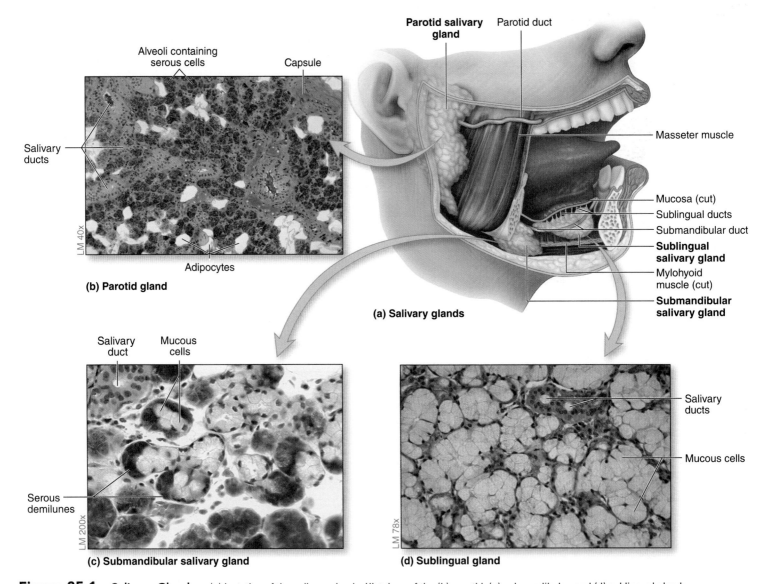

Figure 25.1 Salivary Glands. (a) Location of the salivary glands. Histology of the (b) parotid, (c) submandibular, and (d) sublingual glands.

Chapter Twenty-Five The Digestive System   701

Table 25.1	Salivary Gland Structures		
Structure	**Description and Location**	**Functions**	**Word Origin**
Alveolus	The grape-shaped secretory portion of a gland	NA	*acinus*, grape
Mucous Cells	Cells have flattened nuclei that are located on the basal surface. Mainly located along the tubules of salivary glands.	Secrete mucin	*mucosus*, mucous
Myoepithelial Cells	Cells are not visible in light microscopy. Located around the alveoli and the long axes of the ducts.	Contraction of these cells expels the secretions from salivary glands	*mys*, muscle, + *epithelial*, relating to epithelial tissues
Serous Cells	Cells have round nuclei and contain numerous secretory granules. Located in the alveoli of the parotid gland and in the demilunes of the submandibular and sublingual glands.	Secrete proteins, electrolytes, and salivary amylase	*serous*, having a watery consistency
Serous Demilunes	Crescent- or moon-shaped groups of serous cells located at the periphery of a mucous alveolus	Secrete proteins, electrolytes, and salivary amylase	*serous*, having a watery consistency, + *demilune*, half-moon

Table 25.2	Histological Characteristics of Salivary Glands				
Gland	**Secretory Cells**	**Type of Secretion**	**% of Saliva**	**Opening**	**Word Origin**
Parotid	**Serous:** serous cells occupy ~90% of the gland's volume (the rest is adipose connective tissue)	Mostly water; 25% of dissolved solutes are glycoproteins; high amylase content	25–30%	Empties via the parotid duct opposite the second upper molar	*para*, beside, + *ous*, ear
Sublingual	**Mixed:** 60–70% serous, 30–40% mucus; serous cells are located in serous demilunes	90% of solutes are glycoproteins (the most viscous secretion); low amylase content	3–5%	Empties via multiple ducts into either the submandibular duct or directly into the oral cavity	*sub-*, under, + *lingual*, the tongue
Submandibular	**Mixed:** 80% serous, 20% mucus, contains a few serous demilunes	40–60% of dissolved solutes are glycoproteins; low amylase content; contains lysozyme, an enzyme that inhibits the growth of bacteria	60–70%	Empties via the submandibular ducts between the lingual frenulum and the mandible	*sub-*, under, + *mandible*, the mandible

EXERCISE 25.1

HISTOLOGY OF THE SALIVARY GLANDS

1. Obtain a compound microscope and histology slides of the parotid, submandibular, and sublingual salivary glands.

2. Place the slide of the **parotid gland** on the microscope stage. Bring the tissue sample into focus on low power and then switch to high power.

3. Using figure 25.1*a* and tables 25.1 and 25.2 as guides, identify the following structures on the slide of the parotid gland:

   ☐ adipocytes   ☐ serous cells

4. Place the slide of the **submandibular gland** on the microscope stage. Bring the tissue sample into focus on low power and then switch to high power.

5. In both the submandibular and sublingual salivary glands, the serous cells are located surrounding the mucous cells and are shaped like half moons. Thus, they are referred to as *serous demilunes* (*demi*, half, + *luna*, moon). Scan the slide to locate serous demilunes (figure 25.1*b* and 25.1*c*).

6. Using figure 25.1*c* and tables 25.1 and 25.2 as guides, identify the following structures on the slide of the submandibular gland:

   ☐ mucous cells   ☐ serous demilunes
   ☐ salivary duct

7. Place the slide of the **sublingual gland** on the microscope stage. Bring the tissue sample into focus on low power and then switch to high power.

8. The sublingual gland is similar to the submandibular gland in that it contains serous demilunes. There are very few adipocytes in the gland, so that is one characteristic that can be used to distinguish it from the submandibular gland. It also contains more mucous cells (table 25.2 and figure 25.1*d*).

9. Using figure 25.1*c* and tables 25.1 and 25.2 as guides, identify the following structures on the slide of the sublingual gland:

   ☐ mucous cells   ☐ salivary duct

(continued on next page)

702 Chapter Twenty-Five The Digestive System

(continued from previous page)

10. In the following spaces, sketch the histological appearance of the parotid, submandibular, and sublingual salivary glands. Label mucous cells, serous cells, and serous demilunes in your drawings.

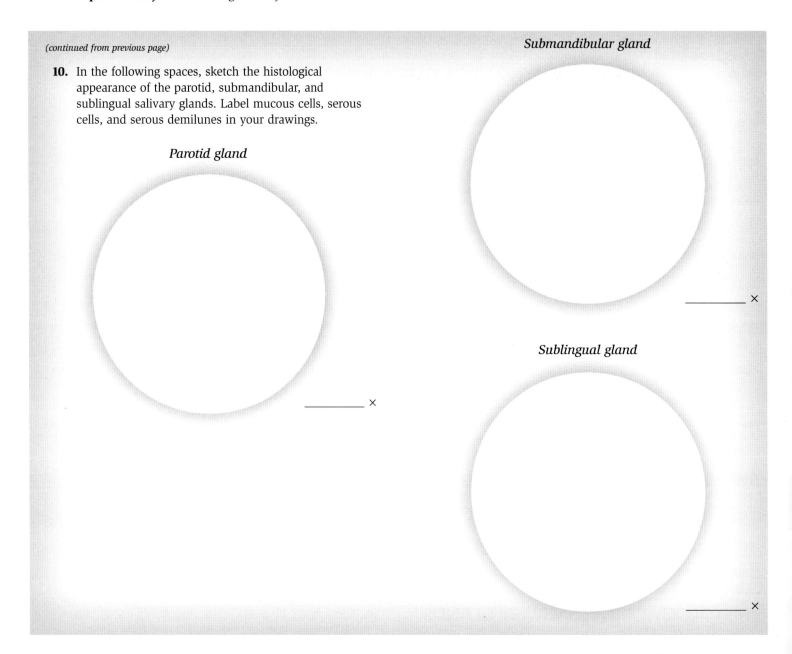

Parotid gland

Submandibular gland

Sublingual gland

The Stomach

The **stomach** is an organ that digests food by mixing it mechanically, through the action of smooth muscle in its walls, and begins to chemically digest food molecules through the action of stomach enzymes and a highly acidic environment within. In this section you will first observe the histology of the stomach wall and then observe specific regional characteristics of the stomach epithelium.

EXERCISE 25.2

WALL LAYERS OF THE STOMACH

The walls of the GI tract are composed of four concentric tunics (layers): The **mucosa, submucosa, muscularis,** and **adventitia/serosa**. **Figure 25.2** shows the tunics that typically compose the GI tract, and **figure 25.3** shows how these wall layers are modified in the stomach. **Table 25.3** describes the types of tissues that are located in each of the layers of the stomach wall.

1. Obtain a histology slide of the stomach (figure 25.3) and place it on the microscope stage. Bring the tissue sample into focus on low power.

2. Using figure 25.3 and table 25.3 as guides, identify the following layers of the stomach wall on the slide:

   ☐ epithelium
   ☐ lamina propria
   ☐ mucosa
   ☐ muscularis
   ☐ muscularis: circular layer
   ☐ muscularis: longitudinal layer
   ☐ muscularis mucosae
   ☐ muscularis: oblique layer
   ☐ serosa
   ☐ submucosa

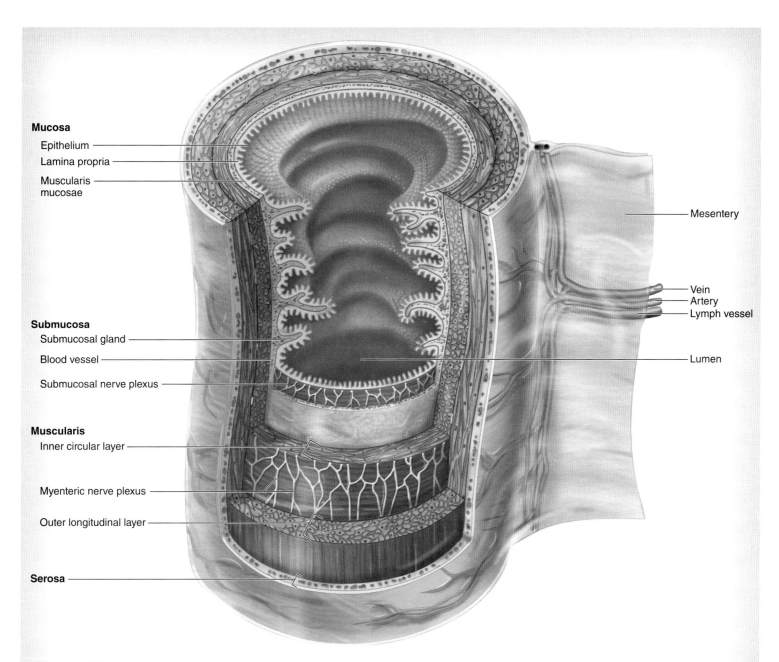

Figure 25.2 Wall Layers of the GI Tract. The four wall layers of the GI tract are the mucosa, submucosa, muscularis, and adventitia/serosa.

3. In the space to the right, sketch the histology of the stomach wall as seen through the microscope. Be sure to label the structures listed in step 2 in your drawing.

_____ ×

(continued on next page)

704 Chapter Twenty-Five *The Digestive System*

(continued from previous page)

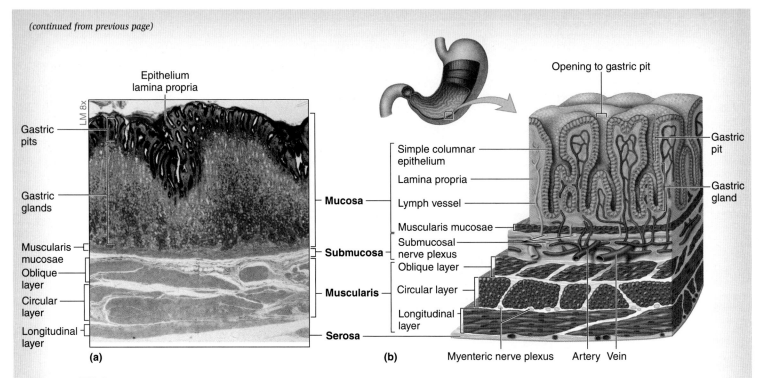

Figure 25.3 **The Stomach Wall.** (a) Histology of the stomach wall. (b) Layers of the stomach wall.

Table 25.3	Wall Layers of the Stomach	
Layer	**Sublayer**	**Characteristics**
Mucosa	Epithelium	Simple columnar epithelium containing five distinct cell types (see table 25.5)
	Lamina propria	Areolar connective tissue that contains blood vessels and nerves. Free lymphocytes and lymphatic nodules are common.
	Muscularis mucosae	Composed of thin layer of smooth muscle
Submucosa	Connective tissue	Composed of areolar and dense irregular connective tissue that contains blood vessels, lymphatic vessels, and the submucosal nerve plexus
Muscularis	Three layers of smooth muscles	Myenteric plexus is located between the layers of smooth muscle
	Inner oblique layer	Contains muscle fibers responsible for causing the "twisting" action of the stomach
	Middle circular layer	Forms the thickest layer of the three layers of muscle. Thickenings of this layer form the inferior esophageal (cardiac) and pyloric sphincters.
	Outer longitudinal layer	Found in the greater and lesser curvatures only
Serosa	NA	Areolar connective tissue with dispersed collagen and elastic fibers; covered by visceral peritoneum (a serous membrane)

EXERCISE 25.3

HISTOLOGY OF THE STOMACH

The epithelium of the stomach is modified to form **gastric pits,** invaginations of the epithelium that contain the openings of **gastric glands (figure 25.4). Epithelial stem cells** (not shown) are located at the junction between the gastric pits and the gastric glands. These cells continuously renew the epithelial lining of the stomach, the epithelium lining the gastric pits, and the epithelial cells that form the gastric glands. The epithelial cells are replaced approximately every three days. Stem cells that will line the gastric pits and the lumen of the stomach migrate up toward the stomach lumen, whereas stem cells that form the gastric glands migrate down towards the muscularis mucosae. The cell types in the gastric pits and glands vary in different regions of the stomach. **Table 25.4** describes the structure of the pits and glands in each of the major regions of the stomach (cardia, fundus/body, and pylorus). **Table 25.5** describes the five cell types that compose the gastric pits and glands, and the types and functions of the secretions produced by each cell.

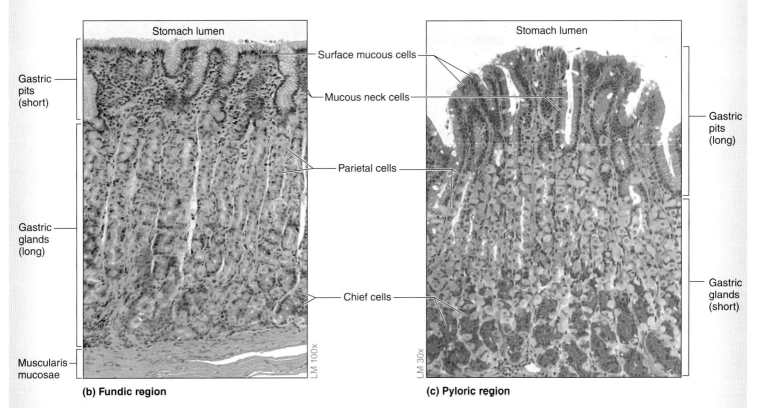

Figure 25.4 Gastric Pits and Gastric Glands. (a) Gastric pits are lined with light-staining, simple columnar mucous cells. The upper regions of the gastric pits contain large, light-staining parietal cells that resemble fried eggs, while the lower regions of the gastric pits contain smaller, dark-staining chief cells. (b) The fundic region of the stomach is characterized by short pits and long glands. (c) The pyloric region of the stomach is characterized by long pits and short glands.

(continued on next page)

Table 25.4 Regional Characteristics of the Gastric Pits and Glands of the Stomach

Region	Pit Structure	Main Cell Type(s)	Gland Structure
Cardia	Short pits with long glands	All mucous	Simple branched, tubular glands
Fundus/Body	Short pits with long glands	Parietal and chief cells with mucous neck cells	Branched, tubular glands
Pylorus	Long pits with short, coiled glands	Mostly mucous	Branched, tubular glands

1. Obtain a histology slide of the stomach and place it on the microscope stage. Bring the tissue sample into focus on low power and then scan the slide to locate the epithelial lining of the stomach (figure 25.4). Once you have the epithelium of the stomach at the center of the field of view, change to high power and bring the tissue sample into focus once again. How would you classify the epithelium that lines the stomach?

   _____

   _____

2. Once you have identified the epithelium lining the stomach wall, locate a gastric pit and gland and move the microscope stage so it is in the center of the field of view.

3. Identify the following structures on the slide:
   - ☐ chief cells
   - ☐ gastric glands
   - ☐ gastric pits
   - ☐ mucous neck cells
   - ☐ parietal cells
   - ☐ surface mucous cells

4. In the space below, sketch the gastric glands and pits as seen through the microscope. Be sure to label the locations of the specialized cell types in your drawing. Can you tell which part of the stomach wall you are looking at based on the structure of the gastric pits and glands? _____

_____ ×

5. *Optional Activity:* **AP|R** 12: Digestive System—Watch the "Stomach" animation for a review of the stomach wall layers and their histology.

Table 25.5 Cell Types in the Gastric Pits and Glands of the Stomach

Cells	Location	Secretions	Action of Secretion
Chief Cells (zymogenic)	Deeper within the gastric glands. Contain numerous eosinophilic (red) granules.	Pepsinogen	A zymogen* that is converted to pepsin when it encounters the acidic environment of the stomach. Pepsin is a protease (it breaks down proteins).
Enteroendocrine Cells	Scattered throughout the gastric glands	Gastrin	Hormone that stimulates chief cells and parietal cells to secrete their products, and stimulates the smooth muscle in the stomach walls to contract
Mucous Neck Cells	Lining the interior of the gastric pits	Mucin	Glycoprotein that protects the mucosa from HCl
Parietal Cells	Superficial within the gastric glands in the fundus and body of the stomach. A few are found in the pylorus. None are found in the cardia.	HCl (hydrochloric acid)	Decreases the pH of the stomach to about 2 (very acidic).
		Intrinsic factor	Necessary for vitamin B_{12} absorption in the small intestine.
Surface Mucous Cells	Covering the ridges between the gastric pits	Mucin	Glycoprotein that protects the mucosa from HCl

*Zymogen is a general term for an inactive protein. Generally, the names of zymogens end in -ogen, as in pepsinogen.

The Small Intestine

After initial digestion in the mouth and stomach, the mixture of the bolus of food and gastric juices is collectively called **chyme.** When this chyme leaves the stomach, it enters the small intestine, where digestion of foodstuffs is completed and nutrients in the food are absorbed into the blood or lymph. The walls of the small intestine contain several modifications that greatly increase its surface area for absorption: circular folds, villi, and microvilli. The mucosal and submucosal layers of the small intestine are thrown into folds called **circular folds** (also called plicae circularis). On the surface of the circular folds, there are folds of mucosa called **villi** (s., *villus*). Each villus is lined with a simple columnar epithelium containing **microvilli** on the apical surface. Intestinal glands (also known as intestinal crypts) are invaginations of the mucosa and are located between the villi. These glands resemble the structure and function of gastric glands. Goblet cells within the epithelium are found in these intestinal glands and produce mucin, which helps protect the epithelial lining and lubricates the passage of chyme through the small intestine. **Figure 25.5** shows the three segments of the small intestine (duodenum, jejunum, and ileum), and **table 25.6** summarizes the distinguishing histological features of each segment.

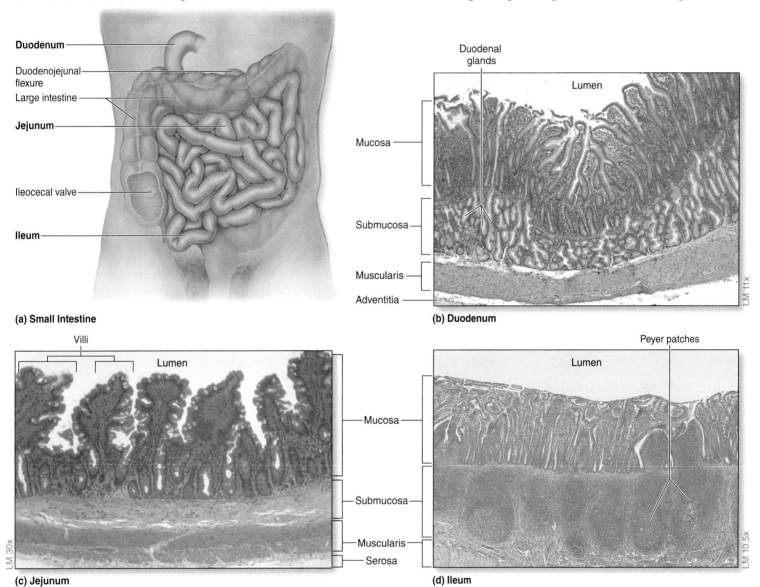

Figure 25.5 **The Small Intestine.** (a) Segments of the small intestine. (b) Histology of the duodenum, demonstrating duodenal glands in the submucosa. (c) Histology of the jejunum, which lacks duodenal glands and Peyer patches. (d) Histology of the ileum, demonstrating Peyer patches in the submucosa.

Table 25.6	Regional Differences in the Small Intestine	
Region	**Distinguishing Histological Characteristics**	**Word Origin**
Duodenum	Contains duodenal glands in the submucosa, which produce an alkaline secretion that protects the duodenum from stomach acids	*duodeno-*, breadth of twelve fingers
Jejunum	Identified by prominent circular folds and a lack of duodenal glands and Peyer patches	*jejunus*, empty
Ileum	Peyer patches are scattered throughout the submucosa. Goblet cells increase in number closer to the iliocecal valve.	*eileo*, to roll up, twist

EXERCISE 25.4

HISTOLOGY OF THE SMALL INTESTINE

1. Obtain a histology slide of the small intestine and place it on the microscope stage. If your laboratory is equipped with slides of each section of the small intestine (duodenum, jejunum, and ileum), be sure to view all three. If not, use figure 25.5 and table 25.6 to decide which part of the small intestine is on the slide.

2. Bring the tissue sample into focus at low power and move the microscope stage until the epithelium is at the center of the field of view. Then change to high power. What type of epithelium lines the small intestine?

   _____

   _____

   What surface modifications are present? _____

   _____

   What is the purpose of these surface modifications?

   _____

3. Using figure 25.5 and table 25.6 as guides, identify the following structures on the slide(s):

   ☐ duodenal glands   ☐ muscularis mucosae
   ☐ epithelium   ☐ Peyer patches
   ☐ lamina propria   ☐ serosa
   ☐ mucosa   ☐ submucosa
   ☐ muscularis   ☐ villi

4. In the spaces to the right, sketch the histology of the duodenum, jejunum, and ileum of the small intestine (if you looked at only one slide, just draw that one). Identify the histological features that will allow you to differentiate between these three portions of the small intestine when viewing histology slides of the small intestine (refer to table 25.6 for reference).

Duodenum

_____ ×

Jejunum

_____ ×

Ileum

_____ ×

The Large Intestine

Chyme leaving the small intestine enters the **large intestine** at the iliocecal junction. A valve, the **iliocecal valve,** controls the passage of chyme from the small intestine to the large intestine. Because the vast majority of nutrients have been absorbed in the small intestine, the function of the large intestine is mainly to absorb water from the chyme that remains, and compact the waste products as **feces** for elimination from the body. Because the feces get more solid and compacted in the large intestine, the epithelium of the large intestine contains many goblet cells, which help lubricate the epithelium to ease the passage of feces through the large intestine.

EXERCISE 25.5

HISTOLOGY OF THE LARGE INTESTINE

1. Obtain a histology slide of the large intestine (colon) and place it on the microscope stage.

2. Bring the tissue sample into focus on low power and identify the epithelial layer **(figure 25.6)**. Move the microscope stage so the epithelium is at the center of the field of view and then change to high power.

3. Using figure 25.6 and your textbook as guides, identify the following structures in the slide of the large intestine:

   ☐ circular muscle layer   ☐ mucosa
   ☐ epithelium   ☐ muscularis
   ☐ goblet cells   ☐ muscularis mucosae
   ☐ lamina propria   ☐ serosa
   ☐ longitudinal muscle   ☐ submucosa
      layer

4. In the space below, sketch the histology of the large intestine as seen through the microscope. Label the four major layers of the wall of the large intestine in your drawing.

_____ ×

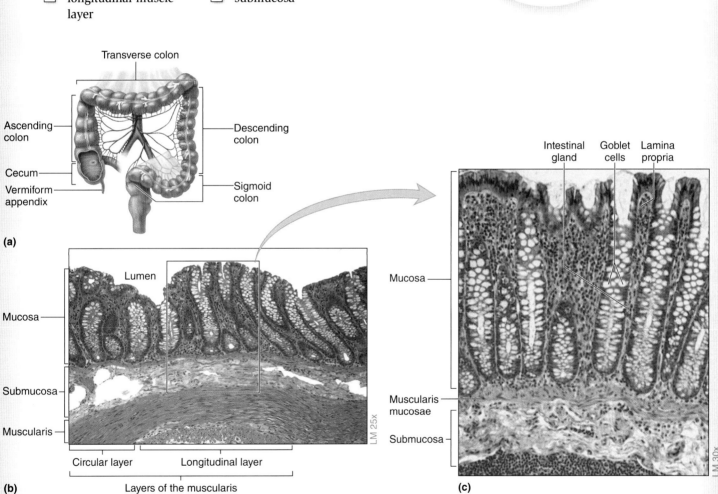

Figure 25.6 **The Large Intestine.** (a) Parts of the large intestine. (b) Histology of the wall of the large intestine. (c) Close-up of the epithelium lining the large intestine; numerous goblet cells are present.

The Liver

The **liver** is the largest accessory organ in the digestive system. Indeed, it is one of the largest organs in the human body and performs numerous vital functions that include producing bile, detoxifying the blood, storing nutrients, and producing plasma proteins. The structural and functional unit of the liver is a **hepatic lobule,** which is a hexagonally-shaped structure consisting of strands of **hepatocytes** (the strands of hepatocytes are **hepatic cords**) radiating from a **central vein** in the middle (**figure 25.7**). In the areas where the outer edges of the hepatic lobules come together there are **portal triads,** which consist of a branch of the hepatic artery, a branch of the hepatic portal vein, and a bile duct. Within the hepatic lobules, in between the hepatic cords, are **hepatic sinusoids:** capillaries that carry blood from the branches of the hepatic artery and hepatic portal vein to the central veins. Along the sinusoids are several macrophage-like cells, **reticuloendothelial** (Kupffer) cells. These cells engulf microorganisms that enter the liver from the portal circulation.

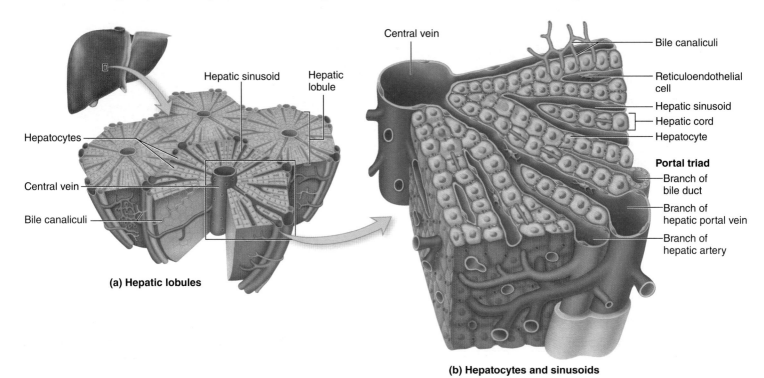

Figure 25.7 Structure of a Hepatic Lobule.

EXERCISE 25.6

HISTOLOGY OF THE LIVER

1. Obtain a histology slide of the liver and place it on the microscope stage. Bring the tissue sample into focus on low power and locate a hepatic lobule (**figure 25.8a**).

2. Move the microscope stage until the hepatic lobule is at the center of the field of view, and then change to high power (**figure 25.8b**). Using figures 25.7 and 25.8 as guides, identify the following structures on the slide of the liver:

   - ☐ branch of bile duct
   - ☐ branch of hepatic artery
   - ☐ branch of hepatic portal vein
   - ☐ central vein
   - ☐ hepatic cords
   - ☐ hepatic lobule
   - ☐ hepatic sinusoids
   - ☐ hepatocyte
   - ☐ portal triad
   - ☐ reticuloendothelial cells

3. In the space below, sketch the liver as seen through the microscope. Be sure to label all of the structures listed in step 2 in your drawing.

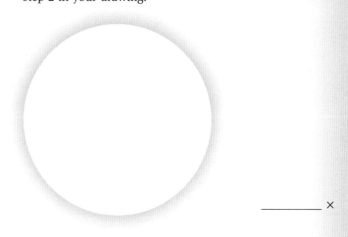

4. *Optional Activity:* **AP|R** 12: Digestive System—Watch the "Liver" animation to help you visualize the organization and structure of a liver lobule.

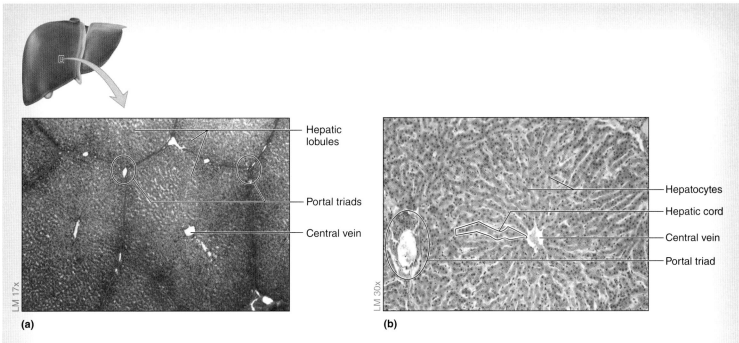

Figure 25.8 The Liver. (a) Low-magnification histology slide demonstrating multiple hepatic lobules with portal triads in the spaces between lobules. (b) Medium-magnification histology slide demonstrating a central vein, hepatocytes arranged into hepatic cords, and a portal triad.

The Pancreas

The **pancreas** is the second-largest accessory organ in the digestive system after the liver. It is both an endocrine and an exocrine gland. The histology of the endocrine part of the pancreas (the pancreatic islets) was covered in chapter 19 ("The Endocrine System"). The **exocrine** portion of the pancreas consists of grape-like bunches of cells called **acini** (s., *acinus*), which are similar in many ways to the cells that compose the salivary glands. The **acinar cells** composing acini produce many substances important for digestion, including digestive enzymes (e.g., pancreatic amylase). Cells of pancreatic ducts produce bicarbonate ion (HCO_3^-). Collectively, these secretions are called **pancreatic juice.** Pancreatic juice is transported from the acinar cells into small ducts that become larger ducts, and eventually dump the secretions into the duodenum via the **main pancreatic duct (figure 25.9).**

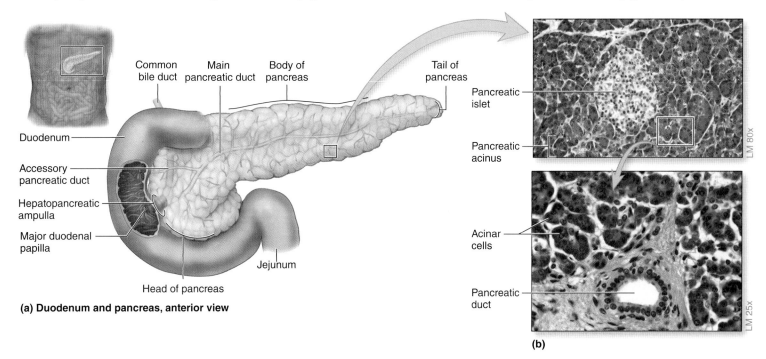

Figure 25.9 The Pancreas. (a) Location of the pancreas. (b) Histology of the pancreas demonstrating acinar cells, which secrete digestive enzymes, and pancreatic islets, which contain the hormone-secreting cells of the pancreas.

Chapter Twenty-Five *The Digestive System*

EXERCISE 25.7

HISTOLOGY OF THE PANCREAS

1. Obtain a histology slide of the **pancreas** and place it on the microscope stage. Bring the tissue sample into focus on low power. Scan the slide until you locate a pancreatic islet surrounded by acinar cells (figure 25.9a).

2. Move the microscope stage until the acinar cells are in the center of the field of view, and then change to high power. Using figure 25.9 as a guide, identify the following structures on the slide:

   ☐ acinar cells   ☐ pancreatic duct
   ☐ pancreatic acinus   ☐ pancreatic islet

3. In the space below, sketch the pancreas as seen through the microscope. Be sure to label all of the structures listed in step 2 in your drawing.

_____ ×

Gross Anatomy

The Oral Cavity, Pharynx, and Esophagus

The oral cavity contains a number of digestive system structures, including the teeth, salivary glands, lips, and tongue. These structures are important for wetting and manipulating food as it enters the GI tract. The resulting bolus (bolus = a wet mass of food) leaves the oral cavity, travels through the oropharynx (*oris*, mouth, + *pharynx*, throat), and enters the esophagus (*oisophagos*, gullet), which transports the bolus to the stomach. The esophagus meets up with the stomach just after it passes through the esophageal hiatus in the diaphragm.

EXERCISE 25.8

GROSS ANATOMY OF THE ORAL CAVITY, PHARYNX, AND ESOPHAGUS

1. Observe a cadaver specimen or a classroom model demonstrating the head and neck.

2. Using your textbook as a guide, identify the structures listed in **figure 25.10** on the cadaver or classroom model. Then label them in figure 25.10.

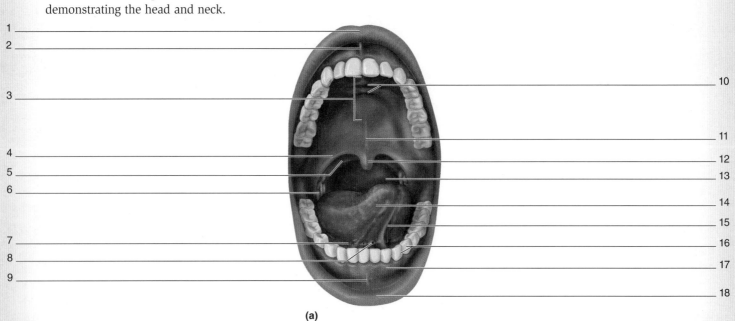

(a)

Figure 25.10 Oral Cavity.

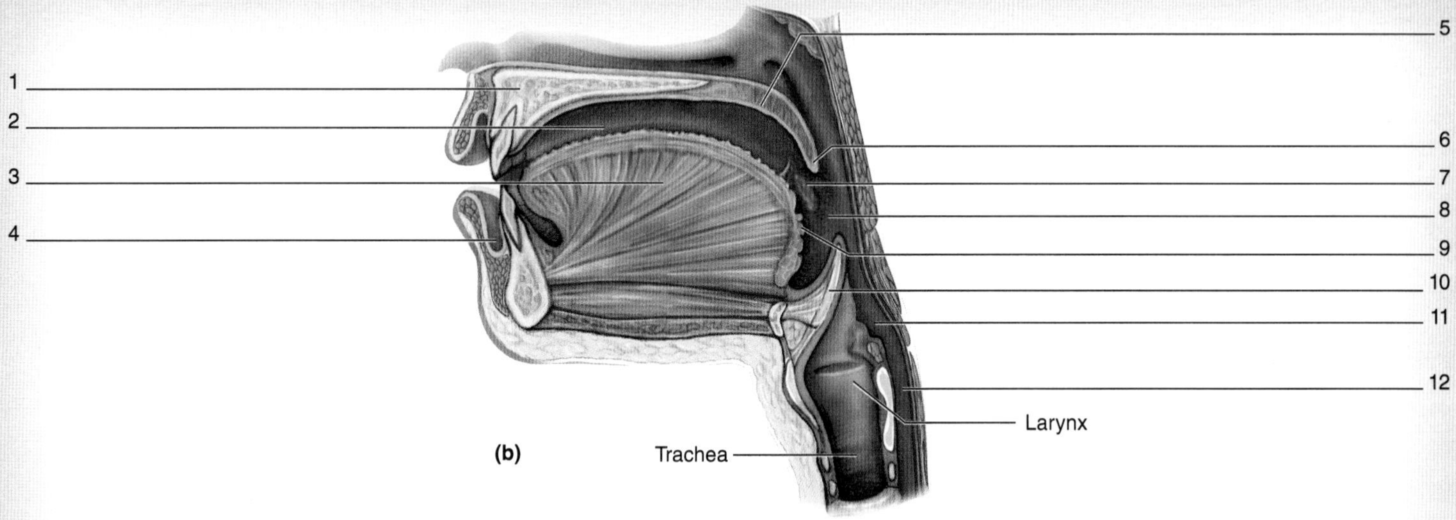

Figure 25.10 Oral Cavity *(continued)*.

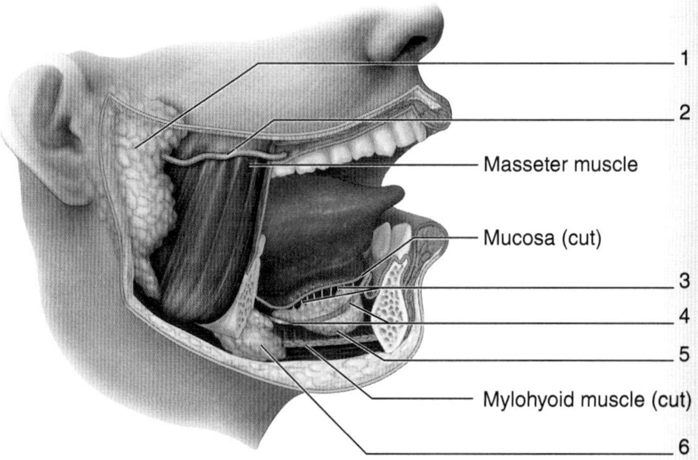

(a) Anterior View of the Oral Cavity

- ☐ fauces
- ☐ gingivae
- ☐ glossopalatine arch
- ☐ hard palate/transverse
- ☐ inferior labial frenulum
- ☐ inferior lip
- ☐ lingual frenulum
- ☐ palatine tonsil
- ☐ pharyngopalatine arch
- ☐ soft palate
- ☐ sublingual duct orifice
- ☐ submandibular duct orifice
- ☐ superior labial frenulum
- ☐ superior lip
- ☐ teeth
- ☐ tongue
- ☐ transverse palatine folds
- ☐ uvula

(b) Midsagittal View of the Oral Cavity and Pharynx

- ☐ epiglottis
- ☐ esophagus
- ☐ hard palate
- ☐ laryngopharynx
- ☐ lingual tonsil
- ☐ oral cavity
- ☐ oropharynx
- ☐ palatine tonsil
- ☐ soft palate
- ☐ tongue
- ☐ uvula
- ☐ vestibule

(c) Salivary Glands

- ☐ parotid duct
- ☐ parotid salivary gland
- ☐ sublingual ducts
- ☐ sublingual salivary gland
- ☐ submandibular duct
- ☐ submandibular salivary gland

The Stomach

The stomach is a large, saclike organ that both stores and breaks down ingested foodstuffs. The stomach is located in the epigastric abdominopelvic region, superficial to the pancreas, and deep to the anterior abdominal wall. **Table 25.7** lists the major features of the stomach and describes their functions.

Table 25.7	Gross Anatomical Features of the Stomach	
Structure	**Description**	**Word Origin**
Body	Main part of the stomach located between the fundus and the pylorus	*body*, the principal mass of a structure
Cardia	Small, narrow, superior portion of the stomach where esophagus enters	*kardia*, heart; relating to the part of the stomach nearest the heart
Cardiac Notch	A deep notch located between the fundus of the stomach and the esophagus	*kardia*, heart; relating to the part of the stomach nearest the heart, + *notch*, indentation
Fundus	The dome-shaped part of the stomach that lies superior to the cardiac notch	*fundus*, bottom
Gastric folds (rugae)	Folds of the mucosal lining of the stomach	*ruga*, a wrinkle
Greater Curvature	The large, inferior convex portion of the stomach. It is one attachment point for the greater omentum.	*greater*, larger
Greater Omentum	A fold of four layers of peritoneum that attaches to the greater curvature of the stomach, drapes over the abdominal contents, and folds back upon itself to attach to the transverse colon	*omentum*, the membrane that encloses the bowels
Inferior Esophageal (cardiac) Sphincter	A physiological sphincter (band of muscle), composed of the part of the diaphragm that surrounds the esophagus. When the diaphragm contracts, it closes off this opening, preventing reflux of stomach contents back into the esophagus. Some circular smooth muscle in the wall of the esophagus also contributes to this sphincter, but its contribution is weak.	*kardia*, heart; relating to the sphincter of the stomach nearest the heart
Lesser Curvature	The small, superior concave portion of the stomach. It is one attachment point of the lesser omentum.	*lesser*, smaller
Lesser Omentum	A fold of four layers of peritoneum that extends between the liver and the lesser curvature of the stomach	*omentum*, the membrane that encloses the bowels
Pyloric Sphincter	An anatomic sphincter (band of muscle), composed of smooth muscle in the wall of the pylorus. It controls passage of chyme from the stomach to the duodenum.	*pyloros*, a gatekeeper, + *sphinkter*, a band
Pylorus	The region of the stomach that opens into the duodenum	*pyloros*, a gatekeeper

EXERCISE 25.9

GROSS ANATOMY OF THE STOMACH

1. Observe a cadaver specimen or a classroom model demonstrating the stomach.

2. Using table 25.7 and your textbook as guides, identify the structures listed in **figure 25.11** on the cadaver or classroom model. Then label them in figure 25.11. (Some answers may be used more than once.)

3. In the space to the right, sketch the stomach. Be sure to label all of the structures listed in figure 25.11 in your drawing.

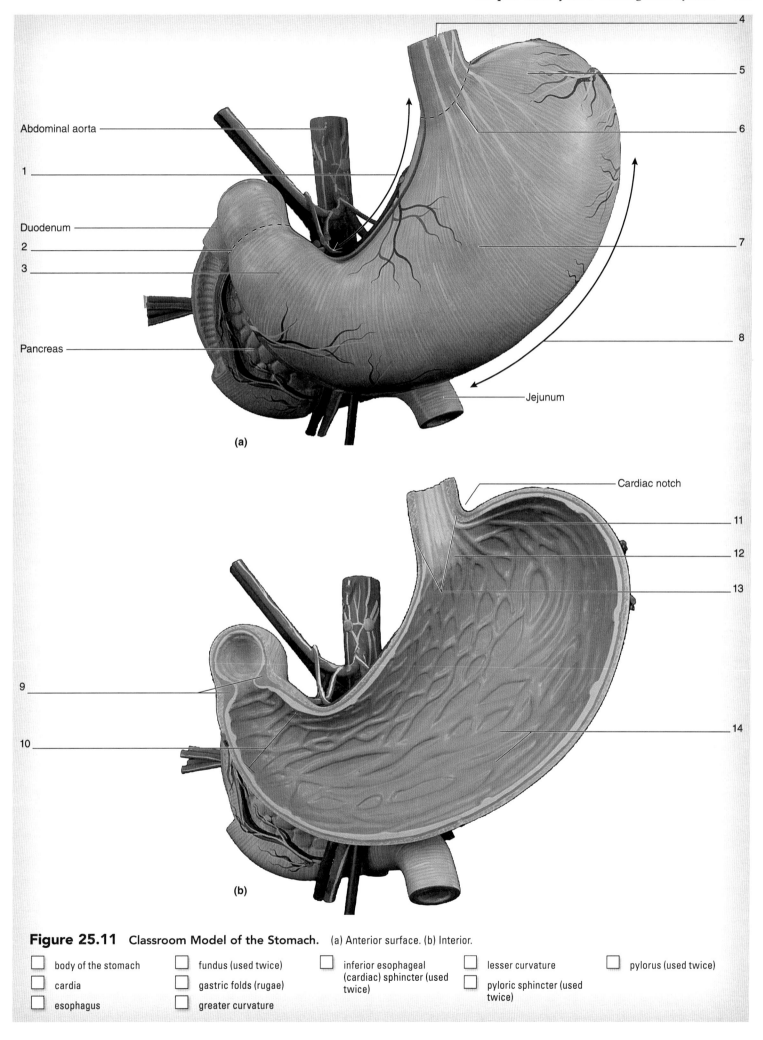

Figure 25.11 Classroom Model of the Stomach. (a) Anterior surface. (b) Interior.

☐ body of the stomach ☐ fundus (used twice) ☐ inferior esophageal (cardiac) sphincter (used twice) ☐ lesser curvature ☐ pylorus (used twice)
☐ cardia ☐ gastric folds (rugae) ☐ pyloric sphincter (used twice)
☐ esophagus ☐ greater curvature

The Duodenum, Liver, Gallbladder, and Pancreas

The **duodenum** (*duodeno-*, breadth of twelve fingers) is the first part of the small intestine. It is C-shaped, and mostly retroperitoneal, which allows it to be anchored to the posterior abdominal wall. This is advantageous because a number of ducts coming from the liver, gallbladder, and pancreas empty their contents into the duodenum. The relationships between the duodenum and the liver, gallbladder, and pancreas are critically important for the process of digestion. The liver produces **bile**, a substance that emulsifies fats, which is temporarily stored and concentrated within the **gallbladder**. The pancreas produces pancreatic juice, which contains digestive enzymes and bicarbonate ion. When these organs (liver, gallbladder, and pancreas) release their secretions into the duodenum, the acidity of the chyme (*chymos*, juice) that has entered the duodenum from the stomach is neutralized and the digestion of proteins and carbohydrates continues. Here the digestion of fats and nucleic acids also begins. **Table 25.8** lists the major features of the liver, gallbladder, and pancreas, and describes each of their functions.

Table 25.8	Gross Anatomical Structures of the Liver, Gallbladder, Pancreas and Their Associated Ducts	
Structure	**Description**	**Word Origin**
Accessory Pancreatic Duct	Excretory duct located in the head of the pancreas. Empties into duodenum at the minor duodenal papilla.	*pankreas*, the sweetbread
Body of the Pancreas	The main portion of the pancreas extending between the head and the tail	*pankreas*, the sweetbread
Caudate Lobe of the Liver	A small lobe of the liver located between the right and left lobes and on the posterior, inferior part of the liver	*caudate*, possessing a tail, + *lobos*, lobe
Common Bile Duct	The bile duct formed from the union of the common hepatic duct and the cystic duct. Empties into the hepatopancreatic ampulla.	*bilis*, a yellow/green fluid produced by the liver
Common Hepatic Duct	The bile duct formed from the union of the right and left hepatic ducts. Drains bile into the common bile duct.	*hepatikos*, liver
Cystic Duct	The bile duct that transports bile from the gallbladder to the common bile duct	*cystic*, relating to the gallbladder
Falciform Ligament	A fold of peritoneum that extends from the diaphragm and anterior abdominal wall to the liver. Its free inferior border contains the round ligament of the liver.	*falx*, sickle, + *forma*, form
Gallbladder	A saclike appendage of the liver that stores and concentrates bile	*gealla*, bile, + *blaedre*, a distensible organ
Head of the Pancreas	The portion of the pancreas that sits in the depression formed by the curvature of the duodenum	*pankreas*, the sweetbread
Hepatopancreatic Ampulla	A duct formed by the joining of the common bile duct and the main pancreatic duct	*hepatikos*, relating to the liver, + *pancreatic*, relating to the pancreas, + *ampulla*, a two-handled bottle
Left Lobe of the Liver	The second largest lobe of the liver. Extends from the falciform ligament toward the midline of the body.	*lobos*, lobe
Main Pancreatic Duct	The main excretory duct of the pancreas. Runs longitudinally in the center of the gland and empties into the duodenum at the major duodenal papilla.	*pankreas*, the sweetbread
Major Duodenal Papilla	A raised "nipplelike" bump located on the posterior wall of the descending part of the duodenum. The hepatopancreatic ampulla empties its contents here.	*major*, great, + *papilla*, a nipple
Minor Duodenal Papilla	A small raised "nipplelike" bump located superior to the major duodenal papilla. Contains the opening of the accessory pancreatic duct.	*minor*, smaller, + *papilla*, a nipple
Porta Hepatis	A depression on the inferomedial part of the liver that contains the hepatic artery, hepatic portal vein, and common bile duct	*porta*, gate, + *hepatikos*, liver
Quadrate Lobe of the Liver	A small lobe of the liver located between the right and left lobes and on the anterior, inferior part of the liver between the gallbladder and the round ligament	*quadratus*, square, + *lobos*, lobe
Right Lobe of the Liver	The largest lobe of the liver, it is on the right side of the abdomen and composes over half of the mass of the liver	*lobos*, lobe
Round Ligament of the Liver (ligamentum teres)	A remnant of the fetal umbilical vein, which connects to the umbilicus. Located within the free edge of the falciform ligament on the anterior abdominal wall.	*ligamentum*, a bandage, + *teres*, round
Tail of the Pancreas	The tapered, right end of the pancreas located near the hilum of the spleen	*pankreas*, the sweetbread

EXERCISE 25.10

GROSS ANATOMY OF THE DUODENUM, LIVER, GALLBLADDER, AND PANCREAS

1. Obtain a classroom model demonstrating the relationships among the duodenum, liver, gallbladder, and pancreas (**figure 25.12**), or view these structures in the superior abdominal cavity of a prosected human cadaver.

2. Using table 25.8 and your textbook as guides, identify structures listed in figure 25.12 on the classroom model or human cadaver. Then label them in figure 25.12.

3. Using table 25.8 and your textbook as guides, sketch the ducts coming from the liver, gallbladder, and pancreas in the space to the right. Show how the ducts merge to eventually empty their contents into the pancreas. Label each duct and organ in your drawing.

4. Obtain a model of the liver (**figure 25.13**), or view the liver from a human cadaver.

5. Using table 25.8 and your textbook as guides, identify structures listed in figure 25.13 on the model of the liver or human cadaver. Then label them in figure 25.13.

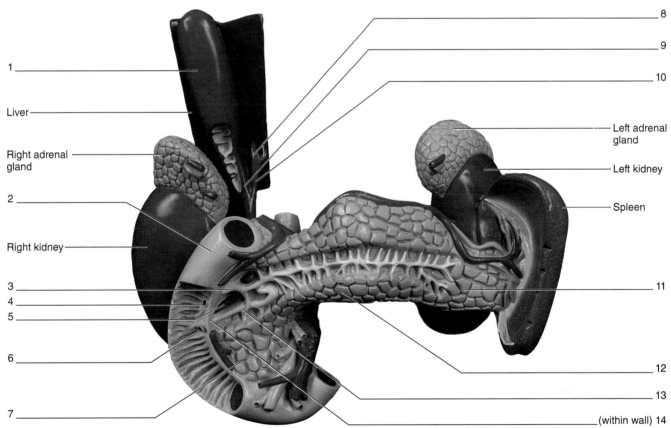

Figure 25.12 Classroom Model of the Duodenum, Liver, Gallbladder, and Pancreas. Anterior view.

- ☐ accessory pancreatic duct
- ☐ body of pancreas
- ☐ common bile duct
- ☐ common hepatic duct
- ☐ cystic duct
- ☐ duodenum
- ☐ gallbladder
- ☐ head of pancreas
- ☐ hepatopancreatic ampulla
- ☐ left and right hepatic ducts
- ☐ main pancreatic duct
- ☐ major duodenal papilla
- ☐ minor duodenal papilla
- ☐ tail of pancreas

(continued on next page)

(continued from previous page)

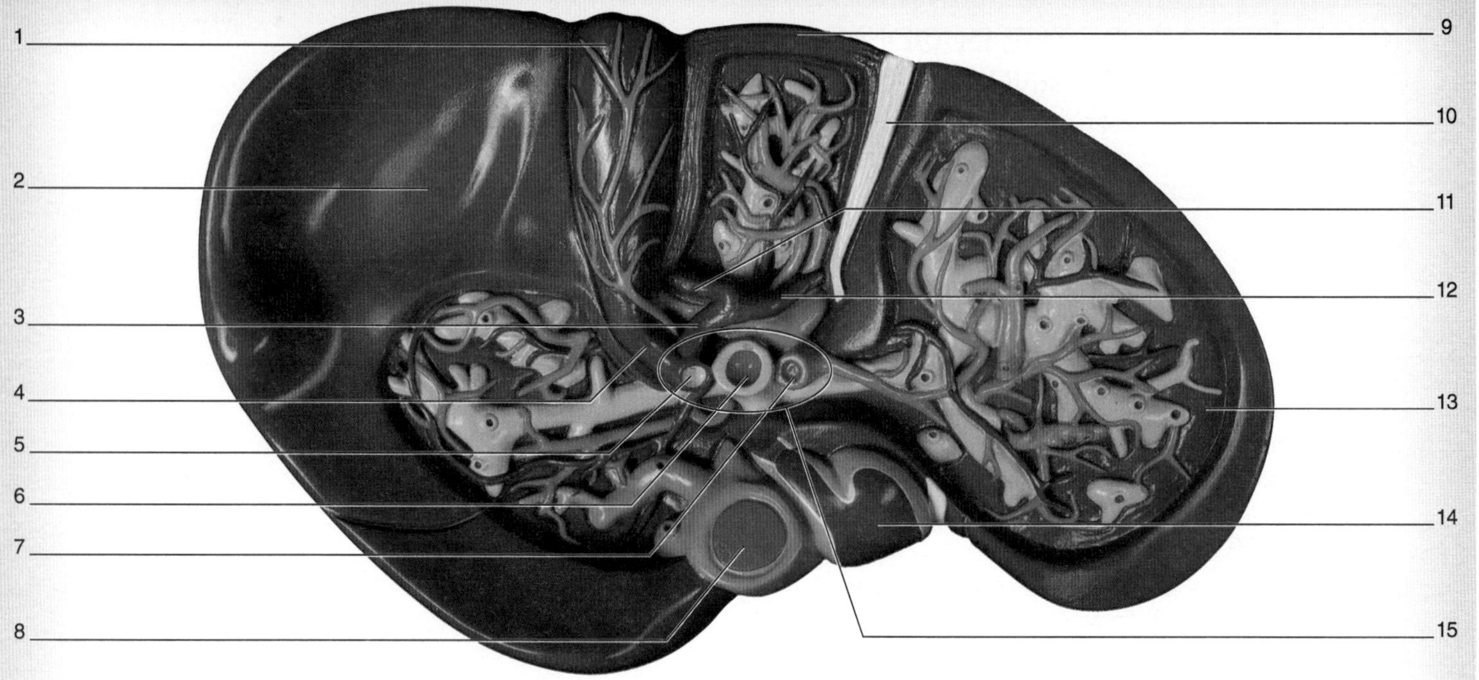

Figure 25.13 Classroom Model of the Liver. Anteroinferior view. In this view, the anterior surface of the liver has been rotated away from your point of view so you can see the structures on the inferior surface of the liver.

☐ caudate lobe of liver   ☐ falciform ligament   ☐ inferior vena cava   ☐ quadrate lobe of liver
☐ common bile duct   ☐ gallbladder   ☐ left hepatic duct   ☐ right hepatic duct
☐ common hepatic duct   ☐ hepatic artery proper   ☐ left lobe of liver   ☐ right lobe of liver
☐ cystic duct   ☐ hepatic portal vein   ☐ porta hepatis

The Jejunum and Ileum of the Small Intestine

The **jejunum** and **ileum** compose the longest part of the small intestine. In general the jejunum is located in the upper left part of the abdominal cavity and the ileum is located in the lower right part of the abdominal cavity. The two segments can be distinguished from each other both histologically and grossly. Distinguishing the two via gross anatomic features involves comparing four anatomic features: circular folds, encroaching fat, arterial arcades, and vasa recta **(figure 25.14)**. **Circular folds** are the mucosal folds found within the lumen of the small intestine. **Encroaching fat** is mesenteric fat that "rides up" upon the wall of the intestine. **Arterial arcades** are arching branches of the mesenteric arteries, and **vasa recta** (*vasa,* vessel, + *rectus,* straight) are straight vessels that come off of the arterial arcades and enter the small intestine proper. **Table 25.9** summarizes the gross anatomical features that distinguish the jejunum from the ileum.

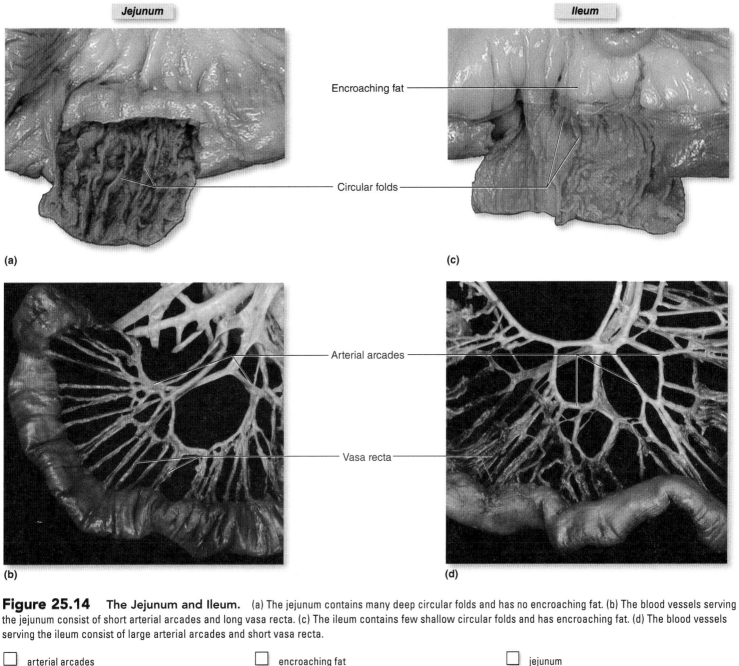

Figure 25.14 The Jejunum and Ileum. (a) The jejunum contains many deep circular folds and has no encroaching fat. (b) The blood vessels serving the jejunum consist of short arterial arcades and long vasa recta. (c) The ileum contains few shallow circular folds and has encroaching fat. (d) The blood vessels serving the ileum consist of large arterial arcades and short vasa recta.

☐ arterial arcades ☐ encroaching fat ☐ jejunum
☐ circular folds ☐ ileum ☐ vasa recta

Table 25.9	Gross Anatomical Differences between the Jejunum and the Ileum			
Part of the Small Intestine	**Circular Folds**	**Encroaching Fat**	**Arterial Arcades**	**Vasa Recta**
Jejunum	Deep, many	No	Fewer, larger	Longer
Ileum	Shallow, few	Yes	More, smaller, stacked upon each other	Shorter

EXERCISE 25.11

GROSS ANATOMY OF THE JEJUNUM AND ILEUM OF THE SMALL INTESTINE

1. Observe a classroom model of the abdominal cavity or the abdominal cavity of a prosected human cadaver in which the small intestine is intact.

2. Using table 25.9 and your textbook as guides, identify the gross structures listed in figure 25.14 on the classroom model of the abdomen or in the abdominal cavity of the human cadaver.

The Large Intestine

The large intestine begins as a large sac called the **cecum,** which is located in the right inferior abdominopelvic quadrant. Exiting the cecum, the **colon** extends along the borders of the abdominal cavity as the ascending colon, transverse colon, and descending colon before entering the pelvic cavity via the sigmoid colon. The sigmoid colon empties into the **rectum,** which is located within the pelvic cavity proper. The rectum ends at the **anus. Table 25.10** summarizes these structures.

Table 25.10	The Cecum, Large Intestine, Rectum, and Anus	
Structure	**Description**	**Word Origin**
Anus	Inferior opening of the GI tract	*anus*, the lower opening of the GI tract
Ascending Colon	Part of the colon that extends from the cecum to the liver	*kolon*, the part of the large intestine from cecum to rectum
Cecum	Blind-ended sac located at the junction between the ileum of the small intestine and the ascending colon	*caecus*, blind
Haustra	Pouches of the colon formed when the taenia coli (longitudinal smooth muscle) contract	*haustus*, to draw up
Ileocecal Valve	Smooth muscle sphincter located where the ileum opens into the cecum	*ileo-*, ileum, + *cecal*, cecum
Left Colic (splenic) Flexure	A curve of the colon medial to the spleen, where the descending colon begins	*splenic*, relating to the spleen, + *flexura*, a bend
Omental (epiploic) Appendices	Small, fatty appendages that hang off of the colon	*omentum*, the membrane that encloses the bowels; *epiploic*, related to the omentum
Rectum	Final portion of the GI tract, located within the pelvic cavity and extending from the sigmoid colon to the anus	*rectus*, straight
Right Colic (hepatic) Flexure	Curve of the colon medial to the liver, where the transverse colon begins	*hepatikos*, relating to the liver, + *flexura*, a bend
Sigmoid Colon	The S-shaped part of the colon that extends from the descending colon to the rectum	*sigma*, the letter S, + *eidos*, resemblance
Taenia Coli	Three small bands of longitudinal smooth muscle of the muscularis externa of the colon; contraction of this muscle creates pouches (haustra) in the colon	*tainia*, a band, + *coli*, colon
Transverse Colon	The part of the colon that extends between the liver and the spleen	*transversus*, crosswise

Chapter Twenty-Five The Digestive System 721

EXERCISE 25.12

GROSS ANATOMY OF THE LARGE INTESTINE

1. Observe either a classroom model of the abdominal cavity that includes the large intestine, or the large intestine of a prosected human cadaver **(figure 25.15)**.

2. Using table 25.10 and your textbook as guides, identify the gross structures listed in figure 25.15 on the classroom model of the abdomen or in the abdominal cavity of the human cadaver.

3. *Optional Activity:* **AP|R 12: Digestive System**—Review the locations and functions of the major organs of the digestive system by watching the "Digestive System Overview" animation.

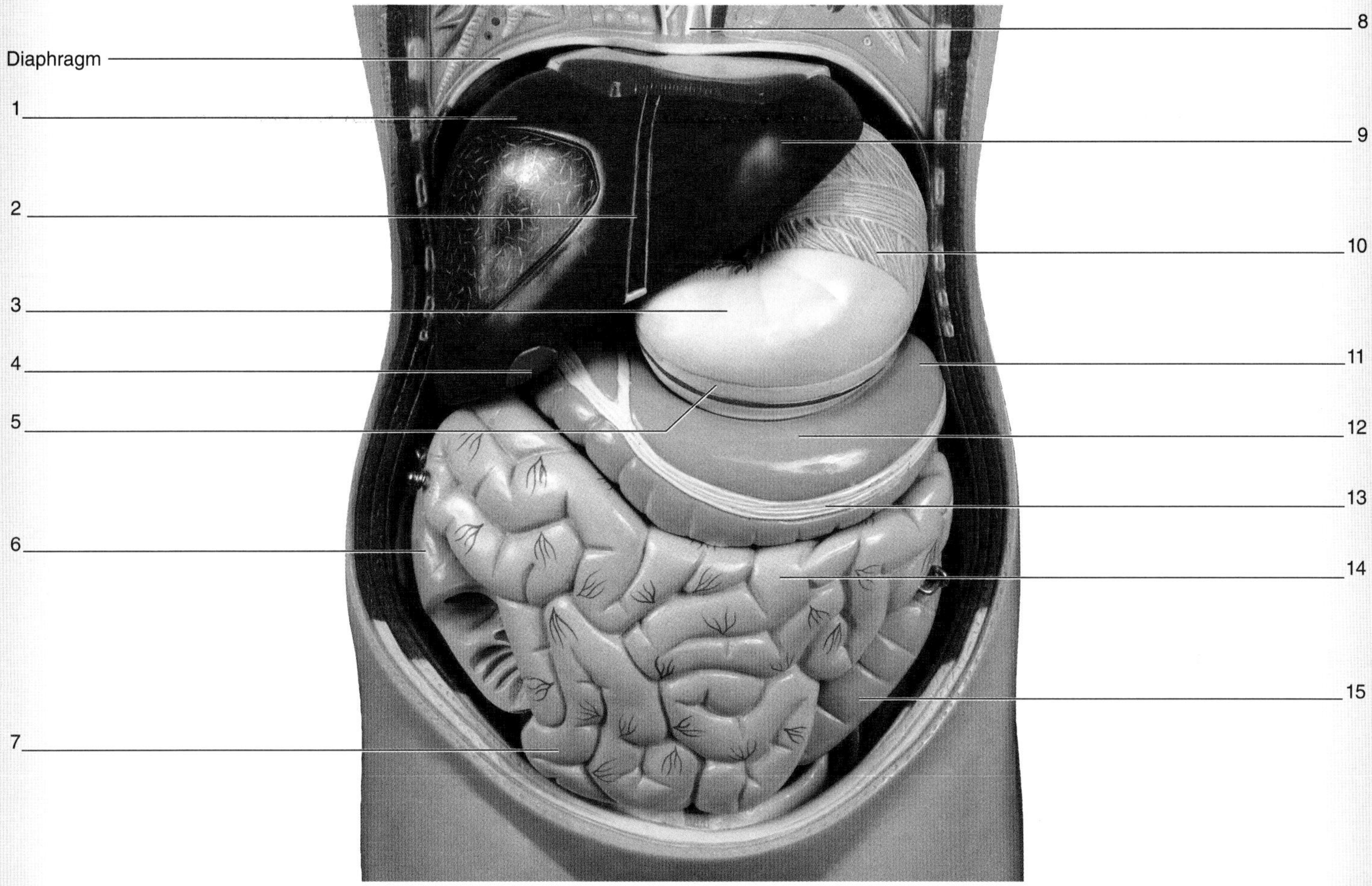

Figure 25.15 Classroom Model of the Abdominal Cavity and Large Intestine. (a) Superficial view. (b) Deep view with the liver, stomach, and small intestine removed.

☐ ascending colon
☐ body of stomach
☐ descending colon
☐ esophagus
☐ falciform ligament
☐ gallbladder
☐ greater curvature of stomach
☐ ileum of small intestine
☐ jejunum of small intestine
☐ left colic (splenic) flexure of colon
☐ left lobe of liver
☐ pylorus of stomach
☐ right lobe of liver
☐ taenia coli
☐ transverse colon

(continued on next page)

(continued from previous page)

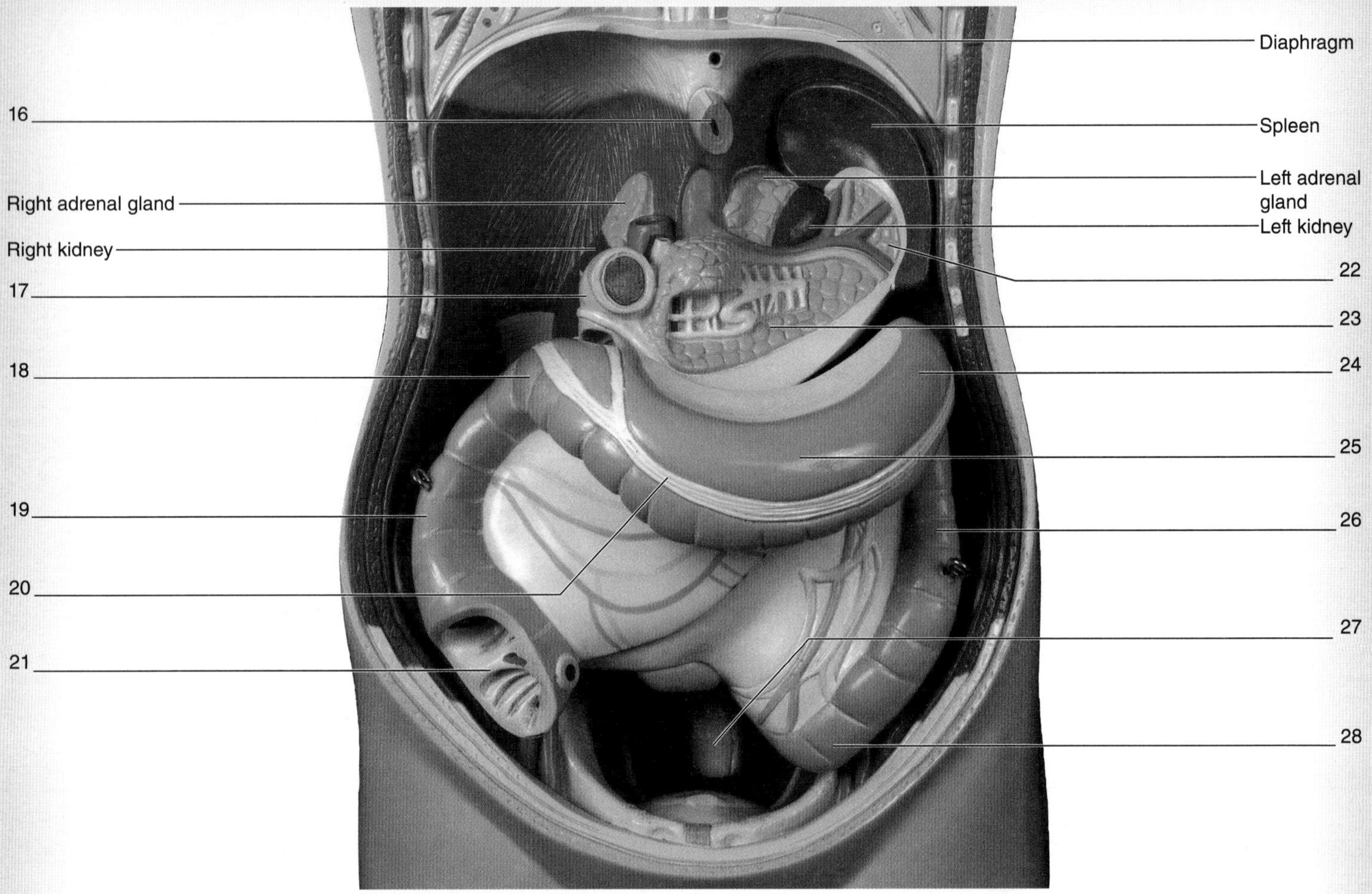

Figure 25.15 Classroom Model of the Abdominal Cavity and Large Intestine *(continued)*.

- ☐ ascending colon
- ☐ body of pancreas
- ☐ cecum
- ☐ descending colon
- ☐ duodenum
- ☐ esophagus
- ☐ left colic (splenic) flexure of colon
- ☐ rectum
- ☐ right colic (hepatic) flexure of colon
- ☐ sigmoid colon
- ☐ taenia coli
- ☐ tail of pancreas
- ☐ transverse colon

Physiology

Digestive Physiology

Previous exercises in this chapter have detailed the anatomy of each of the digestive and accessory organs of the GI tract. In the following laboratory exercises, we will explore some physiological functions of these organs. The digestive system functions to supply the body with nutrients, electrolytes, and water that are obtained from food and drink. Overall, the digestive system provides a conduit to **ingest** food, mechanically and chemically **digest** food, **propel** food along the GI tract, **secrete** products that aid in digestion, **absorb** nutrients, and **eliminate** wastes. Specialized functions of each of the digestive and accessory organs allow these overall processes to take place. The following paragraph summarizes the fate of a food item from the moment it enters the mouth to the moment it exits the anus.

As food enters the mouth, mechanical digestion of the entire food item and chemical digestion of its carbohydrate components begins. Mechanical digestion (mastication) breaks down the food item into smaller pieces, which will allow enzymes to act on the compounds in the food and assist in their degradation (chemical digestion). The act of swallowing moves a bolus of moistened, partially digested food into the esophagus. This bolus is then transported through the esophagus by coordinated contractions of the longitudi-

INTEGRATE

CLINICAL VIEW
Gallstones

Gallstones are hard stones that form from cholesterol and bile deposits in the gallbladder (figure 25.16). The condition of having gallstones is called *cholelithiasis* (*chole*, bile, + *lithos*, stone, + *-iasis*, condition). One of the most serious complications resulting from a gallstone occurs when the stone passes into the cystic duct and makes its way toward the duodenum, but becomes lodged somewhere en route to the duodenum. This blocks the flow of bile from the liver and gallbladder to the duodenum. If it lodges farther down, near the hepatopancreatic ampulla, it can also block the flow of pancreatic juice from the pancreas. Normally some of the pancreatic enzymes (proteolytic enzymes) are not activated until they enter the duodenum. However, when these enzymes build up within the pancreas because of a blockage of flow of pancreatic juice from the pancreas, these enzymes become activated and begin to digest the pancreas itself. This causes *pancreatitis* (inflammation of the pancreas), which can be life threatening.

Figure 25.16 **Gallbladder and Gallstone.** The gallstone is a mass of bile salts that have crystallized and condensed. Note the small crystals that have formed on the larger stone. A stone of this size might not cause problems for the patient if it were too large to enter into the cystic duct.

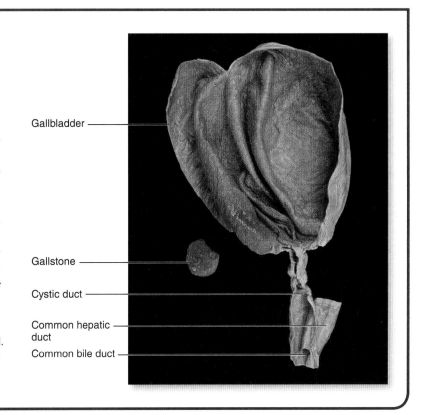

nal and circular layers of muscle that line the esophageal walls (peristalsis). From the esophagus, the bolus enters the stomach, where mechanical digestion (churning) continues and chemical digestion of proteins and fat begins. In addition, the highly acidic environment of the stomach promotes digestion of proteins. From the stomach the acidic, liquid chyme is squeezed (3 mL at a time) into the duodenum of the small intestine. In the duodenum, secretions from accessory digestive organs—the liver, gallbladder, and pancreas—mix with the acidic chyme from the stomach to continue the chemical digestion of proteins, carbohydrates, fats, and nucleic acids. Pancreatic secretions (pancreatic "juice") contain enzymes that are required for protein, carbohydrate, fat, and nucleic acid digestion. Pancreatic secretions also contain bicarbonate ion, which is necessary to neutralize acid. The liver and gallbladder secrete bile into the duodenum as well. Bile is required for emulsification of fats. While all of this chemical digestion begins to take place, the small intestine also continues to mechanically digest the chyme using peristaltic contractions of its smooth muscle.

Table 25.11 Digestion and Absorption of Nutrients

Substance	Monomer Unit	Site of Digestion	Site of Absorption	Mechanism of Absorption	Word Origin
Carbohydrate	Monosaccharides (glucose, galactose, fructose)	Mouth, small intestine	Small intestine	Cotransport with sodium (glucose and galactose); facilitated diffusion (fructose); transport to hepatic portal vein	*carbo-*, carbon + *-hydro*, water
Fat	Fatty acids, monoglycerides, glycerol	Stomach, small intestine	Small intestine	Diffuse into lacteals via chylomicrons; travel via lymph	*faet*, to load
Nucleic Acid	Pentose sugars, nitrogenous bases, phosphates	Small intestine	Small intestine	Active transport; transport to hepatic portal vein	*nucleus*, kernal
Protein	Amino acids	Stomach, small intestine	Small intestine	Cotransport with sodium; transport to hepatic portal vein	*prote-*, primary + *-in*, in

INTEGRATE

LEARNING STRATEGY

Motility in the GI tract is made possible by both an outer longitudinal and an inner circular layer of smooth muscle within the muscularis of the GI tract wall (see figure 25.2). Alternating waves of contraction of these two layers of muscle along the length of the tube moves substances along in a process termed **peristalsis** (*peri-*, around + *stalsis*, compression). Contraction of the outer, longitudinal layer decreases the length of the tube and increases its diameter; contraction of the inner, circular layer increases the length of the tube and decreases the diameter. This coordinated action is the same type of action that an earthworm uses to move about in the dirt. Contraction of the longitudinal layer of the worm's body wall musculature makes the worm short and fat, whereas contraction of the circular layer of the worm's body wall musculature makes the worm long and skinny (much like squeezing a tube of toothpaste).

The signals (action potentials) required for these two layers of muscle to contract come from a network of nerves that lie adjacent to each muscle layer: the submucosal and myenteric plexuses. Collectively, these plexuses comprise the **enteric nervous system**, or "gut brain." While these nerves can operate independently, much like pacemaker cells in the heart, their activity is modulated by the autonomic nervous system.

Following the digestion of carbohydrates, proteins, triglycerides, and nucleic acids into their absorbable form (monosaccharides, amino acids, fatty acids and monoglycerides, and the individual components of a nucleotide), each is transported from the GI tract into the blood or lymph. Consult **table 25.11** for a summary of these processes. The surface area of the small intestine that is available for absorption is immense due to the presence of villi and microvilli. Monosaccharides and amino acids are absorbed into the blood and then immediately enter the hepatic portal circulation (see chapter 20), whereas triglycerides (and other lipids) are packaged into specialized structures (chylomicrons) and absorbed into specialized lymph vessels called **lacteals** (see chapter 23). Water is reabsorbed along the length of the large intestine through the process of osmosis.

Any substances not absorbed remain within the lumen of the large intestine and are moved through the segments of the large intestine and then eliminated as feces. **Table 25.12** summarizes the digestion of macromolecules, including the location and substances secreted in the digestion of carbohydrates, proteins, lipids, and nucleic acids. Use your textbook as a guide as you explore further the processes of digestion, motility, and absorption in the following laboratory exercises.

Table 25.12 Digestion of Macromolecules

Location	Carbohydrates	Proteins	Lipids	Nucleic Acids
Oral Cavity (saliva)	Starch—*salivary amylase*→ partially digested starch	No protein digestion	Lingual lipase added but activated in low pH of stomach	No nucleic acid digestion
Stomach (gastric juice)	No additional enzymes added	Protein—*pepsin*→ polypeptide and peptide fragments	Triglyceride—*lingual lipase*→ monoglyceride and fatty acids (limited amounts) Triglyceride—*gastric lipase*→ monoglyceride and fatty acids (limited amounts)	No nucleic acid digestion
Small Intestine (pancreatic juice secreted into duodenum)	Partially digested starch—*pancreatic amylase*→ oligosaccharides, maltose, and glucose	Protein—*trypsin*→ polypeptide and peptide fragments Protein—*chymotrypsin*→ polypeptide and peptide fragments Protein—*carboxypeptidase*→ amino acids from carboxy-end of peptides	Triglyceride—*pancreatic lipase*→ monoglyceride and fatty acids (within micelles)	DNA—*pancreatic deoxyribonuclease*→ deoxyribonucleotides RNA—*pancreatic ribonuclease*→ ribonucleotides
Small Intestine (brush border enzymes)	Oligosaccharides—*dextrinase* and *glucoamylase*→ maltose, glucose Maltose—*maltase*→ glucose Lactose—*lactase*→ glucose, galactose Sucrose—*sucrase*→ glucose, fructose	Dipeptides—*dipeptidase*→ amino acids Peptides—*aminopeptidase*→ amino acids from amino-end of peptides	No lipid digestion completed by brush border enzymes.	Nucleotides—*phosphatase*→ nucleosides and phosphate Nucleosides—*nucleosidase*→ nitrogenous base and sugar (ribose or deoxyribose)

INTEGRATE

CLINICAL VIEW
Cholera

Diarrhea (*dia-*, through + *-rrhoea*, to flow) is a relatively common condition in which water is either not absorbed or is "pulled" by osmosis into the lumen of the GI tract. When substances such as ions or undigested foodstuffs remain in the lumen of the GI tract, water may be attracted to them, thereby preventing the water's absorption. This is what happens when some pathogens invade the intestinal wall and cause diseases such as cholera, a disease that occurs in areas where sanitary conditions are poor. Cholera is caused by a bacterium, *Vibrio cholerae*, which embeds within the mucosa of the small intestine. Once in place, the bacterium causes large quantities of ions, including sodium, potassium, and chloride, to be pumped into the lumen of the GI tract. The ions attract water via osmosis. The quantity of water within the GI tract becomes so great that the stretch receptors in the intestinal wall are stimulated, which increases gut motility. The result is massive amounts of watery diarrhea. Because dehydration is of great concern in a patient suffering from cholera, the "treatment" is administration of rehydration salts. These salts promote rehydration by taking advantage of the mechanisms of water absorption in the large intestine. That is, the rehydration salts contain both sodium and glucose because sodium and glucose move across the epithelium of the intestinal wall by cotransport. Sodium is the body's greatest source of osmotic pressure, such that where sodium goes (in this case, from the GI tract to the blood), water will follow. Without treatment, a person suffering from cholera may lose up to 20 L of water per day, which can rapidly lead to death.

INTEGRATE

CONCEPT CONNECTION

Acidic and basic solutions play an important role in digestion. The stomach produces acidic chyme due to HCl produced by parietal cells. Acinar cells in the pancreas produce bicarbonate ion to neutralize the acidic chyme once it enters the duodenum of the small intestine. It is interesting that these both involve the same reaction, the *bicarbonate reaction*, which was initially presented in chapter 24: $CO_2 + H_2O \leftrightarrow H_2CO_3 \leftrightarrow H^+ + HCO_3^-$. In the stomach, hydrogen ions move into the lumen of the stomach; in the pancreas, bicarbonate moves into the pancreatic duct to be dumped into the small intestine. The common link is that the source of the hydrogen ion and the bicarbonate is the dissociation of carbonic acid (H_2CO_3). Therefore, both parietal cells and acinar cells contain the enzyme *carbonic anhydrase*. While this reaction is important for digestion, it is also involved in regulating blood pH. As hydrogen ions are pumped to the lumen of the stomach, the bicarbonate moves into the blood. Similarly, as bicarbonate moves to the pancreatic duct, the hydrogen ions move into the blood. This does not become particularly problematic until there is excessive production of acid in the stomach, as occurs during vomiting, or excessive production of bicarbonate in the pancreas, as occurs with diarrhea. When a person vomits a great deal, that person's parietal cells are working extra hard to compensate for the loss of acidic chyme. The action of these cells move H^+ into the lumen of the stomach and release HCO_3^- into the blood. This may raise the blood pH, which can lead to **metabolic alkalosis**. Likewise, when a person suffers a great deal of diarrhea, pancreatic acinar cells work harder to replace the loss of bicarbonate ion. Cells of the small intestine transport HCO_3^- into the lumen of the small intestine and move H^+ into the blood. This may decrease blood pH, which can lead to ***metabolic acidosis***. This concept will be revisited in chapter 26, when you explore the role of the urinary system in regulating acid-base balance.

EXERCISE 25.13

DIGESTIVE ENZYMES

The purpose of this laboratory exercise is to examine the process of chemical digestion. You will observe chemical digestion of carbohydrates by the enzyme amylase, and you will determine the factors that influence the rate of digestion. You will then relate the process of digesting carbohydrates to the digestion of proteins, fats, and nucleic acids by reviewing the breakdown of each into monomer units and the absorption of the monomers into the blood or lymph.

Before you begin, state a hypothesis regarding the effect of temperature on the rate of chemical digestion.

_____

_____

_____

Obtain the Following:
- 0.5% amylase solution
- 0.5% starch solution
- 37°C water bath
- 6 test tubes
- Benedict's solution (or glucose test strips)
- hot plate
- ice
- Lugol's solution (iodine-potassium-iodide)
- porcelain spot plate
- test-tube clamp
- wax pencil

1. Place three test tubes in a rack and label each with a wax pencil (#1, #2, and #3). Fill test tube 1 with 6 mL of 0.5% amylase solution. Fill test tube 2 with 6 mL of 0.5% starch solution. Fill test tube 3 with 5 mL of 0.5% starch solution + 1 mL of 0.5% amylase solution. Shake each tube until the contents are mixed well.

2. To heat the tubes, place the three test tubes in a warm water bath, 37°C (98.6°F), for approximately 10 minutes. Remove the test tubes from the warm water bath.

3. To test each solution for the presence of starch, remove 1 mL of the solution from the test tube and place it in one depression of a spot plate. Add 1 drop of Lugol's solution to the test solution and observe the color. A blue-black color indicates the presence of starch. Record your color observations and conclusions regarding the presence (+) or absence (−) of starch in **table 25.13**.

(continued on next page)

(continued from previous page)

Table 25.13 Data Table for Chemical Digestion Test

Test Tube #	Solution	Lugol's (Color)	Starch Present (+/−)	Benedict's (Color)	Sugar Present (+/−)
1	0.5% Amylase solution				
2	0.5% Starch solution				
3	0.5% Amylase & 0.5% Starch solution				
4	0.5% Amylase at 0°C				
5	0.5% Amylase at 37°C				
6	0.5% Amylase at 100°C				

4. To test each solution for the presence of sugar (monosaccharides and disaccharides), remove 1 mL of solution and place in a clean test tube. Add 1 mL of Benedict's solution, place the tube in boiling water for 2 minutes, and observe the color. Remember to always use a test-tube clamp when lowering the test tube into boiling water. The color may vary depending on the amount of sugar present in the solution. The range in color follows that of the visible light spectrum (red, orange, yellow, green, and blue), where red indicates most and green indicates least sugar present. A solution that is blue in color indicates no sugar present. Record your color observations and conclusions regarding the presence (+) or absence (−) of sugar in table 25.13. Repeat for each of the three solutions. Note that glucose test strips can be used instead of Benedict's solution to test for the presence of sugars.

5. To test the effect of temperature on enzyme activity, place 1 mL of 0.5% amylase solution in each of three clean test tubes and label test tube 1 "cold," test tube 2 "warm," and test tube 3 "hot." Vary the temperature conditions for the solutions by placing test tube 1 in a beaker of ice water (~0°C), test tube 2 in a warm water bath (37°C), and test tube 3 in a beaker of boiling water (100°C). Use a test tube clamp to lower test tube 3 into the boiling water.

6. Remove the three test tubes from their respective water baths, add 5 mL of 0.5% starch solution to each test tube, and mix thoroughly. Note that 0.5% starch solution added to test tube 1 should be chilled to approximately 0°C prior to being added to the amylase solution. Care should be taken when handling and mixing the contents in test tube 3. Return the test tubes to their respective water baths for 10 minutes.

7. Remove the tubes and repeat steps 3 and 4 to test for the presence of starch and sugar, respectively. Record the color of each solution and conclusions regarding the presence (+) or absence (−) of starch and sugar in table 25.13.

8. Describe the effect of increasing temperature on enzymatic activity. _____
_____
_____

9. Chemical digestion is essential for the process of absorption. The process of chemical digestion, as observed in this laboratory exercise, can be applied to not only carbohydrates, but also to proteins, fats, and nucleic acids. Using tables 25.11 and 25.12, and your

textbook as guides, discuss how and where each of the following is chemically digested and absorbed:

a. Carbohydrates: _____
_____
_____

b. Proteins: _____
_____
_____

c. Triglycerides: _____
_____
_____

d. Nucleic Acids: _____
_____
_____

EXERCISE 25.14

Ph.I.L.S. LESSON 37: GLUCOSE TRANSPORT

Starch is broken down into glucose monomers (monosaccharides) during digestion, which are then absorbed into the blood. The movement of these glucose molecules is facilitated by active transport, specifically the sodium/potassium–ATPase pump. The pump moves sodium ions into the interstitial fluid, establishing a concentration gradient with more sodium ions outside the cell than inside the cell. The subsequent movement of sodium ions down its concentration gradient into the cell can be "harnessed" to pump glucose into the epithelial cells against the glucose gradient by symport active transport.

The purpose of this laboratory exercise is to observe the role of the sodium/potassium–ATPase pump in glucose transport across the intestinal wall. An excised segment of a rodent intestinal wall will be everted (turned inside out) and placed in various solutions, and glucose transport will be observed. Ouabain, a substance that blocks the pump, will be added to the solution, and the resulting glucose transport observed.

Before you begin, familiarize yourself with the use of the spectrophotometer to measure absorbance and transmittance. In addition, review the following concepts (use your textbook as a reference):

- The action of amylase
- The anatomical structure of intestinal villi
- The function of the sodium/potassium ATPase in all cells
- The development of a sodium gradient and secondary active transport

State a hypothesis regarding the effect of intraluminal sodium concentration on movement of glucose across the intestinal wall. _____
_____
_____

1. Open Ph.I.L.S. Lesson 37: Glucose Transport (**figure 25.17**).

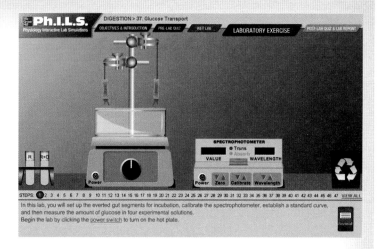

Figure 25.17 Opening Screen for Ph.I.L.S. Lesson 37: Glucose Transport.

2. Click the "power" switch on the hot plate to begin the experiment.

3. Click on the test tube labeled "R," which contains Ringer's solution, on the left side of the screen and drag it to the hot plate just below the section of gut on the left. The gut section will be lowered into the test tube.

4. Repeat the procedure above with the tube labeled "R + O," which contains Ringer's solution and Ouabain, and place the tube on the right side of the hot plate. The section of gut will lower into the solution.

5. Click and drag the "air lines" that have appeared to the opening of each tube. You are now incubating your samples at 37°Celsius with an oxygen air line. The hot plate is moved to the back so that it is out of the way while incubating.

6. Click the "power" switch on the spectrophotometer.

7. Set the wavelength on the spectrophotometer to 450 nm by clicking on the arrows labeled "wavelength."

8. Click and drag the green pipette to the glucose stock solution. When you release the mouse, a pop-up window appears showing a close-up view of the pipette tip.

(continued on next page)

(continued from previous page)

Table 25.14	Serial Glucose Dilution	
Test Tube	Volume of Glucose in Each Tube	Volume of Buffer in Each Tube
50	1 mL	3 mL
100	2 mL	2 mL
150	3 mL	1 mL
Blank	2 mL	3 mL

9. Click the up arrow on the pipette to draw the stock glucose solution into the pipette. Using **table 25.14**, create a serial dilution of glucose with the buffer solution (a serial dilution is the stepwise dilution of a stock solution). Change pipettes between the addition of glucose and buffer by dragging the old pipette to the blue recycle bin. Pick up a new pipette by placing the pipette above the container of clean pipettes.

10. Click and drag the glucose stock to the recycle bin. In its place, a tube of enzyme appears. Add 1 mL of enzyme to each tube (50, 100, and 150) by following the procedure above. (Note the change in gradations on the pipette.) Deposit the pipette in the recycle bin.

11. On the spectrophotometer, use the arrows above the zero on the control panel to set the value to zero.

12. Click the spectrophotometer holder to open it. Click and drag the blank (blk) tube to a position just above the holder and release the mouse. Click the spectrophotometer holder again to close it. Using the arrows of the calibrate section on the control panel of the spectrophotometer, set the value to 100.

13. On the spectrophotometer, change from transmission to absorbance by clicking "absorb" on the control panel. The spectrophotometer is now calibrated.

14. Double click the tube in the spectrophotometer to remove it, and then drag it to the test-tube rack.

15. Click and drag the "50" tube to the spectrophotometer and release the mouse. Close the spectrophotometer holder by clicking on it again.

16. The value in the spectrophotometer "value" window is the absorbance of the sample. Enter that data in **table 25.15**. Repeat the steps above for the remaining tubes.

17. Click on the hot plate to bring it forward. To collect the solutions that have been incubating with the segments of small intestine, click the clamps to raise the section of small intestine. Click the tube labeled "R + O" and place it in the far right space in the test-tube holder. Click and drag the sample labeled "R" to the opening on the left in the test-tube holder.

18. Click and drag the tube labeled "GtR" to just under the small intestine sample on the left side of the hot plate. Click the gray knot at the bottom of the segment to release the contents into the test tube. This tube now contains the solution from the lumen of the gut segment that was placed in Ringer's solution only. Return the "GtR" tube by dragging it back to the test-tube holder.

19. Repeat step 18 for the tube labeled "GtO" and the remaining segment of small intestine. This tube now contains the solution from the lumen of the gut segment that was placed in a solution of Ringer's and Ouabain. Click the hot plate to remove it from the screen.

20. Four clean tubes are now in the middle test-tube rack. Place 1 mL of each solution in the corresponding clean tube (follow the same procedure as above; after you dispense each sample, the pipette will automatically be recycled).

21. Use a clean pipette to add 3 mL of the buffer solution to each tube. Recycle your pipette and add 1 mL of the enzyme solution to each tube.

22. Record the absorbance of each by placing it in the spectrophotometer holder and closing the lid. Click on the "journal" in the lower right corner of the screen. Record your data in **table 25.16**.

Table 25.15	Absorbance Readings for Glucose Dilutions
Glucose Dilution	Absorbance Value
50	
100	
150	

Table 25.16	Absorbance Readings for Test-Tube Solutions
Test-Tube Solution	Absorbance
gtr	
r	
gto	
r+o	

23. Graph the data from tables 25.15 and 25.16 on the graph below (one line for the standard curve and simple data points for the samples that were incubated).

24. Approximate the concentration of glucose in each sample by comparing it to the standard curve. Record the concentration of each sample in **table 25.17**.

25. Make any pertinent observations here. _____
_____
_____

Table 25.17	Concentrations for Test-Tube Solutions
Test-Tube Solution	**Concentration**
gtr	
r	
gto	
r+o	

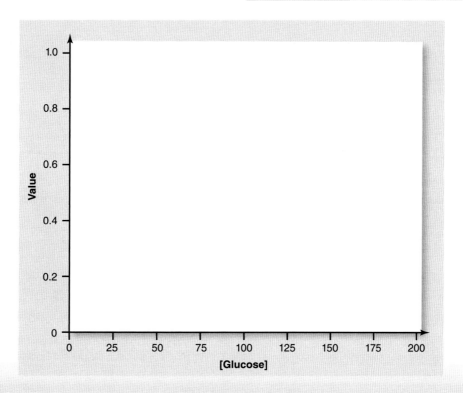

Chapter 25: The Digestive System

Name: _____
Date: _____ Section: _____

POST-LABORATORY WORKSHEET

The ❶ corresponds to the Learning Objective(s) listed in the chapter opener outline.

Do You Know the Basics?

Exercise 25.1: Histology of the Salivary Glands

1. Describe distinguishing histological features of the parotid, submandibular, and sublingual salivary glands. ❶ _____
   _____
   _____

2. Mary likes sour foods, and she decided to eat a slice of lemon. As she bit down on the lemon, she felt an uncomfortable squeezing-type sensation in her cheek as one of her salivary glands emptied its secretions into her mouth. Which salivary gland did she feel? ❷ _____
   _____

Exercise 25.2: Wall Layers of the Stomach

3. Identify the four layers of the stomach wall: ❸

   a. _____

   b. _____

   c. _____

   d. _____

4. Describe the structure and function of the muscularis layer in the stomach. Be sure to address any modifications to this layer that are particular to the stomach. ❹
   _____
   _____

5. The mucosa of the stomach is lined with _____ epithelial tissue. ❺

Exercise 25.3: Histology of the Stomach

6. A student is observing a histology slide of the stomach. He notices in the slide he is observing that there are short gastric pits and long gastric glands, and he can easily identify chief cells, parietal cells, and mucous neck cells. What part of the stomach did the slide most likely come from? ❻ ❼
   _____

Exercise 25.4: Histology of the Small Intestine

7. Describe the three structures that increase the total surface area of the small intestine. ❽
   _____
   _____
   _____

8. _____ cells in the small intestine produce _____ to protect and lubricate the epithelium of the small intestine. ❾

9. What histological feature is unique to the duodenum of the small intestine? ❿
   _____
   _____

10. What histological feature is unique to the ileum of the small intestine? ❿
    _____
    _____

11. The image to the right is a cross section through part of the small intestine.

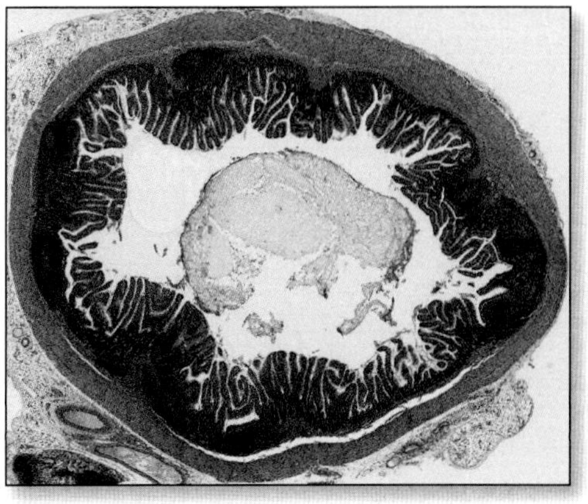

    a. What part of the small intestine is it (duodenum, jejunum, or ileum)?

       _____

    b. In the space below, explain how you came to your answer for part (a). That is, what characteristic(s) did you use to determine which part of the small intestine this sample was taken from?

       _____

       _____

Exercise 25.5: Histology of the Large Intestine

12. What epithelial modification is particularly abundant in the large intestine?

    _____

    _____

13. Describe the wall layers of the large intestine.

    _____

    _____

Exercise 25.6: Histology of the Liver

14. What special cell type is prominent in the epithelium of the large intestine? Why do you think this cell type is particularly abundant in the large intestine?

    _____

    _____

    _____

15. Label the parts of a hepatic lobule in the figure below:

    1 _____
    2 _____
    3 _____
    4 _____
    5 _____
    6 _____
    7 _____
    8 _____
    9 _____
    10 _____
    11 _____
    12 _____
    13 _____

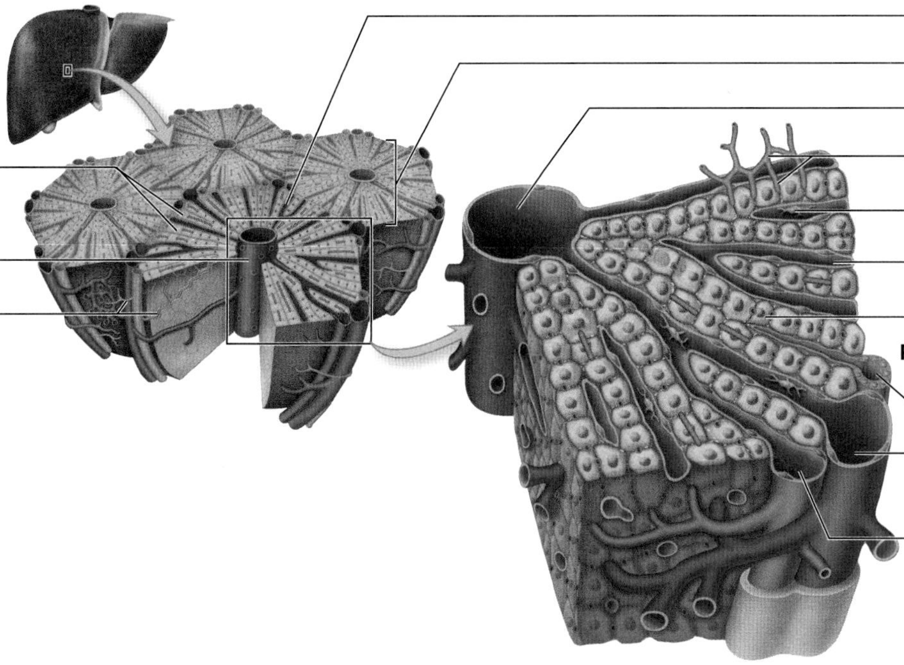

16. Hepatic lobules contain _____, which are capillaries that transport both oxygen-rich blood from the _____ and nutrient-rich blood from the _____ to the central vein.

Exercise 25.7: Histology of the Pancreas

17. Identify two substances produced by the pancreas and released into the duodenum.

    a. _____ (acinar cells)

    b. _____ (cells of pancreatic ducts)

Chapter Twenty-Five *The Digestive System*

18. Exocrine substances produced by _____ cells in the pancreas include _____, which digests carbohydrates, and _____, which neutralizes acidic chyme. Collectively, these substances are known as _____. ⑱

19. Describe the histological structure of the exocrine part of the pancreas. ⑲

    _____

    _____

Exercise 25.8: Gross Anatomy of the Oral Cavity, Pharynx, and Esophagus

20. The esophagus passes through the _____ in the diaphragm prior to entering the stomach. ⑳

Exercise 25.9: Gross Anatomy of the Stomach

21. A patient is suffering from gastroesophageal reflux disease (GERD). When she lies down, she feels a burning sensation in her esophagus caused by the reflux of stomach acids into the esophagus. This occurs because one of the sphincters in her stomach is not working properly. Which sphincter is not functioning properly? _____ Is this sphincter considered an anatomic or physiologic sphincter? _____ What type of muscle composes this sphincter? _____ ㉑

Exercise 25.10: Gross Anatomy of the Duodenum, Liver, Gallbladder, and Pancreas

22. Describe the anatomic relationships among the duodenum, gallbladder, liver, and pancreas. ㉒

    _____

    _____

    _____

23. Identify the four lobes of the liver. ㉓

    a. _____

    b. _____

    c. _____

    d. _____

24. Trace bile from its site of production in the liver through the various ducts to its entry into the duodenum. Assume that it does not enter the gallbladder. ㉔

    _____

    _____

    _____

    _____

Exercise 25.11: Gross Anatomy of the Jejunum and Ileum of the Small Intestine

25. Describe the structure of the following: ㉕

    a. encroaching fat:

    _____

    _____

    b. arterial arcades:

    _____

    _____

    c. vasa recta:

    _____

    _____

26. Describe how to distinguish the jejunum from the ileum of the small intestine, using only gross anatomic features.

_____
_____
_____

Exercise 25.12: Gross Anatomy of the Large Intestine

27. Identify the four parts of the colon.

   a. _____

   b. _____

   c. _____

   d. _____

28. A patient presented to his physician with pain in the left lower quadrant of his abdomen. The source of the pain was an adhesion* between the visceral peritoneum covering part of the patient's colon and the parietal peritoneum lining his anterolateral abdominopelvic wall in that region. What part of the colon was most likely adhered to the abdominopelvic wall?

_____
_____

Exercise 25.13: Digestive Enzymes

29. For each of the following substances, identify the source of secretion and the function of each secretion.

   bile: _____

   pancreatic amylase: _____

   pancreatic lipase: _____

   pepsin: _____

   salivary amylase: _____

30. Describe the effect of temperature on enzyme activity.

_____
_____

31. For each of the following nutrients, identify both the site(s) where the nutrient is digested and the breakdown products of digestion.

   carbohydrate: _____

   triglcyeride: _____

   nucleic acid: _____

   protein: _____

Exercise 25.14: Ph.I.L.S. Lesson 37: Glucose Transport

32. Describe the role of sodium in glucose transport across the wall of the small intestine.

_____
_____
_____

Can You Apply What You've Learned?

33. Using your knowledge of the relationships between ducts draining bile from the liver and gallbladder and the pancreatic ducts, propose a location where a lodged gallstone might cause pancreatitis. _____
_____

*An adhesion (*adhaereo,* to stick to) in the abdominopelvic cavity is an area where two layers of peritoneum are stuck to each other with connective tissue. It usually is the result of some sort of injury or inflammation.

34. Describe why a person suffering from massive bouts of diarrhea is also at risk for electrolyte imbalance. _____

35. Acidic chyme can denature and inactivate enzymes in the small intestine. Describe how an inability to produce bicarbonate in the pancreas can lead to incomplete digestion and issues with absorption. _____

36. Lactose intolerance results from an inability to digest the disaccharide lactose because the enzyme lactase is lacking from the brush border of the small intestine. Lactase is responsible for breaking down lactose into its monomer units, glucose and galactose. Describe why a person suffering from lactose intolerance will commonly suffer from bouts of diarrhea. _____

Can You Synthesize What You've Learned?

37. Weight-loss drugs often work by reducing the amount of fat absorbed from the GI tract into the blood. For instance, Orlistat (Alli®) reduces the activity of lipases in the small intestine. Describe how Orlistat will influence the chemical digestion and absorption of fats, and discuss possible side effects when consuming a high-fat meal. _____

38. Celiac disease is an autoimmune disorder whereby chronic inflammation leads to the destruction of the epithelium of the small intestine. More specifically, the villi in the small intestine begin to atrophy (*a-*, without + *-trophy*, nourishment). Describe how celiac disease will affect chemical digestion and absorption of ingested food. _____

39. *Escherichia coli* (or *E. coli*) is a bacterium that frequently causes severe gastrointestinal symptoms. Ingesting food that is contaminated with *E. coli* often leads to bouts of diarrhea.

   a. Describe what might happen to the pH of the blood of patient infected with *E. coli* (i.e., whether it would increase or decrease) and state the source of the change. _____

   b. Describe how the respiratory rate may change to compensate for the difference described in part a. _____

40. Metformin is a drug frequently prescribed to treat patients who are unable to keep their blood glucose levels under control, such as patients with type II diabetes. Among its actions, metformin works to decrease glucose transport in the GI tract. Describe some potential gastrointestinal side effects that may occur when taking this drug. Describe why this drug may be beneficial for a patient who is unable to utilize insulin for glucose uptake into cells. _____

CHAPTER 26

The Urinary System

OUTLINE AND LEARNING OBJECTIVES

Histology 738

The Kidney 738

EXERCISE 26.1: HISTOLOGY OF THE RENAL CORTEX 740
1. Describe the structure of the nephron
2. List the function of each component of the nephron
3. Describe the type of epithelium found in each part of the nephron, and describe how it is necessary for its function
4. List the components of the renal corpuscle
5. Describe the three structures that compose the filtration membrane
6. Identify the renal cortex and associated structures when viewed through a microscope

EXERCISE 26.2: HISTOLOGY OF THE RENAL MEDULLA 741
7. Identify the renal medulla and associated structures when viewed through a microscope
8. Describe the epithelium that lines the nephron loops and collecting ducts

The Urinary Tract 742

EXERCISE 26.3: HISTOLOGY OF THE URETERS 743
9. Describe the wall-layering pattern of the ureters
10. Identify the layers of the ureter and the type of tissue found in each layer when viewed through a microscope

EXERCISE 26.4: HISTOLOGY OF THE URINARY BLADDER 744
11. Describe the wall-layering pattern of the urinary bladder
12. Identify the layers of the urinary bladder and the type of tissue found in each layer when viewed through a microscope

Gross Anatomy 744

The Kidney 744

EXERCISE 26.5: GROSS ANATOMY OF THE KIDNEY 744
13. Identify structures associated with the kidney on classroom models and/or on preserved kidneys
14. Describe the location of the kidneys with respect to the abdominal cavity and vertebral levels
15. Describe the three layers of the capsule of the kidney
16. Define retroperitoneal

MODULE 13: URINARY SYSTEM

EXERCISE 26.6: BLOOD SUPPLY TO THE KIDNEY 745
17. Identify the blood vessels associated with the kidney on classroom models and/or on preserved kidneys
18. Trace the path of a drop of blood through the kidney, from its entry via the renal artery to its exit via the renal vein
19. Explain how the microscopic structures of the nephron create the gross appearance of the outer cortex and inner medulla of the kidney

EXERCISE 26.7: URINE-DRAINING STRUCTURES WITHIN THE KIDNEY 748
20. Identify the minor and major calyces, renal pelvis, and ureter on a gross specimen of the kidney or on a classroom model of a coronal section of the kidney
21. Trace the path of a drop of filtrate from its exit from the distal convoluted tubule to its entry into the ureter

The Urinary Tract 748

EXERCISE 26.8: GROSS ANATOMY OF THE URETERS 748
22. Trace the path of a drop of urine from its exit from the kidney (via the renal pelvis) to its removal from the body via the urethra
23. Identify the urinary bladder and urethra and their associated structures on classroom models and/or on a cadaver

EXERCISE 26.9: GROSS ANATOMY OF THE URINARY BLADDER AND URETHRA 749
24. Name the three structures that compose the urinary trigone
25. Name the three sections of the male urethra
26. Explain the structural and functional differences between the internal and external urethral sphincters

Physiology 752

Urine Formation 752

EXERCISE 26.10: URINALYSIS 754
27. Describe the process of urine formation
28. Identify normal and abnormal components of urine
29. Conduct a urinalysis using test strips to determine the substances present in urine specimens
30. Measure the specific gravity of urine using a urinometer, and compare results to normal values

Acid-Base Balance 756

EXERCISE 26.11: A CLINICAL CASE STUDY IN ACID-BASE BALANCE 757
31. Describe how the urinary system aids in long-term regulation of acid-base balance
32. Discuss the following compensatory mechanisms responsible for regulating blood pH: buffers, the respiratory system, and the urinary system
33. Apply integrated concepts from the cardiovascular, lymphatic, respiratory, digestive, and urinary systems to a clinical case

INTRODUCTION

The urinary system is responsible for maintaining blood volume and composition. It accomplishes this task through the process of filtration, which involves forcing fluid out of the blood across a membrane called the **filtration membrane**. The fluid thus formed is called **filtrate**. The filtrate flows through the **nephrons**, the structural and functional units of the kidney. Many substances, including over 90% of the water in the filtrate, are **reabsorbed** back into the blood so they are not lost from the body. Other substances, such as **urea** (a breakdown product of protein metabolism), are **secreted** into the filtrate so they can be removed from the body. Ultimately, a small (relative to the volume of blood that is filtered) amount of fluid leaves the kidney as **urine,** which will be transported via the **ureters** to the **urinary bladder** for storage. At a time that is convenient for the individual, the urine is emptied from the bladder and exits the body through the **urethra.**

The structural and functional unit of the kidney is the nephron. Each kidney contains more than 1.25 million nephrons. Remarkably, the kidneys can maintain their function even when 85–90% of their nephrons have been destroyed through disease. However, further losses will result in **kidney failure**. If an individual's kidneys fail, he or she must be placed on **dialysis** (*dialyo,* to separate). This involves filtering the blood using a dialysis machine, or artificial kidney. A patient on dialysis must undergo three to four sessions a week, each session lasting approximately 4 hours. Without dialysis, the individual cannot survive because the balance of fluid, electrolytes, and waste products in the blood cannot be maintained at appropriate levels. The consequences of kidney failure underscore the enormous role the kidneys play in maintaining health.

In these laboratory exercises you will explore the structures of the urinary system with a special emphasis on the kidney. Nearly all functions of the urinary system are functions of the kidney. The remaining structures of this system (ureter, urinary bladder, and urethra) transport or store the urine formed by the kidney. Because the structural and functional units of the kidney—the nephrons—are actively involved in altering the composition of the blood, it is critical to understand the pattern of blood flow through the kidney and how this pattern of blood flow parallels the parts of the nephron. You will accomplish this task by observing the gross anatomy of the kidney and its blood supply, and correlating gross structures with histological observations of the parts of the nephron (renal corpuscles, proximal convoluted tubules, distal convoluted tubules, and nephron loops). Then, you will explore the processes involved in urine formation, identifying normal and abnormal constituents of urine. You will also study the urinary system's role in regulating acid-base balance in the blood. These concepts will be integrated in a clinical case that applies concepts previously covered in multiple chapters. This exercise will enable you to further understand complex physiological systems and their integration to maintain overall homeostasis.

Chapter 26: The Urinary System

PRE-LABORATORY WORKSHEET

Name: _____
Date: _____ Section: _____

Also available at www.connect.mcgraw-hill.com

1. List four functions of the urinary system:

    a. _____

    b. _____

    c. _____

    d. _____

2. List the four components of the urinary system.

    a. _____

    b. _____

    c. _____

    d. _____

3. The structural and functional unit of the kidney is the _____.

4. Describe the structure of transitional epithelium. _____
   _____

5. The three structures that form the boundaries of the trigone within the urinary bladder are as follows:

    a. _____

    b. _____

    c. _____

6. Define *retroperitoneal:* _____
   _____

7. Which of the following is lined with simple cuboidal epithelium with numerous, long, dense microvilli? (circle all that apply)

    a. collecting duct

    b. distal convoluted tubule

    c. renal corpuscle

    d. proximal convoluted tubule

8. The glomerulus of the kidney consists of _____ capillaries.

9. How many layers compose the glomerular capsule of the kidney? _____

10. The renal artery is a branch off of the _____ and the renal vein drains into the _____.

Histology

The Kidney

Each kidney is composed of two major regions: an outer **renal cortex** and an inner **renal medulla (figure 26.1)**. The arrangement of nephrons along the **corticomedullary junction** means that some nephron components fall predominantly in the renal cortex and others in the renal medulla. Thus, the two regions exhibit distinct histological features. The renal cortex contains the renal corpuscles, proximal convoluted tubules (PCTs), distal convoluted tubules (DCTs), and peritubular capillaries. The renal medulla contains the nephron loops, collecting ducts (CDs), and vasa recta. Most structures within the kidney can be identified histologically by recognition of both the region (cortex or medulla) and the type of epithelium lining the structure. **Table 26.1** summarizes the type of epithelium that lines each of the structures and lists the major functions of each structure.

Table 26.1 Histological Features of the Kidney

Structure	Epithelium	Function	Region Where Structure Is Predominantly Located
Glomerulus	Fenestrated endothelium (simple squamous)	Filtration	Cortex
Visceral Layer of Glomerular Capsule	Simple squamous modified to form podocytes	Extensions of podocytes called pedicels contain actin filaments. The membrane-covered openings between the pedicels, called filtration slits, participate in the filtration process.	Cortex
Parietal Layer of Glomerular Capsule	Simple squamous	Forms an outer, impermeable wall of the glomerular capsule	Cortex
Proximal Convoluted Tubule (PCT)	Simple cuboidal with long, dense microvilli. Nuclei are located near the basal surface of the cells.	Reabsorbs glucose, amino acids, Ca^{2+}, PO_4, HCO_3^-, and 80% of the water and NaCl present in the filtrate. Secretes substances like penicillin and toxins after they have undergone modification by the liver. Also secretes organic acids and bases.	Cortex
Descending Limb of the Nephron Loop—Thick Segment	Simple cuboidal	Epithelial cells are impermeable to sodium, but water is drawn out into the interstitial spaces. Thus, the filtrate becomes more concentrated as it moves down the descending nephron loop.	Medulla
Descending Limb of the Nephron Loop—Thin Segment	Simple squamous	Epithelial cells are impermeable to sodium, but water is drawn out into the interstitial spaces. Thus, the filtrate becomes more concentrated as it moves down the descending nephron loop.	Medulla
Ascending Limb of the Nephron Loop—Thin Segment	Simple squamous	Epithelial cells impermeable to water, but sodium passively diffuses out	Medulla
Ascending Limb of the Nephron Loop—Thick Segment	Simple cuboidal. Cells are darker than in the collecting duct.	The epithelial cells are impermeable to water, and they actively transport sodium out of the tubule. Thus, the filtrate becomes less concentrated as it moves up the ascending nephron loop.	Medulla
Distal Convoluted Tubule (DCT)	Simple cuboidal with few, short microvilli. Nuclei are located near the apical surface of the cells.	Secretes H^+ and K^+. Reabsorbs Na^+ and water. Contains the macula densa of the juxtaglomerular apparatus, which is involved with the sensation and regulation of blood pressure.	Cortex
Collecting Tubule and Duct (CD)	Simple cuboidal or simple columnar epithelium. Cells have very precise boundaries. Overall tube diameter is the same as the PCT, but the CD has a larger lumen and no microvilli. Cells are paler than those of the thick segment of the nephron loop.	Concentrates urine under the influence of antidiuretic hormone (ADH). ADH causes CD epithelial cells to transport aquaporins (membrane proteins that transport water) to their apical surface, allowing water to be reabsorbed into the vasa recta.	Medulla

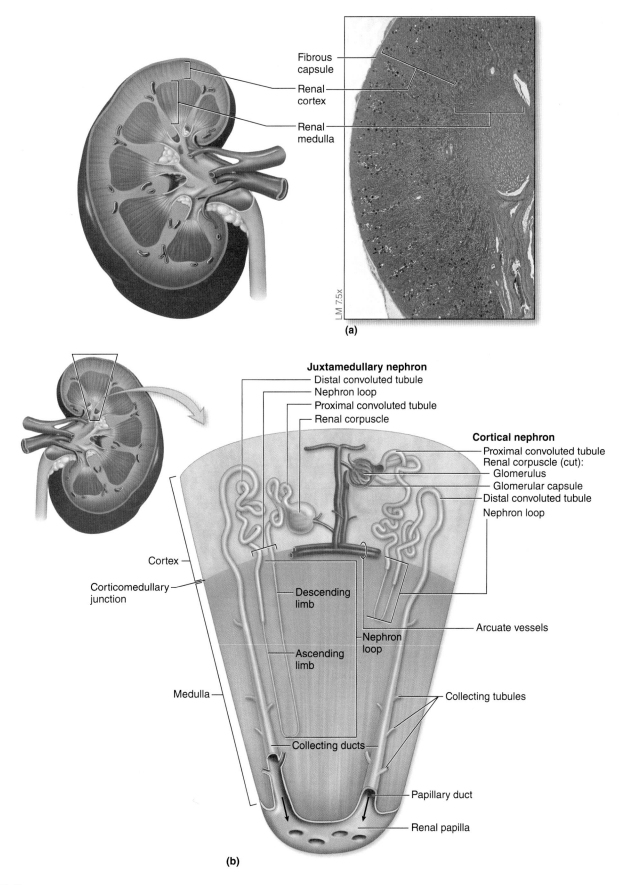

Figure 26.1 Histology of the Kidney. (a) Histological appearance of the kidney at low power demonstrates the fibrous capsule, outer cortex, and inner medulla. (b) Placement of nephron structures within the renal cortex and renal medulla.

EXERCISE 26.1

HISTOLOGY OF THE RENAL CORTEX

The renal cortex contains **renal corpuscles,** which are the site of filtration. Each renal corpuscle is composed of a **glomerulus** (a tuft of capillaries), surrounded by a **glomerular capsule** (also known as Bowman's capsule). The glomerular capsule itself is composed of an inner **visceral layer,** which consists of modified simple squamous epithelial cells called **podocytes,** and an outer **parietal layer,** which consists of unmodified simple squamous epithelium. The renal cortex also contains **proximal convoluted tubules (PCTs),** which are the site of most reabsorption, and **distal convoluted tubules (DCTs),** which are involved in both reabsorption and secretion and compose part of the **juxtaglomerular apparatus.**

EXERCISE 26.1A The Renal Corpuscle

1. Obtain a compound microscope and a histology slide of the kidney and place the slide on the microscope stage.

2. Bring the tissue sample into focus at low power and scan the slide to distinguish between the outer cortex and inner medulla (figure 26.1). Next, move the microscope stage so the renal cortex is in the center of the field of view.

3. Scan the slide in the region of the cortex to locate a circular **renal corpuscle (figure 26.2).** Bring the renal corpuscle into the center of the field of view and then change to high power. Although you won't be able to distinguish the visceral layer of the glomerular capsule from the glomerular capillaries, identify the tissue that contains both the visceral layer of the glomerular capsule and the glomerular capillaries (figure 26.2). Next, identify the parietal layer of the glomerular capsule. What type of epithelium composes the parietal layer of the glomerular capsule?_____

What is the name of the space located between the visceral and parietal layers of the glomerular capsule?

_____

The space you just identified becomes continuous with the lumen of which of the following? (Circle the appropriate response.)

PCT   nephron loop   DCT   collecting duct

4. In the space below, sketch the renal corpuscle as seen through the microscope. Be sure to label the glomerulus (and visceral layer of the glomerular capsule) and the parietal layer of the glomerular capsule on your drawing.

_____ ×

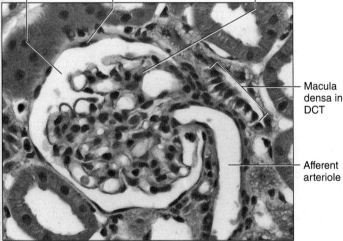

Figure 26.2 The Renal Corpuscle. The renal corpuscle consists of the glomerulus and the glomerular capsule.

5. *Optional Activity:* **AP|R** 13: **Urinary System**—Review the "Kidney—Microscopic Anatomy" animation to visualize the parts of the nephron and their placement in the renal cortex and medulla.

EXERCISE 26.1B Proximal and Distal Convoluted Tubules

1. After completing exercise 26.1A, you should have a slide of the kidney on the microscope stage that has the outer cortex in the center of the field of view. If you did not perform exercise 26.1A, go through steps 1–2 of exercise 26.1A and then continue with this exercise.

2. Scan the slide in the region of the cortex to locate **proximal** and **distal convoluted tubules** (PCTs and DCTs) **(figure 26.3)**. When viewing a histology slide of the cortex of the kidney, you will mostly see cross sections of PCTs and DCTs, with a few renal corpuscles scattered throughout. Although both PCTs and DCTs are lined with simple cuboidal epithelium, the *proximal* convoluted tubules have long, dense microvilli, which make the lumens of the PCTs appear "fuzzy." The distal convoluted tubules have very few, short microvilli, which make the lumens appear to be clear. What does the presence of microvilli indicate about the function of an epithelium? _____

_____

3. In the space below, sketch cross sections of proximal and distal convoluted tubules as seen through the microscope.

_____ ×

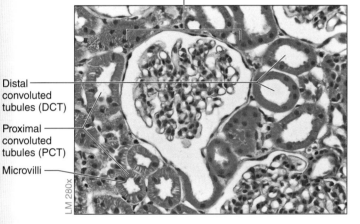

Figure 26.3 The Renal Cortex. The renal cortex contains renal corpuscles, proximal convoluted tubules (PCT), and distal convoluted tubules (DCT). PCTs have "fuzzy" lumens because of the presence of microvilli.

4. Using figures 26.1 to 26.3 and table 26.1 as guides, identify the following structures on the slide of the cortex of the kidney:

- ☐ capsular space
- ☐ distal convoluted tubule
- ☐ microvilli
- ☐ parietal layer of glomerular capsule
- ☐ proximal convoluted tubule
- ☐ renal corpuscle
- ☐ visceral layer of glomerular capsule and glomerular capillaries

EXERCISE 26.2

HISTOLOGY OF THE RENAL MEDULLA

The renal medulla contains **nephron loops** (loops of Henle) and **collecting ducts,** with surrounding capillaries called the **vasa recta.** These structures are all elongated tubules that lie next to each other and function together to concentrate the urine formed by the nephrons.

1. Obtain a compound microscope and a histology slide of the kidney. Place the slide on the microscope stage.

2. Bring the tissue sample into focus at low power and scan the slide to distinguish between the outer cortex and inner medulla. Next, move the microscope stage so the renal medulla is in the center of the field of view (figure 26.1).

3. With the medulla in the center of the field of view, switch to high power and bring the tissue sample into focus once again. In this region of the kidney you will identify thick and thin limbs of the nephron loops, collecting ducts, and possibly vasa recta (table 26.1, figure 26.4). Notice how, in longitudinal section, these structures appear as row after row of cells lined up next to each other. Your main objective in viewing this part of the kidney is to gain an appreciation for the way these structures line up next to each other, which is critical to their function. You should also be able to distinguish thick limbs of the nephron loops from the collecting ducts, based on the diameter of the lumens. Both structures are typically lined with simple cuboidal

(continued on next page)

(continued from previous page)

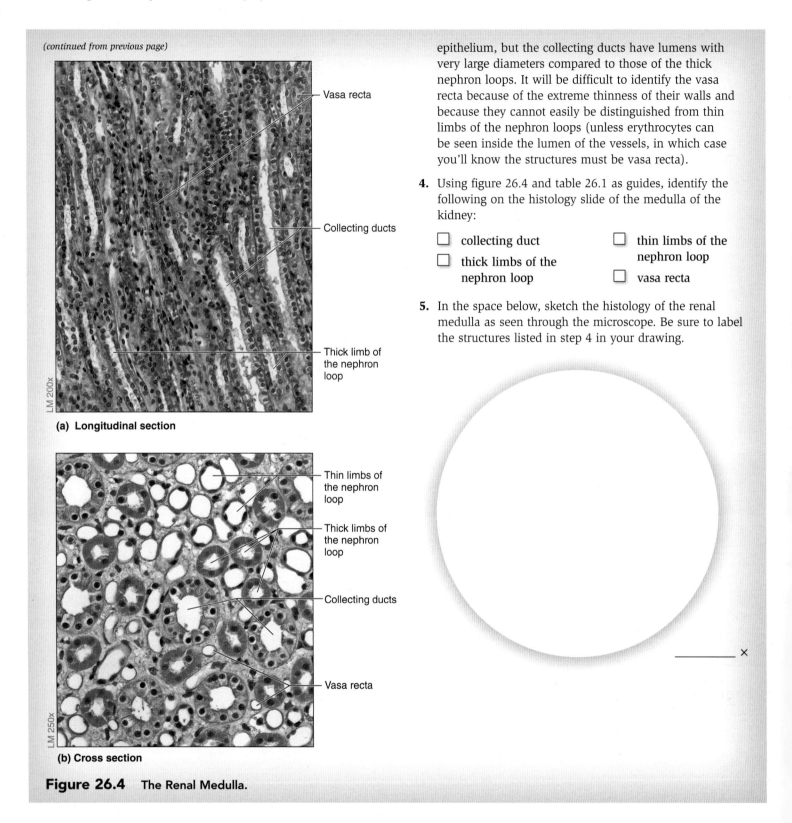

(a) Longitudinal section

(b) Cross section

Figure 26.4 The Renal Medulla.

epithelium, but the collecting ducts have lumens with very large diameters compared to those of the thick nephron loops. It will be difficult to identify the vasa recta because of the extreme thinness of their walls and because they cannot easily be distinguished from thin limbs of the nephron loops (unless erythrocytes can be seen inside the lumen of the vessels, in which case you'll know the structures must be vasa recta).

4. Using figure 26.4 and table 26.1 as guides, identify the following on the histology slide of the medulla of the kidney:

   ☐ collecting duct ☐ thin limbs of the nephron loop
   ☐ thick limbs of the nephron loop ☐ vasa recta

5. In the space below, sketch the histology of the renal medulla as seen through the microscope. Be sure to label the structures listed in step 4 in your drawing.

The Urinary Tract

The **urinary tract** consists of structures that function to transport or store urine: the ureters, urinary bladder, and urethra. In this section you will observe the histological structure of the walls of the ureters and urinary bladder. **Table 26.2** lists the type of epithelium that lines each of the structures of the urinary tract and the calyces of the kidney.

Table 26.2 Urine-Draining Structures

Structure	Epithelium	Number of Cell Layers	Word Origin
Calyces	Transitional	2–3	*calyx*, cup of a flower
Ureter	Transitional	4–5	*oureter*, urinary canal
Urinary Bladder	Transitional	> 6	*urinary*, relating to urine
Urethra	In males, the proximal portions are lined with transitional epithelium and the distal portions are lined with stratified squamous of the external body surface. In females, is primarily stratified squamous.	NA	*ourethra*, canal leading from the bladder

EXERCISE 26.3

HISTOLOGY OF THE URETERS

The **ureters** are long, muscular tubes that transport urine from the hilum of the kidney to the urinary bladder. They are lined with **transitional epithelium,** which has the ability to stretch when urine is being transported through the ureters. Like other tubular structures in the body, the wall of the ureter is composed of multiple layers. The walls of the ureters, however, do not have a submucosa. There are three layers to the walls of the ureters: the mucosa, muscularis, and adventitia **(figure 26.5).** Contraction of smooth muscle in the muscularis layer of the ureters transports urine from the renal pelvis to the urinary bladder. The arrangement of smooth muscle layers in the muscularis of the ureters is just the opposite of that of the digestive tract organs. The inner layer is composed of **longitudinal** smooth muscle, whereas the outer layer is composed of **circular** smooth muscle. In addition, because the ureters are **retroperitoneal** (they lie behind the peritoneal cavity and are not covered by peritoneum), their outer layer is an **adventitia.**

1. Obtain a histology slide demonstrating a cross-sectional view of a ureter. Place it on the microscope stage and bring the tissue sample into focus on low power.

2. Using figure 26.5 as a guide, identify the following structures on the slide of the ureter:

   ☐ adventitia   ☐ outer circular muscle
   ☐ inner longitudinal   ☐ transitional epithelium
       muscle

3. In the space below, sketch a cross section of a ureter as seen through the microscope. Be sure to label all the structures listed in step 2 in your drawing.

Figure 26.5 **The Ureter.** (a) Cross section through entire ureter. (b) Close-up of the mucosa of the ureter.

EXERCISE 26.4

HISTOLOGY OF THE URINARY BLADDER

The wall of the urinary bladder has a more typical arrangement of layers, similar to that of other organs in the body: the mucosa, submucosa, muscularis, and adventitia (with serosa only on the superior surface of the urinary bladder) **(figure 26.6)**. Similar to the ureters, the urinary bladder is lined with **transitional epithelium,** which allows it to stretch as it fills with urine, and its outermost layer is an adventitia because the bladder lies outside the peritoneal cavity. The muscularis layer of the bladder wall is composed of several individual layers of smooth muscle that are collectively referred to as the **detrusor muscle** (*detrudo*, to drive away).

1. Obtain a histology slide of the wall of the urinary bladder. Place it on the microscope stage and observe at low power. Using figure 26.6 as a guide, identify the following structures:

    ☐ adventitia   ☐ serosa
    ☐ mucosa   ☐ submucosa
    ☐ muscularis (detrusor muscle)   ☐ transitional epithelium

2. Move the microscope stage so the epithelium is at the center of the field of view, then change to high power. Look for rounded epithelial cells on the apical surface of the epithelium, which are often binucleate. These cells are sometimes referred to as **dome cells** because of their shape.

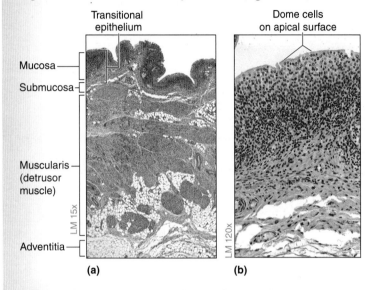

Figure 26.6 Histology of the Urinary Bladder. (a) Section of the wall of the bladder. (b) Close-up of the transitional epithelium.

Gross Anatomy

In this set of exercises you will observe the gross anatomy of the kidneys, ureters, urinary bladder, and urethra, paying particular attention to the location of these structures within the abdominopelvic cavity and their relationships with other organs.

The Kidney

The kidneys are located along the posterior body wall from vertebral levels T_{12}–L_3, with the right kidney slightly lower than the left kidney because of the location of the liver. The kidneys, like the ureters and urinary bladder, are **retroperitoneal** structures (*retro-*, behind, + *peritoneal*, referring to the peritoneum). In the following exercises you will observe the gross anatomy of the kidney, the blood supply to the kidney, and the urine-draining structures within the kidney.

EXERCISE 26.5

GROSS ANATOMY OF THE KIDNEY

1. Each kidney is surrounded and protected by several layers. From innermost to outermost these include the renal capsule, perinephric fat, renal fascia, and paranephric fat. Only the renal capsule typically remains with a kidney that has been removed from a cadaver. The other layers are removed from around the kidney when it is extracted from the body.

2. Obtain a kidney that has been sectioned along a coronal plane, or a classroom model of a coronal section of the kidney **(figure 26.7)**. Whether you are viewing an actual kidney or a model of the kidney, the outermost part of what you are looking at is actually the innermost layer of the capsule: the **fibrous capsule.**

3. Identify the structures listed in figure 26.7 on the kidney or classroom model of the kidney. Then label them in figure 26.7.

Chapter Twenty-Six *The Urinary System* 745

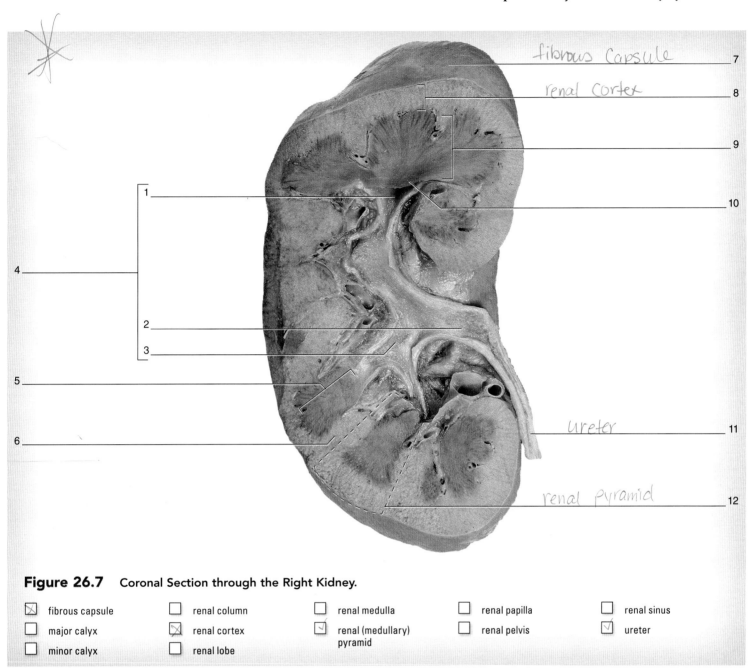

7. fibrous capsule
8. renal cortex
9.
10.
11. ureter
12. renal pyramid

Figure 26.7 Coronal Section through the Right Kidney.

- ☒ fibrous capsule
- ☐ major calyx
- ☐ minor calyx
- ☐ renal column
- ☒ renal cortex
- ☐ renal lobe
- ☐ renal medulla
- ☑ renal (medullary) pyramid
- ☐ renal papilla
- ☐ renal pelvis
- ☐ renal sinus
- ☑ ureter

EXERCISE 26.6

BLOOD SUPPLY TO THE KIDNEY

The kidneys receive approximately 20% to 25% of the blood pumped by the heart each minute. This is not because they have a huge demand for oxygen or nutrients, but instead because their major function is to filter the blood to alter its volume and composition. The kidney accomplishes this through the processes of filtration, reabsorption, and secretion, which all involve transport of fluids and other substances between the functional units of the kidney (nephrons) and the blood. Thus, an understanding of blood flow through the kidney is critical to understanding the function of the kidney.

1. Observe a classroom model of the kidney that demonstrates blood vessels of the kidney.

2. Blood flow through the kidney is unique. **Figure 26.8** diagrams the flow of blood through the kidney. There are three capillary beds in the kidney: the glomerulus (within the renal corpuscle), the peritubular capillaries, and the vasa recta. The **glomerulus,** the first capillary bed, is the site of *filtration,* which is the movement of substances from the blood across the filtration membrane to the capsular space of the renal corpuscle (figure 26.8a,b). Blood enters the glomerulus through an afferent arteriole and exits through an efferent arteriole. The efferent arteriole then leads into the second capillary bed, which can be either the **peritubular capillaries** (within the cortex) or the **vasa recta** (within the medulla). The second capillary bed is the site of

(continued on next page)

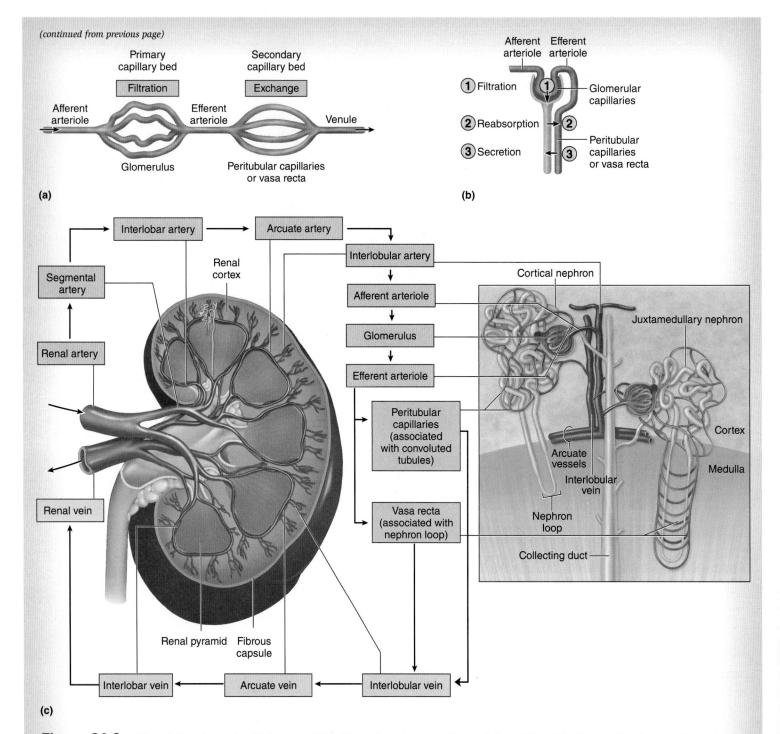

Figure 26.8 Blood Supply to the Kidney. (a) Blood flows through two capillary beds in the kidney. The first capillary bed, the glomerulus, is the site of filtration. The second capillary bed, peritubular capillaries in the cortex or vasa recta in the medulla, is the site of exchange (reabsorption or secretion). (b) The basic renal processes are filtration, reabsorption, and secretion. (c) Blood flow through the kidney.

exchange of fluids, electrolytes, respiratory gases, and nutrients between the tubular portions of the nephron and the blood. Exchange in these capillaries involves either *reabsorption* (movement of substances from the tubular spaces into the blood) or *secretion* (movement of substances from the blood into the tubular spaces). Finally, blood leaving the peritubular capillaries and vasa recta drains into veins that carry the blood back toward the inferior vena cava.

3. Using your textbook as a guide, identify the blood vessels listed in **figure 26.9** on the classroom model of the kidney. Then label them in figure 26.9. What parts of the nephron do the peritubular capillaries surround?

   _____

   What parts of the nephron do the vasa recta surround?

   _____

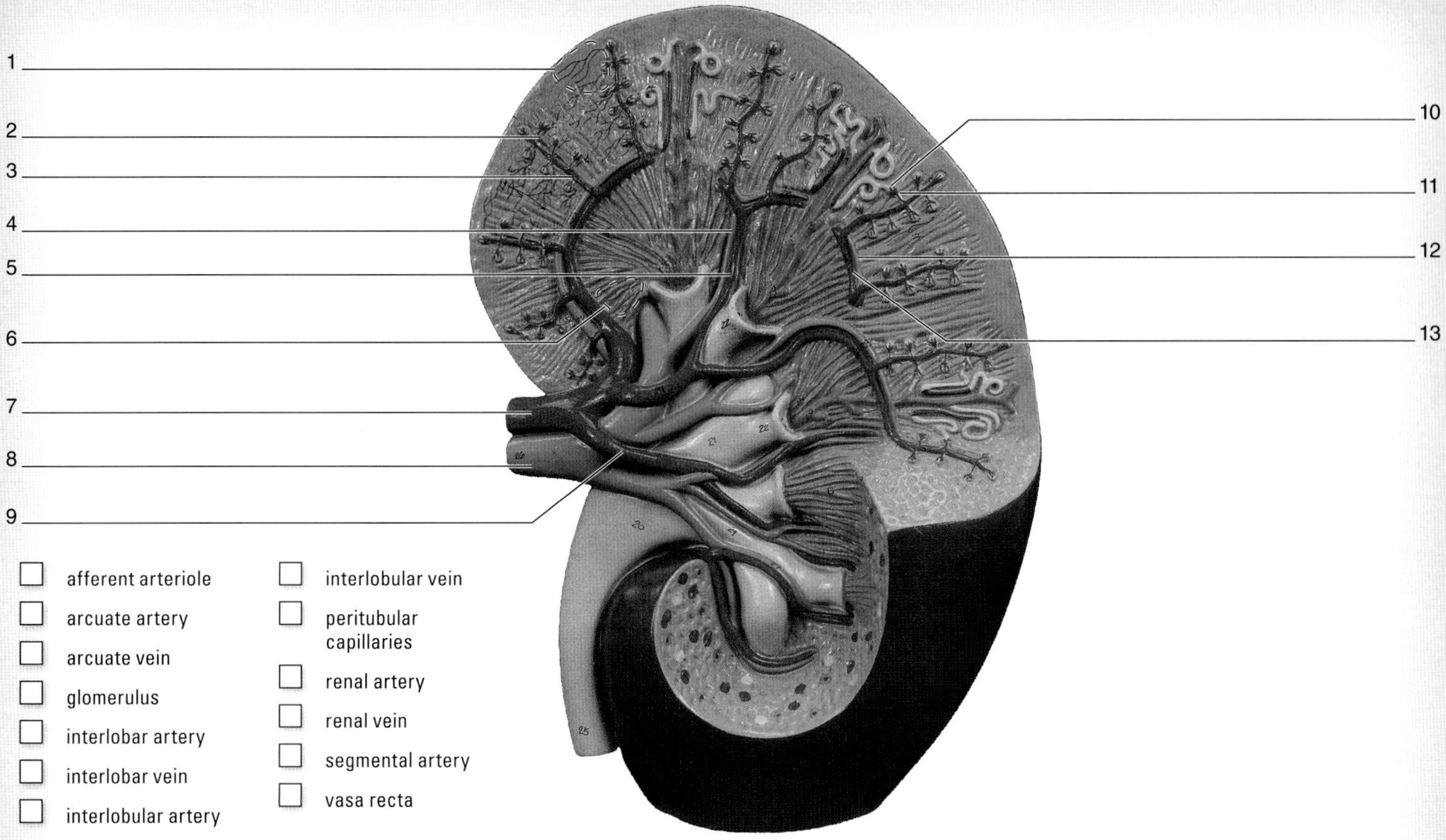

Chapter Twenty-Six The Urinary System   747

1 _____
2 _____
3 _____
4 _____
5 _____
6 _____
7 _____
8 _____
9 _____

10 _____
11 _____
12 _____
13 _____

☐ afferent arteriole
☐ arcuate artery
☐ arcuate vein
☐ glomerulus
☐ interlobar artery
☐ interlobar vein
☐ interlobular artery

☐ interlobular vein
☐ peritubular capillaries
☐ renal artery
☐ renal vein
☐ segmental artery
☐ vasa recta

(a)

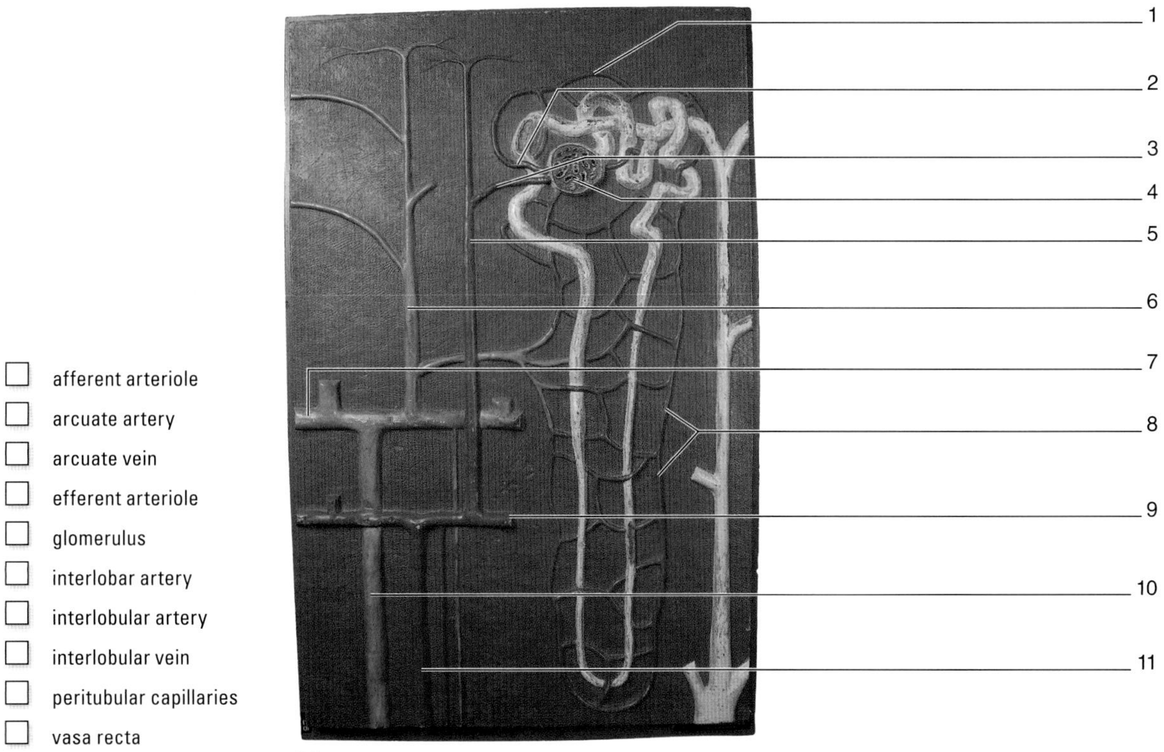

1 _____
2 _____
3 _____
4 _____
5 _____
6 _____
7 _____
8 _____
9 _____
10 _____
11 _____

☐ afferent arteriole
☐ arcuate artery
☐ arcuate vein
☐ efferent arteriole
☐ glomerulus
☐ interlobar artery
☐ interlobular artery
☐ interlobular vein
☐ peritubular capillaries
☐ vasa recta

(b)

Figure 26.9 Models of the Kidney Demonstrating the Blood Supply to the Kidney. (a) Coronal section. (b) Close-up of the renal cortex and renal medulla.

(continued on next page)

4. In the space below, trace the flow of blood, in words, through the kidney starting at the abdominal aorta and ending at the inferior vena cava.

_____

_____

_____

_____

_____

_____

_____

EXERCISE 26.7

URINE-DRAINING STRUCTURES WITHIN THE KIDNEY

Once filtrate has been formed by the nephrons, the fluid is processed as it passes through the nephron tubules (PCT, nephron loop, DCT), collecting tubules, and collecting ducts, and is called **tubular fluid.** Fluid leaving the collecting ducts and entering the papillary ducts is called **urine.** Urine will drain through the minor calyx, major calyx, and renal pelvis within the kidney before entering a ureter.

1. Observe a classroom model of a coronal section of a kidney, or a gross sample of a kidney that has been sectioned along a coronal plane.

2. Using figures 26.7 and 26.9 and table 26.2 as guides, identify the following urine-draining structures of the kidney:

   ☐ collecting duct   ☐ renal papilla
   ☐ papillary duct   ☐ renal pelvis
   ☐ major calyx   ☐ renal pyramid
   ☐ minor calyx   ☐ ureter

3. *Optional Activity:* **AP|R** 13: Urinary System—Watch the "Kidney—Gross Anatomy" and "Urine Formation" animations for an overview of kidney anatomy, vasculature, and function.

The Urinary Tract

The urinary tract consists of the ureters, urinary bladder, and urethra. Within the kidney, urine draining from the calyces begins to collect in the renal pelvis. Every 2 to 3 minutes, peristaltic contractions of the smooth muscle lining the ureters conveys the urine into the ureters, which transport it into the urinary bladder. Urine is then stored within the urinary bladder until a time when it is convenient to allow the urine to exit the body through the urethra. The process of voiding urine is called **micturition.** In this section you will observe the gross anatomy of the structures composing the urinary tract.

EXERCISE 26.8

GROSS ANATOMY OF THE URETERS

The ureters are long, thin, muscular tubes that run from the hilum of the kidney to the posterior, inferior surface of the urinary bladder **(figure 26.10).** Like the kidney, they are located retroperitoneally, behind the peritoneal cavity.

1. Observe a human cadaver or a classroom model of the abdomen.

2. Identify the structures listed in figure 26.10 on the cadaver or classroom model. Then label figure 26.10.

3. Follow the pathway of the ureter from its site of origin near the hilum of the kidney to its entry into the urinary bladder. What major muscle does the ureter cross over to get to the urinary bladder?

_____

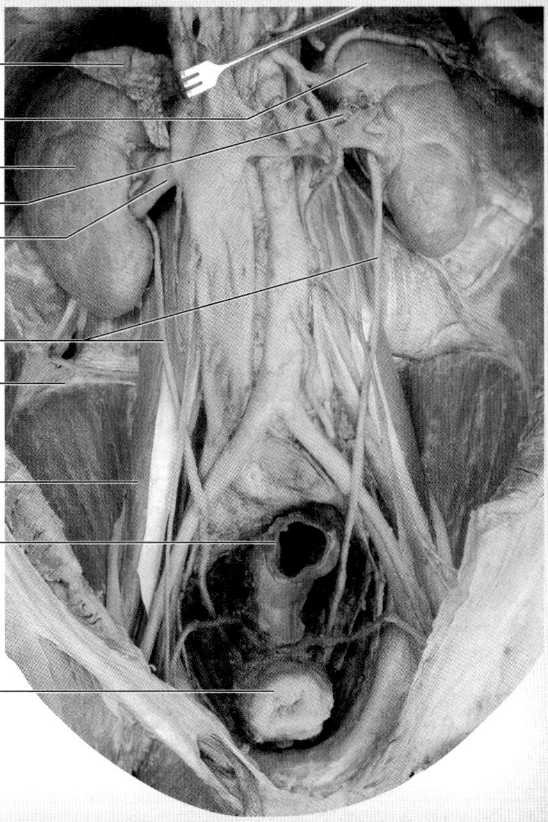

Figure 26.10 Location of the Structures of the Urinary System within the Abdominopelvic Cavity.

- ☐ left kidney
- ☐ right kidney
- ☐ ureters
- ☐ urinary bladder

Labels on figure:
- Adrenal gland
- 1 _____
- 2 _____
- Renal artery
- Renal vein
- 3 _____
- Iliac crest
- Psoas major muscle
- Rectum
- 4 _____

EXERCISE 26.9

GROSS ANATOMY OF THE URINARY BLADDER AND URETHRA

Like the kidneys and ureters, the **urinary bladder** is a retroperitoneal structure. The superior surface of the bladder is covered with peritoneum, but the rest is not. The urinary bladder lies in the pelvic cavity, just posterior to the pubic symphysis. The **urethra** is a passageway extending from the inferior surface of the urinary bladder to the external urethral orifice. It is much longer in males than in females because it extends through the length of the penis. (The urethra also serves as the passageway for sperm during ejaculation.) In both males and females there are both internal and external urethral sphincters. The **internal urethral sphincter** is composed of smooth muscle in the wall of the urinary bladder that encloses the entrance into the urethra, and is under involuntary control. The **external urethral sphincter** is composed of skeletal muscle of the **urogenital diaphragm**, and is under voluntary control.

1. Observe the abdominopelvic cavity of a male human cadaver or a classroom model of the abdominopelvic cavity of a male **(figure 26.11)**. The **trigone** is the area in the floor of the urinary bladder (in both males and females) that is enclosed by imaginary lines that extend between the openings of the two ureters and the urethra. This area is clinically significant because urinary tract infections are common in this area. If the urinary bladder is cut open on the cadaver, or if you can observe a classroom model that demonstrates the interior of the urinary bladder, identify the urinary trigone in the floor of the urinary bladder.

2. Using your textbook as a guide, identify the structures listed in figure 26.11 on the cadaver or classroom model. Then label them in figure 26.11.

3. In the space below, sketch the structures forming the boundaries of the urinary trigone as seen from the interior of the urinary bladder.

(continued on next page)

750 Chapter Twenty-Six *The Urinary System*

(continued from previous page)

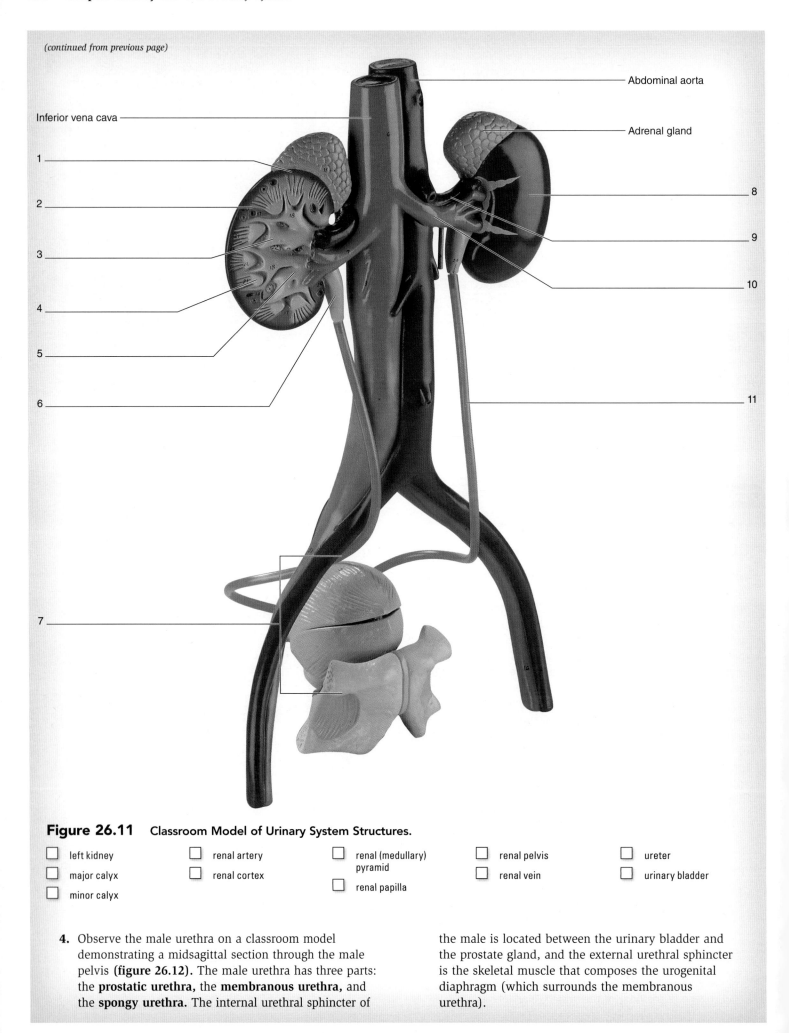

Figure 26.11 **Classroom Model of Urinary System Structures.**

- ☐ left kidney
- ☐ major calyx
- ☐ minor calyx
- ☐ renal artery
- ☐ renal cortex
- ☐ renal (medullary) pyramid
- ☐ renal papilla
- ☐ renal pelvis
- ☐ renal vein
- ☐ ureter
- ☐ urinary bladder

4. Observe the male urethra on a classroom model demonstrating a midsagittal section through the male pelvis **(figure 26.12)**. The male urethra has three parts: the **prostatic urethra,** the **membranous urethra,** and the **spongy urethra.** The internal urethral sphincter of the male is located between the urinary bladder and the prostate gland, and the external urethral sphincter is the skeletal muscle that composes the urogenital diaphragm (which surrounds the membranous urethra).

Chapter Twenty-Six *The Urinary System* 751

5. Using your textbook as a guide, identify the structures listed in figure 26.12 on the classroom model. Then label them in figure 26.12.

6. Observe the female urethra on a classroom model demonstrating a midsagittal section through the female pelvis (**figure 26.13**). The female urethra is short, and

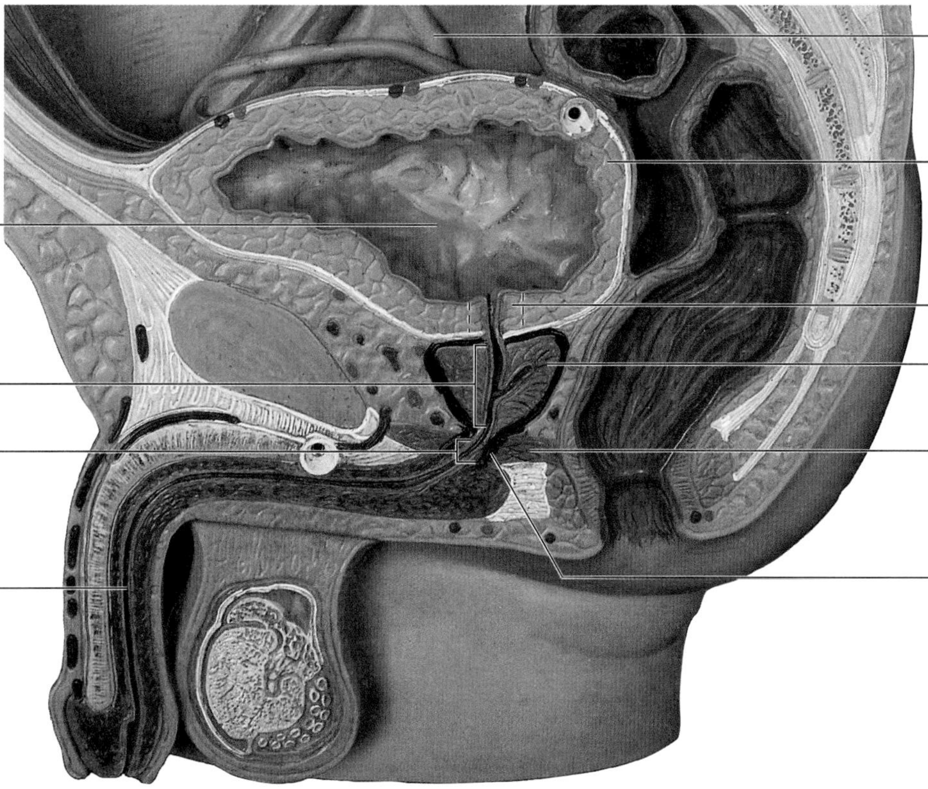

Figure 26.12 Classroom Model of a Midsagittal Section through the Male Pelvis.

- ☐ external urethral sphincter
- ☐ internal urethral sphincter
- ☐ membranous urethra
- ☐ muscularis (detrusor muscle)
- ☐ prostatic urethra
- ☐ spongy urethra
- ☐ ureter
- ☐ urinary bladder

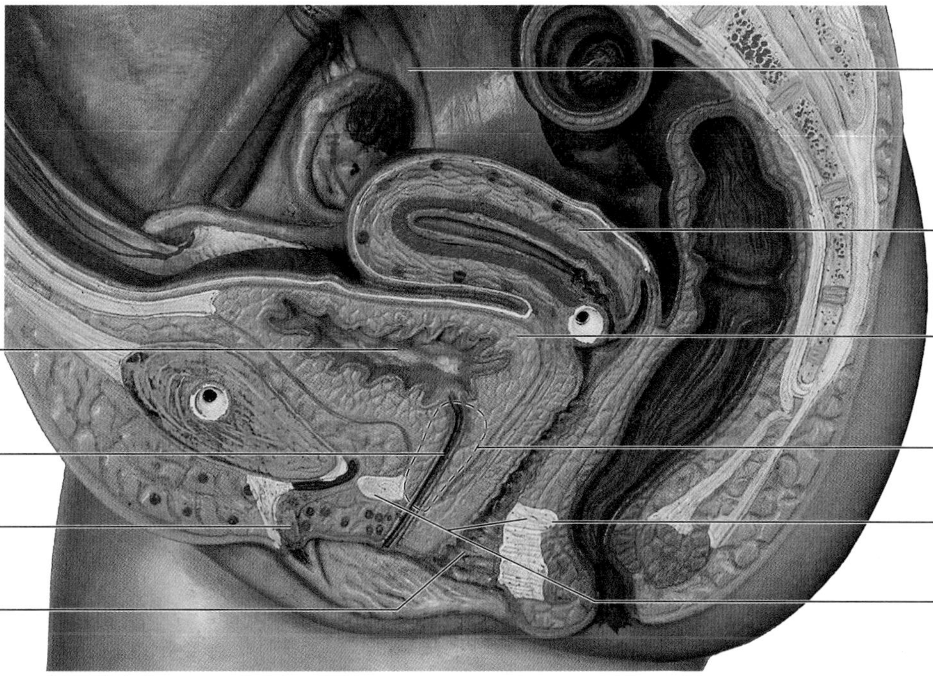

Figure 26.13 Classroom Model of a Midsagittal Section through the Female Pelvis.

- ☐ external urethral sphincter
- ☐ internal urethral sphincter
- ☐ muscularis (detrusor muscle)
- ☐ ureter
- ☐ urethra
- ☐ urinary bladder

(continued on next page)

(continued from previous page)

the external urethral orifice is located between the clitoris and the vagina. The entire urethra is surrounded by the internal urethral sphincter. The external urethral sphincter is skeletal muscle of the urogenital diaphragm, just as it is in the male.

7. Using your textbook as a guide, identify the structures listed in figure 26.13 on the classroom model. Then label them in figure 26.13.

Physiology

Urine Formation

Urine is formed by three main physiological processes that are performed by the nephron: filtration, reabsorption, and secretion. The first process, **filtration,** occurs in the renal corpuscle. Within the nephron, blood enters a specialized capillary bed called a **glomerulus** via the afferent arteriole. The glomerular capillaries are fenestrated capillaries, and are covered by specialized cells called **podocytes**. The podocytes form slits through which blood is filtered. Collectively, the fenestrated epithelial cells, their basement membrane, and the membrane-covered filtration slits formed by podocytes compose the **filtration membrane**. The process of filtration occurs when blood is transported into the glomerular capillaries. Small substances such as water, electrolytes, glucose, and amino acids pass freely across the filtration membrane into the capsular space (Bowman's space). Collectively the fluid and other substances that enter the surrounding capsular space are known as **filtrate**.

The rate at which filtration occurs in the glomerulus, the **glomerular filtration rate** (GFR), is determined by glomerular and capsular hydrostatic pressures and blood colloid osmotic pressure. **Hydrostatic pressure** (*hydro-*, water + *statikos-*, to make stand) is a force per unit area exerted by a fluid in a closed space. For instance, blood flowing through blood vessels exerts a hydrostatic pressure against the vessel wall. You measured "blood pressure" in chapter 22. The **glomerular hydrostatic pressure (HP$_g$)** is the blood pressure within the glomerulus, and therefore is the pressure pushing

INTEGRATE

LEARNING STRATEGY

A unique feature of the glomerulus is that it is a specialized capillary bed that is both supplied *and* drained by arterioles. Constriction or dilation of these arterioles can greatly affect the fluid pressure within the glomerulus, and thus filtration. Specifically, constriction or dilation of the afferent arteriole directly influences GFR and filtration/urine output by the kidneys.

Diameter of the afferent arteriole is self-regulated by a myogenic (*myo*, muscle) response. As systemic blood pressure increases, it increases the stretch on the wall of the afferent arteriole (AA). Smooth muscle in the AA reflexively contracts, thereby constricting the AA and decreasing GFR. The overall goal of the myogenic response is to maintain GFR over a wide range of systemic blood pressures so urinary output does not vary with changes in systemic blood pressure. The concepts and rules governing pressure, flow, and resistance, as described in detail in chapter 22, apply to this situation.

The afferent arteriole is not unlike an outdoor water spigot. When you open the spigot valve, water flows freely. When you close the valve, water merely trickles out. Similarly, when the afferent arteriole dilates, more blood flows into the glomerulus, which increases GFR and filtrate/urine production. When the afferent arteriole constricts, less blood flows into the glomerulus, which decreases GFR and filtrate/urine production. Remember that the blood flowing through the afferent arteriole is representative of the systemic blood pressure. Here is one fun quip to help you easily remember the relationship between blood pressure and GFR: "No BP, no pee pee!"

against the glomerular capillary walls. Thus, glomerular hydrostatic pressure is the primary force that *promotes* filtration **(figure 26.14)**. **Capsular hydrostatic pressure (HP$_c$)** is pressure caused by the presence of filtrate within the capsular (Bowman's) space. Think of this as a fluid pressure pushing back against the the flow of fluid entering the capsular space. Thus, capsular hydrostatic pressure is a pressure that *opposes* filtration. **Blood colloid osmotic pressure (OP$_g$)** is a pressure exerted by proteins such as albumin within the blood plasma. These proteins exert osmotic pressure to "hold on" to water. Thus, blood colloid osmotic pressure is a force that also *opposes* filtration. The **net filtration pressure** is determined by subtracting both the capsular hydrostatic pressure and the blood colloid osmotic pressure from the glomerular hydrostatic pressure (see figure 26.14).

Tubular **reabsorption** involves transport of substances from the filtrate back into the blood. Substances are primarily reabsorbed in the proximal convoluted tubule and include water, glucose, amino acids, and electrolytes. **Table 26.3** provides a quick reference of the parts of the nephron, the substances reabsorbed in each part, and the mechanism of reabsorption for each substance.

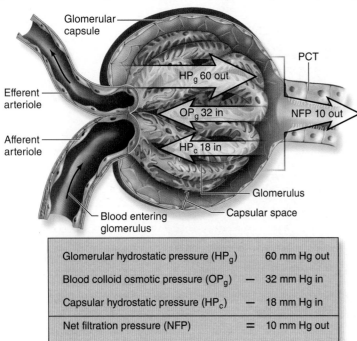

Figure 26.14 Net Filtration Pressure.

INTEGRATE

CONCEPT CONNECTION

Net filtration pressure results from opposing pressures, namely, hydrostatic pressures and colloid osmotic pressure. This is the same mechanism that regulates movement of fluid (blood plasma) from systemic capillaries into the interstitial space on the arterial end of systemic capillaries, and from the interstitial space into post-capillary venules. This process is not 100% efficient, and results in a net loss of fluid from the blood into the interstitial spaces. Recall that this remaining fluid is generally picked up by lymphatic vessels and carried as lymph. Fluid movement from systemic capillaries into the interstitial spaces is driven by hydrostatic pressure within the capillaries. Therefore, hydrostatic pressure "pushes" the fluid into the interstitial space, thereby overcoming the colloid osmotic pressure within the capillaries that tends to want to "hold on" to the water within the blood plasma. The net effect is that fluid flows from the blood capillaries into the tissues. As fluid leaves the capillaries, larger solutes, including albumin (protein), remain in the vessel. On the venous side, hydrostatic pressure within the capillary is quite low because of the net loss of water to the interstitial spaces. However, the colloid osmotic pressure, mainly due to the presence of albumin in the capillaries, is quite high. Therefore, water is drawn back into the vessels via osmosis.

This balance of hydrostatic and colloid osmotic pressures is impacted when there is significant damage to the filtration membrane of the nephron. That is, damage to the filtration membrane may allow larger molecules, including albumin, to enter the filtrate. Because there are no mechanisms for reabsorbing albumin within the nephron, any albumin that is filtered ends up excreted in the urine. In addition, albumin in the filtrate draws more water out of the glomerular capillaries, thus increasing water loss through the kidneys.

The loss of albumin from the blood causes a decrease in the colloid osmotic pressure of systemic capillaries as well. Without the normal ability to "hold on" to water, more water will leave systemic capillaries and enter the interstitial spaces. This results in fluid accumulation in the interstitial space surrounding systemic capillary beds, called **edema.**

Table 26.3 Reabsorption in the Nephron

Structure	Substance Reabsorbed	Mechanism of Reabsorption (Basolateral Membrane)
Proximal Convoluted Tubule	Amino Acids	Facilitated diffusion
	Glucose	
	K^+	Paracellular transport
	Na^+	Active transport
	Water	Osmosis
Descending Limb of Loop of Henle	Water	Osmosis
Ascending Limb of Loop of Henle	Cl^-	Facilitated diffusion
	K^+	
	Na^+	Active transport
Distal Convoluted Tubule	Cl^-	Facilitated diffusion
	K^+	
	Na^+	Active transport
	Water (regulated by ADH)	Osmosis
Collecting Duct	Water (regulated by ADH)	Osmosis + transport through proteins called aquaporins

Table 26.4 Secretion in the Nephron

Structure	Substance Secreted	Mechanism of Secretion
Proximal Convoluted Tubule	H^+	Antiport (with Na^+)
	Drugs and toxins	Active transport
	Ammonia	Simple diffusion
Distal Convoluted Tubule	K^+	Active transport
	H^+	Antiport (with Na^+)

In comparison, tubular **secretion** involves transport of substances from the blood into the filtrate. Numerous substances are secreted including hydrogen ions, drugs, toxins, and ammonia. **Table 26.4** provides a quick reference for the urinary structures, substances secreted, and the site and mechanism of secretion for each substance.

Once the blood has been filtered, and "processed" through tubular reabsorption and tubular secretion, the remaining fluid (now called urine) is drained via the ureters into the urinary bladder where it is stored until it is eliminated from the body. Typically, the composition of urine is slightly acidic, with a pH ranging from 4.5 to 8.0. Given the processes outlined previously, it makes sense that anything that may alter the blood pH might also alter urine pH. For instance, a diet high in protein and low in carbohydrates results in utilization of fats and proteins for energy. The products of fat metabolism may result in the production of ketone bodies, which are acidic and therefore lower the pH of the blood. The increased concentration of hydrogen ions in the blood results in an increased rate of secretion of hydrogen ions into the urine, thereby decreasing the pH of the urine.

Specific gravity is a measure of the density of a substance when compared to that of water. The specific gravity of urine typically ranges from 1.003 to 1.035. Concentrated (darker) urine has a higher specific gravity than does dilute (lighter) urine. Substances that are normally found in the urine include salts (Na^+, Cl^-, K^+, Mg^{2+}, Ca^{2+}, SO_4^{2-}, $H_2PO_4^-$, and NH_4^+), nitrogenous wastes (urea), and toxins. However, substances such as glucose, albumin, red blood cells, white blood cells, or abnormally high levels of the above substances may be cause for concern. For instance, glucose, which is typically reabsorbed via facilitated diffusion in the PCT, may appear in the urine when the carrier molecules are completely saturated. This is often a sign of uncontrolled diabetes mellitus. When larger molecules, such as albumin, red blood cells, or white blood cells appear in the urine, this may be a sign of damage to the glomerular capillaries; such is often the case in chronic hypertension or traumatic injury. The presence of white blood cells may also be

INTEGRATE

CLINICAL VIEW
Diabetes

Most people associate *diabetes* (*diabetes*, siphon) with either type I or type II diabetes *mellitus* (*mellitus*, sweet). However, there are two other forms of diabetes as well: gestational diabetes and diabetes *insipidus*. The common link for all forms of diabetes is the production of copious amounts of diluted urine (**polyuria**). While the etiology of each subtype of diabetes is different, the outcome is the same. These patients are often extremely thirsty with a high urine output. As previously described, diabetes mellitus, whether type I or type II, results from an inability to regulate blood glucose levels. When blood levels of glucose are high, urine production may also be high because water is attracted to the glucose (and sodium) in the filtrate, leading to greater volumes of urine.

During pregnancy, some women may experience high blood glucose levels, a condition called gestational diabetes. To test for gestational diabetes, it is standard practice to give a pregnant woman a glucose tolerance test between 24 and 28 weeks gestation. With this test, the woman drinks a solution with known (high) concentrations of glucose, and her urine is collected 1 hour later. The urine is then tested for the presence of glucose. Those women with elevated levels of glucose in their urine must return for a more prolonged glucose tolerance test that requires collection of both urine and blood samples.

Diabetes *insipidus* is a disease characterized by an inability to produce antidiuretic hormone (ADH), a hormone that targets kidney tubule cells to promote water reabsorption (through aquaporins—*aqua-*, water + *-poros*, passage) in the distal convoluted tubule and collecting ducts. Failure to release ADH means that water reabsorption is dramatically reduced. The excess water is excreted as urine.

indicative of inflammation of the urinary tract. **Table 26.5** identifies substances that are not normally detected in urine, the clinical term associated with each pathology, and possible causes for each condition.

Table 26.5	Abnormal Urine Constituents	
Abnormal Urine Constituent	**Clinical Term**	**Possible Cause**
Bile Pigment	Bilirubinuria	Hepatitis or cirrhosis
Erythrocytes (RBC)	Hematuria	Increased glomerular permeability (hypertension, trauma, toxins)
Glucose	Glucosuria or Glycosuria	Diabetes mellitus
Hemoglobin (Hb)	Hemoglobinuria	Increased glomerular permeability (hypertension, trauma, toxins)
Ketone Bodies	Ketonuria	Diabetes mellitus; diet (low carbohydrate)
Leukocytes (WBC)	Pyuria	Urinary tract infection
Nitrites	Nitrituria	Urinary tract infection
Protein	Proteinuria	Increased glomerular permeability (hypertension, trauma, toxins)

EXERCISE 26.10

URINALYSIS

CAUTION:
The following laboratory exercises may involve the use of human urine samples. These are considered biohazard wastes and should be handled with care. Be sure to take precautionary measures while handling urine samples by wearing nitrile gloves and safety goggles. Careful disposal of biohazard wastes is required. Please consult with your laboratory instructor for further instructions on laboratory policies and procedures for handling and disposing of these substances.

In these laboratory exercises, you will analyze urine samples. You will complete the following steps using your own urine and/or artificial urine. You will observe the following physical characteristics of urine: odor, color, transparency, pH, and specific gravity. You will test for the presence of the following constituents using either combination test strips or individual test strips: glucose, protein, ketones, bilirubin, leukocytes, and hemoglobin. There is also an optional exercise (26.10B) that involves measuring specific gravity of urine with a urinometer.

EXERCISE 26.10A Urinalysis—Testing Characteristincs and Constituents of Urine

1. Obtain a urine sample. If collecting your own urine, be sure to wash your hands prior to collecting the sample. Observe all laboratory safety procedures when handling urine samples. Obtain a clean, transparent collection cup. Be sure to collect a midstream sample (at least 50 mL) to avoid possible contamination by bacteria contained within the urethra. Refrigerate any unused urine immediately.

2. Observe the **odor** of the urine sample. To do this, gently wave your hand over the cup toward your nose. Record your observations in **table 26.6**. Compare your observations with normal values presented in the table. Note that a normal odor may be aromatic but not unpleasant.

3. Observe the **color** of the urine sample. Record your observations in table 26.6, and compare your observations with normal values presented in the table. Note that a normal color is yellow, ranging from light to dark. Abnormal colors may include yellow-brown, due to bile pigments, or red to brown, due to the presence of blood.

4. Observe the **transparency** of the urine. Typically, urine is clear. A slightly cloudy or very cloudy sample may indicate the presence of substances such as bacteria or mucus. Record your observations in table 26.6, and compare your observations with normal values presented in the table.

5. Test the following urine properties or constituents using individual test strips or combination test strips (such as Chemstrip® or Multistix®): pH, specific gravity, glucose, protein, ketones, bilirubin, leukocytes, and hemoglobin. Submerge a dipstick into the urine as directed on the strip container. Remove excess liquid on the strip by touching the strip to the inside rim of the urine container. Remove the strip and wait for the recommended time. Compare the color strip to the available color scale, as directed on the container. Record your observations in table 26.6.

6. Repeat steps 2–5 for an "unknown" sample provided by your instructor.

7. Make note of any pertinent observations here:

_____

_____

_____

Table 26.6	Data Table for Urinalysis Using Test Strips		
Characteristic	**Urine Sample 1**	**Urine Sample 2**	**Normal**
Odor			Aromatic (not unpleasant)
Color			Yellow (light, medium, or dark)
Transparency			Clear
pH			4.5–8.0
Specific Gravity (Exercise 26.10A or 26.10B)			1.003–1.035
Glucose			Negative
Protein			Trace
Ketones			Negative
Bilirubin			Negative
Leukocytes			Negative
Hemoglobin			Negative

(continued on next page)

756 Chapter Twenty-Six *The Urinary System*

(continued from previous page)

EXERCISE 26.10B Urinalysis—Specific Gravity

1. Obtain a urine sample. If collecting your own urine, be sure to wash your hands prior to collecting the sample. Observe all laboratory safety procedures when handling urine samples. Obtain a clean, transparent collection cup. Be sure to collect a midstream sample (at least 50 mL) to avoid possible contamination by bacteria contained within the urethra. Refrigerate any unused urine immediately.

2. Pour the urine into a urinometer, filling it approximately two-thirds full.

3. Lower the specific gravity tube into the urinometer glass vial. Note that the specific gravity tube must float. If it does not float, continue to add urine until the urinometer glass vial floats.

4. Read the specific gravity by aligning the meniscus of the urine with the readings on the floating specific gravity tube **(figure 26.15)**. Record your measurement in table 26.6 and compare this to normal values presented in the table.

5. Repeat steps 2–4 for an "unknown" urine sample provided by your instructor.

6. Make note of any pertinent observations here:

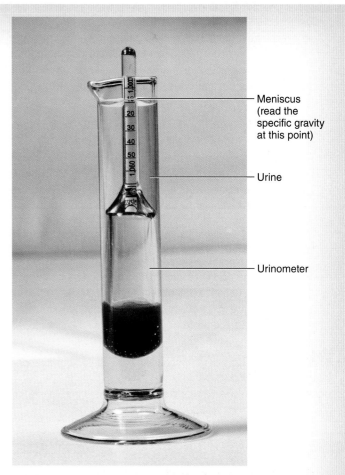

Figure 26.15 Measuring Specific Gravity of Urine Using a Urinometer. Fill the glass tube with enough urine to make the specific gravity tube float. Then, align the meniscus of the urine with the marks on the specific gravity tube to determine the specific gravity.

Acid-Base Balance

Maintaining proper blood pH (7.35–7.45) is a critical homeostatic process that involves both short- and long-term regulation and employs multiple physiological systems. As blood pH constantly changes, the first lines of defense are the buffer systems. The most robust of the chemical buffers is the **carbonic acid** buffering system, which is described in detail in chapter 24. Additional chemical buffers include proteins and phosphates, which act in a similar manner to rapidly buffer against pH changes.

The respiratory system functions as a physiological buffering system. The respiratory system is tightly linked to the carbonic acid buffering system. As discussed in chapter 24, respiratory rates directly impact the amount of carbonic acid in the blood, thereby impacting blood pH. Accumulation of carbon dioxide, whether due to **hypoventilation** or resulting from reduced gas diffusion, leads to the increased formation of carbonic acid and the possible accumulation of hydrogen ions in the blood. This results in a drop in blood pH, or **respiratory acidosis**. Conversely, **hyperventilation** results in low levels of carbon dioxide (blowing off too much carbon dioxide), which may lead to **respiratory alkalosis**. When the source of acid does not involve the respiratory system, it is considered **metabolic**. Clinically, reviewing blood pH as well as carbon dioxide and bicarbonate levels reveals whether or not the patient is in acidosis (pH < 7.35) or in alkalosis (pH > 7.45), and if the source is respiratory (normal P_{CO_2} = 35–45 mm Hg) or metabolic (normal bicarbonate = 22–26 mEq/L blood). **Table 26.7** provides a quick reference guide for classifiying respiratory and metabolic acidosis and alkalosis.

The urinary system also functions as a physiological buffering system. For example, excessive fat metabolism and the loss of bicarbonate due to excessive diarrhea both significantly lower the blood pH. As the blood becomes more acidic, the kidneys secrete excess hydrogen ions and reabsorb more bicarbonate ions in the distal convoluted tubules and collecting tubules, thereby lowering urine pH. Conversely, as the blood becomes more basic, the rate of hydrogen secretion and bicarbonate reabsorption in the kidneys decreases.

Table 26.7	Classification of Respiratory versus Metabolic Acidosis/Alkalosis		
Condition		P_{CO_2}	**Bicarbonate**
Respiratory	Acidosis	> 45 mm Hg	22–26 mEq/L blood (normal range)
	Alkalosis	< 35 mm Hg	
Metabolic	Acidosis	35–45 mm Hg (normal range)	< 22 mEq/L blood
	Alkalosis		> 26 mEq/L blood

EXERCISE 26.11

A CLINICAL CASE STUDY IN ACID-BASE BALANCE

The purpose of this exercise is to apply the concept of acid-base balance to a real scenario, or clinical case. This exercise requires information presented in this chapter, as well as previous chapters. Integrating multiple systems in a case study reinforces a concept that is so important in physiology: homeostasis.

Before you begin, state how both the respiratory system and the urinary system function to maintain homeostatic pH levels in the blood.

_____

_____

_____

_____

Consult with your laboratory instructor as to whether you will be completing this exercise individually, in pairs, or in groups. Review the topics of diffusion and osmosis and the cardiovascular, lymphatic, respiratory, digestive, and urinary systems in your textbook and in previous chapters in this laboratory manual to assist you with this case study.

Case Study:

Robert, now 60 years old, was diagnosed with type I diabetes mellitus at age 10 and has managed the disease with few complications by administering daily insulin injections. Fifteen years ago, Robert complained of extreme fatigue and his health began to decline. Within a few years, Robert was diagnosed with a failing heart, chronic hypertension (high blood pressure), and peripheral artery disease (PAD), in which the peripheral vessels become narrowed or blocked.

Just two months ago, Robert exhibited signs of intermittent confusion followed by bouts of diarrhea and intense feelings of weakness. He was unable to regulate blood glucose levels, and he began experiencing respiratory symptoms. Eventually, his breathing became labored and difficult, particularly in the evenings as he would lie down to sleep. Sometimes, these symptoms would wake him up, even after only 1 or 2 hours of sleep. Robert went to the hospital for evaluation.

Upon examination, doctors heard crackles (see chapter 24) bilaterally in his lungs. Doctors ordered a chest radiograph and a series of laboratory tests (see **table 26.8**, 1st Admission). The results of these tests indicated that Robert was suffering from pneumonitis, or inflammation of the lung tissue. Robert was given a diuretic, which prevented the reabsorption of sodium in kidney tubule cells, and a broad-spectrum antibiotic to fight the infection. He was released from the hospital just two days later.

Robert's condition worsened during the following weeks. The frequency of diarrhea increased, and he was lethargic (lacking energy). Robert had used up his supply of insulin injections when his sister found him unresponsive in his home. Emergency technicians arrived on the scene and transported him to the hospital. They recorded a blood pressure of 90/50 mm Hg (reference = 110/70 mm Hg), a pulse of 90 beats per minute, and oxygen saturation of 90%. Soon, Robert's systolic blood pressure dropped to 70 mm Hg. Robert exhibited poor inspiratory effort, and crackles were heard bilaterally.

Robert was transported to the coronary care unit for further evaluation. Doctors performed an echocardiogram, which revealed some insufficiencies in the atrioventricular and semilunar valves, narrowing of the aorta, and hypertrophy of the left ventricle. Computed tomography (CT) of the chest revealed possible infectious agents in the right upper lobe of the lungs. Doctors performed a series of laboratory tests (see table 26.8, 2nd Admission). Robert's respiratory distress and mental confusion persisted, and he experienced several more bouts of hypotension (low blood pressure). A lumbar puncture revealed elevated glucose and protein levels in the cerebrospinal fluid. Pneumonitis and hypotension persisted, and Robert was found unresponsive just days later. Efforts to resuscitate Robert were unsuccessful.

Answer the following questions regarding Robert's case.

1. Describe how Robert's narrowed aorta might have contributed to chronic respiratory problems, such as pulmonary edema. _____

_____

_____

_____

2. Based on Robert's chest radiograph results, describe how gas exchange across the respiratory membrane might have been affected by his chronic respiratory problems.

_____

_____

_____

_____

3. Justify Robert's blood and urine glucose levels (1st admission) based on the details of his case. Based on this data, estimate the volume of Robert's urinary output at that time. _____

_____

_____

_____

4. Discuss what Robert's glomerular filtration rate revealed about his systemic blood pressure. _____

_____

_____

_____

5. Describe how continued diarrhea might have influenced Robert's blood pH. What compensatory mechanisms could have been initiated? _____

_____

_____

_____

(continued on next page)

(continued from previous page)

6. Describe how chronic respiratory problems might have influenced Robert's blood pH. What compensatory mechanisms could have been initiated? _____

7. Based on the data provided, did Robert suffer from respiratory or metabolic acidosis or alkalosis? Justify your answer. _____

Table 26.8	Data Table for Acid-Base Case Study			
Test	**Variable**	**Reference Range**	**1st Admission**	**2nd Admission**
Arterial Blood Gases	P_{CO_2}	35–45 mm Hg	50	65
	pH	7.35–7.45	7.2	7.1
	P_{O_2}	80–100 mm Hg	120	160
Blood	Bicarbonate (mEq/L)	22–26	25	25
	Chloride (mmol/liter)	100–108	90	105
	Glomerular filtration rate (mL/min)	> 60	20	10
	Glucose (mg/dl)	70–110	680	170
	Hematocrit (%)	42.0–52.0 (men)	35	30
	Hemoglobin (g/dl)	13–18 (men)	10.5	8.5
	Leukocytes (per mm^3)	5000–10,000	9000	10,000
	Potassium (mmol/liter)	3.4–4.8	6	5
	Sodium (mmol/liter)	135–145	130	135
Urine	Albumin (mg/dl)	Negative	1+	2+
	Bacteria	Negative	Moderate	Many
	Color	Yellow	Yellow	Other
	Erythrocytes	0–2	10–25	10–25
	Glucose (mg/dl)	Negative	1000	Negative
	Ketones (mg/dl)	Negative	3+	1+
	Leukocytes	0–2	250	Packed
	pH	4.5–9.0	5	4.5
	Specific gravity	1.003–1.035	1.02	> 1.035

Chapter 26: The Urinary System

Name: _____
Date: _____ Section: _____

POST-LABORATORY WORKSHEET

The ❶ corresponds to the Learning Objective(s) listed in the chapter opener outline.

Do You Know the Basics?

Exercise 26.1: Histology of the Renal Cortex

1. Label the parts of a nephron and the blood supply to the nephron in the figure below: ❶

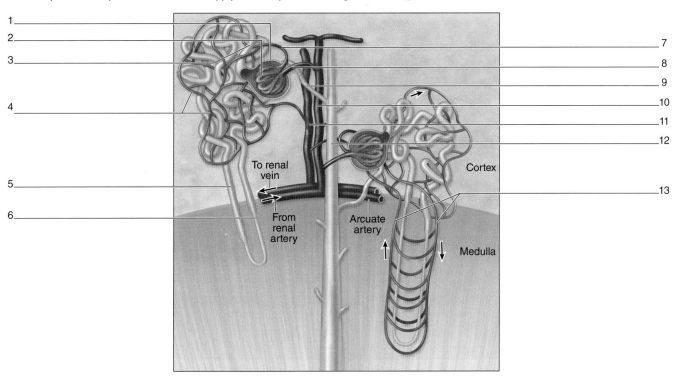

1. _____
2. _____
3. _____
4. _____
5. _____
6. _____
7. _____
8. _____
9. _____
10. _____
11. _____
12. _____
13. _____

2. Fill in the table below with the type of epithelium lining each structure, and a brief description of the function of each structure. ❷ ❸

Structure	Epithelium	Function
Glomerulus		
Visceral Layer of Glomerular Capsule		
Parietal Layer of Glomerular Capsule		
PCT		
Descending Limb of Nephron Loop		
Ascending Limb of Nephron Loop—Thin		
Ascending Limb of Nephron Loop—Thick		
DCT		
Collecting Duct		
Ureter		
Urinary Bladder		
Urethra		

760 Chapter Twenty-Six *The Urinary System*

3. Identify the two components of the renal corpuscle. ❹

   a. _____

   b. _____

4. The filtration membrane contains the following three structures: ❺

   a. _____

   b. _____

   c. _____

Exercise 26.2: Histology of the Renal Medulla

5. Describe the epithelium that lines the nephron loops and collecting ducts. ❽ _____
_____
_____

Exercise 26.3: Histology of the Ureters

6. Describe the histology of the ureter and explain how the ureter transports urine. ❾ _____
_____
_____

7. For each photomicrograph below, identify the organ. Then explain the characteristics you used to determine what organ it is. ❿ ⓬

   (a)

   Organ: _____

   Identifying characteristics: _____
   _____
   _____
   _____
   _____
   _____
   _____

   (b)

   Organ: _____

   Identifying characteristics: _____
   _____
   _____
   _____
   _____
   _____
   _____

   (c)

   Organ: _____

   Identifying characteristics: _____
   _____
   _____
   _____
   _____
   _____
   _____

Chapter Twenty-Six *The Urinary System* 761

Exercise 26.4: Histology of the Urinary Bladder

8. Describe the histology of the urinary bladder. ⓫ _____

_____

Exercise 26.5: Gross Anatomy of the Kidney

9. The layer of dense irregular connective tissue enclosing the kidney is the: _____

_____

_____ . ⓭

10. Which of the two kidneys is located more superiorly than the other (and why)? ⓮ _____

_____

_____

11. A surgeon has the job of removing a patient's kidney. His approach to the kidney will be to first cut through skin and adipose tissue of the lower back, remove a section of the most inferior ribs, and then enter the retroperitoneal space. Once the surgeon has entered the retroperitoneal space, what are the layers of tissue (in order) that the surgeon must cut through to reach the kidney? ⓯ _____

_____

_____

12. Define *retroperitoneal*: ⓰ _____

_____

Exercise 26.6: Blood Supply to the Kidney

13. The _____ supplies oxygen-rich blood to each kidney, whereas the _____ drains blood from each kidney to the inferior vena cava. ⓱

14. Trace the path of a drop of blood through the kidney, from its entry via the renal artery to its exit via the renal vein. ⓲ _____

_____

_____

15. Explain how the microscopic structures of the nephron create the gross appearance of the outer cortex and inner medulla of the kidney. ⓳

_____

_____

_____

Exercise 26.7: Urine-Draining Structures within the Kidney

16. In the space below, trace the path of a drop of filtrate, in words, from its origin in the capsular space to its exit (as urine) from the kidney via the renal

   pelvis. ㉑ _____

_____

Exercise 26.8: Gross Anatomy of the Ureters

17. In the space below, trace the path of a drop of urine, in words, from its exit from the kidney (via the renal pelvis) to its removal from the body through the

   urethra. ㉒ _____

_____

_____

Chapter Twenty-Six The Urinary System

18. Label the figure below.

1 _____

2 _____

Prostate gland _____

3 _____

4 _____

5 _____

6 _____

7 _____

8 _____

9 _____

10 _____

Exercise 26.9: Gross Anatomy of the Urinary Bladder and Urethra

19. What is the detrusor muscle? _____
    _____
    _____

20. What three structures form the boundaries of the urinary trigone?

    a. _____

    b. _____

    c. _____

21. Name the three sections of the male urethra.

    a. _____

    b. _____

    c. _____

22. Explain the structural and functional differences between the internal and external urethral sphincters (i.e., what type of muscle each is composed of and what type of nervous control each sphincter is under). _____
    _____
    _____

Exercise 26.10: Urinalysis

23. The nephron accomplishes the following three processes in the production of urine: ㉗

    a. _____

    b. _____

    c. _____

24. Discuss several abnormal components of urine and identify potential causes of each. ㉘ _____

    _____

    _____

    _____

25. Describe what high specific gravity suggests about the constituents of a urine sample. ㉚ _____

    _____

    _____

26. The urinary system aids in _____-term regulation of acid-base balance. ㉛

Exercise 26.11: A Clinical Case Study in Acid-Base Balance

27. Identify the two physiological compensatory mechanisms employed to regulate acid-base balance in the blood. ㉜

    a. _____

    b. _____

Can You Apply What You've Learned?

28. Describe why muscular contractions are needed to transport urine through the ureters. _____

    _____

    _____

29. Patients with chronic hypertension are at risk for suffering damage to the kidneys. More specifically, increased systemic blood pressure stresses the filtration membrane of the nephron. Describe the specific structures of the nephron that might be damaged in a patient with hypertension. _____

    _____

    _____

    _____

30. Doctors will often first advise patients with hypertension to eat a diet low in sodium. Discuss why this may be beneficial in reducing systemic blood pressure. (Hint: Remember that an increased blood volume will correspond to an increase in systemic blood pressure.) _____

    _____

    _____

    _____

31. Patients in kidney failure require dialysis to perform the processes that are normally accomplished by the nephron. A dialysis membrane does this by mimicking the filtration membrane in the nephron. Describe the essential function typically accomplished by the kidneys that is replaced by the dialysis machine. _____

    _____

32. a. Describe why a person suffering from kidney failure would experience edema, which would be alleviated by dialysis. _____

    _____

    _____

b. What abnormal constituent of urine might be present in a urinalysis of a patient suffering from kidney failure? _____

_____

_____

33. Among its many effects, alcohol inhibits ADH secretion from the posterior pituitary gland. Describe the typical action of this hormone. Then discuss how ingestion of alcohol will affect urine formation. _____

_____

_____

34. A patient with suspected internal bleeding is admitted to the hospital. His systemic blood pressure is 80/50 mm Hg. Predict the patient's GFR given his current systemic blood pressure. Justify your answer. _____

_____

_____

Can You Synthesize What You've Learned?

35. In severe cases of hypertension, an angiotensin-converting enzyme (ACE) inhibitor may be prescribed for a patient. The enzyme ACE converts angiotensin I to angiotensin II. One target of angiotensin II is the adrenal cortex, where angiotensin II stimulates the release of the hormone aldosterone (see chapter 19). Aldosterone increases sodium reabsorption by kidney tubule cells. Describe how an ACE inhibitor will influence blood volume and blood pressure (as it relates to aldosterone). _____

_____

_____

36. In addition to stimulating the release of aldosterone, angiotensin II is a widespread vasoconstrictor. This means that smooth muscle in the walls of blood vessels will be stimulated to contract when angiotensin II is formed.

    a. Describe what will happen to GFR when angiotensin II stimulates the afferent arterioles within the nephron to vasoconstrict. _____

    _____

    _____

    b. Describe what effect this widespread vasoconstriction will have on systemic blood pressure. _____

    _____

37. Susan, a 25-year-old female, contracted food poisoning likely due to undercooked eggs that she recently ate. She has been vomiting violently and cannot keep food or liquids down. Susan sought medical attention. Doctors measured her blood pH at 7.5, P_{CO_2} at 40 mm Hg, P_{O_2} at 95 mm Hg, O_2 saturation at 97%, and bicarbonate at 32 mEq/L.

    a. Classify Susan's case as respiratory acidosis, respiratory alkalosis, metabolic acidosis, or metabolic alkalosis, and justify your answer. _____

    _____

    _____

    b. Describe compensatory mechanisms that may have been initiated to adjust Susan's blood pH back toward normal homeostatic levels. _____

    _____

PART VI | REPRODUCTION

CHAPTER 27

The Reproductive System and Early Development

OUTLINE AND LEARNING OBJECTIVES

Histology 768

Female Reproductive System 768

EXERCISE 27.1: HISTOLOGY OF THE OVARY 768
- ❶ Identify the ovary and associated structures when viewed through a microscope
- ❷ Explain the process of gametogenesis in the female
- ❸ Recognize the stages of ovarian follicle development on histology slides of the ovary

EXERCISE 27.2: HISTOLOGY OF THE UTERINE TUBES 772
- ❹ Identify the uterine tubes and associated structures when viewed through a microscope
- ❺ Describe the wall structure of the uterine tubes
- ❻ Classify the epithelium that lines the uterine tubes

EXERCISE 27.3: HISTOLOGY OF THE UTERINE WALL 774
- ❼ Identify the uterus and layers of the uterine wall when viewed through a microscope
- ❽ Describe the tissues that compose the three layers of the uterine wall

EXERCISE 27.4: HISTOLOGY OF THE VAGINAL WALL 775
- ❾ Identify the vagina and layers of the vaginal wall when viewed through a microscope
- ❿ Classify the epithelium that lines the vaginal wall

Male Reproductive System 776

EXERCISE 27.5: HISTOLOGY OF SEMINIFEROUS TUBULES 776
- ⓫ Identify seminiferous tubules and associated cells when viewed through a microscope
- ⓬ Explain the process of gametogenesis in the male
- ⓭ Recognize the stages of sperm development in histology slides of the testes
- ⓮ Describe the location and function of interstitial cells and sustentacular cells of the testes

EXERCISE 27.6: HISTOLOGY OF THE EPIDIDYMIS 778
- ⓯ Identify the epididymis when viewed through a microscope
- ⓰ Classify the epithelium that lines the epididymis
- ⓱ Describe the structure and function of stereocilia

EXERCISE 27.7: HISTOLOGY OF THE DUCTUS DEFERENS 779
- ⓲ Identify the ductus deferens when viewed through a microscope
- ⓳ Describe the structures that compose the walls of the ductus deferens
- ⓴ Classify the epithelium that lines the ductus deferens

EXERCISE 27.8: HISTOLOGY OF SEMINAL VESICLES 780
- ㉑ Identify the seminal vesicles when viewed through a microscope
- ㉒ Classify the epithelium that lines the seminal vesicles

EXERCISE 27.9: HISTOLOGY OF THE PROSTATE GLAND 781
- ㉓ Identify the prostate gland when viewed through a microscope
- ㉔ Identify corpora aranacea in slides of the prostate gland

EXERCISE 27.10: HISTOLOGY OF THE PENIS 782
- ㉕ Identify the penis when viewed through a microscope
- ㉖ Describe the components that are visible in a cross-sectional view of the penis

Gross Anatomy 783

Female Reproductive System 783

EXERCISE 27.11: GROSS ANATOMY OF THE OVARY, UTERINE TUBES, UTERUS, AND SUSPENSORY LIGAMENTS 783
- ㉗ Describe the structure and function of the ovary, uterine tubes, and uterus
- ㉘ Locate the eight suspensory ligaments of the ovary, uterine tubes, and uterus on classroom models of the female reproductive system
- ㉙ Identify the parts of the uterine tubes and uterus on classroom models of the female reproductive system

EXERCISE 27.12: GROSS ANATOMY OF THE FEMALE BREAST 786
- ㉚ Describe the structure and function of the female breast

Male Reproductive System 788

EXERCISE 27.13: GROSS ANATOMY OF THE SCROTUM, TESTIS, SPERMATIC CORD, AND PENIS 789
- ㉛ Describe the location and function of the epididymis
- ㉜ Describe the location and function of the ductus deferens
- ㉝ Describe the location and function of the seminal vesicles
- ㉞ Describe the location and function of the prostate gland
- ㉟ Identify the three parts of the male urethra on classroom models of the male reproductive system
- ㊱ Describe the relationship between the parts of the male urethra and the male accessory reproductive structures such as the prostate, seminal vesicles, and bulbourethral glands
- ㊲ Explain the contributions of accessory male reproductive glands (seminiferous tubules and prostate) to semen
- ㊳ Describe the components that make up the spermatic cord
- ㊴ Trace the pathway of sperm through the male reproductive tract during ejaculation, starting at the epididymis

 **MODULE 14: REPRODUCTIVE SYSTEM**

| Physiology 791 | Fertilization and Development 797 |

Reproductive Physiology 791

EXERCISE 27.14: A CLINICAL CASE IN REPRODUCTIVE PHYSIOLOGY 795
- **40** Describe the process of meiosis for the production of gametes
- **41** Discuss hormonal regulation of the ovarian and uterine cycles
- **42** Apply concepts from the endocrine, cardiovascular, digestive, urinary, and reproductive systems to a clinical case study

EXERCISE 27.15: EARLY DEVELOPMENT: FERTILIZATION AND ZYGOTE FORMATION 797
- **43** Outline the chronology of events in pre-embryonic development

EXERCISE 27.16: EARLY DEVELOPMENT: EMBRYONIC DEVELOPMENT 798
- **44** Identify the events in embryonic development
- **45** Identify the structures of an embryo

INTRODUCTION

Of all the systems of the human body, the **reproductive system** (*re-*, again, + *productus*, to lead forth) is the only system that is not required for the survival of the human body. Instead, its function is to ensure the survival of the species by allowing for the creation of another human being. Thus far in this course you have discovered much about the amazing complexity and beauty of the human body. We will now turn our attention to the process by which this amazingly complex human comes into being. It requires two separate **sexes:** male and female. Each sex produces **gametes** (sperm in males; ova in females), which are haploid cells that combine to form a **zygote,** the first step in the development of a new human. Male and female gametes are both formed by **meiosis,** which occurs in the **gonads** (*gone*, seed). The human body invests a great deal of energy in the formation of gametes and in the maintenance of characteristics that are meant to attract members of the opposite sex. This much energy expenditure may seem counterintuitive when one thinks about homeostasis as the primary driving force for most organ system functions. However, when you consider the consequences of humans failing to come together for reproducing the species (that is, extinction of the species), the amount of energy invested makes a great deal more sense.

Because the reproductive structures in males and females are different, the structures can seem to be more challenging to learn (because there are more of them). However, all human embryos develop from the same overall plan. Certain embryonic structures degrade in males but not in females. On the other hand, many embryonic structures persist in both males and females—forming apparently very different adult structures. Such structures are called **homologues** (*homos*, the same, + *logos*, relation). Ovaries and testes are examples of homologous structures. As you explore the male and female reproductive systems, pay particular attention to homologous structures. This will not only simplify the task of learning reproductive system structures, it will also help you to appreciate the similarities between the male and female reproductive systems.

In this laboratory session you will explore the histological and gross structure of the male and female reproductive systems, with a focus on the histology of the primary reproductive organs of both females and males, the *ovaries* and *testes*. Observing the histology of the gonads allows us to best see and understand the stages of gametogenesis, which are covered in detail in the physiology section of this chapter. The focus of your observations of the gross anatomy of the reproductive systems will be to understand the regional relationships among gross reproductive structures and structures of other body systems, as well as to understand homologous relationships between male and female reproductive structures. This study of anatomical structures will be related to hormonal regulation of gametogenesis for both males and females. You will then apply these concepts to a clinical case involving infertility. Finally, you will explore the processes involved in early development, from fertilization to embryonic development.

Chapter 27: The Reproductive System and Early Development

PRE-LABORATORY WORKSHEET

Name: _____
Date: _____ Section: _____

Also available at www.connect.mcgraw-hill.com

1. List three functions of the reproductive system.

    a. _____

    b. _____

    c. _____

2. The female gonad is the _____, while the male gonad is the _____.

3. The gamete produced in females is the _____, while the gamete produced in males is the _____.

4. In the space below, make a drawing depicting the stages of meiosis.

Stages of Meiosis

Histology

Female Reproductive System

The female reproductive system has two important functions: produce the female gametes (ova) and support and protect the developing embryo/fetus that is formed when an ovum becomes fertilized by a sperm. Components of the female reproductive system include the ovaries, uterine tubes, uterus, vagina, and mammary glands. Exercises 27.1 to 27.4 explore the internal detail and wall structures of the ovary, uterine tube, uterus, and vagina. Mammary glands will be covered in the gross anatomy portion of this chapter.

EXERCISE 27.1

HISTOLOGY OF THE OVARY

The **ovary** is the primary reproductive organ in the female and functions in the production of the ova (eggs) and the female sex steroid hormones **estrogen** and **progesterone**. The ovaries contain several **ovarian follicles,** all at various stages of development. Each follicle contains a developing **oocyte** **(table 27.1)**. The structure and function of the ovary is best viewed at the histological level so you can appreciate the characteristics of ovarian follicles at each stage of development.

1. Obtain a histology slide of an ovary and place it on the microscope stage.

2. Bring the tissue sample into focus on low power and identify the outer **cortex** and inner **medulla** of the ovary **(figure 27.1)**. Next, move the microscope stage so the ovarian cortex is at the center of the field of view and then change to high power.

3. Observe the outermost region of the ovarian cortex. Locate the two layers of tissue that compose the outer coverings of the ovary: the germinal epithelium and the tunica albuginea **(figure 27.2)**. The **germinal epithelium** (*germen*, a sprout) is a simple cuboidal epithelium that forms the outermost covering of the ovary. The name *germinal* refers to the fact that scientists once thought that the germ cells (oocytes) were formed from this epithelial layer. Most ovarian cancers arise in the germinal epithelium. Deep to the germinal epithelium is the **tunica albuginea** (*tunica*, a coat, + *albugineus*, white spot), which is composed of dense irregular connective tissue.

4. Focus on the cortex and identify follicles in each stage of follicular development listed in **table 27.2**.

5. Using tables 27.1 and 27.2 and figures 27.1 and 27.2 as guides, locate the following structures on the slide of the ovary:

   ☐ corpus albicans ☐ secondary follicle
   ☐ corpus luteum ☐ secondary oocyte
   ☐ germinal epithelium ☐ tunica albuginea
   ☐ primary follicle ☐ vesicular (Graafian) follicle
   ☐ primary oocyte
   ☐ primordial follicle

Table 27.1 Developmental Stages of an Oocyte

Cell Name/ Oocyte Stage	Description and Function	Stage of Mitosis/ Meiosis	Ploidy	Word Origin
Oogonia	Primitive germ cells that undergo mitosis in the second to fifth months of embryonic life form oogonia that will subsequently develop into primary oocytes. Approximately 70% of the primary oocytes will degenerate in a process called atresia.	Formed by mitosis; undergo mitosis to form primary oocytes	Diploid (2n)	*oon*, egg, + *gonia*, generation
Primary Oocyte	Oogonia has replicated DNA and has begun the process of meiosis and arrest in prophase of meiosis I. Primary oocytes will not undergo subsequent meiotic divisions unless the follicle is stimulated to mature by follicle-stimulating hormone (FSH) and luteinizing hormone (LH).	Arrested in Prophase I of meiosis	Diploid (2n)	*primary*, first, + *oon*, egg, + *kytos*, cell
Secondary Oocyte	Under the influence of FSH and LH, primary oocytes complete meiosis I to become secondary oocytes, which will not undergo subsequent meiotic divisions unless fertilization occurs	Arrested in Metaphase II of meiosis	Haploid (1n)	*secondary*, second, + *oon*, egg, + *kytos*, cell
Definitive Oocyte	Upon fertilization, the secondary oocyte undergoes the second meiotic division to become an ovum	Formed after secondary oocyte is fertilized and completes meiosis	Haploid (1n)	*oon*, egg, + *kytos*, cell

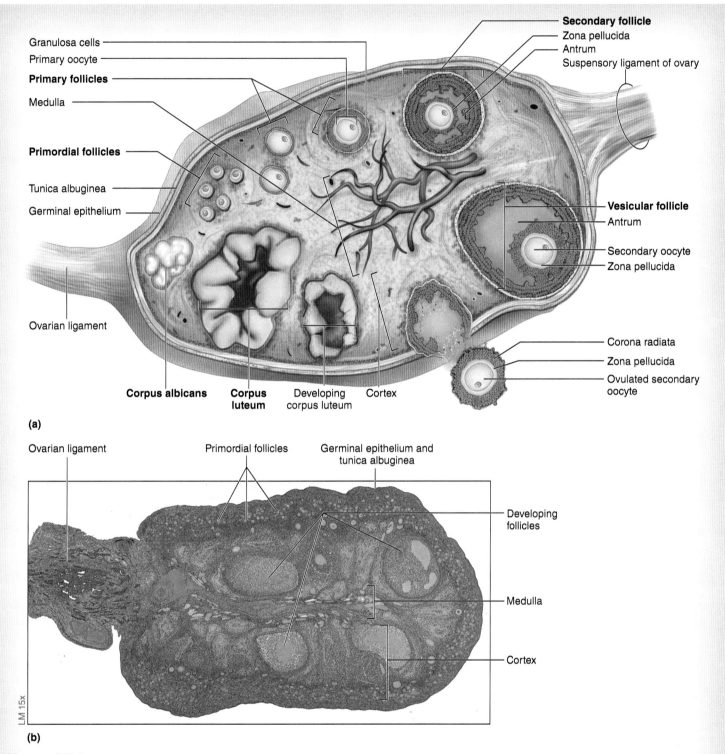

Figure 27.1   The Ovary.   (a) The ovary is the primary reproductive organ of the female and produces the female gametes, the ova. (b) Histological structure of the ovary demonstrating the outer cortex containing follicles in various stages of development, and the inner medulla, which consists mainly of blood vessels and nerves.

6. Next move the microscope stage so a vesicular follicle is in the center of the field of view. A **vesicular (Graafian) follicle (figure 27.3)** is a mature follicle that is ready to be released during ovulation, after stimulation by a surge in luteinizing hormone (LH) secreted by the anterior pituitary gland. **Table 27.3** lists the histological features of a vesicular follicle. Using table 27.3 and figure 27.3 as guides, identify the following parts of the vesicular follicle:

- ☐ antrum
- ☐ corona radiata
- ☐ cumulus oophorus
- ☐ granulosa cells
- ☐ secondary oocyte
- ☐ theca externa
- ☐ theca interna
- ☐ zona pellucida

(continued on next page)

(continued from previous page)

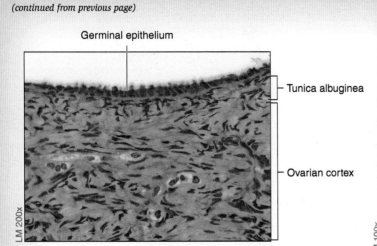

Figure 27.2 Germinal Epithelium and Tunica Albuginea of the Ovary. The germinal epithelium is the most common origin site for ovarian cancers. The tunica albuginea is a "white coat" of dense irregular connective tissue that surrounds the entire ovary internal to the germinal epithelium.

7. In the spaces to the right, sketch primordial, primary, secondary, and vesicular follicles as seen through the microscope.

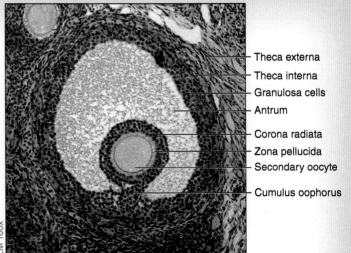

Figure 27.3 Vesicular (Graafian) Follicle. A vesicular follicle contains a secondary oocyte and a single antrum.

Primordial follicle

_____ ×

Table 27.2	Developmental Stages of Ovarian Follicles		
Photograph			
Follicle Stage	Primordial Follicle	Primary Follicle	Secondary (Maturing) Follicle
Description and Function	Contains a primary oocyte surrounded by a single layer of flattened follicular cells	Contains a primary oocyte surrounded by one or more layers of cuboidal follicular cells	Contains a full-size primary oocyte surrounded by an extracellular glycoprotein coat (the zona pellucida), which separates it from a peripheral layer of follicular cells. Contains one or more fluid-filled antra. Thecal cells develop into internal and external layers.
Oocyte Stage	Primary oocyte	Primary oocyte	Primary oocyte
Word Origins	*primus*, first, + *ordior*, to begin, + *folliculus*, a small sac	*primus*, first, + *ordior*, to begin, + *folliculus*, a small sac	*secunda*, second, + *ordior*, to begin, + *folliculus*, a small sac

Chapter Twenty-Seven *The Reproductive System and Early Development*

Table 27.3 — Components of a Vesicular Follicle

Structure	Description and Function	Word Origin
Antrum	The fluid-filled space in the center of the follicle	*antron*, a cave
Corona Radiata	A single layer of columnar cells derived from the cumulus oophorus that attach to the zona pellucida of the oocyte	*corona*, a crown, + *radiatus*, to shine
Cumulus Oophorus	A "mound" of granulosa cells that supports and surrounds the ovum within the follicle	*cumulus*, a heap, + *oophoron*, ovary
Granulosa Cells	Epithelial cells lining the follicle that will become the luteal cells of the corpus luteum after ovulation. They secrete the liquor folliculi, which is the fluid that fills the antrum.	*granulum*, a small grain
Theca Externa	The external fibrous layer of a well-developed vesicular follicle. The cells and fibers are arranged in concentric layers.	*theca*, a box, + *externus*, external
Theca Interna	The inner cellular layer of the vesicular follicle; these cells secrete androgen that is converted to estrogen by granulosa cells	*theca*, a box, + *internus*, internal
Zona Pellucida	A thick coat of glycoproteins that surrounds the oocyte	*zona*, zone, + *pellucidus*, clear

Primary follicle   *Secondary follicle*   *Vesicular follicle*

_____ ×   _____ ×   _____ ×

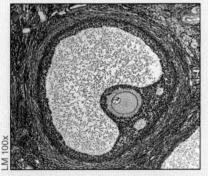

Vesicular (Graafian) Follicle

Has nearly the same structure and function as a secondary follicle, but contains a secondary oocyte and only a single large antrum

Secondary oocyte

vesicular, a blister, + *folliculus*, a small sac

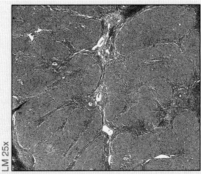

Corpus Luteum

A "yellow body" that gets its color from the steroid hormones it secretes, which are lipids. After ovulation, the theca interna cells enlarge and continue to secrete the steroid hormones estrogen and progesterone. In the center of the corpus luteum is a large blood clot.

NA

corpus, body, + *luteus*, yellow

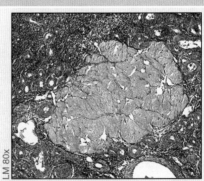

Corpus Albicans

If fertilization does not occur, the corpus luteum stops secreting hormones after two weeks and becomes a smaller, *inactive* "white body" consisting mainly of scar tissue

NA

corpus, body, + *albus*, white

(continued on next page)

(continued from previous page)

8. Using table 27.2 as a guide, locate a corpus luteum and corpus albicans on the slide.

9. In the spaces below, sketch a corpus luteum and a corpus albicans as seen through the microscope.

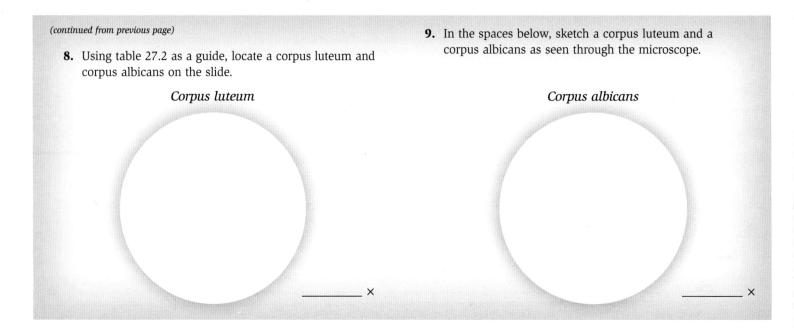

Corpus luteum ____×

Corpus albicans ____×

EXERCISE 27.2

HISTOLOGY OF THE UTERINE TUBES

The **uterine tubes** (fallopian tubes, or oviducts) are long muscular tubes that transport the ovum from the ovary to the uterus. The wall of the uterine tube is composed of three layers: mucosa, muscularis, and serosa.

1. Obtain a histology slide demonstrating a cross-sectional view of a uterine tube.

2. Place the slide on the microscope stage and bring the tissue sample into focus on low power. Note the highly folded **mucosa** of the tube **(figure 27.4)**.

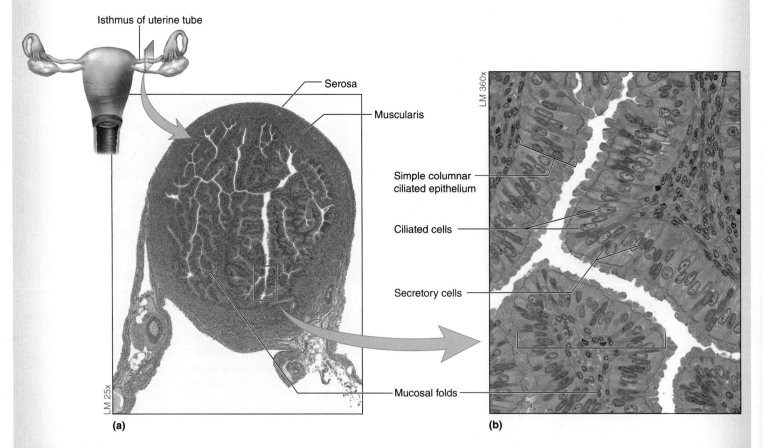

Figure 27.4 Histology of the Uterine Tubes. (a) Cross section through the isthmus of the uterine tubes demonstrating the many folds of the mucosa. (b) The epithelial cells lining the uterine tubes contain two cell types: ciliated cells and secretory cells. Ciliated cells have light nuclei, whereas secretory cells have dark nuclei.

3. Move the microscope stage until you have an area of folded mucosa at the center of the field of view and then switch to high power. The epithelium of the uterine tube contains two cell types, **ciliated epithelial cells** and **secretory cells**. Table 27.4 describes the histological components of the uterine tubes. Focus on the epithelial cells lining the uterine tube, and distinguish ciliated cells from secretory cells (figure 27.4*b*).

4. Using table 27.4 and figure 27.4 as guides, identify the following structures on the slide of the uterine tube:

   ☐ ciliated cells   ☐ secretory cells
   ☐ mucosal folds   ☐ serosa
   ☐ muscularis

5. In the space below, sketch a cross section of the uterine tube, as seen through the microscope. Be sure to label the structures listed in step 4 in your drawing.

_____ ×

Table 27.4	Components of the Uterine Tube	
Uterine Tube Structure	**Description and Function**	**Word Origin**
Ampulla	The wide part of the uterine tube located medial to the infundibulum. Its mucosa is highly folded and lined with simple ciliated columnar epithelium and secretory cells. Fertilization most commonly occurs in the ampulla.	*ampulla*, a two-handled bottle
Ciliated Cells	Simple ciliated columnar epithelial cells; cilia beat toward the uterus to transport the ovum to the uterus. Nuclei stain lighter than those of the secretory cells.	*cilium*, eyelash
Infundibulum	The funnel-like expansion of the ovarian end of the uterine tube. Fingerlike extensions of the infundibulum are fimbriae (*fimbria*, fringe).	*infundibulum*, a funnel
Isthmus	The narrow part of the uterine tube located right next to the uterus	*isthmos*, a constriction
Mucosa	The inner layer of the wall of the uterine tube. It is highly folded in the infundibulum and ampulla of the uterine tube. Contains ciliated columnar epithelium and a layer of areolar connective tissue.	*mucosus*, mucous
Muscularis	The middle layer of the wall of the uterine tube. It consists of smooth muscle whose contraction assists in transporting the ovum toward the uterus.	*musculus*, mouse (referring to muscle)
Secretory Cells	Nonciliated columnar epithelial cells that promote the activation of spermatozoa (capacitation) and provide nourishment for the ovum. Nuclei stain darker than those of the ciliated cells.	*secretus*, to separate
Serosa	The outer layer of the uterine tube. It consists of simple squamous epithelium (a fold of peritoneum).	*serosus*, serous
Uterine Part (Interstitial Segment)	The part of the uterine tube that penetrates the wall of the uterus	*intra–*, within, + *muralis*, wall (*inter*, between, + *sisto*, to stand)

EXERCISE 27.3

HISTOLOGY OF THE UTERINE WALL

The **uterus** is a muscular structure whose primary function is to house the developing fetus.

1. Obtain a histology slide that includes a portion of the uterine wall. Place it on the microscope stage and bring the tissue sample into focus on low power.

2. The uterine wall is composed of three layers: endometrium, myometrium, and perimetrium. The inner lining of the uterus, the **endometrium,** goes through its own cycle of growth during the ovarian cycle. Each time a woman's ovary undergoes a single ovarian cycle, the endometrium becomes prepared for the possibility that a fertilized egg will become implanted. If no implantation occurs, the innermost layer of the endometrium, the **functional layer,** sloughs off in the process of **menstruation** (*menstruus*, monthly). **Table 27.5** describes the layers of the uterine wall and the components that make up each layer. **Table 27.6** summarizes three phases of the menstrual cycle and demonstrates the appearance of the endometrium during each phase.

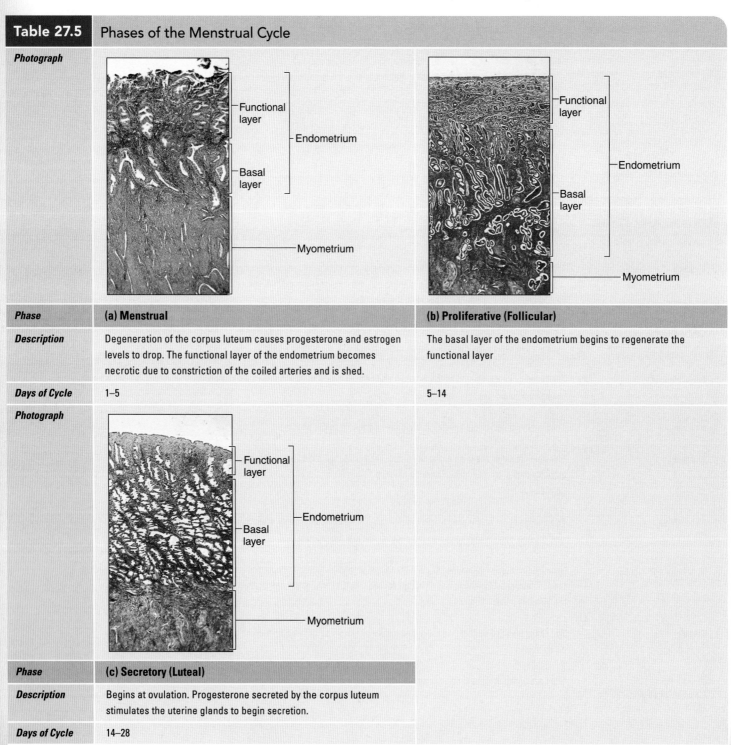

Table 27.5 Phases of the Menstrual Cycle

	(a) Menstrual	(b) Proliferative (Follicular)
Description	Degeneration of the corpus luteum causes progesterone and estrogen levels to drop. The functional layer of the endometrium becomes necrotic due to constriction of the coiled arteries and is shed.	The basal layer of the endometrium begins to regenerate the functional layer
Days of Cycle	1–5	5–14

	(c) Secretory (Luteal)
Description	Begins at ovulation. Progesterone secreted by the corpus luteum stimulates the uterine glands to begin secretion.
Days of Cycle	14–28

3. Using tables 27.5 and 27.6 as guides, identify the following structures on the slide of the uterus:
   - ☐ endometrium (basal layer)
   - ☐ endometrium (functional layer)
   - ☐ myometrium
   - ☐ perimetrium

4. Move the microscope stage so the functional layer of the endometrium is in the center of the field of view. Switch to high power and identify **uterine glands.**

Table 27.6	Wall Layers of the Uterus	
Wall Layer	**Description and Function**	**Word Origin**
Endometrium	The mucous membrane composing the inner layer of the uterine wall. Consists of simple columnar epithelium and a lamina propria with simple tubular uterine glands. The structure, thickness, and state of the endometrium undergoes marked changes during the menstrual cycle.	*endon*, within, + *metra*, uterus
Functional Layer (Stratum Functionalis)	The apical layer of the endometrium. Most of this layer is shed during menstruation.	*stratum*, a layer, + *functus*, to perform
Basal Layer (Stratum Basalis)	The basal layer of the endometrium. It undergoes minimal changes during the menstrual cycle and serves as the basis for regrowth of the more apical stratum functionalis.	*stratum*, a layer, + *basalis*, basal
Myometrium	The muscular wall of the uterus composed of three layers of smooth muscle	*mys*, muscle, + *metra*, uterus
Perimetrium	The outermost covering of the uterus; a serous membrane formed from peritoneum	*peri-*, around, + *metra*, uterus

EXERCISE 27.4

HISTOLOGY OF THE VAGINAL WALL

The **vagina** (*vagina*, sheath) is a muscular tube that receives the penis during intercourse and serves as the birth canal during parturition (*parturio*, to be in labor).

1. Obtain a histology slide demonstrating a portion of the vaginal wall. Place it on the microscope stage and bring the tissue sample into focus on low power. The walls of the vagina are composed of a mucosa and muscularis; the mucosa is lined with a nonkeratinized stratified squamous epithelium (**figure 27.5**).

2. Using figure 27.5 as a guide, identify the following structures on the slide of the vagina:
   - ☐ lamina propria
   - ☐ mucosa
   - ☐ muscularis
   - ☐ nonkeratinized stratified squamous epithelium

3. *Optional Activity:* **AP|R** 14: Reproductive System— Watch the "Female Reproductive System Overview" animation to review the female reproductive organs and their histological features.

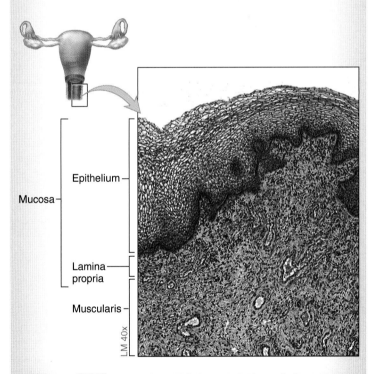

Figure 27.5 Histology of the Vaginal Epithelium. The epithelium is nonkeratinized stratified squamous.

Male Reproductive System

The male reproductive system consists of the testes, epididymis, ductus deferens, seminal vesicles, prostate gland, bulbourethral glands, and penis. Unlike the female reproductive system, which must support and nourish a developing fetus, the primary functions of the male reproductive system are to produce the male gamete (sperm) and provide nourishment for the sperm and a mechanism for the sperm to be delivered to the female reproductive tract. Testes also produce testosterone hormone. Like female reproductive organs, male reproductive structures have unique histological features that will help you identify them through the microscope.

EXERCISE 27.5

HISTOLOGY OF THE SEMINIFEROUS TUBULES

1. Obtain a slide demonstrating seminiferous tubules and place it on the microscope stage. The **testes** are the primary reproductive organ in the male, and function in the production of sperm and the male sex steroid hormone **testosterone**. Within each testis are several hundred, coiled **seminiferous tubules** (figure 27.6), which are the site of sperm production, or **spermatogenesis** (*sperma*, seed, + *genesis*, origin). Within each tubule, **spermatogonia** undergo successive meiotic divisions as they move from the basal to the apical surface of the tubule epithelium. **Table 27.7** describes the developmental stages of the sperm, and **table 27.8** describes the structure and function of the accessory cell types located within the testes.

2. Observe the slide on low power and identify several cross sections of the seminiferous tubules. Move the microscope stage so that one tubule is at the center of the field of view, and change to high power.

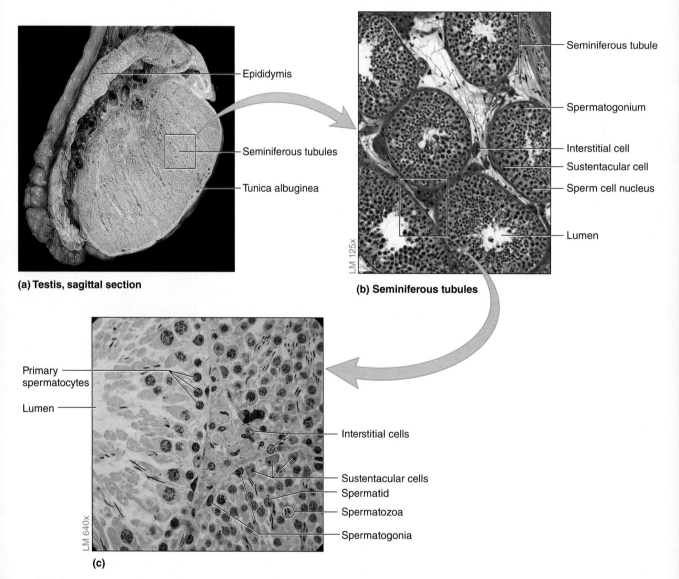

Figure 27.6 Testes and Seminiferous Tubules. (a) Testes and epididymis showing seminiferous tubules within the testes and tunica albuginea surrounding the testes. (b) Medium-power histological appearance of seminiferous tubules. (c) High-power histological view of seminiferous tubules demonstrating stages of spermatogenesis and sustentacular cells.

Table 27.7 Developmental Stages of Sperm

Cell Name	Description and Function	Stage of Mitosis/Meiosis	Ploidy	Word Origin
Spermatogonia	Primitive male gametes that are formed by mitosis from male germline stem cells	Formed by mitosis; undergo mitosis to form primary spermatocytes	Diploid (2n)	*sperma*, seed, + *gonia*, generation
Primary Spermatocyte	Male gamete that has replicated its DNA in preparation for meiosis. It will subsequently undergo meiosis I to form two secondary spermatocytes.	DNA is replicated in prophase of meiosis I	Diploid (2n)	*primary*, first, + *sperma*, seed, + *kytos*, cell
Secondary Spermatocyte	Male gamete that has completed meiosis I and will subsequently undergo meiosis II to form two spermatids	Formed after the first meiotic division	Haploid (1n)	*secondary*, second, + *sperma*, seed, + *kytos*, cell
Spermatid	Male gamete that has completed meiosis, but has not undergone spermiogenesis, the process of becoming a mature spermatozoon	Formed after the second meiotic division	Haploid (1n)	*sperma*, seed, + *-id*, a young specimen
Spermatozoon (Sperm)	A mature male gamete that has undergone spermiogenesis	Formed by maturation of spermatids	Haploid (1n)	*sperma*, seed, + *zoon*, animal

Table 27.8 Accessory Cells of the Testis

Cell Name	Description and Function	Word Origin
Interstitial (Leydig) Cells	Produce the steroid hormone testosterone, which is required for proper sperm development and for the development of male secondary sex characteristics	*inter-*, between, + *sisto*, to stand
Sustentacular (Sertoli) Cells	Surround multiple developing spermatocytes to provide them with support and nourishment. Phagocytize excess cytoplasm from developing spermatocytes. Form the blood-testis barrier, which protects the developing spermatocytes from antigens that circulate in the blood. Produce androgen-binding protein (ABP), which concentrates testosterone around the developing spermatocytes.	*sustento*, to hold upright

3. Using tables 27.7 and 27.8 and figure 27.6 as guides, identify the following structures:

   ☐ interstitial cell ☐ spermatids
   ☐ primary spermatocyte ☐ spermatogonia
   ☐ seminiferous tubule ☐ sustentacular cell

4. In the space to the right, sketch a seminiferous tubule as seen through the microscope. Be sure to label the structures listed in step 3 in your drawing.

_____ ×

5. *Optional Activity:* **AP|R** 14: Reproductive System—Watch the "Spermatogenesis" animation to visualize the formation of sperm in the seminiferous tubules.

EXERCISE 27.6

HISTOLOGY OF THE EPIDIDYMIS

After spermatozoa are formed within the seminiferous tubules, they are transported through the efferent ductules, straight tubules, and rete testis to enter the long, coiled tube that lies upon the testis, the **epididymis** (*epi-*, upon, + *didymos*, a twin [related to *didymoi*, testes]). The epididymis is the first part of the duct system in the male reproductive tract. Here the spermozoa undergo the process of maturation and are stored until ejaculation takes place. One of the most characteristic features of the epididymis is the presence of thousands of spermatozoa in the lumen of the tube **(figure 27.7)**. **Table 27.9** describes the structure and function of the male accessory reproductive structures, including the epididymis.

1. Obtain a slide of the epididymis and place it on the microscope stage. Bring the tissue sample into focus on low power and identify several cross sections of the epididymis. Move the microscope stage so that one part of the lumen of the epididymis is at the center of the field of view, and then change to high power. Notice the sperm inside the lumen of the tubule, and how they do not come right up against the apical surface of the columnar epithelial cells. This is because of the stereocilia on the apical surface of the epithelial cells. **Stereocilia** (*stereo*, solid, + *cilium*, eyelid) are single, long microvilli (ironically, they are not cilia at all!) that increase the surface area of the epithelial cells for the purpose of secreting substances that nourish the sperm and absorbing substances from the sperm as they undergo maturation.

2. Using table 27.9 and figure 27.7 as guides, identify the following structures on the slide of the epididymis:

   ☐ pseudostratified ciliated columnar epithelial cells
   ☐ sperm cells
   ☐ stereocilia

3. In the space below, sketch the epididymis as seen through the microscope. Be sure to label the structures listed in step 2 in your drawing.

_____ ×

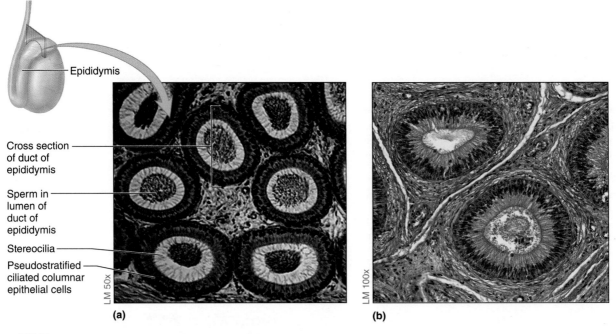

Figure 27.7 Cross Section through the Epididymis. The epididymis is a long, coiled tube. In these sections you can see several cross sections through the duct. (a) Note the numerous spermatozoa filling the lumen. (b) At higher magnification you can better view the pseudostratified columnar epithelium with stereocilia.

Table 27.9	Male Accessory Reproductive Structures		
Structure	**Epithelium**	**Function**	**Word Origin**
Epididymis	Pseudostratified columnar epithelium with cilia	Site of storage and maturation of sperm. Walls contain some smooth muscle that will propel sperm into the ductus deferens during ejaculation.	*epi-*, upon, + *didymos*, a twin (related to *didymoi*, testes)
Ductus Deferens (Vas Deferens)	Pseudostratified columnar epithelium with stereocilia	A thick, muscular tube whose walls undergo peristaltic contractions during ejaculation to propel sperm into the urethra	*ductus*, to lead, + *defero*, to carry away
Prostate Gland	Glandular epithelium resembling pseudostratified or simple columnar	Accessory reproductive gland that contributes approximately 25% of the volume of semen. The fluid produced is rich in Vitamin C (citric acid) and enzymes (such as PSA, prostate-specific antigen) that are important for proper sperm function.	*prostates*, one standing before
Seminal Vesicles	Pseudostratified columnar	Accessory reproductive gland that contributes approximately 60% of the volume of semen. The fluid produced is rich in fructose, which is a source of energy for the sperm. They also produce prostaglandins, which are important in promoting sperm motility and may stimulate uterine contractions.	*seminal*, relating to sperm, + *vesicula*, a blister
Bulbourethral Glands	Glandular epithelium consisting of simple and pseudostratified columnar	Accessory reproductive gland that produces a mucuslike lubricating substance during sexual arousal that lubricates the urethra and neutralizes the acidity of the urine, thus preparing the way for the spermatozoa to pass	*bulbus*, bulb, + *urethral*, relating to the urethra

EXERCISE 27.7

HISTOLOGY OF THE DUCTUS DEFERENS

The **ductus deferens** (vas deferens) is a continuation of the epididymis and is the route by which sperm travel from the epididymis to the urethra during ejaculation (*ejaculo*, to shoot out). It is a highly muscular tube lined with pseudostratified columnar epithelium with cilia **(figure 27.8)**.

1. Obtain a slide demonstrating the ductus deferens and place it on the microscope stage. Bring the tissue sample into focus on low power and identify the cross section of the ductus deferens (figure 27.8). Move the microscope stage so the lumen of the ductus deferens is at the center of the field of view, and then change to high power.

2. Using table 27.9 and figure 27.8 as guides, identify the following structures on the slide of the ductus deferens:

   - ☐ cilia
   - ☐ lumen
   - ☐ mucosa
   - ☐ pseudostratified ciliated columnar epithelial cells
   - ☐ smooth muscle

3. In the space below, sketch a cross section of the ductus deferens as seen through the microscope. Be sure to label the structures listed in step 2 in your drawing.

(continued on next page)

780 Chapter Twenty-Seven *The Reproductive System and Early Development*

(continued from previous page)

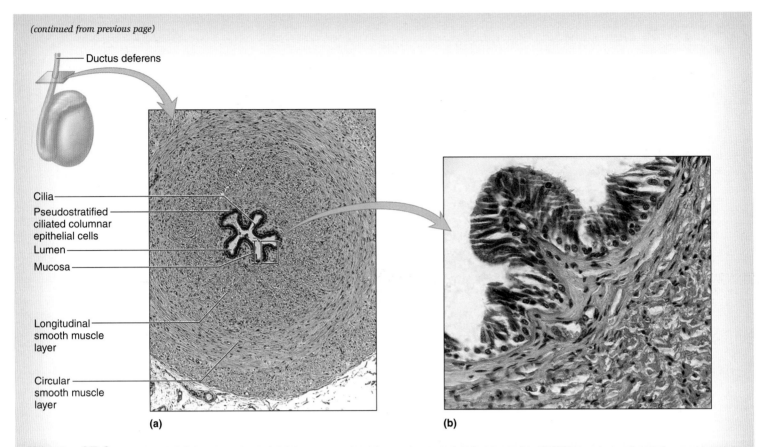

Figure 27.8 Ductus Deferens. (a) Cross section through the entire ductus deferens, demonstrating thick layers of smooth muscle surrounding the lumen. (b) Close-up view of the pseudostratified ciliated columnar epithelium.

EXERCISE 27.8

HISTOLOGY OF THE SEMINAL VESICLES

As each ductus deferens approaches the posterior wall of the bladder, each comes together with ducts from a **seminal vesicle (figure 27.9)**. These accessory reproductive glands produce substances (including fructose, which is a source of energy for the sperm) that are an important component of **semen**, which is the fluid expelled from the penis during ejaculation (table 27.9).

1. Obtain a slide of the seminal vesicles and place it on the microscope stage. Bring the tissue sample into focus on low power. Identify cross sections of the seminal vesicles. Move the microscope stage so the lumen of a portion of the seminal vesicle is at the center of the field of view, and then change to high power.

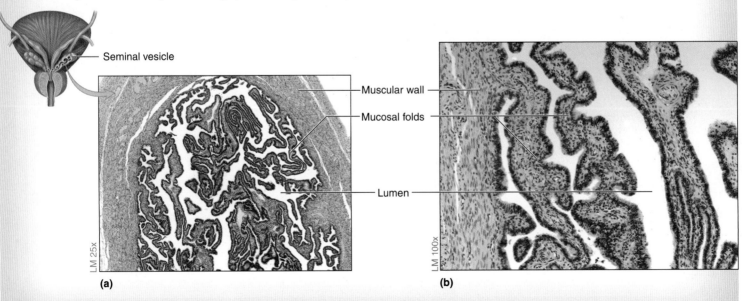

Figure 27.9 The Seminal Vesicles. (a) Low magnification. (b) Medium magnification.

2. Using table 27.9 and figure 27.9 as guides, identify the following structures on the slide of the seminal vesicles:

   ☐ lumen   ☐ muscular wall
   ☐ mucosal folds

3. In the space to the right, sketch the seminal vesicles as seen through the microscope. Be sure to label the structures listed in step 2 in your drawing.

_____ ×

EXERCISE 27.9

HISTOLOGY OF THE PROSTATE GLAND

The **prostate** gland is an accessory reproductive gland located immediately inferior to the urinary bladder. Like the seminal vesicles, the prostate produces substances that will become part of semen and that are important for proper sperm function.

1. Obtain a slide demonstrating the prostate gland and place it on the microscope stage. Bring the tissue sample into focus on low power and identify the prostatic urethra and the prostate gland itself **(figure 27.10)**. Move the microscope stage so a part of the gland relatively far away from the urethra is at the center of the field of view, and then change to a higher power.

2. Using table 27.9 and figure 27.10 as guides, identify the following structures on the slide of the prostate:

   ☐ tubuloalveolar glands

3. As men age, calcifications often form in the prostate. Such calcifications are called **prostatic calculi** (corpora aranacea) (figure 27.10). Scan the slide and see if any of these are present in the specimen you are viewing.

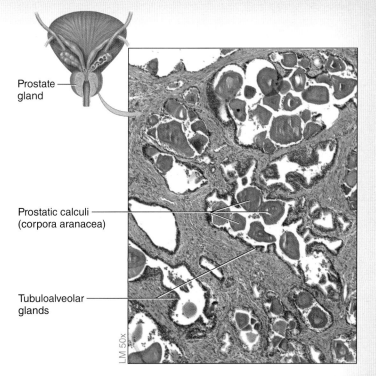

Figure 27.10 The Prostate Gland. Note the many prostatic calculi (corpora aranacea) seen at high magnification.

EXERCISE 27.10

HISTOLOGY OF THE PENIS

The **penis** is the male copulatory organ and is homologous to the female clitoris. It is composed of three erectile bodies, the paired **corpora cavernosa** and the single **corpus spongiosum**, which contains the male urethra **(figure 27.11)**. In the center of each corpus cavernosum is a central artery that supplies blood to the penis. The erectile tissues of the corpus cavernosa are the equivalent of the deep venous drainage accompanying the central arteries. During an erection, the veins draining the penis become constricted and venous blood fills up the erectile tissues.

1. Obtain a slide demonstrating a cross section of the penis and place it on the microscope stage. Observe at low power.

2. Using figure 27.11 as a guide, identify the following structures on the slide of the penis:

   ☐ central artery   ☐ dorsal vein
   ☐ corpus cavernosum   ☐ superficial fascia
   ☐ corpus spongiosum   (connective tissue)
   ☐ dorsal surface   ☐ urethra
                                    ☐ ventral surface

3. In the space below, sketch a cross section of the penis as seen through the microscope. Be sure to label the structures listed in step 2 in your drawing.

_____ ×

4. *Optional Activity:* **AP|R** 14: Reproductive System— Watch the "Male Reproductive System Overview" animation to review the male reproductive structures and their relationships.

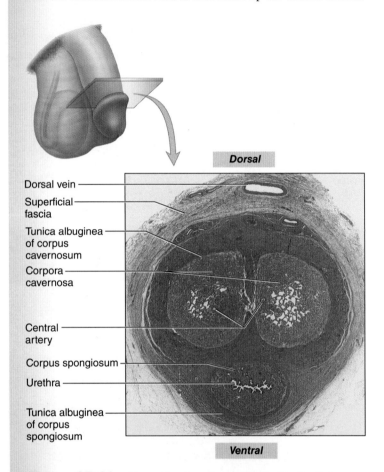

Figure 27.11 Cross Section through the Shaft of the Penis. The penis contains two paired corpora cavernosa dorsally, and a single corpus spongiosum ventrally. Note the male urethra in the center of the corpus spongiosum.

Gross Anatomy

Female Reproductive System

The internal detail, wall structures, and functions of most of the female reproductive organs were covered in the histology section of this chapter. In this section, you will observe the gross structure of these organs to gain an appreciation of their location within the female pelvic cavity, and their relationships with each other. You will also observe the numerous suspensory ligaments that anchor the ovaries, uterine tubes, and uterus in place within the pelvic cavity, and you will observe the primary anatomical features of the female breast.

EXERCISE 27.11

GROSS ANATOMY OF THE OVARY, UTERINE TUBES, UTERUS, AND SUSPENSORY LIGAMENTS

The ovary, uterine tubes, and uterus are all contained within the pelvic cavity of the female and are held in place by a number of suspensory ligaments, many of which form from folds of the peritoneum as it drapes over these structures. **Table 27.10** summarizes the structure, function, and location of the suspensory ligaments.

1. Observe a female human cadaver or classroom models of the female pelvis and female reproductive organs. Also observe a classroom model demonstrating a sagittal section through a female pelvis.

2. Using table 27.10 and your textbook as guides, identify the structures listed in **figures 27.12** and **27.13** on the cadaver or on classroom models. Then label them in figures 27.12 and 27.13. (Some answers may be used more than once.)

3. *Optional Activity:* **AP|R** 14: Reproductive System—In the Quiz section, select the "Structures: Female" option to test yourself on gross anatomy of female structures.

Table 27.10	Suspensory Ligaments of the Ovary, Uterine Tubes, and Uterus	
Suspensory Ligament	**Description and Function**	**Word Origin**
Broad Ligament	Fold of peritoneum draped over the superior surface of the uterus. Portions of the broad ligament form the mesovarium and the mesosalpinx.	*broad*, wide, + *ligamentum*, a bandage
Mesosalpinx	The mesentery of the uterine tube, which is formed as a fold of the most superior part of the broad ligament	*meso-*, a mesentery-like structure, + *salpinx*, trumpet (tube)
Mesovarium	The mesentery of the ovary, which is formed as a posterior extension of the broad ligament	*meso-*, a mesentery-like structure, + *ovarium*, ovary
Ovarian Ligament	Ligament contained within folds of the broad ligament. It extends from the medial part of the ovary to the superolateral surface of the body of the uterus.	*ovarium*, ovary, + *ligamentum*, a bandage
Round Ligament of the Uterus	Ligament attached to the superolateral surface of the body of the uterus. It extends laterally to the deep inguinal ring, passes through the inguinal canal, and attaches to the skin of the labia majora.	*ligamentum*, a bandage, + *uterus*, uterus
Suspensory Ligament of the Ovary	A fold of peritoneum draping over the ovarian artery and vein superolateral to the ovary. Anchors the ovary to the lateral body wall.	*suspensio-*, to hang up, + *ligamentum*, a bandage
Transverse Cervical (Cardinal) Ligament	Ligament extending laterally from the cervix and vagina, connecting them to the pelvic wall	*transverse-*, across, + *cervix*, neck, + *ligamentum*, a bandage
Uterosacral Ligaments	Ligament connecting the inferior part of the uterus to the sacrum posteriorly	*utero-*, the uterus, + *sacral*, the sacrum

(continued on next page)

784 Chapter Twenty-Seven *The Reproductive System and Early Development*

(continued from previous page)

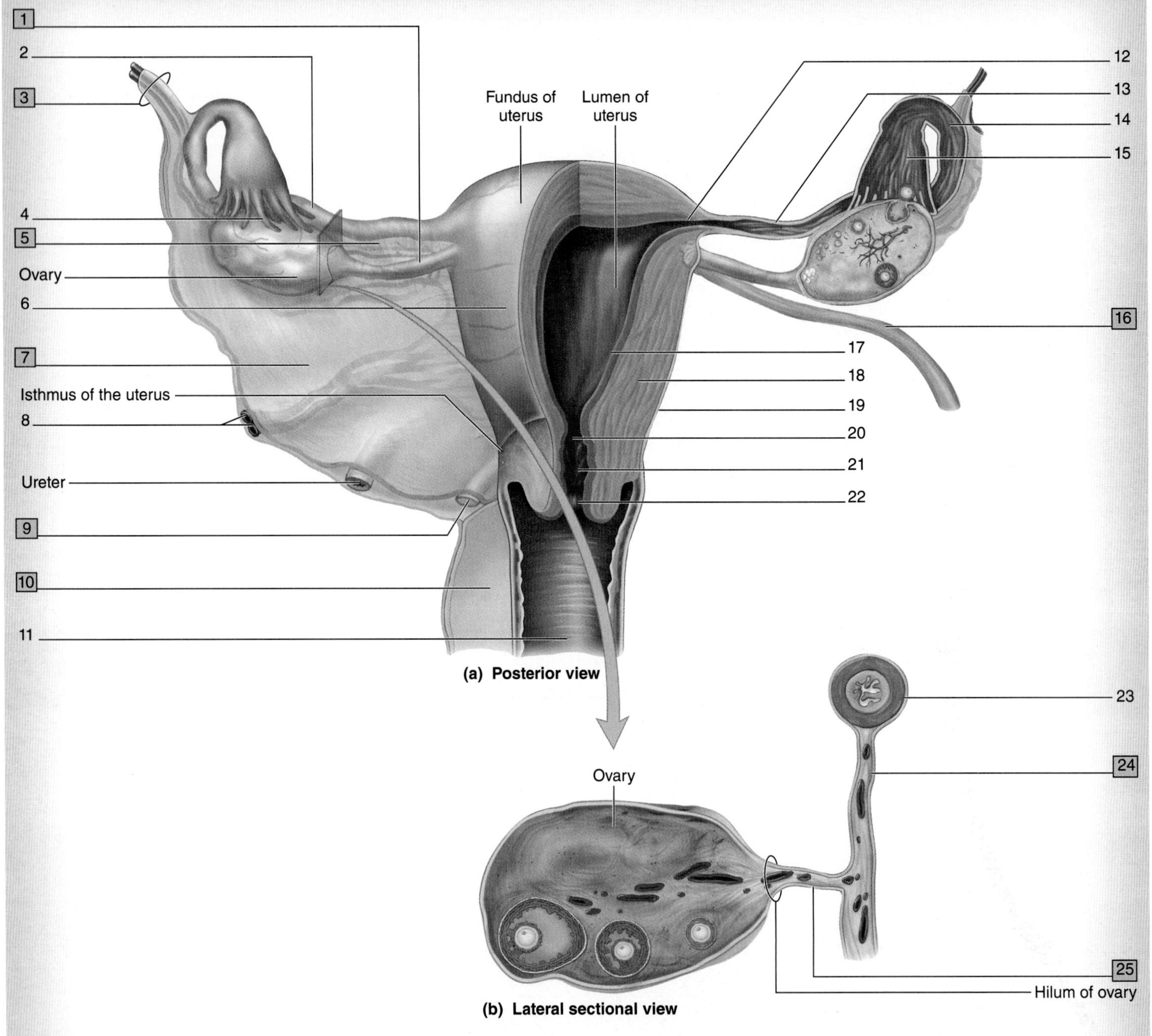

(a) Posterior view

(b) Lateral sectional view

Figure 27.12 Suspensory Ligaments of the Ovary, Uterine Tubes, and Uterus as Seen from a Posterior View. Numbers indicating suspensory ligaments are highlighted in green. Numbers of lines to reproductive and other associated structures have no highlight. (a) Posterior view. (b) Lateral view.

- ☐ ampulla of uterine tube
- ☐ body of uterus
- ☐ broad ligament
- ☐ cervical canal
- ☐ endometrium
- ☐ external os
- ☐ fimbria
- ☐ infundibulum of uterine tube
- ☐ internal os
- ☐ isthmus of uterine tube
- ☐ mesosalpinx (used twice)
- ☐ mesovarium
- ☐ myometrium
- ☐ ovarian ligament
- ☐ perimetrium
- ☐ round ligament of the uterus
- ☐ suspensory ligament of the ovary
- ☐ transverse cervical ligament
- ☐ uterine blood vessels
- ☐ uterine part of uterine tube
- ☐ uterine tube (used twice)
- ☐ uterosacral ligament
- ☐ vagina

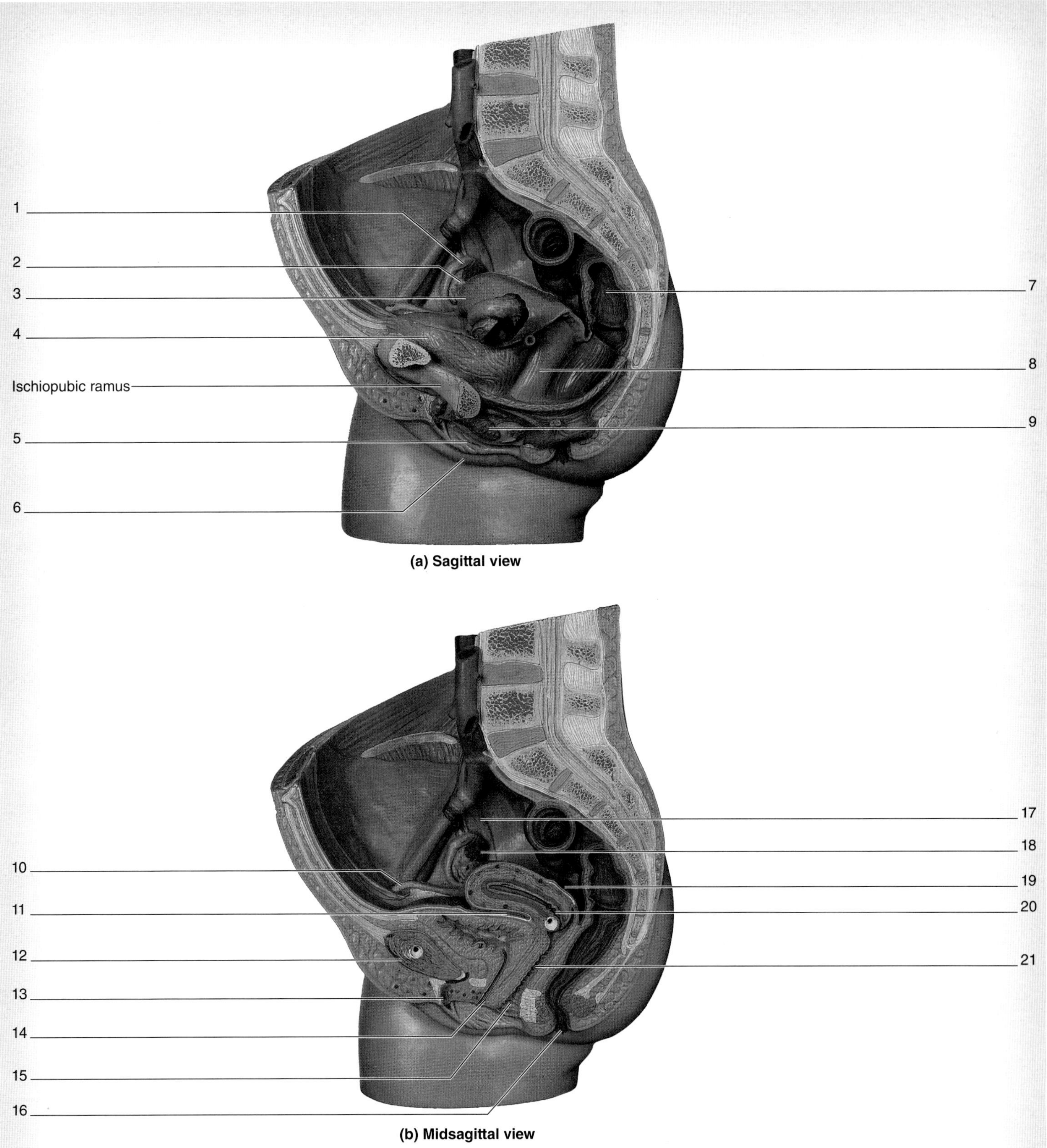

Chapter Twenty-Seven *The Reproductive System and Early Development* 785

1 _____
2 _____
3 _____
4 _____
Ischiopubic ramus _____
5 _____
6 _____

7 _____
8 _____
9 _____

(a) Sagittal view

10 _____
11 _____
12 _____
13 _____
14 _____
15 _____
16 _____

17 _____
18 _____
19 _____
20 _____
21 _____

(b) Midsagittal view

Figure 27.13 **Classroom Model of the Female Pelvic Cavity.** (a) Midsagittal view with pelvic structures intact. (b) Midsagittal view.

- ☐ anus
- ☐ bulb of the vestibule
- ☐ cervix of uterus
- ☐ clitoris
- ☐ external urethral orifice
- ☐ fimbria of uterine tube
- ☐ labia majora
- ☐ labia minora
- ☐ ovary
- ☐ pubic symphysis
- ☐ rectouterine pouch
- ☐ rectum
- ☐ round ligament of the uterus
- ☐ ureter
- ☐ urinary bladder
- ☐ uterine tube
- ☐ uterus
- ☐ vagina (used twice)
- ☐ vaginal orifice
- ☐ vesicouterine pouch

> **EXERCISE 27.12**

GROSS ANATOMY OF THE FEMALE BREAST

The female breast consists largely of fatty tissue and suspensory ligaments. Imbedded within are numerous modified sweat glands, the **mammary glands** (*mamma*, breast), which are compound tubuloalveolar exocrine glands. These glands enlarge greatly during pregnancy, enabling them to produce milk to nourish the new baby. **Table 27.11** describes the structures that compose the female breast.

1. Observe the breast of a prosected female human cadaver or a classroom model of the female breast.
2. Using table 27.11 and your textbook as guides, identify the structures listed in **figure 27.14** on the cadaver or the classroom model of the female breast. Then label them in figure 27.14. (Some answers may be used more than once.)

Table 27.11	The Female Breast	
Structure	**Description and Function**	**Word Origin**
Alveoli	Secretory units of the mammary glands, which produce milk	*alveolus*, a concave vessel, a bowl
Areola	The pigmented area of skin surrounding the nipple	*areola*, area
Areolar Glands	Sebaceous glands deep to the skin of the areola; produce sebum, which keeps the skin of the areola moist, particularly during lactation	*areolar*, relating to the areola of the breast
Lactiferous Ducts	Ducts that form from the confluence of small ducts draining milk from the alveoli and lobules	*lacto-*, milk + *ductus*, to lead
Lactiferous Sinuses	10–20 large channels that form from the confluence of several lactiferous ducts; the spaces where milk is stored prior to release from the nipple	*lacto-*, milk + *sinus*, cavity
Lobes	Large subdivisions of the mammary glands	*lobos*, lobe
Lobules	Smaller subdivisions of the mammary glands, which contain the alveoli	*lobulus*, a small lobe
Nipple	A cylindrical projection in the center of the breast that contains the openings of the lactiferous ducts	*neb*, beak or nose
Suspensory Ligaments	Bands of connective tissue that anchor the breast skin and tissue to the deep fascia overlying the pectoralis major muscle	*suspensio*, to hang up, + *ligamentum*, a band

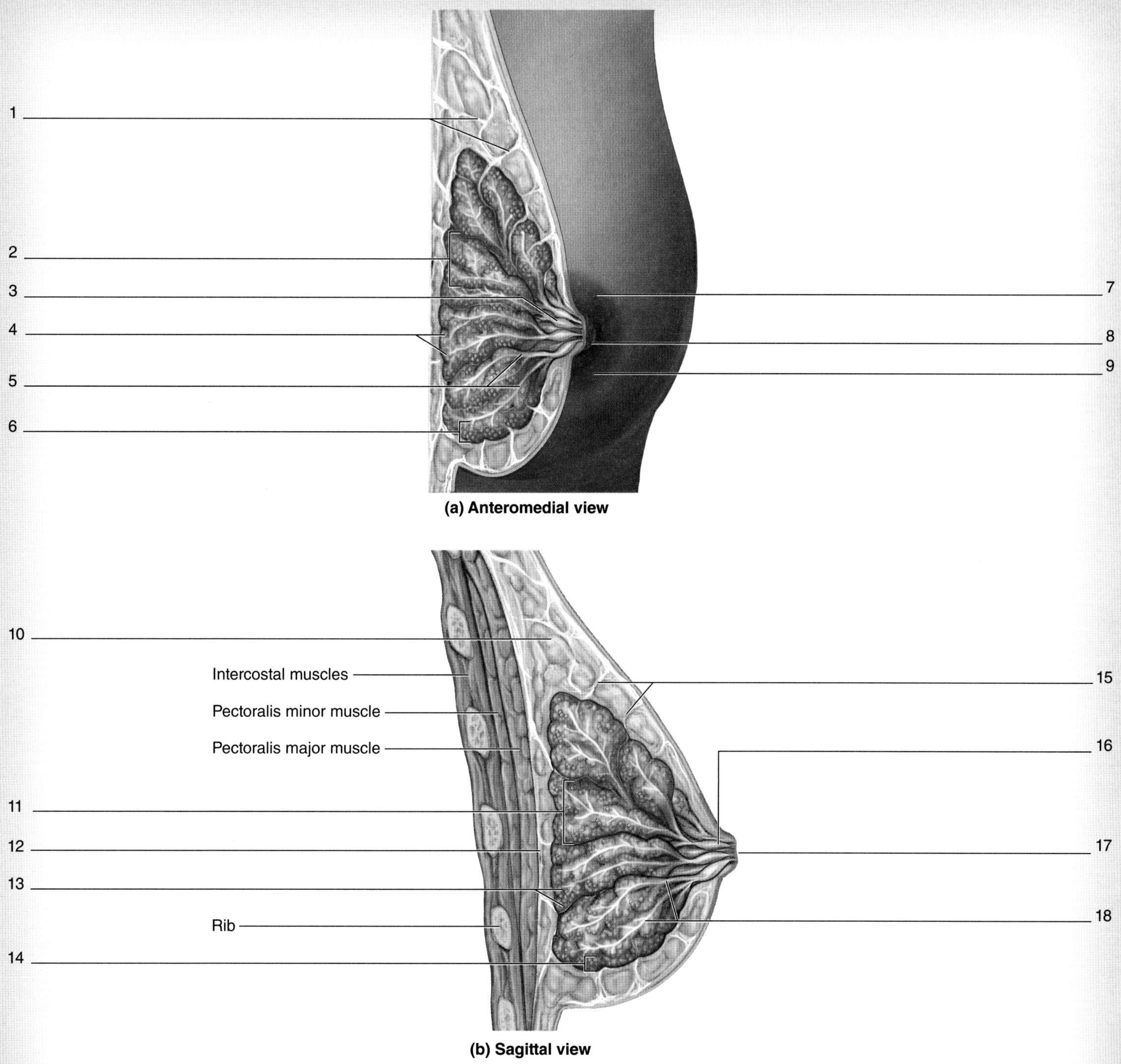

Chapter Twenty-Seven *The Reproductive System and Early Development* 787

(a) Anteromedial view

1 _____
2 _____
3 _____
4 _____
5 _____
6 _____
7 _____
8 _____
9 _____

(b) Sagittal view

Intercostal muscles
Pectoralis minor muscle
Pectoralis major muscle
Rib

10 _____
11 _____
12 _____
13 _____
14 _____
15 _____
16 _____
17 _____
18 _____

Figure 27.14 **The Female Breast.** (a) Anterior view. (b) Sagittal view.

- [ ] adipose tissue
- [ ] alveoli (used twice)
- [ ] areola
- [ ] areolar gland
- [ ] deep fascia
- [ ] lactiferous ducts (used twice)
- [ ] lactiferous sinus (used twice)
- [ ] lobe (used twice)
- [ ] lobule (used twice)
- [ ] nipple (used twice)
- [ ] suspensory ligaments (used twice)

(continued on next page)

INTEGRATE

CLINICAL VIEW
Breast Cancer and Breast Self-Examination (BSE)

Incidence

In females, breast cancer is the second most commonly diagnosed cancer after skin cancer, and the second most common cause of death from cancer after lung cancer. Due to increased screening for breast cancers, currently only about 20% of patients diagnosed with breast cancer are likely to die from breast cancer. Luckily, breast cancer is a highly curable disease if caught early, and there are several screening mechanisms that can be used to identify early lesions: clinical breast exam (CBE), mammography, and breast self-exam (BSE). Of these, breast self-exam is one of the most common ways that breast cancers are first noticed.

What Is Breast Cancer?

The most common forms of breast cancers are carcinomas, which are tumors that originate in epithelial tissues. In the breast, the mammary glands and ducts consist of modified epithelial tissue, and most breast cancers originate in these tissues. The most common types of breast cancer are ductal carcinoma in situ (DCIS) and lobular carcinoma in situ (LCIS), which are tumors that originate in cells lining the ducts (DCIS) or in the glands themselves (LCIS). These tumors remain limited to the epithelial tissue (*in situ* means a tumor has not crossed the basement membrane of the epithelium). More serious forms of cancer are invasive carcinomas, which are tumors that have spread beyond the basement membrane of the epithelial tissues.

Screening
Clinical Breast Exam (CBE)

All women of reproductive age are encouraged to make annual visits to their gynecologist for a clinical breast exam and a pap smear to test for early cervical changes that are risk factors for cervical cancer. The physician performing the CBE is palpating for abnormal lumps in the breast. However, the physician can also help teach a woman how to perform BSE so she can monitor the condition of her breasts monthly instead of just during yearly visits. The vast majority of breast cancers are discovered by the patient herself.

Mammography

A mammogram is an x-ray of the breast that is used to look for areas of extra density or calcifications, which can be early signs of breast cancer. Most women are advised to start having mammograms at age 40, but those with a family history of the disease are often advised to have their first mammogram at age 35. Mammography is good at locating very small tumors (especially DCIS), which may not be palpable on self-examinations or clinical examinations.

Breast Self-Exam (BSE)

Women should perform a breast self-exam at least once a month to look for any unusual bumps, lumps, or thickening of breast tissue. Upon palpation, a typical breast feels a bit lumpy, with some firmer areas and some softer areas. The firmer/harder areas can be thickenings of the suspensory ligaments, or fibrocystic changes to the breast. The softer areas are typically just adipose tissue. There are a number of benign (noncancerous) changes to the breast that can make breasts feel "lumpy." Specifically, fibrocystic changes in the breast consist of fluid-filled cysts and are often surrounded by dense fibrous tissue. This tissue, though often large and dense, is typically mobile (moves around easily), and often mirrors itself on the opposite breast. That is, if you find such changes in the lower-right quadrant of the right breast, you might also find it on the lower-right quadrant of the left breast. On the contrary, breast cancers tend to feel much firmer, like a kernel of unpopped popcorn, and they are immobile (they do not move when you touch them). They are generally painless and sometimes cause dimpling of the skin overlying the tumor.

The following are three procedures for performing a breast self-exam:

1. In the shower: Use soapy hands to palpate the breast while raising the arm on the same side as the breast you are palpating. Starting at the periphery of the breast, use two or three fingers to make small circles that progressively move toward the nipple. Squeeze the nipple to look for discharge. Be sure to palpate all the way up into the axilla, as the majority of tumors arise in the outer/upper quadrant toward the axillary region of the breast.

2. Lying down: Use the same procedure as in the shower. It is good to do an exam lying down because it makes the breasts lie flat, which may make it easier to feel certain lumps.

3. Standing in front of a mirror: Place your hands on your hips, tilt your elbows forward (anterior), and look for any indentations in the skin of the breast or "orange peel"–looking skin, which can be indicative of an underlying tumor.

If you discover anything that concerns you, make an appointment with your gynecologist as soon as possible to discuss your findings so the physician can help determine if what you felt was something benign or something that needs further exploration. Always remember that breast cancer is *highly curable* when caught early.

Male Reproductive System

The **testes,** the primary male reproductive organs, function in the production of sperm and testosterone. Sperm produced in the testes are stored in the **epididymis** and transported via the **ductus deferens** to the male urethra during ejaculation. The **seminal vesicles** and **prostate gland** are accessory reproductive glands that produce substances that nourish and protect the sperm and also compose the vast majority of semen. The **bulbourethral glands** are small glands located within the urogenital diaphragm that produce a mucouslike substance during sexual arousal that neutralizes the acidity of the male urethra and provides lubrication to ease the passage of semen during ejaculation.

In exercises 27.5 to 27.10 you examined the histological appearance of many of these structures. Now we will observe the male reproductive structures at the gross level.

EXERCISE 27.13

GROSS ANATOMY OF THE SCROTUM, TESTIS, SPERMATIC CORD, AND PENIS

1. Observe a prosected male human cadaver or classroom models of the male reproductive organs.

2. Using your textbook as a guide, identify the structures listed in **figures 27.15** and **27.16** on the cadaver or on classroom models. Then label them in figures 27.15 and 27.16.

3. *Optional Activity:* **AP|R** **14: Reproductive System**—In the Quiz section, select the "Structures: Male" option to test your knowledge of male reproductive gross anatomy.

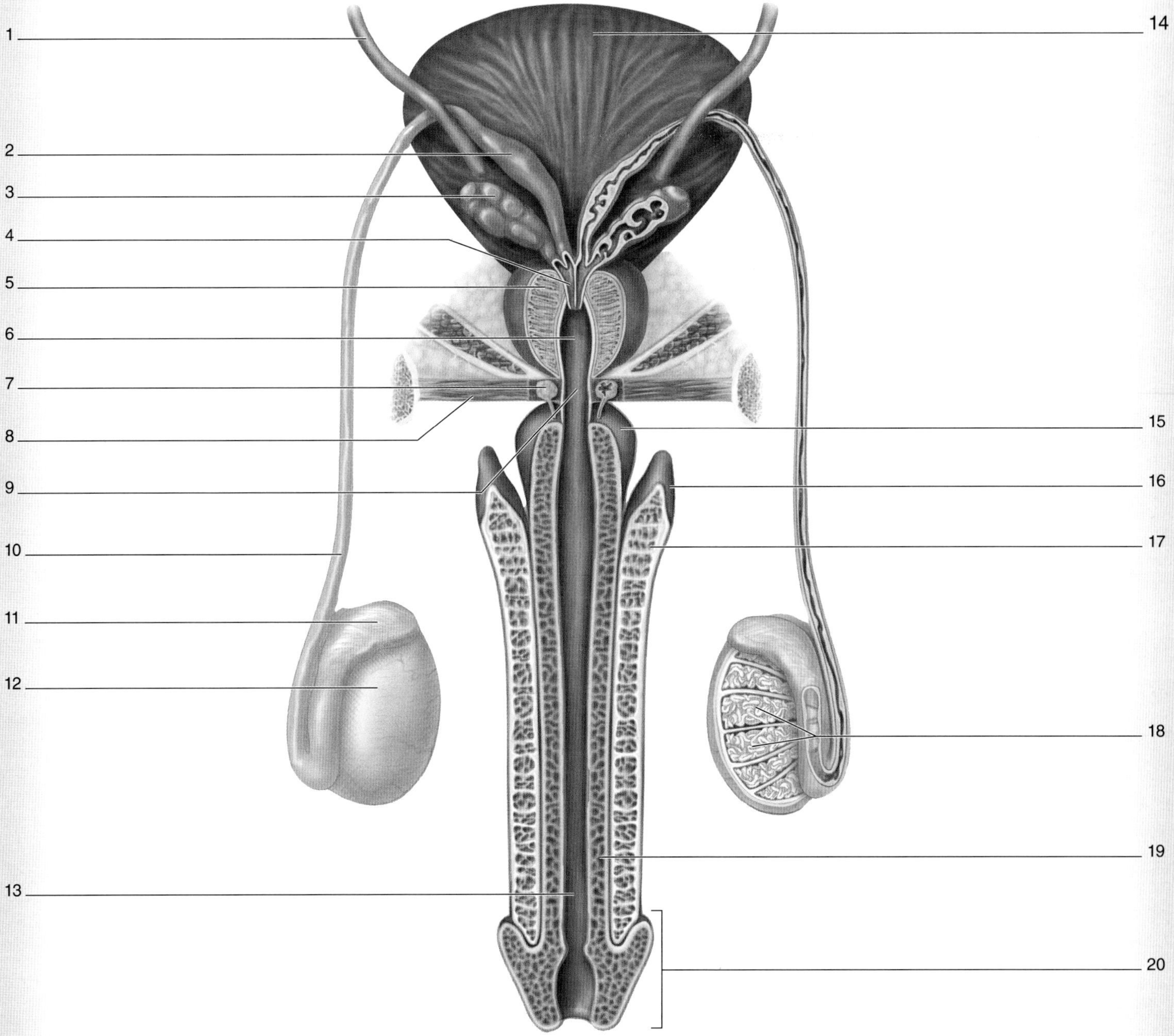

Figure 27.15 Male Reproductive Tract Structures. Posterior view.

- ☐ ampulla of ductus deferens
- ☐ bulb of penis
- ☐ bulbourethral gland
- ☐ corpus cavernosum
- ☐ corpus spongiosum
- ☐ crus of penis
- ☐ ductus deferens
- ☐ ejaculatory duct
- ☐ epididymis
- ☐ glans penis
- ☐ membranous urethra
- ☐ prostate gland
- ☐ prostatic urethra
- ☐ seminal vesicle
- ☐ seminiferous tubules
- ☐ spongy urethra
- ☐ testis
- ☐ ureter
- ☐ urinary bladder
- ☐ urogenital diaphragm

(continued on next page)

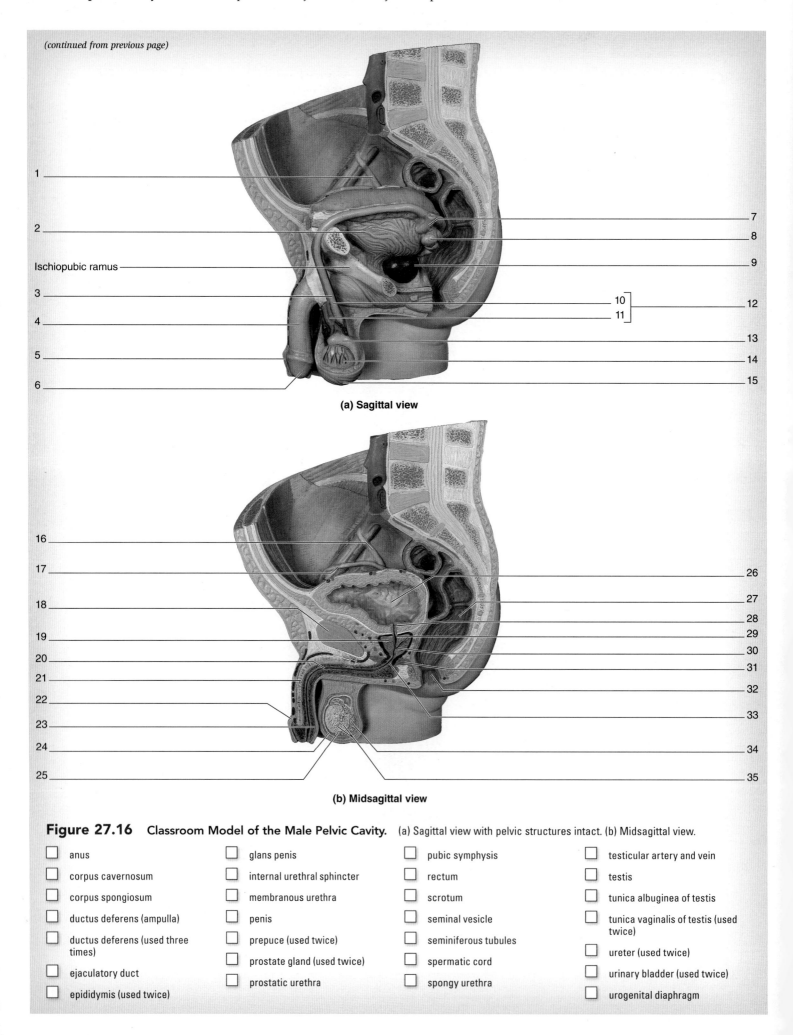

Figure 27.16 Classroom Model of the Male Pelvic Cavity. (a) Sagittal view with pelvic structures intact. (b) Midsagittal view.

- ☐ anus
- ☐ corpus cavernosum
- ☐ corpus spongiosum
- ☐ ductus deferens (ampulla)
- ☐ ductus deferens (used three times)
- ☐ ejaculatory duct
- ☐ epididymis (used twice)
- ☐ glans penis
- ☐ internal urethral sphincter
- ☐ membranous urethra
- ☐ penis
- ☐ prepuce (used twice)
- ☐ prostate gland (used twice)
- ☐ prostatic urethra
- ☐ pubic symphysis
- ☐ rectum
- ☐ scrotum
- ☐ seminal vesicle
- ☐ seminiferous tubules
- ☐ spermatic cord
- ☐ spongy urethra
- ☐ testicular artery and vein
- ☐ testis
- ☐ tunica albuginea of testis
- ☐ tunica vaginalis of testis (used twice)
- ☐ ureter (used twice)
- ☐ urinary bladder (used twice)
- ☐ urogenital diaphragm

Physiology

Reproductive Physiology

Gametogenesis, the production of the **gametes** (sperm and secondary oocytes) is accomplished by meiosis. **Meiosis** is similar in many ways to mitosis, but there are two nuclear divisions (I and II), which result in a reduction of chromosome number from a diploid number (2n, or 46) to a haploid number (n, or 23). Another advantage of meiosis is the phenomenon of **synapsis** (crossing over) that occurs when homologous chromosomes are adjacent to each other. Synapsis is a process that introduces variability in the genetic code, which ensures that daughter cells are genetically different from parent cells.

Table 27.12 depicts and describes each phase of meiosis and the defining features of each phase.

Human somatic cells contain 23 pairs of chromosomes, or **homologous pairs.** These include 22 pairs of autosomal chromosomes, and one pair of sex chromosomes. Human gametes (egg and sperm) contain 22 autosomal chromosomes and one sex chromosome, an X chromosome for oocytes and an X or Y chromosome for sperm. Meiosis yields four genetically different, haploid daughter cells. The process of **fertilization** brings haploid sperm together with a haploid secondary oocyte. When a sperm cell has fertilized an oocyte, the resulting cell (called a zygote) is once again diploid.

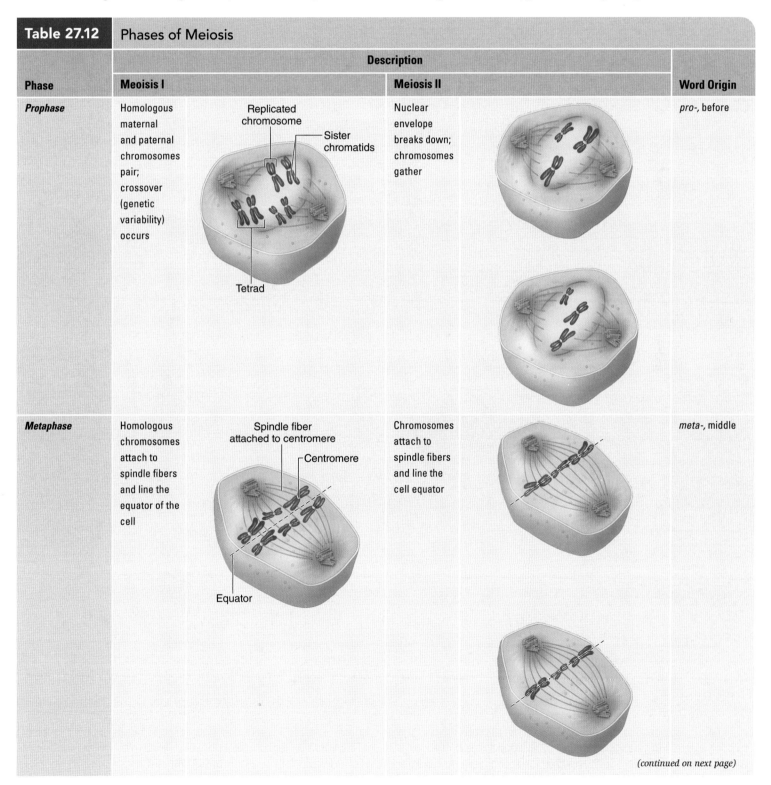

Table 27.12 Phases of Meiosis

Phase	Description — Meoisis I	Description — Meiosis II	Word Origin
Prophase	Homologous maternal and paternal chromosomes pair; crossover (genetic variability) occurs	Nuclear envelope breaks down; chromosomes gather	*pro-*, before
Metaphase	Homologous chromosomes attach to spindle fibers and line the equator of the cell	Chromosomes attach to spindle fibers and line the cell equator	*meta-*, middle

(continued on next page)

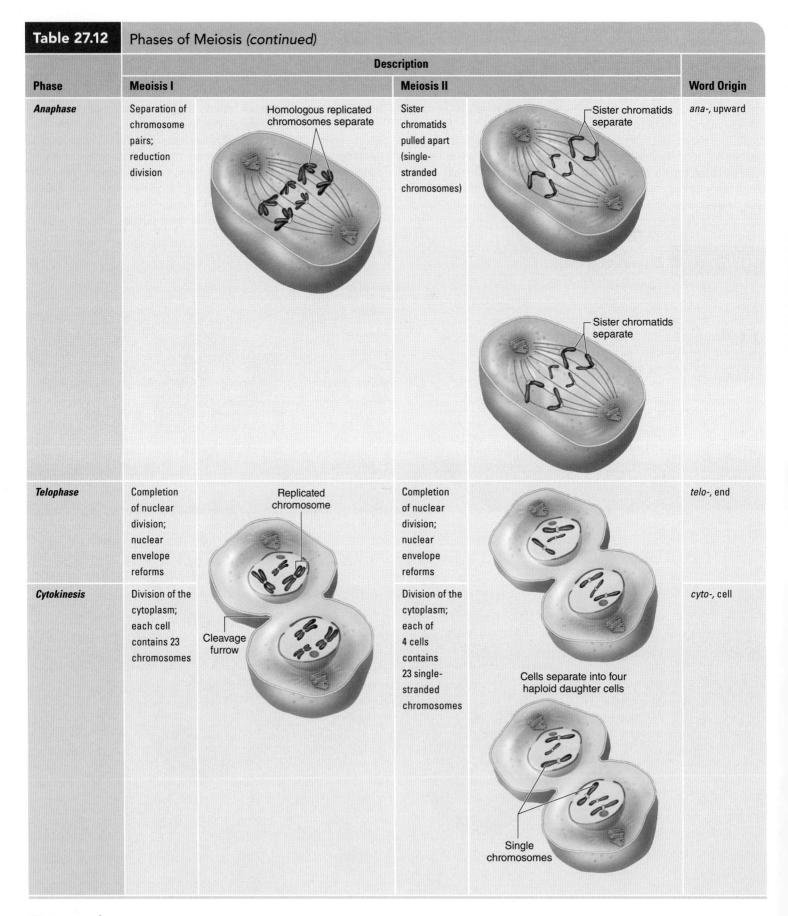

Table 27.12 Phases of Meiosis (continued)

Phase	Meiosis I	Meiosis II	Word Origin
Anaphase	Separation of chromosome pairs; reduction division	Sister chromatids pulled apart (single-stranded chromosomes)	ana-, upward
Telophase	Completion of nuclear division; nuclear envelope reforms	Completion of nuclear division; nuclear envelope reforms	telo-, end
Cytokinesis	Division of the cytoplasm; each cell contains 23 chromosomes	Division of the cytoplasm; each of 4 cells contains 23 single-stranded chromosomes	cyto-, cell

Oogenesis

The process of producing a mature egg (ovum) in females requires both mitotic and meiotic division. Primordial germ cells, or oogonia (singular, *oogonium*), divide by mitosis. Primary oocytes start the process of meiosis. **Oogenesis** involves forming secondary oocytes from primary oocytes. However, rather than proceeding rapidly through the two phases of meiosis, primary and secondary oocytes experience periods of "arrest" where meiosis is stopped until hormones or events trigger continuation of meiosis. These "arrested" oocytes are housed within ovarian follicles. **Primordial follicles** contain primary oocytes, which are available to be selected each month once a female reaches puberty, the time at which monthly hormone fluctuations begin. Only a few primordial follicles will mature into primary follicles, and fewer still will become secondary follicles. At ovulation a large vesicular follicle bursts and releases a secondary oocyte from the follicle's fluid-filled sac. The secondary oocyte will then be drawn into the uterine tubes to potentially be fertilized by sperm. The final phases of meiosis occur after fertilization (see table 27.12).

The result of fertilization is formation of a **zygote,** which will subsequently undergo mitosis to form a mass of cells that may implant in the uterine wall. Development *in utero* continues, as described in exercise 27.16. **Figure 27.17** depicts the overall process of oogenesis, including identification of the various phases and arrests of meiosis, corresponding stages of the ovarian follicles, and histological images of each stage.

Hormones are responsible for regulating the monthly changes that occur both in the ovaries (ovarian cycle) and in the uterus (uterine cycle). The overall purpose is to produce viable eggs (ova) for fertilization and to provide an attractive place for implantation of the fertilized egg. The release of gonadotropin-releasing hormone (GnRH) from the hypothalamus stimulates the release of luteinizing hormone (LH) and follicle-stimulating hormone (FSH) from the anterior pituitary gland. It is LH and FSH that stimulate the development of the follicles during the **follicular phase** of the ovarian cycle. As the

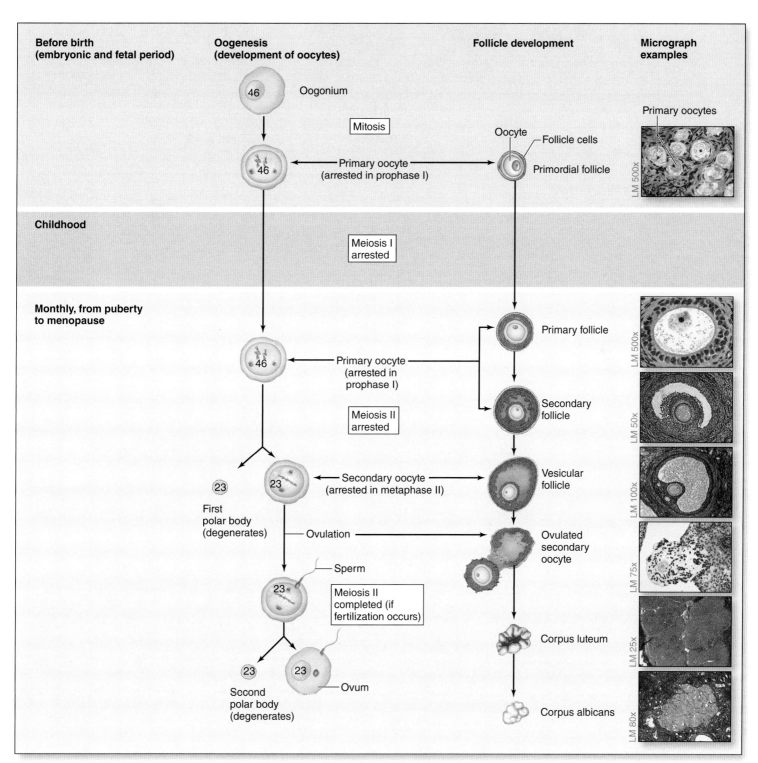

Figure 27.17 Oogenesis.

follicles develop, the granulosa cells surrounding the follicle secrete estrogen. Estrogen stimulates the cells in the uterine endometrium to proliferate during the **proliferative phase** of the uterine cycle.

Once ovulation occurs, the ruptured vesicular follicle becomes the corpus luteum. This structure begins producing progesterone and estrogen throughout the **luteal phase** of the ovarian cycle. Increasing levels of progesterone result in thickening and increased vascularization of the endometrium. In addition, the corpus luteum maintains elevated progesterone and estrogen levels in the event fertilization occurs, but prior to the formation of the placenta, the structure responsible for maintaining a viable pregnancy. Human chorionic gonadotropin (hCG), the hormone secreted by the developing embryo, prevents the degradation of the corpus luteum in the event that fertilization and implantation occur. If fertilization and implantation do not occur, the corpus luteum degenerates into the corpus albicans. This degeneration also decreases levels of progesterone

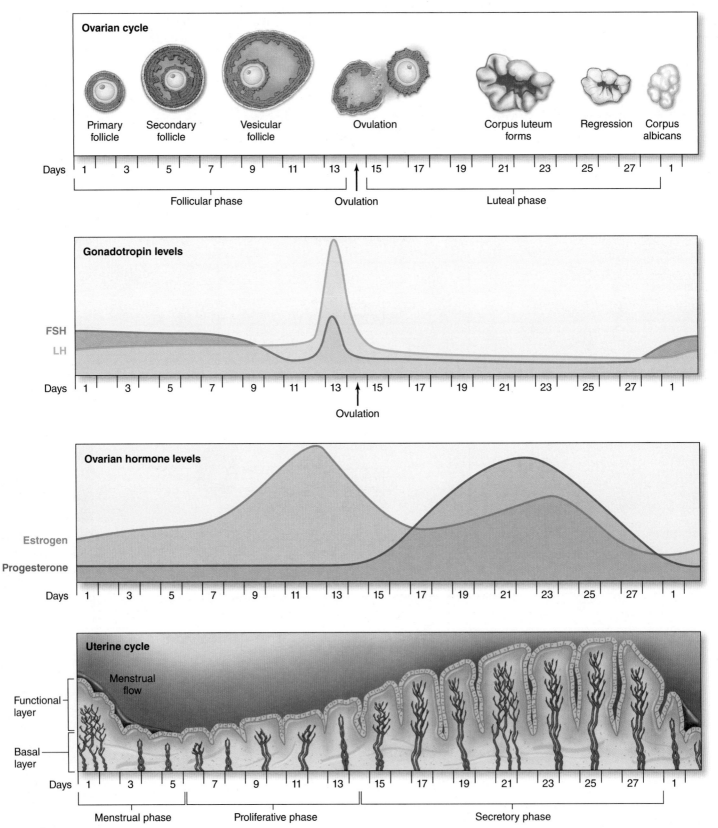

Figure 27.18 Ovarian and Uterine Cycles.

and estrogen. As the progesterone levels drop, blood vessels (spiral arterioles) in the uterine wall constrict, and the endometrial layer is shed for menses. Typically, these changes occur in 28-day cycles, as depicted in **figure 27.18.** However, cycles can vary in length. Research suggests that the duration of the luteal phase remains consistent at roughly 14 days, whereas the duration of the foliicular phase may vary.

Spermatogenesis

The production of sperm, **spermatogenesis,** in males is similar to oogenesis in females. It is interesting that the overall process is regulated by the same hormones (GnRH, LH, and FSH) that regulate oogenesis in females, with the exception of males secreting testosterone in place of estrogen. GnRH from the hypothalamus stimulates the release of LH and FSH from the anterior pituitary gland. LH stimulates interstitial cells in the seminiferous tubules of the testes to release testosterone and FSH stimulates sustentacular cells to release androgen binding protein (ABP) and inhibin. The elevated levels of testosterone are stimulatory for sperm production, yet are inhibitory for GnRH production by the hypothalamus. The process begins as spermatogonia, or primordial germ cells, actively undergo mitosis and remain at the basement membrane of the seminiferous tubules. These spermatogonia divide to form primary spermatocytes, which are diploid cells. Primary spermatocytes then undergo meiosis. Unlike females, males have the ability to continuously form new primary spermatocytes for gamete formation. After meosis I, the spermatocyte is considered secondary and haploid, with only 23 chromosomes. Once the secondary spermatocyte undergoes meiosis II, it is considered a spermatid and is located closer to the lumen of the seminferous tubule. Each of the four spermatids produced then undergoes the maturation process, or **spermiogenesis.** This process involves the development of a head, or chromosome-containing acrosome, a midpiece containing abundant mitochondria, and a tail (flagellum) of the functional spermatozoon (plural, *spermatozoa*). The spermatozoon continues the maturation process in the epididymis before becoming a viable sperm cell.

INTEGRATE

CONCEPT CONNECTION

The estrogen released by follicular cells acts directly on the hypothalamus and the anterior pituitary gland to regulate the amount of gonadotropin-releasing hormone (GnRH) released from the hypothalamus, and of luteinizing hormone (LH) and follicle-stimulating hormone (FSH) released from the anterior pituitary gland. Prior to ovulation, the increasing estrogen is stimulatory (positive feedback) for the release of GnRH, LH, and FSH. However, once ovulation occurs, estrogen is inhibitory (negative feedback) to the release of these hormones. It is a peak in LH that allows ovulation to occur; therefore, this switch from positive to negative feedback limits the number of eggs released each month. Incidentally, fertility drugs will often override these feedback mechanisms, thereby releasing multiple secondary oocytes with the chance for multiples (twins, triplets, or more). The release of these hormones was first introduced in the chapter on the endocrine system, chapter 19. The concepts of negative and positive feedback appear throughout the book, and are critical to understanding homeostasis.

EXERCISE 27.14

A CLINICAL CASE IN REPRODUCTIVE PHYSIOLOGY

The purpose of this exercise is to apply the concept of reproductive physiology to a real scenario, or clinical case study. This exercise requires information presented in this chapter, as well as in previous chapters. Integrating multiple systems in a case study reinforces a concept that is very important in physiology: homeostasis.

Before you begin, state your hypothesis regarding how the ovarian cycle may be influenced by increasing levels of testosterone. _____
_____
_____

Consult with your laboratory instructor as to whether you will be completing this exercise individually, in pairs, or in groups. Review the topics of the cardiovascular, lymphatic, respiratory, digestive, urinary, and reproductive systems in your textbook and in previous chapters in this laboratory manual as needed for assistance with the case study.

Case Study:
Elizabeth, a 32-year-old newlywed, was eager to start a family. She was career focused, as was her husband, Frank, and both thrived under high-stress conditions. Their jobs required long hours, with little time to cook or clean at home. For that reason, they were constantly eating on the go, often getting takeout meals that were, admittedly, not the healthiest. Despite this hectic schedule, the couple was ready to have a child of their own. So, they began trying to conceive. Both Elizabeth and Frank were optimistic that they would be able to conceive despite having many friends that struggled with infertility. No women in Elizabeth's family had struggled before, so she had no real concerns.

Frank and Elizabeth's attempts to conceive were met with disappointment. They "tried" for 1 year with no luck, and ultimately sought medical advice, because Elizabeth was concerned that her age would soon make conception more difficult. Both Frank and Elizabeth visited a fertility clinic to investigate their case further. Doctors first tested Frank's sperm count and sperm viability, because this is the simplest test to conduct. There was no evidence of inadequate numbers of sperm or their inability to fertilize an egg. Frank and Elizabeth were relieved, yet still frustrated that they were unsure as to the cause of their infertility.

Doctors then made note of Elizabeth's physical health. At 5 feet, 0 inches, Elizabeth weighed 175 pounds. In addition to being overweight, she also suffered from chronic hypertension and an elevated heart rate. Doctors suggested a change in lifestyle to reduce stress, which they hoped would help Elizabeth lose weight and improve her cardiovascular health. Doctors also inquired about Elizabeth's reproductive health. Elizabeth admitted that she had been prescribed birth control pills at

(continued on next page)

(continued from previous page)

15 years old to help regulate her irregular (unpredictable) periods. Since stopping the pill just 1 year ago, her periods were again irregular, with cycle durations ranging from 31 to 51 days. Menses would last 7 days with no remarkable pain or discomfort. She also noticed that she had an increased incidence of facial hair and acne, something that was quite unusual for her. Based on these symptoms, doctors ordered blood samples **(table 27.13)** to test for a condition called polycystic ovarian syndrome (PCOS).

PCOS is an endocrine disorder, characterized by elevated levels of testosterone, that is a major cause of infertility in females. Patients are often diabetic (type II), and exhibit signs of insulin resistance. Patients tend to be overweight and experience irregular periods due to a disruption in the positive and negative feedback mechanisms that regulate the ovarian and uterine cycles. Often, patients with PCOS do not ovulate, thereby preventing fertilization and implantation. Instead of ovulating, partially developed follicles may persist as "cysts" in the ovary. Clinically, ultrasound is used to observe ovarian cysts, and a diagnosis of PCOS requires that one ovary has more than twelve follicles ranging 2–9 mm in diameter. It is proposed that regulation of blood glucose levels and weight loss may allow ovulation to resume, although the exact mechanisms are unclear. In the event that dietary changes are unsuccessful, drugs can be administered to regulate blood glucose levels as well as stimulate ovulation.

Table 27.13	Elizabeth's Blood Test Results	
Blood Test	**Result**	**Reference Range**
Cholesterol (mg/dl)	218	< 200
FSH (mIU/L)	6	3–20
Glucose (mg/dl)	141	70–110
Insulin (µU/mL)	167	0–20
LH (mIU/L)	18	3–20
Testosterone (pg/mL)	15	1.1–6.3
Triglycerides (mg/dl)	291	40–150

Answer the following questions regarding Elizabeth's case:

a. Doctors often prescribe metformin, a drug that decreases glucose transport in the GI tract, for patients suffering from PCOS. Describe how this would be beneficial, given the symptoms of the disease.

b. Elizabeth's blood samples were taken on day 3 of the menstrual cycle. Describe the relative levels of estrogen, progesterone, LH, and FSH during this phase of the cycle.

c. How might these hormone levels be different if samples were collected on day 14 (assume a 28-day cycle)?

d. Describe the potential consequences of elevated LH when compared to FSH levels.

e. Discuss why elevated testosterone may lead to anovulation. (Hint: Testosterone exhibits similar feedback mechanisms in males and females).

f. Do you think that Elizabeth is suffering from PCOS?

Fertilization and Development

The time period between **fertilization** and birth is considered the **prenatal period.** Fertilization yields a diploid cell, or **zygote,** that undergoes multiple mitotic divisions, known as **cleavage** events. Two cells become four cells, and then eight cells, until there is a mass of cells, called a **blastocyst,** which is ready for implantation in the uterine wall. This cleavage process takes approximately 2 weeks and constitutes the **pre-embryonic period.** Implantation marks the official start of gestation, a period that lasts 38 weeks in all. Weeks 3 through 8 constitute the **embryonic period,** a time in which the organism is called an **embryo**. During this time, major organs and organ systems develop. In the ninth week of development, the organism is classified as a **fetus.** The **fetal period** lasts for the remaining 30 weeks of gestation. During the fetal period, organs and systems continue to grow and develop in preparation for birth and to sustain life separate from the mother's womb. The following laboratory exercises are designed to have you further explore the events of the pre-embryonic and embryonic periods.

EXERCISE 27.15

EARLY DEVELOPMENT: FERTILIZATION AND ZYGOTE FORMATION

The following laboratory exercise is designed to reinforce your understanding of the changes that occur during the pre-embryonic period.

1. Using your textbook as a guide, complete **table 27.14** below, depicting the chronology of events in pre-embryonic development.

Table 27.14	Pre-Embryonic Period		
Developmental Stage	**Time of Occurence**	**Location**	**Events**
	Ovum pronucleus / Sperm pronucleus — 120 μm		
	Nucleus — 120 μm		
	4-cell stage / 8-cell stage — 120 μm		
	Morula — 120 μm		

(continued on next page)

798 Chapter Twenty-Seven *The Reproductive System and Early Development*

(continued from previous page)

Table 27.14	Pre-Embryonic Period *(continued)*		
Developmental Stage	Time of Occurence	Location	Events
	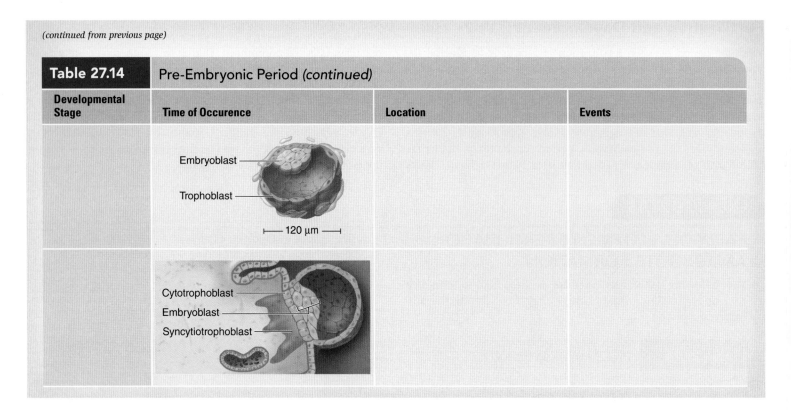		

EXERCISE 27.16

EARLY DEVELOPMENT: EMBRYONIC DEVELOPMENT

The following laboratory exercise is designed to reinforce your understanding of the changes that occur in an organism during the embryonic period.

1. Using your textbook as a guide, label the structures of the 3-week-old embryo identified in **figure 27.19**.

2. Using your textbook as a guide, complete **table 27.15** depicting the chronology of events in embryonic development.

3. Using your textbook as a guide, label the structures of the 4-week-old embryo identified in **figure 27.20**.

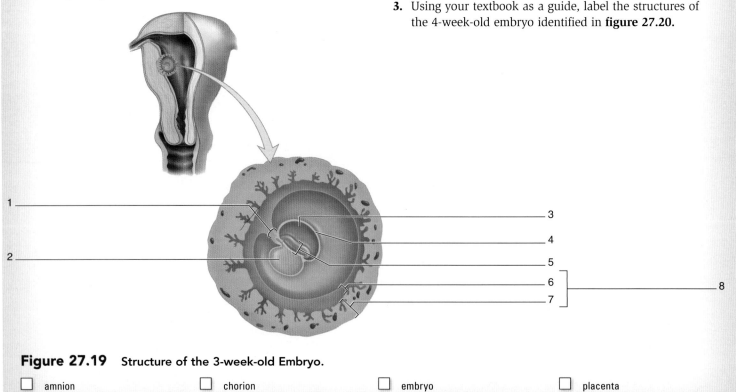

Figure 27.19 Structure of the 3-week-old Embryo.

- ☐ amnion
- ☐ chorion
- ☐ embryo
- ☐ placenta
- ☐ amniotic cavity
- ☐ connecting stalk
- ☐ functional layer of uterus
- ☐ yolk sac

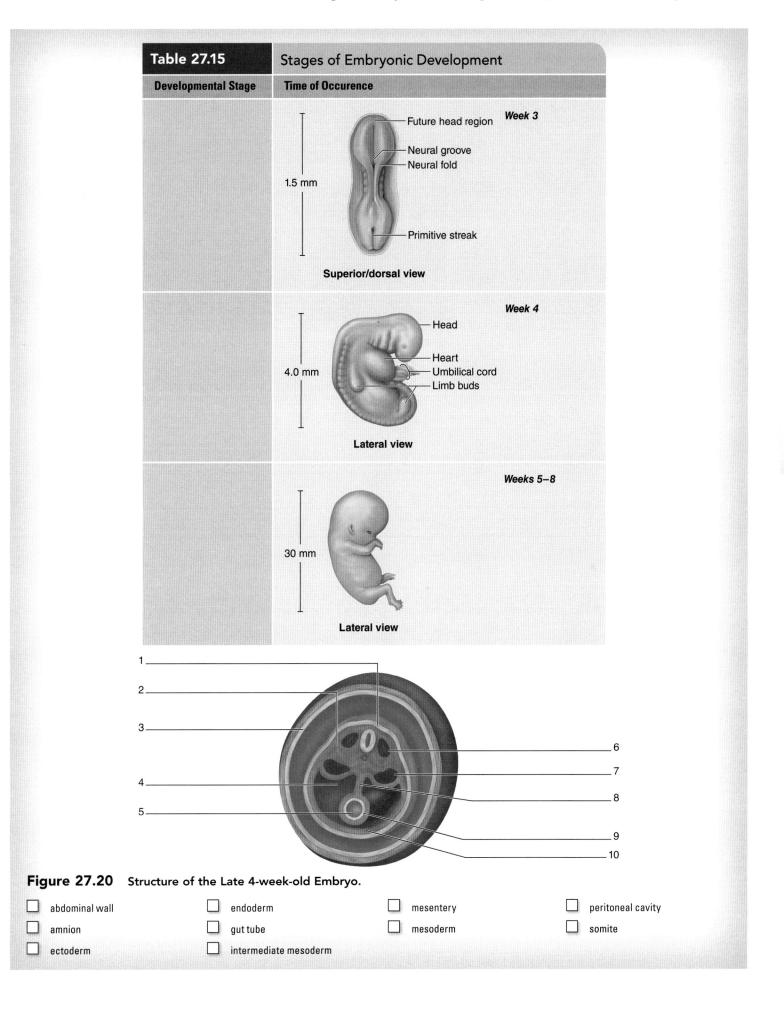

Figure 27.20 Structure of the Late 4-week-old Embryo.

☐ abdominal wall ☐ endoderm ☐ mesentery ☐ peritoneal cavity
☐ amnion ☐ gut tube ☐ mesoderm ☐ somite
☐ ectoderm ☐ intermediate mesoderm

Chapter 27: The Reproductive System

Name: _____
Date: _____ Section: _____

POST-LABORATORY WORKSHEET

The ❶ corresponds to the Learning Objective(s) listed in the chapter opener outline.

Do You Know the Basics?

Exercise 27.1: Histology of the Ovary

1. Identify the structure shown in the photomicrograph below. Describe the characteristic features of the structure that allowed you to come to your answer. ❶

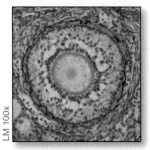

Structure: _____

Description: _____

_____

_____

2. Briefly describe the process of gametogenesis in females. ❷

_____

_____

_____

3. Compare and contrast the following: ❸

   a. primary follicle versus primary oocyte

   _____

   _____

   _____

   b. secondary follicle versus secondary oocyte

   _____

   _____

   _____

4. How can you distinguish histologically between a secondary follicle and a vesicular (Graafian) follicle? ❸

_____

_____

_____

5. The figure below demonstrates a _____ follicle, which contains a _____ oocyte. The oocyte is in the _____ stage of meiosis. ❸

Exercise 27.2: Histology of the Uterine Tubes

6. Name the three layers that compose the wall of the uterine tube. ❺

    a. _____

    b. _____

    c. _____

7. What type of epithelium lines the uterine tube? ❻ _____

Exercise 27.3: Histology of the Uterine Wall

8. The two layers of the uterine endometrium are the _____ layer and the _____ layer. The _____ layer is shed during menstruation. ❽

Exercise 27.4: Histology of the Vaginal Wall

9. What type of epithelium lines the vaginal wall? ❿ _____

Exercise 27.5: Histology of the Seminiferous Tubules

10. Briefly describe the process of gametogenesis in males. ⓬

    _____

    _____

    _____

11. Describe histological features of the following structures: ⓭

    a. interstitial cell _____

    b. primary spermatocyte _____

    c. secondary spermatocyte _____

    d. seminiferous tubule _____

    e. spermatids _____

    f. spermatogonia _____

    g. sustentacular cell _____

12. List three functions of sustentacular (Sertoli) cells of the testis. ⓮

    a. _____

    b. _____

    c. _____

Chapter Twenty-Seven The Reproductive System and Early Development

Exercise 27.6: Histology of the Epididymis

13. What type of epithelium lines the epididymis? 16

_____

_____

14. What is the function of stereocilia within the epididymis? 17

_____

_____

_____

15. Identify the structure shown in the photomicrograph below. Then describe the characteristic features of the structure that allowed you to come to your answer. 18

Structure: _____

Description: _____

_____

_____

Exercise 27.7: Histology of the Ductus Deferens

16. Describe the structures that compose the walls of the ductus deferens. 19

_____

_____

17. What type of epithelium lines the ductus deferens? 20

_____

_____

Exercise 27.8: Histology of the Seminal Vesicles

18. What type of epithelium lines the seminal vesicles? 22 _____

Exercise 27.9: Histology of the Prostate Gland

19. Describe distinguishing features of the prostate gland when viewed through a microscope. 23 _____

_____

_____

Exercise 27.10: Histology of the Penis

20. Identify the three erectile bodies in the penis. 26

   a. _____

   b. _____

   c. _____

Exercise 27.11: Gross Anatomy of the Ovary, Uterine Tubes, Uterus, and Suspensory Ligaments

21. Put the following terms in the correct order to describe the pathway that an ovum takes as it travels from the ovary to the uterus in the female.

    ampulla of uterine tube

    infundibulum of uterine tube

    isthmus of uterine tube

    ovarian cortex

    uterine part of uterine tube

    uterus

    _____
    _____
    _____

22. Does the inguinal canal exist in females? _____ If so, what structure(s) run through it? _____
    _____
    _____

23. The peritoneal cavity in the female is considered an "open" cavity. Based on your knowledge of the female reproductive organs, explain why.
    _____
    _____
    _____

24. You are a nurse and your job is to insert a catheter into the urethra of a female patient.

    a. Where will you look for the opening to the female urethra (be specific)?
    _____
    _____

    b. Approximately how long must the catheter be in order to reach into the bladder of the female?
    _____
    _____

Exercise 27.12: Gross Anatomy of the Female Breast

25. What type of glands are mammary glands?
    _____
    _____

Exercise 27.13: Gross Anatomy of the Scrotum, Testis, Spermatic Cord, and Penis

26. Label the following diagram of the testis. 31 32

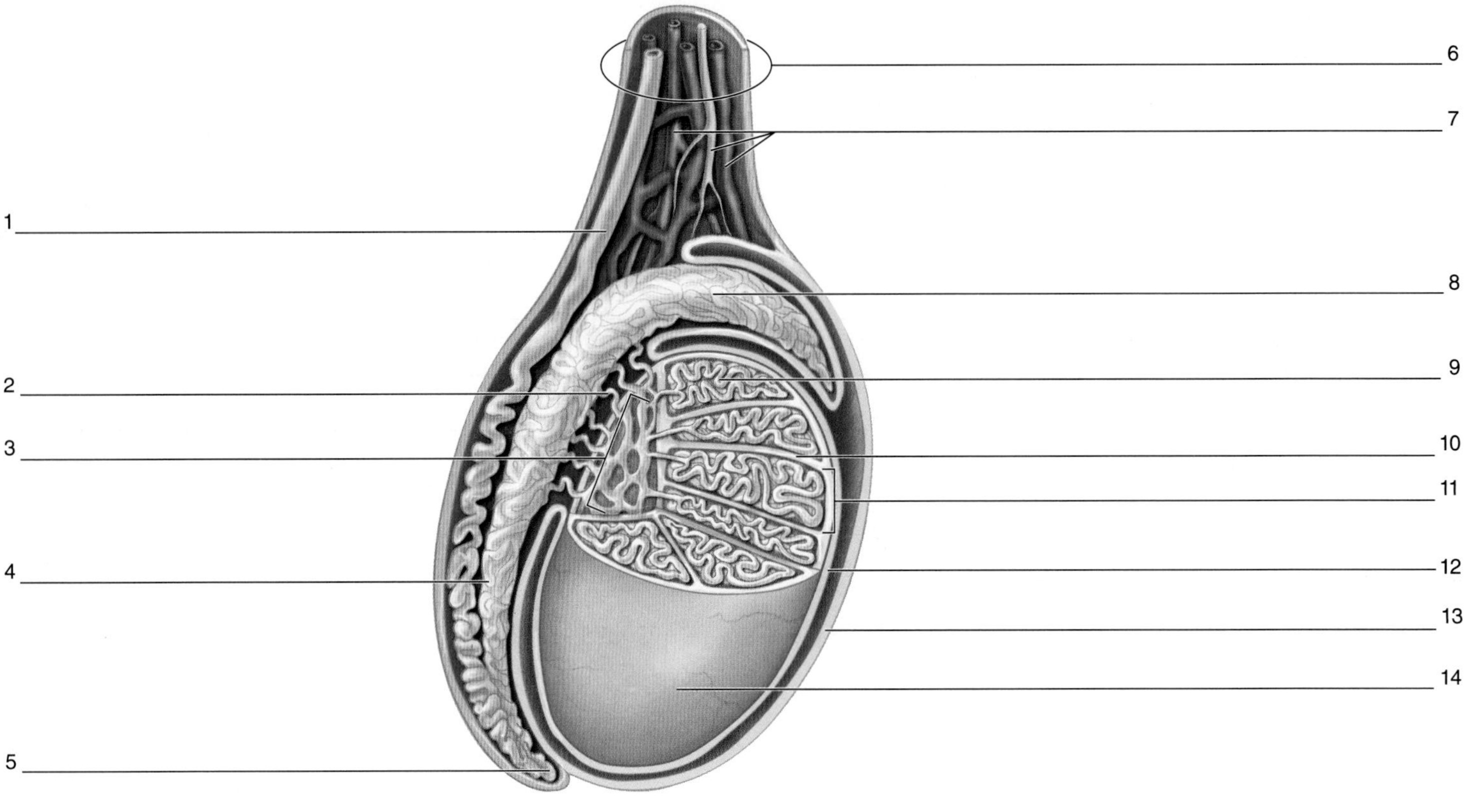
1
2
3
4
5
6
7
8
9
10
11
12
13
14

27. What is the function of the epididymis? 31

28. Identify one substance produced by the seminal vesicles that is an important component of semen. 33

29. Describe the location of the prostate gland. 34

30. Provide the homologous female and male reproductive structures for each blank in the table. 36

Female Structure	Male Homologue
Ovary	
	Penis
Labia Majora	
	Bulbourethral Glands

31. Based on your observations of male reproductive structures (and using table 27.9 as a guide), list the main constituents of semen and the relative percentages of each constituent.

32. Describe the components that make up the spermatic cord.

33. Trace the pathway a spermatozoon travels, starting at the epididymis, as it travels through the male reproductive tract during ejaculation. Name all accessory reproductive glands encountered along the way.

Exercise 27.14: A Clinical Case in Reproductive Physiology

34. Put the following terms in the correct sequence to describe the correct maturation sequence for a developing sperm cell.

    primary spermatocyte

    secondary spermatocyte

    spermatid

    spermatogonia

    spermatozoon

35. Identify the two phases of the ovarian cycle.
    a. _____
    b. _____

36. Identify the two phases of the uterine cycle.
    a. _____
    b. _____

37. What is the function of the corpus luteum?

Exercise 27.15: Early Development: Fertilization and Zygote Formation

38. The diploid cell that forms as a result of fertilization is called a _____.

39. The structure that implants in the uterine wall is called a _____.

Exercise 27.16: Early Development: Embryonic Development

40. Identify the weeks of gestation during which embryonic development occurs. _____

41. Identify the three germ layers present in an embryo.

    a. _____

    b. _____

    c. _____

Can You Apply What You've Learned?

42. Unlike in the epididymis, it is highly unlikely that sperm cells are visible inside the lumen of the ductus deferens when viewed histologically. Why do you think this is? _____

_____

_____

43. The homologous structure to the male penis is the female clitoris. The clitoris consists of two paired erectile tissues, the corpora cavernosa, but it does not contain a corpus spongiosum. What structure is the clitoris "missing" as compared to the penis? _____

_____

44. A woman that is trying to conceive may purchase ovulation kits in order to test for the release of the secondary oocyte. An ovulation kit requires placing a dipstick in a collected urine sample. A positive test indicates that ovulation will happen typically within 48 hours. For which hormone is this test sensitive?

_____

45. One cause of miscarriage, or loss of pregnancy, is low levels of progesterone in the initial stages of development.

    a. Describe what mechanisms are in place to ensure adequate levels of progesterone in the event that fertilization occurs. _____

    _____

    b. What long-term mechanisms are in place to provide a continuous supply of progesterone (to maintain pregnancy)? _____

    _____

Can You Synthesize What You've Learned?

46. Viagra®, a drug known for its ability to enhance erection in males, is a drug that was first prescribed to treat hypertension. Using the concepts of pressure, flow, and resistance, describe why a drug used to counteract high blood pressure may also lead to an erection in males. _____

_____

_____

_____

47. Birth control pills work to override the hormone cycles in order to prevent a secondary oocyte from being released. Based on your knowledge of hormonal regulation of the ovarian cycle, design a drug that could be used to prevent ovulation. Be specific about what hormones would be present in the female and why. Would you expect the levels of these hormones to change throughout the 28-day cycle? _____

_____

_____

_____

ANSWER KEYS

This appendix includes answers labeling exercises, table completion activities, and calculation questions found in each chapter. Answers to in-exercise questions prompting students to record their personal observations and drawings are not provided due to their variability.

Chapter 1

Exercise 1.3, #1
Practice:
a 0.5 cm
b 0.74 ounces
c 0.5 L
d 12.7 cm
e 9.9 ft
f 11.5 L
g 3000 mL

Figure 1.2 Identification of Common Dissection Instruments
1 forceps
2 scalpel blades
3 scalpel (disposable)
4 scalpel blade handle (#3)
5 scalpel blade handle (#4)
6 scissors (curved)
7 scissors (pointed)
8 blunt probe
9 dissecting pins
10 dissecting needle
11 blunt probe
12 hemostat

Exercise 1.5
1 A
2 B
3 A
4 A
5 B
6 B
7 A
8 C
9 A

Chapter 2

Chapter Activities

Figure 2.2 Sections Through a Human Brain
1 coronal
2 midsagittal
3 transverse
4 sagittal

Figure 2.3 Posterior View of an Individual with Three Reference Locations Marked
1 thoracic
2 antebrachial
3 femoral

Figure 2.4 Regional Terms
1 oral
2 cervical
3 axillary
4 brachial
5 antebrachial
6 carpal
7 digital
8 femoral
9 crural
10 frontal
11 orbital
12 buccal
13 mental
14 mammary
15 pelvic
16 inguinal
17 tarsal
18 otic
19 vertebral
20 sacral
21 sural
22 calcaneal
23 occipital
24 lumbar
25 perineal
26 popliteal

Figure 2.5 Body Cavities
1 posterior aspect
2 ventral cavity
3 cranial cavity
4 vertebral canal
5 thoracic cavity
6 abdominopelvic cavity
7 diaphragm
8 abdominal cavity
9 pelvic cavity
10 thoracic cavity
11 abdominopelvic cavity
12 mediastinum
13 pleural cavity
14 pericardial cavity
15 abdominal cavity
16 pelvic cavity

Exercise 2.5, #2

Organ	Quadrant(s)	Region(s)
Left kidney	Left upper	Left hypochondriac, left lumbar
Liver	Right upper, left upper	Right hypochondriac, right lumbar, epigastric
Pancreas	Right and left upper	Epigastric, left hypochondriac
Small intestine	Right and left lower	Umbilical, hypogastric, right and left lumbar and right and left iliac
Spleen	Left upper	Left hypochondriac
Stomach	Right and left upper	Epigastric, left hypochondriac
Urinary bladder	Right and left lower	Hypogastric

Chapter 3

No Activities

Chapter 4

Concept Connection (p. 73)
Molecular weight of $KMnO_4$ = 158
Molecular weight of $C_{16}H_{18}ClN_3S$ = 423

Chapters 5-6

No Activities

Chapter 7

Chapter Activities

Figure 7.6 Identifying Bones Based on Structure
1 flat
2 short
3 long
4 irregular
5 long
6 irregular
7 irregular

Figure 7.10 The Human Skeleton
(a) Anterior View
1 skull
2 mandible
3 clavicle
4 scapula
5 sternum
6 humerus
7 rib
8 vertebra
9 ilium
10 sacrum
11 coccyx
12 ischium
13 radius
14 ulna
15 pubis
16 carpals
17 metacarpals
18 phalanges
    (of the hand)
19 femur
20 patella
21 fibula
22 tibia
23 tarsals
24 metatarsals
25 phalanges
    (of the foot)

(b) Posterior View
1 skull
2 mandible
3 scapula
4 humerus
5 rib
6 vertebra
7 ilium
8 sacrum
9 coccyx
10 ischium
11 radius
12 ulna
13 pubis
14 carpals
15 metacarpals
16 phalanges
    (of the hand)
17 femur
18 fibula
19 tibia
20 tarsals
21 metatarsals
22 phalanges
    (of the foot)

Chapter 8

Chapter Activities

Figure 8.1 Anterior View of the Skull
1 frontal bone
2 parietal bone
3 glabella
4 sphenoid bone
5 superior orbital fissure
6 lacrimal bone
7 nasal bone
8 zygomatic bone
9 infraorbital foramen
10 maxilla
11 superciliary arch
12 supraorbital foramen (notch)
13 supraorbital margin
14 temporal bone
15 optic canal
16 inferior orbital fissure
17 perpendicular plate of ethmoid
18 vomer
19 inferior nasal concha
20 alveolar processes
21 mandible
22 mental foramen
23 mental protuberance

Figure 8.2 The Orbit
1 nasal bone
2 supraorbital foramen (notch)
3 ethmoid bone
4 frontal bone
5 zygomatic process of frontal bone
6 optic foramen
7 superior orbital fissure
8 lacrimal bone
9 inferior orbital fissure
10 zygomatic bone
11 infraorbital foramen
12 maxilla

Figure 8.3 The Nasal Cavity
1. crista galli (of ethmoid bone)
2. cribriform plate of ethmoid bone
3. sella turcica
4. sphenoid sinus
5. palatine bone
6. sphenoid bone
7. frontal sinus
8. nasal bone
9. superior nasal concha
10. middle nasal concha
11. lacrimal bone
12. inferior nasal concha
13. maxilla
14. perpendicular plate of ethmoid bone
15. vomer

Figure 8.4 The Mandible
1. condylar process
2. coronoid process
3. mandibular foramen
4. condylar process
5. mandibular notch
6. mylohyoid line
7. ramus
8. alveolar process
9. mental foramen
10. angle
11. body
12. mental protuberance

Figure 8.5 Lateral View of the Skull
1. parietal eminence
2. parietal bone
3. inferior temporal line
4. squamous suture
5. lambdoid suture
6. squamous part of temporal bone
7. occipital bone
8. external acoustic meatus
9. mastoid process
10. styloid process
11. head of mandible
12. zygomatic process of temporal bone
13. temporal process of zygomatic bone
14. body of mandible
15. coronal suture
16. frontal bone
17. superior temporal line
18. pterion
19. greater wing of sphenoid bone
20. nasal bone
21. lacrimal bone
22. ethmoid bone
23. lacrimal groove
24. zygomatic bone
25. maxilla
26. mental foramen
27. mental protuberance

Figure 8.6 Posterior View of the Skull
1. sagittal suture
2. parietal foramina
3. parietal bone
4. parietal eminence
5. lambdoid suture
6. sutural (wormian) bone
7. occipital bone
8. temporal bone
9. external occipital protuberance
10. mastoid process

Figure 8.7 Superior View of the Skull
1. frontal bone
2. coronal suture
3. sagittal suture
4. parietal bone
5. parietal foramina
6. lambdoid suture
7. sutural (wormian) bone
8. occipital bone

Figure 8.8 Inferior View of the Skull
1. maxilla
2. palatine bone
3. vomer
4. sphenoid bone
5. foramen ovale
6. foramen spinosum
7. foramen lacerum
8. jugular foramen
9. carotid canal
10. inferior nuchal line
11. occipital bone
12. incisive foramen
13. temporal process of zygomatic bone
14. zygomatic process of temporal bone
15. mandibular fossa
16. styloid process
17. temporal bone
18. mastoid process
19. occipital condyle
20. hypoglossal canal
21. basilar region of occipital bone
22. foramen magnum
23. external occipital crest
24. superior nuchal line
25. external occipital protuberance

Figure 8.9 Cranial Fossae
1. anterior cranial fossa
2. middle cranial fossa
3. posterior cranial fossa

Figure 8.11 Superior View of the Cranial Floor
1. frontal sinus
2. frontal bone
3. optic canal
4. lesser wing of sphenoid bone
5. foramen rotundum
6. greater wing of sphenoid bone
7. temporal bone
8. petrous part of temporal bone
9. parietal bone
10. internal acoustic meatus
11. foramen magnum
12. occipital bone
13. internal occipital protuberance
14. frontal crest
15. crista galli
16. cribriform plate of the ethmoid bone
17. sella turcica
18. foramen lacerum
19. foramen ovale
20. foramen spinosum
21. jugular foramen
22. hypoglossal canal
23. groove for sigmoid sinus
24. basilar part of occipital bone
25. groove for transverse sinus
26. internal occipital crest
27. anterior clinoid process
28. hypophyseal fossa
29. posterior clinoid process

Figure 8.12 The Hyoid Bone
1. greater cornu
2. lesser cornu
3. body

Figure 8.13 The Fetal Skull
1. frontal bone
2. parietal bone
3. sphenoidal fontanel
4. sphenoid bone
5. mandible
6. temporal bone
7. mastoid fontanel
8. occipital bone
9. frontal bone
10. anterior fontanelle
11. parietal bone

Figure 8.14 Lateral View of the Vertebral Column
1. cervical vertebrae
   Number of vertebrae: 7
2. thoracic vertebrae
   Number of vertebrae: 12
3. lumbar vertebrae
   Number of vertebrae: 5
4. sacrum
   Number of vertebrae: 5
5. coccygeal vertebrae
   Number of vertebrae: 4
6. cervical curvature
7. thoracic curvature
8. lumbar curvature
9. sacral curvature

Figure 8.15 A Typical Vertebra
1. lamina
2. transverse process
3. superior articular process
4. body
5. spinous process
6. vertebral arch
7. transverse process
8. spinous process
9. vertebral (spinal) foramen
10. transverse process
11. superior articular process
12. body
13. inferior articular process

Figure 8.16 Cervical Vertebra
1. body
2. transverse process
3. superior articular process (and facet)
4. transverse foramen
5. pedicle
6. vertebral (spinal) foramen
7. lamina
8. spinous process
9. spinous process
10. superior articular process (and facet)
11. transverse foramen
12. body
13. inferior articular process (and facet)

Figure 8.17 The Atlas (C1)
1. anterior tubercle
2. anterior arch
3. superior articular facet
4. transverse process
5. transverse foramen
6. posterior tubercle
7. posterior arch
8. articular facet for dens
9. vertebral foramen

Figure 8.18 The Axis (C2)
1. dens (odontoid process)
2. superior articular process
3. transverse foramen
4. transverse process
5. pedicle
6. vertebral (spinal) foramen
7. lamina
8. spinous process

Figure 8.19 Thoracic Vertebra
1. body
2. vertebral (spinal) foramen
3. superior articular process
4. spinous process
5. superior costal facet
6. pedicle
7. costal facet
8. transverse process
9. lamina
10. costal facet
11. spinous process
12. inferior articular process
13. superior articular process
14. superior costal facet
15. body
16. inferior costal facet

Figure 8.20 Lumbar Vertebra
1. body
2. vertebral (spinal) foramen
3. superior articular process (and facet)
4. spinous process
5. pedicle
6. transverse process
7. lamina
8. superior articular process (and facet)
9. transverse process
10. spinous process
11. inferior articular process (and facet)
12. body

Figure 8.21 Sacrum and Coccyx
1. superior articular process
2. ala
3. sacral promontory
4. anterior sacral foramina
5. transverse ridges
6. coccygeal cornu
7. coccyx
8. sacral canal
9. superior articular facet
10. median sacral crest
11. auricular surface
12. posterior sacral foramina
13. sacral hiatus
14. coccygeal cornu
15. coccyx

Figure 8.22 The Sternum
1. suprasternal notch
2. manubrium
3. sternal angle
4. body
5. second rib
6. xiphoid process

Figure 8.23 A Typical Rib
1. superior articular facet
2. inferior articular facet
3. shaft
4. head
5. neck
6. tubercle
7. articular facet for transverse process
8. angle
9. costal groove

Chapter 9

Chapter Activities

Figure 9.1 The Clavicle
1. acromial end (lateral)
2. sternal end (medial)
3. acromial end
4. conoid tubercle
5. sternal end
6. costal tuberosity

Figure 9.2 The Right Scapula
1. acromion
2. coracoid process
3. glenoid cavity
4. lateral (axillary) border
5. inferior angle
6. suprascapular notch
7. spine
8. superior border
9. superior angle
10. subscapular fossa
11. medial (vertebral) border
12. acromion
13. spine
14. glenoid cavity
15. infraglenoid tubercle
16. lateral (axillary) border
17. superior angle
18. supraglenoid tubercle
19. coracoid process
20. subscapular fossa
21. inferior angle
22. coracoid process
23. suprascapular notch
24. superior border
25. supraspinous fossa
26. spine
27. infraspinous fossa
28. medial (vertebral) border
29. acromion
30. glenoid cavity
31. lateral (axillary) border
32. inferior angle

Figure 9.3 The Right Humerus
1. anatomical neck
2. greater tubercle
3. lesser tubercle
4. intertubercular sulcus
5. surgical neck
6. deltoid tuberosity
7. shaft
8. coronoid fossa
9. radial fossa
10. lateral epicondyle
11. capitulum
12. trochlea
13. head
14. medial epicondyle
15. head
16. greater tubercle
17. anatomical neck
18. surgical neck
19. deltoid tuberosity
20. medial epicondyle
21. olecranon fossa
22. lateral epicondyle
23. trochlea
24. lateral epicondyle
25. medial epicondyle
26. trochlea
27. supracondylar ridges
28. ulna
29. radius
30. humerus
31. medial epicondyle
32. lateral epicondyle
33. radius
34. ulna

Figure 9.4 The Radius
1. head
2. neck
3. radial tuberosity
4. shaft
5. styloid process of radius
6. ulnar notch

Figure 9.5 The Ulna
1. trochlear notch
2. coronoid process
3. radial notch
4. tuberosity of ulna
5. shaft of ulna
6. styloid process of ulna
7. olecranon
8. coronoid process

Exercise 9.2D, #3

Carpal Bone	Word Origin	Bone Shape/Appearance
Scaphoid	*skaphe*, boat	Shaped like a boat
Lunate	*luna*, moon	Moon-shaped
Triquetrum	*triquetrus*, 3-cornered	Triangular
Pisiform	*pisum*, pea	Pea-shaped
Trapezium	*trapezion*, a table	Table-shaped
Trapezoid	*trapezion*, a table	Table-shaped
Capitate	*caput*, head	Head-shaped
Hamate	*hamus*, a hook	Hook-shaped

Figure 9.7 The Metacarpals and Phalanges
1. phalanges
2. metacarpals
3. carpals
4. distal phalanx
5. middle phalanx
6. proximal phalanx
7. distal phalanx of pollex
8. proximal phalanx of pollex
9. metacarpal V
10. metacarpal IV
11. metacarpal III
12. metacarpal I
13. metacarpal II

Figure 9.8 Surface Anatomy of Upper Limb
1. clavicle
2. acromion of scapula
3. deltoid tuberosity
4. medial epicondyle of humerus
5. suprasternal notch
6. styloid process of ulna
7. olecranon
8. styloid process of radius
9. spine of scapula

Figure 9.9 The Right Os Coxae
1. ala
2. anterior gluteal line
3. posterior gluteal line
4. posterior superior iliac spine
5. posterior inferior iliac spine
6. greater sciatic notch
7. body of ischium
8. ischial spine
9. lesser sciatic notch
10. ischial tuberosity
11. iliac crest
12. anterior superior iliac spine
13. inferior gluteal line
14. anterior inferior iliac spine
15. lunate surface
16. acetabulum
17. superior pubic ramus
18. pubic crest
19. pubic tubercle
20. inferior pubic ramus
21. obturator foramen
22. ramus of ischium
23. iliac fossa
24. anterior superior iliac spine
25. anterior inferior iliac spine
26. arcuate line
27. pectineal line
28. superior pubic ramus
29. pubic tubercle
30. symphysial surface of pubic bone
31. obturator foramen
32. inferior pubic ramus
33. iliac crest
34. posterior superior iliac spine
35. auricular surface
36. posterior inferior iliac spine
37. greater sciatic notch
38. ischial spine
39. lesser sciatic notch
40. body of ischium
41. ischial tuberosity
42. ramus of ischium

Figure 9.11 The Right Femur
1. head
2. greater trochanter
3. fovea
4. neck
5. intertrochanteric crest
6. lesser trochanter
7. shaft
8. head
9. shaft
10. patellar surface
11. intercondylar fossa
12. lateral epicondyle
13. medial epicondyle
14. head
15. greater trochanter
16. neck
17. intertrochanteric crest
18. lesser trochanter
19. shaft
20. adductor tubercle
21. lateral epicondyle
22. medial epicondyle
23. patellar surface
24. lateral condyle
25. medial condyle
26. head
27. fovea
28. greater trochanter
29. neck
30. intertrochanteric crest
31. lesser trochanter
32. pectineal line
33. gluteal tuberosity
34. linea aspera
35. medial supracondylar line
36. lateral supracondylar line
37. lateral epicondyle
38. popliteal surface
39. adductor tubercle
40. medial epicondyle
41. medial condyle
42. intercondylar fossa
43. lateral condyle

Figure 9.12 The Tibia
1. lateral condyle
2. intercondylar eminence
3. medial condyle
4. tibial tuberosity
5. anterior border
6. medial malleolus
7. medial condyle
8. intercondylar eminence
9. lateral condyle
10. fibular articular facet
11. medial malleolus

Figure 9.13 The Right Fibula
1. head
2. neck
3. shaft
4. lateral malleolus

Figure 9.14 The Tarsals
1. phalanges
2. metatarsals
3. medial cuneiform
4. navicular
5. intermediate cuneiform
6. lateral cuneiform
7. cuboid
8. talus
9. calcaneus

Exercise 9.5D, #5

Tarsal Bone	Word Origin	Bone Shape/Appearance
Talus	*talus*, ankle	Convex, triangular
Calcaneus	*calcaneous*, heel	Elongated
Navicular	*navis*, ship	Shaped like a boat
Medial Cuneiform	*cuneus*, wedge	Wedge-shaped
Intermediate Cuneiform	*cuneus*, wedge	Wedge-shaped
Lateral Cuneiform	*cuneus*, wedge	Wedge-shaped
Cuboid	*kybos*, cube	Cube-shaped

Figure 9.15   The Metatarsals and Phalanges
1. metatarsal II
2. metatarsal III
3. metatarsal IV
4. metatarsal V
5. lateral cuneiform
6. cuboid
7. distal phalanx
8. middle phalanx
9. proximal phalanx
10. metatarsal I
11. medial cuneiform
12. intermediate cuneiform
13. navicular
14. talus
15. calcaneus

Figure 9.16   Surface Anatomy of the Lower Limb
1. iliac crest
2. anterior superior iliac spine
3. sacrum
4. coccyx
5. greater trochanter of femur
6. ischial tuberosity
7. lateral epicondyle of femur
8. head of fibula
9. lateral malleolus
10. medial epicondyle of femur
11. patella
12. tibial tuberosity
13. shaft of tibia
14. medial malleolus
15. calcaneus
16. metatarsals
17. phalanges

Chapter 10

Chapter Activities
Figure 10.1   Fibrous Joints
1. syndesmosis
2. gomphosis
3. suture

Figure 10.2   Cartilaginous Joints
1. synchondrosis
2. symphysis
3. synchondrosis
4. symphysis

Figure 10.3   Diagram of a Representative Synovial Joint
1. periosteum
2. yellow bone marrow
3. fibrous layer of articular capsule
4. synovial membrane
5. synovial (joint) cavity
6. articular capsule
7. articular cartilage
8. ligament

Figure 10.4   Structural Classifications of Synovial Joints
1. saddle
2. hinge
3. plane
4. ball-and-socket
5. condylar
6. pivot

Figure 10.5   A Representative Synovial Joint: Right Knee Joint
1. posterior cruciate ligament
2. lateral condyle of femur
3. lateral meniscus
4. fibular collateral ligament
5. anterior cruciate ligament
6. fibula
7. medial condyle of femur
8. medial meniscus
9. tibial collateral ligament
10. tibia
11. anterior cruciate ligament
12. medial condyle of femur
13. medial meniscus
14. posterior cruciate ligament
15. tibial collateral ligament
16. tibia
17. femur
18. lateral condyle of femur
19. fibular collateral ligament
20. lateral meniscus
21. fibula
22. femur
23. quadriceps femoris tendon
24. suprapatellar bursa
25. patella
26. prepatellar bursa
27. articular cartilage
28. patellar ligament
29. infrapatellar bursa
30. tibia

Chapter 11

Chapter Activities
Figure 11.7   Muscles of the Human Body

Muscle Number	Muscle Name	Architecture
1	Orbicularis oculi	Circular
2	Deltoid	Multipennate
3	Pectoralis major	Convergent
4	Sartorius	Parallel
5	Rectus femoris	Bipennate
6	Trapezius	Convergent
7	Triceps brachii	Bipennate
8	Extensors of wrist/hand	Parallel
9	Gastrocnemius	Bipennate (individual heads) Multipennate (entire muscle)

Chapter 12

Chapter Activities
Figure 12.1   Muscles of Facial Expression
1. epicranius (occipitofrontalis)
2. epicranial aponeurosis (connective tissue)
3. frontal belly of occipitofrontalis
4. procerus
5. orbicularis oculi
6. levator labii superioris
7. zygomaticus minor
8. zygomaticus major
9. depressor anguli oris
10. depressor labii inferioris
11. platysma
12. corrugator supercilii
13. nasalis
14. levator anguli oris
15. risorius
16. orbicularis oris
17. mentalis
18. epicranial aponeurosis (connective tissue)
19. frontal belly of occipitofrontalis
20. epicranius (occipitofrontalis)
21. buccinator
22. orbicularis oculi
23. levator labii superioris
24. zygomaticus minor
25. zygomaticus major
26. orbicularis oris
27. levator anguli oris
28. depressor labii inferioris
29. depressor anguli oris
30. platysma

Figure 12.2   Muscles of Mastication
1. temporalis
2. masseter
3. temporalis
4. lateral pterygoid
5. medial pterygoid

Figure 12.3   Extrinsic Eye Muscles
1. superior rectus
2. lateral rectus
3. inferior oblique
4. inferior rectus
5. superior oblique (tendon)
6. medial rectus
7. lateral rectus
8. inferior oblique

Figure 12.4   Muscles That Move the Tongue
1. palatoglossus
2. styloglossus
3. hyoglossus
4. genioglossus

Figure 12.5   Muscles of the Pharynx
1. tensor veli palatini
2. levator veli palatini
3. superior constrictor
4. stylopharyngeus
5. middle constrictor
6. inferior constrictor

Concept Connection (p. 299)
The right sternocleidomastoid rotates the neck to the left. The right splenius capitis rotates the neck to the right. In summary: To rotate your neck to the right, you use the sternal head of the sternocleidomastoid on the left side of the neck, and the splenius capitis muscle on the right side of the neck.

Figure 12.6   Muscles of the Head and Neck
1. mylohyoid
2. stylohyoid
3. digastric (anterior belly)
4. digastric (posterior belly)
5. omohyoid (superior belly)
6. sternohyoid
7. sternocleidomastoid
8. thyrohyoid
9. sternothyroid
10. scalenes
11. splenius capitis
12. sternocleidomastoid
13. scalenes
14. omohyoid (inferior belly)
15. digastric (posterior belly)
16. mylohyoid
17. digastric (anterior belly)
18. sternothyroid
19. omohyoid (superior belly)
20. sternohyoid
21. semispinalis capitis
22. sternocleidomastoid
23. splenius capitis
24. rectus capitis posterior minor
25. rectus capitis posterior major
26. superior oblique
27. inferior oblique

Figure 12.7 Muscles of the Vertebral Column
1. splenius capitis
2. splenius cervicis
3. iliocostalis
4. longissimus
5. spinalis
6. semispinalis capitis
7. semispinalis cervicis
8. semispinalis thoracis
9. multifidus
10. quadratus lumborum

Figure 12.8 Muscles of Respiration
1. external intercostals
2. internal intercostals
3. transverse thoracis
4. diaphragm
5. internal intercostals
6. external intercostals
7. diaphragm

Figure 12.10 Diaphragm
1. caval opening (for inferior vena cava)
2. aortic opening (hiatus)
3. esophageal opening (hiatus)

Figure 12.11 Muscles of the Abdominal Wall
1. tendinous intersections
2. rectus abdominis
3. transversus abdominis
4. internal oblique
5. external oblique

Chapter 13

Chapter Activities

Figure 13.1 Muscles That Move the Pectoral Girdle and Glenohumeral Joint
1. trapezius
2. deltoid
3. pectoralis major
4. biceps brachii (long head)
5. biceps brachii (short head)
6. subscapularis
7. coracobrachialis
8. pectoralis minor
9. serratus anterior
10. trapezius
11. rhomboid minor
12. rhomboid major
13. deltoid
14. rhomboid major
15. latissimus dorsi
16. levator scapulae
17. supraspinatus
18. infraspinatus
19. teres minor
20. teres major
21. serratus anterior

Figure 13.2 Anterior (Flexor) Compartment of the Arm
1. coracobrachialis
2. biceps brachii (short head)
3. biceps brachii (long head)
4. tendon of the long head of biceps brachii
5. coracobrachialis
6. brachialis

Figure 13.3 Posterior (Extensor) Compartment of the Arm
1. lateral head of triceps brachii
2. long head of triceps brachii

Figure 13.4 Anterior (Flexor) Muscles of the Forearm
1. pronator teres
2. brachioradialis
3. flexor retinaculum
4. flexor carpi radialis
5. palmaris longus
6. flexor carpi ulnaris
7. flexor digitorum superficialis
8. palmar aponeurosis
9. supinator
10. flexor pollicis longus
11. pronator quadratus
12. flexor digitorum profundus

Figure 13.7 Posterior (Extensor) Compartment of the Forearm
1. anconeus
2. extensor carpi ulnaris
3. extensor digiti minimi
4. extensor retinaculum
5. extensor digitorum tendons
6. extensor carpi radialis longus
7. extensor carpi radialis brevis
8. extensor digitorum
9. abductor pollicis longus
10. extensor pollicis brevis
11. extensor pollicis longus
12. extensor indicis
13. supinator
14. abductor pollicis longus
15. extensor pollicis brevis

Figure 13.8 Intrinsic Muscles of the Hand
1. lateral lumbricals
2. flexor digiti minimi brevis
3. abductor digiti minimi
4. first dorsal interosseous
5. medial lumbrical
6. adductor pollicis
7. flexor pollicis brevis
8. abductor pollicis brevis
9. thenar group
10. hypothenar group

Figure 13.9 Muscles That Move the Pelvic Girdle
1. gluteus maximus
2. piriformis
3. superior gemellus
4. obturator internus
5. inferior gemellus
6. gluteus medius
7. gluteus minimus
8. gluteus medius
9. quadratus femoris
10. iliacus
11. psoas
12. iliopsoas
13. tensor fasciae latae
14. iliotibial tract

Figure 13.11 Anterior Compartment of the Thigh
1. inguinal ligament
2. tensor fasciae latae
3. iliotibial tract
4. rectus femoris
5. vastus lateralis
6. pectineus
7. adductor longus
8. gracilis
9. sartorius
10. vastus medialis
11. quadriceps tendon

Figure 13.12 Medial Compartment of the Thigh
1. pectineus
2. adductor brevis
3. adductor longus
4. gracilis

Figure 13.13 Posterior Compartment of the Thigh
1. semimembranosus
2. semitendinosus
3. biceps femoris (long head)
4. biceps femoris (short head)
5. adductor magnus

Figure 13.14 Anterior Compartment of the Leg
1. extensor digitorum longus
2. extensor hallucis longus
3. fibularis tertius tendon
4. tibialis anterior

Figure 13.15 Lateral View of the Leg
1. gastrocnemius
2. soleus
3. fibularis longus
4. fibularis brevis
5. fibularis tertius
6. extensor digitorum brevis
7. fibularis tertius tendon
8. tibialis anterior
9. extensor digitorum longus
10. extensor hallucis longus
11. extensor hallucis brevis
12. extensor hallucis longus tendon
13. extensor digitorum longus tendons

Figure 13.16 Posterior Compartment of the Leg
1. plantaris
2. popliteus
3. tibialis posterior
4. flexor digitorum longus
5. flexor hallucis longus
6. calcaneal tendon

Figure 13.17 Intrinsic Muscles of the Foot
1. flexor digitorum brevis
2. abductor hallucis
3. abductor digiti minimi
4. lumbricals
5. tendon of flexor hallucis longus
6. tendons of flexor digitorum longus
7. quadratus plantae
8. adductor hallucis
9. flexor hallucis brevis
10. flexor digiti minimi brevis
11. quadratus plantae
12. abductor digiti minimi
13. abductor hallucis
14. plantar interossei
15. dorsal interossei

Chapter 14

No Activities

Chapter 15

Chapter Activities

Figure 15.2 Meningeal Structures

(a) Coronal Section
1. superior sagittal sinus
2. falx cerebri
3. pia mater
4. subarachnoid space
5. arachnoid mater
6. falx cerebri
7. dura mater (periosteal layer)
8. dura mater
9. superior sagittal sinus
10. arachnoid villi
11. dura mater (meningeal layer)

(b) Midsagittal Section
1. falx cerebri
2. diaphragma sellae
3. superior sagittal sinus
4. inferior sagittal sinus
5. tentorium cerebelli
6. straight sinus
7. confluence of sinuses
8. falx cerebelli

Figure 15.4 Cast of the Ventricles of the Brain
1. lateral ventricles
2. interventricular foramen
3. third ventricle
4. cerebral aqueduct
5. fourth ventricle

Figure 15.5 Superior View of the Brain
1. frontal lobe
2. precentral gyrus
3. central sulcus
4. postcentral gyrus
5. longitudinal fissure
6. parietal lobe
7. occipital lobe

Figure 15.6 Lateral View of the Brain
1. frontal lobe
2. parietal lobe
3. central sulcus
4. precentral gyrus
5. postcentral gyrus
6. lateral sulcus
7. temporal lobe
8. pons
9. medulla oblongata
10. occipital lobe
11. transverse fissure
12. cerebellum
13. spinal cord

Appendix Answer Keys

Figure 15.7 Inferior View of the Brain
1. frontal lobe
2. infundibulum
3. mammillary bodies
4. temporal lobe
5. pons
6. occipital lobe
7. olfactory bulb
8. olfactory tract
9. optic chiasm
10. optic nerve
11. optic tract
12. midbrain
13. cerebellum
14. medulla oblongata
15. spinal cord

Figure 15.8 Midsagittal View of the Brain
1. frontal lobe
2. cingulate gyrus
3. corpus callosum
4. septum pellucidum
5. interthalamic adhesion
6. third ventricle
7. hypothalamus
8. midbrain
9. infundibulum
10. cerebral peduncle
11. temporal lobe
12. pons
13. medulla oblongata
14. spinal cord
15. central sulcus
16. parietal lobe
17. thalamus
18. parieto-occipital sulcus
19. occipital lobe
20. pineal body (gland)
21. tectal plate (corpora quadrigemina)
22. mammillary body
23. cerebral aqueduct
24. fourth ventricle
25. cerebellum

Exercise 15.7, #4

Point of Exit	Cranial Nerve
Midbrain	III, IV
Pons	V, VI, VII, part of VIII
Medulla oblongata	part of VIII, IX, X, XI, XII

Chapter 16

Chapter Activities

Figure 16.2 Regional Gross Anatomy of the Spinal Cord
1. cervical plexus
2. cervical enlargement
3. brachial plexus
4. lumbosacral enlargement
5. L1 vertebra
6. conus medullaris
7. lumbar plexus
8. sacral plexus
9. filum terminale
10. cauda equina
11. brain (cerebellum)
12. posterior rootlets
13. posterior median sulcus
14. posterior roots
15. anterior roots
16. posterior median sulcus
17. pia mater
18. denticulate ligament
19. dura mater
20. conus medullaris
21. dura mater
22. cauda equina
23. filum terminale
24. posterior root ganglion

Figure 16.3 Model of a Spinal Cord in Cross Section
1. posterior funiculus
2. posterior horn
3. lateral horn
4. posterior rootlets
5. posterior root ganglion
6. spinal nerve
7. anterior root
8. anterior horn
9. posterior median sulcus
10. anterior rootlets
11. anterior median fissure

Figure 16.5 The Phrenic Nerves in the Thoracic Cavity
1. right phrenic nerve
2. diaphragm
3. left phrenic nerve

Figure 16.7 Major Nerves of the Brachial Plexus
1. lateral cord
2. posterior cord
3. medial cord
4. musculocutaneous nerve
5. axillary nerve
6. radial nerve
7. median nerve
8. ulnar nerve
9. superior trunk
10. middle trunk
11. inferior trunk

Figure 16.8 Nerves of the Lumbosacral Plexus Within the Gluteal Region
1. inferior gluteal nerve
2. posterior femoral cutaneous nerve
3. pudendal nerve
4. superior gluteal nerve
5. sciatic nerve

Figure 16.9 Nerves of the Lumbosacral Plexus
1. obturator nerve
2. femoral nerve
3. tibial nerve
4. common fibular nerve

Chapter 17

Chapter Activities

Figure 17.2 Components of a Reflex
1. control center
2. sensory neuron
3. motor neuron
4. effector
5. receptor

Figure 17.4 Overview of the Parasympathetic Division of the ANS
1. oculomotor nerve (CN III)
2. facial nerve (CN VII)
3. glossopharyngeal nerve (CN IX)
4. vagus nerve (CN X)
5. cardiac plexus
6. abdominal aortic plexus
7. pelvic splanchnic nerves

Figure 17.5 Overview of the Sympathetic Division of the ANS
1. sympathetic trunk ganglia (paravertebral)
2. adrenal medulla
3. prevertebral ganglia
4. sympathetic trunk

Chapter 18

Chapter Activities

Figure 18.12 External Eye Structures
1. eyelashes
2. medial canthus
3. eyebrow
4. superior eyelid
5. pupil
6. sclera
7. iris
8. inferior eyelid

Chapter 19

Chapter Activities

Figure 19.9 Labeling Major Endocrine Glands of the Body
1. parathyroid glands
2. thyroid gland
3. thymus
4. adrenal cortex
5. adrenal medulla
6. testes
7. pineal gland
8. hypothalamus
9. pituitary gland
10. pancreas
11. ovaries
12. hypothalamus
13. anterior pituitary gland
14. pineal gland
15. posterior pituitary gland

Chapter 20

No Activities

Chapter 21

Chapter Activities

Figure 21.1 Cardiac Muscle Tissue
1. striations
2. intercalated disc
3. nucleus
4. cardiac muscle cell

Figure 21.3 Location of Heart within the Thoracic Cavity
1. right lung/pleural cavity
2. diaphragm
3. mediastinum
4. left lung/pleural cavity
5. heart

Figure 21.4 The Pericardial Sac
1. fibrous pericardium
2. parietal layer of serous pericardium
3. pericardial cavity
4. visceral layer of serous pericardium (epicardium)
5. myocardium
6. endocardium
7. pericardial cavity
8. myocardium
9. endocardium
10. visceral layer of serous pericardium (epicardium)
11. fibrous pericardium
12. diaphragm

Figure 21.12 Circulation to and from the Heart Wall
1. right coronary artery
2. marginal artery
3. small cardiac vein
4. left coronary artery
5. circumflex artery
6. great cardiac vein
7. anterior interventricular artery
8. coronary sinus
9. right coronary artery
10. posterior interventricular artery
11. middle cardiac vein

Chapter 22

Chapter Activities

Exercise 22.7, #4
right ventricle → pulmonary semilunar valve → pulmonary trunk → pulmonary arteries → lungs → pulmonary veins → left atrium

Figure 22.6   Pulmonary Circuit
1. right atrium
2. right AV valve
3. right ventricle
4. pulmonary semilunar valve
5. pulmonary trunk
6. pulmonary arteries
7. branch of pulmonary artery
8. pulmonary capillaries
9. branch of pulmonary vein
10. pulmonary veins
11. left atrium
12. left AV valve
13. left ventricle
14. aortic semilunar valve
15. aorta

Figure 22.7   Circulation to the Head and Neck
(a) Arterial Supply
1. internal carotid artery
2. external carotid artery
3. common carotid artery
4. vertebral artery
5. thyrocervical trunk
6. subclavian artery
7. superficial temporal artery
8. occipital artery
9. facial artery
10. brachiocephalic trunk (artery)

(b) Venous Drainage
1. vertebral vein
2. external jugular vein
3. internal jugular vein
4. subclavian vein
5. right brachiocephalic vein
6. superficial temporal vein
7. facial vein
8. superior thyroid vein

Figure 22.8   Circulation from the Aortic Arch to the Anterior Part of the Right Parietal Bone and Back to the Superior Vena Cava
1. brachiocephalic trunk (artery)
2. right common carotid artery
3. right external carotid artery
4. right superficial temporal artery
5. right superficial temporal vein
6. right internal jugular vein
7. right brachiocephalic vein

Figure 22.9   Circulation to the Brain
(a) Arterial Supply
1. middle cerebral artery
2. internal carotid artery
3. posterior cerebral artery
4. anterior communicating artery
5. anterior cerebral artery
6. internal carotid artery
7. posterior communicating artery
8. posterior cerebral artery
9. basilar artery
10. vertebral artery

(b) Venous Drainage
1. straight sinus
2. transverse sinus
3. sigmoid sinus
4. internal jugular vein
5. superior sagittal sinus
6. inferior sagittal sinus
7. cavernous sinus

Figure 22.10   Circulation from the Aortic Arch to the Right Parietal Lobe of the Brain and Back to the Right Brachiocephalic Vein
1. brachiocephalic trunk (artery)
2. right common carotid artery
  a. Alternate 2: right subclavian artery
3. right internal carotid artery
  b. Alternate 3: right vertebral artery
4. middle cerebral artery
  c. Alternate 4: basilar artery
  d. Alternate 5: posterior cerebral artery
  e. Alternate 6: posterior communicating artery
5. superior sagittal sinus
6. transverse sinus
7. sigmoid sinus
8. right internal jugular vein
9. right brachiocephalic vein

Figure 22.11   Circulation to the Thoracic and Abdominal Walls
(a) Arterial Supply
1. right subclavian artery
2. internal thoracic artery
3. brachiocephalic trunk
4. internal thoracic artery
5. anterior intercostal arteries
6. posterior intercostal arteries
7. superior epigastric artery
8. descending abdominal aorta
9. right lumbar artery
10. inferior epigastric artery
11. left subclavian artery
12. posterior intercostal arteries (1–2)
13. aortic arch
14. posterior intercostal arteries (3–11)
15. descending thoracic aorta
16. left common iliac vein

(b) Venous Drainage
1. right subclavian vein
2. right brachiocephalic vein
3. superior vena cava
4. anterior intercostal veins
5. azygos vein
6. internal thoracic vein
7. right posterior intercostal vein
8. inferior vena cava
9. right superior epigastric vein
10. right lumbar veins
11. right inferior epigastric vein
12. left subclavian vein
13. left brachiocephalic vein
14. accessory hemiazygos vein
15. left posterior intercostal vein
16. hemiazygos vein
17. inferior vena cava
18. left common iliac vein
19. median sacral vein

Figure 22.12   Circulation from the Left Ventricle of the Heart to the Right Kidney and Back to the Right Atrium of the Heart
1. ascending aorta
2. aortic arch
3. descending thoracic aorta
4. descending abdominal aorta
5. right renal artery
6. right renal vein
7. inferior vena cava

Figure 22.13   Arterial Supply to Abdominal Organs
(a) Arterial Supply to the Stomach, Spleen, Pancreas, Duodenum, and Liver
1. celiac trunk
2. common hepatic artery
3. hepatic artery proper
4. left hepatic artery
5. right hepatic artery
6. gastroduodenal artery
7. right gastric artery
8. left gastric artery
9. splenic artery
10. descending abdominal aorta

(b) Arterial Supply to the Small and Large Intestines
1. middle colic artery
2. right colic artery
3. ileocolic artery
4. celiac trunk
5. superior mesenteric artery
6. left colic artery
7. descending abdominal aorta
8. inferior mesenteric artery
9. sigmoid arteries
10. superior rectal artery

Figure 22.14   The Hepatic Portal System
1. inferior vena cava
2. hepatic veins
3. hepatic portal vein
4. superior mesenteric vein
5. gastric veins
6. splenic vein
7. inferior mesenteric vein

Figure 22.15   Circulation from the Abdominal Aorta to the Spleen and Back to the Right Atrium of the Heart
1. abdominal aorta
2. celiac trunk
3. splenic artery
4. splenic vein
5. hepatic portal vein
6. hepatic veins
7. inferior vena cava

Figure 22.16   Circulation from the Abdominal Aorta to the Duodenum and Back to the Right Atrium of the Heart
1. abdominal aorta
2. celiac trunk
3. common hepatic artery
4. gastroduodenal artery
5. superior mesenteric vein
6. hepatic portal vein
7. hepatic veins
8. inferior vena cava

Figure 22.17   Circulation from the Abdominal Aorta to the Sigmoid Colon and Back to the Right Atrium of the Heart
1. abdominal aorta
2. inferior mesenteric artery
3. sigmoid arteries
4. inferior mesenteric vein
5. splenic vein
6. hepatic portal vein
7. hepatic veins
8. inferior vena cava

Figure 22.18   Circulation to the Upper Limb
(a) Arterial Supply
1. brachiocephalic trunk (artery)
2. subclavian artery
3. axillary artery
4. deep brachial artery
5. brachial artery
6. ulnar artery
7. radial artery
8. deep palmar arch
9. superficial palmar arch
10. digital arteries

(b) Venous Drainage
1. brachiocephalic vein
2. subclavian vein
3. axillary vein
4. cephalic vein
5. basilic vein
6. brachial veins
7. median cubital vein
8. radial veins
9. cephalic vein
10. deep palmar venous arch
11. superficial palmar venous arch
12. dorsal venous network
13. digital veins
14. ulnar veins

Figure 22.19   Circulation from the Aortic Arch to the Anterior Surface of the Index Finger and Back along a Superficial Route to the Superior Vena Cava
1. brachiocephalic trunk (artery)
2. subclavian artery
3. axillary artery
4. brachial artery
5. radial artery
6. superficial palmar arch
7. digital artery
8. digital vein
9. superficial palmar venous arch
10. cephalic vein
  a. Alternate 11: median cubital vein
  b. Alternate 12: basilic vein
  c. Alternate 13: axillary vein
11. subclavian vein
12. brachiocephalic vein
13. superior vena cava

Figure 22.20 Circulation from the Aortic Arch to the Capitate Bone of the Wrist and Back along a Deep Route to the Superior Vena Cava
1. brachiocephalic trunk (artery)
2. subclavian artery
3. axillary artery
4. brachial artery
5. ulnar artery
6. deep palmar arch
7. deep palmar venous arch
8. ulnar veins
9. brachial vein
10. axillary vein
11. subclavian vein
12. brachiocephalic vein
13. superior vena cava

Figure 22.21 Circulation to the Lower Limb
(a) Arterial Supply
1. common iliac artery
2. internal iliac artery
3. external iliac artery
4. femoral artery
5. deep femoral artery
6. anterior tibial artery
7. fibular artery
8. dorsalis pedis artery
9. popliteal artery
10. posterior tibial artery
11. lateral plantar artery
12. medial plantar artery
13. plantar arterial arch
14. digital arteries

(b) Venous Drainage
1. common iliac vein
2. external iliac vein
3. internal iliac vein
4. deep femoral vein
5. femoral vein
6. great saphenous vein
7. anterior tibial veins
8. fibular veins
9. dorsal venous arch
10. popliteal vein
11. small saphenous vein
12. posterior tibial veins
13. small saphenous vein
14. lateral plantar vein
15. medial plantar vein
16. digital veins

Figure 22.22 Circulation from the Abdominal Aorta to the Dorsal Surface of the Big Toe and Back along a Superficial Route to the Inferior Vena Cava
1. abdominal aorta
2. common iliac artery
3. external iliac artery
4. femoral artery
5. popliteal artery
6. anterior tibial artery
7. dorsalis pedis artery
8. digital artery
9. digital vein
10. dorsal venous arch
11. great saphenous vein
12. femoral vein
13. external iliac vein
14. common iliac vein
15. inferior vena cava

Figure 22.23 Circulation from the Abdominal Aorta to the Cuboid Bone of the Foot and Back along a Deep Route to the Inferior Vena Cava
1. abdominal aorta
2. common iliac artery
3. external iliac artery
4. femoral artery
5. popliteal artery
6. posterior tibial artery
7. lateral plantar artery
8. lateral plantar vein
9. posterior tibial vein
10. popliteal vein
11. femoral vein
12. external iliac vein
13. common iliac vein
14. inferior vena cava

Exercise 22.14, #2
left ventricle → aorta → common iliac arteries → internal iliac arteries → umbilical arteries → placenta → umbilical vein → ductus venosus → inferior vena cava → right atrium

Table 22.4	Fetal Cardiovascular Structures and Associated Postnatal Structures	
Fetal Cardiovascular Structure	Postnatal Structure	Function of Fetal Cardiovascular Structure
Ductus arteriosus	Ligamentum arteriosum	Shunt blood from the pulmonary trunk to the aorta, thereby bypassing the pulmonary circuit and nonfunctional lungs
Ductus venosus	Ligamentum venosum	Carry oxygenated blood from umbilical vein to the inferior vena cava
Foramen ovale	Fossa ovalis	Shunt blood from the right atrium to the left atrium, thereby bypassing the nonfunctional lungs
Umbilical arteries	Medial umbilical ligaments	Carry deoxygenated blood from the fetus to the placenta so it can obtain oxygen from the mother's blood
Umbilical vein	Round ligament of the liver (ligamentum teres)	Carry oxygenated blood from the placenta to the fetal circulation

Figure 22.24 The Fetal Circulation
1. superior vena cava
2. lung
3. right atrium
4. liver
5. ductus venosus
6. inferior vena cava
7. umbilical vein
8. aortic arch
9. ductus arteriosus
10. pulmonary artery
11. pulmonary veins
12. foramen ovale
13. right ventricle
14. heart
15. descending abdominal aorta
16. common iliac artery
17. umbilical arteries
18. umbilical cord
19. placenta

Chapter 23

Chapter Activities
Figure 23.11 Major Lymph Vessels of the Body
1. right lymphatic duct
2. thoracic duct
3. lymph nodes
4. cisterna chyli

Figure 23.12 Mucosa-Associated Lymphatic Tissue (MALT)
1. pharyngeal tonsils
2. palatine tonsils
3. lingual tonsils
4. vermiform appendix

Figure 23.13 Lymph Node and Its Components
1. lymphatic nodules
2. afferent lymphatic vessels
3. capsule
4. medullary sinus
5. medullary cords
6. medulla
7. trabeculae
8. lymphatic nodules
9. peritrabecular space
10. subcapsular space
11. hilum
12. efferent lymphatic vessel

Figure 23.14 Gross Anatomy of the Spleen
1. splenic artery
2. hilum of the spleen
3. spleen
4. splenic vein

Chapter 24

Chapter Activities
Figure 24.7 The Upper Respiratory Tract: Midsagittal View
1. frontal sinus
2. superior nasal concha
3. middle nasal concha
4. inferior nasal concha
5. vestibule
6. epiglottis
7. thyroid cartilage
8. cricoid cartilage
9. sphenoidal sinus
10. superior meatus
11. middle meatus
12. inferior meatus
13. nasopharynx
14. oropharynx
15. laryngopharynx
16. lumen of larynx

Figure 24.8 Classroom Model of the Larynx
1. epiglottis
2. hyoid bone
3. thyroid cartilage
4. laryngeal prominence
5. tracheal "C" ring
6. epiglottis
7. hyoid bone
8. thyrohyoid membrane
9. cuneiform cartilage
10. corniculate cartilage
11. thyroid cartilage
12. arytenoid cartilage
13. cricoid cartilage
14. trachealis muscle
15. trachea
16. epiglottis
17. corniculate cartilage
18. false vocal cords
19. true vocal cords
20. cricoid cartilage
21. hyoid bone
22. thyrohyoid membrane
23. thyroid cartilage

Figure 24.9 The Pleural Cavities
1. visceral pleura
2. left lung
3. right lung
4. mediastinum
5. diaphragm
6. parietal pleura

Figure 24.10 The Right Lung
1. superior lobe
2. horizontal fissure
3. oblique fissure
4. middle lobe
5. inferior lobe
6. base
7. horizontal fissure
8. middle lobe
9. oblique fissure
10. apex
11. superior lobe
12. oblique fissure
13. pulmonary arteries
14. primary bronchi
15. pulmonary veins
16. inferior lobe
17. base

Figure 24.11 The Left Lung
1. apex
2. superior lobe
3. oblique fissure
4. cardiac notch
5. inferior lobe
6. base
7. superior lobe
8. pulmonary artery
9. primary bronchi
10. pulmonary veins
11. inferior lobe
12. base
13. oblique fissure
14. cardiac impression
15. cardiac notch

Figure 24.12 The Bronchial Tree
1. right main bronchus
2. right segmental bronchi
3. right lobar bronchi
4. trachea
5. left main bronchus
6. left segmental bronchi
7. left lobar bronchi
8. carina (internal structure)

Chapter 25

Chapter Activities

Figure 25.10 Oral Cavity
(a) Anterior View of the Oral Cavity
1. superior lip
2. superior labial frenulum
3. hard palate
4. glossopalatine arch
5. pharyngopalatine arch
6. palatine tonsil
7. sublingual duct orifice
8. submandibular duct orifice
9. inferior labial frenulum
10. transverse palatine folds
11. soft palate
12. uvula
13. fauces
14. tongue
15. lingual frenulum
16. teeth
17. gingivae
18. inferior lip

(b) Midsagittal View of the Oral Cavity and Pharynx
1. hard palate
2. oral cavity
3. tongue
4. vestibule
5. soft palate
6. uvula
7. palatine tonsil
8. oropharynx
9. lingual tonsil
10. epiglottis
11. laryngopharynx
12. esophagus

(c) Salivary Glands
1. parotid salivary gland
2. parotid duct
3. sublingual ducts
4. submandibular duct
5. sublingual salivary gland
6. submandibular salivary gland

Figure 25.11 Classroom Model of the Stomach
1. lesser curvature
2. pyloric sphincter
3. pylorus
4. esophagus
5. fundus
6. inferior esophageal (cardiac) sphincter
7. body of the stomach
8. greater curvature
9. pyloric sphincter
10. pylorus
11. fundus
12. cardia
13. inferior esophageal (cardiac) sphincter
14. gastric folds (rugae)

Figure 25.12 Classroom Model of the Duodenum, Liver, Gallbladder, and Pancreas
1. gallbladder
2. duodenum
3. accessory pancreatic duct
4. minor duodenal papilla
5. common bile duct
6. major duodenal papilla
7. head of pancreas
8. left and right hepatic ducts
9. cystic duct
10. common hepatic duct
11. tail of pancreas
12. body of pancreas
13. main pancreatic duct
14. hepatopancreatic ampulla

Figure 25.13 Classroom Model of the Liver
1. gallbladder
2. right lobe of liver
3. common hepatic duct
4. cystic duct
5. common bile duct
6. hepatic portal vein
7. hepatic artery proper
8. inferior vena cava
9. quadrate lobe of liver
10. falciform ligament
11. right hepatic duct
12. left hepatic duct
13. left lobe of liver
14. caudate lobe of liver
15. porta hepatis

Figure 25.15 Classroom Model of the Abdominal Cavity and Large Intestine
1. right lobe of liver
2. falciform ligament
3. pylorus of stomach
4. gallbladder
5. greater curvature of stomach
6. ascending colon
7. ileum of small intestine
8. esophagus
9. left lobe of liver
10. body of stomach
11. left colic (splenic) flexure
12. transverse colon
13. taenia coli
14. jejunum of small intestine
15. descending colon
16. esophagus
17. duodenum
18. right colic (hepatic) flexure of colon
19. ascending colon
20. taenia coli
21. cecum
22. tail of pancreas
23. body of pancreas
24. left colic (splenic) flexure of colon
25. transverse colon
26. descending colon
27. rectum
28. sigmoid colon

Chapter 26

Chapter Activities

Figure 26.7 Coronal Section through the Right Kidney
1. minor calyx
2. renal pelvis
3. major calyx
4. renal (medullary) pyramid
5. renal column
6. fibrous capsule
7. renal cortex
8. renal medulla
9. renal papilla
10. renal sinus
11. renal lobe
12. ureter

Figure 26.9 Models of the Kidney Demonstrating the Blood Supply to the Kidney
(a) Coronal Section
1. peritubular capillaries
2. interlobular vein
3. interlobular artery
4. interlobar vein
5. interlobar artery
6. vasa recta
7. renal artery
8. renal vein
9. segmental artery
10. glomerulus
11. afferent arteriole
12. arcuate vein
13. arcuate artery

(b) Close-up of the Renal Cortex and Renal Medulla
1. peritubular capillaries
2. efferent arteriole
3. afferent arteriole
4. interlobular artery
5. interlobular vein
6. arcuate vein
7. vasa recta
8. arcuate artery
9. interlobar vein
10. interlobar artery

Exercise 26.6, #4
abdominal aorta → renal artery → segmental artery → interlobar artery → arcuate artery → interlobular artery → afferent arteriole → glomerulus → efferent arteriole → peritubular capillaries or vasa recta → interlobular vein → arcuate vein → interlobar vein → renal vein → inferior vena cava

Figure 26.10 Location of the Structures of the Urinary System within the Abdominopelvic Cavity
1. left kidney
2. right kidney
3. ureter
4. urinary bladder

Figure 26.11 Classroom Model of Urinary System Structures
1. renal cortex
2. renal (medullary) pyramid
3. major calyx
4. renal papilla
5. minor calyx
6. renal pelvis
7. urinary bladder
8. left kidney
9. renal artery
10. renal vein
11. ureter

Figure 26.12 Classroom Model of a Midsagittal Section through the Male Pelvis
1. urinary bladder
2. prostatic urethra
3. membranous urethra
4. spongy urethra
5. ureter
6. muscularis (detrusor muscle)
7. internal urethral sphincter
8. external urethral sphincter

Figure 26.13 Classroom Model of a Midsagittal Section through the Female Pelvis
1. urinary bladder
2. urethra
3. ureter
4. muscularis (detrusor muscle)
5. internal urethral sphincter
6. external urethral sphincter

Chapter 27

Chapter Activities

Figure 27.12 Suspensory Ligaments of the Ovary, Uterine Tubes, and Uterus as Seen from a Posterior View
1. ovarian ligament
2. uterine tube
3. suspensory ligament of the ovary
4. fimbria
5. mesosalpinx
6. body of uterus
7. broad ligament
8. uterine blood vessels
9. uterosacral ligament
10. transverse cervical ligament
11. vagina
12. uterine part of uterine tube
13. isthmus of uterine tube
14. ampulla of uterine tube
15. infundibulum of uterine tube
16. round ligament of the uterus
17. endometrium
18. myometrium
19. perimetrium
20. internal os
21. cervical canal
22. external os
23. uterine tube
24. mesosalpinx
25. mesovarium

Figure 27.13 Classroom Model of the Female Pelvic Cavity
1. uterine tube
2. ovary
3. uterus
4. urinary bladder
5. labia minora
6. labia majora
7. rectum
8. vagina
9. bulb of the vestibule
10. round ligament of the uterus
11. vesicouterine pouch
12. pubic symphysis
13. clitoris
14. external urethral orifice
15. vaginal orifice
16. anus
17. ureter
18. fimbria of uterine tube
19. rectouterine pouch
20. cervix of uterus
21. vagina

Figure 27.14 The Female Breast
1. suspensory ligaments
2. lobe
3. lactiferous sinus
4. alveoli
5. lactiferous ducts
6. lobule
7. areolar gland
8. nipple
9. areola
10. adipose tissue
11. lobe
12. deep fascia
13. alveoli
14. lobule
15. suspensory ligaments
16. lactiferous sinus
17. nipple
18. lactiferous ducts

Figure 27.15 Male Reproductive Tract Structures
1. ureter
2. ampulla of ductus deferens
3. seminal vesicle
4. ejaculatory duct
5. prostate gland
6. prostatic urethra
7. bulbourethral gland
8. urogenital diaphragm
9. membranous urethra
10. ductus deferens
11. epididymis
12. testis
13. spongy urethra
14. urinary bladder
15. bulb of penis
16. crus of penis
17. corpus cavernosum
18. seminiferous tubules
19. corpus spongiosum
20. glans penis

Figure 27.16 Classroom Model of the Male Pelvic Cavity
1. ureter
2. urinary bladder
3. ductus deferens
4. penis
5. prepuce
6. glans penis
7. ductus deferens (ampulla)
8. seminal vesicle
9. prostate gland
10. ductus deferens
11. testicular artery and vein
12. spermatic cord
13. epididymis
14. testis
15. scrotum
16. ureter
17. ductus deferens
18. pubic symphysis
19. prostate gland
20. spongy urethra
21. corpus cavernosum
22. prepuce
23. corpus spongiosum
24. tunica albuginea of testis
25. tunica vaginalis of testis
26. urinary bladder
27. rectum
28. internal urethral sphincter
29. prostatic urethra
30. ejaculatory duct
31. urogenital diaphragm
32. anus
33. membranous urethra
34. epididymis
35. seminiferous tubules

Figure 27.19 Structure of the 3-week-old Embryo
1. connecting stalk
2. yolk sac
3. amniotic cavity
4. amnion
5. embryo
6. chorion
7. functional layer of uterus
8. placenta

Figure 27.20 Structure of the Late 4-week-old Embryo
1. ectoderm
2. mesoderm
3. amnion
4. peritoneal cavity
5. gut tube
6. somite
7. intermediate mesoderm
8. mesentery
9. endoderm
10. abdominal wall

credits

PHOTOGRAPHS

Chapter 1
All Photos: © Christine Eckel

Chapter 2
Figure 2.2 (top left, top right): © The McGraw-Hill Companies, Inc./Photo and Dissection by Christine Eckel; 2.2 (bottom left, bottom right): © Christine Eckel; p. 41 (left, right): © Geoff Brightling/Dorling Kindersley Media Library

Chapter 3
Figure 3.1–3.2: © Christine Eckel; 3.4a–c: © The McGraw-Hill Companies, Inc./Photo by Dr. Alvin Telser; 3.6, pp. 54–55: © Christine Eckel

Chapter 4
Figure 4.1a–e: © Christine Eckel; 4.2: © Ed Reschke; p. 64 (left): © Michael Abbey/Photo Researchers, Inc.; p. 64 (right), p. 65 (middle, right): © Carolina Biological Supply Company/Phototake; p. 65 (right): © Michael Abbey/Photo Researchers, Inc.; 4.3: © The Science Source/Photo by Christine Eckel; 4.6–4.14: © Christine Eckel

Chapter 5
Figure 5.1: © Ward's Natural Science/Photo by Christine Eckel; p. 96 (left, middle left): © Victor P. Eroschenko; p. 96 (middle right, right): © The McGraw-Hill Companies, Inc./Photo by Dr. Alvin Telser; p. 97 (left): © The McGraw-Hill Companies, Inc./Dennis Strete, photographer; p. 97 (middle left): © Dr. Richard Kessel/Visuals Unlimited; p. 97 (middle right): © Ed Reschke/Peter Arnold, Inc./Photolibrary Group; p. 97 (right): © Victor P. Eroschenko; 5.3: © The McGraw-Hill Companies, Inc./Photo by Dr. Alvin Telser; 5.4: © Ed Reschke; 5.5: © Victor P. Eroschenko; 5.6: © The McGraw-Hill Companies, Inc./Photo by Dr. Alvin Telser; 5.7a: © Dr. Gladden Willis/Visuals Unlimited; 5.7b: © Victor P. Eroschenko; 5.7c: © Ed Reschke; 5.8: © The McGraw-Hill Companies, Inc./Photo by Dr. Alvin Telser; 5.9: © Dr. Frederick Skvara/Visuals Unlimited; p.104 (left, middle): © The McGraw-Hill Companies, Inc./Photo by Dr. Alvin Telser; p.104 (right): © Ed Reschke; 5.10: © Victor P. Eroschenko; 5.12–5.14: © The McGraw-Hill Companies, Inc./Photo by Dr. Alvin Telser; 5.15: © Ed Reschke; 5.16–5.19: © The McGraw-Hill Companies, Inc./Christine Eckel, photographer; 5.20: © Ed Reschke/Getty Images; 5.21: © The McGraw-Hill Companies, Inc./Photo by Dr. Alvin Telser; 5.22: © Ed Reschke; 5.23–5.27a: © The McGraw-Hill Companies, Inc./Photo by Dr. Alvin Telser; 5.27b: © Christine Eckel; 5.28a: © Ed Reschke/Peter Arnold, Inc./Photolibrary Group; 5.28b: © Biophoto Associates/Photo Researchers, Inc.; p.119 (top): © The McGraw-Hill Companies, Inc./Photo by Dr. Alvin Telser; p.119 (bottom): © The McGraw-Hill Companies, Inc./Dennis Strete, photographer; p.122a: © The McGraw-Hill Companies, Inc./Christine Eckel, photographer; p. 122b–c: © Christine Eckel; p. 122d–f, p. 123g–h: © The McGraw-Hill Companies, Inc./Christine Eckel, photographer

Chapter 6
Figure 6.1: © Astrid & Hanns-Frieder Michler/Photo Researchers, Inc.; 6.2: © Tom Caceci; 6.3a: © James Stevenson/Photo Researchers, Inc.; 6.3b: © Christine Eckel; 6.4–6.6: © Ed Reschke; 6.7–6.8: © Biophoto Associates/Photo Researchers, Inc.; 6.9: © Fred Hossler/Custom Medical Stock Photo; 6.10: © Science VU/Visuals Unlimited; 6.11: © Professor Peter Cull/Photo Researchers, Inc.; p. 141: © Christine Eckel

Chapter 7
Figure 7.1: © Manfred Kage/Peter Arnold, Inc./Photolibrary Group; 7.2: © The McGraw-Hill Companies, Inc./Photo by Dr. Alvin Telser; 7.3a: © Ed Reschke; 7.3b: © Dr. Michael Klein/Peter Arnold, Inc./Photolibrary Group; 7.4: © Biophoto Associates/Photo Researchers, Inc.; 7.6: © Christine Eckel; 7.7a: © The McGraw-Hill Companies, Inc./Photo by Christine Eckel; 7.7b: © Ralph T. Hutchings/Visuals Unlimited; 7.8–7.10: © Christine Eckel; p. 160 (top): © Leonard Lessin/SPL/Photo Researchers, Inc.; p. 160 (bottom), p. 161: © Christine Eckel

Chapter 8
Figure 8.1: © The McGraw-Hill Companies, Inc./Photo by Christine Eckel; 8.2–8.3: © Christine Eckel; 8.4–8.8: © The McGraw-Hill Companies, Inc./Photo by Christine Eckel; 8.10: © Andy Crawford/Dorling Kindersley Media Library; 8.11a–b: © Christine Eckel; 8.12: © Ralph T. Hutchings/Visuals Unlimited; 8.13–8.22: © The McGraw-Hill Companies, Inc./Photo by Christine Eckel; 8.23, p. 196: © Christine Eckel

Chapter 9
Figure 9.1a–b: © Christine Eckel; 9.2–9.3: © The McGraw-Hill Companies, Inc./Photo by Christine Eckel; 9.4–9.6: © Christine Eckel; 9.7–9.9: © The McGraw-Hill Companies, Inc./Photo by Christine Eckel; 9.10 (top left, top right): © David Hunt/Smithsonian Institution; 9.10 (bottom left, bottom right): © L. Basset/Visuals Unlimited; 9.11a–d: © The McGraw-Hill Companies, Inc./Photo by Christine Eckel; 9.12–9.13: © Christine Eckel; 9.14–9.15: © The McGraw-Hill Companies, Inc./Photo by Christine Eckel; 9.16a–c: © The McGraw-Hill Companies, Inc./Photo by JW Ramsey

Chapter 10
Figure 10.5a–b: © The McGraw-Hill Companies, Inc./Photo by Christine Eckel; 10.6a–c: © Christine Eckel

Chapter 11
Figure 11.1e: © James Dennis/Phototake; 11.2: © Dr. David M. Phillips/Visuals Unlimited; 11.3: © Rick Ash; 11.4c: © Dr. Thomas Caceci, Virginia-Maryland Regional College of Veterinary Medicine; 11.5a–b: © The McGraw-Hill Companies, Inc./Photo by Dr. Alvin Telser; 11.6: © Ed Reschke/Peter Arnold, Inc./Photolibrary Group; 11.13: © Christine Eckel

Chapter 12
Figure 12.1a–b: © The McGraw-Hill Companies, Inc./Photo by Christine Eckel; 12.3a–b: © Denoyer-Geppert/Photo by MH/Christine Eckel; 12.6–12.11: © The McGraw-Hill Companies, Inc./Photo by Christine Eckel; 12.13: © Christine Eckel

Chapter 13
Figure 13.1–13.4a: © The McGraw-Hill Companies, Inc./Photo by Christine Eckel; 13.4b: © Christine Eckel; 13.5: © The McGraw-Hill Companies, Inc./Photo by JW Ramsey; 13.6: © Christine Eckel; 13.7a: © The McGraw-Hill Companies, Inc./Photo by Christine Eckel; 13.8: © Christine Eckel; 13.9a: © The McGraw-Hill Companies, Inc./Photo by Christine Eckel; 13.9b: © Christine Eckel; 13.11: © The McGraw-Hill Companies, Inc./Photo by Christine Eckel; 13.12: © Christine Eckel; 13.13–13.16: © The McGraw-Hill Companies, Inc./Photo by Christine Eckel

Chapter 14
Figure 14.1: © Dr. Robert Calentine/Visuals Unlimited; 14.2–14.9: © Rick Ash

Chapter 15
Figure 15.2b: © The McGraw-Hill Companies, Inc./Photo by Christine Eckel; 15.4: 3BScientific Model VH410; 15.5a: © Laerdal Medical/Photo by Christine Eckel; 15.5b–15.8: © The McGraw-Hill Companies, Inc./Photo and Dissection by Christine Eckel; 15.9–15.17, pp. 421–422: © Christine Eckel

Chapter 16
Figure 16.1a: © Ed Reschke; 16.1b: © Lester V. Bergman/Corbis; 16.1c: © Ed Reschke; 16.1d: © The McGraw-Hill Companies, Inc./Photo by Dr. Alvin Telser; 16.2 (top): © The McGraw-Hill Companies, Inc./Photo and Dissection by Christine Eckel; 16.2 (middle): © Christine Eckel; 16.2 (bottom): © The McGraw-Hill Companies, Inc./Photo and Dissection by Christine Eckel; 16.3: © Copyright by Denoyer-Geppert; 16.5–16.9: © The McGraw-Hill Companies, Inc./Photo and Dissection by Christine Eckel

Chapter 17
All photos: © Jill Braaten

Chapter 18
Figure 18.1: © The McGraw-Hill Companies, Inc./Photo by Dr. Alvin Telser; 18.2: © Carolina Biological Supply Company/Phototake; 18.3b: © Dr. Gladden Willis/Visuals Unlimited; 18.3c: © Christine Eckel; 18.3d: © The McGraw-Hill Companies, Inc./Photo by Dr. Alvin Telser; 18.3e: © Dr. John D. Cunningham/Visuals Unlimited; 18.4b: © The McGraw-Hill Companies, Inc./Photo by Dr. Alvin Telser; 18.5b: © Dr. John D. Cunningham/Visuals Unlimited; 18.6: © The McGraw-Hill Companies, Inc./Photo by Dr. Alvin Telser; 18.7a: © Gene Cox/Photo Researchers, Inc.; 18.7b: © Victor P. Eroschenko; 18.8c: © Biophoto Associates/Photo Researchers, Inc.; 18.9: © Science VU/Visuals Unlimited; 18.10, 18.11a: Photo by Christine Eckel; 18.11b: Photo by MH/JW Ramsey; 18.12: © The McGraw-Hill Companies, Inc./Photo by JW Ramsey; 18.13–18.15: © Christine Eckel; 18.16a: © 3BScientific Model E10/Photo by Christine Eckel; 18.16b: © Denoyer-Geppert; 18.18–18.20: © Christine Eckel; 18.21: © The McGraw-Hill Companies, Inc./Rick Brady, photographer; 18.23–18.26: © The McGraw-Hill Companies

Chapter 19

Figure 19.2a: © The McGraw-Hill Companies, Inc./Photo by Dr. Alvin Telser; 19.2b: © Educational Images/Custom Medical Stock Photo; 19.3: © Victor P. Eroschenko; 19.4: © Christine Eckel; 19.5: © The McGraw-Hill Companies, Inc./Photo by Dr. Alvin Telser; 19.6a: © Victor P. Eroschenko; 19.6b: © The McGraw-Hill Companies, Inc./Photo by Dr. Alvin Telser; 19.6c: © Victor P. Eroschenko; 19.7b: © Carolina Biological Supply Company/Phototake; 19.7c, 19.8 (right): © The McGraw-Hill Companies, Inc./Photo by Dr. Alvin Telser; p. 527 (left-right): © The McGraw-Hill Companies, Inc./Christine Eckel, photographer

Chapter 20

Figure 20.1, 20.3: © Christine Eckel; 20.4a–g: © The McGraw-Hill Companies, Inc./Photo by Dr. Alvin Telser; 20.5a: © Dr. Dorothea Zucker-Franklin/Phototake; 20.8–20.10a: © Christine Eckel; 20.10b-c: © McGraw-Hill Companies/Womack Photography Ltd.; 20.11a–c, p. 546: © Christine Eckel

Chapter 21

Figure 21.1: © Dennis Drenner/Visuals Unlimited; 21.2: © Rick Ash; 21.3: © 3BScientific Model G15; 21.4a: © Christine Eckel; 21.5: © The McGraw-Hill Companies, Inc./Photo and Dissection by Christine Eckel; 21.6a: © Christine Eckel; 21.6b: © Denoyer-Geppert/Photo by Christine Eckel; 21.7a: © Christine Eckel; 21.7b: © Denoyer-Geppert/Photo by Christine Eckel; 21.8, 21.9a: © Christine Eckel; 21.9b: © Denoyer-Geppert/Photo by Christine Eckel; 21.10–21.15: © Christine Eckel; 21.20a–c: © doc-stock GmbH/Phototake; p.583 (top, bottom): © The McGraw-Hill Companies, Inc./Photo and Dissection by Christine Eckel

Chapter 22

Figure 22.1–22.3: © Christine Eckel; 22.4: © The McGraw-Hill Companies, Inc./Photo by Dr. Alvin Telser; 22.5: © Christine Eckel; p. 598 (left, right): © Dr. Don W. Fawcett/Visuals Unlimited; p. 599: Image and © Dr. H. Jastrow from Dr. Jastrow's electron microscopic atlas http://www.drjastrow.de

Chapter 23

Figure 23.1a: © Ed Reschke; 23.1b: © The McGraw-Hill Companies, Inc./Photo by Dr. Alvin Telser; 23.2: © Christine Eckel; 23.3b: © The McGraw-Hill Companies, Inc./Photo by Dr. Alvin Telser; 23.3c: © Biophoto Associates/Photo Researchers, Inc.; 23.3d: © The McGraw-Hill Companies, Inc./Photo by Dr. Alvin Telser; 23.4a–b: © The McGraw-Hill Companies, Inc./Christine Eckel, photographer; 23.5: © The McGraw-Hill Companies, Inc./Photo by Dr. Alvin Telser; 23.6a–c: © The McGraw-Hill Companies, Inc./Christine Eckel, photographer; 23.9a–b: © The McGraw-Hill Companies, Inc./Photo by Dr. Alvin Telser; 23.9c: © Christine Eckel; 23.10b–c: © The McGraw-Hill Companies, Inc./Photo by Dr. Alvin Telser; 23.14: © The McGraw-Hill Companies, Inc./Photo and Dissection by Christine Eckel

Chapter 24

Figure 24.2: © Victor P. Eroschenko; 24.3a: © Science VU/Visuals Unlimited; 24.3b–c: © Christine Eckel; 24.4a–c: © The McGraw-Hill Companies, Inc./Christine Eckel, photographer; 24.5a–b: © The McGraw-Hill Companies, Inc./Photo by Dr. Alvin Telser; 24.7: © The McGraw-Hill Companies, Inc./Photo and Dissection by Christine Eckel; 24.8a–c: © 3BScientific Model G21/Photo by Christine Eckel; 24.9: © Denoyer-Geppert/Photo by Christine Eckel; 24.12b: © 3BScientific Model G23; 24.14: © McGraw-Hill Companies/J. Womack Photography; p. 690: Christine Eckel

Chapter 25

Figure 25.1b: © Carolina Biological Supply Company/Phototake; 25.1c: © Dr. Frederick Skvara/Visuals Unlimited; 25.1d: © The McGraw-Hill Companies, Inc./Photo by Dr. Alvin Telser; 25.3a: © Victor P. Eroschenko; 25.4b: © Dr. John D. Cunningham/Visuals Unlimited; 25.4c, 25.5b: © Victor P. Eroschenko; 25.5c–d: © Carolina Biological Supply Company/Phototake; 25.6b–25.9b (top): © Victor P. Eroschenko; 25.9b (bottom): © Dr. Alvin Telser/Visuals Unlimited; 25.11a–b: © Photo by Christine Eckel; 25.12–25.13: © Denoyer-Geppert/Photo by Christine Eckel; 25.14a: © Christine Eckel; 25.14b: Courtesy of David A. Morton and Chris Steadman, University of Utah School of Medicine; 25.14c: © Christine Eckel; 25.14d: Courtesy of David A. Morton and Chris Steadman, University of Utah School of Medicine; 25.15a–b: © Denoyer-Geppert/Photo by Christine Eckel; 25.16: © Christine Eckel; p. 731: © The McGraw-Hill Companies, Inc./Christine Eckel, photographer

Chapter 26

Figure 26.1–26.6: © The McGraw-Hill Companies, Inc./Photo by Dr. Alvin Telser; 26.7: © Ralph T. Hutchings/Visuals Unlimited; 26.9a: © Denoyer-Geppert/Photo by Christine Eckel; 26.9b–26.10: © Christine Eckel; 26.11: © Copyright by Denoyer-Geppert; 26.12: © 3BScientific Model H11/Photo by Christine Eckel; 26.13: © 3BScientific Model H10/Photo by Christine Eckel; 26.15: © Christine Eckel; p. 760 (top): © The McGraw-Hill Companies, Inc./Christine Eckel, photographer; p. 760 (middle, bottom): © Christine Eckel

Chapter 27

Figure 27.1–27.2: © The McGraw-Hill Companies, Inc./Photo by Dr. Alvin Telser; 27.3: © Christine Eckel; p. 770 (bottom left, bottom middle): © The McGraw-Hill Companies, Inc./Photo by Dr. Alvin Telser; p. 770 (bottom right): © Ed Reschke/Getty Images; p. 771 (left): © Ed Reschke/Peter Arnold, Inc./Photolibrary Group; p. 771 (middle, right), 27.4a–b: © The McGraw-Hill Companies, Inc./Photo by Dr. Alvin Telser; p.774 (left): © Educational Images/Custom Medical Stock Photo; p. 774 (right), p. 775 (left): © Biophoto Associates/Photo Researchers, Inc.; 27.5: © The McGraw-Hill Companies, Inc./Photo by Dr. Alvin Telser; 27.6a: From *Anatomy & Physiology Revealed*, © The McGraw-Hill Companies, Inc./The University of Toledo, photography and dissection; 27.6b: © The McGraw-Hill Companies, Inc./Photo by Dr. Alvin Telser; 27.6c: © Christine Eckel; 27.7a: © Ed Reschke/Getty Images; 27.7b: © The McGraw-Hill Companies, Inc./Christine Eckel, photographer; 27.8a: © The McGraw-Hill Companies, Inc./Photo by Dr. Alvin Telser; 27.8b: © The McGraw-Hill Companies, Inc./Christine Eckel, photographer; 27.9–27.10: © The McGraw-Hill Companies, Inc./Photo by Dr. Alvin Telser; 27.11: © The McGraw-Hill Companies, Inc./Christine Eckel, photographer; 27.13a–b: © 3BScientific Model H10/Photo by Christine Eckel; 27.16a–b: © 3BScientific Model H11/Photo by Christine Eckel; 27.17 (1, 2): © The McGraw-Hill Companies, Inc./Al Telser, photographer; 27.17 (3): © Ed Reschke/Getty Images; 27.17 (4): © Ed Reschke/Peter Arnold, Inc./Photolibrary Group; 27.17 (5): © Dr. Francisco Gaytan; 27.17 (6–7): © The McGraw-Hill Companies, Inc./Al Telser, photographer; p. 800: © Christine Eckel; p. 801: © The McGraw-Hill Companies, Inc./Christine Eckel, photographer; p. 802: © Christine Eckel

index

A bands, 254, 255f, 256, 256f, 260f, 261, 272
Abdominal, as term, 35f
Abdominal aorta, 521f, 607f, 609f, 610f, 612f, 613f, 621f, 622f, 624f, 715f, 750f
Abdominal cavity
   blood vessels, 609–13, 609f–613f
   innervation, 416, 416f
Abdominal muscles, **310–12,** 310t–311t, 311f
Abdominal wall, 799f
   blood vessels, 606–9, 607f, 608f
   muscles, 310–12, 310t–311t, 311f–313f
Abdominis
   definition of, 263t
   rectus, 310t, 311f, 312f, 313f
   transversus, 310t, 311f, 312, 312f, 313f
Abdominopelvic cavity, 749, 750f, 751f
Abdominopelvic quadrants, 37–38, 37f–38f
Abdominopelvic regions, 37–38, 37f–38f
Abducens nerve (CN VI), 296f, 398t
   disorders, 410t
   functions, 399f, 409t, 413, 413f, 498
   of sheep, 405f
Abduction
   defined, 242t
   hip, **336**–38
Abductor, definition of, 263t
Abductor digiti minimi, 334t, 335f, 347t, 349f
Abductor hallucis, 347t, 349f
Abductor pollicis brevis, 334t, 335f
Abductor pollicis longus, 332t, 333f, 335f
ABP. *See* Androgen binding protein
Absolute refractory period, **371,** 371f
Absorption of nutrients, **722,** 724, 724t
Accessory hemiazygos vein, 608f
Accessory nerve (CN XI), 300t, 398t, 399f, 436f
   disorders, common, 410t
   function, 409t, 417, 417f
   sheep, 405f
Accessory pancreatic duct, 711f, 716t, 717f
Accommodation, near point, 498
Acetabulum, 214t, 216f
Achilles (Calcaneal) tendon, 346, 346f
Acid-base balance. *See also* pH
   case study, 757–58
   and digestion, 725
   regulation of, 681, 756
   respiratory system and, 681
Acidophils, 512, 513t, 514f, 515
Acidosis, **681**
   metabolic, **725, 756,** 756t
   respiratory, **756,** 756t
Acinar cells, **711,** 711f, 725
Acini, pancreatic, **520,** 521f, **711,** 711f, 725
ACL (anterior cruciate ligament), 243, 243t, 244f–245f
Acoustic (auditory) meatus
   external, 170t, 176f, **179,** 487f, 488t, 489, **490,** 490f
   internal, 170t, **179,** 180f, 415f
Acromial end, 202t, 203f
Acromial process, **213,** 213f
Acromioclavicular joint, **213**
Acromion, 202t, 204f
ACTH. *See* Adrenocorticotropic hormone
Actin, **254,** 270, 270f, 271f, 276, 555f, 579
Action potentials, 68, 117
   and cardiac muscle contraction, 569
   compound, 372–73, 372f, 374
   conduction velocity, temperature and, 374–75, 374f
   definition of, **371**
   generation of, 371, 371f
   propagation of, 371, 371f
   refractory period, absolute, 371, 371f, 375–76
   refractory period, relative, 371, 371f, 375–76
Adaptation of receptors
   olfactory, 496
   sensory, **491, 493**
Adaptive immune response, 539
Adduction, 242t
Adductor, definition of, 263t
Adductor brevis, 267f, 341f, 341t, 342
Adductor gracilis, 342
Adductor hallucis, **347,** 349f
   oblique head, 348t
   transverse head, 348t

Adductor longus, 267f, **339,** 340f, 341f, 341t, 342
Adductor magnus, 267f, 341f, 341t, 342, 343f
Adductor pollicis, 334t, 335, 335f
Adductor tubercle, 218t, 220f
Adenohypophysis. *See* Anterior pituitary
Adenoids (pharyngeal tonsils), 646, 646f, 646t, 655f
ADH (antidiuretic hormone; vasopressin), 393, 513t, 514, 753t, 754
Adipocytes, 105t, 107, 107f, 701
Adipose tissue, 469f, 480f, 787f
   as connective tissue, 105t, 137, 137f
   identification of, 107, 107f
   in medullary cavity, 155f, 156
ADP, and muscle contraction, 271f
Adrenal cortex, **517**–19, 518f, 519t, 522f–523f, 523
Adrenal glands, 510, 510f, 522f–523f, 523, 717f, 721f–722f, 749f, 750f
   histology, 517–19, 518f, 519t
   hormones, 517, 519t
Adrenal medulla, **517**–19, 518f, 519t, 522f–523f, 523
   innervation of, 456, 457f, 458f–459f
Adrenocorticotropic hormone (ACTH), 512, 513t
Adventitia
   digestive tract, **702,** 707f
   respiratory tract, 664, 664f, 667f, 667t, 668
   urinary tract, **743,** 743f, 744, 744f
Afferent arteriole, 740f, 746f, 747f, 752
Afferent lymphatic vessels, 648, 649, 649f, 650f, 650t, 651, 656f
Agar, solute diffusion in, 69–**70,** 69f, 70f, 71t
Agglutination, 538, 544, 544f
Agonist (prime mover), **245, 299, 328**
Air sacs. *See* Alveoli
Ala, 189f, 214t, 216f
Albumin, 753
Aldosterone, 519t
Alkalosis, 681
   metabolic, **725, 756,** 756t
   respiratory, **756,** 756t
Alpha cells, 520t
ALS. *See* Amyotrophic lateral sclerosis
Alveolar ducts, 666, 670, 670f, 670t, 671
Alveolar gas exchange, 679, **680**–81
Alveolar macrophages, 670f, 671, 671t
Alveolar nerves, 413f
Alveolar processes, 169t, 172t, 175f, **176,** 237f
Alveolar sacs, 601f, 670, 670f, 670t, 671f
Alveoli
   breast, 786t, 787f
   respiratory, 601f, 666, **669,** 670, 670f, 670t, 671t, 679, 679f
   salivary glands, 700f, 701t
   type I cells, 670, 670f, 671t
   type II cells, 670f, 671t
Aminopeptidase, 724t
Ammonia, tubular secretion, 753t
Amnion, 798f, 799f
Amniotic cavity, 798f
Amphiarthrosis, 233, 234t, 236, 237
Ampulla
   of ductus deferens, 789f
   of uterine tube, 773t, 784f
Amylase, 725–27, 726t
Amyotrophic lateral sclerosis (Lou Gehrig's Disease; ALS), 270
Anaphase
   in meiosis, 791t–792t
   in mitosis, 64f–65f, 67
Anastomoses, **613,** 613f
Anatomical neck, of humerus, 205t, 207f, 208f
Anatomical terminology, 31–38
   abdominopelvic quadrants, 37–38, 37f–38f
   abdominopelvic regions, 37–38, 37f–38f
   anatomic planes, 31–32, 31f, 31t
   anatomic position, 31, 31f
   body cavities, 36, 36f
   body membranes, 36, 36f
   directional terms, 33, 33t
   regional terms, 34–35, 34t
Anatomic planes, 31–32, 31f, 31t
Anatomic position, 31, 31f
Anatomy
   etymology of term, **28**

sectional, **31**
Anconeus, 333f
Androgen binding protein (ABP), 795
Androgens, 519t
Anemia, 538
Angle
   of mandible, 169t
   of rib, 190t, 192f
Animal cell, 61–67
   classroom model, observation of, 66f, 67
   human cheek cell, observation of, 63–65, 63f, 64f
   parts, structure and function of, 61t–62t
Ansa cervivalis, 417f, 436f
Antagonist, **245,** 299
Antebrachial, as term, 35f
Anterior, definition of, 33t, 263t
Anterior arch, of atlas, **186,** 187f
Anterior cavity, of eye, 482t, 484, 484f, **486**
Anterior cerebral artery, 604f
Anterior chamber, of eye, 482t, 484, 484f, **486**
Anterior communicating artery, 604f
Anterior compartments, muscles of, **326**
   arm, 326–28, 327f, 327t
   forearm, 329–31, 329t, 330f, 331f
   leg, 344, 344f, 344t
   thigh, 339–41, 339f, 340f, 340t
Anterior cruciate ligament (ACL), 243, 243t, 244f–245f
Anterior funiculus, 430, 431f, 432t
Anterior horn cell, 359f, **360,** 360f
Anterior horns, of spinal cord, 430, **430,** 431, 431f, 432t, 435f, 459
Anterior humeral circumflex artery, 614f
Anterior iliac spines, 214t, 216f, 312
Anterior intercostal arteries, 607f
Anterior intercostal vein, 608f
Anterior interventricular artery, 557f, 563f, 564f
Anterior interventricular sulcus, 566, 566f
Anterior pituitary gland (adenohypophysis)
   gross anatomy, 522f–523f, 523
   histology, **512**–15, 513t, 514f
   hormones, 393, 512, 513t, 769, 793, 795
Anterior ramus, **435,** 436, 436f, 438f
Anterior (motor) root, of spinal cord, 433t, 434f, 435f
Anterior rootlets, 434f, 435f
Anterior sacral foramina, 189f
Anterior scalenes, 436f
Anterior superior iliac spine, **227,** 228f
Anterior tibial artery, 619f, 621f
Anterior tibial vein, 620f
Anterior tubercle (atlas), 186, 187f
Antibodies, formation of, 539
Anticubital, as term, 35f
Antidiuretic hormone (ADH; vasopressin), 393, 513t, 514, 753t, **754**
Antigen-antibody complexes, 539
Antigens, spleen and, 654
Anti-inflammatory drugs, uses of, 34
Antiserum, for snake venom, **657**
Antrum, 769, 769f, 770f, 771t
Anulus fibrosus, **238,** 238f
Anus, 698, 720, 720f, 785f, 790f
Aorta, 556f, **560,** 561f, 562, 563f, 566–67, 566f, 567f, 568, 601f
   abdominal, 521f, 607f, 609f, 610f, 612f, 613f, 621f, 622f, 624f, 715f, 750f
   ascending, 557f, 609f
   connective tissue in, 109, 109f
   thoracic, 607f, 609f
Aortic arch, 557f, 603f, 605f, 607f, 609f, 624f
Aortic semilunar valve, **560,** 561f, 562–63, 563f, 601f
Apex
   of heart, **557,** 557f, 566, 566f, 567f, 568
   of lung, 676f, 677f
Apical surface, of epithelial tissue, 96, 99, 99f
Apocrine sweat glands, 131, 131t, **135,** 135f
Aponeurosis, **310,** 310t, 313f
   epicranial, 294f
   of internal and external oblique, **312,** 312f
   palmar, 330f, **331**
   plantar, **347**
Apoptosis, 651
Appendicular muscles, **320**
   arm, 326–28, 327f, 327t
   foot, 347, 347t–348t, 349f

forearm, 329–33
hand, 334–35, 334t, 335f
hip, 336–39, 336f, 337f, 338f
learning strategies, 302, 326, 328t, 335, 341, 342, 347
leg, 344–46
thigh, 339–43
Appendicular skeleton, **157,** 165, 202–28
   limb, lower, 218–27, 218t–219t
   limb, upper, 205–13, 205t–206t
   pectoral girdle, 202–5, 202t
   pelvic girdle, 214–18, 214t–215t
Appendix, 610f, 645, 645f, **648,** 648f, 655f, 709f
Aquaporins, 754
Aqueous humor, 482t, **486**
Arachnoid mater, 384, 384f, 385t, 387f–388f, 433, 435f
   in sheep, 404f, **406,** 406f
Arachnoid trabeculae, 384f
Arachnoid villi (granulations), **384,** 384f, 387f–388f
Arches, of atlas, 186, **186,** 187f
Arcuate artery, 746f, 747f
Arcuate line, 214t, 216f
Arcuate vein, 746f, 747f
Arcuate vessels, 739f, 746f
Areola, 786t, 787f
Areolar connective tissue, 105t, 106, 106f, 108–9, 108f, 130
Areolar glands, 786t, 787f
Arm, **326**
   compartments, fascial, 266, 266t, 267f, 326–28
   of microscope, 46f, 47t, 52f
   muscles, 326–28, 327f, 327t, 328f, 328t
Arrector pili muscles, **133,** 134, 137f
Arterial arcades, **718,** 719f
Arteriole, 595, **595,** 595f, 601f
   afferent, 740f, 746f, 747f, 752
   definition of, 590, **590**
   diameter of, and blood vessel resistance, 625
   efferent, 746f, 747f
   types of, 594
   vessel wall, 592t
Artery, 592t. *See also* Arteriole
   definition of, **590**
   elastic, 592t, 594f
   muscular, 592t, **594**–95, 595f
   pulmonary circuit, 600–601, 601f
   systemic circuit, 601–22
   vessel wall, 592t, 593, 593f
Arthrology, definition of, **233**
Articular capsule, 239f, 239t
Articular cartilage, 155f, 156, 239f, 239t, 244f–245f
Articular facets, **185, 186,** 187f, 189f, 190t, 192f, 219t, 222f
Articular processes, 183t, **185,** 185f, 187, 187f, 188, 188f, 189f
Articulations
   cartilaginous joints, 237–38, 237f, 238f
   classification of, 233, 234t
   definition of 233
   fibrous joints, 236, 236t, 237f
   synovial joints, 239–45
Arytenoid cartilage, 673t, 674f
Arytenoid muscles, 674f
Ascending aorta, 557f, 609f
Ascending colon, 610f, 611f, 709f, 720, 720t, 721f–722f
Astrocytes, **362,** 362–63, 362t, 363f
Astronauts, bone loss in, 158
Atherosclerotic plaques, 606
Atlantoaxial joint, **187**
Atlanto-occipital joint, **186**
Atlas, 183t, **186–87,** 187f, 241f, 434f
ATP, and muscle contraction, 270, 271f, 273
Atrial fibrillation, **572,** 572f
Atrioventricular (AV) node, 569, 569t, 570, 570f
Atrioventricular (AV) valve, **578**
   left, **560,** 561f, 562, 567f, 568, 601f
   right, 558f, **559,** 559f, 560, 563f, 567f, 568, 601f
Atrium, cardiac
   cardiac conduction system, 569–70
   fetal, 624f
   left, **560,** 562, 601f

I-1

Atrium, cardiac (continued)
right, 557f, 558, **558**, 558f, 559f, 560, 564f, 601f
in sheep, 566, **566**, 566f, 567f, 568
Auditory canal, 179
Auditory meatus. See Acoustic (auditory) meatus
Auditory nerve, internal, 415, 415f
Auditory ossicles, 487f, 488t, **490**, 490f
Auditory tube (pharyngotympanic tube; eustachian tube), 487f, 488t, 489, **490**, 490f, 655f, 672f
Auricle (pinna), 487f, 488t, 489, **490**, 557f, 558, 559f, 561f, 564f, 566, **566**, 566f, 567f, 568
Auricular artery, posterior, 602f
Auricular nerve, great, 436f
Auricular processes, 183t
Auricular surface, 189f, 214t, 216f
Auricular vein, posterior, 602f
Auscultation
heart sounds, 578-79, 578f, 626
respiratory sounds, 682
Auscultatory method, **625**
Autonomic nervous system (ANS), **450**, 450f, **452**, 453-59
and enteric nervous system, 724
parasympathetic division, **453**-55, 453f, 454f, 455t
sympathetic division, **453**, 455-56, 455t, 457f, 458f-459f
Autonomic reflexes, **450**, **452**, 461-62
Autosomal chromosomes, 791
Avascular tissue
defined, **96**
epithelial tissue as, 96
AV block (heart block), 575-76
AV bundle (Bundle of His), **569**, 569f, 570f
AV node. See Atrioventricular (AV) node
Axial, as term, **165**
Axial muscles
abdominal wall, 310-12, 310t-311t, 311f-313f
head, 292-99
neck, 299-303
respiration muscles, 307-9, 307t, 308f, 309f
vertebral column, 303, 303t-306t, 306f
Axial skeleton, **157**, 165, 169-92
skull, 169-82, 169t-170t, 172t
thoracic cage, 190-92, 190t
vertebral column, 183-89, 183t
Axillary, defined, 34t, 35f
Axillary artery, 438f, 440f, 614f, 616f, 617f
Axillary nerve, 437f, 437t, 438f, 439f, 440f
Axillary skin, **135**, 135f
Axillary vein, 615f, 616f, 617f
Axis, 183t, 186-**87**, 187f, 241f
Axon(s), 118, 118f, **359**, 359t, 364, 364f
myelinated, 364, 364f, 365, 366f, 371
optic projection, 411f
peripheral nerves, 365-66, 366f
unmyelinated, 358
Axonal branches, of somatic motor neuron, 258f, **259**
Axon hillock, **359**, 359t
Azygos vein, 608f, 655

Babinski reflex, 461
Balance, and stretch reflexes, 459
Ball and socket joint, 240, 240t, 241f
Barany test, **499**, 501, 501f
Basal cells
of olfactory epithelium, 472, 472f, 473t, 665f, 665t, 666
of taste bud, 471f, 471t
of trachea, 666f, 667f, 668
Basal layer, of uterine lining, 794f
Basal surface, of epithelial tissue, 96, 99, 99f, **103**
Base
of heart, **557**, 557f
of lung, 676f, 677f
of microscope, 46f, 47, 47t, 52f
Basement membrane, **96**, **103**, 598, 598t, 665f, 670, 680
Basilar artery, 604f, 605f
Basilar membrane, cochlea, 476, **476**, 477f, 477t, 490, 490f
Basilar part, 180f
Basilic vein, 590, 615f, 616f
Basket cells. See Purkinje cell
Basophilic stains, 63
Basophils, 512, 513t, 514f, 515, 531t
identification of, 533-35, 534f, 537
Benedict's solution, 75, 726, 726t
Beta cells, 520t

B-flat mnemonic, 512
Bicarbonate ion, 681, 711, 716, 723, 725, 756, 756t
Bicarbonate reaction, 725
Biceps, definition of, 262t
Biceps brachii, 267f, 324f, 325f, 327f, 327t, 328
Biceps femoris, 267f, 342, 342t, 343, 343f, 443f
Bifid processes, 186
Bile, **716**, 723
Bile canaliculi, 710, 710f
Bile duct, 521f, 710, 710f, 716t, 717f, 718f, 723, 723f
Bile pigment, in urine, 754t
Bilirubin, in urine, 755, 755t
Bilirubinuria, 754t
Biohazard bags, **14**, 14f
BIOPAC lessons
blood pressure, 628-31, 628f
pulmonary function tests, 683-86
Bipennate architecture, in skeletal muscle, 264t
Bipolar cell layer, retina, 474, 474f, 475f, 476
Bitter, as basic taste, 494, 494t
Black bile, 529, 530, 530t
Bladder, innervation of, 454f
Blade remover, **16**, 18f
Blastocyst, **797**
Blood. See also Erythrocytes; Leukocytes; Platelets (thrombocytes)
as ancient Greek humor, 529, 530t
calcium levels, 151, 510
cell types in, 113-14, 114f, 530t, 533-36, 534f, 537
centrifugal separation of, 438f, 538
cholesterol levels, 439, 542, 545
coagulation, 538
coagulation time, 538, 542-43
as connective tissue, 113-14, 114f
diagnostic tests, 538-45
filtration of, 752
glucose levels, 2, 510, 754
glucose transport in, 727-29
handling of, 533
hemopoiesis, 156, 227, 536
pH of, 681, 686-87, 725, 756, 756t
substances transported in, 538
transfusions, 544
viscosity, and blood vessel resistance, **625**
Blood colloid osmotic pressure (OP_g), **752**, 752f, 753
Bloodletting, 529
Blood pressure, 625, **625**, 627-31
Blood sample, collection of, 538, 538f
Blood smear, human, making of, 533-34, 533f
Blood type, determination of, 538-39, 544, 544f
Blood vessel(s). See also Artery; Vein
fetal, 623, 623t, 624f
learning strategies, 600
in medullary cavity, 155f, 156
pulmonary circuit, 600-601, 601f
resistance in, 625
respiratory tract, 664f
in sheep brain, **406**, 406f
systemic circuit, 601-22
uterine, 784f
wall structure, 592-93, 592t, 593f
Blunt dissection, 14, 22, 22f
Blunt probe, 11t, 19, 20f, 22, 22f
B-lymphocytes, 645, 645f, 653, 657
Body
of ischium, 214t
of nail, 136, 136f
of pancreas, 711f, 716t, 717f, 721f-722f
of sternum, 190t, **191**, 191f, 721f-722f
of stomach, 714t, 715f
of uterus, 784f
of vertebra, **185**, 185f, 186, 186f, 188, 188f
Body cavities, 36, 36f
Body membranes, 36, 36f
Body temperature
and action potential conduction velocity, 374-75, 374f
and diffusion rate, 71-73, 72t
and digestion rate, 725-27, 726t
regulation of, 524-25
Bolus, 722-23
Bone(s). See also Compact bone; Spongy (cancellous) bone
calcium storage in, 151
classification of, 152-53, 152f, 152t
as connective tissue, 145, 151
as dynamic, metabolically active tissue, 145, 155
endochondral ossification, 150-51, 150f

formation and remodeling, 150-51, 150f, 151, 158
matrix, 151
phosphate storage in, 151
slides, types of, 147
structure and function of, 110, 110t
types of, 110
Bone cell(s), types, 147t
Bone marrow
red, 153-54, 154f, 156, 227, 536-37, 537f
transplantation of, 227, 537
yellow, 153, 154f, 155, 155f, 227, 239f
Bone marrow cavity, 149f, 150
Bone tissue, 147-51, 147t, 536, 537f
Bony matrix (lamellae), 110, 110t
Border, anterior, of tibia, 219t, 222f
Bowman's (glomerular) capsule, 738t, 739f, **740**, 740f
Bowman's space (capsular space), 740f, 752
Boyle's Gas Law, 679
Brachial, defined, 34t, 35f
Brachial artery, 590, 614f, 616f, 617f, 627
Brachialis, 267f, 326, 327f, 327t, **328**
Brachial plexus, 307f, 434f, **437**-38, 437f, 438f, 439f, 440f, 459
Brachial vein, 590, 615f, 617f
Brachii, definition of, 263t
Brachiocephalic trunk (artery), 602f, 603f, 605f, 607f, 614f, 616f, 617f
Brachiocephalic veins, 602f, 603f, 605f, 608f, 615f, 616f, 617f
Brachioradialis, 330f, 332t, 333f
Bradycardia, 570
Brain, human, 390-98, 391t-393t, 394f-397f. See also Cranial nerves; specific parts
blood vessels, 603-6, 604f, 605f
gliomas, 362
meninges, 384-**87**, 384f-385f, 385t-386t, 387f-388f
structures, learning strategies for, 381-82, 389, 390, 400
tissues, delicacy of, 401
ventricles, 388-89, 388f, 389f, 390t
Brain, sheep, 401-8, 401f-407f
tissues, delicacy of, 401
Branches, of brachial plexus, **437**, 437f
Branching cells, in cardiac muscle, 260f, 261
Breast
cancer of, 788
gross anatomy, 786-87, 786t, 878f
self-examination (BSE), 788
Brevis, definition of, 262t
Broad ligament, 783t
Bronchi, 666, **668**-69, 668f, 669f, 673, 676, 676f, 677, 677f, 678f, 678t, 679
Bronchial arteries, 676, 677, 677f
Bronchial sounds, 682
Bronchial tree, 678-79, 678f, 678t
Bronchial veins, 676, 677, 677f
Bronchioles, 666, **668**-69, 668f, 669, 669t, 670, **670**, 670f, 671f, 673, 679f
Bronchopulmonary segments, **678**
Bronchoscopy, 679
Brush border, 97t
BSE (breast self-examination), 788
Buccal, defined, 34t, 35f
Buccinator, 292t, 294f
Buffer systems, and acid-base balance, 756
Buffy coat, 438f, 538
Bulb
of hair follicle, **133**, 133f, 134, 134f, 134t
of penis, 789f
of vestibule, 785f
Bulbourethral glands, 779t, **788**, 789f
Bundle of His (AV bundle), **569**, 569f, 570f
Bursae, 239t, 243, 243t, 244f-245f

Cadavers
emotional responses to, 2
proper treatment of, 2
tissues, disposal of, 14
Calcaneal, as term, 35f
Calcaneal (Achilles) tendon, 346, 346f
Calcaneus, 219t, 224, 224f, 225, 226f, **227**, 228f
Calcified cartilage, zone of, 150f, 151
Calcitonin, 510, **516**
Calcium
blood levels, 151, 510
and muscle contraction, 270, 270f, 278
Calcium channels, voltage-gated, and cardiac refractory period, 571
Calvaria, **169**
Calyx, 743f, 745f, 748, 750f
Canaliculus, 110t, 113, 113f, **148**, 148f
Canals, in bone, 110t

Cancellous bone. See Spongy (cancellous) bone
Cancer
breast, 788
carcinomas, 114, 788
cervical, 788
ovarian, 768, 770f
sarcomas, 114
skin, 129, 129f
Canthus, medial, 485f
Capillaries, 597-99, 598t-599t
in lungs, 670, 671t
vessel wall, 592t
Capillary beds, **590**, 609, 752
in kidneys, 745-46, 746f
Capillary endothelium, 681
Capillary tube, heparinized, 540, 540f, 542, 543f
Capillary tufts, sheep brain, 401f, 403, 403f, 404
Capitate, 206t, 211, 211f, 617, 617f
Capitis (longissimus, semispinalis, splenius), 299, 300t, 301f-302f, 304t, 305t, 306f
Capitulum, 205t, 207f, 207f-208f
Capsular hydrostatic pressure (HP_c), **752**, 752f, 753
Capsular space (Bowman's space), 740f, 752
Capsule. See also Glomerular capsule; Internal capsule
of adrenal gland, **517**, 519
of kidney, 739f, 744, 745f, 746f
of lymph node, **648**, 649, 649f, 650f, 650t, 656f
of parotid salivary gland, 700f
of spleen, 653, 653f, 654, 654t
of thymus, 652f
Carbohydrates, digestion of, 723-24, 723t, 724t, 725-27, 726t
Carbon dioxide
and blood pH, 681, 756, 756t
and gas exchange, 679, 680
and gas transport, 681
Carbonic acid, 725, **756**
Carbonic anhydrase, 681, 725
Carboxypeptidase, 724t
Carcinomas, 114, 788
Cardia, 714t, 715f
Cardiac cycle, **578**, 626
Cardiac impression, **677**, 677f
Cardiac muscle, **252**
contraction of, 555, 579-80
histology, 554, 554f
identification and classification, 116, 116f
intrinsic conduction system, 569-70, 569t, 570f
refractory period, 571
Starling's Law, and, 579-80
structure and function, 114, 115t
Cardiac muscle cell, 554f
Cardiac muscle tissue, **260**-61, 260f
appearance, 254t, 260, 260f
contraction, regulation of, 277
functions, 254t
vs. skeletal muscle, 277
Cardiac notch, 677f, 714t, 715f
Cardiac output, **578**
Cardiac plexus, 454f, 458f-459f
Cardiac veins, 564f, 565, 565t
Cardiovascular disease, risk levels, determining, 539
Cardiovascular system
endothelium in, 99
lymphatic system and, **641**
Carina, 678f, 679
Carotid arteries, **601**, 602f, **603**, 603f, 604f, 605f, 607f
Carotid canal, 170t, **178**, 178f
Carotid sinus, 603f
Carpal, defined, 34t, 35f
Carpals, 153f, 157f, 206t, **211**, 211f, 212
Carpal tunnel, **438**
Carpal tunnel syndrome, 428, **438**
Cartilage
articular, 155f, 156, 239f, 239t, 244f-245f
arytenoid, 673t, 674f
corniculate, 673t, 674f
costal, 192
cricoid, 672f, 674f
cuneiform, 219t, 224, 224f, 225, 226f, 673t, 674f
elastic, 110t, 111f, 112-13, 112f, 490, 594, 594f
of epiglottis, 673t, 674f
fibrocartilage, 110t, 111f, 112, 112f, 237, 238, 238f

I-2

hyaline, 110t, 111–12, 111f, 155f, 156, 237, 238, 668f, 669
  respiratory tract, 664, 664f, 668f, 669, 669t
  hypertrophic, zone of, 150f, 151
  of larynx, 673t
  proliferating, zone of, 150f, 151
  of respiratory tract, 666
  resting, zone of, 150f, 151
  septal, 174f
  structure and function, 110, 110t
  thyroid, 672f, 673t
  trachea, 667f, 667t, 668
  types, 110, 110t
Cartilaginous joints, **237**–38, 237t, 238f
Catalase, 62t
Catecholamines, **517**, 519t
Cauda equina, 430, 431f, 433t, 434f
Caudal, defined, 33t
Caudate lobe, liver, 716t, 717f, 718f
Caval opening, 309, 309f
Cavernous sinus, 604f
CBE (clinical breast exam), 788
Cecum, 610f, 648, 648f, 655f, 709f, **720**, 720t, 721f–722f
Celiac ganglion, 457f
Celiac trunk, 610f, 612f
Cell(s). See also Nucleus; *specific types*
  animal, structure and function, 61–67
  as basic unit of life, 57
  meiosis, **766**, **791**, 791t–792t, 792–93, 795
  membrane of. See Plasma membrane
  mitosis, 64f–65f, 65–66, 792
  multinuclear, 57
  recognition of, 57–58
Cell body
  of anterior horn cell, 360, 360f
  of neuron, 117t, 118, 118f, 358, **359**, 359t
Cell cycle, 65
Cellular processes, of astrocytes, 362, 363f
Cellular tissue
  defined, 96
  epithelial tissue as, 96
Celsius, conversion to/from Fahrenheit, 9
Central artery
  penis, 782, 782f
  spleen, 653f, 654, **654**, 654t
Central canal
  in bone (haversian canal), 110t, 113, 113f, 148–49, 148f
  in spinal cord, **388**, 388f, 430, 431f, 432t, 435f
  sheep, 406, 407f
Central nervous system. See also Brain; Spinal cord
  development of, 388
  glial cells in, 362–63
Central sulcus, human, 391t, **394**, 394f, **395**, 395f, 397f
Central tendon of diaphragm, **309**
Central vein, **710**, 710f, 711f
Centrifuge, separation of whole blood sample by, 438f, 538
Centriole, 61t, 66f, 67
Centromere, 791t–792t
Centrosome, 61t
Cephalic, defined, 33t, 34t, 35f
Cephalic vein, 590, 615f, 616f
Cerebellum
  human, 391t, 395f, 396f, 397f, 434f
  Purkinje cells in, 361, 361f
  sheep, 402f, 404, 404f, 406, 406f, 407f
Cerebral aqueduct, **388**, 388f, 389f, 390t
  human, 391f, 397f
  sheep, 406, 407f
Cerebral arterial circle, **603**, 604f
Cerebral arteries, 604f, 605f
Cerebral cortex, 384f, 498
  sheep, 407, 407f
Cerebral peduncle
  human, 391f, 397f
  sheep, 404, 404f, 406, 407, 407f, **408**
Cerebrospinal fluid (CSF)
  circulation of, 406
  transport of, 384
Cerebrum, 497
  functions, 497
  pyramidal cells in, 360–61, 361f
  sheep, 406, 407f
Cervical, defined, 34t, 35f
Cervical canal, 784f
Cervical cancer, 788
Cervical curvature, 184f
Cervical enlargement, **430**, 434f
Cervical nerve, transverse, 436f
Cervical plexus, 307t, 434f, 436, 436f

Cervical region, 183
Cervical spinal nerves, 300t–301t
Cervical vertebrae, **183**, 183t, 184f, **186**, 186f, 431f, 432t
Cervicis (splenius cervicis), 304t
Cervix, 785f
Cheek cell, human, 63–65, 63f, 64f
Chemical digestion, 722–23
Chemicals, hazardous, safe use of, 13, 13t
Chief (principal) cells
  gastric, 705f, 706, 706t
  parathyroid, **516**, 516f, 517t
Childbirth, pelvic structure and, 215, 218
Chin, 182t
Chloride, tubular reabsorption, 753t
Choanae, 672f
Cholelithiasis, 723, 723f
Cholera, 725
Cholesterol
  blood levels, 439, 542, 545
  good and bad forms of, 539
Chondroblasts, 110, 111–113, 111f
Chondrocytes, 110, 110t, 111f, 112, 112f, 113, 150f
Chordae tendineae, **559**, 559f, 560, 561f, 562, 567, 567f, 568
Chorda tympani, 413f, 414f
Chorion, 798f
Choroid, 474, 474f, 475t, 484, 484f, 486, **486**, 486f
Choroid layer, 482t
Choroid plexus, 363, 363f
  sheep, 407f, **408**
Chromaffin cells, 518f, 519, 519t
Chromatin, 61t, 62t, 64f–65f, 66f, 67
Chromatophilic substance (Nissl bodies), 61t, 117, 118, 118f, **359**, 359t, 360, 360f
Chromophobes, 512, 513t, 514f, 515
Chromosomes, in mitosis, 64f–65f
Chylomicrons, 723t, 724
Chyme, **707**, 708, 716, 723
Chymotrypsin, 724t
Cilia
  epithelial cell types with, 98f, 100, 100f, 101f, 102
  respiratory tract, 669
  structure and function, 97t, 104
  trachea, 666t, 667f, 668
Ciliary body, 482t, 484, 484f, 486f, **486**, 486f
Ciliary ganglion, 412f, 454f
Ciliary muscle, 482t, 484f
Ciliary process, 484f
Cingulate gyrus, human, 391t, 397f
Circadian rhythms, regulation of, 515
Circular architecture, in skeletal muscle, 264t
Circular folds (plicae circularis), **707**, **718**, 719f
Circular layer of muscularis, 702, 703f, 704f, 704f, 709, 709f
Circular smooth muscle, **743**, 743f, 780f
Circulatory system, fetal, 623, 623f, 624f
Circumduction, 242t
Circumferential lamellae, **148**–49, 148f
Circumflex artery, 563f, 564f
Circumvallate papillae, 414f, 470f, 470t, 471, 471f, 472
Cirrhosis, 613
Cisterna chyli, 655f
Clavicle, 157f, 202t, **203**, 203f, **213**, 213f
Clavicular head, of sternocleidomastoid, **299**
Clavicular notch, 190t
Cleavage, **797**
Cleavage furrow, 64f–65f, 791t–792t
Clinical breast exam (CBE), 788
Clinoid processes
  anterior, 180f
  posterior, **179**, 180f
Clitoris, 785f
Clotting proteins, 113
Coagulation, **538**
Coagulation time, **538**, 542–43
Coccygeal cornu, 189f
Coccygeal region, 183
Coccygeal vertebrae, 184f
Coccyx, 157f, **183**, 183t, 189, 189f, 228f
Cochlea, 415f, **476**, 477f, 487f, 489, 489t, 498
Cochlear duct (membranous labyrinth), 477f, 478f, 490f
Cochlear duct (scala media), **476**, 477f, 478t
Cochlear implants, 498
Cochlear nerve, 415, 415f, 476, 477f, 478f, 487f
Colic arteries, 610
Colic flexures, 720t, 721f–722f
Colick, 530
Collagen fibers, 104, 104t, 105t, 106, 106f, 107, 107f, 108–9, 108f, 109f, 112, 112f

in blood vessels, 594
in bone, 151
in meniscus, 156, 156f
in reticular tissue, 651
in scar tissue, 562
in tendons and ligaments, 156
in trachea, 664f
Collecting ducts, 738, 738t, 740, **741**, 742f, 746f, 748, 753t
Collecting tubules, 738t, 739f, 748, 756
Colliculi (superior and inferior), 393t, 406, 406f, 407f
Colloid, **516**, 516f
Colon, **720**
  ascending, 610f, 611f, 709f, 720, 720t, 721f–722f
  descending, 610, 611f, 709f, 720, 720t, 721f–722f
  innervation, 454f
  sigmoid, 610f, 709f, 720, 720t, 721f–722f
  transverse, 610f, 709f, 720, 720t, 721f–722f
Columnar epithelium. See also Pseudostratified columnar epithelium
  in respiratory tract, 104
  simple, 98f, 100, 100f, 647, 647f, 668f, 705f, 707, 772f
    ciliated, 669, 669t
  stratified, 98f
  structure and function, 96t
  types, identification of, 98f, 100, 100f, 101f, 102–3, 102f
Commissures
  gray, 430, 431f, 432t
  white, 430, 431f, 433t
Common bile duct, 716t, 717f, 718f, 723, 723f
Common carotid arteries, 602f, 603f, 605f, 607f
Common fibular nerve, 441f, **442**, 443f
Common hepatic artery, 610f, 612f
Common hepatic duct, 716t, 717f, 718f, 723, 723f
Common iliac artery, 607f, 610f, 619f, 621f, 622f, 624f
Common iliac vein, 608f, 620f, 621f, 622f
Communicating arteries, 604f, 605f
Compact bone, 110, 110t, 111f, 113, 113f, 148–49, 148f, **153**–54, 154f, 155, 155f
Compartments, fascial, **266**
  arm, 266, 266t, 267f, 326–28
  forearm, 266, 266t, 329–33
  leg, 266, 266t, 344–46
  thigh, 266, 266t, 267f, 339–43
Compound action potential (CAP), 372–73, 372f, 374
Compound microscope, 45–51
  care of, 45, 45f
  cleanup procedures, 51, 51f
  depth of field, 47t, **51**, 51f
  field of view, 47t, **49**–50
  focusing of, 48–49
  magnification, 45, 49
  parts of, 46, 46f, 47t
  specimen size, estimation of, 50–51, 50f
  troubleshooting, 48
  working distance, 47t, **49**, 49f
Concentric lamellae, **148**–49
Conchae, 172t
Conclusion(s), in scientific method, 5, 5f
Condenser, of microscope, 46f, 47t
Conducting zone, **664**, 669, 673
Conduction deafness, 499
Conduction velocity, of action potential, **371**
Conductive tissues, muscle as, **252**
Condylar (ellipsoid) joint, 240t, 241f
Condylar process, 175f
Condyle
  femoral, 218, 220f, 221f, 243, 244f–245f
  occipital, 170t, 178f, 186
  tibial, 219t, 222f, 243
Cones, 475t, **497**
Confluence of sinuses, 385f, 386f, 387f–388f
  sheep, 401, 401f
Confounding variables, **4**
Connecting stalk, 798f
Connective tissue, 104–14
  embryonic, 105, 105f
  fiber types, 104, 104t
  fluid, **104**, 113–14, 114f
  of hair follicle, 133t, 134, 134f
  identification of, 106–14, 106f, 110f
  proper, 104–9, 105t
  of skeletal muscle, 257, 257f
  staining of, 108
  structure, 104
  supporting, 104, 110–13, 110t
  types of, 104

Conoid tubercle, 200, 202t, 203f
Constrictor
  definition of, 263t
  inferior, 298t, 299f
  middle, 298t, 299f
  superior, 298t, 299f
Continuous capillaries, **597**, 598t
Contractile tissues, muscle as, **252**
Contraction, 252
Control center
  in homeostasis, 2
  in reflexes, **450**, **452**, 452f, 452t
Control value, in scientific experiment, **4**
Conus medullaris, 433t, 434f
Convergent architecture, in skeletal muscle, 264t
Convoluted tubules
  distal (DCTs), 738, 738t, 739f, **740**, **741**, 741f, 748, 753t, 756
  proximal (PCTs), 736, 738, 738t, 739f, **740**, **741**, 741f, 748, 752, 753, 753t
Coracobrachialis, 324t, 325f, 326, 327f, 327t
Coracoid process, 200, 202t, 204f
Cord(s), of brachial plexus, **437**–38, 437t, 438f, 440f
Cornea, 481t, 484, 484f, 485, 485f
Corneal reflex, 413, 461, 461f
Corniculate cartilage, 673t, 674f
Cornu, 183t, 189, 189f
Coronal plane, 31, 31f, 31t
Coronal suture, 172t, 176f, 177f
Corona radiata, 769, 769f, 770f, 771t
Coronary arteries, 557f, 562, **562**, 563f, 563t, 564f
Coronary circulation, 562–65, 563f, 564f, 565t
Coronary sinus, 558f, **559**, 559f, 560, 564f, 565, 565t
Coronary sulcus, 565, 566, 566f
Coronoid fossa, 205t, 207f–208f
Coronoid process, 169t, **175**, 175f, 206f, 210f
Corpora arenacea
  pineal sand, **515**, 515f
  prostatic calculi, **781**, 781f
Corpora quadrigemina (tectal plate), human, 391t, 397f
Corpus albicans, 768, 769f, 770t–771t, 772, 794, 794f
Corpus callosum, 406, 407f
  human, 391t, 397f
  sheep, 407, 407f, **408**
Corpus cavernosum, **782**, 782f, 789f, 790f
Corpus luteum, 768, 769f, 770t–771t, 772, 774f–775f, 794, 794f
Corpus spongiosum, **782**, 782f, 789f, 790f
Corrugator supercilii, 292t, 294f
Cortex
  of hair follicle, 133t, 134, 134f
  of kidney, 746f
  of lymph node, **648**, 649, 649f, 650f, 650t, 656f
  of ovary, **768**, 769f, 770
  of thymus, 652, 652f
Cortical nephron, 739f, 746f
Corticomedullary junction, **738**
Corticosteroids, **517**, 523
Corticotropin-releasing hormone (CRH), 513t
Cortisol, 519t
Costal cartilage, 192
Costal facets, 183t, **188**, 188f
Costal groove, 190t, 192f
Costal tuberosity, 202t, 203f
Costocervical trunk, 607f
Coughing reflex, 416
Counteraction, in homeostasis, **2**
Cow
  bone, dissection, 155–56, 155f, 156f
  eye, dissection, 485–87, 485f, 486f, 487f
Coxal, as term, 35f
Cranial, defined, 33t, 35f
Cranial bones, **169**, 169t
Cranial dural septae, **384**, 385t, 387f–388f
  sheep, 403
Cranial floor, 179, 179f, 180f
Cranial fossa, 179, 179f, 180f
Cranial meninges, 384–87, 384f–385f, 385t–386t, 387f–388f
Cranial nerves, 398–400, 398t–399t, 399f
  I (olfactory), 398t, 409t, **472**, 472f
    disorders, common, 410t
    functions, 409t, 411, 411f
  II (optic), 398t, 399f, 412f
    disorders, common, 410t
    function, 409t, 411, 411f, 475, 497
    gross anatomy, 396f, 481t, 484, 484f, 485, 485f, **486**, 486f

I-3

Cranial nerves, II (optic) (continued)
   histology, **474**, 474f
   and pupillary reflexes, 462
   sheep, 401f, 402f, 404, 404f, 405f
  III (oculomotor), 296t, 398t, 399f
   disorders, common, 410t
   functions, 409t, 412, 412f, 498
   and parasympathetic division, 453f, 454f, 455
   and pupillary reflexes, 462
   sheep, 405f
  IV (trochlear), 296t, 398t, 399f, 400
   disorders, common, 410t
   functions, 409t, 412, 412f, 498
  V (trigeminal), **295**, 295t, 298t, 300t, 303, 398t, 399f
   and corneal reflex, 461
   disorders, common, 410t
   functions, 409t, 413, 413f
   sheep, 401f, 402, 402f, 403, 403f, 405f
   V1 (ophthalmic branch), 409t, 413, 413t
   V2 (maxillary branch), 409t, 413, 413t
   V3 (mandibular branch), 409t, 413, 413f
  VI (abducens), 296t, 398t
   disorders, 410t
   functions, 399f, 409t, 413, 413f, 498
   of sheep, 405f
  VII (facial), **292**, 292t–293t, 300t, 398t, 399f
   and corneal reflex, 461
   disorders, common, 410t
   functions, 409t, 414, 414f
   and parasympathetic division, 453f, 454f, 455
   sheep, 405f
   and stroke, 303
  VIII (vestibulocochlear), 398t, 399f, 478t
   cochlear branch, 477f, 490, 490f
   and cochlear implants, 498
   disorders, common, 410t
   functions, 409t, 415, 415f, 501
   sheep, 405f
   vestibular branch, 498
  IX (glossopharyngeal), 298t, 398t, 399f, 414f
   disorders, common, 410t
   function, 409t, 414f, 416, 416f
   and parasympathetic division, 453f, 454f, 455
  X (vagus), 297t, 298t, 398t, 399f, 417f, 673
   disorders, common, 410t
   function, 409t, 416, 416f
   and parasympathetic division, 453f, 454f, 455
  XI (accessory), 300t, 398t, 399f, 436f
   disorders, common, 410t
   function, 409t, 417, 417f
   sheep, 405f
  XII (hypoglossal), 297t, 300t, 301t, 303, 398t, 399f, 436f
   disorders, common, 410t
   functions, 409t, 417, 417f
  disorders, common, 410t
  functions of, 409t, 410–17
  learning strategies for, 400, 408
  modalities within, 408t
  sheep, 405, 405f
Cranial reflexes, **450**, **452**, 461–62
Craniosacral (parasympathetic) division, 453–55, 453f, 454f, 455t
Cranium, 157f
CRH. See Corticotropin-releasing hormone
Cribriform foramina, 169t, 172t, **472**, 472f, 473t
Cribriform plate, 169t, 172t, **173**, 174f, 175, 180f, 472f, 473t
Cricoarytenoid muscle, posterior, 674f
Cricoid cartilage, 672f, 674f
Cricothyroid muscle, 674f
Crista galli, 169t, 174f, 180f
Crossbridge cycle, 270, 271f, 276, 278, 555
Crossing over (synapsis), **791**, 791t–792t
Cruciate ligaments, **243**, 243f, 244f–245f
Crural, as term, 35t
Crus of penis, 789f
Crypts, **646**–47, 646f, 646t
Cubital vein, median, 615f, 616f
Cuboid, 219t, 224, 224f, 225, 226f, 622f
Cuboidal epithelium, 516, 516f, 517f
  structure and function, 96t
  types, identification of, 98f, 99, 99f, 102–3, 102f
Cumulus oophorus, 769, 769f, 770f, 771t
Cuneiform cartilage, 219t, 224, 224f, 225, 226f, 673t, 674f
Current, **370**
Curvatures, spinal, 184, 184f

Cusps
  of aortic semilunar valve, 563f
  of atrioventricular (AV) valves, 558f, 559, 559f, 560, 561f, 563f
  of pulmonary semilunar valves, 559, 559f, 563f
Cutaneous nerves, 442f, 443f
Cutaneous receptors, and plantar reflex, 460
Cuticle (eponychium), 136, 136f, 136t
Cuticle layer, of hair follicle, 133t, 134, 134f
Cystic duct, 716t, 717f, 718f, 723, 723f
Cystic vein, 611f
Cytokinesis, 64f–65f, 65, 791t–792t
Cytoplasm, 66f, 67
  animal cell, 64f
  in cytokinesis, 65
  staining of, 57
  structure and function, 61t
Cytoskeleton, 61t
Cytosol, 61t
Cytotrophoblast, 797t–798t

DAB/PAD mnemonic, 335, 347
Dalfampridine, 376
Dalton's Law of Partial Pressures, **680**
Data analysis, in scientific method, **5**, 5t
Data collection, in scientific method, **4**–5
Data presentation, 7–8, 7f
Daughter cells, 64f–65f, 65, 791t–792t
DCIS (ductal carcinoma in situ), 788
DCTs. See Distal convoluted tubules
Deafness, types of, **499**
Decalcified bone preparations, **147**, 148, 148f
Deep, defined, 33t
Deep brachial artery, 614f
Deep fascia, **257**, 787f
Deep femoral vein, 620f
Deep fibular nerve, 441t, 443
Deep palmar arch, 614f, 617f
Deep palmar venous arch, 615f, 617f
Deflection waves, 570, 570f, 573
Delayed-onset muscle soreness (DOMS), 320
Delta cells, 520t
Deltoid, as term, 35f, 261t, 262t
Deltoid muscles, 264, 265f, 324t, 325f
Deltoid tuberosity, 205f, 207f–208f, **213**, 213f
Demyelinating diseases, 376
Dendrite, 118, 118f, 358, **359**, 359t, 472f
Dens (odontoid process)
  of atlas, 186, 241f
  of axis, **187**
Dense connective tissue, 104, 105t
  appearance vs. smooth muscle tissue, 260
  irregular, 105t, 108–9, 108f, 130, 130f, 137, 137f, 768
  regular, 105t, 108, 108f
Denticulate ligaments, **433**, 433f, 434f
Deoxyribonuclease, 724t
Dependent variable, **4**
Depolarization, of excitable membrane, 371, 371f
Depression (joint movement), 242t
Depressor, definition of, 263t
Depressor anguli oris, 293t, 294f
Depressor labii inferioris, 293t, 294f
Depth of field, of microscope, 47, **51**, 51f
Dermal papillae, **128**, 128f, 130, 130f, 132, **133**, 134, 134f, **468**, 468f, 480, 480f
Dermis, 108–9, 108f, **125**, 130–36, 133
  glands in, 131–32, 131t, 133–34, 133f, 135, 135f
  layers, 130–31, 130f, **468**, 468f
  sensory receptors in, 131–32, 132f, 480f
  structure and function, 132, 137, 137f
Descending abdominal aorta, 607f, 609f, 610f, 624f
Descending colon, 610f, 611f, 709f, 720, 720f, 721f–722f
Descending thoracic aorta, 607f, 609f
Desmosomes, **261**, 671t
Desquamation, **127t**
Detrusor muscle, **744**, 744f, 750f, 751f
Developing long bone, 150–51, 150f
Development
  embryonic period, **797**, 798, 798f, 799f, 799t
  fetal period, 797
  pre-embryonic period, 797, 797t–798t
  prenatal period, 797
Diabetes insipidus, 754
Diabetes mellitus, 753, 754, 757–58, 796
Dialysate, 83
Dialysis, **736**
Diaphragm
  blood vessels, 612f
  central tendon of, **309**

gross anatomy, 308f, **309**, 556f, 608f, 610f, 611f, 656f, 675f, 721f–722f
  innervation, 307t, 436, 436f
  in respiration, 307t, **662**, 679
Diaphragma sellae, 385t, 387f–388f, **396**
  sheep, 401f, **402**, 402f, 403, 403f, 404
Diaphragmatic surface, **557**, 557f, 653f
Diaphysis, 150, **153**–54, 154f, 155f, 156, 238f
Diarrhea, 725
Diarthrosis, 233, 234t, 239, **241**, 242f
Diastasis symphysis pubis, 218
Diastole, **578**, **625**
Diastolic pressure, **625**
Diencephalon, 388, 388f, 390t
Diffusion, 67–76, 81–83, 82f
  across semipermeable membrane, 68, 68f
  facilitated, 523
  membrane permeability and, 75–76, 76f
  osmosis, 77–79, 77f, 78f, 79t, 81–83, 82f
  process, 67–68, 67f
  simple squamous epithelium and, 99
  solute molecular weights and, 73–74, 74f, 75t
  solvent viscosity and, 69–70, 69f, 70f, 71f
  temperature and, 71–73, 72f
Digastric, definition of, 262t
Digastric muscle, 300t, 301f–302f
Digestion, 716, **722**–25, 723f, 724f
  acid-base balance and, 725
Digital, definition of, 34t, 35f
Digital arteries, 614f, 616f, 619f, 621f
Digital veins, 615f, 616f, 620f, 621f
Dilator, definition of, 263t
Diopter adjustment ring, of microscope, 46f, 47t, 52f
Dipeptidase, 724t
Directional terms, 33, 33t
  for muscles, 301
Disease(s), demyelinating, 376
Dissecting microscope, 52–53, 52f
Dissection
  blunt, 14, 22, 22f
  care and prudence in, 14, 19, 22
  cow bone, 155–56, 155f, 156f
  cow eye, 485–87, 485f, 486f, 487f
  instruments, 11, 11t, 12f, 12t
  protecting underlying tissues, 19, 20f
  safety, 10
  with scissors, 20–21, 21f
  sharp, **14**, 19, 19f, 20f
  sheep brain, 401–8, 401f–407f
  sheep heart, 567–69, 567f, 568f
Distal, defined, 33t, 34
Distal convoluted tubules (DCTs), 738, 738t, **740**, **741**, 741f, 748, 753t, 756
Distal phalanx, 136t, 212f, **331**
Distal supracondylar line, 220f
Divisions, of brachial plexus trunks, **437**, 437t, 438f
Dome cells, 103, 103f, **744**, 744f
DOMS (delayed-onset muscle soreness), 320
Donor, of blood, 544
Dorsal interossei, 334t, 335, 335f, 347, 348t, 349f
Dorsal rami, 300t, 303t, 304t–306t
Dorsal root ganglia, **364**, 365f, 430, 431f, 433f, 434f
Dorsal vein, penis, 782f, 782f
Dorsal venous arch, 620f
Dorsi, definition of, 263t
Dorsiflexion, 242t
Dorsum, of foot, 35f
Dorsum sellae, **179**, 180f
Ductal carcinoma in situ (DCIS), 788
Ductus arteriosus, 623t, 624f
Ductus deferens (vas deferens), 779–80, 779t, 780f, **789**, 789f, 790f
Ductus venosus, 623, 623t, 624f
Duodenal glands, 707f, 708
Duodenal papilla, 521f, 711, 711f, 716f, 717f
Duodenojejunal flexure, 707f
Duodenum, 521f, 611f, 612f, 715f
  blood vessels, 610f
  in digestion, 723, 725
  gross anatomy, **716**–18, 716f, 717f, 721f–722f
  histology, 707f, 707f, 707t
  pancreas and, 711, 711f
Dural venous sinuses, **384**, 385f, 386t, 387f–388f, 401, 401f, 402f, 604f, 605f, 606
Dura mater, **384**, 385t, 433, 434f, 435f
  meningeal layer, 384, 384f, 385f, 387f–388f
  periosteal layer, 384, 384f, 385f, 387f–388f
  sheep, 401, 401f, 402f, 404

Ear
  external, 487, 487f, 488t
  functions of, 487
  gross anatomy, 487–90, 487f, 488t–489t, 490f
  inner, 487, 487f, 489t
  middle, 487, 487f, 488t
  soundwave pathways, 490, 490f
  structure and function, 476–79, 477f, 477t–478t
Eardrum (tympanic membrane), 415f, 487f, 488t, 489, **490**, 490f
Eccrine (merocrine) sweat gland, 102–3, 102f, 130f, **131**, 131t, 137, 137f, 469f, 480f
ECG (electrocardiography), 570, 629
  abnormal, 572f, 572f, 575–76, 577
  and cardiac cycle, 626
  deflection waves. normal, 570, 570f
  and electrical axis of heart, 573, 573f
  standard, 574
  standard limb leads for, 570, 570f
Ectoderm, of embryo, 799f
Edema, **641**, 753
EDV (end diastolic volume), 578
EEG (Electroencephalogram), 398
Effector
  in homeostasis, 2
  in reflexes, **450**, **452**, 452f, 452t
Efferent arteriole, 746f, 747f
Efferent ductules, 778
Efferent lymphatic vessels, **648**, 650f, 650t, 651, 656f
Einthoven's Law, **570**
Einthoven's triangle, 570
Ejaculation, 779, 788
Ejaculatory duct, 789f, 790f
Elastic cartilage, 110t, 111f, 112–13, 112f, 490, 594, 594f
Elastic connective tissue, 105t, 109, 109f
Elastic fibers, 104, 104t, 105t, 106, 106f, 108, 109, 109f, 112–13, 112f, 594, 596, 596f
Elastic lamina, 594–95, 595f
Elastic tissues, muscle as, **252**
Electrocardiography, with ECG, 574
Electroencephalogram (EEG), 398
Electromyograms, 279–80, 279f
Eleidin, 127t
Elevation, 242t
Elimination of wastes, **722**
Embolism, 606
Embryo, **797**, 798f
Embryoblast, 797t–798t
Embryonic connective tissue, 105, 105f
Embryonic period, **797**, 798, 798f, 799f, 799t
Emphysema, 680
Encroaching fat, on jejunum and ileum, **718**, 719f
End diastolic volume (EDV), **578**
Endocardium, **555**, 555f, 555t, 558
Endochondral ossification, **150**–51, 150f
Endocrine system
  classical glands of, 510, 510f, 522f–523f, 523
  functions, 510
  histology, 512–24
  hypothalamus and, 393, 510f, 512, 513t, 514, 522f–523f, 523
  ovaries and, 510, 510f, 522f–523f, 523
  pancreas and, **520**–22, 520t, 521f, 522f–523f, 523
Endoderm, 799f
Endolymph, 476, 477f, 477t, 489, 490
Endometrium, 774, **774**, 774t–775t, 784f, 794
Endomysium, 255f, **257**, 257f, 258f
Endoneurium, 365–66, 366f
Endoplasmic reticulum (ER)
  of neuron, 117, 118f
  rough, 61t, 66f, 67, **360**
  smooth, 61t, 66f, 67
  structure and function, 61t
Endosteum, 147t
Endothelium, 98, **555**
  capillaries, 597, 598, 598t–599t, 670, 680, 681
  defined, 99
End systolic volume (ESV), **578**
Energy
  kinetic, **370**
  potential, **370**
Enteric nervous system, **724**
Enteroendocrine cells, 706
Entropy, 67–68
Enzymes, digestive, 700, 711, 716, 723, 724t, 725–27, 726t
Eosin (E), 57, 94
Eosinophil(s), 531t, 533–35, 534f, 537

I-4

Eosinophilic stains, 63
Ependymal cells, **362,** 362t, **363,** 363f
Epicardium, **555,** 555f, 555t, 556, 556f, 558, 566, 566f
Epicondyle. *See* Lateral epicondyle; Medial epicondyle
Epicranial aponeurosis, 294f
Epicranius (occipitofrontalis), 293t, 294f
Epidermal-melanin unit, 129, 129f
Epidermal ridges, **128,** 128f, 130f, 137, 137f
Epidermis, 108–9, 108f, **125, 127**–29, 130f, 133, **468,** 468f, 480f
  cell types in, 127
  layers of, 127–28, 127t, 128f
  structure and function, 132, 137, 137f
Epididymis
  anatomy, 779t, 789f, 790f
  function, **788,** 795
  histology, 776f, **778**–79, 778f, 779t
Epidural (anesthetic procedure), **432**
Epidural space, 386t, 432, 435f
Epigastric arteries, 607f
Epigastric region, 37, 37f
Epigastric veins, 608f
Epiglottis, 672f, 673t, 674f, 712f–713f
Epilepsy, **398**
Epimysium, 255f, **257,** 257f
Epinephrine, 517, 519t
Epineurium, **365**–66, 366f
Epiphyseal line, 154, 154f, 155f
Epiphyseal (growth) plate, **150**–51, 150f, 238f
Epiphysis, 150, **153**–54, 154f, 155f, 156, 238f
Epiploic (omental) appendices, 720t
Epithalamus, human, 391t
Epithelial cells
  ciliated, 772f, **773,** 773t
  of scalp, 133
  shape of, 96t, 97
  stem cells, **705**
  surface modifications, 97t, 104
Epithelial tissue, 96–104
  apical surface of, 96, 99, 99f
  basal surface of, 99, 99f. 96, 103
  basement membrane, 96, 103
  cell surface modifications, 97t, 104
  identification and classification of, 96, 96t, 97t, 98–104, 98f
  simple, **97**
  specialized cells, 97t
  stratified, **97**
  structure of, 96
Epithelium
  germinal, 768, **768,** 769f, 770f
  intestinal, 709, 709f
  lymphatic tissues, 646–47, 646f, 646t, 647, 647f, 648, 648f
  male reproductive structures, 779t
  olfactory, 175, 427f, 472, 473t, **665–66,** 665f, 665t
  renal, 738, 738t
  respiratory, 664, 664f, 666, 667f, 668, 669t, 670t, 680
  small intestine, 707f, 708
  stomach wall, 702, 703f, 704f, 704t
  trachea, 666t
  uterine tube, 772f, 773
  vaginal wall, 775, 775f
Eponychium (cuticle), 136f, 136t
Equator, cellular, 791t–792t
Equatorial plate, 64f–65f
Equilibrium
  solution in, 68, 68f
  testing of, 498–501
Equipment, laboratory
  dissection instruments, 11, 11t, 12f, 12t
  safety equipment, 10
Erector spinae, **303,** 304t, 306f
ERV (expiratory reserve volume), 679, 679t, 680f, 684f, 685, 685t, 689, 689t
Erythrocytes (red blood cells), **113,** 114, 114f
  cell count, 538
  centrifugal separation of, 438f, 538
  characteristics of, 530t
  in gas transport, 681
  identification of, 533–35, 534f, 537
  and respiration, **662**
  and tonicity, 81–83, 82f
  in urine, 753, 754t
Esophageal opening, 309, 309f
Esophageal plexus, 454f
Esophageal sphincter, inferior, 714t, 715f
Esophageal veins, 613
Esophagus, 299f
  blood vessels, 610f
  connective tissue in, 111–12, 111f
  in digestion, 722–23
  epithelial tissue, 101, 101f
  gross anatomy, 655, 672f, 712, 712f–713f, 715f, 721f–722f
  histology, 666, 667f
  innervation, 454f
Estrogen, **768,** 774t–775t, 793–95, 794f
ESV (end systolic volume), **578**
Ethanol, 13, 13t
Ethmoid bone, 153f, 169, 173, 173f, 174f, 175, 176f, 182t
Eustachian tube (pharyngotympanic tube; auditory tube), 487f, 488t, 489, **490,** 490f, 655f, 672f
Eversion, 242t
Exchange. *See* Gas exchange
Excitable tissue
  definition of, 371
  muscle as, **252,** 371
  nervous tissue as, **356, 371**
Exercise, and respiration changes, 688–89
Exocrine glands, pancreas as, **520, 711**–12, 711f
Exocytosis, 131t
Experimental design, in scientific method, **4**
Expiration, **662, 679**
Expiratory reserve volume (ERV), 679, 679t, 680f, 684f, 685, 685t, 689, 689t
Expression, facial, muscles of, 292, 292t–293t, 294f
Extensible tissues, muscle as, **252**
Extension, 242t
Extensor, definition of, 263t
Extensor carpi radialis brevis, 332t, 333f
Extensor carpi radialis longus, 332t, 333f
Extensor carpi ulnaris, 332t, 333f
Extensor compartment, **326**
Extensor digiti minimi, 332t, 333f
Extensor digitorum, 264, 265f, 332t, 333f
Extensor digitorum brevis, 345f, 347t
Extensor digitorum longus, 344f, 344t, 345f
Extensor digitorum longus tendon, 345f
Extensor hallucis brevis, 345f, 347t
Extensor hallucis longus, 344f, 344t, 345f
Extensor hallucis longus tendon, 345f
Extensor indicis, 332t, 333f
Extensor pollicis brevis, 332t, 333f
Extensor pollicis longus, 332t, 333f
Extensor retinaculum, 333f
External acoustic (auditory) meatus, 170t, 176f, **179,** 487f, 488t, 489, **490,** 490f
External auditory canal, **179**
External carotid arteries, **601,** 602f, 603f
External iliac artery, 607f, 619f, 621f, 622f
External iliac vein, 608f, 620f, 621f, 622f
External intercostal membrane, 308, **308**
External intercostal muscles, 307t, **308,** 308f, 311f, **662,** 679
External jugular vein, **601,** 602f
External oblique, **310,** 310t, 311f, 312, 312f, 313f
External occipital crest, 178f
External occipital protuberance, 170t, 177f, 178f
External os, 784f
External urethral sphincter, **749,** 750, 751f, 752
Externus, definition of, 263t
Extracellular matrix, of bone, 110
Extraocular muscles, 296, 481t, 485, 485f
Extrinsic (extraocular) muscles, **296,** 481t, 485, 485f
Extrinsic regulation of muscle contraction, 277
Extrinsic tongue muscles, 297, 297t, 298f
Eye
  cow, 485–87, 485f, 486f, 487f
  external structures, 481t, 485f
  gross anatomy, 481–85
  internal structures, 482t–483t
  movement, control of, 498
  muscles, 252, 296, 296t, 297f, 412
Eyebrow, 485f
Eyelash, 485f
Eyelid, 485f
Eye piece, of microscope, 46f, 47t, 52f
Eyewear, protective, 10

Facial artery, 602f
Facial bones, **169,** 169t
Facial muscles
  of expression, 292, 292t–293t, 294f
  of mastication, 295, 295f, 295t
Facial nerve (CN VII), **292,** 292t–293t, 300t, 398t, 399f
  and corneal reflex, 461
  disorders, common, 410t
  functions, 409t, 414, 414f
  and parasympathetic division, 453f, 454f, 455
  sheep, 405f
  and stroke, 303
Facial vein, 602f, 604f
Facilitated diffusion, **523**
Fahrenheit, conversion to/from Celsius, 9
Falciform ligament, 716t, 717f, 718f, 721f–722f
Fallopian tubes (uterine tubes; oviducts)
  gross anatomy, 783–85, 784f, 785f
  histology, 100, 100f, 772–73, 772f, 773f
Falx cerebelli, 386t, 387f–388f
  sheep, 403
Falx cerebri, 384f, 385f, 386t, 387f–388f
  sheep, 402f
Fascia
  chicken, 19, 19f
  deep, **257,** 787f
Fascial compartments. *See* Compartments, fascial
Fascicles, **254,** 255f, **257,** 264, **365**–66, 366f
Fat
  digestion of, 723–24, 723t, 724t, 727
  encroaching, on jejunum and ileum, **718,** 719f
  subcutaneous, 130f
Fatigue, muscle fiber type and, 268, 268t, 269, 269t
Fauces, 712f–713f
F cells, 520t
Feces, **708**
Female
  pelvis, **215**–18, 217f
  reproductive system
    components, 768
    functions, 768
    gross anatomy, 783–888
    histology, 768–75
  urethra, 751–52, 751f
Femoral, defined, 34t, 35f
Femoral artery, 619f, 621f, 622f
  deep, 619f
Femoral circumflex arteries, 619f
Femoral circumflex veins, 620f
Femoral condyles, 218t, 220f, 243, 244f–245f
Femoral cutaneous nerves, 442f, 443f
Femoral nerve, 441t, 443, 443f
Femoral triangle, **339,** 340f
Femoral vein, 620f, 621f, 622f
Femoris, definition of, 263t
Femur, 152f, 157f, 267f
  fracture of, 246
  gross anatomy, 218t, **220,** 220f–221f
  histology, 154
  joints, 244f–245f
Fenestrated capillary, **597,** 598t–599t
Fenestrations, **597,** 598f
Fertility drugs, 795
Fertilization, **791,** 793, 797
Fetal period, **797**
Fetus, **797**
  blood vessels, 623, 623t, 624f
  development, 797
  skull, 181, 182f
FEV (forced expiration volume), **680**
Fibrillation, 572, 572f
Fibrin, **113,** 538
Fibrinogen, **113**
Fibroblast, 104t, **106,** 106f, 108–9, 108f, 109, 109f, 260
Fibrocartilage, 110t, 111f, 112, 112f, **237,** 238, 238f
Fibrous capsule, renal, 739f, **744,** 745f, 746f
Fibrous joints, 236, 236t, 237f
Fibrous pericardium, **556,** 556f
Fibrous sheaths, 331f
Fibrous tunic, 481t
Fibula, 157f, 219t, **223,** 223f, 244f–245f
Fibular artery, 619f
Fibular articular facet, 219t, 222f
Fibular collateral ligament, 243t, 244f–245f
Fibularis brevis, **344,** 345f, 345t
Fibularis longus, 344, **344,** 345f, 345t
Fibularis tertius, 344t, 345f
Fibularis tertius tendon, 344f, 345f
Fibular nerves, 441t, **442,** 443, 443f
Fibular veins, 620f
Field of view, of microscope, 47t, **49**–50
Fight or flight response, 453, 453f, 455–56
Filiform papillae, 470f, 470t, 471
Filtrate, **79, 736, 752**
Filtration, **79**–81, 80f
  by nephrons, **752**
Filtration membrane, **736, 752,** 753
Filum terminale, 433, 433t, 434f
Fimbria, 784f, 785f
Finger pulse, 626
Fixation, **13**
Flat bone, classification of, 152f, 152t
Flexion, 242t, 252
Flexor, definition of, 263t
Flexor carpi radialis, 329, 329t, 330f, 331f
Flexor carpi ulnaris, 329, 329t, 330f, 331f
Flexor compartment, **326**
Flexor digiti minimi brevis, 334t, 335f, 348t, 349f
Flexor digitorum brevis, 347t, 349f
Flexor digitorum longus, 346f, 346t, 349f
Flexor digitorum profundus, 329t, 330f, **331,** 331f
Flexor digitorum superficialis, 329t, 330f, **331,** 331f
Flexor hallucis brevis, 348t, 349f
Flexor hallucis longus, 346f, 346t, 349f
Flexor pollicis brevis, 334t, 335f
Flexor pollicis longus, 329t, 330f
Flexor retinaculum, **329**–31, 330f
Fluid connective tissue, **104,** 113–14, 114f
Focus adjustment knobs, of microscope, 46f, 47t, 52f
Foliate papillae, 470f, 470t, **471**
Folic acid, 189
Follicle, lymphatic, **645,** 645f–650f, 646–47, 648, 649, 650t, 656f
Follicle, ovarian, **768,** 769f, 792–93
  developmental stages, 770t–771t
  primary, 768, 769f, 770t–771t, 794f
  primordial, 768, 769f, 792–93
  secondary, 768, 769f, 770t–771t, 794f
  vesicular (Graafian), 768, 769, 769f, 770f, 770t–771t, 793, 794f
Follicle, hair, **133,** 133f, 134, 134f, 135, 135f, 137, 137f
Follicle-stimulating hormone (FSH), 512, 513t, 793, 795
Follicular cells, 516, 516f, 517t, **524,** 795
Follicular phase, of ovarian cycle, **793**–94, 794f
Fontanelles, **181,** 182f
Foot muscles, intrinsic, 347, 347t–348t, 349f
Foramen, 178
Foramen lacerum, 170t, **178,** 178f, 180f
Foramen magnum, 170t, 178f, 180f, 417f
Foramen ovale, 170t, 178f, 180f, **558**–59, 558f, 623t, 624f
Foramen rotundum, 170t, 180f
Foramen spinosum, 170t, 178f, 180f
Forced expiration volume (FEV), **680**
Forced vital capacity (FVC), **680**
Forceps, 11t, 19, 19f, **20,** 21f, **22,** 22f
Forearm, **326**
  compartments, fascial, 266, 267f, 329–33
  muscles, 329–32, 329t, 330f, 331f, 332t, 333f
Forehead, bones, 182t
Formalin, **13,** 13t
Fornix
  human, 391t
  sheep, 406, 407, 407f, **408**
Fossa
  coronoid, 205t, 207f–208f
  cranial, 179, 179f, 180f
  glenoid, 202t, 203
  hypophyseal, 179, 180f
  iliac, 214t, 216f
  incisive foramen, 169t, 172t, 178f
  infraspinous, 202t, 204f
  intercondylar, 218t, 220f
  lacrimal, 172t
  mandibular, 170t, 178f
  olecranon, 205t, 207f–208f
  ovalis, 558, **558,** 558f, 560, **560,** 561f, 562
  radial, 205t, 207f–208f
  subscapular, 202t, 204f
  supraspinous, 202t, 204f
  tuberculum sellae, 179, 180f
Fourth ventricle, 388f, 389f, 390t
  human, 391t, 397f
  sheep, 406, 407f
Fovea, 218t, 220f
Fovea centralis, **474,** 474f, 475t, 482t, 484, 484f
Free nerve ending, **132,** 132t, 137, 468t, **479,** 480, 480f
Frenulum (inferior and lingual), 712f–713f
Frontal, as term, 35f
Frontal bone, 152f, 169, 169t, 171f, 173, 173f, 176f, 177f, 180f, 182f, 182t
Frontal crest, 180f
Frontal lobe
  human, 391t, 394f, 395f, 396f, 397f
  sheep, 404, 404f
Frontal process, 170t, 172t, 173f
Frontal sinus, 169t, 174f, 180f, 672f
FSH. *See* Follicle-stimulating hormone

Functional layer, of uterus, 794f, 798f
Functional residual capacity, 679, 679t, 680f, 684f, 685, 685t
Fundus
  stomach, 705f, 706t, 714t, 715f
  uterus, 784f
Fungiform papillae, 470f, 470t, 471
Funiculi, 430, 431f, 432t, 435f
Funny bone, 438
FVC (forced vital capacity), 680

Gag reflex, 416
GAGS. See Glycosaminoglycans
Gallbladder, 716
  blood vessels, 610f, 611f
  functions, 716, 723
  gross anatomy, 716–18, 716t, 717f, 718f, 721f–722f
  innervation, 454f
Gallstones, 723, 723f
Gametes, 766, 766, 791
Gametogenesis, 791
Ganglia, 358, 364
  satellite cells in, 364–65, 365f
Ganglion cell layer, retina, 474, 474f, 475, 475t, 476
Gap junctions, 261
Gas exchange, 681
  alveolar, 679, 680–81
  systemic, 679, 681
Gas transport, 679, 681
Gastric arteries, 610f
Gastric folds (rugae), 714t, 715f
Gastric glands, 704f, 705–6, 705f, 706t
Gastric lipase, 724t
Gastric pits, 704f, 705–6, 705f, 706t
Gastric veins, 611f
Gastrin, 705f, 706t
Gastrocnemius, 264, 265f, 274, 276, 345f, 346, 346t, 443f
Gastroduodenal artery, 610f, 612f
Gastroepiploic arteries, 610f
Gastroepiploic veins, 611f
Gastrointestinal (GI) tract, 698
  digestion in, 722–25, 723t, 724t
  motility in, 724
  walls, 698, 702–6, 703f, 704f, 704t
Gemellus, 336t, 337f, 338
General senses
  definition of, 468
  functional properties of, 491–93
  gross anatomy, 479–90
  histology, 468–69
Geniculate ganglion, 414f
Genioglossus, 297t, 298f
Geniohyoid muscle, 300t
Genitalia. See Reproductive system
Genitofemoral nerve, 443f
Germinal center, 645, 645f, 649f, 650f
Germinal epithelium, 768, 769f, 770f
Gestational diabetes, 754
GFR (glomerular filtration rate), 752
GH (growth hormone), 512, 513t
GHIH (growth-hormone-inhibiting hormone), 513t
GHRH (growth-hormone-releasing hormone), 513t
Gingivae, 712f–713f
GI tract. See Gastrointestinal (GI) tract
Glabella, 171f, 172t
Glabrous skin. See Thick skin
Glands, in dermis, 131–32, 131t, 133–34, 133f, 135, 135f
Glandular epithelium, 779t
Glans penis, 789f, 790f
Glassy membrane, of hair follicle, 133t, 134, 134f
Glenohumeral joint, muscles, 323–26, 324t, 325f
Glenoid cavity, 204f
Glenoid fossa, 202t, 203
Glial cells
  histology, 118, 118f, 360, 361f, 362–65, 362f, 515
  structure and function, 117, 117t, 356
Gliomas, 362
Glomerular capillaries, 740f, 752, 753
Glomerular capsule (Bowman's capsule), 738t, 739f, 740, 740f
Glomerular filtration rate (GFR), 752
Glomerular hydrostatic pressure (HP$_g$), 752, 752f, 753
Glomerulus
  function, 752, 752
  gross anatomy, 745, 746f, 747f

histology, 738t, 739f, 740, 740f
Glossopharyngeal nerves (CN IX), 298t, 398t, 399f, 414f
  disorders, common, 410t
  function, 409t, 414f, 416, 416f
  and parasympathetic division, 453f, 454f, 455
Glottis, 673t
Gloves, 10
Glucagon, 510, 520t
Glucocorticoids, 519t
Glucose
  blood levels, 2, 510, 754
  blood transport, 727–29
  facilitated diffusion across cell membrane, 523
  tubular reabsorption of, 753t
  in urine, 753, 754t, 755, 755t
Glucose test strips, 75
Glucosuria, 754t
Gluteal, defined, 34t, 35f
Gluteal lines, 214t, 216f
Gluteal muscles, 336–38, 336t, 337f, 338f
Gluteal nerves, 441t, 442, 442f
Gluteal tuberosity, 218t, 220f
Gluteus maximus, 336t, 337f, 338, 442f
Gluteus medius, 336–38, 336t, 337f, 339, 442f
Gluteus minimus, 336–38, 336t, 337f, 339, 442f
Glycerinated muscle tissue, 270, 270f, 271f, 273
Glycerine (glycerol), 13, 13t
Glycogenolysis, 524
Glycosaminoglycans (GAGs), 110
Glycosuria, 754t
GnRH. See Gonadotropin-releasing hormone
Goblet cells, 100, 495
  epithelial cell types with, 98f, 100, 100f, 101f, 102
  intestinal, 707, 708, 709, 709f
  structure and function, 97t, 104
  of trachea, 666f, 667f, 668
Golgi apparatus, 62t, 66f, 67
Golgi tendon organs, 350
Gomphosis, 236, 236t, 237f
Gonadocorticoids, 519t
Gonadotropin, 794f
Gonadotropin-releasing hormone (GnRH), 513t, 793, 795
Gonads, 766
Goose bumps, 133
GPA mnemonic, 512
Graafian (vesicular) follicle, 768, 769, 769f, 770f, 770t–771t, 793, 794f
Gracilis, 267f, 340f, 341, 341f, 341t, 342
  as term, 261t, 262t
Graded potentials, 371, 371f
Grand mal seizure, 398
Granular layer, of Purkinje cell, 361, 361f
Granulosa cells, 769, 769f, 770f, 771t
Graph(s), 7
Gray commissure, 430, 430, 431f, 432t
Gray matter, 356, 358, 358f, 360, 360f, 384f, 430, 431f, 432t, 459
Greater curvature, of stomach, 714t, 715f, 721f–722f
Greater omentum, 714t, 715f
Greater sciatic notch, 214t, 216f
Greater trochanter of femur, 218t, 220f, 227, 228f, 337f
Greater tubercle (humerus), 205t, 207f–208f
Greater wing, 170t, 173f, 176f, 180f
Great saphenous vein, 620f, 621f
Greece, classical, four bodily humors in, 529, 530t
Gross anatomy, defined, 1
Ground bone preparations, 147, 148, 148f
Ground substance, in supporting connective tissue, 110, 110t
Growth hormone (GH), 512, 513t
Growth-hormone-inhibiting hormone (GHIH), 513t
Growth-hormone-releasing hormone (GHRH), 513t
Growth (epiphyseal) plate, 150–51, 150f, 238f
Gustation (taste), 470–72, 470f, 470t, 471f
  basic taste sensations, 494, 494t
  olfaction and, 470, 495–96
Gustatory cells, 471, 471f, 471t
Gut tube, 799f
Gynglymoid (hinge) joint, 240t, 241f
Gyrus (gyri)
  human, 393, 394, 394f, 395f
  sheep brain, 406, 406, 406f

Hair cells
  semicircular canals, 498

spiral organ, 477f, 478t, 490, 490f
  vestibule, 498
Hair follicles, 133, 133f, 134, 134f, 135, 135f, 137, 137f
Hair shaft, 480f
Hallux, 219t, 226, 621f
Hamate, 206t, 211, 211f
Hamstring muscles, 245, 342–43
Hand, muscles, intrinsic, 334–35, 334f, 335f
Hard palate, 672f, 712f–713f
Haustra, 720t
Haversian (central) canal, 110t, 113, 113f, 148–49, 148f
Haversian systems. See Osteon
Hazardous chemicals, safe use of, 13, 13t
Hazardous waste
  defined, 14
  disposal of, 14, 14t, 15
hCG. See Human chorionic gonadotropin
HDLs. See High density lipoproteins
Head
  blood vessels, 601–3, 602f, 603f
  muscles, 292–99
Head, of bone
  femur, 218t, 220f
  fibula, 219t, 227, 228f
  humerus, 205t, 207f–208f
  mandible, 175, 175f, 176f
  metatarsals, 219t
  radius, 206t, 209f
  rib, 190t, 192f
Head, of microscope, 46f, 47t, 52f
Head, of pancreas, 711f, 716f, 717f
Head locking screw, of microscope, 52f
Hearing, 476–79, 477f, 477t–478t
  testing, 499
Heart. See also entries under cardiac
  arrhythmias, treatment of, 376
  coronary circulation, 562–65, 563f, 564f, 565t
  electrical abnormalities, 572, 572f, 575–76
  electrical axis of, 573, 573f
  electrical conduction system, 569–70, 569f, 570f
  fetal, 624f
  human, gross anatomy, 557–62, 557f
  innervation of, 454t, 456
  location within thoracic cavity, 556f
  refractory period, 571
  sheep, gross anatomy, 566–69, 566f–568f
  sound of (auscultation), 578–79, 578f, 626
  wall, histology, 555, 555f, 555t, 558
Heart attack. See Myocardial infarction
Heart block (AV block), 575–76
Heart disease, deaths from, 552
Helicotrema, 478t, 490f
Hematocrit
  and blood vessel resistance, 625
  measurement of, 538, 540, 540f
  normal levels, 542
Hematoxylin (H), 57, 94
Hematuria, 754t
Heme group, 681
Hemiazygos vein, 608f
Hemocytoblasts, 227
Hemodialysis, 83
Hemoglobin, 538, 681
  oxygen binding, pH and, 686–87
  in urine, 754t, 755, 755t
Hemoglobin content, 541–42, 541f, 542
Hemoglobinometer, 541, 541f
Hemoglobinuria, 754t
Hemopoiesis, 156, 227, 536
Hemorrhagic stroke, 303
Hemorrhoids, 613
Hemostasis, 538
Hemostat, 11t, 16, 17f, 18f, 19
Henry's Law, 680
Hepatic artery, 610f, 612f, 710, 710f, 718f
Hepatic colic flexures, 720t, 721f–722f
Hepatic cords, 710, 710f, 711f
Hepatic ducts, 716t, 717f, 718f, 723, 723f
Hepatic lobule, 710, 710f, 711f
Hepatic portal system, 609, 611, 611f, 613, 723f, 724
Hepatic portal vein, 609, 611f, 612f, 613f, 710, 710f, 718f
Hepatic sinusoids, 710, 710f
Hepatic veins, 608f, 611, 611f, 612f, 613
Hepatocytes, 597, 599t, 710, 710f, 711f
Hepatopancreatic ampulla, 711, 711f, 716t, 717f
Hidden border, of nail, 136, 136f, 136t
High density lipoproteins (HDLs), 539, 542
Hilum
  lung, 675, 676f, 677, 677f

lymph node, 648, 649, 649f, 650f, 650t, 656f
ovary, 784f
spleen, 654t, 656f
Hinge (ginglymoid) joint, 240t, 241f
Hip
  abduction, 336–38
  muscles, 336–39, 336t, 337f, 338f
  replacement surgery, 246, 246f
Histology, 1, 57, 93
Histone proteins, 64f–65f
Holocrine secretion, 131t
Homeostasis
  acid-base balance, 681, 756
  and autonomic nervous system, 453f
  blood calcium levels, 151
  blood pH, 681
  hypothalamus and, 393
  overview of, 2
  potassium levels, 367
  sodium levels, 369–70
Homologous chromosomes, 791, 791t–792t
Homologous structures, in reproductive system, 766
Horizontal fissure, 676f
Horizontal plate, 174f
Hormone(s)
  adrenal, 517, 519t
  in bone formation and remodeling, 151
  and childbirth, 218
  pancreatic, 510, 520t
  parathyroid, 151, 510, 516, 517t
  pineal, 515
  pituitary, 393, 512, 513t, 514, 795
  sex, 510
  steroid, 517
  stomach, 510
  thymus, 510, 651
  thyroid, 510, 516, 517t, 524–25
Horn cell, anterior, 359t, 360, 360f
HP$_c$ (capsular hydrostatic pressure), 752, 752f, 753
HP$_g$ (glomerular hydrostatic pressure), 752, 752f, 753
Human cadavers
  emotional responses to, 2
  proper treatment of, 2
  tissues, disposal of, 14
Human chorionic gonadotropin (hCG), 794
Humectants, 13, 13t
Humeral circumflex artery, 614f
Humerus, 154, 157f, 205t, 207, 207f–208f, 213, 213f, 267f
Humors, in classical Greek medicine, 529, 530t
Hyaline cartilage, 110t, 111–12, 111f
  bone, 155f, 156
  in joints, 237, 238
  respiratory tract, 664, 664f, 668f, 669, 669t
Hydrochloric acid, 705f, 706t, 725
Hydrogen ions
  and acid-base balance, 756
  tubular secretion and reabsorption, 753t
Hydrogen peroxide, 62t
Hydrostatic pressure, 79–81, 752
Hydroxyapatite, 151
Hyoglossus, 297t, 298f
Hyoid bone, 181, 181f, 298f, 674f
Hypercapnia, 681
Hyperpolarization, of excitable membrane, 371, 371f
Hypertension, 753
Hypertrophic cartilage, zone of, 150f, 151
Hyperventilation, 681, 756
Hypocapnia, 681
Hypochondriac regions, 37, 37f
Hypodermis (subcutaneous layer), 130f, 137, 137f, 480f
Hypoglossal canal, 170t, 178f, 180f
Hypoglossal nerve (CN XII), 297t, 300t, 301t, 303, 398t, 399f, 436f
  disorders, common, 410t
  functions, 409t, 417, 417f
Hyponychium, of nail, 136, 136f, 136t
Hypophyseal fossa, 179, 180f
Hypothalamo-hypophyseal axis, 393
Hypothalamus, 391t, 397
  as endocrine organ, 393, 510f, 512, 513t, 514, 522f–523f, 523
  and homeostasis, 393
  in oogenesis, 793, 795
  sheep, 407, 407f
Hypothenar eminence, 334
Hypothenar group, 334, 334f, 335f
Hypothesis, in scientific method, 4
Hypotonic solution, 65
Hypoventilation, 681, 756

I-6

Hypoxia, **562**
H zone, 255f, 272

I bands, 254, 255f, 256, 256f, 260f, 261, 272
Ileocecal junction, 708
Ileocecal valve, 707f, **708**, 720t
Ileocolic artery, 610f
Ileum
  gross anatomy, 610f, 655f, **718**–19, 719f, 719t, 721f–722f
  histology, 647, 647f, 707, 707f, 707t
Iliac arteries, 607f, 610f, 619f, 621f, 622f, 624f
Iliac crest, 154, 214t, 216f, **227**, 228f, 246, 337f, 537, 749f
Iliac fossa, 214t, 216f
Iliac regions, 37, 37f
Iliac spines, 214t, 216f, 227, 228f, 312
Iliac tuberosity, 214t
Iliacus, 336t, 337f, **339**
Iliac veins, 608f, 620f, 621f, 622f
Iliocostalis, 304t, 306f
Iliohypogastric nerve, 443f
Ilioinguinal nerve, 443f
Iliopsoas, 336, 336t, 337f, **339**, 340f
Iliotibial tract, 267f, 337f, 338, 340f
Ilium, 157f, **214**, 214t
Illuminator lens adjustment, of microscope, 52f
Immune system, lymphatic system and, **641**
Immunity, passive, **657**
Impressions, in lungs, **675**–76, 677, 677f
Incisive foramen (fossa), 169t, 172f, 178f
Incus, 487f, 488f, 489, 490, 490f
Independent variable, **4**
Inferior, defined, 33t, 34
Inferior alveolar nerve, 413f
Inferior angle, of scapula, 202t, 204f
Inferior articular facets, **185**
Inferior articular process, 183t, **185**, 185f, 187, 188, 188f
Inferior colliculus, 391t, 406, 406f
Inferior costal facets, 183t, **188**, 188f
Inferior gemellus, 336t, 337f, 338
Inferior gluteal nerve, 441t, **442**, 442f
Inferior lobe, lung, 676f, 677f, 678f
Inferior mesenteric artery, 610f, 613f
Inferior mesenteric ganglion, 457f
Inferior mesenteric vein, **609**, 611f, 613f
Inferior nasal concha, 169, 169t, 171f, 174f, 182t
Inferior nuchal line, 178f
Inferior oblique, 296t, 297f, 412f, 484, 484f
Inferior orbital fissure, 170t, 171f, 172t, 173f
Inferior petrosal sinus, 604f
Inferior phrenic artery, 607f
Inferior pubic ramus, 215t, 216f
Inferior rectus, 296t, 297f, 412f, 484, 484f
Inferior sagittal sinus, 386t, 387f–388f, 604f
Inferior temporal line, 176f
Inferior trunk, brachial plexus, **437**, 437t, 438f, 440f
Inferior vena cava, 309, 521f, **558**, 558f, 559f, 560, 566f, 567, 608f, 609f, 611, 611f, 612f, 613, 613f, 621f, 622f, 624f, 656f, 718f, 750f
Infraglenoid tubercle, 202t, 204f, 328f
Infrahyoid muscles, 300t–301t
Infraorbital foramen, 169t, 171f, 172t, 173f
Infrapatellar bursa, 243f, 244f–245f
Infraspinatus, 324f, 325f, **326**
  as term, 263t
Infraspinous fossa, 202t, 204f
Infundibulum (pituitary stalk)
  human, 179, 391t, **396**, 396f, 397f, 773t, 784f
  sheep, 405f
Ingestion, **722**
Inguinal, defined, 34t, 35f
Inguinal canal, 310t, **312**, 313f
Inguinal ligament, 311t, **312**, 313f, **339**, 340f, 607f, 608f, 619f, 621f
Inguinal ring, superficial, **312**, 313f
Inhibin, 513t, 795
Inner circular layer, smooth muscle cells, 259, 259f, 260
Innervation, reciprocal, 245
Inspiration, **662**, **679**
Inspiratory capacity, 679, 679t, 680f, 684f, 685, 685t
Inspiratory reserve volume (IRV), 679, 679t, 680f, 684f, 685, 685t, 689, 689t
Insulin, 510, 520t
Integument. *See also* Dermis; Epidermis; Scalp
  defined, **125**
  model, 137, 137f
  nails, 136, 136f, 136t
Interatrial septum, **560**, 561f
Intercalated discs, 115, 115f, 116, 116f, 260f, **261**, 554f

Intercellular clefts, **597**
Intercondylar eminence, 219t, 222f
Intercondylar fossa, 218t, 220f
Intercostal arteries, 607f
Intercostal membranes, **308**
Intercostal muscles, 307t, 308, **308**, 308f, 311f, **662**, 679
Intercostal nerves, 307t, **435**
Intercostal veins, 608f
Interior scalenes, 307t
Interlobar artery, 746f, 747f
Interlobar vein, 746f, 747f
Interlobular artery, 746f, 747f
Interlobular vein, 746f, 747f
Intermediate filament, 61t
Intermediate mass (interthalamic adhesion), 391t, **397**, 397f
Intermediate mesoderm, 799f
Internal acoustic (auditory) meatus, 170t, **179**, 180f, 415f
Internal auditory canal, **179**, **415**
Internal capsule, sheep brain, 407, 407f, **408**
Internal carotid arteries, 603, **603**, 604f, 605f
Internal iliac arteries, 607f, 619f
Internal iliac vein, 608f, 620f
Internal intercostal membrane, **308**
Internal intercostal muscles, 307t, **308**, 308f
Internal jugular vein, **384**, 385f, 602f, **603**, 603f, 604f, 605f
Internal oblique, **310**, 310t, 311f, 312, 312f
Internal occipital crest, 180f
Internal occipital protuberance, 180f
Internal os, 784f
Internal thoracic vein, 602f
Internal urethral sphincter, **749**, 750, 751f, 752, 790f
Internus, definition of, 263t
Interosseous, definition of, 263t
Interosseous membrane, **236**, 237f
Interosseous muscles, 334–**35**, 334t, 335f, 347, 348f, 349f
Interphase, 64f–65f, 65, 67
Interspinales, 306t
Interstitial cells, 776f, 777
Interstitial lamellae, 148–49
Interstitial segment, 773t
Interthalamic adhesion (intermediate mass), 391t, **397**, 397f
Intertransversarii, 306t
Intertrochanteric crest, 218t, 220f
Intertrochanteric line, 218t, 220f
Intertubercular sulcus, 205f, 207f–208f
Interventricular arteries, 557f, 563f, 564f
Interventricular foramen, 389f
Interventricular septum, 562f
Interventricular sulci, 557f, 566, 566f
Intervertebral disc, 112, 112f, **238**
Intervertebral foramen, 183t, **185**
Intestinal arteries, 610f
Intestinal blood vessels, 610f, 611f
Intestinal crypts (intestinal glands), 707, 709f
Intestinal veins, 611f
Intramembranous ossification, **236**
Intrinsic factor, 706t
Intrinsic muscles
  foot, 347, 347t–348t, 349f
  hand, **334**–35, 334t, 335f
  tongue, 297, 297t, 298f
Intrinsic regulation of muscle contraction, 277
Inversion, 242t
Iodine, as indicator solution, 75
Iodine potassium iodide (Lugol's solution), 76
Iris, 482t, 484, 484f, 485f
Iris diaphragm lever, of microscope, 46f, 47
Irregular bone, classification of, 152f, 152t
Irregular dense connective tissue, 105f, 108–9, 108f, 130, 130f, 137, 137f, 768
IRV (inspiratory reserve volume), 679, 679t, 680f, 684f, 685, 685t, 689, 689t
Ischemia, **562**, 606
Ischemic stroke, 303
Ischial body, 216f
Ischial spine, 214t, 215, 216f
Ischial tuberosity, 214t, 216f, **227**, 228f
Ischiopubic ramus, 790f
Ischium, 157f, **214**, 214t, 216f
Islets of Langerhans (pancreatic islets)
  histology, 520–22, 520t, 521f, 711, 711f
  hormones, 520t
Isthmus
  thyroid, **516**
  uterine tube, 773t, 784f
IT band. *See* Iliotibial tract

Jejunum, 707, 707f, 707t, **718**–19, 719f, 719t, 721f–722f
Joint. *See* Articulations
Jugular foramen, 170t, 178f, 180f, 417f
Jugular veins, 384, 385f, **601**, 602f, **603**, 603f, 604f, 605f
Juxtaglomerular apparatus, **740**
Juxtamedullary nephron, 739f, 746f

Keratin, **127**, 127t
Keratinization, 97t
Keratinized stratified squamous epithelium, 97t, 98f, 127
Keratinocyte, **127**, 127t, 137, 137f
Ketone bodies, 753
  in urine, 754t, 755, 755t
Ketonuria, 754t
Kidney(s), 656f, 721f–722f, 749f, 750f
  blood supply to, 745–48, 746f, 747f
  dialysis, **736**
  failure, **736**
  function, 745–46
  gross anatomy, 744–45, 745f
  hemodialysis, 83
  histology, 99–100, 99f, 738–42, 738t, 739f
  innervation, 454f
  urine-draining structures in, 748
  urine formation, 752–54
Kinetic energy, **370**
Knee joint, 243, 243t, 244f–245f
  muscles, 341
Korotkoff sounds, **625**, 629
Kupffer (reticuloendothelial) cells, **710**, 710f

Labial frenulum, 712f–713f
Labia majora, 785f
Labia minora, 785f
Laboratory equipment
  dissection instruments, 11, 11t, 12f, 12t
  safety equipment, 10
Laboratory waste, disposal of, 14, 14t, 15
Lacrimal bones, 169, 169t, 173, 173f, 174f, 176f, 182t
Lacrimal caruncle, 481t
Lacrimal fossa, 172t
Lacrimal gland, 414f, 454f, 481t, 484, 484f
Lacrimal groove, 169t, 173f, 176f
Lacrimal puncta, 481t
Lacrimal sac, 481t
Lacteals, 723t, **724**
Lactiferous ducts, 786t, 787f
Lactiferous sinuses, 786t, 787f
Lacuna(e)
  in bone, 110, 110t, 113, 113f, 147t, **148**–49, 148f, 149f, 150, 150f
  in cartilage, 110, 110t, 111f, 112–13, 112f
Lambdoid suture, 176f, **177**, 177f
Lamellae (bony matrix), 110, 110t, 148, 148f
Lamellated (Pacinian) corpuscle, 130f, **132**, 132t, 137, 137f, 468f, **469**, 469f, 480, 480f
Lamina, 183t, **185**, 185f, 187
Lamina propria
  intestines, 707f, 708, 709, 709f
  olfactory epithelium, 472f
  respiratory tract, **664**, 664f, 666t, 667f, 668
  stomach, 702, 703f, 704f, 704t
  vaginal wall, 775, 775f
Large bronchi, **668**–69, 668f, 669t
Large intestine, 707f
  gross anatomy, 720–21, 720t, 721f–722f
  histology, **708**–9, 709f
  physiology, **724**
Laryngeal prominence, 674f
Laryngopharynx, 672f, 712f–713f
Larynx, 666, 672f, **673**, 674, 674f, 678f
  innervation, 416, 416f
  muscles, 673
Lateral, defined, 33t
Lateral (axillary) border, 202t, 204f
Lateral chord, of brachial plexus, **437**–38, 438f
Lateral compartment muscles, leg, **344**, 345f, 345t
Lateral condyle
  femoral, 218t, 220f, 243, 244f–245f
  tibial, 219t, 222f, 243
Lateral epicondyle
  of femur, 218t, 220f, **227**, 228f
  of humerus, 205t, 207f–208f
Lateral funiculus, 430, 431f, 432t
Lateral horns, of spinal cord, 430, **430**, 431f, 432t, 435f, 459
Lateral lumbrical, 335f
Lateral malleolus, 219t, 223f, **227**, 228f
Lateral mass, **187**
Lateral plantar artery, 619f, 622f

Lateral plantar vein, 620f, 622f
Lateral pterygoid, 295f, 295t
Lateral pterygoid plate, 178f
Lateral rectus, 296t, 297f, 413, 413f, 484, 484f
Lateral rotators, of hip, 336, 338
Lateral sulcus, human, 392t, 394, **395**, 395f
Lateral supracondylar line, 218t, 220f
Lateral sural cutaneous nerve, 443f
Lateral ventricles, **388**, 388f, 389, 389f, 390t
  sheep, 407, 407f
Latissimus, definition of, 262t
Latissimus dorsi, 324t, 325f
Law of entropy, **67**
LCIS (lobular carcinoma in situ), 788
LDLs. *See* Low density lipoproteins
Leather, as connective tissue, 130
Left bundle branch, **569**, 569t, 570f
Left subclavian artery, 607f
Left subclavian vein, 608f
Leg
  compartments, fascial, 266, 267f, 344–46
  muscles, 344–46, 344f–346f, 344t–346t
  nerves of, 443, 443f
Lens, of eye, 482t, 484, 484f, **486**, 486f, 487f
Lesser curvature, stomach, 714t, 715f
Lesser omentum, 714f, 715f
Lesser sciatic notch, 214t, 216f
Lesser trochanter, 218t, 221f, **339**
Lesser tubercle (humerus), 205f, 207f–208f
Lesser wing (sphenoid bone), 170t, **179**, 180f
Leukemia, 537, 538
Leukocytes (white blood cells), **113**, 114, 114f, 653
  abnormal, 537
  cell count, 538, 539–40, 539f
  centrifugal separation of, 438f, 538
  characteristics of, 530t
  production of, 537
  types of, 531t
  in urine, 753–54, 754t, 755, 755t
Leukopoiesis, 537
Levator, definition of, 263t
Levator anguli oris, 293t, 294f
Levator labii superioris, 293t, 294f
Levator palpebrae superioris, 296t, 484f
Levator scapulae, 301t–302t, 323t, 325f, **326**
Levator veli palatini, 298t, 299f
LH (luteinizing hormone), 512, 513t, 769, 793, 795
Life, definition of, 57
Ligament
  attachment to periosteum, 155f, 156
  connective tissue in, 108, 108f
  knee, 243, 243t, 244f–245f
  synovial joint, 239f
Ligaments, cruciate, 243, 243t
Ligamentum arteriosum, 557f, **560**, 561f
Ligamentum teres, 716t, 717f
Light control, of microscope, 46f, 47t
Limb buds, 799f
Limbic system, **397**, 470
Limbs, lower
  blood vessels, 618–22, 619f–622f
  bones, 218–27, 218t–219t
  compartments, fascial, 266, 266t, 267f
  muscles, 339–49
  nerves of, 443, 443f
Limbs, upper
  blood vessels, 614–17, 614f–617f
  bones, 205–13, 205t–206t
  compartments, fascial, 266, 266t, 267f
  muscles, 326–35
Linea alba, **310**, 311f, 311t, 312f, 313f
Linea aspera, 218t, 221f
Linea terminalis, 214t
Lingual frenulum, 712f–713f
Lingual lipase, 724t
Lingual nerve, 413f, 414f
Lingual tonsils, **646**, 646f, 646t, 655f, 672f, 712f–713f
Lingual vein, 602f
Lipid droplet, 107f, 597, 599t
Lipolysis, **524**
Lips, 712f–713f
Liver, 656f
  blood vessels, 610f, 611f, 612f, 613f
  cirrhosis, **613**
  fetal, 624f
  functions, 609–11, 710, 716, 723
  gross anatomy, 716–18, 716t, 717f, 718f
  histology, 710, 710f, 711f
  innervation, 454f
Lobar (secondary) bronchi, 678f, 678t, 679
Lobe(s)
  kidney, 745f

I-7

Lobe(s) (continued)
  liver, 716t, 717t, 718f, 721f–722f
  lung, 676f, 677f, **678**, 678f
  mammary gland, 786t, 787f
  thymus, **651**, 652f
Lobular carcinoma in situ (LCIS), 788
Lobules
  breast, 787f
  liver, **710**, 710f, 711f
  lungs, **678**
  mammary gland, 786t
  thymus, **651**, 652, 652f
Locomotion, gluteal muscles in, 338f
Long bone
  classification of, 152f, 152t
  developing, **150**, 152f
  endochondral ossification, 150–51, 150f
  structure of, 153–56, 154f, 155f
Long head of biceps brachii, **328**
Longissimus, 262t, 304t, 306f
Longissimus capitis, 300t, 301f–302f
Longitudinal fissure
  human, 392t, **394**, 394f
  sheep, 404, 404f, 406, 406f, 407, 407f
Longitudinal layer of muscularis, 702, 703f, 704f, 704t, 709, 709f
Longitudinal smooth muscle, **743**, 743f, 780f
Long thoracis nerve, **438**, 438f, 439t
Longus, definition of, 262t
Loops of Henle. See Nephron loops
Loose connective tissue, 19, 19f, 104, 105t
Lou Gehrig's Disease (Amyotrophic lateral sclerosis), 270
Low density lipoproteins (LDLs), 439, 542
Lower respiratory tract
  gross anatomy, 673–79
  histology, 666–69
Lugol's solution, 76, 725, 726t
Lumbar, defined, 34t, 35f
Lumbar arteries, right, 607f
Lumbar curvature, 184f
Lumbar enlargement, **430**, 434f
Lumbar plexus, 434f, **440**. See also Lumbosacral plexus
Lumbar puncture (spinal tap), **432**
Lumbar regions, 37, 37f, **183**
Lumbar veins, left, 608f
Lumbar vertebrae, **183**, 183t, 184f, 188, 188f
  characteristics, 432t
  histology, 431f
  injuries to, 427
Lumbosacral plexus, **435**, 440–43, 441t, 442f, 443f
Lumbrical, definition of, 261t, 262t
Lumbricals, 334–**35**, 334t, 335, 335f, 347t, 349f
Lumen, **98**
Lunate, 206t, 211, 211f
Lunate surface, 214t, 216f
Lung(s)
  bronchial tree, 678–79, 678f, 678t
  fetal, 624f
  gross anatomy, 675–77, 676f, 677f
  histology, 669–71, 670f, 670t, 671t
  innervation, 454f
  left, 675f, 677, 677f, 678f, 679
  right, 675f, 676, 676f, 678f, 679
Lung capacity
  measurement of, 679–80, 679t, 680f, 683–86, 684f, 685, 685t, 689, 689t
  total, 679, 679t, 680f, 684f, 685, 685t
Lunula, 136t
Luteal phase, of uterine cycle, **794**, 794f, 795
Luteinizing hormone (LH), 512, 513t, 769, 793, 795
Lymph, 641
Lymphatic ducts, 655f
Lymphatic fluid, 649f
Lymphatic nodule (follicle), **645**, 645f–650f, 646–47, **648**, 649, 650t, 656f
Lymphatic system
  and cardiovascular system, **641**
  gross anatomy, 655–56, 655f, 656f
  histology, 644–48
  and immune system, 641
  organs, 648–54
Lymphatic tissue, in GI tract, 698
Lymphatic vessels
  gross anatomy, 655, 655f, 656f
  histology, 593f, 644, 644f, **648**, 648, 649, 649f, 650f, 650t, 651
Lymph nodes, 648–51, 649f, 650f, 650t, 655, 655f, 656f
Lymphocytes
  characteristics, 531t, 533–36, 534f, 537
  function, 539, 651

histology, 107, 107f, 652f, 653
Lysosomes, 62t, 66f, 67
Lysozymes, 700

Macroglia, 362
Macrophages, 645, 645f, 653
  alveolar, 670f, 671, 671t
Macula densa, 740f
Macula lutea, 483t
Magnesium, and muscle contraction, 270
Magnetic resonance imaging (MRI), 356
Main (primary) bronchi, 676f, 677f, 678f, 678t
Main pancreatic duct, **711**, 711f, 716t, 717f
Major, definition of, 262t
Major calyx, 748, 750f
Major duodenal papilla, 711, 711f
Male
  pelvis, **215**–18, 217f
  reproductive system
    components, 776, 779t
    functions, 776
    gross anatomy, 788–90
    histology, 776–82
  urethra, 750–51, 751f, 782, 782f, 788
Malleolus
  lateral, 219t, 223f, **227**, 228f
  medial, 219t, **227**, 228f
Malleus, 487f, 488t, 489, 490, 490f
MALT (mucosa-associated lymphatic tissue), **645**, 645f, 645t, 655, 655f
Mammary, as term, 35f
Mammary glands, **786**
Mammillary bodies
  human, 392t, **396**, 396f, 397f
  sheep, **405**, 405f, 406, 407f
Mammography, 788
Mandible, 157f, 169, 169t, 171f, **175**–76, 175f, 182f, 182t
Mandibular branch of trigeminal nerve (V₃), 409t, 413, 413f
Mandibular condyle, **175**
Mandibular foramen, 169t, 175f, **176**
Mandibular fossa, 170t, 178f
Mandibular notch, 175f
Manubrium, 190t, 191, **191**, 191f
MAP (mean arterial blood pressure), **625**
Marginal artery, 557f, 563f, 564f
Marginal sinuses, 604f
Masseter, **295**, 295t, 295f, 700f, 712f–713f
Mastication, muscles of, 295, 295t, 295f
Mastoid fontanelle, 182f
Mastoid foramen, 178f
Mastoid process, 170t, 176f, 177f, 178f, **299**
Maxilla, 169, 169t, 171f, 173, 173f, 174f, 176f, 178f, 182f
Maxillary artery, 602f
Maxillary bones (palatine processes), 169t, 172t, **173**, 174f
Maxillary branch of trigeminal nerve (V₂), 409t, 413, 413t
Maxillary process, 170t
Maxillary vein, 602f
Maximal stimulus, in muscle contraction, 274
Mean, **5**, 5t
Mean arterial blood pressure (MAP), **625**
Measurement systems, 8–10
  conversions, 8, 9, 9t
  temperature scales, 9
Meatus, acoustic
  external, 170t, 176f, 179, 487f, 488t, 489, 490, 490f
  internal, 170t, 179, 180f, 415f
Meatus, nasal cavity, 672f
Mechanical digestion, 722–23
Mechanical stage, of microscope, 46f, 47t
Medial, defined, 33t
Medial (vertebral) border, 202t, 204f
Medial chord, of brachial plexus 437–38, 438f
Medial compartment muscles, thigh, **341**–42, 341f, 341t
Medial condyle
  femur, 218t, 221f, 243, 244f–245f
  tibia, 243
Medial epicondyle
  femur, 218t, 221f, **227**, 228f
  humerus, 205t, 207f–208f, **213**, 213f
Medial lumbricals, 335f
Medial malleolus, 219t, **227**, 228f
Medial plantar artery, 619f
Medial plantar vein, 620f
Medial pterygoid, 295f, 295t
Medial pterygoid plate, 178f
Medial rectus, 296f, 297f, 412f, 484, 484f
Medial supracondylar line, 218t, 221f
Medial sural cutaneous nerve, 443f

Median, **5**
Median fissure, anterior, 430, 431f, 432t, 435f
Median nerve, 437f, **438**, 438f, 439t, 440f
Median sacral crest, 189f
Median sulcus, posterior, 430, 431f, 433f, 435f
Mediastinum, **556**, 556f, 651, 651f, **675**, 675f
Medical imaging, and sectional anatomy, 31
Medulla
  kidney, 746f
  lymph node, **648**, 649, 649f, 650f, 650t, 656f
  ovary, **768**, 769f
  thymus, 652, 652f
Medulla oblongata, 415f, 417f
  human, 392t, 395f, 396f, 397f
  sheep, 401f, 402f, 404, 404f, 406, 407f
Medullary cavity, 153–54, 154f, 155, 155f, 156
Medullary cords, **649**, 649f, 650f, 650t, 656f
Medullary sinuses, **649**, 649f, 650f, 650t, 651, 656f
Megakaryocytes, 113, **536**–37, 537f
Meiosis, **766**, 791, 791t–792t, 792–93, 795
Meissner (tactile) corpuscle, **130**, 130f, **132**, 132t, 137, 137f, 468–69, **468–69**, 468f, 468t, 480, 480f
Melanin, 127, 127t, **129**, 129f
Melanocytes, **129**, 129f
Melanoma, 129, 129f
Melatonin, 515
Membrane permeability, and diffusion, 75–76, 76f
Membrane potential, 366–70
  depolarization and repolarization, 371, 371f
  membrane transport mechanisms and, 373
  potassium ions and, 367–68, 367f
  sodium ions and, 369–70
Membrane transport mechanisms, 58, 67–83
  diffusion, 67–76, 81–83, 82f
  and membrane potential, 373
  osmosis, 77–79, 81–83, 82f
Membrane voltage, **370**
Membranous labyrinth (cochlear duct), 477f, 490f
Membranous urethra, **750**, 751f, 789f, 790f
Memory, in adaptive immune response, 539
Meniere's disease, 501
Meningeal layer, of dura mater, **384**, 384f, 385t, 387f–388f
Meningeal spaces, 386t
Meninges, cranial, 384–**87**, 384f–385f, 385t–386t, 387f–388f
Meniscus, **156**, 156f, 239f, 239t, 243, 243t, 244f–245f
Menstruation, **774**, 774t–775t, 794f, 795
Mental, defined, 34t, 35f
Mental foramen, 169t, 171f, 172t, 175f, **176**, 176f
Mentalis, 293t, 294f
Mental nerve, 413f
Mental protuberance, 169t, 171f, 172t, 175f, **176**, 176f
Merkel (tactile) cells, 127, **132**, 132t
Merkel (tactile) disc, 468t, **479**, 480
Merocrine (eccrine) sweat gland, 102–3, 102f, 130f, **131**, 131t, 137, 137f, 469f, 480f
Mesencephalon, 388, 388f, 390t
Mesenchymal cells, **105**, 105f
Mesenchyme, **104**, 105, 105f, 147t
Mesenteric arteries, 610f, 613f
Mesenteric ganglia, 457f
Mesenteric veins, 609, 611f, 612f, 613f
Mesentery, 703f, 799f
Mesoderm, 799f
Mesosalpinx, 783t
Mesothelium, **99**, 557
Mesovarium, 783t
Metabolic acidosis, **725**, 756, 756t
Metabolic alkalosis, **725**, 756, 756t
Metabolism, thyroid hormone and, 524–25
Metacarpals, 153f, 157f, 206t, 211f, **212**, 212f
Metaphase
  in meiosis, 791t–792t
  in mitosis, 64f–65f, 67
Metaphysis, 150–51, **153**, 154f, 155f
Metastasis, 129
Metatarsals, 157f, 219t, 224f, **226**, **227**, 228f
Metencephalon, 388, 388f, 390t
Methylene blue, 63, 63f, 73–74, 75t
Metopic suture, 172t
Metric system, **8**, 8t
  conversions, 8, 9, 9t
MHC protein, 651
MI. See Myocardial infarction
Microfilament, 61t
Microglia, **362**, 362t
Microhematocrit centrifuge, 540, 540f

Micrometer (mm), 50
Microscope(s). See also Slides
  care of, 45, 45f
  cleanup procedures, 51, 51f
  compound, parts of, 46, 46f, 47t
  depth of field, 47t, **51**, 51f
  dissecting (stereomicroscope), parts of, 52–53, 52f
  field of view, 47t, **49–50**
  focusing of, 48–49
  good technique, importance of, 43
  magnification, 45, 49, 52
  oil immersion objective, 534, 534f, 535, 539
  specimen size, estimation of, 50–51, 50f
  troubleshooting, 48
  working distance, 47t, **49**, 49f
Microtome, 94
Microtubule, 61t
Microvilli
  of convoluted tubules, 741, 741f
  epithelial cell types with, **100**, 100f
  gustatory, 471f
  in small intestine, **707**, 724
  structure and function, 97t, 104
Midbrain, human, 396f, 397f
Midclavicular line, **308**
Middle cerebral artery, 604f, 605f
Middle lobe, lung, 676f, 678f
Middle nasal concha, 169t, 174f
Middle phalanx, **331**
Middle scalenes, 307t
Middle trunk, brachial plexus, **437**, 437t, 438f, 440f
Midpalmar group, **334**–35, 334t
Midsagittal plane, 31f, 31t
Mineralocorticoids, 519t
Minor, definition of, 262t
Minor calyx, 748, 750f
Mirror, of microscope, 52f
Mitochondrion, 62t, 66f, 67
Mitosis, 64f–65f, 65–66, 792
Mitotic spindle, 64f–65f
M line, 255f
Mnemonic devices
  cranial nerves, 400, 408
  defined, **400**
  femoral triangle structures (NAVEL), 341
  interossei (DAB/PAD), 335, 347
  pituitary hormones (GPA and B-flat), 512
Moist epithelium, stratified squamous, 101
Molecular layer, of Purkinje cell, 361, 361f
Molecular weight
  calculation of, 73
  and diffusion, 73–74, 74f, 75f
Monocytes, 531t
  identification of, 533–36, 534f, 537
Morula, 797t–798t
Motility, **724**
Motor end plate, 258f
Motor information, and nerve function, **408**
Motor neurons
  ALS and, 270
  ANS and, 459
  neuromuscular junctions, 257–59, 258f
  in reflexes, **450**, **452**, 452f, 452t
Motor units, **257**, 258f, **268**
  and muscle fatigue, 269
  recruitment of, 268, 274
MRI. See Magnetic resonance imaging
Mucin, 495, 700, 705f, 706f, 707
Mucosa
  ductus deferens, 779, 780f
  intestines, 707, 707f, 708, 709, 709f
  mouth, 700f
  olfactory tract, **665**, 665f
  oral cavity, 712f–713f
  respiratory tract, **664**, 664f, 665, 665f, 666f, 667f, 668
  stomach, 702, **702**, 703f, 704f, 704t
  urinary tract, 743, 743f, 744, 744f
  uterine tubes, 772, 772f, 773t
  vaginal wall, 775, 775f
Mucosa-associated lymphatic tissue (MALT), **645**, 645f, 645t, 655, 655f
Mucosal folds
  of seminal vesicle, 780f, 781
  in uterine tube, 772f, 773
Mucous cells, **700**, 700f, 701, 701t, 702, 705f, 706, 706t
Mucous neck cell, 705f, 706, 706t
Mucus, 104, 495, 664f, 700
Multifidus muscles, **303**, 305t
Multipennate architecture, in skeletal muscle, 264t
Multiple sclerosis, 376

I-8

Muscle(s). *See also* Cardiac muscle; Skeletal muscle; Smooth muscle
abdominal wall, 310–12, 310t–311t, 311f–313f
action, learning strategy for, 241
agonists, **245, 299, 328**
antagonists, **245,** 299
arm, 326–28, 327f, 327t
contraction of, 555
delayed-onset muscle soreness (DOMS), 320
directional terms for, 301
eye, 252, 296, 296t, 297f
face, 292–95
facial expression, 292, 292t–293t, 294f
foot, 347, 347t–348t, 349f
forearm, 329–33
glenohumeral joint, 323–26, 324t, 325f
hand, 334–35, 334t, 335f
head, 292–99
hip, 336–39, 336f, 337f, 338f
larynx, 673
learning strategies for, 252, 301, 320, 326, 328t, 335, 341, 342, 347
leg, 344–46
mastication, 295, 295f, 295t
neck, 299–300, 300t–301t, 301f–302f
pectoral girdle, 323–26, 323t, 325f
pharynx, 298–99, 298t, 299f
respiration, 307–9, 307t, 308f, 309f, 662, 679
synergists, **245**
thigh, 339–43
tongue, 297, 297f, 298f
types of, 252
vertebral column, 303, 303t–306t, 306f
Muscle cells, excitability of, 68
Muscle fibers
skeletal, **254,** 255f, 256, 256f, 257, 257f, 258f, 272, 272f
types of, 268, 268t, 269, 269t
smooth, 259, 259f
Muscle spindles, 350, 459, 460
Muscle tissue, 114–17
identification and classification, 116–17
properties, 252
striation, 116, 116f, **254,** 254f, 260, 260f, 272, 277, 555, 555f
types of, 114, 115f
Muscle twitch, **274**–75, 275f, 275t, 278–80, 278f, 279f
Muscularis
GI tract, **702,** 703f, 704f, 704t, 707f, 708, 709, 709f, 724
urinary tract, 743, 743f, 744, 744f, 750f
uterine tubes, 772, 772f, 773, 773f
vaginal wall, 775, 775f
Muscularis mucosae, 702, 703f, 704f, 704t, 705, 708, 709, 709f
Muscular system. *See* Appendicular muscles; Axial muscles; Muscle(s)
Muscular wall, of seminal vesicle, 780f, 781
Musculophrenic artery, 607f
Musculophrenic vein, 608f
Musculoskeletal nerve, **437,** 437t, 438f, 439t, 440f
Musculoskeletal system, muscle groups, major, 265–66, 265f, 266t, 267f
Myelencephalon, 388, 388f
Myelin, 358, 371
Myelination, 364, 364f, 365, 366f, 371, **371**
Myelin sheath, 364, 364f, 365, 366f
Myenteric nerve plexus, 703f, 704f, 704t, 724
Mylohyoid line, 175f
Mylohyoid muscle, 300t, 301f–302f, 700f, 712f–713f
Myoblasts, 254
Myocardial infarction (MI), **562**
Myocardium, **555,** 555f, 555t, 556f, 558
Myoepithelial cells, 701f
Myofibrils, **254,** 255f, 258f, 260
Myofilaments, **254,** 255f, 258f
Myogenic response, 752
Myometrium, 774, 774t–775t, 784f
Myosin, **254,** 270, 270f, 271f, 276, 555, 579

Nail(s), **136,** 136f, 136t
Nasal, as term, 35f
Nasal bones, 169, 170t, 171f, **173,** 173f, 174f, 176f, 182t
Nasal cavity, 173–75, 174f, 182t, 411f, 664, 672f
Nasal conchae, 169, 169t, 171f, 174f, 182t, 672f
Nasal epithelium, 411
Nasalis, 293t, 294f
Nasal septum, **173,** 174f, 182t
Nasal spine, 172t
Nasolacrimal duct, 481t

Nasopharynx, 672f
NAVEL mnemonic, 341
Navicular, 219t, 224, 224f, 225, 226f
Neck
blood vessels, 601–3, 602f, 603f
muscles, 299–300, 300t–301t, 301f–302f
Neck, of bone
femur, 218f
fibula, 219t
radius, 206t, 209f
rib, 190t, 192f
Necrotic tissue, **562,** 606
Needles, dissection, 11t
Negative feedback, in homeostasis, **2**
Nephron(s), 736, **736,** 752, 753t
Nephron loops (loops of Henle)
function, 740, 748, 752, 753t
gross anatomy, 746f
histology, 738, 738f, 739f, 741–42, 742f
Nephron tubules. *See* Distal convoluted tubules; Nephron loops; Proximal convoluted tubules
Nerve(s). *See also* Action potentials; Cranial nerves; Spinal nerves
in peripheral nervous system, 365–66, 366f
visceral, 408t
as white matter, **358**
Nerve block, pudendal, 442
Nerve deafness, 499
Nervous system
autonomic, 450, 450f, 453–59, 724
central, 362–63, 388 (*See also* Brain; Spinal cord)
enteric, 724
peripheral, 364–66, 366f, 435–44 (*See also* Cranial nerves; Spinal nerves)
somatic, 450, 450f, 452, 452f, 452t
tracts in, **358**
Nervous tissue. *See also* Gray matter; White matter
cell types, 117, 117t, 356
excitability of, 356, 371
Net filtration pressure, **752,** 752f, 753
Neural fold, 799f
Neural groove, 799f
Neural tube disorders, 189
Neural tunic, **474**
Neurofibril node (node of Ranvier), 359, **364,** 364f
Neuroglia. *See* Glial cells
Neurohypophysis. *See* Posterior pituitary gland
Neurolemmocytes (Schwann cells), 362, 362t, 364, 364f
Neuromuscular junction, **257**–59, 258f
Neuron(s), 359–61, **360**
bipolar, **359**
excitability of, 68
and gray matter, 358
identification of, 118, 118f
membrane potential, resting, 366–70
multipolar, **359,** 359t
structure and function, **117,** 117t, 356, 359, 359t
unipolar, **359,** 364–65, 365f
Neutrophils, 531t, 533–35, 534f, 537
Nipple, 786f, 787f
Nissl bodies (chromatophilic substance), 61t, 117, 118, 118f, **359,** 359t, 360, 360f
Nissl stain, 360
Nitrites, in urine, 754t
Nitriuria, 754t
Node of Ranvier (neurofibril node), 359, **364,** 364f
Nonkeratinized stratified squamous epithelium, 98f, 101, 101f, 775, 775f
Norepinephrine, 517, 519t
Nose, 182t, 664
Nosepiece, of microscope, 46f, 47t, 52f
Nostril, 672f
Nuchal lines, 178f
Nuclear envelope, 61t, 62t, 66f, 67
Nuclear pore, 61t, 62t, 66f, 67
Nuclei, central nervous system, **358**
Nucleic acids, digestion of, 723–24, 723t, 724t, 727
Nucleolus, 61t, 62t, 64f, 66f, 67
of anterior horn cell, 360, 360f
staining of, 57
structure and function, 62t
Nucleosidase, 724t
Nucleus, 66f, 67
animal cell, 64f
cardiac muscle, 260f, 261
in mitosis, 64f–65f
skeletal muscle cells, 256, 256f, 257, 272

smooth muscle cells, 259, 259f, 260
staining of, 57, 58, 63
structure and function, 61t, 62t
Nucleus pulposus, **238,** 238f
Nutrient artery, **156**
Nutrients, absorption of, **722,** 724, 724t
Nystagmus, 501

Obicularis oculi, 264, 265f
Objective lens, of microscope, 46f, 47t, 52f
Oblique(s), 296t, 297f, **310,** 310t, 311f, 312, 312f, 313f, 412, 412f, **484,** 484f
Oblique arytenoid muscle, 674f
Oblique fissure, 676f, 677f
Oblique layer of muscularis, 702, 704f, 704t
Oblique plane, 31f
Observation, in scientific method, **4**
Obstructive pulmonary diseases, 680
Obturator artery, 619f
Obturator foramen, 214t, 215, 216f, 217f
Obturator internus, 336f, 337f, 338
Obturator nerve, 441t, 443, 443f
Occipital, defined, 35f
Occipital artery, 602f
Occipital bone, 169, 170t, 176f, 177f, 178f, 180f, 182f
Occipital condyles, 170t, 178f, **186**
Occipital crests, 178f
Occipital lobe, 392t, 394f, 395f, 396f, 397f, **474,** 497
Occipital nerve, lesser, 436f
Occipital protuberances, 170t, 177f, 178f, 180f
Occipital sinus, 386f, 604f
Occipitofrontalis (epicranius), 293t, 294f
Ocular lens, of microscope, 46f, 47t
Oculomotor nerve (CN III), 296t, 398t, 399f
disorders, common, 410t
functions, 409t, 412, 412f, 498
and parasympathetic division, 453f, 454f, 455
and pupillary reflexes, 462
sheep, 405f
Odontoid process (dens)
of atlas, 186, 241f
of axis, **187**
Oil immersion objective, 534, 534f, 535, 539
Olecranon, 206t, 210f, **213,** 213f
Olecranon fossa, 205f, 207f–208f
Olecranon process, 328f
Olfaction, 175
adaptation, 496
gustation and, 470, 495–96
histology, 472–73, 472f, 473t
Olfactory bulb
human, 392t, 396f, 399f, **411,** 411f, **472,** 472f, 473t
sheep, 401f, 402f, 403, 403f, 404, 404f, 405f
Olfactory epithelium, 175, **472,** 472f, 473t, 665–66, 665f, 665t
Olfactory foramina, **411**
Olfactory glands, 665f, 665t, 666
Olfactory hairs, 472f, 473, 473t
Olfactory mucosa, 665–66, 665f, 665t
Olfactory nerve (CN I), 398t, 409t, **472,** 472f
disorders, common, 410t
function, 409t, 411, 411f
Olfactory receptor cells, 175, **472,** 472f, 473t, 665f, 665t, 666
Olfactory tract
human, 392t, 396f, 399f, **411,** 411f, **472,** 472f, 473t
sheep, 403, 403f, 404, 404f, 405f
Oligodendrocytes, **362,** 362t
Omental (epiploic) appendices, 720t
Omohyoid muscle, 300t, 301f–302f
Oocyte, **768**
definitive, 768t
developmental stages, 768t
primary, 768, 768t, 769f, 792–93
secondary, 768, 768t, 769, 769f, 770f, 791, 792–93, 795
Oogenesis, 792–95, 793f
Oogonia, 768t, 792
Open scissors technique, 20–21, 21f, 22, 22f
OP_g (blood colloid osmotic pressure), 752, 752f, 753
Ophthalmic branch of trigeminal nerve (V_1), 409t, 413, 413t
Ophthalmic veins, 604f
Opponens digiti minimi, 334t
Opponens pollicis, 334t
Opposition, 242t, **334**
Optic canal, 171f, 172t, 180f
Optic chiasm, **411,** 411f, 604f
human, 392t, **396,** 396f, 399f

sheep, 401f, 403, 403f, 404, 404f, 405f, 406, 407f
Optic disc (blind spot), 411, **475,** 483t, 484, 484f, **486,** 486f, 497
Optic foramen, 170t, 173f
Optic nerve (CN II), 398t, 399f, 412f
disorders, common, 410t
function, 409t, **411,** 411f, 475, 497
gross anatomy, 396f, 481t, 484, 484f, 485, 485f, **486,** 486f
histology, **474,** 474f
and pupillary reflexes, 462
sheep, 401f, 402f, 404, 404f, 405f
Optic tract, 396f, 411f
Oral, as term, 35f
Oral cavity, 182t, 672f, 698, 712, 712f–713f
digestion in, 723t, 724t
Ora serrata, 483t
Orbicularis oculi, 293t, 294f, 461
Orbicularis oris, 293t, 294f
Orbit, 173, 173f, 182t
Orbital, defined, 34t, 35f
Orbital fat pad, 481t, 485, 485f
Orbital fissures, 170t, 171f, 172t, 173f
Organelle(s)
in cytokinesis, 65
recognition of, 57–58
structure and function, of, 61t–62t
Organ of Corti (spiral organ), **476,** 477f, 478t, 489f, 490f
Oris, definition of, 263t
Oropharynx, 672f, 712f–713f
Ortic opening, 309, 309f
Os (external and internal), 784f
Osmosis, 77–79, 77f, 78f, 79t, 81–83, 82f
Ossicles, 489
Ossification
intramembranous, **236**
zone of, 150f, 151
Osteoarthritis, 246
Osteoblast, 110, 147t, **149**–50, 149f, 151, 158
Osteoclast, 147t, **149**–50, 149f, 151, 158
Osteocyte, 113, 113f, 147t, 148–49, 148f, 149f, 150
Osteoid, 151
Osteon, 110, 110t, 113, 113f, **148**–49, 148f
Osteoprogenitor cells, 110, 147t, **148**–49, 148f, 150f
Otic, defined, 35f
Otic ganglion, 454f
Ouabain, 727–28
Outer longitudinal layer, smooth muscle cells, 259, 259f, 260
Oval window, 477f, 487f, 488f, 489, **489,** 490, 490f
Ovarian cycle, 774, 793–94, 794f
Ovarian follicle, **768,** 769f, 792–93. *See also* Lymphatic nodule
developmental stages, 770t–771t
primary, 768, 769f, 770f–771f, 794f
primordial, 768, 769f, 792–93
secondary, 768, 769f, 770f–771f, 794f
vesicular (Graafian), 768, 769, 769f, 770f, 770t–771t, 793, 794f
Ovarian ligament, 769f, 783t
Ovary, 784f, 785f
cancer of, 768, 770f
cysts, 796
as endocrine organ, 510, 510f, 522f–523f, 523
gross anatomy, 783–85, 784f, 785f
histology, 768–72, 768f, 768t
Oviducts (fallopian tubes; uterine tubes)
gross anatomy, 783–85, 784f, 785f
histology, 100, 100f, 772–73, 772f, 773f
Ovulation, 794f
Ovum (ova), 766, 768
Ovum pronucleus, 797t–798t
Oxygen
consumption, TH and, **524**
and gas exchange, 679, 680
and gas transport, 681
Oxyphil cells, 516, 516f, 517f
Oxytocin, 2, 393, 513t, 514

Pacemaker
artificial, 569–70
intrinsic, 569
Pacinian (lamellated) corpuscle, 130f, **132,** 132t, 137, 137f, 468t, 469, 469f, 480, 480f
Packed cell volume (PCV), 538, 540
Palate, 655f
hard, 672f, 712f–713f
innervation, 416, 416f
soft, 672f, 712f–713f

I-9

Palatine bones, 169, 170t, **173**, 174f, 178f, 182t
Palatine folds, transverse, 712f–713f
Palatine foramina, 178f
Palatine processes (maxillary bones), 169t, 172t, **173**, 174f
Palatine tonsils, **646**, 646f, 646t, 655f, 672f, 712f–713f
Palatoglossal arch, 712f–713f
Palatoglossus, 297t, 298f
Palatopharygeal arch, 712f–713f
Palatopharyngeus, 298t
Palmar, as term, 35f
Palmar aponeurosis, 330f, **331**
Palmar arches, 614f, 616f, 617f
Palmaris longus, 329–31, 329t, 330f, 331f
Palmar venous arches, 615f, 616f, 617f
Pancreas, **520**, 711
  in digestion, 723
  endocrine, **520**–22, 520t, 521f, 522f–523f, 523
  exocrine, **520**, **711**–12, 711f
  functions, 520, 716
  gross anatomy, 510f, 611f, 656f, 715f, 716–18, 716t, 717f, 721f–722f
  histology, 711–12, 711f
  hormones, 510, 520t
  innervation, 454f
Pancreatic acini, **520**, 521f, **711**, 711f, 725
Pancreatic amylase, 711
Pancreatic ducts, 521f, 711, 711f, 716t, 717f
Pancreatic islets (islets of Langerhans)
  histology, **520**–22, 520t, 521f, 711, 711f
  hormones, 520t
Pancreatic juice, **711**, 716, 723
Pancreatic lipase, 724t
Pancreatic polypeptide, 520t
Pancreatitis, 723
Papillae
  dermal, **128**, 128f, 130, 130f, 132, **133**, 134, 134f, **468**, 468f, 480, 480f
  duodenal, 521f, 711, 711f, 716t, 717f
  foliate, 470f, 470t, **471**
  renal, 739f, 745f, 748, 750f
  of tongue, 470f, 470t, 471
  vallate, 470f, 470t, **471**, 471f, 472
Papillary dermis, **130**, 130f, 137, 137f, 468, 468f, 480, 480f
Papillary duct, 739f, 748
Papillary muscles, 558f, **559**, 559f, 560, 561f, 562, 567, 567f
Pap smear, 788
Parafollicular cells, **516**, 516f, 517t
Parallel architecture, in skeletal muscle, **264**, 264t
Paranasal sinuses, 664, 672f
Paranephric fat, 744
Parasympathetic (craniosacral) division, **453**–55, 453f, 454f, 455t
Parasympathetic ganglia, 455, 455t
Parathormone. See Parathyroid hormone
Parathyroid gland, 510f, **516**, 522f–523f, 523
  and blood calcium levels, 151
  histology, 516, 517t
  hormones, 151, 510, 516, 517t
Parathyroid hormone (PTH), 151, 510, **516**
Parietal bone, 169, 170t, 171f, 176f, 177f, 180f, 182f, 603f
Parietal cell, 705f, 706, 706f, 725
Parietal eminence, 176f, 177f
Parietal foramina, 177f
Parietal layer
  of glomerular capsule, 738t, 740, 740f
  of renal cortex, **740**, 740f
  of serous membranes, 34, **556**, 556f, 557
Parietal lobe, 392t, 394f, 395f, 397f, 605f
Parietal pericardium, 566, **566**
Parietal pleura, **675**, 675f
Parieto-occipital sulcus, 392t, 397f
Parotid duct, 700f, 712f–713f
Parotid salivary gland, 414f, 454f, 700–702, 700f, **701**, 701f, 712f–713f
Paroxisomes, 62t
Pars distalis, 514f
Pars intermedia, 514f
Pars nervosa, 514f
Particle size, and filtration, 79–81
Passive immunity, **657**
Patella, 153f, **227**, 228f, 243, 244f–245f, 345f
Patellar, as term, 35f
Patellar ligament, 243t, 244f–245f, 460
Patellar reflex, 350, 444, 460, 460f
Patellar surface, 218t, 221f
Patellofemoral joint, **243**
Patent foramen ovale, 558–59
PCOS (polycystic ovarian syndrome), 796

PCTs. See Proximal convoluted tubules
PCV (packed cell volume), 538, 540
Pectinate muscle, **558**, 558f, 560
Pectineal line, 215t, 216f, 218t, 221f
Pectineus, 340f, 341f, 341t, 342
Pectoral, as term, 35f
Pectoral girdle, **323**
  bones, **157**, **202**–5, 202t, 213
  muscles, 323–26, 323t, 325f
Pectoralis, as term, 263t
Pectoralis major, 264, 265f, 324t, 325f
Pectoralis minor, 323t, 324t, 325f, **326**
Pedicle, 183f, **185**, 185f, 187f
Pelvic, as term, 35f
Pelvic girdle, 157, **214**–18, 214t–215t, 227
Pelvic splanchnic nerves, 453f
Pelvic viscera, innervation of, 455
Pelvis, male and female, **215**–18, 217f
Penile urethra, 789f, 790f
Penis, 790f
  gross anatomy, 789, 789f, 790f
  histology, 782, 782f
  innervation, 454f
Pennate architecture, in skeletal muscle, **264**, 264t
Pepsin, 724t
Pepsinogen, 705f, 706f
Perforating (Volkmann) canals, 110t, **148**–49
Pericardial cavity, 556–57, 556f
Pericardial fluid, **557**
Pericardial sac, 555, **556**, 556f
Pericarditis, 34
Pericardium, 555, **556**, 556f
Perichondrium, 110, 110t, 111–13, 111f, 664f
Perilymph, 476, 477f, 478f, 489, **490**
Perimetrium, 774
Perimysium, 255f, **257**, 257f
Perineal, as term, 35f
Perineum, **312**
Perineurium, 365–66, 366f
Periodontal membrane, **236**, 237f
Periosteal dura, 387f–388f
Periosteal layer, of dura mater, **384**, 384f, 385f, 387f–388f
Periosteum, 110, 147t, **148**–49, 155f, 156, 239f, 384f
Peripheral nervous system, 435–44. See also Cranial nerves; Spinal nerves
  glial cells in, 364–65
  nerves in, histology, 365–66, 366f
Peristalsis, 723, **724**
Peritoneal cavity, 799f
Peritoneum, 312f, 749, 783
Peritonitis, 34
Peritrabecular spaces, 649, 650f, 650t, 656f
Peritubular capillaries, 738, **745**–46, 746f, 747f
Peroneus brevis, **344**
Peroneus longus, **344**
Perpendicular plate (ethmoid bone), 169t, 171f, 172t, **173**, 174f
Pes, as term, 35f
Petrosal nerve, greater, 414f
Petrosal sinuses, 604f
Petrous part of temporal bone, 170t, **179**, 180f, 487, 487f
Peyer patches, **645**, 645t, **647**, 647f, 707f, 708
pH. See also Acid-base balance
  of blood, 681, 686–87, 725, 756, 756t
  of urine, 753, 755
Phalanges (phalanx)
  foot, 157f, 219f, 224, 224f, 225, **226**, 226f, **227**, 228f
  hand, 157f, 206t, **212**, 212f, 331
Pharyngeal artery, ascending, 602f
Pharyngeal tonsils, **646**, 646f, 646t, 655f, 672f
Pharyngeal vein, 602f
Pharyngotympanic tube (auditory tube; eustachian tube), 487f, 488t, 489, **490**, 490f, 655f, 672f
Pharynx, 298–99, 298t, 299f, 664, 672f
  gross anatomy, 712, 712f–713f
  muscles, **298**–**99**, 298t, 299f
Phenol, **13**, 13t
Ph.I.L.S lessons, 278–79, 278f
  action potential conduction velocity, 374–75, 374f
  action potential refractory periods, 375–76
  action potentials, compound, 372–73, 372f
  ECG, abnormal, 577
  ECG and finger pulse, 626
  ECG and heart block, 575–76
  EMG and twitch amplitude, 279–80, 279f
  glucose transport, 727–29
  heart electrical axis, 573, 573f

heart refractory period, 571
heart sounds, 578–79
membrane resting potential, 367–68, 367f, 369–70
muscle length-tension relationship, 276–77, 276t
muscle stimulation, 274–75, 274f, 275t
osmosis and diffusion, 81–83, 82f
pH and oxygen binding, 686–87
Starling's law, 579–80
thyroid gland and metabolic rate, 524–25
Phlebotomist, 538
Phlegm, 529, 530t
Phonation, 673
Phosphatase, 724t
Phosphate, bone storage of, 151
Photoreceptors, 497. See also Cones; Rod(s)
Phrenic arteries, 607f
Phrenic nerves, 307t, **436**, 436f
Physiology, as term, **1**, **28**
Pia mater, **384**, 384f, 385t, 387f–388f, 433, 434f, 435f
  sheep, **406**
Pigmentation, of skin, 129, 129f
Pigmented layer, retina, 474, 474f
PIH (prolactin-inhibiting hormone), 513t
Pineal body (gland), 510f, 522f–523f, 523
  histology, **515**, 515f
  human, 392t, 397f
  sheep, 406, 406f, 407f
Pinealocytes, **515**, 515f
Pineal sand (corpora arenacea), **515**, 515f
Pinna (auricle), 487f, 488t, 489, **490**, 557f, 558, 559f, 561f, 564f, 566, 566f, 567f, 568
Pins, dissecting, 11t
Piriformis, 336f, 337f, **338**, 339
Piriformis syndrome, 339
Pisiform, 206t, **211**, 211f
Pituicytes, **514**, 514f, 515
Pituitary gland, 396, 510f, **512**, 604f. See also Anterior pituitary gland; Posterior pituitary gland
  functions, 512
  gross anatomy, 522f–523f, 523
  histology, **512**–15, 513t, 514f
  hormones, 512
  sheep, 401f, **402**–6, 403f, 404f, 407f
Pituitary stalk (infundibulum)
  human, 179, 391t, **396**, 396f, 397f, 773t, 784f
  sheep, 405f
Pivot joint, 240t, 241f
Placenta, 623, 624f, 794, 798f
Plane (gliding) joint, 240t, 241f
Plantar, defined, 34f
Plantar aponeurosis, **347**
Plantar arterial arch, 619f
Plantar arteries, 619f, 622f
Plantar flexion, 242t
Plantar interossei, 334t, 335, 347, 348t, 349f
Plantar musculature, 347, 347t–348t, 349f
Plantar reflex, 460, 460f
Plantar veins, 620f, 622f
Plasma, **113**
  centrifugal separation of, 438f, 538
Plasma membrane, 66f, 67. See also Membrane transport mechanisms
  animal cell, 64f
  and facilitated diffusion, 523
  proteins, 68, 366, 523
  selective permeability of, 58, 68
  staining of, 57
  structure and function, 62t
Platelet plug formation, 538
Platelets (thrombocytes), **113**, 114, 114f, **536**, 537f
  centrifugal separation of, 438f, 538
  characteristics of, 530t
  identification of, 533–35, 534f, 537
  number and function, assessment of, 538
  precursor cells, 536
Platysma, 293f, 294f, 301f–302f
Pleura
  parietal, **675**, 675f
  visceral, **675**, 675f
Pleural cavities, 556f, **675**, 675f
Pleurisy, 34
Plexus(es), **435**
  abdominal aortic, 454f
  brachial, 307t, 434f, 437–38, 437t, 438f, 439f, 440f, 459
  cardiac, 454f, 458f–459f
  cervical, 307t, 434f, 436, 436f
  choroid, 363, 363f, 407f, 408
  esophageal, 454f

hypogastric, 454f
lumbar, 434f, 440
lumbosacral, **435**, 440–43, 441t, 442f, 443f
myenteric, 703f, 704f, 704t, 724
pulmonary, 454f
sacral, 434f, 440
submucosal, 703f, 704f, 704t, 724
Plicae circulares (circular folds), **707**, **718**, 719f
Pluck, definition of, 676
Pneumothorax, 675
Podocytes, **740**, 752
Pointed scissors, dissection with, **20**–21, 21f, **22**, 22f
Polarity, in epithelial tissue, **96**
Pollex, 206t, **212**, 212f, 334
Polycystic ovarian syndrome (PCOS), 796
Polyurea, **754**
Pons, 414f, 415f
  human, 392t, 395f, 396f, 397f
  sheep, 404, 404f, 406, 407, 407f
Popliteal, as term, 35f
Popliteal artery, 619f, 621f, 622f
Popliteal surface, 218t, 221f
Popliteal vein, 620f, 621f, 622f
Popliteus, 346f, 346t
Porta hepatis, 716t, 717f, 718f
Portal triads, **710**, 710f, 711f
Portal veins, 609
Positive feedback, in homeostasis, 2
Postcapillary venule, 592t, 596
Postcentral gyrus, 393t, 394, 394f, 395f
Posterior, definition of, 33t, 263t
Posterior arch, of atlas, **186**, 187f
Posterior cavity (vitreous chamber), 483t, 484, 484f, 486
Posterior cerebral artery, 604f
Posterior chamber, of eye, 483t, 484, 484f, **486**
Posterior chord, of brachial plexus, 437, 438f
Posterior clinoid processes, **179**
Posterior communicating artery, 604f, 605f
Posterior compartments, muscles of, **326**
  arm, 328, 328f, 328t
  forearm, 332, 332t, 333f
  leg, 346, 346f, 346t
  thigh, 342–43, 342t, 343f
Posterior cruciate ligaments, 243t, 244f–245f
Posterior funiculus, 430, 431f, 432t, 435f
Posterior horns, of spinal cord, 430, 431, 431f, 433t, 435f
Posterior humeral circumflex artery, 614f
Posterior iliac spines, 214t, 216f
Posterior intercostal arteries, 607f
Posterior intercostal vein, 608f
Posterior interventricular artery, 563t, 564f
Posterior interventricular sulcus, 566, 566f
Posterior median sulcus, 434f
Posterior pituitary gland (neurohypophysis)
  gross anatomy, 522f–523f, 523
  histology, **512**–15, 513t, 514f
  hormones, 393, 513t, 514
Posterior ramus, **435**
Posterior root, of spinal cord, 430, 431f, 434f
Posterior root ganglia, **364**, 365f, 430, 431f, 433t, 434f
Posterior rootlets, 434f, 435f
Posterior sacral foramina, 189f
Posterior scalenes, 307t
Posterior tibial artery, 619f, 622f
Posterior tibial vein, 620f, 622f
Posterior tubercle (atlas), **186**, 187f
Postganglionic fibers, 455, 458f–459f
Posture, and stretch reflexes, 459
Potassium channel blockers, 376
Potassium channels, voltage-gated, 371, 571
Potassium ions
  and action potential generation, 371
  leak channels, 366, 376
  and membrane potential, 366, 367–68, 367f
  tubular secretion and reabsorption, 753t
Potassium permanganate, 69–74, 69f, 70f, 75f
Potential energy, **370**
Power switch, of microscope, 46f, 47t
P-Q segment, 570, 571f
Precentral gyrus, 393t, 394, 394f, 395f
Pre-embryonic period, **797**, 797t–798t
Preganglionic fibers, 453f, 454f, 458f–459f
Prenatal period, **797**
Prepatellar bursa, 243t, 244f–245f
Prepuce, 790f
Preservatives, **13**, 13t
Prevertebral ganglia, 457f
Primary brain vesicles, **388**

I-10

Primary curvatures, 184
Primary oocyte, 768, 768t, 769f, 792–93
Primary ovarian follicle, 768, 769f, 770t–771t, 794f
Primary spermatocytes, 776f, 777, 777t, 795
Prime mover (agonist), **245, 299, 328**
Primitive streak, 799f
Principal cells. *See* Chief (principal) cells
P-R interval, **570,** 571f, 575–76
PRL. *See* Prolactin
Probe, blunt, 11t, 19, 20f, 22, 22f
Procerus, 293t, 294f
Profundus, definition of, 263t
Progesterone, **768,** 774t–775t, 794–95, 794f
Prolactin (PRL), 512, 513t
Prolactin-inhibiting hormone (PIH), 513t
Proliferating cartilage, zone of, 150f, 151
Proliferative phase of uterine cycle, **794,** 794f
Pronation, 242t
Pronator, definition of, 263t
Pronator quadratus, 329t, 330f
Pronator teres, 329, 329t, 330f, 331f
Prophase
  in meiosis, 791t–792t
  in mitosis, 64f–65f, 67
Proplatelets, 536
Proprioceptors, 350
Propulsion of food, **722**
Propylthiouracil (PTU), 524–25
Prosencephalon, 388
Prostate gland
  function, **788**
  gross anatomy, 750f, 789f, 790f
  histology, 779t, **781,** 781f
Prostatic calculi (corpora arenacea), **781,** 781f
Prostatic urethra, **750,** 751f, 789f, 790f
Protective equipment, 10
Protein(s)
  clotting, 113
  digestion of, 723–24, 723t, 724t, 727
  in urine, 754t, 755, 755t
Protein pumps, in plasma membrane, 68, 366, 523
Proteinuria, 754t
Proteoglycans, 151
Protraction, 242t
Proximal, defined, 33t, 34
Proximal convoluted tubules (PCTs), 736, 738, 738t, **740, 741,** 741f, 748, 752, 753, 753t
Proximal phalanx, 212f
Pseudostratified columnar epithelium, 669t
  ciliated, 668f, 669, 778, 778f, 779, 779t, 780f
  in respiratory tract, 104
  with stereocilia, 779t
  types, identification of, 98f, 101f, **102**
Psoas, 337f
Psoas major, 336t, **339,** 749f
Pterion, 176f
Pterygoid(s), 295f, 295t
Pterygoid plates, 178f
Pterygopalatine ganglion, 414f, 454f
PTH (parathyroid hormone), 151, 510, **516**
PTU (Propylthiouracil), 524–25
Puberty, hormones regulating, 515
Pubic, as term, 35f
Pubic crest, 215t, 216f
Pubic rami, 215t, 216f
Pubic symphysis, 218, **238,** 238f, 785f, 790f
Pubic tubercle, 215t, 216f
Pubis, 157f, **214,** 215t, 217f
Pudendal nerve, 441f, **442,** 442f
Pudendal nerve block, **442**
Pulmonary arteries, 557f, 561f, 601f, 624f, 668f, 669, 676, 677, 677f
Pulmonary capillaries, 601f, 680
Pulmonary circuit, 600–601, 601f
Pulmonary diseases, 680
Pulmonary fibrosis, 680
Pulmonary function tests, 683–86
Pulmonary plexus, 454f
Pulmonary semilunar valve, **559,** 559f, 560, 563f, 601f
Pulmonary surfactant, **671t**
Pulmonary trunk, 557f, **559,** 559f, 560, 561f, 563f, 566–67, 566f, 601f
Pulmonary veins, 557f, **560,** 561f, 562, 566f, 567, 601f, 624f, 676, 677, 677f
Pulmonary ventilation, **679**
Pulse, monitoring, 625, 626
Pulse points, **625**
Pulse pressure, **625**
Pupil, 483f, 485f
Pupillary reflexes, 462, 462f
Pupillary sphincter, 462
Purkinje cell(s), 359t, **361,** 361f

Purkinje cell layer, 361, 361f
Purkinje fibers, 569, 569t, 570f
P waves, **570,** 571f, 572, 573
Pyloric sphincter, 714t, 715f
Pylorus, 705f, 706t, 714t, 715f, 721f–722f
Pyramidal cells, 359t, **360–61,** 361f
Pyuria, 754t

QRS complex, **570,** 571f, 572, 573, 575
Q-T interval, **570,** 571f
Quadrate lobe, of liver, 716t, 717f, 718f
Quadratus femoris, 336t, 337f, 338
Quadratus lumborum, 305t, 306f
Quadratus plantae, 347t, 349f
Quadriceps, 245
Quadriceps, as term, 262t
Quadriceps femoris muscle group, 340t
Quadriceps femoris tendon, 243t, 244f–245f
Quadriceps tendon, 340f

Radial artery, 614f, 616f
Radial fossa, 205t
Radial nerve, **437,** 437t, 438f, 439t, 440f, 459
Radial notch, 206t, 210f
Radial veins, 615f
Radicular fila (rootlets), 433t
Radioulnar joint, 236, 237f
Radius, 157f, 206t, 207f–208f, **209,** 209f
Ramus
  of ischium, 214t, 216f
  of mandible, 169t, 172t, 175f
  of spinal nerve, **435,** 435, 436, 436f, 438f
Range, of data, **5,** 5t
Reabsorption, of filtrate, **736,** 752, 753t
Receptive field, **491**
  size of, **491**–92, 491f, 492f
Receptors
  adaptation of, 491, 496
  in homeostasis, **2**
  in reflexes, 450, 452, 452f, 452t
Reciprocol innervation, 245
Recruitment, for action potential, 372
Rectal artery, superior, 610f
Rectal veins, varicosities of, 613
Rectouterine pouch, 785f
Rectum, **720,** 720t, 721f–722f, 749f, 785f, 790f
  blood vessels, 610f
  innervation, 454f
Rectus, as term, 262t
Rectus abdominis, **310,** 310t, 311f, 312f, 313f
Rectus femoris, 264, 265f, 267f, 339, 340f, 340t
Rectus muscles, 296t, 297f, 412f, 484, 484f
Rectus sheath, 311f, 311t, **312,** 312f
Red blood cell(s). *See* Erythrocytes
Red blood cell count, 538
Red bone marrow, 153–54, 154f, **156,** 227
  identification of megakaryocytes in, 536–37, 537f
Red pulp, **653,** 653f, 654, 654t
Reflex arc of somatic nervous system, 452, 452f, 452t
Reflexes
  autonomic, 450, 452, 461–62
  Babinski, 461
  corneal, 413, 461, 461f
  coughing, 350
  definition of, **449**
  gag, 350
  muscle spindles and, 350
  patellar, 350, 444, 460, 460f
  plantar, 460, 460f
  pupillary, 462, 462f
  reflex arc, components of, 450, 452, 452f, 452t
  somatic, 450, 452, 459–61
  speed of, 449–50
  spinal, 444, 450, 452
  stretch, 350, 444, 459
  types of, 450, 452
Refractory period
  absolute, **371,** 371f, 375–76
  relative, **371,** 371f, 375–76
Regional terms, 34–35, 34t
Regression, in ovarian cycle, 794f
Regular dense connective tissue, 105t, 108, 108f
Rejection, of blood transfusion, 544
Relative refractory period, **371,** 371f
Relaxin, 218
Renal artery, 609f, 746f, 747f, 749f, 750f
Renal capsule, 739f, 744, 745f, 746f
Renal column, 745f
Renal corpuscles, 738, 739f, **740,** 740f, 741f, 752
Renal cortex
  gross anatomy, 745f, 746f, 750f
  histology, **738,** 739f, 740–41, 740f, 741f

Renal fascia, 744
Renal lobe, 745f
Renal medulla, **738,** 739f, 741–42, 742f, 745f
Renal papilla, 739f, 745f, 748, 750f
Renal pelvis, 745f, 748, 750f
Renal pyramid, 745f, 746f, 748, 750f
Renal sinus, 745f
Renal vein, 609f, 746f, 747f, 749f, 750f
Repolarization, of excitable membrane, 371, 371f
Reposition, 242t
Reproductive system
  clinical case, 795–96
  female
    components, 768
    functions, 768
    gross anatomy, 783–888
    histology, 768–75
  functions of, 766
  genitalia, innervation, 442
  homologous structures in, 766
  male
    components, 776, 779t
    functions, 776
    gross anatomy, 788–90
    histology, 776–82
    physiology, 791–96
Residual volume (RV), 679, 679t, 680f, 684f, 685, 685t
Resistance, in blood vessels, 625
Respiration
  and acid-base balance, 756, 756t
  auscultation of, 682
  and blood pH homeostasis, 681
  exercise-induced changes in, 688–89
  mechanics of, 682
  muscles of, 307–9, 307t, 308f, 309f, 662, 679
  overview, 662
  physiology of, 679–89
  pulmonary diseases, 680
Respiratory acidosis, **756,** 756t
Respiratory alkalosis, **756,** 756t
Respiratory membrane, **670,** 671t
  rate of diffusion across, **680**–81
Respiratory tract
  conducting zone, 664, 669, 673
  epithelial tissue in, 104
  lower
    gross anatomy, 673–79
    histology, 666–69
  respiratory zone, 664, 669, 673
  upper
    gross anatomy, 671–72
    histology, 664–66
Respiratory volumes, measurement of, 679–80, 679f, 680f, 683–86, 684f, 685t, 688–89
Respiratory zone, 664, 669, 673
Response, in homeostasis, **2**
Resting cartilage, zone of, 150f, 151
Resting membrane potential, 366–70
  depolarization and repolarization, 371, 371f
  potassium ions and, 367–68, 367f
  sodium ions and, 369–70
Restrictive pulmonary diseases, 680
Rete testis, 778
Reticular connective tissue, 105t, 107, 107f, 651
Reticular dermis, 480, 480f
Reticular fibers, 104, 104t, 105t, 107, 107f, 651
Reticular layer, of dermis, **130,** 130f, 137, 137f, 468, 468f
Reticuloendothelial (Kupffer) cells, **710,** 710f
Retina, **411,** 474–76, 474f, 475t, 483t, **486,** 486f
Retraction, 242t
Retroperitoneal, definition of, **743**
Rh blood type, 538
Rheumatoid arthritis (RA), 523
Rh factor, 538
Rhombencephalon, 388
Rhomboid, as term, 262t
Rhomboids, 323t, 325f, **326**
Ribonuclease, 724t
Ribosomes, 62t, 66f, 67
Ribs, 157f, 190, 190t, 192, 192f
  true, **192**
Right brachial artery, 627
Right bundle branch, **569,** 569t, 570f
Right renal artery, 609f
Right renal vein, 609f
Right subclavian artery, 605f, 607f
Right subclavian vein, 608f
Right vertebral artery, 605f
Rima glottidis, 673t
Rinne test, **499,** 499f

Risorius, 293t, 294f
Rod(s), 475t, **497**
Rod and cone layer, 474, 474f, 475t, 476
Romberg test, **499,** 500, 500f
Root, of hair, **133,** 133f, 134
Root ganglia, posterior, **364,** 365f, 430, 431f, 433f, 434f, 435f
Rootlets (radicular fila), 433t
Root sheath, of hair, 133t, 134, 134f
Rotation, 242t
Rotator cuff, muscles of, **326**
Rotatore muscles, 303, 305t
Rough endoplasmic reticulum, 61t, 66f, 67, **360**
Round ligament
  liver, 716t, 717f
  uterus, **312,** 783t, 785f
Round window, 477f, 489, 490, 490f
Rugae (gastric folds), 714t, 715f
RV (residual volume), 679, 679t, 680f, 684f, 685, 685t
R wave, 573, 573f, 575

SA. *See* Sinoatrial (SA) node
Saccule, 477f, 489, 489t
Sacral, as term, 35f
Sacral arteries, median, 607f
Sacral canal, 189f
Sacral crest, 183t
Sacral curvature, 184f
Sacral foramina, 183t, 189f
Sacral hiatus, 183t, 189f
Sacral plexus, 434f, **440.** *See also* Lumbosacral plexus
Sacral promontory, 183t, 189f
Sacral region, **183,** 183t
Sacral vein, median, 608f
Sacral vertebrae, 431f, 432f
Sacrotuberous ligament, 337f, **338**
Sacrum, 157f, **183,** 183t, 184f, **189,** 189f, 228f
Saddle (sellar) joint, 240t, 241f
Safety
  equipment, 10
  hazardous chemicals, 13, 13t
  laboratory waste disposal, 14, 14t, 15
  urine samples, handling of, 754
Sagittal (parasagittal) plane, 31, 31f, 31t
Sagittal sinus
  inferior, 386f, 387f–388f, 604f
  superior, 384f, 386f, 387f–388f, 401, 401f, 604f, 605f
Sagittal suture, **177,** 177f
Saliva, **700**
Salivary amylase, **700**
Salivary ducts, 700f, 701
Salivary glands
  gross anatomy, 712f–713f
  histology, **700**–702, 700f, 701t
  innervation, 454f, 455
Salt, as basic taste, 494, 494t
Salt(s), and muscle contraction, 273
Sample size, statistical tests for, 4
Saphenous vein, 620f, 621f
Sarcolemma, 254, 258f
Sarcomas, 114
Sarcomeres, **254,** 255f, 256, 260, 271f, 277, 555, 579
Sarcoplasmic reticulum, 278
Sartorius, 264, 265f, 267f, **339,** 340f, 340t
Satellite cells, **256,** 362, 362t, **364**–65, 365f
Scala media (cochlear duct), 476, 477f, 478f, 490
Scala tympani, 476, 477f, 478f, 490f
Scala vestibuli, 476, 477f, 478f, 490f
Scalenes, 301f–302f, 308f
  anterior, 436f
  interior, 307t
  middle, 307t
  posterior, 307t
Scalene tubercle, 190t
Scalp, 133–34, 133f
Scalpel, 12t, **16,** 16f, 19, 19f, 20f
Scalpel blade, 12t, 16, **16,** 17f–18f
Scalpel blade handle, 12t, **16,** 17f, 19, 20f, 22
Scaphoid, 206t, 211, 211f
Scapula, 157f, 202t, **203,** 204f, 299
Scapular line, **308**
Scar tissue, after myocardial infarction, 562, 562f
Schwann cells. *See* Neurolemmocytes
Sciatica, **339,** 428
Sciatic nerve, 337f, **338,** 339, 441t, **442,** 442f
  frog, action potentials in, 372–75
  tibial division, **442,** 443
Sciatic notches, 214t, 215, 216f, 217f
Science, defined, 4

I-11

Scientific method, steps of, **4**–6
Scientist, defined, **4**
Scissors
　description and use, 12t
　dissection with, 20–21, 21f, 22, 22f
Sclera, **296,** 474, 474f, 475t, 481t, **485,** 485f
Scleral venous sinus, 484f
Scrotum, 789, 789f, 790f
Sebaceous glands, 131t, **133,** 133f, 134, 135f, 137, 137f, 480f
Sebum, **133**
Secondary brain vesicles, **388,** 388f, **389**
Secondary curvatures, **184**
Secondary oocyte, 768, 768t, 769, 769f, 770f, 791, 792–93, 795
Secondary ovarian follicle, 768, 769f, 770f–771f, 794f
Secondary spermatocytes, 777t
Second law of thermodynamics, **67**
Secretion
　apocrine, 135
　in digestion, **722**
　renal, **736, 753,** 753t
Secretory cells, 772f, **773,** 773t
Sectional anatomy, **31**
Segmental artery, 746f, 747f
Segmental (tertiary) bronchi, 678f, 678t
Seizure, epileptic, **398**
Selection, of T-lymphocytes, **651**
Selective permeability, of plasma membrane, **58,** 68, 366
Sella turcica, 170t, 174f, **179,** 180f, 396
Semen, **780**
Semicircular canals, 415f, 487f, 489, 489t, **498**
Semilunar valve, **578**
　aortic, 560, 561f, 562–63, 563f, 601f
　pulmonary, 559, 559f, 560, 563f, 601f
Semimembranosus, 267f, 342, 342t, **343,** 343f
Seminal vesicles
　function, **788**
　gross anatomy, 789f, 790f
　histology, 779t, **780**–81, 780f
Seminiferous tubules, **776**–77, 776f, 789f, 790f, 795
Semipermeable membrane
　diffusion across, 68, 68f
　permeability, and diffusion rate, 75–76, 76f
Semispinalis capitis, 300t, 301f–302f, 305f, 306f
Semispinalis cervicis, 306f
Semispinalis muscles, **303,** 305f
Semispinalis thoracis, 305t, 306f
Semitendinosus, 267f, 342, 342t, **343,** 343f, 443f
Senses, general
　definition of, **468**
　functional properties of, 491–93
　gross anatomy, 479–90
　histology, 468–69
Senses, special
　definition of, 368, **468**
　functional properties of, 494–501
　gross anatomy, 481–90
　histology, **470**–79
Sensory adaptation, **491, 493**
Sensory information, and nerve function, **408**
Sensory neurons, in reflexes, **450, 452,** 452f, 452t
Sensory receptors
　for blood pH, 681
　density of, **491**
　in dermis, 131–32, 132t
　receptive field size, **491**–92, 491f, 492f
　skin, 468–69, 468f, 468t, 469f, 479–80, 480f, 491–92, 491f, 492f
Septae, of thymus, **651,** 652f
Septal cartilage, 174f
Septum pellucidum, human, 393t, 397f
Serous cells, **700,** 700f, 701, 701t, 702
Serous demilunes, 700f, 701, 701t, 702
Serous fluid, 34, 675
Serous membranes/serosa
　definition of, **34**
　GI tract, **702,** 703f, 704f, 704t, 709
　inflammation of, 34
　parietal layer of, 34, **556,** 556f, 557
　pericardial, 555, 556, 556f
　structure and function, 34
　uterine tubes, 772, 772f, **773,** 773f
　visceral layer of, 34, 555, **556,** 556f, 557
Serratus anterior, 308f, 323f, 325f, **326, 438**
Serratus posterior inferior, 306f
Sertoli (sustentacular) cells, 776f, **777,** 777t
Set point, in homeostasis, **2**
Sex chromosomes, 791
Sexes, **766**

Sex hormones, 510
Shaft, of hair, **133,** 133f, 134
Sharp dissection, **14,** 19, 19f, 20f
Sharps containers, **14,** 14f
Sheep
　brain structures, 401–8, 401f–407f
　heart, anatomy, 566–69, 566f–568f
Short bone, classification of, 152f, 152t
Short head of biceps brachii, **328**
Shunts, in fetal circulatory system, **623**
Sigmoid arteries, 610f, 613f
Sigmoid colon, 610f, 709f, 720, 720t, 721f–722f
Sigmoid sinus, 180f, **384,** 385f, 386f, 604f, 605f
Silver stain, 362
Simple columnar epithelium, 98f, 100, 100f, 647, 647f, 668f, 705f, 707, 772f
　ciliated, 669, 669f
Simple cuboidal epithelium, 98f, 99, 99f, 741–42, 768
　ciliated, 670t
Simple epithelial tissue, **97**
Simple squamous epithelium, 98–99, 98f, 669, 670t, 671t, 740
Sinoatrial (SA) node, 569, 569t, 570, 570f
Sinuses
　coronary, 558f, 559, 560, 564f, 565, 565t
　dural venous, **384,** 385f, 386t, 387f–388f, 401, 401f, 402f, 604f, 605f, **606**
　lactiveous, 786f, 787f
　medullary, 649, 649f, 650f, 650t, 651, 656f
　paranasal, 664, 672f
　renal, 745f
　scleral venous, 484f
　skull, 169f, 174f, 180f
Sinusoid capillaries, 597, **597,** 599t
Sinusoids, splenic, 653f, 654, 654t
Sister chromatids, 64f–65f, 791f–792f
SITS acronym, 326
Size of microscope specimen, estimation of, 50–51, 50f
Skeletal muscle, **252**
　architecture of, 264, 264t
　blood vessels, innervation of, 456
　components of, 145
　connective tissues covering, 257, 257f
　contraction of, 270, 270f, 271f
　fiber types, 268, 268t, 269, 269t
　force generation in, 268–80
　identification and classification, 116, 116f
　learning strategies for, 261
　length-tension relationship, 276–77, 276t
　naming methods for, 261, 261t–263t
　neuromuscular junctions, 257–59, 258f
　sliding filament theory and, 276–77, 276t
　structure and function, 114, 115t
Skeletal muscle tissue
　appearance, 254, 254t
　cells, 254
　contraction, regulation of, 277
　functions, 254t
　muscle fibers, 272, 272f
　structure, 254, 255f, 256, 256f
　vs. cardiac muscle, 277
Skeleton/skeletal system. See also Bone(s)
　appendicular, 165, 202–28
　axial, 165, 169–92
　human, 157–58, 157f, 158f
　learning strategies for, 157, 165–66, 179, 186, 212
　limb
　　lower, 218–27, 218t–219t
　　upper, 205–13, 205t–206t
　pectoral girdle, 202–5, 202t
　pelvic girdle, 214–18, 214t–215t
　skull bones, 169–82, 169t–170t, 172t
　thoracic cage, 190–92, 190t
　vertebral column, 183–89, 183t
Skin
　cancer of, 129, 129f
　components of, 125, **127**
　connective tissue in, 108–9, 108f
　defined, **125**
　facial, wrinkling of, 292
　pigmentation, 129, 129f
　sensory receptors, 468–69, 468f, 468t, 469f, 479–80, 480f, 491–92, 491f, 492f
　thick, 127, **128**
　thin, 127
Skull, 157f
　bones, 169–82, 169t–170t, 172t
　fetal, 181, 182f
　sutures, 236, 237f
Slides
　of bone, types of, 147
　care of, 45

histology, 94, 94f
　stains, 57. 94, 63, 108
　viewing of, 48–49, 51, 97
　wet mount preparation, 63, 63f
Sliding filament theory, 276–77, 276t, 579
Small bronchi, **668**–69, 668f, 669t
Small intestine. See also Duodenum; Ileum; Jejunum
　blood vessels, 611f
　in digestion, 723, 723t, 724t
　epithelial tissue in, 100, 100f
　histology, 707–8, 707f, 707t
　innervation, 454f, 456
Small saphenous vein, 620f
Smooth endoplasmic reticulum, 61t, 66f, 67
Smooth muscle, **252**
　in blood vessels, 595, 595f, 596, 596f
　in bronchi, 668f, 669, 669t
　in ductus deferens, 779f, 780f
　in GI tract, 704f, 704t, 724
　in respiratory tract, 669
　structure and function, 114, 115t
　in urinary tract, 743, 743f, 744, 744f
Smooth muscle tissue, 259–60, 259f
　appearance, 254t, 259, 259f, 260
　cells, 259, 259f
　functions, 254t
　identification and classification, 117, 117f
Snake venom, antiserum for, 657
Snellen eye chart, 497, 497f
Sodium channels, voltage-gated, and action potential generation, 371
Sodium ions
　and action potential generation, 371
　leak channels, 366, 373
　and membrane potential, 366
　tubular secretion and reabsorption, 753t
Sodium/potassium pumps, 366, 727
Soft palate, 672f, 712f–713f
Soleus, 345f, **346,** 346t
Solubility coefficient, of gas, 680
Solute
　defined, **67**
　diffusion rate, factors affecting, 69–76
　membrane-impermeable, 77
Solvent
　defined, **67**
　viscosity, and diffusion rate, 69–70, 69f, 70f, 71t
Somatic motor neurons
　ALS and, 270
　anterior horn cells, 360, 360f
　function of, 408t
　location of, 459
　neuromuscular junction, 257–59, 258f
Somatic nervous system (SNS), **450,** 450f, 452, 452f, 452t
Somatic reflexes, **450, 452,** 459–61
Somatic sense receptors, 468–69
Somatic senses, **468**
Somatic sensory nerves, function of, 408t
Somatic sensory neurons, in posterior root ganglia, 364, 365f
Somatostatin, 520t
Somite, 799f
Sour, as basic taste, 494, 494t
Special senses
　definition of, 368, **468**
　functional properties of, 494–501
　gross anatomy, 481–90
　histology, 470–79
Special sensory nerves, function of, 408t
Specific gravity, of urine, **753,** 756t
Specificity, in adaptive immune response, 539
Spectrophotometer, 686, 727–28
Sperm, 766, 776f, 778, 778f, 788, 791
Spermatic cord, 312, **312,** 313f, 789, 789f, 790f
Spermatids, 776f, **777,** 777t, 795
Spermatocytes, 776f, 777, 777t, 795
Spermatogenesis, **776,** 777t, **795**
Spermatogonia, **776,** 776f, 777, 777t, 795
Spermatozoa, 776f, 795
Spermatozoon, 777t
Sperm pronucleus, 797t–798t
Sphenoidal fontanelle, 182f
Sphenoid bone, 169, 170t, 171f, 173, 173f, 174f, 176f, 178f, 179, 182f, 182t
Sphenoid sinus, 174f, 672f
Sphygmomanometer, **625,** 627–28
Spina bifida, 189
Spinal cord
　central canal of, 388, 388f
　gross anatomy, 433–35, 433t, 434f, 435f
　histology, 430, 431f, 432t–433t
　human, 395f, 396f, 397f

　injuries to, 427–28
　learning strategies for, 428, 431
　nervous tissues in, 358, 358f, 360, 360f
　organization, 430, 431f, 432t–433t
　regional characteristics, 432t
　sheep, 401f, 404, 404f, 406, 406f, 407f
Spinal (vertebral) foramen, 183t, **185,** 185f, 186, 186f, 187f, 188, 188f
Spinal ganglion, satellite cells in, 364–65, 365f
Spinalis, 304f, 306f
Spinal nerves, 433t, **435**–44, 435f
　brachial plexus, 307t, 434f, 437–38, 437t, 438f, 439t, 440f, 459
　cervical plexus, 307t, 434f, 436, 436f
　flow of information in, 444
　learning strategies for, 428, 436, 437, 440, 442
　lumbar, 310t
　lumbosacral plexus, 440–43, 441t, 442f, 443f
　thoracic, 300t, 310t
Spinal reflexes, 444, **450,** 452
Spinal tap (lumbar puncture), 432
Spindle fibers, 791t–792t
Spine, of scapula, 202t, 204f, **213**
Spinous processes, 183t, **185,** 185f, 186, 186f, 187f, 188, 188f
Spiral ganglion, 476, 478t
Spiral organ (Organ of Corti), **476,** 477f, 478t, 489t, 490f
Spirometry, **679,** 680, 683–86
Splanchnic nerves, pelvic, 454f, 458f–459f
Spleen
　blood vessels, 610f, 611f, 612f
　gross anatomy, 521f, 656f, 721f–722f
　histology, 653–54, 653f, 654f
　innervation, 454f
Splenic artery, 610f, 612f, **653,** 653f, 654, 656f
Splenic colic flexures, 720t, 721f–722f
Splenic sinusoids, 653f, 654, 654t
Splenic vein, **609,** 611f, 612f, 613f, **653,** 653f, 656f
Splenius capitis, 299, 300t, 301f–302f, 306f
Splenius cervicis, 301f–302f, 304t, 306f
Splenius muscles, 303t
Spongy (cancellous) bone, 110, 111f, **149**–50, 149f, **153**–54, 154f, 155f, 156
Spongy urethra, **750,** 751f
Squamous epithelial cells, 101, 101f
　structure and function, 96t
　types, identification of, 98–99, 98f, 101, 101f
Squamous part of temporal bone, 170t, 176f, **179**
Squamous suture, 172t, 176f
Stage plate, of microscope, 52f
Stains, 57, 94, 63, 108
Standard deviation, **5,** 5t
Stapedius, 488t, 489, **490**
Stapes, 487f, 488t, 489, 489t, 490, 490f
Starling's Law of the Heart, 579–80
Statistical tests, for sample size, 4
Stem cells
　epithelial, **705**
　hemocytoblasts, 227
　in stomach, 705
Stenoclavicular joint, **203,** 203f
Stereocilia, **778,** 778f, 779t
Stereomicroscope, 52–53, 52f
Sternal, as term, 35f
Sternal angle, 190t, **191,** 191f
Sternal end, 202t, 203f
Sternal head, of sternocleidomastoid, **299**
Sternoclavicular joint, **213,** 213f
Sternocleidomastoid muscle, **299**–300, 300t, 301f–302f
　innervation, 417, 417f
Sternocostal joints, 238, 238f
Sternohyoid muscle, 301f–302f, 301t
Sternothyroid muscle, 301f–302f, 301t
Sternum, 154, 157f, 190, 190t, 191, 191f, 308f, 309f
Steroid hormones, 517
Stethoscope
　in auscultation of respiratory sounds, 682
　in auscultatory blood pressure measurement, 625, 627–28
Stimulus, in homeostasis, **2**
Stomach, **702**
　blood vessels, 611f
　in digestion, 723, 723t, 724t
　gross anatomy, 714–15, 714f, 715f
　histology, 705–6, 705f, 706t
　hormones secreted by, 510
　innervation, 454f
　wall layers, 702–3, 703f, 704f, 704t
Straight sinus, 385f, 386t, 387f–388f, 604f

I-12

Straight tubules, 778
Stratified epithelium, **97,** 98f, 101, 101f, 102–3, 102f
 cuboidal, 102–3, 102f
 squamous, 108–9, 108f, 471f, 646f, 743t
 squamous keratinized, 97t, 98f, 127
 squamous moist, 101
 squamous nonkeratinized, 98f, 101, 101f, 775, 775f
Stratum basale, 127t, 128f, 129, 129f, 137, 137f, 480, 480f
Stratum corneum, 127t, 128f, 137, 137f, 480, 480f
Stratum granulosum, 127t, 128f, 137, 137f, 480, 480f
Stratum lucidum, 127t, 128f, 137, 137f, 480, 480f
Stratum spinosum, 127t, 128f, 129, 129f, 137, 137f, 480, 480f
Stretch reflex, 350, 444, **459**
Striations
 cardiac muscle, 116, 116f, 254, 254t, 260, 260f, 277, 554f, 555
 skeletal muscle, 116, 116f, 254, 254t, 272, 277
Stroke, **303,** 606
Stroke volume, **578**
S-T segment, 570, 571f
Styloglossus, 297t, 298f
Stylohyoid, 300t, 301f–302f
Styloid process
 of radius, 206f, 209f, **213,** 213f
 of temporal bone, 170t, 176f, 178f, 298f
 of ulna, 206f, 210f, **213,** 213f
Stylomastoid foramen, 178f, 414f
Stylopharyngeus, 298t
Subarachnoid space, **384,** 384f, 386t, 387f–388f, 432, 435f
Subcapsular spaces, 649, 649f, 650f, 650t, 656f
Subclavian arteries, 438f, 602f, 605f, 607f, 614f, 616f, 617f
Subclavian veins, 602f, 608f, 615f, 616f, 617f
Subcostal nerve, 443f
Subcutaneous layer (hypodermis), 130f, 137, 137f, 480f
Subdural space, 384f, 386t
Subepithelial tissues, of GI tract, 698
Sublingual ducts, 700f, 712f–713f
Sublingual salivary gland, 454f, 700–702, 700f, **701,** 701t, 712f–713f
Submandibular ducts, 700f, 712f–713f
Submandibular ganglion, 414f, 454f
Submandibular salivary gland, 454f, 700–702, 700f, 701t, 712f–713f
Submucosa
 GI tract, 702, **702,** 703f, 704f, 704t, 707, 707f, 708, 709, 709f
 respiratory tract, **664,** 664f, 666, 666f, 667f, 668
 urinary tract, 744, 744f
Submucosal glands, 664f, 666t, 667f, 668, 703f
Submucosal nerve plexus, 703f, 704f, 704t, 724
Subpubic angles, 215
Subscapularis, 324t, 325f, **326**
Subscapular artery, 614f
Subscapular fossa, 202t, 204f
Substage lamp, of microscope, 46f, 47f
Sulci, 384f
 sheep, 406, **406,** 406f
Superciliary arches, 169t, 171f, 172t
Superficial, defined, 33t
Superficial fascia, penis, 782, 782f
Superficial fibular nerve, 441t, 443
Superficial inguinal ring, **312**
Superficialis, definition of, 263t
Superficial palmar arch, 616f
Superficial palmar venous arch, 615f, 616f
Superior, defined, 33t, 34
Superior alveolar nerve, 413f
Superior angle, of scapula, 202t, 204f
Superior articular facets, **185, 186**
Superior articular process, 183t, **185,** 185f, 187, 187f, 188, 188f, 189f
Superior border, of scapula, 202t, 204f
Superior colliculus
 human, 393t
 sheep, 406, 406f, 407f
Superior costal facets, 183t, **188,** 188f
Superior gemellus, 336t, 337f, 338
Superior gluteal nerve, 441t, **442,** 442f
Superior lobe, lung, 676f, 677f, 678f
Superior mediastinum, 651, 651f
Superior mesenteric artery, 610f, 611f
Superior mesenteric ganglion, 457f
Superior mesenteric vein, **609,** 612f

Superior nasal concha, 169t, 174f
Superior nuchal line, 178f
Superior oblique, 296t, 297f, 412, 412f, 484f
Superior orbital fissure, 170t, 171f, 172t, 173f
Superior palmar arch, 614f
Superior petrosal sinus, 604f
Superior phrenic artery, 607f
Superior pubic ramus, 215f, 216f
Superior rectus, 296f, 297f, 412f
Superior sagittal sinus, **384,** 386t, 387f–388f, 604f, 605f
 sheep, 401, 401f
Superior temporal line, 176f
Superior trunk, brachial plexus, **437,** 437t, 438f, 440f
Superior vena cava, 557f, **558,** 558f, 559f, 560, 566f, 567, 603f, 605f, 608f, 616f, 617f, 624f
Supination, 242t
Supinator, 263t, **328,** 329t, 330f, 332t, 333f
Supporting cells (sustentacular cells)
 of cochlea, 476, 477f, 478t
 of olfactory epithelium, 472, 472f, 473t, 665f, 665t, 666
 of taste bud, 471, 471f, 471t
Supporting connective tissue, **104,** 110–13, 110t
Supraclavicular nerves, 436f
Supracondylar lines, 218f, 220f, 221f
Supracondylar ridges, 205t, 207f–208f
Supraglenoid tubercle, 202t, 204f, 328t
Suprahyoid artery, 602f
Suprahyoid muscles, 300t
Supramaximal stimuli, in muscle contraction, 274
Supraorbital foramen, 169t, 171f, 172t, 173f
Supraorbital margin, 171f, 172t, 173f
Suprapatellar bursa, 243t, 244f–245f
Suprascapular notch, 202t, 204f
Supraspinatus, 263t, 324t, 325f, **326**
Supraspinous fossa, 202t, 204f
Suprasternal notch, 190t, **191,** 191f
Supreme intercostal vein, 608f
Sural, as term, 35f
Sural cutaneous nerve, 443f
Surface antigen D, 538
Surgical neck, of humerus, 205t, 207f–208f
Suspensory ligaments
 breast, 786, 786f, 787f
 eye, 483t, 484, **486,** 486f
 ovary, uterine tubes, and uterus, 783–85, 784f, 785f
Sustentacular (Sertoli) cells, 776f, 777, 777t
Sutural (Wormian) bone, 177f
Sutures, **169,** 177, **236,** 236f, 237f
Swallowing, muscles for, 298–99
Sweat glands
 apocrine, 131, 131t, **135,** 135f
 eccrine (merocrine), 102–3, 102f, 130f, **131,** 131t, 137, 137f, 469f, 480f
Sweet, as basic taste, 494, 494t
Sympathetic division, **453,** 455–56, 455t, 457f, 458f–459f
Sympathetic ganglia, 455, 455t
Sympathetic trunk, 457f
Symphysial surface, pubic bone, 215t, 216f
Symphysis, **237,** 237t, 238, 238f
Synapsis (crossing over), **791,** 791t–792t
Synaptic cleft, 258f
Synaptic knob, 258f, 259, 359t
Synaptic vesicle, 258f
Synarthrosis, 233, 234t, 236, 237
Synchondrosis, 237, 237t, 238, 238f
Syncytiotrophoblast, 797t–798t
Syndesmosis, **236,** 236t, 237f
Synergist, 299
 biceps brachii as, **328**
 of trapezius, **324–26**
Synovial cavity, 239f, 239t
Synovial fluid, 239f, 239t
Synovial joints
 classifications, 240, 240t
 knee joint, 243, 243t, 244f–245f
 movement, 241, 242f
 structure, **239,** 239f, 239t
Synovial membrane, 239f, 239t
Systemic circuit, **601**–22
Systemic gas exchange, 679
Systole, **578, 625**
Systolic pressure, **625**

Table(s), conventions for, **7,** 7t
Tachycardia, 570
Tactile (Merkel) cells, 127, **132,** 132t
Tactile (Meissner) corpuscle, **130,** 130f, **132,** 132t, 137, 137f, **468**–69, 468f, 468t, 480, 480f

Tactile (Merkel) disc, 468t, **479,** 480
Tactile localization, **492,** 492f
Tactile menisci, 479
Taenia coli, 720t, 721f–722f
Tail, of pancreas, 711f, 716t, 717f, 721f–722f
Tallquist method, 542
Talus, 219t, 224, 224f, 225, 226f
Tapetum lucidum, 483t, **486,** 486f
Tarsal, as term, 35f
Tarsals, 152f, 157f, 219t, **224**–25, 224f, 226f
Taste buds. See also Gustation
 function, 494, 494t
 histology, **471**–72, 471t
 innervation of, 414f
Taste pores, 471, 471t
Tectal plate (corpora quadrigemina), human, 391f, 397f
Tectorial membrane, 476, 477f, 478t, 490f
Teeth, 712f–713f
Telencephalon, 388, 388f, 390t
Telodendria, 359t
Telophase
 in meiosis, 791t–792t
 in mitosis, 64f–65f, 67
Temperature
 and action potential conduction velocity, 374–75, 374f
 and diffusion rate, 71–73, 72t
 and digestion rate, 725–27, 726t
 regulation of, 524–25
Temporal artery, superficial, 602f, 603f
Temporal bone, 153f, 169, 170t, 176, 177f, 178f, 179, 180f, 182f, 487, 487f
Temporalis, **295,** 295f, 295t
Temporal lines, 176f
Temporal lobe
 human, 393t, 395f, 396f, 397f
 sheep, 404, 404f
Temporal process, 170t, 176, 176f, 178f
Temporal vein, superficial, 602f, 603f
Temporomandibular joint, **175**
Tendinous intersections, abdominal, 311f, 311t, **312,** 313f
Tendon(s)
 attachment to periosteum, 155f, 156
 connective tissue in, 108, 108f
 grafting of, 346
 knee, 243t, 244f–245f
 leg, 344f, 345f, 346, 346f
Tendon sheaths, 239t, 243, 244f–245f
Tensor fascia latae, 336f, 337f, **338,** 340f
Tensor tympani, 487f, 488t, 489, **490**
Tensor veli palatini, 298t, 299f
Tentorium cerebelli, 385f, 386f, 387f–388f
 sheep, 402f, 403
Teres, definition of, 262t
Teres major, 324t, 325f
Teres minor, 324t, 325f, **326**
Tertiary (segmental) bronchi, 678f, 678t
Testes, **788**
 as endocrine organ, 510, 510f, 522f–523f, 523
 gross anatomy, 789, 789f, 790f
 histology, **776**–77, 776f, 777t
Testicular artery and vein, 790f
Testosterone, **776,** 795, 796
Tetanus, 278–79, 278f
Tetraiodothyronine (T_4; thyroxine), **524**
TH (thyroid hormone), **516,** 524–25
Thalamus, 474, 498
 human, 393t, 397, 397f
 lateral geniculate nucleus of, 411f
 sheep, 406, 407, 407f
Theca externa, 769, 769f, 770f, 771f
Theca interna, 769, 769f, 770f, 771f
Thenar eminence, **334**
Thenar group, **334,** 334t, 335f
Thick skin, 127, **128**
Thigh
 fascial compartments, 266, 266f, 267f, 339–43
 muscles, 339–43, 340f, 340t–342t, 341f, 343f
 nerves, 443, 443f
Thin skin, 127
Third ventricle, 388f, 389f, 390t, 397
 human, 393t, 397f
 sheep, 406, 407f
Thoracic aorta, descending, 607f, 609f
Thoracic artery, internal, 602f, 607f
Thoracic cage, 190–92, 190t, 191
 muscles of, 307–9, 307t, 309f
Thoracic cavity, **556,** 556f
 innervation, 416, 416f
 and respiration, 679
Thoracic curvature, 184f
Thoracic duct, 655, 655f

Thoracic region, **183**
Thoracic spinal nerves, 300t
Thoracic vein, internal, 608f
Thoracic vertebrae, **183,** 183t, 184f, **185,** 188, 188f
 characteristics, 432t
 histology, 431f
Thoracic wall, blood vessels, 606–9, 607f, 608f
Thoracis muscle, 305t, 308, 308f
Thorax, points of reference on, 309
Threshold stimulus
 for action potential, 371, 371f, 372
 in muscle contraction, 274
Thrombocytes. See Platelets
Thymic corpuscles, **651**–52, 652f
Thymopoietin, 651
Thymosin, 510, **651**
Thymulin, 651
Thymus, 510, 510f, 651–52, 651f
Thyrocervical trunk, 602f, 607f
Thyroglobulins, **516**
Thyrohyoid, 301f–302f
Thyrohyoid membrane, 674f
Thyrohyoid muscle, 301f–302f, 301t, 674f
Thyroid artery, superior, 602f
Thyroid cartilage, 298f, 299, 672f, 673t, 674f
Thyroid follicles, **516,** 516f
Thyroid gland, 510, 510f
 and blood calcium levels, 151
 gross anatomy, 522f–523f, 523, 651f, 674f
 histology, 516, 516f, 517f
 hormones, 510, 516, 517t, 524–25
Thyroid hormone (TH), **516,** 524–25
Thyroid-stimulating hormone (TSH), 512, 513t, **524**
Thyroid vein, superior, 602f
Thyrotropin-releasing hormone (TRH), 313t, **524**
Thyroxine (tetraiodothyronine; T_4), **524**
Tibia, 157f, 219t, **222**–23, 222f, 244f–245f
Tibial arteries, 619f, 621f, 622f
Tibial collateral ligament, 243t, 244f–245f
Tibial condyles, 219t, 222f, 243
Tibialis anterior, 344, 344f, 344t, 345f
Tibialis posterior, 346f, 346t
Tibial nerve, 441t, **442,** 443, 443f
Tibial tuberosity, 219t, 222f, 227, 228f
Tibial veins, 620f, 622f
Tibiofemoral joint, **243**
Tidal volume (TV), 679, 679t, 680f, 684f, 685, 685t, 688, 688t, 689t
Tissue(s), defined, **57**
Tissue forceps, **19,** 19f, 20, 21f
T-lymphocytes, 645, 645f, 651, 653, 657
Tongue, 672f, 712f–713f
 influences on, 417, 417f
 innervation of, 414f
 muscles, **297,** 297t, 298f
 papillae, 470f, 470t, 471
 taste buds, 471–72, 471t
Tonicity, erythrocytes and, 81–83, 82f
Tonometer, 686
Tonsils, **645,** 645f, 646–47, 646f, 646t, 655f, 672f
Tooth, root, 237f
Total peripheral resistance (TPR), 625
Touch, receptive field, size of, 429f, 491–92, 491f
TPR (total peripheral resistance), 625
Trabeculae
 bone, 149–50, 149f, 155f, **156**
 lymph node, **648,** 649, 649f, 650f, 650t, 656f
 spleen, 653, 653f, 654, 654t
 thymus, 652, 652f
Trabeculae carneae, 558f, **559,** 559f, 560, 561f, 562, 567, 567f, 568
Trabecular arteries, 653f, **654**
Trachea
 connective tissue in, 111–12, 111f
 epithelial tissue in, 101f, 102
 gross anatomy, 651f, 672f, 673, 674f, 678f
 histology, **666**–68, 666t–667t, 667f
 innervation, 454f
Tracheal cartilage, 667f, 667t, 668
Trachealis muscle, 667f, 667t, 668, 674f
Tracheal ring, 664f, 674f
Tract(s), nervous system, **358**
Transitional epithelium
 identification of, 98f, 103, 103f
 structure and function, 96t
 in urinary tract, **743,** 743f, 743t, **744,** 744f
Transverse arytenoid muscle, 674f
Transverse cervical (cardinal) ligament, 783t
Transverse colon, 610f, 709f, 720, 720t, 721f–722f

I-13

Transverse fissure
　human, 393t, 395f
　sheep, 404, 404f, 406, 406f
Transverse foramen, 183t, **186,** 186f, 187f
Transverse plane, 31, 31f, 31t
Transverse processes, 183t, **185,** 185f, 186, 186f, 187f, 189
Transverse ridges, 189f
Transverse sinuses, 180f, 385f, 386f, 604f, 605f
　sheep, 401, 401f, 402f
Transverse thoracis, 308, **308,** 308f
Transversospinal muscles, **303,** 305f, 306f
Transversus abdominis, **310,** 310t, 311f, 312, 312f, 313f
Transversus thoracis, 307t, 308
Trapezium, 206t, 211, 211f
Trapezius, 264, 265f, 299, 301f–302f, 323t, **324,** 325f
　innervation, 417, 417f
　as term, 262t
Trapezoid, 206t, 211, 211f
Tray, dissecting, description and use, 11t
T-regulatory cells, 652
TRH (thyrotropin-releasing hormone), 313t, **524**
Triceps, definition of, 262t
Triceps brachii, 264, 265f, 267f, 324t, 328f, 328t, 459
Triceps surae, **346**
Trigeminal nerve (CN V), **295,** 295f, 298t, 300t, 303, 398t, 399f
　V1 (ophthalmic branch), 409t, 413, 413f
　V2 (maxillary branch), 409t, 413, 413f
　V3 (mandibular branch), 409t, 413, 413f
　and corneal reflex, 461
　disorders, common, 410t
　functions, 409t, 413, 413f
　sheep, 401f, **402,** 402f, 403, 403f, 405f
Trigone, **749,** 750f
Triiodothyronine (T₃), **524**
Triquetrum, 206t, 211, 211f
Trochanter, femur, 218f, 220f, 221f, 227, 228f, 337f, 339
Trochlea, 205t, 207f–208f
Trochlear nerve (CN IV), 296t, 398t, 399f, 400
　disorders, common, 410t
　functions, 409t, 412, 412f, 498
Trochlear notch, 206t, 210f
Trophoblast, 797t–798t
Tropomyosin, 270, 270f, 271f, 555
Troponin, 270, 270f, 271f, 555
Troponin/tropomyosin complex, 270, 270f, 271f
True rib, **192**
True vocal cords, 674f
Trunk(s), of brachial plexus, **437,** 437t, 438f, 440f
Trypsin, 724t
TSH. *See* Thyroid-stimulating hormone
TSH (thyroid-stimulating hormone), 512, 513t, **524**
Tubercle
　atlas, 186, 187f
　humerus, 205t, 207f–208f
　rib, **188,** 190t, 192f
Tuberculum sellae fossa, **179,** 180f
Tuberosity
　costal, 202t, 203f
　deltoid, 205t, 207f–208f, 213, 213f
　gluteal, 218t, 220f
　iliac, 214t
　ischial, 214t, 216f, 227, 228f
　radial, 206t, 209f
　tibial, 219t, 222f, 227, 228f
　of ulna, 206t, 210f
Tubular fluid, **748**
Tubular reabsorption, **736,** 752, 753t
Tubular secretion, **753,** 753t
Tubuloalveolar glands, 781, 781f
Tumors, brain, 362
Tunic(s)
　of blood vessel walls, 592–93, 592t, 593f
　of lymphatic vessel, 644
　of stomach wall, 702–3, 703f, 704f, 704t
Tunica albuginea, 768, **768,** 769f, 770f, 776f, 790f
Tunica externa, 592–93, 592t, 593f, 594, 594f, 595, 595f, 596, 596f
Tunica intima, 592–93, 592t, 593f, 594, 594f, 595, 595f, 596, 596f, 597, 644
Tunica media, 592–93, 592t, 593f, 594, 594f, 595, 595f, 596, 596f
Tunica vaginalis, 790f

TV (tidal volume), 679, 679t, 680f, 684t, 685, 685t, 688, 688t, 689t
T waves, **570,** 571f, 572, 573
Two-point discrimination test, 491, 491f
Tympanic cavity, 415f
Tympanic membrane (eardrum), 415f, 487f, 488t, 489, **489, 490,** 490f
Type I muscle fibers, 268, 268t, 269, 269t
Type IIa muscle fibers, 268, 268t, 269, 269t
Type IIb muscle fibers, 268, 268t, 269, 269t

Ulna, 153f, 157f, 206t, 207f–208f, **210**–11, 210f
Ulnar artery, 614f, 617f
Ulnar nerve, **213, 437**–38, 437f, 438f, 439t, 440f
Ulnar notch, 209f
Ulnar veins, 615f, 617f
Umami, as basic taste, 494, 494t
Umbilical, defined, 34t
Umbilical arteries, 623t, 624f
Umbilical cord, 624f, 799f
Umbilical region, 37, 37f
Umbilical veins, 623t, 624f
　varicosities of, 613
Umbilicus, 311f, 313f
Umbrella cells, 103
Unipennate architecture, in skeletal muscle, 264t
Unit(s). *See* Measurement
Upper respiratory tract
　gross anatomy, 671–72
　histology, 664–66
Urea, 736
Ureters
　function, 753
　gross anatomy, 736, 745f, **748,** 748–52, 749f, 750f, 751f, 785f, 789f, 790f
　histology, 743, 743f, 743t
　innervation, 454f
Urethra
　female, 743t, 751–52, 751f
　function, **736**
　histology, 743t
　male, 743t, 750–51, 751f, 782, 782f, 788
Urethral orifice, external, 785f
Urethral sphincters, **749,** 750, 751f, 752, 790f
Urinalysis, 754–56, 755t
Urinary bladder
　function, **736,** 753
　gross anatomy, **749**–52, 749f, 750f, 751f, 785f, 789f, 790f
　histology, 103, 103f, 743t, 744, 744f
Urinary system
　and acid-base balance, 756, 756t
　function of, 736
Urinary tract
　gross anatomy, 748–52
　histology, **742**–44, 743f
　infections of, 749
Urine, **736, 748,** 753
　abnormal constituents, 753–54, 754t
　color of, **755,** 755t
　normal constituents, 753
　odor of, 755, 755t
　pH of, 753, 755
　safe handling of, 754
　specific gravity, 753, 756t
　transparency of, **755,** 755t
Urinometer, 756, 756t
Urogenital diaphragm, **749,** 788, 789f, 790f
Uterine contraction, positive feedback in, 2
Uterine cycle, 793, 794–95, 794f
Uterine glands, **774**
Uterine part of uterine tube, 773t
Uterine tubes (fallopian tubes; oviducts)
　gross anatomy, 783–85, 784f, 785f
　histology, 100, 100f, **772**–73, 772f, 773f
Uterine wall, 774, 774t–775t
Uterosacral ligament, 783t
Uterus
　functional layer of, 794f, 798f
　gross anatomy, 783–85, 784f, 785f
　innervation, 454f
　round ligament of, **312**
Utricle, 477f, 489, 489t
Uvula, 416, 672f, 712f–713f

Vagina, **775,** 784f, 785f
　innervation, 454f
　wall, histology, 775, 775f
Vaginal orifice, 785f
Vagus nerve (CN X), 297t, 298t, 398t, 399f, 417f, 673

disorders, common, 410t
function, 409t, 416, 416f
and parasympathetic division, 453f, 454f, 455
Vallate (circumvallate) papillae, 414f, 470f, 470f, **471,** 471f, 472
Valve(s), vein, 596, 596f
Valve, of lymphatic vessel, 650
Valve cusp, of lymphatic vessel, 644, 644f
Variable(s), in scientific experiment, **4**
Varicosities, **613**
Vasa recta, **718,** 719f, 738, **741**–42, 742f, **745**–46, 746f, 747f
Vasa vasorum, 593f, 594, 594f, 596, 596f
Vascular spasm, in hemostasis, 538
Vascular tunic, 483t, 484
Vas deferens (ductus deferens), 779–80, 779t, 780f, 788
Vasopressin (antidiuretic hormone), 393, 513t, 514, 753t, 754
Vastus intermedius, 267f, 340t
Vastus lateralis, 267f, 338, 340f, 340t
Vastus medialis, 267f, 340f, 340t
Vein, 596, 596f
　definition of, **590**
　epithelial cells in, 98–99, 98f
　individual variation in patterns of, 590
　overview of, 590
　pulmonary circuit, 600–601, 601f
　systemic circuit, 601–22
　vessel wall, 592t, 593, 593f
Veli palatini
　levator, 298t, 299f
　tensor, 298t, 299f
Vena cava
　inferior, 309, 521f, **558,** 558f, 559f, 560, 566f, 567, 608f, 609f, 611, 611f, 612f, 613, 613f, 621f, 622f, 624f, 656f, 718f, 750f
　superior, 557f, **558,** 558f, 559f, 560, 566f, 567, 603f, 605f, 608f, 616f, 617f, 624f
Venom, antiserum for, 657
Venous arch, dorsal, 621f
Venous network, dorsal, 615f
Ventilation, mechanics of, 682
Ventricles, brain, 388–89, 388f, 389f, 390t
　cerebral aqueduct, 388, 388f, 389f, 390t, 391t, 397
　fourth ventricle, 388f, 389f, 390t, 391t, 397f, 406, 407f
　lateral ventricles, **388,** 388f, 389, 389f, 390t, 407, 407f
　third ventricle, 388f, 389f, 390t, 393t, **397,** 397f, 407, 407f
Ventricles, heart
　cardiac conduction system, 569–70
　fetal, 624f
　left, 557f, **558,** 559f, 560, 561f, 562, 562f, 563f, 601f
　right, 557f, **558,** 558f, 559, 559f, 562f, 563f, 564f, 601f
　sheep, 566, 566f, 567f
Ventricular contraction, premature, 572, 572f
Ventricular fibrillation, 572, 572f
Ventricular space, 363, 363f
Venule, **590,** 595f, **596,** 601f
　postcapillary, 592t, **596**
Vermiform appendix, 610f, **645,** 645f, **648,** 648f, 655f, 709f
Vertebrae, 152f, 153f, 157f, 183–89, 183t
Vertebral, defined, 34t, 35f
Vertebral arch, **185,** 185f
Vertebral arteries, **186,** 602f, **603,** 604f, 607f
Vertebral column, 183–89, 183t
　muscles, **303,** 303t–306t, 306f
Vertebral (spinal) foramen, 183t, **185,** 185f, 186, 186f, 187f, 188, 188f
Vertebral vein, 602f
Vertebra prominens, 183t
Vertigo, 501
Vesicles, brain
　primary, **388**
　secondary, **388,** 388f, **389**
Vesicouterine pouch, 785f
Vesicular (Graafian) follicle, 768, **769,** 769f, 770f, 770t–771t, 793, 794f
Vesicular sounds, 682
Vestibular folds, 673t
Vestibular ligaments, 673t
Vestibular membrane, 476, 477f, 478f, 490f
Vestibular nerve, **415,** 415f, 487f
Vestibule, 477f, 487f, 489t, **498,** 499, 672f, 712f–713f, 785f

Vestibulocochlear nerve (CN VIII), 398t, 399f, 478t
　cochlear branch, 477f, 490, 490f
　and cochlear implants, 498
　disorders, common, 410t
　functions, 409t, 415, 415f, 501
　sheep, 405f
　vestibular branch, 498
Vibrio cholerae, 725
Villi, intestinal, **707,** 707f, 708, 724
Visceral layer
　of glomerular capsule, 738t, **740,** 740f
　of serous membranes, 34, 557
　of serous pericardium, 555, **556,** 556f
Visceral motor nerves, 408t
Visceral muscle. *See* Smooth muscle
Visceral organs, innervation, 416, 416f, 454f, 455
Visceral pericardium, 555, **566,** 566f
Visceral pleura, **675,** 675f
Visceral sensory nerves, function of, 408t
Visceral sensory receptors, 468
Visceral surface, 653f
Viscosity of blood, and blood vessel resistance, **625**
Vision
　functional tests, 497–98, 497f
　retina structure and function, 474–76, 474f, 475t
Visual acuity test, 497, 497f
Visual cortex, 411f
Vital capacity, 679, 679t, 680f, 684t, 685, 685t, 689, 689t
Vitreous chamber (posterior cavity), 483t, 484, 484f
Vitreous humor, 483t, **486,** 486f
Vocal cords, false, 674f
Vocal folds, **673,** 673t
Vocal ligaments, 673, 673t
Volkmann (perforating) canals, 110t, **148**–49
Volt, 370
Voltage-gated channels
　and action potential generation, 371, 371f, 373
　and cardiac refractory period, 571
Vomer, 169, 170t, 171f, **173,** 174f, 178f, 182t
Vomiting, and acid-base balance, 725

Water
　osmosis, 77–79, 77f, 78f, 79f, 81–83, 82f
　solute diffusion in, 69–70, 69f, 70f, 71t
Weber test, **499,** 499f
Wet mount slide, preparation of, 63, 63f
White blood cell(s). *See* Leukocytes
White blood cell count, 538, 539–40, 539f
White commissure, 430, 431f, 433f
Whitefish, embryo, mitosis in, 66
White matter, **358,** 358f, 360, 360f, **361,** 361f, 384f, 430, 431f, 432t, 433t
White pulp, **653,** 653f, 654, 654t
Word etymology, value of knowing, 27–28
Working distance, of microscope, 47t, **49,** 49f
Wormian (sutural) bone, 177f
Wright's stain, 533

Xiphisternal joint, 190t
Xiphoid, 190t
Xiphoid process, **191,** 191f

Yellow bile, 529, 530t
Yellow bone marrow, **153,** 154f, 155, 155f, 227, 239f
Yolk sac, 798f

Z discs, **254,** 255f, 256, 260, 272
Zona fasciculata, 517–19, 518f, 519t
Zona glomerulosa, 517–19, 518f, 519t
Zona pellucida, 769, 769f, 770f, 771t
Zona reticularis, 517–19, 518f, 519t
Zone of calcified cartilage, 150f, 151
Zone of hypertrophic cartilage, 150f, 151
Zone of ossification, 150f, 151
Zone of proliferating cartilage, 150f, 151
Zone of resting cartilage, 150f, 151
Zoom knob, of microscope, 52f
Zygomatic arch, **176,** 176f
Zygomatic bone, 169, 170t, 173, 173f, 176, 176f, 182t
Zygomatic process, 170t, 173f, 176, 176f, 178f
Zygomaticus, 293t, 294f
Zygote, **766,** 791, **793, 797**